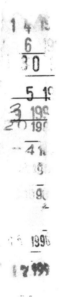

A Textbook of Quantitative Inorganic Analysis
including Elementary Instrumental Analysis

VOGEL'S TEXTBOOK OF QUANTITATIVE INORGANIC ANALYSIS

including
ELEMENTARY INSTRUMENTAL ANALYSIS

Fourth edition
Revised by the following members of
The School of Chemistry
Thames (formerly Woolwich) Polytechnic London, S.E.18

J. Bassett, M.Sc, C.Chem., F.R.I.C.
Senior Lecturer in Inorganic Chemistry

R. C. Denney, B.Sc., Ph.D., C.Chem., F.R.I.C.
Senior Lecturer in Analytical Chemistry

G. H. Jeffery, B.Sc., Ph.D., C.Chem., F.R.I.C.
Formerly Principal Lecturer
and Deputy Head of the School of Chemistry

J. Mendham, M.Sc., C.Chem., M.R.I.C.
Senior Lecturer in Analytical Chemistry

Longman
Scientific &
Technical

Copublished in the United States with
John Wiley & Sons, Inc., New York

Longman Scientific & Technical
Longman Group UK Limited,
Longman House, Burnt Mill, Harlow,
Essex CM20 2JE, England
and Associated Companies throughout the world

Copublished in the United States with
John Wiley & Sons, Inc., 605 Third Avenue, New York, NY 10158

First published 1939
New impressions 1941, 1942
New impression with corrections 1943
New impressions 1944, 1945, 1946, 1947, 1948
Second edition 1951
New impressions 1953, 1955, 1957, 1958, 1959, 1960
Third edition (published under the title
A TEXT-BOOK OF QUANTITATIVE INORGANIC ANALYSIS
INCLUDING ELEMENTARY INSTRUMENTAL ANALYSIS) *1961*
New impression with corrections 1962
New impressions 1964, 1968, 1969, 1971, 1974, 1975
Fourth edition 1978
New impression with minor corrections 1979
New impressions 1981, 1983, 1985, 1986, 1987

British Library Cataloguing in Publication Data
(Available)

ISBN 0-582-46321-1

Library of Congress Cataloging in Publication Data

Vogel, Arthur
  A textbook of quantitative inorganic analysis, including
elementary instrumental analysis.

  Includes bibliographies and index.
  1. Chemistry, Analytic—Quantitative.   2. Chemistry,
Inorganic.   I. Bassett, John, 1924–   II. Title.
QD101.2.V63   1978   546   77–5545
ISBN 0–470–20608–X (USA only)

Printed and bound in Great Britain by
William Clowes Limited, Beccles and London

CONTENTS

Chapter III   Common apparatus and basic techniques   58

Chapter VIII Paper, thin layer and column chromatography

Chapter IX Gas chromatography

PART D TITRIMETRY AND GRAVIMETRY

Chapter X   Titrimetric analysis

PART E ELECTROANALYTICAL METHODS

Chapter XII Electro-gravimetry

Chapter XIII Coulometry

Chapter XIV Potentiometry

Chapter XV Conductometric titrations 615

Chapter XVI Voltammetry 632

Chapter XVII   Amperometry 672

PART F   SPECTROANALYTICAL METHODS

Chapter XVIII   Colorimetry and spectrophotometry

Chapter XIX Fluorimetry — 773

Chapter XX Nephelometry and turbidimetry

Chapter XXI Emission spectrography

Chapter XXII Flame spectrometry

PART G   THERMAL METHODS

Chapter XXIII   Thermal analysis   849

Thermometric titrations   864

Appendices   869

Index   901

FOREWORD

SI units have been used throughout this book, but with the acceptance of 'litre' as a special name for the cubic decimetre we have the introduction of a non-SI term.

In this book the recommended convention has been adopted, namely that concentration data of high precision are expressed in terms of the dm^3, and only data of moderate accuracy are expressed in terms of the litre.

Concentrations of solutions are usually expressed in terms of moles per cubic decimetre: a molar solution (M) has one mole of solute per dm^3.

For some purposes, however, it is more convenient to work in terms of *equivalents* rather than moles (see Chapter X); a normal solution (N) has one equivalent of solute per dm^3.

PREFACE TO THE FOURTH EDITION

Successive editions of Arthur I. Vogel's books, and especially his *Textbook of Quantitative Inorganic Analysis*, have become accepted internationally as standard texts in colleges and laboratories. Only his untimely death in 1966 prevented him from extending and improving the various volumes which have become so familiar to generations of undergraduates.

In carrying out the revision necessary for this Fourth Edition we have been very conscious of the fact that we have been revising a book which possesses the character of one particular person. Because of this we have sought to retain that character throughout the reorganisation and introduction of new material. At the same time we have made a number of changes which are intended to make the individual sections and chapters as self-contained as possible. We have also chosen to emphasise more strongly the importance of statistics and sampling to the analytical chemist and have introduced a chapter on thermal analysis.

Some of the traditional methods of analysis which occupied substantial sections of earlier editions have received fairly heavy pruning in order to create sufficient space for the enlargement and introduction of sections dealing with instrumental methods. It has been difficult to decide what to delete and what to retain, but in those areas in which the third edition devoted space to several titrimetric or gravimetric procedures for individual elements we have reduced the number of entries to those of widest application.

The chapter on flame photometry has been rewritten to allow for the inclusion of a more substantial section dealing with the development of atomic absorption spectroscopy. At the same time the other chapters on spectroanalytical methods have been substantially reorganised and extended. The chapter on infrared spectrophotometry that was included in the third edition has been deleted in view of the limited application of this technique for quantitative inorganic analysis.

The opportunity has been taken to rearrange the chapters dealing with electroanalytical methods, and polarography is now included under the wider name 'voltammetry'. We have also extended the section on separative techniques to include a chapter on gas chromatography because of its general application in analytical chemistry, and in the context of this book for its use with certain volatile inorganic compounds.

Most of the important data in the appendices have been retained, but we have deleted both the chemical factors and the five figure logarithms as we have found that these are rarely used. A table giving a wide range of reagents suitable for the determination of metals has been introduced (by permission of Hopkin and

Williams Ltd.) listing a range of reagents suitable for determinations of metals. Where necessary the other tables have also been revised.

The work now occupies twenty-three chapters divided into seven parts. In this edition we have included a number of references to the sources of the new material that has been incorporated into the book. SI units have been employed throughout and the chemical nomenclature up-dated.

We should at this stage like to express our sincere appreciation to the many companies and publishers who have been so tolerant and understanding in providing us with information, diagrams and photographs for the new edition. That we had more than we could use is an indication of the great interest they have taken in this production.

We are all indebted to our wives in many different ways for their encouragement and assistance during the many months we have spent writing, revising, checking and finally proof reading. That we have actually finished the task is greatly due to their help throughout. Acknowledgement is also made of the helpful discussions with colleagues and of the assistance given by members of the laboratory staff of the School of Chemistry.

In conclusion we would like to say how pleased we are that we have been given this opportunity to carry on the work of a man who did so much to promote high standards in analytical chemistry. We hope that our efforts in producing the Fourth Edition of this book will in themselves serve as part of the memorial to the work of Arthur I. Vogel.

J. Bassett   R. C. Denney   G. H. Jeffery   J. Mendham

Thames Polytechnic, Woolwich, London, S.E.18.

PREFACE TO FIRST EDITION

In writing this book, the author had as his primary object the provision of a complete up-to-date text-book of quantitative inorganic analysis, both theory and practice, at a moderate price to meet the requirements of University and College students of all grades. It is believed that the material contained therein is sufficiently comprehensive to cover the syllabuses of all examinations in which quantitative inorganic analysis plays a part. The elementary student has been provided for, and those sections devoted to his needs have been treated in considerable detail. The volume should therefore be of value to the student throughout the whole of his career. The book will be suitable *inter alia* for students preparing for the various Intermediate B.Sc. and Higher School Certificate Examinations, the Ordinary and Higher National Certificates in Chemistry, the Honours and Special B.Sc. of the Universities, the Associateship of the Institute of Chemistry, and other examinations of equivalent standard. It is hoped, also, that the wide range of subjects discussed within its covers will result in the volume having a special appeal to practising analytical chemists and to all those workers in industry and research who have occasion to utilise methods of inorganic quantitative analysis.

The kind reception accorded to the author's *Text Book of Qualitative Chemical Analysis* by teachers and reviewers seems to indicate that the general arrangement of that book has met with approval. The companion volume on *Quantitative Inorganic Analysis* follows essentially similar lines. Chapter I is devoted to the theoretical basis of quantitative inorganic analysis, Chapter II to the experimental technique of quantitative analysis, Chapter III to volumetric analysis, Chapter IV to gravimetric analysis (including electro-analysis), Chapter V to colorimetric analysis, and Chapter VI to gas analysis; a comprehensive Appendix has been added, which contains much useful matter for the practising analytical chemist. The experimental side is based essentially upon the writer's experience with large classes of students of various grades. Most of the determinations have been tested out in the laboratory in collaboration with the author's colleagues and senior students, and in some cases this has resulted in slight modifications of the details given by the original authors. Particular emphasis has been laid upon recent developments in experimental technique. Frequently the source of certain apparatus or chemicals has been given in the text; this is not intended to convey the impression that these materials cannot be obtained from other sources, but merely to indicate that the author's own experience is confined to the particular products mentioned.

The ground covered by the book can best be judged by perusal of the Table of Contents. An attempt has been made to strike a balance between the classical and modern procedures, and to present the subject of analytical chemistry as it is to-day. The theoretical aspect has been stressed throughout, and numerous cross-references are given to Chapter I (the theoretical basis of quantitative inorganic analysis).

No references to the original literature are given in the text. This is because the introduction of such references would have considerably increased the size and therefore the price of the book. However, a discussion on the literature of analytical chemistry is given in the Appendix (Section **A, 3**). With the aid of the various volumes mentioned therein—which should be available in all libraries of analytical chemistry—and the Collective Indexes of *Chemical Abstracts* or of *British Chemical Abstracts*, little difficulty will, in general, be experienced in finding the original sources of most of the determinations described in the book.

In the preparation of this volume, the author has utilised pertinent material wherever it was to be found. While it is impossible to acknowledge every source individually (see, for example, Section **A, 3**), mention must, however, be made of Hillebrand and Lundell's *Applied Inorganic Analysis* (1929) and of Mitchell and Ward's *Modern Methods in Quantitative Chemical Analysis* (1932). In con-clusion, the writer wishes to express his thanks: to Dr. G. H. Jeffery, A.I.C., for reading the galley proofs and making numerous helpful suggestions; to Mr. A. S. Nickelson, B.Sc., for reading some of the galley proofs; to his laboratory steward, Mr. F. Mathie, for preparing a number of the diagrams, including most of those in Chapter VI, and for his assistance in other ways; to Messrs. A. Gallenkamp and Co., Ltd., of London, E.C.2, and to Messrs. Fisher Scientific Co, of Pittsburgh, Pa., for providing a number of diagrams and blocks;* and to Mr. F. W. Clifford, F.L.A., Librarian to the Chemical Society, and his able assistants for their help in the task of searching the extensive literature.

Any suggestions for improving the book will be gratefully received by the author.

Woolwich Polytechnic, London, S.E.18.   June, 1939.

---

* Acknowledgment to other firms and individuals is made in the body of the text.

PART A

FUNDAMENTALS OF QUANTITATIVE INORGANIC ANALYSIS

CHAPTER I   INTRODUCTION

I, 1. CHEMICAL ANALYSIS. 'The resolution of a chemical compound into its proximate or ultimate parts; the determination of its elements or of the foreign substances it may contain': thus reads a dictionary definition.

This definition outlines in very broad terms the scope of analytical chemistry. When a completely unknown sample is presented to an analyst, the first requirement is usually to ascertain what substances are present in it. This fundamental problem may sometimes be encountered in the modified form of deciding what impurities are present in a given sample, or perhaps of confirming that certain specified impurities are absent. The solution of such problems lies within the province of **qualitative analysis** and is outside the scope of the present volume.

Having ascertained the nature of the constituents of a given sample, the analyst is then frequently called upon to determine how much of each component, or of specified components, is present. Such determinations lie within the realm of **quantitative analysis**, and to supply the required information a variety of techniques is available.

I, 2. SAMPLING. The results obtained for the proportion of a certain constituent in a given sample may form the basis of assessing the value of a large consignment of the commodity from which the sample was drawn. In such cases it is absolutely essential to be certain that the sample used for analysis is truly representative of the whole. When dealing with a homogeneous liquid, sampling presents few problems, but if the material under consideration is a solid mixture, then it is necessary to combine a number of portions to ensure that a representative sample is finally selected for analysis. The analyst must therefore be acquainted with the normal standard sampling procedures employed for different types of materials.

I, 3. TYPES OF ANALYSIS. With an appropriate sample available, attention must now be given to the question of the most suitable technique or techniques to be employed for the required determinations. One of the major decisions to be made by an analyst is the choice of the most effective procedure for a given analysis, and in order to arrive at the correct decision, not only must he be familiar with the practical details of the various techniques and of the theoretical principles upon which they are based, he must also be conversant with the conditions under which each method is reliable, must be aware of possible interferences which may arise, and must be capable of devising means of circumventing such problems. He will also be concerned with questions regarding the accuracy and the precision to be expected from given methods and, in addition, must not overlook such factors as time and costing. The most accurate method for a certain determination may prove to be lengthy or to involve the use of expensive reagents, and in the interests of economy it may be necessary to choose a method which, although somewhat less exact, yields results of sufficient accuracy in a reasonable time.

Important factors which must be taken into account when selecting an appropriate method of analysis include (*a*) the nature of the information which is sought, (*b*) the size of sample available and the proportion of the constituent to be determined, and (*c*) the purpose for which the analytical data are required.

The nature of the information sought: this may consist of a requirement for very detailed data, or alternatively, results of a general character may suffice. With respect to the information which is furnished, different types of chemical analysis may be classified as follows:

(i) *proximate analysis*, in which the amount of each element in a sample is determined with no concern as to the actual compounds present;

(ii) *partial analysis*, which deals with the determination of selected constituents in the sample;

(iii) *trace constituent analysis* is a specialised instance of partial analysis in which we are concerned with the determination of specified components present in very minute quantity;

(iv) *complete analysis*, when the proportion of each component of the sample is determined.

On the basis of sample size, analytical methods are often classified as:

Macro, the determination of quantities of 0.1 g or more;

Semi-micro, dealing with quantities ranging from 0.01 g to 0.1 g;

Micro, for quantities not exceeding 0.001 g. The term semi-micro is not very apt, referring as it does to quantities larger than micro and it has been proposed that it should be replaced by the term *meso*.

A *major constituent* is one present in excess of 1 per cent, a *minor constituent* is one constituting from 0.01 to 1 per cent of the sample, and a *trace constituent* is one present to an extent of less than 0.01 per cent of the sample.

The purpose for which the analytical data are required may perhaps be related to process control and quality control. In such circumstances the objective is checking that raw materials and finished products conform to specification, and it may also be concerned with monitoring various stages in a manufacturing process. For this kind of determination, methods must be employed which are quick and which can be readily adapted for routine work: in this area instrumental methods have an important role to play, and in certain cases may lend themselves to automation. On the other hand, the problem may be one which requires detailed consideration and which may be regarded as being more in the nature of a research topic.

I, 4. USE OF LITERATURE. Faced with a research-type problem the analyst will frequently be dealing with a situation which is outside his normal experience and it will be necessary to seek guidance from published data. This will involve consultation of multi-volume reference works such as Kolthoff and Elving, *Treatise on Analytical Chemistry*; Wilson and Wilson, *Comprehensive*

Analytical Chemistry; Fresenius and Jander, *Handbuch der analytischen Chemie*; of a compendium of methods such as Meites, *Handbook of Analytical Chemistry*; or of specialised monographs dealing with particular techniques or types of material. It may be necessary to seek for more up-to-date information than that available in the books which have been consulted and this will necessitate making use of review publications (e.g. Annual Reports of the Chemical Society; Selected Annual Reviews of the Society for Analytical Chemistry), and of abstracts (e.g. Analytical Abstracts; Chemical Abstracts), and referring to journals devoted to analytical chemistry and to specific techniques.*

Such a literature survey may lead to the compilation of a list of possible procedures and the ultimate selection must then be made in the light of the criteria previously enunciated, and with special consideration being given to questions of possible interferences and to the equipment available.

I, 5. COMMON TECHNIQUES. The main techniques employed in quantitative inorganic analysis are based upon (*a*) the quantitative performance of suitable chemical reactions and either measuring the amount of reagent needed to complete the reaction, or ascertaining the amount of reaction product obtained; (*b*) appropriate electrical measurements (e.g. potentiometry); (*c*) the measurement of certain optical properties (e.g. absorption spectra); or (*d*), in some cases, a combination of optical or electrical measurements and quantitative chemical reaction (e.g. amperometric titration).

The quantitative execution of chemical reactions is the basis of the traditional or 'classical' methods of chemical analysis: gravimetry, titrimetry and volumetry. In *gravimetric analysis* the substance being determined is converted into an insoluble precipitate which is collected and weighed, or in the special case of *electrogravimetry*, electrolysis is carried out and the material deposited on one of the electrodes is weighed.

In *titrimetric analysis* (hitherto often termed volumetric analysis), the substance to be determined is allowed to react with an appropriate reagent added as a standard solution, and the volume of solution needed for complete reaction is determined. The common types of reaction which find use in titrimetry are (*a*) neutralisation (acid-base) reactions; (*b*) complex-forming reactions; (*c*) precipitation reactions; (*d*) oxidation-reduction reactions.

Volumetry is concerned with measuring the volume of gas evolved or absorbed in a chemical reaction.

Electrical methods of analysis (apart from electrogravimetry referred to above) involve the measurement of current, voltage or resistance in relation to the concentration of a certain species in solution. Techniques which can be included under this general heading are (i) *voltammetry* (measurement of current at a micro-electrode at a specified voltage); (ii) *coulometry* (measurement of current and time needed to complete an electrochemical reaction or to generate sufficient material to react completely with a specified reagent); (iii) *potentiometry* (measurement of the potential of an electrode in equilibrium with an ion to be determined); (iv) *conductimetry* (measurement of the electrical conductivity of a solution).

Optical methods of analysis are dependent either upon (i) the absorption of

---

* Selected bibliographies are given in Section I, 12 and at the conclusion of each chapter.

radiant energy and the measurement of the amount of energy of a particular wavelength absorbed by the sample, or (ii) the emission of radiant energy and measurement of the amount of energy of a particular wavelength emitted. *Absorption methods* are usually classified according to the wavelength involved as (*a*) *visible spectrophotometry* (colorimetry), (*b*) *ultraviolet spectrophotometry*, and (*c*) *infrared spectrophotometry* (this region in fact has few applications in quantitative inorganic analysis).

Atomic absorption spectroscopy involves vapourising the specimen, often by spraying a solution of the sample into a flame, and then studying the absorption of radiation from an electric lamp producing the spectrum of the element to be determined.

Although not strictly absorption methods in the sense in which the term is usually employed, *turbidimetric* and *nephelometric methods* which involve measuring the amount of light stopped or scattered by a suspension may also be referred to at this point.

Emission methods involve submission of the sample to heat or electrical treatment so that atoms are raised to excited states causing them to emit energy: it is the intensity of this emitted energy which is measured. The common techniques are:

(*a*) *emission spectrography*, where the sample is subjected to an electric arc or spark and the light emitted (which may extend into the ultraviolet region) is examined;

(*b*) *flame photometry*, in which a solution of the sample is injected into a flame and the light emitted is investigated;

(*c*) *fluorimetry*, in which a suitable substance in solution (commonly a metal-fluorescent reagent complex) is excited by irradiation with visible or ultraviolet radiation, and the characteristic emitted radiation is then examined.

I, 6. OTHER TECHNIQUES. In addition to the main general methods of quantitative inorganic analysis outlined above there are also certain specialised techniques, amongst which are: X-ray fluorescence, methods based on the measurements of radioactivity, and the so-called kinetic methods.

In the *X-ray fluorescence* method a metallic sample or a specimen of a rock is subjected to irradiation by a powerful beam of X-rays of short wavelength. This beam may displace an electron from the innermost electron shell of an atom, and to replace the lost electron, another electron may jump from one of the outer shells and in so doing release energy in the form of X-rays. The resultant 'secondary' or 'fluorescent' X-ray radiation will be emitted at wavelengths characteristic of the atom involved, and the intensity of the radiation can be used to assess the amount of the element giving rise to it present in the sample. This is an example of a number of so-called non-destructive methods of testing which lie outside the scope of this book. A number of texts dealing with various non-destructive techniques have been included in the bibliography under Section **I, 12**.

Methods based on the measurement of *radioactivity* belong to the realm of radiochemistry and may involve measurement of the intensity of the radiation from a naturally radioactive material, or alternatively, measurement of induced radioactivity arising from the exposure of the sample under investigation to a neutron source (activation analysis).

Kinetic methods of quantitative analysis are based upon the fact that the speed of a given chemical reaction may frequently be increased by the addition of a small amount of a catalyst, and within limits, the rate of the catalysed reaction will be governed by the amount of the catalyst present. If a calibration curve is prepared showing variation of reaction rate with amount of catalyst used, then measurement of reaction rate will make it possible to determine how much catalyst has been added in a certain instance, and this provides a sensitive method for determining sub-microgram amounts of appropriate substances.

I, 7. INSTRUMENTAL METHODS. The methods dependent upon measurement of an electrical property, and those based upon determination of the extent to which radiation is absorbed or upon assessment of the intensity of emitted radiation, all require the use of a suitable instrument, e.g. polarograph, spectrophotometer, etc., and in consequence such methods are referred to as *instrumental methods*. Instrumental methods are usually much faster than purely chemical procedures, are normally applicable at concentrations far too small to be amenable to determination by classical methods, and find wide application in industry. In many cases a recorder can be attached to the instrument so that absorption curves, polarograms, titration curves, etc., can be plotted automatically, and in fact, by the incorporation of appropriate servo-mechanisms, the whole analytical process may, in suitable cases, be completely automated.

Despite the advantages possessed by instrumental methods in many directions, their widespread adoption has not rendered the purely chemical or 'classical' methods obsolete; the situation is influenced by three main factors.
1. The apparatus required for classical procedures is cheap and readily available in all laboratories, but many instruments are expensive and their use will only be justified if numerous samples have to be analysed.
2. With most instrumental methods it is necessary to carry out a calibration operation using a sample of material of known composition as reference substance: the exact analytical data for this standard must be established by alternative procedures which will normally mean by the use of classical chemical methods.
3. Whilst an instrumental method is ideally suited to the performance of a large number of routine determinations, for an occasional, non-routine analysis, it is often simpler to use a classical method than to go to the trouble of preparing requisite standards and carrying out the calibration of an instrument.

Clearly, instrumental and classical methods must be regarded as supplementing each other.

I, 8. TIME AND MONEY, ACCURACY AND RANGE. Salient information relating to the more common quantitative techniques is presented below. The methods are arranged in columns according to the cost of the main equipment involved:

Cheap, less than £100;
Moderate, in the range £200–£1000;
High, in the range £1000–£4000;
Expensive, in excess of £4000.

This division is obviously somewhat arbitrary, and the cost of an instrument for a given technique varies widely according to the degree of sophistication offered.

The columns have been divided to give an indication of the time required for each technique. At the head of each column are the 'fast' procedures, that is, those which can normally be completed in ten to fifteen minutes, whilst the methods grouped at the foot of each column are the 'slow' procedures, those which normally require more than one hour for completion. Speed of operation is very much a personal factor, and it must be emphasised that the time required for preparing the final solution for analysis (including the removal of interferences) has not been taken into account, nor has the time needed for instrument calibration or for standardisation procedures.

Other information included is an indication by means of the letters H, M, L of the accuracy normally to be expected from a given method in the hands of a competent analyst:

H   high accuracy, results better than 1 per cent;
M   moderate accuracy, results in the range 1 per cent–5 per cent;
L   low accuracy, results not better than 5 per cent.

The numeral following the letter showing the accuracy provides an indication of the concentration range in which the method can be satisfactorily employed:

1   g—cg per cubic decimetre;
2   g—mg per cubic decimetre;
3   g—μg per cubic decimetre;
4   dg—mg per cubic decimetre;
5   mg—μg per cubic decimetre;
6   g—mg actual weight;
7   dg—mg actual weight (macro analysis);
8   p.p.m.—p.p.b.;
9   1 per cent—0.001 per cent;
10 1 per cent—p.p.b.

I, 9. INTERFERENCES. Whatever the method finally chosen for the required determination, it should, ideally, be a *specific method*; that is to say, it should be capable of measuring the amount of desired substance accurately, no matter what other substances may be present. In practice few analytical procedures attain this ideal, but many methods are *selective*; in other words, they can be used to determine any of a small group of ions in the presence of certain specified ions. In many instances the desired selectivity is achieved by carrying out the procedure under carefully controlled conditions, particularly with reference to the pH of the solution.

Frequently, however, there are substances present that prevent direct measurement of the amount of a given ion; these are referred to as *interferences*, and the selection of methods for separating the interferences from the substance to be determined are as important as the choice of the method of determination. Typical separation procedures include the following:

(*a*) *Selective precipitation.* The addition of appropriate reagents may convert interfering ions into precipitates which can be filtered off, careful pH control is often necessary in order to achieve a clean separation, and it must be borne in mind that precipitates tend to adsorb substances from solution and care must be taken to ensure that as little as possible of the substance to be determined is lost in this way.

(*b*) *Masking.* A complexing agent is added, and if the resultant complexes are sufficiently stable they will fail to react with reagents added in a subsequent

Conspectus of common quantitative analytical methods

	Cheap	Moderate	High	Expensive
FAST	Titrimetry (H1)	Conductivity (M1)	Atomic absorption (H4)	Emission spectrography (direct reading) (L8) X-ray fluorescence (M9)
		Flame photometry (M5)		
		←- -→		
		Coulometry (H4)		
		←- →		
		Potentiometry (M2)		
		←- →		
		Polarography (M4)		
		←- →		
		Spectrophotometry (M4)		
		←- →		
		Fluorimetry (M5)		
SLOW	Gravimetry (H7)	Electro-gravimetry (H7)		Differential thermal analysis (M6)
	Kinetic methods (L3)			Activation analysis (M10) Emission spectrometry (L8)

operation: this may be a titrimetric procedure or a gravimetric precipitation method.

(*c*) *Selective oxidation* (*reduction*). The sample is treated with a selective oxidising or reducing agent which will react with some of the ions present: the resultant change in oxidation state will often facilitate separation. For example, to precipitate iron as hydroxide, the solution is always oxidised so that iron(III) hydroxide is precipitated: this precipitates at a lower pH than does iron(II) hydroxide and the latter could be contaminated with the hydroxides of many bivalent metals.

(*d*) *Solvent extraction.* When metal ions are converted into chelate compounds by treatment with suitable organic reagents, the resulting complexes are soluble in organic solvents and can thus be extracted from the aqueous solution. Many ion-association complexes containing bulky ions which are largely organic in character (e.g. the tetraphenylarsonium ion $(C_6H_5)_4As^+$) are soluble in organic solvents and can thus be utilised to extract appropriate metal ions from aqueous solution. Such treatment may be used to isolate the ion which is to be determined, or alternatively, to remove interfering substances.

(*e*) *Ion exchange.* Ion exchange materials are insoluble substances containing ions which are capable of replacement by ions from a solution containing electrolytes. The phosphate ion is an interference encountered in many analyses involving the determination of metals; in other than acidic solutions the phosphates of most metals are precipitated. If, however, the solution is passed through a column of an anion exchange resin in the chloride form, then phosphate ions are replaced by chloride ions. Equally, the determination of

9

phosphates is difficult in the presence of a variety of metallic ions, but if the solution is passed through a column of a cation exchange resin in the protonated form, then the interfering cations are replaced by hydrogen ions.

(*f*) *Chromatography.* The term chromatography is applied to a separation technique in which the components of a solution are made to travel down a column at different rates, the column being packed with a suitable finely divided solid termed the *stationary phase*, for which such diverse materials as cellulose powder, silica gel and alumina are employed. Having introduced the test solution to the top of the column, an appropriate solvent (the *mobile phase*) is allowed to flow slowly through the column. In *adsorption chromatography* the solutes are adsorbed on the column material and are then eluted by the mobile phase: the less easily adsorbed components are eluted first and the more readily adsorbed components are eluted more slowly, thus effecting separation. In *partition chromatography* the solutes are partitioned between the mobile phase and a film of liquid (commonly water) firmly adsorbed on the surface of the stationary phase. A typical example is the separation of cobalt from nickel in solution in concentrated hydrochloric acid: the stationary phase is cellulose powder, the mobile phase, acetone containing hydrochloric acid; the cobalt is eluted whilst the nickel remains on the column. If compounds of adequate voltatility are selected (some metal chelates are suitable) then *gas chromatography* may be carried out in which the mobile phase is a current of gas, e.g. nitrogen. It is frequently possible to dispense with a column and to use the adsorbent spread as a thin layer on a glass plate (*thin layer chromatography*) and in some cases a roll or a sheet of filter paper without any added adsorbent may be used (*paper chromatography*): these techniques are especially useful for handling small amounts of material.

I, 10.   ACCURACY AND PRECISION.   The best method or methods of dealing with interferences having finally been decided upon and the most appropriate method of determination chosen, the analysis will be carried out in duplicate and preferably in triplicate. The results must then be evaluated to decide on the best value to report, and to attempt to establish the probable limits of error of this value. The analyst will thus be concerned with the question of *precision*, that is, the agreement between a set of results for the same quantity, and also with *accuracy*, that is, with the difference between the measured value and the true value for the quantity which has been determined. *Statistical methods* will be used to demonstrate the degree of reliability of the results obtained.

I, 11.   SUMMARY.   Summarising, the following steps are necessary when confronted with an unfamiliar quantitative determination.
(i)   Sampling.
(ii)   Literature survey and selection of possible methods of determination.
(iii) Consideration of interferences and procedures for their removal.
   Pooling the information gathered under headings (ii) and (iii), a final selection will be made of the method of determination and of the procedure for eliminating interferences.
(iv)   Dissolution of sample.
(v)   Removal or suppression of interferences.
(vi)   Performance of the determination.
(vii) Statistical analysis of the results.

I, 12. Selected bibliography

The following selection of reference books, journals, and review articles is not intended to be exhaustive. The books listed include the widely known general textbooks and reference works devoted to quantitative chemical analysis, and to books giving general accounts of various topics referred to in this chapter. Succeeding chapters carry individual bibliographies which include the more important books dealing with specialised techniques.

A. General reference books

1. ASTM Standards. American Society for Testing Materials, Philadelphia, 1964.
2. D. Abbott and R. S. Andrews (1970). *An Introduction to Chromatography*, 2nd edn. London; Longman.
3. J. A. Barnard and A. Chayen (1965). *Modern Methods of Chemical Analysis*. London; McGraw-Hill.
4. R. Belcher and A. J. Nutten (1970). *Quantitative Inorganic Analysis*. 3rd edn. London; Butterworth.
5. R. Belcher and C. L. Wilson (1964). *New Methods of Analytical Chemistry*. 2nd edn. London; Chapman and Hall.
6 E. W. Berg (1963). *Physical and Chemical Methods of Separation*. New York; McGraw-Hill.
7. W. G. Berl (1960). *Physical Methods in Chemical Analysis*. 2nd edn. New York; Academic Press.
8. D. Betteridge and H. E. Hallam (1972). *Modern Analytical Methods*. London; The Chemical Society.
9. D. R. Browning (1969). *Electrometric Methods*. London; McGraw-Hill.
10. G. Charlot (1960). *Les méthodes de la Chimie Analytique: analyse quantitative minérale*. 4th edn. Paris; Masson et cie.
11. J. A. Dean (1969). *Chemical Separation Methods*. New York; Van Nostrand.
12. G. W. Ewing (1968). *Instrumental Methods of Chemical Analysis*. 3rd edn. New York; McGraw-Hill.
13. G. W. Ewing (1971). *Topics in Chemical Instrumentation*. Easton; Chemical Education Pub. Co.
14. W. Fresenius and G. Jander (from 1944). *Handbuch der Analytischen Chemie, Dritter Teil*. Berlin; Springer-Verlag.
15. N. H. Furman and F. G. Welcher (1962). *Standard Methods of Chemical Analysis*. 6th edn. Princeton; Van Nostrand.
16. W. F. Hildebrand, G. E. J. Lundell, H. A. Bright and J.'A. Hoffman (1953). *Applied Inorganic Analysis*, 2nd edn. New York; Wiley.
17. K. Kodoma (1963). *Methods of Quantitative Inorganic Analysis: an encyclopaedia of gravimetric, titrimetric and colorimetric methods*. New York; Wiley.
18. I. M. Kolthoff and P. J. Elving (from 1959). *Treatise on Analytical Chemistry*. New York; Wiley.
19. P. Kruger (1971). *Principles of Activation Analysis*. New York; Wiley-Interscience.
20. N. A. Lange (1966). *Handbook of Chemistry*. 10th edn. New York; McGraw-Hill.
21. H. A. Liebhafsky, H. G. Pfeiffer, E. H. Winslow, P. D. Zemany, and S. S. Liebhafsky (1972). *X-ray Absorption and Emission in Analytical Chemistry*. New York; Wiley-Interscience.
22. L. Meites (1963). *Handbook of Analytical Chemistry*. New York; McGraw-Hill.
23. W. F. Pickering (1971). *Modern Analytical Chemistry*. New York; Marcel Dekker.
24. D. Samuelson (1963). *Ion-exchange Separations in Analytical Chemistry*. New York; Wiley.
25. A. Seidell and W. F. Linke (1958). *Solubilities of Inorganic and Metal-organic Compounds*. 4th edn. Princeton, Van Nostrand.

26. S. Siggia (1968). *Survey of Analytical Chemistry.* New York; McGraw-Hill.
27. L. G. Sillen and A. E. Martell (1964). *Stability Constants of Metal Ion Complexes.* London; The Chemical Society.
28. D. A. Skoog and D. M. West (1970). *Fundamentals of Analytical Chemistry.* 2nd edn. London; Holt, Rinehart and Winston.
29. D. A. Skoog and D. M. West (1971). *Principles of Instrumental Analysis.* New York; Holt, Rinehart and Winston.
30. F. D. Snell and C. L. Hilton (from 1969). *Encyclopaedia of Industrial Chemical Analysis.* New York; Wiley.
31. C. R. N. Strouts, H. N. Wilson and R. T. Parry-Jones (1962). *Chemical Analysis: the working tools.* 2nd edn. Oxford; Oxford University Press.
32. R. C. Weast (1972). *Handbook of Chemistry and Physics.* 53rd edn. Cleveland; Chemical Rubber Publishing Co.
33. R. K. Webster (1960). *Methods in Geochemistry.* London; Interscience.
34. F. J. Welcher (1947). *Organic Analytical Reagents.* Princeton; Van Nostrand.
35. T. S. West (1972). *Analytical Chemistry (Parts 1, 2).* MTP Series I, Vols. 12, 13. London; Butterworths.
36. H. H. Willard, L. L. Merritt and J. A. Dean (1974). *Instrumental Methods of Analysis.* 5th edn. New York; Van Nostrand.
37. C. L. Wilson and D. W. Wilson (from 1959). *Comprehensive Analytical Chemistry.* Amsterdam; Elsevier.
38. J. G. Dick (1973). *Analytical Chemistry.* New York; McGraw-Hill Book Co.
39. J. S. Fritz and G. H. Schenck (1974). *Quantitative Analytical Chemistry.* 3rd edn. Boston; Allyn and Bacon.
40. D. G. Peters, J. M. Hayes and G. M. Hieftje (1974). *Chemical Separations and Measurements.* Philadelphia; W. B. Saunders Co.
41. W. E. Harris and B. Kratochvil (1974). *Chemical Separations and Measurements.* Philadelphia; W. B. Saunders Co.
          (This is complementary to No. 40.)
42. H. A. Strobel (1973). *Chemical Instrumentation—A Systematic Approach to Instrumental Analysis.* 2nd edn. Reading, Mass.; Addison-Wesley Publishing Co.
43. D. J. Pietrzyk and C. W. Frank (1974). *Analytical Chemistry: An Introduction.* New York; Academic Press Inc.
44. H. F. Walton and J. Reyes (1973). *Modern Chemical Analysis and Instrumentation.* New York; Marcel Dekker Inc.
45. T. H. Gouw (1972). *Guide to Modern Methods of Instrumental Analysis.* New York; J. Wiley and Sons Inc.
46. J. W. Robinson (1973). *Undergraduate Instrumental Analysis.* 2nd edn. New York; Marcel Dekker Inc.
47. H. A. Laitinen and W. E. Harris (1975). *Chemical Analysis.* 2nd edn. New York; McGraw-Hill.
48. D. de Soete, R. Gijbels and J. Hoste (1972). *Neutron Activation Analysis.* Chichester, Wiley.
49. P. Kruger (1971). *Principles of Activation Analysis.* New York; Wiley-Interscience.
50. G. D. Chase and J. L. Rabinowitz (1962). *Principles of Radioisotope Methodology.* 2nd edn. Minneapolis; Burgess.
51. H. A. C. McKay (1971). *Principles of Radiochemistry.* London; Butterworths.
52. C. A. Anderson (ed.) (1973). *Microprobe Analysis.* New York; Wiley-Interscience.
53. H. A. Liebhafsky, H. G. Pfeiffer, E. H. Winslow and P. D. Zemany (1972). *X-rays, Electrons and Analytical Chemistry.* Chichester; Wiley.
54. R. Jenkins and J. L. de Vries (1970). *Practical X-ray Spectrometry.* London; Macmillan.

B. Journals, abstracts and reviews

1. *Advances in Analytical Chemistry and Instrumentation.*
2. *Analytical Abstracts.*
3. *Analytica Chimica Acta.*
4. *Analytical Chemistry* (includes *Annual Review* in April issue).
5. *Annual Reports of the Chemical Society*, London.
6. *Chemical Abstracts.*
7. *Chemia Analityczna.*
8. *Chemical Titles.*
9. *Chimie Analytique.*
10. *Current Chemical Papers.*
11. *Journal of Analytical Chemistry of the USSR (Zhurnal analitischeskoi Khimii).*
12. *Selected Annual Reviews of the Analytical Sciences.*
13. *Spectrochimica Acta.*
14. *Talanta.*
15. *The Analyst.*
16. *Zeitschrift für analytische chemie.*
17. *Journal of Electroanalytical Chemistry.*
18. *Journal of the Polargraphic Society.*
19. *Journal of Scientific Instruments.*
20. *Mikrochimica Acta.*

CHAPTER II

FUNDAMENTAL THEORETICAL PRINCIPLES

II, 1. ELECTROLYTIC DISSOCIATION. Many of the reactions of qualitative and quantitative analysis take place in aqueous solution. It is therefore necessary to have a general knowledge of the conditions which exist in such solutions. It is assumed that the reader is familiar with the broad concepts of the simple theory of electrolytic dissociation.

Ionisation of acids and bases in solution. An acid may be defined as a substance which, when dissolved in water, undergoes dissociation with the formation of hydrogen ions as the only positive ions:

$$HCl \rightleftharpoons H^+ + Cl^-$$
$$HNO_3 \rightleftharpoons H^+ + NO_3^-$$

Actually the hydrogen ion H^+ (or proton) does not exist in the free state in aqueous solution; each hydrogen ion combines with one molecule of water to form the *hydroxonium* ion H_3O^+. The hydroxonium ion is a hydrated proton. The above equations are therefore more accurately written:

$$HCl + H_2O \rightleftharpoons H_3O^+ + Cl^-$$
$$HNO_3 + H_2O \rightleftharpoons H_3O^+ + NO_3^-$$

The ionisation may be attributed to the great tendency of the free hydrogen ions H^+ to combine with water molecules to form hydroxonium ions. Hydrochloric and nitric acids are almost completely dissociated in aqueous solution in accordance with the above equations; this is readily demonstrated by freezing-point measurements and by other methods.

Polyprotic (polybasic) acids ionise in stages. In sulphuric acid, one hydrogen atom is almost completely ionised:

$$H_2SO_4 \rightleftharpoons H^+ + HSO_4^-$$
$$\text{or } H_2SO_4 + H_2O \rightleftharpoons H_3O^+ + HSO_4^-$$

The second hydrogen atom is only partially ionised, except in very dilute solution:

$$HSO_4^- \rightleftharpoons H^+ + SO_4^{2-}$$
$$\text{or } HSO_4^- + H_2O \rightleftharpoons H_3O^+ + SO_4^{2-}$$

Phosphoric acid also ionises in stages:

$$H_3PO_4 \rightleftharpoons H^+ + H_2PO_4^- \rightleftharpoons 2H^+ + HPO_4^{2-} \rightleftharpoons 3H^+ + PO_4^{3-}$$

or $H_3PO_4 + H_2O \rightleftharpoons H_3O^+ + H_2PO_4^-$
$ H_2PO_4^- + H_2O \rightleftharpoons H_3O^+ + HPO_4^{2-}$
$ HPO_4^{2-} + H_2O \rightleftharpoons H_3O^+ + PO_4^{3-}$

The successive stages of ionisation are known as the primary, secondary, and tertiary ionisations respectively. As already mentioned, these do not take place to the same degree. The primary ionisation is always greater than the secondary, and the secondary very much greater than the tertiary.

Acids of the type of acetic acid CH_3COOH give an almost normal freezing-point depression in aqueous solution; the extent of dissociation is accordingly small. It is usual, therefore, to distinguish between acids which are completely or almost completely ionised in solution and those which are only slightly ionised. The former are termed **strong acids** (examples: hydrochloric, hydrobromic, hydriodic, iodic, nitric, and perchloric acids, primary ionisation of sulphuric acid), and the latter are called **weak acids** (examples: nitrous acid, acetic acid, carbonic acid, boric acid, phosphorous acid, phosphoric acid, hydrocyanic acid, and hydrogen sulphide). There is, however, no sharp division between the two classes.

A **base** may be defined as a substance which, when dissolved in water, undergoes dissociation with the formation of hydroxide ions OH^- as the only negative ions. Thus sodium hydroxide, potassium hydroxide, and the hydroxides of certain bivalent metals are almost completely dissociated in aqueous solution:

$NaOH \rightarrow Na^+ + OH^-$
$Ba(OH)_2 \rightarrow Ba^{++} + 2OH^-$

These are **strong bases**. Aqueous ammonia solution, however, is a **weak base**. Only a small concentration of hydroxide ions is produced in aqueous solution:

$NH_3 + H_2O \rightleftharpoons NH_4^+ + OH^-$

General concept of acid and bases. The Brønsted theory. The simple Arrhenius concept given in the preceding paragraphs suffices for many of the requirements of quantitative inorganic analysis in aqueous solution. It is, however, desirable to have some knowledge of the general theory of acids and bases proposed by J. N. Brønsted in 1923, since this is applicable to all solvents. According to Brønsted, an **acid** is a species having a tendency to lose a proton, and a **base** is a species having a tendency to add on a proton. This may be represented as:

Acid $\rightleftharpoons$ Proton + Conjugate Base
$A \rightleftharpoons H^+ + B$ \hfill (a)

It must be emphasised that the symbol H^+ (p^+ is sometimes used) represents the proton and not the 'hydrogen ion' of variable nature existing in different solvents (OH_3^+, NH_4^+, $CH_3CO_2H_2^+$, $C_2H_5OH_2^+$, etc.); the definition is therefore independent of solvent. The above equation represents a hypothetical scheme for defining A and B and not a reaction which can actually occur. Acids need not be neutral molecules (e.g., HCl, H_2SO_4, CH_3CO_2H), but may also be anions (e.g., HSO_4^-, $H_2PO_4^-$, $HOOC\cdot COO^-$) and cations (e.g., NH_4^+, $C_6H_5NH_3^+$, $Fe(H_2O)_6^{3+}$). The same is true of bases where the three classes can be illustrated by NH_3, $C_6H_5NH_2$, H_2O; CH_3COO^-, OH^-, HPO_4^{2-}, $OC_2H_5^-$; $Fe(H_2O)_5(OH)^{2+}$.

Since the free proton cannot exist in solution in measureable concentration, reaction does not take place unless a base is added to accept the proton from the acid. By combining the equations

$$A_1 \rightleftharpoons B_1 + H^+ \quad \text{and} \quad B_2 + H^+ \rightleftharpoons A_2,$$

we obtain $A_1 + B_2 \rightleftharpoons A_2 + B_1$ (b)

A_1–B_1 and A_2–B_2 are two conjugate acid–base pairs. This is the most important expression for reactions involving acids and bases; it represents the transfer of a proton from A_1 to B_2 or from A_2 to B_1. The stronger the acid A_1 and the weaker A_2, the more complete will be the reaction (b). The stronger acid loses its proton more readily than the weaker; similarly, the stronger base accepts a proton more readily than does the weaker base. It is evident that the base or acid conjugate to a strong acid or a strong base is always weak, whereas the base or acid conjugate to a weak acid or weak base is always strong.

In aqueous solution a Brønsted acid A

$$A + H_2O \rightleftharpoons H_3O^+ + B$$

is strong when the above equilibrium is virtually complete to the right so that [A] is almost zero. A strong base is one for which [B], the equilibrium concentration of base other than hydroxide ion, is almost zero.

Acids may thus be arranged in series according to their relative combining tendencies with a base, which for aqueous solutions (in which we are largely interested) is water:

$$HCl + H_2O \rightleftharpoons H_3O^+ + Cl^-$$

  Acid$_1$ Base$_2$ Acid$_2$ Base$_1$

This process is essentially complete for all typical 'strong' (i.e., highly ionised) acids, such as HCl, HBr, HI, HNO_3, and $HClO_4$. In contrast with the 'strong' acids, the reactions of a typical 'weak' or slightly ionised acid, such as acetic acid or propionic acid, proceeds only slightly to the right in the equation:

$$CH_3COOH + H_2O \rightleftharpoons H_3O^+ + CH_3COO^-$$

  Acid$_1$ Base$_2$ Acid$_2$ Base$_1$

The typical strong acid of the water system is the hydrated proton H_3O^+, and the role of the conjugate base is minor if it is a sufficiently weak base, e.g., Cl^-, Br^-, and ClO_4^-. The conjugate bases have strengths that vary inversely as the strengths of the respective acids. It can easily be shown that the basic ionisation constant of the conjugate base $K_{B,conj.}$ is equal to $K_w/K_{A,conj.}$, where K_w is the ionic product of water.

Scheme (b) includes reactions formerly described by a variety of names, such as dissociation, neutralisation, hydrolysis and buffer action (see below). One acid–base pair may involve the solvent (in water H_3O^+–H_2O or H_2O–OH^-), showing that ions such as H_3O^+ and OH^- are in principle only particular examples of an extended class of acids and bases, though, of course, they do occupy a particularly important place in practice. It follows that the properties of an acid or base may be greatly influenced by the nature of the solvent employed.

Another definition of acids and bases is due to G. N. Lewis (1938). From the experimental point of view Lewis regards all substances which exhibit 'typical' acid–base properties (neutralisation, replacement, effect on indicators, catalysis), irrespective of their chemical nature and mode of action, as acids or bases. He relates the properties of acids to the acceptance of electron pairs, and bases as donors of electron pairs, to form covalent bonds regardless of whether protons are involved. On the experimental side Lewis' definition brings together a wide range of qualitative phenomena, e.g., solutions of BF_3, BCl_3, $AlCl_3$, or SO_2 in an inert solvent cause colour changes in indicators similar to those produced by hydrochloric acid, and these changes are reversed by bases so that titrations can be carried out. Compounds of the type of BF_3 are usually described as **Lewis acids** or **electron acceptors**. The Lewis bases (e.g., ammonia, pyridine) are virtually identical with the Brønsted bases. The great disadvantage of the Lewis definition of acids is that, unlike proton-transfer reactions, it is incapable of general quantitative treatment.

Salts. The structure of numerous salts in the solid state has been investigated by means of X-rays and by other methods, and it has been shown that they are composed of charged atoms or groups of atoms held together in a crystal lattice. When these salts are dissolved in a solvent of high dielectric constant such as water, or are heated to the melting point, the crystal forces are weakened and the substances dissociate into the pre-existing charged particles or ions, so that the resultant liquids are good conductors of electricity. There are, however, some exceptions: feebly ionised salts (**weak electrolytes**) are exemplified by the cyanides, thiocyanates, and halides of mercury and cadmium, and by lead acetate.

The theoretical implications of the theory of complete ionisation, due to Debye, Hückel, and Onsager, have been fully worked out by these authors. In particular, they have been able to account for the increasing equivalent conductance with decreasing concentration over the concentration range 0–$0.002M$. For full details the reader must be referred to textbooks of physical chemistry.

It is important to realise that whilst complete ionisation occurs with strong electrolytes, this does not mean that the *effective concentrations* of the ions are the same at all concentrations, for if this were the case, the osmotic properties of aqueous solutions could not be accounted for. The variation of osmotic properties with dilution is ascribed to changes in the *activity* of the ions; these are dependent upon the electrical forces between the ions. Expressions for the variations of the activity or of related quantities, applicable to dilute solutions, have also been deduced by the Debye–Hückel–Onsager theory. Further consideration of the concept of activity will be found in Section **II, 3**.

II, 2. THE LAW OF MASS ACTION. Guldberg and Waage in 1867 clearly stated the **law of mass action** (sometimes termed the law of chemical equilibrium) in the form: the velocity of a chemical reaction is proportional to the product of the active masses of the reacting substances. For the present we shall interpret 'active mass' by concentration and express it in mols per cubic decimetre. By applying the law to homogeneous systems, i.e., to systems in which all the reacting molecules are present in one phase, for example in solution, we can arrive at a mathematical expression for the condition of equilibrium in a reversible reaction.

Let us consider first the simple reversible reaction at constant temperature:

$$A + B \rightleftharpoons C + D$$

The velocity with which A and B react is proportional to their concentrations, or

$$v_1 = k_1 \times [A] \times [B]$$

where k_1 is a constant known as the velocity coefficient, and the square brackets (see p. 30 footnote) denote the molecular concentrations of the substances enclosed within the brackets. Similarly, the velocity with which the reverse reaction occurs is given by:

$$v_2 = k_2 \times [C] \times [D]$$

At equilibrium, the velocities of the reverse and the forward reactions will be equal (the equilibrium is a dynamic and not a static one) and therefore $v_1 = v_2$,

or $\quad k_1 \times [A] \times [B] = k_2 \times [C] \times [D]$

or $\qquad \dfrac{[C] \times [D]}{[A] \times [B]} = \dfrac{k_1}{k_2} = K$

K is the **equilibrium constant** of the reaction at the given temperature.

The expression may be generalised. For a reversible reaction represented by:

$$p_1 A_1 + p_2 A_2 + p_3 A_3 + \ldots \rightleftharpoons q_1 B_1 + q_2 B_2 + q_3 B_3 + \ldots$$

where p_1, p_2, p_3 and q_1, q_2, q_3 are the number of molecules of reacting substances, the condition for equilibrium is given by the expression:

$$\frac{[B]^{q_1} \times [B_2]^{q_2} \times [B_3]^{q_3} \ldots}{[A_1]^{p_1} \times [A_2]^{p_2} \times [A_3]^{p_3} \ldots} = K$$

This result may be expressed in words: when equilibrium is reached in a reversible reaction, at constant temperature, the product of the molecular concentrations of the resultants (the substances on the right-hand side of the equation) divided by the product of the molecular concentrations of the reactants (the substances on the left-hand side of the equation), each concentration being raised to a power equal to the number of molecules of that substance taking part in the reaction, is constant.

II, 3. ACTIVITY AND ACTIVITY COEFFICIENT. In our deduction of the law of mass action it was assumed that the effective concentrations or active masses of the components could be expressed by the stoichiometric concentrations. According to modern thermodynamics, this is not strictly true. The rigorous equilibrium equation for, say, a binary electrolyte:

$$AB \rightleftharpoons A^+ + B^-$$

is $\qquad \dfrac{(a_{A^+} \times a_{B^-})}{a_{AB}} = K_a$

where a_{A^+}, a_{B^-}, and a_{AB} represent the **activities** of A^+, B^-, and AB respectively, and K_a is the **true** or **thermodynamic dissociation constant**. The concept of **activity**, a thermodynamic quantity, is due to G. N. Lewis. The quantity is related to the concentration by a factor, termed the **activity coefficient**:

activity = concentration × activity coefficient

Thus at any concentration

$$a_{A^+} = y_{A^+} \cdot [A^+], \ a_{B^-} = y_{B^-} \cdot [B^-], \text{ and } a_{AB} = y_{AB} \cdot [AB]$$

where y refers to the activity coefficients,* and the square brackets to the concentrations. Substituting in the above equation, we obtain:

$$\frac{y_{A^+} . [A^+] \times y_{B^-} . [B^-]}{y_{AB} . [AB]} = \frac{[A^+] . [B^-]}{[AB]} \times \frac{y_{A^+} \times y_{B^-}}{y_{AB}} = K_a$$

This is the rigorously correct expression for the law of mass action as applied to weak electrolytes.

The activity coefficient varies with the concentration. For ions it also varies with the valency, and is the same for all *dilute* solutions having the same **ionic strength**, the latter being a measure of the electrical field existing in the solution. The term ionic strength, designated by the symbol I, is defined as equal to one half of the sum of the products of the concentration of each ion multiplied by the square of its valency, or $I = 0.5\Sigma c_i z_i^2$, where c_i is the ionic concentration in mols per cubic decimetre of solution and z_i is the valency of the ion concerned. An example will make this clear. The ionic strength of $0.1M$-HNO_3 solution containing $0.2M$-$Ba(NO_3)_2$ is given by:

$$0.5 \{0.1 \text{ (for } H^+) + 0.1 \text{ (for } NO_3^-)$$
$$+ 0.2 \times 2^2 \text{ (for } Ba^{2+}) + (0.2 \times 2) \text{ (for } NO_3^-)\} = 0.5\{1.4\} = 0.7.$$

The activity coefficient depends upon the *total* ionic strength of the solution in a manner which is discussed in Section **II, 8**. The activity coefficients of un-ionised molecules do not differ considerably from unity and for weak electrolytes in which the ionic concentration and therefore the ionic strength is small, the error introduced by neglecting the difference between the actual values of the activity coefficient of the ions, y_{A^+} and y_{B^-}, and unity is small (< 5 per cent). Hence for weak electrolytes, the true or thermodynamic expression reduces to $[A^+] \times [B^-]/[AB] = K$, and the constants obtained by the use of simple concentrations will be accurate to 2–5 per cent; such values are sufficiently precise for many of the calculations related to quantitative analysis.

II, 4. ACID–BASE EQUILIBRIA IN WATER. Let us consider the dissociation of a weak electrolyte, such as acetic acid, in dilute aqueous solution:

$$CH_3COOH + H_2O \rightleftharpoons H_3O^+ + CH_3COO^-$$

This will be written for simplicity in the conventional manner:

$$CH_3COOH \rightleftharpoons H^+ + CH_3COO^-$$

where H^+ represents the hydrated hydrogen ion. Applying the law of mass action, we have:

$$[CH_3COO^-] \times [H^+]/[CH_3COOH] = K$$

K is the equilibrium constant at a particular temperature and is usually known as the **ionisation constant** or **dissociation constant**. If one mol of the electrolyte is

---

* The symbol used is dependent upon the method of expressing the concentration of the solution. The recommendations of the IUPAC Commission on Symbols, Terminology and Units (1969) are as follows: concentration in mols per cubic decimetre (molarity), activity coefficient represented by y, concentration in mols per kilogram (molality), activity coefficient represented by γ, concentration expressed as mole fraction, activity coefficient represented by f.

dissolved in V cubic decimetres of solution ($V = 1/c$, where c is the concentration in mols per dm^3), and if α is the degree of ionisation at equilibrium, then the amount of un-ionised electrolyte will be $(1 - \alpha)$ mols, and the amount of each of the ions will be α mols. The concentration of un-ionised acetic acid will therefore be $(1 - \alpha)/V$, and the concentration of each of the ions α/V. Substituting in the equilibrium equation, we obtain the expression:

$$\alpha^2/(1 - \alpha)\,V = K \quad \text{or} \quad \alpha^2 c/(1 - \alpha) = K.$$

This is known as **Ostwald's dilution law**.

Interionic effects are, however, not negligible even for weak acids and the activity coefficient product must be introduced into the expression for the ionisation constant:

$$K = \frac{\alpha^2 c}{(1 - \alpha)} \cdot \frac{y_{H^+} \cdot y_{A^-}}{y_{HA}}$$

where the y's refer to activity coefficients and A^- to CH_3COO^-.

Reference must be made to textbooks of physical chemistry (see Bibliography at the end of the chapter) for details of the methods used to evaluate true dissociation constants of acids.

From the point of view of quantitative inorganic analysis, sufficiently accurate values for the ionisation constants of weak monoprotic acids may be obtained by using the classical Ostwald dilution law expression: the resulting 'constant' is sometimes called the 'concentration dissociation constant'.

II, 5. STRENGTHS OF ACIDS AND BASES. The Brønsted expression for acid–base equilibria (see Section **II, 1**)

$$A_1 + B_2 \rightleftharpoons A_2 + B_1 \tag{a}$$

leads, upon application of the law of mass action, to the expression:

$$K = \frac{[A_2]\,[B_1]}{[A_1]\,[B_2]} \tag{1}$$

where the constant K depends on the temperature and the nature of the solvent. This expression is strictly valid only for extremely dilute solutions: when ions are present the electrostatic forces between them have appreciable effects on the properties of their solutions, and deviations are apparent from ideal laws (which are assumed in the derivation of the mass-action law by thermodynamic or kinetic methods); the deviations from the ideal laws are usually expressed in terms of activities or activity coefficients For our purpose, the deviations due to interionic attractions and ionic activities will be regarded as small for small ionic concentrations and the equations will be regarded as holding in the same form at higher concentrations, provided that the total ionic concentration does not vary much in a given set of experiments.

To use the above expression for measuring the *strength of an acid*, a standard acid–base pair, say A_2–B_2, must be chosen, and it is usually convenient to refer acid–base strength to the *solvent*. In water the acid–base pair H_3O^+–H_2O is taken as the standard. The equilibrium defining acids is therefore:

$$A + H_2O \rightleftharpoons B + H_3O^+ \tag{b}$$

and the constant

$$K' = \frac{[B][H_3O^+]}{[A][H_2O]} \tag{2}$$

gives the strength of A, that of the ion H_3O^+ being taken as unity. Equation (b) represents what is usually described as the dissociation of the acid A in water, and the constant K' is closely related to the dissociation constant of A in water as usually defined and differing only in the inclusion of the term $[H_2O]$ in the denominator. The latter term represents the 'concentration' of water molecules in liquid water (55.5 mols per cubic decimetre on the ordinary volume concentration scale). When dealing with dilute solutions, the value of $[H_2O]$ may be regarded as constant, and equation (2) may be expressed as:

$$K_a = \frac{[B][H^+]}{[A]} \tag{3}$$

by writing H^+ for H_3O^+ and remembering that the hydrated proton is meant. This equation defines the strength of the acid A. If A is an uncharged molecule (e.g., a weak organic acid), B is the anion derived from it by the loss of a proton, and (3) is the usual expression for the ionisation constant. If A is an anion such as $H_2PO_4^-$, the dissociation constant $[HPO_4^{2-}][H^+]/[H_2PO_4^-]$ is usually referred to as the second dissociation constant of phosphoric acid. If A is a cation acid, for example the ammonium ion, which interacts with water as shown by the equation

$$NH_4^+ + H_2O \rightleftharpoons NH_3 + H_3O^+$$

the acid strength is given by $[NH_3][H^+]/[NH_4^+]$.

On the above basis it is, in principle, unnecessary to treat the strength of bases separately from acids, since any protolytic reaction involving an acid must also involve its conjugate base. The basic properties of ammonia and various amines in water are readily understood on the Brønsted concept.

$$H_2O = H^+ + OH^-$$
$$\frac{NH_3 + H^+ = NH_4^+}{NH_3 + H_2O \rightleftharpoons NH_4^+ + OH^-}$$

The basic dissociation constant K_b is given by:

$$K_b = \frac{[NH_4^+][OH^-]}{[NH_3]} \tag{4}$$

where $[NH_3]$ represents the total concentration of ammonia, irrespective of whether it is present as free NH_3 or as NH_4OH; no reliable evidence is available as to the actual existence of NH_4OH. Since $[H^+][OH^-] = K_w$ (the ionic product of water), we have

$$K_b = K_w/K_a$$

The values of K_a and K_b for different acids and bases vary through many powers of ten. It is often convenient to use the dissociation constant exponent pK defined by

$$pK = \log_{10} 1/K = -\log_{10} K$$

the larger the pK_a value the weaker the acid and the stronger the base.

21

For very weak or slightly ionised electrolytes, the expression $\alpha^2/(1-\alpha)V = K$ reduces to $\alpha^2 = KV$ or $\alpha = \sqrt{KV}$, since α may be neglected in comparison with unity. Hence for any two weak acids or bases at a given dilution V (in dm^3), we have $\alpha_1 = \sqrt{K_1 V}$ and $\alpha_2 = \sqrt{K_2 V}$, or $\alpha_1/\alpha_2 = \sqrt{K_1}/\sqrt{K_2}$. Expressed in words, for any two weak or slightly dissociated electrolytes at equal dilutions, the degrees of dissociation are proportional to the square roots of their ionisation constants. Some values for the dissociation constants at 25 °C for weak acids and bases are collected in Appendix XI.

II, 6.   DISSOCIATION OF POLYPROTIC (POLYBASIC) ACIDS.

When a polyprotic acid is dissolved in water, the various hydrogen atoms undergo ionisation to different extents. For a **diprotic acid** H_2A, the primary and secondary dissociations can be represented by the equations:

$$H_2A \rightleftharpoons H^+ + HA^- \tag{a}$$
$$HA^- \rightleftharpoons H^+ + A^{2-} \tag{b}$$

If the acid is a weak electrolyte, the law of mass action may be applied, and the following expressions obtained:

$$[H^+] \times [HA^-]/[H_2A] = K_1 \tag{1}$$
$$[H^+] \times [A^{2-}]/[HA^-] = K_2 \tag{2}$$

K_1 and K_2 are known as the **primary** and **secondary dissociation constants** respectively. Each stage of the dissociation process has its own ionisation constant, and the magnitudes of these constants give a measure of the extent to which each ionisation has proceeded at any given concentration. The greater the value of K_1 relative to K_2, the smaller will be the secondary dissociation, and the greater must be the dilution before the latter becomes appreciable. It is thereore possible that a diprotic (or polyprotic) acid may behave, so far as dissociation is concerned, as a monoprotic acid. This is indeed characteristic of many polyprotic acids.

A **triprotic acid** H_3A (e.g., orthophosphoric acid) will similarly yield three dissociation constants, K_1, K_2, and K_3, which may be computed in an analogous manner:

$$H_3A \rightleftharpoons H^+ + H_2A^- \tag{c}$$
$$H_2A^- \rightleftharpoons H^+ + HA^{2-} \tag{d}$$
$$HA^{2-} \rightleftharpoons H^+ + A^{3-} \tag{e}$$

We can now apply some of the theoretical considerations to actual examples encountered in analysis.

Example 1.   To calculate the concentrations of HS^- and S^{2-} in a solution of hydrogen sulphide.

A saturated aqueous solution of hydrogen sulphide at 25 °C, at atmospheric pressure, is approximately $0.1M$, and for H_2S the primary and secondary dissociation constants may be taken as 1.0×10^{-7} mol dm^{-3} and 1×10^{-14} mol dm^{-3} respectively.

In the solution the following equilibria are involved:

$$H_2S + H_2O \rightleftharpoons HS^- + H_3O^+; \quad K_1 = [H^+][HS^-]/[H_2S] \tag{e}$$
$$HS^- + H_2O \rightleftharpoons S^{2-} + H_3O^+; \quad K_2 = [H^+][S^{2-}]/[HS^-] \tag{f}$$
$$H_2O \rightleftharpoons H^+ + OH^- \tag{g}$$

Electroneutrality requires that the total cation concentration must equal total anion concentration and hence, taking account of valencies,

$$[H^+] = [HS^-] + 2[S^{2-}] + [OH^-] \qquad (h)$$

but since in fact we are dealing with an acid solution, $[H^+] > 10^{-7} > [OH^-]$ and we can simplify equation (h) to read

$$[H^+] = [HS^-] + 2[S^{2-}] \qquad (j)$$

The 0.1 mol H_2S is present partly as undissociated H_2S and partly as the ions HS^- and S^{2-}, and it follows that

$$[H_2S] + [HS^-] + [S^{2-}] = 0.1 \qquad (k)$$

The very small value of K_2 indicates that the secondary dissociation and therefore $[S^{2-}]$ is extremely minute, and ignoring $[S^{2-}]$ in equation (j) we are left with the result

$$[H^+] \approx [HS^-] \qquad (l)$$

Since K_1 is also small, $[H^+] \ll [H_2S]$ and so equation (k) can be reduced to

$$[H_2S] \approx 0.1 \qquad (m)$$

Using these results in the expression for K_1 we find

$$[H^+]^2/0.1 = 1 \times 10^{-7}; \quad [H^+] = [HS^-] = 1.0 \times 10^{-4} \text{ mol dm}^{-3}.$$

From equation (f) it then follows that

$$(1.0 \times 10^{-4})[S^{2-}]/(1.0 \times 10^{-4}) = 1 \times 10^{-14}$$

and
$$[S^{2-}] = 1 \times 10^{-14} \text{ mol dm}^{-3}.$$

If we multiply K_1 by K_2 we find $[S^{2-}] = 1 \times 10^{-21}/[H^+]^2$.

Thus the concentration of the sulphide ion is inversely proportional to the square of the hydrogen-ion concentration, i.e., if we, say, double $[H^+]$ by the addition of a strong acid, the $[S^{2-}]$ will be reduced to $\frac{1}{2}^2$ or $\frac{1}{4}$ of its original value.

II, 7. COMMON ION EFFECT. The concentration of a particular ion in an ionic reaction can be increased by the addition of a compound which produces that ion upon dissociation. The particular ion is thus derived from the compound already in solution and also from the added reagent, hence the name **common ion**. We shall confine our attention to the case in which the original compound is a weak electrolyte in order that the law of mass action may be applicable. The result is usually that there is a higher concentration of this ion in solution than that derived from the original compound alone, and new equilibrium conditions will be produced. Examples of the calculation of the common ion effect are given below. In general, it may be stated that if the total concentration of the common ion is only slightly greater than that which the original compound alone would furnish, the effect is small; if, however, the concentration of the common ion is very much increased (e.g., by the addition of a completely dissociated salt), the effect is very great, and may be of considerable practical importance. Indeed, the common ion effect provides a valuable method for controlling the concentration of the ions furnished by a weak electrolyte.

Example 2. To calculate the sulphide-ion concentration in a 0.25M-hydrochloric acid solution saturated with hydrogen sulphide.

This concentration has been chosen since it is that at which the sulphides of certain heavy metals are precipitated. The total concentration of hydrogen sulphide may be assumed to be approximately the same as in aqueous solution, i.e., $0.1M$; the $[H^+]$ will be equal to that of the completely dissociated HCl, i.e., $0.25M$, but the $[S^{2-}]$ will be reduced below 1×10^{-14}.

Substituting in equations (a) and (b) (Section **II, 6**), we find:

$$[HS^-] = \frac{K_1 \times [H_2S]}{[H^+]} = \frac{1.0 \times 10^{-7} \times 0.1}{0.25} = 4.0 \times 10^{-8} \text{ mol dm}^{-3}$$

$$[S^{2-}] = \frac{K_2 \times [HS^-]}{[H^+]} = \frac{(1 \times 10^{-14}) \times (4 \times 10^{-8})}{0.25} = 1.6 \times 10^{-21} \text{ mol dm}^{-3}$$

Thus by changing the acidity from $1.0 \times 10^{-4}M$ (that present in saturated H_2S water) to $0.25M$, the sulphide-ion concentration is reduced from 1×10^{-14} to 1.6×10^{-21}.

Example 3.   What effect has the addition of 0.1 mol of anhydrous sodium acetate to 1 dm^3 of $0.1M$-acetic acid upon the degree of dissociation of the acid?

The dissociation constant of acetic acid at 25 $°C$ is 1.82×10^{-5} mol dm^{-3} and the degree of ionisation α in $0.1M$ solution may be computed by solving the quadratic equation:

$$\frac{[H^+] \times [C_2H_3O_2{}^-]}{[H \cdot C_2H_3O_2]} = \frac{\alpha^2 c}{(1-\alpha)} = 1.82 \times 10^{-5}.$$

For our purpose it is sufficiently accurate to neglect α in $(1-\alpha)$ since α is small:

$$\therefore \alpha = \sqrt{K/c} = \sqrt{1.82 \times 10^{-4}} = 0.0135$$

Hence in $0.1M$-acetic acid,

$$[H^+] = 0.00135, [C_2H_3O_2{}^-] = 0.00135, \text{ and } [H \cdot C_2H_3O_2] = 0.0986 \text{ mol dm}^{-3}.$$

The concentrations of sodium and acetate ions produced by the addition of the completely dissociated sodium acetate are:

$$[Na^+] = 0.1, \text{ and } [C_2H_3O_2{}^-] = 0.1 \text{ mol dm}^{-3} \text{ respectively.}$$

The acetate ions from the salt will tend to decrease the ionisation of the acetic acid, and consequently the acetate-ion concentration derived from it. Hence we may write $[C_2H_3O_2{}^-] = 0.1$ for the solution, and if α' is the new degree of ionisation, $[H^+] = \alpha'c = 0.1\alpha'$, and $[H \cdot C_2H_3O_2] = (1-\alpha')c = 0.1$, since α' is negligibly small.

Substituting in the mass-action equation:

$$\frac{[H^+] \times [C_2H_3O_2{}^-]}{[H \cdot C_2H_3O_2]} = \frac{0.1\alpha' \times 0.1}{0.1} = 1.82 \times 10^{-5}$$

or $\qquad\qquad \alpha' = 1.8 \times 10^{-4}$

$$[H^+] = \alpha'c = 1.8 \times 10^{-5} \text{ mol dm}^{-3}.$$

The addition of a tenth of a mole of sodium acetate to a $0.1M$ solution of acetic acid has decreased the degree of ionisation from 1.35 to 0.018 per cent, and the hydrogen-ion concentration from 0.00135 to 0.000018 mol dm^{-3}.

Example 4.   What effect has the addition of 0.5 mol of ammonium chloride to 1 dm^3 of 0.1M-aqueous ammonia solution upon the degree of dissociation of the base?

(Dissociation constant of NH_3 in water $= 1.8 \times 10^{-5}$ mol dm^{-3})

In 0.1M-ammonia solution $\alpha = \sqrt{1.8 \times 10^{-5}/0.1} = 0.0135$. Hence $[OH^-] = 0.00135$, $[NH_4{}^+] = 0.00135$, and $[NH_3] = 0.0986$ mol dm^{-3}. Let α' be the degree of ionisation in the presence of the added ammonium chloride. Then $[OH^-] = \alpha'c = 0.1\alpha'$, and $[NH_3] = (1-\alpha')c = 0.1$, since α' may be taken as negligibly small. The addition of the completely ionised ammonium chloride will, of necessity, decrease the $[NH_4{}^+]$ derived from the base and increase $[NH_3]$, and as a first approximation $[NH_4{}^+] = 0.5$.

Substituting in the equation:

$$\frac{[NH_4{}^+] \times [OH^-]}{[NH_3]} = \frac{0.5 \times 0.1\alpha'}{0.1} = 1.8 \times 10^{-5}$$

$\alpha' = 3.6 \times 10^{-5}$ and $[OH^-] = 3.6 \times 10^{-6}$ mol dm^{-3}.

The addition of half a mole of ammonium chloride to a 0.1M solution of aqueous ammonia has decreased the degree of ionisation from 1.35 to 0.0036 per cent, and the hydroxide-ion concentration from 0.00135 to 0.0000036 mol dm^{-3}.

II, 8. SOLUBILITY PRODUCT. For sparingly soluble salts (i.e., those of which the solubility is less than 0.01 mol per dm^3) it is an experimental fact that the product of the total molecular concentrations of the ions is a constant at constant temperature. This product K_S is termed the **solubility product.** For a binary electrolyte:

$$AB \rightleftharpoons A^+ + B^-$$

$$K_{s(AB)} = [A^+] \times [B^-]$$

In general, for an electrolyte A_pB_q, which ionises into pA^{q+} and qB^{p-} ions:

$$A_pB_q \rightleftharpoons pA^{q+} + qB^{p-}$$

$$K_{s(A_pB_q)} = [A^{q+}]^p \times [B^{p-}]^q$$

A plausible deduction of the solubility product relation is the following. When excess of a sparingly soluble electrolyte, say silver chloride, is shaken up with water, some of it passes into solution to form a saturated solution of the salt and the reaction appears to cease. The following equilibrium is actually present (the silver chloride is completely ionised in solution):

$$AgCl \text{ (solid)} \rightleftharpoons Ag^+ + Cl^-$$

The velocity of the forward reaction depends only upon the temperature and at any given temperature:

$$v_1 = k_1$$

where k_1 is a constant. The velocity of the reverse reaction is proportional to the activity of each of the reactants; hence at any given temperature:

$$v_2 = k_2 \times a_{Ag^+} \times a_{Cl^-}$$

where k_2 is another constant. At equilibrium the two velocities are equal, i.e.

$$k_1 = k_2 \times a_{Ag^+} \times a_{Cl^-}$$

or $a_{Ag^+} \times a_{Cl^-} = k_1/k_2 = K_{s(AgCl)}$

In the very dilute solutions with which we are concerned, the activities may be taken as practically equal to the concentrations so that $[Ag^+] \times [Cl^-]$ = const.

It is important to note that the solubility product relation applies with sufficient accuracy for purposes of quantitative analysis only to saturated solutions of slightly soluble electrolytes and with *small* additions of other salts. In the presence of moderate concentrations of salts, the ionic concentration, and therefore the ionic strength of the solution, will increase. This will, in general, lower the activity coefficients of both ions, and consequently the ionic concentrations (and therefore the solubility) must increase in order to maintain the solubility product constant. This effect, which is most marked when the added electrolyte does not possess an ion in common with the sparingly soluble salt, we may term the **salt effect**. It can be shown on the basis of the Debye-Hückel–Onsager theory that for aqueous solutions at 25 °C:

$$\log y_i = -\frac{0.505z_i^2 . I^{0.5}}{1 + 3.3 \times 10^7 a . I^{0.5}}$$

where y_i is the activity coefficient of the ion, z_i is the valency of the ion concerned, I is the ionic strength of the solution (Section **II, 3**), and a is the average 'effective diameter' of all the ions in the solution. For very dilute solutions ($I^{0.5} < 0.1$) the second term of the denominator is negligible and the equation reduces to:

$$\log y_i = -0.505z_i^2 . I^{0.5}$$

For more concentrated solutions ($I^{0.5} > 0.3$) an additional term BI is added to the equation; B is an empirical constant. For a more detailed treatment of the influence of salts upon solubility and solubility product, the reader is referred to textbooks of electrochemistry.

It will be clear from the above short discussion that two factors may come into play when a solution of a salt containing a common ion is added to a saturated solution of a slightly soluble salt. At moderate concentrations of the added salt, the solubility will generally decrease, but with higher concentrations of the soluble salt, when the ionic strength of the solution increases considerably and the activity coefficients of the ions decrease, the solubility may actually increase. This is one of the reasons why a very large excess of the precipitating agent is avoided in quantitative analysis.

A few examples may help the reader to fully understand the subject. The concentrations are expressed in mols per dm^3 for the calculation of solubility products.

Example 5.   The solubility of silver chloride is 0.0015 g per dm^3. Calculate the solubility product.

The molecular weight of silver chloride is 143.5. The solubility is therefore $0.0015/143.5 = 1.05 \times 10^{-5}$ mol per dm^3. In a saturated solution, 1 mole of AgCl will give 1 mole each of Ag^+ and Cl^-. Hence $[Ag^+] = 1.05 \times 10^{-5}$ and $[Cl^-] = 1.05 \times 10^{-5}$ mol dm^{-3}.

$K_{s(AgCl)} = [Ag^+] \times [Cl^-] = (1.05 \times 10^{-5}) \times (1.05 \times 10^{-5}) = 1.1 \times 10^{-10}$ mol^2 dm^{-6}.

Example 6.   Calculate the solubility product of silver chromate, given that its solubility is 2.5×10^{-2} g per dm^3.

$$Ag_2CrO_4 \rightleftharpoons 2Ag^+ + CrO_4^{--}$$

The molecular weight of Ag_2CrO_4 is 332, hence the solubility $= 2.5 \times 10^{-2}/332$ $= 7.5 \times 10^{-5}$ mol dm^{-3}.

Now 1 mole of Ag_2CrO_4 gives 2 moles of Ag^+ and 1 mole of CrO_4^{2-}; therefore

$$
\begin{aligned}
K_{s(Ag_2CrO_4)} &= [Ag^+]^2 \times [CrO_4^{2-}] \\
&= (2 \times 7.5 \times 10^{-5})^2 \times (7.5 \times 10^{-5}) \\
&= 1.7 \times 10^{-12} \text{ mol}^3 \text{ dm}^{-9}.
\end{aligned}
$$

Example 7.   The solubility product of magnesium hydroxide is 3.4×10^{-11} mol^3 dm^{-9}. Calculate its solubility in grams per dm^3.

$$Mg(OH)_2 \rightleftharpoons Mg^{2+} + 2OH^-$$

$$[Mg^{2+}] \times [OH^-]^2 = 3.4 \times 10^{-11}$$

The molecular weight of magnesium hydroxide is 58. Each mole of magnesium hydroxide, when dissolved, yields 1 mole of magnesium ions and 2 moles of hydroxyl ions. If the solubility is x mol dm^{-3}, $[Mg^{2+}] = x$ and $[OH^-] = 2x$. Substituting these values in the solubility product expression:

$$x \times (2x)^2 = 3.4 \times 10^{-11}$$
or
$$x = 2.0 \times 10^{-4} \text{ mol dm}^{-3}$$
$$= 2.0 \times 10^{-4} \times 58 = 1.2 \times 10^{-2} \text{ g dm}^{-3}$$

The great importance of the solubility product concept lies in its bearing upon precipitation from solution, which is, of course, one of the principal operations of quantitative analysis. The solubility product is the ultimate value which is attained by the ionic concentration product when equilibrium has been established between the solid phase of a difficultly soluble salt and the solution. If the experimental conditions are such that the ionic concentration product is different from the solubility product, then the system will attempt to adjust itself in such a manner that the ionic and solubility products are equal in value. Thus, if, for a given electrolyte, the product of the concentrations of the ions in solution is arbitrarily made to exceed the solubility product, as, for example, by the addition of a salt with a common ion, the adjustment of the system to equilibrium results in precipitation of the solid salt, provided supersaturation conditions are excluded. If the ionic concentration product is less than the solubility product or can arbitrarily be made so, as, for example, by complex salt formation or by the formation of weak electrolytes, then a further quantity of solute can pass into solution until the solubility product is attained, or, if this is not possible, until all the solute has dissolved.

II, 9. QUANTITATIVE EFFECTS OF A COMMON ION.   An important application of the solubility product principle is to the calculation of the solubility of sparingly soluble salts in solutions of salts with a common ion. Thus the solubility of a salt MA in the presence of a relatively large amount of the common M^+ ions,* supplied by a second salt MB, follows from the definition of solubility products:

$$[M^+] \times [A^-] = K_{s(MA)}$$
$$\text{or} \qquad [A^-] = K_{s(MA)}/[M^+] \qquad\qquad (1)$$

The solubility of the salt is represented by the $[A^-]$ which it furnishes in solution. It is clear that the addition of a common ion will *decrease* the solubility of the salt.

Example 8.   Calculate the solubility of silver chloride in (a) $0.001 M$- and (b) $0.01 M$-sodium chloride solutions respectively ($K_{s(AgCl)} = 1.1 \times 10^{-10}$ mol^2 dm^{-6}).

In a saturated solution of silver chloride $[Cl^-] = \sqrt{1.1 \times 10^{-10}} = 1.05 \times 10^{-5}$ mol dm^{-3}; this may be neglected in comparison with the excess of Cl^- ions added.

For (a) $[Cl^-] = 1 \times 10^{-3}, [Ag^+] = 1.1 \times 10^{-10}/1 \times 10^{-3}$
$\qquad\qquad = 1.1 \times 10^{-7}$ mol dm^{-3}

For (b) $[Cl^-] = 1 \times 10^{-2}, [Ag^+] = 1.1 \times 10^{-10}/1 \times 10^{-2}$
$\qquad\qquad = 1.1 \times 10^{-8}$ mol dm^{-3}

Thus the solubility is decreased 100 times in $0.001 M$-sodium chloride and 1000 times in $0.01 M$-sodium chloride. Similar results are obtained for $0.001 M$- and $0.01 M$-silver nitrate solution.

Example 9.   Calculate the solubilities of silver chromate in 0.001 *M*- and 0.01 *M*- silver nitrate solutions, and in 0.001 *M*- and 0.01 *M*-potassium chromate solutions (Ag_2CrO_4: $K_s = 1.7 \times 10^{-12}$ mol^3 dm^{-9}, solubility in water = 7.5×10^{-5} mol dm^{-3}).

$$[Ag^+]^2 \times [CrO_4{}^{2-}] = 1.7 \times 10^{-12}$$
$$\text{or} \qquad [CrO_4{}^{2-}] = 1.7 \times 10^{-12}/[Ag^+]^2$$

For $0.001 M$-silver nitrate solution: $[Ag^+] = 1 \times 10^{-3}$

$$\therefore [CrO_4{}^{2-}] = 1.7 \times 10^{-12}/1 \times 10^{-6} = 1.7 \times 10^{-6} \text{mol dm}^{-3}.$$

For $0.01 M$-silver nitrate solution: $[Ag^+] = 1 \times 10^{-2}$

$$[CrO_4{}^{2-}] = 1.7 \times 10^{-12}/1 \times 10^{-4} = 1.7 \times 10^{-8} \text{mol dm}^{-3}.$$

The solubility product equation gives:

$$[Ag^+] = \sqrt{1.7 \times 10^{-12}/[CrO_4{}^{2-}]}$$
For $[CrO_4{}^{2-}] = 0.001, [Ag^+] = \sqrt{1.7 \times 10^{-12}/1 \times 10^{-3}}$
$\qquad\qquad = 4.1 \times 10^{-5}$ mol dm^{-3}.
For $[CrO_4{}^{2-}] = 0.01, [Ag^+] = \sqrt{1.7 \times 10^{-12}/1 \times 10^{-2}}$
$\qquad\qquad = 1.3 \times 10^{-5}$ mol dm^{-3}.

This decrease in solubility by the common-ion effect is of fundamental importance in gravimetric analysis. By the addition of a suitable excess of a precipitating agent, the solubility of a precipitate is usually decreased to so small a value that the loss from solubility influences is negligible. Let us consider a specific case—the determination of silver as silver chloride. Here the chloride

---

* This enables us to neglect the concentration of M^+ ions supplied by the sparingly soluble salt itself, and thus to simplify the calculation.

solution is added to the solution of the silver salt. If an exactly equivalent amount is added, the resultant saturated solution of silver chloride will contain 0.0015 g per dm^3 (*Example* 5). If 0.2 g of silver chloride is produced and the volume of the solution and washings is 500 cm^3, the loss, owing to solubility, will be 0.00075 g or 0.38 per cent of the weight of the salt; the analysis would then be 0.38 per cent too low. By using an excess of the precipitant, say, to a concentration of 0.01M, the solubility of the silver chloride is reduced to 1.5×10^{-5} g per dm^3 (*Example* 8), and the loss will be $1.5 \times 10^{-5} \times 0.5 \times 100/0.2 = 0.0038$ per cent. Silver chloride is therefore very suitable for the quantitative determination of silver with high accuracy.

It should, however, be noted that as the concentration of the excess of precipitant increases, so too does the ionic strength of the solution. This leads to a decrease in activity coefficient values with the result that to maintain the value of K_s *more* of the precipitate will dissolve. In other words there is a limit to the amount of precipitant which can be safely added in excess. Also, addition of excess precipitant may sometimes result in the formation of soluble complexes causing some precipitate to dissolve.

II, 10. FRACTIONAL PRECIPITATION. We have thus far considered the solubility product principle in connection with the precipitation of one sparingly soluble salt. We shall now extend our studies to the case where two slightly soluble salts may be formed. For simplicity, we shall study the situation which arises when a precipitating agent is added to a solution containing two anions, both of which form slightly soluble salts with the same cation, e.g., when silver nitrate solution is added to a solution containing both chloride and iodide ions. The questions which arise are: which salt will be precipitated first, and how completely will the first salt be precipitated before the second ion begins to react with the reagent?

The solubility products of silver chloride and silver iodide are respectively 1.2×10^{-10} mol^2 dm^{-6} and 1.7×10^{-16} mol^2 dm^{-6}; i.e.,

$$[Ag^+] \times [Cl^-] = 1.2 \times 10^{-10} \tag{1}$$
$$[Ag^+] \times [I^-] = 1.7 \times 10^{-16} \tag{2}$$

It is evident that silver iodide, being less soluble, will be precipitated first since its solubility product will be first exceeded. Silver chloride will be precipitated when the Ag^+ ion concentration is greater than

$$\frac{K_{s(AgCl)}}{[Cl^-]} = \frac{1.2 \times 10^{-10}}{[Cl^-]}$$

and then both salts will be precipitated simultaneously. When silver chloride commences to precipitate, silver ions will be in equilibrium with both salts, and equations (1) and (2) will be simultaneously satisfied, or

$$[Ag^+] = \frac{K_{s(AgI)}}{[I^-]} = \frac{K_{s(AgCl)}}{[Cl^-]} \tag{3}$$

$$\text{and} \quad \frac{[I^-]}{[Cl^-]} = \frac{K_{s(AgI)}}{K_{s(AgCl)}} = \frac{1.7 \times 10^{-16}}{1.2 \times 10^{-10}} = 1.4 \times 10^{-6} \tag{4}$$

Hence when the concentration of the iodide ion is about one millionth part of the chloride-ion concentration, silver chloride will be precipitated. If the initial concentration of both chloride and iodide ions is $0.1 M$, then silver chloride will be precipitated when

$$[I^-] = 0.1 \times 1.4 \times 10^{-6} = 1.4 \times 10^{-7} M = 1.8 \times 10^{-5} \text{ g dm}^{-3}.$$

Thus an almost complete separation is theoretically possible. The separation is feasible in practice if the point at which the iodide precipitation is complete can be detected. This may be done: (a) by the use of an adsorption indicator (see Section X, 30C), or (b) by a potentiometric method with a silver electrode (see Chapter XIV).

For a mixture of bromide and iodide:

$$\frac{[I^-]}{[Br^-]} = \frac{K_{s(AgI)}}{K_{s(AgBr)}} = \frac{1.7 \times 10^{-16}}{3.5 \times 10^{-13}} = \frac{1}{2.0 \times 10^3}$$

Precipitation of silver bromide will occur when the concentration of the bromide ion in the solution is 2.0×10^3 times that of the iodide concentration. The separation is therefore not quite so complete as in the case of chloride and iodide, but can nevertheless be effected with fair accuracy with the aid of adsorption indicators (Section X, 30C).

II, 11. COMPLEX IONS. The increase in solubility of a precipitate upon the addition of excess of the precipitating agent is frequently due to the formation of a complex ion. A **complex ion** is formed by the union of a simple ion with either other ions of opposite charge or with neutral molecules. Let us examine a few examples in detail.

When potassium cyanide solution is added to a solution of silver nitrate, a white precipitate of silver cyanide is first formed because the solubility product of silver cyanide:

$$[Ag^+] \times [CN^-] = K_{s(AgCN)} \tag{1}$$

is exceeded. The reaction is expressed:

$$CN^- + Ag^+ = AgCN$$

The precipitate dissolves upon the addition of excess of potassium cyanide, the complex ion $[Ag(CN)_2]^-$ being produced:

$AgCN \text{ (solid)} + CN^- \text{ (excess)} \rightleftharpoons [Ag(CN)_2]^-$ *
(or $AgCN + KCN = K[Ag(CN)_2]$—a soluble complex salt)

This complex ion dissociates to give silver ions, since the addition of sulphide ions yields a precipitate of silver sulphide (solubility product $1.6 \times 10^{-49} \text{ mol}^3 \text{ dm}^{-9}$), and also silver is deposited from the complex cyanide solution upon electrolysis. The complex ion thus dissociates in accordance with the equation:

$$[Ag(CN)_2]^- \rightleftharpoons Ag^+ + 2CN^-$$

---

* Square brackets are commonly used for two purposes: to denote concentrations and also to include the whole of a complexion; for the latter purpose curly brackets (braces) are sometimes used. With careful scrutiny there should be no confusion regarding the sense in which the square brackets are used: with complexes there will be no charge signs *inside* the brackets.

Applying the law of mass action, we obtain the dissociation constant of the complex ion:

$$\frac{[Ag^+] \times [CN^-]^2}{[\{Ag(CN)_2\}^-]} = K_{diss.} \tag{2}$$

which has a value of 1.0×10^{-21} mol^2 dm^{-6} at the ordinary temperature. By inspection of this expression, and bearing in mind that excess of cyanide ion is present, it is evident that the silver ion concentration must be very small, so small in fact that the solubility product of silver cyanide is not exceeded.

The inverse of equation (2) gives us the **stability constant** or **formation constant** of the complex ion:

$$K = \frac{[\{Ag(CN)_2^-\}]}{[Ag^+] \times [CN^-]^2} = 10^{21} \ mol^{-2} \ dm^6 \tag{3}$$

Consider now a somewhat different type of complex-ion formation, viz., the production of a complex ion with constituents other than the common ion present in the solution. This is exemplified by the solubility of silver chloride in ammonia solution. The reaction is:

$$AgCl + 2NH_3 \rightleftharpoons [Ag(NH_3)_2]^+ + Cl^-$$

Here again, electrolysis, or treatment with hydrogen sulphide, shows that silver ions are present in solution. The dissociation of the complex ion is represented by:

$$[Ag(NH_3)_2]^+ \rightleftharpoons Ag^+ + 2NH_3$$

and the dissociation constant is given by:

$$K_{diss.} = \frac{[Ag^+] \times [NH_3]^2}{[\{Ag(NH_3)_2\}^+]} = 6.8 \times 10^{-8} \ mol^2 dm^{-6}.$$

The stability constant $K = 1/K_{diss.} = 1.5 \times 10^7 \ mol^{-2} dm^6.$

The magnitude of the dissociation constant clearly shows that only a very small silver ion concentration is produced by the dissociation of the complex ion.

The stability of complex ions varies within very wide limits. It is quantitatively expressed by means of the **stability constant**. The more stable the complex, the greater is the stability constant, i.e., the smaller is the tendency of the complex ion to dissociate into its constituent ions. When the complex ion is very stable, e.g., the hexacyanoferrate(II) ion $[Fe(CN)_6]^{4-}$, the ordinary ionic reactions of the components are not shown.

The application of complex-ion formation in chemical separations depends upon the fact that one component may be transformed into a complex ion which is no longer precipitable with the precipitating agent, whereas another component is precipitated. One example may be mentioned here. This is concerned with the separation of cadmium and copper. Excess of potassium cyanide solution is added to the solution containing the two salts when the complex ions $[Cd(CN)_4]^{2-}$ and $[Cu(CN)_4]^{3-}$ respectively are formed. Upon passing hydrogen sulphide into the solution containing excess of CN^- ions, a precipitate of cadmium sulphide is produced. Despite the higher solubility product of CdS (1.4×10^{-28} mol^2 dm^{-6} as against 6.5×10^{-45} mol^2 dm^{-6} for copper sulphide), the former is precipitated because the complex cyanocuprate(I) ion has a greater stability constant (2×10^{27} mol^{-4} dm^{12} as compared with 7 $\times 10^{10}$ mol^{-4} dm^{12} for the cadmium compound).

For a further discussion of complex ions and stability constants, see Chapter X (Sections **19–21**).

II, 12. EFFECT OF ACIDS UPON THE SOLUBILITY OF A PRE-CIPITATE.

For sparingly soluble salts of a strong acid the effect of the addition of an acid will be similar to that of any other indifferent electrolyte but if the sparingly soluble salt MA is the salt of a weak acid HA, then acids will, in general, have a solvent effect upon it. Let us suppose that hydrochloric acid is added to an aqueous suspension of such a salt. The following equilibrium will be established:

$$M^+ + A^- + H^+ \rightleftharpoons HA + M^+$$

If the dissociation constant of the acid HA is very small, the anion A^- will be removed from the solution to form the undissociated acid HA. Consequently more of the salt will pass into solution to replace the anions removed in this way, and this process will continue until equilibrium is established (i.e., until $[M^+] \times [A^-]$ has become equal to the solubility product of MA) or, if sufficient hydrochloric acid is present, until the sparingly soluble salt has dissolved completely. Similar reasoning may be applied to salts of acids, such as phosphoric acid ($K_1 = 7.5 \times 10^{-3}$ mol dm^{-3}; $K_2 = 6.2 \times 10^{-8}$ mol dm^{-3}; $K_3 = 5 \times 10^{-13}$ mol dm^{-3}), oxalic acid ($K_1 = 5.9 \times 10^{-2}$ mol dm^{-3}; $K_2 = 6.4 \times 10^{-5}$ mol dm^{-3}), and arsenic acid. Thus the solubility of, say, silver orthophosphate in dilute nitric acid is due to the removal of the $PO_4{}^{3-}$ ion as $HPO_4{}^{2-}$ and/or $H_2PO_4{}^-$:

$$PO_4{}^{3-} + H^+ \rightleftharpoons HPO_4{}^{2-}$$
$$HPO_4{}^{2-} + H^+ \rightleftharpoons H_2PO_4{}^-$$

With salts of weak acids, such as carbonic ($K_1 = 4.3 \times 10^{-7}$ mol dm^{-3}; $K_2 = 5.6 \times 10^{-11}$ mol dm^{-3}), sulphurous ($K_1 = 1.7 \times 10^{-2}$ mol dm^{-3}; $K_2 = 1.0 \times 10^{-7}$ mol dm^{-3}), and nitrous ($K_1 = 4.6 \times 10^{-4}$ mol dm^{-3}) acids, an additional factor contributing to the increased solubility is the actual disappearance of the acid from solution either spontaneously or on gentle warming. An explanation is thus provided for the well-known solubility of the sparingly soluble sulphites, carbonates, oxalates, phosphates, arsenites, arsenates, cyanides (with the exception of silver cyanide, which is actually a salt of the strong acid $H[Ag(CN)_2]$), fluorides, acetates, and salts of other organic acids in strong acids.

The sparingly soluble sulphates (e.g., those of barium, strontium, and lead) also exhibit increased solubility in acids as a consequence of the weakness of the second stage of ionisation of sulphuric acid ($K_2 = 1.2 \times 10^{-2}$ mol dm^{-3}):

$$SO_4{}^{2-} + H^+ \rightleftharpoons HSO_4{}^-$$

Since, however, K_2 is comparatively large, the solvent effect is relatively small; this is why, in the quantitative separation of barium sulphate, precipitation may be carried out in slightly acid solution in order to obtain a more easily filterable precipitate and to reduce coprecipitation (Section **XI, 5**).

The precipitation of substances within a controlled range of pH is discussed in Section **XI, 10**.

II, 13. EFFECT OF TEMPERATURE UPON THE SOLUBILITY OF A PRECIPITATE. The solubility of the precipitates encountered in quantitative analysis increases with rise of temperature. With some substances the influence of temperature is small, but with others it is quite appreciable. Thus the solubility of silver chloride at 10 and 100 °C is 1.72 and 21.1 mg dm^{-3} respectively, whilst that of barium sulphate at these two temperatures is 2.2 and 3.9 mg dm^{-3} respectively. In many instances, the common-ion effect reduces the solubility to so small a value that the temperatures effect, which is otherwise appreciable, becomes very small. Wherever possible it is advantageous to filter while the solution is hot; the rate of filtration is increased, as is also the solubility of foreign substances, thus rendering their removal from the precipitate more complete. The double phosphates of ammonium with magnesium, maganese or zinc, as well as lead sulphate and silver chloride, are usually filtered at the laboratory temperature to avoid solubility losses.

II, 14. EFFECT OF THE SOLVENT UPON THE SOLUBILITY OF A PRECIPITATE. The solubility of most inorganic compounds is reduced by the addition of organic solvents, such as methanol, ethanol, and propan-1-ol, acetone, etc. For example, the addition of about 20 per cent by volume of ethanol renders the solubility of lead sulphate practically negligible, thus permitting quantitative separation. Similarly calcium sulphate separates quantitatively from 50 per cent ethanol. Other examples of the influence of solvent will be found in Chapter XI.

II, 15. THE IONIC PRODUCT OF WATER. Kohlrausch and Heydweiller (1894) found that the most highly purified water that can be obtained possesses a small but definite conductivity. Water must therefore be slightly ionised in accordance with the equation:

$$H_2O \rightleftharpoons H^+ + OH^- *$$

Applying the law of mass action to this equation, we obtain, for any given temperature:

$$\frac{a_{H^+} \times a_{OH^-}}{a_{H_2O}} = \frac{[H^+].[OH^-]}{[H_2O]} \times \frac{y_{H^+} \cdot y_{OH^-}}{y_{H_2O}} = \text{a constant}$$

Since water is only slightly ionised, the ionic concentrations will be small, and their activity coefficients may be regarded as unity; the activity of the unionised molecules may also be taken as unity. The expression thus becomes:

$$\frac{[H^+] \times [OH^-]}{[H_2O]} = \text{a constant}$$

In pure water or in dilute aqueous solutions, the concentration of the undissociated water may be considered constant. Hence:

$$[H^+] \times [OH^-] = K_w$$

---

* Strictly speaking the hydrogen ion H$^+$ exists in water as the hydroxonium ion H$_3$O$^+$ (Section **II, 1**). The electrolytic dissociation of water should therefore be written:

$$2H_2O \rightleftharpoons H_3O^+ + OH^-$$

For the sake of simplicity, the more familiar symbol H$^+$ will be retained.

where K_w is the **ionic product of water**. It must be pointed out that the assumption that the activity coefficients of the ions are unity and that the activity coefficient of water is constant applies strictly to pure water and to very dilute solutions (ionic strength <0.01); in more concentrated solutions, i.e., in solutions of appreciable ionic strength, the electrical environment affects the activity coefficients of the ions (compare Section **II, 8**) and also the activity of the un-ionised water. The ionic product of water will then not be constant, but will depend upon the ionic environment. It is, however, difficult to determine the activity coefficients, except under specially selected conditions, so that in practice the ionic product K_w, although not strictly constant, is employed.

The ionic product varies with the temperature, but under ordinary experimental conditions (at about 25 °C) its value may be taken as 1×10^{-14} with concentrations expressed in mol dm^{-3}. This is sensibly constant in dilute aqueous solutions. If the product of [H$^+$] and [OH$^-$] in aqueous solution momentarily exceeds this value, the excess ions will immediately combine to form water. Similarly, if the product of the two ionic concentrations is momentarily less than 10^{-14}, more water molecules will dissociate until the equilibrium value is attained.

The hydrogen- and hydroxide-ion concentrations are equal in pure water; therefore $[H^+] = [OH^-] = \sqrt{K_w} = 10^{-7}$ mol dm^{-3} at about 25 °C. A solution in which the hydrogen- and hydroxide-ion concentrations are equal is termed an exactly **neutral solution**. If [H$^+$] is greater than 10^{-7}, the solution is **acid**, and if less than 10^{-7}, the solution is **alkaline** (or **basic**). It follows that at ordinary temperatures [OH$^-$] is greater than 10^{-7} in alkaline solution and less than this value in acid solution.

In all cases the reaction of the solution can be quantitatively expressed by the magnitude of the hydrogen-ion (or hydroxonium-ion) concentration, or, less frequently, of the hydroxide-ion concentration, since the following simple relations between [H$^+$] and [OH$^-$] exist:

$$[H^+] = \frac{K_w}{[OH^-]}, \text{ and } [OH^-] = \frac{K_w}{[H^+]}$$

The variation of K_w with temperature is shown in Table II, 1.

Table II, 1.   Ionic Product of Water at Various Temperatures

Temp. (°C)	$K_w \times 10^{14}$	Temp. (°C)	$K_w \times 10^{14}$
0°	0.12	35°	2.09
5°	0.19	40°	2.92
10°	0.29	45°	4.02
15°	0.45	50°	5.47
20°	0.68	55°	7.30
25°	1.01	60°	9.61
30°	1.47		

II, 16.   THE HYDROGEN-ION EXPONENT. For many purposes, especially when dealing with small concentrations, it is cumbersome to express concentrations of hydrogen and hydroxyl ions in terms of mols per cubic decimetre. A very convenient method was proposed by S. P. L. Sørensen (1909).

He introduced the **hydrogen-ion exponent** pH defined by the relationships:

$$pH = \log_{10} 1/[H^+] = -\log_{10}[H^+], \text{ or } [H^+] = 10^{-pH}$$

The quantity pH is thus the logarithm (to the base 10) of the reciprocal of the hydrogen-ion concentration, or is equal to the logarithm of the hydrogen-ion concentration with negative sign. This method has the advantage that all states of acidity and alkalinity between those of solutions containing on the one hand, 1 mol dm^{-3} of hydrogen ions, and on the other hand, 1 mol dm^{-3} of hydroxide ions, can be expressed by a series of positive numbers between 0 and 14. Thus a neutral solution with $[H^+] = 10^{-7}$ has a pH of 7; a solution with a hydrogen ion concentration of 1 mol dm^{-3} has a pH of 0 ($[H^+] = 10^0$); and a solution with a hydroxide ion concentration of 1 mol dm^{-3} has $[H^+] = K_w/[OH^-] = 10^{-14}/10^0 = 10^{-14}$, and possesses a pH of 14. A neutral solution is therefore one in which pH = 7, an acid solution one in which pH < 7, and an alkaline solution one in which pH > 7. An alternative definition for a neutral solution, applicable to all temperatures, is one in which the hydrogen-ion and hydroxide-ion concentrations are equal. In an acid solution the hydrogen-ion concentration exceeds the hydroxide-ion concentration, whilst in an alkaline or basic solution, the hydroxide-ion concentration is greater.

Example 10.   (i) Find the pH of a solution in which $[H^+] = 4.0 \times 10^{-5}$ mol dm^{-3}.

$$\begin{aligned} pH = \log_{10} 1/[H^+] &= \log 1 - \log[H^+] \\ &= \log 1 - \log 4.0 \times 10^{-5} \\ &= 0 - \bar{5}.602 \\ &= \underline{4.398} \end{aligned}$$

(ii) Find the hydrogen ion concentration corresponding to pH = 5.643.

$$pH = \log_{10} 1/[H^+] = \log 1 - \log[H^+] = 5.643$$
$$\therefore \log[H^+] = -5.643$$

This must be written in the usual form containing a negative characteristic and a positive mantissa:

$$\log[H^+] = -5.643 = \bar{6}.357$$

By reference to tables of antilogarithms we find $[H^+] = 2.28 \times 10^{-6}$ mol dm^{-3}.
    (iii) Calculate the pH of a 0.901 M solution of acetic acid in which the degree of dissociation is 12.6 per cent. The hydrogen ion concentration of the solution is 0.125×0.01

$$= 1.25 \times 10^{-3} \text{ mol } dm^{-3}.$$
$$\begin{aligned} pH = \log_{10} 1/[H^+] &= \log 1 - \log[H^+] \\ &= 0 - \bar{3}.097 \\ &= \underline{2.903.} \end{aligned}$$

The hydroxide-ion concentration may be expressed in a similar way:

$$pOH = -\log_{10}[OH^-] = \log_{10} 1/[OH^-], \text{ or } [OH^-] = 10^{-pOH}$$

If we write the equation:

$$[H^+] \times [OH^-] = K_w = 10^{-14}$$

in the form:

$$\log[H^+]+\log[OH^-]=\log K_w = -14$$
$$\text{then} \quad pH+pOH = pK_w = 14$$

This relationship should hold for all dilute solutions at about 25 °C.

Fig. II, 1 will serve as a useful mnemonic for the relation between $[H^+]$, pH, $[OH^-]$, and pOH in acid and alkaline solution.

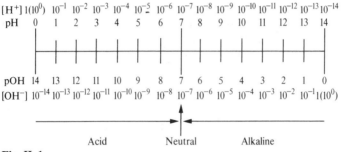

Fig. II, 1

The logarithmic or exponential method has also been found useful for expressing other small quantities which arise in quantitative analysis. These include: (i) dissociation constants (Section **II, 4**), (ii) other ionic concentrations, and (iii) solubility products (Section **II, 8**).

(i)   For any acid with a dissociation constant of K_a:

$$pK_a = \log 1/K_a = -\log K_a$$

Similarly for any base with dissociation constant K_b:

$$pK_b = \log 1/K_b = -\log K_b$$

(ii)   For any ion I of concentration $[I]$:

$$pI = \log 1/[I] = -\log[I]$$

Thus, for $[Na^+] = 8 \times 10^{-5}, pNa = 4.1$.

(iii)   For a salt with a solubility product K_s:

$$pK_s = \log 1/K_s = -\log K_s.$$

II, 17.   THE HYDROLYSIS OF SALTS.   Salts may be divided into four main classes:
1.   those derived from strong acids and strong bases, e.g., potassium chloride;
2.   those derived from weak acids and strong bases, e.g., sodium acetate;
3.   those derived from strong acids and weak bases, e.g., ammonium chloride; and
4.   those derived from weak acids and weak bases, e.g., ammonium formate or aluminium acetate.

When any of these is dissolved in water, the solution, as is well known, is not always neutral in reaction. Interaction may occur with the ions of water, and the resulting solution will be neutral, acid, or alkaline according to the nature of the salt.

With an aqueous solution of a salt of class (1), neither the anions have any tendency to combine with the hydrogen ions nor the cations with the hydroxide ions of water, since the related acids and bases are strong electrolytes. The equilibrium between the hydrogen and hydroxide ions in water:

$$H_2O \rightleftharpoons H^+ + OH^- \tag{a}$$

is therefore not disturbed and the solution remains neutral.

Consider, however, a salt MA derived from a weak acid HA and a strong base BOH {class (2)}. The salt is completely dissociated in aqueous solution:

$$MA \longrightarrow M^+ + A^-$$

A very small concentration of hydrogen and hydroxide ions, originating from the small but finite ionisation of water, will be initially present. HA is a weak acid, that is, it is dissociated only to a small degree; the concentration of A^- ions which can exist in equilibrium with H^+ ions is accordingly small. In order to maintain the equilibrium, the large initial concentration of A^- ions must be reduced by combination with H^+ ions to form undissociated HA:

$$H^+ + A^- \rightleftharpoons HA \tag{b}$$

The hydrogen ions required for this reaction can be obtained only from the further dissociation of the water; this dissociation produces simultaneously an equivalent quantity of hydroxyl ions. The hydrogen ions are utilised in the formation of HA, consequently the hydroxide ion concentration of the solution will increase and the solution will react alkaline. The net result is that the anions of the salt react with the hydrogen ions of the water, yielding the weak acid HA, and there is an increase in the concentration of hydroxide ions over that present in water.

It is usual in writing equations involving equilibria between completely dissociated and slightly dissociated or sparingly soluble substances to employ the ions of the former and the molecules of the latter. The reaction is thefore written:

$$A^- + H_2O \rightleftharpoons OH^- + HA \tag{c}$$

This equation can also be obtained by combining (a) and (b), since both equilibria must co-exist. This interaction between the ion (or ions) of a salt and the ions of water is called **hydrolysis**.

Let us now study the salt of a strong acid and a weak base {class (3)}. Here the initial high concentration of cations M^+ will be reduced by combination with the hydroxide ions of water to form the little dissociated base MOH until the equilibrium:

$$M^+ + OH^- \rightleftharpoons MOH \tag{d}$$

is attained. The hydrogen-ion concentration of the solution will thus be increased, and the solution will react acid. The hydrolysis is here represented by:

$$M^+ + H_2O \rightleftharpoons MOH + H^+ \tag{e}$$

For salts of class (4), in which both the acid and the base are weak, two reactions will occur simultaneously

$$M^+ + H_2O \rightleftharpoons MOH + H^+ \tag{f}$$
$$A^- + H_2O \rightleftharpoons HA + OH^- \tag{g}$$

The reaction of the solution will clearly depend upon the relative dissociation constants of the acid and the base. If they are equal in strength, the solution will be neutral; if $K_a > K_b$, it will be acid, and if $K_b > K_a$, it will be alkaline.

Having considered all the possible cases, we are now in a position to give a more general definition of hydrolysis. **Hydrolysis** is the interaction between an ion (or ions) of a salt and the ions of water with the production of (a) a weak acid or a weak base, or (b) of both a weak acid and a weak base.

The phenomenon of salt hydrolysis may be regarded as a simple application of the general Brønsted equation

$$A_1 + B_2 \rightleftharpoons A_2 + B_1$$

Thus the equation for the hydrolysis of ammonium salts

$$NH_4^+ + H_2O \rightleftharpoons NH_3 + H_3O^+$$

is really identical with the expression used to define the strength of the ammonium ion as a Brønsted acid (see Section **II, 5**) and the constant K_a for NH_4^+ is in fact what is usually termed the hydrolysis constant of an ammonium salt.

The hydrolysis of the sodium salt of a weak acid can be treated similarly. Thus for a solution of sodium acetate

$$CH_3COO^- + H_2O \rightleftharpoons CH_3COOH + OH^-,$$

the hydrolysis constant is

$$[CH_3COOH][OH^-]/[CH_3COO^-] = K_h = K_w/K_a$$

where K_a is the dissociation constant of acetic acid.

II, 18. HYDROLYSIS CONSTANT AND DEGREE OF HYDRO-LYSIS. Case 1. Salt of a weak acid and a strong base.

The equilibrium in a solution of a salt MA may be represented by:

$$A^- + H_2O \rightleftharpoons OH^- + HA$$

Applying the law of mass action, we obtain:

$$\frac{a_{OH^-} \times a_{HA}}{a_{A^-}} = \frac{[OH^-].[HA]}{[A^-]} \times \frac{y_{OH^-} \cdot y_{HA}}{y_{A^-}} = K_h \tag{1}$$

where K_h is the **hydrolysis constant.** The solution is assumed to be dilute so that the activity of the unionised water may be taken as constant, and the approximation that the activity coefficient of the un-ionised acid is unity and that both ions have the same activity coefficient may be introduced. Equation (1) then reduces to:

$$K_h = \frac{[OH^-] \times [HA]}{[A^-]} \tag{2}$$

This is often written in the form:

$$K_h = \frac{[Base] \times [Acid]}{[Unhydrolysed\,Salt]}$$

the free strong base and the unhydrolysed salt are completely dissociated and the acid is very little dissociated.

The **degree of hydrolysis** is the fraction of each mole hydrolysed at equilibrium. Let 1 mol of salt be dissolved in V dm^3 of solution, and let x be the degree of hydrolysis. The concentrations in mols dm^{-3} are:

$$A^- + H_2O \rightleftharpoons OH^- + HA$$
$$(1-x)/V \qquad\qquad x/V \qquad x/V$$

Substituting these values in (2):

$$K_h = \frac{[OH^-] \times [HA]}{[A^-]} = \frac{x/V \times x/V}{(1-x)/V} = \frac{x^2}{(1-x)V}.$$

This expression enables us to calculate the degree of hydrolysis at the dilution V; it is evident that as V increases, the degree of hydrolysis x must increase.

The two equilibria:

$$H_2O \rightleftharpoons H^+ + OH^-$$
$$\text{and} \quad HA \rightleftharpoons H^+ + A^-$$

must co-exist with the hydrolytic equilibrium:

$$A^- + H_2O \rightleftharpoons HA + OH^-$$

Hence the two relationships:

$$[H^+] \times [OH^-] = K_w$$
$$\text{and} \quad [H^+] \times [A^-]/[HA] = K_a$$

must hold in the same solution as:

$$[OH^-] \times [HA]/[A^-] = K_h$$

$$\text{But} \quad \frac{K_w}{K_a} = \frac{[H^+] \times [OH^-] \times [HA]}{[H^+] \times [A^-]} = \frac{[OH^-] \times [HA]}{[A^-]} = K_h$$

therefore $\quad K_w/K_a = K_h$

or $\qquad\qquad pK_h = pK_w - pK_a$

The hydrolysis constant is thus related to the ionic product of the water and the ionisation constant of the acid. Since K_a varies slightly and K_w varies considerably with temperature, K_h and consequently the degree of hydrolysis will be largely influenced by changes of temperature.

The hydrogen-ion concentration of a solution of a hydrolysed salt can be readily computed. The amounts of HA and of OH$^-$ ions formed as a result of hydrolysis are equal, therefore in a solution of the pure salt in water [HA] = [OH$^-$]. If the concentration of the salt is c mol dm^{-3}, then:

$$\frac{[HA] \times [OH^-]}{[A^-]} = \frac{[OH^-]^2}{c} = K_h = \frac{K_w}{K_a}$$

and $\qquad\qquad [OH^-] = \sqrt{c \cdot K_w/K_a}$

or $\qquad [H^+] = \sqrt{K_w \cdot K_a/c}$, since $[H^+] = K_w/[OH^-]$

and $\quad pH = \frac{1}{2}pK_w + \frac{1}{2}pK_a + \frac{1}{2}\log c$

To be consistent we should use $pc = -\log c$.

$$pH = \tfrac{1}{2}pK_w + \tfrac{1}{2}pK_a - \tfrac{1}{2}pc \tag{3}$$

Equation (3) can be employed for the calculation of the pH of a solution of a salt of a weak acid and a strong base. Thus the pH of a solution of sodium benzoate $(0.05 \text{ mol dm}^{-3})$ is given by:

$$pH = 7.0 + 2.10 - \tfrac{1}{2}(1.30) = 8.45$$

(Benzoic acid: $K_a = 6.37 \times 10^{-5} \text{ mol dm}^{-3}; pK_a = 4.20$)

Such a calculation will provide useful information as to the indicator which should be employed in the titration of a weak acid and a strong base (see Section **X, 13**).

Example 11.   Calculate: (i) the hydrolysis constant, (ii) the degree of hydrolysis, and (iii) the hydrogen-ion concentration of a solution of sodium acetate $(0.01 \text{ mol dm}^{-3})$ at the laboratory temperature.

$$K_h = \frac{K_w}{K_a} = \frac{1.0 \times 10^{-14}}{1.82 \times 10^{-5}} = 5.5 \times 10^{-10}$$

The degree of hydrolysis x is given by:

$$K_h = \frac{x^2}{(1-x)V}$$

Substituting for K_h and $V\,(=1/c)$, we obtain:

$$5.5 \times 10^{-10} = \frac{x^2 \times 0.01}{(1-x)}$$

Solving this quadratic equation for x, $x = 0.000235$ or 0.0235 per cent.

$$C_2H_3O_2^- + H_2O \rightleftharpoons H \cdot C_2H_3O_2 + OH^-$$

$(1-x)$ moles x moles x moles

If the solution were completely hydrolysed, the concentration of acetic acid produced would be 0.01 mol dm^{-3}. But the degree of hydrolysis is 0.0235 per cent, therefore the concentration of acetic acid is $2.35 \times 10^{-6} \text{ mol dm}^{-3}$. This is also equal to the hydroxide-ion concentration produced, i.e., $pOH = 5.63$.

$$pH = 14.0 - 5.63 = 8.37$$

The pH may also be computed from equation (3):

$$pH = \tfrac{1}{2}pK_w + \tfrac{1}{2}pK_a - \tfrac{1}{2}pc$$
$$= 7.0 + 2.37 - \tfrac{1}{2}(2) = 8.37.$$

Case 2.   Salt of a strong acid and a weak base.
The hydrolytic equilibrium is represented by:

$$M^+ + H_2O \rightleftharpoons MOH + H^+$$

By applying the law of mass action along the lines of **Case 1**, the following equations are obtained:

$$K_h = \frac{[\text{H}^+] \times [\text{MOH}]}{[\text{M}^+]} = \frac{[\text{Acid}] \times [\text{Base}]}{[\text{Unhydrolysed Salt}]} = \frac{K_w}{K_b}$$

$$= \frac{x^2}{(1-x)V}$$

K_b is the dissociation constant of the base. Furthermore, since [MOH] and $[\text{H}^+]$ are equal:

$$K_h = \frac{[\text{H}^+] \times [\text{MOH}]}{[\text{M}^+]} = \frac{[\text{H}^+]^2}{c} = \frac{K_w}{K_b}$$

$$[\text{H}^+] = \sqrt{c \cdot K_w/K_b},$$

or $\quad \text{pH} = \tfrac{1}{2}pK_w - \tfrac{1}{2}pK_b + \tfrac{1}{2}pc$ \hfill (4)

Equation (4) may be applied to the calculation of the pH of solutions of salts of strong acids and weak bases. Thus the pH of a solution of ammonium chloride $(0.2 \text{ mol dm}^{-3})$ is:

$$\text{pH} = 7.0 - 2.37 + \tfrac{1}{2}(0.70) = 4.98$$

(Ammonia in water: $K_b = 1.85 \times 10^{-5} \text{ mol dm}^{-3}$; $pK_b = 4.74$)

Case 3.   Salt of a weak acid and a weak base.
The hydrolytic equilibrium is expressed by the equation:

$$\text{M}^+ + \text{A}^- + \text{H}_2\text{O} \rightleftharpoons \text{MOH} + \text{HA}$$

Applying the law of mass action and taking the activity of un-ionised water as unity, we have:

$$K_h = \frac{a_{\text{MOH}} \times a_{\text{HA}}}{a_{\text{M}^+} \times a_{\text{A}^-}} = \frac{[\text{MOH}] \cdot [\text{HA}]}{[\text{M}^+] \cdot [\text{A}^-]} \times \frac{y_{\text{MOH}} \cdot y_{\text{HA}}}{y_{\text{M}^+} \cdot y_{\text{A}^-}}$$

By the usual approximations, that is, by assuming that the activity coefficients of the un-ionised molecules and, less justifiably, of the ions are unity, the following approximate equation is obtained:

$$K_h = \frac{[\text{MOH}] \times [\text{HA}]}{[\text{M}^+] \times [\text{A}^-]}$$

$$= \frac{[\text{Base}] \times [\text{Acid}]}{[\text{Unhydrolysed Salt}]^2}.$$

If x is the degree of hydrolysis of 1 mol of the salt dissolved in V dm^3 of solution, then the individual concentrations are:

$$[\text{MOH}] = [\text{HA}] = x/V; \quad [\text{M}^+] = [\text{A}^-] = (1-x)/V$$

leading to the result

$$K_h = \frac{x/V \cdot x/V}{(1-x)/V \cdot (1-x)/V} = \frac{x^2}{(1-x)^2}$$

The degree of hydrolysis and consequently the pH is independent of the concentration of the solution.*

It may be readily shown that:

$$K_h = K_w/K_a \times K_b$$
$$\text{or} \quad pK_h = pK_w - pK_a - pK_b$$

This expression enables us to compute the value of the degree of hydrolysis x from the dissociation constants of the acid and the base.

The hydrogen-ion concentration of the hydrolysed solution is calculated in the following manner:

$$[H^+] = K_a \times \frac{[HA]}{[A^-]} = K_a \times \frac{x/V}{(1-x)/V} = K_a \times \frac{x}{(1-x)}$$

But $x/(1-x) = \sqrt{K_h}$

$$\text{hence} \quad [H^+] = K_a \cdot \sqrt{K_h} = \sqrt{K_w \times K_a/K_b}$$

$$\text{or} \qquad pH = \tfrac{1}{2}pK_w + \tfrac{1}{2}pK_a - \tfrac{1}{2}pK_b \tag{5}$$

If the ionisation constants of the acid and the base are equal, that is, $K_a = K_b, pH = \tfrac{1}{2}pK_w = 7.0$, and the solution is neutral, although hydrolysis may be considerable. If $K_a > K_b$, pH < 7 and the solution is acid, but when $K_b > K_a$, pH > 7 and the solution reacts alkaline.

The pH of a solution of ammonium acetate is given by:

$$pH = 7.0 + 2.37 - 2.37 = 7.0$$

i.e., the solution is approximately neutral. On the other hand, for a dilute solution of ammonium formate:

$$pH = 7.0 + 1.88 - 2.37 = 6.51$$
$$(\text{Formic acid}: K_a = 1.77 \times 10^{-4} \text{ mol dm}^{-3}; pK_a = 3.75)$$

i.e., the solution reacts slightly acid.

II, 19.   BUFFER SOLUTIONS.   A solution of hydrochloric acid (0.0001 mol dm^{-3}) should have a pH equal to 4, but the solution is extremely sensitive to traces of alkali from the glass of the containing vessel and to ammonia from the air. Likewise a solution of sodium hydroxide (0.0001 mol dm^{-3}), which should have a pH of 10, is sensitive to traces of carbon dioxide from the atmosphere. Aqueous solutions of potassium chloride and of ammonium acetate have a pH of about 7. The addition to 1 dm^3 of these solutions of 1 cm^3 of a solution of hydrochloric acid (1 mol dm^{-3}) results in a change of pH to 3 in the former case and in very little change in the latter. The resistance of a solution to changes in hydrogen ion concentration upon the addition of small amounts of acid or alkali is termed **buffer action**; a solution which possesses such properties is known as a **buffer solution**. It is said to possess 'reserve acidity' and 'reserve alkalinity'. Buffer solutions usually consist of solutions containing a mixture of a weak acid HA and

---

* This applies only if the original assumptions as to activity coefficients are justified. In solutions of appreciable ionic strength, the activity coefficients of the ions will vary with the total ionic strength.

its sodium or potassium salt (A^-), or of a weak base B and its salt (BH^+). A buffer, then, is usually a mixture of an acid and its conjugate base. In order to understand buffer action, let us study first the equilibrium between a weak acid and its salt. The dissociation of a weak acid is given by:

$$HA \rightleftharpoons H^+ + A^-$$

and its magnitude is controlled by the value of the dissociation constant K_a:

$$\frac{a_{H^+} \times a_{A^-}}{a_{HA}} = K_a, \quad \text{or} \quad a_{H^+} = \frac{a_{HA}}{a_{A^-}} \times K_a \tag{1}$$

The expression may be approximated by writing concentrations for activities

$$[H^+] = \frac{[HA]}{[A^-]} \times K_a \tag{2}$$

This equilibrium applies to a mixture of an acid HA and its salt, say MA. If the concentration of the acid be c_a and that of the salt be c_s, then the concentration of the undissociated portion of the acid is $c_a - [H^+]$. The solution is electrically neutral, hence $[A^-] = c_s + [H^+]$ (the salt is completely dissociated). Substituting these values in the equilibrium equation (2), we have:

$$[H^+] = \frac{c_a - [H^+]}{c_s + [H^+]} \times K_a \tag{3}$$

This is a quadratic equation for $[H^+]$ and may be solved in the usual manner. It can, however, be simplified by introducing the following further approximations. In a mixture of a weak acid and its salt, the dissociation of the acid is repressed by the common ion effect, and $[H^+]$ may be taken as negligibly small by comparison with c_a and c_s. Equation (3) then reduces to:

$$[H^+] = \frac{c_a}{c_s} . K_a, \quad \text{or} \quad [H^+] = \frac{[Acid]}{[Salt]} \times K_a \tag{4}$$

$$\text{or} \quad pH = pK_a + \log \frac{[Salt]}{[Acid]} \tag{5}$$

The equations can be readily expressed in a somewhat more general form when applied to a Brønsted acid A and its conjugate base B:

$$A \rightleftharpoons H^+ + B$$

(e.g., CH_3COOH and CH_3COO^-, etc.). The expression for pH is:

$$pH = pK_a + \log \frac{[B]}{[A]}$$

where $\quad K_a = [H^+][B]/[A]$.

Similarly for a mixture of a weak base of dissociation constant K_b and its salt with a strong acid:

$$[OH^-] = \frac{[Base]}{[Salt]} \times K_b \tag{6}$$

$$\text{or} \quad pOH = pK_b + \log \frac{[Salt]}{[Base]} \tag{7}$$

Let us confine our attention to the case in which the concentrations of the acid and its salt are equal, i.e., of a half-neutralised acid. Then $pH = pK_a$. Thus the pH of a half-neutralised solution of a weak acid is equal to the negative logarithm of the dissociation constant of the acid. For acetic acid, $K_a = 1.82 \times 10^{-5}$ mol dm^{-3}, $pK_a = 4.74$; a half-neutralised solution of, say, $0.1M$-acetic acid will have a pH of 4.74. If we add a small concentration of H^+ ions to such a solution, the former will combine with the acetate ions to form undissociated acetic acid:

$$H^+ + CH_3COO^- \rightleftharpoons CH_3COOH$$

Similarly, if a small concentration of hydroxide ions be added, the latter will combine with the hydrogen ions arising from the dissociation of the acetic acid and form un-ionised water; the equilibrium will be disturbed, and more acetic acid will dissociate to replace the hydrogen ions removed in this way. In either case, the concentration of the acetic acid and acetate ion (or salt) will not be appreciably changed. It follows from equation (5) that the pH of the solution will not be materially affected.

Example 12.   Calculate the pH of the solution produced by adding 10 cm^3 of 1 M hydrochloric acid to 1 dm^3 of a solution which is 0.1 M in acetic acid and $0.1M$ in sodium acetate ($K_a = 1.82 \times 10^{-5}$ mol dm^{-3}).

The pH of the acetic acid–sodium acetate buffer solution is given by the equation:

$$pH = pK_a + \log \frac{[\text{Salt}]}{[\text{Acid}]} = 4.74 + 0.0 = 4.74$$

The hydrogen ions from the hydrochloric acid react with acetate ions forming practically undissociated acetic acid, and neglecting the change in volume from 1000 cm^3 to 1010 cm^3 we can say

$$CH_3COO^- = 0.1 - 0.01 = 0.09$$
$$CH_3COOH = 0.1 + 0.01 = 0.11$$
and $$pH = 4.74 + \log 0.09/0.11 = 4.74 - 0.09 = \underline{4.65}.$$

Thus the pH of the acetic acid–sodium acetate buffer solution is only altered by 0.09 pH unit on the addition of the hydrochloric acid. The same volume of hydrochloric acid added to one litre of water (pH = 7) would lead to a solution with pH = $-\log(0.01) = 2$; a change of 5 pH units. This example serves to illustrate the regulation of pH exercised by buffer solutions.

A solution containing equal concentrations of acid and its salt, or a half-neutralised solution of the acid, has the maximum buffer capacity. Other mixtures also possess considerable buffer capacity, but the pH will differ slightly from that of the half-neutralised acid. Thus in a quarter-neutralised solution of acid, $[\text{Acid}] = 3\,[\text{Salt}]$:

$$pH = pK_a + \log\tfrac{1}{3} = pK_a + \bar{1}.52$$
$$= pK_a - 0.48$$

For a three-quarter-neutralised acid, $[\text{Salt}] = 3\,[\text{Acid}]$:

$$pH = pK_a + \log 3$$
$$= pK_a + 0.48$$

In general, we may state that the buffering capacity is maintained for mixtures within the range 1 acid: 10 salt and 10 acid: 1 salt. The approximate pH range of a weak acid buffer is:

$$pH = pK_a \pm 1$$

The concentration of the acid is usually of the order $0.05-0.2$ mol dm^{-3}. Similar remarks apply to weak bases.

The preparation of a buffer solution of a definite pH is a simple process if the acid (or base) of appropriate dissociation constant is found: small variations in pH are obtained by variations in the ratio of the acid to the salt concentration. One example is given in Table II, 2.

Table II, 2. pH of Acetic Acid–Sodium Acetate Buffer Mixtures
10 cm^3 mixtures of x cm^3 of 0.2M-acetic acid and y cm^3 of 0.2M-sodium acetate

Acetic Acid (x cm^3)	Sodium Acetate (y cm^3)	pH
9.5	0.5	3.42
9.0	1.0	3.72
8.0	2.0	4.05
7.0	3.0	4.27
6.0	4.0	4.45
5.0	5.0	4.63
4.0	6.0	4.80
3.0	7.0	4.99
2.0	8.0	5.23
1.0	9.0	5.57
0.5	9.5	5.89

Before leaving the subject of buffer solutions, it is necessary to draw attention to a possible erroneous deduction from equation (5), namely, that the hydrogen-ion concentration of a buffer solution is dependent only upon the ratio of the concentrations of acid and salt and upon K_a, and not upon the actual concentrations; otherwise expressed, that the pH of such a buffer mixture should not change upon dilution with water. This is approximately although not strictly true. In deducing equation (2), concentrations have been substituted for activities, a step which is not entirely justifiable except in dilute solutions. The exact expression controlling buffer action is:

$$a_{H^+} = \frac{a_{HA}}{a_{A^-}} \times K_a = \frac{c_a \cdot y_a}{c_s \cdot y_{A^-}} \times K_a \qquad (8)$$

The activity coefficient y_a of the undissociated acid is approximately unity in dilute aqueous solution. Expression (8) thus becomes:

$$a_{H^+} = \frac{[\text{Acid}]}{[\text{Salt}] \times y_{A^-}} \times K_a \qquad (9)$$

or $pH = pK_a + \log[\text{Salt}]/[\text{Acid}] + \log y_{A^-}$ (10)

According to the Debye–Hückel–Onsager theory, the activity coefficient of an ion y_i in aqueous solution at 25 °C is given by:

$$\log y_i = -\frac{0.505 z_i^2 I^{0.5}}{1 + A I^{0.5}} + BI$$

where z_i is the valency of the ion, I is the ionic strength of the solution, and A and B are constants. This may be written in the form

$$\log y_i = -0.505z_i^2 I^{0.5} + CI$$

where C is another constant approximately equal to $0.505z_i^2 A + B$ and usually has a value varying between 0.2 and 1.5. Substituting for y_{A^-} in (10), we obtain:

$$pH = pK_a + \log[\text{Salt}]/[\text{Acid}] - 0.505z_i^2 I^{0.5} + CI \tag{11}$$

The activity coefficient of the ion y_{A^-} generally increases with decrease of concentration, so that when a buffer solution is diluted, y_{A^-} increases and consequently a_{H^+} will increase {equation (9)}. For most practical purposes the change in pH is small, but for exact work it must be taken into account. The addition of salts to buffer mixtures results in a change of the ionic strength of the solution; this will affect the pH of the solution {equation (11)}. Indeed, in all buffer solutions a correction should, strictly speaking, be applied for the ionic strength of the solution.

Buffer mixtures are not confined to mixtures of monoprotic acids or monoacid bases and their salts. We may employ a mixture of salts of a polyprotic acid, e.g., NaH_2PO_4 and Na_2HPO_4. The salt NaH_2PO_4 is completely dissociated:

$$NaH_2PO_4 \rightleftharpoons Na^+ + H_2PO_4^-$$

The ion $H_2PO_4^-$ acts as a monoprotic acid:

$$H_2PO_4^- \rightleftharpoons H^+ + HPO_4^{2-}$$

for which $K(\equiv K_2$ for phosphoric acid) is 6.2×10^{-8} mol dm^{-3}. The addition of the salt Na_2HPO_4 is analogous to the addition of, say, acetate ions to a solution of acetic acid, since the tertiary ionisation of phosphoric acid ($HPO_4^{2-} \rightleftharpoons H^+ + PO_4^{3-}$) is small ($K_3 = 5 \times 10^{-13}$ mol dm^{-3}). The mixture of NaH_2PO_4 and Na_2HPO_4 is therefore an effective buffer over the range pH 7.2 ± 1.0 ($= pK \pm 1$). It will be noted that this is a mixture of a Brønsted acid and its conjugate base.

Buffer solutions find many applications in quantitative inorganic analysis, e.g., many precipitations are only quantitative under carefully controlled conditions of pH, as are also many compleximetric titrations: numerous examples of their use will be found throughout the book.

II, 20. ELECTRODE POTENTIALS.

When a metal is immersed in a solution containing its own ions, say, zinc in zinc sulphate solution, a potential difference is established between the metal and the solution. The potential difference E for an electrode reaction

$$M^{n+} + ne \rightleftharpoons M$$

is given by the expression:

$$E = E^{\ominus} + \frac{RT}{nF} \ln a_{M^{n+}} \tag{1}$$

where R is the gas constant, T is the absolute temperature, F the Faraday constant, n the valency of the ions, $a_{M^{n+}}$ the activity of the ions in the solution, and $E^{\ominus}$ is a constant dependent upon the metal. Equation (1) can be simplified by introducing the known values of R and F, and converting natural logarithms to base 10 by multiplying by 2.3026; it then becomes:

$$E = E^{\ominus} + \frac{0.0001983T}{n} \log a_{M^{n+}}$$

For a temperature of 25 °C ($T = 298K$):

$$E = E^{\ominus} + \frac{0.0591}{n} \log a_{M^{n+}} \tag{2}$$

For most purposes in quantitative analysis, it is sufficiently accurate to replace $a_{M^{n+}}$ by $c_{M^{n+}}$, the ion concentration (in moles per dm^3):

$$E = E^{\ominus} + \frac{0.0591}{n} \log c_{M^{n+}} \tag{3}$$

The latter is a form of the **Nernst equation**.

If in equation (2), $a_{M^{n+}}$ is put equal to unity, E is equal to $E^{\ominus}$. $E^{\ominus}$ is called the **standard electrode potential of the metal**.

In order to determine the potential difference between an electrode and a solution, it is necessary to have another electrode and solution of accurately known potential difference. The two electrodes can then be combined to form a voltaic cell, the e.m.f. of which can be directly measured. The e.m.f. of the cell is the arithmetical sum or difference of the electrode potentials (depending upon the sign of these two potentials); the value of the unknown potential can then be calculated. The primary reference electrode is the **normal or standard hydrogen electrode** (see also Section **XIV, 2**). This consists of a piece of platinum foil, coated electrolytically with platinum black, and immersed in a solution of hydrochloric acid containing hydrogen ions at unit activity. (This corresponds to 1.8M-hydrochloric acid at 25 °C.) Hydrogen gas at a pressure of one atmosphere is passed over the platinum foil through the side tube C (Fig. II, 2) and

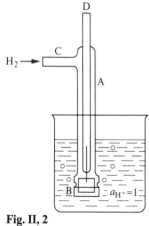

Fig. II, 2

escapes through the small holes B in the surrounding glass tube A. Because of the periodic formation of bubbles, the level of the liquid inside the tube fluctuates, and a part of the foil is alternately exposed to the solution and to hydrogen. The lower end of the foil is continuously immersed in the solution to avoid interruption of the electric current. Connection between the platinum foil and an external circuit is made with mercury in D. The platinum black has the remarkable property of adsorbing large quantities of hydrogen, and it permits the change from the gaseous to the ionic form and the reverse process to occur without hindrance; it therefore behaves as though it were composed entirely of hydrogen, that is, as a hydrogen electrode. Under fixed conditions, viz., hydrogen gas at atmospheric pressure and unit activity of hydrogen ions in the solution in contact with the electrode, the hydrogen electrode possesses a definite potential. By convention, the potential of the standard hydrogen electrode is equal to zero at all temperatures. Upon connecting the standard hydrogen electrode with a metal electrode (a metal in contact with a solution of its ions of unit activity) by means of a salt (say, potassium chloride) bridge, the **standard electrode potential** may be determined. the cell is usually written as

$$Pt,H_2 \,|\, H^+ (a = 1) \,\|\, M^{n+} (a = 1) \,|\, M$$

In this scheme, a single vertical line represents a metal–electrolyte boundary at which a potential difference is taken into account: a double vertical line represents a liquid junction at which the potential is to be disregarded or is considered to be eliminated by a salt bridge.

When we speak of the electrode potential of a zinc electrode, we mean the e.m.f. of the cell:

$$Pt,H_2 \,|\, H^+ (a = 1) \,\|\, Zn^{2+} \,|\, Zn$$

or the e.m.f. of the half-cell $Zn^{2+} \,|\, Zn$. The cell reaction is:

$$H_2 + Zn^{2+} \longrightarrow 2H^+ (a = 1) + Zn$$

and the half-cell reaction is written as:

$$Zn^{2+} + 2e \rightleftharpoons Zn$$

The electrode potential of the Fe^{3+}, $Fe^{2+} \,|\, Pt$ electrode is the e.m.f. of the cell:

$$Pt,H_2 \,|\, H^+ (a = 1) \,\|\, Fe^{3+}, Fe^{2+} \,|\, Pt$$

or the e.m.f. of the half-cell $Fe^{3+}, Fe^{2+} \,|\, Pt$. The cell reaction is:

$$\tfrac{1}{2}H_2 + Fe^{3+} \longrightarrow H^+ (a = 1) + Fe^{2+}$$

and the half-cell reaction is written:

$$Fe^{3+} + e \rightleftharpoons Fe^{2+}$$

The convention is adopted of writing all half-cell reactions as reductions:

$$M^{n+} + ne \rightleftharpoons M$$

e.g., $\quad Zn^{2+} + 2e \rightleftharpoons Zn \qquad\qquad\qquad E^\ominus = -0.76 \text{ volt}$

When the activity of the ion M^{n+} is equal to unity (approximately true for a $1M$ solution), the electrode potential E is equal to the standard potential $E^\ominus$. Some important standard electrode potentials referred to the standard hydrogen electrode at 25 °C (in aqueous solution) are collected in Table II, 3.

Table II, 3. Standard Electrode Potentials at 25 °C

Electrode reaction	$E^\ominus$ (volts)	Electrode reaction	$E^\ominus$ (volts)
$Li^+ + e = Li$	-3.045	$Tl^+ + e = Tl$	-0.336
$K^+ + e = K$	-2.925	$Co^{2+} + 2e = Co$	-0.277
$Ba^{2+} + 2e = Ba$	-2.90	$Ni^{2+} + 2e = Ni$	-0.25
$Sr^{2+} + 2e = Sr$	-2.89	$Sn^{2+} + 2e = Sn$	-0.136
$Ca^{2+} + 2e = Ca$	-2.87	$Pb^{2+} + 2e = Pb$	-0.126
$Na^+ + e = Na$	-2.714	$2H^+ + 2e = H_2$	0.000
$Mg^{2+} + 2e = Mg$	-2.37	$Cu^{2+} + 2e = Cu$	$+0.337$
$Al^{3+} + 3e = Al$	-1.66	$Hg^{2+} + 2e = Hg$	$+0.789$
$Mn^{2+} + 2e = Mn$	-1.18	$Ag^+ + e = Ag$	$+0.799$
$Zn^{2+} + 2e = Zn$	-0.763	$Pd^{2+} + 2e = Pd$	$+0.987$
$Fe^{2+} + 2e = Fe$	-0.440	$Pt^{2+} + 2e = Pt$	$+1.2$
$Cd^{2+} + 2e = Cd$	-0.403	$Au^{3+} + 3e = Au$	$+1.50$

It may be noted that the standard hydrogen electrode is rather difficult to manipulate. In practice, electrode potentials on the hydrogen scale are usually determined indirectly by measuring the e.m.f. of a cell formed from the electrode in question and a convenient reference electrode whose potential with respect to the hydrogen electrode is accurately known. The reference electrodes generally used are the calomel electrode and the silver–silver chloride electrode (see Sections **XIV, 3–4**).

When metals are arranged in the order of their standard electrode potentials, the so-called **electrochemical series** of the metals is obtained. The greater the negative value of the potential, the greater is the tendency of the metal to pass into the ionic state. A metal will normally displace any other metal below it in the series from solutions of its salts. Thus magnesium, aluminium, zinc, or iron will displace copper from solutions of its salts; lead will displace copper, mercury, or silver; copper will displace silver.

The standard electrode potential is a quantitative measure of the readiness of the element to lose electrons. It is therefore a measure of the strength of the element as a reducing agent in aqueous solution; the more negative the potential of the element, the more powerful is its action as a reductant.

It must be emphasised that standard electrode potential values relate to an *equilibrium* condition between the metal electrode and the solution. Potentials determined under, or calculated for, such conditions are often referred to as *reversible electrode potentials*, and it must be remembered that the Nernst equation is only strictly applicable under such conditions.

II, 21. CONCENTRATION CELLS. An electrode potential varies with the concentration of the ions in the solution. Hence two electrodes of the same metal, but immersed in solutions containing different concentrations of its ions, may form a cell. Such a cell is termed a **concentration cell**. The e.m.f. of the cell will be the algebraic difference of the two potentials, if a salt bridge be inserted to eliminate the liquid–liquid junction potential. It may be calculated as follows. At 25 °C:

$$E = \frac{0.0591}{n}\log c_1 + E^{\ominus} - \left(\frac{0.0591}{n}\log c_2 + E^{\ominus}\right)$$

$$= \frac{0.0591}{n}\log \frac{c_1}{c_2}, \text{ where } c_1 > c_2$$

As an example we may consider the cell:

$$\overset{-}{Ag}\left|\begin{matrix} AgNO_3 \text{ aq.} \\ [Ag^+] = 0.00475M \end{matrix}\right|\left|\begin{matrix} AgNO_3 \text{ aq.} \\ [Ag^+] = 0.043M \end{matrix}\right| \overset{+}{Ag}$$

$$\underset{E_2}{\longleftarrow} \qquad\qquad\qquad \underset{E_1}{\longrightarrow}$$

Assuming that there is no potential difference at the liquid junction:

$$E = E_1 - E_2 = \frac{0.0591}{1}\log\frac{0.043}{0.00475} = 0.056 \text{ volt}$$

II, 22. CALCULATION OF THE e.m.f. OF A VOLTAIC CELL. An interesting application of electrode potentials is to the calculation of the e.m.f. of a voltaic cell. One of the simplest of galvanic cells is the Daniell cell. It consists of a rod of zinc dipping into zinc sulphate solution and a strip of copper in copper sulphate solution; the two solutions are generally separated by placing one inside a porous pot and the other in the surrounding vessel. The cell may be represented as:

$$Zn \,|\, ZnSO_4 \,aq. \,\|\, CuSO_4 aq. |Cu$$

At the zinc electrode, zinc ions pass into solution, leaving an equivalent negative charge on the metal. Copper ions are deposited at the copper electrode, rendering it positively charged. By completing the external circuit, the current (electrons) passes from the zinc to the copper. The chemical reactions in the cell are as follows:

(a) zinc electrode, $Zn \rightleftharpoons Zn^{2+} + 2e$;
(b) copper electrode, $Cu^{2+} + 2e \rightleftharpoons Cu$

The net chemical reaction is:

$$Zn + Cu^{2+} = Zn^{2+} + Cu$$

The potential difference at each electrode may be calculated by the formula given above, and the e.m.f. of the cell is the algebraic difference of the two potentials, the correct sign being applied to each.

As an example we may calculate the e.m.f. of the Daniell cell with molar concentrations of zinc ions and copper (II) ions:

$$E = E_{(Cu)}^{\ominus} - E_{(Zn)}^{\ominus} = +0.34 - (-0.76) = 1.10 \; volts$$

The small potential difference produced at the contact between the two solutions (the so-called liquid–junction potential) is neglected.

II, 23. OXIDATION–REDUCTION CELLS.

Reduction is accompanied by a gain of electrons, and oxidation by a loss of electrons. In a system containing both an oxidising agent and its reduction product, there will be an equilibrium between them and electrons. If an inert electrode, such as platinum, is placed in a redox system, for example, one containing Fe(III) and Fe(II) ions, it will assume a definite potential indicative of the position of equilibrium. If the system tends to act as an oxidising agent, then $Fe^{3+} \longrightarrow Fe^{2+}$ and it will take electrons from the platinum, leaving the latter positively charged; if, however, the system has reducing properties ($Fe^{2+} \longrightarrow Fe^{3+}$), electrons will be given up to the metal, which will then acquire a negative charge. The magnitude of the potential will thus be a measure of the oxidising or reducing properties of the system.

To obtain comparative values of the 'strengths' of oxidising agents, it is necessary, as in the case of the electrode potentials of the metals, to measure under standard experimental conditions the potential difference between the platinum and the solution relative to a standard of reference. The primary standard is the standard or normal hydrogen electrode (Section II, 20) and its potential is taken as zero. The standard experimental conditions for the redox system are those in which the ratio of the activity of the oxidant to that of the reductant is unity. Thus for the $Fe^{3+}-Fe^{2+}$ electrode, the redox cell would be:

$$Pt,H_2 \left| H^+ (a = 1) \right\| \left. \begin{matrix} Fe^{3+} \,(a = 1) \\ Fe^{2+} \,(a = 1) \end{matrix} \right| Pt$$

The potential measured in this way is called the **standard reduction potential**. A selection of standard reduction potentials is given in Table II, 4.

The standard potentials enable us to predict which ions will oxidise or reduce other ions at unit activity (or molar concentration). The most powerful oxidising agents are those at the upper end of the table, and the most powerful reducing

Table II, 4.   Standard reduction potentials at 25°C

Half-reaction	$E^{\ominus}$, volts
$F_2 + 2e \rightleftharpoons 2F^-$	+2.65
$S_2O_8^{2-} + 2e \rightleftharpoons 2SO_4^{2-}$	+2.01
$Co^{3+} + e \rightleftharpoons Co^{2+}$	+1.82
$Pb^{4+} + 2e \rightleftharpoons Pb^{2+}$	+1.70
$MnO_4^- + 4H^+ + 3e \rightleftharpoons MnO_2 + 2H_2O$	+1.69
$Ce^{4+} + e \rightleftharpoons Ce^{3+}$ (nitrate medium)	+1.61
$BrO_3^- + 6H^+ + 5e \rightleftharpoons \frac{1}{2}Br_2 + 3H_2O$	+1.52
$MnO_4^- + 8H^+ + 5e \rightleftharpoons Mn^{2+} + 4H_2O$	+1.52
$Ce^{4+} + e \rightleftharpoons Ce^{3+}$ (sulphate medium)	+1.44
$Cl_2 + 2e \rightleftharpoons 2Cl^-$	+1.36
$Cr_2O_7^{2-} + 14H^+ + 6e \rightleftharpoons 2Cr^{3+} + 7H_2O$	+1.33
$Tl^{3+} + 2e \rightleftharpoons Tl^+$	+1.25
$MnO_2 + 4H^+ + 2e \rightleftharpoons Mn^{2+} + 2H_2O$	+1.23
$O_2 + 4H^+ + 4e \rightleftharpoons 2H_2O$	+1.23
$IO_3^- + 6H^+ + 5e \rightleftharpoons \frac{1}{2}I_2 + 3H_2O$	+1.20
$Br_2 + 2e \rightleftharpoons 2Br^-$	+1.07
$HNO_2 + H^+ + e \rightleftharpoons NO + H_2O$	+1.00
$NO_3^- + 4H^+ + 3e \rightleftharpoons NO + 2H_2O$	+0.96
$2Hg^{2+} + 2e \rightleftharpoons Hg_2^{2+}$	+0.92
$ClO^- + H_2O + 2e \rightleftharpoons Cl^- + 2OH^-$	+0.89
$Cu^{2+} + I^- + e \rightleftharpoons CuI$	+0.86
$Hg_2^+ + 2e \rightleftharpoons 2Hg$	+0.79
$Fe^{3+} + e \rightleftharpoons Fe^{2+}$	+0.77
$BrO^- + H_2O + 2e \rightleftharpoons Br^- + 2OH^-$	+0.76
$BrO_3^- + 3H_2O + 6e \rightleftharpoons Br^- + 6OH^-$	+0.61
$MnO_4^{2-} + 2H_2O + 2e \rightleftharpoons MnO_2 + 4OH^-$	+0.60
$MnO_4^- + e \rightleftharpoons MnO_4^{2-}$	+0.56
$H_3AsO_4 + 2H^+ + 2e \rightleftharpoons H_3AsO_3 + H_2O$	+0.56
$Cu^{2+} + Cl^- + e \rightleftharpoons CuCl$	+0.54
$I_2 + 2e \rightleftharpoons 2I^-$	+0.54
$IO^- + H_2O + 2e \rightleftharpoons I^- + 2OH^-$	+0.49
$[Fe(CN)_6]^{3-} + e \rightleftharpoons [Fe(CN)_6]^{4-}$	+0.36
$UO_2^{2+} + 4H^+ + 2e \rightleftharpoons U^{4+} + 2H_2O$	+0.33
$IO_3^- + 3H_2O + 6e \rightleftharpoons I^- + 6OH^-$	+0.26
$Cu^{2+} + e \rightleftharpoons Cu^+$	+0.15
$Sn^{4+} + 2e \rightleftharpoons Sn^{2+}$	+0.15
$TiO^{2+} + 2H^+ + e \rightleftharpoons Ti^{3+} + H_2O$	+0.1
$S_4O_6^{2-} + 2e \rightleftharpoons 2S_2O_3^{2-}$	+0.08
$2H^+ + 2e \rightleftharpoons H_2$	0.00
$V^{3+} + e \rightleftharpoons V^{2+}$	−0.26
$Cr^{3+} + e \rightleftharpoons Cr^{2+}$	−0.41
$Bi(OH)_3 + 3e \rightleftharpoons Bi + 3OH^-$	−0.44
$Fe(OH)_3 + e \rightleftharpoons Fe(OH)_2 + OH^-$	−0.56
$U^{4+} + e \rightleftharpoons U^{3+}$	−0.61
$AsO_4^{3-} + 3H_2O + 2e \rightleftharpoons H_2AsO_3^- + 4OH^-$	−0.67
$[Sn(OH)_6]^{2-} + 2e \rightleftharpoons [HSnO_2]^- + H_2O + 3OH^-$	−0.90
$[Zn(OH)_4]^{2-} + 2e \rightleftharpoons Zn + 4OH^-$	−1.22
$[H_2AlO_3]^- + H_2O + 3e \rightleftharpoons Al + 4OH^-$	−2.35

agents at the lower end. Thus permanganate ion can oxidise Cl^-, Br^-, I^-, Fe^{2+} and $[Fe(CN)_6]^{4-}$; Fe^{3+} can oxidise H_3AsO_3 and I^- but not $Cr_2O_7^{2-}$ or Cl^-. It must be emphasised that for many oxidants the pH of the medium is of great importance, since they are generally used in acidic media. Thus in measuring the standard potential of the $MnO_4^- - Mn^{2+}$ system; $MnO_4^- + 8H^+ + 5e = Mn^{2+} + 4H_2O$ it is necessary to state that the hydrogen-ion activity is unity; this leads to $E^\ominus = +1.52$ volts. Similarly, the value of $E^\ominus$ for the $Cr_2O_7^{2-} - Cr^{3+}$ system is $+1.33$ volts. This means that the $MnO_4^- - Mn^{2+}$ system is a better oxidising agent than the $Cr_2O_7^{2-} - Cr^{3+}$ system. Since the standard potentials for $Cl_2 - 2Cl^-$ and $Fe^{3+} - Fe^{2+}$ systems are $+1.36$ and 0.77 volt respectively, permanganate and dichromate will oxidise Fe(II) ions but only permanganate will oxidise chloride ions; this explains why dichromate but not permanganate (except under very special conditions) can be used for the titration of Fe(II) in hydrochloric acid solution. Standard potentials do not give any information as to the speed of the reaction: in some cases a catalyst is necessary in order that the reaction may proceed with reasonable velocity.

Standard potentials are determined with full consideration of activity effects, and are really limiting values. They are rarely, if ever, observed directly in a potentiometric measurement. In practice, measured potentials determined under defined conditions (formal potentials) are very useful for predicting the possibilities of redox processes. Further details are given in Section **X, 32**.

II, 24. CALCULATION OF THE STANDARD REDUCTION PO-TENTIAL. A reversible oxidation–reduction system may be written in the form (*oxidant* = substance in oxidised state, *reductant* = substance in reduced state):

$$Oxidant + ne \rightleftharpoons Reductant$$
$$\text{or} \qquad Ox + ne \rightleftharpoons Red$$

The electrode potential which is established when an inert or unattackable electrode is immersed in a solution containing both oxidant and reductant is given by the expression:

$$E_T = E^\ominus + \frac{RT}{nF} \ln \frac{a_{Ox}}{a_{Red}}$$

where E_T is the observed potential of the redox electrode at temperature T relative to the standard or normal hydrogen electrode taken as zero potential, $E^\ominus$ is the standard reduction potential,* n the number of electrons gained by the oxidant in being converted into the reductant, and a_{Ox} and a_{Red} are the activities of the oxidant and reductant respectively.

Since activities are often difficult to determine directly, they may be replaced by concentrations; the error thereby introduced is usually of no great importance. The equation therefore becomes:

$$E_T = E^\ominus + \frac{RT}{nF} \ln \frac{c_{Ox}}{c_{Red}}$$

---

* $E^\ominus$ is the value of E_T at unit activities of the oxidant and reductant. If both activities are variable, e.g., Fe^{3+} and Fe^{2+}, $E^\ominus$ corresponds to an activity ratio of unity.

Substituting the known values of R and F, and changing from natural to common logarithms, we have for a temperature of 25 °C ($T = 298$K):

$$E_{25°} = E^{\ominus} + \frac{0.0591}{n} \log \frac{[Ox]}{[Red]}$$

If the concentrations (or, more accurately, the activities) of the oxidant and reductant are equal, $E_{25°} = E^{\ominus}$, i.e., the standard reduction potential. It follows from this expression that, for example, a ten-fold change in the ratio of the concentrations of the oxidant to the reductant will produce a change in the potential of the system of $0.0591/n$ volts.

II, 25. EQUILIBRIUM CONSTANTS OF OXIDATION–REDUCTION REACTIONS.

The general equation for the reaction at an oxidation–reduction electrode may be written:

$$pA + qB + rC \ldots\ldots + ne \rightleftharpoons sX + tY + uZ + \ldots\ldots$$

The potential is given by:

$$E = E^{\ominus} + \frac{RT}{nF} \ln \frac{a_A^p \cdot a_B^q \cdot a_C^r \ldots\ldots}{a_X^s \cdot a_Y^t \cdot a_Z^u \ldots\ldots}$$

where a refers to activities, and n to the number of electrons involved in the oxidation–reduction reaction. This expression reduces to the following for a temperature of 25 °C (concentrations are substituted for activities to permit ease of application in practice):

$$E = E^{\ominus} + \frac{0.0591}{n} \log \frac{c_A^p \cdot c_B^q \cdot c_C^r \ldots\ldots}{c_X^s \cdot c_Y^t \cdot c_Z^u \ldots\ldots}$$

It is, of course, possible to calculate the influence of the change of concentration of certain constituents of the system by the use of the latter equation. Consider, for example, the permanganate reaction:

$$MnO_4^- + 8H^+ + 5e \rightleftharpoons Mn^{2+} + 4H_2O$$

$$E = E^{\ominus} + \frac{0.0591}{5} \log \frac{[MnO_4^-] \times [H^+]^8}{[Mn^{2+}]} \text{ (at 25 °C)}$$

The concentration (or activity) of the water is taken as constant, since it is assumed that the reaction takes place in dilute solution, and the concentration of the water does not change appreciably as the result of the reaction. The equation may be written in the form:

$$E = E^{\ominus} + \frac{0.0591}{5} \log \frac{[MnO_4^-]}{[Mn^{2+}]} + \frac{0.0591}{5} \log [H^+]^8$$

This enables us to calculate the effect of change in the ratio $[MnO_4^-]/[Mn^{2+}]$ at any hydrogen-ion concentration, other factors being maintained constant. In this system, however, difficulties are experienced in the calculation owing to the fact that the reduction products of the permanganate ion vary at different hydrogen-ion concentrations. In other cases no such difficulties arise, and the calculation may be employed with confidence. Thus in the reaction:

$$H_3AsO_4 + 2H^+ + 2e \rightleftharpoons H_3AsO_3 + H_2O$$

$$E = E^{\ominus} + \frac{0.0591}{2} \log \frac{[H_3AsO_4] \times [H^+]^2}{[H_3AsO_3]} \text{ (at 25 °C)}$$

or $\quad E = E^{\ominus} + \frac{0.0591}{2} \log \frac{[H_3AsO_4]}{[H_3AsO_3]} + \frac{0.0591}{2} \log [H^+]^2$

We are now in a position to calculate the equilibrium constants of oxidation–reduction reactions, and thus to determine whether such reactions can find application in quantitative analysis. Let us consider first the simple reaction:

$$Cl_2 + 2Fe^{2+} \rightleftharpoons 2Cl^- + 2Fe^{3+}$$

The equilibrium constant is given by:

$$\frac{[Cl^-]^2 \times [Fe^{3+}]^2}{[Cl_2] \times [Fe^{2+}]^2} = K$$

The reaction may be regarded as taking place in a voltaic cell, the two half-cells being a $Cl_2, 2Cl^-$ system and a Fe^{3+}, Fe^{2+} system. The reaction is allowed to proceed to equilibrium; the total voltage or e.m.f. of the cell will then be zero, i.e., the potentials of the two electrodes will be equal:

$$E^{\ominus}_{Cl_2, 2Cl^-} + \frac{0.059}{2} \log \frac{[Cl_2]}{[Cl^-]^2} = E^{\ominus}_{Fe^{3+}, Fe^{2+}} + \frac{0.059}{1} \log \frac{[Fe^{3+}]}{[Fe^{2+}]}$$

Now $E^{\ominus}_{Cl_2, 2Cl^-} = 1.36$ volts and $E^{\ominus}_{Fe^{3+}, Fe^{2+}} = 0.75$ volt, hence

$$\log \frac{[Fe^{3+}]^2 \times [Cl^-]^2}{[Fe^{2+}]^2 \times [Cl_2]} = \frac{0.61}{0.02965} = 20.67 = \log K$$

or $\qquad\qquad\qquad K = 4.7 \times 10^{20}$

The large value of the equilibrium constant signifies that the reaction will proceed from left to right almost to completion, i.e., an iron(II) salt is almost completely oxidised by chlorine.

Consider now the more complex reaction:

$$MnO_4^- + 5Fe^{2+} + 8H^+ \rightleftharpoons Mn^{2+} + 5Fe^{3+} + 4H_2O$$

The equilibrium constant K is given by:

$$K = \frac{[Mn^{2+}] \times [Fe^{3+}]^5}{[MnO_4^-] \times [Fe^{2+}]^5 \times [H^+]^8}$$

The term $4H_2O$ is omitted, since the reaction is carried out in dilute solution, and the water concentration may be assumed constant. The hydrogen-ion concentration is taken as molar. The complete reaction may be divided into two half-cell reactions corresponding to the partial equations:

$$MnO_4^- + 8H^+ + 5e \rightleftharpoons Mn^{2+} + 4H_2O \tag{1}$$

and $\qquad\qquad\qquad Fe^{2+} \rightleftharpoons Fe^{3+} + e \tag{2}$

For (1) as an oxidation–reduction electrode, we have:

$$E = E^{\ominus} + \frac{0.059}{5} \log \frac{[MnO_4^-] \times [H^+]^8}{[Mn^{2+}]}$$

$$= 1.52 + \frac{0.059}{5} \log \frac{[MnO_4^-] \times [H^+]^8}{[Mn^{2+}]}$$

The partial equation (2) may be multiplied by 5 in order to balance (1) electrically:

$$5Fe^{2+} \rightleftharpoons 5Fe^{3+} + 5e \tag{2'}$$

For (2′) as an oxidation–reduction electrode:

$$E = E^{\ominus} + \frac{0.059}{5} \log \frac{[Fe^{3+}]^5}{[Fe^{2+}]^5}$$

$$= 0.77 + \frac{0.059}{5} \log \frac{[Fe^{3+}]^5}{[Fe^{2+}]^5}$$

Combining the two electrodes into a cell, the e.m.f. will be zero when equilibrium is attained, i.e.,

$$1.52 + \frac{0.059}{5} \log \frac{[MnO_4^-] \times [H^+]^8}{[Mn^{2+}]} = 0.77 + \frac{0.059}{5} \log \frac{[Fe^{3+}]^5}{[Fe^{2+}]^5}$$

or $\quad \log \dfrac{[Mn^{2+}] \times [Fe^{3+}]^5}{[MnO_4^-] \times [Fe^{2+}]^5 \times [H^+]^8} = \dfrac{5(1.52-0.77)}{0.059} = 63.5$

$$K = \frac{[Mn^{2+}] \times [Fe^{3+}]^5}{[MnO_4^-] \times [Fe^{3+}]^5 \times [H^+]^8} = 3 \times 10^{63}$$

This result clearly indicates that the reaction proceeds virtually to completion. It is a simple matter to calculate the residual Fe(II) concentration in any particular case. Thus suppose we titrate 10 cm³ of 0.1N-potassium permanganate with an approximately 0.1N solution of iron(II) ions in the presence of molar concentration of hydrogen ions. Let the volume of the solution at the equivalence point be 100 cm³. Then $[Fe^{3+}] = 0.01\,N$, since it is known that the reaction is practically complete, $[Mn^{2+}] = \frac{1}{5} \times [Fe^{3+}] = 0.002N$, and $[Fe^{2+}] = x$. Let the excess of permanganate solution at the end point be one drop or 0.05 cm³; its concentration will be $0.05 \times 0.1/100 = 5 \times 10^{-5}N = [MnO_4^-]$. Substituting these values in the equation:

$$K = \frac{(2 \times 10^{-3}) \times (1 \times 10^{-2})^5}{(5 \times 10^{-5}) \times x^5 \times 1^8} = 3 \times 10^{63}$$

or $\quad x = [Fe^{2+}] = 5 \times 10^{-15} N$

It is clear from what has already been stated that standard reduction potentials may be employed to determine whether redox reactions are sufficiently complete for their possible use in quantitative analysis. It must be emphasised, however, that these calculations provide no information as to the speed of the reaction, upon which the application of that reaction in practice will ultimately depend. This question must form the basis of a separate experimental study, which may include the investigation of the influence of temperature, variations of pH and of the concentrations of the reactants, and the influence of catalysts. Thus, theoretically, potassium permanganate should quantitatively oxidise oxalic acid in aqueous solution. It is found, however, that the reaction is extremely slow at the ordinary temperature, but is more rapid at about 80 °C, and also increases in velocity when a little manganese(II) ion has been formed, the latter apparently acting as a catalyst.

It is of interest to consider the calculation of the equilibrium constant of the general redox reaction, viz.:

$$a\,Ox_I + b\,Red_{II} \rightleftharpoons b\,Ox_{II} + a\,Red_I$$

The complete reaction may be regarded as composed of two oxidation–reduction electrodes. aOx_I, $a\,Red_I$ and $b\,Ox_{II}$, $b\,Red_{II}$ combined together into a cell; at equilibrium, the potentials of both electrodes are the same:

$$E_1 = E_1{}^\ominus + \frac{0.0591}{n}\log\frac{[Ox_I]^a}{[Red_I]^a}$$

$$E_2 = E_2{}^\ominus + \frac{0.0591}{n}\log\frac{[Ox_{II}]^b}{[Red_{II}]^b}$$

At equilibrium, $E_1 = E_2$, hence:

$$E_1{}^\ominus + \frac{0.0591}{n}\log\frac{[Ox_I]^a}{[Red_I]^a} = E_2{}^\ominus + \frac{0.0591}{n}\log\frac{[Ox_{II}]^b}{[Red_{II}]^b}$$

$$\text{or}\quad \log\frac{[Ox_{II}]^b \times [Red_I]^a}{[Red_{II}]^b \times [Ox_I]^a} = \log K = \frac{n}{0.0591}(E_1{}^\ominus - E_2{}^\ominus)$$

This equation may be employed to compute the equilibrium constant of any redox reaction, provided the two standard potentials $E_1{}^\ominus$ and $E_2{}^\ominus$ are known; from the value of K thus obtained, the feasibility of the reaction in analysis may be ascertained.

It can readily be shown that the concentrations at the equivalence point, when equivalent quantities of the two substances Ox_I and Red_{II} are allowed to react, are given by:

$$\frac{[Red_I]}{[Ox_I]} = \frac{[Ox_{II}]}{[Red_{II}]} = {}^{a+b}\!\sqrt{K}$$

This expression enables us to calculate the exact concentration at the equivalence point in any redox reaction of the general type given above, and therefore the feasibility of a titration in quantitative analysis.

II, 26. Selected bibliography
A full discussion of the topics considered in this Chapter will be found in textbooks of physical chemistry, and many of the topics are featured in textbooks of electrochemistry: a selection of books for further reading follows.

1. J. N. Butler (1964). *Solubility and pH Calculations*. Reading, Mass.; Addison Wesley Publishing Co.
2. J. N. Butler (1964). *Ionic Equilibrium*. Reading, Mass.; Addison Wesley Publishing Co.
3. C. W. Davies (1967). *Electrochemistry*. London; George Newnes Ltd.

4. R. B. Fischer and D. G. Peters (1969). *A Brief Introduction to Quantitative Chemical Analysis*. 3rd edn. Philadelphia; W. B. Saunders Co.
5. H. A. Flaschka, A. J. Barnard and P. E. Sturrock (1969). *Quantitative Analytical Chemistry*. Vol. I. New York; Barnes and Noble.
6. S. Glasstone (1942). *Introduction to Electrochemistry*. New York; Van Nostrand Co. Inc.
7. L. F. Hamilton, S. G. Simpson and D. W. Ellis (1969). *Calculations of Analytical Chemistry*. New York; McGraw-Hill Inc.
8. I. M. Kolthoff and P. J. Elving (1959). *Treatise on Analytical Chemistry. Part I, Theory and Practice*. Vol. I. New York; Interscience Publishers Inc.
9. H. A. Laitinen (1960). *Chemical Analysis. An Advanced Text and Reference*. New York; McGraw-Hill & Co.
10. W. J. Moore (1972). *Physical Chemistry*. 5th edn. London; Longmans.
11. F. Steel (1970). *Grundlagen der Analytischen Chemie*. 5th edn. Weinheim; Verlag Chemie.
12. W. F. Sheehan (1970). *Physical Chemistry*. 2nd edn. Boston; Allyn and Bacon.
13. D. A. Skoog and D. M. West (1970). *Fundamentals of Analytical Chemistry*. 2nd edn. London; Holt, Rinehart and Winston.
14. T. B. Smith (1940). *Analytical Processes*. 2nd edn. London; Arnold.
15. C. L. Wilson and D. W. Wilson (1959). *Comprehensive Analytical Chemistry*. Vol. IA. Amsterdam; Elsevier Publishing Co.
16. J. G. Dick (1973). *Analytical Chemistry*. New York; McGraw-Hill Book Co.

CHAPTER III

COMMON APPARATUS AND BASIC TECHNIQUES

III INTRODUCTION. In this Chapter the more important basic techniques and the apparatus commonly used in analytical operations will be described, and it is essential that the beginner should become familiar with these procedures, and also acquire dexterity in handling the various pieces of apparatus. The habit of clean, orderly working must also be cultivated, and observance of the following points will be helpful in this direction.

1. The bench must be kept clean and a bench-cloth must be available so that any spillages of solid or liquid chemicals (solutions) can be removed immediately.

2. All glassware must be scrupulously clean (see Section **III, 14**), and if it has been standing for any length of time, must be rinsed with distilled or de-ionised water before use. The outsides of vessels may be dried with a lint-free glass-cloth which is reserved exclusively for this purpose, and which is frequently laundered, but the cloth should not be used on the insides of the vessels.

3. Under no circumstances should the working surface of the bench become cluttered with apparatus. All the apparatus associated with some particular operation should be grouped together on the bench; this is most essential to avoid confusion when duplicate determinations are in progress. Apparatus for which no further immediate use is envisaged should be returned to the locker, but if it will be needed at a later stage, it may be placed at the back of the bench.

4. If a solution, precipitate, filtrate, etc., is set aside for subsequent treatment, the container must be labelled so that the contents can be readily identified, and the vessel must be suitably covered to prevent contamination of the contents by dust: in this context, bark corks are usually unsuitable; they invariably tend to shed some dust. For temporary labelling, a 'Chinagraph' pencil or a felt tip pen which will write directly on to glass is preferable to the gummed labels which are used when more permanent labelling is required.

5. Reagent bottles must never be allowed to accumulate on the bench; they must be replaced on the reagent shelves immediately after use.

6. It should be regarded as normal practice that all determinations are performed in duplicate.

7. A stiff covered notebook of A4 size must be provided for **recording experimental observations as they are made**. A double page should be devoted to each determination, the title of which, together with the date, must be clearly indicated. One of the two pages must be reserved for the experimental observations, and the other should be used for a brief description of the procedure followed, but with a full account of any special features associated with the determination. In most cases it will be found convenient to divide the page on which the experimental observations are to be recorded into two halves by a vertical line: the observations relating to duplicate determinations can then be recorded side by side.

The record must conclude with the calculations and the results with appropriate comments upon the degree of accuracy achieved.

BALANCES

III, 2. THE ANALYTICAL BALANCE. One of the most important tools of the analytical chemist is the balance, and it is essential that the underlying principles of the theory and construction of this fundamental item of equipment should be understood.

The conventional free-swinging, equal-arm chemical balance is now rarely used, but to appreciate the developments which have been incorporated in present-day analytical balances (see Section **III, 6**) it is necessary to discuss the

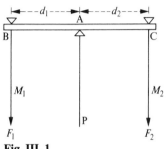

Fig. III, 1

essential features of the simple balance. Such a balance may be regarded as a rigid beam BC having a central fulcrum (A) and two arms of equal length; the two ends of the beam carry prism edges upon which the balance pans are supported by means of a suitable suspension (Fig.III, 1). Let us suppose a body of mass M_1 is placed on the left-hand pan of the balance; the pointer P (attached to the beam) will be deflected to the right. To restore the pointer to its original position, bodies of known mass, termed 'weights', are added to the right-hand pan. When equilibrium is restored, the principle of the lever requires that the following relation holds:

$$F_1 \times d_1 = F_2 \times d_2$$

where F_1 and F_2 are the forces acting upon the left-hand and right-hand prism edges respectively, and d_1 and d_2 are the respective distances of these from the central prism edge. Since the balance has equal arms, $d_1 = d_2$ and $F_1 = F_2$. Now, the origin of the forces F_1 and F_2 lies in the attraction of gravity on the bodies in the left-hand and right-hand pans respectively, or, otherwise expressed,

$$F_1 = M_1g \text{ and } F_2 = M_2g$$

where M_1 and M_2 are the masses (or quantities of matter) in the left-hand and right-hand pans respectively, and g is the acceleration due to gravity. Strictly speaking, F_1 and F_2 are the true **weights** in the two pans. But:

$$\frac{F_1}{F_2} = \frac{M_1g}{M_2g} = \frac{M_1}{M_2}$$

i.e., the ratio of the forces with which the bodies in the two pans are attracted by gravity is equal to the ratio of the two masses. In quantitative analysis we are interested only in the amount of matter in the body, i.e., in its mass: this is independent of g. At any given place, the weights are proportional to the masses. It has become customary to employ the term 'weight' synonymously with the mass, and it is in this sense that 'weight' is employed in quantitative analysis. The analytical balance, strictly speaking, determines mass and not weight.

Although in the above discussion it was considered that the weight M_2 was made equal to the weight M_1 so that the pointer P was returned to its original position, it will in fact suffice if M_2 is adjusted to be close to, but slightly less than M_1, and then the deflection of the pointer from the null position measured. Using known differences between M_1 and M_2, the deflection of the pointer can be calibrated with reference to weight, and so in any instance the additional weight which needs to be added to M_2 can be deduced from the deflection: M_1 is then equal to M_2 plus the weight corresponding to the observed deflection of the pointer. The object to be weighed is of course always placed in the left hand pan of the balance.

III, 3. ESSENTIAL FEATURES OF THE ANALYTICAL BALANCE. The beam must be light but at the same time sufficiently rigid so that it does not bend under load; aluminium alloys are commonly used as the construction material, but a recent innovation is the use of titanium. The suspension points (the fulcrum A and the pan-carrying knife edges at B and C) are constructed of synthetic sapphire, and each knife edge bears upon a flat plate of the same material; this arrangement provides a very hard-wearing, almost frictionless, suspension system. It is however essential that the knife edges be protected from undue wear, and from injury during the transfer of weights or objects to or from the pans, and the balance is therefore provided with a device whereby the knife edges and the planes are slightly separated when the beam is brought to rest.

A pointer is attached to the centre of the beam and moves over a scale at the bottom of the pillar on which the beam is supported; a small moveable weight is attached to the pointer and this can be used to adjust the centre of gravity of the balance. The beam is also provided with adjusting screws, commonly at its ends, so that slight inequalities between the two sides of the beam (including the attached pans), can be compensated for. The beam carries a scale divided into 100 equal parts so that adjustments of weight of less than 10 mg can be readily made by moving a **rider** along the scale. The **rider** consists of a small piece of gold or platinum wire, bent into such a shape that it will straddle and rest upon the scale; movement of the rider is carried out by means of a special hook and carriage attachment.

Balances are usually classified according to their capacity and the following ranges are commonly recognised: 'Analytical', capacity 150–200 g, precision (see Section **III, 4**) 0.1 mg; 'Semi-micro', capacity 75–100 g, precision 0.01 mg; 'Micro', capacity 10–30 g, precision 0.001 mg.

III, 4. THE REQUIREMENTS OF A GOOD BALANCE. The requirements of a good balance may be summarised as follows.

(*a*) *The balance must have good reproducibility or precision.* The **precision** may be expressed in terms of the standard deviation of a series of weighings of the same mass, and is a function of the quality of the balance, of the environment in which it is used, and the skill of the operator. For good precision the environment must be vibration-free and the balance must be protected from air currents and from corrosive fumes. In terms of construction, the requirements are that (i) the arms of the beam are of equal length, (ii) the beam is rigid and does not bend appreciably under load, (iii) the three bearing edges lie in the same plane and are parallel to each other.

(*b*) *The balance must be stable, that is, the beam must return to the horizontal position after swinging.* This is attained by proper adjustment of the centre of gravity.

(*c*) *The balance must be sensitive, that is, 0.1 mg should be readily detectable with average loads.* We may define the **sensitivity of a balance** as the angular deflection α of the beam when a known small weight is added. It can be shown that the angle α is determined by the excess of weight w producing the deflection α, the length, d, of the balance arm, the weight, W, of the beam, and the distance, h, between the centre of gravity and the point of support of the beam. Expressed mathematically (see Ref. 1),

$$\tan \alpha = \frac{wd}{Wh}$$

The angular deflection of the beam is equal to the angular deflection of the pointer, and the latter is directly proportional to the number of divisions between the two points of rest on the scale at the foot of the beam. This leads directly to the usual definition of sensitivity, viz., the **sensitivity** of a balance is the number of scale divisions that the rest point (or equilibrium point) is displaced by a weight of 1 mg.

In an ideal balance, free from friction and with a perfectly rigid beam, the sensitivity would be independent of the load. Most balances, however, exhibit a decreasing sensitivity with increasing load, and this change of sensitivity provides a good criterion as to the maximum safe load that a balance can carry. The criterion is: no greater load should ever be placed upon the balance pans than the load at which the sensitivity becomes 40 per cent of its maximum value.

III, 5. WEIGHTS, REFERENCE MASSES. The determination of the mass of an object with an equal-arm balance necessitates the use of a series of reference masses termed **weights**. For scientific work, the international metric system of weights and measures is employed. The fundamental standard of mass is the **international prototype kilogram**, which is a mass of platinum–iridium alloy made in 1887 and deposited in the International Bureau of Weights and Measures near Paris. Authentic copies of the standard are kept by the appropriate responsible authorities* in the various countries of the world; these copies are employed for the comparison of secondary standards, which are used in the calibration of weights for scientific work. The unit of mass that is almost universally employed in laboratory work, however, is the **gram**, which may be defined as the one-thousandth part of the mass of the international prototype kilogram.

An ordinary set of analytical weights contains the following: grams, 100, 50, 30, 20, 10, 5, 3, 2, 1: milligrams, 500–100 and 50–10 in the same 5, 3, 2, 1 sequence. Other sequences, such as 5, 2, 1, 1, or 5, 2, 2, 1, are also encountered, but the provision of duplicate weights is not recommended. The set of weights will also contain a rider (see Section **III, 3**) for weights below 10 mg. The weights

---

*The National Physical Laboratory (NPL) in Great Britain, the National Bureau of Standards (NBS) in USA, etc.

from 1 g upwards are constructed from a non-magnetic nickel – chromium alloy (80 % Ni, 20 % Cr), or from austenitic stainless steel; plated brass is sometimes used but is less satisfactory. The fractional weights are made from the same alloys, or from a non-tarnishable metal such as gold or platinum. For handling the weights, a pair of forceps, preferably ivory-tipped, are provided and the weights are stored in a box with suitably shaped compartments.

A set of weights should be calibrated before it is adopted for laboratory use and it is usually advisable to recalibrate at yearly intervals: the calibration is carried out by comparison of each piece of the set against the corresponding piece from a standard set which has been calibrated by a nationally recognised institution, e.g., in Great Britain, the National Physical Laboratory. If a set of standard weights is not available, then the weights in one set may be inter-calibrated: for details of this procedure see T. W. Richards (Ref. 3).

The National Physical Laboratory at Teddington recognises only one grade of weights, 'Class A', in which the following tolerances are permitted: 100 g, 0.5 mg; 50 g, 0.25 mg; 30 g, 0.15 mg; 20 g, 0.10 mg; 10 g–100 mg, 0.05 mg; 50–10 mg, 0.02 mg.

The National Bureau of Standards at Washington recognises the following classes of precision weights:

Class M. For use as reference standards, for work of the highest precision, and where a high degree of constancy over a period of time is required.

Class S. For use as working reference standards or as high precision analytical weights.

Class S-1. Precision analytical weights for routine analytical work.

Class J. Microweight standards for microbalances.

It must be emphasised again (cf. Section **III, 2**) that the 'weights' which have been discussed are strictly masses: some laboratory suppliers do now list 'boxes of masses', but most analysts will undoubtedly still refer to 'weights'.

III, 6. TWO-KNIFE SINGLE-PAN BALANCE. The simple balance described in Section **III, 2** has been subject to many modifications; these have been chiefly directed towards improving the speed of weighing, but other beneficial results have also accrued. An interesting account of the development of the analytical balance has been given by J. T. Stock (Ref. 2). Important developments culminating in the introduction of the two-knife single-pan balance by E. Mettler in 1946 are discussed briefly in the following paragraphs.

A. Aperiodic balances. In the aperiodic balance an air-damping device is attached to the beam: this consists of pistons attached to the beam and operating in stationary cylinders closed at one end and with the minimum clearance between piston and cylinder. The resulting damping effect brings the beam to rest 10–15 seconds after release, and if the object to be weighed is slightly heavier than the weights deployed in the pan, the beam will be tilted up, and the pointer will be deflected to the right. The displacement from the null position can be measured by means of a graticule attached either to the right-hand end of the beam, or to the foot of the pointer, and illuminated by a suitably placed lamp. The scale in the graticule can be calibrated to measure the weight corresponding to the deflection of the pointer, and with this arrangement the rider can be dispensed with, and all weights below 10 mg can be read on the optical scale: an appropriate optical system is incorporated so that the scale reading is displayed on a suitably placed ground-glass screen. The principle can be extended, and the optical scale enlarged, to read up to 100 mg or even up to 1000 mg with consequent elimination of some or all the fractional weights; in such cases some form of vernier or micrometer is included to facilitate the reading of the fourth decimal place (0.1 mg). The sensitivity of the balance is unaffected by the damping effect but clearly the optical scale will only be accurate for loads within the range in which the sensitivity is constant.

B. Dial-controlled weight loading.   Instead of having to transfer weights individually from the box to the balance pan, it is a great improvement to have the weights suspended within the balance case in such a position that a series of hooked levers can place them on (or remove them from) a carrier bar attached to the right-hand arm of the balance beam. For ease in handling, the weights are fabricated in ring or other convenient form, and the operating levers are controlled by knobs with engraved dials on the outside of the balance case: these dials are rotated to add or remove appropriate rings from the carrier bar. A common arrangement is to provide two dials, one covering the range 0.1–0.9 g and the other the range 1–9 g: the weights from 10 g upwards are contained in a special holder located within the balance case, and of course, the weight below 0.1 g is determined from an optical scale as described under *A*. Alternatively, the optical scale may be used to cover the range 0–1000 mg and the two weight-loading dials are then used respectively for the 1–9 g and 10–90 g ranges: additional weights are then no longer required.

C. Controlled release mechanisms.   In order to maintain the sensitivity-load relationship and the precision of weighing, wear on the knife edges of the balance must be kept to a minimum. This is achieved by controlling the rate of impact of the knife edges upon the supporting planes when the balance is released. In the 'Releas-o-matic' device used by Messrs Oertling, the beam movement was controlled pneumatically by a graphite piston moving within a graphite cylinder: the use of graphite obviates lubrication problems and the device always operates smoothly. The rate of movement of the piston is controlled by the rate at which air enters the cylinder through an adjustable needle valve. Some manufacturers use a small synchronous motor coupled to a gear unit which releases the beam at a steady, slow rate.

D. Preweighing devices.   The weighing operation can be greatly speeded up if the weight of the object to be weighed is already known approximately. Accordingly, most manufacturers now incorporate in their balances a 'pre-weighing' facility. In some systems, the beam release lever is turned in one direction so that the beam is only partially released, and weights are then added until the beam is almost balanced. The release lever is then turned back to the rest position, and still moving in the same direction, to the completely free position; the optical scale then comes into operation and the full weight is obtained.

With Oertling and Sartorius balances, operation of the pre-weigh lever leads to the beam becoming supported on a spring which is fitted with an optical read-out system giving the approximate weight of the object (see Fig. III, 4). The correct weights may then be selected on the weight-loading dials and the final weight is obtained by releasing the beam in the normal manner.

In the **two-knife single-pan balance** the foregoing improvements are incorporated in an instrument in which the balance beam is unsymmetrical, one balance pan with its suspension is replaced by a counterpoise, and the dial-operated weights are suspended from a carrier attached to the remaining pan support: these changes are illustrated in Fig. III, 2.

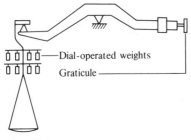

—Dial-operated weights
Graticule—

Fig. III, 2

In this system, all the ring weights are permanently in position on the carrier when the balance is at rest, and if the beam is then released, the weights exactly balance the counterpoise and the balance is at the zero point. When an object to be weighed is placed, upon the pan, weights must bè *removed* from the carrier to compensate for the weight of the object. The weighing is then completed by allowing the beam to take up its rest position and reading, on an optical scale, the displacement of the beam. Weighing is thus accomplished by **substitution**: the weight of the object on the balance pan is substituted for the weights removed from the carrier. This method of weighing possesses certain advantages over the normal procedure involved with a conventional two-pan balance:

(*a*) the load on the beam is maintained constant and therefore the sensitivity of the balance remains constant;

(*b*) there are no errors arising from any inequality in the two arms of the balance;

(*c*) accuracy is enhanced by the elimination of one knife edge: it is easier to achieve co-planarity of two knife edges than of three {cf. Section **III, 4** (*a*), (iii)}.

A minor disadvantage is that with the beam always under maximum load there tends to be increased wear of the knife edges; however, by good design, and by making use of controlled release, this defect can be minimised, and in any case, is far outweighed by the advantages of the system. At the present time, this type of balance may be considered as the standard analytical balance, and two-arm

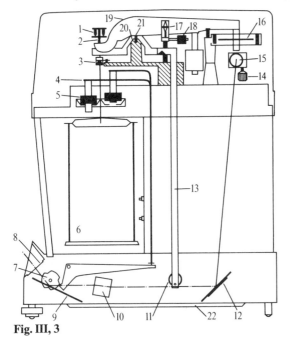

1 Compensating stirrup
2 Front knife edge
3 Overhead pan brake
4 Weight carriage
5 Built-in weights
6 Weighing pan
7 Weight control mechanism
8 Optical projection screen
9 Micrometer mirror
10 Micrometer prism
11 Arrestment cam
12 Zero point mirror
13 Arrestment rod
14 Objective focus control
15 Optical scale and objective
16 Air damper
17 Sensitivity adjustment
18 Macro zero adjustment
19 Beam
20 Main bearing
21 Main knife
22 Voltage selector edge

Fig. III, 3

balances must be regarded as obsolete. The essential features of a typical substitution balance (the Sartorius 2400 range) are shown in Fig. III, 3. The principle of the pre-weighing system of this balance is shown in Fig. III, 4. In the pre-weigh position, all the built-in weights are removed from the carrier, the beam bears down on the supporting spring, and the weight of the object is read off on the scale shown on the right of the diagram. An actual balance of this type (Oertling Model R41) is shown in Fig. III, 5. In this instrument, the readings of the dials controlling the weights removed from the carrier are displayed in windows which are adjacent to those provided for the optical scale, and thus the weight of the object can be read immediately as a complete digital display. A taring device is also incorporated in the balance.

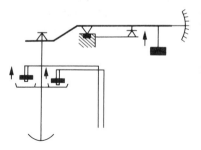

Fig. III, 4

Fig. III, 5

The weights included in such a balance are made from non-magnetic stainless steel, and are adjusted to within NPL Class A limits, so that for most purposes, calibration is unnecessary. The weights can of course be readily checked against a standard set of weights, and it is good practice to make such a comparison at appropriate intervals. A method for inter-calibration of the weights attached to one balance has been described by Lashof and MacCurdy (Ref. 4).

III, 7. TOP LOADING BALANCES. Originally developed to enable masses of up to about 1 kg to be weighed rapidly with an accuracy of about 0.1 g, the top loading balance has now become an indispensable item of laboratory equipment. Slightly different models will cater for loads of up to 5 kg with an accuracy of about 0.1 g on the one hand, or with loads of up to 200 g with an accuracy of 0.01 g or even 0.001 g: with these more sensitive models it is essential that the balance be provided with some form of screen to shield the pan from draughts. A typical balance of this type (Oertling Model TD30) is shown in Fig. III, 6.

Fig. III, 6

In such a balance weighing up to 1 kg, weights moving up in steps of 100 g are controlled by a knob at the side of the balance, and the weight selected appears as a digit on an optical read-out panel on the front of the instrument. The remaining weights (tens, units and tenths of a gram) then appear on the same read-out panel and are obtained from a magnified image of a graticule attached to the balance beam. Such balances incorporate a special knife suspension system, and the speed of weighing is partly attained with the aid of a magnetic damping device to reduce the oscillations of the beam. A taring device is frequently fitted, and this is very useful when it is required to weigh out a given quantity of material.

III, 8. ELECTRONIC BALANCES. A recent development in the design of two-knife single pan balances is the replacement of the optical read-out system used for the fractional weights by an electrically operated measuring system. The principle of the system incorporated in the Mettler HE20 electronic analytical balance is shown in Fig. III, 7. Dial-operated weights (tens and units) are added or removed in accordance with the usual substitution principle; the weights selected are detected electrically (9) and the appropriate figures displayed on a

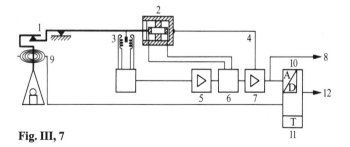

Fig. III, 7

digital display unit (10). The fractional part of the weight is determined with the aid of an electrical sensor (3) which responds to the deflection of the beam, and modulates an amplifier (5), thus generating a current in the 'compensation system' (2). The compensation system current gives rise to a magnetic field which serves to maintain the beam in equilibrium: the magnitude of the current necessary to produce the requisite magnetic field is proportional to the weight which is being determined. This information is likewise transmitted to the display unit (10). Connection (12) can also be made to a print-out unit so that a permanent record of the weight can be recorded automatically.

The 'Gravitron' is a novel form of top-loading balance introduced by International Electronics Ltd. It contains no moving parts, and the weight of an object placed on top of the balance is obtained by means of a transducer in conjunction with a solid-state electrical circuit; the weight is indicated on a neon-tube digital display.

III, 9. OTHER TYPES OF BALANCES. Amongst other types of balances, brief reference will be made to the following.

A. Torsion suspension balances. In balances of this type, the normal beam is replaced by two comparatively slender, parallel beams, held in position by metal bands which are pulled taut around metal supports. Three supports are employed: one at each end of the beam assembly which carry the balance pans, and one in the centre which supports the complete assembly. There are no knife edges to wear, and as the beam does not have to be arrested, weighing is speeded up.

B. Torsion balances. If a suitable fine wire or fibre is stretched taut, and a horizontal beam attached to its mid-point, then if a weight is placed at one end of the beam, thus pulling it downwards, the suspension wire experiences a torque which causes it to twist. By means of a graduated dial attached to one end of the suspension wire, the latter can be rotated manually until the beam is returned to the horizontal position. The dial can be calibrated against known weights so that it reads the weight directly.

This is the principle used in decimicrobalances such as the Oertling Q01, in which the torsion system is made from a quartz fibre, and which weighs to 0.0001 mg.

C. Electrobalances. This term is applied to balances such as the Cahn electrobalance, in which an electromagnetic force is used to counteract the deflection of the balance beam caused by adding a weight to one side. At the fulcrum of the beam, and at right angles to it, a wire coil is attached. This coil is mounted between the poles of a permanent magnet, one situated above, and the other below the coil. If a current is passed through the coil the resultant electromagnetic interaction applies a torque to the beam. The current is adjusted by means of a potentiometer until the beam is restored to the null position, the potentiometer dial being calibrated to read weights directly; a sensitivity of $0.1–0.02\ \mu g$ can be achieved.

III, 10. CARE AND USE OF ANALYTICAL BALANCES. No matter what type of analytical balance is employed (single pan or two pan), due attention must be paid to the manner in which it is used. The following points should be carefully observed.

1. The balance should be placed upon a firm foundation which is as free from mechanical vibration as possible. The ideal foundation is a concrete or stone slab resting upon brick piers, which are either sunk into the ground or, if this is not practicable, into the concrete floor or sub-floor of the laboratory. If this is not possible, the balance should be set up on a stout table or shelf and protected, when necessary, by sheets of shock-absorbing media, such as cork mats or sheet rubber on which the balance is placed: anti-vibration tables, designed for balances, are available from most laboratory suppliers. It is best to keep the balance in a room separate from the laboratory in order to protect it from fumes, and it should be located in a draught-free position away from direct sunlight.

The balance must be level. This adjustment may be made with the aid of the levelling-screws and spirit levels on the base of the instrument.

2. When not in use, the balance beam should be raised so as to protect the knife edges and bearing planes. The doors of the balance should be kept closed whenever possible.

3. To release the balance, the beam should be lowered very gently.

4. Objects to be weighed must be allowed to attain the temperature of the balance before weighing is attempted, otherwise the air currents produced inside the balance case may introduce errors. If the object has been heated, sufficient time must be allowed for cooling. The time required to attain the balance-room temperature varies with the size, etc., of the object, but as a rule 30–40 minutes is sufficient.

5. The object to be weighed should always be placed in the centre of the pan; the same remark applies to the weights if a two pan balance is used.

6. The weights used with a non-dial loading balance must be handled only with the forceps provided.

7. When objects are being added to or removed from the pan, the beam arrest must be raised so as to protect the knife edges from injury. A similar remark applies to alteration of the weights on a single pan balance, unless it is in the 'pre-weigh' position.

8. As soon as all external weights have been added, the balance case must be closed. Hence with a fully dial-operated single pan balance, the case will be closed as soon as the object has been placed on the pan.

9. No chemicals or objects which might injure the balance pans should ever be placed directly upon them. Substances must be weighed in suitable containers, such as small beakers, weighing bottles or crucibles, or upon watch glasses.

Liquids and volatile or hygroscopic solids must be weighed in tightly closed vessels, such as stoppered weighing bottles.

10. The balance must not be overloaded (see Section **III, 4** (*c*)).

11. Nothing must be left on the pan when the weighing has been completed. If any substance is spilled accidentally upon the pan or upon the floor of the balance case, it must be removed at once. The pans should be lightly brushed periodically with a camel-hair brush to remove dust which may have collected.

12. A beginner should never attempt to adjust a balance: help should be sought from an experienced operator.

In the actual weighing process, the exact sequence of operations will be partly dependent upon the make of balance in use and the arrangement of the controls, but with a single pan balance it will include the following steps.

1. Sit opposite the centre of the balance.
2. Brush the pan lightly with the camel hair brush to remove any dust.
3. Carefully release the beam and check that the empty balance gives a zero reading: if necessary the requisite adjustment should be made.
4. With the beam at rest, place the object, which must be at or near room temperature, on the pan, and close the balance case.
5. Set the balance to the 'preweigh' position and from the scale reading select the appropriate gram weights with the weight loading dials.
6. Release the beam fully and record the final weight: with some balances this may necessitate the adjustment of a vernier control to enable the fourth place (0.1 mg) to be read.
7. When weighings are completed, arrest the beam, return the weight dials to zero, remove the object which has been weighed, clear up any accidental spillages, and close the balance case.

III, 11. ERRORS IN WEIGHING. The chief sources of error are the following:
1. Change in the condition of the containing vessel or of the substance between successive weighings.
2. Effect of the buoyancy of the air upon the object and the weights.
3. Inaccuracy of the weights.

1. The first source of error is occasioned by change in weight of the containing vessel: (*a*) by absorption or loss of moisture, (*b*) by electrification of the surface caused by rubbing, and (*c*) by its temperature being different from that of the balance case. These errors may be largely eliminated by wiping the vessel gently with a linen cloth, and allowing it to stand at least 30 minutes in the balance room before weighing. The electrification, *which may cause a comparatively large error, particularly if both the atmosphere and the cloth are dry*, is slowly dissipated on standing; it may be removed by placing a piece of pitchblende or similar *feebly* radioactive material in the balance case to ionise the air. Hygroscopic, efflorescent, and volatile substances must be weighed in completely closed vessels. Substances which have been heated in an air oven or ignited in a crucible are generally allowed to cool in a desiccator containing a suitable drying agent. The time of cooling in a desiccator cannot be exactly specified, since it will depend upon the temperature and upon the size of the crucible as well as upon the material of which it is composed. Platinum vessels require a shorter time than those of porcelain, glass, or silica. It has been customary to leave platinum crucibles in the desiccator for 20–25 minutes, and crucibles of other materials for

30–35 minutes before being weighed. It is advisable to cover crucibles and other open vessels.

2. When a substance is immersed in a fluid, its true weight is diminished by the weight of the fluid which it displaces. If the object and the weights have the same density, and consequently the same volume, no error will be introduced on this account. If, however, as is usually the case, the density of the object is different from that of the weights, the volumes of air displaced by each will be different. If the substance has a lower density than the weights, as is usual in analysis, the former will displace a greater volume of air than the latter, and it will therefore weigh less in air than in a vacuum. Conversely, if a denser material (e.g., one of the precious metals) is weighed, the weight in a vacuum will be less than the apparent weight in air.

Consider the weighing of 1 litre of water, first *in vacuo*, and then in air. It is assumed that the flask containing the water is tared by an exactly similar flask, that the temperature of the air is 20 °C and the barometric pressure is 760 mm of mercury. The weight of 1 litre of water *in vacuo* at 20 °C and 760 mm is 998.23 g. If the water is weighed in air, it will be found that 998.23 g are too heavy. We can readily calculate the difference. The weight of 1 litre of air displaced by the water is 1.20 g. Assuming the weights to have a density of 8.0, they will displace $998.23/8.0 = 124.8$, or $124.8 \times 1.20/1000 = 0.15$ g of air. The net difference in weight will therefore be $1.20 - 0.15 = 1.05$ g. Hence the weight in air of 1 litre of water under the experimental conditions named is $998.23 - 1.05 = 997.18$ g, a difference of 0.1 per cent from the weight *in vacuo*.

Let us now extend our enquiry to the case of a solid, such as potassium chloride, under the above conditions. The density of potassium chloride is 1.99. If 2 g of the salt are weighed, the apparent loss in weight (= weight of air displaced) is $2 \times 0.0012/1.99 = 0.0012$ g. The apparent loss in weight for the weights is $2 \times 0.0012/8.0 = 0.00030$ g. Hence 2 g of potassium chloride will weigh $0.0012 - 0.00030 = 0.00090$ g less in air than *in vacuo*, a difference of 0.05 per cent.

It must be pointed out that for most analytical purposes where it is desired to express the results in the form of a percentage, the ratio of the weights in air, so far as solids are concerned, will give a result which is practically the same as that which would be given by the weights *in vacuo*. Hence no buoyancy correction is necessary in these cases. However, where absolute weights are required, as in the calibration of graduated glassware, corrections for the buoyancy of the air must be made (compare Section **III, 15**).

Let us now consider the general case. It is evident that the weight of an object *in vacuo* is equal to the weight in air *plus* the weight of air displaced by the object *minus* the weight of air displaced by the weights. It can easily be shown that if W_v = weight *in vacuo*, W_a = apparent weight in air, d_a = density of air, d_w = density of the weights, and d_b = density of the body, then:

$$W_v = W_a + d_a \left(\frac{W_v}{d_b} - \frac{W_a}{d_w} \right)$$

The density of the air will depend upon the humidity, the temperature, and the pressure. For an average relative humidity (50 per cent) and average conditions of temperature and pressure in a laboratory, the density of the air will rarely fall outside the limits 0.0011 and 0.0013 g cm^{-3}. It is therefore permissible for analytical purposes to take the weight of 1 cm^3 of air as 0.0012 g.

Since the difference between W_v and W_a does not usually exceed 1 to 2 parts per thousand, we may write:

$$W_v = W_a + d_a \left(\frac{W_a}{d_b} - \frac{W_a}{d_w} \right)$$

$$= W_a + W_a \left\{ 0.0012 \left(\frac{1}{d_b} - \frac{1}{8.0} \right) \right\} = W_a + k W_a / 1000$$

where

$$k = 1.20 \left(\frac{1}{d_b} - \frac{1}{8.0} \right)$$

The values of k for $d_a = 0.0012$ and $d_w = 8.0$ have been calculated and are collected in Table III, 1. If a substance of density d_b weighs W_a grams in air, then $W_a . k$ *milligrams* are to be added to the weight in air in order to obtain the weight *in vacuo*. The correction is positive if the substance has a density lower than 8.0 (stainless steel), and negative if the density of the substance is greater than 8.0.

Table III, 1   Reductions of weighings made in air with weights of density 8.0 to *vacuo*

d_b	k	d_b	k	d_b	k
0.5	+2.25	1.9	+0.48	11.0	−0.04
0.6	+1.85	2.0	+0.45	12.0	−0.05
0.7	+1.56	2.5	+0.33	13.0	−0.06
0.8	+1.35	3.0	+0.25	14.0	−0.06
0.9	+1.18	3.5	+0.19	15.0	−0.07
1.0	+1.05	4.0	+0.15	16.0	−0.07
1.1	+0.94	4.5	+0.12	17.0	−0.08
1.2	+0.85	5.0	+0.09	18.0	−0.08
1.3	+0.77	5.5	+0.07	19.0	−0.09
1.4	+0.71	6.0	+0.05	20.0	−0.09
1.5	+0.65	7.0	+0.02	21.0	−0.10
1.6	+0.60	8.0	±0.00	22.0	−0.10
1.7	+0.56	9.0	−0.02	23.0	−0.10
1.8	+0.52	10.0	−0.03	24.0	−0.10

3. Accuracy of the weights can be ensured by periodical checks against a standard set of weights.

GRADUATED GLASSWARE

III, 12. UNITS OF VOLUME. For scientific purposes the convenient unit to employ for measuring reasonably large volumes of liquids is the cubic decimetre (dm^3), or, for smaller volumes, the cubic centimetre (cm^3). For many years the fundamental unit employed was the *litre*, based upon the volume occupied by one kilogram of water at $4\,°C$ (the temperature of maximum density of water): the relationship between the litre as thus defined and the cubic decimetre was established as

1 litre $= 1.000028\ dm^3$

or   1 millilitre $= 1.000028\ cm^3$.

In 1964 the *Conférence Générale des Poids et des Mésures* (*CGPM*) decided to accept the term **litre** as a special name for the cubic decimetre, and to discard the original definition of the litre. It is suggested (Ref. 5) that the litre should not be used to express results of high precision, and that the use of the term millilitre should not be encouraged. From the point of view of the analyst, however, the litre and the millilitre are sufficiently precise for the requirements of titrimetric analysis, and by virtue of established usage the millilitre will for many years be regarded as synonymous with the cubic centimetre.

III, 13. TEMPERATURE STANDARD. The capacity of a glass vessel varies with the temperature, and it is therefore necessary to define the temperature at which its capacity is intended to be correct. A temperature of 20 °C has been almost universally adopted. A subsidiary standard temperature of 27 °C is accepted by the British Standards Institution, for use in tropical climates where the ambient temperature is consistently above 20 °C. The US Bureau of Standards, Washington, in compliance with the view held by some chemists that 25 °C more nearly approximates to the average laboratory temperature in the United States, will calibrate glass volumetric apparatus marked either 20 °C or 25 °C.

If we take the coefficient of cubical expansion of soda glass as about 0.000025 and of borosilicate glass about 0.000010 per 1 °C Table III, 2 gives the correction to be added when the sign is +, or subtracted when the sign is −, to or from the capacity of a 1000-cm^3 flask correct at 20 °C in order to obtain the capacity at other temperatures.

Table III, 2   Temperature corrections for 1000-cm^3 glass flask for the expansion of glass (standard temperature, 20 °C)

Temperature (°C)	Correction (cm^3)	
	Soda glass	Borosilicate glass
5	−0.39	−0.15
10	−0.26	−0.10
15	−0.13	−0.05
20	0.00	0.00
25	+0.13	+0.05
30	+0.26	+0.10

In the use of graduated glassware for measurement of the volume of liquids, the expansion of the liquid must also be taken into consideration if temperature corrections are to be made. Table III, 3 gives the corrections to be added or subtracted in order to obtain the volume occupied at 20 °C by a volume of water which at the tabulated temperature is contained in an accurate 1000-cm^3 flask having a standard temperature of 20 °C. It will be seen that the allowance for the expansion of water is considerably greater than that for the expansion of the glass. For dilute (e.g., 0.1 M) aqueous solutions, the corrections can be regarded as approximately the same as for water, but with more concentrated solutions the correction increases, and for non-aqueous solutions the corrections can be quite large (Ref. 6a).

Table III, 3   Temperature corrections for volumes of water measured in a 1000-cm^3 glass flask (standard temperature, 20°C)

Temperature (°C)	Correction (cm^3)	
	Soda glass	Borosilicate glass
5	+1.37	+1.61
10	+1.24	+1.40
15	+0.77	+0.84
20	0.00	0.00
25	−1.03	−1.11
30	−2.31	−2.46

III, 14. GRADUATED APPARATUS. The most commonly used apparatus in titrimetric (volumetric) analysis are graduated flasks, burettes, and pipettes. Graduated cylinders and weight pipettes are less widely employed. Each of these will be described in turn.

Graduated apparatus for quantitative analysis is generally made to specification limits, particularly with regard to the accuracy of calibration. In Great Britain there are two grades of apparatus available, designated Class A and Class B by the British Standards Institution. The tolerance limits are closer for Class A apparatus, and such apparatus is intended for use in work of the highest accuracy: Class B apparatus is employed in routine work. In the United States, specifications for only one grade are available from the National Bureau of Standards at Washington, and these are equivalent to the British Class A.

Cleaning of glass apparatus.   Before describing graduated apparatus in detail, reference must be made to the important fact that all such glassware must be perfectly clean and free from grease, otherwise the results will be unreliable. One test for cleanliness of glass apparatus is that on being filled with distilled water and the water withdrawn, only an unbroken film of water remains. If the water collects in drops, the vessel is dirty and must be cleaned. Various methods are available for cleaning glassware.

Many commercially available detergents are suitable for this purpose, and some manufacturers market special formulations for cleaning laboratory glassware; some of these, e.g., 'Decon 90' made by Decon Laboratories of Portslade, are claimed to be specially effective in removing contamination due to radioactive materials.

'**Teepol**' is a relatively mild and inexpensive detergent which may be used for cleaning glassware. The laboratory stock solution may consist of a 10 per cent solution in distilled water. For cleaning a burette, 2 cm^3 of the stock solution diluted with 40 cm^3 of distilled water is poured into the burette, allowed to stand for $\frac{1}{2}$ to 1 minute, the detergent run off, the burette rinsed thrice with tap water, and then several times with distilled water. A 25 cm^3 pipette may be similarly cleaned using 1 cm^3 of the stock solution diluted with 25–30 cm^3 of distilled water.

A method which is frequently used consists in filling the apparatus with 'cleaning mixture', a nearly saturated solution of powdered sodium or potassium dichromate in concentrated sulphuric acid, and allowing it to stand for several hours, preferably overnight; the acid is then poured off, the apparatus thoroughly rinsed with distilled water, and allowed to drain until dry. [It may be mentioned that potassium dichromate is not very soluble in concentrated sulphuric acid (about 5 g per litre), whereas sodium dichromate $Na_2Cr_2O_7,2H_2O$ is much more soluble (about 70 g per litre); for this reason, as well as the fact that it is much cheaper, the latter is usually preferred for the preparation of 'cleaning mixture'. From time to time it is advisable to filter the sodium dichromate–sulphuric acid mixture through a little glass wool placed in the apex of a glass funnel: small particles or sludge, which are often present and may block the tips of burettes, are thus removed.] A more efficient cleaning liquid is a mixture of concentrated sulphuric acid and fuming nitric acid; this may be used if the vessel is very greasy and dirty, but must be handled with extreme caution.

A very effective de-greasing agent, which it is claimed is much quicker acting than 'cleaning mixture', is obtained by dissolving 100 g of potassium hydroxide in 50 cm³ of water, and after cooling, making up to 1 litre with industrial methylated spirit (Ref. 6b).

III, 15. GRADUATED FLASKS. A graduated flask (known alternatively as a volumetric flask or a measuring flask), is a flat-bottomed, pear-shaped vessel with a long narrow neck. A thin line etched around the neck indicates the volume that it holds at a certain definite temperature, usually 20 °C (both the capacity and temperature are clearly marked on the flask); the flask is then said to be graduated *to contain*. Flasks with one mark are always taken *to contain* the volume specified. A flask may also be marked *to deliver* a specified volume of liquid under certain definite conditions; these are, however, not suitable for exact work and are not widely used. Vessels intended to contain definite volumes of liquid are marked C or TC or In, while those intended to deliver definite volumes are marked D or TD.

The mark extends completely around the neck in order to avoid errors due to parallax when making the final adjustment; the lower edge of the meniscus should be tangential to the graduation mark, and both the front and the back of the mark should be seen as a single line. The neck is made narrow so that a small change in volume will have a large effect upon the height of the meniscus: the error in adjustment of the meniscus is accordingly small.

The flasks should be fabricated in accordance with BS 1792 and the opening should be ground to standard (interchangeable) specifications and fitted with an interchangeable glass or plastic (commonly polypropylene) stopper. They should conform to either Class A or Class B specification; examples of permitted tolerances for the latter Grade are as follows:

Flask size	5	25	100	250	1000	cm³
Tolerance	0.04	0.06	0.15	0.30	0.80	cm³

For Class A flasks the tolerances are approximately halved: such flasks may be purchased with a works calibration certificate, or with a British Standard Test (BST) Certificate.

Graduated flasks are available in the following capacities: 1, 2, 5, 10, 20, 50, 100, 200, 250, 500, 1000, 2000 and 5000 cm³. They are employed in making up standard solutions to a given volume; they can also be used for obtaining, with the aid of pipettes, aliquot portions of a solution of the substance to be analysed.

Calibration. For most analytical purposes flasks of Class A standard may be used without calibration, but for the highest accuracy, all flasks (unless carrying a *recent* BST Certificate) should be calibrated; this involves determining the weight of water held by the flask when it is filled to the mark. For this purpose a large balance which will accommodate the largest flask to be calibrated (say one litre) is required: a top pan balance of suitable loading and sensitivity characteristics may be used.

The flask is first thoroughly cleaned and dried, and after standing in the balance room for an hour is stoppered and weighed. A small filter funnel, the stem of which has been drawn out so that it reaches below the graduation mark of the flask, is then inserted into the neck and de-ionised (distilled) water, which has also been standing in the balance room for an hour, is added slowly until the

Table III, 4   Weight of water to give one litre at 20 °C*

Flask of soda glass, coefficient of cubical expansion, 0.000025/°C

Temp. (°C)	Weight (g)	Volume of 1 g of water (cm³)	Temp. (°C)	Weight (g)	Volume of 1 g of water (cm³)
10	998.39	1.0016	23	996.60	1.0034
11	998.32	1.0017	24	996.38	1.0036
12	998.23	1.0018	25	996.17	1.0038_5
13	998.14	1.0018_5	26	995.93	1.0041
14	998.04	1.0019	27	995.69	1.0043
15	997.93	1.0021	28	995.44	1.0046
16	997.80	1.0022	29	995.18	1.0048
17	997.66	1.0023	30	994.91	1.0051
18	997.51	1.0025	31	994.64	1.0054
19	997.35	1.0026	32	994.35	1.0057
20	997.18	1.0028	33	994.06	1.0060
21	997.00	1.0030	34	993.75	1.0063
22	996.80	1.0032	35	993.45	1.0066

Flask of borosilicate glass, coefficient of cubical expansion, 0.000010/°C

Temp. (°C)	Weight (g)	Volume of 1 g of water (cm³)	Temp. (°C)	Weight (g)	Volume of 1 g of water (cm³)
15	998.00	1.0020	24	996.33	1.0037
16	997.86	1.0021	25	996.09	1.0039
17	997.71	1.0023	26	995.85	1.0042
18	997.54	1.0025	27	995.49	1.0045
19	997.37	1.0026	28	995.32	1.0047
20	997.18	1.0028	29	995.05	1.0050
21	996.98	1.0030	30	994.76	1.0053
22	996.78	1.0032	31	994.47	1.0056
23	996.56	1.0034_5	32	994.17	1.0059

Note. For the calibration of flasks of capacity other than 1 litre, the corresponding multiple or sub-multiple of the above values is taken.

* The above figures refer to the apparent weight in grams in air against brass weights, density 8.4 g cm⁻³. The modern basis for density of weights of 8.0 g cm⁻³ will result in a difference of 7.1 parts per million: this will obviously not affect the table to the significant figures quoted.

More elaborate tables will be found in BS 1797:1968 (Tables for Use in the Calibration of Volumetric Glassware).

mark is reached. The funnel is then carefully removed, taking care not to wet the neck of the flask above the mark, and then, using a dropping tube, water is added dropwise until the meniscus stands on the graduation mark. The stopper is replaced, the flask reweighed, and the temperature of the water noted.

The true volume of the water filling the flask to the graduation mark can be calculated with the aid of Table III, 4. The values in the table have been obtained by making allowance for (a) the difference in volume of the glass vessel at the calibration temperature and at 20 °C, (b) the density of water at the temperature of the calibration, and (c) the effect of buoyancy of the air upon the water and the brass weights. The figures apply to an atmospheric pressure of 760 mm of mercury and a relative humidity of the air of 50 per cent; the usual deviation from these figures will affect the buoyancy correction (compare Section **III, 11**) only slightly and can be neglected for most purposes.

III, 16. PIPETTES. Pipettes are of two kinds: (i) those which have one mark and *deliver* a small, constant volume of liquid under certain specified conditions (**transfer pipettes**); (ii) those in which the stems are graduated and are employed to deliver various small volumes at discretion (**graduated or measuring pipettes**). The **transfer pipette** consists of a cylindrical bulb joined at both ends to narrower tubing: a calibration mark is etched around the upper (suction) tube, while the lower (delivery) tube is drawn out to a fine tip. The graduated or measuring pipette is usually intended for the delivery of pre-determined variable volumes of liquid: it does not find wide use in accurate work for which a burette is generally preferred. Transfer pipettes are constructed with capacities of 1, 2, 5, 10, 20, 25, 50 and 100 cm³; those of 10, 25 and 50 cm³ capacity are most frequently employed in macro work. They should conform to BS 1583 and should carry a colour code ring at the suction end to identify the capacity (BS 3996): as a safety measure an additional bulb is often incorporated above the graduation mark. They may be fabricated from lime-soda or Pyrex glass, and some high-grade pipettes are manufactured in Corex glass (Corning Glass Works, USA). This is glass which has been subjected to an ion exchange process which strengthens the glass and also leads to greater surface hardness, thus giving a product which is resistant to scratching and chipping. Pipettes are available to Class A and Class B specifications: for the latter Grade typical tolerance values are:

Pipette capacity	5	10	25	50	100	cm³
Tolerance	0.01	0.04	0.06	0.08	0.12 cm³	

whilst for Class A, the tolerances are approximately halved.

In using such pipettes, they are first rinsed with the liquid, then filled by suction to about 1–2 cm above the mark, and the upper end of the pipette is closed with the tip of the dry index finger (Fig. III, 8); any adhering liquid is wiped from the outside of the lower stem. The liquid is allowed to run out slowly by slightly relaxing the pressure of the finger and by carefully rotating the pipette until the bottom of the meniscus just reaches the graduation mark; the pipette must be held vertically so that the mark is at the same level as the eye. Any drops adhering to the tip are removed by stroking against a

Fig. III, 8

glass surface. The liquid is then allowed to run into the receiving vessel, the tip of the pipette touching the wall of the vessel. When the continous discharge has ceased, the jet is held in contact with the side of the vessel for 15 seconds (**draining time**). At the end of the draining time, the tip of the pipette is removed from contact with the wall of the receptacle; the liquid remaining in the jet of the pipette must not be removed either by blowing or by other means.

A pipette will not deliver constant volumes of liquid if discharged too rapidly. The orifice must be of such size that the time of outflow is about 20 seconds for a 10-cm^3 pipette, 30 seconds for a 25-cm^3 pipette, and 35 seconds for a 50-cm^3 pipette.

Various devices are available for handling corrosive or toxic liquids with transfer pipettes. Some attachments (e.g. the **Griffin pipette filler**) consist of a rubber or plastic bulb with glass ball valves operated between finger and thumb: these control the entry and expulsion of air to and from the bulb, and thus the flow of liquid into and out of the pipette. In other devices, a piston-control is attached to the suction end of the pipette. With the **'Exelo' safety pipette**, the suction end of the pipette fits snugly into a hollow barrel with an air vent at the top. With the barrel pushed right down, the tip of the pipette is placed into the liquid, the vent closed by the fore-finger, and by pulling the barrel slowly upwards, liquid is sucked into the pipette until it is above the graduation mark; the pipette is then controlled by finger pressure on the air vent and operated as a conventional pipette.

Calibration.   Class A pipettes are usually satisfactory for most analytical purposes and may be purchased with a BST Certificate. When calibration of a pipette is necessary, the following procedure should be used.

The pipette must first be thoroughly cleaned using one of the cleaning agents referred to in Section **III, 14**. If it is necessary to soak the pipette for an extended period of time, it may be left standing in the cleaning solution contained in a tall jar: a chromatography jar or a tall measuring cylinder are suitable. Alternatively, attach a short piece of rubber tubing and a pinch clip to the upper end of the pipette, and after filling completely with the cleaning solution, close the pinch clip and clamp the pipette in a vertical position with the jet dipping into cleaning solution contained in a beaker. After this treatment, wash the pipette thoroughly with tap water and finally with distilled water.

The pipette is then filled with distilled water, which has been standing in the balance room for at least an hour, to a short distance above the mark. Water is run out until the meniscus is exactly on the mark, and the out-flow is then stopped. The drop adhering to the jet is removed by bringing the surface of some water contained in a beaker in contact with the jet, and then removing it without jerking. The pipette is then allowed to discharge into a clean, weighed stoppered flask (or a large weighing bottle) and held so that the jet of the pipette is in contact with the side of the vessel (it will be necessary to incline slightly either the pipette or the vessel). The pipette is allowed to drain for 15 seconds after the outflow has ceased, the jet still being in contact with the side of the vessel. At the end of the draining time the receiving vessel is removed from contact with the tip of the pipette, thus removing any drop adhering to the outside of the pipette and ensuring that the drop remaining in the end is always of the same size. To determine the instant at which the outflow ceases, the motion of the water surface down the delivery tube of the pipette is observed, and the delivery time is considered to be complete when the meniscus comes to rest slightly above the end of the delivery tube. The draining time of 15 seconds is counted from this moment. The receiving vessel is weighed, and the temperature of the water noted. The capacity of the pipette is then calculated with the aid of Table III, 4. At least two determinations should be made.

Graduated pipettes consist of straight, fairly narrow tubes with no central bulb, and are also constructed to a standard specification (BS 700); they are likewise colour coded in accordance with BS 3996. Three different types are available:

Type 1 delivers a measured volume from a top zero to a selected graduation mark;

Type 2 delivers a measured volume from a selected graduation mark to the jet: i.e. the zero is at the jet;

Type 3 calibrated to *contain* a given capacity from the jet to a selected graduation mark, and thus to *remove* a selected volume of solution.

Automatic pipettes. The Dafert pipette (Fig. III, 9) is an automatic version of a transfer pipette. One side of the two-way tap is connected to a reservoir

containing the solution to be dispensed, and when the tap is in the appropriate position, solution fills the pipette completely, excess solution draining away through the overflow chamber. The pipette now contains a definite volume of solution which is delivered to the receiver by appropriate manipulation of the tap. These pipettes, which are constructed to conform to BS 1132, are available in a range of sizes from 5–100 cm^3 and are useful in routine work.

Autodispensers are also useful for measuring definite volumes of solutions on a routine basis. solution is forced out of a container by depressing a syringe plunger: the movement of the plunger and hence the volume of liquid dispensed, is controlled by means of a moveable clamp. the plunger is spring loaded, so that when released,

Fig. III, 9   it returns to its original position and is immediately ready for operation again.

Tilting pipettes, which are attached to a reagent bottle, are only suitable for delivering approximate volumes of solution.

III, 17. BURETTES.   Burettes are long cylindrical tubes of uniform bore throughout the graduated length, terminating at the lower end in a glass stop-cock and a jet; in cheaper varieties, the stopcock may be replaced by a rubber pinch valve incorporating a glass sphere. A diaphragm-type plastic burette tap is marketed: this can be fitted to an ordinary burette and provides a delicate control of the outflow of liquid. The merits claimed include: (*a*) the tap cannot stick, because the liquid in the burette cannot come into contact with the threaded part of the tap; (*b*) no lubricant is generally required; (*c*) there is no contact between ground glass surfaces; and (*d*) burettes and taps can be readily replaced. Burette taps made of polytetrafluoroethylene (PTFE or Teflon) are also available; these have the great advantage that no lubricant is required.

It is sometimes advantageous to employ a burette with an extended jet which is bent twice at right angles so that the tip of the jet is displaced by some 7.5–10 cm from the body of the burette. Insertion of the tip of the burette into complicated assemblies of apparatus is thus faciliated, and there is a further advantage, that if heated solutions have to be titrated the body of the burette is kept away from the source of heat. Burettes fitted with two-way stopcocks are useful for attachment to reservoirs of stock solutions.

As with other graduated glassware, burettes are produced to both Class A and Class B specifications in accordance with the appropriate standard (BS 846), and Class A burettes may be purchased with BST Certificate. All Class A and some Class B burettes have graduation marks which completely encircle the burette; this is a very important feature for the avoidance of parallax errors in reading the burette. Typical values for the tolerances permitted for Class A burettes are:

Total capacity	5	10	50	100	cm^3
Tolerance	0.02	0.02	0.06	0.10 cm^3	

for Class B, these values are approximately doubled. In addition to the volume requirements, limits are also imposed on the length of the graduated part of the burette and on the drainage time.

When in use, a burette must be firmly supported on a stand, and various types of burette holders are available for this purpose. The use of an ordinary laboratory clamp is not recommended: the ideal type of holder permits the burette to be read without the need of removing it from the stand, and amongst holders which the authors have found to be particularly satisfactory are the Fisher burette holder, in both the original and the cheaper students' version, and the Gallenkamp burette holder.

Lubricants for glass stopcocks. The object of lubricating the stopcock of a burette is to prevent sticking or 'freezing' and to ensure smoothness in action. The simplest lubricant is pure Vaseline, but this is rather soft, and, unless used sparingly, portions of the grease may readily become trapped at the point where the jet is joined to the barrel of the stopcock, and lead to blocking of the jet. Various products are available commercially (e.g., Gallenkamp rubber grease) which are better suited to the lubrication of burette stopcocks. *Silicone-containing lubricants should be avoided* since they tend to 'creep' with consequent contamination of the walls of the burette.

To lubricate the stopcock, the plug is removed from the barrel and two thin streaks of lubricant are applied to the length of the plug on lines roughly midway between the ends of the bore of the plug. Upon replacing in the barrel and turning the tap a few times, a uniform thin film of grease is distributed round the ground joint. A spring or some other form of retainer may be subsequently attached to the key to lessen the chance of it becoming dislodged when in use.

Reference is again made to the Teflon stopcocks and to the diaphragm type of burette tap which do not require lubrication.

The mode of use of a burette is as follows. If necessary, the burette is thoroughly cleaned using one of the cleaning agents described in Section **III, 14**, and is then well rinsed with distilled water. The plug of the stopcock is removed from the barrel, and after wiping the plug and the inside of the barrel dry, the stopcock is lubricated as described in the preceding paragraph. Using a small funnel, about 10 cm^3 of the solution to be used are introduced into the burette, and then after removing the funnel, the burette is tilted and rotated so that the solution flows over the whole of the internal surface; the liquid is then discharged through the stopcock. After repeating the rinsing process, the burette is clamped *vertically* in the burette holder and then filled with the solution to a little above the zero mark. The funnel is removed, and the liquid discharged through the stopcock until the lowest point of the liquid meniscus just touches the zero mark; the jet is inspected to ensure that all air bubbles have been removed and that it is completely full of liquid. To read the position of the meniscus, the eye must be at the same level as the meniscus, in order to avoid errors due to parallax. In the best type of burette, the graduations are carried completely round the tube for each cm³ and half-way round for the other graduation marks: parallax is thus easily avoided. To aid the eye in reading the position of the meniscus a piece of white paper or cardboard, the lower half of which is blackened either by painting with dull black paint or by pasting a piece of dull black paper upon it, is employed. When this is placed so that the sharp dividing line is 1–2 mm below the meniscus, the bottom of the meniscus appears to be darkened and is sharply outlined against the white background; the level of the liquid can then be accurately read. A variety of 'burette readers' are available from laboratory supply houses, and a home-made device which is claimed to be particularly effective has been described by Woodward and Redman (Ref. 6c). For all ordinary purposes readings are made to 0.05 cm³, but for precision work, readings should be made to 0.01–0.02 cm³, using a lens to assist the estimation of the subdivisions.

To deliver liquid from a burette into a conical flask or other similar receptacle, place the fingers of the left hand behind the burette and the thumb in front, and hold the tap between the thumb and the fore and middle fingers (Fig. III, 10). In

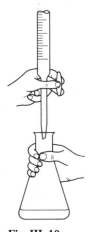

this way, there is no tendency to pull the plug out of the barrel of the stopcock, and the operation is under complete control. Any drop adhering to the jet after the liquid has been discharged is removed by bringing the side of the receiving vessel into contact with the jet. During the delivery of the liquid, the flask may be gently rotated with the right hand to ensure that the added liquid is well mixed with any existing contents of the flask.

Calibration of a burette. If it is necessary to calibrate a burette, it is essential to establish that it is satisfactory with regard to (a) leakage, and (b) delivery time, before undertaking the actual calibration process. The burette must naturally be subjected to a thorough cleaning and rinsing procedure, and then to test for leakage, the plug is removed from the barrel of the stopcock and both parts of the stopcock are carefully cleaned of all grease; after wetting well with de-ionised water, the stopcock is reassembled. The burette is placed in the holder,

Fig. III, 10 filled with distilled (de-ionised) water, adjusted to the zero mark, and any drop of water adhering to the jet removed with a piece of filter paper. The burette is then allowed to stand for ten minutes, and if the meniscus has not fallen by more than one half of a scale division, the burette may be regarded as satisfactory as far as leakage is concerned.

To test the delivery time, again separate the components of the stopcock, dry, grease and reassemble, then fill the burette to the zero mark with distilled water, and place in the holder. Adjust the position of the burette so that the jet comes inside the neck of a conical flask standing on the base of the burette stand, but does not touch the side of the flask. Open the stopcock fully, and note the time taken for the meniscus to reach the lowest graduation mark of the burette: this should agree closely with the time marked on the burette, and in any case, must fall within the limits laid down by BS 846.

If the burette passes these two tests, the calibration may be proceeded with. Fill the burette with distilled water which has been allowed to stand in the balance room for at least an hour to acquire room temperature: ideally, this should be as near to 20 °C as possible. Weigh a clean, dry stoppered flask of about 100 cm³ capacity, then after adjusting the burette to the zero mark and removing any drop adhering to the jet, place the flask in position under the jet, open the stopcock fully and allow water to flow into the flask. As the meniscus approaches the desired calibration point on the burette, reduce the rate of flow until eventually it is discharging dropwise, and adjust the meniscus exactly to the required mark. Do not wait for drainage, but remove any drop adhering to the jet by touching the neck of the flask against the jet, then re-stopper and reweigh the flask. Repeat this procedure for each graduation to be tested; for a 50 cm^3 burette, this will usually be every 5 cm^3. Note the temperature of the water, and then, using Table III, 4, the volume delivered at each point is calculated from the weight of water collected. The results are most conveniently used by plotting a calibration curve for the burette.

III, 18.   WEIGHT BURETTES.   For work demanding the highest possible accuracy in transferring various quantities of liquids, weight burettes are employed. As their name implies, they are weighed before and after a transfer of liquid. A very useful form is shown diagrammatically in Fig. III, 11(*a*). There are two ground-glass caps, the lower one is closed, whilst the upper one is provided with a capillary opening; the loss by evaporation is accordingly negligible. For hygroscopic liquids, a small ground-glass cap is fitted to the top of the capillary tube. The burette is roughly graduated in 5-cm^3 intervals. The titre thus obtained is in terms of weight loss of the burette, and for this reason the titrants are prepared on a weight/weight basis rather than a weight/volume basis. The errors associated with the use of a volumetric burette, such as those of drainage, reading, and change in temperature, are obviated, and weight burettes are especially useful when dealing with non-aqueous solutions or with viscous liquids. The advantages of weight titrations are discussed in Ref. 7.

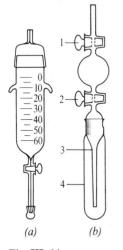

An alternative form of weight burette due to Redman (Ref. 6d) consists of a glass bulb, flattened on one side so that it will stand on a balance pan. Above the flattened side is the stopcock-controlled discharge jet, and a filling orifice which is closed with a glass stopper. The stopper and short neck into which it fits are pierced with holes, by alignment of which air can be admitted, thus permitting discharge of the contents of the burette through the delivery jet.

The Lunge–Rey pipette is shown in Fig. III, 11(*b*). There is a small central bulb (5–10 cm^3 capacity) closed by two stopcocks *1* and *2*; the pipette *3* below the stopcock has a capacity of about 2 cm^3, and is fitted with a ground-on test-tube *4*. This pipette is of particular value for the weighing out of corrosive and fuming liquids.

(*a*)   (*b*)

Fig. III, 11

III, 19.   PISTON BURETTES.   In piston burettes, the delivery of the liquid is controlled by movement of a tightly fitting plunger within a graduated tube of uniform bore. They are particularly useful when the piston is coupled to a motor drive, and in this form serve as the basis of automatic titrators such as the instruments supplied *inter alia* by Mettler Ltd, Metrohm Ltd, Radiometer Ltd. These instruments can provide automatic plotting of titration curves, and provision is made for a variable rate of delivery as the end-point is approached so that there is no danger of overshooting the end-point.

III, 20. GRADUATED (MEASURING) CYLINDERS. These are graduated vessels available in capacities from 2 to 2000 cm^3. Since the area of the surface of the liquid is much greater than in a graduated flask, the accuracy is not very high. Graduated cylinders cannot therefore be employed for work demanding even a moderate degree of accuracy. They are, however, useful where only rough measurements are required.

WATER FOR LABORATORY USE

III, 21. PURIFIED WATER. From the earliest days of quantitative chemical measurements it has been recognised that some form of purification is required for water which is to be employed in analytical operations, and with increasingly lower limits of detection being attained in instrumental methods of analysis, correspondingly higher standards of purity are imposed upon the water used for preparing solutions. Standards have now been laid down for water to be used in laboratories (Ref. 8), which prescribe limits for non-volatile residue, for residue remaining after ignition, for pH and for conductivity. The British Standard 3978 gives the limit for non-volatile residue as 5 mg l^{-1}, for residue after ignition as 2 mg l^{-1}, for pH, 5.0–7.5, and for conductivity, 10 megohm^{-1} per centimetre.

For many years the sole method of purification available was by distillation, and **distilled water** was universally employed for laboratory purposes. The modern water-still is usually made of glass, is heated electrically, and provision is made for interrupting the current in the event of failure of the cooling water, or of the boiler-feed supply; the current is also cut off when the receiver is full.

Pure water can also be obtained by allowing tap water to percolate through a mixture of ion-exchange resins: a strong acid resin which will remove cations from the water and replace them by hydrogen ions, and a strong base resin (OH$^-$ form) which will remove anions. A number of units are commercially available (Permutit, Elgastat, etc.) for the production of **de-ionised water**, and the usual practice is to monitor the quality of the product by means of a conductivity meter. The resins are usually supplied in an interchangeable cartridge, so that maintenance is reduced to a minimum. A mixed-bed ion-exchange column fed with distilled water is capable of producing water with the very low conductivity of about 0.2×10^{-6} ohm^{-1} cm^{-1}, but in spite of this very low conductivity, the water may contain traces of organic impurities which can be detected by means of a spectrofluorimeter. For most purposes however the traces of organic material present in de-ionised water can be ignored, and it may be used in most situations where distilled water is acceptable.

An alternative method of purifying water is by **reverse osmosis**. Under normal conditions, if an aqueous solution is separated by a semi-permeable membrane from pure water, osmosis will lead to water entering the solution to dilute it. If however, sufficient pressure is applied to the solution, i.e. a pressure in excess of its osmotic pressure, then water will flow through the membrane *from* the solution; the process of reverse osmosis is taking place.

This principle has been adapted in the Milli-Q3 system of the Millipore Corporation (Bedford, Massachusetts) as a method of purifying tap water. The tap water, at a pressure of 3–5 atmospheres, is passed through a tube containing the semi-permeable membrane. The permeate which is collected usually still contains traces of inorganic material and is therefore not suitable for operations requiring very pure water, but it will serve for many laboratory purposes, and is very suitable for further purification by ion-exchange treatment. In the Milli-Q2 system, water produced by reverse osmosis is passed first through a bed of activated charcoal which removes organic contaminants, and is then passed through a mixed bed ion-exchange column; the resultant effluent will then meet the most stringent requirements.

III, 22. WASH BOTTLES. A wash bottle is a flat-bottomed flask fitted up to deliver a fine stream of distilled water or other liquid for use in the transfer and washing of precipitates. A convenient size is a 500–750 cm^3 flask of Pyrex or other resistance glass; it should be fitted up as shown in Fig. III, 12. A rubber bung is used, and the glass tubes above the bung should be in the same straight

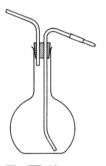

line and lie in the same plane. The jet should deliver a fine stream of water; a suitable diameter of the orifice is 1 mm. All glass tubing must be rounded in the Bunsen flame after cutting. Thick string, foam rubber, thin sheet cork, or other insulating material, held in place by copper wire, should be wrapped round the neck of the flask in order to protect the hand when hot water is used. Asbestos paper is best applied wet and allowed to dry overnight: there is sufficient adhesive material in the paper to make it cling tightly. In order to protect the mouth from scalding by the back rush of steam through the mouth-piece when the blowing is stopped, it is convenient to use a three-holed rubber stopper; a short piece of glass tubing open at both ends is inserted in the third hole. The thumb is kept over this tube whilst the water is being blown out, and is removed immediately before the mouth pressure is released. All-glass wash bottles, fitted with ground-glass joints, can be purchased. They should be used with organic solvents that attack rubber.

Fig. III, 12

A **polythene wash bottle** is available commercially and is inexpensive. It is fitted with a plastic cap and also with a plastic jet, and has flexible sides. The bottle can be held in the hand; application of slight pressure by squeezing gives an easily controllable jet of water. It is more or less unbreakable and is inert to many wash liquids. A polythene wash bottle should be used only for cool liquids.

Polythene wash bottles are sometimes charged with wash liquids other than water. Attention must be drawn to the fact that the components of some wash solutions may pass into the polythene and may be released into the space in the bottle when it is set aside: repeated fillings and rinsings may be required to remove the chemicals from the bottle. It is safer to label the wash bottle and to reserve it for the special wash liquid. Such wash solutions include a weakly acid solution saturated with hydrogen sulphide, dilute aqueous ammonia, saturated bromine water, and dilute nitric acid.

GENERAL APPARATUS

III, 23.   GLASSWARE, CERAMICS, PLASTIC WARE.   In the following sections, a brief account of general laboratory apparatus relevant to quantitative analysis will be given. The commonest materials of construction of such apparatus are glass, porcelain, fused silica, and various plastics; the merits and disadvantages of these are considered below.

Glassware.   In order to avoid the introduction of impurities during analysis, apparatus of resistance glass should be employed. For most purposes Pyrex glass (a borosilicate glass) is to be preferred. Resistance glass is very slightly affected by all solutions, but, in general, attack by acid solutions is less than that by pure water or by alkaline solutions; for this reason the latter should be acidified whenever possible, if they must be kept in glass for any length of time. Attention should also be given to watch, clock, and cover glasses; these should also be of resistance glass. As a rule, glassware should not be heated with a naked flame; a wire gauze, preferably with an asbestos centre, should be interposed between the flame and the glass vessel.

For special purposes, Corning Vycor glass (96 per cent silica) may be used. It has great resistance to heat and equally great resistance to thermal shock, and is unusually stable to acids (except hydrofluoric acid), water, and various solutions.

The most satisfactory **beakers** for general use are those provided with a spout. The advantages of this form are: (*a*) convenience of pouring, (*b*) the spout forms a convenient place at which a stirring rod may protrude from a covered beaker, and (*c*) the spout forms an outlet for steam or escaping gas when the beaker is covered with an ordinary clock glass. The size of a beaker must be selected with due regard to the volume of the liquid which it is to contain. The most useful sizes are from 250 to 600 cm^3.

Conical (or Erlenmeyer's) **flasks** of 200–500-cm^3 capacity find many applications, for example, in titrations.

Funnels should enclose an angle of 60°. The most useful sizes for quantitative analysis are those with diameters of 5.5, 7 and 9 cm. The stem should have an internal diameter of about 4 mm and should not be more than 15 cm long. For filling burettes and transferring solids to graduated flasks, a short-stem, wide-necked funnel is useful.

Porcelain apparatus.   Porcelain is generally employed for operations in which hot liquids are to remain in contact with the vessel for prolonged periods. It is usually considered to be more resistant to solutions, particularly alkaline solutions, than glass, although this will depend primarily upon the quality of the glaze. Shallow porcelain basins with lips are employed for evaporations. **Casseroles** are lipped, flat-bottomed porcelain dishes provided with handles; they are more convenient to use than dishes.

Porcelain **crucibles** are very frequently utilised for igniting precipitates and heating small quantities of solids because of their cheapness and their ability to withstand high temperatures without appreciable change. Some reactions, such as fusion with sodium carbonate or other alkaline substances, and also evaporations with hydrofluoric acid cannot be carried out in porcelain crucibles owing to the resultant chemical attack. A slight attack of the porcelain also takes place with pyrosulphate fusions.

Fused-silica apparatus.   Two varieties of silica apparatus are available commercially, the translucent and the transparent grades. The former is much cheaper and can usually be employed instead of the transparent variety. The advantages of silica ware are: (a) its great resistance to heat shock because of its very small coefficient of expansion, (b) it is not attacked by acids at a high temperature, except by hydrofluoric acid and phosphoric acid, and (c) it is more resistant to pyrosulphate fusions than is porcelain. The chief disadvantages of silica are: (a) it is attacked by alkaline solutions and particularly by fused alkalis and carbonates, (b) it is more brittle than ordinary glass, and (c) it requires a much longer time for heating and cooling than does, say, platinum apparatus.

Corning Vycor apparatus (96 per cent silica glass) possesses most of the merits of fused silica and is transparent. The smallest Vycor crucible has a capacity of 30 cm^3, but pure silica crucibles as small as 5 cm^3 are produced.

Plastic apparatus.   Plastic materials are widely used for a variety of items of common laboratory equipment such as aspirators, beakers, bottles, Buchner funnels and flasks, centrifuge tubes, conical flasks, filter crucibles, filter funnels, measuring cylinders, scoops, spatulas, stoppers, tubing, weighing bottles, etc.; such products are often cheaper than their glass counterparts, and are frequently less fragile. Although inert towards many chemicals, there are some limitations on the use of plastic apparatus, not the least of which is the generally rather low maximum temperature to which it may be exposed: salient properties of the commonly used plastic materials are summarised in Table III, 5.

Table III, 5   Plastics used for laboratory apparatus

Material	Appearance[a]	Highest temperature (°C)	Chemical reagents[b] Acids		Alkalis		Attacking organic solvents[c]
			Weak	Strong	Weak	Strong	
Polythene (L.D.)	TL	80–90	R	R*	V	R	1, 2
Polythene (H.D.)	TL–O	100–110	V	R*	V	V	2
Polypropylene	T–TL	120–130	V	R*	V	V	2
TPX (Polymethylpentene)	T	170–180	V	R*	V	V	1, 2
Polystyrene	T	85	V	R*	V	V	Most
PTFE (Teflon)	O	250–300	V	V	V	V	V
Polycarbonate	T	120–130	R	A	F	A	Most
PVC (Polyvinylchloride)	T–O	50–70	R	R*	R	R	2, 3, 4
Nylon	TL–O	120	R	A	R	F	V

(a) O = opaque; T = transparent; TL = translucent.
(b) A = attacked; F = fairly resistant; R = resistant; R* = generally resistant but attacked by oxidising mixtures; V = very resistant.
(c) 1 = hydrocarbons; 2 = chlorohydrocarbons; 3 = ketones; 4 = cyclic ethers; V = very resistant.

Attention is drawn to the extremely inert character of Teflon, which is so lacking in reactivity that it is used as the liner in pressure digestion vessels in which substances are decomposed by heating with hydrofluoric acid, or with concentrated nitric acid (see Section **III, 35**).

III, 24.   METAL APPARATUS.   Crucibles and basins required for special purposes are often fabricated from various metals, amongst which platinum holds pride of place by virtue of its general resistance to chemical attack.

Platinum. Platinum is used mainly for crucibles, dishes and electrodes; it has a very high melting point (1773 °C), but the pure metal is too soft for general use, and is therefore always hardened with small quantities of rhodium, iridium, or gold. These alloys are slightly volatile at temperatures above 1100 °C, but retain most of the advantageous properties of pure platinum, such as resistance to most chemical reagents, including molten alkali carbonates and hydrofluoric acid (the exceptions are dealt with below), excellent conductivity of heat, and extremely small adsorption of water vapour. A 25-cm^3 platinum crucible has an area of 80–100 cm^2 and, in consequence, the error due to volatility may be appreciable if the crucible is made of an alloy of high iridium content. The magnitude of this loss will be evident from the following table, which gives the approximate loss in weight of crucibles expressed in mg/100 cm^2/hour at the temperature indicated:

Temp. (°C)	Pure Pt	99 % Pt – 1 % Ir	97.5 % Pt – 2.5 % Ir
900	0.00	0.00	0.00
1000	0.08	0.30	0.57
1200	0.81	1.2	2.5

An alloy consisting of 95 per cent platinum and 5 per cent gold (e.g. Engelhard alloy 7070) is referred to as a 'non-wetting' alloy and it is claimed that fusion samples are readily removed from crucibles composed of this alloy.

Platinum crucibles should be supported, when heated, upon a platinum triangle. If the latter is not available, a silica triangle may be used. Nichrome and other metal triangles should be avoided; pipe-clay triangles may contain enough iron to damage the platinum. Hot platinum crucibles must always be handled with platinum-tipped crucible tongs; unprotected brass or iron tongs produce stains on the crucible. Platinum vessels must not be exposed to a luminous flame, nor may they be allowed to come into contact with the inner cone of a gas flame (see Fig. III, 13); this may result in the disintegration of the surface of the metal, causing it to become brittle, owing, probably, to the formation of a carbide of platinum.

It must be appreciated that at high temperatures platinum permits the flame gases to diffuse through it, and this may cause the reduction of some substances not otherwise affected. Hence if a covered crucible is heated by a gas flame there is a reducing atmosphere in the crucible: in an open crucible diffusion into the air is so rapid that this effect is not appreciable. Thus if iron(III) oxide is heated in a covered crucible, it is partly reduced to metallic iron, which alloys with the platinum; sodium sulphate is similarly partly reduced to the sulphide. It is, advisable, therefore, in the ignition of iron compounds or sulphates to place the crucible in a slanting position with free access of air.

Platinum apparatus may be used without significant loss for:

1. Fusions with (*a*) sodium carbonate or fusion mixture, (*b*) borax and lithium metaborate, (*c*) alkali bifluorides, and (*d*) alkali hydrogensulphates (slight attack in the last case above 700 °C, which is diminished by the addition of ammonium sulphate).

2. Evaporations with (*a*) hydrofluoric acid, (*b*) hydrochloric acid in the absence of oxidising agents which yield nascent chlorine, and (*c*) concentrated sulphuric acid (a slight attack may occur).

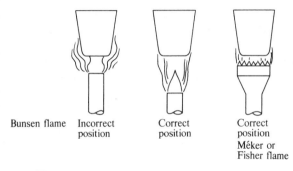

Bunsen flame   Incorrect   Correct   Correct
                position   position   position
                                        Méker or
                                        Fisher flame

Fig. III, 13

3. Ignition of (*a*) barium sulphate and sulphates of metals which are not readily reducible, (*b*) the carbonates, oxalates, etc., of calcium, barium and strontium, and (*c*) oxides which are not readily reducible, e.g., CaO, SrO, Al$_2$O$_3$, Cr$_2$O$_3$, Mn$_3$O$_4$, TiO$_2$, ZrO$_2$, ThO$_2$, MoO$_3$, and WO$_3$. (BaO, or compounds which yield BaO on heating, attack platinum.)

Platinum is *attacked* under the following conditions, and such operations must not be conducted in platinum vessels:

1. Heating with the following liquids: (*a*) aqua regia, (*b*) hydrochloric acid and oxidising agents, (*c*) liquid mixtures which evolve bromine or iodine, and (*d*) concentrated phosphoric acid (slight, but appreciable action after prolonged heating).

2. Heating with the following solids, their fusions, or vapours: (*a*) oxides, peroxides, hydroxides, nitrates, nitrites, sulphides, cyanides, hexacyano-ferrate(III), and hexacyanoferrate(II) of the alkali and alkaline-earth metals (except oxides and hydroxides of calcium and strontium); (*b*) molten lead, silver, copper, zinc, bismuth, tin, or gold, or mixtures which form these metals upon reduction; (*c*) phosphorus, arsenic, antimony, or silicon, or mixtures which form these elements upon reduction, particularly phosphates, arsenates, and silicates in the presence of reducing agents; (*d*) sulphur (slight action), selenium, and tellurium; (*e*) volatile halides (including iron(III) chloride), especially those which decompose readily; (*f*) all sulphides or mixtures containing sulphur and a carbonate or hydroxide; and (*g*) substances of unknown composition: (*h*) heating in an atmosphere containing chlorine, sulphur dioxide, or ammonia, whereby the surface is rendered porous.

Solid carbon, however produced, presents a hazard. It may be burnt off at low temperatures, with free access to air, without harm to the crucible, but it should never be ignited strongly. Precipitates in filter paper should be treated in a similar manner; strong ignition is only permissible after *all* the carbon has been removed. Ashing in the presence of carbonaceous matter should not be conducted in a platinum crucible, since metallic elements which may be present will attack the platinum under reducing conditions.

Cleaning and preservation of platinum ware.   All platinum apparatus (crucibles, dishes, etc.) should be kept clean, polished, and in proper shape. If, say, a platinum crucible becomes stained, a little sodium carbonate should be fused in the crucible, the molten solid poured out on to a dry stone or iron slab, the residual solid dissolved out with water, and the vessel then digested with concentrated hydrochloric acid: this treatment may be repeated, if necessary. If fusion with sodium carbonate is without effect, potassium hydrogen sulphate may be substituted; a slight attack of the platinum will occur. Disodium tetraborate may also be used. In some cases, the use of hydrofluoric acid or potassium hydrogen fluoride may be necessary. Iron stains may be removed by heating the covered crucible with a gram or two of A.R. ammonium chloride and applying the full heat of a burner for 2–3 minutes.

All platinum vessels must be handled with care to prevent deformation and denting. Platinum crucibles must on no account be squeezed with the object of loosening the solidified cake after a fusion. Box-wood formers can be purchased for crucibles and dishes; these are invaluable for re-shaping dented or deformed platinum ware.

Platinum-clad stainless steel laboratory ware is available for the evaporation of solutions of corrosive chemicals. These vessels have all the corrosion-resistance properties of platinum up to about 550 °C. The main features are: (i) much lower cost than similar apparatus of platinum; (ii) the overall thickness is about four times that of similar all-platinum apparatus, thus leading to greater mechanical strength; and (iii) less susceptible to damage by handling with tongs, etc.

Silver apparatus. The chief uses of silver crucibles and dishes in the laboratory are in the evaporation of alkaline solutions and for fusions with caustic alkalis; in the latter case, the silver is slightly attacked. Gold vessels (m.p. 1050 °C) are more resistant than silver to fused alkalis. Silver melts at 960 °C, and care should therefore be taken when it is heated over a bare flame.

Nickel ware. Crucibles and dishes of nickel are employed for fusions with alkalis and with sodium peroxide. In the peroxide fusion a little nickel is introduced, but this is usually not objectionable. No metal entirely withstands the action of fused sodium peroxide. Nickel oxidises in air, hence nickel apparatus cannot be used for operations involving weighing.

Iron ware. Iron crucibles may be substituted for those of nickel in sodium peroxide fusions. They are not so durable, but are much cheaper.

Stainless-steel ware. Beakers, crucibles, dishes, funnels, etc., of stainless steel are available commercially and have obvious uses in the laboratory. They will not rust, are tough, strong, and highly resistant to denting and scratching.

III, 25. HEATING APPARATUS. Various methods of heating are required in the analytical laboratory ranging from gas burners, electric hot plates and ovens to muffle furnaces.

Burners. The ordinary **Bunsen burner** is widely employed for the attainment of moderately high temperatures. The maximum temperature is attained by adjusting the regulator so as to admit rather more air than is required to produce a non-luminous flame; too much air gives a noisy flame, which is unsuitable.

Owing to the differing combustion characteristics and calorific values of the various gaseous fuels which are commonly available (town gas, natural gas, liquefied petroleum (bottled) gas), slight variations in dimensions, including jet size and aeration controls, are necessary: for maximum efficiency it is essential that, unless the burner is of the 'All Gases' type which can be adjusted, the burner should be the one intended for the available gas supply.

An improvement in design has been effected in burners in which both the gas and air supply can be regulated. The flow of gas is controlled at the base of the burner by means of a screw which operates a needle valve; the supply of air is regulated by screwing the tube of the burner up or down and thus allowing more or less air to enter through the holes at the base. The **Pittsburgh universal burner** (sometimes termed a **Tirril burner**) is of this type. A temperature of 1050–1150 °C in a covered platinum crucible or 600–700 °C in a covered porcelain crucible can be attained with these burners.

With a **Meker burner** a temperature of 1100–1200 °C is said to be reached in a covered platinum crucible and 800–900 °C in a covered porcelain crucible. The volume of air passing through a fully aerated ordinary Bunsen burner is about 2.5 times the volume of the gas (town gas); this is not sufficient for the complete combustion of the gas, but if attempts are made to increase the aeration, the flame 'strikes back' and burns at the bottom of the tube. In the Meker burner the holes for the admission of air are large enough to pass sufficient air for the complete combustion of the gas, and the tube is narrowest near the base and widens out near the top, thus resulting in a more perfect mixing of the gas with air; a nickel grid is fitted into the top of the burner in order to prevent the flame striking back. The gas burns in many small flames, with the top of each inner reducing cone about 1 mm above the top of the burner. The numerous small flames combine to give a very hot and highly concentrated flame, which is oxidising in character except below the tips of the tiny flames; the maximum temperature is attained just a little above the small flames, i.e., about 2–3 mm from the top of the burner. The burner is used for the ignition of precipitates that require a high temperature for conversion into a weighable form, and also for some fusions.

The **'Amal' burner** attempts to combine the chief features of the improved Bunsen burner and the Meker burner. The flame can be turned down very low without flashing back, and it also furnishes a very hot flame.

The so-called **'electric Bunsen burner'** is an electric heating unit designed so that the heat is directed by radiation and convection into a small volume. In one form, horizontal radiation from a vertically mounted tubular heating element is concentrated over a very small area by reflection from a polished, anodised parabolic reflector of pure aluminium: the glowing element may be fitted with a quartz sleeve to protect it against spillage. Attachments are available for heating crucibles, for conversion into a hot plate, and also into a small water bath. In another form, a replaceable heating element (conically shaped and in a refractory casing) is mounted in the top of a cylindrical housing, which forms the supporting case. The housing is provided with air-circulation holes in the lower part; a concentric gap between the heating element and the housing prevents any undue temperature rise. A variable transformer or an energy regulator controls the electric energy supplied, and hence the temperature. The attractive features include a concentrated source of heat, cleanliness, absence of smell, use in any position, and independence of draughts.

An **immersion heater ('red rod')**, consisting of a radiant heater encased in a silica sheath, is useful for the direct heating of most acids and other liquids (except hydrofluoric acid and concentrated caustic alkalis). Infrared radiation passes through the silica sheath with little absorption, so that a large proportion of heat is transferred to the liquid by radiation. The heater is almost unaffected by violent thermal shock due to the low coefficient of thermal expansion of the silica.

Steam or water baths. Boiling water or steam baths are employed for heating solutions just below boiling, for slow evaporation of liquids to reduce their volumes, for digestion of precipitates, etc. The simplest form is a lipped beaker in which water is boiled, the vessel being supported on the rim. Some of those available commercially have a small number of openings on the top, and these are fitted with a series of copper or stainless-steel rings: vessels of various sizes can be heated either on the surface of the bath or partially immersed. The bath is partly filled with water and heated by steam or electricity. Electrically heated water baths should preferably be provided with a constant-level device, thus eliminating the danger of running dry and consequent overheating; all should be fitted with a cut-out switch to prevent overheating if the water supply should fail.

Hot plates. The electrically-heated hot plate, preferably provided with three controls—'Low', 'Medium' and 'High'—is of great value in the analytical laboratory. The heating elements and the internal wiring should be totally enclosed; this protects them from fumes or spilled liquids. Electric hot plates with 'stepless' controls are also marketed; these permit of much greater selection of surface temperatures to be made. A combined electric hot plate and magnetic stirrer is described in Section **III, 27**.

Electric ovens. The most convenient type is an electrically heated, thermostatically controlled drying oven having a temperature range from room temperature to about 250–300 °C; the temperature can be controlled to within ± 1–2 °C. They are used principally for drying precipitates or solids at comparatively low controlled temperatures, and have virtually superseded the steam oven.

A recent introduction is the 'Mercury 450' **microwave oven** (marketed by Baird and Tatlock Ltd) which is particularly valuable for determining the moisture content of materials.

Muffle furnaces. An electrically heated furnace of muffle form should be available in every well-equipped laboratory. The maximum temperature should be about 1200 °C. If possible, a thermo-couple and indicating pyrometer should be provided; otherwise the ammeter in the circuit should be calibrated, and a chart constructed showing ammeter and corresponding temperature readings. Gas-heated muffle furnaces are marketed; these may give temperatures up to about 1200°C.

Air baths. For drying solids and precipitates at temperatures up to 250 °C in which acid or other corrosive vapours are evolved, an electric oven should not be used. An air bath may be constructed from a cylindrical metal (copper, iron, or nickel) vessel, wrapped with asbestos cloth (of about $\frac{1}{8}$ in. thickness) and held in position by copper wire ligatures. The bottom of the vessel may be pierced with numerous holes and covered with a circular asbestos board. A silica triangle, the legs of which are appropriately bent, is inserted inside the bath for supporting an evaporating dish, crucible, etc. The whole is heated by a Bunsen flame, which is shielded from draughts. The insulating layer of air prevents bumping by reducing the rate at which heat reaches the contents of the inner dish or crucible. An air bath of similar construction but with special heat-resistant glass sides may also be used; this possesses the obvious advantage of visibility inside the air bath.

Infrared lamps and heaters. Infrared lamps with internal reflectors are available commercially and are valuable for evaporating solutions. The lamp may be mounted immediately above the liquid to be heated: the evaporation takes place rapidly, without spattering and also without creeping. Units are obtainable which permit the application of heat to both the top and bottom of a number of crucibles, dishes, etc., at the same time; this assembly can char filter papers in crucibles quite rapidly, and the filter paper does not catch fire.

Crucible and beaker tongs. Crucible tongs should be made of solid nickel, nickel steel, or other rustless ferro-alloy. For handling hot platinum crucibles or dishes, platinum-tipped tongs must be used.

Beaker tongs are available for handling beakers (Griffin form) of 100–2000 cm³ capacity. They are made of stainless steel and have woven asbestos mittens. The tongs have stainless-steel jaws covered with asbestos sleeves and also die-cast aluminium grips. An adjustable screw with locknut limits the span of the jaws and enables the user to adjust the jaw span to suit the container size.

III, 26. DESICCATORS AND DRY BOXES. It is usually necessary to ensure that substances which have been dried by heating (e.g. in an oven, or by ignition), are not unduly exposed to the atmosphere, otherwise they will absorb moisture more or less rapidly. In many cases, storage in the dry atmosphere of a desiccator, allied to minimum exposure to the atmosphere during subsequent operations will be sufficient to prevent appreciable absorption of water vapour. Some substances however are so sensitive to atmospheric moisture that all handling must be carried out in a 'dry box'.

Fig. III, 14

A desiccator is a covered glass container designed for the storage of objects in a dry atmosphere. A common form of desiccator (Scheibler pattern) is shown in Fig. III, 14, it is usually charged with some drying agent, such as anhydrous calcium chloride (largely used in elementary work), silica gel, activated alumina, or anhydrous calcium sulphate ('Drierite'). Silica gel, alumina and calcium sulphate can be obtained which have been impregnated with a cobalt salt so that they are self-indicating: the colour changes from blue to pink when the desiccant is exhausted. The spent material can be regenerated by heating in an electric oven at 150–180 °C (silica gel); 200–230 °C (activated alumina): 230–250 °C (Drierite) and it is therefore convenient to place these drying agents in a shallow dish which is situated at the bottom of the desiccator, and which can be easily removed for baking as required.

The action of desiccants can be considered from two points of view. The amount of moisture that remains in a closed space, containing incompletely consumed desiccant, is related to the vapour pressure of the latter, i.e. the vapour pressure is a measure of the extent to which the desiccant can remove moisture,

Table III, 6   Comparative efficiency of drying agents

Drying agent	Residual water per litre of air in mg	Drying agent	Residual water per litre of air in mg
$CuSO_4$	2.8	KOH (sticks)	0.014
$CaCl_2$ (gran. 'anhyd.' tech.)	1.5	Al_2O_3	0.005
$ZnCl_2$ (sticks)	1.0	$CaSO_4$	0.005
NaOH (sticks)	0.8	H_2SO_4	0.003
H_2SO_4 (95 %)	0.3	$Mg(ClO_4)_2$	0.002
Silica gel	0.03	BaO	0.0007
$Mg(ClO_4)_2, 2H_2O$	0.03	P_2O_5	0.00002

and therefore its efficiency. A second factor is the weight of water that can be removed per unit weight of desiccant, i.e., the drying capacity. In general, substances that form hydrates have higher vapour pressures but also have greater drying capacities. It must be remembered that a substance cannot be dried by a desiccant the vapour pressure of which is greater than that of the substance itself.

The relative efficiencies of various drying agents will be evident from the data presented in Table III, 6. These were determined by aspirating properly conditioned air through U-tubes charged with the desiccants; they are applicable, strictly, to the use of these desiccants in absorption tubes, but the figures may reasonably be applied as a guide for the selection of desiccants for desiccators. It would appear from the table that a hygroscopic material such as ignited alumina should not be allowed to cool in a covered vessel over 'anhydrous' calcium chloride; anhydrous magnesium perchlorate or phosphorus pentoxide is satisfactory.

There is however much controversy regarding the effectiveness of desiccators. If the lid is briefly removed from a desiccator then it may take as long as two hours to remove the atmospheric moisture thus introduced, and to re-establish the dry atmosphere: during this period, a hygroscopic substance may actually gain in weight whilst in the desiccator. It is therefore advisable that any substance which is to be weighed should be kept in a vessel with as tightly fitting a lid as possible whilst it is in the desiccator.

The problem of the cooling of hot vessels within a desiccator is also important. A crucible which has been strongly ignited and immediately transferred to a desiccator may not have attained room temperature even after one hour. The situation can be improved by allowing the crucible to cool for a few minutes before transferring to the desiccator, and then a cooling time of 20–25 minutes is usually adequate. The inclusion in the desiccator of a metal block (e.g. aluminium), upon which the crucible may be stood, is also helpful in ensuring the attainment of temperature equilibrium.

If a Scheibler-type desiccator is employed as a cooling receptacle for weighing vessels it may be provided with a porcelain plate on feet, which contains apertures for crucibles, etc.: the porcelain plate should be wedged into the sides, if necessary, with cork or some other material. For small desiccators, a silica triangle, with wire ends suitably bent, may be used. The ground edge of the desiccator should be lightly coated with white Vaseline or a special grease in order to make it air tight: too much grease may permit the lid to slide.

When a hot object, such as a crucible, is placed in a desiccator, about 5–10 seconds should elapse for the air to become heated and expand before putting the cover in place. When re-opening, the cover should be slid open very gradually in order to prevent any sudden inrush of air due to the partial vacuum which exists owing to the cooling of the expanded gas content of the desiccator, and thus prevent the precipitate being blown out of the crucible.

A desiccator is frequently also employed for the thorough drying of solids for analysis and for other purposes. Its efficient operation depends upon the condition of the desiccant; the latter should therefore be renewed at frequent intervals, particularly if its drying capacity is low. For dealing with large quantities of solid a vacuum desiccator is advisable.

Convenient types of **'vacuum' desiccators** are illustrated in Fig. III, 15. Large surfaces of the solid can be exposed; the desiccator may be evacuated, and drying is thus much more rapid than in the ordinary Scheibler type. These desiccators

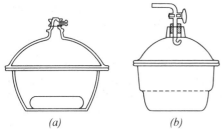

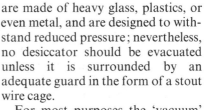

(a)   (b)

Fig. III, 15

are made of heavy glass, plastics, or even metal, and are designed to withstand reduced pressure; nevertheless, no desiccator should be evacuated unless it is surrounded by an adequate guard in the form of a stout wire cage.

For most purposes the 'vacuum' produced by an efficient water pump (20–30 mm mercury) will suffice; a guard tube containing desiccant should be inserted between the pump and the desiccator. The sample to be dried should be covered with a watch or clock glass, so that no mechanical loss ensues as a result of the removal or admission of air. Air must be admitted slowly into an exhausted desiccator: if the substance is very hygroscopic, a drying train should be attached to the stopcock. In order to maintain a satisfactory vacuum within the desiccator, the flanges on both the lid and the base must be well lubricated with Vaseline or other suitable grease. In some desiccators an elastomer ring is incorporated in a groove in the flange of the lower component of the desiccator: when the pressure is reduced, the ring is compressed by the lid of the desiccator, and an airtight seal is produced without the need for any grease. The same desiccants are used as with an ordinary desiccator.

For the efficient drying of small quantities of materials, the **'drying pistol'** (Fig. III, 16) may be used. The substance is placed in a porcelain boat which is inserted into the heating tube B, and is heated by the vapour of the liquid boiling in A:

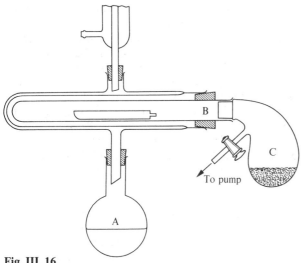

Fig. III, 16

water is often suitable for this purpose. B is joined to the vessel C containing a desiccant, and also carrying a connection to a vacuum pump. Electrically heated forms of drying pistol are also available and these can be controlled to operate over a wide range of temperatures.

Dry boxes (Glove boxes), which are especially intended for the manipulation of materials which are very sensitive to atmospheric moisture (or to oxygen), consist of a plastic or metal box provided with a window (of glass or clear plastic) on the upper side, and sometimes also on the side walls. A pair of rubber or plastic gloves are fitted through air-tight seals through the front side of the box, and by placing the hands and forearms into the gloves, manipulations may be carried out inside the box. One end of the box is fitted with an air-lock so that apparatus and materials can be introduced into the box without disturbing the atmosphere inside. A tray of desiccant placed inside the box will maintain a dry atmosphere, but to counter the unavoidable leakages in such a system, it is advisable to supply a slow current of dry air to the box; inlet and outlet taps are provided to control this operation. If the box is flushed out before use with an inert gas (e.g. nitrogen), and a slow stream of the gas is maintained whilst the box is in use, materials which are sensitive to oxygen can be safely handled. For a detailed discussion of the construction and uses of glove boxes (see Ref. 9).

III, 27.   STIRRING APPARATUS.   Many operations involving solutions of reagents require the thorough mixing of two or more reactants, and apparatus suitable for this purpose ranges from a simple glass stirring rod to electrically operated stirrers.

Stirring rods.   These are made from glass rod 3–5 mm in diameter, cut into suitable lengths. Both ends should be rounded by heating in the Bunsen or blowpipe flame. The length of the stirring rod should be suitable for the size and the shape of the vessel for which it is employed, e.g., for use with a beaker provided with a spout, it should project 3–5 cm beyond the lip when in a resting position.

A short piece of Teflon or of rubber tubing (or a rubber cap) is fitted tightly over one end of a stirring rod of convenient size. This is the so-called **policeman**; it is used for detaching particles of a precipitate adhering to the side of a vesssel which cannot be removed by a stream of water from a wash bottle; it should not, as a rule, be employed for stirring, nor should it be allowed to remain in a solution.

Boiling rods.   Boiling liquids and liquids in which a gas, such as hydrogen sulphide, sulphur dioxide, etc., has to be removed by boiling can be prevented from super-heating and 'bumping' by the use of a boiling rod (Fig. III, 17). This consists of a piece of glass tubing closed at one end and sealed approximately 1 cm from the other end; the latter end is immersed in the liquid. When the rod is removed, the liquid in the open end must be shaken out and the rod rinsed with a jet of water from a wash bottle. This device should not be used in solutions which contain a precipitate.

Fig. III, 17

Stirring may be conveniently effected with the so-called **magnetic stirrer**. A rotating field of magnetic force is employed to induce variable-speed stirring action within either closed or open vessels. The stirring is accomplished with the aid of a small cylinder of iron sealed in Pyrex glass, polythene, or Teflon, which is caused to rotate by a rotating magnet. A stirrer, fitted with an electric hot plate, is depicted in sectional diagram in Fig. III, 18. A speed control is provided, together with a dial to indicate the setting.

The usual type of **glass paddle stirrer** is also widely used in conjunction with an electric motor fitted with either a transformer-type, or a solid state speed controller. The stirrer may be either connected directly to the motor shaft or to a

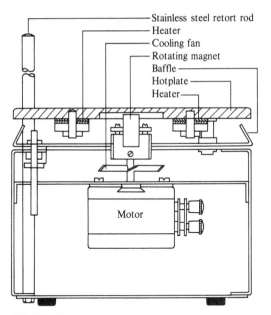

Stainless steel retort rod
Heater
Cooling fan
Rotating magnet
Baffle
Hotplate
Heater

Fig. III, 18

spindle actuated by a gear box which forms an integral part of the motor housing; by these means, wide variation in stirrer speed can be achieved.

Under some circumstances, e.g. the dissolution of a sparingly soluble solid, it may be more advantageous to make use of a **mechanical shaker**. Various models are available, ranging from 'wrist action shakers' which will accommodate small to moderate size flasks, to those equipped with a comparatively powerful electric motor and capable of shaking the contents of large bottles, such as 'Winchester quarts', vigorously.

III, 28. FILTRATION APPARATUS. The simplest apparatus used for filtration is the **filter funnel** fitted with a **filter paper**. The funnel should have an angle as close to 60° as possible, and a long stem (15 cm) to promote rapid filtration. Filter papers are made in varying grades of porosity, and one appropriate to the type of material to be filtered must be chosen (see Section **III, 38**).

In the majority of quantitative determinations involving the collection and weighing of a precipitate, it is convenient to be able to collect the precipitate in a crucible in which it can be weighed directly, and various forms of **filter crucible** have been devised for this purpose. The first of these was the **Gooch crucible** which may be encountered in porcelain, in silica, and (rarely) in platinum: the porcelain variety is most common, and the term 'Gooch crucible' is generally understood to refer to the porcelain crucible.

The **Gooch crucible** consists of a tall form crucible with the base pierced with a number of small holes. The holes are covered by a pad of asbestos, produced by sucking a slurry of asbestos fibres in water through the crucible under reduced pressure: the exact procedure for preparation of the crucible will be described later (Section **III, 40**). The asbestos employed must be carefully selected and purified. **NOTE** The normal recommended precautions **must** be taken when handling asbestos (Ref. 24).

Sintered glass crucibles are made of resistance glass and have a porous disc of sintered ground glass fused into the body of the crucible. The filter disc is made in varying porosities as indicated by numbers from 0 (the coarsest) to 5 (the finest); the range of pore diameter for the various grades is as follows:

Porosity	0	1	2	3	4	5
Pore diameter (μm)	200–250	100–120	40–50	20–30	5–10	1–2

Porosity 3 is suitable for precipitates of moderate particle size, and porosity 4 for fine precipitates such as barium sulphate. These crucibles should not be heated above about 200 °C.

Silica crucibles of similar pattern are also available, and, although expensive, have certain advantages in thermal stability.

Filter crucibles with porous filter base are available in porcelain (porosity 4), in silica (porosities 1, 2, 3, 4), and in alumina (coarse, medium and fine porosities): these have the advantage as compared with sintered crucibles, of being capable of being heated to much higher temperatures. Nevertheless, the heating must be gradual otherwise the crucible may crack at the join between porous base and glazed side.

For filtering large quantities of material, a **Buchner funnel** is usually employed; alternatively, one of the modified funnels shown diagrammatically in Fig. III, 19 may be used. Here (a) is the ordinary **porcelain Buchner funnel**; (b) is the **'slit sieve' glass funnel**. In both cases, one or, better, two good-quality filter papers are placed on the plate; the glass type is preferable since it is transparent and it is easy to see whether the funnel is perfectly clean. (c) is a Pyrex funnel with a **sintered glass plate**; no filter paper is required so that strongly acidic and weakly alkaline solutions can be readily filtered with this funnel. In all cases the funnel of appropriate size is fitted into a filter flask (d), and the filtration conducted under the diminished pressure provided by a filter pump or vacuum line.

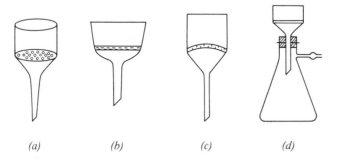

(a)   (b)   (c)   (d)

Fig. III, 19

One of the disadvantages of the porcelain Buchner funnel is that, being of one-piece construction, the filter plate cannot be removed for thorough cleaning and it is difficult to see whether the whole of the plate is clean on both sides. In a modern polythene version, the funnel is made in two sections which can be unscrewed thus permitting inspection of both sides of the plate.

The **Hartley funnel**, shown in Fig. III, 20, consists of three detachable parts: an

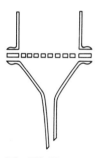

Fig. III, 20

upper and lower part, both of which have flanges extending beyond the actual filtering area, and a detachable plate ground on both sides to fit the flanges on the lower and upper portions. A filter paper is used which covers the whole surface of the ground area; creeping of the material to be filtered underneath the edges of the filter paper is thus avoided. In use, the filter paper is first wetted with the appropriate solvent, placed on the filter disc, and then the flanged ring placed on the paper. Generally the weight of the ring renders the joint leak-proof, but small clips can be used to keep the three parts together; it is advisable to apply suction before the funnel is filled with liquid.

In some circumstances, separation of solid from a liquid is better achieved by use of a **centrifuge** than by filtration, and a small, electrically driven centrifuge is a useful piece of equipment for an analytical laboratory. It may be employed for removing the mother liquor from recrystallised salts, for collecting difficultly filterable precipitates, and for the washing of certain precipitates by decantation. It is particularly useful when small quantities of solids are involved; centrifuging, followed by decantation and re-centrifuging, avoids transference losses and yields the solid phase in a compact form. Another valuable application is for the separation of two immiscible phases.

III, 29. WEIGHING BOTTLES. Most chemicals are weighed *by difference* by placing the material inside a stoppered weighing bottle which is then weighed. The requisite amount of substance is shaken out into a suitable vessel (beaker or flask), and the weight of substance taken is determined by reweighing the weighing bottle. In this way, the substance dispensed receives the minimum exposure to the atmosphere during the actual weighing process; a feature of some importance if the material is hygroscopic.

The most convenient form of weighing bottle is one fitted with an *external* cap and made of glass, polythene or polycarbonate. A weighing bottle with an *internally* fitting stopper is not recommended; there is always the danger that small particles may lodge at the upper end of the bottle and be lost when the stopper is pressed into place.

If the substance is unaffected by exposure to the air, it may be weighed on a watch glass, or in a disposable plastic container. The weighing funnel (Fig. III, 21) is very useful, particularly when the solid is to be transferred to a flask: having weighed the solid into the scoop-shaped end which is flattened so that it will stand

Fig. III, 21

on the balance pan, the narrow end is inserted into the neck of the flask and the solid washed into the flask with a stream of water from a wash bottle.

Woodward and Redman (Ref. 6e) have described a specially designed weighing bottle which will accommodate a small platinum crucible: when a substance has been ignited in the crucible, the crucible is transferred to the weighing bottle and subsequently weighed in this. This device obviates the need for a desiccator.

If the substance to be weighed is a liquid, a Lunge–Rey pipette (Fig. III, 11; Section **III, 18**) may be used. Alternatively, the liquid is placed in a weighing bottle fitted with a cap carrying a dropping tube.

Reagents and Standard Solutions

III, 30.   REAGENTS. The purest reagents available should be used for quantitative analysis; the analytical reagent quality (AR) is generally employed. In Great Britain AR ('AnalaR') chemicals from BDH Chemicals Ltd, and from Hopkin and Williams Ltd conform to the specifications given in their handbook *'AnalaR' Standards for Laboratory Chemicals.* In the USA the American Chemical Society committee on Analytical Reagents has established standards for certain reagents, and manufacturers supply reagents which are labelled 'Conforms to ACS Specifications'. In addition, certain manufacturers market chemicals of high purity, and each package of these analysed chemicals has a label giving the manufacturer's limits of certain impurities.

With the increasingly lower limits of detection being achieved in various types of instrumental analysis, there is an ever growing demand for reagents of correspondingly improved specification, and some manufacturers are now offering a range of specially purified reagents, e.g., the BDH Chemicals 'Aristar' chemicals; Hopkin and Williams PVS (Purified for Volumetric Standardisation) range.

In some instances, where a reagent of the requisite purity is not available, it may be advisable to weigh out a suitable portion of the appropriate *pure* metal [e.g. the Johnson, Matthey 'Specpure' range, or the MARZ grade metals supplied by the Materials Research Corporation, USA (Materials Research Co, London, England)], and to dissolve this in the appropriate acid.

It must be remembered that the label on a bottle is not an infallible guarantee of the purity of a chemical, for the following reasons:

(a) Some impurities may not have been tested for by the manufacturer.

(b) The reagent may have been contaminated after its receipt from the manufacturers either by the stopper having been left open for some time, with the consequent exposure of the contents to the laboratory atmosphere or by the accidental return of an unused portion of the reagent to the bottle.

(c) In the case of a solid reagent, it may not be sufficiently dry. This may be due either to insufficient drying by the manufacturers or to leakage through the stoppers during storage, or to both of these causes.

However, if the analytical reagents are purchased from a manufacturing firm of repute, and instructions given (a) that no bottle is to be opened for a longer time than is absolutely necessary, and (b) that no reagent is to be returned to the bottle after it has been removed, the likelihood of any errors arising from some of the above possible causes is considerably reduced. Liquid reagents should be poured from the bottle; a pipette should never be inserted into the reagent bottle. Particular care should be taken to avoid contamination of the stopper of the reagent bottle. When a liquid is poured from a bottle, the stopper should never be placed on the shelf or on the working bench; it may be placed upon a clean watchglass, and many chemists cultivate the habit of holding the stopper between the thumb and fingers of one hand. The stopper should be returned to the bottle immediately after the reagent has been removed, and all reagent bottles should be kept scrupulously clean, particularly round the neck or mouth of the bottle.

If there is any doubt as to the purity of the reagents used, they should be tested by standard methods for the impurities that might cause errors in the determinations. It may be mentioned that not all chemicals employed in quantitative analysis are available in the form of analytical reagents; the purest commercially available products should, if necessary, be purified by known methods: see below. The exact mode of drying, if required, will vary with the reagent; details are given for specific reagents in the text.

III, 31. PURIFICATION OF SUBSTANCES. If a reagent of adequate purity for a particular determination is not available, then the purest available product must be purified: this is most commonly done by **recrystallisation** from water. A known weight of the solid is dissolved in a volume of water sufficient to give a saturated or nearly saturated solution at the boiling point: a beaker, conical flask or porcelain dish may be used. The hot solution is filtered through a fluted filter paper placed in a short-stemmed funnel, and the filtrate collected in a beaker: this process will remove insoluble material which is usually present. If the substance crystallises out in the funnel, it should be filtered through a hot-water funnel. The clear hot filtrate is cooled rapidly by immersion in a dish of cold water or in a mixture of ice and water, according to the solubility of the solid; the solution is constantly stirred in order to promote the formation of small crystals, which occlude less mother liquor than larger crystals. The solid is then separated from the mother liquor by filtration, using one of the Buchner-type funnels shown in Fig. III, 19 and Fig. III, 20 (Section **III, 28**). When all the liquid has been filtered, the solid is pressed down on the funnel with a wide glass stopper, sucked as dry as possible, and then washed with small portions of the original solvent to remove the adhering mother liquor. The recrystallised solid is dried upon clock glasses at or above the laboratory temperature according to the nature of the material; care must of course be taken to exclude dust. The dried solid is preserved in glass-stoppered bottles. It should be noted that unless great care is taken when the solid is removed from the funnel, there is danger of introducing fibres from the filter paper, or small particles of glass from the glass filter disc: scraping of the filter paper or of the filter disc must be avoided.

Some solids are either too soluble, or the solubility does not vary sufficiently with temperature, in a given solvent for direct crystallisation to be practicable. In many cases, the solid can be precipitated from, say, a concentrated aqueous solution by the addition of a liquid, miscible with water, in which it is less soluble. Ethanol, in which many inorganic compounds are almost insoluble, is generally used. Care must be taken that the amount of ethanol or other solvent added is not so large that the impurities are also precipitated. Potassium hydrogencarbonate and antimony potassium tartrate may be purified by this method.

Sublimation. This process is employed to separate volatile substances from non-volatile impurities. Iodine, arsenic trioxide and ammonium chloride can be purified in this way. The substance is placed in a porcelain dish or casserole; the latter is gently heated with a small flame and the vapour condensed upon a cool surface, such as a large inverted glass funnel containing a plug of glass wool at the apex, or, preferably, a flask containing cold water.

Pure iodine is required in analysis, and details for its purification will not be out of place here. Grind together 10 g of iodine with 4 g of potassium iodide (any chlorine or bromine present will thus be retained as the non-volatile potassium salts) and transfer the mixture to a casserole: place a flask through which a gentle stream of cold water is circulating, on the casserole (Fig. III, 22). Heat very gently

Fig. III, 22

until sufficient iodine has sublimed on to the bottom of the flask, allow to cool, and remove the flask with the iodine adhering to it. Pass a rapid stream of ice-cold water through the flask; this will cause the glass to contract somewhat and the whole of the crust can then be removed by scraping with a clean glass rod and is collected on a clock glass. Break up the large pieces, and repeat the sublimation without the addition of potassium iodide. Remove the second sublimate as before, and grind the iodine in a glass mortar. Dry in a desiccator containing calcium chloride; no grease whatever should be exposed on the inside, since iodine vapour attacks grease forming hydrogen iodide.

Zone refining is a purification technique originally developed for the refinement of certain metals, and which is applicable to all substances of reasonably low melting point which are stable at the melting temperature. In a zone refining apparatus, the substance to be purified is packed into a column of glass or stainless steel, which may vary in length from six inches (semimicro apparatus) to three feet. An electric ring heater which heats a narrow band of the column is allowed to fall slowly by a motor-controlled drive, from the top to the bottom of the column. The heater is set to produce a molten zone of material at a temperature 2–3 °C above the melting point of the substance, which travels slowly down the tube with the heater. Since impurities normally lower the melting point of a substance, it follows that the impurities tend to flow down the column in step with the heater, and thus to become concentrated in the lower part of the tube. The process may be repeated a number of times (the apparatus may be programmed to reproduce automatically a given number of cycles), until the required degree of purification has been achieved. The benzoic acid PVS grade marketed by Hopkin and Williams is purified by this technique (see Ref. 10).

III, 32.   PREPARATION AND STORAGE OF STANDARD SOLUTIONS.

In any analytical laboratory it is essential to maintain stocks of solutions of various reagents: some of these will be of accurately known concentration (i.e. standard solutions) and correct storage of such solutions is imperative.

According to BS 4245 solutions should be classified as:
1. reagent solutions which are of approximate concentration;
2. standard solutions which have a known concentration of some chemical;
3. standard reference solutions which have a known concentration of a primary standard substance (Section **X, 6**);
4. standard titrimetric solutions which have a known concentration (determined either by weighing or by standardisation), of a substance other than a primary standard.

The IUPAC Commission on Analytical Nomenclature (Ref. 11), refer to 3 and 4 respectively as Primary Standard Solutions and Secondary Standard Solutions.

For **reagent solutions** as defined above (i.e. 1), it is usually sufficient to weigh out approximately the amount of material required, using a watch glass or a plastic weighing container, and then to add this to the required volume of solvent which has been measured with a measuring cylinder.

To prepare a **standard solution** the following procedure is followed. A short-stemmed funnel is inserted into the neck of a graduated flask of the appropriate size. A suitable amount of the chemical is placed in a weighing bottle which is weighed, and then the required amount of substance is transferred from the weighing bottle to the funnel, taking care that no particles are lost. After the weighing bottle has been re-weighed, the substance in the funnel is washed down with a stream of the liquid. The funnel is thoroughly washed, inside and out, and then removed from the flask; the contents of the flask are dissolved, if necessary, by shaking or swirling the liquid, and then made up to the mark: for the final adjustment of volume, a dropping tube drawn out to form a very fine jet is employed.

If a watch glass is employed for weighing out the sample, the contents are transferred as completely as possible to the funnel, and then a wash bottle is used to remove the last traces of the substance from the watch glass. If the weighing scoop (Fig. III, 21; Section **III, 29**) is used, then of course a funnel is not needed provided that the flask is of such a size that the end of the scoop is an easy fit in the neck.

If the substance is not readily soluble in water, it is advisable to add the material from the weighing bottle or the watch glass to a beaker, followed by distilled water; the beaker and its contents are then heated gently with stirring until the solid has dissolved. After allowing the resulting concentrated solution to cool a little, it is transferred through the short-stemmed funnel to the graduated flask, the beaker is rinsed thoroughly with several portions of distilled water, adding these washings to the flask, and then finally the solution is made up to the mark: it may be necessary to allow the flask to stand for a while before making the final adjustment to the mark to ensure that the solution is at room temperature. *Under no circumstances may the graduated flask be heated.*

In some circumstances it may be considered preferable to prepare the standard solution by making use of one of the concentrated volumetric solutions supplied in sealed ampoules by various manufacturers (BDH Chemicals 'CVS' solutions, Hopkin and Williams 'Convol' solutions, May and Baker 'Volucon' solutions, etc.) which only require dilution in a graduated flask to produce a standard solution.

Solutions which are comparatively stable and unaffected by exposure to air may be stored in litre, or in 'Winchester quart' bottles; for work requiring the highest accuracy, Pyrex, or other resistance glass, bottles fitted with ground-glass stoppers should be employed, the solvent action of the solution being thus considerably reduced. It is however necessary to use a rubber bung instead of a glass stopper for alkaline solutions, and in many instances a polythene container (e.g. an aspirator) may well replace glass vessels. It should be noted however that for some solutions as, for example, iodine and silver nitrate, glass containers only may be used, and in both these cases the bottle should be made of dark (brown) glass: solutions of EDTA (Section **X, 50**) are best stored in polythene containers.

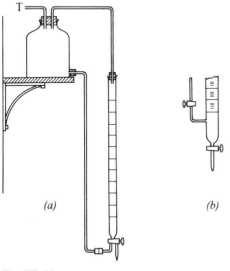

(a)   *(b)*

Fig. III, 23

The bottle should be clean and dry: a little of the stock solution is introduced, the bottle well rinsed with this solution, drained, and the remainder of the solution poured in and the bottle immediately stoppered. If the bottle is not dry, but has recently been thoroughly rinsed with distilled water, it may be rinsed successively with three small portions of the solution and drained well after each rinsing; this procedure is, however, less satisfactory than that employing a clean and dry vessel. Immediately after the solution has been transferred to the stock bottle, it should be labelled with: (1) the name of the solution, (2) its concentration, (3) the date of preparation, and (4) the initials of the person who prepared the solution, together with any other relevant data. Unless the bottle is completely filled, internal evaporation and condensation will cause drops of water to form on the upper part of the inside of the vessel. For this reason, the bottle must be thoroughly shaken before removing the stopper.

For expressing concentrations of reagents, the molar system is universally applicable, i.e., the number of moles of solute present in $1\,dm^3$ of solution. Concentrations may also be expressed in terms of normality if no ambiguity is likely to arise (see Section **X, 3**), and in fact BS 2445 (1968) recommends that concentrations should usually be expressed in normalities.

Solutions liable to be affected by access of air (e.g., alkali hydroxides which absorb carbon dioxide; iron(II) and titanium(III) which are oxidised) may be stored in the apparatus shown diagrammatically in Fig. III, 23 (*a*). The burette has a three-way tap which enables it either to be filled from the stock bottle or to be emptied. If such a burette is not available, one with a side-tube and stopcock (*b*) will serve equally well. The tube T is permanently connected with a source of hydrogen (e.g., from a Kipp's apparatus) if the solution is oxidised upon exposure to air, or to a soda–lime or sodium hydroxide–asbestos guard tube, if it contains caustic alkali. In the latter case, particularly if soda glass vessels are used, the solution may become contaminated with silicates owing to the attack of the alkali on the glass: it is better to employ a storage vessel of resistance glass or, preferably, of polythene.

A more compact apparatus is shown in Fig. III, 24. A is a large storage bottle of 10–15 litres capacity. B is a 50-cm^3 burette provided with an automatic filling device at C (the point of the drawn-out tube is adjusted to be exactly at the zero mark of the burette), D is the burette-bottle clamp, E is a two-holed rubber stopper, F is a ground-glass tension joint, a rubber tube is connected to a source of hydrogen (for example, a Kipp's apparatus) and to the T-joint below L, H is a Bunsen valve, and J is hydrogen. The burette is filled by closing tap K and passing hydrogen through the rubber tube attached to the T-piece (below tap L) with tap L closed; taps L and K are opened, and the excess of liquid allowed to siphon back.

Two other apparatus for the storage of standard solutions are shown in Fig. III, 25. Fig. III, 25(*b*) is self-explanatory. The solution is contained in the storage bottle A, and the 50-cm^3 burette is fitted into this by means of a

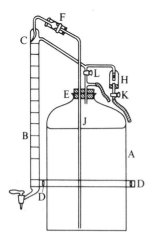

Fig. III, 24

ground-glass joint B. To fill the burette, tap C is opened and the liquid pumped into the burette by means of the small bellows E. F is a small guard tube; this is filled with soda-lime or 'carbosorb' when caustic alkali is contained in the storage bottle. Bottles with a capacity up to 2 litres are provided with standard ground-

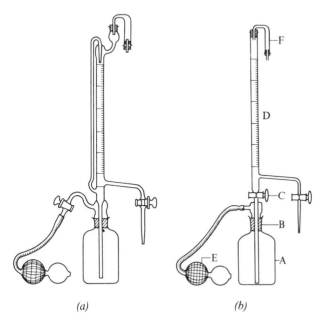

(a)　　　　　　*(b)*

Fig. III, 25

glass joints; large bottles, up to 15 litres capacity, can also be obtained. Fig. III, 25(*a*) portrays a similar apparatus, but with an automatic filling device. The solution is pumped into the burette and enters it through a glass tube which terminates in a capillary exactly at the zero mark; immediately the pressure is released, the solution above the zero mark is automatically siphoned back into the storage bottle.

The Dafert pipette (Fig. III, 9; Section **III, 16**) is a convenient apparatus for dispensing fixed volumes of a standard solution, as are also the various **liquid dispensers** which are available.

SOME BASIC TECHNIQUES

III, 33. PREPARATION OF THE SUBSTANCE FOR ANALYSIS. Presented with a large quantity of a material to be analysed, the analyst is immediately confronted with the problem of selecting a representative sample for the analytical investigations. It may well be that the material is in such large pieces that comminution is necessary in order to produce a specimen suitable for handling in the laboratory. These important factors are considered in Chapter V

(Sections **V, 2**; **V, 3**), and as explained therein, the material is usually dried at 105–110 °C before analysis.

III, 34.   WEIGHING THE SAMPLE.   If necessary refer to Section **III, 11** dealing with the operation of a chemical balance, and to Sections **III, 29** and **III, 26** which are concerned with the use and function of weighing bottles and desiccators respectively.

The material, prepared as above, is usually transferred to a weighing bottle which is stoppered and stored in a desiccator. Samples of appropriate size are withdrawn from the weighing bottle as required, the bottle being weighed before and after the withdrawal, so that the weight of substance is obtained by difference.

Attention is drawn to the Mettler vibro-spatula which is a useful adjunct to the weighing out of powders. The spatula is connected to the electric mains, and the powder is placed on the blade of the spatula. When the current is switched on by pressing the button 1, the blade is caused to vibrate and to deposit solid gradually into the beaker or other container over which it is held: the intensity of the vibration may be adjusted by means of the knurled head 2 (Fig. III, 26).

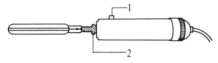

Fig. III, 26

III, 35.   SOLUTION OF THE SAMPLE.   Whilst many substances can be dissolved directly in water or in dilute acids, materials such as minerals, refractories, and alloys must usually be treated with a variety of reagents in order to discover a suitable solvent: in such cases the preliminary qualitative analysis will have revealed the best procedure to adopt. Each case must be considered on its merits; no attempt at generalisation will therefore be made. We can, however, discuss the experimental technique of the simple process of solution of a substance in water or in acids, and also the method of treatment of insoluble substances.

For a substance which dissolves readily, the sample is weighed out into a beaker, and the beaker immediately covered with a clock glass of suitable size (its diameter should not be more than about 1 cm larger than that of the beaker) with its convex side facing downwards. The beaker should have a spout in order to provide an outlet for the escape of steam or gas. The solvent is then added by pouring it carefully down a glass rod, the lower end of which rests against the wall of the beaker; the clock glass is displaced somewhat during this process. If a gas is evolved during the addition of the solvent (e.g., acids with carbonates, metals, alloys, etc.), the beaker must be kept covered as far as possible during the addition. The reagent is then best added by means of a pipette or by means of a funnel with a bent stem inserted beneath the clock glass at the spout of the beaker; loss by spirting or as spray is thus prevented. When the evolution of gas has ceased and the substance has completely dissolved, the under side of the clock glass is well rinsed with a stream of water from a wash bottle, care being taken that the washings fall on to the side of the beaker and not directly into the solution. If warming is necessary, it is usually best to carry out the dissolution in a conical (Erlenmeyer) flask with a small funnel in the mouth; loss of liquid by spirting is thus prevented and the escape of gas is not hindered.

It may often be necessary to reduce the volume of the solution, or sometimes to evaporate completely to dryness. Wide and shallow vessels are most suitable, since a large surface is thus exposed and evaporation is thereby accelerated. We may employ shallow beakers of resistance glass, Pyrex evaporating dishes, porcelain basins or casseroles, silica or platinum basins; the material selected will depend upon the extent of attack of the hot liquid upon it and upon the constituents being determined in the subsequent analysis. Evaporations should be carried out on the steam bath or upon a low-temperature hot plate; slow evaporation is preferable to vigorous boiling, since the latter may lead to some mechanical loss in spite of the precautions to be mentioned below. During evaporations, the vessel must be covered by a Pyrex clock glass of slightly larger diameter than the vessel, and supported either on a large all-glass triangle or upon three small U-rods of Pyrex glass hanging over the rim of the container. Needless to say, at the end of the evaporation the sides of the vessel, the lower side of the clock glass and the triangle and glass hooks (if employed) should be rinsed with distilled water into the vessel.

For evaporation at the boiling point either a conical flask with a short Pyrex funnel in the mouth or a round-bottomed flask inclined at an angle of about 45° may be employed; in the latter the drops of liquid, etc., thrown up by the ebullition or by effervescence will be retained by striking the inside of the flask, while gas and vapour will escape freely.

Substances which are insoluble or only partially soluble in acids are brought into solution by fusion with the appropriate reagent. The most commonly used fusion reagents, or **fluxes** as they are called, are anhydrous sodium carbonate either alone or, less frequently, mixed with potassium nitrate or sodium peroxide; potassium, or sodium pyrosulphate; sodium peroxide; sodium or potassium hydroxide. Of recent years *anhydrous* lithium metaborate has found favour as a flux, especially for materials containing silica (Ref. 14); when the resulting fused mass is dissolved in dilute acids, no separation of silica takes place as it does when a sodium carbonate melt is similarly treated. Other advantages claimed for lithium metaborate are:

1. no gases are evolved during the fusion or during the dissolution of the melt, and hence there is no danger of losses due to spitting;
2. fusions with lithium metaborate are usually quicker (15 minutes will often suffice), and can be performed at a lower temperature than with other fluxes;
3. the loss of platinum from the crucible is less during a lithium metaborate fusion than with a sodium carbonate fusion;
4. many elements can be determined directly in the acid solution of the melt without the need for tedious separations.

It is claimed that in some circumstances the fusion may be performed in a high purity graphite crucible but this is not always satisfactory (Ref. 15). Naturally, the flux employed will depend upon the nature of the insoluble substance. The vessel in which fusion is effected must be carefully chosen; platinum crucibles are employed for sodium carbonate, lithium metaborate and potassium pyrosulphate; nickel or silver crucibles for sodium or potassium hydroxide; nickel, gold, silver, or iron crucibles for sodium carbonate and/or sodium peroxide; nickel crucibles for sodium carbonate and potassium nitrate (platinum is slightly attacked). To carry out the fusion, a layer of the flux is placed at the bottom of the crucible, and then the intimate mixture of the flux and the finely-divided substance added; the crucible should be not more than about half-full, and should, generally, be kept covered during the whole process. The crucible is very gradually heated at first, and the temperature slowly raised to the required temperature. The final temperature should not be higher than is actually necessary; any possible further attack of the flux upon the crucible is thus avoided. When the fusion, which usually takes 30–60 minutes, has been completed, the crucible is grasped by means of the crucible tongs and gently rotated and tilted so that the molten material distributes itself around the walls of the container and solidifies there as a thin layer. This procedure greatly facilitates the subsequent detachment and solution of the fused mass. When cold, the crucible is placed in a casserole, porcelain dish, platinum basin, or Pyrex beaker (according to the nature of the flux) and covered with water. Acid is added, if necessary, and the vessel is covered with a clock glass, the temperature raised to 95–100 °C, and maintained until solution is achieved.

Many of the substances which require fusion treatment to render them soluble will in fact dissolve in mineral acids if the digestion with acid is carried out under pressure, and consequently at higher temperatures than those normally achieved. Such drastic treatment requires a container capable of withstanding the requisite pressure, and also resistant to chemical attack. A satisfactory solution of this problem was achieved by Bernas (Ref. 12), who devised a stainless steel pressure vessel, capacity $50 \, cm^3$, fitted with a Teflon liner. **Acid digestion vessels** made on this principle are now available from Uniseal Decomposition Vessels Ltd, Haifa, Israel (British agents, S. and J. Juniper and Co, Harlow, Essex), and also from the Parr Instrument Co, USA; they may be heated to temperatures of 150–180 °C, will withstand pressures of 80–90 atmospheres, and under these conditions decomposition of refractory materials may be accomplished in 45 minutes. Apart from the saving in time which is achieved, and the fact that the use of expensive platinum ware is obviated, other advantages of the method are that no losses can occur during the treatment, and the resulting solution is free from the heavy loading of alkali metals which follows the usual fusion procedures. A full discussion of decomposition techniques is given in Reference 16.

III, 36. PRECIPITATION. The conditions for precipitation are given in Section **XI, 6**. Precipitations are usually carried out in resistance-glass beakers, and the solution of the precipitant is added slowly (for example, by means of a pipette, burette, or tap funnel) and with efficient stirring of the suitably diluted solution. The addition must always be made without splashing; this is best achieved by allowing the solution of the reagent to flow down the side of the beaker or precipitating vessel. Only a moderate excess of the reagent is generally required; a very large excess may lead to increasing solubility (compare Section **II, 8**) or contamination of the precipitate. After the precipitate has settled, a few drops of the precipitant should always be added to determine whether further precipitation occurs. As a general rule, precipitates are not filtered off immediately after they have been formed; most precipitates, with the exception of those which are definitely colloidal, such as iron (III) hydroxide, require more or less digestion (Section **XI, 5**) to complete the precipitation and make all particles of filterable size. In some cases digestion is carried out by setting the beaker aside and leaving the precipitate in contact with the mother liquor at room temperature for 12–24 hours; in others, where a higher temperature is permissible, digestion is usually effected near the boiling point of the solution. Hot plates, water baths, or even a low flame if no bumping occurs, are employed for the latter purpose; in all cases the beaker should be covered with a clock glass with the convex side turned down. If the solubility of the precipitate is appreciable, it may be necessary to allow the solution to attain room temperature before filtration.

III, 37.   FILTRATION.   This operation is the separation of the precipitate from the mother liquor, the object being to get the precipitate and the filtering medium quantitatively free from the solution. The media employed for filtration are: (1) filter paper; (2) filter mats of purified asbestos (Gooch crucibles) or of platinum (Munroe crucibles); (3) porous fritted plates of resistance glass, e.g., Pyrex (sintered glass filtering crucibles), of silica (Vitreosil filtering crucibles), or of porcelain (porcelain filtering crucibles): see Section **III, 28**.

The choice of the filtering medium will be controlled by the nature of the precipitate (filter paper is especially suitable for gelatinous precipitates) and also by the question of cost. The limitations of the various filtering media are given in the account which follows.

III, 38.   FILTER PAPERS.   Quantitative filter papers must have a very small ash content; this is achieved during manufacture by washing with hydrochloric and hydrofluoric acids. The sizes generally used are circles of 7.0, 9.0, 11.0, and 12.5 cm diameter, those of 9.0 and 11.0 cm being most widely employed. The ash of a 11-cm circle should not exceed 0.0001 g; if the ash exceeds this value, it should be deducted from the weight of the ignited residue. Manufacturers give values for the average ash per paper: the value may also be determined, if desired, by igniting several filter papers in a crucible. Quantitative filter paper is made of various degrees of porosity. The filter paper used must be of such texture as to retain the smallest particles of precipitate and yet permit of rapid filtration. Three textures are generally made, one for very fine precipitates, a second for the average precipitate which contains medium-sized particles, and a third for gelatinous precipitates and coarse particles. The speed of filtration is slow for the first, fast for the third, and medium for the second. 'Hardened' filter papers are made by further treatment of quantitative filter papers with acid; these have an extremely small ash, a much greater mechanical strength when wet, and are more resistant to acids and alkalis: they should be used in all quantitative work.

The characteristics of the Whatman series of hardened ashless filter papers, manufactured by E. W. Balston, are shown in Table III, 7: for further details consult Ref. 13.

Table III, 7   'Whatman' quantitative filter papers

Filter paper	Fast speed. Retains coarse particles	Medium speed. Retains medium-sized particles	Slow speed. Retains fine particles
Double acid washed, hardened	No. 541	No. 540	No. 542
Ash (mg)*	0.08	0.08	0.09

* Value for 12.5 cm circle.

The size of the filter paper selected for a particular operation is determined by the bulk of the precipitate, and not by the volume of the liquid to be filtered. The entire precipitate should occupy about a third of the capacity of the filter at the end of the filtration. The funnel should match the filter paper in size; the folded paper should extend to within 1–2 cm of the top of the funnel, but never closer than 1 cm.

A funnel with an angle as nearly 60° as possible should be employed; the stem should have a length of about 15 cm in order to promote rapid filtration. The filter paper must be carefully fitted into the funnel so that the upper portion beds tightly against the glass. Some analysts recommend that the filter paper should rest completely against the wall of the funnel; this is really unnecessary, since a filter paper which adheres snugly to the funnel over the upper half only will permit more rapid filtration. To prepare the filter paper for use, the dry paper is usually folded exactly in half and exactly again in quarters. The folded paper is then opened so that a 60° cone is formed with three thicknesses of paper on the one side and a single thickness on the other; the paper is then adjusted to fit the funnel. The paper is placed in the funnel, moistened thoroughly with water, pressed down tightly to the sides of the funnel, and then filled with water. If the paper fits properly, the stem of the funnel will remain filled with liquid during the filtration. Another method of folding the filter, which is preferable to that just described, consists in folding the paper across a diameter and then once again so that the two halves of the first crease do not quite coincide (the two extreme edges should enclose an angle of 3–4° for a 60° funnel); the corner of the fold should be torn off to a depth of about one-third of the radius of the paper. When this filter is opened and placed in the funnel, it should fit the walls tightly at the upper half: if it does not fit properly, the angle of the second fold must be adjusted until it does. Here also, the ultimate test of a proper fit is that the stem of the funnel remains filled with liquid throughout the filtration.

To carry out a filtration, the funnel containing the properly fitted paper is placed in a funnel stand (or is supported vertically in some other way) and a clean beaker placed so that the stem of the funnel just touches the side; this will prevent splashing. The liquid to be filtered is then poured down a glass rod into the filter, directing the liquid against the side of the filter and not into the apex (Fig. III, 27): the lower end of the stirring rod should be very close to, but should not quite touch, the filter paper on the side having three thicknesses of paper. The paper is never filled completely with the solution; the level of the liquid should not rise closer than to within 5–10 mm from the top of the paper. A precipitate which tends to remain in the bottom of the beaker should be removed by holding the glass rod across the beaker, tilting the beaker, and directing a jet of water from a wash bottle so that the precipitate is rinsed into the filter funnel. This procedure may also be adopted to transfer the last traces of the precipitate in the beaker to the filter. Any precipitate which adheres firmly to the side of the beaker or to the stirring rod may be removed with a rubber-tipped rod or 'policeman' (Section **III, 27**).

Filtration by suction is rarely necessary: with gelatinous and some finely divided precipitates, the suction will

Fig. III, 27

draw the particles into the pores of the paper, and the speed of filtration will actually be reduced rather than increased. If suction is used with filter paper, it is necessary to support the paper in a perforated cone made of platinum ('filter cone') or in a Whatman filter cone (hardened, No. 51).

III, 39. FILTER PULP. Dittrich first recommended the use of filter paper pulp as an aid in the filtration and washing of gelatinous or slimy precipitates, which tend to clog the pores of ordinary filter paper. The macerated paper may be prepared by vigorously shaking an ordinary quantitative filter paper, torn into small pieces, with hot distilled water in a stoppered conical flask until it is disintegrated to a pulp. Ashless grade filter clippings (ash not exceeding 0.1 per cent) are marketed (Whatman) for the preparation of pure filter pulp by dispersing in distilled water.

Filter pulp tablets are marketed ready for use; and are easily disintegrated in distilled water. Whatman 'accelerators' are small discs, each weighing ca. 0.4 g and giving an ash of about 0.00006 g. Whatman 'ashless tablets' are larger, weighing 2.4 g each, and giving an ash of about 0.0003 g. Either one or two 'accelerators' or a quarter of a 'tablet' is used in the average precipitation. The bulk of the filter pulp should be approximately equal to that of the gelatinous precipitate. The filter pulp is added *after* the precipitate has formed and *immediately before* filtration. When the bulk of the precipitate and paper is large, it is usually advisable to support the filter paper on a Whatman filter cone and drain well on the pump upon completion of the washing. It is best not to dry the filter and precipitate completely, as a hard mass may be formed, which is difficult to insert in the crucible in the subsequent ignition. While still slightly moist it should be transferred to the crucible, and the drying completed in the crucible by heating over a very small non-luminous flame.

III, 40. GOOCH CRUCIBLES. The characteristics of Gooch crucibles have been described in Section **III, 28**. In use, the crucible is supported in a special holder, known as a Gooch funnel, by means of a wide rubber tube (Fig. III, 28);

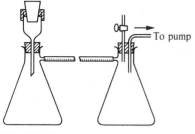

the bottom of the crucible should be quite free from the side of the funnel and from the rubber gasket, the latter in order to be sure that the filtrate does not come into contact with the rubber. The Gooch funnel passes through a one-holed rubber bung into a large filter flask of about 750 cm^3 capacity. The tip of the funnel must project below the side arm of the filter flask so that any risk that the liquid may be sucked out of the filter flask may be avoided. The filter flask should be coupled with another flask of similar capacity, and the latter connected to a water filter pump; if the water in the pump should 'suck back', it will first enter the empty flask and the filtrate will not be contaminated. It is advisable also to have some sort of pressure regulator to limit the maximum pressure under which filtration is conducted. A simple method is to insert a glass tap in the second filter flask, as in Fig. III, 28; alternatively, a glass T-piece may be introduced between the receiver and the pump, and one arm closed either by a glass tap or by a piece of heavy rubber

Fig. III, 28

tubing ('pressure' tubing) carrying a screw clip.

The rubber sleeve for fitting the Gooch crucible into the Gooch funnel may be replaced with advantage by either:

(a) a solid rubber ring, shaped to hold the crucible, which fits into a 7.5-cm filter funnel;

(b) a solid rubber washer; the wide part of the washer merely rests on top of the Gooch funnel, and the inside part conforms to the shape of, and supports the crucible.

The Gooch crucible should be of suitable size. One about 4 cm in height, with a capacity of 25 cm^3 and perforations about 0.5–0.8 mm in diameter, will be found serviceable for most purposes. The crucible is first placed in the suction-filtering apparatus, and then half to two-thirds filled with the suspension of asbestos in water. The whole is allowed to stand for 2–3 minutes in order to allow the larger particles to settle to the bottom, and then suction is applied gently. When the water has passed through, the pump is turned on full, and the mat sucked down tight. The final uniform pad of asbestos should have a thickness of 2–3 mm; it is possible to tell approximately when the mat has the correct thickness by holding the crucible up to the light and looking through it, when the outline of the holes should be barely visible. If the pad is too thin, more asbestos must be added and the process repeated. The asbestos pad is now thoroughly washed with distilled water under the maximum suction of the pump until no fine fibres pass into the filtrate. It should be mentioned that some analysts prefer to place a perforated porcelain plate ('Witt' plate) upon the asbestos mat to prevent its dislodgement: a little more of the suspension is poured in to furnish enough asbestos to barely cover the plate and to hold it in place. This procedure is unnecessary if it be remembered that no liquid may be poured into the crucible unless suction is being applied. Liquids should be poured gently on to the centre of the mat down a stirring rod: a jet of water from a wash bottle should never be directed into the prepared crucible. If these precautions are taken there is little danger that the mat will become torn and allow the precipitate to pass through.

The crucible is placed on a small ignition dish or saucer or upon a shallow-form Vitreosil capsule and dried to constant weight at the same temperature as that which will be subsequently used in drying the precipitate. For temperatures up to about 250 °C a thermostatically controlled electric oven should be used. For higher temperatures, the crucible may be heated in an electrically heated muffle furnace. In all cases the crucible is allowed to cool in a desiccator before weighing. The asbestos normally used for Gooch crucibles tends to lose weight above about 280 °C, hence it is recommended that precipitates which require heating above about 250 °C should not be collected in Gooch crucibles. Porous filtering crucibles (see below) may be employed.

III, 41. CRUCIBLES WITH PERMANENT POROUS PLATES. Reference has already been made to these crucibles and to crucibles with a porous base in Section III, 28. They possess an obvious advantage over Gooch crucibles in that no preparation of a filter mat is necessary, and there is none of the possible instability associated with filter mats, i.e., the possibility that the mat may become dislodged during the filtering process. They are used exactly as described for Gooch crucibles, and a Gooch funnel is used to support the crucible during the filtration process.

Care must be taken with porous base crucibles, as with sintered glass crucibles, to avoid attempting the filtration of materials that may clog the filter plate. A new crucible should be washed with concentrated hydrochloric acid and then with distilled water. The crucibles are chemically inert and are resistant to all solutions which do not attack silica; they are attacked by hydrofluoric acid, fluorides, and strongly alkaline solutions.

Crucibles fitted with permanent porous plates are cleaned by shaking out as much of the solid as possible, and then dissolving out the remainder of the solid with a suitable solvent.

A hot $0.1M$ solution of the tetrasodium salt of ethylenediamine tetra-acetic acid is an excellent solvent for many of the precipitates (except metallic sulphides and hexacyanoferrates(III)) encountered in analysis. These include barium sulphate, calcium oxalate, calcium phosphate, calcium oxide, lead carbonate, lead iodate, lead oxalate, and ammonium magnesium phosphate: for calcium and barium precipitates the solution must be alkaline (pH >10). The crucible may either be completely immersed in the hot reagent or the latter may be drawn by suction through the crucible.

III, 42. WASHING OF PRECIPITATES. Most precipitates are produced in the presence of one or more soluble compounds. Since the latter are frequently not volatile at the temperature at which the precipitate is ultimately dried, it is necessary to wash the precipitate to remove such material as completely as possible. The minimum volume of the washing liquid required to remove the objectionable matter should be used, since no precipitate is absolutely insoluble. Qualitative tests for the removal of the impurities should be made on small volumes of the filtered washing solution. Furthermore, it is better to wash with a number of small portions of the washing liquid, which are well drained between each washing, than with one or two large portions, or by adding fresh portions of the washing liquid whilst solution still remains on the filter (see Section **XI, 8**).

The ideal washing liquid should comply as far as possible with the following conditions:
1. it should have no solvent action upon the precipitate, but dissolve foreign substances easily;
2. it should have no dispersive action on the precipitate;
3. it should form no volatile or insoluble product with the precipitate;
4. it should be easily volatile at the temperature of drying of the precipitate;
5. it should contain no substance which is likely to interfere with subsequent determinations in the filtrate.

In general, pure water should not be used unless it is certain that it will not dissolve appreciable amounts of the precipitate or peptise it. If the precipitate is appreciably soluble in water, a common ion is usually added, since any electrolyte is less soluble in a dilute solution containing one of its ions than it is in pure water (Section **II, 9**); as an example the washing of calcium oxalate with dilute ammonium oxalate solution may be cited. If the precipitate tends to become colloidal and pass through the filter paper (this is frequently observed with gelatinous or flocculent precipitates), a wash solution containing an electrolyte must be employed (compare Section **XI, 3**). The nature of the electrolyte is immaterial, provided it has no action upon the precipitate during washing and is volatilised during the final heating. Ammonium salts are usually selected for this purpose: thus ammonium nitrate solution is employed for washing iron(III) hydroxide. In some cases it is possible to select a solution which will both reduce the solubility of the precipitate and prevent peptisation; for example, the use of dilute nitric acid with silver chloride. Some precipitates tend to oxidise during washing; in such instances the precipitate cannot be allowed to run dry, and a special washing solution which re-converts the oxidised compounds into the original condition must be employed, e.g., acidulated hydrogen sulphide water for copper sulphide. Gelatinous precipitates, like aluminium hydroxide, require more washing than crystalline ones, such as calcium oxalate. With gelatinous precipitates there is also a danger of channel formation if the wash liquid is allowed to drain completely; these precipitates should be washed as far as possible by decantation.

III, 43. TECHNIQUE OF FILTRATION. When the proper filtering medium (filter paper, Gooch crucible, etc.) has been prepared, as much as possible of the supernatant liquid is poured off by directing the stream of liquid against a glass rod held against the lip of the beaker (compare Fig. III, 27, Section **III, 38**) without disturbing the precipitate. The precautions already mentioned against filling a filter paper too full must be observed. In most cases, particularly if the precipitate settles rapidly or is gelatinous, **washing by decantation** may be employed. Twenty to fifty cm^3 of a suitable wash liquid is added to the residue in the beaker, the solid stirred up and allowed to settle. If the solubility of the precipitate allows, the solution should be heated, since, *inter alia*, the rate of filtration will thus be increased. When the supernatant liquid is clear, as much as possible of the clear liquid is decanted through the filtering medium. This process is repeated three to five times (or as many times as is necessary) before the precipitate is transferred to the filter. The main bulk of the precipitate is first transferred by mixing with the wash solution and pouring off the suspension, the process being repeated until most of the solid has been removed from the beaker. The precipitate adhering to the sides and the bottom of the beaker is removed as follows. The beaker is grasped in the left hand, and the stirring rod is held firmly against the top of the beaker with the index finger and should project 2–3 cm beyond the lip; the wash bottle is controlled by the right hand. The beaker is inclined and a stream of water (or wash liquid) is directed against the precipitate to dislodge it and wash it against the rod into the filter or filtering crucible. After the above treatment there will generally be small amounts of the precipitate adhering to the walls of the beaker. These are removed by rubbing with a 'policeman'; when all the particles have been dislodged, the 'policeman' is rinsed with the wash liquid, and the remaining precipitate transferred to the filter or filtering crucible.

Where the precipitate is washed on the filter, in the last stages the washing solution is directed along the rim and then gradually towards the apex of the cone. In all cases, tests for the completeness of washing must be made by collecting a small sample of the washing solution after it is estimated that most of the impurities have been removed, and applying an appropriate qualitative test. Where filtration is carried out under suction, a small test-tube may be attached to the bottom of the Gooch funnel by means of a wire.

III, 44. DRYING AND IGNITION OF PRECIPITATES. After a precipitate has been filtered and washed, it must be brought to a constant composition before it can be weighed. The further treatment will depend both upon the nature of the precipitate and upon that of the filtering medium; this treatment consists in drying or igniting the precipitate. Which of the latter two terms is employed depends upon the temperature at which the precipitate is heated. There is, however, no definite temperature below or above which the precipitate is said to be dried or ignited respectively. The meaning will be adequately conveyed for our purpose if we designate *drying* when the temperature is below 250 °C (the maximum temperature which is readily reached in the usual thermostatically controlled, electric drying-oven), and *ignition* above 250 °C up to, say, 1200 °C. Precipitates that are to be dried should be collected on filter paper, or in Gooch, sintered glass, or porcelain filtering crucibles. Precipitates that are to be ignited are collected on filter paper, porcelain filtering crucibles, or silica filtering crucibles. Ignition is simply effected by placing in a special ignition dish or in a larger nickel or platinum crucible, and heating with the appropriate burner; alternatively, these crucibles (and, indeed, any type of crucible) may be placed in an electrically heated muffle furnace, which is equipped with a pyrometer and a means for controlling the temperature.

Attention is directed to the information provided by thermogravimetric analysis (see Chapter XXIII) concerning the range of temperature to which a precipitate should be heated for a particular composition. In general, thermal gravimetric curves seem to suggest that in the past precipitates were heated for too long a period and at too high a temperature. It must, however, be borne in mind that in some cases the thermal gravimetric curve is influenced by the experimental conditions of precipitation, and even if a horizontal curve is not obtained, it is possible that a suitable weighing form may be available over a certain temperature range. Nevertheless, thermograms do provide valuable data concerning the range of temperature over which a precipitate has a constant composition under the conditions that the thermogravimetric analysis was made; these, at the very least, provide a guide for the temperature at which a precipitate should be dried and heated for quantitative work, but due regard must be paid to the general chemical properties of the weighing form.

Although precipitates which require ignition will usually be collected in porcelain or silica filtering crucibles, there may be some occasions where filter paper has been used, and it is therefore necessary to describe the method to be adopted in such cases. The exact technique will depend upon whether the precipitate may be safely ignited in contact with the filter paper or not. It must be remembered that some precipitates, such as barium sulphate, may be reduced or changed in contact with filter paper or its decomposition products.

A. Incineration of the filter paper in the presence of the precipitate. A silica crucible is first ignited to constant weight (i.e., to within 0.0002 g) at the same temperature as that to which the precipitate is ultimately heated. The well-drained filter paper and precipitate are carefully detached from the funnel; the filter paper is folded so as to completely enclose the precipitate, care being taken not to tear the paper. The packet is then placed point down in the weighed crucible, which is supported on a pipe-clay, or better, a silica triangle resting on a ring stand as in Fig. III, 29. The crucible is slightly inclined, as shown in the diagram, and partially covered with the lid, which should rest partly on the triangle. A *very small flame* is then placed under the crucible lid; drying thus proceeds quickly and without undue risk. When the moisture has been expelled, the flame is increased slightly so as to *slowly* carbonise the paper. The paper should not be allowed to inflame, as this may cause a mechanical expulsion of fine

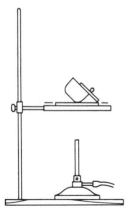

Fig. III, 29

particles of the precipitate owing to the rapid escape of the products of combustion: if, by chance, it does catch fire, the flame should be extinguished by momentarily placing the cover on the mouth of the crucible with the aid of a pair of crucible tongs. When the paper has completely carbonised and vapours are no longer evolved, the flame is moved to the back (bottom) of the crucible and the carbon slowly burned off whilst the flame is gradually increased.* After all the carbon has been burned away, the crucible is covered completely (if desired, the crucible may be placed in a vertical position for this purpose) and heated to the required temperature by means of a Bunsen or Meker flame. Usually it takes about 20 minutes to char the paper, and 30–60 minutes to complete the ignition.

When the ignition is ended, the flame is removed and, after 1–2 minutes, the crucible and lid are placed in a desiccator containing a suitable desiccant (Section **III, 26**), and allowed to cool for 25–30 minutes. The crucible and lid are then weighed: The crucible and contents are then ignited at the same temperature for 10–20 minutes, allowed to cool in a desiccator as before, and weighed again. The ignition is repeated until constant weight is attained. Crucibles should always be handled with clean crucible tongs and preferably with platinum-tipped tongs.

It is important to note that 'heating to constant weight' has no real significance unless the periods of heating, cooling of the *covered* crucible, and weighing are duplicated. There is some doubt as to whether cooling in a desiccator containing a desiccant is really successful in all cases in preventing some moisture being absorbed by the crucible and contents; this possible error is minimised by covering the crucible and weighing the precipitate as soon as it acquires the laboratory temperature. The empty crucible and lid should, of course, be subjected to the same treatment.

B. Incineration of the filter paper apart from the precipitate. This method is employed in all those cases where the ignited substance is reduced by the burning paper; for example, barium sulphate, lead sulphate, bismuth oxide, copper oxide, etc. The funnel containing the precipitate is covered by a piece of qualitative filter paper upon which is written the formula of the precipitate and the name of the owner; the paper is made secure by crumpling its edges over the rim of the funnel so that they will engage the outer conical portion of the funnel. The funnel is placed in the steam oven, or in a drying oven maintained at 100–105 °C, for 1–2 hours or until completely dry. A sheet of glazed paper about 25 cm square (white or black, to contrast with the colour of the precipitate) is placed on the bench away from all draughts. The dried filter is removed from the funnel, and as much as possible of the precipitate is removed from the paper and allowed to drop on a clock glass resting upon the glazed paper. This is readily done by very gently rubbing the sides of the filter paper together, when the bulk of the precipitate becomes detached and drops upon the clock glass. Any small

---

* If the carbon on the lid is oxidised only slowly, the cover may be heated separately in a flame. It is, of course, held in clean crucible tongs.

particles of the precipitate which may have fallen upon the glazed paper are brushed into the crucible with a small camel-hair brush. The clock glass containing the precipitate is then covered with a larger clock glass or with a beaker. The filter paper is now carefully folded and placed inside a weighed porcelain or silica crucible. The crucible is placed on a triangle and the filter paper incinerated as detailed above. The crucible is allowed to cool, and the filter ash subjected to a suitable chemical treatment in order to convert any reduced or changed material into the form finally desired. The cold crucible is then placed upon the glazed paper and the main part of the precipitate carefully transferred from the clock glass to the crucible. A small camel-hair brush will assist in the transfer. Finally, the precipitate is brought to constant weight by heating to the necessary temperature as detailed under **A**.

III, 45. PERFORATED SCREENS FOR CRUCIBLES. It is often important to exclude flame gases from the interior of a crucible during an ignition, e.g., in the ignition of iron(III) oxide. For this purpose we may employ a vitreosil plate, about 10 cm square, in which a round opening is cut large enough to admit the crucible to two-thirds of its depth. The plate is held at an angle of about 30° from the horizontal by means of a clamp; alternatively, but less satisfactorily, it may be suspended on a tripod. Asbestos board may also be employed, but this has the disadvantage that fibres may adhere to the crucible: this difficulty is less likely to occur with 'Uralite'.

III, 46. THE SCHÖNIGER OXYGEN FLASK METHOD FOR ELEMENTAL ANALYSIS. One of the most useful methods available for micro-analysis is that developed by Schöniger in 1955 (Refs. 17 and 18). It is based upon the procedure for the combustion of organic materials in an atmosphere of oxygen originally introduced by Hempel (Ref. 19) for determining sulphur in coal.

Although the procedure is used to analyse organic substances it is included in this book as the elements of the organic materials combusted are, in fact, determined in their inorganic forms using many of the titrimetric or spectrophotomeric methods described in later sections.

A number of reviews of the oxygen flask method have been published (Refs. 20 and 21) giving considerable details of all aspects of the subject.

In outline the procedure consists of carefully weighing about 5–10 mg of sample onto a shaped piece of paper (Fig. III, 30c) which is folded in such a way that the tail (wick) is free. This is then placed in a platinum basket or carrier suspended from the ground glass stopper of a $500 \, cm^3$ or 1 litre flask. The flask, containing a few cm^3 of absorbing solution (e.g. aqueous sodium hydroxide), is filled with oxygen and then sealed with the stopper with the platinum basket attached.

The wick of the sample paper can either be ignited before the stopper is placed in the flask neck, or better still ignited by remote electrical control, or by an infrared lamp. In any case combustion is rapid and usually complete within 5–10 seconds. After standing for a few minutes until any combustion cloud has disappeared, the flask is shaken for 2–3 minutes to ensure that complete absorption has taken place. The solution can then be treated by a method appropriate to the element being determined.

Organic sulphur is converted to sulphur trioxide and sulphur dioxide by the

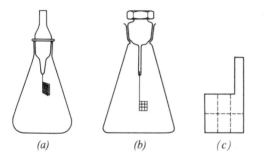

(a)   (b)   (c)

**Fig. III, 30   Conventional flasks for microdeterminations: (a) airleak design, (b)
stopper design (c) filter paper for wrapping sample
Reproduced by permission from A. M. G. Macdonald (1965). In
Advances in Analytical Chemistry and Instrumentation. (Ed. C. N.
Reilley), Vol 4, p. 75. New York; Interscience.**

combustion, absorbed in hydrogen peroxide, and the sulphur determined as
sulphate.

The combustion products of organic halides are usually absorbed in sodium
hydroxide containing some hydrogen peroxide. The resulting solutions may be
analysed by a range of available procedures. For chlorides the method most
commonly used is that of argentimetric potentiometric titration (Ref. 22) (see
Section **XIV, 25**), whilst for bromides a mercurimetric titration (Ref. 23) is
comparable with the argentimetric method.

Phosphorus from organophosphorus compounds, which are combusted to
give mainly orthophosphate, can be absorbed by either sulphuric acid or nitric
acid and readily determined spectrophotometrically either by the molybdenum
blue method or as the phosphovanadomolybdate (Section **XVIII, 33**).

Procedures have also been devised for the determination of metallic constituents. Thus, mercury is absorbed in nitric acid and titrated with sodium
diethyldithiocarbamate, whilst zinc is absorbed in hydrochloric acid and
determined by an EDTA titration (see Section **X, 67**).

III, 47.   References

1. National Physical Laboratory (1956). *Balances, Weights and Precise Laboratory
   Weighing.* London; Her Majesty's Stationery Office.
2. (a) J. T. Stock (1969). *The Development of the Chemical Balance.* London; Her
   Majesty's Stationery Office.
   (b) J. T. Stock (1973). *Analytical Chemistry*, **45**, 947A.
3. (a) T. W. Richards (1900). *J. Am. Chem. Soc.*, **22**, 144.
   (b) A. I. Vogel (1961). *Quantitative Inorganic Analysis.* 3rd edn. London; Longmans.
4. T. W. Lashof and L. B. MacCurdy (1954). *Analytical Chemistry*, **26**, 707.
5. M. L. McGlashan (1971). *Physico-Chemical Quantities and Units.* 2nd edn (p. 45).
   London; Royal Institute of Chemistry.
6. C. Woodward and H. N. Redman (1973). *High Precision Titrimetry.* London; Society
   for Analytical Chemistry.
   (a), p. 14. (b), p. 5. (c), p. 11. (d), p. 10. and p. 12.
7. Technical Information Bulletin (1967). *Gravimetric Titrimetry—A Review of the
   Literature.* Princeton; Mettler Instrument Corporation.

8. (a) BS 3978 1966: *Water for Laboratory Use*. London; British Standards Institution.
(b) D1193–70 (1970). *Standard Specification for Reagent Water*. Easton, Md.; American Society for Testing Materials.
(c) *Reagent Chemicals; Supplement 1* (1969). Washington, DC; American Chemical Society Publications.

9. D. F. Shriver (1969). *The Manipulation of Air-Sensitive Materials*. New York; McGraw-Hill Book Co.

10. R. G. Bates and E. Wichers (1957). *J. Res. Natl. Bur. Stand.*, **59**, 9.

11. E. B. Sandell and T. S. West (1969). 'Recommended Nomenclature for Titrimetric Analysis.' *Pure and Applied Chemistry*, **18**, 429.

12. B. Bernas (1968). 'A New Method for Decomposition and Comprehensive Analysis of Silicates.' *Anal. Chem.*, **40**, 1682.

13. J. C. Meakin and M. C. Pratt (1972). *Manual of Laboratory Filtration*. Maidstone; E. W. Balston Ltd.

14. C. O. Ingamells (1964). 'Rapid Chemical Analysis of Silicate Rocks.' *Talanta*, **11**, 665.

15. H. Bennett and G. J. Oliver (1971). 'Loss of Cobalt and Iron from Lithium Borate Fusions in Graphite Crucibles.' *Analyst*, **96**, 427.

16. J. Dolezal, P. Povondra and Z. Sulcek (1968). *Decomposition Techniques in Inorganic Analysis*. London; Iliffe Books Ltd.

17. W. Schöniger (1955). *Mikrochim. Acta*, 123.

18. W. Schöniger (1956). *Mikrochim. Acta*, 869.

19. W. Hempel (1892). *Z. Angew. Chem.*, **13**, 393.

20. A. M. G. Macdonald (1965), in C. N. Reilley (ed.). *Advances in Analytical Chemistry and Instrumentation*. Vol. 4, p. 75, New York; Interscience.

21. A. M. G. Macdonald (1961). *Analyst*, **86**, 3.

22. Analytical Methods Committee (1963). *Analyst*, **88**, 415.

23. R. C. Denney and P. A. Smith (1974). *Analyst*, **99**, 166.

24. Asbestos Regulations (1969), Department of Employment Health Booklets and Technical Data. H.M.S.O., London.

III, 48. Selected bibliography

Many of the books detailed in Section **I, 12** contain descriptions of simple apparatus and basic techniques and are relevant to the present Chapter. Attention is particularly directed to:

(a) Kolthoff and Elving. *Treatise on Analytical Chemistry* (Vols. 7, 9, 10).

(b) Strouts, Wilson and Parry Jones. *Chemical Analysis: the Working Tools*.

(c) Wilson and Wilson. *Comprehensive Analytical Chemistry* (Vol. IA).

1. W. M. MacNevin (1951). *The Analytical Balance. Its Care and Use*. Handbook Publishers, Sandusky, Ohio.

2. National Physical Laboratory (1956). *Balances, Weights and Precise Laboratory Weighing*. (Notes on Applied Science, No. 7.) Her Majesty's Stationery Office, London.

3. P. H. Bigg (1959). 'Weight-in-air Basis of Adjustment of Precision Weights.' *Journal of Scientific Instruments*, **36**, 359.

4. British Standards Institution, London.
BS 501, 554: 1952. *Report on Metric Units of Volume and Standard Temperature of Volumetric Glassware*.
BS 604: 1952. *Graduated Measuring Cylinders*.
BS 676: 1953. *Flasks with Graduated Necks*.
BS 700: 1962. *Graduated Pipettes and One-Mark Cylindrical Pipettes*.
BS 846: 1962. *Burettes and Bulb Burettes*.
BS 1132: 1966. *Automatic Pipettes*.
BS 1583: 1961. *One-Mark Bulb Pipettes*.
BS 1792: 1960. *One-Mark Graduated Flasks*.

BS 1797: 1968. *Tables for Use in the Calibration of Volumetric Glassware.*

BS 2058: 1961. *Weighing Pipettes.*

BS 1752: 1963. *Laboratory sintered or fritted filters.*

BS 2648: 1955. *Performance requirements for electrically-heated laboratory drying ovens.*

BS 3423: 1962. *Recommendations for the design of glass vacuum desiccators.*

BS 3978: 1966. *Water for laboratory use.*

BS 3996: 1966. *Colour coding for one-mark and graduated pipettes.*

BS 4244: 1967. *Porcelain and silica crucibles.*

BS 2445: 1968. *Recommendations for solutions used in chemical analysis. Terminology, presentation and concentration.*

5. R. F. Hirsch. 'Modern Laboratory Balances. Part I.' *J. Chem. Ed.,* **44**, A1023, 1967. 'Part II.' *J. Chem. Ed.,* **45**, A7, (1968).

6. *'AnalaR' Standards for Laboratory Chemicals* (1967). 6th edn. London; AnalaR Standards Ltd.

PART B   **ERRORS AND SAMPLING**

CHAPTER IV ERRORS AND STATISTICS

IV, 1. LIMITATIONS OF ANALYTICAL METHODS. The function of the analyst is to obtain a result as near to the true value as possible by the correct application of the analytical procedure employed. The level of confidence that the analyst may enjoy in his results will be very small unless he has knowledge of the accuracy and precision of the method used as well as being aware of the sources of error which may be introduced. Quantitative analysis is not simply a case of taking a sample, carrying out a single determination and then claiming that the value obtained is irrefutable. It also requires a sound knowledge of the chemistry involved, the possibilities of interferences from other ions, elements and compounds as well as the statistical distribution of values. The purpose of this chapter is to explain some of the terms employed and to outline the statistical methods that may be applied.

IV, 2. ACCURACY. The accuracy of a determination may be defined as the concordance between it and the true or most probable value. For analytical methods there are two possible ways of determining the accuracy; the so-called absolute method and the comparative method.

Absolute method. A synthetic sample containing known amounts of the constituents in question is used. Known amounts of a constituent can be obtained by weighing out pure elements or compounds of known stoichiometric composition. These substances, primary standards, may be available commercially or they may be prepared by the analyst and subjected to rigorous purification by recrystallisation, etc. The substances must be of known purity. The test of the accuracy of the method under consideration is carried out by taking varying amounts of the constituent and proceeding according to specified instructions. The amount of the constituent must be varied, because the determinate errors in the procedure may be a function of the amount used. The difference between the mean of an adequate number of results and the amount of the constituent actually present, usually expressed as parts per thousand, is a measure of the accuracy of the method in the absence of foreign substances.

The constituent in question will usually have to be determined in the presence of other substances, and it will therefore be necessary to know the effect of these upon the determination. This will require testing the influence of a large number of elements, each in varying amounts—a major undertaking. The scope of such tests may be limited by considering the determination of the component in a specified range of concentration in a material whose composition is more or less fixed both with respect to the elements which may be present and their relative amounts. It is desirable, however, to study the effect of as many foreign elements as feasible. In practice, it is frequently found that separations will be required before a determination can be made in the presence of varying elements; the accuracy of the method is likely to be largely controlled by the separations involved.

Comparative method.   Sometimes, as in the analysis of a mineral, it may be impossible to prepare solid synthetic samples of the desired composition. It is then necessary to resort to standard samples of the material in question (mineral, ore, alloy, etc.) in which the content of the constituent sought has been determined by one or more supposedly 'accurate' methods of analysis. This comparative method, involving secondary standards, is obviously not altogether satisfactory from the theoretical standpoint, but is nevertheless very useful in applied analysis. Standard samples are issued by the US National Bureau of Standards, Washington, and by the Bureau of Analysed Samples, Middlesbrough.

If several fundamentally different methods of analysis for a given constituent are available, e.g., gravimetric, titrimetric, spectrophotometric, or spectrographic, the agreement between at least two methods of essentially different character can usually be accepted as indicating the absence of an appreciable determinate error in either (a determinate error is one which can be evaluated experimentally or theoretically).

IV, 3.   PRECISION.   Precision may be defined as the concordance of a series of measurements of the same quantity. The mean deviation or the relative mean deviation is a measure of precision. In quantitative analysis the precision of measurements rarely exceeds 1 to 2 parts per thousand.

Accuracy expresses the correctness of a measurement, and precision the reproducibility of a measurement. Precision always accompanies accuracy, but a high degree of precision does not imply accuracy. This may be illustrated by an example.

Example.   A substance was known to contain 49.06 ± 0.02 per cent of a given constituent A. The results obtained by two observers using the same substance and the same general technique were:

Observer (1).   49.01; 49.21; 49.08.   Mean = 49.10 per cent.
Relative mean error = $(49.10 - 49.06)/49.06 = 0.08$ per cent.
Relative mean deviation* = $[(0.09 + 0.11 + 0.02)/3] \times 100/49.10$
$\qquad\qquad\qquad\qquad = 0.15$ per cent.
Observer (2).   49.40; 49.44; 49.42.   Mean = 49.42 per cent.
Relative mean error = $(49.42 - 49.06)/49.06 = 0.73$ per cent.
Relative mean deviation* = $[(0.02 + 0.02 + 0.00)/3] \times 100/49.42$
$\qquad\qquad\qquad\qquad = 0.03$ per cent.

The analyses of observer (1) were therefore accurate and precise; those of observer (2) were unusually precise, but less accurate than those of observer (1). Some small source of constant error appears to be present in the results of (2).

---

* See Section **IV, 7.**

IV, 4. CLASSIFICATION OF ERRORS. The errors which affect an experimental result may be conveniently divided into those of the determinate and the indeterminate kind.

Determinate or constant errors. These are errors which can be avoided, or whose magnitude can be determined. The most important of these are:

1. Operational and personal errors. These are due to factors for which the individual analyst is responsible and are not connected with the method or procedure: they form part of the 'personal equation' of an observer. The errors are mostly physical in nature and occur when sound analytical technique is not followed. Examples are: mechanical loss of materials in various steps of an analysis, underwashing or overwashing of precipitates, ignition of precipitates at incorrect temperatures, insufficient cooling of crucibles before weighing, allowing hygroscopic materials to absorb moisture before or during weighing, and use of reagents containing harmful impurities.

Personal errors may arise from the constitutional inability of an individual to make certain observations accurately. Thus some persons are unable to judge colour changes sharply in visual titrations, which may result in a slight overstepping of the end-point.

2. Instrumental and reagent errors. These arise from the faulty construction of balances, the use of uncalibrated or improperly calibrated weights, graduated glassware, and other instruments; the attack of reagents upon glassware, porcelain, etc., resulting in the introduction of foreign materials; volatilisation of platinum at very high temperatures; and the use of reagents containing impurities.

3. Errors of method. These originate from incorrect sampling and from incompleteness of a reaction. In gravimetric analysis errors may arise owing to appreciable solubility of precipitates, co-precipitation, and post-precipitation, decomposition, or volatilisation of weighing forms on ignition, and precipitation of substances other than the intended ones. In titrimetric analysis errors may occur owing to failure of reactions to proceed to completion, occurrence of induced and side reactions, reaction of substances other than the constituent being determined, and a difference between the observed end point and the stoichiometric end point of a reaction.

4. Additive and proportional errors. The absolute value of an additive error is independent of the amount of the constituent present in the determination. Examples of additive errors are loss in weight of a crucible in which a precipitate is ignited, and errors in weights. The presence of this error is revealed by taking samples of different weights.

The absolute value of a proportional error depends upon the amount of the constituent. Thus a proportional error may arise from an impurity in a standard substance, which leads to an incorrect value for the normality of a standard solution. Other proportional errors may not vary linearly with the amount of the constituent, but will at least exhibit an increase with the amount of constituent present. One example is the ignition of aluminium oxide: at 1200 °C the aluminium oxide is anhydrous and virtually non-hygroscopic; ignition of various weights at an appreciably lower temperature will show a proportional type of error.

Indeterminate or accidental errors. These errors manifest themselves by the slight variations that occur in successive measurements made by the same observer with the greatest care under as nearly identical conditions as possible.

They are due to causes over which the analyst has no control, and which, in general, are so intangible that they are incapable of analysis. If a *sufficiently large number of observations* is taken, it can be shown that these errors lie on a curve of the form shown in Fig. IV, 1 (Section **IV, 8**). An inspection of this error curve shows: (*a*) small errors occur more frequently than large ones; (*b*) large errors occur relatively infrequently; and (*c*) positive and negative errors of the same numerical magnitude are equally likely to occur.

IV, 5. MINIMISATION OF ERRORS. Determinate errors can often be materially reduced by one of the following methods:

1. Calibration of apparatus and application of corrections. All instruments (weights, flasks, burettes, pipettes, etc.) should be calibrated, and the appropriate corrections applied to the original measurements. In some cases where an error cannot be eliminated, it is possible to apply a correction for the effect that it produces; thus an impurity in a weighed precipitate may be determined and its weight deducted.

2. Running a blank determination. This consists in carrying out a separate determination, the sample being omitted, under exactly the same experimental conditions as are employed in the actual analysis of the sample. The object is to find out the effect of the impurities introduced through the reagents and vessels, or to determine the excess of standard solution necessary to establish the end-point under the conditions met with in the titration of the unknown sample. A large blank correction is undesirable, because the exact value then becomes uncertain and the precision of the analysis is reduced.

3. Running a control determination. This consists in carrying out a determination under as nearly as possible identical experimental conditions upon a quantity of a standard substance which contains the same weight of the constituent as is contained in the unknown sample. The weight of the constituent in the unknown can then be calculated from the relation:

$$\frac{\text{Result found for standard}}{\text{Result found for unknown}} = \frac{\text{Weight of constituent in standard}}{x}$$

where x is the weight of the constituent in the unknown.

In this connection it must be pointed out that standard samples which have been analysed by a number of skilled analysts are commercially available. These include certain primary standards (sodium oxalate, potassium hydrogenphthalate, arsenic(III) oxide, and benzoic acid) and ores, ceramic materials, irons, steels, steel-making alloys, and non-ferrous alloys. All of these are obtainable from the US Bureau of Standards, Department of Commerce, Washington, DC. Many of these are also available as the 'British Chemical Standards' and are supplied by the Bureau of Analysed Samples, Ltd, Newham Hall, Middlesbrough, England.

4. Use of independent methods of analysis. In some instances the accuracy of a result may be established by carrying out the analysis in an entirely different manner. Thus iron may first be determined gravimetrically by precipitation as iron(III) hydroxide after removing the interfering elements, followed by ignition of the precipitate to iron(III) oxide. It may then be determined titrimetrically by reduction to the iron(II) state, and titration with a standard solution of an oxidising agent, such as potassium dichromate or cerium(IV) sulphate. Another example that may be mentioned is the determination of the strength of a hydrochloric acid solution both by titration with a standard solution of a strong base and by precipitation and weighing as silver chloride. If the results obtained by the two radically different methods are concordant, it is highly probable that the values are correct within small limits of error.

5. Running of parallel determinations. These serve as a check on the result of a single determination and indicate only the precision of the analysis. The values obtained for constituents which are present in not too small an amount should not vary among themselves by more than three parts per thousand. If larger variations are shown, the determinations must be repeated until satisfactory concordance is obtained. Duplicate, and at most triplicate, determinations should suffice. It must be emphasised that good agreement between duplicate and triplicate determinations does not justify the conclusion that the result is correct; a constant error may be present. The agreement merely shows that the accidental errors, or variations of the determinate errors, are the same, or nearly the same, in the parallel determinations.

6. Standard addition. A known amount of the constituent being determined is added to the sample, which is then analysed for the total amount of constituent present. The difference between the analytical results for samples with and without the added constituent gives the recovery of the amount of added constituent. If the recovery is satisfactory our confidence in the accuracy of the procedure is enhanced. The method is usually applied to physico-chemical procedures such as polarography and spectrophotometry.

7. Internal standards. This procedure is of particular value in spectroscopic and chromatographic determinations. It involves adding a fixed amount of a reference material (the internal standard) to a series of known concentrations of the material to be measured. The ratios of the physical value (absorption or peak size) of the internal standard and the series of known concentrations is plotted against the concentration values. This should give a straight line. Any unknown concentration can then be determined by adding the same quantity of internal standard and finding where the ratio obtained falls on the concentration scale.

8. Amplification methods. In determinations in which a very small amount of material is to be measured this may be beyond the limits of the apparatus available. In these circumstances if the small amount of material can be reacted in such a way that every molecule produces two or more molecules of some other measurable material the amplification of the quantity may then be within the scope of the apparatus or method available.

9. Isotopic dilution. A known amount of the element being determined, containing a radioactive isotope, is mixed with the sample and the element is isolated in a pure form (usually as a compound), which is weighed or otherwise determined. The radioactivity of the isolated element is measured and compared with that of the added element: the weight of the element in the sample can then be calculated.

IV, 6. SIGNIFICANT FIGURES AND COMPUTATIONS. The term digit denotes any one of the ten numerals, including the zero. A significant figure is a digit which denotes the amount of the quantity in the place in which it stands. The digit zero is a significant figure except when it is the first figure in a number. Thus in the quantities 1.2680 g and 1.0062 g the zero is significant, but in the quantity 0.0025 kg the zeros are not significant figures; they serve only to locate the decimal point and can be omitted by proper choice of units, e.g., 2.5 g. The first two numbers contain five significant figures, but 0.0025 contains only two significant figures.

Observed quantities should be noted with *one uncertain figure retained*. Thus in most analyses weights are determined to the nearest tenth of a milligram, e.g., 2.1546 g. This means that the weight is less than 2.1547 g and more than 2.1545 g. A weight of 2.150 g would signify that it has been determined to the nearest milligram, and that the weight is nearer to 2.150 g than it is to either 2.151 g or 2.149 g. The digits of a number which are needed to express the precision of the measurement from which the number was derived are known as significant figures.

There are a number of rules for computations with which the student should be familiar.

1. Retain as many significant figures in a result or in any data as will give only one uncertain figure. Thus a volume which is known to be between 20.5 cm³ and 20.7 cm³ should be written as 20.6 cm³, but not as 20.60 cm³, since the latter would indicate that the value lies between 20.59 cm³ and 20.61 cm³. Also, if a weight, to the nearest 0.1 mg, is 5.2600 g, it should not be written as 5.260 g or 5.26 g, since in the latter case an accuracy of a centigram is indicated and in the former a milligram.

2. In rounding off quantities to the correct number of significant figures, add one to the last figure retained if the following figure (which has been rejected) is 5 or over. Thus the average of 0.2628, 0.2623, and 0.2626 is 0.2626 (0.2625_7).

3. In addition or subtraction, there should be in each number only as many significant figures as there are in the least accurately known number. Thus the addition

$$168.11 + 7.045 + 0.6832$$

should be written

$$168.11 + 7.05 + 0.68 = 175.84$$

The sum or difference of two or more quantities cannot be more precise than the quantity having the largest uncertainty.

4. In multiplication or division, retain in each factor one more significant figure than is contained in the factor having the largest uncertainty. The percentage precision of a product or quotient cannot be greater than the percentage precision of the least precise factor entering into the calculation. Thus the multiplication

$$1.26 \times 1.236 \times 0.6834 \times 24.8652$$

should be carried out using the values

$$1.26 \times 1.236 \times 0.683 \times 24.87$$

and the result expressed to three significant figures.

Where a large number of multiplications and divisions are to be made, the use of logarithms is recommended. Four-figure logarithm tables are sufficiently precise if interpolation is used.

A 25 cm slide rule is accurate to about 0.25 per cent, and is useful in checking calculations. The Otis King calculator has an accuracy of about four times that of the 25 cm slide rule, and the rotary scales are 170 cm long; it is of convenient size for the pocket, and is very useful in the analytical laboratory.

With the advent of many reasonably priced electronic pocket calculators statistical calculations are now easy to carry out and the saving in time achieved very quickly covers the initial financial outlay. Apart from normal arithmetic functions a suitable calculator for statistical work should give squares and square roots, possess a floating decimal point and at least a six digit display.

For processing large amounts of data, and retrieval or comparison with previous results, many hours of work are saved by use of computers. Although computer programming is outside the scope of this book it should be pointed out that standard programs now exist in ALGOL, COBOL, FORTRAN IV, etc., for calculating statistical functions and carrying out the more involved mathematical determinations such as that for binary mixtures by ultraviolet/visible spectroscopy (Section **XVIII, 38**).

IV, 7. MEAN (AVERAGE) DEVIATION. STANDARD DEVIATION.
When a quantity is measured with the greatest exactness that the instrument, method, and observer are capable of, it is found that the results of successive determinations differ among themselves to a greater or lesser extent. The average value is accepted as the most probable. This may not always be the true value. In some cases the difference may be small, in others it may be large; the reliability of the result depends upon the magnitude of this difference. It is therefore of interest to enquire briefly into the factors which affect and control the trustworthiness of chemical analysis.

The **absolute error** of a determination is the difference between the observed or measured value and the true or most probable value of the quantity measured. The absolute error is a measure of the **accuracy** of the measurement.

The **relative error** is the absolute error divided by the true or most probable value; it is usually expressed in terms of percentage or in parts per thousand. The true or absolute value of a quantity cannot be established experimentally, so that the observed result must be compared with the most probable value. With pure substances the quantity will ultimately depend upon the atomic weights of the constituent elements. Determinations of the atomic weights have been made with the utmost care, and the accuracy obtained usually far exceeds that attained in ordinary quantitative analysis; the analyst must accordingly accept their reliability. With natural or industrial products, we must accept provisionally the results obtained by analysts of repute using carefully tested methods. If several analysts determine the same constituent in the same sample by different methods, the most probable value, which is usually the average, can be deduced from their results. In both cases the establishment of the most probable value involves the application of statistical methods and the concept of precision.

The agreement between a series of results is measured by computing their **mean deviation**. This is evaluated by determining the arithmetical mean of the results, then calculating the deviation of each individual measurement from the mean, and finally dividing the sum of the deviations, regardless of sign, by the number of measurements. The **relative mean deviation** is the mean deviation divided by the mean. This may be expressed in terms of percentage or in parts per thousand. An example will make this clear.

In analytical chemistry one of the most common statistical terms employed is the **standard deviation** of a population of observations. This is also called the root mean square deviation as it is the square root of the mean of the sum of the squares of the differences between the values and the mean of those values (this is expressed mathematically below) and is of particular value in connection with the normal distribution, Section **IV, 8**.

Example. The percentages of a constituent A in a compound AB were found to be 48.32, 48.36, 48.23, 48.11, and 48.38 per cent. Calculate the mean deviation and the relative mean deviation.

Results	*Deviations*
48.32	0.04
48.36	0.08
48.23	0.05
48.11	0.17
48.38	0.10
5)241.40	5)0.44

Mean $= 48.28$ *Mean deviation $= 0.09$*

Relative mean deviation $= 0.09 \times 100/48.28 = 0.19$ per cent
$= 1.9$ parts per thousand

If we consider a series of n observations arranged in ascending order of magnitude:

$$x_1, x_2, x_3, \ldots x_{n-1}, x_n,$$

the **arithmetic mean** (often called simply the **mean**) is given by:

$$\bar{x} = \frac{x_1 + x_2 + x_3 \ldots + \ldots + x_{n-1} + x_n}{n}$$

The spread of the values is measured most efficiently by the **standard deviation** s defined by:

$$s = \sqrt{\frac{(x_1 - \bar{x})^2 + (x_2 - \bar{x})^2 + \ldots (x_n - \bar{x})^2}{n-1}}$$

In this equation the denominator is $(n-1)$ rather than n when the number of values is small.

The equation may also be written as:

$$s = \sqrt{\frac{\Sigma(x - \bar{x})^2}{n-1}}$$

The square of the standard deviation is called the **variance**. A more accurate measure of the precision, known as the **coefficient of variation** (C.V.), is given by:

$$\text{C.V.} = \frac{s \times 100}{\bar{x}}$$

Example. Analyses of a sample of iron ore gave the following percentage values for the iron content: 7.08, 7.21, 7.12, 7.09, 7.16, 7.14, 7.07, 7.14, 7.18, 7.11. Calculate the mean, standard deviation and coefficient of variation for the values.

results (x)	$x - \bar{x}$	$(x - \bar{x})^2$
7.08	−0.05	0.0025
7.21	0.08	0.0064
7.12	−0.01	0.0001
7.09	−0.04	0.0016
7.16	0.03	0.0009
7.14	0.01	0.0001
7.07	−0.06	0.0036
7.14	0.01	0.0001
7.18	0.05	0.0025
7.11	−0.02	0.0004

total 71.30 $\Sigma = 0.0182$

mean ($\bar{x}$) 7.13

$$s = \sqrt{\frac{0.0182}{9}}$$

$$= \sqrt{0.0020}$$
$$= \pm 0.045$$
$$\text{C.V.} = \frac{0.045 \times 100}{7.13} = 0.63$$

IV, 8.   NORMAL (GAUSSIAN) DISTRIBUTION.   Continuous data, of the type resulting from a number of analyses of an individual chemical sample, fall within a range of values that satisfy the Normal (or Gaussian) distribution. This is a bell-shaped curve that is symmetrical about the mean as shown in Fig. IV, 1.

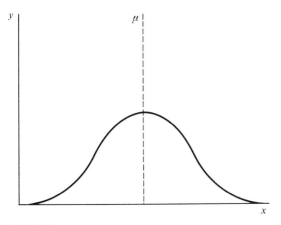

Fig. IV, 1

The curve satisfies the equation:

$$y = \frac{1}{\sigma\sqrt{2\pi}}\, e^{-\frac{(x-\mu)^2}{2\sigma^2}}$$

With this type of distribution about 68 per cent of all values will fall within one standard deviation on either side of the mean, 95 per cent will fall within two standard deviations, 99.7 per cent within three standard deviations, and 99.994 within four standard deviations.

It is important to know that the Greek letters σ and μ refer to the standard deviation and mean respectively of a total population; whilst the Roman letters s and $\bar{x}$ are used for samples of populations.

IV, 9.   COMPARISON OF RESULTS.   Statistical figures obtained from a set of results are of limited value by themselves. It is only by comparing them with the true value or with other sets of data that it is possible to determine whether the analytical procedure has been accurate or precise or if it is superior to another method. There are three common methods for testing results; (*a*) Student's *t*-test, (*b*) the Variance ratio test (*F*-test) and (*c*) the Chi square distribution.

These methods of test require a knowledge of what is known as the number of degrees of freedom. In statistical terms this is the number of independent values necessary to determine the statistical quantity. Thus a sample of n values has n degrees of freedom, whilst the sum $\Sigma(x - \bar{x})^2$ is considered to have $n - 1$ degrees of freedom, as for any defined value of $\bar{x}$ only $n - 1$ values can be freely assigned, the nth being automatically defined from the other values.

(*a*) Student's *t*-test. This is a test (Ref. 1) used for small samples; its purpose is to compare the mean from a sample with some standard value and to express some level of confidence in the significance of the comparison. It is also used to test the difference between the means of two sets of data $\bar{x}_1$ and $\bar{x}_2$.

The value of t is obtained from the equation:

$$t = \frac{(\bar{x} - \mu)\sqrt{n}}{s} \tag{i}$$

where μ is the true value.

It is then related to a set of *t*-tables (Appendix XV) in which the probability (P) of the *t*-value falling within certain limits is expressed, either as a percentage or as a function of unity, relative to the number of degrees of freedom.

Example.   If $\bar{x}$ the mean of 12 determinations $= 8.37$, and μ the true value $= 7.91$, say whether or not this result is significant if the standard deviation is 0.17.

From equation (i)

$$t = \frac{(8.37 - 7.91)\sqrt{12}}{0.17}$$
$$= 9.4$$

From *t* tables for *eleven* degrees of freedom (one less than those used in the calculation)

for $P =$	0.10(10%)	0.05(5%)	0.01(1%)
$t =$	1.80	2.20	3.11

and as the calculated value for t is 9.4 the result is highly significant. The t table tells us that the probability of obtaining the difference of 0.46 between the experimental and true result is less than 1 in 100. This implies that some particular bias exists in the laboratory procedure.

Had the calculated value for t been less than 1.80 then there would have been no significance in the results and no apparent bias in the laboratory procedure, as the tables would have indicated a probability of greater than 1 in 10 of obtaining that value. It should be pointed out that these values refer to what is known as a double-sided, or two-tailed, distribution because it concerns probabilities of values both less and greater than the mean. In some calculations an analyst may only be interested in one of these two cases, and under these conditions the t-test becomes single-tailed so that the probability from the tables is halved.

(b) *F-test.* This is used to compare the precision of two sets of data (Ref. 2); for example, the results of two different analytical methods or the results from two different laboratories. It is calculated from the equation:

$$F = \frac{s_A^2}{s_B^2} \qquad\qquad (ii)$$

N.B. the larger value of s is always used as the numerator so that the value of F is always greater than unity. The value obtained for F is then checked for its significance against values in the F table calculated from a Gaussian distribution (Appendix XVI) corresponding to the number of degrees of freedom for the two sets of data.

Example. The standard deviation from one set of 11 determinations $s_A = 0.210$, and the standard deviation from another 13 determinations was $s_B = 0.641$. Is there any significant difference between the precision of these two sets of results?

From equation (ii)

$$F = \frac{(0.641)^2}{(0.210)^2} = \frac{0.411}{0.044}$$
$$F = 9.4$$

Using the F tables, we look for the values corresponding to 12 degrees of freedom for s_B and 10 degrees of freedom for s_A. These give us three values:

for $P = 0.10 \quad 0.05 \quad 0.01$
$F = 2.19 \quad 2.75 \quad 4.30$

The first value (2.19) corresponds to 10 per cent probability, the second value (2.75) to 5 per cent probability and the third value (4.30) to 1 per cent probability.

Under these conditions there is less than 1 chance in 100 that these precisions are similar. To put it another way, the difference between the two sets of data is highly significant.

Had the value of F turned out to be less than 2.19 then it would have been possible to say that there was no significant difference between the precisions.

(c) Chi square test (χ^2). This is used to determine whether or not a set of data differs significantly from a theoretical or defined distribution (Ref. 3); that is, whether the observed frequencies of an occurrence correspond to the predicted frequencies. Chi square is calculated from the equation:

$$\chi^2 = \sum \frac{(O - E)^2}{E} \qquad\qquad (iii)$$

where O is the observed frequency and E the expected frequency.

Thus if a coin is tossed 100 times and the tails come up 25 times we would wish to ascertain if there is any real indication of bias.

Normally we would expect an equal chance of obtaining heads or tails, in this case 50 heads and 50 tails.

From equation (iii)

$$\chi^2 = \frac{(25-50)^2}{50} + \frac{(75-50)^2}{50}$$
$$= \frac{2 \times 625}{50}$$
$$= 25$$

For one degree of freedom the χ^2 tables give the following values (Appendix XVIII):

1 % level 6.63
0.1 % level 10.83

The value of 25 obtained in the above calculation is well beyond 10.83 and we can say that there is a significant bias in the spinning of the coin.

IV, 10.   THE NUMBER OF PARALLEL DETERMINATIONS.   To avoid unnecessary time and expenditure an analyst needs some guide to the number of repetitive determinations he needs to carry out to obtain a suitably reliable result. He will be aware that the greater the number he carries out the greater the reliability, but at the same time will know that after a certain number of determinations any improvement in precision and accuracy is very small.

Although rather involved statistical methods exist for establishing the number of parallel determinations, a reasonably good assessment can be made by establishing the variation of the value for the absolute error Δ obtained for an increasing number of determinations.

$$\Delta = \frac{ts}{n}$$

The value for t is taken from the 95 per cent confidence limit column of the t tables for $n-1$ degrees of freedom.

The values for Δ are used to calculate the reliability interval L from the equation:

$$L = \frac{100\Delta}{z}\%$$

where z is the approximate percentage level of the unknown being determined. The number of replicate analyses is assessed from the magnitude of the change in L with the number of determinations.

Example.   Ascertain the number of replicate analyses desirable (a) for the determination of approximately 2 per cent Cl^- in a material if the standard deviation for determinations is 0.051, (b) for approximately 20 per cent Cl^- if the standard deviation of determinations is 0.093.

(a) For 2 per cent Cl^- :

Number of determinations	$\Delta = \dfrac{ts}{n}$	$L = \dfrac{100\Delta}{z}\%$	Difference
2	$12.7 \times 0.051 \times 0.71 = 0.4599$	22.99	
3	$4.3 \times 0.051 \times 0.58 = 0.1272$	6.36	16.63
4	$3.2 \times 0.051 \times 0.50 = 0.0816$	4.08	2.28
5	$2.8 \times 0.051 \times 0.45 = 0.0642$	3.21	0.87
6	$2.6 \times 0.051 \times 0.41 = 0.0544$	2.72	0.49

(b) For 20 per cent Cl^- :

Number of determinations	$\Delta = \dfrac{ts}{n}$	$L = \dfrac{100\Delta}{z}\%$	Difference
2	$12.7 \times 0.093 \times 0.71 = 0.838$	4.19	
3	$4.3 \times 0.093 \times 0.58 = 0.232$	1.16	3.03
4	$3.2 \times 0.093 \times 0.50 = 0.148$	0.74	0.42
5	$2.8 \times 0.093 \times 0.45 = 0.117$	0.59	0.15
6	$2.6 \times 0.093 \times 0.41 = 0.099$	0.49	0.10

In (a) the reliability interval is greatly improved by carrying out a third analysis. This is less the case with (b) as the reliability interval is already narrow. In this second case no substantial improvement is gained by carrying out more than two analyses.

This subject is dealt with in more detail by Eckschlager (Ref. 4), and Shewell (Ref. 5) has discussed other factors which influence the value of parallel determinations.

IV, 11. THE VALUE OF STATISTICS. Correctly used, statistics are an essential tool to the analyst. They can, in particular, prevent him from making hasty judgements on the basis of limited information. It has only been possible at this stage to give a brief resume of the main statistical techniques that may be applied. The reader is advised to make himself fully conversant with these methods by obtaining one of the many excellent statistics texts now available.

IV, 12. References

1. 'Student' (1908). (W. G. Gosset), *Biometrika*. **6**, 1.
2. J. Mandel (1964). *The Statistical Analysis of Experimental Data*, New York; Interscience.
3. C. J. Brookes, I. G. Betteley and S. M. Loxston (1966). *Mathematics and Statistics for Chemists*, New York; John Wiley, p. 304.
4. K. Eckschlager (1969). *Errors, Measurements and Results in Chemical Analysis*, London; Van Nostrand Reinhold.
5. C. T. Shewell (1959). *Analytical Chemistry*, **31**, No. 5, 21A.

IV, 13. Selected bibliography

1. G. T. Wernimont (1949). 'Statistics Applied to Analysis', *Analytical Chemistry*, **21**, 115.
2. R. B. Dean and W. J. Dixon (1951). 'Simplified Statistics for Small Numbers of Observations', *Analytical Chemistry*, **23**, 636.
3. D. R. Read (1951). 'Statistical Methods with Special Reference to Analytical Chemistry', *Royal Institute of Chemistry, Lectures, Monographs and Reports*, No. 1.
4. I. M. Kolthoff and P. J. Elving (Ed.) (1950). *Treatise on Analytical Chemistry.* Part I. *Theory and Practice.* Vol. 1. Chapter 2. *Errors in Chemical Analysis.* Ch. 3. *Accuracy and Precision*, New York; Interscience Publishers.
5. C. R. N. Strouts, J. H. Gilfillan and H. N. Wilson (1955). *Analytical Chemistry. The Working Tools.* Volume II. Chapter 28. *The Application of Statistical Methods to Chemical Analysis*, Oxford; Oxford University Press.
6. C. W. Wilson and D. W. Wilson (Ed.) (1959). *Comprehensive Analytical Chemistry.* Vol. IA. *Classical Analysis.* Ch. 4. *Statistics.* Amsterdam; Elsevier Publishing Company.
7. R. A. Fisher and F. Yates (1953). *Statistical Tables for Biological, Agricultural and Medical Research.* 4th edn., Edinburgh; Oliver and Boyd.
8. D. J. Finney (1953). *Experimental Design and its Statistical Basis.* Cambridge; Cambridge University Press.
9. C. A. Bennett and N. L. Franklin (1954). *Statistical Analysis in Chemistry and the Chemical Industry.* New York; John Wiley.
10. W. J. Dixon and F. J. Massey, Jr. (1957). *Introduction to Statistical Analysis.* 2nd edn. New York; McGraw-Hill Book Co.
11. O. L. Davies (1957). *Statistical Methods in Research and Production, with Special Reference to the Chemical Industry.* 3rd edn. Edinburgh; Oliver and Boyd.
12. H. A. Strobel (1960). *Chemical Instrumentation. A Systematic Approach to Instrumental Analysis.* Ch. 2. *Errors of Measurement.* Reading, Mass.; Addison-Wesley Publishing Co.
13. D. A. Pantony (1961). *A Chemist's Introduction to Statistics, Theory of Error, and Design of Experiment.* Lecture Series, No. 2. Royal Institute of Chemistry, London.
14. J. Mandel and F. J. Linnig (1956, 1958). 'Statistical Methods in Chemistry', *Analytical Chemistry*, **28**, 770; **30**, 739.
15. B. N. Nilson (1960). 'Statistical Methods in Chemistry', *Analytical Chemistry*, **32**, 161R.
16. J. D. Hinchen (1969). *Practical Statistics for Chemical Research.* London; Methuen and Co.
17. M. J. Moroney (1965). *Facts from Figures.* 3rd edn., revised. Penguin Books, Harmondsworth.
18. H. L. Youmans (1973). *Statistics for Chemistry.* Columbus, Ohio; Merrill Publishing Co.

CHAPTER V **SAMPLING**

V, 1. THE BASIS OF SAMPLING. The purpose of analysis is to determine the quality or composition of a material; and for the analytical results obtained to have any validity or meaning it is essential that adequate sampling procedures be adopted. Sampling is the process of extracting from a large quantity of material a small portion which is truly representative of the composition of the whole material.

Sampling methods fall into three main groups:
1. those in which all the material is examined;
2. casual sampling on an *ad hoc* basis;
3. methods in which portions of the material are selected based upon statistical probabilities.

Procedure 1 is normally impracticable, as the majority of methods employed are destructive, and in any case the amount of material to be examined is frequently excessive. Even for a sample of manageable size the analysis would be very time consuming, require large quantities of reagents, and would monopolise instruments for long periods.

Sampling according to 2 is totally unscientific and can lead to decisions being taken on inadequate information. In this case, as the taking of samples is entirely casual, any true form of analytical control or supervision is impossible.

For these reasons the only reliable basis for sampling must be a mathematical one using statistical probabilities. This means that although not every item or every part of the sample is analysed the limitations of the selection are carefully calculated and known in advance. Having calculated the degree of acceptable risk or margin of variation, the sampling plan is then chosen that will give the maximum information and control that is compatible with a rapid turn over of samples. For this reason, in the case of sampling from batches the selection of individual samples is carried out according to special random tables (Ref. 1) which ensure that personal factors do not influence the choice.

V, 2. SAMPLING AND PHYSICAL STATE. Many of the problems occurring during sampling arise from the physical nature of the materials to be studied (Ref. 2). Although gases and liquids can, and do, present difficulties the greatest problems of adequate sampling undoubtedly arise with solids.

1. Gases. Few problems arise over homogeneity of gas mixtures where the storage vessel is not subjected to temperature or pressure variations. Difficulties may arise if precautions are not taken to clear valves, taps and connecting lines of any other gas prior to passage of the sample. Similarly care must be taken that no gaseous components will react with the sampling and analytical devices.

2. *Liquids.* In most cases general stirring or mixing is sufficient to ensure homogeneity prior to sampling. Where separate phases exist it is necessary to determine the relative volumes of each phase in order to compare correctly the composition of one phase with the other. The phases should in any case be individually sampled as it is not possible to obtain a representative sample of the combined materials even after vigorously shaking the separate phases together.

3. *Solids.* It is with solids that real difficulties over homogeneity arise. Even materials that superficially have every appearance of being homogeneous in fact may have localised concentrations of impurities and vary in composition. The procedure adopted to obtain as representative a sample as possible will depend greatly upon the type of solid. This process is of great importance since, if it is not satisfactorily done, the labour and time spent in making a careful analysis of the sample may be completely wasted. If the material is more or less homogeneous, sampling is comparatively simple. If, however, the material is bulky and heterogeneous, sampling must be carried out with great care, and the method will vary somewhat with the nature of the bulk solid.

The underlying principle of the sampling of material in bulk, say, of a truck load of coal or iron ore, is to select a large number of portions in a systematic manner from different parts of the bulk and then to combine them. This large sample of the total weight is crushed mechanically, if necessary, and then shovelled into a conical pile. Every shovelful must fall upon the apex of the cone and the operator must walk around the cone as he shovels; this ensures a comparatively even distribution. The top of the cone is then flattened out and divided into quarters. Opposite quarters of the pile are then removed, mixed to form a smaller conical pile, and again quartered. This process is repeated, further crushing being carried out if necessary, until a sample of suitable weight (say, 200–300 g) is obtained.

If the quantity of material is of the order of 2–3 kilos or less, intermixing may be accomplished by the method known as 'tabling'. The finely divided material is spread on the centre of a large sheet of oilcloth or similar material. Each corner is pulled in succession over its diagonal partner, the lifting being reduced to a minimum; the particles are thus caused to roll over and over on themselves, and the lower portions are constantly brought to the top of the mass and thorough intermixing ensues. The sample may then be rolled to the centre of the cloth, spread out, and quartered as before. The process is repeated until a sufficiently small sample is obtained. The final sample for the laboratory, which is usually between 25 and 200 g in weight, is placed in an air-tight bottle. This method produces what is known as the 'average sample' and any analysis on it should always be compared with those of a second sample of the same material obtained by the identical routine.

Mechanical methods also exist for dividing up particulate material into suitably sized samples. Samples obtained by these means are usually representative of the bulk material within limits of less than ± 1 per cent, and are based upon the requirements established by the British Standards Institution. Sample dividers* exist with capacities of up to $10\,dm^3$ and operate either by means of a

---

* Available from A. Gallenkamp and Co Ltd, P.O. Box 290, Technico House, Christopher St, London, EC2R 2ER; Glen Creston, 37 The Broadway, Stanmore, Middlesex, and The Pascall Engineering Co Ltd, Gatwick Road, Crawley, Sussex, RH10 2 RS.

series of rapidly rotating sample jars under the outlet of a loading funnel, or by a rotary cascade from which the samples are fed into a series of separate compartments. Sample dividers can lead to a great deal of time saving in laboratories dealing with bulk quantities of powders or minerals.

The sampling of metals and alloys may be effected by drilling holes through a representative ingot at selected points; *all* the material from the holes is collected, mixed, and a sample of suitable size used for analysis. Turnings or scrapings from the outside are not suitable as these frequently possess superficial impurities from the castings or moulds.

In some instances in which grinding presents problems it is possible to obtain a suitable homogeneous sample by dissolving a portion of the material in an appropriate solvent.

Before analysis the representative solid sample is usually dried at 105–110 °C, or at some higher specified temperature if necessary, to constant weight. The results of the analysis are then reported on the 'dry' basis, viz., on a material dried at a specified temperature. The loss in weight on drying may be determined, and the results may be reported, if desired, on the original 'moist' basis; these figures will only possess real significance if the material is not appreciably hygroscopic and no chemical changes, other than the loss of water, take place on drying.

In a course of systematic quantitative analysis, such as that with which we are chiefly concerned in the present book, the unknowns supplied for analysis are usually portions of carefully analysed samples which have been finely ground until uniform.

It should be borne in mind that although it is possible to generalise on sampling procedures all industries have their own established methods for obtaining a record of the quantity and/or quality of their products. The sampling procedures for tobacco leaves will obviously differ from those used for bales of cotton or for coal. But although the types of samples differ considerably the actual analytical methods later used are of general application.

V, 3.   CRUSHING AND GRINDING.   If the material is hard (e.g., a sample of rock), it is first broken into small pieces on a hard steel plate with a hardened hammer. The loss of fragments is prevented by covering the plate with a steel ring, or in some other manner. The small lumps may be broken in a **'percussion' mortar** (also known as a 'diamond' mortar) (Fig. V, 1). The mortar and pestle are

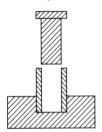

Fig. V, 1

constructed entirely of hard tool steel. One or two small pieces are placed in the mortar, and the pestle inserted into position; the latter is struck lightly with a hammer until the pieces have been reduced to a coarse powder. The whole of the hard substance may be treated in this manner. The coarse powder is then *ground* in an agate mortar in small quantities at a time. A mortar of mullite is claimed to be superior to one of agate: mullite is a homogeneous ceramic material that is harder, more resistant to abrasion and less porous than agate. A synthetic sapphire mortar and pestle (composed essentially of a specially prepared form of pure aluminium oxide) is marketed; it is extremely hard (comparable with tungsten carbide) and will grind materials not readily reduced in ceramic or metal mortars. Mechanical (motor-driven) mortars are available commercially.

V, 4. HAZARDS IN SAMPLING. The handling of many materials is fraught with hazards (Ref. 3) and this is no less so when sampling materials in preparation for chemical analysis. The sampler must always wear adequate protective clothing and if possible have detailed prior knowledge of the material being sampled. When dangers from toxicity exist the necessary antidotes and treatment procedures should be available and established before sampling commences (Ref. 4). In no instances should naked flames be allowed anywhere near the sampling area.

Apart from the toxic nature of many gases the additional hazards are those of excessive release of gas due to pressure changes, spontaneous ignition of inflammable gases and sudden vaporisation of liquified gases.

With liquids dangers frequently arise from easily volatised and readily inflammable liquids. In all cases precautions should be greater than under normal circumstances due to the unpredictable nature and conditions of taking samples. The sampler must always be prepared for the unexpected, as can arise, for example, if a container has built up excess pressure, or if the wrong liquid has been packed. Toxic and unknown liquids should never be sucked along tubes or into pipettes by mouth.

Even the sampling of solids must not be casually undertaken, and the operator should always use a face mask as a protection until it is established that the powdered material is not hazardous.

It should be borne in mind that sampling of radioactive substances is a specialist operation at all times and should only be carried out under strictly controlled conditions within restricted areas. In almost all instances the operator must be protected against the radioactive emanations from the substance he is sampling.

Correct sampling of materials is therefore of importance in two main respects; first to obtain a representative portion of the material for analysis, and secondly to prevent the occurrence of accidents when sampling hazardous materials.

V, 5. References

1. J. Murdoch and J. A. Barnes (1970). *Statistical Tables for Science, Engineering and Management*. 2nd edn. London; Macmillan, pp. 30–33.
2. C. R. N. Strouts, J. H. Gilfillan and H. N. Wilson (1955). *Analytical Chemistry. The Working Tools*. Vol. I. Ch. 3, *Sampling*. London; Oxford University Press.
3. N. Irving Sax (Ed.) (1968). *Dangerous Properties of Industrial Materials*. 3rd edn. Reinhold, New York.
4. G. D. Muir. (Ed.) (1971). *Hazards in the Chemical Laboratory*. London; Royal Institute of Chemistry.

V, 6. Selected bibliography

1. C. L. Wilson and D. W. Wilson (Ed.) (1959). *Comprehensive Analytical Chemistry*. Vol. IA, *Classical Analysis*, Ch. II.3., *Sampling*. Amsterdam; Elsevier Publishing Co.
2. H. A. Laitinen (1960). *Chemical Analysis. An Advanced Text and Reference*. Ch. 27, *Sampling*. New York; McGraw-Hill BookCo.
3. N. V. Steere (Ed.) (1967). *Handbook of Laboratory Safety*. Cleveland, Ohio; Chemical Rubber Co.

PART C   SEPARATIVE TECHNIQUES

PART C   **INTRODUCTION**

An ideal analytical method would enable a species to be determined directly in various matrices. Few, if any, analytical measurements are, however, wholly specific for a single species so that a major problem in quantitative analysis is the elimination of interferences. The following two general methods are available for dealing with substances that interfere in an analytical measurement:
determination of uranium as 8-hydroxyquinolate.

1. Masking of the potential interference(s) so as to prevent it contributing to the measurement step (Ref. 1); this is commonly effected by the introduction of a complexing agent that reacts selectively with the interfering substance, e.g., the masking of iron(III) and aluminium by EDTA in the solvent extraction and
2. The isolation of the species to be determined in a separate phase from the interfering species, by means of one of the various separative techniques, e.g., ion exchange, gas–liquid and liquid–liquid chromatography, solvent extraction. Such separations are, of course, based upon equilibrium processes so that complete separation of the interference from the species required is never possible. In practice the aim of the separation procedure will be to lower the concentration of the interference to a tolerable level while at the same time ensuring that any losses of the desired constituent are smaller than the allowable error in the analysis. The level of interfering substance which can be tolerated will be dependent upon the relative sensitivity of the final analytical measurement for the interfering and required constituents. If the measurement is much less affected by the interfering substance (i.e. lower sensitivity), then a partial separation may be adequate, but if the sensitivity of the method for the two constituents is about the same then a virtually complete separation may be required.

In trace analysis, where the ratio of required minor constituent to major component may be as small as 10^{-6} or 10^{-7}, the separation procedure may also effect a useful preconcentration of the trace constituent, thus providing an adequate amount of the substance for the measurement to be employed.

The aim of the present section is to survey and illustrate the application of the chief separative techniques in inorganic quantitative analysis. Wherever possible the use of an appropriate instrumental method is indicated for the final analytical measurement, as well as titrimetric or other classical methods.

No attempt has been made to deal with the general theory of separatory methods since this is already adequately considered in various analytical chemistry texts (Refs. 2, 3 and 4).

References

1. D. D. Perrin (1975). 'Selection of Masking Agents for Use in Analytical Chemistry.' *CRC crit. Rev. analyt. Chem.* **5**, 85.
2. L. B. Rogers (1961). Principles of Separations', in *Treatise on Analytical Chemistry. Part 1* (ed. J. M. Kolthoff and P. J. Elving). Vol. 2, New York; Interscience.
3. B. L. Karger, L. R. Snyder and C. Horvath (1973). *An Introduction to Separation Science.* New York; Wiley.
4. J. A. Dean (1969). *Chemical Separation Methods.* New York; Van Nostrand Reinhold Co.

CHAPTER VI   SOLVENT EXTRACTION

VI, 1.   GENERAL DISCUSSION.   Liquid–liquid extraction is a technique in which a solution (usually aqueous) is brought into contact with a second solvent (usually organic), essentially immiscible with the first, in order to bring about a transfer of one or more solutes into the second solvent. The separations that can be performed are simple, clean, rapid, and convenient. In many cases separation may be effected by shaking in a separatory funnel for a few minutes. The technique is equally applicable to trace level and large amounts of materials.* We are concerned largely with samples in aqueous solution, and the production of chelate and of ion association extraction systems.

To understand the fundamental principles of extraction, the various terms used for expressing the effectiveness of a separation must first be considered. For a solute A distributed between two immiscible phases a and b, the Nernst distribution (or partition) law states that, provided its molecular state is the same in both liquids and that the temperature is constant:

$$\frac{\text{Concentration of solute in solvent } a}{\text{Concentration of solute in solvent } b} = \frac{[A]_a}{[A]_b} = K_D$$

where K_D is a constant known as the **distribution** (or partition) **coefficient**. The law, as stated, is not thermodynamically rigorous (e.g., it takes no account of the activities of the various species, and for this reason would be expected to apply only in very dilute solutions, where the ratio of the activities approaches unity), but is a useful approximation. The law in its simple form does not apply when the distributing species undergoes dissociation or association in either phase. In the practical applications of solvent extraction we are interested primarily in the fraction of the total solute in one or other phase, quite regardless of its mode of dissociation, association, or interaction with other dissolved species. It is convenient to introduce the term distribution ratio D (or extraction coefficient E):

$$D = (C_A)_a/(C_A)_b$$

where the symbol C_A denotes the concentration of A in all its forms as determined analytically.

---

* In addition to the examples in this chapter, other (and more complex) examples will be found in Chapter XVIII, e.g., Section **XVIII, 28**.

A problem often encountered in practice is to determine what is the most efficient method for removing a substance quantitatively from solution. It can be shown that if V cm^3 of, say, an aqueous solution containing x_0 g of a solute be extracted n times with v-cm^3 portions of a given solvent, then the weight of solute x_n remaining in the water layer is given by the expression:

$$x_n = x_0 \left(\frac{DV}{DV+v} \right)^n$$

where D is the distribution ratio between water and the given solvent. It follows, therefore, that the best method of extraction with a given volume of extracting liquid is to employ several fractions of the liquid rather than to utilise the whole quantity in a single extraction.

Let us take a particular example. Let us suppose that we shake 50 cm^3 of water containing 0.1 g of iodine with 25 cm^3 of carbon tetrachloride. The distribution coefficient of iodine between water and carbon tetrachloride at the ordinary laboratory temperature is 1/85, i.e., at equilibrium the iodine concentration in the aqueous layer is 1/85th of that in the carbon tetrachloride layer. We will compute the weight of iodine remaining in the aqueous layer after one extraction with 25 cm^3 and also after, say, three extractions with 8.33 cm^3 of the solvent by application of the above formula. The former can be simply computed as follows. If x_1 g of iodine remains in the 50 cm^3 of water, its concentration is $x_1/50$ g cm^{-3}; the concentration in the carbon tetrachloride layer will be $(0.1 - x_1)/25$ g cm^{-3}. Hence:

$$\frac{x_1/50}{(0.1 - x_1)/25} = \frac{1}{85}, \text{ or } x_1 = 0.00230 \text{ g}$$

The concentration in the aqueous layer after three extractions with 8.33 cm^3 of carbon tetrachloride is given by:

$$x_3 = 0.1 \left(\frac{(1/85) \times 50}{(50/85) + 8.33} \right)^3 = 0.0000145 \text{ g}$$

The extraction may therefore be regarded as virtually complete.

If we confine our attention to the distribution of a solute A between water and an organic solvent, we may write the percentage extraction $E_\%$ as:

$$E_\% = \frac{100[A]_o V_o}{[A_o]V_o + [A_w]V_w} = \frac{100D}{D + (V_w/V_o)}$$

where V_o and V_w represent the volumes of the organic and aqueous phases respectively. Thus the percentage of extraction varies with the volume ratio of the two phases and the distribution coefficient.

If the solution contains two solutes A and B it often happens that under the conditions favouring the complete extraction of A, some B is extracted as well. The effectiveness of separation increases with the magnitude of the **separation coefficient or factor** β, which is related to the individual distribution ratios as follows:

$$\beta = \frac{[A]_o/[B]_o}{[A]_w/[B]_w} = \frac{[A]_o/[A]_w}{[B]_o/[B]_w} = \frac{D_A}{D_B}$$

If $D_A = 10$ and $D_B = 0.1$, a single extraction will remove 90.9 per cent of A and 9.1 per cent of B (ratio 10:1); a second extraction of the same aqueous phase will bring the total amount of A extracted up to 99.2 per cent, but increases that of B to 17.4 per cent (ratio 5.7:1). More complete extraction of A thus involves an increased contamination by B. Clearly, when one of the distribution ratios is relatively large and the other very small, almost complete separation can be quickly and easily achieved. If the separation factor is large but the smaller distribution ratio is of sufficient magnitude that extraction of both components occurs, it is necessary to resort to special techniques to suppress the extraction of the unwanted component.

VI, 2. FACTORS FAVOURING SOLVENT EXTRACTION. It is well known that hydrated inorganic salts tend to be more soluble in water than in organic solvents such as benzene, chloroform, etc., whereas organic substances tend to be more soluble in organic solvents than in water unless they incorporate a sufficient number of hydroxyl, sulphonic, or other hydrophilic groupings. In solvent extraction analysis for metals we are concerned with methods by which the water solubility of inorganic cations may be masked by interaction with appropriate (largely organic) reagents; this will in effect remove some or all of the water molecules associated with the metal ion to which the water solubility is due.

Ionic compounds would not be expected to extract into organic solvents from aqueous solution because of the large loss in electrostatic solvation energy which would occur. The most obvious way to make an aqueous ionic species extractable is to neutralise its charge. This can be done by formation of a neutral metal chelate complex or by ion association; the larger and more hydrophobic the resulting molecular species the better will be its extraction.

In chelation complexes (sometimes called inner complexes when uncharged) the central metal ion co-ordinates with a polyfunctional organic base to form a stable ring compound, e.g., copper(II) 'acetylacetonate' or iron(III) 'cupferrate':

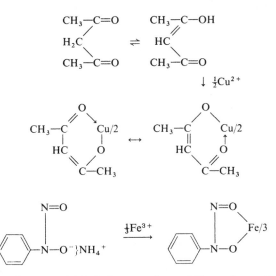

The factors which influence the stability of metal ion complexes are discussed in Section **X, 20**, but it is appropriate to emphasise here the significance of the chelate effect and to list the features of the ligand which affect chelate formation:

(i) *The basic strength of the chelating group.* The stability of the chelate complexes formed by a given metal ion generally increase with increasing basic strength of the chelating agent, as measured by the pK_a values.

(ii) *The nature of the donor atoms in the chelating agent.* Ligands which contain donor atoms of the soft base type form their most stable complexes with the relatively small group of class (b) metal ions (i.e., soft acids) and are thus more selective reagents. This is illustrated by the reagent diphenylthiocarbazone (dithizone) used for the solvent extraction of metal ions such as Pd^{2+}, Ag^+, Hg^{2+}, Cu^{2+}, Bi^{3+}, Pb^{2+} and Zn^{2+}.

(iii) *Ring size.* Five- or six-membered conjugated chelate rings are most stable since these have minimum strain. The functional groups of the ligand must be so situated that they permit the formation of a stable ring.

(iv) *Resonance and steric effects.* The stability of chelate structures is enhanced by contributions of resonance structures of the chelate ring: thus copper acetylacetonate (see formula above) has greater stability than the copper chelate of salicylaldoxime. A good example of steric hindrance is 2,9-dimethylphenanthroline (neocuproine), which does not give a complex with iron(II) as does the unsubstituted phenanthroline; this hindrance is at a minimum in the tetrahedral grouping of the reagent molecules about a univalent tetra-coordinated ion such as that of copper(I). A nearly specific reagent for copper is thus available.

The choice of a satisfactory chelating agent for a particular separation should, of course, take all the above factors into account. The critical influence of pH on the solvent extraction of metal chelates is shown in the following section.

VI, 3. QUANTITATIVE TREATMENT OF SOLVENT EXTRACTION EQUILIBRIA.

The solvent extraction of a neutral metal chelate complex formed from the chelating agent HR according to the equation

$$M^{n+} + nR^- \rightleftharpoons MR_n$$

may be treated quantitatively on the basis of the following assumptions: (*a*) that the reagent and the metal complex exist as simple unassociated molecules in both phases; (*b*) solvation plays no significant part in the extraction process; and (*c*) the solutes are uncharged molecules and their concentrations are generally so low that the behaviour of their solutions departs little from ideality. The dissociation of the chelating agent HR in the aqueous phase is represented by the equation

$$HR \rightleftharpoons H^+ + R^-$$

The various equilibria involved in the solvent extraction process are expressed in terms of the following thermodynamic constants:

Dissociation constant of complex, $K_c = [M^{n+}]_w[R^-]_w^n/[MR_n]_w$
Dissociation constant of reagent, $K_r = [H^+]_w[R^-]_w/[HR]_w$
Partition coefficient of complex, $p_c = [MR_n]_w/[MR_n]_o$
Partition coefficient of reagent, $p_r = [HR]_w/[HR]_o$

where the subscripts c and r refer to complex and reagent, and w and o to aqueous and organic phase respectively.

The distribution ratio, i.e., the ratio of the amount of metal extracted as complex into the organic phase to that remaining in all forms in the aqueous phase, is given by

$$D = [MR_n]_o / \{[MR_n]_w + [M^{n+}]_w\}$$

which can be shown (Ref. 1) to reduce to

$$D = K[HR]_o^n / [H^+]_w^n$$

where $K = (K_r p_r)^n / K_c p_c$

If the reagent concentration remains virtually constant

$$D = K^* / [H^+]_w^n \text{ where } K^* = K[HR]_o^n$$

and the percentage of solute extracted, E, is given by

$$\log E - \log(100 - E) = \log D$$
$$= \log K^* + n\text{pH}$$

The distribution of the metal in a given system is a function of the pH alone. The equation represents a family of sigmoid curves when E is plotted against pH, with the position of each along the pH axis depending only on the magnitude of K^* and the slope of each uniquely depending upon n. Some theoretical extraction curves for divalent metals showing how the position of the curves depends upon the magnitude of K^* are depicted in Fig. VI, 1; Fig. VI, 2, illustrates how the slope depends upon n. It is evident that a ten-fold change in reagent concentration is exactly offset by a ten-fold change in hydrogen-ion concentration, i.e., by a change of a single unit of pH: such a change of pH is much easier to effect in practice. If $\text{pH}_{\frac{1}{2}}$ is defined as the pH value at 50 per cent extraction ($E_\% = 50$) we see from the above equation that

$$\text{pH}_{\frac{1}{2}} = -\frac{1}{n}\log K^*$$

The difference in $\text{pH}_{\frac{1}{2}}$ values of two metal ions in a specific system is a measure of the separatability of the two ions. If the $\text{pH}_{\frac{1}{2}}$ values are sufficiently far apart, then excellent separation can be achieved by controlling the pH of extraction. It is often helpful to plot the extraction curves of metal chelates. If one takes as the criterion of a successful single-stage separation of two metals by pH control a 99 per cent extraction of one with a maximum of 1 per cent extraction of the other,

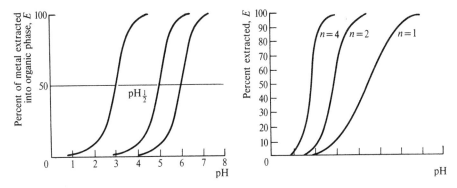

Fig. VI, 1   Fig. VI, 2

for bivalent metals a difference of two pH units would be necessary between the two $pH_\frac{1}{2}$ values; the difference is less for tervalent metals. Some figures for the extraction of metal dithizonates in chloroform are:

Metal Ion	Cu(II)	Hg(II)	Ag	Sn(II)	Co	Ni, Zn	Pb
Optimum pH of Extraction	1	1–2	1–2	6–9	7–9	8	8.5–11

If the pH is controlled by a buffer solution, then those metals with $pH_\frac{1}{2}$ values in this region, together with all metals having smaller $pH_\frac{1}{2}$ values, will be extracted. The $pH_\frac{1}{2}$ values may be altered (and the selectivity of the extraction thus increased) by the use of a competitive complexing agent or of masking agents. Thus in the separation of mercury and copper by extraction with dithizone in carbon tetrachloride at pH 2 the addition of EDTA forms a water-soluble complex which completely masks the copper but does not affect the mercury extraction. Cyanides raise the $pH_\frac{1}{2}$ values of mercury, copper, zinc, and cadmium in dithizone extraction with carbon tetrachloride.

VI, 4. ION-ASSOCIATION COMPLEXES. In ion-association complexes the inorganic ion associates with oppositely charged ions to form a neutral extractable species. Such complexes may form clusters with increasing concentration which are larger than just simple ion pairs, particularly in organic solvents of low dielectric constant. The following types of ion-association complexes may be recognised:

(i) those formed from a reagent yielding a large organic ion, e.g., the tetraphenylarsonium, $(C_6H_5)_4As^+$, and tetrabutylammonium, $(n\text{-}C_4H_9)_4N^+$ ions, which form large ion aggregates or clusters with suitable oppositely charged ions, e.g., the perrhenate ion, ReO_4^-. These large and bulky ions do not have a primary hydration shell and cause disruption of the hydrogen-bonded water structure; the larger the ion the greater the amount of disruption and the greater the tendency for the ion association species to be pushed into the organic phase.

These large ion extraction systems lack specificity since any relatively large unhydrated univalent cation will extract any such large univalent anion. On the other hand polyvalent ions, because of their greater hydration energy, are not so easily extracted and good separations are possible between MnO_4^-, ReO_4^- or TcO_4^- and CrO_4^{2-}, MoO_4^{2-} or WO_4^{2-}, for example.

(ii) those involving a cationic or anionic chelate complex of a metal ion. Thus chelating agents having two uncharged donor atoms, such as 1:10-phenanthroline, form cationic chelate complexes which are large and hydrocarbon like. Tris-(phenanthroline) iron(II) perchlorate extracts fairly well into chloroform, and extraction is virtually complete using large anions such as long chain alkyl sulphonate ions in place of ClO_4^-.

Dagnall and West (Ref. 7) have described the formation and extraction of a blue *ternary* complex, Ag(I)-1,10-phenanthroline-Bromopyrogallol Red (BPR), as the basis of a highly sensitive spectrophotometric procedure for the determination of traces of silver (Section **VI, 15**). The reaction mechanism for the formation of the blue complex in aqueous solution was investigated by photometric and potentiometric methods and these studies led to the conclusion that the complex is an ion-association system, $(Ag(phen)_2)_2BPR^{2-}$, i.e., involving a cationic chelate complex of a metal ion (Ag^+) associated with an anionic counter-ion derived from the dyestuff (BPR). Ternary complexes have been reviewed by Babko (Ref. 8).

Types (i) and (ii) represent extraction systems involving coordinately unsolvated large ions and differ in this important respect from type (iii).

(iii) those in which solvent molecules are directly involved in formation of the ion-association complex. Most of the solvents (ethers, esters, ketones and alcohols) which participate in this way contain donor oxygen atoms and the coordinating ability of the solvent is of vital significance. The coordinated solvent molecules facilitate the solvent extraction of salts such as chlorides and nitrates by contributing both to the size of the cation and the resemblance of the complex to the solvent.

A class of solvents which shows very marked solvating properties for inorganic compounds comprises the esters of orthophosphoric acid. The functional group in these molecules is the semipolar phosphoryl group, $\geqslant P^+ - O^-$, which has a basic oxygen atom with good steric availability. A typical compound is tri-*n*-butyl phosphate (TBP) which has been widely used in solvent extraction on both the laboratory and industrial scale; of particular note is the use of TBP for the extraction of uranyl nitrate and its separation from fission products.

The mode of extraction in these 'oxonium' systems may be illustrated by considering the ether extraction of iron(III) from strong hydrochloric acid solution. In the aqueous phase chloride ions replace the water molecules coordinated to the Fe^{3+} ion, yielding the tetrahedral $FeCl_4^-$ ion. It is recognised that the hydrated hydronium ion, $H_3O^+(H_2O)_3$ or $H_9O_4^+$, normally pairs with the complex halo-anions, but in the presence of the organic solvent, solvent molecules enter the aqueous phase and compete with water for positions in the solvation shell of the proton. On this basis the primary species extracted into the ether (R_2O) phase is considered to be $[H_3O(R_2O)_3^+, FeCl_4^-]$ although aggregation of this species may occur in solvents of low dielectric constant.

VI, 5. EXTRACTION REAGENTS.

Many complexes of metals in aqueous solution are coloured: when extracted with an organic solvent, the coloured extract may be used directly for the determination of the concentration of the metal by colorimetric or, preferably, spectrophotometric techniques (see Chapter XVIII). These techniques are particularly applicable with many chelate complexes, although a number of coloured inorganic complexes, such as 'molybdenum blue' and iodobismuthite ion, can be treated in this way. In this section we shall discuss a limited number of chelating and extraction reagents,* as well as some organic solvents with special interest as to their selective extraction properties.

1. **Acetylacetone** (Pentane-2,4-dione) $CH_3CO\cdot CH_2\cdot COCH_3$. Acetylacetone is a colourless mobile liquid, b.p. 139 °C, which is sparingly soluble in water (0.17 g per cm³ at 25 °C) and miscible with many organic solvents. It is useful both as a solution (in carbon tetrachloride, chloroform, benzene, xylene, etc.) and as the pure liquid. The compound is a β-diketone and forms well-defined chelates with over sixty metals. Many of the chelates (acetylacetonates) are soluble in organic solvents, and the solubility is of the order of grams per dm³, unlike that of most analytically used chelates, so that macro- as well as micro-

---

* The formulae of the compounds not given in the text will be found in Section **XI, 11**.

scale separations are possible. The selectivity can be increased by using EDTA as a masking agent. The use of acetylacetone as both solvent and extractant {e.g., for Al, Be, Ce, Co(III), Ga, In, Fe, U(VI), etc.} offers several advantages over its use in solution in carbon tetrachloride, etc.: extraction may be carried out at a lower pH than otherwise feasible because of the higher reagent concentration; and often the solubility of the chelate is greater in acetylacetone than in many organic solvents. The solvent generally used is carbon tetrachloride; the organic layer is heavier than water.

An interesting application is the separation of cobalt and nickel: neither Co(II) nor Ni(II) form extractable chelates, but Co(III) chelate is extractable; extraction is therefore possible following oxidation.

2. **Thenoyltrifluoroacetone,** (TTA), $C_4H_3S \cdot CO \cdot CH_2 \cdot COCF_3$. This is a crystalline solid, m.p. 43 °C; it is, of course, a β-diketone, and the trifluoromethyl group increases the acidity of the enol form so that extractions at low pH values are feasible. The reactivity of TTA is similar to that of acetylacetone: it is generally used as a 0.1–0.5M solution in benzene or toluene. The difference in extraction behaviour of hafnium and zirconium, and also among lanthanoids and actinoids, is especially noteworthy.

3. **8-Hydroxyquinoline (oxine).** Oxine is a versatile organic reagent and forms chelates with many metallic ions. The divalent and trivalent metal chelates have the general formulae $M(C_9H_6ON)_2$ and $M(C_9H_6ON)_3$; the oxinates of the higher-valent metals may differ somewhat in composition, e.g., $Ce(C_9H_6ON)_4$; $Th(C_9H_6ON)_4 \cdot (C_9H_7ON)$; $WO_2(C_9H_6ON)_2$; $MoO_2(C_9H_6ON)_2$; $U_3O_6(C_9H_6ON)_6 \cdot (C_9H_7ON)$. Oxine is generally used as a 1 per cent (0.07M) solution in chloroform, but concentrations as high as 10 per cent are advantageous in some cases (e.g., for strontium).

8-Hydroxyquinoline, having both a phenolic hydroxyl group and a basic nitrogen atom, is amphoteric in aqueous solution; it is completely extracted from aqueous solution by chloroform at pH < 5 and pH > 9; the distribution coefficient of the neutral compound between chloroform and water is 720 at 18 °C. The usefulness of this sensitive reagent has been extended by the use of masking agents (cyanide, EDTA, citrate, tartrate, etc.) and by control of pH.

4. **Dimethylglyoxime.** The complexes with nickel and with palladium are soluble in chloroform. The optimum pH range for extraction of the nickel complex is 4–12 in the presence of tartrate and 7–12 in the presence of citrate (solubility 35–50 μg Ni per cm^3 at room temperature); if the amount of cobalt exceeds 5 mg some cobalt may be extracted from alkaline solution. Palladium(II) may be extracted out of *ca.* 1M-sulphuric acid solution.

5. **1-Nitroso-2-naphthol.** The reagent forms extractable complexes (chloroform) with Co(III) in an acid medium and with Fe(II) in a basic medium.

6. **Cupferron (ammonium salt of *N*-nitroso-*N*-phenylhydroxylamine).** The reagent is used in cold aqueous solution (about 6 per cent). Metal cupferrates are soluble in diethyl ether and in chloroform, and so the reagent finds wide application in solvent-extraction separation schemes. Thus Fe(III), Ti, and Cu may be extracted from 1.2M HCl solution by chloroform: numerous other elements may be extracted largely in acidic solution.

7. **Diphenylthiocarbazone** (dithizone), $C_6H_5 \cdot N{=}N \cdot CS \cdot NH \cdot NH \cdot C_6H_5$. The compound is insoluble in water and dilute mineral acids, and is readily soluble in dilute aqueous ammonia. It is used in dilute solution in chloroform or carbon tetrachloride. Dithizone is an important selective reagent for quantitative

determinations of metals: colorimetric (and, of course, spectrophotometric) analyses are based upon the intense green colour of the reagent and the contrasting colours of the metal dithizonates in organic solvents. The selectivity is improved by the control of pH and the use of masking agents, such as cyanide, thiocyanate, thiosulphate, and EDTA.

8. **Sodium diethyldithiocarbamate,** $\{(C_2H_5)_2N{\cdot}CS{\cdot}S\}^- Na^+$. This reagent is generally used as a 2 per cent aqueous solution; it decomposes rapidly in solutions of low pH. It is an effective extraction reagent for over twenty metals into various organic solvents, such as chloroform, carbon tetrachloride, and ethanol. The selectivity is enhanced by the control of pH and the addition of masking agents.

9. **Toluene-3,4-dithiol (dithiol),** $CH_3{\cdot}C_6H_3{\cdot}(SH)_2$. This compound is a solid, m.p. $31°$, which forms complexes in acid solution with Mo(VI), W(VI), and Re(VI) that are extractable by chloroform or pentyl acetate. Control of the pH or the use of citric acid permits the selective extraction of molybdenum.

10. **Tri-*n*-butyl phosphate,** $(n\text{-}C_4H_9)_3PO_4$. This solvent is useful for the extraction of metal thiocyanate complexes, of nitrates from nitric acid solution (e.g., cerium, thallium, and uranium), of chloride complexes, and of acetic acid from aqueous solution. In the analysis of steel, iron(III) may be removed as the soluble 'iron(III) thiocyanate'. The solvent is non-volatile, non-inflammable, and rapid in its action.

11. **Tri-*n*-octylphosphine oxide,** $(n\text{-}C_8H_{17})_3PO$. This compound (TOPO) dissolved in cyclohexane $(0.1M)$ is an excellent extraction solvent. Thus the distribution ratio of U(VI) is of the order of 10^5 times greater for TOPO than for tri-*n*-butyl phosphate. The following elements are completely extracted from $1M$-hydrochloric acid: Cr(VI) as $H_2Cr_2O_7,2TOPO$; Zr(IV) as $ZrCl_4,2TOPO$; Ti(IV); U(VI) as $UO_2(NO_3)_2,2TOPO$; Fe(III); Mo(VI) and Sn(IV). If the hydrochloric acid concentration is increased to $7M$, Sb(III), Ga(III) and V(IV) are completely extracted.

VI, 6. SOME PRACTICAL CONSIDERATIONS. Solvent extraction is generally employed in analysis to separate a solute (or solutes) of interest from substances which interfere in the ultimate quantitative analysis of the material; sometimes the interfering solutes are extracted selectively. Solvent extraction is also used to concentrate a species which in aqueous solution is too dilute to be analysed.

The **choice of solvent for extraction** is governed by the following considerations:

(i) A high distribution ratio for the solute and a low distribution ratio for undesirable impurities.

(ii) Low solubility in the aqueous phase.

(iii) Sufficiently low viscosity and sufficient density difference from the aqueous phase to avoid the formation of emulsions.

(iv) Low toxicity and inflammability.

(v) Ease of recovery of solute from the solvent for subsequent analytical processing. Thus the b.p. of the solvent and the ease of stripping by chemical reagents merits attention when a choice is possible.

Sometimes mixed solvents may be used to improve the above properties. Salting-out agents may also improve extractability.

Extraction. Extraction may be accomplished in either a batch operation or

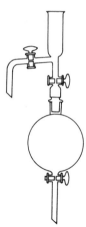

Fig. VI, 3

a continuous operation. Batch extraction, the simplest and most widely used method, is employed where a large distribution ratio for the desired separation is readily obtainable. A small number of batch extractions readily remove the desired component completely. It may be carried out in a simple separatory funnel. For solvents lighter than water the modified separatory funnel (Fig. VI, 3), designed to simplify the removal of the lighter phase, may be used. After equilibration, the lighter (e.g., ethereal) and aqueous layers are displaced upwards by introducing mercury through the stopcock at the bottom of the bulb with the aid of a subsidiary mercury levelling bulb. The stopcocks should be well ground so that a lubricant is not required; if a lubricant is used at all it should preferably be a silicone-type grease.

The two layers are shaken in a separatory funnel until equilibrium is attained, after which the two layers are allowed to settle completely before sampling. The extraction and sampling should be performed at constant temperature, since the distribution ratio as well as the volumes of the solvent are influenced by temperature changes. It must be borne in mind that too violent agitation of the extraction mixture often serves no useful purpose: simple repeated inversions of the vessel suffice to give equilibrium in a relatively few inversions. If droplets of aqueous phase are entrained in the organic extract it is possible to remove them by filtering the extract through a dry filter paper: the filter paper should be washed several times with fresh organic solvent.

When the distribution ratio is low continuous methods of extraction are used. This procedure makes use of a continuous flow of immiscible solvent through the solution; if the solvent is volatile, it is recycled by distillation and condensation and is dispersed in the aqueous phase by means of a sintered glass disc or equivalent device. Apparatus are available for effecting such continuous extractions with automatic return of the volatilised solvent (see Selected Bibliography at the end of this chapter).

Stripping. Stripping is the removal of the extracted solute from the organic phase for further preparation for the detailed analysis. In many colorimetric procedures involving an extraction process the concentration of the desired solute is determined directly in the organic phase by measuring the absorbance of a known volume of the solution of the coloured complex.

Where other methods of analysis are to be employed, or where further separation steps are necessary, the solute must be removed from the organic phase to a more suitable medium. If the organic solvent is volatile (e.g., diethyl ether) the simplest procedure is to add a small volume of water and evaporate the solvent on a water bath; care should be taken to avoid loss of a volatile solute during the evaporation. Sometimes adjustment of the pH of the solution, change in valence state, or the use of competitive water-soluble complexing reagents may be employed to prevent loss of the solute. When the extracting solvent is non-volatile the solute is removed from the solvent by chemical means, e.g., by shaking the solvent with a volume of water containing acids or other reagents, whereby the extractable complex is decomposed. The metal ions are then quantitatively back-extracted into the aqueous phase.

Impurities present in the organic phase may sometimes be removed by

backwashing. The organic extract when shaken with one or more small portions of a fresh aqueous phase containing the optimum reagent concentration and of correct pH will result in the redistribution of the impurities in favour of the aqueous phase, since their distribution ratios are low: most of the desired element will remain in the organic layer.

Completion of the analysis. Having separated a particular element or substance by solvent extraction, the final step involves the quantitative determination of the element or substance of interest. Simple colorimetric or, better, spectrophotometric methods may be applied directly to the solvent extract utilising the absorption bands of the complex in the ultraviolet or visible region. A typical example is the determination of nickel as dimethylglyoximate in chloroform by measuring the absorption of the complex at 366 nm.

With ion-association complexes, improved results can often be obtained by developing a chelate complex after extraction. An example is the extraction of uranyl nitrate from nitric acid into tributyl phosphate and the subsequent addition of dibenzoylmethane to the solvent to form a soluble coloured chelate.

If direct analysis of the solvent extract is impracticable the element is usually backwashed into an aqueous phase which can be analysed by standard methods.

Further techniques which may be applied directly to the solvent extract are flame spectrophotometry and atomic absorption spectrophotometry (AAS). An example of the former technique is the determination of copper as the salicylaldoxime complex in chloroform; the organic extract is sprayed directly into an oxy-acetylene flame and the spectral emission of copper at 324.7 nm is measured. The direct use of the solvent extract in AAS may be advantageous since the presence of the organic solvent generally enhances the sensitivity of the method (Chapter XXII).

Automation of solvent extraction. Although automatic methods of analysis do not fall within the scope of the present text, it is appropriate to emphasise here that solvent extraction methods offer considerable scope for automation (Refs 2, 3, 4). Details of the use of automatic analysers are best obtained by referring to the manufacturers' manuals (e.g., Ref. 5) as this is a rapidly expanding field with many changes occurring.

SOME APPLICATIONS*

VI, 7. DETERMINATION OF BERYLLIUM AS THE ACETYL-ACETONE COMPLEX. *Discussion.* Beryllium forms an acetylacetone complex, which is soluble in chloroform, and yields an absorption maximum at 295 nm. The excess of acetylacetone in the chloroform solution may be removed by rapid washing with $0.1M$-sodium hydroxide solution. It is advisable to treat the solution containing up to 10 μg of Be with up to 10 cm^3 of 2 per cent EDTA solution: the latter will mask up to 1 mg of Fe, Al, Cr, Zn, Cu, Pb, Ag, Ce, and U.

Procedure. Prepare a solution containing 10 μg of beryllium in 50 cm^3: use A.R. beryllium sulphate, $BeSO_4,4H_2O$. To 50.0 cm^3 of this solution contained in a beaker, add dilute hydrochloric acid until the pH is 1.0, and then introduce 10.0

---

* Further, and more complex, applications will be found in Chapter XVIII.

cm^3 of 2 per cent EDTA solution. Adjust the pH to 7 by the addition of 0.1M-sodium hydroxide solution. Add 5.0 cm^3 of 1 per cent aqueous acetylacetone and readjust the pH to 7–8. After standing for 5 minutes, extract the colourless beryllium complex with three 10-cm^3 portions of chloroform. Wash the chloroform extract rapidly with two 50-cm^3 portions of 0.1M-sodium hydroxide in order to remove the excess of acetylacetone. To determine the absorbance at 295 nm (in the ultraviolet region of the spectrum) it may be necessary to dilute the extract with chloroform. Measure the absorbance using 1.0-cm absorption cells against a blank.

Repeat the determination with a solution containing 100 μg of iron(III) and of aluminium ion; the absorbance is unaffected.

VI, 8. DETERMINATION OF BORON USING FERROIN. *Discussion.* The method is based upon the complexation of boron as the bis(salicylato)borate(III) anion (A), (Borodisalicylate), and the solvent extraction into chloroform of the ion-association complex formed with the ferroin.

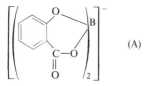

(A)

The intensity of the colour of the extract due to ferroin is observed spectrophotometrically and may be related by calibration to the boron content of the sample.

The method has been applied to the determination of boron in river water and sewage (Ref. 6), the chief sources of interference being copper(II) and zinc ions, and anionic detergents. The latter interfere by forming ion-association complexes with ferroin which are extracted by chloroform; this property may, however, be utilised for the joint determination of boron and anionic detergents by the one procedure. The basis of this joint determination is that the ferroin–anionic detergent complex may be immediately extracted into chloroform, whereas the formation of the borodisalicylate anion from boric acid and salicylate requires a reaction time of one hour prior to extraction using ferroin. The absorbance of the chloroform extract obtained after zero minutes thus gives a measure of the anionic detergent concentration, whereas the absorbance of the extract after a one hour reaction period corresponds to the amount of boron plus anionic detergent present. Interference due to copper(II) ions may be eliminated by masking with EDTA.

Reagents. Sulphuric acid solution, 0.05M.

Sodium hydroxide solution, 0.1M.

Sodium salicylate solution, 10 per cent w/v.

EDTA solution, 1 per cent w/v; use the disodium salt of EDTA.

Ferroin solution, $2.5 \times 10^{-2}M$. Dissolve 0.695 g of iron(II) sulphate heptahydrate and 1.485 g of 1,10-phenanthroline hydrate in 100 cm^3 of distilled water.

Boric acid solution, $2.5 \times 10^{-4}M$. Dissolve 61.8 mg of boric acid in 1 dm^3 of distilled water; dilute 250 cm^3 of this solution to 1 dm^3 to give the standard boric acid solution.

Use analytical reagent grade materials whenever possible and store the solutions in polythene bottles.

Procedure. (a) **Zero minutes reaction time.** Neutralise a measured volume of the sample containing 1–2 mg dm^{-3} of boron with sodium hydroxide or sulphuric acid ($0.05M$) to a pH of 5.5 (use a pH meter). Note the change in volume and hence calculate the volume correction factor to be applied to the final result. Measure 100 cm^3 of the neutralised sample solution into a flask, add 10 cm^3 of 10 per cent sodium salicylate solution and 17.5 cm^3 of $0.05M$ sulphuric acid, and mix the solutions thoroughly. Adjust the pH of the solution to pH 6 to 7 with $0.1M$ sodium hydroxide and transfer the solution immediately to a separating funnel; wash the flask with 20 cm^3 of distilled water and add the washings to the rest of the solution. Add by pipette 1 cm^3 of 1 per cent EDTA solution and 1 cm^3 of 2.5 $\times 10^{-2}M$ ferroin solution and again throughly mix the solution. Add 50 cm^3 of chloroform and shake the funnel for thirty seconds to mix the phases thoroughly. Allow the layers to separate and transfer the chloroform layer to another separating funnel. Wash the chloroform by shaking it vigorously for thirty seconds with 100 cm^3 of water and repeat this process with a second 100 cm^3 of water. Filter the chloroform phase through cotton-wool and measure the absorbance, A_o, against pure chloroform at 516 nm in a 1 cm cell (this 1 cm cell reading is used to calculate the boron concentration on the basis of equation (1), but if the zero minutes reading is to be used for determination of anionic detergent concentration a 2 cm cell reading is more suitable).

(b) **One hour reaction time.** Measure a second 100 cm^3 of neutralised sample solution into a flask, add 10 cm^3 of 10 per cent sodium salicylate solution and 17.5 cm^3 of $0.05M$ sulphuric acid solution. Mix the solutions thoroughly, allow the mixture to stand for one hour and adjust the pH of the solution to 6–7 with $0.1M$ sodium hydroxide. Now proceed as previously described, under (a), to obtain the absorbance, A_{1h}. The absorbance, A, to be used in the calculation of the boron concentration is obtained from the following equation:

$$A = A_{1h} - (A_o - A_{(blank)o}) \tag{1}$$

$A_{(blank)o}$ is determined by repeating procedure (a), i.e., zero minutes, using 100 cm^3 of distilled water in place of the sample solution.

Calculate the amount of boron present by reference to a calibration graph of absorbance against boron concentration (mg dm^{-3}). Multiply the result obtained by the appropriate volume correction factor arising from neutralisation of the sample.

Calibration. Take 5, 10, 25, 50, 75 and 100 cm^3 of the standard boric acid solution ($2.5 \times 10^{-4}M$) and make each up to 100 cm^3 with distilled water; this yields a boron concentration range up to 2.70 mg dm^{-3}. Continue with each solution as described under Procedure (b), i.e., one hour reaction time, except that the initial neutralisation of the boron solution to pH 5.5 is not necessary. Construct a calibration graph of absorbance at 516 nm against boron concentration, mg dm^{-3}. For maximum accuracy, the calibration should be carried out immediately prior to the analysis of samples.

VI, 9. DETERMINATION OF COPPER AS THE DIETHYLDITHIO-CARBAMATE COMPLEX. *Discussion.* Sodium diethyldithiocarbamate (B) reacts with a slightly acidic or ammoniacal solution of copper(II) in low

concentration to produce a brown colloidal suspension of the copper(II) diethyldithiocarbamate. The suspension may be extracted with an organic solvent (chloroform, carbon tetrachloride or butyl acetate) and the coloured extract analysed spectrophotometrically at 560 nm (butyl acetate) or 435 nm (chloroform or carbon tetrachloride).

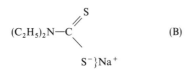

(B)

Many of the heavy metals give slightly soluble products (some white, some coloured) with the reagent, most of which are soluble in the organic solvents mentioned. The selectivity of the reagent may be improved by the use of masking agents, particularly EDTA.

The reagent decomposes rapidly in solutions of low pH.

Procedure.   Dissolve 0.0393 g of A.R. copper(II) sulphate pentahydrate in 1 dm^3 of water in a graduated flask. Pipette 10.0 cm^3 of this solution (containing about 100 μg Cu) into a beaker, add 5.0 cm^3 of 25 per cent aqueous citric acid solution, render slightly alkaline with dilute ammonia solution and boil off the excess of ammonia; alternatively, adjust to pH 8.5 using a pH meter. Add 15.0 cm^3 of 4 per cent EDTA solution and cool to room temperature. Transfer to a separatory funnel, add 10 cm^3 of 0.2 per cent aqueous sodium diethyldithio-carbamate solution, and shake for 45 seconds. A yellow-brown colour develops in the solution. Pipette 20 cm^3 of butyl acetate into the funnel and shake for 30 seconds. The organic layer acquires a yellow colour. Cool, shake for 15 seconds and allow the phases to separate. Remove the lower aqueous layer; add 20 cm^3 of 5 per cent sulphuric acid (v/v), shake for 15 seconds, cool, and separate the organic phase. Determine the absorbance at 560 nm in 1.0-cm absorption cells against a blank. All the copper is removed in one extraction.

Repeat the experiment in the presence of 1 mg of iron(III); no interference can be detected.

VI, 10.   DETERMINATION OF COPPER AS THE 'NEO-CUPROIN' COMPLEX.   *Discussion.*   'Neo-cuproin' (2,9-dimethyl-1,10-phenanthroline) can, under certain conditions, behave as an almost specific reagent for copper(I). The complex is soluble in chloroform and absorbs at 457 nm. It may be applied to the determination of copper in cast iron, alloy steels, lead–tin solder, and various metals.

Procedure.   To 10.0 cm^3 of the solution containing up to 200 μg of copper in a separatory funnel, add 5.0 cm^3 of 10 per cent hydroxylammonium chloride solution to reduce Cu(II) to Cu(I), and 10 cm^3 of a 30 per cent sodium citrate solution to complex any other metals which may be present. Add ammonia solution until the pH is about 4 (Congo red paper), followed by 10 cm^3 of a 0.1 per cent solution of 'neo-cuproin' in absolute ethanol. Shake for about 30 seconds with 10 cm^3 of chloroform and allow the layers to separate. Repeat the extraction with a further 5 cm^3 of chloroform. Measure the absorbance at 457 nm against a blank on the reagents which have been treated similarly to the sample.

VI, 11. DETERMINATION OF IRON BY CHLORIDE EXTRAC-
TION. *Discussion.* The extraction of iron(III) chloride from hydrochloric acid with diethyl ether (probably as the solvated complex $H[FeCl_4]$) has long been known, but the amount of metal extracted depends upon the concentration of the acid and passes through a maximum at about $6M$-hydrochloric acid.

Elements that extract well as chloride complexes include Sb(V), As(III), Ga(III), Ge(IV), Tl(III), Hg(II), Mo(VI), Pt(II), and Au(III). Elements which are partially extracted include Sb(III), As(V), V(V), Co(II), Sn(II), and Sn(IV). Many solvents with donor oxygen atoms, including di-isopropyl ether, $\beta\beta'$-dichlorodiethylether, ethyl acetate, butyl acetate, and pentyl acetate, have been employed. In most cases the optimum extraction depends upon the acid concentration.

The extraction of large amounts of iron is conveniently made with iso-butyl acetate: this solvent has the merit of low volatility and of almost negligible temperature rise during the extraction (unlike diethyl ether).

To gain experience in the procedure, experimental details are given for the extraction of iron(III) in hydrochloric acid solution with diethyl ether.

Procedure. Weigh out 16.486 g of A.R. hydrated ammonium iron(III) sulphate and dissolve it in 250 cm^3 of $6M$-hydrochloric acid in a graduated flask. Extract 25.0 cm^3 of the iron(III) solution (which contains 200 mg of Fe) with three 25-cm^3 portions of pure diethyl ether (1): shake gently for 3 minutes during each extraction. Combine the three ether extracts and strip the iron from the ether by shaking with 25 cm^3 of water: approximately 99.9 per cent of the iron is removed by this method. Boil off any ether remaining in the aqueous extract on a water-bath (caution!), and determine the iron by titration with standard $0.1N$-potassium dichromate after previous reduction to the iron(II) state. The iron recovered should not be less than 99.6 per cent (2).

Notes. 1. The factors of importance in the diethyl ether extraction of iron are:

(*a*) The iron must be in the iron(III) state, since iron(II) chloride is not extracted.

(*b*) The hydrochloric acid concentration must be close to $6M$.

(*c*) The extraction should be carried out in subdued light, since ether photochemically reduces iron(III).

(*d*) The ether should be free from ethanol and peroxides because these reduce iron(III) chloride.

(*e*) The concentration of anions other than chloride should be kept low.

(*f*) Heat is generated by the mixing of the ether and the hydrochloric acid–iron(III) chloride solution so that cooling of the mixture under the tap or in ice is essential.

2. The procedure may be adapted to the **determination of iron in an iron ore or a steel**. The broad details are as follows. Dissolve a 0.5 g sample, accurately weighed, in 25 cm^3 of $6M$-hydrochloric acid and 4 cm^3 of concentrated nitric acid by heating the mixture on a water bath. Evaporate the solution to dryness and then dissolve the residue in 15 cm^3 of 1:1 hydrochloric acid. Transfer the solution to a continuous extractor and rinse the vessel with a little $6M$-hydrochloric acid. Extract the solution with diethyl ether or with peroxide-free di-isopropyl ether until the ether layer above the solution is colourless. Transfer the ethereal solution of the iron(III) chloride to a separatory funnel, strip the iron from the ethereal solution by two or three washings with an equal volume of water. Determine the iron content as above.

VI, 12. DETERMINATION OF IRON AS THE 8-HYDROXY-QUINOLATE.* *Discussion.* Iron(III) (50–200 μg) can be extracted from aqueous solution with a 1 per cent solution of 8-hydroxyquinoline in chloroform by double extraction when the pH of the aqueous solution is between 2 and 10. At a pH of 2–2.5 nickel, cobalt, cerium(III), and aluminium do not interfere. Iron(III) oxinate is dark-coloured in chloroform and absorbs at 470 nm.

Procedure. Weigh out 0.0226 g of A.R. hydrated ammonium iron(III) sulphate and dissolve it in 1 dm^3 of water in a graduated flask; 50 cm^3 of this solution contain 100 μg of iron. Place 50.0 cm^3 of the solution in a 100-cm^3 separatory funnel, add 10 cm^3 of a 1 per cent oxine (A.R.) solution in chloroform and shake for 1 minute. Separate the chloroform layer. Transfer a portion of the latter to a 1.0-cm absorption cell. Determine the absorbance at 470 nm in a spectrophotometer, using the solvent as a blank or reference. Repeat the extraction with a further 10 cm^3 of 1 per cent oxine solution in chloroform, and measure the absorbance to confirm that all the iron was extracted.

Repeat the experiment using 50.0 cm^3 of the iron(III) solution in the presence of 100 μg of aluminium ion and 100 μg of nickel ion at pH 2.0 (use a pH meter to adjust the acidity) and measure the absorbance. Confirm that an effective separation has been achieved.

Note. Some typical results are given below. Absorbance after first extraction 0.605; after second extraction 0.004; in presence of 100 μg Al and 100 μg Ni the absorbance obtained is 0.602.

VI, 13. DETERMINATION OF LEAD BY THE DITHIZONE METHOD.† *Discussion.* Diphenylthiocarbazone (dithizone) behaves in solution as a tautomeric mixture of (C) and (D):

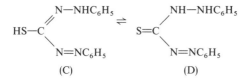

(C)   (D)

It functions as a monoprotic acid (pK_a = 4.7) up to a pH of about 12; the acid proton is that of the thiol group in (C). 'Primary' metal dithizonates are formed according to the reaction:

$$M^{n+} + nH_2Dz \rightleftharpoons M(HDz)_n + nH^+$$

Some metals, notably copper, silver, gold, mercury, bismuth, and palladium, form a second complex (which we may term 'secondary' dithizonates) at a higher pH range or with a deficiency of the reagent:

$$2M(HDz)_n \rightleftharpoons M_2Dz_n + nH_2Dz$$

---

* In this and the succeeding determinations of minute quantities of the various elements involving solvent extraction and spectrophotometric analysis of the solvent extracts, some simple experiments with known microgram quantities of the elements will be described. These will illustrate the principles involved; they can be readily adapted to the determination of unknown solutions by utilising a calibration graph (absorbance—concentration in μg).

† This experiment is not recommended for elementary students or students having little experience of analytical work.

In general, the 'primary' dithizonates are of greater analytical utility than the 'secondary' dithizonates, which are less stable and less soluble in organic solvents.

Dithizone is a violet-black solid which is insoluble in water, soluble in dilute ammonia solution, and also soluble in chloroform and in carbon tetrachloride to yield green solutions. It is an excellent reagent for the determination of small (microgram) quantities of many metals, and can be made selective for certain metals by resorting to one or more of the following devices:

(*a*) Adjusting the pH of the solution to be extracted. Thus from dilute acid solution (0.1–0.5*N*) silver, mercury, copper, and palladium can be separated from other metals; bismuth can be extracted from a weakly acidic medium; lead and zinc from a neutral or faintly alkaline medium; cadmium from a strongly basic solution containing citrate or tartrate.

(*b*) Adding a complex-forming agent or masking agent, e.g., cyanide, thiocyanate, thiosulphate, or EDTA.

It must be emphasised that dithizone is an extremely sensitive reagent and is applicable to quantities of metals of the order of micrograms. Only the purest dithizone (e.g., A.R.) may be used, since the reagent tends to oxidise to diphenylthiocarbadiazone $S=C(N=NC_6H_5)_2$: the latter does not react with metals, is insoluble in ammonia solution, and dissolves in organic solvents to give yellow or brown solutions. Reagents for use in dithizone methods of analysis must be of the highest purity (e.g., A.R.). De-ionised water and redistilled acids are recommended: ammonia solution should be prepared by passing ammonia gas into water. Weakly basic and neutral solutions can frequently be freed from reacting heavy metals by extracting them with a fairly strong solution of dithizone in chloroform until a green extract is obtained. Vessels (of Pyrex) should be rinsed with dilute acid before use. Blanks must always be run.

Only one example of the use of dithizone in solvent extraction will be given in order to illustrate the general technique involved.

Procedure. Dissolve 0.0067 g of pure lead chloride in 1 dm^3 of water in a graduated flask. To 10.0 cm^3 of this solution (containing about 50 μg of lead) contained in a 250 cm^3 separatory funnel, add 75 cm^3 of ammonia–cyanide–sulphite mixture (1), adjust the pH of the solution to 9.5 (pH meter) by the cautious addition of hydrochloric acid*, then add 7.5 cm^3 of a 0.005 per cent solution of dithizone in chloroform (2), followed by 17.5 cm^3 of chloroform. Shake for 1 minute, and allow the phases to separate. Determine the absorbance at 510 nm against a blank solution in a 1.0 cm absorption cell. A further extraction of the same solution gives zero absorption indicative of the complete extraction of the lead. Almost the same absorbance is obtained in the presence of 100 μg of copper ion and 100 μg of zinc ion.

Notes. 1. This solution is prepared by diluting 35 cm^3 of concentrated ammonia solution (sp. gr. 0.88) and 3.0 cm^3 of 10 per cent potassium cyanide solution (**caution**) to 100 cm^3, and then dissolving 0.15 g of sodium sulphite in the solution.

2. One cm^3 of this solution is equivalent to about 20 μg of lead.

---

* It is essential that the pH of the mixture does not fall below 9.5, even temporarily, as there is always the possibility that HCN could be liberated.

VI, 14.   DETERMINATION OF MOLYBDENUM BY THE THIOCY-ANATE METHOD. *Discussion.*

Molybdenum(VI) in acid solution when treated with tin(II) chloride (best in the presence of a little iron(II) ion) is converted largely into molybdenum(V): this forms a complex with thiocyanate ion, probably largely $Mo(SCN)_5$, which is red in colour. The latter may be extracted with solvents possessing donor oxygen atoms (3-methylbutanol is preferred). The colour depends upon the acid concentration (optimum concentration $1 M$) and the concentration of the thiocyanate ion ($\not< 1$ per cent, but colour intensity is constant in the range 2–10 per cent); it is little influenced by excess of tin(II) chloride. The molybdenum complex has maximum absorption at 465 nm.

Reagents. *Standard molybdenum solution.* Dissolve 0.184 g of A.R. ammonium molybdate $(NH_4)_6[Mo_7O_{24}]4H_2O$ in 1 dm^3 of distilled water in a graduated flask: this gives a 0.001 per cent Mo solution containing 10 μg Mo per cm^3. Alternatively, dissolve 0.150 g of A.R. molybdenum trioxide in a few cm^3 of dilute sodium hydroxide solution, dilute with water to about 100 dm^3, render slightly acidic with dilute hydrochloric acid, and then dilute to 1 dm^3 with water in a graduated flask: this is a 0.0100 per cent solution. It can be diluted to 0.001 per cent with 0.1 M-hydrochloric acid.

Ammonium Iron sulphate solution. Dissolve 10 g of the A.R. salt in 100 cm^3 of very dilute sulphuric acid.

Tin(II) chloride solution. Dissolve 10 g of A.R. tin(II) chloride dihydrate in 100 cm^3 of 1 M-hydrochloric acid.

Potassium thiocyanate solution. Prepare a 10 per cent aqueous solution from the A.R. salt.

Procedure. Construct a calibration curve by placing 1.0, 2.0, 3.0, 4.0, and 5.0 cm^3 of the 0.001 per cent Mo solution (containing 10 μg, 20 μg, 30 μg, 40 μg, and 50 μg Mo) severally in 50-cm^3 separatory funnels and diluting each with an equal volume of water. Add to each funnel 2.0 cm^3 of concentrated hydrochloric acid, 1.0 cm^3 of the ammonium iron(II) sulphate solution, and 3.0 cm^3 of the potassium thiocyanate solution; shake gently and then introduce 3.0 cm^3 of the tin(II) chloride solution. Add water to bring the total volume in each separatory funnel to 25 cm^3 and mix. Pipette 10.0 cm^3 of redistilled 3-methylbutanol into each funnel and shake individually for 30 seconds. Allow the phases to separate, and carefully run out the lower aqueous layer. Remove the glass stopper and pour the alcoholic extract through a small plug of purified glass wool in a small funnel and collect the organic extract in a 1.0-cm absorption cell. Measure the absorbance at 465 nm in a spectrophotometer against a 3-methylbutanol blank. Plot absorbance against μg of Mo. A straight line is obtained over the range 0–50 μg Mo: Beer's law is obeyed.

Determine the concentration of Mo in unknown samples supplied and containing less than 50 μg Mo per 10 cm^3: use the calibration curve, and subject the unknown to the same treatment as the standard solutions.

The above procedure may be adapted to the **determination of molybdenum in steel**. Dissolve a 1.00-g sample of the steel (accurately weighed) in 5 cm^3 of 1:1 hydrochloric acid and 15 cm^3 of 70 per cent perchloric acid. Heat the solution until dense fumes are evolved and then for 6–7 minutes longer. Cool, add 20 cm^3 of water, and warm to dissolve all salts. Dilute the resulting cooled solution to volume in a 1-dm^3 flask. Pipette 10.0 cm^3 of the diluted solution into a 50-cm^3 separatory funnel, add 3 cm^3 of the tin(II) chloride solution, and continue as detailed above. Measure the absorbance of the extract at 465 nm with a spectrophotometer, and compare this value with that obtained with known amounts of molybdenum. Use the calibration curve prepared with equal amounts of iron and varying quantities of molybdenum. If preferred, a mixture of 3-methylbutanol and carbon tetrachloride, which is heavier than water, can be used as extractant.

Note. Under the above conditions of determination the following elements interfere in the amount specified when the amount of Mo is 10 μg (error greater than 3 per cent): V, 0.4 mg, yellow colour (interference prevented by washing extract with tin(II) chloride solution); Cr(VI), 2 mg, purple colour; W(VI), 0.15 mg, yellow colour; Co, 12 mg, slight green colour; Cu, 5 mg; Pb, 10 mg; Ti(III), 30 mg (in presence of sodium fluoride).

VI, 15. DETERMINATION OF NICKEL AS THE DIMETHYLGLY-OXIME COMPLEX. *Discussion.*

Nickel (200–400 μg) forms the red dimethylglyoxime complex in a slightly alkaline medium; it is only slightly soluble in chloroform (35–50 μg Ni cm^{-3}). The optimum pH range for extraction of the nickel complex is 7–12 in the presence of citrate. The nickel complex absorbs at 366 nm and also at 465–470 nm.

Procedure. Weigh out 0.135 g of pure ammonium nickel sulphate (NiSO$_4$,(NH$_4$)$_2$SO$_4$,6H$_2$O) and dissolve it in 1 dm^3 of water in a graduated flask. Transfer 10.0 cm^3 of this solution (Ni content about 200 μg) to a beaker containing 90 cm^3 of water, add 5.0 g of A.R. citric acid, and then dilute ammonia solution until the pH is 7.5. Cool and transfer to a separatory funnel, add 20 cm^3 of dimethylglyoxime solution (1) and, after standing for a minute or two, 12 cm^3 of chloroform. Shake for 1 minute, allow the phases to settle out, separate the red chloroform layer, and determine the absorbance at 366 nm in a 1.0-cm absorption cell against a blank. Extract with a further 12 cm^3 of chloroform and measure the absorbance of the extract at 366 nm; very little nickel will be found.

Repeat the experiment in the presence of 500 μg of iron(III) and 500 μg of aluminium ion; no interference will be detected.

Note. 1. The dimethylglyoxime reagent is prepared by dissolving 0.50 g of A.R. dimethylglyoxime in 250 cm^3 of ammonia solution and diluting to 500 cm^3 with water.

Note. 2. Cobalt forms a brown soluble dimethylglyoxime complex which is very slightly extracted by chloroform; the amount is only significant if large amounts of Co ($>$2–3 mg) are present. If Co is suspected it is best to wash the organic extract with *ca.* 0.5M-ammonia solution: enough reagent must be added to react with the Co and leave an excess for the Ni. Large amounts of cobalt may be removed by oxidising with hydrogen peroxide, complexing with ammonium thiocyanate (as a 60 per cent aqueous solution), and extracting the compound with a pentyl alcohol–diethylether (3:1) mixture. Copper(II) is extracted to a small extent, and is removed from the extract by shaking with 0.5M-ammonia solution. Copper in considerable amounts is not extracted if it is complexed with thiosulphate at pH 6.5. Much Mn tends to inhibit the extraction of Ni; this difficulty is overcome by the addition of hydroxylammonium chloride. Iron(III) does not interfere.

VI, 16.   DETERMINATION OF SILVER BY EXTRACTION AS ITS ION-ASSOCIATION   COMPLEX   WITH   1,10-PHENANTHROLINE   AND BROMOPYROGALLOL RED. *Discussion.* Silver can be extracted from a nearly neutral aqueous solution into nitrobenzene as the blue ternary ion-association complex formed between silver(I) ions, 1,10-phenanthroline and bromopyrogallol red. The method is highly selective in the presence of EDTA, bromide and mercury(II) ions as masking agents and only thiosulphate appears to interfere (Ref. 7).

Reagents. Silver nitrate solution, $10^{-4}M$. Prepare by dilution of a standard $0.1M$ silver nitrate solution.

1,10-Phenanthroline solution. Dissolve 49.60 mg of analytical grade 1,10-phenanthroline in distilled water and dilute to 250 cm^3.

Ammonium acetate solution, 20 per cent. Dissolve 20 g of the analytical grade salt in distilled water and dilute to 100 cm^3.

Bromopyrogallol red solution, $10^{-4}M$. Dissolve 14.0 mg of bromopyrogallol red and 2.5 g of ammonium acetate in distilled water and dilute to 250 cm^3. This solution should be discarded after five days.

EDTA solution, $10^{-1}M$. Dissolve 3.7225 g of analytical grade disodium salt in distilled water and dilute to 100 cm^3.

Sodium nitrate solution, $1M$. Dissolve 8.5 g of analytical grade sodium nitrate in distilled water and dilute to 100 cm^3.

Nitrobenzene, analytical grade.

Sodium hydroxide, analytical grade pellets.

Procedure. (*a*) **Calibration.** Pipette successively 1, 2, 3, 4 and 5 cm^3 of $10^{-4}M$ silver nitrate solution, 1 cm^3 of 20 per cent ammonium acetate solution, 5 cm^3 of $10^{-3}M$ 1,10-phenanthroline solution, 1 cm^3 of $10^{-1}M$ EDTA solution and 1 cm^3 of 1M sodium nitrate solution into five 100 cm^3 separating funnels. Add sufficient distilled water to give the same volume of solution in each funnel, then add 20 cm^3 of nitrobenzene and shake by continuous inversion for one minute. Allow about ten minutes for the layers to separate, then transfer the lower organic layers to different 100 cm^3 separating funnels and add to the latter 25 cm^3 of $10^{-4}M$ bromopyrogallol red solution. Again shake by continuous inversion for one minute and allow about thirty minutes for the layers to separate. Run the lower nitrobenzene layers into clean, dry 100 cm^3 beakers and swirl each beaker until all cloudiness disappears (Note 1). Finally transfer the solutions to 1 cm cells and measure the absorbance at 590 nm against a blank carried through the same procedure but containing no silver. Plot a calibration curve of absorbance against silver content (μg).

$$1 \text{ cm}^3 \text{ of } 10^{-4}M \text{ AgNO}_3 = 10.788 \ \mu\text{g of Ag}$$

(*b*) **Determination.** To an aliquot of the silver(I) solution containing between 10 and 50 μg of silver, add sufficient EDTA to complex all those cations present which form an EDTA complex. If gold is present ($\not> 250 \ \mu$g) it is masked by adding sufficient bromide ion to form the $AuBr_4^-$ complex. Cyanide, thiocyanate or iodide ions are masked by adding sufficient mercury(II) ions to complex these anions followed by sufficient EDTA to complex any excess mercury(II). Add 1 cm^3 of 20 per cent ammonium acetate solution, etc., and proceed as described under Calibration.

Note. 1. More rapid clarification of the nitrobenzene extract is obtained if the beakers contain about 5 pellets of sodium hydroxide. The latter is, however, a source of instability of the colour system and its use is therefore not recommended.

VI, 17. DETERMINATION OF URANIUM AS THE 8-HYDROXY-QUINOLATE. *Discussion.*

Uranium(VI) may be determined as the 8-hydroxyquinolate in concentrations up to 900 μg at a pH of 8.8 in the presence of a little EDTA; the yellow oxinate complex absorbs at 400 nm. Many interfering elements (e.g., iron and aluminium but not titanium) may be masked by increasing the quantity of EDTA present in the solution.

Simple experimental details follow: these are designed to give the student experience in the method. Absorbances are read against a blank solution using 1.0 cm absorption cells.

Procedure. Weigh out 0.106 g of A.R. uranyl nitrate $UO_2(NO_3)_2,6H_2O$ and dissolve it in 1 dm^3 of water in a graduated flask. Mix 10.0 cm^3 of the solution (containing about 500 μg of U) with 5.0 cm^3 of 0.02M-EDTA, adjust the pH to 8.8 and dilute to 100 cm^3. Extract with two 10-cm^3 portions of 1.0 per cent oxine (A.R.) solution in chloroform, and measure the absorbance after each extraction.

Note. Some typical results with a spectrophotometer were: absorbance at first extraction, 0.510; at second extraction, 0.005.

Repeat the experiment in the presence of 500 μg of iron (as A.R. iron(III) alum) and 500 μg of aluminium (as A.R. ammonium alum): it will be found that the absorbance has increased, suggesting interference from these elements. Increase the volume of 0.02M-EDTA solution by 5-cm^3 portions until the absorbance is identical with that of the original uranium solution; about 20 cm^3 will be required.

VI, 18. References

1. H. Irving and R. J. P. Williams (1949). 'Metal Complexes and Partition Equilibria', *J. Chem. Soc.*, 1841.
2. J. Dunbar (1963), in *Proceedings of Technicon Symposium on Automated Analytical Chemistry,* 130.
3. J. M. Carter and G. Nickless (1970). 'A Solvent-extraction Technique with the Technicon AutoAnalyser', *Analyst,* **95**, 148.
4. F. Trowell (1969). 'Automated Solvent Extraction', *Laboratory Practice,* **18**, 44.
5. 'Automating Manual Methods using Technicon AutoAnalyser 11 System Techniques', *Manual TN1-0170-01.* New York; Technicon Instruments Corporation, 1972.
6. J. Bassett and P. J. Matthews (1974). 'A Spectrophotometric Method for the Determination of Boron in Water by the use of Ferroin', *Analyst,* **99**, 1.
7. R. M. Dagnall and T. A. West (1964). 'A Selective Extraction System for Trace Amounts of Silver', *Talanta,* **11**, 1627.
8. A. K. Babko (1968). *Talanta,* **15**, 721.

VI, 19. Selected bibliography

1. F. D. Snell and C. T. Snell. *Colorimetric Methods of Analysis.* Vol. II. Inorganic (1949). Vol. IIA (1959). New York; Van Nostrand.
2. H. Irving (1951). 'Solvent Extraction and its Applications to Inorganic Analysis', *Quarterly Reviews,* **5**, 200.
3. H. Freiser (1952). 'The Stability of Metal Chelates in Relation to their Use in Analysis', *Analyst,* **77**, 830.

4. G. Charlot and D. Bézier (translated by R. C. Murray) (1957). *Quantitative Inorganic Analysis*. Chapter XIII. Reactions in the Presence of Two Immiscible Solvents. Separation by Extraction. London; Methuen.

5. G. H. Morrison and H. Freiser (1957). *Solvent Extraction in Analytical Chemistry*. New York; John Wiley.

6. E. B. Sandell (1959). *Colorimetric Determination of Traces of Metals*. 3rd edn. New York; Interscience Publishers.

7. L. Alders (1959). *Liquid-Liquid Extraction, Theory and Laboratory Practice*. 2nd edn. Amsterdam; Elsevier Publishing Co.

8. L. C. Craig (1956). 'Extraction', *Analytical Chemistry*, **28**, 723.

9. G. H. Morrison and H. Freiser. 'Extraction', *Analytical Chemistry*, 1958, **30**, 633; 1960, **32**, 37R.

10. A. K. De, S. M. Khopkar and R. A. Chalmers (1970). *Solvent Extraction of Metals*. London; Van Nostrand Reinhold Co.

11. J. Stary (1964). *The Solvent Extraction of Metal Chelates*. Oxford; Pergamon Press.

12. Y. Marcus (ed.) (1971). *Solvent Extraction Reviews*. Vol. I. New York; Marcel Dekker Inc. Ltd.

13. H. Irving and R. J. P. Williams (1961). 'Liquid-Liquid Extraction', in *Treatise on Analytical Chemistry*, (ed. 1. Kolthoff and P. Elving) Part I, Vol. 3. New York; Interscience.

14. R. M. Diamond and D. G. Tuck (1960). 'Extraction of Inorganic Compounds into Organic Solvents', in F. A. Cotton (ed.). *Progress in Inorganic Chemistry* Vol. 2, New York; Interscience.

15. Yu. A. Zolotov (1970). (trans. J. Schmorak). *Extraction of Chelate Compounds*. Ann Arbor; Ann Arbor Science Publishers Inc.

16. A. S. Kertes and Y. Marcus (ed.) (1970). *Solvent Extraction Research. Proceedings of the Fifth International Conference on Solvent Extraction Chemistry*. Chichester; John Wiley and Sons, Sussex.

CHAPTER VII ION EXCHANGE

VII, 1. GENERAL DISCUSSION. The term ion exchange is generally understood to mean the exchange of ions of like sign between a solution and a solid highly insoluble body in contact with it. The solid (ion exchanger) must, of course, contain ions of its own, and for the exchange to proceed sufficiently rapidly and extensively to be of practical value, the solid must have an open, permeable molecular structure so that ions and solvent molecules can move freely in and out. Many substances, both natural (e.g., certain clay minerals) and artificial, have ion exchanging properties, but for analytical work synthetic organic ion exchangers are chiefly of interest, although some inorganic materials, e.g., zirconyl phosphate and ammonium 12-molybdophosphate, also possess useful ion exchange capabilities and have specialised applications (Ref. 1). All ion exchangers of value in analysis have several properties in common: they are almost insoluble in water and in organic solvents, and they contain active or counter ions that will exchange reversibly with other ions in a surrounding solution without any appreciable physical change occurring in the material. The ion exchanger is of complex nature and is, in fact, polymeric. The polymer carries an electric charge that is exactly neutralised by the charges on the counter ions. These active ions are cations in a **cation exchanger** and anions in an **anion exchanger**. Thus a cation exchanger consists of a polymeric anion and active cations, while an anion exchanger is a polymeric cation with active anions.

A widely used cation exchange resin is that obtained by the copolymerisation

of styrene $\left(\bigcirc\!\!\!\!\!\!\diagdown -CH\!=\!CH_2 \right)$ and a small proportion of divinylbenzene

$\left(CH_2\!=\!CH-\bigcirc\!\!\!\!\!\!\diagdown -CH\!=\!CH_2 \right)$, followed by sulphonation; it may be

represented as

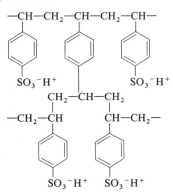

The formula enables us to visualise a typical cation exchange resin. It consists of a polymeric skeleton, held together by linkings crossing from one polymer chain to the next: the ion exchange groups are carried on this skeleton. The physical properties are largely determined by the degree of cross-linking. This cannot be determined directly in the resin itself: it is often specified as the mole per cent of the cross-linking agent in the mixture polymerised. Thus 'polystyrene sulphonic acid, 5 per cent DVB' refers to a resin containing nominally 1 mole in 20 of divinylbenzene: the true degree of cross-linking probably differs somewhat from the nominal value, but the latter is nevertheless useful for grading resins. Highly cross-linked resins are generally more brittle, harder, and more impervious than the lightly cross-linked materials; the preference of a resin for one ion over another is influenced by the degree of cross linking. The solid granules of resin swell when placed in water, but the swelling is limited by the cross-linking. In the above example the divinylbenzene units 'weld' the polystyrene chains together and prevent it from swelling indefinitely and dispersing. The resulting structure is a vast sponge-like network with negatively charged sulphonate ions attached firmly to the framework. These fixed negative charges are balanced by an equivalent number of cations: hydrogen ions in the hydrogen form of the resin and sodium ions in the sodium form of the resin, etc. These ions move freely within the water-filled pores and are sometimes called mobile ions; they are the ions which exchange with other ions. When a cation exchanger containing mobile ions C^+ is brought into contact with a solution containing cations A^+ the latter diffuse into the resin structure and cations C^+ diffuse out until equilibrium is attained. The solid and the solution then contain both cations C^+ and A^+ in numbers depending upon the position of equilibrium. The same mechanism operates for the exchange of anions in an anion exchanger.

Anion exchangers are likewise cross-linked, high-molecular-weight polymers. Their basic character is due to the presence of amino, substituted amino, or quaternary ammonium groups. The polymers containing quaternary ammonium groups are strong bases; those with amino or substituted amino groups possess weak basic properties. A widely used anion exchange resin is prepared by copolymerisation of styrene and a little divinylbenzene, followed by chloro-methylation (introduction of the $-CH_2Cl$ grouping, say, in the free para position) and interaction with a base such as trimethylamine. A hypothetical formulation of such a polystyrene anion exchange resin is given on p. 167.

Numerous types of both cation and anion exchange resins have been prepared, but only a few can be mentioned here. Cation exchange resins include that prepared by the copolymerisation of methacrylic acid $CH_2=C(CH_3)-COOH$ with glycol *bis*methacrylate

$$CH_2=C(CH_3)-COOCH_2$$
$$CH_2=C(CH_3)-COOCH_2$$

(as the cross-linking agent); this contains free $-COOH$ groups and has weak acidic properties. Weak cation exchange resins containing free $-COOH$ and $-OH$ groups have also been synthesised. Anion exchange resins containing primary, secondary, or tertiary amino groups possess weakly basic properties.

We may define a **cation exchange resin** as a high molecular weight, cross-linked polymer containing sulphonic, carboxylic, phenolic, etc., groups as an integral part of the resin and an equivalent amount of cations: an **anion exchange resin** is a polymer containing amine (or quaternary ammonium) groups as integral parts of the polymer lattice

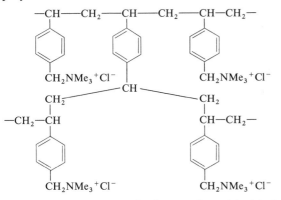

and an equivalent amount of anions such as chloride, hydroxyl, or sulphate ions. The fundamental requirements of a useful resin are:
1. The resin must be sufficiently cross-linked to have only a negligible solubility.
2. The resin must be sufficiently hydrophilic to permit diffusion of ions through the structure at a finite and usable rate.
3. The resin must contain a sufficient number of accessible ionic exchange groups and must be chemically stable.
4. The swollen resin must be denser than water.

Some of the commercially available ion exchange resins are collected in Table VII, 1. These resins, produced by different manufacturers, are often inter-changeable, and similar types will generally behave in a similar manner. The reader will find the table useful if it is desired to repeat original work carried out with a resin which, for some reason, is not immediately available.

VII, 2. ACTION OF ION EXCHANGE RESINS. Cation exchange resins* contain free cations which can be exchanged for cations in solution (soln).

$$(Res.A^-)B^+ + C^+ \text{ (solution)} \rightleftharpoons (Res.A^-)C^+ + B^+ \text{ (soln)} \tag{1}$$

If the experimental conditions are such that the equilibrium is completely displaced from left to right the ion C^+ is completely fixed on the cation exchanger. If the solution contains several ions (C^+, D^+, and E^+) the exchanger may show different affinities for them, thus making separations possible. A typical example is the displacement of sodium ions in a sulphonate resin by calcium ions:

$$2(Res.SO_3^-)Na^+ + Ca^{2+}(soln) \rightleftharpoons (Res.SO_3)^-{}_2Ca^{2+} + 2Na^+ (soln) \tag{2}$$

The reaction is reversible; by passing a solution containing sodium ions through the product, the calcium ions may be removed from the resin and the original sodium form regenerated. Similarly, by passing a solution of a neutral salt

---

* These will be represented by $(Res.A^-)B^+$, where Res. is the basic polymer of the resin, A^- is the anion attached to the polymeric framework, B^+ is the active or mobile cation: thus a sulphonated polystyrene resin in the hydrogen form would be written as $(Res.SO_3^-)H^+$. A similar nomenclature will be employed for anion exchange resins, e.g., $(Res.NMe_3^+)Cl^-$.

Table VII, 1. Comparable ion exchange materials

Type	Permutit Co Ltd, London, England	Rohm & Hass Co, USA	Dow Chemical Co, USA	Merck, Bayer, West Germany	Diamond Alkali Co, USA	Bio-Rad Labs Bromley, England
Strong acid cation exchangers	Zerolit 215 Zerolit 225	Amberlite 100 Amberlite 120 Amberlite 200	Dowex 30 Dowex 50	Lewatit S 1020 Lewatit S 1080	Duolite C 10 Duolite C 20 Duolite C 25	Bio-Rex 40 AG 50 W
Weak acid cation exchangers	Zerolit 226	Amberlite 50	Dowex CCR–1	Lewatit CP 3050	Duolite CS–100 Duolite CS–101	Bio-Rex 70
Strong base anion exchangers	Zerolit FF	Amberlite 400 Amberlite 410	Dowex 1 Dowex 2	Lewatit M 5020 Lewatit M 5080	Duolite A–42	AG 1 AG 21 K
Weak base anion exchangers	Zerolit H Zerolit G	Amberlite 68 Amberlite 45	Dowex 3	Lewatit MP 7080	Duolite A1–A7 Duolite A–14	AG 3
Chelating resins			Dowex A–1			Chelex 100

through the hydrogen form of a sulphonic resin, an equivalent quantity of the corresponding acid is produced by the following typical reaction:

$$(Res.SO_3^-)H^+ + Na^+Cl^- \text{ (soln)} \rightleftharpoons (Res.SO_3^-)Na^+ + H^+Cl^- \text{ (soln)}$$

$$(3)$$

For the strongly acidic cation exchange resins, such as the cross-linked polystyrene sulphonic acid resins, the exchange capacity is virtually independent of the pH of the solution. For weak acid cation exchangers, such as those containing the carboxylate group, ionisation occurs to an appreciable extent only in alkaline solution, i.e., in their salt form; consequently the carboxylic resins have very little action in solutions below pH 7. These carboxylic exchangers in the hydrogen form will absorb strong bases from solution:

$$(Res.COO^-)H^+ + Na^+OH^- \text{ (soln)} \rightleftharpoons (Res.COO^-)Na^+ + H_2O$$

$$(4)$$

but will have little action upon, say, sodium chloride; hydrolysis of the salt form of the resin occurs so that the base may not be completely absorbed even if an excess of resin is present.

Strongly basic anion exchange resins, e.g., a cross-linked polystyrene containing quaternary ammonium groups, are largely ionised in both the hydroxide and the salt forms. Some of their typical reactions may be represented as:

$$2(Res.NMe_3^+)Cl^- + SO_4^{2-} \text{ (soln)} \rightleftharpoons (Res.NMe_3^+)_2SO_4^{2-} + 2Cl^- \text{ (soln)}$$

$$(5)$$

$$(Res.NMe_3^+)Cl^- + OH^- \text{ (soln)} \rightleftharpoons (Res.NMe_3^+)OH^- + Cl^- \text{ (soln)}$$

$$(6)$$

$$(Res.NMe_3^+)OH^- + H^+Cl^- \text{ (soln)} \rightleftharpoons (Res.NMe_3^+)Cl^- + H_2O$$

$$(7)$$

These resins are similar to the sulphonate cation exchange resins in their activity, and their action is largely independent of pH. Weakly basic ion exchange resins contain little of the hydroxide form in basic solution. The equilibrium of, say,

$$(Res.NMe_2) + H_2O \rightleftharpoons (Res.NHMe_2)^+OH^-$$

$$(8)$$

is mainly to the left and the resin is largely in the amine form. This may also be expressed by stating that in basic solution the free base Res.NHMe$_2$·OH is very little ionised. In acidic solution, however, they behave like the strongly basic ion exchange resins, yielding the highly ionised salt form:

$$(Res.NMe_2) + H^+Cl^- \rightleftharpoons (Res.NHMe_2^+)Cl^-$$

$$(9)$$

They can be used in acid solution for the exchange of anions, for example:

$$(Res.NHMe_2^+)Cl^- + NO_3^- \text{ (soln)} \rightleftharpoons (Res.NHMe_2^+)NO_3^- + Cl^- \text{ (soln)}$$

$$(10)$$

Basic resins in the salt form are readily regenerated with alkali.

Ion exchange equilibria. The ion exchange process, involving the replacement of the exchangeable ions A_R in the resin by ions of like charge B_S from a solution, may be written:

$$A_R + B_S \rightleftharpoons B_R + A_S$$

The process is a reversible one. The extent to which one ion is absorbed in preference to another is of fundamental importance: it will determine the readiness with which two or more substances, which form ions of like charge, can be separated by ion exchange and also the ease with which the ions can subsequently be removed from the resin. The factors determining the distribution of ions between an ion exchange resin and a solution include:

(i) *Nature of exchanging ions.* (a) At low aqueous concentrations and at ordinary temperatures the extent of exchange increases with increasing valency of the exchanging ion, i.e.,

$$Na^+ < Ca^{2+} < Al^{3+} < Th^{4+}.$$

(b) Under similar conditions and constant valence, for univalent ions the extent of exchange increases with decrease in size of the hydrated cation $Li^+ < H^+ < Na^+ < NH_4^+ < K^+ < Rb^+ < Cs^+$, while for divalent ions the ionic size is an important factor but the incomplete dissociation of salts of bivalent metals also plays a part

$$Cd^{2+} < Be^{2+} < Mn^{2+} < Mg^{2+} = Zn^{2+} < Cu^{2+}$$
$$= Ni^{2+} < Co^{2+} < Ca^{2+} < Sr^{2+} < Pb^{2+} < Ba^{2+}.$$

(c) With strongly basic anion exchange resins, the extent of exchange for univalent anions varies with the size of hydrated ion in a similar manner to that indicated for cations. In dilute solution polyvalent anions are generally absorbed preferentially.

(d) When a cation in solution is being exchanged for an ion of different valency the relative affinity of the higher valent ion increases in direct proportion to the dilution. Thus to exchange a higher valent ion on the exchanger for one of lower valency in solution, exchange will be favoured by increasing the concentration, while if the lower valent ion is in the exchanger and the higher valent ion is in solution, exchange will be favoured by high dilutions.

(ii) *Nature of ion exchange resin.* The absorption* of ions will depend upon the nature of the functional groups in the resin. It will also depend upon the degree of cross-linking: as the degree of cross-linking is increased, resins become more selective towards ions of different sizes (the volume of the ion is assumed to include the water of hydration), the ion with the smaller hydrated volume will usually be absorbed preferentially.

Ion exchange capacity. The total ion exchange capacity of a resin is dependent upon the total number of ion-active groups per unit weight of material, and the greater the number of ions, the greater will be the capacity. The *total ion exchange capacity* is usually expressed as milli-equivalents per gram of exchanger: it may be regarded as an equivalent weight, the latter being the reciprocal of the former, i.e., meq. per g $= 1000$/equivalent weight. The capacities of the weakly acidic and weakly basic ion exchangers are functions of pH, the former reaching moderately constant values at pH above about 9 and the latter at pH below about 5. Values for the total exchange capacities, expressed as meq. per g of dry resin, for a few typical resins are: Zerolit 225 (Na form), 4.5–5; Zerolit 226

---

* The term absorption is used whenever ions or other solutes are taken up by an ion exchanger. It does not imply any specific types of forces responsible for this uptake.

(H form), 9–10; Zerolit FF (Cl form), 4.0; Zerolit G (Cl form), 4.0. The total exchange capacity expressed as meq. cm^{-3} of the wet resin is about $\frac{1}{3}$–$\frac{1}{2}$ of the meq. g^{-1} of the dry resin. These figures are useful in estimating very approximately the quantity of resin required in a determination: an adequate excess must be employed, since the 'break through' capacity is often much less than the total capacity of the resin. In most cases a 100 per cent excess is satisfactory.

The exchange capacity of a cation exchange resin may be measured in the laboratory by determining the number of milligram equivalents of sodium ion which are absorbed by 1 g of the dry resin in the hydrogen form. Similarly, the exchange capacity of a strongly basic anion exchange resin is evaluated by measuring the amount of chloride ion taken up by 1 g of dry resin in the hydroxide form.

Changing the ionic form. Some widely used resins. It is frequently necessary to convert a resin completely from one ionic form to another. This should be done after regeneration, if this is being practised to 'clean' the resin (e.g., if the 'standard' grade of ion exchanger is used). An excess of a suitable salt solution should be run through a column of the resin. Ready conversion will occur if the ion to be introduced into the resin has a higher, or only a slightly lower, affinity than that actually on the resin. When replacing an ion of lower valency on the exchanger by one of higher valency, the conversion is assisted by using a dilute solution of replacing salt (preferably as low as 0.01M), while to substitute a higher valent ion in the exchanger by one of lower valency, a comparatively concentrated solution should be used (say, a 1M solution).

Strongly acidic cation exchangers are usually supplied in the hydrogen or sodium forms, and strongly basic anion exchangers in the chloride or hydroxide forms; the chloride form is preferred to the free base form, since the latter readily absorbs carbon dioxide from the atmosphere and becomes partly converted into the carbonate form. Weakly acidic cation exchangers are generally supplied in the hydrogen form, while weakly basic anion exchange resins are available in the hydroxide or chloride forms.

Strongly acidic cation exchangers (polystyrene sulphonic acid resins)— Zerolit 225, Amberlite 120, etc. These resins are usually marketed in the sodium form,* and to convert them into the hydrogen form (which, it may be noted, are also available commercially) the following procedure may be used.

The resin, after regeneration (see Section **VII, 7**) if the 'standard' grade is used, is treated with 2M- or with 10 per cent hydrochloric acid: one bed volume of the acid is passed through the column in 10–15 minutes. The effluent should then be strongly acid to methyl orange indicator; if it is not, further acid must be used (about three bed volumes may be required). The excess of acid is drained to almost bed level and the remaining acid washed away with distilled or de-ionised water, the volume required being about six times that of the bed. This operation occupies about 20 minutes: it is complete when the final 100 cm^3 of effluent requires less than 1 cm^3 of 0.02M sodium hydroxide to neutralise its acidity using methyl orange as indicator. The resin can now be employed for the exchange of its hydrogen ions for cations present in a given solution. Tests on the treated effluent show that its acidity, due to the exchange, rises to a maximum, which is

---

* The resin is supplied in moist condition, and should not be allowed to dry out; particle fracture may occur after repeated drying and re-wetting.

maintained until the capacity is exhausted when the acidity of the treated solution falls. Regeneration is then necessary and is performed, after back-washing, with $2M$ hydrochloric acid as before.

Weakly acidic cation exchangers (polymethylacrylic acid, etc., resins)—Zerolit 226, Amberlite 50, etc. These resins are usually supplied in the hydrogen form. They are readily changed into the sodium form by treatment with M-sodium hydroxide; an increase in volume of 80–100 per cent may be expected, the swelling is reversible and does not appear to cause any damage to the bead structure. Below a pH of about 3.5, the hydrogen form exists almost entirely in the little-ionised carboxylic acid form. Exchange with metal ions will occur in solution only when these are associated in solution with anions of weak acids, i.e., pH values above about 4.

The exhausted resin is more easily regenerated than the strongly acidic exchangers: about 1.5 bed volumes of $1M$-hydrochloric acid will usually suffice.

Strongly basic anion exchangers (polystyrene quaternary ammonium resins—Zerolit FF, Amberlite 400, etc. These resins are usually supplied in the chloride form. For conversion into the hydroxide form, treatment with $1M$-sodium hydroxide is employed, the volume used depending upon the extent of conversion desired; two bed volumes are satisfactory for most purposes. The rinsing of the resin free from alkali should be done with de-ionised water free from carbon dioxide to avoid converting the resin into the carbonate form; about 2 dm^3 of such water will suffice per 100 g of resin. An increase in volume of about 20 per cent occurs in the conversion of the resin from the chloride to the hydroxide form.

Weakly basic anion exchangers (polystyrene tertiary amine resins)—Zerolit G, Amberlite 45, etc. These resins are generally supplied in the free base (hydroxide) form. The salt form may be prepared by treating the resin with about four bed volumes of the appropriate acid (e.g., $1M$-hydrochloric acid) and rinsing with water to remove the excess of acid; the final effluent will not be exactly neutral, since hydrolysis occurs slowly, resulting in slightly acidic effluents. As with cation exchange, quantitative anion exchange will occur only if the anion in the resin has a lower affinity for the resin than the anion to be exchanged in the solution. When the resin is exhausted, regeneration can be accomplished by treatment with excess of $1M$-sodium hydroxide, followed by washing with de-ionised water until the effluent is neutral. If ammonia solution is used for regeneration the amount of washing required is reduced.

VII, 3. ION EXCHANGE CHROMATOGRAPHY. If a mixture of two or more different cations, A, B, etc., is passed through an ion exchange column, and if the quantities of these ions are small compared with the total capacity of the column for ions, then it may be possible to recover the absorbed ions separately and consecutively by using a suitable regenerating (or eluting) solution. If cation A is held more firmly by the exchange resin than cation B, all the B present will flow out of the bottom of the column before any of A is liberated, provided that the column is long enough and other experimental factors are favourable for the particular separation. This separation technique is sometimes called **ion exchange chromatography**. Its most spectacular success has been the separation of the lanthanoids and also of other cations of very similar properties (e.g., Hf and Zr; Nb and Ta; Na and K).

The process of removing absorbed ions is known as elution, the solution

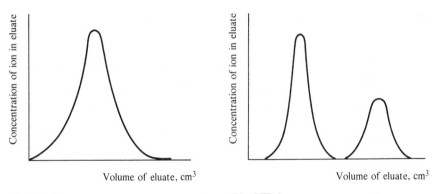

Fig. VII, 1   Fig. VII, 2

employed for elution is termed the eluant, and the solution resulting from elution is called the eluate. The liquid entering the ion exchange column may be termed the influent and the liquid leaving the column is conveniently called the effluent. If a solution of a suitable eluant is passed through a column charged with an ion A the course of the reaction may be followed by analysing continuously the effluent solution. If the concentration of A in successive portions of the eluate is plotted against the volume of the eluate, an elution curve is obtained such as is shown in Fig. VII, 1. It will be seen that practically all the A is contained in a certain volume of liquid and also that the concentration of A passes through a maximum.

If the ion exchange column is loaded with several ions of similar charge, B, C, etc., elution curves may be obtained for each ion by the use of appropriate eluants. If the elution curves are sufficiently far apart, as in Fig. VII, 2, a quantitative separation is possible; only an incomplete separation is obtained if the elution curves overlap. Ideally the curves should approach a Gaussian (normal) distribution (Section **IV, 8**) and excessive departure from this distribution may indicate faulty technique and/or column operating conditions.

The rate at which two constituents separate in the column is determined by the ratio of the two corresponding **distribution coefficients**, where the distribution coefficient is given by the equation

$$K_d = \frac{\text{amount of solute on resin}}{\text{weight of resin, g}} \div \frac{\text{amount of solute in solution}}{\text{volume of solution, cm}^3}$$

The distribution coefficient can be determined by batch experiments in which a small known quantity of resin is shaken with a solution containing a known concentration of the solute, followed by analysis of the two phases after equilibrium has been attained. The **separation factor**, α, is used as a measure of the chromatographic separation possible and is given by the equation,

$$\alpha = K_{d_1}/K_{d_2}$$

where K_{d_1} and K_{d_2} are the distribution coefficients of the two constituents. The greater the deviation of α from unity the easier will be the separation. For normal laboratory practice, a useful guide is that quantitative separation should be achieved if α is above 1.2 or less than 0.8.

An important relationship exists between the weight distribution coefficient and the volume of eluant (V_{max}) required to reach the maximum concentration of an eluted ion in the effluent. This is given by the equation:

$$V_{max} = K_d V_o + V_o$$

where V_o is the volume of liquid in the interstices between the individual resin beads. If the latter are spheres of uniform size and close-packed in the column, V_o is approximately 0.4 of the total bed volume, V_b. The void fraction V_o/V_b of the column may, however, be determined experimentally or calculated from density data (Ref. 2).

The **volume distribution coefficient** is also a useful parameter for chromatographic calculations and is defined as

$$D_v = \frac{\text{amount of ion in 1 cm}^3 \text{ of resin bed}}{\text{amount of ion in 1 cm}^3 \text{ of interstitial volume}}$$

It is related to the weight distribution coefficient by

$$D_v = K_d \beta$$

where β is the void fraction of the settled column.
It is also related to V_{max} by the equation

$$V_{max} = V_b(D_v + \beta)$$

It should be remembered that the relationships given above are strictly applicable only when the loading of the column is less than 5 per cent of its capacity.

The application of these parameters may be illustrated by the following example.

Example. A mixture of *ca.* 0.05 meq each of chloride and bromide ions is to be separated on an anion exchange column of length 10 cm and 1 cm^2 cross section, using 0.035M potassium nitrate as the eluant. The distribution coefficients (K_d) for the chloride and bromide ions respectively are 29 and 65.

$$\text{Separation factor } \alpha = \frac{65}{29}$$

$$= 2.24$$

This value indicates that a satisfactory separation could be achieved, and this is confirmed by calculation of the V_{max} values for the appearance of the chloride and bromide peaks.

From the column dimensions, the bed volume is

$$V_b = 10 \text{ cm} \times 1.0 \text{ cm}^2 = 10.0 \text{ cm}^3$$

and the void volume (assuming $\beta = 0.4$) is

$$V_o = 0.4 \times 10.0 \text{ cm}^3 = 4 \text{ cm}^3$$

Hence for the chloride peak,

$$V_{max} = K_d V_o + V_o = (29 \times 4) + 4 = 120 \text{ cm}^3$$

and for bromide,

$$V_{max} = (65 \times 4) + 4 = 264 \text{ cm}^3$$

The relatively large values of V_{max} indicate, however, that the separation will be lengthy and the elution bands broad, particularly for the bromide band. The use of a more concentrated solution of eluant significantly reduces the values of V_{max} and the elution bands become much sharper. Thus the distribution coefficient for bromide using a 0.35M potassium nitrate solution is 6.5 and using the same column, $V_{max} = (6.5 \times 4) + 4 = 30 \text{ cm}^3$.

In many cases the efficient separation of a mixture by ion exchange chromatography requires that the eluant concentration be changed during the course of the elution. This may be done in a stepwise manner or by a continuous change in concentration as in gradient elution; the latter procedure can be carried out using simple laboratory equipment. A comprehensive discussion of the technique and of gradient elution devices is given in the review by L. R. Snyder (Ref. 3).

The scope of separations by ion exchange chromatography may be extended by using for fixation or for elution a solution capable of complexing the ions exchanged. The formation of complexes may assist separations by diminishing the concentrations of free ions, and also by producing complexes of different stabilities, thus leading to significantly different behaviour with selected eluants.

The results of ion exchange separations may be influenced by varying the pH, the solvent or eluant, the temperature, the nature of the ion exchange resin, the particle size, the rate of flow of eluant, and the length of the column.

VII, 4. ION EXCHANGE IN ORGANIC AND AQUEOUS–ORGANIC SOLVENTS.

Investigations in aqueous systems have established many of the fundamental principles of ion exchange as well as providing useful applications. The scope of the ion exchange process has, however, been extended during the last decade or so by the use of both organic and mixed aqueous–organic solvent systems (Ref. 4 and 5).

The organic solvents generally used are oxo-compounds of the alcohol, ketone and carboxylic acid types, generally having dielectric constants below 40. Cations and anions should, therefore, pair more strongly in such solvent systems than in water and this factor may in itself be expected to alter selectivities for the resin. In addition to influencing these purely electrostatic forces, the presence of the organic solvent may enhance the tendency of a cation to complex with anionic or other ligands thus modifying its ion exchange behaviour. In mixed aqueous–organic solvents the magnitude of such effects will clearly be dependent on the proportion of organic solvent present.

As already indicated, ion exchange resins are osmotic systems which swell owing to solvent being drawn into the resin. Where mixed solvent systems are used the possibility of preferential osmosis occurs and it has been shown that strongly acid cation and strongly basic anion resin phases tend to be predominantly aqueous with the ambient solution predominantly organic. This effect (preferential water sorption by the resin) increases as the dielectric constant of the organic solvent decreases.

An interesting consequence of selective sorption is that conditions for partition chromatography arise which may enhance the normal ion exchange separation factors. This aspect has been utilised by Korkisch (Ref. 6) for separation of inorganic ions by the so-called 'Combined Ion Exchange–Solvent Extraction Method' (CISE), and is illustrated by experiment **VII, 16.**

VII, 5. CHELATING ION EXCHANGE RESINS.

The use of complexing

agents in solution in order to enhance the efficiency of separation of cation mixtures (e.g. lanthanoids) using conventional cation or anion exchange resins is well established. An alternative mode of application of complex formation is, however, the use of chelating resins which are ion exchangers in which various chelating groups (e.g., dimethylglyoxime and iminodiacetic acid) have been incorporated and are attached to the resin matrix.

An important feature of chelating ion exchangers is the greater selectivity which they offer compared with the conventional type of ion exchanger. The affinity of a particular metal ion for a certain chelating resin depends mainly on the nature of the chelating group, and the selective behaviour of the resin is largely based on the different stabilities of the metal complexes formed on the resin under various pH conditions. It may be noted that the binding energy in these resins is of the order of 60–105 kJ mole^{-1}, whereas in ordinary ion exchangers the strength of the electrostatic binding is only about 8–13 kJ mole^{-1}.

The exchange process in a chelating resin is generally slower than in the ordinary type of exchanger, the rate apparently being controlled by a particle diffusion mechanism.

According to Gregor *et al.* (Ref. 7) the following properties are required for a chelating agent which is to be incorporated as a functional group into an ion exchange resin:

1. the chelating agent should yield, either alone or with a cross-linking substance, a resin gel of sufficient stability or be capable of incorporation into a polymer matrix;
2. the chelating group must have sufficient chemical stability, so that during the synthesis of the resin its functional structure is not changed by polymerisation or any other reaction;
3. the steric structure of the chelating group should be compact so that the formation of the chelate rings with cations will not be hindered by the resin matrix;
4. the specific arrangements of the ligand groups should be preserved in the resin. This is particularly necessary since the complexing agents forming sufficiently stable complexes are usually at least tridentate.

These considerations indicate that many chelating agents could not be incorporated into a resin without loss of their selective complexing abilities. Ligands which do not form $1:1$ complexes (e.g., 8-quinolinol) would be unsuitable, as also would molecules such as EDTA, which are insufficiently compact. In the latter case, it is improbable that the chelate configurations occurring in aqueous solution could be maintained in a cross-linked polymer. The closely related iminodiacetic acid group does, however, meet the requirements described, being compact and forming $1:1$ complexes with metal cations.

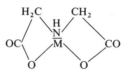

The basicity of the nitrogen atom can be influenced by whether the imino group is attached directly to a benzene nucleus or whether a methylene group is interposed.

Although chelating resins containing various ligand donor atoms have been synthesised, the iminodiacetic acid resins (N and O donor atoms) undoubtedly form the largest group (Refs. 8 and 9). The resin based on iminodiacetic acid in a styrene–divinylbenzene matrix is available commercially under the trade names of Dowex Chelating Resin A-1 and Chelex 100, and its chemical and physical properties have been fully investigated.

The starting material for the synthesis of this chelating resin is chloromethylated styrene-divinylbenzene which undergoes an amination reaction and is then treated with monochloracetic acid:

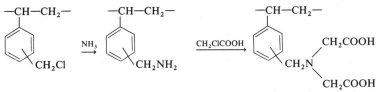

The selectivity of this type of exchange resin is illustrated by Chelex 100 which shows unusually high preference for copper, iron and other heavy metals (i.e., metals which form complexes having high stability constants with this type of ligand) over such cations as sodium, potassium and calcium; it is also much more selective for the alkaline earths than for the alkali metal cations. The resin's high affinity for these ions makes it very useful for removing, concentrating or analysing traces of them in solutions, even when large amounts of sodium and potassium are present.

VII, 6. LIQUID ION EXCHANGERS. The ion exchange processes involving exchange resins occur between a solid and liquid phase whereas in the case of liquid ion exchangers the process takes place between two immiscible solutions. Liquid ion exchangers consist of high molecular weight acids and bases which possess low solubility in water and high solubility in water-immiscible solvents. Thus a solution of a base insoluble in water, in a solvent which is water-immiscible, can be used as an anion exchanger; similarly a solution of an acid insoluble in water can act as a cation exchanger for ions in aqueous solution. A comprehensive list of liquid ion exchangers has been given by Coleman et al. (Ref. 10).

The liquid anion exchangers at present available are based largely on primary, secondary and tertiary aliphatic amines, e.g., the exchangers Amberlite LA.1 [N-dodecenyl(trialkylmethyl)amine] and Amberlite LA.2 [N-lauryl(trialkylmethyl)amine], both secondary amines. These anion exchange liquids are best employed as solutions (ca. 2.5 to 12.5% v/v) in an inert organic solvent such as benzene, toluene, kerosene, petroleum ether, cyclohexane, octane, etc.

The liquid exchangers Amberlite LA.1 and LA.2 may be used to remove acids from solution

$$R'R''NH + HX \longrightarrow R'R''NH_2X$$

or in a salt form for various ion exchange processes

$$R'R''NH_2Cl + NaNO_3 \longrightarrow R'R''NH_2NO_3 + NaCl$$

Examples of liquid cation exchangers are alkyl and dialkyl phosphoric acids, alkyl sulphonic acids and carboxylic acids, although only two appear to have been used to any extent, viz., di-(2-ethylhexyl)orthophosphoric acid and dinonylnaphthalene sulphonic acid.

The operation of liquid ion exchangers involves the selective transfer of a solute between an aqueous phase and an immiscible organic phase containing the liquid exchangers. Thus high molecular weight amines in acid solution yield large cations capable of forming extractable species (e.g., ion pairs) with various anions. The technique employed for separations using liquid ion exchangers is thus identical to that used in solvent extraction separations and these exchangers thus offer many of the advantages of both ion exchange and solvent extraction. There are, however, certain difficulties and disadvantages associated with their use which it is important to appreciate in order to make effective use of liquid ion exchangers.

Probably the chief difficulty which arises is that due to the formation of emulsions between the organic and aqueous phases. This makes separation of the phases difficult and sometimes impossible. It is clearly important to select liquid exchangers having low surface activity and to use conditions which will minimise the formation of stable emulsions (see Section **VI, 6**).

Another disadvantage in the use of liquid ion exchangers is that it is frequently necessary to back-extract the required species from the organic phase into an aqueous phase prior to completing the determination. The organic phase may, however, sometimes be used directly for determination of the extracted species, in particular by aspirating directly into a flame and estimating extracted metal ions by flame photometry or atomic absorption spectroscopy.

The extraction of metals by liquid amines has been widely investigated and depends on the formation of anionic complexes of the metals in aqueous solution. Such applications are illustrated by the use of Amberlite LA.1 for extraction of zirconium and hafnium from hydrochloric acid solutions, and the use of liquid amines for extraction of uranium from sulphuric acid solutions (Refs. 11 and 12).

Exhausted liquid ion exchangers may be regenerated in an analogous manner to ion exchange resins, e.g., Amberlite LA.1 saturated with nitrate ions can be converted to the chloride form by treatment with excess sodium chloride solution.

APPLICATIONS IN ANALYTICAL CHEMISTRY

VII, 7. EXPERIMENTAL TECHNIQUES. The simplest apparatus for ion exchange work in analysis consists of a burette provided with a glass-wool plug or sintered glass disc (porosity 0 or 1) at the lower end. Another simple column is shown in Fig. VII, 3, (a); the ion exchange resin is supported on a glass-wool plug or sintered-glass disc. A glass-wool pad may be placed at the top of the bed of resin and the eluting agent is added from a tap funnel supported above the column. The siphon overflow tube, attached to the column by a short length of rubber or PVC tubing, ensures that the level of the liquid does not fall below the top of the resin bed, so that the latter is always wholly immersed in the liquid. The ratio of the height of the column to the diameter is not very critical but is usually 10 or 20 to 1. Another form of column is depicted in Fig. VII, 3, (b) (not drawn to scale): a convenient size is 30 cm long, the lower portion of about 10 mm and the

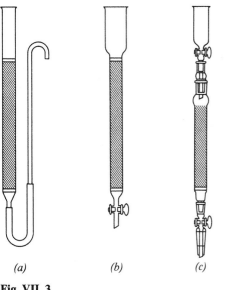

(a)   *(b)*   *(c)*

Fig. VII, 3

upper portion of about 25 mm internal diameter. A commercially available column, fitted with ground-glass joints, is illustrated in Fig. VII, 3, (*c*).

The ion exchange resin should be of small particle size, so as to provide a large surface of contact; it should, however, not be so fine as to produce a very slow flow rate. For most analytical work 50–100-mesh or 100–200-mesh materials are satisfactory. In all cases the diameter of the resin bead should be less than one-tenth of that of the column. Resins of medium and high cross-linking rarely show any further changes in volume, and only if subjected to large changes of ionic strength will any appreciable volume change occur. Resins of low cross-linking may change in volume appreciably even with small variations of ionic strength, and this may result in channelling and possible blocking of the column; these effects limit the use of these materials. To obtain satisfactory separations, it is essential that the solutions should pass through the column in a uniform manner. The resin particles should be packed uniformly in the column: the resin bed should be free from air bubbles so that there is no channelling.

To prepare a well-packed column, a supply of exchange resin of narrow size range is desirable. An ion exchange resin swells if the dry solid is immersed in water; no attempt should therefore be made to set up a column by pouring the dry resin into a tube and then adding water, since the expansion will probably shatter the tube. The resin should be stirred with water in an open beaker for several minutes, any fine particles removed by decantation, and the resin slurry transferred portionwise to the tube previously filled with water. The tube may be tapped gently to prevent the formation of air bubbles. To ensure the removal of entrained air bubbles, of any remaining fine particles, and also to ensure an even distribution of resin granules, it is advisable to 'backwash' the resin column before use, i.e., a stream of good quality distilled water or of de-ionised water is run up through the bed from the bottom at a sufficient flow rate to loosen and suspend the exchanger granules. The enlarged upper portion of the exchange tube shown in Fig. VII, 3, *b* or *c*, will hold the resin suspension during washing. If a tube of uniform bore is used the volume of resin employed must be suitably adjusted or else a tube attached by a rubber bung to the top of the column; the tube dips into an open filter flask, the side arm of which acts as the overflow and is connected by rubber tubing to waste. When the wash water is clear the flow of water is stopped and the resin is allowed to settle in the tube. The excess of water is drained off; the water level must never fall below the surface of the resin, or else channelling will occur, with consequent incomplete contact between the resin and solutions used in subsequent operations. The apparatus with a side-arm outlet (Fig. VII, 3, *a*) has an advantage in this respect in that the resin will not run dry even if left unattended, since the outlet is above the surface of the resin.

Ion exchange resins (standard grades) as received from the manufacturers may contain unwanted ionic impurities and sometimes traces of water-soluble intermediates or incompletely polymerised material; these must be washed out before use. This is best done by passing $2M$-hydrochloric acid and $2M$-sodium hydroxide alternately through the column, with distilled-water rinsings in between, and then washing with water until the effluent is neutral and salt free. 'Analytical grade' and/or 'chromatographic grade' ion exchange resins that have undergone this preliminary washing are available commercially.

For analytical work the exchange resin of 'analytical' grade (Amberlite) or of 'chromatographic' grade (Permutit; Amberlite, etc.) of a particle size of 100–200 mesh is preferred. However, for student work, the 'standard' grade of resin of 50–100 or 15–50 mesh, which is less expensive, is generally satisfactory. The 'standard' grade of resin must, however, be conditioned before use. Cation exchange resins must be soaked in a beaker in about twice the volume of $2M$-hydrochloric acid for 30–60 minutes with occasional stirring; the fine particles are removed by decantation or by back-washing in a column with distilled or de-ionised water until the supernatant liquid is clear. Anion exchange resins may be washed with water in a beaker until the colour of the decanted wash liquid reaches a minimum intensity; they may then be transferred to a wide glass column and cycled between $1M$-hydrochloric acid and $1M$-alkali. Sodium hydroxide is used for strongly basic resins, and ammonia (preferably) or sodium carbonate for weakly basic resins. For all resins the final treatment should be with a solution leading to the resin in the desired ionic form.

A 50-cm^3 or 100-cm^3 burette, with Pyrex glass-wool plug or sintered-glass disc at the lower end, can generally be used for the determinations described below: alternatively, the column with side arm (Fig. VII, 3, *a*) is equally convenient in practice for student use. Reference will be made to the Permutit resins; the equivalent Amberlite or other resin (see Table VII, I in Section **VII, 1**) may of course be used.

VII, 8. DETERMINATION OF THE CAPACITY OF AN ION EXCHANGE RESIN (COLUMN METHOD). Cation exchange resin.

Dry the purified resin (e.g., Zerolit 225 in the hydrogen form) by placing it in an evaporating dish, cover with a clock glass supported on two glass rods to provide protection from dust while giving access to the air, and leave in a warm place (25–35 °C) until the resin is completely free-running (2–3 days). The capacity of the resulting resin remains constant over a long period if kept in a closed bottle. Drying at higher temperatures (say, 100 °C) is not recommended, owing to possible fracture of the resin beads.

Partly fill a small column, 15 cm × 1 cm (Fig. VII, 3, *a*) with distilled water, taking care to displace any trapped air from beneath the sintered-glass disc. Weigh out accurately about 0.5 g of the air-dried resin in a glass scoop and transfer it with the aid of a small camel-hair brush through a dry funnel into the column. Add sufficient distilled water to cover the resin. Dislodge any air bubbles that stick to the resin beads by applying an intermittent pressure to the rubber tubing, thus causing the level of the liquid in the column to rise and fall slightly. Adjust the level of the outlet tube so that the liquid in the column will drain to a level about 1 cm above the resin beads.

Fill a 250-cm^3 separatory funnel with *ca.* 0.25M-sodium sulphate solution. Allow this solution to drip into the column at a rate of about 2 cm^3 per minute, and collect the effluent in a 500-cm^3 conical flask. When all the solution has passed through the column, titrate the effluent with standard 0.1M-sodium hydroxide using phenolphthalein as indicator.

The reaction may be represented as:

$$2R^-H^+ + 2Na^+ \rightleftharpoons 2R^-Na^+ + 2H^+$$

and proceeds to completion because of the large excess and large volume of sodium sulphate solution passed through the column.

The capacity of the resin in milli-equivalents per gram is given by av/W, where a is the molarity of the sodium hydroxide solution, v is the volume in cm^3, and W is the weight (g) of the resin.

Anion exchange resin. Proceed as in the previous experiment using 1.0 g, accurately weighed, of the air-dried strongly basic anion exchanger (e.g., Zerolit FF, chloride form). Fill the 250-cm^3 separatory funnel with *ca.* 0.25M-sodium nitrate solution, and allow this solution to drop into the column at the rate of about 2 cm^3 per minute. Collect the effluent in a 500-cm^3 conical flask, and titrate with standard 0.1M-silver nitrate using potassium chromate as indicator.

The reaction which occurs may be written as:

$$R^+Cl^- + NO_3^- \rightleftharpoons R^+NO_3^- + Cl^-$$

The capacity of the resin expressed as milli-equivalents per gram is given by bv/W, where v cm^3 of bM AgNO$_3$ are required by W g of the resin.

VII, 9. SEPARATION OF ZINC AND MAGNESIUM ON AN ANION EXCHANGER. *Theory.*

Several metal ions (e.g., those of Fe, Al, Zn, Co, Mn, etc.) can be absorbed from hydrochloric acid solutions on anion exchange resins owing to the formation of negatively charged chloro complexes. Each metal is absorbed over a well-defined range of pH, and this property can be used as the basis of a method of separation. Zinc is absorbed from 2M-acid, while magnesium (and aluminium) are not; thus by passing a mixture of zinc and magnesium through a column of anion exchange resin a separation is effected. The zinc is subsequently eluted with dilute nitric acid.

Procedure. Prepare a column of the anion exchange resin using about 15 g of Zerolit FF in the chloride form (Section **VII, 7**). The column should be made up in 2M hydrochloric acid.

Prepare standard zinc (about 2.5 mg Zn/cm^3) and magnesium (about 1.5 mg Mg/cm^3) ion solutions by dissolving accurately weighed quantities of A.R. zinc shot and magnesium (for Grignard reaction) in 2M-hydrochloric acid and diluting each to volume in a 250-cm^3 graduated flask. Pipette 10.0 cm^3 of the zinc ion solution and 10.0 cm^3 of the magnesium ion solution into a small separatory funnel supported in the top of the ion exchange column, and mix the solutions. Allow the mixed solution to flow through the column at a rate of about 5 cm^3 per minute. Wash the funnel and column with 50 cm^3 of 2M-hydrochloric acid: do not permit the level of the liquid to fall below the top of the resin column. Collect all the effluent in a conical flask; this contains all the magnesium. Now change the receiver. Elute the zinc with 30 cm^3 of water, followed by 80 cm^3 of *ca.* 0.25M-nitric acid. Determine the magnesium and the zinc in the respective eluates by neutralisation with sodium hydroxide solution, followed by titration with standard EDTA solution using a buffer solution of pH = 10 and Solochrome Black indicator (Sections **X, 64, 67**).

The following results were obtained in a typical experiment:

Weight of zinc taken = 25.62 mg, found = 25.60 mg
Weight of magnesium taken = 14.95 mg, found = 14.89 mg

Magnesium may conveniently be determined by atomic absorption spectroscopy (Section **XXII, 22**) if a smaller amount (*ca.* 4 mg) is used for the separation. Collect the magnesium effluent in a 1 dm^3 graduated flask, dilute to the mark with de-ionised water and aspirate the solution into the flame of an atomic absorption spectrometer. Calibrate the instrument using standard magnesium solutions covering the range 2 to 8 p.p.m.

VII, 10. SEPARATION OF CHLORIDE AND BROMIDE ON AN ANION EXCHANGER. *Theory.* The anion exchange resin, originally in the chloride form, is converted into the nitrate form by washing with sodium nitrate solution. A concentrated solution of the chloride and bromide mixture is introduced at the top of the column. The halide ions exchange rapidly with the nitrate ions in the resin, forming a band at the top of the column. Chloride ion is more rapidly eluted from this band than bromide ion by sodium nitrate solution, so that a separation is possible. The progress of elution of the halides is followed by titrating fractions of the effluents with standard silver nitrate solution.

Procedure. Prepare an anion exchange column (Section **VII, 7**) using about 40 g of Zerolit FF (chloride form). The ion exchange tube may be 16 cm long and about 12 mm internal diameter. Wash the column with 0.6*M*-sodium nitrate until the effluent contains no chloride ion (silver nitrate test) and then wash with 50 cm^3 of 0.3*M*-sodium nitrate.

Weigh out accurately about 0.10 g of A.R. sodium chloride and about 0.20 g of A.R. potassium bromide, dissolve in about 2.0 cm^3 of water and transfer quantitatively to the top of the column with the aid of 0.3*M*-sodium nitrate. Pass 0.3*M*-sodium nitrate through the column at a flow rate of about 1 cm^3 per minute and collect the effluent in 10-cm^3 fractions. Transfer each fraction in turn to a conical flask, dilute with an equal volume of water, add 2 drops of 0.2*M*-potassium chromate solution and titrate with standard 0.02*M*-silver nitrate.

Before commencing the elution titrate 10.0 cm^3 of the 0.3*M*-sodium nitrate with the standard silver nitrate solution, and retain the product of the blank titration for comparing with the colour in the actual titrations of the eluates. When the titre of the eluate falls almost to zero (i.e., nearly equal to the blank titration)—*ca.* 150 cm^3 of effluent—elute the column with 0.6*M*-sodium nitrate. Titrate as before until no more bromide is detected (titre almost zero). A new blank titration must be made with 10.0 cm^3 of the 0.6*M*-sodium nitrate.

Plot a graph of the total effluent collected against the concentration of halide in each fraction (millimols per litre). The sum of the titres using 0.3*M*-sodium nitrate eluant (less blank for *each* titration) corresponds to the chloride, and the parallel figure with 0.6*M*-sodium nitrate corresponds to the bromide recovery.

A typical experiment gave the following results:

Weight of sodium chloride used = 0.1012 g ≡ 61.37 mg Cl$^-$
Weight of potassium bromide used = 0.1934 g ≡ 129.87 mg Br$^-$
Concentration of silver nitrate solution = 0.01936*M*

Cl$^-$: total titres (less blanks) = 89.54 cm^3 ≡ 61.47 mg
Br$^-$: total titres (less blanks) = 83.65 cm^3 ≡ 129.4 mg

VII, 11. DETERMINATION OF THE TOTAL CATION CONCENTRATION IN WATER.

Theory. The following procedure is a rapid one for the determination of the total cations present in water, particularly that used for industrial ion exchange plant, but may be used for all samples of water, including tap-water. When water containing dissolved ionised solids is passed through a cation exchanger in the hydrogen form all cations are removed and replaced by hydrogen ions. By this means any alkalinity present in the water is destroyed, and the neutral salts present in solution are converted into the corresponding mineral acids. The effluent is titrated with 0.02M-sodium hydroxide using screened methyl orange as indicator.

Procedure. Prepare a 25–30-cm column of Zerolit 225 in a 14–16-mm chromatographic tube (Section **VII, 7**). Pass 250 cm^3 of 2M-hydrochloric acid through the tube during about 30 minutes; rinse the column with distilled water until the effluent is just alkaline to screened methyl orange or until a 10-cm^3 portion of the effluent does not require more than one drop of 0.02M-sodium hydroxide to give an alkaline reaction to bromothymol blue indicator. The resin is now ready for use: the level of the water should never be permitted to drop below the upper surface of the resin in the column. Pass 50.0 cm^3 of the sample of water under test through the column at a rate of 3–4 cm^3 per minute, and discard the effluent. Now pass two 100.0-cm^3 portions through the column at the same rate, collect the effluents separately, and titrate each with standard 0.02M-sodium hydroxide using screened methyl orange as indicator. After the determination has been completed, pass 100–150 cm^3 of distilled or de-ionised water through the column.

From the results of the titration calculate the milli-equivalents of calcium present in the water. It may be expressed, if desired, as the equivalent mineral acidity (E.M.A.) in terms of mg CaCO$_3$ per dm^3 of water (i.e., parts per million of CaCO$_3$). In general, if the titre is A cm^3 of sodium hydroxide of molarity B for an aliquot volume of V cm^3, the E.M.A. is given by $(AB \times 50 \times 1000)/V$.

Commercial samples of water are frequently alkaline due to the presence of hydrogen carbonates, carbonates, or hydroxides. The alkalinity is determined by titrating a 100.0-cm^3 sample with 0.02M-hydrochloric acid using screened methyl orange as indicator (or to a pH of 3.8). To obtain the total cation content in terms of CaCO$_3$, the total methyl orange alkalinity is added to the E.M.A.

VII, 12. SEPARATION OF COBALT AND NICKEL ON AN ANION EXCHANGER.

Theory. The separation is based upon the fact that cobalt, but not nickel, forms a monovalent complex anion (probably [CoCl$_3$]$^-$) in 9M-hydrochloric acid, and this anion is rapidly extracted from the solution by a strongly basic anion exchanger, such as Zerolit FF. The nickel is not retained by the resin, presumably because of the instability of the anionic chloro complex, and can be washed out of the column with 9M-hydrochloric acid. Upon washing the column with water, the cobalt complex is decomposed and passes out in the effluent as cobalt(II) chloride. The nickel and cobalt in the respective effluents may be determined, after evaporation of the excess of hydrochloric acid, by titration with EDTA.

Reagents. *Anion exchange column.* Prepare an anion exchange column using 25–30 g of Zerolit FF (chloride form). Mix the resin with about 100 cm³ of water in a measuring cylinder and shake for a few minutes: decant the liquid as soon as the larger particles have settled. The volume of the resin should be about 25 cm³. Stir the resin with distilled water, allow to settle, and decant the supernatant liquid: repeat the process until the supernatant liquid is clear. Transfer the resin slurry to a burette containing a plug of glass wool until a column of well-packed resin about 22 cm long is obtained; alternatively, use an ion exchange tube (see Fig. VII, 3, *a*). Wash the resin in the column once with water. Do not allow the level of the liquid in the column to fall below the upper surface of the resin: the level should preferably be about 1 cm above it.

Cobalt-ion solution. Dissolve 5.0 g A.R. hydrated cobalt(II) chloride in 9M-hydrochloric acid and dilute to 250 cm³ with 9M-hydrochloric acid.

Nickel-ion solution. Dissolve 2.5 g pure nickel carbonate in 9M-hydrochloric acid and dilute to 250-cm³ with 9M-hydrochloric acid.

Procedure. Pass 50 cm³ of 9M-hydrochloric acid through the column and drain to almost bed level. Mix 10.0 cm³ of each of the cobalt and nickel-ion solutions in a small beaker, transfer 10.0 cm³ of the mixed solution with the aid of a pipette to the top of the resin column, and lower this solution to the upper part of the column with a little 9M-hydrochloric acid. Pass 100 cm³ of 9M-hydrochloric acid through the column in order to elute the nickel; collect the eluate in a 400-cm³ beaker. Concentrate the eluate to a small volume on a wire gauze (FUME CUPBOARD!) in order to remove the excess of acid. Neutralise the resulting solution with A.R. potassium hydroxide, dilute to 100 cm³ with distilled water, add 10 cm³ of buffer solution (prepared by mixing equal volumes of 1M-NH$_4$Cl and 1M-aqueous ammonia), about 15 drops of Bromopyrogallol Red indicator solution* and titrate with standard 0.02M-EDTA until the colour changes from blue to wine red {see Section **X, 58**(*b*)}. Perform a similar titration with 5.00 cm³ of the original nickel-ion solution.

After the nickel has been eluted from the column, pass 150 cm³ of water at the rate of 4–5 cm³ per minute through the resin to decompose the anionic cobalt chloro complex, and collect the effluent in a small beaker. Concentrate the effluent to a small volume, partly neutralise with A.R. potassium hydroxide, and adjust the pH of the solution to about 6 by the addition of powdered hexamine. Add a few milligrams of Xylenol Orange indicator,† warm to about 60 °C, and then titrate with standard 0.02M-EDTA (slowly near the end-point) until the colour changes from red to orange-yellow. Perform a similar titration upon 5.00 cm³ of the original cobalt-ion solution.

Compare the amounts of nickel and cobalt recovered with those actually used. Some typical results are given below.

5.00 cm³ of the original nickel-ion solution required 20.15 cm³ of 0.02M-EDTA. The nickel-ion solution recovered after passage through the column required 20.05 cm³ of 0.02M-EDTA.

5.00 cm³ of the original cobalt-ion solution required 21.45 cm³ of 0.02M-EDTA. The cobalt-ion solution recovered from the column required 21.35 cm³ of 0.02M-EDTA.

---

* Full details of this indicator are given in Section **X, 28**.

† This indicator is used as a solid mixture; details are given in Section **X, 28**.

VII, 13.   SEPARATION OF CADMIUM AND ZINC ON AN ANION EXCHANGER.   *Theory.*   Cadmium and zinc form negatively charged chloro complexes which are absorbed by a strongly basic anion exchange resin, such as Zerolit FF. The maximum absorption of cadmium and zinc is obtained in 0.12M-hydrochloric acid containing 100 g of sodium chloride per dm^3. The zinc is eluted quantitatively by a 2M-sodium hydroxide solution containing 20 g of sodium chloride per dm^3, while the cadmium is retained on the resin. Finally, the cadmium is eluted with 1M-nitric acid. The zinc and cadmium in their respective effluents may be determined by titration with standard EDTA.

Elements such as Fe(III), Mn, Al, Bi, Ni, Co, Cr, Cu, Ti, the alkaline-earth metals, and the lanthanoids are not absorbed on the resin in the HCl–NaCl medium.

Reagents.   *Anion exchange column.*   Prepare an anion exchange column using 25–30 g of Zerolit FF (chloride form) following the experimental details given in Section **VII, 12**. Allow the resin to settle in 0.5M-hydrochloric acid. Transfer the resin slurry to the column: after settling, the resin column should be about 20 cm in length if a 50-cm^3 burette is used.

Reagent I.   This consists of 0.12M-hydrochloric acid containing 100 g of A.R. sodium chloride per dm^3.

Reagent II.   This consists of 2M-sodium hydroxide containing 20 g of A.R. sodium chloride per dm^3.

Zinc-ion solution.   Dissolve about 7.0 g of A.R. zinc sulphate heptahydrate in 25 cm^3 of Reagent I.

Cadmium-ion solution.   Dissolve about 6.0 g of A.R. crystallised cadmium sulphate in 25 cm^3 of Reagent I.

EDTA solution, 0.01M.   See Section **X, 50**.

Buffer solution, pH = 10.   Dissolve 7.0 g of A.R. ammonium chloride and 57 cm^3 of concentrated ammonia solution (sp. gr. 0.88) in water and dilute to 100 cm^3.

Solochrome Black indicator mixture.   Triturate 0.20 g of the solid dyestuff with 50 g of A.R. potassium chloride.

Xylenol Orange indicator.   Triturate 0.20 g of the solid dyestuff with 50 g of A.R. potassium chloride (or nitrate). This solid mixture is used because solutions of Xylenol Orange are not very stable.

Nitric acid, ca. 1M.

Procedure.   Wash the anion exchange column with two 20-cm^3 portions of Reagent I; drain the solution to about 0.5 cm above the top of the resin. Mix thoroughly equal volumes (2.00 cm^3 each) of the zinc- and cadmium-ion solutions and transfer by means of a pipette 2.00 cm^3 of the mixed solution to the top of the resin column. Allow the solution to drain to within about 0.5 cm of the top of the resin and wash down the tube above the resin with a little of Reagent I. Pass 150 cm^3 of Reagent II through the column at a flow rate of about 4 cm^3 per minute and collect the eluate (containing the zinc) in a 250-cm^3 graduated flask; dilute to volume with water. Wash the resin with about 50 cm^3 of water to remove most of the sodium hydroxide solution. Now place a 250-cm^3 graduated flask in position as receiver and pass 150 cm^3 of 1M-nitric acid through the column at a rate of about 4 cm^3 per minute; the cadmium will be eluted. Dilute the effluent to 250 cm^3 with distilled water.

The resin may be regenerated by passing Reagent I through the column, and can then be used again for analysis of another Zn–Cd sample.

Analyses. (*a*) Original zinc-ion solution. Dilute 2.00 cm^3 (pipette) to 100 cm^3 in a graduated flask. Pipette 10.0 cm^3 of the diluted solution into a 250-cm^3 conical flask, add *ca*. 90 cm^3 of water, 2 cm^3 of the buffer solution, and sufficient of the Solochrome Black indicator mixture to impart a pronounced red colour to the solution. Titrate with standard 0.01M-EDTA to a pure blue colour (see Section **X, 61**).

(*b*) Zinc-ion eluate. Pipette 50.0 cm^3 of the solution into a 250-cm^3 conical flask, neutralise with hydrochloric acid, and dilute to about 100 cm^3 with water. Add 2 cm^3 of the buffer mixture, then a little Solochrome Black indicator powder, and titrate with standard 0.01M-EDTA until the colour changes from red to pure blue.

(*c*) Original cadmium-ion solution. Dilute 2.00 cm^3 (pipette) to 100 cm^3 in a graduated flask. Pipette 10.0 cm^3 of the diluted solution into a 250-cm^3 conical flask, add *ca*. 40 cm^3 of water, followed by solid hexamine and a few milligrams of Xylenol Orange indicator. If the pH is correct (5–6) the solution will have a pronounced red colour (see Section **X, 61**). Titrate with standard 0.01M-EDTA until the colour changes from red to clear orange-yellow.*

(*d*) Cadmium-ion eluate. Pipette 50.0 cm^3 of the solution into a conical flask, and partially neutralise (to pH 3–4) with aqueous sodium hydroxide. Add solid hexamine (to give a pH of 5–6) and a little Xylenol Orange indicator. Titrate with standard 0.01M-EDTA to a colour change from red to clear orange-yellow.

Some typical results are given below.

0.200 cm^3 of original Zn^{2+} solution required 17.50 cm^3 of 0.01038M-EDTA
∴ Weight of Zn^{2+} per cm^3 = 17.50 × 5 × 0.01038 × 65.38 = 59.35 mg
50.0 cm^3 of Zn^{2+} eluate ≡ 17.45 cm^3 of 0.01038M-EDTA.
∴ Zn^{2+} recovered = 5 × 17.45 × 0.01038 × 65.38 = 59.21 mg
0.200 cm^3 of original Cd^{2+} solution required 19.27 cm^3 of 0.01038M-EDTA.
∴ Weight of Cd^{2+} per cm^3 = 5 × 19.27 × 0.01038 × 112.4 = 112.4 mg
50.0 cm^3 of Cd^{2+} eluate ≡ 19.35 cm^3 of 0.01038M-EDTA.
∴ Cd^{2+} recovered = 5 × 19.35 × 0.01038 × 112.4 = 112.8 mg

VII, 14.   DETERMINATION OF FLUORIDE WITH THE AID OF A CATION EXCHANGER.

Theory. Soluble metallic fluorides may be analysed by passing an aqueous solution through a cation exchange column in a polythene tube, collecting the liberated hydrofluoric acid in a polythene beaker, and titrating it with standard sodium hydroxide solution.

The student may determine the fluoride content of sodium fluoride to gain experience in the determination.

Procedure.   Obtain a polythene tube, 125 cm long and 12 mm internal diameter, provided with a nozzle at the lower end. Fill the tube above the nozzle with short lengths of polythene tubing (15 mm × 2 mm) stacked vertically to provide a support for the resin. Attach a short length of thin-walled PVC tubing to the jet outlet, and then attach a length of about 10 cm polythene tubing (6 mm internal and 10 mm external diameter) to the latter. Attach a pinch-cock or screw clip to the thin-walled PVC tubing; this will enable the flow of liquid to be stopped at will. Charge the column in the usual manner with Zerolit 225,

---

* The solution may also be titrated at pH = 10 using Solochrome Black as indicator.

hydrogen form (volume about 15 cm^3); leave the column full of water to just above the bed of resin. Prepare a *ca.* 0.1*M*-sodium fluoride solution, using an accurately weighed amount of the dry A.R. salt. Pass 25.0 cm^3 of this solution through the column followed by 4 × 15 cm^3 of boiled-out distilled water (or deionised water) and collect the effluent in a polythene beaker. Maintain a rate of flow of about 4 cm^3 per minute. Titrate the total effluent with standard 0.1*M*-sodium hydroxide, using phenol red or phenolphthalein as indicator.

Calculate the fluoride content of the sample of sodium fluoride.

VII, 15.   DETERMINATION OF SULPHUR IN IRON PYRITES WITH THE AID OF A CATION EXCHANGER.   *Theory.*   The sample of iron pyrites is dissolved in a mixture of concentrated nitric and hydrochloric acids. After dilution (and filtration, if necessary), the solution is passed through a cation exchanger (sulphonic acid type) in the hydrogen form. The effluent contains hydrogen ion as the only cation. The sulphate is determined by precipitation as barium sulphate. The barium sulphate may either be weighed or dissolved in excess of standard EDTA solution and the excess titrated with standard magnesium chloride solution (Section **X, 75**).

Reagents.   *Barium chloride solution, ca. 0.05*M.   Prepare from the A.R. solid.

*EDTA solution, 0.05*M.   See Section **X, 50**.

*Magnesium chloride, 0.05*M.   Prepare from pure magnesium (Section **X, 62**).

Buffer solution, pH = 10.   Add 7.0 g of A.R. ammonium chloride to 57 cm^3 of concentrated ammonia solution (sp. gr. 0.88) and dilute to 100 cm^3 with water.

Solochrome Black indicator.   See Section **X, 28**.

Procedure.   Weigh out accurately about 0.50 g of iron pyrites* and treat it with 10 cm^3 of a mixture of 3 volumes of concentrated nitric acid and 1 volume of concentrated hydrochloric acid in a 250-cm^3 beaker. Allow the reaction to proceed at room temperature for 30 minutes, then warm the covered beaker on a steam bath until all reaction appears to cease, remove the clock-glass cover, and evaporate the solution to dryness. Treat the residue with 5 cm^3 of concentrated hydrochloric acid and evaporate to dryness again. Dissolve the residue in 1–2 cm^3 of warm concentrated hydrochloric acid, dilute to about 100 cm^3 with hot water, and filter through a sintered-glass crucible (G3). Wash the residue with hot water, and combine the washings with the filtrate.

Percolate the combined solutions through a 25-cm column (contained in a burette or tube with overflow as in Fig. VII, 3, *a*) of a cation exchange resin (e.g., Zerolit 225) in the hydrogen form; pass water through the column until the effluent is neutral. Maintain a flow rate of about 3 cm^3 per minute and collect the effluent in a 500-cm^3 graduated flask. Finally, dilute the solution to the mark with water. Pipette 25.0 cm^3 of the solution into a 250-cm^3 beaker, dilute to 50 cm^3, heat to boiling, and add a slight excess of 0.05*M*-barium chloride (about 12 cm^3) with stirring. Keep on a steam bath for 1 hour. Filter through a filter-paper disc (Whatman No. 542) supported on a Gooch porcelain crucible, and wash the precipitate with cold water. Transfer the precipitate and filter paper back to the

---

* Iron Pyrites, No. 44G (one of the British Chemical Standards) may be used for practice in this determination.

original beaker. Introduce 35.0 cm^3 of standard 0.05M-EDTA into the beaker, followed by 5 cm^3 of concentrated ammonia solution: boil gently until the precipitate dissolves (about 10 minutes). Dilute the clear solution to 100 cm^3, add 4 cm^3 of buffer solution and a few drops of Solochrome Black indicator. Titrate the excess of EDTA with standard 0.05M-magnesium chloride until the colour changes from blue to wine red.

Calculate the percentage of sulphur in the sample of iron pyrites.

Some typical results are given below.

Weight of iron pyrites = 0.5001 g
EDTA solution = 0.04697M. MgCl$_2$ solution = 0.05160M.
Volume of EDTA solution = 35.00 cm^3
Mean titre of excess of EDTA = 26.75 cm^3 of MgCl$_2$ solution.
∴ S present in 25.00 cm^3 of solution =
    $\{(35.00 \times 0.04967) - (26.75 \times 0.05160)\} \times 32.06 = 0.3582 \times 32.06$ mg
∴ Per cent of S in iron pyrites = $(0.3582 \times 32.06 \times 20 \times 100)/500.1 = 45.9_3$.

The analysed Ridsdale sample contained 46.1 per cent S.

VII, 16. SEPARATION OF COBALT AND URANIUM FROM MIXED AQUEOUS–ORGANIC SOLVENT USING A CATION EXCHANGE RESIN.

Theory. The increased selectivity of ion exchange resins which may be achieved by the use of mixed aqueous–organic solvent systems is illustrated by the separation of uranyl ion from cobalt(II) using a strong acid cation exchange resin. It has been shown (Ref. 6) that uranium, as UO_2^{2+}, in a mixture composed of 90 per cent tetrahydrofuran and 10 per cent 6M nitric acid (v/v) has a much lower distribution coefficient than have most other di- and even higher-valent ions. This forms the basis of the separation of small amounts of cobalt from relatively large amounts of uranium. The low distribution coefficient of UO_2^{2+} is probably due to the formation of an anionic nitrato complex $UO_2(NO_3)_3^-$ which may be eluted from the cation exchange resin as an ion association complex with tetrahydrofuran.

$$\begin{array}{c} CH_2-CH_2 \\ | \quad\quad >O + HNO_3 \\ CH_2-CH_2 \end{array} \rightleftharpoons \left[\begin{array}{c} CH_2-CH_2 \\ | \quad\quad >O-H \\ CH_2-CH_2 \end{array}\right]^+ NO_3^-$$

$$(THFH)^+$$

$$(THFH)^+NO_3^- + UO_2(NO_3)_2 \rightleftharpoons (THFH)^+(UO_2(NO_3)_3)^-$$

Cobalt is retained on the resin and can subsequently be eluted with a mixture of 90 per cent tetrahydrofuran and 10 per cent 6M-hydrochloric acid.

Reagents. *Mixed solvent (A).* 90 per cent tetrahydrofuran + 10 per cent 6M-nitric acid. Prepare from pure reagents.

Mixed solvent (B). 90 per cent tetrahydrofuran + 10 per cent 6M-hydrochloric acid. Prepare from pure reagents.

Cation exchange resin. Zerolit 225 (H$^+$ form).

Sample solution. Dissolve 2 g uranyl nitrate hexahydrate and 5 mg cobalt(II) in 10 cm^3 of mixed solvent (A).

Procedure. Equilibrate the resin (*ca.* 5 g) by allowing it to stand in the mixed solvent(A), *ca.* 20 cm^3, for about 30 minutes. Prepare a small column (5 cm × 1.0 cm) of the resin and introduce the sample solution on to the top of the resin column. Elute uranium with the mixed solvent(A) at a flow rate of about 2 cm^3

min^{-1} until the eluate is no longer yellow (50 cm^3 of mixed solvent should be sufficient).

With the mixed solvent(B) (*ca.* 100 cm^3) elute the cobalt using a similar flow rate. Collect the blue eluate in a 100 cm^3 graduated flask and make up to the mark with more mixed solvent(B).

Measure the absorbance of the cobalt solution against the mixed solvent(B) at 675 nm. Calibrate the spectrophotometer using solutions of 2, 4, 6, 8, 10 mg of cobalt(II) in 100 cm^3 of mixed solvent(B).

VII, 17. DETERMINATION OF URANIUM WITH THE AID OF A LIQUID ANION EXCHANGER. *Theory.*

The formation of an anionic sulphato complex by uranium(VI) in relatively dilute sulphuric acid solution provides the basis for separation of uranium from solutions containing high concentrations of iron salts. Uranium is extracted from a sulphuric acid solution using a chloroform solution of the liquid anion exchanger, Amberlite LA.1. Back extraction of uranium with sodium carbonate solution gives an alkaline solution which reacts with hydrogen peroxide to give yellow peruranate. This selective reaction enables uranium (10–100 mg) to be determined spectrophotometrically by measuring the absorbance of the solution at 410 nm (Ref. 13).

Reagents. *Sodium carbonate solution,* 100 g dm^{-3}. Prepare from A.R. solid and de-ionised water.

Hydrogen peroxide, 20 vols.

Liquid anion exchanger solution. Dissolve 4 cm^3 of Amberlite LA.1 in pure chloroform and make up to 100 cm^3 with this solvent.

Uranium solution. Prepare a standard uranium solution by dissolving 0.524 g A.R. uranyl nitrate hexahydrate, $UO_2(NO_3)_2.6H_2O$, in 250 cm^3 0.5M-sulphuric acid.

Procedure. Add increments of 10, 20, 30, 40 cm^3 of the standard uranium solution to beakers each containing 90 cm^3 of sodium carbonate solution. Heat each solution and boil for about 5 minutes, cool and dilute to 200 cm^3 in a graduated flask. Transfer a 25 cm^3 aliquot in each case to a 50 cm^3 graduated flask, add 5 cm^3 of hydrogen peroxide (20 vol.) and dilute the solution with water to the mark. Measure the absorbances of the solutions at 410 nm using a 2 cm cell (it is advisable to de-gas the solutions by shaking them thoroughly prior to measuring their absorbances). Prepare a calibration graph of absorbance against uranium concentration.

Prepare a sample solution containing approximately 25 mg of uranium(VI) and 500 mg iron(III), (5 g ammonium iron(III) sulphate may conveniently be used for this second ion), in 50 cm^3 of 1M sulphuric acid and dilute to 100 cm^3 with water. Transfer the solution to a separating funnel (250 cm^3), add 30 cm^3 of anion exchanger solution and shake for 30 seconds. Allow the two phases to separate and run the lower chloroform layer into a clean beaker. Repeat the extraction of the aqueous phase with two further 30 cm^3 portions of anion exchanger solution and combine the three chloroform extracts. Wash out the separating funnel and return the combined extracts to it. Add 30 cm^3 of sodium carbonate solution and shake for 30 seconds. Allow the two phases to separate, run the lower chloroform layer back into the original beaker and transfer the sodium carbonate extract to a clean beaker. Return the chloroform solution to the funnel, add another 30 cm^3 of sodium carbonate solution and repeat the process. Make a third sodium carbonate extraction and combine the sodium carbonate extracts.

Heat the combined solutions to boiling and boil for about 5 minutes, cool and dilute to 200 cm^3 in a graduated flask. Transfer a 25 cm^3 aliquot (if necessary filter the solution through a dry Whatman No. 541 paper) to a 50 cm^3 graduated flask, add 5 cm^3 of hydrogen peroxide (20 vol.) and dilute the solution with water to the mark. Measure the absorbance of the solution against water at 410 nm using a 2 cm cell. Compare the value obtained with the calibration graph previously prepared.

VII, 18. CONCENTRATION OF COPPER(II) IONS FROM A BRINE SOLUTION USING A CHELATING ION EXCHANGE RESIN. *Theory.* Conventional anion and cation exchange resins appear to be of limited use for concentrating trace metals from saline solutions such as sea water. The introduction of chelating resins, particularly those based on iminodiacetic acid, makes it possible to concentrate trace metals from brine solutions and separate them from the major components of the solution. Thus the elements cadmium, copper, cobalt, nickel and zinc are selectively retained by the resin Chelex-100 and can be recovered subsequently for determination by atomic absorption spectrophotometry (Ref. 14). To enhance the sensitivity of the AAS procedure the eluate is evaporated to dryness and the residue dissolved in 90 per cent aqueous acetone.* The use of the chelating resin offers the advantage over concentration by solvent extraction that, in principle, there is no limit to the volume of sample which can be used.

Reagents. *Standard copper(II) solutions.* Dissolve 100 mg of spectroscopically pure copper metal in a slight excess of nitric acid and dilute to 1 dm^3 in a graduated flask with de-ionised water. Pipette a 10 cm^3 aliquot into a 100 cm^3 graduated flask and make up to the mark with acetone (A.R.); the resultant solution contains 10 μg of copper per cm^3. Use this stock solution to prepare a series of standard solutions containing 1.0–5.0 μg of copper per cm^3, each solution being 90 per cent with respect to acetone.

Sample solution. Prepare a sample solution containing 100 μg of copper(II) in 1 dm^3 of 0.5M-sodium chloride solution in a graduated flask.

Ion exchange column. Prepare the Chelex-100 resin (100–500 mesh) by digesting it with excess (about 2–3 bed-volumes) of 2M-nitric acid at room temperature. Repeat this process twice and then transfer sufficient resin to fill a 1.0 cm diameter column to a depth of 8 cm. Wash the resin column with several bed-volumes of de-ionised water.

Procedure. Allow the whole of the sample solution (1 dm^3) to flow through the resin column at a rate not exceeding 5 cm^3 min^{-1}. Wash the column with 250 cm^3 of de-ionised water and reject the washings. Elute the copper(II) ions with 30 cm^3 of 2M-nitric acid, place the eluate in a small conical flask (100 cm^3, preferably silica) and evaporate carefully to dryness on a hot plate (use a low temperature setting). Dissolve the residue in 1 cm^3 of 0.1M-nitric acid introduced by pipette and then add 9 cm^3 of acetone. Determine copper in the resulting solution using an atomic absorption spectrophotometer which has been calibrated using the standard copper(II) solutions.

Note. All glass and silica apparatus to be used should be allowed to stand

---

* In the illustrative experiment described here, copper(II) ions in a brine solution are concentrated from 0.1 p.p.m. to about 3.3 p.p.m. prior to determination by atomic absorption spectrophotometry.

overnight filled with a 1:1 mixture of concentrated nitric and sulphuric acids and then thoroughly rinsed with de-ionised water. This treatment effectively removes traces of metal ions.

VII, 19. References

1. C. B. Amphlett (1964). *Inorganic Ion Exchangers*. Amsterdam; Elsevier.
2. J. Inczédy (1966). *Analytical Applications of Ion Exchangers*. Pergamon Press. 1st English edition.
3. L. R. Snyder (1965). *Chromatog. Rev.*, **7**, 1.
4. G. J. Moody and J. D. R. Thomas (1968). *Analyst*, **93**, 557.
5. W. R. Heumann (1971). 'Ion exchange in non-aqueous and mixed media', *CRC crit. Rev. analyt. Chem.*, **2**, 425.
6. J. Korkisch (1966). 'Combined ion exchange-solvent extraction (CISE): A novel separation technique for inorganic ions', *Separation Science*, **1** (2, 3), 159.
7. H. P. Gregor *et al.* (1952). *Ind. Eng. Chem.*, **44**, 2834.
8. E. Blasius and B. Brozio (1967). 'Chelating Ion-exchange Resins', in (ed. H. A. Flashka and A. J. Barnard), *Chelates in Analytical Chemistry*, Vol. 1, p. 49. New York; Marcel Dekker.
9. G. Schmuckler (1965). 'Chelating Resins—Their Analytical Properties and Applications', *Talanta*, **12**, 281.
10. C. F. Coleman *et al.* (1962). *Talanta*, **9**, 297.
11. H. Green (1964). 'Recent Uses of Liquid Ion Exchangers in Inorganic Analysis', *Talanta*, **11**, 1561.
12. H. Green (1973). 'Use of Liquid Ion-exchangers in Inorganic Analysis', *Talanta*, **20**, 139.
13. H. Green (1964). 'Determination of Uranium in Cast Iron', *BCIRA Journal*, **12**, 632.
14. J. P. Riley and D. Taylor (1968). 'Chelating Resins for the Concentration of Trace Elements from Sea Water and Their Analytical Use in Conjunction with Atomic Absorption Spectrophotometry', *Anal. Chim. Acta*, **40**, 479.

VII, 20. Selected bibliography

1. O. Samuelson (1962). *Ion Exchanger Separations in Analytical Chemistry*. New York; John Wiley.
2. G. H. Osborn (1961). *Synthetic Ion-Exchangers*. 2nd edn. London; Chapman and Hall.
3. E. Lederer and M. Lederer (1957). *Chromatography*. Division II. 'Ion Exchange Chromatography.' Amsterdam; Elsevier.
4. L. Meites and H. C. Thomas (1958). *Advanced Analytical Chemistry*. Ch. II. Ion Exchange and Chromatographic Methods. New York; McGraw-Hill Book Co.
5. R. Kunin (1958). *Ion Exchange Resins*. 2nd edn. New York; John Wiley.
6. J. E. Salmon and D. K. Hale (1959). *Ion Exchange. A Laboratory Manual*. London; Butterworths.
7. G. W. Ewing (1960). *Instrumental Methods of Chemical Analysis*. Ch. 19. Ion Exchange. 2nd edn. New York; McGraw-Hill Book Co.
8. *Ion Exchange Resins* (1971). 5th Edition, 3rd Impression (revised). Poole, Dorset; The British Drug Houses Ltd.
9. W. Rieman and R. Sargent (1961). *Ion Exchange*, in W. G. Berl. *Physical Methods in Chemical Analysis*. Vol. 4. New York; Academic Press.
10. R. Kunin, F. X. McGarvey, and A. Farren (1956). 'Ion Exchange', *Analytical Chemistry*, **28**, 729.
11. R. Kunin, F. X. McGarvey, and D. Zobian (1958). 'Ion Exchange', *Analytical Chemistry*, **30**, 681.
12. R. Kunin (1960). 'Ion Exchange', *Analytical Chemistry*, **32**, 67R.

13. J. A. Marinsky and Y. Marcus (eds.) (1973). 'Ion Exchange and Solvent Extraction—A Series of Advances', New York; Marcel Dekker.
14. F. Helfferich (1962). 'Ion Exchange', New York; McGraw-Hill.
15. M. Qureshi *et al.* (1972). 'Recent Progress in Ion-exchange Studies on Insoluble Salts of Polybasic Metals', *Separation Science*, **7**, 615.
16. W. Riemann and H. F. Walton (1970). *Ion Exchange in Analytical Chemistry*. Oxford; Pergamon Press.
17. J. Inczédy (1972). 'Use of ion exchangers in analytical Chemistry', *Rev. analyt. Chem.*, **1**, 157.

CHAPTER VIII
PAPER, THIN LAYER AND COLUMN CHROMATOGRAPHY

VIII, 1. GENERAL INTRODUCTION. Chromatography has been defined as primarily a separation process which is used for the separation of essentially molecular mixtures. It depends upon the redistribution of the molecules of the mixture between two or more phases. The various types of chromatography include adsorption chromatography, fluid partition chromatography, and ion exchange. The main systems employed in partition chromatography are: gas partition (see Chapter IX), liquid partition employing fixed beds (i.e. column chromatography), thin layer and paper chromatography. In each case distribution takes place between a 'stationary' sorbed 'liquid' phase and a mobile fluid in intimate contact with it. In liquid partition chromatography, a mobile liquid phase flows over an essentially stationary liquid phase sorbed on a support; in paper chromatography the support is paper or treated paper, whereas in thin layer chromatography the adsorbent is coated on a glass plate or plastic foil. We shall deal only with selected aspects of partition chromatography upon cellulose with particular reference to inorganic analysis.

The simplified technique for paper chromatography will be evident from the following description of a classical experiment designed to separate a mixture of amino acids. A strip of Whatman No. 1 filter paper, about 25–30 cm long and 1.5 cm wide, is marked lightly with a pencil line about 5 cm from one end. The mixture, containing 5–15 micrograms (μg) each of glycine, alanine, valine, and leucine in 2–4 microlitres (μl) of total solution, is spotted from a capillary pipette on to a marked spot A in the middle of the pencil line (Fig. VIII, 1a). The solvent is allowed to evaporate. The developer is prepared by shaking together liquified phenol and water in about equal quantities in a separatory funnel for several minutes; the phases are allowed to separate cleanly, and the upper aqueous layer is drawn off into the dish, which is placed in the bottom of the large gas jar (Fig. VIII, 1b—not drawn to scale). The paper is hung in the gas jar with the upper end held in the glass trough. The liquid is introduced to saturate the air in the gas jar with water and phenol in the ratio that will be in equilibrium with the developer; after a while the paper becomes conditioned to the reagents. (This conditioning of the paper is not usually necessary in inorganic chromatography.) The developer, which is the lower phenolic phase, is now introduced into the glass trough and the gas jar is closed. The developer moves by capillary action into the paper and, aided by gravity, passes down over the mixture at A, and development proceeds. The bulk mobile phase is phenol containing water, and the thin stationary phase is water (containing some phenol) sorbed on the paper. After

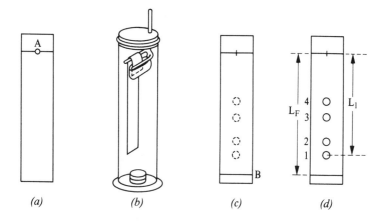

(a)   *(b)*   *(c)*   *(d)*

Fig. VIII, 1

the front of the developer has moved to almost the lower edge of the paper, the gas jar is opened, the paper removed, and the position of the front B marked immediately (Fig. VIII, 1c). The paper is then allowed to dry. The amino acids are colourless and, in order to reveal the position of the zones, a colour reaction with ninhydrin is used. The paper is sprayed with a 0.1 per cent solution of ninhydrin in butanol and warmed for a short time to accelerate the reaction with the amino acids. The amino acid zones appear as red-purple spots (Fig. VIII, 1d). The resulting chromatogram is described and the zones are characterised by R_F values. The R_F value is defined by the relation:

$$R_F = \frac{\text{distance (cm) from starting line to centre of zone}}{\text{distance (cm) from starting line to solvent front}}$$

The R_F value measures the velocity of movement of the zone relative to that of the developer front. The measurement is made by measuring the distances from the starting line (centre of initial mixed zone) to the developer front and the centre of density of each zone: thus for zone 1, $R_F = L_1/L_F$ (Fig. VIII, 1d). The R_F values will identify the amino acids, and the intensity of the zone may be used as a measure of the concentration by comparison with standard spots.

Chromatography on cellulose is basically a solvent-extraction type process; the materials to be separated undergo partition between the aqueous phase held in the inert cellulose matric and the organic solvent used as the mobile phase. Those components of the mixture to be separated which are most readily soluble. in the organic mobile phase will have R_F values near, or equal to, unity. Those components which have a lower solubility in the organic phase will have R_F values near to zero. The R_F value is characteristic of a particular species in any given type of separation, and is sometimes used for the qualitative identification of the unknown species. In simple cellulose chromatography the mechanism is largely partition in type; adsorption processes play only a small part (e.g., due to the presence of small quantities of carboxyl groups in the paper), but this effect is not normally apparent when strongly acidic solvents are employed.

In inorganic separations on unmodified cellulose two main groups of factors appear to govern the mobility of different elements. First, those factors which increase their solubility in the organic phase and thus lead to high R_F values.

Inorganic salts are not usually soluble in organic solvents and such solubility is often indicative of complex formation. With solvents containing donor oxygen atoms in the presence of a small amount of hydrochloric acid, those metals which form chloro complexes move readily, probably owing to the solubility of the free chloro-complex acid in the organic solvent. Thus iron(III) is readily mobile with solvents containing hydrochloric acid while nickel(II) does not move appreciably: this behaviour may be compared with the different absorption of these metals on to anion exchange resins from hydrochloric acid solution. The second group of factors governing mobility are those which tend to cause low R_F values, i.e., retain the components of the mixture in the stationary aqueous phase of the cellulose. This behaviour is apparent with metals if anions are present in the mixture which form strong water-soluble complexes or which give an insoluble precipitate. Such interference may often be overcome by prior elimination of the offending anion (by precipitation) or by the addition of an otherwise inert species which complexes the interfering anion more strongly than do the components of the mixture to be separated.

Reference should be made to **modified cellulose** obtained by the introduction of diethylaminoethyl groups, of carboxyl groups and of phosphate groups into the cellulose matrix. Such substituted celluloses possess the advantages of normal ion exchange materials with the additional merit that they can be obtained in sheet form, thus enabling standard paper-strip methods to be used. Cellulose phosphate appears to be an excellent selective exchanger; thus thorium is taken up by cellulose phosphate from $4N$-acid, and so rendering possible the recovery of this element from sulphuric acid solutions of monazite. Iron(III) ions and uranyl ions behave similarly, and can be eluted only by the use of a complexing agent such as ammonium carbonate.

Cellulose phosphate is essentially cellulose dihydrogen phosphate, and is a bifunctional exchanger containing both strong acid (thus simulating strongly acidic cation exchange resins) and very weak acid groups. Carboxymethyl-cellulose may be regarded as a weakly acidic cation exchanger which functions most readily at a pH above 4–5. Diethylaminoethyl-cellulose resembles in base strength the corresponding tertiary-amino ion exchange resins; it does not function in strongly alkaline solutions. As in most other cellulose derivatives, the majority of the substituted groups are located accessibly close to the surface of the exchanger, thus facilitating the exchange with large molecules more readily than an orthodox resin exchanger. The modified cellulose ion exchangers are marketed* in both paper sheet and in powder form.

There are three main methods of conducting inorganic chromatographic separations on cellulose, viz., by the use of (a) glass plates or plastic foils coated with thin layers of cellulose, (b) paper strips, and (c) columns of cellulose. Thin layer and paper strip methods are essentially micro-analytical in character and cannot usually be employed with more than 100 μg of sample. For quantitative work the plate or strip may be sprayed with a reagent which forms coloured complexes with the elements present; the quantities of these elements can then be estimated by visual comparison of the spots produced with those obtained using standard solutions of the elements under identical conditions. Alternatively the portions of the cellulose thin layer, or paper strip, containing the spots may be

---

* By the manufacturers of Whatman filter papers, W. & R. Balston and Co Ltd.

suitably treated to extract the separated elements for determination by spectrophotometric or other appropriate instrumental methods.

The cellulose column method is generally employed for macro-scale work, it being usual to condition the column by prior equilibration with the solvent to be used for the separation. The most convenient procedure is to use conditions under which the element to be determined is eluted first from the column; it may then be determined by a standard method, either titrimetric or instrumental.

It will be appropriate at this point, following the above general introduction to the topic, to enlarge upon the two particular aspects of inorganic thin layer chromatography and of the development of high performance liquid chromatography.

VIII, 2. THIN LAYER CHROMATOGRAPHY. The technique of thin layer chromatography (TLC) uses an adsorbent coated on a glass plate as the stationary phase and development of the chromatogram takes place as the mobile phase percolates through the adsorbent. Thin layer chromatography has, as is well known, distinct advantages over paper chromatography because of its convenience and rapidity, its greater sharpness of separation and its high sensitivity. A brief description of experimental procedure will be given here with particular reference to the separation of cations.

Preparation of the plate. In thin layer chromatography a variety of coating materials are available, although silica gel is used more often than other materials. The separation of cations on silica gel is not, however, always satisfactory as many cations have similar R_F values and remain grouped together on this adsorbent. Cellulose powder is recommended as an adsorbent for separation of cations by TLC even though separations may be slower than those obtained on silica gel. The use of cellulose powder may be regarded as a substitute for paper chromatography and data obtained for inorganic paper chromatography are generally applicable to inorganic TLC on cellulose.

Thin layers of cellulose can be made by spreading an aqueous slurry of cellulose powder using one of the commercially available applicators.* It is most important that the glass plates to which the thin layer is to be applied should be thoroughly clean and this can be accomplished by washing the plates in a concentrated sodium carbonate solution followed by thorough rinsing with distilled water.

The aqueous slurry of cellulose powder is prepared by mixing about 15 g powder in 90 cm^3 of distilled water and dispering the powder for about 1 min using a mechanical mixer. The cellulose powder used for inorganic TLC is of a special microcrystalline nature.† In partition chromatography, unactivated plates are used and cellulose layers thus require no activation by heating. The coated plate may be dried overnight at room temperature.

Ready-to-use thin layers, prepared with the most widely used adsorbents, are now available, e.g. as precoated glass plates and plastic foils. Plastic sheets precoated with cellulose (which may also incorporate fluorescent material) are

---

* Available from Shandon Scientific Company Ltd, Camlab (Glass) Ltd, and from Griffin and George Ltd.

† Supplied by E. Merck Laboratory Chemicals (distributed in the UK by Anderman and Co Ltd).

marketed* and are very convenient for inorganic TLC work as they can be cut to the required size.

Sample application.   The sample solution to be applied should contain between 0.1 and 10 mg of the cation per cm^3 and may be neutral or dilute acid; about 1 μl of solution is applied with a microsyringe or micropipette near one end of the chromatoplate (about 1.5–2.0 cm from the edge of the plate) and the latter air dried. Equilibration of the chromatoplates is not necessary and development of the plate can start immediately after it is dried.

Development of plates.   The chromatogram is usually developed by the ascending technique in which the plate is immersed in the developing solvent (redistilled or chromatographic grade solvent should be used) to a depth of 0.5 cm. The tank or chamber used is preferably lined with sheets of filter paper which dip into the solvent in the base of the chamber; this ensures that the chamber is saturated with solvent vapour (Fig. VIII, 2). Development is allowed to proceed until the solvent front has travelled the required distance (usually 10–15 cm), the plate is then removed from the chamber and the solvent front immediately marked with a pencil line.

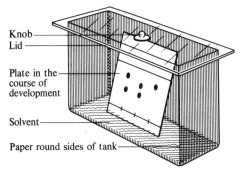

Knob
Lid
Plate in the course of development
Solvent
Paper round sides of tank

Fig. VIII, 2   Reproduced from D. Abbott and R. S. Andrews (1965). *An Introduction to Chromatography.* **London; Longman.**

The positions of the separated solutes can be located by various methods. Coloured substances can be seen directly when viewed against the stationary phase while colourless species may usually be detected by spraying the plate with an appropriate reagent which produces coloured areas in the regions which they occupy. Some compounds fluoresce in ultraviolet light and may be located in this way. Alternatively if fluorescing material is incorporated in the adsorbent the solute can be observed as a dark spot on a fluorescent background when viewed under ultraviolet light. (When locating zones by this method the eyes should be protected by wearing special protective goggles or spectacles.) The spots located by this method can be delineated by marking with a needle.

Thin layer chromatography using microscope slides.   The separation of cations may be very conveniently achieved using cellulose-coated microscope slides (or precoated cellulose sheets cut into pieces about the size of microscope slides). Very small spots (diameter *ca.* 1 mm) of sample solution are introduced onto the plate, e.g., by using a pointed paper strip saturated with the solution.

---

* Manufactured by the Eastman Kodak Co.

The plate is air dried and developed in a small bottle until the solvent front has travelled 2–3 cm. Separation is achieved in 5–10 minutes and the separated cations may be located by the usual procedures.

Quantitative inorganic thin layer chromatography.   Quantitative analysis of separated constituents on thin layer plates is generally carried out by measurement of the photodensity and area of the spot, i.e. by photodensitometry of the plate.* This type of procedure requires comparison with spots obtained using known amounts of standard mixtures which must be chromatographically examined on the same plate as the sample.

An alternative procedure involves removal of the separated components from the plate by scraping off the relevant portion of the adsorbent. The component is eluted (extracted) from the adsorbent with a suitable solvent and determined by an appropriate physical technique, e.g., by ultraviolet, visible, or fluorescence spectrophotometry.

VIII, 3.   HIGH PERFORMANCE LIQUID CHROMATOGRAPHY.   No account of contemporary chromatographic techniques would be complete without mention of high performance liquid chromatography (HPLC). Classical column liquid chromatography is a well established separation procedure in which the mobile liquid phase flows slowly through the column by means of gravity. The method is generally characterised by low column efficiencies and long separation times. Since about 1969, however, there has been a very marked revival of interest in the technique of liquid column chromatography due to the development by Kirkland and Huber of high pressure systems operating at pressures up to $2.07 \times 10^7 \, Nm^{-2}$ (3000 p.s.i.) (Ref. 1). In this method small diameter columns (1–3 mm) with support particle sizes in the region of 30 μm are used and the eluent is pumped through the column at a high flow rate (*ca.* 1 to 5 cm^3 min^{-1}). Separations by this method may be effected much more rapidly (about 100 times faster) than by the use of conventional liquid chromatography.

Although the currently available commercial equipment† is rather expensive, HPLC has already shown itself to have wide application in organic chemistry. The development of inorganic applications is likely, e.g., in the field of ion exchange chromatography where pellicular resins, that is, resins produced as thin coatings on the surface of spherical glass beads (20–50 micrometres diameter), are now available commercially under the trade name 'Zipax'. The beads have a porous surface about 2 micrometres thick which serves to bond the resin coating.

The application of high performance ion exchange chromatography for the separation of inorganic compounds is illustrated by the separation of transplutonium elements; the method is based on sequential elution of the cations from a cation exchange column with an anionic complexing agent. Exposure of the resin to radiation from the radioisotopes is reduced since a satisfactory separation is achieved in a much shorter time, and radiation induced damage to the resin is minimised (Ref. 2).

---

* A low cost densitometer using fibre optics has been developed by Kontes, Instrument Group, Spruce Street, Vineland, New Jersey, USA, 08360.

† Reeve Angel Ltd now market an economical familiarisation kit for HPLC which can be operated up to $2.4 \times 10^6 \, Nm^{-2}$ (350 p.s.i.). More advanced equipment is supplied by Waters Assocs., Perkin Elmer Ltd, etc.

VIII, 4. SEPARATION OF NICKEL, MANGANESE, COBALT AND ZINC AND DETERMINATION OF R_F VALUES. *Discussion.* This experiment illustrates the separation of the above four metals using either thin layer or paper strip techniques. A small measured sample of the solution (e.g. from an Agla micrometer syringe or a micropipette) is placed near one end of the thin layer plate or the paper strip and the chromatogram developed using a mixed acetone–hydrochloric acid solvent. After the solvent front has moved a suitable distance the plate or paper strip is removed from the tank or jar, the solvent evaporated after marking the solvent front and the metal ions rendered visible in the form of coloured spots or bands by spraying with a suitable reagent. The experiment permits the evaluation of R_F values which are approximately Ni 0.1, Mn 0.25, Co 0.55, Zn 0.9. For quantitative work the bands can be cut from the paper strip, the metals extracted and determined; alternatively and more conveniently a reflectance densitometer may be used (see footnote to page 198). The procedures used for quantitative analysis by thin layer chromatography have already been described.

   Reagents. *Nickel-, manganese-, cobalt-, and zinc-ion solution.* Prepare this from the following A.R. salts: $(NH_4)_2SO_4, NiSO_4, 6H_2O$; $MnSO_4, 4H_2O$; $CoCl_2, 6H_2O$ or $(NH_4)_2SO_4, CoSO_4, 6H_2O$; and $ZnSO_4, 7H_2O$. Dissolve appropriate quantities in $2M$-hydrochloric acid to give solutions containing 10 µg each of Ni^{2+}, Co^{2+}, Mn^{2+}, and Zn^{2+} in 0.01 cm³ of solution.

   Acetone–HCl solvent. Mix 43.5 cm³ of A.R. acetone, 4 cm³ of concentrated hydrochloric acid (sp. gr. 1.18), and 2.5 cm³ of water.

   Spraying reagent PACF/R (trisodium pentacyanoammine-ferrate/rubeanic acid). Dissolve 0.70 g of the pentacyanoammine-ferrate in 20 cm³ of water and pour the resulting solution into a solution of 0.25 g of rubeanic acid in 10 cm³ of ethanol. Shake the mixture for 15 minutes and filter. The filtered solution is ready for use. The reagent should be prepared on the day it is to be used.

   Preparation of $Na_3[Fe(CN)_5NH_3]6H_2O$. Weigh out 10 g of finely-ground, sodium nitroprusside (disodium pentacyanonitrosyl ferrate) into a small conical flask, and add 24 cm³ of concentrated ammonia solution, sp. gr. 0.88. Shake well and loosen the cork so that gas can escape; when all the solid has dissolved and gas evolution has begun, place the flask in a refrigerator at $-7°C$ for 48 hours. Warm to room temperature, add a little more concentrated ammonia solution. sp. gr. 0.88, filter at the pump, remove as much liquid as possible, and then wash the precipitate with a small volume of methanol. Rapidly remove as much as possible of the methanol by suction, transfer the product to a glass dish in a desiccator over calcium chloride, and keep it in the dark. The yield is 5.4 g.

   Acetic acid, ca. *0.2*M. Dilute 11 cm³ of glacial acetic acid to 1 litre with water.

   Procedure A. **Separation using paper strips.** Pour 25 cm³ of the acetone–HCl solvent into the tall gas jar or 2-litre measuring cylinder (Fig. VIII, 1) and an equal volume into the upper solvent container.* Allow to stand for at least 15 minutes before commencing the separation. Fold a 30-cm strip of Whatman's No. 1 paper (2.5 cm wide) about 3 cm from one end. Draw a thin pencil line about 2.5 cm from the fold and mark the mid point. Pipette 0.05 cm³ of

---

* The external dimensions of the Pyrex boat are about $40 \times 25 \times 35$ mm (internal capacity ca. 30 cm³), the two arms are of 6-mm rod and are spaced about 10 mm from the sides of the boat, and the vertical rod is 6 mm.

the test solution on the strip; hold the paper vertically and draw the tip of the micro-pipette (or syringe) along the pencil line and near the mid point. Prepare a second strip in the same way and allow both to dry in the air for about 30 minutes. Hang the strips in the jar with the folded ends in the upper solvent container, and allow the solvent to diffuse within 2.5 cm of the bottom of the paper strips (about 3 hours). Remove the strips, allow the solvent to evaporate, and expose to ammonia vapour to neutralise the acid. Spray with PACF/R reagent; use an atomiser. Remove the excess of the reagent by dipping the paper strips into dilute acetic acid (0.2M). The metals are visible as coloured bands: Ni, blue; Mn, blue; Co, brown; and Zn, red.

Determine the R_F values as described in Section **VIII, 1**.

Procedure B. **Separation by thin layer chromatography.** Prepare a cellulose coated glass plate, or more conveniently cut out a strip (5 × 20 cm) from a plastic sheet precoated with cellulose. Draw a thin pencil line about 2 cm from one end of the strip (or plate), apply a 10 μl sample of the metal ion solution to the centre of the line using a microsyringe and allow to air dry. Place the strip upright in the developing tank so that the sample end is immersed in the developing solvent to a depth of 0.5 cm; the solvent should be introduced into the tank at least 15 minutes before commencing the separation. Allow the development to proceed until the solvent front has travelled about 15 cm (this takes approximately 30 minutes), remove the strip (or plate) and allow about 10 minutes for the solvent to evaporate. Neutralise excess acid by exposing the strip to ammonia vapour for about 10 minutes and spray with pentacyanoammineferrate/rubeanic acid reagent using an atomiser. Remove excess reagent as described under Procedure A; the metals are visible as similarly coloured spots and the R_F values can be determined.

VIII, 5. SEPARATION OF NICKEL, COPPER, COBALT, AND ZINC.

This provides an alternative experiment to that of the previous Section.

Reagents. *Nickel-, copper-, cobalt-, and zinc-ion solution.* Prepare this from the following A.R. salts: $(NH_4)_2SO_4.NiSO_4,6H_2O$; $CuSO_4,5H_2O$; $CoCl_2,6H_2O$ or $(NH_4)_2SO_4CoSO_4,6H_2O$; and $ZnSO_4,7H_2O$. Appropriate quantities are dissolved in distilled water to give a solution containing about 1 mg of each metal ion in 1 cm^3 of solution.

Solvent. Prepare a mixture of acetone, ethyl acetate, and 6M-hydrochloric acid in the ratio of 9:9:2 by volume from the A.R. reagents.

Spraying reagent. Prepare a 0.1 per cent solution of rubeanic acid in ethanol.

Procedure. Apply 10 μl of the solution of metal ions to either a paper strip or cellulose coated plate or foil. Allow to air dry and continue the separation as described under either Procedure A or B (previous section). After development, remove the paper strip or the plate (foil), allow the solvent to evaporate, and expose it to ammonia vapour to neutralise the acid. Spray the paper strip (both sides) or thin layer plate (foil) with the rubeanic acid reagent. Nickel is rendered visible as a blue-purple band, cobalt as a yellow-orange band, and copper as an olive-green band. The zinc is visible as a pink band by spraying with a dilute solution of dithizone in chloroform.

Evaluate the R_F values of the four ions by the method described in Section **VIII, 1**.

VIII, 6. SEMI-QUANTITATIVE SEPARATION OF COPPER, COBALT, AND NICKEL ON SLOTTED PAPER STRIPS. *Discussion.* A specially cut

rectangular sheet (21.3 cm × 11 cm) of Whatman's filter paper, provided with eleven slots (3 mm × 9 cm) cut into the paper parallel to the short side so as to leave twelve strips 1.5 cm wide joined at the top and bottom (Fig. VIII, 3), is

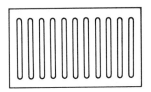

Fig. VIII, 3

employed*. This permits the separation of a number of samples simultaneously with very simple apparatus and is especially valuable for semi-quantitative trace metal analysis: the technique is useful for geochemical prospecting. The metals are detected by spraying the strips with suitable reagents, and the amounts present are detected by visual comparison with standards. Thus copper, cobalt, and nickel may be determined after a single separation.

An aliquot of the solution is applied to one end of each strip of paper so that, by leaving the two end strips vacant, ten sample solutions may be placed on the sheet. The volume of test solution used is about 0.01 cm³, and it is applied with the aid of a capillary pipette so that it spreads right across the strip to form a thin rectangular patch. The sheet is bent so that it forms a cylinder and clipped at the upper end with a paper clip, preferably of polythene. After suitable drying, the sheet is placed vertically in a covered beaker containing the solvent, the depth of which must not exceed 1 cm. The solvent is allowed to diffuse up the strip and reaches the level of the top of the slots in 10–35 minutes, the time depending upon the solvent. The sheet is removed just before the solvent reaches this level, and the positions of the metals on the strip are located by various treatments of the sheet, according to the metals to be determined. For the chromatographic separation of copper, cobalt, and nickel from each other and from other elements, a mixture of pure ethyl methyl ketone (75 parts), concentrated hydrochloric acid (15 parts), and water (10 parts, v/v) is satisfactory. With this solvent, the approximate R_F values are: copper, 0.65; cobalt, 0.45; and nickel, 0.10.

Reagents. *Solvent.* Mix 37.5 cm³ of pure ethyl methyl ketone, 5 cm³ of water and 7.5 cm³ of concentrated hydrochloric acid (sp. gr. 1.18).

Copper solution. Dissolve 1.12 g of A.R. copper(II) chloride dihydrate in 50 cm³ of concentrated hydrochloric acid (sp. gr. 1.18) and 5 cm³ of concentrated nitric acid (sp. gr. 1.42), and dilute to 100 cm³ with water.

Cobalt solution. Dissolve 1.70 g of A.R. hydrated cobalt chloride in 50 cm³ of concentrated hydrochloric acid and 5 cm³ of concentrated nitric acid, and dilute to 100 cm³ with water.

Nickel solution. Dissolve 1.86 g of A.R. nickel sulphate hexahydrate in 50 cm³ of concentrated hydrochloric acid and 5 cm³ of concentrated nitric acid and dilute to 100 cm³ with water.

All the above solutions contain about 0.42 g of each metal in 100 cm³ of solution. Prepare a standard mixed Cu, Co, Ni solution containing *ca.* 0.1 g of each metal in 100 cm³ by placing 25.0 cm³ of each of the above solutions in a 100-cm³ graduated flask and diluting to the mark with a mixed acid solution (50 cm³ of concentrated hydrochloric acid + 5 cm³ of concentrated nitric acid + 45 cm³

---

* This is marketed as Whatman Pattern F-SF 2925 CRL.

of water). Use this standard solution to prepare a series of solutions as follows:

Solution No.	1	2	3	4	5	6	7	8	9	10
Standard Cu, Co. Ni solution, cm^3	10	9	8	7	6	5	4	3	7.5	4.5
Mixed acid solution, cm^3	0	1	2	3	4	5	6	7	2.5	5.5

In order to gain experience in the determination, solutions *1* to *8* will be regarded as standards, and solutions *9* and *10* as unknowns.

Rubeanic acid solution. Dissolve 0.1 g of rubeanic acid in 60 cm^3 of ethanol and dilute to 100 cm^3 with water.

Procedure.   Pour 25 cm^3 of the ketone solvent into a 600-cm^3 Pyrex beaker and cover it with a Petri dish. By means of a graduated micropipette add 0.01 cm^3 of each of the solutions Nos. *1* to *10* evenly along lines about 2 cm from one side of the sheet of slotted paper (Whatman C.R.L. pattern): each paper contains twelve strips, but do not use the two outer strips. Form the sheet into a cylinder and secure the two upper ends together with a paper clip. Place the cylinder with the sample spots lowermost in a clean, dry beaker in a boiling water bath; remove the paper cylinder after 3 minutes and stand it immediately in the solvent beaker, replacing the Petri dish cover. Allow the solvent to diffuse to the top of the strips (time: up to 60 minutes), remove the cylinder from the beaker, and allow it to dry in the air for 5 minutes. Place the paper cylinder inside a second 600-cm^3 beaker, fitted with a Petri dish cover and containing a small (30-cm^3) beaker charged with concentrated ammonia solution; leave for 2 minutes. Take the cylinder from the beaker, lay the sheet flat on plate glass, and spray* it lightly on both sides with the rubeanic acid solution. Allow the sheet to dry in the air. Compare (visually) the copper (grey-green), cobalt (orange), and nickel (blue) bands for solutions *9* and *10* with those produced by the standards, and in this way estimate their Cu, Co, and Ni contents.

VIII, 7.   SEPARATION OF IRON AND ALUMINIUM ON A CELLULOSE COLUMN.   *Theory.*

The ions of iron(III) and of aluminium may be separated on a cellulose column using A.R. acetone containing 2 per cent of hydrochloric acid as solvent; the iron(III) forms a chloro complex and is readily mobile in the solvent, probably as the metal chloro-complex acid. The aluminium remaining in the cellulose column may then be eluted with 1M-hydrochloric acid. The acetone–hydrochloric acid solvent may be used to separate a number of binary mixtures: Ni, Pb, Al, Cr, Ti, Zr, and Th (immobile) from Co, Cu, Cd, Fe, Zn, Mn, V, and Mo (mobile).

Reagents.   *Iron(III) solution.* Dissolve sufficient A.R. ammonium iron(III) sulphate, accurately weighed, in 8M-hydrochloric acid to give a solution containing about 25 mg of iron(III) per cm^3.

Aluminium-ion solution. Dissolve sufficient A.R. aluminium ammonium sulphate, accurately weighed, in 8M-hydrochloric acid to give a solution containing about 25 mg of aluminium per cm^3.

Standard EDTA solution, ca. *0.02*M. See Section **X, 50**.

Standard zinc sulphate solution, ca. *0.02*M. Dissolve pure zinc in dilute sulphuric acid.

Variamine Blue B indicator. Prepare a 1 per cent aqueous solution.

---

* An excellent spray is marketed by Shandon Scientific Co Ltd, by Baird and Tatlock (London) Ltd, and by Microchemical Specialities Co of Berkeley, 10, California.

Solochrome Black indicator. See Section **X, 28**.

Water repellent for glassware. A 2 per cent solution of dichlorodimethylsilane in carbon tetrachloride is satisfactory.

Organic extraction solvent. Add 6 cm³ of concentrated hydrochloric acid to 300 cm³ of A.R. acetone. (The latter usually contains up to about 0.7 per cent of water.)

Whatman's 'ashless tablets'. For preparation of cellulose pulp.

Preparation of cellulose column. Prepare cellulose pulp* by boiling Whatman 'ashless tablets' with an aqueous solution containing 5 cm³ of concentrated nitric acid (sp. gr. 1.42) per 100 cm³ for 2–3 minutes; decant, remove the excess of acid by washing with water. Then wash with ethanol, followed by diethyl ether.

Preparation of glass extraction tube. Use a Pyrex tube, 30–40 cm long, about 16 mm internal diameter, and drawn out at the lower end to an internal diameter of about 3 mm. It is necessary to treat the inside surface of the glass extraction tube with a water-repellent material to prevent 'wall' effects due to creep of the aqueous test solution at the top of the column down between the glass tube and the cellulose. Shake a small amount of the water-repellent solution in the clean tube until the whole of the surface has been treated, remove the excess after a few minutes, and wash the tube with ethanol, followed by a little water. In the absence of fluoride, the 'silicone'-treated tube retains its water-repellent properties for a large number of separations.

Procedure. Place a large glass bead (*ca.* 6 mm diameter) at the bottom of the glass column (to act as a support for the cellulose); fit the narrow end of the tube with a short length of rubber tubing, a screw clip, and a short glass tube.† Fill the column to about two-thirds of its length with the organic extraction solvent. Introduce some cellulose pulp or powder. With the aid of a glass rod, which has one end flattened to form a plunger of slightly smaller diameter than that of the glass column, beat the cellulose up to form a smooth slurry. Allow the solvent to run out until most of the cellulose has settled; add more solvent and cellulose pulp (or powder) and beat the cellulose up with the glass plunger. Repeat the process until a 'settled' column of about 12 cm length results. Press the cellulose gently down until the rate of flow of solvent through the column decreases to about 2 cm³ per second. Maintain the solvent level above that of the 'settled' cellulose. Finally, run 50 cm³ of the solvent through the column to remove any traces of iron present in the cellulose: close the screw clip when the level of the liquid is just above that of the cellulose column.

Mix equal volumes (say, 5.00 cm³) of the prepared iron and aluminium solutions and pipette 2.00 cm³ of the resulting solution into a 100-cm³ beaker. Add sufficient dry cellulose pulp (or powder) to form a friable mass when mixed, and transfer the wad to the top of the column; rinse the beaker with 5 cm³ of the organic solvent and transfer to the column. Gently beat the wad of cellulose with the glass plunger and pack it down to form a continuous part of the column. Rest the glass plunger in the sample beaker. Detach the screw clip: collect the eluate in a 350-cm³ conical flask. Introduce 150 cm³ of the solvent into the column in 5-cm³ portions; use each portion to rinse the beaker which contained the sample. Add 50 cm³ of water to the eluate; this contains all the iron.

---

* Whatman Cellulose Powder, Standard Grade, may be used directly, and is more convenient.
† Alternatively, a glass stopcock may be sealed to the narrow end of the glass column.

Replace the receiver by another 350-cm^3 conical flask. Elute the aluminium from the column with 100 cm^3 of 1M-hydrochloric acid, added in 5-cm^3 portions. Remove the acetone from both eluates by evaporation on a water bath.

Analysis of eluates. (*a*) Iron solution. Add a few cm^3 of 20-volume hydrogen peroxide to oxidise any iron(II) present to iron(III); boil for 10–15 minutes to destroy the excess of hydrogen peroxide. Cool, dilute to 100 cm^3, add dilute ammonia solution until the pH is 2–3 (use Congo Red paper or a pH meter), followed by 5–6 drops of Variamine Blue B indicator. Titrate with standard EDTA solution until the colour changes from blue-violet to yellow.

(*b*) Aluminium solution. Dilute the eluate to 250 cm^3 in a graduated flask. To 50.0 cm^3 of this solution add 25.0 cm^3 of standard EDTA solution, adjust the pH to between 7 and 8 (use a pH meter or phenol red paper), and introduce a few drops of Solochrome Black indicator. Titrate *rapidly* with standard zinc sulphate solution until the colour changes from blue to wine red.

Standardisation of iron(III) solution with EDTA. Dilute 2.00 cm^3 of the iron(III) solution to 100 cm^3 with water, add dilute ammonia solution to the first perceptible colour change of Congo Red paper (pH 2–3), followed by 5 drops of Variamine Blue B indicator. Titrate with standard EDTA solution until the colour changes from blue-violet to yellow; the colour is grey just before the end point.

Standardisation of aluminium-ion solution with EDTA. Dilute 2.00 cm^3 of the aluminium-ion solution to 100 cm^3 in a graduated flask. To 10.00 cm^3 of the diluted solution add 25.0 cm^3 of standard EDTA solution, followed by 1M-ammonia solution to a pH of 7–8. Introduce a few drops of Solochrome Black indicator and titrate rapidly with standard zinc sulphate solution to the blue to wine-red end point. It may be necessary to add a drop of ammonia solution to maintain the pH above 7, otherwise the blue colour tends to fade.

Calculate the weight of iron(III) and of aluminium ion in the volume of solution employed.

Some typical results are given below.

2.00 cm^3 of the iron(III) solution required 40.70 cm^3 of 0.01964M-EDTA;
∴ 2.00 cm^3 of solution contains 40.70 × 0.01964 × 55.85 = 44.67 mg Fe.
In aluminium-ion titration, excess of EDTA required 17.65 cm^3 of 0.01920M-ZnSO$_4$.
∴ 2.00 cm^3 of the aluminium-ion solution contains
10 × {(25.00 × 0.01964) − (17.65 × 0.01920)} × 26.98 = 41.01 mg Al.
Iron recovered from 2.00 cm^3 of mixed solution = 20.25 cm^3 of 0.01964M-EDTA = 20.25 × 0.01964 × 55.85 = 22.13 mg.
Aluminium recovered from 2.00 cm^3 of mixed solution: 25.00 cm^3 of 0.01964M-EDTA added; excess of EDTA ≡ 17.70 cm^3 of 0.01920M-ZnSO$_4$;
∴ Eluate contains
5 × {(25.00 × 0.01964) − (17.70 × 0.01920)} × 26.98 = 20.40 mg Al.

Alternatively the metal ions may be determined by atomic absorption spectrophotometry. Suitable solutions are obtained by collecting each eluate (after removal of acetone) in a 1 dm^3 graduated flask and diluting to the mark with de-ionised water. The instrument should be calibrated with standard Fe^{3+} and Al^{3+} solutions covering the 0–50 p.p.m. concentration range.

VIII, 8. SEPARATION OF COBALT AND NICKEL ON A CELLULOSE COLUMN. *Theory.* See Section **VIII, 7**.

Reagents. *Cobalt-ion solution.* Dissolve about 2.0 g of A.R. hydrated cobalt chloride in 25 cm^3 of 8M-hydrochloric acid.

Nickel-ion solution. Dissolve about 11.0 g of A.R. nickel sulphate hepta-hydrate in 25 cm^3 of 8M-hydrochloric acid.

Standard EDTA solution, ca. *0.02*M. See Section **X, 50**.

Xylenol Orange indicator. Prepare a 0.5 per cent aqueous solution.

Bromopyrogallol Red indicator. Prepare a 0.05 per cent solution in ethanol.

Buffer solution. Mix equal volumes of 1 M-ammonium chloride and 1 M-aqueous ammonia solutions.

Organic extraction solvent. Add 6 cm^3 of concentrated hydrochloric acid to 300 cm^3 of A.R. acetone.

Procedure. Prepare a cellulose column about 15 cm long as detailed in Section **VIII, 7**, use Whatman Cellulose Powder, Standard Grade.

Mix exactly equal volumes (say, 5.00 cm^3 each) of the cobalt- and nickel-ion solutions. Pipette 2.00 cm^3 of the mixed solution into a 100-cm^3 beaker. Add sufficient dry cellulose powder to the solution to form a friable mass and transfer it to the top of the column as completely as possible; rinse the beaker with 5 cm^3 of the organic extraction solvent and add to the column. Gently beat the wad of cellulose with the glass plunger, and pack it down to form a continuous part of the column. Introduce 125 cm^3 of the extraction solvent into the column in 5-cm^3 portions; use each portion first to rinse the beaker which contained the sample. Collect the eluate in a 350-cm^3 conical flask; the eluate contains the cobalt as the deep blue chloro complex.

Now elute the nickel from the cellulose by passing 100 cm^3 of 1M-hydrochloric acid through the column; add the acid in 5-cm^3 portions. Collect the eluate in a 250-cm^3 conical flask; this now contains the nickel.

Analysis of eluates. (*a*) Cobalt solution. Add 50 cm^3 of water to the eluate and remove the acetone by volatilisation on a steam bath. Boil the solution for several minutes, cool and dilute to about 100 cm^3 with water. Adjust the pH to 6 by the addition of solid hexamine, add a few drops of Xylenol Orange indicator, and titrate with standard 0.02M-EDTA until the colour changes from red to orange yellow (compare Section **X, 61**).

(*b*) Nickel solution. Heat the eluate on a steam bath to remove acetone, cool and dilute to 250 cm^3 in a graduated flask. Transfer 50.0 cm^3 of the diluted solution to a 250-cm^3 conical flask, dilute to about 100 cm^3 with water, nearly neutralise with A.R. sodium hydroxide, add 10 cm^3 of the buffer solution, 10 drops of Bromopyrogallol Red indicator, and titrate, with standard 0.02M-EDTA until the colour changes from blue to red (compare Section **X, 58**).

Standardisation of cobalt solution with EDTA. Dilute 2.00 cm^3 of the prepared cobalt solution to 100 cm^3 with water, adjust the pH to 6 by the addition of solid hexamine, add a few drops of Xylenol Orange indicator, and titrate with the standard EDTA solution until the colour changes from red to orange yellow.

Standardisation of the nickel solution with EDTA. Dilute 2.00 cm^3 of the prepared nickel solution to 100 cm^3 with water in a graduated flask. Pipette 10.0 cm^3 of the diluted solution into a conical flask, dilute to 100 cm^3, add 10 cm^3 of the buffer solution, 10 drops of Bromopyrogallol Red indicator. Titrate with standard 0.02M-EDTA to a colour change from blue to red.

Calculate the percentage of cobalt and nickel recovered in the separation and

also the weights of these elements in the volume of solution employed.
Some typical results are given below.

$2.00 \, cm^3$ of cobalt-ion solution required $31.85 \, cm^3$ of $0.01975M$-EDTA;

$\therefore 2.00 \, cm^3$ of solution contained $31.85 \times 0.01957 \times 58.94 = 36.73 \, mg$ Co.

$0.200 \, cm^3$ of nickel-ion solution required $15.03 \, cm^3$ of $0.01957 \, M$-EDTA;

$\therefore 2.00 \, cm^3$ of solution contained $10 \times 15.03 \times 0.01957 \times 58.69 = 172.63 \, mg$ Ni.

Cobalt recovered from $2.00 \, cm^3$ of the mixed solution $\equiv 15.95 \, cm^3$ of $0.01957M$-EDTA;

$\therefore$ Weight of cobalt in $2.00 \, cm^3$ of the mixed solution $= 15.95 \times 0.01957 \times 58.94 = 18.39 \, mg$

Nickel recovered from $2.00/5 \, cm^3$ of the mixed solution $\equiv 14.95 \, cm^3$ of $0.01957M$-EDTA:

Weight of nickel in $2.00 \quad cm^3$ of the mixed solution $= 5 \times 14.95 \times 0.01957 \times 58.69 = 85.86 \, mg$.

VIII, 9. SEPARATION OF COPPER AND NICKEL ON A CELLULOSE COLUMN. *Theory.* See Section **VIII, 7**.

Reagents. *Copper-ion solution.* Dissolve about 2.8 g of A.R. copper(II) sulphate pentahydrate in $50 \, cm^3$ of $8M$-hydrochloric acid.

Nickel-ion solution. Dissolve 11.0 g of A.R. hydrated nickel sulphate in $25 \, cm^3$ $8M$-hydrochloric acid.

Standard EDTA solution, ca. *0.02*M. See Section **X, 50**.

Fast sulphon Black F (C.I. No. 26990) *indicator.* Prepare a 0.5 per cent aqueous solution.

Bromopyrogallol Red indicator. Prepare a 0.05 per cent solution in ethanol.

Organic extraction solvent. Add $6.0 \, cm^3$ of concentrated hydrochloric acid to $300 \, cm^3$ of A. R. acetone.

Procedure. Prepare a cellulose column about 15 cm long as described in Section **VIII, 7**; use Whatman Cellulose Powder, Standard Grade.

Mix exactly equal volumes (say, $5.00 \, cm^3$ each) of the copper- and nickel-ions solutions. Use $2.00 \, cm^3$ of the mixed solution for the separation exactly as detailed in Section **VIII, 8**. Elute the copper from the cellulose column as an orange brown solution with $125 \, cm^3$ of the organic extraction solvent, and the nickel with $100 \, cm^3$ of $1M$-hydrochloric acid.

Analysis of eluates. (*a*) Copper solution. Add $50 \, cm^3$ of water to the eluate, neutralise the solution with A.R. sodium hydroxide, and remove the acetone by evaporation on a steam bath. Render the solution faintly acid with dilute hydrochloric acid and boil for a few minutes. Cool the greenish-yellow solution, add $50 \, cm^3$ of water, followed by $10 \, cm^3$ of concentrated ammonia solution and 10 drops of Fast Sulphon Black F indicator. Titrate with standard $0.02M$-EDTA until the colour changes from blue-violet to bright green (compare Section **X, 56**).

(*b*) Nickel solution. Proceed as in section **VIII, 8**.

Standardisation of the copper solution with EDTA. Pipette $2.00 \, cm^3$ of the copper solution into a 250-cm^3 conical flask, dilute to $100 \, cm^3$ with water, and just neutralise wth A.R. sodium hydroxide. Add $10 \quad cm^3$ of concentrated ammonia solution and 10 drops of Fast Sulphon Black F indicator. Titrate with standard $0.02M$-EDTA to a colour change from blue-violet to bright green.

Standardisation of the nickel solution with EDTA. See Section **VIII, 8**.

Calculate the percentage of copper and nickel recovered in the separation and also the weights of these elements in the volume of solution employed.

Some typical results are given below.

2.00 cm^3 of the copper-ion required 29.97 cm^3 of 0.01971M-EDTA;

∴ 2.00 cm^3 of solution contained 29.97 × 0.01971 × 63.54 = 37.53 mg Cu.

0.200 cm^3 of nickel-ion solution required 15.03 cm^3 of 0.01971M-EDTA;

∴ 2.00 cm^3 of the solution contained 10 × 15.03 × 0.01971 × 58.69 = 173.86 mg Ni.

Copper recovered from 2.00 cm^3 of mixed solution ≡ 14.90 cm^3 of 0.01971M-EDTA.

∴ Weight of copper in 2.00 cm^3 of the mixed solution = 14.90 × 0.01971 × 63.54 = 18.66 mg.

Nickel recovered from 2.00/5 cm^3 of the mixed solution ≡ 14.95 cm^3 of 0.01971M-EDTA.

∴ Weight of nickel in 2.00 cm^3 of the mixed solution = 5 × 14.95 × 0.01971 × 58.69 = 86.47 mg.

VIII, 10.   References

1. J. J. Kirkland (ed.) (1971). *Modern Practice of Liquid Chromatography*. New York; Wiley-Interscience.
2. D. O. Campbell and S. R. Buxton (1970). *Ind. Eng. Chem. Process Design Develop.*, **9**, 89; **9**, 95.

VIII, 11.   Selected bibliography

1. W. G. Berl (1951). *Physical Methods of Analysis*. Vol. II. Chromatographic Analysis. New York; Academic Press.
2. C. R. N. Strouts, J. H. Gilfillan, and H. N. Wilson (1955). *Analytical Chemistry, The Working Tools*. Vol. II. Ch. 27. Chromatography. Oxford: Clarendon Press.
3. F. H. Pollard and J. F. W. McOmie (1953). *Chromatographic Methods of Inorganic Analysis*. London; Butterworths.
4. H. G. Cassidy (1957). *Fundamentals of Chromatography*, in A. Weissberger. *Technique of Organic Chemistry*. Vol. 10. New York; Interscience.
5. E. Lederer and M. Lederer (1957). *Chromatography*. Amsterdam; Elsevier Publishing Co.
6. G. W. Ewing (1968). *Instrumental Methods of Chemical Analysis*. Ch. 18. Chromatography. 3rd edn. New York; McGraw-Hill.
7. R. C. Brimley and F. C. Barrett (1953). *Practical Chromatography*. New York; Reinhold.
8. I. M. Hais and K. Macek (eds.) (1964). *Paper Chromatography*. 3rd edn. New York; Academic Press.
9. J. G. Kirchner (1967). *Thin-layer Chromatography*. New York; Interscience.
10. K. Randerath (1968). *Thin Layer Chromatography*. 2nd edn. New York; Academic Press.
11. E. Stahl (ed.) (1964). *Thin Layer Chromatography—A Laboratory Handbook*. New York; Academic Press.
12. F. W. H. M. Merkus (1970). 'Progress in Inorganic Thin-Layer Chromatography'. in *Progress in Separation and Purification*, ed. E. S. Perry and C. J. Van Oss, Vol. 3, p. 233, New York; Wiley.
13. F. H. Pollard, K. W. C. Burton and D. Lyons (1964). 'Thin-layer Chromatography in Inorganic Chemistry', *Lab. Practice*, **13**, 505.
14. J. P. Garel (1965). 'Thin-layer Chromatography III. Application in Inorganic Chemistry', *Bull. Soc. chim. France*, 1899.
15. R. C. Denney (1976). *A Dictionary of Chromatography*. London; Macmillan.

16. J. Michal (1974). *Inorganic Chromatographic Analysis.* London; Van Nostrand Reinhold.
17. L. R. Snyder and J. J. Kirkland (1973). *Introduction to Modern Liquid Chromatography.* Chichester, Sussex; John Wiley and Sons.
18. J. C. Touchsone (ed.) (1973). *Quantitative Thin-Layer Chromatography.* Chichester, Sussex; John Wiley and Sons.

CHAPTER IX GAS CHROMATOGRAPHY

IX, 1. INTRODUCTION. Gas chromatography is a process by which a mixture is separated into its constituents by a moving gas phase passing over a stationary sorbent. The technique is thus similar to liquid–liquid chromatography except that the mobile liquid phase is replaced by a moving gas phase. Gas chromatography is divided into two major categories: **gas–liquid chromatography** (GLC), where separation occurs by partitioning a sample between a mobile gas phase and a thin layer of non-volatile liquid coated on an inert support, and **gas–solid chromatography** (GSC), which employs a solid of large surface area as the stationary phase. The present chapter deals with gas–liquid chromatography and some of its applications in the field of inorganic analysis, particularly in the gas chromatography of metal chelates. Before considering these applications, however, it is appropriate to describe briefly the apparatus used in, and some of the basic principles of, gas chromatography. For more detailed accounts of these topics the texts listed in the bibliography at the end of this chapter should be consulted.

IX, 2. APPARATUS. A gas chromatograph {see block diagram Fig. IX, 1(*a*)} consists essentially of the following parts:

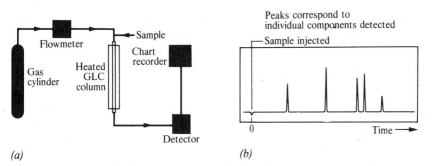

(a) *(b)*

Fig. IX, 1   (*a*) **Block diagram of a gas chromatograph,** (*b*) **Typical chart record**
   Reproduced by permission from R. C. Denney (1970). *The Truth About Breath Tests.* **London; Nelson.**

(1) **a supply of carrier gas from a high-pressure cylinder.** The carrier gas used is either helium, nitrogen, hydrogen or argon, the choice of gas depending on factors such as availability, purity required, consumption and the type of

detector employed. Thus helium is preferred when thermal conductivity detectors are employed because of its high thermal conductivity relative to that of the vapours of most organic compounds. Associated with this high pressure supply of carrier gas are the attendant pressure regulators and flow meters to control and monitor the carrier gas flow; the operating efficiency of the apparatus is very dependent on the maintenance of a constant flow of carrier gas.

(2) sample injection system. Liquid samples are introduced using a microsyringe with hypodermic needle. The latter is inserted through a self-sealing silicone rubber septum and the sample injected smoothly into a heated metal block at the head of the column. Manipulation of the syringe may be regarded as an art developed with practice and the aim must be to introduce the sample in a reproducible manner. The temperature of the sample port should be such that the liquid is rapidly vaporised but without either decomposing or fractionating the sample; a useful rule of thumb is to set the sample port temperature approximately to the boiling point of the least volatile component. For greatest efficiency, the smallest possible sample size (1 to 10 μl) consistent with detector sensitivity should be used.

(3) the column. The actual separation of sample components is effected in the column where the nature of the solid support, type and amount of liquid phase, method of packing, length and temperature are important factors in obtaining the desired resolution. Analytical columns are usually prepared with 2 to 6 mm internal diameter glass tubing or 3 to 10 mm outer diameter metal tubing which is normally coiled for compactness. Glass columns must be used if any of the sample components are decomposed by contact with metal.

The material chosen as the inert support should be of uniform granular size and have good handling characteristics (i.e. be strong enough not to break down in handling) and be capable of being packed into a uniform bed in a column. The surface area of the material should be large so as to promote distribution of the liquid phase as a film and ensure the rapid attainment of equilibrium between the stationary and mobile phases. The most commonly used supports (e.g. Celite) are made from diatomaceous materials which can hold liquid phases in amounts exceeding 20 per cent without becoming too sticky to flow freely and can be easily packed.

Commercial preparations of these supports are available in narrow mesh-range fractions; to obtain particles of uniform size the material should be sieved to the desired particle size range and repeatedly water floated to remove fine particles which contribute to excessive pressure drop in the final column. To a good approximation the height equivalent to a theoretical plate (see Ref. 12) is proportional to the average particle diameter so that theoretically the smallest possible particles should be preferred in terms of column efficiency. Decreasing particle size will, however, rapidly increase the necessary applied gas pressure to achieve flow through the column and in practice the best choice is 80/100 mesh for a 3 mm i.d. column. It may be noted here that for effective packing of any column the internal diameter of the tubing should be at least eight times the diameter of the solid support particles.

The selection of the most suitable liquid phase for a particular separation is crucial. Liquid phases can be broadly classified as follows:

1. Non-polar hydrocarbon-type liquid phases, e.g., paraffin oil (Nujol), squalane, Apiezon L grease and silicone gum rubber; the latter is used for high temperature work (upper limit $\sim 400\,°C$).

2. Compounds of intermediate polarity which possess a polar or polarisable group attached to a large non-polar skeleton, e.g., esters of high molecular weight alcohols such as dinonyl phthalate.

3. Polar compounds containing a relatively large proportion of polar groups, e.g. the carbowaxes (polyglycols).

4. Hydrogen bonding class, i.e. polar liquid phases such as glycol, glycerol, hydroxyacids etc., which possess an appreciable number of hydrogen atoms available for hydrogen bonding.

Studies with metal chelates have shown the best results to be obtained with the silicones, e.g., SE-30 silicone gum rubber.

The column packing is prepared by adding the correct amount of liquid phase dissolved in a suitable solvent (e.g., acetone or dichloromethane) to a weighed quantity of the solid support in a suitable dish. The volatile solvent is removed either by spontaneous evaporation or careful heating, the mixture being gently agitated to ensure a uniform distribution of the liquid phase in the support. Final traces of the solvent may be removed under vacuum and the column packing re-sieved to remove any fines produced during the preparation. The relative amount of stationary liquid phase in the column packing is usually expressed on the basis of the per cent by weight of liquid phase present, e.g., 15 per cent loading indicates that 100 g column packing contains 15 g of liquid phase on 85 g of inert support. The solid support should remain free flowing after being coated with the liquid phase.

(4) the detector. The function of the detector, which is situated at the exit of the separation column, is to sense and measure the small amounts of the separated components present in the carrier gas stream leaving the column. The output from the detector is fed to a recorder which produces a pen-trace called a **chromatogram** (Fig. IX, 1(b)). The choice of detector will depend on factors such as the concentration level to be measured and the nature of the separated components. Thus for work in the microgram range the detector most generally used is probably the thermal conductivity cell, while for work ranging down to the picogram level (e.g., ultra-trace analysis of metals), more sensitive detectors such as those based on ionisation phenomena are required. The detectors most widely used in the gas chromatography of metal chelates are the thermal conductivity, flame ionisation and electron capture detectors, and a brief description of these will be given.

Thermal conductivity detector. Thermal conductivity cells or katharo-meters are the most widely used detectors in gas chromatography. These detectors employ a heated metal filament or a thermistor (a semiconductor of fused metal oxides) to sense changes in the thermal conductivity of the carrier gas stream. Helium and hydrogen are the best carrier gases to use in conjunction with this type of detector since their thermal conductivities are much higher than any other gases; on safety grounds helium is preferred because of its inertness.

In the detector two pairs of matched filaments are arranged in a Wheatstone bridge circuit; two filaments in opposite arms of the bridge are surrounded by the carrier gas only, while the other two filaments are surrounded by the effluent from the chromatographic column. This type of thermal conductivity cell is illustrated in Fig. IX, 2(a) with two gas channels through the cell; a sample channel and a reference channel. When pure carrier gas passes over both the reference and sample filaments the bridge is balanced, but when a vapour emerges from the column the rate of cooling of the sample filaments changes and

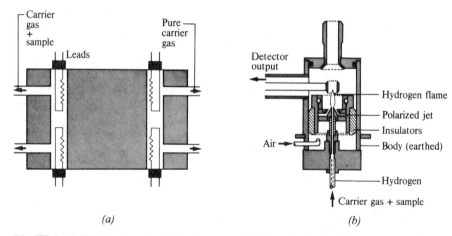

Fig. IX, 2   (a) **Thermal conductivity detector**   (b) **Flame ionization detector**

the bridge becomes unbalanced. The extent of this imbalance is a measure of the concentration of vapour in the carrier gas at that instant, and the out-of-balance signal is fed to a recorder thus producing the chromatogram. The differential technique used is thus based on the measurement of the difference in thermal conductivity between the carrier gas and the carrier gas/sample mixture.

Thermal conductivity detectors have been employed in gas chromatographic studies in metal chelates, e.g., for the quantitative determination of mixtures of beryllium, aluminium, gallium and indium trifluoroacetylacetonates (Ref. 1).

The use of these detectors in studies on metal halides may, however, pose corrosion problems which are not significant with metal chelates, owing to the reactivity of the metal halides.

Flame ionisation detector.   The basis of this detector is that the effluent from the column is mixed with hydrogen and burned in air to produce a flame which has sufficient energy to ionise solute molecules having low ionisation potentials. The ions produced are collected at electrodes and the resulting ion current measured; the burner jet is the negative electrode while the anode is usually a wire or grid extending into the tip of the flame. This is shown diagrammatically in Fig. IX, 2(b).

The combustion of mixtures of hydrogen and air produces very few ions so that with only the carrier gas and hydrogen burning an essentially constant signal is obtained. When, however, carbon-containing compounds are present ionisation occurs and there is a large increase in the electrical conductivity of the flame. Because the sample is destroyed in the flame a stream-splitting device is employed when further examination of the eluate is necessary; this device is inserted between the column and detector and allows the bulk of the sample to by-pass the detector.

The flame ionisation detector is obviously suitable in the study of metal chelates formed with organic ligands and has been used for quantitative work involving chelates of acetylacetone and fluorinated derivatives. It is found that the introduction of fluorine atoms into the chelates diminishes the response of the flame detector, which is opposite to the effect observed with the electron capture detector (Refs. 2 and 3).

Electron capture detector.   Gases at near atmospheric pressure are normally very good electrical insulators but if they become ionised, e.g., by exposure to α- or β-radiation from a radioactive source, they will conduct an electric current. Most ionisation detectors are based on measurement of the increase in current (above that due to the background ionisation of the carrier gas) which occurs when a more readily ionised molecule appears in the gas stream. The principles of operation of the various ionisation detectors have been reviewed by J. E. Lovelock (Ref. 4). The electron capture detector differs from other ionisation detectors in that it exploits the recombination phenomenon, being based on electron capture by compounds having an affinity for free electrons; the detector thus measures a decrease rather than an increase in current.

A β-ray source (commonly a foil containing ^{3}H or ^{63}Ni) is used to generate 'slow' electrons by ionisation of the carrier gas (nitrogen preferred) flowing through the detector. These slow electrons migrate to the anode under a fixed potential and give rise to a steady base-line current. When an electron-capturing gas (i.e., eluate molecules) emerges from the column and reacts with an electron, the net result is the replacement of an electron by a negative ion of much greater mass with a corresponding reduction in current flow.

The electron capture detector is very sensitive to certain molecules such as halogen-containing compounds but insensitive to others such as hydrocarbons. The response of the detector is clearly related to the electron affinity of the solute molecules and, not surprisingly, it exhibits high sensitivity to fluorinated β-diketonate complexes. The electron affinity of metal chelates appears to be a function both of the nature of the metal ion and the extent of the halogenation in the ligand. The detector is of great value in detecting ultra-trace amounts of metals (Ref. 5).

Element selective detectors.   Many samples, e.g. those originating from environmental studies, contain so many constituent compounds that the gas chromatogram obtained is a complex array of peaks. For the analytical chemist, who may be interested in only a few of the compounds present, the replacement of the essentially non-selective type of detector (i.e. thermal conductivity, flame ionisation, etc.) by a system which responds selectively to some property of certain of the eluted species may overcome this problem.

The most common selective detectors in use at present respond to the presence of a characteristic element or group in the eluted compound. The electron capture detector, dealt with in detail above, comes in this category of selective detectors. Similarly, the flame photometric detector, which is a modified form of the flame ionisation detector (F.I.D.), has been specially developed for the detection of phosphorus and sulphur compounds to the extent that its response is 10 000 times greater for compounds containing these elements than for hydrocarbons. Another form of the F.I.D. is the thermionic detector which employs a hydrogen flame burning at a jet with an alkali metal salt tip and has a selective sensitivity for compounds containing halogens, nitrogen, phosphorus and sulphur (Ref. 12). A particularly high degree of specific molecular identification can also be achieved using on-line mass spectrometry or Fourier transform infrared spectrometry, although these are normally employed for organic compounds. The principles and applications of element selective detectors have been reviewed (Ref. 6).

The element specificity of atomic absorption spectrometry has also been used in conjunction with gas chromatography to separate and determine organo-metallic compounds of similar chemical composition, e.g., lead alkyls in petroleum; here lead is determined by AAS for each compound as it passes from the gas chromatograph (Ref. 7).

IX, 3.   PROGRAMMED-TEMPERATURE GAS CHROMATOGRAPHY.

Gas chromatograms are usually obtained with the column kept at a constant temperature. Two important disadvantages result from this isothermal mode of operation:

1. Early peaks are sharp and closely spaced (i.e. resolution is relatively poor in this region of the chromatogram), whereas late peaks tend to be low, broad and widely spaced (i.e., resolution is excessive).
2. Compounds of high boiling point are often undetected, particularly in the study of mixtures of unknown composition and wide boiling point range; the solubilities of the higher-boiling substances in the stationary phase are so large that they are almost completely immobilised at the inlet to the column especially where the latter is operated at a relatively low temperature.

The above consequences of isothermal operation may be largely avoided by using the technique of programmed temperature gas chromatography (PTGC) in which the temperature of the whole column is raised during the sample analysis (the variation of temperature with time may be linear or non-linear according to the separation to be effected). An alternative technique is chromathermography in which a fixed temperature gradient is maintained down the column, the column inlet being kept at the highest temperature and the outlet at the lowest.

Programmed-temperature gas chromatography permits the separation of compounds of a very wide boiling range more rapidly than by isothermal operation of the column. The peaks on the chromatogram are also sharper and more uniform in shape so that, using PTGC, peak heights may be used to obtain accurate quantitative analysis.

IX, 4.   QUANTITATIVE ANALYSIS BY GLC.   The quantitative de-termination of a component in gas chromatography using differential-type detectors of the type previously described is based upon measurement of the recorded peak area or peak height; the latter is more suitable in the case of small peaks, or peaks with narrow band width. In order that these quantities may be related to the amount of solute in the sample two conditions must prevail:

(a) the response of the detector-recorder system must be linear with respect to the concentration of the solute;
(b) factors such as the rate of carrier gas flow, column temperature, etc., must be kept constant or the effect of variation must be eliminated, e.g., by use of the internal standard method.

Peak area is commonly used as a quantitative measure of a particular component in the sample and can be measured by one of the following techniques:

1. Planimetry.   The planimeter is a mechanical device which enables the peak area to be measured by tracing the perimeter of the peak. The method is slow but can give accurate results with experience in manipulation of the planimeter. Accuracy and precision, however, decrease as peak area diminishes.

2. Geometrical methods.   In the so-called triangulation methods, tangents are drawn to the inflection points of the elution peak and these two lines, together

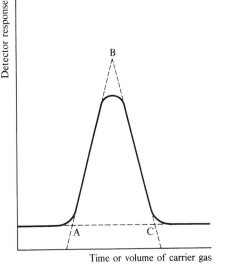

Fig. IX, 3   Measurement of peak area by triangulation

with the base line form a triangle (Fig. IX, 3); the area of the latter is calculated as one-half the product of the base length times the peak height, the value obtained being about 97 per cent of the actual area under the chromatographic peak when this is Gaussian in shape (Ref. 8).

The area may also be computed as the product of the peak height times the width at half the peak height, i.e., *by the height × width at half height method.* Since the exact location of the tangents (required for the triangulation method) to the curve is not easily determined it is in general more accurate to use the method based on width at half-height.

3. Integration by weighing. The chromatographic peak is carefully cut out of the chart and the paper weighed on an analytical balance. The accuracy of the method is clearly dependent upon the constancy of the thickness and moisture content of the chart paper, and it is usually preferable (unless an automatic integrator is available) to use geometrical methods.

4. Automatic integration. Integrators of this type may be divided into two groups, viz., the mechanical type such as the ball and disc integrator, and the more complex electronic type such as the digital integrator. These devices are designed for attachment to the detector/recorder system so that integration of the area may be carried out simultaneously with the recording of the chromatogram. Electronic integrators give the best precision but are very expensive.

5. Data evaluation. It is, of course, necessary to correlate peak area with the amount or concentration of a particular solute in the sample; this is usually done by construction of a calibration graph of peak area versus amount of solute. The calibration determinations must be carried out under conditions which are as similar as possible to those used in the chromatographic study of the sample.

Quantitative analysis using the internal standard method. The height and area of chromatographic peaks are affected not only by the amount of sample but also by fluctuations of the carrier gas flow rate, the column and detector temperatures etc., i.e., by variations of those factors which influence the sensitivity and response of the detector. The effect of such variations can be eliminated by use of the internal standard method in which a known amount of a reference substance is added to the sample to be analysed before injection into the column. The requirements for an effective internal standard (Section **IV, 5**) have been specified as follows (Ref. 9):

(*a*) it should give a completely resolved peak, but should be eluted close to the components to be measured;

(*b*) its peak height or peak area should be similar in magnitude to those of the components to be measured;

(*c*) it should be chemically similar to but not present in the original sample.

The procedure comprises the addition of a constant amount of internal standard to a fixed volume of several synthetic mixtures which contain varying known amounts of the component to be determined. The resulting mixtures are chromatographed and a calibration curve is constructed of the per cent of component in the mixtures against the ratio of component peak area/standard peak area. The analysis of the unknown mixture is carried out by addition of the same amount of internal standard to the specified volume of the mixture; from the observed ratio of peak areas the solute concentration is read off using the calibration curve.

IX, 5.   GAS CHROMATOGRAPHY OF METAL CHELATES.

Although inorganic compounds are generally not so volatile as are organic compounds, gas chromatography has been applied in the study of certain inorganic compounds which possess the requisite properties. If gas chromatography is to be used for metal separation and quantitative analysis, the types of compounds which can be used are limited to those that can be readily formed in virtually quantitative and easily reproducible yield. This feature, together with the requirements of sufficient volatility and thermal stability necessary for successful gas chromatography, make neutral metal chelates the most favourable compounds for use in metal analysis. β-Diketone ligands, e.g., acetylacetone and the fluorinated derivatives, trifluoroacetylacetone (TFA) and hexafluoroacetylacetone (HFA) form stable, volatile chelates with aluminium, beryllium, chromium(III) and a number of other metal ions; it is thus possible to chromatograph a wide range of metals as their β-diketone chelates.

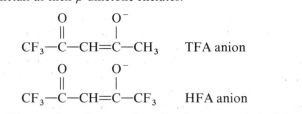

The number of reported applications to analytical determinations at the trace level appear to be few, probably the best known being the determination of beryllium in various samples. The method generally involves the formation of the volatile beryllium trifluoroacetylacetonate chelate, its solvent extraction into benzene with subsequent separation and analysis by gas chromatography (Ref. 10).

A number of important requirements must be met if gas chromatography is to be successfully applied to metal analysis, and these will be briefly considered.

1. Ease of formation of the metal compound. Chelating agents of the β-diketone type form complexes of high solvolytic stability by simple reactions with metal ions. The reactions may take place in aqueous or non-aqueous media and are pH dependent, a feature which provides a measure of selectivity when mixtures of metal ions are being investigated. The solubility of these chelates in a number of organic solvents permits their solvent extraction after preparation in aqueous media, again providing enhanced selectivity, and yielding suitable samples for injection into the chromatographic column.

2. Volatility. The most important requirement is that the metal compound must be sufficiently volatile to be chromatographed in the gas phase. It is, of course, not essential for the column to be operated at a temperature above the boiling point of the compound and satisfactory elution will usually occur even if the compound only possesses a vapour pressure of the order of a few mm at the column temperature. The following types of metal compounds are volatile at reasonably low temperatures: metal alkyls and alkoxides, metal carbonyls, certain metal hydrides, metal cyclopentadienyls and related complexes with π-acceptor ligands, and neutral metal chelates such as those formed with β-diketone ligands. When other factors, such as thermal stability of the compound, are considered the choice of metal compound is limited to two major groups— metal halides and metal β-diketone chelates. An interesting feature of the latter group is the markedly greater volatility of the fluorocarbon chelates (TFA and HFA chelates) as compared with the corresponding acetylacetonates. Thus chromium(III) hexafluoroacetylacetonate is eluted rapidly at quite low column temperatures ranging down to 30 °C, whereas for a similar elution of chromium(III) acetylacetonate the column temperature required must be about 150 °C. The possibility of operating at lower column temperatures by using the fluorocarbon chelates is clearly important in minimising any tendency to thermal decomposition.

3. Thermal stability. An important criterion for quantitative work is that the compounds should possess sufficient thermal stability to enable them to be eluted without degradation; it may be possible, however, to obtain quantitative results even when thermal decomposition occurs provided that it is only slight and that the extent of the decomposition is reproducible under the given conditions.

Thermal degradation of the sample is indicated by the following observations:
(*a*) the presence of a residue in the injection port, although this may also be due to incomplete vapourisation of the sample;
(*b*) appearance of spurious chromatographic peaks;
(*c*) discoloration of the column packing material.

A more complete picture of the composition of the eluted material corresponding to each gas chromatographic peak will, however, be obtained by collecting and identifying the eluted material. A glass U-tube cooled in solid carbon dioxide provides an adequate means of trapping metal chelates in the effluent; the U-tube is connected to the exit port of the column with a short length of Teflon tubing. Identification of small effluent samples may be achieved using melting point determination, but physical techniques such as visible and ultraviolet spectroscopy, infrared spectroscopy, etc., can often provide a more detailed analysis of the material.

4. Solvolytic stability. In addition to being thermally stable, the metal compounds must possess solvolytic stability particularly in relation to the liquid stationary phase in the column. If the molecules of the liquid phase function effectively as ligands, solvolysis may occur through ligand substitution in the metal complex. The compounds should also not react chemically with the solid stationary support or the materials of construction of the column. The reactivity of the metal halides gives rise to a number of difficulties:

(a) the halides are easily hydrolysed and special precautions must be taken to remove traces of moisture from the carrier gas;

(b) at the elevated temperatures in the column the halides react with many of the liquid stationary phases so that careful choice of the latter is required;

(c) metal surfaces in the flow system are often attacked and corroded.

Thus, despite the volatility of a number of metal halides their usefulness in gas chromatography is limited by these considerations and the use of metal chelates offers greater scope in metal analysis.

5. Health hazards.   The effluent emerging from the gas chromatographic instrument, if not trapped, will, of course, diffuse into the laboratory atmosphere and may constitute a health hazard. This may happen if metals such as lead, mercury and zinc, which act as cumulative poisons, are allowed to pass into the laboratory atmosphere as volatile metal complexes. To prevent such atmospheric contamination the effluent stream should either be passed to a fume hood or through a cold trap to remove the volatile metal compound. Appropriate precautions must, of course, also be taken if the sample contains radioactive material or when radioisotopes are employed in ionisation detectors.

IX, 6. DETERMINATION OF ALUMINIUM BY GAS CHROMATOGRAPHIC ANALYSIS OF ITS TRIS(ACETYLACETONATO) COMPLEX.

The purpose of this experiment is to illustrate the application of gas chromatographic analysis to the quantitative determination of trace amounts of metals as their chelate complexes. The procedure described for the determination of aluminium may be adapted for the separation and determination of aluminium and chromium(III) as their acetylacetonate (Ref. 3).

Sample.   The solvent extraction of aluminium from aqueous solution using acetylacetone (Ref. 11) can provide a suitable sample solution for gas chromatographic analysis.

Take 5 cm^3 of a solution containing about 15 mg of aluminium and adjust the pH to between 4 and 6. Equilibrate the solution for 10 minutes with two successive 5 cm^3 portions of a solution made up of equal volumes of acetylacetone (pure, redistilled) and A.R. chloroform. Combine the organic extracts. Fluoride ion causes serious interference to the extraction and must be previously removed.

Introduce a 0.30 μl portion of the solvent extract into the gas chromatograph. It is found that solutions of concentrations greater than 0.3M are unsuitable as they deposit solid and thus cause a blockage of the 1 μl microsyringe used for the injection of the sample. The syringe is flushed several times with the sample solution, filled with the sample to the required volume, excess liquid wiped from the tip of the needle and the sample injected into the chromatograph.

Apparatus.   A suitable instrument is the Pye 104 Chromatograph equipped with a flame ionisation detector, with an Autolab 6300 Digital Integrator linked to a Westrex Teletype for printout. The use of a digital integrator is particularly convenient for quantitative determinations, but other methods of measuring peak area may be used (Section **IX, 4**).

Pure nitrogen (white spot), at a flow rate of $40 \, cm^3 \, min^{-1}$, is used as carrier gas. The dimensions of the glass column are 1.6 m length and 6 mm o.d., and is packed with 5 per cent by weight SE-30 on Chromosorb W as the stationary phase. The column is maintained at a temperature of 165 °C.

Procedure. Extract a series of aqueous aluminium solutions containing 5 to 25 mg aluminium in $5 \, cm^3$, using the procedure described above under Sample. Calibrate the apparatus by injecting $0.30 \, \mu l$ of each extract into the column and recording the peak area on the chromatogram. Plot a graph of peak area against concentration.

Determine aluminium (present as its acetylacetonate) in the sample solution by injecting $0.30 \, \mu l$ into the column. Record the peak area obtained and read off the aluminium concentration from the calibration graph.

IX, 7. References

1. J. E. Schwarberg, R. W. Moshier and J. H. Walsh (1964). *Talanta*, **11**, 1213.
2. D. K. Albert (1964). 'Comparison of Electron Capture and Hydrogen Flame Detectors for Gas Chromatographic Determination of Trace Amounts of Metal Chelates', *Anal. Chem.*, **36**, 2034.
3. R. D. Hill and H. Gesser (1963). 'An Investigation into the Quantitative Gas Chromatographic Analysis of Metal Chelates using a Hydrogen Flame Ionisation Detector', *J. Gas Chromatography*, **1**, 11.
4. J. E. Lovelock (1961). 'Ionisation Methods for Analysis of Gases and Vapors', *Anal. Chem.*, **33**, 162.
5. W. D. Ross, R. E. Sievers and G. Wheeler Jr. (1965). 'Quantitative Ultratrace Analysis of Mixtures of Metal Chelates by Gas Chromatography', *Anal. Chem.*, **37**, 598.
6. D. F. S. Natusch and T. M. Thorpe (1973). 'Element Selective Detectors in Gas Chromatography', *Anal. Chem.*, **45**, 1184A.
7. P. R. Ballinger and I. M. Whittemore (1968). *Proceedings of the American Chemical Society, Div. of Petroleum Chemistry*, **13**, [3], 133.
8. L. Condal-Bosch (1964). 'Some Problems of Quantitative Analysis in Gas Chromatography', *J. Chem. Educ.*, **41**, A235.
9. D. Harvey and D. E. Chalkley (1955). *Fuel*, **34**, 191.
10. R. S. Barratt (1973). 'Analytical Applications of Gas Chromatography of Metal Chelates', *Proc. Soc. Analyt. Chem.*, **10**, 167.
11. I. M. Kolthoff and P. J. Elving (ed.) (1966). *Treatise on Analytical Chemistry*. Part II, Vol. 4, p. 392. New York; Interscience.
12. R. C. Denney (1976). *A Dictionary of Chromatography*. London; Macmillan.

IX, 8. Selected bibliography

1. S. Dal Nogare and R. S. Juvet Jr. (1962). *Gas–Liquid Chromatography*. New York; Interscience.
2. H. Purnell (1962). *Gas Chromatography*. New York; Wiley.
3. A. I. M. Keulemans (1959). *Gas Chromatography*. 2nd edn. New York; Reinhold.
4. R. W. Moshier and R. E. Sievers (1965). *Gas Chromatography of Metal Chelates*. Oxford; Pergamon Press.
5. W. W. Brandt (1963). 'Gas Chromatography', in *Technique of Inorganic Chemistry*, ed. Jonassen and Weissberger, Vol. 3, p. 1. New York; Interscience.
6. J. Trenchant (ed.) (1969). *Practical Manual of Gas Chromatography*. Amsterdam and London; Elsevier.

PART D   **TITRIMETRY AND GRAVIMETRY**

CHAPTER X   TITRIMETRIC ANALYSIS

A.   Theoretical Considerations

X, 1. TITRIMETRIC ANALYSIS. The term titrimetric analysis refers to quantitative chemical analysis carried out by determining the volume of a solution of accurately known concentration which is required to react quantitatively with the solution of the substance to be determined. The solution of accurately known strength is called the **standard solution**; see Section **X, 3**. The weight of the substance to be determined is calculated from the volume of the standard solution used and the known laws of stoichiometry.

The standard solution is usually added from a **burette**. The process of adding the standard solution until the reaction is just complete is termed a **titration**, and the substance to be determined is **titrated**. The point at which this occurs is called the **equivalence point** or the **theoretical** (or **stoichiometric**) **end-point**. The completion of the titration should, as a rule, be detectable by some change, unmistakable to the eye, produced by the standard solution itself (e.g., potassium permanganate) or, more usually, by the addition of an auxiliary reagent, known as an **indicator**. After the reaction between the substance and the standard solution is practically complete, the indicator should give a clear visual change (either a colour change or the formation of a turbidity) in the liquid being titrated. The point at which this occurs is called the **end-point of the titration**. In the ideal titration the visible end point will coincide with the stoichiometric or theoretical end-point. In practice, however, a very small difference usually occurs; this represents the **titration error**. The indicator and the experimental conditions should be so selected that the difference between the visible end-point and the equivalence point is as small as possible.

The term volumetric analysis was formerly used, but it has now been replaced by **titrimetric analysis**, since it is considered that the latter expresses the process of titration rather better, and the former may be confused with measurements of volume, such as those involving gases. The reagent of known concentration is called the **titrant** and the substance being titrated is termed the **titrand**. The alternative name has not been extended to apparatus used in the various operations: thus the terms volumetric glassware and volumetric flasks are still retained, but it is probably better to employ the expressions graduated glassware and graduated flasks which are used throughout this book.

For use in titrimetric analysis a reaction must fulfil the following conditions:

1. There must be a simple reaction which can be expressed by a chemical equation; the substance to be determined should react completely with the reagent in stoichiometric or equivalent proportions.

2. The reaction should be practically instantaneous or proceed with very great speed. (Most ionic reactions satisfy this condition.) In some cases the addition of a *catalyst* increases the speed of a reaction.

3. There must be a marked change in free energy leading to alteration in some physical or chemical property of the solution at the equivalence point.

4. An indicator should be available which, by a change in physical properties (colour or formation of a precipitate), should sharply define the end point of the reaction. [If no visible indicator is available for the detection of the equivalence point, the latter can often be determined by following during the course of the titration: (a) the potential between an indicator electrode and a reference electrode (**potentiometric titration**, see Chapter XIV); (b) the change in electrical conductivity of the solution (**conductometric titration**, see Chapter XV); (c) the current which passes through the titration cell between an indicator electrode (e.g., the dropping mercury electrode) and a depolarised reference electrode (e.g., the saturated calomel electrode) at a suitable applied e.m.f. (**amperometric titration**, see Chapter XVII); or (d) the change in absorbance of the solution (**spectrophotometric titration**, see Section **XVIII, 39**).]

Titrimetric methods are, as a rule, susceptible of high precision (1 part in 1000) and possess several advantages, wherever applicable, over gravimetric methods. They need simpler apparatus, and are, generally, quickly performed; tedious and difficult separations can often be avoided. The following are required for titrimetric analysis: (i) calibrated measuring vessels, including burettes, pipettes, and measuring flasks (see Chapter III); (ii) substances of known purity for the preparation of standard solutions; (iii) a visual indicator or an instrumental method for detecting the completion of the reaction.

X, 2. CLASSIFICATION OF REACTIONS IN TITRIMETRIC ANALYSIS.
The reactions employed in titrimetric analysis fall into two main classes:

(a) Those in which no change in oxidation state occurs; these are dependent upon the combination of ions.

(b) Oxidation–reduction reactions; these involve a change of oxidation state or, otherwise expressed, a transfer of electrons.

For purposes of convenience, however, these two types of reactions are divided into four main classes:

1. Neutralisation reactions, or acidimetry and alkalimetry. These include the titration of free bases, or those formed from salts of weak acids by hydrolysis, with a standard acid (**acidimetry**), and the titration of free acids, or those formed by the hydrolysis of salts of weak bases, with a standard base (**alkalimetry**). These reactions involve the combination of hydrogen and hydroxide ions to form water.

2. Complex formation reactions. These depend upon the combination of ions, other than hydrogen or hydroxide ions, to form a soluble, slightly dissociated ion or compound, as in the titration of a solution of a cyanide with silver nitrate ($2CN^- + Ag^+ \rightleftharpoons [Ag(CN)_2]^-$) or of chloride ion with mercury(II) nitrate solution ($2Cl^- + Hg^{2+} \rightleftharpoons HgCl_2$).

Ethylenediaminetetra–acetic acid, largely as the disodium salt EDTA, is a very important reagent for complex formation titrations and, indeed, EDTA has beome one of the most important reagents used in titrimetric analysis. The use of metal ion indicators has greatly enhanced its value in titrimetry. The subject is discussed fully later in this chapter (Part A.2).

3. Precipitation reactions. These depend upon the combination of ions to form a simple precipitate as in the titration of silver ion with a solution of a chloride (Section **X, 29**). No change in oxidation state occurs.

4. Oxidation–reduction reactions. Under this heading are included all reactions involving change in oxidation number or transfer of electrons (Section **X, 3**) among the reacting substances. The standard solutions are either oxidising or reducing agents. The principal oxidising agents are potassium permanganate, potassium dichromate, cerium(IV) sulphate, iodine, potassium iodate, and potassium bromate. Frequently used reducing agents are iron(II) and tin(II) compounds, sodium thiosulphate, arsenic(III) oxide, mercury(I) nitrate, vanadium(II) chloride or sulphate, chromium(II) chloride or sulphate, and titanium(III) chloride or sulphate.

X, 3. STANDARD SOLUTIONS. A **standard solution** is one which contains a known weight of the reagent in a definite volume of solution, and for many years concentrations were expressed in terms of molarity (i.e., number of moles per litre) and normality (i.e., number of equivalents per litre). With the adoption by the International Union of Pure and Applied Chemistry of the mole as a base unit of quantity with the definition,

> 'The mole is the amount of substance which contains as many elementary units as there are atoms in 0.012 kilogram of carbon-12. The elementary unit must be specified and may be an atom, a molecule, an ion, a radical, an electron or other particle or a specified group of such particles',

the mole is no longer a unit of mass, but is one of amount of substance, and terms such as gram-molecule, gram-ion, etc., are obsolete.

With the introduction of this definition came proposals that the terms 'molarity' (the number of moles of solute per litre of solution), equivalent weight, and normality (the number of equivalents of solute per litre of solution), should be abandoned. However, experience has shown that there are certain practical advantages in retaining the use of the terms **equivalent** and **normal solution**, and the latest IUPAC recommendations (Ref. 1) suggest the following definitions:

> 'The **equivalent** of a substance is that amount of it which, in a specified reaction, combines with, releases or replaces that amount of hydrogen which is combined with 3 grams of carbon-12 in methane $^{12}CH_4$.'

In this definition, the amount of hydrogen referred to may be replaced by the equivalent amount of electricity or by one equivalent of any other substance, but the reaction to which the definition is applied must be clearly specified.

Although the terms mole and equivalent as now defined refer to an amount of substance, each definition does in fact refer to a specified mass of carbon-12, and hence we can say for example

1 mole of Hg_2Cl_2 has a mass of 0.47208 kg
1 mole of $Na_2CO_3 \cdot 10H_2O$ has a mass of 0.286004 kg
1 mole of H_2SO_4 has a mass of 0.098078 kg
1 equivalent of $Na_2CO_3 \cdot 10H_2O$ has a mass of 0.143002 kg
1 equivalent of H_2SO_4 has a mass of 0.049039 kg

and it is therefore quite permissible to refer to weighing out one mole of a certain reagent, because this refers to a definite mass of the substance.

A **normal solution** is defined as a solution containing one equivalent of a defined species per dm^3 according to the specified reaction, and a **molar solution** as one containing one mole of a defined species per dm^3, i.e., a concentration of 1 mol dm^{-3}.

As already explained (Section **III, 12**), the term litre is accepted as a special name for the cubic decimetre, but with the suggestion that the litre should not be used to express results of high precision (Ref. 2) the recommendations of Ref. 1 can be summarised as follows:

1. Wherever possible, concentrations should be expressed in terms of moles per cubic decimetre ($mol \, dm^{-3}$ or $mol \, 1^{-1}$).
2. The symbol M to signify $mol \, dm^{-3}$ should be retained, but the term molarity should be discontinued.
3. The use of the term equivalent, defined as above, and given in the appropriate SI unit should be retained, as should likewise the term normality based on the redefined equivalent.

The above definition of normal solution utilises the term 'equivalent'. This quantity varies with the type of reaction, and, since it is difficult to give a clear definition of 'equivalent' which will cover all reactions, it is proposed to discuss this subject in some detail below. It often happens that the same compound possesses different equivalents in different chemical reactions. The situation may therefore arise in which a solution has normal concentration when employed for one purpose, and a different normality when used in another chemical reaction.

Neutralisation reactions. The **equivalent of an acid** is that mass of it which contains 1.008 (more accurately 1.0078) g of replaceable hydrogen. The equivalent of a monoprotic acid, such as hydrochloric, hydrobromic, hydriodic, nitric, perchloric, or acetic acid, is identical with the mole. A normal solution of a monoprotic acid will therefore contain 1 mole per dm^3 of solution. The equivalent of a diprotic acid (e.g., sulphuric or oxalic acid), or of a triprotic acid (e.g., phosphoric acid) is likewise $\frac{1}{2}$ and $\frac{1}{3}$ respectively of the mole.

The **equivalent of a base** is that mass of it which contains one replaceable hydroxyl group, i.e., 17.008 g of ionisable hydroxyl; 17.008 g of hydroxyl are equivalent to 1.008 g of hydrogen. The equivalents of sodium hydroxide and potassium hydroxide are the mole, of calcium hydroxide, strontium hydroxide, and barium hydroxide half a mole.

Salts of strong bases and weak acids possess alkaline reactions in aqueous solution because of hydrolysis (Section **II, 17**). A mole of sodium carbonate, with methyl orange as indicator, reacts with 2 moles of hydrochloric acid to form 2 moles of sodium chloride; hence its equivalent is half a mole. Sodium tetraborate, under similar conditions, also reacts with 2 moles of hydrochloric acid, and its equivalent is, likewise, half a mole.

Complex formation and precipitation reactions. Here the equivalent is the mass of the substance which contains or reacts with 1 mole of a univalent cation M^+ (which is equivalent to 1.008 g of hydrogen), $\frac{1}{2}$ mole of a bivalent cation M^{2+}, $\frac{1}{3}$ mole of a trivalent cation M^{3+}, etc. For the cation, the equivalent is the mole divided by the valency. For a reagent which reacts with this cation, the equivalent is the mass of it which reacts with one equivalent of the cation. The **equivalent of a salt** in a precipitation reaction is the mole divided by the total valency of the *reacting* ion. Thus the equivalent of silver nitrate in the titration of chloride ion is the mole.

In a complex formation reaction the equivalent is most simply deduced by writing down the ionic equation of the reaction. For example, the equivalent of potassium cyanide in the titration with silver ions is 2 moles, since the reaction is:

$$2CN^- + Ag^+ \rightleftharpoons [Ag(CN)_2]^-$$

In the titration of zinc ion with potassium hexacyanoferrate(II) solution:

$$3Zn^{2+} + 2K_4Fe(CN)_6 = 6K^+ + K_2Zn_3[Fe(CN)_6]_2$$

the equivalent of the hexacyanoferrate(II) is one-third of the mole. For other examples of complex formation reactions, see Sections **X, 19–27**; it is apparent that in many complexation reactions it is preferable to work in moles rather than equivalents.

Oxidation–reduction reactions. The equivalent of an oxidising or reducing agent is most simply defined as that mass of the reagent which reacts with or contains 1.008 g of available hydrogen or 8.000 g of available oxygen. By 'available' is meant capable of being utilised in oxidation or reduction. The amount of available oxygen may be indicated by writing the *hypothetical* equation, e.g.,

$$2KMnO_4 = K_2O + 2MnO + 5O$$

i.e., in acid solution $2KMnO_4$ gives up 5 atoms of available oxygen, which is taken up by the reducing agent, hence its equivalent is $2KMnO_4/10$. For potassium dichromate in acid solution, the hypothetical equation is:

$$K_2Cr_2O_7 = K_2O + Cr_2O_3 + 3O$$

The equivalent is $K_2Cr_2O_7/6$. This elementary treatment is limited in application, but is useful for beginners.

A more general and fundamental view is obtained by a consideration of: (*a*) the number of electrons involved in the partial ionic equation representing the reaction, and (*b*) the change in the 'oxidation number' of a significant element in the oxidant or reductant. Both methods will be considered in some detail.

In quantitative analysis we are chiefly concerned with reactions which take place in solution, i.e., ionic reactions. We shall therefore limit our discussion of oxidation–reduction to such reactions. The oxidation of iron(II) chloride by chlorine in aqueous solution may be written:

$$2FeCl_2 + Cl_2 = 2FeCl_3$$

or may be expressed ionically:

$$2Fe^{2+} + Cl_2 = 2Fe^{3+} + 2Cl^-$$

The ion Fe^{2+} is converted into ion Fe^{3+} (oxidation), and the neutral chlorine molecule into negatively charged chloride ions Cl^- (reduction); the conversion of Fe^{2+} into Fe^{3+} requires the loss of one electron, and the transformation of the neutral chlorine molecule into chloride ions necessitates the gain of two electrons. This leads to the view that, for reactions in solution, oxidation is a process involving a loss of electrons, as in

$$Fe^{2+} - e = Fe^{3+}$$

and reduction is the process resulting in a gain of electrons, as in

$$Cl_2 + 2e = 2Cl^-$$

In the actual oxidation–reduction process electrons are transferred from the reducing agent to the oxidising agent. This leads to the following definitions. **Oxidation** is the process which results in the loss of one or more electrons by atoms or ions. **Reduction** is the process which results in the gain of one or more electrons by atoms or ions. An **oxidising agent** is one that gains electrons and is reduced; a **reducing agent** is one that loses electrons and is oxidised.

In all oxidation–reduction processes (or **redox** processes) there will be a reactant undergoing oxidation and one undergoing reduction, since the two reactions are complementary to one another and occur simultaneously—one cannot take place without the other. The reagent suffering oxidation is termed the reducing agent or **reductant**, and the reagent undergoing reduction is called the oxidising agent or **oxidant**. The study of the electron changes in the oxidant and reductant forms the basis of the **ion-electron method** for balancing ionic equations. The equation is accordingly first divided into two balanced, partial equations representing the oxidation and reduction respectively. It must be remembered that the reactions take place in aqueous solution so that in addition to the ions supplied by the oxidant and reductant, the molecules of water H_2O, hydrogen ions H^+, and hydroxide ions OH^- are also present, and may be utilised in balancing the partial ionic equation. The unit change in oxidation or reduction is a charge of one electron, which will be denoted by e. To appreciate the principles involved, let us consider first the reaction between iron(III) chloride and tin(II) chloride in aqueous solution. The partial ionic equation for the reduction is:

$$Fe^{3+} \longrightarrow Fe^{2+} \tag{1}$$

and for the oxidation is:

$$Sn^{2+} \longrightarrow Sn^{4+} \tag{2}$$

The equations must be balanced not only with regard to the number and kind of atoms, but also electrically, that is, the net electric charge on each side must be the same. Equation (1) can be balanced by adding one electron to the left-hand side:

$$Fe^{3+} + e \rightleftharpoons Fe^{2+} \tag{1'}$$

and equation (2) by adding two electrons to the right-hand side:

$$Sn^{2+} \rightleftharpoons Sn^{4+} + 2e \tag{2'}$$

These partial equations must then be multiplied by coefficients which result in the number of electrons utilised in one reaction being equal to those liberated in the other. Thus equation (1') must be multiplied by two, and we have:

$$2Fe^{3+} + 2e \rightleftharpoons 2Fe^{2+} \tag{1''}$$
$$Sn^{2+} \rightleftharpoons Sn^{4+} + 2e \tag{2''}$$

Adding (1'') and (2''), we obtain:

$$2Fe^{3+} + Sn^{2+} + 2e \rightleftharpoons 2Fe^{2+} + Sn^{4+} + 2e$$

and by cancelling the electrons common to both sides, the simple ionic equation is obtained:

$$2Fe^{3+} + Sn^{2+} = 2Fe^{2+} + Sn^{4+}$$

The following facts must be borne in mind. All strong electrolytes are completely dissociated; hence only the ions actually taking part or resulting from the reaction need appear in the equation. Substances which are only slightly ionised, such as water, or which are sparingly soluble and thus yield only a small concentration of ions, e.g., silver chloride and barium sulphate, are, in general, written as molecular formulae because they are present mainly in the undissociated state.

The complete rules for the application of the ion-electron method may be expressed as follows:

(a) ascertain the products of the reaction;
(b) set up a partial equation for the oxidising agent;
(c) set up a partial equation for the reducing agent in the same way;
(d) multiply each partial equation by a factor so that when the two are added the electrons just compensate each other;
(e) add the partial equations and cancel out substances which appear on both sides of the equation.

A few examples follow.

Reaction I: the reduction of potassium permanganate by iron(II) sulphate in the presence of dilute sulphuric acid.

The first partial equation (reduction) is:

$$MnO_4^- \longrightarrow Mn^{2+}$$

To balance atomically, $8H^+$ is required:

$$MnO_4^- + 8H^+ \longrightarrow Mn^{2+} + 4H_2O$$

and to balance it electrically $5e$ is needed on the left-hand side:

$$MnO_4^- + 8H^+ + 5e \rightleftharpoons Mn^{2+} + 4H_2O$$

The second partial equation (oxidation) is:

$$Fe^{2+} \longrightarrow Fe^{3+}$$

To balance this electrically one electron must be added to the right-hand side or subtracted from the left-hand side:

$$Fe^{2+} \rightleftharpoons Fe^{3+} + e$$

Now the gain and loss of electrons must be equal. One permanganate ion utilises 5 electrons, and one iron(II) ion liberates 1 electron; hence the two partial equations must apply in the ratio of 1:5.

$$
\begin{array}{l}
MnO_4^- + 8H^+ + 5e \rightleftharpoons Mn^{2+} + 4H_2O \\
5(Fe^{2+} \rightleftharpoons Fe^{3+} + e) \\
\hline
MnO_4^- + 8H^+ + 5Fe^{2+} = Mn^{2+} + 5Fe^{3+} + 4H_2O
\end{array}
$$

or

Reaction II: the interaction of potassium dichromate and potassium iodide in the presence of dilute sulphuric acid.

$$Cr_2O_7^{2-} \longrightarrow Cr^{3+}$$
$$Cr_2O_7^{2-} + 14H^+ \longrightarrow 2Cr^{3+} + 7H_2O$$

To balance electrically, add $6e$ to the left-hand side:

$$Cr_2O_7^{2-} + 14H^+ + 6e \rightleftharpoons 2Cr^{3+} + 7H_2O$$

The various stages in the deduction of the second partial equation are

$$I^- \longrightarrow I_2$$
$$2I^- \longrightarrow I_2$$
$$2I^- \rightleftharpoons I_2 + 2e$$

One dichromate ion uses $6e$, and two iodide ions liberate $2e$; hence the two partial equations apply in the ratio of $1:3$:

$$Cr_2O_7^{2+} + 14H^+ + 6e \rightleftharpoons 2Cr^{3+} + 7H_2O$$
$$3(2I^- \rightleftharpoons I_2 + 2e)$$

or $\quad Cr_2O_7^{2-} + 14H^+ + 6I^- = 2Cr^{3+} + 7H_2O + 3I_2$

We can now apply our knowledge of partial ionic equations to the subject of equivalents. The standard oxidation–reduction process is $H \rightleftharpoons H^+ + e$, where e represents an electron per atom, or the Avogadro's number of electrons per mole. If we know the change in the number of electrons per ion in any oxidation–reduction reaction, the equivalent may be calculated. The **equivalent of an oxidant or a reductant** is the mole divided by the number of electrons which 1 mole of the substance gains or loses in the reaction, e.g.:

$$MnO_4^- + 8H^+ + 5e \rightleftharpoons Mn^{2+} + 4H_2O. \quad \text{Eq.} = MnO_4^-/5 = KMnO_4/5.$$

$$Cr_2O_7^{2-} + 14H^+ + 6e \rightleftharpoons 2Cr^{3+} + 7H_2O. \quad \text{Eq.} = Cr_2O_7^{2-}/6 = K_2Cr_2O_7/6.$$

$$Fe^{2+} \rightleftharpoons Fe^{3+} + e. \qquad\qquad\qquad \text{Eq.} = Fe^{2+}/1 = FeSO_4/1.$$

$$C_2O_4^{2-} \rightleftharpoons 2CO_2 + 2e. \qquad\qquad \text{Eq.} = C_2O_4^{2-}/2 = H_2C_2O_4/2.$$

$$SO_3^{2-} + H_2O \rightleftharpoons SO_4^{2-} + 2H^+ + 2e. \qquad \text{Eq.} = SO_3^{2-}/2 = Na_2SO_3/2.$$

For convenience of reference the partial ionic equations for a number of oxidising and reducing agents are collected in Table X, 1.

Table X, 1. Ionic equations for use in the calculation of the equivalents of oxidising and reducing agents

Substance	Partial ionic equation
OXIDANTS	
Potassium permanganate (acid)	$MnO_4^- + 8H^+ + 5e \rightleftharpoons Mn^{2+} + 4H_2O$
Potassium permanganate (neutral)	$MnO_4^- + 2H_2O + 3e \rightleftharpoons MnO_2 + 4OH^-$
Potassium permanganate (strongly alkaline)	$MnO_4^- + e \rightleftharpoons MnO_4^{2-}$
Cerium(IV) sulphate	$Ce^{4+} + e \rightleftharpoons Ce^{3+}$
Potassium dichromate	$Cr_2O_7^{2-} + 14H^+ + 6e \rightleftharpoons 2Cr^{3+} + 7H_2O$
Chlorine	$Cl_2 + 2e \rightleftharpoons 2Cl^-$
Bromine	$Br_2 + 2e \rightleftharpoons 2Br^-$
Iodine	$I_2 + 2e \rightleftharpoons 2I^-$
Iron(III) chloride	$Fe^{3+} + e \rightleftharpoons Fe^{2+}$
Potassium bromate	$BrO_3^- + 6H^+ + 6e \rightleftharpoons Br^- + 3H_2O$
Potassium iodate	$IO_3^- + 6H^+ + 6e \rightleftharpoons I^- + 3H_2O$
Sodium hypochlorite	$ClO^- + H_2O + 2e \rightleftharpoons Cl^- + 2OH^-$
Hydrogen peroxide	$H_2O_2 + 2H^+ + 2e \rightleftharpoons 2H_2O$

Substance	Partial ionic equation
Manganese dioxide	$MnO_2 + 4H^+ + 2e \rightleftharpoons Mn^{2+} + 2H_2O$
Sodium bismuthate	$BiO_3^- + 6H^+ + 2e \rightleftharpoons Bi^{3+} + 3H_2O$
Nitric acid (conc.)	$NO_3^- + 2H^+ + e \rightleftharpoons NO_2 + H_2O$
Nitric acid (dilute)	$NO_3^- + 4H^+ + 3e \rightleftharpoons NO + 2H_2O$

REDUCTANTS	
Hydrogen	$H_2 \rightleftharpoons 2H^+ + 2e$
Zinc	$Zn \rightleftharpoons Zn^{2+} + 2e$
Hydrogen sulphide	$H_2S \rightleftharpoons 2H^+ + S + 2e$
Hydrogen iodide	$2HI \rightleftharpoons I_2 + 2H^+ + 2e$
Oxalic acid	$C_2O_4^{2-} \rightleftharpoons 2CO_2 + 2e$
Iron(II) sulphate	$Fe^{2+} \rightleftharpoons Fe^{3+} + e$
Sulphurous acid	$H_2SO_3 + H_2O \rightleftharpoons SO_4^{2-} + 4H^+ + 2e$
Sodium thiosulphate	$2S_2O_3^{2-} \rightleftharpoons S_4O_6^{2-} + 2e$
Titanium(III) sulphate	$Ti^{3+} \rightleftharpoons Ti^{4+} + e$
Tin(II) chloride	$Sn^{2+} \rightleftharpoons Sn^{4+} + 2e$
Tin(II) chloride (in presence of hydrochloric acid)	$Sn^{2+} + 6Cl^- \rightleftharpoons SnCl_6^{2-} + 2e$
Hydrogen peroxide	$H_2O_2 \rightleftharpoons 2H^+ + O_2 + 2e$

The other procedure which is of value in the calculation of the equivalents of substances is the '**oxidation number**' **method**. This is a development of the view that oxidation and reduction are attended by changes in valency and was originally developed from an examination of the formulae of the initial and final compounds in a reaction. The **oxidation number** (this will be abbreviated to O.N.) of an element is a number which, applied to that element in a particular compound, indicates the amount of oxidation or reduction which is required to convert one atom of the element from the *free state* to that in the compound. If oxidation is necessary to effect the change, the oxidation number is positive, and if reduction is necessary, the oxidation number is negative.

The following rules apply to the determination of oxidation numbers:
1. The O.N. of the free or uncombined element is zero.
2. The O.N. of hydrogen (except in certain hydrides) has a value of $+1$.
3. The O.N. of oxygen (except in peroxides) is -2.
4. The O.N. of a metal in combination (except in hydrides) is usually positive.
5. The O.N. of a radical or ion is that of its electrovalency with the correct sign attached, i.e., is equal to its electrical charge.
6. The O.N. of a compound is always zero, and is determined by the sum of the oxidation numbers of the individual atoms each multiplied by the number of atoms of the element in the molecule.

The **equivalent of an oxidising agent** is determined by the change in oxidation number which the reduced element experiences. It is that quantity of oxidant which involves a change of one unit in the oxidation number. Thus in the normal reduction of potassium permanganate in the presence of dilute sulphuric acid to a Mn(II) salt:

$$\overset{+1 \ +7 \ -8}{K \ MnO_4} \longrightarrow \overset{+2 \ +6 \ -8}{Mn \ S \ O_4}$$

the change in the oxidation number of the manganese is from $+7$ to $+2$. The equivalent of potassium permanganate is therefore $\frac{1}{5}$ mole. Similarly for the reduction of potassium dichromate in acid solution:

$$\overset{+2\ +12\ -14}{K_2Cr_2O_7} \longrightarrow \overset{+6\ -6}{Cr_2(SO_4)_3}$$

the change in oxidation number of *two* atoms of chromium is from $+12$ to $+6$, or by 6 units of reduction. The equivalent of potassium dichromate is accordingly $\frac{1}{6}$ mole. In order to find the equivalent of an oxidising agent, we divide the mole by the change in oxidation number *per molecule* which some key element in the substance undergoes.

The **equivalent of a reducing agent** is similarly determined by the change in oxidation number which the oxidised element suffers. Consider the conversion of iron(II) into iron(III) sulphate:

$$\overset{+2\ -2}{2(FeSO_4)} \longrightarrow \overset{+6\ -6}{Fe_2(SO_4)_3}$$

Here the change in oxidation number *per atom* of iron is from $+2$ to $+3$, or by 1 unit of oxidation, hence the equivalent of iron(II) sulphate is 1 mole. Another important reaction is the oxidation of oxalic acid to carbon dioxide and water:

$$\overset{+2\ +6\ -8}{H_2C_2O_4} \longrightarrow \overset{+4\ -4}{2CO_2}$$

The change in oxidation number of two atoms of carbon is from $+6$ to $+8$, or by 2 units of oxidation. The equivalent of oxalic acid is therefore $\frac{1}{2}$ mole.

In general, it may be stated:

(i) The equivalent of an element taking part in an oxidation–reduction (redox) reaction is the atomic mass divided by the change in oxidation number.

(ii) When an atom in any complex molecule suffers a change in oxidation number (oxidation or reduction), the equivalent of the substance is the mole divided by the change in oxidation number of the oxidised or reduced element. If more than one atom of the reactive element is present, the mole is divided by the total change in oxidation number.

A useful summary of common oxidising and reducing agents, together with the various transformations which they undergo is given in Table X, 2.

Table X, 2

Substance	Radical or element involved	O.N. of 'Effec- tive' element	Reduction product	New O.N.	De- crease in O.N.	Gain in elec- trons
COMMON OXIDISING AGENTS						
KMnO$_4$ (acid)	MnO$_4^-$	$+7$	Mn^{2+}	$+2$	5	5
KMnO$_4$ (neutral)	MnO$_4^-$	$+7$	MnO$_2$ or Mn^{4+}	$+4$	3	3
KMnO$_4$ (strongly alkaline)	MnO$_4^-$	$+7$	MnO$_4^{2-}$	$+6$	1	1
K$_2$Cr$_2$O$_7$	Cr$_2$O$_7^{2-}$	$+6$	Cr^{3+}	$+3$	3	3
HNO$_3$ (dil.)	NO$_3^-$	$+5$	NO	$+2$	3	3
HNO$_3$ (conc.)	NO$_3^-$	$+5$	NO$_2$	$+4$	1	1
Cl$_2$	Cl	0	Cl$^-$	-1	1	1

Substance	Radical or element involved	O.N. of 'Effective' element	Reduction product	New O.N.	Decrease in O.N.	Gain in electrons
COMMON OXIDISING AGENTS						
Br_2	Br	0	Br^-	-1	1	1
I_2	I	0	I^-	-1	1	1
$3HCl:1HNO_3$	Cl	0	Cl^-	-1	1	1
H_2O_2	O_2	-1	O^{2-}	-2	1	1
Na_2O_2	O_2	-1	O^{2-}	-2	1	1
$KClO_3$	ClO_3^-	$+5$	Cl^-	-1	6	6
$KBrO_3$	BrO_3^-	$+5$	Br^-	-1	6	6
KIO_3	IO_3^-	$+5$	I^-	-1	6	6
NaOCl	OCl^-	$+1$	Cl^-	-1	2	2
$FeCl_3$	Fe^{3+}	$+3$	Fe^{2+}	$+2$	1	1
$Ce(SO_4)_2$	Ce^{4+}	$+4$	Ce^{3+}	$+3$	1	1
COMMON REDUCING AGENTS						
H_2SO_3 or Na_2SO_3	SO_3^{2-}	$+4$	SO_4^{2-}	$+6$	2	2
H_2S	S^{2-}	-2	S°	0	2	2
HI	I^-	-1	I°	0	1	1
$SnCl_2$	Sn^{2+}	$+2$	Sn^{4+}	$+4$	2	2
Metals, e.g., Zn	Zn	0	Zn^{2+}	$+2$	2	2
Hydrogen	H	0	H^+	$+1$	1	1
$FeSO_4$ (or any iron(II) salt)	Fe^{2+}	$+2$	Fe^{3+}	$+3$	1	1
Na_3AsO_3	AsO_3^{3-}	$+3$	AsO_4^{3-}	$+5$	2	2
$H_2C_2O_4$	$C_2O_4^{2-}$	$+3$	CO_2	$+4$	1	1
$Ti_2(SO_4)_3$	Ti^{3+}	$+3$	Ti^{4+}	$+4$	1	1

We are now in a position to understand more clearly why the equivalents of some substances vary with the reaction. We will consider two familiar examples by way of illustration. A normal solution of iron(II) sulphate $FeSO_4,7H_2O$ will have an equivalent of 1 mole when employed as a reductant, and $\frac{1}{2}$ mole when employed as a precipitant with aqueous ammonia. A solution of iron(II) sulphate which is normal as a precipitant will be half normal as a reductant. Potassium tetroxalate $KHC_2O_4,H_2C_2O_4,2H_2O$ contains three replaceable hydrogen atoms; its equivalent in neutralisation reactions is therefore $\frac{1}{3}$ mole:

$$KHC_2O_4,H_2C_2O_4,2H_2O + 3KOH = 2K_2C_2O_4 + 5H_2O$$

As a reducing agent, a mole contains $2C_2O_4^{2-}$, and the equivalent is accordingly $\frac{1}{4}$ mole:

$$C_2O_4^{2-} - 2e = 2CO_2$$

A solution of the salt which is $3N$ as an acid is $4N$ as a reducing agent.

When a sequence of reactions is involved in a chemical process the reaction which determines the equivalent is the one in which the standard solution is actually used. Thus if sodium nitrate is reduced to ammonia with Devarda's alloy and the ammonia is titrated with standard acid, the equivalent of the sodium nitrate is not determined by the reduction but by the reaction between ammonia and the acid. Since the equivalent of ammonia NH_3 is 1 mole, that of sodium nitrate $NaNO_3$ is also 1 mole, because 1 mole of $NaNO_3$ yields 1 mole of NH_3.

X, 4.   ADVANTAGES OF THE USE OF THE EQUIVALENT SYSTEM.

The most important advantage of the equivalent system is that the calculations of titrimetric analysis are rendered very simple, since at the end point the number of equivalents of the substance titrated is equal to the number of equivalents of the standard solution employed. We may write:

$$\text{Normality} = \frac{\text{Number of equivalents}}{\text{Number of dm}^3}$$

$$= \frac{\text{Number of milli-equivalents}}{\text{Number of cm}^3}$$

Hence: number of milli-equivalents = number of cm^3 × normality. If the volumes of solutions of two different substances A and B which exactly react with one another are V_A cm^3 and V_B cm^3 respectively, then these volumes severally contain the same number of equivalents or milli-equivalents of A and B. Thus:

$$V_A \times \text{normality}_A = V_B \times \text{normality}_B \qquad (1)$$

In practice V_A, V_B, and normality$_A$ (the standard solution) are known, hence normality$_B$ (the unknown solution) can be readily calculated.

Example 1.   How many cm^3 of 0.2N-hydrochloric acid are required to neutralise 25.0 cm^3 of 0.1N-sodium hydroxide?

Substituting in equation (1), we obtain:

$$x \times 0.2 = 25.0 \times 0.1, \text{ whence } x = 12.5 \ cm^3$$

Example 2.   How many cm^3 of N-hydrochloric acid are required to precipitate completely 1 g of silver nitrate?

The equivalent of $AgNO_3$ in a precipitation reaction is 1 mole or 169.89 g.

Hence 1 g of $AgNO_3$ = 1 × 1000/169.89 = 5.886 milli-equivalents.

Now number of milli-equivalents of HCl = number of milli-equivalents of $AgNO_3$:

$$x \times 1 = 5.886, \text{ whence } x = 5.90 \ cm^3$$

Example 3.   25 cm^3 of an iron(II) sulphate solution react completely with 30.0 cm^3 of 0.125N-potassium permanganate. Calculate the strength of the iron solution in grams of $FeSO_4$ per dm³.

A normal solution of $FeSO_4$ as a reductant contains 1 mole per dm³ or 151.90 g per dm³ (Table X, 2). Let the normality of the iron solution be n_A. Then:

$$25 \times n_A = 30 \times 0.125$$

or $\qquad n_A = 30 \times 0.125/25 = 0.150N$

Hence the solution will contain 0.150 × 151.90 = 22.78 g $FeSO_4$ per dm³.

Example 4.   What volume of 0.127N reagent is required for the preparation of 1000 cm^3 of 0.1N solution?

$$V_A \times \text{normality}_A = V_B \times \text{normality}_B$$

$$V_A \times 0.127 = 1000 \times 0.1$$

or $\qquad V_A = 1000 \times 0.1/0.127 = 787.4 \ cm^3$

Hence it is necessary to dilute 787.4 cm³ of 0.127N solution to 1 dm³. Strictly speaking it is not correct to add 212.6 cm³ of water, because there is usually a volume change on mixing. This change is so small, however, that diluted solutions are often prepared by the addition of the calculated amount of water to a measured volume of standard reagent.

X, 5. PREPARATION OF STANDARD SOLUTIONS. If a reagent is available in the pure state, a solution of definite normality is prepared simply by weighing out an equivalent, or a definite fraction or multiple thereof, dissolving it in the solvent, usually water, and making up the solution to a known volume. It is not really essential to weigh out the equivalent (or a multiple or sub-multiple thereof); in practice it is often more convenient to prepare the solution a *little* more concentrated than is ultimately required, and then to dilute it with distilled water until the desired normality is obtained. If N_1 is the required normality, V_1 the volume after dilution, N_2 the normality originally obtained, and V_2 the original volume taken, $N_1 V_1 = N_2 V_2$, or $V_1 = N_2 V_2 / N_1$. The volume of water to be added to the volume V_2 is $(V_1 - V_2)$ cm³ (compare *Example* 4 in Section **X, 4**). The following is a list of some of the substances which can be obtained in a state of high purity and are therefore suitable for the preparation of standard solutions: sodium carbonate, potassium hydrogenphthalate, benzoic acid, sodium tetraborate, sulphamic acid, potassium hydrogen iodate, sodium oxalate, silver, silver nitrate, sodium chloride, potassium chloride, iodine, potassium bromate, potassium iodate, potassium dichromate, and arsenic(III) oxide.

When the reagent is not available in the pure form as in the cases of most alkali hydroxides, some inorganic acids and various deliquescent substances, solutions of the approximate normality required are first prepared. These are then standardised by titration against a solution of a pure substance of known concentration. It is generally best to standardise a solution by a reaction of the same type as that for which the solution is to be employed, and as nearly as possible under identical experimental conditions. The titration error and other errors are thus considerably reduced or are made to cancel out. This indirect method is employed for the preparation, *inter alia*, of solutions of most acids (for hydrochloric acid, the constant-boiling-point mixture of definite composition can be weighed out directly, if desired), sodium, potassium and barium hydroxides, potassium permanganate, ammonium and potassium thiocyanates, and sodium thiosulphate.

X, 6. PRIMARY STANDARD SUBSTANCES. A primary standard substance should satisfy the following requirements:

1. It must be easy to obtain, to purify, to dry (preferably at 110–120 °C), and to preserve in a pure state. (This requirement is not usually met by hydrated substances, since it is difficult to remove surface moisture completely without effecting partial decomposition.)

2. The substance should be unaltered in air during weighing; this condition implies that it should not be hygroscopic, nor oxidised by air, nor affected by carbon dioxide. The standard should maintain its composition unchanged during storage.

3. The substance should be capable of being tested for impurities by qualitative and other tests of known sensitivity. (The total amount of impurities should not, in general, exceed 0.01–0.02 per cent.)

4. It should have a high equivalent so that the weighing errors may be negligible. (The precision in weighing is ordinarily 0.1–0.2 mg; for an accuracy of 1 part in 1000, it is necessary to employ samples weighing at least *ca.* 0.2 g.)

5. The substance should be readily soluble under the conditions in which it is employed.

6. The reaction with the standard solution should be stoichiometric and practically instantaneous. The titration error should be negligible, or easy to determine accurately by experiment.

In practice, an ideal primary standard is difficult to obtain, and a compromise between the above ideal requirements is usually necessary. The substances commonly employed as primary standards are: Acid-base reactions—sodium carbonate Na_2CO_3, sodium tetraborate $Na_2B_4O_7$, potassium hydrogen-phthalate $KH(C_8H_4O_4)$, constant-boiling-point hydrochloric acid, potassium hydrogen iodate $KH(IO_3)_2$, benzoic acid $H(C_7H_5O_2)$.

Complex formation reactions—silver, silver nitrate, sodium chloride, various metals (e.g., zinc, magnesium, copper, and spectroscopically pure manganese) and salts, depending upon the reaction used.

Precipitation reactions—silver, silver nitrate, sodium chloride, potassium chloride, and potassium bromide (prepared from potassium bromate).

Oxidation–reduction reactions—potassium dichromate $K_2Cr_2O_7$, potassium bromate $KBrO_3$, potassium iodate KIO_3, potassium hydrogen iodate $KH(IO_3)_2$, iodine I_2, sodium oxalate $Na_2C_2O_4$, arsenic(III) oxide As_2O_3, and pure iron.

Hydrated salts, as a rule, do not make good standards because of the difficulty of efficient drying. However, those salts which do not effloresce, such as sodium tetraborate $Na_2B_4O_7,10H_2O$, and copper sulphate $CuSO_4,5H_2O$, are found by experiment to be satisfactory secondary standards. (See Ref. 11.)

A secondary standard is a substance which may be used for standardisations, and whose content of the active substance has been found by comparison against a primary standard.

A. 1 THEORY OF ACID-BASE TITRATIONS

X, 7. NEUTRALISATION INDICATORS. The object of titrating, say, an alkaline solution with a standard solution of an acid is the determination of the amount of acid which is exactly equivalent chemically to the amount of base present. The point at which this is reached is the **equivalence point, stoichiometric point**, or **theoretical end-point**; an aqueous solution of the corresponding salt results. If both the acid and base are strong electrolytes, the resultant solution will be neutral and have a pH of 7 (Section **II, 16**); but if either the acid or the base is a weak electrolyte, the salt will be hydrolysed to a certain degree, and the solution at the equivalence point will be either slightly alkaline or slightly acid. The exact pH of the solution at the equivalence point can readily be calculated from the ionisation constant of the weak acid or the weak base and the concentration of the solution (see Section **II, 18**). For any actual titration the correct end point will be characterised by a definite value of the hydrogen-ion concentration of the solution, the value depending upon the nature of the acid and the base and the concentration of the solution.

A large number of substances are available, called **neutralisation** or **acid-base indicators**, which possess different colours according to the hydrogen-ion concentration of the solution. The chief characteristic of these indicators is that the change from a predominantly 'acid' colour to a predominantly 'alkaline' colour is not sudden and abrupt, but takes place within a small interval of pH (usually about two pH units) termed the **colour-change interval** of the indicator. The position of the colour-change interval in the pH scale varies widely with different indicators. For most acid-base titrations we can therefore select an indicator which exhibits a distinct colour change at a pH close to that obtaining at the equivalence point.

The first useful theory of indicator action was suggested by W. Ostwald. All indicators in general use are very weak organic acids or bases. Ostwald considered that the undissociated indicator acid (HIn) or base (InOH) had a different colour from that of its ion. The equilibria in aqueous solution may be written:

$$HIn \rightleftharpoons H^+ + In^- \tag{1}$$

$$\text{and} \quad InOH \rightleftharpoons OH^- + In^+ \tag{1'}$$

$$\underset{\substack{\text{unionised} \\ \text{colour}}}{} \qquad \underset{\substack{\text{ionised} \\ \text{colour}}}{}$$

If the indicator is a free amine or substituted amine, the equilibrium is:

$$In + H_2O \rightleftharpoons OH^- + HIn^+ \tag{1''}$$

Let us consider an indicator which is a weak acid. In acid solution, i.e., in the presence of excess of H^+ ions, the ionisation will be depressed (common-ion effect) and the concentration of In^- will be very small; the colour will therefore be that of the unionised form. If the medium is alkaline, the decrease of $[H^+]$ will result in the further ionisation of the indicator; $[In^-]$ increases and the colour of the ionised form becomes apparent. By applying the law of mass action, we obtain:

$$\frac{a_{H^+} \times a_{In^-}}{a_{HIn}} = \frac{[H^+] \times [In^-]}{[HIn]} \times \frac{y_{H^+} \cdot y_{In^-}}{y_{HIn}} = K_{in} \tag{2}$$

$$\text{and} \quad [H^+] = \frac{[HIn]}{[In^-]} \times K_{in} \times \frac{y_{HIn}}{y_{H^+} \cdot y_{In^-}}$$

$$= \frac{[\text{Un-ionised form}]}{[\text{Ionised form}]} \times K_{in} \times \frac{y_{HIn}}{y_{H^+} \cdot y_{In^-}} \tag{3}$$

where K_{in} is the **ionisation constant of the indicator**. If the activity coefficients are assumed to be unity—a not entirely justifiable assumption, as will be evident from the ensuing discussion—equation (3) reduces to the simplified 'concentration form':

$$[H^+] = \frac{[HIn]}{[In^-]} \times K_{in} = \frac{[\text{Un-ionised form}]}{[\text{Ionised form}]} \times K_{in} \tag{3'}$$

The actual colour of the indicator, which depends upon the ratio of the concentrations of the ionised and un-ionised forms, is thus directly related to the hydrogen-ion concentration. Equation (3') (the simplified or 'classical' form) may be written logarithmically:

$$pH = \log \frac{[In^-]}{[HIn]} + pK_{in} \tag{4}$$

For an indicator which is a weak base an exactly analogous expression to (3') may be deduced, which in its simplified form is:

$$[OH^-] = \frac{[InOH]}{[In^+]} \times K_{in} \tag{5}$$

where K_{in} is now the corresponding base dissociation constant. This may be written:

$$[H^+] = \frac{K_w \cdot [In^+]}{K_{in} \cdot [InOH]} \tag{6}$$

since $\quad K_w = [H^+] \times [OH^-]$ (approximately)

The simple Ostwald theory of the colour change of indicators requires revision, but the modified views of indicator action lead to equations similar to the above. The colour changes are believed to be due to structural changes, including the production of quinonoid and resonance forms; these may be illustrated by reference to phenolphthalein, the changes of which are characteristic of all phthalein indicators: see the formulae (I–IV) given below. In the presence of dilute alkali the lactone ring in (I) opens to yield (II), and the triphenylcarbinol structure (II) undergoes loss of water to produce the resonating ion (III) which is red. If phenolphthalein is treated with excess of concentrated alcoholic alkali the red colour first produced disappears owing to the formation of (IV).

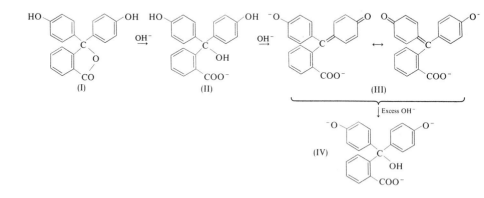

The Brønsted concept of acids and bases makes it unnecessary to distinguish between acid and base indicators: emphasis is placed upon the charge types of the acid and alkaline forms of the indicator. The equilibrium between the acidic form In_A and the basic form In_B may be expressed as:

$$In_A \rightleftharpoons H^+ + In_B \qquad (7)$$

and the equilibrium constant as:

$$\frac{a_{H^+} \times a_{In_B}}{a_{In_A}} = K_{In} \qquad (8)$$

The colour of an indicator, as perceived by the eye, is determined by the ratio of the concentrations of the acidic and basic forms. This is given by:

$$\frac{[In_A]}{[In_B]} = \frac{a_{H^+} \times y_{In_B}}{K_{In} \times y_{In_A}} \qquad (9)$$

where y_{In_A} and y_{In_B} are the activity coefficients of the acidic and basic forms of the indicator. Equation (9) may be written in the logarithmic form:

$$pH = -\log a_{H^+} = pK_{In} + \log\frac{[In_B]}{[In_A]} + \log\frac{y_{In_B}}{y_{In_A}} \qquad (10)$$

The pH will depend upon the ionic strength of the solution (which is, of course, related to the activity coefficient—see Section **II, 3**). Hence when making a colour comparison for the determination of the pH of a solution not only must the indicator concentration be the same in the two solutions but the ionic strength must also be equal or approximately equal. The equation incidentally provides an explanation of the so-called salt and solvent effects which are observed with indicators. The colour-change equilibrium at any particular ionic strength (constant activity–coefficient term) can be expressed by the modified equation:

$$pH = pK'_{In} + \log\frac{[In_B]}{[In_A]} \qquad (11)$$

where pK'_{In} is termed the **apparent indicator constant**.

The value of the ratio $[In_B]/[In_A]$ (i.e., [Basic form]/[Acidic form]) can be determined by a visual colour comparison or, more accurately, by a spectrophotometric method. Both forms of the indicator are present at any hydrogen-ion concentration. It must be realised, however, that the human eye has a limited ability to detect either of two colours when one of them predominates. Experience shows that the solution will appear to have the 'acid' colour, i.e., of In_A, when the ratio of $[In_A]$ to $[In_B]$ is above approximately 10, and the 'alkaline' colour, i.e., of In_B, when the ratio of $[In_B]$ to $[In_A]$ is above approximately 10. Thus only the 'acid' colour will be visible when $[In_A]/[In_B] > 10$, the corresponding limit of pH given by equation (11) is:

$$pH = pK'_{In} - 1$$

only the alkaline colour will be visible when $[In_B]/[In_A] > 10$, and the corresponding limit of pH is:

$$pH = pK'_{In} + 1$$

The colour-change interval is accordingly $pH = pK'_{In} \pm 1$, i.e., over approximately two pH units. Within this range the indicator will appear to change from one colour to the other. The change will be gradual, since it depends upon the ratio of the concentrations of the two coloured forms (acidic form and basic form).

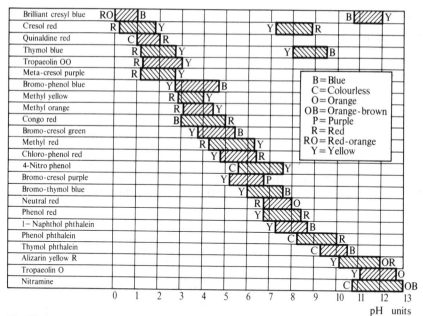

Fig. X, 1

When the pH of the solution is equal to the apparent dissociation constant of the indicator pK'_{In}, the ratio $[In_A]$ to $[In_B]$ becomes equal to 1, and the indicator will have a colour due to an equal mixture of the 'acid' and 'alkaline' forms. This is sometimes known as the *middle tint* of the indicator. This applies strictly only if the two colours are of equal intensity. If one form is more intensely coloured than the other or if the eye is more sensitive to one colour than the other, then the middle tint will be slightly displaced along the pH range of the indicator.

Table X, 3.   Colour changes and pH range of certain indicators

Indicator	Chemical name	pH range	Colour in Acid Solution	Colour in Alkaline Solution	pK'_{In}
Brilliant cresyl blue (acid)	Amino-diethylamino-methyl-diphenazonium chloride	0.0–1.0	Red-orange	Blue	—
Cresol red (acid)	1-Cresolsulphone-phthalein	0.2–1.8	Red	Yellow	—
Quinaldine red	1-(*p*-Dimethyl-amino-phenyl-ethylene)-quinoline ethiodide	1.5–2.5	Colourless	Red	—
Thymol blue (acid)	Thymol-sulphone-phthalein	1.2–2.8	Red	Yellow	1.7
m-Cresol purple	*m*-Cresolsulphone-phthalein	1.2–2.8	Red	Yellow	—
Pentamethoxy red	2,4,2′,4′2″-Pentamethoxy triphenyl carbinol	1.2–3.2	Red-violet	Colourless	—

Indicator	Chemical name	pH range	Colour in Acid Solution	Colour in Alkaline Solution	pK'_{In}
Tropaeolin OO	Diphenylamino-*p*-benzene-sodium sulphonate	1.3–3.0	Red	Yellow	—
Bromo-phenol blue	Tetrabromophenol-sulphone-phthalein	3.0–4.6	Yellow	Blue	4.1
Methyl yellow	Dimethylamino-azo-benzene	2.9–4.0	Red	Yellow	3.3
Ethyl orange		3.0–4.5	Red	Orange	—
Methyl orange	Dimethylamino-azo-benzene sodium sulphonate	3.1–4.4	Red	Orange	3.7
Congo red	Diphenyl-bis-azo-1-naphthylamine-4-sulphonic acid	3.0–5.0	Blue	Red	—
Bromo-cresol green	Tetrabromo-*m*-cresol-sulphone-phthalein	3.8–5.4	Yellow	Blue	4.7
Methyl red	*o*-Carboxybenzene-azodi-methyl-aniline	4.2–6.3	Red	Yellow	5.0
Ethyl red		4.5–6.5	Red	Orange	—
Propyl red		4.6–6.6	Red	Yellow	—
Chlorophenol red	Dichloro-phenol-sulphone-phthalein	4.8–6.4	Yellow	Red	6.1
4-Nitrophenol	4-Nitrophenol	5.6–7.6	Colourless	Yellow	7.1
Bromocresol purple	Dibromo-*o*-cresol-sulphone-phthalein	5.2–6.8	Yellow	Purple	6.1
Bromophenol red	Dibromo-phenol-sulphone-phthalein	5.2–6.8	Yellow	Red	
Azolitmin (litmus)	—	5.0–8.0	Red	Blue	—
Bromo-thymol blue	Dibromo-thymol-sulphone-phthalein	6.0–7.6	Yellow	Blue	7.1
Neutral red	Amino-dimethyl-amino-tolu-phenazonium chloride	6.8–8.0	Red	Orange	—
Phenol red	Phenol-sulphone-phthalein	6.8–8.4	Yellow	Red	7.8
Cresol red (base)	1-Cresolsulphone-phthalein	7.2–8.8	Yellow	Red	8.2
1-Naphthol phthalein	1-Naphtholphthalein	7.3–8.7	Yellow	Blue	8.4
m-Cresol purple	*m*-Cresolsulphone-phthalein	7.6—9.2	Yellow	Purple	—
Thymol blue (base)	Thymol-sulphone-phthalein	8.0–9.6	Yellow	Blue	8.9
o-Cresol-phthalein	Di-*o*-cresol-phthalide	8.2–9.8	Colourless	Red	—
Phenol-phthalein	Phenolphthalein	8.3–10.0	Colourless	Red	9.6
Thymolphthalein	Thymolphthalein	8.3–10.5	Colourless	Blue	9.3
Alizarin yellow R	*p*-Nitrobenzene-azo-salicylic acid	10.1–12.0	Yellow	Orange red	—
Brilliant cresyl blue (base)	Amino-diethylamino-methyl-diphenazonium chloride	10.8–12.0	Blue	Yellow	—
Tropaeolin O	*p*-Sulphobenzene-azo-resorcinol	11.1–12.7	Yellow	Orange	—
Nitramine	2,4,6-Trinitro-phenyl-methyl-nitroamine	10.8–13.0	Colourless	Orange-brown	—

Table X, 3 contains a selected list of indicators suitable for titrimetric analysis and also for the colorimetric determination of pH. The colour-change intervals of most of the various indicators listed in the table are represented graphically in Fig. X, 1.

It is necessary to draw attention to the pH of various types of water which may be encountered in quantitative analysis. Water in equilibrium with the normal atmosphere containing 0.03 per cent by volume of carbon dioxide has a pH of about 5.7; very carefully prepared conductivity water has a pH close to 7; water saturated with carbon dioxide under a pressure of one atmosphere has a pH of about 3.7 at 25 °C. The analyst may therefore be dealing, according to the conditions that prevail in the laboratory, with water having a pH between the two extremes pH 3.7 and pH 7. Hence for indicators which show their alkaline colours at pH values above 4.5, the effect of carbon dioxide introduced during a titration, either from the atmosphere or from the titrating solutions, must be seriously considered. This subject is discussed again later (Section **X, 12**).

X, 8. PREPARATION OF INDICATOR SOLUTIONS. As a rule the stock solutions of the indicators contain 0.5–1 g of indicator per litre of solvent. If the substance is soluble in water, e.g., a sodium salt, water is the solvent; in most other cases 70–90 per cent ethanol is employed. It should now be stated that the synthetic indicators, particularly the sulphonephthaleins and phthaleins which exhibit brilliant colour changes, may be used with confidence in all those cases where the older ones, largely natural products, were formerly employed.

Methyl orange. This indicator is encountered in commerce either as the free acid or as the sodium salt.

Dissolve 0.5 g of the free acid in 1 litre of water, and filter the cold solution if a precipitate separates.

Dissolve 0.5 g of the sodium salt in 1 litre of water, add 15.2 cm^3 of 0.1M-hydrochloric acid, and filter, if necessary, when cold.

Methyl red. Dissolve 1 g of the free acid in 1 litre of hot water, or dissolve in 600 cm^3 of ethanol and dilute with 400 cm^3 of water.

Phenolphthalein. Dissolve 5 g of the reagent in 500 cm^3 of ethanol and add 500 cm^3 of water with constant stirring. Filter, if a precipitate forms.

Alternatively, dissolve 1 g of the dry indicator in 60 cm^3 of 2-ethoxyethanol (cellosolve), b.p. 135 °C, and dilute to 100 cm^3 with distilled water: the loss by evaporation is less with this preparation.

Thymolphthalein. Dissolve 0.4 g of the reagent in 600 cm^3 of ethanol and add 400 cm^3 of water with stirring.

1-Naphtholphthalein. Dissolve 1 g of the indicator in 500 cm^3 of ethanol and dilute with 500 cm^3 of water.

Sulphonephthaleins. These indicators are usually supplied in the acid form. They are rendered water-soluble by adding sufficient sodium hydroxide to neutralise the potential sulphonic acid group. One gram of the indicator is triturated in a clean glass mortar with the appropriate quantity of 0.1M-sodium hydroxide solution, and then diluted with water to 1 litre. The following volumes of 0.1M-sodium hydroxide are required for 1 g of the indicators: bromo-phenol blue, 15.0 cm^3; bromo-cresol green, 14.4 cm^3; bromo-cresol purple, 18.6 cm^3; chloro-phenol red, 23.6 cm^3; bromo-thymol blue, 16.0 cm^3; phenol red, 28.4 cm^3; thymol blue, 21.5 cm^3; cresol red, 26.2 cm^3; meta-cresol purple, 26.2 cm^3.

Quinaldine Red. Dissolve 1 g in 100 cm^3 of 80 per cent ethanol.

Methyl yellow, neutral red, and Congo red.   Dissolve 1 g of the indicator in 1 litre of 80 per cent ethanol. Congo red may also be dissolved in water.

4-Nitrophenol.   Dissolve 2 g of the solid in 1 litre of water.

Alizarin yellow R.   Dissolve 0.5 g of the indicator in 1 litre of 80 per cent ethanol.

Tropaeolin O and Tropaeolin OO.   Dissolve 1 g of the solid in 1 litre of water.

Many of the indicator solutions are available commercially already prepared for use. These should be bought from the actual chemical manufacturers, who will usually supply full details as to the method of preparation, concentration of the solution, etc.

X, 9.   MIXED INDICATORS.   For some purposes it is desirable to have a sharp colour change over a narrow and selected range of pH; this is not easily seen with an ordinary acid–base indicator, since the colour change extends over two units of pH. The result may, however, be achieved by the use of a suitable mixture of indicators; these are generally selected so that their pK'_{In} values are close together and the overlapping colours are complementary at an intermediate pH value. A few examples will be given in some detail.

(*a*) A mixture of equal parts of neutral red (0.1 per cent solution in ethanol) and methylene blue (0.1 per cent solution in ethanol) gives a sharp colour change from violet-blue to green in passing from acid to alkaline solution at pH 7. This indicator may be employed to titrate acetic acid with ammonia solution or vice versa. Both acid and base are approximately of the same strength, hence the equivalence point will be at a pH of *ca.* 7 (Section **X, 15**); owing to the extended hydrolysis and the flat nature of the titration curve, the titration cannot be performed except with an indicator of very narrow range.

(*b*) A mixture of phenolphthalein (3 parts of a 0.1 per cent solution in ethanol) and 1-naphtholphthalein (1 part of a 0.1 per cent solution in ethanol) passes from pale rose to violet at pH = 8.9. The mixed indicator is suitable for the titration of phosphoric acid to the diprotic stage ($K_2 = 6.3 \times 10^{-8}$; equivalence point at pH = *ca.* 8.7).

(*c*) A mixture of thymol blue (3 parts of a 0.1 per cent aqueous solution of the sodium salt) and cresol red (1 part of a 0.1 per cent aqueous solution of the sodium salt) changes from yellow to violet at pH = 8.3. It has been recommended for the titration of carbonate to the hydrogen-carbonate stage.

Other examples are included in Table X, 4. The abbreviations p. = part. w. = water, e = ethanol, Na = Na salt, are used.

The colour change of a single indicator may also be improved by the addition of a pH-sensitive dyestuff to produce the complement of one of the indicator colours. A typical example is the addition of xylene cyanol FF to methyl orange (1.0 g of methyl orange and 1.4 g of xylene cyanol FF in 500 cm^3 of 50 per cent ethanol): here the colour change from the alkaline to the acid side is green → grey → magenta, the middle (grey) stage being at pH = 3.8. The above is an example of a **screened indicator**, and the mixed indicator solution is sometimes known as 'screened' methyl orange. Another example is the addition of methyl green (2 parts of a 0.1 per cent solution in ethanol) to phenolphthalein (1 part of a 0.1 per cent solution in ethanol); the former complements the red-violet basic colour of the latter, and at a pH of 8.4–8.8 the colour change is from grey to pale blue.

Table X, 4. Some mixed indicators

Indicator mixture	pH	Colour change	Composition
Bromocresol green; methyl orange	4.3	Orange $\longrightarrow$ blue-green	1 p. 0.1% (Na) in w.; 1 p. 0.2% in w.
Bromocresol green; chlorophenol red	6.1	Pale green $\longrightarrow$ blue violet	1 p. 0.1% (Na) in w.; 1 p. 0.1% (Na) in w.
Bromothymol blue; neutral red	7.2	Rose pink $\longrightarrow$ green	1 p. 0.1% in e.; 1 p. 0.1% in e.
Bromothymol blue; phenol red	7.5	Yellow $\longrightarrow$ violet	1 p. 0.1% (Na) in w.; 1 p. 0.1% (Na) in w.
Thymol blue; cresol red	8.3	Yellow $\longrightarrow$ violet	3 p. 0.1% (Na) in w.; 1 p. 0.1% (Na) in w.
Thymol blue; phenolphthalein	9.0	Yellow $\longrightarrow$ violet	1 p. 0.1% in 50% e.; 3 p. 0.1% in 50% e.
Thymolphthalein; phenolphthalein	9.9	Colourless $\longrightarrow$ violet	1 p. 0.1% in e.; 1 p. 0.1% in w.

X, 10. UNIVERSAL OR MULTIPLE RANGE INDICATORS. By suitably mixing certain indicators changes in colour may occur over a considerable portion of the pH range. Such mixtures are usually called **'universal indicators'**. They are not suitable for quantitative titrations, but may be employed for the determination of the approximate pH of a solution by the colorimetric method. One such universal indicator is prepared thus: dissolve 0.1 g of phenolphthalein, 0.2 g of methyl red, 0.3 g of methyl yellow, 0.4 g of bromothymol blue, and 0.5 g of thymol blue in 500 cm^3 of absolute ethanol, and add sodium hydroxide solution until the colour is yellow. The colour changes are as follows: pH 2, red; pH 4, orange; pH 6, yellow; pH 8, green; pH 10, blue.

Another recipe for the preparation of a universal indicator follows. Dissolve 0.05 g of methyl orange, 0.15 g of methyl red, 0.3 g of bromothymol blue, and 0.35 g of phenolphthalein in 1 litre of 66 per cent ethanol. The colour changes are: pH up to 3, red; pH 4, orange-red; pH 5, orange; pH 6, yellow; pH 7, yellowish-green; pH 8, greenish-blue; pH 9, blue; pH 10, violet; pH 11, reddish-violet. Several 'universal indicators' are available commercially as solutions and as test papers.

X, 11. NEUTRALISATION CURVES. An insight into the mechanism of neutralisation processes is obtained by studying the changes in the hydrogen-ion concentration during the course of the appropriate titration. The change in pH in the neighbourhood of the equivalence point is of the greatest importance, as it enables us to select an indicator which will give the smallest titration error. The curve obtained by plotting pH as ordinates against the percentage of acid neutralised (or the number of cm^3 of alkali added) as abscissae is known as the **neutralisation** (or, more generally, the **titration**) **curve.** This may be evaluated experimentally by determination of the pH at various stages during the titration by a potentiometric method (Sections **XIV, 16; 25**), or it may be computed with the aid of the theoretical principles that we have already studied. We shall, for the present, adopt the latter method.

X, 12. NEUTRALISATION OF A STRONG ACID AND A STRONG BASE. We shall assume that both the acid and the base are completely dissociated and that the activity coefficients of the ions are unity in order to

Table X, 5. pH during titration of 100 cm^3 of HCl with NaOH of equal concentration

Cm^3 of NaOH added	M solution pH	0.1M solution pH	0.01M solution pH
0	0.0	1.0	2.0
50	0.5	1.5	2.5
75	0.8	1.8	2.8
90	1.3	2.3	3.3
98	2.0	3.0	4.0
99	2.3	3.3	4.3
99.5	2.6	3.6	4.6
99.8	3.0	4.0	5.0
99.9	3.3	4.3	5.3
100.0	7.0	7.0	7.0
100.1	10.7	9.7	8.7
100.2	11.0	10.0	9.0
100.5	11.4	10.4	9.4
101	11.7	10.7	9.7
102	12.0	11.0	10.0
110	12.7	11.7	10.7
125	13.0	12.0	11.0
150	13.3	12.3	11.3
200	13.5	12.5	11.5

calculate the change of pH during the course of the neutralisation of the strong acid and the strong base, or vice versa, at the laboratory temperature. For simplicity of calculation we shall start with 100 cm^3 of, say, M-hydrochloric acid and add M-sodium hydroxide solution. The pH of M-hydrochloric acid is 0. When 50 cm^3 of the M base have been added, 50 cm^3 of un-neutralised M acid will be present in a total volume of 150 cm^3.

$[H^+]$ will therefore be $50 \times 1/150 = 3.33 \times 10^{-1}$, or pH $= 0.48$

for 75 cm^3 of base, $[H^+] = 25 \times 1/175 = 1.43 \times 10^{-1}$, pH $= 0.84$

for 90 cm^3 of base, $[H^+] = 10 \times 1/190 = 5.26 \times 10^{-2}$, pH $= 1.3$

for 98 cm^3 of base, $[H^+] = 2 \times 1/198 = 1.01 \times 10^{-2}$, pH $= 2.0$

for 99 cm^3 of base, $[H^+] = 1 \times 1/199 = 5.03 \times 10^{-3}$, pH $= 2.3$

for 99.9 cm^3 of base, $[H^+] = 0.1 \times 1/199.9 = 5.00 \times 10^{-4}$, pH $= 3.3$

Upon the addition of 100 cm^3 of base, the pH will change sharply to 7, i.e., the theoretical equivalence point provided carbon dioxide is absent; the resulting solution is simply one of sodium chloride.

With 100.1 cm^3 of base, $[OH^-] = 0.1/200.1 = 5.00 \times 10^{-4}$,
    pOH $= 3.3$ and pH $= 10.7$
With 101 cm^3 of base, $[OH^-] = 1/201 = 5.00 \times 10^{-3}$, pOH $= 2.3$,
    and pH $= 11.7$

These results show that as the titration proceeds, initially the pH rises slowly, but between the addition of 99.9 and 100.1 cm^3 of alkali, the pH of the solution rises from 3.3 to 10.7, i.e., in the vicinity of the equivalence point the rate of change of pH of the solution is very rapid.

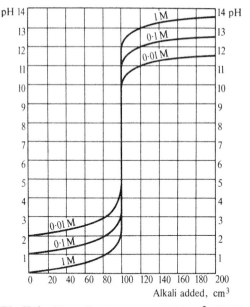

Fig. X, 2 **Neutralisation curves of 100 cm³ of HCl with NaOH of same concentration (calculated)**

The complete results, extended to 200 cm³ of alkali, are collected in Table X, 5; this also includes the figures for 0.1M and 0.01M solutions of acid and base respectively. The additions of alkali have been extended in all three cases to 200 cm³; it is evident that the range from 200 to 100 cm³ and beyond represents the reverse titration of 100 cm³ of alkali with the acid in the presence of the non-hydrolysed sodium chloride solution. The data in the table are presented graphically in Fig. X, 2.

In quantitative analysis we are especially interested in the changes of pH near the equivalence point. This part of Fig. X, 2 is accordingly shown on a larger scale in Fig. X, 3, on which are also indicated the colour-change intervals of some of the common indicators.

With M solutions, it is evident that any indicator with an effective range between pH 3 and 10.5 may be used. The colour change will be sharp and the titration error negligible.

With 0.1M solutions, the ideal pH range for an indicator is limited to 4.5–9.5. Methyl orange will exist chiefly in the alkaline form when 99.8 cm³ of alkali have been added, and the titration error will be 0.2 per cent, which is negligibly small for most practical purposes; it is therefore advisable to add sodium hydroxide solution until the indicator is present completely in the alkaline form. The titration error is also negligibly small with phenolphthalein.

With 0.01M solutions, the ideal pH range is still further limited to 5.5–8.5; such indicators as methyl red, bromothymol blue, or phenol red will be suitable. The titration error for methyl orange will be 1–2 per cent.

The above considerations apply to solutions which do not contain carbon dioxide. In practice, carbon dioxide is usually present (compare Section **X, 7**); it may be derived from the small quantity of carbonate in the sodium hydroxide and/or from the atmosphere. The gas is in equilibrium with carbonic acid, of

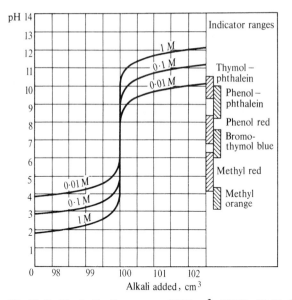

Fig. X, 3 **Neutralisation curves of 100 cm³ of HCl with NaOH of same concentration in vicinity of equivalence point (calculated)**

which both stages of ionisation are weak. This will introduce a small error when indicators of high pH range (above pH 5) are used, e.g., phenolphthalein or thymolphthalein. More acid indicators, such as methyl orange and methyl yellow, are unaffected by carbonic acid. Kolthoff has calculated that the difference in the amounts of sodium hydroxide solution used with methyl orange and phenolphthalein is not greater than 0.15–0.2 cm³ of 0.1M-sodium hydroxide when 100 cm³ of 0.1M-hydrochloric acid are titrated. A method of eliminating this error, other than that of selecting an indicator with a pH range below pH 5, is to boil the solution while still acid and to continue the titration with the cold solution. Boiling the solution is particularly efficacious when titrating dilute (e.g., 0.01M) solutions.

X, 13. NEUTRALISATION OF A WEAK ACID WITH A STRONG BASE. We shall confine our attention to 0.1M solutions; other concentrations can be treated analogously. Let us study the neutralisation of 100 cm³ of 0.1M-acetic acid with 0.1M-sodium hydroxide solution. The pH of the solution at the equivalence point is given by (Section **II, 18**):

$$pH = \tfrac{1}{2}pK_w + \tfrac{1}{2}pK_a - \tfrac{1}{2}pc$$
$$= 7 + 2.37 - \tfrac{1}{2}(1.3) = 8.72$$

For other concentrations, we may employ the approximate mass action expression:

$$[H^+] \times [CH_3COO^-]/[CH_3COOH] = K_a \tag{1}$$

or $[H^+] = [CH_3COOH] \times K_a/[CH_3COO^-]$

or $pH = \log[Salt]/[Acid] + pK_a \tag{2}$

247

The concentration of the salt (and of the acid) at any point is calculated from the volume of alkali added, due allowance being made for the total volume of the solution.

The initial pH of $0.1M$-acetic acid is computed from equation (1); the dissociation of the acid is relatively so small that it may be neglected in expressing the concentration of acetic acid. Hence from equation (1):

$$[H^+] \times [CH_3COO^-]/[CH_3COOH] = 1.82 \times 10^{-5}$$

or $\quad [H^+]^2/0.1 = 1.82 \times 10^{-5}$

or $\qquad [H^+] = \sqrt{1.82 \times 10^{-6}} = 1.35 \times 10^{-3}$

or $\qquad pH = 2.87$

When 50 cm^3 of $0.1M$-alkali have been added,

$$[Salt] = 50 \times 0.1/150 = 3.33 \times 10^{-2}$$
$$\text{and} \quad [Acid] = 50 \times 0.1/150 = 3.33 \times 10^{-2}$$
$$pH = \log(3.33 \times 10^{-2}/3.33 \times 10^{-2}) + 4.74 = 4.74$$

The pH values at other points on the titration curve are similarly calculated. After the equivalence point has been passed, the solution contains excess of OH$^-$ ions which will repress the hydrolysis of the salt; the pH may be assumed, with sufficient accuracy for our purpose, to be that due to the excess of base present, so that in this region the titration curve will almost coincide with that for $0.1M$-hydrochloric acid (Fig. X, 2 and Table X, 5). All the results are collected in Table X, 6, and are depicted graphically in Fig. X, 4. The results for the titration of

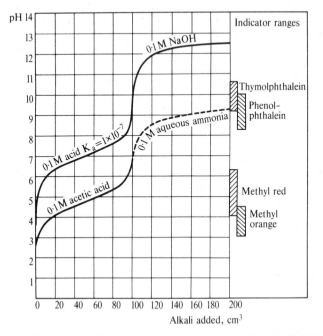

Fig. X, 4 Neutralisation curves of $0.1M$-acetic acid and of $0.1M$-acid ($K_a = 1 \times 10^{-7}$) with $0.1M$-sodium hydroxide (calculated)

Table X, 6.   Neutralisation of 100 cm^3 of 0.1M-acetic acid ($K_a = 1.82 \times 10^{-5}$) and of 100 cm^3 of 0.1$M$-HA ($K_a = 1 \times 10^{-7}$) with 0.1$M$-sodium hydroxide

Cm^3 of 0.1M-NaOH used	0.1M-acetic acid pH	0.1M-HA ($K_a = 1 \times 10^{-7}$) pH
0	2.9	4.0
10	3.8	6.0
25	4.3	6.5
50	4.7	7.0
90	5.7	8.0
99.0	6.7	9.0
99.5	7.0	9.3
99.8	7.4	9.7
99.9	7.7	9.8
100.0	8.7	9.9
100.2	10.0	10.0
100.5	10.4	10.4
101	10.7	10.7
110	11.7	11.7
125	12.0	12.0
150	12.3	12.3
200	12.5	12.5

100 cm^3 of 0.1M solution of a weaker acid ($K_a = 1 \times 10^{-7}$) with 0.1M-sodium hydroxide at the laboratory temperature are also included.

For 0.1M-acetic acid and 0.1M-sodium hydroxide, it is evident from the titration curve that neither methyl orange nor methyl red can be used as indicators. The equivalence point is at pH 8.7, and it is necessary to use an indicator with a pH range on the slightly alkaline side, such as phenolphthalein, thymolphthalein, or thymol blue (pH range, as base, 8.0–9.6). For the acid with $K_a = 10^{-7}$, the equivalence point is at pH = 10, but here the rate of change of pH in the neighbourhood of the stoichiometric point is very much less pronounced, owing to considerable hydrolysis. Phenolphthalein will commence to change colour after about 92 cm^3 of alkali have been added, and this change will occur to the equivalence point; thus the end-point will not be sharp and the titration error will be appreciable. With thymolphthalein, however, the colour change covers the pH range 9.3–10.5; this indicator may be used, the end-point will be more sharp than for phenolphthalein, but nevertheless somewhat gradual, and the titration error will be about 0.2 per cent. Acids that have dissociation constants less than 10^{-7} cannot be satisfactorily titrated in 0.1M solution with a simple indicator.

In general, it may be stated that weak acids ($K_a > 5 \times 10^{-6}$) should be titrated with phenolphthalein, thymolphthalein, or thymol blue as indicators.

X, 14.   NEUTRALISATION OF A WEAK BASE WITH A STRONG ACID.   We may illustrate this case by the titration of 100 cm^3 of 0.1M-aqueous ammonia ($K_b = 1.8 \times 10^{-5}$) with 0.1M-hydrochloric acid at the ordinary laboratory temperature. The pH of the solution at the equivalence point is given by the equation (Section **II, 18**):

$$\text{pH} = \tfrac{1}{2}pK_w - \tfrac{1}{2}pK_b + \tfrac{1}{2}pc$$
$$= 7 - 2.37 + \tfrac{1}{2}(1.3) = 5.28$$

For other concentrations, the pH may be calculated with sufficient accuracy as follows (compare previous section):

$$[NH_4^+] \times [OH^-]/[NH_3] = K_b \tag{1}$$

or

$$[OH^-] = [NH_3] \times K_b/[NH_4^+] \tag{2}$$

or

$$pOH = \log [Salt]/[Base] + pK_b \tag{3}$$

or

$$pH = pK_w - pK_6 - \log [Salt]/[Base] \tag{4}$$

After the equivalence point has been reached, the solution contains excess of H^+ ions, hydrolysis of the salt will be repressed, and the subsequent pH changes may be assumed, with sufficient accuracy for our purpose, to be those due to the excess of acid present.

The results computed in the above manner are represented graphically in Fig. X, 5; the results for the titration of 100 cm^3 of a 0.1M solution of a weaker base ($K_b = 1 \times 10^{-7}$) are also included.

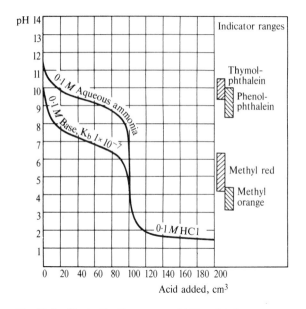

Fig. X, 5   Neutralisation curves of 100 cm^3 0.1M-aqueous ammonia ($K_a = 1.8 \times 10^{-5}$) and of 0.1M-base ($K_b = 1 \times 10^{-7}$) with 0.1M-hydrochloric acid.

It is clear that neither thymolphthalein nor phenolphthalein can be employed in the titration of 0.1M-aqueous ammonia. The equivalence point is at pH 5.3, and it is necessary to use an indicator with a pH range on the slightly acid side (3–6.5), such as methyl orange, methyl red, bromophenol blue, or bromocresol green. The last-named indicators may be utilised for the titration of all weak bases ($K_b > 5 \times 10^{-6}$) with strong acids.

For the weak base ($K_b = 1 \times 10^{-7}$), bromo-phenol blue or methyl orange may be used; no sharp colour change will be obtained with bromo-cresol green or with methyl red, and the titration error will be considerable.

X, 15. NEUTRALISATION OF A WEAK ACID WITH A WEAK BASE. This case is exemplified by the titration of 100 cm^3 of 0.1M-acetic acid ($K_a = 1.8 \times 10^{-5}$) with 0.1M-aqueous ammonia ($K_b = 1.8 \times 10^{-5}$). The pH at the equivalence point is given by (Section **II, 18**):

$$pH = \tfrac{1}{2}pK_w + \tfrac{1}{2}pK_a - \tfrac{1}{2}pK_b$$
$$= 7.0 + 2.37 - 2.37 = 7.0$$

The neutralisation curve up to the equivalence point is almost identical with that using 0.1M-sodium hydroxide as the base; beyond this point the titration is virtually the addition of 0.1M-aqueous ammonia solution to 0.1M-ammonium acetate solution and equation (4) (Section **X, 14**) is applicable to the calculation of the pH. The titration curve for the neutralisation of 100 cm^3 of 0.1M-acetic acid with 0.1M-aqueous ammonia at the laboratory temperature is shown by the dotted line in Fig. X, 4. The chief feature of the curve is that the change of pH near the equivalence point and, indeed, during the whole of the neutralisation curve is very gradual. There is no sudden change in pH, and hence no sharp end-point can be found with any simple indicator. A mixed indicator, which exhibits a sharp colour change over a very limited pH range, may sometimes be found which is suitable. Thus for acetic acid–ammonia solution titrations, neutral red–methylene blue indicator may be used (see Section **X, 9**), but on the whole, it is best to avoid the use of indicators in titrations involving *both* a weak acid and a weak base.

X, 16. NEUTRALISATION OF A POLYPROTIC ACID WITH A STRONG BASE. The shape of the titration curve will depend upon the relative magnitudes of the various dissociation constants. It is assumed that titrations take place at the ordinary laboratory temperature in solutions of concentration of 0.1M or stronger. For a diprotic acid, if the difference between the primary and secondary dissociation constants is very large ($K_1/K_2 > 10\,000$), the solution behaves like a mixture of two acids with constants K_1 and K_2 respectively; the considerations given previously may be applied. Thus for sulphurous acid, $K_1 = 1.7 \times 10^{-2}$ and $K_2 = 1.0 \times 10^{-7}$, it is evident that there will be a sharp change of pH near the first equivalence point, but for the second stage the change will be less pronounced, yet just sufficient for the use of, say, thymolphthalein as indicator (see Fig. X, 4). For carbonic acid, however, for which $K_1 = 4.3 \times 10^{-7}$ and $K_2 = 5.6 \times 10^{-11}$, only the first stage will be just discernible in the neutralisation curve (see Fig. X, 4); the second stage is far too weak to exhibit any point of inflexion and there is no suitable indicator available for direct titration. As indicator for the primary stage, thymol blue may be used (see Section **X, 13**), although a mixture of thymol blue (3 parts) and cresol red (1 part) (see Section **X, 9**) is more satisfactory; with phenolphthalein the colour change will be somewhat gradual and the titration error may be several per cent.

It can be shown that the pH at the first equivalence point for a diprotic acid is given by

$$[H^+] = \sqrt{\frac{K_1 K_2 c}{K_1 + c}}$$

Provided that the first stage of the acid is weak and that K_1 can be neglected by comparison with c, the concentration of salt present, this expression reduces to

$[H^+] = \sqrt{K_1 K_2}$, or $pH = \frac{1}{2}pK_1 + \frac{1}{2}pK_2$.

With a knowledge of the pH at the stoichiometric point and also of the course of the neutralisation curve, it should be an easy matter to select the appropriate indicator for the titration of any diprotic acid for which K_1/K_2 is at least 10^4. For many diprotic acids, however, the two dissociation constants are too close together and it is not possible to differentiate between the two stages. If K_2 is not less than about 10^{-7}, all the replaceable hydrogen may be titrated, e.g., sulphuric acid (primary stage—a strong acid), oxalic acid, malonic, succinic, and tartaric acids.

Similar remarks apply to triprotic acids. These may be illustrated by reference to orthophosphoric acid, for which $K_1 = 7.5 \times 10^{-3}$, $K_2 = 6.2 \times 10^{-8}$, and $K_3 = 5 \times 10^{-13}$. Here $K_1/K_2 = 1.2 \times 10^5$ and $K_2/K_3 = 1.2 \times 10^5$, so that the acid will behave as a mixture of three monoprotic acids with the dissociation constants given above. Neutralisation proceeds almost completely to the end of the primary stage before the secondary stage is appreciably affected, and the secondary stage proceeds almost to completion before the tertiary stage is apparent. The pH at the first equivalence point is given approximately by $(\frac{1}{2}pK_1 + \frac{1}{2}pK_2) = 4.6$, and at the second equivalence point by $(\frac{1}{2}pK_2 + \frac{1}{2}pK_3) = 9.7$; in the very weak third stage, the curve is very flat and no indicator is available for direct titration. The third equivalence point may be computed approximately from the equation (Section **X, 13**):

$$pH = \frac{1}{2}pK_w + \frac{1}{2}pK_a - \frac{1}{2}pc$$
$$= 7.0 + 6.15 - \frac{1}{2}(1.6)$$
$$= 12.35 \text{ for } 0.1M \text{ } H_3PO_4.$$

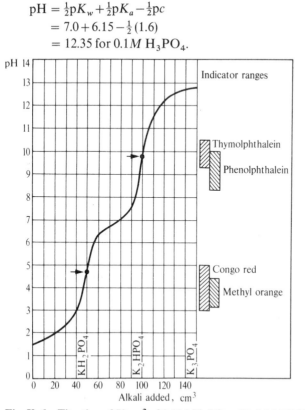

Fig. **X, 6**   Titration of 50 cm^3 of $0.1M$-H_3PO_4 with $0.1M$-KOH

For the primary stage (phosphoric acid as a monoprotic acid), methyl orange, bromo-cresol green, or Congo red may be used as indicators. The secondary stage of phosphoric acid is very weak (see acid $K_a = 1 \times 10^{-7}$ in Fig. X, 4) and the only suitable simple indicator is thymolphthalein (see Section **X, 14**); with phenolphthalein the error may be several per cent. A mixed indicator composed of phenolphthalein (3 parts) and 1-naphtholphthalein (1 part) is very satisfactory for the determination of the end-point of phosphoric acid as a diprotic acid (see Section **X, 9**). The experimental neutralisation curve of 50 cm^3 of 0.1M-orthophosphoric acid with 0.1M-potassium hydroxide, determined by potentiometric titration, is shown in Fig. X, 6.

There are a number of triprotic acids, e.g., citric acid with $K_1 = 9.2 \times 10^{-4}$, $K_2 = 2.7 \times 10^{-5}$, $K_3 = 1.3 \times 10^{-6}$, the three dissociation constants of which are too close together for the three stages to be differentiated easily. If $K_3 > ca. \ 10^{-7}$, all the replaceable hydrogen may be titrated; the indicator will be determined by the value of K_3.

X, 17. TITRATION OF ANIONS OF WEAK ACIDS (BRØNSTED BASES) WITH STRONG ACIDS. 'DISPLACEMENT TITRATIONS'.

So far we have dealt with titrations involving a strong base, the hydroxide ion, but titrations are also possible with weaker bases (Brønsted bases), such as the carbonate ion, the borate ion, the acetate ion, etc. Formerly titrations involving these ions were regarded as titrations of solutions of hydrolysed salts, and the net result was that the weak acid was displaced by the stronger acid. Thus in the titration of sodium acetate solution with hydrochloric acid the following equilibria were considered:

$$CH_3.COO^- + H_2O \rightleftharpoons CH_3.COOH + OH^- \text{ (hydrolysis)}$$

$$H^+ + OH^- = H_2O \text{ (strong acid reacts with } OH^- \text{ produced by}$$
hydrolysis).

The net result thus appeared to be:

$$H^+ + CH_3COO^- = CH_3.COOH$$

or $$CH_3.COONa + HCl = CH_3.COOH + NaCl$$

i.e., the weak acetic acid was apparently displaced by the strong hydrochloric acid, and the process was referred to as a **displacement titration**. On the Brønsted theory the so-called titration of solutions of hydrolysed salts is merely the titration of a weak Brønsted base with a strong (highly ionised) acid. When the anion of a weak acid is titrated with a strong acid the titration curve is identical with that observed in the reverse titration of a weak acid itself with a strong base (compare Section **X, 13**).

A few examples encountered in practice will now be considered.

Titration of borate ion with a strong acid. The titration of the tetraborate ion with hydrochloric acid is similar. The net result of the displacement titration is given by:

$$B_4O_7{}^{2-} + 2H^+ + 5H_2O = 4H_3BO_3$$

Boric acid behaves as a weak monoprotic acid with a dissociation constant of

6.4×10^{-10}. The pH at the equivalence point in the titration of $0.2M$-sodium tetraborate with $0.2M$-hydrochloric acid is that due to $0.1M$-boric acid, i.e., 5.6. Further addition of hydrochloric acid will cause a sharp decrease of pH and any indicator covering the pH range 3.7–5.1 (and slightly beyond this) may be used; suitable indicators are bromo-cresol green, methyl orange, bromo-phenol blue, and methyl red.

Titration of carbonate ion with a strong acid. A solution of sodium carbonate may be titrated to the hydrogen-carbonate stage (i.e., with one equivalent of acid), when the net reaction is:

$$CO_3^{2-} + H^+ = HCO_3^-$$

The equivalence point for the primary stage of ionisation of carbonic acid is at $pH = (\frac{1}{2}pK_1 + \frac{1}{2}pK_2) = 8.3$, and we have seen (Section **X, 14**) that *inter alia* thymol blue and, less satisfactorily, phenolphthalein, or a mixed indicator (Section **X, 19**) may be employed to detect the end point.

Sodium carbonate solution may also be titrated until all the carbonic acid is displaced (two equivalents of acid). The net reaction is then:

$$CO_3^{2-} + 2H^+ = H_2CO_3$$

The same end point is reached by titrating sodium hydrogen-carbonate solution with hydrochloric acid:

$$HCO_3^- + H^+ = H_2CO_3$$

The end point with $100 \, \text{cm}^3$ of $0.2M$-sodium hydrogen-carbonate and $0.2M$-hydrochloric acid may be deduced as follows from the known dissociation constant and concentration of the weak acid. The end point will obviously occur when $100 \, \text{cm}^3$ of hydrochloric acid have been added, i.e., the solution now has a total volume of $200 \, \text{cm}^3$. Consequently since the carbonic acid liberated from the sodium hydrogencarbonate (0.02 moles) is now contained in a volume of $200 \, \text{cm}^3$, its concentration is $0.1M$. K_1 for carbonic acid has a value of 4.3×10^{-7}, and hence we can say:

$$[H^+] \times [HCO_3^-]/[H_2CO_3] = K_1 = 4.3 \times 10^{-7}$$

and since

$$[H^+] = [HCO_3^-]$$
$$[H^+] = \sqrt{4.3 \times 10^{-7} \times 0.1} = 2.07 \times 10^{-4}$$

The pH at the equivalence point is thus approximately 3.7; the secondary ionisation and the loss of carbonic acid, due to escape of carbon dioxide, have been neglected. Suitable indicators are therefore methyl yellow, methyl orange, Congo red, and bromo-phenol blue. The experimental titration curve, determined with the hydrogen electrode, for $100 \, \text{cm}^3$ of $0.1M$-sodium carbonate and $0.1M$-hydrochloric acid is shown in Fig. X, 7.

Cations of weak bases (Brønsted acids; such as the anilinium ion $C_6H_5NH_3^+$) may be titrated with strong bases, and the treatment is similar. These were formerly regarded as salts of weak bases (e.g., aniline, $K_b = 4.0 \times 10^{-10}$) and strong acids: an example is aniline hydrochloride or anilinium chloride.

X, 18. CHOICE OF INDICATORS IN NEUTRALISATION RE-ACTIONS.

As a general rule it may be stated that for a titration to be feasible, there should be a change of approximately two units of pH at or near the stoichiometric point produced by the addition of a small volume of the reagent. The pH at the equivalence point may be computed by means of the equations given in Section **II, 18** (see also below), the pH at either side of the equivalence point (0.1–1 cm^3) may be calculated as described in the preceding sections, and the difference will indicate whether the change is large enough to permit a sharp end point to be observed. Alternatively, the pH change on both sides of the equivalence point may be noted from the neutralisation curve determined by potentiometric titration (Sections **XIX, 16; 25**). If the pH change is satisfactory, an indicator should be selected that changes at or near the equivalence point.

For convenience of reference, we shall summarise the conclusions already deduced from theoretical principles.

Strong acid and strong base. For 0.1M solutions or stronger, any indicator may be used which has a range between the limits pH 4.5 and pH 9.5. With 0.01M solutions, the pH range is somewhat smaller (5.5–8.5). If carbon dioxide is present, the solution should either be boiled whilst still acid and the solution titrated when cold, or an indicator with a range below pH 5 be employed.

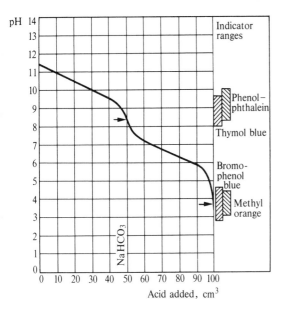

Fig. X, 7   Titration of 100 cm^3 of 0.1M-Na$_2$CO$_3$ with 0.1M-HCl

Weak acid and a strong base. The pH at the equivalence point is calculated from the equation:

$$pH = \tfrac{1}{2}pK_w + \tfrac{1}{2}pK_a - \tfrac{1}{2}pc$$

The pH range for acids with $K_a > 10^{-5}$ is 7–10.5; for weaker acids ($K_a > 10^{-6}$) the range is reduced (8–10). The pH range 8–10.5 will cover most of the examples likely to be encountered; this permits of the use of thymol blue, thymolphthalein, or phenolphthalein.

Weak base and strong acid.   The pH at the equivalence point is computed from the equation:

$$pH = \tfrac{1}{2}pK_w - \tfrac{1}{2}pK_b + \tfrac{1}{2}pc$$

The pH range for bases with $K_b > 10^{-5}$ is 3–7, and for weaker bases $(K_b > 10^{-6})$ 3–5. Suitable indicators will be methyl red, methyl orange, methyl yellow, bromo-cresol green, and bromo-phenol blue.

Weak acid and weak base.   There is no sharp rise in the neutralisation curve and, generally, no simple indicator can be used. The titration should therefore be avoided, if possible. The approximate pH at the equivalence point can be computed from the equation:

$$pH = \tfrac{1}{2}pK_w + \tfrac{1}{2}pK_a - \tfrac{1}{2}pK_b$$

It is sometimes possible to employ a mixed indicator (see Section **X, 9**) which exhibits a colour change over a very limited pH range, for example, neutral red–methylene blue for ammonia solution and acetic acid.

Polyprotic acids (or mixtures of acids, with dissociation constants K_1, K_2, and K_3) **and strong bases.**   The first stoichiometric end point is given approximately by:

$$pH = \tfrac{1}{2}(pK_1 + pK_2)$$

The second stoichiometric end point is given approximately by:

$$pH = \tfrac{1}{2}(pK_2 + pK_3)$$

Anion of a weak acid titrated with a strong acid.   The pH at the equivalence point is given by:

$$pH = \tfrac{1}{2}pK_w - \tfrac{1}{2}pK_a - \tfrac{1}{2}pc$$

Cation of a weak base titrated with a strong base.   The pH at the stoichiometric end-point is given by:

$$pH = \tfrac{1}{2}pK_w - \tfrac{1}{2}pK_b - \tfrac{1}{2}pc$$

As a general rule, wherever an indicator does not give a sharp end point, it is advisable to prepare an equal volume of a comparison solution containing the same quantity of indicator and of the final products and other components of the titration as in the solution under test, and to titrate to the colour shade thus obtained.

In cases where it proves impossible to find a suitable indicator (and this will occur if for example we have to deal with strongly coloured solutions) then titration may be possible by an electrometric method such as conductometric, potentiometric or amperometric titration; see Chapters XIII–XVII. In some instances, spectrophotometric titration (Chapter XVIII) may be feasible. It should also be noted, that if it is possible to work in a non-aqueous solution rather than in water, then acidic and basic properties may be altered according to the solvent chosen, and titrations which are difficult in aqueous solution may then become easy to perform. This procedure, although widely used for the analysis of organic materials, is of very limited application with inorganic substances and will not therefore be discussed further; for details Ref. 3 may be consulted.

A. 2 THEORY OF COMPLEXATION TITRATIONS

X, 19. INTRODUCTION. A complexation reaction with a metal ion involves the replacement of one or more of the co-ordinated solvent molecules by other nucleophilic groups. The groups bound to the central ion are called **ligands** and in aqueous solution the reaction can be represented by the equation:

$$M(H_2O)_n + L = M(H_2O)_{(n-1)}L + H_2O.$$

Here the ligand (L) can be either a neutral molecule or a charged ion, and successive replacement of water molecules by other ligand groups can occur until the complex ML_n is formed; n is the co-ordination number of the metal ion and represents the maximum number of monodentate ligands that can be bound to it.

Ligands may be conveniently classified on the basis of the number of points of attachment to the metal ion. Thus simple ligands, such as halide ions or the molecules H_2O or NH_3, are **monodentate**, i.e., the ligand is bound to the metal ion at only one point by the donation of a lone pair of electrons to the metal. When, however, the ligand molecule or ion has two atoms, each of which has a lone pair of electrons, then the molecule has two donor atoms and it may be possible to form two coordinate bonds with the same metal ion; such a ligand is said to be **bidentate** and may be exemplified by consideration of the tris(ethylenediamine)cobalt(III) complex, $[Co(en)_3]^{3+}$. In this 6-coordinate octahedral complex of cobalt(III), each of the bidentate ethylenediamine molecules is bound to the metal ion through the lone pair electrons of the two nitrogen atoms. This results in the formation of three 5-membered rings, each including the metal ion; the process of ring formation is called **chelation**.

Multidentate ligands contain more than two coordinating atoms per molecule, e.g., 1,2-diaminoethanetetra-acetic acid (ethylenediaminetetra-acetic acid, EDTA) which has two donor nitrogen atoms and four donor oxygen atoms in the molecule can be hexadentate.

In the foregoing it has been assumed that the complex species does not contain more than one metal ion, but under appropriate conditions, a **binuclear complex**, i.e., one containing two metal ions, or even a **polynuclear complex** containing more than two metal ions may be formed. Thus interaction between Zn^{2+} and Cl^- ions may result in the formation of binuclear complexes, e.g., $[Zn_2Cl_6]^{2-}$ in addition to simple species such as $ZnCl_3^-$ and $ZnCl_4^{2-}$. The formation of bi- and poly-nuclear complexes will clearly be favoured by a high concentration of the metal ion; if the latter is present as a trace constituent of a solution, poly-nuclear complexes are unlikely to be formed.

X, 20. STABILITY OF COMPLEXES. The thermodynamic stability of a species is a measure of the extent to which this species will be formed from other species under certain conditions, provided that the system is allowed to reach equilibrium. Consider a metal ion M in solution together with a monodentate ligand L, then the system may be described by the following step-wise equilibria, in which, for convenience, coordinated water molecules are not shown:

$$M + L \rightleftharpoons ML; \quad K_1 = [ML]/[M][L]$$
$$ML + L \rightleftharpoons ML_2; \quad K_2 = [ML_2]/[ML][L]$$
$$ML_{(n-1)} + L \rightleftharpoons ML_n; \quad K_n = [ML_n]/[ML_{(n-1)}][L]$$

257

The equilibrium constants K_1, K_2, ... K_n are referred to as **step-wise stability constants**.

An alternative way of expressing the equilibria is as follows:

$$M + L \rightleftharpoons ML; \quad \beta_1 = [ML]/[M][L]$$

$$M + 2L \rightleftharpoons ML_2; \quad \beta_2 = [ML_2]/[M][L]^2$$

$$M + nL \rightleftharpoons ML_n; \quad \beta_n = [ML_n]/[M][L]^n$$

The equilibrium constants β_1, β_2 ... β_n are called the **overall stability constants** and are related to the stepwise stability constants by the general expression

$$\beta_n = K_1 \times K_2 \times ... K_n.$$

In the above equilibria it has been assumed that no insoluble products are formed nor any polynuclear species.

A knowledge of stability constant values is of considerable importance in analytical chemistry, since they provide information about the concentrations of the various complexes formed by a metal in specified equilibrium mixtures; this is invaluable in the study of complexometry, and of various analytical separation procedures such as solvent extraction, ion exchange, and chromatography (Refs. 20 and 21).

X, 21. FACTORS INFLUENCING THE STABILITY OF COMPLEXES.

The stability of a complex will obviously be related to (a) the complexing ability of the metal ion involved, and (b) to characteristics of the ligand, and it is important to examine these factors briefly.

(a) **Complexing ability of metals.** The relative complexing ability of metals is conveniently described in terms of the **Schwarzenbach classification** which is broadly based upon the division of metals into **Class A** and **Class B** Lewis acids, i.e., electron acceptors. Class A metals are distinguished by an order of affinity (in aqueous solution) towards the halogens $F^- \gg Cl^- > Br^- > I^-$, and form their most stable complexes with the first member of each group of donor atoms in the Periodic Table (i.e., nitrogen, oxygen and fluorine). Class B metals coordinate much more readily with I^- than with F^- in aqueous solution, and form their most stable complexes with the second (or heavier) donor atom from each group (i.e., P, S, Cl). The Schwarzenbach classification defines three categories of metal ion acceptors:

1. Cations with noble gas configurations. The alkali metals, alkaline earths and aluminium belong to this group which exhibit Class A acceptor properties. Electrostatic forces predominate in complex formation, so interactions between small ions of high charge are particularly strong and lead to stable complexes. Thus fluoro-complexes are particularly stable, water is more strongly bound than ammonia which has a smaller dipole moment, and cyanide ions have little tendency to form complexes since they only exist in alkaline solutions where they cannot compete successfully with hydroxyl ions.

2. Cations with completely filled d sub-shells. Typical of this group are copper(I), silver(I) and gold(I) which exhibit Class B acceptor properties. These ions have high polarising power and the bonds formed in their complexes have appreciable covalent character. Complexes are more stable the more noble the metal and the less electronegative the donor atom of the ligand; thus cadmium(II)

and mercury(II) form strong complexes with I^- and CN^- ions, but weak complexes with F^-.

3. Transition metal ions with incomplete d sub-shells. In this group both Class A and Class B tendencies can be distinguished. The elements with Class B characteristics form a roughly triangular group within the Periodic Table, with the apex at copper and the base extending from rhenium to bismuth. To the left of this group, elements in their higher oxidation states tend to exhibit Class A properties, while to the right of the group, the higher oxidation states of a given element have a greater Class B character.

The concept of **Hard and Soft acids and bases** is useful in characterising the behaviour of Class A and Class B acceptors. A **soft base** may be defined as one in which the donor atom is of high polarisability and of low electronegativity, is easily oxidised, or is associated with vacant, low-lying orbitals. These terms describe, in different ways, a base in which the donor atom electrons are not tightly held, but are easily distorted or removed. **Hard bases** have the opposite properties, i.e., the donor atom is of low polarisability and high electronegativity, is difficult to reduce, and is associated with vacant orbitals of high energy which are inaccessible.

On this basis, it is seen that Class A acceptors prefer to bind hard bases, e.g., with nitrogen, oxygen and fluorine donor atons, whilst Class B acceptors prefer to bind to the softer bases, e.g., P, As, S, Se, Cl, Br, I donor atoms. Examination of the Class A acceptors shows them to have the following distinguishing features; small size, high positive oxidation state, and the absence of outer electrons which are easily excited to higher states. These are all factors which lead to low polarisability, and such acceptors are called **hard acids**. Class B acceptors, however, have one or more of the following properties: low positive or zero oxidation state, large size, and several easily excited outer electrons (for metals these are the d electrons). These are all factors which lead to high polarisability, and Class B acids may be called **soft acids**.

A general principle may now be stated which permits correlation of the complexing ability of metals: 'Hard acids prefer to associate with hard bases and soft acids with soft bases'. This statement must not, however, be regarded as exclusive, i.e., under appropriate conditions soft acids may complex with hard bases or hard acids with soft bases.

(b) **Characteristics of the ligand.** Among the characteristics of the ligand which are generally recognised as influencing the stability of complexes in which it is involved are (i) the basic strength of the ligand, (ii) its chelating properties (if any), and (iii) steric effects. From the point of view of the analytical applications of complexes, the chelating effect is of paramount importance and therefore merits particular attention.

The term 'chelate effect' refers to the fact that a chelated complex, i.e., one formed by a bidentate or a multidentate ligand, is more stable than the *corresponding* complex with monodentate ligands: the greater the number of points of attachment of ligand to the metal ion, the greater the stability of the complex. Thus the complexes formed by the nickel(II) ion with (a) the monodentate NH_3 molecule, (b) the bidentate ethylenediamine (1,2-diaminoethane), and (c) the hexadentate ligand 'penten' $\{(H_2N \cdot CH_2 \cdot CH_2)_2N \cdot CH_2 \cdot CH_2 \cdot N (CH_2 \cdot CH_2 \cdot NH_2)_2\}$ show an overall stability constant value for the ammonia complex of 3.1×10^8, which is increased by a factor of about 10^{10} for the complex of ligand (b), and is approximately ten times greater still for the third complex.

The chelate effect can often be attributed to the increase in entropy which accompanies chelation; in this context the displacement of water molecules from the hydrated ion must be borne in mind.

The most common steric effect is that of inhibition of complex formation owing to the presence of a large group either attached to, or in close proximity to, the donor atom.

A further factor which must also be taken into consideration from the point of view of the analytical applications of complexes and of complex-formation reactions is the rate of reaction: to be analytically useful it is usually required that the reaction be rapid. An important classification of complexes is based upon the rate at which they undergo substitution reactions and leads to the two groups of **labile** and **inert complexes**. The term labile complex is applied to those cases where nucleophilic substitution is complete within the time required for mixing the reagents. Thus, for example, when excess of aqueous ammonia is added to an aqueous solution of copper(II) sulphate the change in colour from pale to deep blue is instantaneous; the rapid replacement of water molecules by ammonia indicates that the Cu(II) ion forms kinetically labile complexes. The term inert is applied to those complexes which undergo slow substitution reactions, i.e., reactions with half-times of the order of hours or even days at room temperature. Thus the Cr(III) ion forms kinetically inert complexes, so that the replacement of water molecules coordinated to Cr(III) by other ligands, is a very slow process at room temperature.

Kinetic inertness or lability is influenced by many factors, but the following general observations form a convenient guide to the behaviour of the complexes of various elements.
(i) Main group elements usually form labile complexes.
(ii) With the exception of Cr(III) and Co(III), most first-row transition elements form labile complexes.
(iii) Second- and third-row transition elements tend to form inert complexes.

For a full discussion of the topics introduced in this Section a textbook of Inorganic Chemistry (e.g., Ref. 4), or one dealing with complexes (e.g., Ref. 5), should be consulted.

X, 22. A SIMPLE COMPLEXATION TITRATION.

A simple example of the application of a complexation reaction to a titration procedure is the titration of cyanide with silver nitrate solutions, a method first proposed by Liebig. When a solution of silver nitrate is added to a solution containing cyanide ions (e.g., an alkali cyanide) a white precipitate is formed when the two liquids first come into contact with one another, but on stirring it re-dissolves owing to the formation of a stable complex cyanide, the alkali salt of which is soluble:

$$Ag^+ + 2CN^- \rightleftharpoons [Ag(CN)_2]^-$$

When the above reaction is complete, further addition of silver nitrate solution yields the insoluble silver cyanoargentate (sometimes termed insoluble silver cyanide); the end-point of the reaction is therefore indicated by the formation of a permanent precipitate or turbidity.

The only difficulty in obtaining a sharp end-point lies in the fact that silver cyanide, precipitated by local excess concentration of silver ion somewhat prior to the equivalence point, is very slow to redissolve and the titration is time-consuming. In the Déniges modification, iodide ion (usually as KI, *ca.* 0.01*M*) is used as the indicator and aqueous ammonia (*ca.* $0.2M$) is introduced to dissolve the silver cyanide.

The iodide ion and ammonia solution are added before the titration is commenced; the formation of silver iodide (as a turbidity) will indicate the end point:

$$[Ag(NH_3)_2]^+ + I^- \rightleftharpoons AgI + 2NH_3$$

During the titration any silver iodide which would tend to form will be kept in solution by the excess of cyanide ion always present until the equivalence point is reached:

$$AgI + 2CN^- \rightleftharpoons [Ag(CN)_2]^- + I^-$$

The method may also be applied to the analysis of silver halides by dissolution in excess of cyanide solution and back-titration with standard silver nitrate. It can also be utilised indirectly for the determination of several metals, notably nickel, cobalt, and zinc, which form stable stoichiometric complexes with cyanide ion. Thus if a Ni(II) salt in ammoniacal solution is heated with excess of cyanide ion, the $[Ni(CN)_4]^{2-}$ ion is formed quantitatively; since it is more stable than the $[Ag(CN)_2]^-$ ion, the excess of cyanide may be determined by the Liebig–Dénigès method. The metal ion determinations are, however, more conveniently made by titration with EDTA: see the following Sections.

X, 23. COMPLEXONES. The formation of a single complex species rather than the stepwise production of such species will clearly simplify complexometric titrations and facilitate the detection of end-points. Schwarzenbach (Ref. 6) realised that the acetate ion is able to form acetato complexes of low stability with nearly all polyvalent cations, and that if this property could be reinforced by the chelate effect, then much stronger complexes would be formed by most metal cations. He found that the aminopolycarboxylic acids are excellent complexing agents: the most important of these is 1,2-diaminoethanetetra-acetic acid (ethylenediaminetetra-acetic acid). The formula (I) is preferred to (II), since it has been shown from

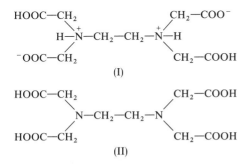

measurements of the dissociation constants that two hydrogen atoms are probably held in the form of zwitterions. The values of pK are respectively pK_1 = 2.0, pK_2 = 2.7, pK_3 = 6.2, and pK_4 = 10.3 at 20 °C; these values suggest that it behaves as a dicarboxylic acid with two strongly acidic groups and that there are two ammonium protons of which the first ionises in the pH region of about 6.3 and the second at a pH of about 11.5. Various trivial names (see Ref. 19) are used for ethylenediaminetetra-acetic acid and its sodium salts, and these include Trilon B, Complexone III, Sequestrene, Versene, and Chelaton 3; the disodium salt is most widely employed in titrimetric analysis. To avoid the constant use of the long name, the abbreviation EDTA is utilised for the disodium salt.

Other complexing agents (complexones) which are sometimes used include (a) nitrilotriacetic acid (III) (NITA or NTA or Complexone I; this has $pK_1 = 1.9$, $pK_2 = 2.5$, and $pK_3 = 9.7$), (b) trans-1,2-diaminocyclohexane-N,N,N',N',-tetra-acetic acid (IV): this should presumably be formulated as a zwitterion structure like (I); the abbreviated name is CDTA, DCyTA or DCTA or Complexone IV),

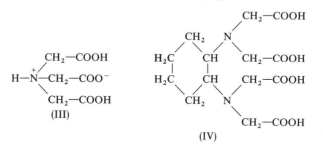

(III)

(IV)

(c) 2,2'-ethylenedioxybis{ethyliminodi(acetic acid)} (V) also known as ethylene glycolbis(2-aminoethyl ether)N,N,N',N'-tetra-acetic acid (EGTA), and (d) triethylenetetramine-N,N,N',N'',N''',N'''-hexa-acetic acid (TTHA), (VI),

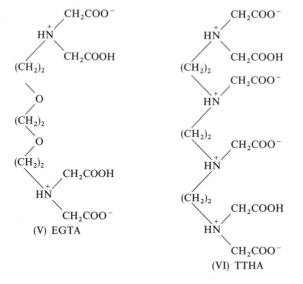

(V) EGTA

(VI) TTHA

CDTA often forms stronger metal complexes than does EDTA and thus finds applications in analysis, but the metal complexes are formed rather more slowly than with EDTA so that the end point of the titration tends to be drawn out with the former reagent. EGTA finds analytical application mainly in the determination of calcium in a mixture of calcium and magnesium and is probably superior to EDTA in the calcium/magnesium water-hardness titration (Section X, 63). TTHA forms 1:2 complexes with many trivalent cations and with some

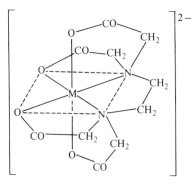

Fig. X, 8

divalent metals, and Přibil and Veselý (Ref. 7) have devised procedures for determining the components of mixtures of certain ions without the use of masking agents (see Section **X, 27**).

However, EDTA has the widest general application in analysis because of its powerful complexing action and commercial availability. The spatial structure of its anion, which has six donor atoms, enables it to satisfy the coordination number of six frequently encountered among the metal ions and to form strain-less five-membered rings on chelation. The resulting complexes have similar structures but differ from one another in the charge they carry. One such structure suggested for the complex with a divalent ion is shown in Fig. X, 8; this structure shows the complex ion exhibiting the maximum chelating power as a hexa-dentate ligand, but this may not be true for all metal-EDTA complexes.

To simplify the following discussion EDTA is assigned the formula H_4Y: the disodium salt is therefore Na_2H_2Y and affords the complex-forming ion H_2Y^{2-} in aqueous solution; it reacts with all metals in a 1:1 ratio. The reactions with cations, e.g., M^{2+}, may be written as:

$$M^{2+} + H_2Y^{2-} \rightleftharpoons MY^{2-} + 2H^+ \tag{1}$$

For other cations, the reactions may be expressed as:

$$M^{3+} + H_2Y^{2-} \rightleftharpoons MY^- + 2H^+ \tag{2}$$

$$M^{4+} + H_2Y^{2-} \rightleftharpoons MY + 2H^+ \tag{3}$$

$$\text{or} \quad M^{n+} + H_2Y^{2-} \rightleftharpoons (MY)^{(n-4)+} + 2H^+ \tag{4}$$

One mole of the complex-forming H_2Y^{2-} reacts in all cases with one mole of the metal ion and in each case, also, two moles of hydrogen ion are formed. It is apparent from equation (4) that the dissociation of the complex will be governed by the pH of the solution; lowering the pH will decrease the stability of the metal-EDTA complex. The more stable the complex, the lower the pH at which an EDTA titration of the metal ion in question may be carried out. Table X, 7 indicates minimum pH values for the existence of EDTA complexes of some selected metals.

Table X, 7. Stability with respect to pH of some metal-EDTA complexes

Minimum pH at which complexes exist	Selected metals
1–3	Zr^{4+}; Hf^{4+}; Th^{4+}; Bi^{3+}; Fe^{3+}
4–6	Pb^{2+}; Cu^{2+}; Zn^{2+}; Co^{2+}; Ni^{2+}; Mn^{2+}; Fe^{2+}; Al^{3+}; Cd^{2+}; Sn^{2+}
8–10	Ca^{2+}; Sr^{2+}; Ba^{2+}; Mg^{2+}

It is thus seen that, in general, EDTA complexes with divalent metal ions are stable in alkaline or slightly acidic solution, whilst complexes with tri- and tetra-valent metal ions may exist in solutions of much higher acidity.

X, 24.   STABILITY CONSTANTS OF EDTA COMPLEXES.   The stability of a complex is characterised by the stability constant (or formation constant) K:

$$M^{n+} + Y^{4-} \rightleftharpoons (MY)^{(n-4)+} \tag{5}$$

$$K = [(MY)^{(n-4)+}]/[M^{n+}][Y^{4-}] \tag{6}$$

Some values for the stability constants (expressed as $\log K$) of metal–EDTA complexes are collected in Table X, 8: these apply to a medium of ionic strength I $= 0.1$ at $20\,°C$.

Table X, 8.   Stability constants of metal–EDTA complexes

Mg^{2+}	8.7	Zn^{2+}	16.7	La^{3+}	15.7
Ca^{2+}	10.7	Cd^{2+}	16.6	Lu^{3+}	20.0
Sr^{2+}	8.6	Hg^{2+}	21.9	Sc^{3+}	23.1
Ba^{2+}	7.8	Pb^{2+}	18.0	Ga^{3+}	20.5
Mn^{2+}	13.8	Al^{3+}	16.3	In^{3+}	24.9
Fe^{2+}	14.3	Fe^{3+}	25.1	Th^{4+}	23.2
Co^{2+}	16.3	Y^{3+}	18.2	Ag^{+}	7.3
Ni^{2+}	18.6	Cr^{3+}	24.0	Li^{+}	2.8
Cu^{2+}	18.8	Ce^{3+}	15.9	Na^{+}	1.7

In equation (6) only the fully ionised form of EDTA, i.e., the ion Y^{4-}, has been taken into account, but at low pH values the species HY^{3-}, H_2Y^{2-}, H_3Y^- and even undissociated H_4Y may well be present; in other words, only a part of the EDTA uncombined with metal may be present as Y^{4-}. Further, in equation (6) the metal ion M^{n+} is assumed to be uncomplexed, i.e., in aqueous solution it is simply present as the hydrated ion. If, however, the solution also contains substances other than EDTA which can complex with the metal ion, then the whole of this ion uncombined with EDTA may no longer be present as the simple hydrated ion. Thus, in practice, the stability of metal-EDTA complexes may be altered by (a) variation in pH and (b) by the presence of other complexing agents. The stability constant of the EDTA complex will then be different from the value recorded for a specified pH in pure aqueous solution; the value recorded for the new conditions is termed the **apparent** or **conditional stability constant**. It is clearly necessary to examine the effect of these two factors in some detail.

(a) **pH effect.**   The apparent stability constant at a given pH may be calculated from the ratio K/α, where α is the ratio of the total uncombined EDTA (in all forms) to the form Y^{4-}. Thus K_H, the apparent stability constant for the metal-EDTA complex at a given pH, can be calculated from the expression

$$\log K_H = \log K - \log \alpha \tag{7}$$

The factor α can be calculated from the known dissociation constants of EDTA, and since the proportions of the various ionic species derived from EDTA will be dependent upon the pH of the solution, α will also vary with pH; a plot of $\log \alpha$ against pH shows a variation of $\log \alpha = 18$ at pH $= 1$ to $\log \alpha = 0$ at pH $= 12$: such a curve is very useful for dealing with calculations of apparent stability constants. Thus, for example, from Table X, 8 log K of the EDTA complex of the Pb^{2+} ion is 18.0 and from a graph of $\log \alpha$ against pH, it is found that at a pH $= 5.0$, $\log \alpha = 7$. Hence from equation (7), at a pH of 5.0 the lead–EDTA complex has an apparent dissociation constant given by:

$$\log K_H = 18.0 - 7.0 = 11.0.$$

Carrying out a similar calculation for the EDTA complex of the Mg^{2+} ion ($\log K = 8.7$), for the same pH (5.0), it is found:

$$\log K_H(\text{Mg(II)–EDTA}) = 8.7 - 7.0 = 1.7.$$

These results imply that at the specified pH the magnesium complex is appreciably dissociated, whereas the lead complex is stable, and clearly titration of an Mg(II) solution with EDTA at this pH will be unsatisfactory, but titration of the lead solution under the same conditions will be quite feasible. In practice, for a metal ion to be titrated with EDTA at a stipulated pH the value of $\log K_H$ should be greater than 8 when a metallochromic indicator is used.

As indicated by the data quoted in the previous Section, the value of $\log \alpha$ is small at high pH values, and it therefore follows that the larger values of $\log K_H$ are found with increasing pH. However, by increasing the pH of the solution the tendency to form slightly soluble metallic hydroxides is enhanced owing to the reaction:

$$(MY)^{(n-4)+} + nOH^- \rightleftharpoons M(OH)_n + Y^{4-}.$$

The extent of hydrolysis of $(MY)^{(n-4)+}$ depends upon the characteristics of the metal ion, and is largely controlled by the solubility product of the metallic hydroxide and, of course, the stability constant of the complex. Thus iron(III) is precipitated as hydroxide ($K_{sol} = 1 \times 10^{-36}$) in basic solution, but nickel(II), for which the relevant solubility product is 6.5×10^{-18}, remains complexed. Clearly the use of excess EDTA will tend to reduce the effect of hydrolysis in basic solutions. It follows that for each metal ion there exists an optimum pH which will give rise to a maximum value for the apparent stability constant.

(b) **The effect of other complexing agents.** If another complexing agent (say NH_3) is also present in the solution, then in equation (6) $[M^{n+}]$ will be reduced owing to complexation of the metal ions with ammonia molecules. It is convenient to indicate this reduction in effective concentration by introducing a factor β, defined as the ratio of the sum of the concentrations of all forms of the metal ion not complexed with EDTA to the concentration of the simple (hydrated) ion. The apparent stability constant of the metal–EDTA complex, taking into account the effects of both pH and the presence of other complexing agents, is then given by:

$$\log K_{HZ} = \log K - \log \alpha - \log \beta. \tag{8}$$

X, 25. TITRATION CURVES. If, in the titration of a strong acid, pH is plotted against the volume of the solution of the strong base added, a point of inflexion occurs at the equivalence point (compare Section **X, 12**). Similarly, in the EDTA titration, if pM (negative logarithm of the 'free' metal ion concentration: $pM = -\log[M^{n+}]$) is plotted against the volume of EDTA solution added, a point of inflexion occurs at the equivalence point; in some instances this sudden increase may exceed 10 pM units. The general shape of titration curves obtained

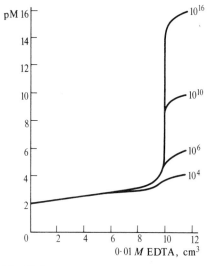

Fig. X, 9

by titrating $10.0 \, \text{cm}^3$ of a $0.01 M$ solution of a metal ion M with a $0.01 M$-EDTA solution is shown in Fig X, 9. The apparent stability constants of various metal–EDTA complexes are indicated at the extreme right of the curves. It is evident that the greater the stability constant, the sharper is the end-point provided the pH is maintained constant.

In acid-base titrations the end-point is generally detected by a pH-sensitive indicator. In the EDTA titration a metal ion-sensitive indicator (abbreviated to *metal indicator* or metal ion indicator) is often employed to detect changes of pM. Such indicators (which contain types of chelate groupings and generally possess resonance systems typical of dyestuffs) form complexes with specific metal ions; these differ in colour from the free indicator and, in consequence, a sudden colour change occurs at the equivalence point. The end-point of the titration can also be evaluated by other methods, which include potentiometric, amperometric, conductometric, and spectrophotometric techniques.

X, 26. TYPES OF EDTA TITRATIONS. The most important procedures for the titration of metal ions with EDTA are:

A. Direct titration. The solution containing the metal ion to be determined is buffered to the desired pH (e.g., to pH = 10 with $NH_4{}^+$–aq. NH_3) and titrated directly with the standard EDTA solution. It may be necessary to prevent precipitation of the hydroxide of the metal (or a basic salt) by the addition of some auxiliary complexing agent, such as tartrate or citrate or triethanolamine. At the equivalence point the magnitude of the concentration of the metal ion being determined decreases abruptly. This is generally determined by the change in colour of a metal indicator which responds to changes in pM: the end point may also be determined by amperometric, conductometric, spectrophotometric, or, in some cases, by potentiometric methods.

B. Back-titration. Many metals cannot, for various reasons, be titrated directly: thus they may precipitate from the solution in the pH range necessary for the titration, or they may form inert complexes, or a suitable metal indicator is not available. In such cases an excess of standard EDTA solution is added, the resulting solution is buffered to the desired pH, and the excess of the reagent is back-titrated with a standard metal ion solution; a solution of zinc chloride or sulphate or of magnesium chloride or sulphate is often used for this purpose. The end-point is detected with the aid of the metal indicator which responds to the metal ion introduced in the back-titration.

C. Replacement or substitution titration. Substitution titrations may be used for metal ions that do not react (or react unsatisfactorily) with a metal indicator, or for metal ions which form EDTA complexes that are more stable than those of other metals such as magnesium and calcium. The metal cation M^{n+} to be determined may be treated with the magnesium complex of EDTA, when the following reaction occurs:

$$M^{n+} + MgY^{2-} \rightleftharpoons (MY)^{(n-4)+} + Mg^{2+}$$

The amount of magnesium ion set free is equivalent to the cation present and can be titrated with a standard solution of EDTA and a suitable metal indicator.

An interesting application is the titration of calcium. In the direct titration of calcium ions Solochrome Black (Eriochrome Black T) gives a poor end-point; if magnesium is present, it is displaced from its EDTA complex by calcium and an improved end-point results (compare Section **X, 55**).

D. Alkalimetric titration.   When a solution of disodium ethylenediamine-tetra-acetate, Na_2H_2Y, is added to a solution containing metallic ions, complexes are formed with the liberation of two equivalents of hydrogen ion:

$$M^{n+} + H_2Y^{2-} \rightleftharpoons (MY)^{(n-4)+} + 2H^+$$

The hydrogen ions thus set free can be titrated with a standard solution of sodium hydroxide using an acid–base indicator or a potentiometric end-point; alternatively, an iodate–iodide mixture is added as well as the EDTA solution and the liberated iodine is titrated with a standard thiosulphate solution.

The solution of the metal to be determined must be accurately neutralised before titration; this is often a difficult matter on account of the hydrolysis of many salts, and constitutes a weak feature of alkalimetric titration.

E. Miscellaneous Methods.   Exchange reactions between the tetracyano-nickelate(II) ion $[Ni(CN)_4]^{2-}$ (the potassium salt is readily prepared) and the element to be determined, whereby nickel ions are set free, have a limited application. Thus silver and gold, which themselves cannot be titrated complexometrically, can be determined in this way.

$$[Ni(CN)_4]^{2-} + 2Ag^+ \rightleftharpoons 2[Ag(CN)_2]^- + Ni^{2+}$$

These reactions take place with sparingly soluble silver salts, and hence provide a method for the determination of the halide ions Cl^-, Br^-, I^-, and the thiocyanate ion SCN^-. The anion is first precipitated as the silver salt, the latter dissolved in a solution of $[Ni(CN)_4]^{2-}$, and the equivalent amount of nickel thereby set free is determined by rapid titration with EDTA using an appropriate indicator (Murexide, Bromopyrogallol Red).

Sulphate may be determined by precipitation as barium sulphate or as lead sulphate, the precipitate is dissolved in an excess of standard EDTA solution, and the excess of EDTA is back-titrated with a standard magnesium or zinc solution using Solochrome Black (Eriochrome Black T) as indicator.

Phosphate may be determined by precipitating as $Mg(NH_4)PO_4,6H_2O$, dissolving the precipitate in dilute hydrochloric acid, adding an excess of standard EDTA solution, buffering at pH = 10, and back-titrating with standard magnesium ion solution in the presence of Solochrome Black.

X, 27. TITRATION OF MIXTURES, SELECTIVITY, MASKING AND DEMASKING AGENTS.   EDTA is a very unselective reagent because it complexes with numerous di-, tri-, and tetra-valent cations. When a solution containing two cations which complex with EDTA is titrated without the addition of a complex-forming indicator, and if a titration error of 0.1 per cent is permissible, then the ratio of the stability constants of the EDTA complexes of the two metals M and N must be such that $K_M/K_N \geq 10^6$ if N is not to interfere with

the titration of M. Strictly, of course, the constants K_M and K_N considered in the above expression should be the apparent stability constants of the complexes. If complex-forming indicators are used, then for a similar titration error $K_M/K_N \geq 10^8$.

The following procedures will help to increase the selectivity:

(a) **By suitable control of the pH of the solution.** This, of course, makes use of the different stabilities of metal–EDTA complexes. Thus bismuth and thorium can be titrated in an acidic solution (pH = 2) with Xylenol Orange or Methylthymol Blue as indicator and most divalent cations do not interfere. A mixture of bismuth and lead ions can be successfully titrated by first titrating the bismuth at pH 2 with Xylenol Orange as indicator, and then adding hexamine to raise the pH to about 5, and titrating the lead (see Section **X, 72**).

(b) **By the use of masking agents.** **Masking** may be defined as the process in which a substance, without physical separation of it or its reaction products, is so transformed that it does not enter into a particular reaction. **Demasking** is the process in which the masked substance regains its ability to enter into a particular reaction.

By the use of masking agents, some of the cations in a mixture can often be 'masked' so that they can no longer react with EDTA or with the indicator. An effective masking agent is the cyanide ion; this forms stable cyanide complexes with the cations of Cd, Zn, Hg(II), Cu, Co, Ni, Ag, and the platinum metals, but not with the alkaline earths, manganese, and lead:

$$M^{2+} + 4CN^- \longrightarrow [M(CN)_4]^{2-}$$

It is therefore possible to determine cations such as Ca^{2+}, Mg^{2+}, Pb^{2+}, and Mn^{2+} in the presence of the above-mentioned metals by masking with an excess of potassium or sodium cyanide. A small amount of iron may be masked by cyanide if it is first reduced to the iron(II) state by the addition of ascorbic acid. Titanium(IV), iron(III), and aluminium can be masked with triethanolamine; mercury with iodide ions; and aluminium, iron(III), titanium(IV), and tin(II) with ammonium fluoride (the cations of the alkaline-earth metals yield slightly soluble fluorides).

Sometimes the metal may be transformed into a different oxidation state: thus copper(II) may be reduced in acid solution by hydroxylamine or ascorbic acid. After rendering ammoniacal, nickel or cobalt can be titrated using, for example, murexide as indicator without interference from the copper, which is now present as Cu(I). Iron(III) can often be similarly masked by reduction with ascorbic acid.

(c) **The cyanide complexes** of zinc and cadmium may be **demasked** with formaldehyde–acetic acid solution or, better, with chloral hydrate:

$$[Zn(CN)_4]^{2-} + 4H^+ + 4HCHO \longrightarrow Zn^{2+} + 4HO \cdot CH_2 \cdot CN$$

The use of masking and selective demasking agents permits the successive titration of many metals. Thus a solution containing Mg, Zn, and Cu can be titrated as follows:

(i) Add excess of standard EDTA and back-titrate with standard Mg solution using Solochrome Black (Eriochrome Black T) as indicator. This gives the sum of all the metals present.

(ii) Treat an aliquot portion with excess of KCN and titrate as before. This gives Mg only.

(iii) Add excess of chloral hydrate (or of formaldehyde–acetic acid solution,

3:1) to the titrated solution in order to liberate the Zn from the cyanide complex, and titrate until the indicator turns blue. This gives the Zn only. The Cu content may then be found by difference.

(d) **Classical separations** may be applied if these are not tedious; thus the following precipitates may be used for separations in which, after being redissolved, the cations can be determined complexometrically: CaC_2O_4, nickel dimethylglyoximate, $Mg(NH_4)PO_4,6H_2O$, and CuSCN.

(e) **Solvent extraction** is occasionally of value. Thus zinc can be separated from copper and lead by adding excess of ammonium thiocyanate solution and extracting the resulting zinc thiocyanate with 4-methyl pentan-2-one (isobutyl methyl ketone); the extract is diluted with water and the zinc content determined with EDTA solution.

(f) **The indicator** chosen should be one for which the formation of the metal-indicator complex is sufficiently rapid to permit establishment of the end-point without undue waiting, and should preferably be reversible.

(g) **Anions**, such as orthophosphate, which can interfere in complexometric titrations may be removed using ion exchange resins. For the use of ion exchange resins in the separation of cations and their subsequent EDTA titration, see Chapter VII.

(h) **'Kinetic masking'** is a special case in which a metal ion does not effectively enter into the complexation reaction because of its kinetic inertness (see Section **X, 21**). Thus the slow reaction of chromium(III) with EDTA makes it possible to titrate other metal ions which react rapidly, without interference, from Cr(III); this is illustrated by the determination of iron(III) and chromium(III) in a mixture (Section **X, 68**).

X, 28. METAL ION INDICATORS. General properties. The success of an EDTA titration depends upon the precise determination of the end-point. The most common procedure utilises metal ion indicators. The requisites of a metal ion indicator for use in the visual detection of end-points include:

(*a*) The colour reaction must be such that before the end-point, when nearly all the metal ion is complexed with EDTA, the solution is strongly coloured.

(*b*) The colour reaction should be specific or at least selective.

(*c*) The metal indicator-complex must possess sufficient stability, otherwise, because of dissociation, a sharp colour change is not obtained. The metal indicator–complex must, however, be less stable than the metal–EDTA complex to ensure that, at the end-point, EDTA removes metal ions from the metal indicator–complex. The change in equilibrium from the metal indicator–complex to the metal–EDTA complex should be sharp and rapid.

(*d*) The colour contrast between the free indicator and the metal indicator–complex should be such as to be readily observed.

(*e*) The indicator must be very sensitive to metal ions (i.e., to pM) so that the colour change occurs as near to the equivalence point as possible.

(*f*) The above requirements must be fulfilled within the pH range at which the titration is performed.

Dyestuffs which form complexes with specific metal cations can serve as indicators of pM values; 1:1-complexes (metal: dyestuff = 1:1) are common, but 1:2-complexes and 2:1-complexes also occur. The metal ion indicators, like EDTA itself, are chelating agents; this implies that the dyestuff molecule possesses several ligand atoms suitably disposed for coordination with a metal atom. They can, of course, equally take up protons, which also produces a colour change; metal ion indicators are therefore not only pM but also pH indicators.

Theory of the visual use of metal ion indicators. Our discussion will be confined to the more common 1:1-complexes. The use of a metal ion indicator in an EDTA titration may be written as:

$$M\text{--}In + EDTA \longrightarrow M\text{--}EDTA + In$$

This reaction will proceed if the metal indicator–complex M–In is less stable than the metal–EDTA complex M–EDTA. The former dissociates to a limited extent, and during the titration the free metal ions are progressively complexed by the EDTA until ultimately the metal is displaced from the complex M–In to leave the free indicator (In). The stability of the metal indicator–complex may be expressed in terms of the *formation constant* (or *indicator constant*) K_{In}:

$$K_{In} = [M\text{--}In]/[M][In]$$

The indicator colour change is affected by the hydrogen-ion concentration of the solution, and no account of this has been taken in the above expression for the formation constant. Thus Solochrome Black (Eriochrome Black T), which may be written as H_2In^-, exhibits the following acid–base behaviour:

$$H_2In^- \underset{5.3-7.3}{\overset{pH}{\rightleftharpoons}} HIn^{2-} \underset{10.5-12.5}{\overset{pH}{\rightleftharpoons}} In^{3-}$$
$$\text{Red} \qquad\qquad \text{Blue} \qquad\qquad \text{Yellow-orange}$$

In the pH range 7–11, in which the dye itself exhibits a blue colour, many metal ions form red complexes; these colours are extremely sensitive, as is shown, for example, by the fact that 10^{-6}–10^{-7} molar solutions of magnesium ion give a distinct red colour with the indicator. From the practical viewpoint, it is more convenient to define the *apparent indicator constant* K'_{In}, which varies with pH, as:

$$K'_{In} = [MIn^-]/[M^{n+}][In]$$

where [MIn⁻] = concentration of metal ion–indicator complex,
   [M^{n+}] = concentration of metallic ion, and
   [In] = concentration of indicator not complexed with metallic ion.

(This, for the above indicator, is equal to $[H_2In^-]+[HIn^{2-}]+[In^{3-}]$.)
The equation may be expressed as:

$$\log K'_{In} = pM + \log[MIn^-]/[In];$$

$\log K'_{In}$ gives the value of pM when half the total indicator is present as the metal ion complex. Some values for $\log K'_{In}$ for CaIn⁻ and MgIn⁻ respectively (where H_2In^- is the anion of Solochrome Black (Eriochrome Black T) are: 0.8 and 2.4 at pH = 7; 1.9 and 3.4 at pH = 8; 2.8 and 4.4 at pH = 9; 3.8 and 5.4 at pH = 10; 4.7 and 6.3 at pH = 11; 5.3 and 6.8 at pH = 12. For a small titration error K'_{In} should be large ($> 10^4$), the ratio of the apparent stability constant of the metal–EDTA complex K'_{MY} to that of the metal indicator–complex K'_{In} should be large ($> 10^4$), and the ratio of the indicator concentration to the metal ion concentration should be small ($< 10^{-2}$).

The visual metallochromic indicators discussed above form by far the most important group of indicators for EDTA titrations and the operations subsequently described will be confined to the use of indicators of this type; nevertheless there are certain other substances which can be used as indicators (see Ref. 8).

Some examples of metal ion indicators.   Numerous compounds have been proposed for use as pM indicators; a selected few of these will be described. Where applicable, Colour Index (C.I.) references are given (Ref. 9). It has been pointed out by West (Ref. 8), that apart from a few miscellaneous compounds, the important visual metallochromic indicators fall into three main groups: (*a*) hydroxyazo compounds, (*b*) phenolic compounds and hydroxy-substituted triphenylmethane compounds, (*c*) compounds containing an aminomethyldicarboxymethyl group: many of these are also triphenylmethane compounds.

Note.   In view of the varying stability of solutions of these indicators, and the possible variation in sharpness of the end-point with the age of the solution, it is generally advisable (if the stability of the indicator solution is suspect), to dilute the solid indicator with 100–200 parts of A.R. potassium (or sodium) chloride, nitrate or sulphate (potassium nitrate is usually preferred) and grind the mixture well in a glass mortar. The resultant mixture is usually stable indefinitely if kept dry and in a tightly stoppered bottle.

Murexide.   (C.I. 56085)   This is the ammonium salt of purpuric acid, and its anion has the structure (I). It is of interest because it was probably the first metal ion indicator to be employed in the EDTA titration. Murexide solutions are reddish-violet up to pH $= 9\,(H_4D^-)$, violet from pH 9 to pH 11 (H_3D^{2-}), and blue-violet (or blue) above pH 11 (H_2D^{3-}). These colour changes are probably due to the progressive displacement of protons from the imido groups; since there are four such groups, murexide may be represented as H_4D^-. Only two of these four acidic hydrogens can be removed by adding an alkali hydroxide, so that only two pK values need be considered; these are $pK_4 = 9.2\,(H_2D^- \longrightarrow H_3D^{2-})$ and $pK_3 = 10.5\,(H_3D^{2-} \longrightarrow H_2D^{3-})$. The anion H_4D^- can also take up a proton to yield the yellow and unstable purpuric acid, but this requires a pH of about 0.

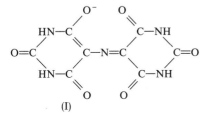

(I)

Murexide forms complexes with many metal ions: only those with Cu, Ni, Co, Ca and the lanthanoids are sufficiently stable to find application in analysis. Their colours in alkaline solution are orange (copper), yellow (nickel and cobalt), and red (calcium); the colours vary somewhat with the pH of the solution.

Murexide may be employed for the direct EDTA titration of calcium at pH $= 11$; the colour change at the end point is from red to blue-violet, but is far from ideal. The colour change in the direct titration of nickel at pH 10–11 is from yellow to blue-violet.

Aqueous solutions of murexide are unstable and must be prepared each day. The indicator solution may be prepared by suspending 0.5 g of the powdered dyestuff in water, shaking thoroughly, and allowing the undissolved portion to settle. The saturated supernatant liquid is used for titrations. Every day the old supernatant liquid is decanted and the residue treated with water as before to provide a fresh solution of the indicator. Alternatively, one may prepare a mixture of the indicator with pure sodium chloride in the ratio 1:500, and employ 0.2–0.4 g in each titration. A screened indicator, consisting of 0.2 g of murexide, 0.5 g of Naphthol Green B, and 100 g of pure sodium chloride ground together to form a uniformly coloured mixture has been proposed; about 0.2 g of the mixture is suitable for 100 cm³ of the sample solution. The colour change for calcium is from olive-green, through grey, to a sudden blue.

Solochrome Black (Eriochrome Black T). This substance is sodium 1-(1-hydroxy-2-naphthylazo)-6-nitro-2-naphthol-4-sulphonate(II); and has the Colour Index reference C.I. 14645. In strongly acidic solutions the dye tends to polymerise to a red-brown product, and consequently the indicator is rarely applied in the EDTA titration of solutions more acidic than pH = 6.5.

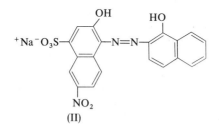

(II)

The sulphonic acid group gives up its proton long before the pH range of 7–12, which is of immediate interest for metal-ion indicator use. Only the dissociation of the two hydrogen atoms of the phenolic groups need therefore be considered, and so the dyestuff may be represented by the formula H_2D^-. The two pK values for these hydrogen atoms are 6.3 and 11.5 respectively. Below pH = 5.5, the solution of Solochrome Black (Eriochrome Black T) is red (due to H_2D^-), between pH 7 and 11 it is blue (due to HD^{2-}), and above pH = 11.5 it is yellowish-orange (due to D^{3-}). In the pH range 7–11 the addition of metallic salts produces a brilliant change in colour from blue to red:

$$M^{2+} + HD^{2-} \text{ (blue)} \longrightarrow MD^- \text{ (red)} + H^+$$

This colour change can be observed with the ions of Mg, Mn, Zn, Cd, Hg, Pb, Cu, Al, Fe, Ti, Co, Ni, and the Pt metals. To maintain the pH constant (*ca.* 10) a buffer mixture is added, and most of the above metals must be kept in solution with the aid of a weak complexing reagent such as ammonia or tartrate. The cations of Cu, Co, Ni, Al, Fe(III), Ti(IV), and certain of the Pt metals form such stable indicator complexes that the dyestuff can no longer be liberated by adding EDTA: direct titration of these ions using Solochrome Black (Eriochrome Black T) as indicator is therefore impracticable, and the metallic ions are said to 'block' the indicator. However, with Cu, Co, Ni, and Al a back-titration can be carried out, for the rate of reaction of their EDTA complexes with the indicator is extremely slow and it is possible to titrate the excess of EDTA with standard zinc or magnesium ion solution.

Cu, Ni, Co, Cr, Fe, or Al, even in traces, must be absent when conducting a direct titration of the other metals listed above; if the metal ion to be titrated does not react with the cyanide ion or with triethanolamine, these substances can be used as masking reagents. It has been stated that the addition of 0.5–1 cm³ of

0.001 M-o-phenanthroline prior to the EDTA titration eliminates the 'blocking effect' of these metals with Solochrome Black (Eriochrome Black T) and also with Xylenol Orange (see below).

The *indicator solution* is prepared by dissolving 0.2 g of the dyestuff in 15 cm³ of triethanolamine with the addition of 5 cm³ of absolute ethanol to reduce the viscosity; the reagent is stable for several months. A 0.4 per cent solution of the pure dyestuff in methanol remains serviceable for at least a month.

It may be noted that the dyestuff in which the nitro group is absent, viz., sodium 1-(1-hydroxy-2-naphthylazo)-2-naphthol-4-sulphonate (**Solochrome Black 6B; Eriochrome Blue-Black B;** Colour Index No. 14640) is superior as far as the solution stability is concerned, and the colour change is sharper with Mg and certain other metals (Zn and Pb excepted). The colour change is from red to blue: the indicator may be screened with a little 0.5 per cent aqueous tartrazine solution, when the resulting end-point is from scarlet (or orange red) to apple green. A 0.5 per cent ethanolic solution of the dyestuff is stable for at least two months.

Patton and Reeder's indicator. The indicator is 2-hydroxy-1-(2-hydroxy-4-sulpho-1-naphthylazo)-3-naphthoic acid (III); the name may be abbreviated to HHSNNA. Its main use is in the direct titration of calcium, particularly in the presence of magnesium. A sharp colour change from wine red to pure blue is obtained when calcium ions are titrated with EDTA at pH values between 12 and 14. Interferences are similar to those observed with Solochrome Black (Eriochrome Black T), and can be obviated similarly. This indicator may be used as an alternative to murexide for the determination of calcium.

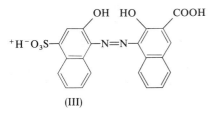

(III)

The dyestuff is thoroughly mixed with 100 times its weight of sodium sulphate, and 1 g of the mixture is used in each titration. The indicator is not very stable in alkaline solution.

Solochrome Dark Blue or Calcon. (C.I. 15705). This is sometimes referred to as Eriochrome Blue Black RC; it is in fact sodium 1-(2-hydroxy-1-naphthylazo)-2-naphthol-4-sulphonate, as shown in formula (IV). The dyestuff has two ionisable phenolic hydrogen atoms; the protons ionise stepwise with pK's of 7.4 and 13.5 respectively. An important application of the indicator is in the complexometric titration of calcium in the presence of magnesium; this must

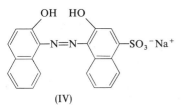

(IV)

be carried out at a pH of about 12.3 (obtained, for example, with a diethylamine buffer: 5 cm³/100 cm³ of solution) in order to avoid the interference of magnesium. Under these conditions magnesium is precipitated quantitatively as the hydroxide. The colour change is from pink to pure blue.

The *indicator solution* is prepared by dissolving 0.2 g of the dyestuff in 50 cm³ of methanol.

Calmagite. This indicator, 1-(1-hydroxyl-4-methyl-2-phenylazo)-2-naphthol-4-sulphonic acid (V), has the same colour change as Solochrome Black (Eriochrome Black T), but the colour change is somewhat clearer and sharper. An important advantage is that aqueous solutions of the indicator are stable almost indefinitely. It may be substituted for Solochrome Black (Eriochrome Black T) without change in the experimental procedures for the titration of calcium plus magnesium (see Sections **X, 55, 64**).

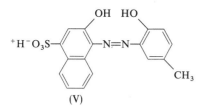

(V)

Calmagite functions as an acid–base indicator:

$$\underset{\substack{\text{Bright} \\ \text{red}}}{H_3D} \overset{\text{low pH}}{\rightleftharpoons} \underset{\substack{\text{Bright} \\ \text{red}}}{H_2D^-} \overset{\text{pH 7.1–9.1}}{\rightleftharpoons} \underset{\substack{\text{Clear} \\ \text{blue}}}{HD^{2-}} \overset{\text{pH 11.4–13.3}}{\rightleftharpoons} \underset{\substack{\text{Reddish-} \\ \text{orange}}}{D^{3-}}$$

The hydrogen of the sulphonic acid group plays no part in the functioning of the dye as a metal ion indicator. The acid properties of the hydroxyl groups are expressed by $pK_1 = 8.14$ and $pK_2 = 12.35$. The blue colour of Calmagite at pH $= 10$ is changed to red by the addition of magnesium ions, the change being reversible:

$$\underset{\text{Clear blue}}{HD^{2-}} \overset{Mg^{2+}}{\rightleftharpoons} \underset{\text{Red}}{MgD^{2-}}$$

This is the basis of the indicator action in the EDTA titration. The pH $= 10$ is attained by the use of an aqueous ammonia–ammonium chloride buffer mixture.

The combining ratio between calcium or magnesium and the indicator is 1:1; the magnesium compound is the more stable. Calmagite is similar to Solochrome Black in that small amounts of copper, iron, and aluminium interfere seriously in the titration of calcium and magnesium, and similar masking agents may be used. Potassium hydroxide should be employed for the neutralisation of large amounts of acid since sodium ions in high concentration cause difficulty.

The *indicator solution* is prepared by dissolving 0.05 g of Calmagite in 100 cm³ of water. It is stable for at least 12 months when stored in a polythene bottle out of sunlight.

Calcichrome. This indicator, cyclotris-7-(1-azo-8-hydroxynaphthalene-3,6-disulphonic acid) (VI), is unusual in having a cyclic structure, and is very selective for calcium. It is in fact not very suitable as an indicator for EDTA titrations because the colour change is not particularly sharp, but if EDTA is replaced by CDTA (see Section **X, 23**), then the indicator gives good results for calcium in the presence of large amounts of barium and small amounts of strontium (Ref. 10).

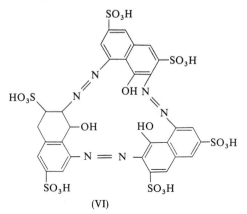

(VI)

Fast Sulphon Black F. (C.I. 26990). This dyestuff is the sodium salt of 1-hydroxy-8-(2-hydroxynaphthylazo)-2-(sulphonaphthylazo)-3,6-disulphonic acid (VII). The colour reaction seems virtually specific for copper ions. In ammoniacal solution it forms complexes with only copper and nickel; the presence of ammonia or pyridine is required for colour formation. In the direct titration of copper in ammoniacal solution the colour change at the end-point is from magenta or (depending upon the concentration of copper(II) ions) pale blue to bright green. The indicator action with nickel is poor. Metal ions, such as those of Cd, Pb, Ni, Zn, Ca, and Ba, may be titrated using this indicator by the prior addition of a reasonable excess of standard copper(II) solution.

The *indicator solution* consists of a 0.5 per cent aqueous solution.

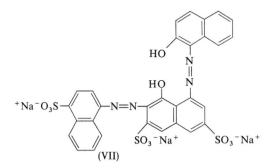

(VII)

Catechol Violet. This indicator, also termed Pyrocatechol Violet, is catechol sulphonphthalein (VIII). It also possesses acid–base indicator properties (H_4D). An aqueous solution of Catechol Violet is coloured yellow; at a pH below 1.5, the colour is red; it is yellow between pH = 2 and 6 (anion H_3D^-), at pH = 7 it is violet (anion H_2D^{2-}), and above pH = 10 the colour is blue (anion D^{4-}). The colour change is ascribed to the progressive ionisation of the hydroxyl groups. The blue, strongly alkaline solutions are unstable and lose their colour fairly rapidly, probably owing to atmospheric oxidation.

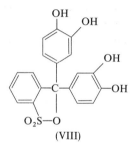

(VIII)

Catechol violet forms coloured compounds (usually blue or green-blue) with many metals; the most stable of these complexes are formed in the pH range 2–6, so that there is a sharp colour change from yellow to blue when certain cations (e.g., of bismuth and thorium) are added to the indicator solution. Complexes of the indicator with ions of divalent metals, such as Cu, Zn, Cd, Ni, and Co, do not form until the pH is about 7, so that on adding these metal ions to the indicator there is only a change from violet to blue, which is less easy to detect. The determination of copper in the presence of a pyridine buffer is, however, fairly satisfactory; the colour change is from blue to green or yellowish-green.

The *indicator solution* is prepared by dissolving 0.1 g of the dyestuff in 100 cm^3 of water. The solution is stable for several weeks.

Bromopyrogallol Red.   This metal ion indicator is dibromopyrogallol sulphonphthalein (IX) and is more resistant to oxidation than Catechol Violet; it also possesses acid–base indicator properties. The indicator is coloured orange-yellow in strongly acidic solution, claret red in nearly neutral solution, and violet to blue in basic solution. The dyestuff forms coloured complexes with many cations. It is valuable for the determination, *inter alia*, of bismuth (pH = 2–3, nitric acid solution; end-point blue to claret red).

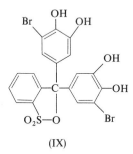

(IX)

The *indicator solution* is prepared by dissolving 0.05 g of the solid reagent in 100 cm^3 of 50 per cent ethanol.

Xylenol Orange.   This indicator, prepared by the condensation of *o*-cresolsulphonephthalein (Cresol Red) with formaldehyde and iminodiacetic acid, is 3,3′-*bis*[*N,N*-di(carboxymethyl)-aminomethyl]-*o*-cresolsulphonephthalein(X). This dyestuff retains the acid–base properties of Cresol Red and displays metal indicator properties even in acid solution (pH = 3–5). Acidic solutions of the indicator are coloured lemon-yellow and those of the metal complexes intensely red.

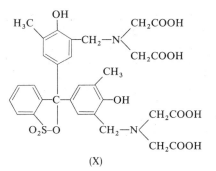

(X)

Direct EDTA titrations of Bi, Th, Zn, Cd, Pb, Co, etc., are readily carried out and the colour change is sharp. Iron(III) and, to a lesser extent, aluminium interfere. By appropriate pH adjustment certain pairs of metals may be titrated successfully in a single sample solution. Thus bismuth may be titrated at pH = 1–2, and zinc or lead after adjustment to pH = 5 by addition of hexamine.

The indicator solution is prepared by dissolving 0.5 g of Xylenol Orange in 100 cm³ of water.

Thymolphthalein Complexone. (Thymolphthalexone). This is thymolphthalein di(methylimine diacetic acid) (XI); it contains a stable lactone ring and reacts only in an alkaline medium. The indicator may be used for the titration of calcium; the colour change is from blue to colourless (or a slight pink). Manganese and also nickel may be determined by adding an excess of standard EDTA solution, and titrating the excess with standard calcium chloride solution; the colour change is from very pale blue to deep blue.

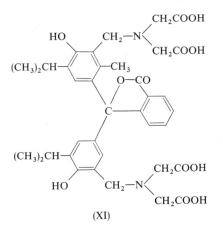

(XI)

The *indicator solution* consists of a 0.5 per cent solution in ethanol. Alternatively, a finely ground mixture (1:100) with A.R. potassium nitrate may be used.

Methylthymol Blue (Methylthymol Blue Complexone). This compound (XII) is very similar in structure to the preceding one from which it is derived by replacement of the lactone grouping by a sulphonic acid group. By contrast, however, it will function in both acidic and alkaline media, ranging from pH = 0 under which condition bismuth may be titrated with a colour change from blue to yellow, to pH = 12, where the alkaline earths may be titrated with a colour change from blue to colourless. At intermediate pH values a wide variety of bivalent metal ions may be titrated; of particular interest is its use as an indicator for the titration of Hg(II), an ion for which very few indicators are available. It is also suitable for determining calcium in the presence of magnesium provided that the proportion of the latter is not too high, and is therefore of value in determining the hardness of water. The indicator does not keep well in solution and is used as a solid: 1 part to 100 of potassium nitrate.

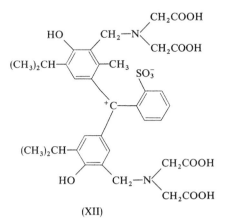

(XII)

Zincon is 1-(2-hydroxy-5-sulphophenyl)-3-phenyl-5-(2-carboxyphenyl)-formazan (XIII) which is a specific indicator for zinc at pH 9–10. Its most important use, however, is as indicator for titration of calcium in the presence of magnesium, using the complexone EGTA (Section **X, 23**); the magnesium–EGTA complex is relatively weak and does not interfere with the calcium titration. Calcium and magnesium do not give coloured complexes with the indicator, and the procedure is to add a little of the zinc complex of EGTA. The titration is carried out in a buffer at pH 10, and under these conditions calcium ions decompose the Zn–EGTA complex, liberating zinc ions which give a blue colour with the indicator. As soon as all the calcium has been titrated, excess EGTA reconverts the zinc ions to the EGTA complex, and the solution acquires the orange colour of the metal-free indicator.

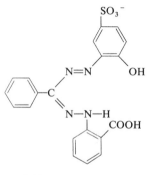

(XIII)

0.1M-potassium iodide with 0.1M-silver nitrate are included in the same table ($K_{sol.AgI} = 1.7 \times 10^{-16}$).

It will be seen by inspecting the silver-ion exponents in the neighbourhood of the equivalence point (say, between 99.8 and 100.2 cm³) that there is a marked change in the silver-ion concentration, although the change is more pronounced for silver iodide than for silver chloride, since the solubility product of the latter is about 10^6 larger than for the former. This is shown more clearly in the titration curve in Fig. X, 10, which represents the changes of pAg⁺ in the range between 10 per cent before and 10 per cent after the stoichiometric point in the titration of 0.1M-chloride and 0.1M-iodide respectively with 0.1M-silver nitrate. An almost identical curve is obtained by potentiometric titration using a silver electrode (see Section **XIV, 25**); the pAg⁺ values may be computed from the e.m.f. figures exactly as in the calculation of pH.

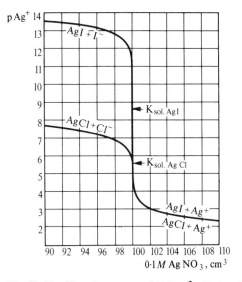

Fig. X, 10   Titration curves of 100 cm³ of 0.1M-NaCl and of 100 cm³ of 0.1M-KI respectively with 0.1M-AgNO₃ (calculated)

X, 30.   DETERMINATION OF END-POINTS IN PRECIPITATION REACTIONS.   Many methods are utilised in determining end-points in these reactions, but only the most important will be mentioned here.

A. Formation of a coloured precipitate.   This may be illustrated by the Mohr procedure for the determination of chloride and bromide. In the titration of a neutral solution of, say, chloride ions with silver nitrate solution, a small quantity of potassium chromate solution is added to serve as indicator. At the end-point the chromate ions combine with silver ions to form the sparingly soluble, red silver chromate.

The theory of the process is as follows. We have here a case of fractional precipitation (Section **II, 10**), the two sparingly soluble salts being silver chloride (K_{sol} 1.2 × 10⁻¹⁰) and silver chromate (K_{sol} 1.7 × 10⁻¹²). Let us consider an actual example encountered in practice, viz., the titration of, say, 0.1M-sodium chloride with 0.1M-silver nitrate in the presence of a few cm³ of dilute potassium chromate solution. Silver chloride is the less soluble salt and, furthermore, the initial chloride concentration is high, hence silver chloride will be precipitated. At the first point where red silver chromate is just precipitated, we shall have both salts in equilibrium with the solution, hence:

$$[Ag^+] \times [Cl^-] = K_{sol. AgCl} = 1.2 \times 10^{-10}$$

$$[Ag^+]^2 \times [CrO_4{}^{2-}] = K_{sol. Ag_2CrO_4} = 1.7 \times 10^{-12}$$

$$[Ag^+] = \frac{K_{sol. AgCl}}{[Cl^-]} = \sqrt{\frac{K_{sol. Ag_2CrO_4}}{[CrO_4{}^{2-}]}}$$

$$\frac{[Cl^-]}{\sqrt{[CrO_4{}^{2-}]}} = \frac{K_{sol. AgCl}}{\sqrt{K_{sol. Ag_2CrO_4}}} = \frac{1.2 \times 10^{-10}}{\sqrt{1.7 \times 10^{-12}}} = 9.2 \times 10^{-5}$$

At the equivalence point $[Cl^-] = \sqrt{K_{sol. AgCl}} = 1.1 \times 10^{-5}$. If silver chromate is to precipitate at this chloride-ion concentration:

$$[CrO_4{}^{2-}] = \left(\frac{[Cl^-]}{9.2 \times 10^{-5}}\right)^2 = \left(\frac{1.1 \times 10^{-5}}{9.2 \times 10^{-5}}\right)^2 = 1.4 \times 10^{-2}$$

or the potassium chromate solution should be $0.014M$. It should be noted that a slight excess of silver nitrate solution must be added before the red colour of silver chromate is visible. In practice, a more dilute solution ($0.003–0.005M$) of potassium chromate is generally used, since a chromate solution of concentration $0.01–0.02M$ imparts a distinct deep orange colour to the solution, which renders the detection of the first appearance of silver chromate somewhat difficult. We can readily calculate the error thereby introduced, for if $[CrO_4{}^{2-}] = $ (say) 0.003, silver chromate will be precipitated when:

$$[Ag^+] = \sqrt{\frac{K_{sol. Ag_2CrO_4}}{CrO_4{}^{2-}}} = \sqrt{\frac{1.7 \times 10^{-12}}{3 \times 10^{-3}}} = 2.4 \times 10^{-5}$$

If the theoretical concentration of indicator is used:

$$[Ag^+] = \sqrt{\frac{1.7 \times 10^{-12}}{1.4 \times 10^{-2}}} = 1.1 \times 10^{-5}$$

The difference is 1.3×10^{-5} equivalent dm^{-3}. If the volume of the solution at the equivalence point is 150 cm^3, then this corresponds to $1.3 \times 10^{-5} \times 150 \times 10^4/1000 = 0.02$ cm^3 of $0.1M$-silver nitrate. This is the theoretical titration error, and is therefore negligible. In actual practice another factor must be considered, viz., the small excess of silver nitrate solution which must be added before the eye can detect the colour change in the solution; this is of the order of one drop or *ca.* 0.05 cm^3 of $0.1M$-silver nitrate.

The titration error will increase with increasing dilution of the solution being titrated and is quite appreciable (*ca.* 0.4* per cent) in dilute, say, $0.01M$, solutions when the chromate concentration is of the order $0.003–0.005M$. This is most simply allowed for by means of an indicator blank determination, e.g., by measuring the volume of standard silver nitrate solution required to give a

---

* The errors for $0.1M$- and $0.01M$-bromide may be calculated to be 0.04 and 0.4 per cent respectively.

perceptible coloration when added to distilled water containing the same quantity of indicator as is employed in the titration. This volume is subtracted from the volume of standard solution used.

It must be mentioned that the titration should be carried out in neutral solution or in very faintly alkaline solution, i.e., within the pH range 6.5–9. In acid solution, the following reaction occurs:

$$2CrO_4{}^{2-} + 2H^+ \rightleftharpoons 2HCrO_4{}^- \rightleftharpoons Cr_2O_7{}^{2-} + H_2O$$

$HCrO_4{}^-$ is a weak acid, consequently the chromate-ion concentration is reduced and the solubility product of silver chromate may not be exceeded. In markedly alkaline solutions, silver hydroxide ($K_{sol.}$ 2.3×10^{-8}) might be precipitated. A simple method of making an acid solution neutral is to add an excess of pure calcium carbonate or sodium hydrogen carbonate. An alkaline solution may be acidified with acetic acid and then a slight excess of calcium carbonate is added. The solubility product of silver chromate increases with rising temperature; the titration should therefore be performed at room temperature. By using a mixture of potassium chromate and potassium dichromate in proportions such as to give a neutral solution, the danger of accidentally raising the pH of an unbuffered solution beyond the acceptable limits is minimised; the mixed indicator has a buffering effect and adjusts the pH of the solution to 7.0 ± 0.1. In the presence of ammonium salts, the pH must not exceed 7.2 because of the effect of appreciable concentrations of ammonia upon the solubility of silver salts. Titration of iodide and of thiocyanate is not successful because silver iodide and silver thiocyanate adsorb chromate ions so strongly that a false and somewhat indistinct end-point is obtained.

B. Formation of a soluble coloured compound. This procedure is exemplified by the method of Volhard for the titration of silver in the presence of free nitric acid with standard potassium or ammonium thiocyanate solution. The indicator is a solution of iron(III) nitrate or of iron(III) ammonium sulphate. The addition of the thiocyanate solution produces first a precipitate of silver thiocyanate ($K_{sol.}$ 7.1×10^{-13}):

$$Ag^+ + SCN^- \rightleftharpoons AgSCN$$

When this reaction is complete, the slightest excess of thiocyanate produces a reddish-brown coloration, due to the formation of a complex ion:*

$$Fe^{3+} + SCN^- \rightleftharpoons [FeSCN]^{2+}$$

This method may be applied to the determination of chlorides, bromides, and iodides in acid solution. Excess of standard silver nitrate solution is added, and the excess is back-titrated with standard thiocyanate solution. For the chloride estimation, we have the following two equilibria during the titration of excess of silver ions:

$$Ag^+ + Cl^- \rightleftharpoons AgCl$$
$$Ag^+ + SCN^- \rightleftharpoons AgSCN$$

---

* This is the complex formed when the ratio of thiocyanate ion to iron(III) ion is low; higher complexes, $[Fe(SCN)_2]^+$, etc., are important only at higher concentrations of thiocyanate ion.

The two sparingly soluble salts will be in equilibrium with the solution, hence:

$$\frac{[Cl^-]}{[SCN^-]} = \frac{K_{sol.\,AgCl}}{K_{sol.\,AgSCN}} = \frac{1.2 \times 10^{-10}}{7.1 \times 10^{-13}} = 169$$

When the excess of silver has reacted, the thiocyanate may react with the silver chloride, since silver thiocyanate is the less soluble salt until the ratio $[Cl^-]/(SCN^-)$ in the solution is 169:

$$AgCl + SCN^- \rightleftharpoons AgSCN + Cl^-$$

This will take place before reaction occurs with the iron(III) ions in the solution, and there will consequently be a considerable titration error. It is therefore absolutely necessary to prevent the reaction between the thiocyanate and the silver chloride. This may be effected in several ways, of which the first is probably the most reliable:

(i) The silver chloride is filtered off before back-titrating. Since at this stage the precipitate will be contaminated with adsorbed silver ions, the suspension should be boiled for a few minutes to coagulate the silver chloride and thus remove most of the adsorbed silver ions from its surface before filtration. The cold filtrate is titrated.

(ii) After the addition of the silver nitrate, potassium nitrate is added as coagulant, the suspension is boiled for about 3 minutes, cooled and then titrated immediately. Desorption of silver ions occurs and, on cooling, re-adsorption is largely prevented by the presence of potassium nitrate.

(iii) An immiscible liquid is added to 'coat' the silver chloride particles and thereby protect them from interaction with the thiocyanate. The most successful liquid is nitrobenzene (*ca.* 1.0 cm^3 for each 50 mg of chloride): the suspension is well shaken to coagulate the precipitate before back-titration.

With bromides, we have the equilibrium:

$$\frac{[Br^-]}{[SCN^-]} = \frac{K_{sol.\,AgBr}}{K_{sol.\,AgSCN}} = \frac{3.5 \times 10^{-13}}{7.1 \times 10^{-13}} = 0.5$$

The titration error is small, and no difficulties arise in the determination of the end-point. Silver iodide ($K_{sol.}$ 1.7×10^{-16}) is less soluble than the bromide; the titration error is negligible, but the iron(III) indicator should not be added until excess of silver is present, since the dissolved iodide reacts with Fe^{3+} ions:

$$2Fe^{3+} + 2I^- \rightleftharpoons 2Fe^{2+} + I_2$$

C. Use of adsorption indicators. K. Fajans introduced a useful type of indicator for precipitation reactions as a result of his studies on the nature of adsorption The action of these indicators is due to the fact that at the equivalence point the indicator is adsorbed by the precipitate, and during the process of adsorption a change occurs in the indicator which leads to a substance of different colour; they have therefore been termed **adsorption indicators**. The substances employed are either acid dyes, such as those of the fluorescein series, e.g., fluorescein and eosin which are utilised as the sodium salts, or basic dyes, such as those of the rhodamine series (e.g., rhodamine 6G), which are applied as the halogen salts.

The theory of the action of these indicators is based upon the properties of colloids, Section **XI, 3**. When a chloride solution is titrated with a solution of

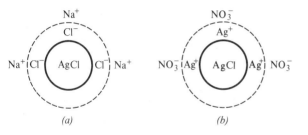

Fig. X, 11 (*a*) **AgCl precipitated in the presence of excess of Cl⁻**
(*b*) **AgCl precipitated in the presence of excess of Ag⁺**

silver nitrate, the precipitated silver chloride adsorbs chloride ions (a precipitate has a tendency to adsorb its own ions); this may be termed the primary adsorbed layer, and it will hold by secondary adsorption oppositely charged ions present in solution (shown diagrammatically in Fig. X, 11, *a*). As soon as the stoichiometric point is reached, silver ions are present in excess; these will now be primarily adsorbed, and nitrate ions will be held by secondary adsorption (Fig. X, 11, *b*). If fluorescein is also present in the solution, the negative fluorescein ion, which is much more strongly adsorbed than the nitrate ion, is immediately adsorbed, and will reveal its presence on the precipitate, not by its own colour, which is that of the solution, but by the formation of a pink complex of silver and a modified fluorescein ion on the surface with the first trace of excess of silver ions. An alternative view is that during the adsorption of the fluorescein ion a rearrangement of the structure of the ion occurs with the formation of a coloured substance. It is important to notice that the colour change takes place at the *surface* of the precipitate. If chloride is now added, the suspension will remain pink until chloride ions are present in excess, the adsorbed silver will then be converted into silver chloride, which will then primarily adsorb chloride ions. The fluorescein ions secondarily adsorbed will pass back into solution, to which they impart a greenish-yellow colour.

The following conditions will govern the choice of a suitable adsorption indicator:

(*a*) The precipitate should separate as far as possible in the colloidal condition. Large quantities of neutral salts, particularly of multivalent ions, should be avoided owing to their coagulating effect. The solution should not be too dilute, as the amount of precipitate formed will be small and the colour change far from sharp with certain indicators.

(*b*) The indicator ion must be of opposite charge to the ion of the precipitating agent.

(*c*) The indicator ion should not be adsorbed before the particular compound has been completely precipitated, but it should be strongly adsorbed immediately after the equivalence point. The indicator ion should not be too strongly adsorbed by the precipitate; if this occurs, e.g., eosin (tetrabromo-fluorescein) in the chloride–silver titration, the adsorption of the indicator ion may be a primary process and will take place before the equivalence point.

A disadvantage of adsorption indicators is that silver halides are sensitised to the action of light by a layer of adsorbed dyestuff. For this reason, titrations should be carried out with a minimum exposure to sunlight. When using adsorption indicators, only 2×10^{-4} to 3×10^{-3} mol of dye per mol of silver halide is added; this small concentration is used so that an appreciable fraction of the added indicator is actually adsorbed on the precipitate.

For the titration of chlorides, fluorescein may be used. This indicator is a very weak acid ($K_a = ca.\ 1 \times 10^{-8}$), hence even a small amount of other acids reduces the already minute ionisation, thus rendering the detection of the end-point (which depends essentially upon the adsorption of the free anion) either impossible or difficult to observe. The optimum pH range is between 7 and 10. Dichlorofluorescein is a stronger acid and may be utilised in slightly acid solutions of pH greater than 4.4; this indicator has the further advantage that it is applicable in more dilute solutions.

Eosin (tetrabromofluorescein) is a stronger acid than dichlorofluorescein and can be used down to a pH of 1–2; the colour change is sharpest in an acetic acid solution (pH $<$ 3). Eosin is so strongly adsorbed on silver halides that it cannot be used for chloride titrations; this is because the eosin ion can compete with chloride ion before the equivalence point and thereby gives a premature indication of the end-point. With the more strongly adsorbing ions, Br^-, I^- and SCN^-, the competition is not serious and a very sharp end-point is obtained in the titration of these ions, even in dilute solutions. The colour on the precipitate is magenta. Rose Bengal (dichlorotetraiodofluorescein) and dimethyldiiodofluorescein) have been recommended for the titration of iodides.

Many other dyestuffs have been recommended as adsorption indicators, not only for the titration of halides but also for other ions. Thus cyanide ion may be titrated with standard silver nitrate solution using diphenylcarbazide as adsorption indicator (see Section **X, 22**): the precipitate is pale violet at the end-point. A selection of adsorption indicators, their properties and uses is given in Table X, 10.

D. Turbidity method. The appearance of a turbidity is sometimes utilised to mark the end-point of a reaction, as in Liebig's method for cyanides (see Section **X, 22**). A method which should be included here is the **turbidity procedure for the determination of silver with chloride**, first introduced by Gay Lussac. A standard solution of sodium chloride is titrated with a solution of silver nitrate or *vice versa*. Under certain conditions the addition of an indicator is unnecessary, because the presence of a turbidity caused by the addition of a few drops of one of the solutions to the other will show that the end-point has not been reached. The titration is continued until the addition of the appropriate solution produces no turbidity. Accurate results are obtained.

The procedure may be illustrated by the following simple experiment, which is a modification of the Gay Lussac–Stas method. The sodium chloride solution is added to the silver solution in the presence of free nitric acid and a small quantity of pure barium nitrate (the latter to assist coagulation of the precipitate). Weigh out accurately about 0.40 g of silver nitrate into a well-stoppered 200 cm³ bottle. Add about 100 cm³ of water, a few drops of concentrated nitric acid, and a small crystal of barium nitrate. Titrate with standard 0.1M sodium chloride by adding 20 cm³ at once, stoppering the bottle, and shaking it vigorously until the precipitate of silver chloride has coagulated and settled, leaving a clear solution. The volume of sodium chloride solution taken should leave the silver still in excess. Continue to add the chloride solution, 1 cm³ at a time, stoppering and shaking after each addition, until no turbidity is produced: note the total volume

Table X, 10. Selected adsorption indicators: properties and uses

Indicator	Use	Colour change at end-point*	Further data of interest
Fluorescein	Cl^-, Br^-, I^-, with Ag^+	Yellowish green $\rightarrow$ pink	Solution must be neutral or weakly basic
Dichloro-(R)-fluorescein	Cl^-, Br^-, BO_3^- with Ag^+	Yellowish green $\rightarrow$ red	Useful pH range 4.4–7
Tetrabromo-(R)-fluorescein (eosin)	Br^-, I^-, SCN^- with Ag^+	Pink $\rightarrow$ reddish-violet	Best in acetic acid solution; useful down to pH 1–2
Dichloro-(P)-tetraiodo-(R)-fluorescein (Rose Bengal)	I^- in presence of Cl^- with Ag^+	Red $\rightarrow$ purple	Accurate if $(NH_4)_2CO_3$ added
Di-iodo-(R)-dimethyl-(R)-fluorescein	I^- with Ag^+	Orange-red $\rightarrow$ blue-red	Useful pH range 4–7
Tartrazine	Ag^+ with I^- or SCN^-; I^- + Cl^- with excess Ag^+, back-titration with I^-	Colourless *solution* $\rightarrow$ green *solution*	Sharp colour change in I^- + Cl^- back-titration
Sodium alizarin sulphonate (alizarin red S)	$[Fe(CN)_6]^{4-}$, $[MoO_4]^{2-}$ with Pb^{2+}	Yellow $\rightarrow$ pink	Neutral solution
Rhodamine 6G	Ag^+ with Br^-	Orange-pink $\rightarrow$ reddish-violet	Best in dilute (up to 0.3M) HNO_3
Phenosafranine	Cl^-, Br^- with Ag^+ Ag^+ with Br^-	Red ppt. $\rightarrow$ blue ppt. Blue ppt. $\rightarrow$ red ppt.	Sharp, reversible colour change on ppt., but only if NO_3^- is present. Tolerance up to 0.2M-HNO_3

* The colour change is as indicator passes from solution to precipitate, unless otherwise stated.

287

of sodium chloride solution. Repeat the determination, using a fresh sample of silver nitrate of about the same weight, and run in initially that voume of the $0.1M$ sodium chloride, less 1 cm^3, which the first titration has indicated will be required, and thereafter add the chloride solution dropwise (i.e., in about 0.05 cm^3 portions). It will be found that the end-point can be determined within one drop.

A detailed account of modern nephelometric techniques is given in Chapter XX (Nephelometry and Turbidimetry).

A. 4   THEORY OF OXIDATION-REDUCTION TITRATIONS

X, 31.   CHANGE OF THE ELECTRODE POTENTIAL DURING THE TITRATION OF A REDUCTANT WITH AN OXIDANT.   In Sections **X, 11–16** it has been shown how to calculate the change in pH during acid–base titrations, and how the titration curves thus obtained can be used (*a*) to ascertain the most suitable indicator to be used in a given titration, and (*b*) to compute the titration error. Similar procedures may be carried out for oxidation–reduction titrations, and we will consider first a simple case which involves only the valency change of ions, and is theoretically independent of the hydrogen-ion concentration. A suitable example, for purposes of illustration, is the titration of 100 cm^3 of $0.1N$-iron(II) with $0.1N$-cerium(IV) in the presence of dilute sulphuric acid:

$$Ce^{4+} + Fe^{2+} \rightleftharpoons Ce^{3+} + Fe^{3+}$$

The quantity corresponding to $[H^+]$ in acid–base titrations is the ratio $[Ox]/[Red]$. We are concerned here with two systems, the Fe^{3+}/Fe^{2+} ion electrode (1), and the Ce^{4+}/Ce^{3+} ion electrode (2).

For (1) at 25 °C:

$$E_1 = E_1^{\ominus} + \frac{0.0591}{1} \log \frac{[Fe^{3+}]}{[Fe^{2+}]} = +0.75 + 0.0591 \log \frac{[Fe^{3+}]}{[Fe^{2+}]}$$

For (2), at 25 °C:

$$E_2 = E_2^{\ominus} + \frac{0.0591}{1} \log \frac{[Ce^{4+}]}{[Ce^{3+}]} = +1.45 + 0.0591 \log \frac{[Ce^{4+}]}{[Ce^{3+}]}$$

The equilibrium constant of the reaction is given by (Section **II, 25**):

$$\log K = \log \frac{[Ce^{3+}] \times [Fe^{3+}]}{[Ce^{4+}] \times [Fe^{2+}]}$$

$$= \frac{1}{0.0591}(1.45 - 0.75)$$

$$= 11.84$$

or $K = 7 \times 10^{11}$

The reaction is therefore virtually complete.

During the addition of the cerium(IV) solution up to the equivalence point, its only effect will be to oxidise the iron(II) (since K is large) and consequently change the ratio $[Fe^{3+}]/[Fe^{2+}]$. When 10 cm³ of the oxidising agent have been added, $[Fe^{3+}]/[Fe^{2+}] = 10/90$ (approx.), and $E_1 = 0.75 + 0.0591 \log 10/90 = 0.75 - 0.056 = 0.69$ volt:

with 50 cm³ of the oxidising agent, $E_1 = E_2^\ominus = 0.75$ volt
with 90 cm³, $E_1 = 0.75 + 0.0591 \log 90/10 = 0.81$ volt
with 99 cm³, $E_1 = 0.75 = 0.0591 \log 99/1 = 0.87$ volt
with 99.9 cm³, $E_1 = 0.75 + 0.0591 \log 99.9/0.1 = 0.93$ volt

At the equivalence point (100.0 cm³) $[Fe^{3+}] = [Ce^{3+}]$ and $[Ce^{4+}] = [Fe^{2+}]$, and the electrode potential is given by:*

$$\frac{E_1^\ominus + E_2^\ominus}{2} = \frac{0.75 + 1.45}{2} = 1.10 \text{ volts}$$

The subsequent addition of cerium(IV) solution will merely increase the ratio $[Ce^{4+}]/[Ce^{3+}]$. Thus:

with 100.1 cm³, $E_1 = 1.45 + 0.0591 \log 0.1/100 = 1.27$ volts
with 101 cm³, $E_1 = 1.45 + 0.0591 \log 1/100 = 1.33$ volts
with 110 cm³, $E_1 = 1.45 + 0.0591 \log 10/100 = 1.39$ volts
with 190 cm³, $E_1 = 1.45 + 0.0591 \log 90/100 = 1.45$ volts

These results are shown in Fig. X, 12.

It is of interest to calculate the iron(II) concentration in the neighbourhood of the equivalence point. When 99.9 cm³ of the cerium(IV) solution have been added, $[Fe^{2+}] = 0.1 \times 0.1/199.9 = 5 \times 10^{-5}$, or $pFe^{2+} = 4.3$. The concentration

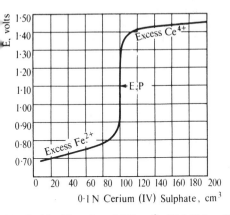

Fig. X, 12   **Titration of 100 cm³ of 0.1 M-iron(II) with 0.1 M-cerium sulphate (calculated)**

---

* For a deduction of this expression and a discussion of the approximations involved, see a textbook of electrochemistry. It can similarly be shown that for the reaction:

$$a \, Ox_I + b \, Red_{II} \rightleftharpoons b \, Ox_{II} + a \, Red_I$$

the potential at the equivalence point is given by:

$$E_0 = \frac{bE_1^\ominus + aE_2^\ominus}{a+b}$$

where $E_1^\ominus$ refers to Ox_I, Red_I, and $E_2^\ominus$ to Ox_{II}, Red_{II}.

at the equivalence point is given by (Section **II, 25**):

$$[Fe^{3+}]/[Fe^{2+}] = \sqrt{K} = \sqrt{7 \times 10^{11}} = 8.4 \times 10^5$$

Now $[Fe^{3+}] = 0.05N$, hence $[Fe^{2+}] = 5 \times 10^{-2}/8.5 \times 10^5 = 6 \times 10^{-8}N$, or $pFe^{2+} = 7.2$. Upon the addition of 100.1 cm³ of cerium(IV) solution, the reduction potential (*vide supra*) is 1.27 volts. The $[Fe^{3+}]$ is practically unchanged at $5 \times 10^{-2}N$, and we may calculate $[Fe^{2+}]$ with sufficient accuracy for our purpose from the equations:

$$E = E_1^{\ominus} + 0.0591 \log \frac{[Fe^{3+}]}{[Fe^{2+}]}$$

$$1.27 = 0.75 + 0.0591 \log \frac{5 \times 10^{-2}}{[Fe^{2+}]}$$

$$[Fe^{2+}] = 1 \times 10^{-10}$$

or $pFe^{2+} = 10$

Thus pFe^{2+} changes from 4.3 to 10 between 0.1 per cent before and 0.1 per cent after the stoichiometric end-point. These quantities are of importance in connection with the use of indicators for the detection of the equivalence point.

It is evident that the abrupt change of the potential in the neighbourhood of the equivalence point is dependent upon the standard potentials of the two oxidation–reduction systems that are involved, and therefore upon the equilibrium constant of the reaction; it is independent of the concentrations unless these are extremely small. The change in redox potential for a number of typical oxidation–reduction systems is exhibited graphically in Fig. X, 13. For the MnO_4^-, Mn^{2+} system and others which are dependent upon the pH of the solution, the hydrogen-ion concentration is assumed to be molar: lower acidities give lower potentials. The value at 50 per cent oxidised form will, of course, correspond to the standard redox potential. As an indication of the application of the curves, we may take the titration of iron(II) with potassium dichromate. The titration curve would follow that of the Fe(II)/Fe(III) system until the end-point was reached, then it would rise steeply and continue along the curve for the $Cr_2O_7^{2+}/Cr^{3+}$ system: the potential at the equivalence point can be computed as already described.

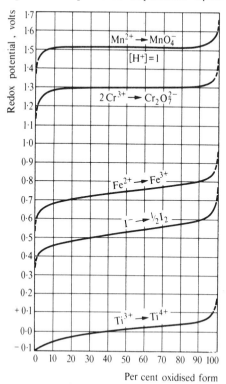

Fig. X, 13   Variation of Redox Potentials with Oxidant/Reductant Ratio

It is possible to titrate two substances by the same titrant provided that the standard potentials of the substances being titrated, and their oxidation or reduction products, differ by about 0.2 V. Stepwise titration curves are obtained in the titration of mixtures or of substances having several oxidation states. Thus the titration of a solution containing Cr(VI), Fe(III) and V(V) by an acid titanium(III) chloride solution is an example of such a mixture: in the first step Cr(VI) is reduced to Cr(III) and V(V) to V(IV); in the second step Fe(III) is reduced to Fe(II); in the third step V(IV) is reduced to V(III); chromium is evaluated by difference of the volumes of titrant used in the first and third steps. Another example is the titration of a mixture of Fe(II) and V(IV) sulphates with Ce(IV) sulphate in dilute sulphuric acid: in the first step Fe(II) is oxidised to Fe(III) and in the second 'jump' V(IV) is oxidised to V(V), the latter change is accelerated by heating the solution after oxidation of the Fe(II) ion is complete. The titration of a substance having several oxidation states is exemplified by the stepwise reduction by acid chromium(II) chloride of Cu(II) ion to the monovalent state and then to the metal.

X, 32. FORMAL POTENTIALS. Standard potentials $E^{\ominus}$ are evaluated with full regard to activity effects and with all ions present in simple form: they are really limiting or ideal values and are rarely observed in a potentiometric measurement. In practice, the solutions may be quite concentrated and frequently contain other electrolytes; under these conditions the activities of the pertinent species are much smaller than the concentrations, and consequently the use of the latter may lead to unreliable conclusions. Also, the actual active species present (see example below) may differ from those to which the ideal standard potentials apply. For these reasons 'formal potentials' have been proposed to supplement standard potentials. The formal potential is the potential observed experimentally in a solution containing equal numbers of moles of the oxidised and reduced substances together with other specified substances at specified concentrations. It is found that formal potentials vary appreciably, for example, with the nature and concentration of the acid that is present. The formal potential incorporates in one figure the effects resulting from variation of activity coefficients with ionic strength, acid–base dissociation, complexation, liquid–junction potentials, etc., and thus have a real practical value. Formal potentials do not have the theoretical significance of standard potentials, but they are observed values in actual potentiometric measurements. In dilute solutions they usually obey the Nernst equation fairly closely in the form:

$$E = E^{\ominus\prime} + \frac{0.0591}{n} \log \frac{[\text{Ox}]}{[\text{Red}]} \text{ at } 25\,^{\circ}\text{C}$$

where $E^{\ominus\prime}$ is the formal potential and corresponds to the value of E at *unit* concentrations of oxidant and reductant, and the quantities in square brackets refer to molecular concentrations. It is useful to determine and to tabulate $E^{\ominus\prime}$ with equivalent amounts of various oxidants and their conjugate reductants at various concentrations of different acids. If one is dealing with solutions whose composition is identical with or similar to that to which the formal potential pertains, more trustworthy conclusions can be derived from formal potentials than from standard potentials.

To illustrate how the use of standard potentials may occasionally lead to erroneous conclusions, let us consider the hexacyanoferrate(II)–hexacyano-ferrate(III) and the iodide–iodine systems. The standard potentials are:

$$[Fe(CN)_6]^{3-} + e \rightleftharpoons [Fe(CN)_6]^{4-}; \quad E^{\ominus} = +0.36 \text{ volt}$$
$$I_2 + 2e \rightleftharpoons 2I^- \qquad ; \quad E^{\ominus} = +0.54 \text{ volt}$$

It would be expected that iodine would quantitatively oxidise hexa-cyanoferrate(II) ions:

$$2[Fe(CN)_6]^{4-} + I_2 = 2[Fe(CN)_6]^{3-} + 2I^-$$

In actual fact $[Fe(CN)_6]^{4-}$ ion oxidises iodide ion quantitatively in media containing about $1M$-hydrochloric, sulphuric, or perchloric acid. This is because in solutions of low pH, protonation occurs and the species derived from $H_4Fe(CN)_6$ are weaker than those derived from $H_3Fe(CN)_6$; the activity of the $[Fe(CN)_6]^{4-}$ ion is decreased to a greater extent than that of the $[Fe(CN)_6]^{3-}$ ion, and therefore the reduction potential is increased. The actual redox potential of a solution containing equal concentrations of both cyanoferrates in $1M$-HCl, H_2SO_4 or $HClO_4$ is $+0.71$ volt, a value that is greater than the potential of the iodine–iodide couple.

Some results of formal potential measurements may now be mentioned. If there is no great difference in complexation of either the oxidant or its conjugate reductant in various acids, the formal potentials lie close together in these acids. Thus for the Fe(II)–Fe(III) system $E^{\ominus} = +0.77$ volt, $E^{\ominus\prime} = +0.73$ volt in $1M$-$HClO_4$, $+0.70$ volt in $1M$-HCl, $+0.68$ volt in $1M$-H_2SO_4, and $+0.61$ volt in $0.5M$-$H_3PO_4 + 1M$-H_2SO_4. It would seem that complexation is least in perchloric acid and greatest in phosphoric acid.

For the Ce(III)–Ce(IV) system $E^{\ominus\prime} = +1.44$ volts in $1M$-H_2SO_4, $+1.61$ volts in $1M$-HNO_3, and $+1.70$ volts in $1M$-$HClO_4$. Perchloric acid solutions of cerium(IV) perchlorate, although unstable on standing, react rapidly and quantitatively with many inorganic compounds and have greater oxidising power than cerium(IV) sulphate–sulphuric acid or cerium(IV) nitrate–nitric acid solutions.

X, 33. DETECTION OF THE END-POINT IN OXIDATION–REDUCTION TITRATIONS. A. Internal oxidation–reduction indicators.

We have already seen (Sections **X, 10–16**) that acid–base indicators are employed to mark the sudden change in pH during acid–base titrations. Similarly an oxidation–reduction indicator should mark the sudden change in the oxidation potential in the neighbourhood of the equivalence point in an oxidation–reduction titration. The ideal oxidation–reduction indicator will be one with an oxidation potential intermediate between that of the solution titrated and that of the titrant, and which exhibits a sharp, readily detectable colour change.

An oxidation–reduction indicator (redox indicator) is a compound which exhibits different colours in the oxidised and reduced forms:

$$In_{Ox} + ne \rightleftharpoons In_{Red}$$

The oxidation and reduction should be reversible. At a potential E the ratio of the concentrations of the two forms is given by the Nernst equation:

$$E = E_{In}^{\ominus} + \frac{RT}{nF} \ln a_{In\,ox}/a_{In\,red}$$

$$E \approx E_{\text{In}}^{\ominus} + \frac{RT}{nF} \ln \frac{[\text{In}_{\text{Ox}}]}{[\text{In}_{\text{Red}}]}$$

where $E_{\text{In}}^{\ominus}$ is the standard (strictly the formal) potential of the indicator. If the colour intensities of the two forms are comparable a practical estimate of the colour-change interval corresponds to the change in the ratio $[\text{In}_{\text{Ox}}]/[\text{In}_{\text{Red}}]$ from 10 to $\frac{1}{10}$; this leads to an interval of potential of:

$$E = E_{\text{In}}^{\ominus} \pm \frac{0.0591}{n} \text{ volts at } 25\,^{\circ}\text{C}$$

If the colour intensities of the two forms differ considerably the intermediate colour is attained at a potential somewhat removed from $E_{\text{In}}^{\ominus}$, but the error is unlikely to exceed 0.06 volt. For a sharp colour change at the end-point, $E_{\text{In}}^{\ominus}$ should differ by about at least 0.15 volt from the standard (formal) potentials of the other systems involved in the reaction.

One of the best oxidation–reduction indicators is the 1,10-phenanthroline-iron(II) complex. The base 1,10-phenanthroline combines readily in solution with iron(II) salts in the molecular ratio 3 base : 1 iron(II) ion forming the intensely red 1,10-phenanthroline-iron(II) complex ion; with strong oxidising agents the iron(III) complex ion is formed, which has a pale blue colour. The colour change is a very striking one:

$$[\text{Fe}(\text{C}_{12}\text{H}_8\text{N}_2)_3]^{3+} + e \rightleftharpoons [\text{Fe}(\text{C}_{12}\text{H}_8\text{N}_2)_3]^{2+}$$

pale blue   *deep red*

The standard redox potential is 1.14 volts; the formal potential is 1.06 volts in $1M$-hydrochloric acid solution. The colour change, however, occurs at about 1.12 volts, because the colour of the reduced form (deep red) is so much more intense than that of the oxidised form (pale blue). The indicator is of great value in the titration of iron(II) salts and other substances with cerium(IV) sulphate solutions. It is prepared by dissolving 1,10-phenanthroline hydrate (molecular weight = 198.1) in the calculated quantity of $0.02M$ acid-free iron(II) sulphate, and is therefore 1,10-phenanthroline-iron(II) complex sulphate (known as *ferroin*). One drop is usually sufficient in a titration: this is equivalent to less than 0.01 cm^3 of $0.1N$ oxidising agent, and hence the indicator blank is negligible at this or higher concentrations.

It has been shown (Section **X, 31**) that the potential at the equivalence point is the mean of the two standard redox potentials. In Fig. X, 12, the curve shows the variation of the potential during the titration of $0.1N$-iron(II) ion with $0.1N$-cerium(IV) solution, and the equivalence point is at 1.10 volts. Ferroin changes from deep red to pale blue at a redox potential of 1.12 volts: the indicator will therefore be present in the red form. After the addition of, say, a 0.1 per cent excess of cerium(IV) sulphate solution the potential rises to 1.27 volts, and the indicator is oxidised to the pale blue form. It is evident that the titration error is negligibly small.

The standard or formal potential of ferroin can be modified considerably by the introduction of various substituents in the 1,10-phenanthroline nucleus. The most important substituted ferroin is 5-nitro-1,10-phenanthroline iron(II) sulphate (nitroferroin) and 4,7-dimethyl-1,10-phenanthroline iron(II) sulphate (dimethylferroin). The former ($E^{\ominus} = 1.25$ volts) is especially suitable for titrations using Ce(IV) in nitric or perchloric acid solution where the formal potential of the oxidant is high. The 4,7-dimethylferroin has a sufficiently low formal potential ($E^{\ominus}$ = 0.88 volt) to render it useful for the titration of Fe(II) with dichromate in 0.5M-sulphuric acid.

Mention should be made of one of the earliest internal indicators. This is a 1 per cent solution of diphenylamine in concentrated sulphuric acid, and was introduced by Knop for the titration of iron(II) with potassium dichromate solution. An intense blue-violet coloration is produced at the end-point. The addition of phosphoric acid is desirable, for it lowers the formal potential of the Fe(III)–Fe(II) system so that the equivalence point potential coincides more nearly with that of the indicator. The action of diphenylamine (I) as an indicator depends upon its oxidation first into colourless diphenylbenzidine (II), which is the real indicator and is reversibly further oxidised to diphenylbenzidine violet (III).

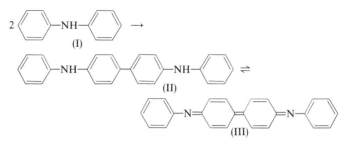

Diphenylbenzidine violet undergoes further oxidation if it is allowed to stand with excess of dichromate solution; this further oxidation is irreversible, and red or yellow products of unknown composition are produced.

A solution of diphenylbenzidine in concentrated sulphuric acid acts similarly to diphenylamine. The reduction potential of the system II, III is 0.76 volt in 0.5–

Table X, 11. Some oxidation–reduction indicators

Indicator	Colour change		Formal potential (volts) at pH = 0
	Oxidised form	Reduced form	
5-Nitro-1,10-phenanthroline iron(II) sulphate (nitroferroin)	Pale blue	Red	1.25
1,10-Phenthroline iron(II) sulphate (ferroin)	Pale blue	Red	1.06
2,2′-Bipyridyl iron(II) sulphate	Faint blue	Red	1.02
5,6-Dimethylferroin	Pale blue	Red	0.97
N-phenylanthranilic acid,	Purple red	Colourless	0.89
4,7-Dimethyl-1, 10-phenanthroline iron(II) sulphate (4,7-dimethylferroin)	Pale blue	Red	0.88
Diphenylaminesulphonic acid	Red-violet	Colourless	0.85
Diphenylbenzidine	Violet	Colourless	0.76
Diphenylamine	Violet	Colourless	0.76
3,3′-Dimethylnaphthidine	Purplish-red	Colourless	0.71
Starch-I_3^-, KI	Blue	Colourless	0.53
Methylene blue	Blue	Colourless	0.52

$1M$-sulphuric acid. It is therefore evident that a lowering of the potential of the Fe(III)–(Fe(II) system is desirable, as already mentioned, in order to obtain a sharp colour change. The disadvantage of diphenylamine and of diphenylbenzidine is their slight solubility in water. This has been overcome by the use of the soluble barium or sodium diphenylaminesulphonate, which are employed in 0.2 per cent aqueous solution. The redox potential $(E_{\text{In}}^{\ominus})$ is slightly higher (0.85 volt in $0.5M$-sulphuric acid), and the oxidised form has a reddish-violet colour resembling that of potassium permanganate, but the colour slowly disappears on standing; the presence of phosphoric acid is desirable in order to lower the redox potential of the system.

A list of selected redox indicators, together with their colour changes and reduction potentials in an acidic medium, are given in Table X, 11.

At this stage reference may be made to **potential mediators**, i.e., substances which undergo reversible oxidation–reduction and reach equilibrium *rapidly*. If we have a mixture of two ions, say M^{2+} and M^+, which reaches equilibrium slowly with an inert electrode, and a very small quantity of a cerium(IV) salt is added, then the reaction:

$$M^+ + Ce^{4+} \longrightarrow M^{2+} + Ce^{3+}$$

takes place until the tendency of M^+ to be oxidised to M^{2+} is exactly balanced by the tendency of Ce^{3+} to be oxidised to Ce^{4+}, that is, until the M^{2+}, M^+ and Ce^{4+}, Ce^{3+} potentials are equal. A platinum or other inert electrode rapidly attains equilibrium with the Ce(III) and Ce(IV) ions, and will soon register a stable potential which is also that due to the $M^{2+} + e \rightleftharpoons M^+$ system. If the potential mediator is employed in small amount, then a negligible quantity of M^+ is converted into M^{2+} when equilibrium is reached, and the measured potential may be regarded as that of the original system. Potential mediators are, of course, useful in the measurement of the oxidation–reduction potentials of redox systems; in this connection mention may be made of the use of potassium iodide ($\equiv$ iodide–iodine system) in the arsenate–arsenite system in acid solution. It is evident that redox indicators (e.g., 1,10-phenanthroline iron(II) ion) may act as potential mediators.

B. The reagent may serve as its own indicator. This is well illustrated by potassium permanganate; here, however, sensitive internal indicators (1,10-permanganate will impart a visible pink coloration to several hundred cm³ of solution, even in the presence of slightly coloured ions, such as iron(III). The colour of cerium(IV) sulphate and of iodine solutions have also been employed in the detection of end-points, but the colour change is not so marked as for potassium permanganate; here, however, sensitive internal indicators (ortho-phenanthroline iron(II) ion or N-phenylanthranilic acid and starch respectively) are available.

This method has the drawback that an excess of oxidising agent is always present at the end-point. For work of the highest accuracy, the indicator blank may be determined and allowed for, or the error may be considerably reduced by performing the standardisation and determination under similar experimental conditions.

C. External indicators. The best-known example of an external indicator in a redox process is the spot-test method for the titration of iron(II) with standard potassium dichromate solution. Near the equivalence point, drops of the solution are removed and brought into contact with dilute, freshly prepared potassium hexacyanoferrate(III) solution on a spot plate. The end-point is reached when the drop first fails to give a blue coloration. Another example is provided by the titration of zinc ions with standard potassium hexacyano-ferrate(II) solution; here a solution of uranyl acetate or nitrate is the external indicator, and titration is continued until a drop of the solution just imparts a brown colour to the indicator. External indicators are virtually superseded by the more satisfactory internal oxidation–reduction indicators: thus in the first example 1,10-phenanthroline iron(II) ion or N-phenylanthranilic acid is suitable, whilst for the second 3, 3'-dimethylnaphthidine may be used.

    D. Potentiometric methods. This is a procedure which depends upon measurement of the e.m.f. between a reference electrode and an indicator (redox) electrode at suitable intervals during the titration, i.e., a **potentiometric titration** is carried out. The procedure is discussed fully in Chapter XIV and suffice at this stage to point out that the procedure is applicable not only to those cases where suitable indicators are available, but also to those cases, e.g., coloured or very dilute solutions, where the indicator method is inapplicable, or of limited accuracy.

B EXPERIMENTAL DETAILS

B. 1 AQUEOUS ACID-BASE TITRATIONS

Acidimetry and Alkalimetry

X, 34. PREPARATION OF A STANDARD ACID. *Discussion.* Two acids, namely hydrochloric acid and sulphuric acid, are widely employed in the preparation of standard solutions of acids. Both of these are commercially available as concentrated solutions; concentrated hydrochloric acid is about $10.5–12M$, and concentrated sulphuric acid is about $18M$. By suitable dilution, solutions of any desired *approximate* strength may be readily prepared. Hydrochloric acid is generally preferred, since most chlorides are soluble in water. Sulphuric acid forms insoluble salts with calcium and barium hydroxides; for titration of hot liquids or for determinations which require boiling for some time with excess of acid, standard sulphuric acid is, however, preferable. Nitric acid is rarely employed, because it almost invariably contains a little nitrous acid, which has a destructive action upon many indicators.

    For the present, we shall confine our attention to the preparation of standard solutions of hydrochloric acid. Two methods are available. The first utilises the experimental fact that aqueous solutions of hydrochloric acid lose either hydrogen chloride or water upon boiling, according as to whether they are stronger or weaker than the constant-boiling-point mixture, until they attain a practically constant composition (constant-boiling-point mixture), which depends upon the prevailing pressure. The composition of this constant-boiling mixture and its dependence upon pressure have been determined with great accuracy by Foulk and Hollingsworth. The relevant data are collected in Table X, 12.

Table X, 12. Composition of constant-boiling-point hydrochloric acid

Pressure (mm of Hg)	Per cent HCl in acid (vac. wt.)	Grams of acid, weighed in air, containing 36.47 g of HCl
780	20.173	180.621
770	20.197	180.407
760	20.221	180.193
750	20.245	179.979
740	20.269	179.766
730	20.293	179.555

The constant-boiling-point acid is neither hygroscopic nor appreciably volatile, and its concentration remains unchanged if kept in a well-stoppered vessel out of direct sunlight. This acid may be employed directly in the preparation of a solution of hydrochloric acid of known concentration.

In the second method a solution of the approximate strength required is prepared, and this is standardised against some standard alkaline substance, such as sodium tetraborate or anhydrous sodium carbonate; standard potassium iodate or pure silver may also be used (see Sections **X, 84, 130**). If a solution of an exact normality is required, a solution of an approximate strength somewhat greater than that desired is first prepared; this is suitably diluted with water after standardisation (for a typical calculation, see Section **X, 4**).

The student should read the theoretical sections **X, 1–18**, before embarking upon the experimental work.

X, 35. PREPARATION OF CONSTANT-BOILING-POINT HYDRO-CHLORIC ACID.

Mix 400 cm^3 of pure concentrated hydrochloric acid with 250–400 cm^3 of distilled water so that the specific gravity of the resultant acid is 1.10 (test with a hydrometer). Insert a thermometer in the neck of a 1-litre Pyrex distillation flask so that the bulb is just opposite the side tube, and attach a condenser to the side tube; use an all-glass apparatus. Place 500 cm^3 of the diluted acid in the flask, distil the liquid at a rate of about 3–4 cm^3 per minute and collect the distillate in a small Pyrex flask. From time to time pour the distillate into a 500-cm^3 measuring-cylinder. When 375 cm^3 has been collected in the measuring-cylinder, collect a further 50 cm^3 in the small Pyrex flask; watch the thermometer to see that the temperature remains constant. Remove the receiver and stopper it; this contains the pure constant-boiling-point acid. Note the barometric pressure to the nearest mm at intervals during the distillation and take the mean value. Interpolate the concentration of the acid from Table X, 12.

X, 36. DIRECT PREPARATION OF 0.1M-HYDROCHLORIC ACID FROM THE CONSTANT-BOILING-POINT ACID.

Clean and dry a small, stoppered conical flask; a glass-stoppered flask is preferable. After weighing, do not handle the flask directly with the fingers; handle it with the aid of a tissue or a linen cloth. Add the calculated quantity of constant-boiling-point acid required for the preparation of 1 dm^3 of 0.1M-acid (see Table X, 12) with the aid of a pipette; make the final adjustment with a dropper pipette. Reweigh the flask to 0.001 g after replacing the stopper. Add an equal volume of water to prevent loss of acid, and transfer the contents to a 1 dm^3 graduated flask. Wash out the weighing flask several times with distilled water and add the washings to the original solution. Make up to the mark with distilled water. Insert the stopper and mix the solution thoroughly by shaking and inverting the flask repeatedly.

Note. Unless a solution of exact concentration is required, it is not necessary to weigh out the exact quantity of constant-boiling acid; the concentration may be calculated from the weight of acid used. Thus, if 18.305 g of acid, prepared at 760 mm, was diluted to 1 dm^3, its concentration would be $18.305/180.193 = 0.10158M$ dm^{-3}.

X, 37. PREPARATION OF APPROXIMATELY 0.1M-HYDROCHLORIC ACID AND STANDARDISATION.

Measure out by means of a graduated cylinder or a burette 9 cm^3 of pure concentrated hydrochloric acid; pour the acid into a litre measuring-cylinder containing about 500 cm^3 of distilled water. Make up to the litre mark with distilled water and thoroughly mix by shaking. This will give a solution approximately 0.1M (1).

Note. 1. If 1M-hydrochloric acid is required, use 90 cm^3 of the concentrated acid. If 0.01M-acid is required, dilute two 50-cm^3 portions of the approximately 0.1M-acid, removed with a 50-cm^3 pipette, in a graduated flask to 1 litre.

Approximately (0.05M)-sulphuric acid is similarly prepared from 3 cm^3 of pure concentrated sulphuric acid.

Two excellent methods (utilising acid-base indicators) are available for standardisation. The first is widely employed, but the second is more convenient, less time-consuming, and equally accurate.

A. Standardisation with anhydrous sodium carbonate. *Pure sodium carbonate.* Analytical reagent-quality sodium carbonate of 99.9 per cent purity is obtainable commercially. This contains a little moisture and must be dehydrated by heating at 260–270 °C for half an hour and allowed to cool in a desiccator before use. Alternatively, pure sodium carbonate may be prepared by heating A.R. sodium hydrogencarbonate to 260–270 °C for 60–90 minutes; the temperature must not be allowed to exceed 270 °C, for above this temperature the sodium carbonate may lose carbon dioxide. It has been recommended (Ref. 11) that the A.R. sodium hydrogencarbonate be decomposed by adding to hot (85 °C) water, and then the hydrated sodium carbonate which crystallises out is filtered off and dehydrated by heating, first at 100 °C, and finally at 260–270 °C.

In all cases the crucible is allowed to cool in a desiccator, and, before it is quite cold, the solid is transferred to a warm, dry, glass-stoppered tube or bottle, out of which, when cold, it may be weighed rapidly as required. It is important to remember that anhydrous sodium carbonate is hygroscopic and exhibits a tendency to change into the monohydrate.

Procedure. Weigh out accurately from a weighing bottle about 0.2 g of the pure sodium carbonate into a 250-cm^3 conical flask (1), dissolve it in 50—75 cm^3 of water, and add 2 drops of methyl orange indicator (2) or preferably of methyl orange-indigo carmine indicator (Section **X, 9**), which gives a very much more satisfactory end-point.* Rinse a clean burette three times with 5-cm^3 portions of

---

* This indicator is prepared by dissolving 1 g of methyl orange and 2.5 g of *purified* indigo carmine in 1 litre of distilled water, and filtering the solution. The colour change on passing from alkaline to acid solution is from green to magenta with a neutral-grey colour at pH of about 4.

the acid; fill the burette to a point 2–3 cm above the zero mark and open the stopcock momentarily in order to fill the jet with liquid. Examine the jet to see that no air bubbles are enclosed. If there are, more liquid must be run out until the jet is completely filled. Re-fill, if necessary, to bring the level above the zero mark; then slowly run out the liquid until the level is between the 0.0 and 0.5-cm^3 marks. Read the position of the meniscus to 0.01 cm^3 (Section **III, 17**). Place the conical flask containing the sodium carbonate solution upon a piece of unglazed white paper or a white tile beneath the burette, and run in the acid slowly from the burette. During the addition of the acid, the flask must be constantly rotated with one hand whilst the other hand controls the stopcock. Continue the addition until the methyl orange becomes a very faint yellow or the green colour commences to become paler, when the methyl orange–indigo carmine indicator is used. Wash the walls of the flask down with a little distilled water from a wash bottle, and continue the titration very carefully by adding the acid dropwise until the colour of the methyl orange becomes orange or a faint pink, or the colour of the mixed indicator is a neutral grey. This marks the end point of the titration, and the burette-reading should be taken and recorded in a note-book. The procedure is repeated with two or three other portions of sodium carbonate. The first (or preliminary) titration will indicate the location of the true end point within 0.5 cm^3. With experience and care, subsequent titrations can be carried out very accurately, and should yield concordant results. From the weights of sodium carbonate and the volumes of hydrochloric acid employed, the strength of the acid may be computed for each titration. The arithmetical mean is used to calculate the strength of the solution.

Notes. 1. For *elementary students*, an approximately 0.1N solution of sodium carbonate may be prepared by weighing out accurately about 1.3 g of pure sodium carbonate in a weighing bottle or in a small beaker, transferring it to a 250-cm^3 graduated flask, dissolving it in water (Section **III, 32**), and making up to the mark. The flask is well shaken, then 25.00 cm^3 portions are withdrawn with a pipette and titrated with the acid as described above. Individual titrations should not differ by more than 0.1 cm^3. Record the results as in Section **X, 42**.

2. To obtain the most accurate results, a comparison solution, saturated with carbon dioxide and containing the same concentration of sodium chloride (the colour of methyl orange in a saturated aqueous solution of carbon dioxide is sensitive to the concentration of sodium chloride) and indicator as the titrated solution at the end-point, should be used.

The mixed indicator, **bromocresol green–dimethyl yellow**, may be used with advantage. The indicator consists of 4 parts of a 0.2 per cent ethanolic solution of bromocresol green and 1 part of a 0.2 per cent ethanolic solution of dimethyl yellow: about 8 drops are used for 100 cm^3 of solution. The colour change is from blue to greenish yellow at pH = 4.0–4.1; the colour is yellow at pH = 3.9.

Methyl red may be used as an indicator provided the carbon dioxide in solution is expelled at the end point by boiling. This indicator gives a red colour with high concentrations of carbon dioxide such as are produced during titrations involving carbonates. Add the standard hydrochloric acid to the cold sodium carbonate solution containing 3 drops of 0.1 per cent methyl red until the indicator changes colour. Boil the solution gently (preferably with a small funnel in the mouth of the flask) for 2 minutes to expel carbon dioxide: the original colour of the indicator will return. Repeat the process until the colour no longer changes on boiling. Generally, boiling and cooling must be repeated twice. Care must be taken to avoid loss of liquid by spattering during boiling. The colour change is more easily perceived than with methyl orange.

Calculation of normality. The normality may be computed from the equation:

$$Na_2CO_3 + 2HCl = 2NaCl + CO_2 + H_2O$$

but the best method is to derive the normality entirely in terms of the primary standard substance, here, sodium carbonate. The equivalent (Section **X, 3**) of sodium carbonate is 52.997 or 53.00 g. If the weight of the sodium carbonate is divided by the number of cm^3 of hydrochloric acid to which it is equivalent, as found by titration, we have the weight of primary standard equivalent to 1 cm^3 of the acid. Thus if 0.2500 g of sodium carbonate is required for the neutralisation of 45.00 cm^3 of hydrochloric acid, 1 cm^3 of the acid would be equivalent to 0.2500/45.00 = 0.005556 g of sodium carbonate. The milli-equivalent or the weight in 1 cm^3 of N-sodium carbonate solution is 0.05300 g. Hence the normality of the acid is 0.005556/0.05300 = 0.1048N.

Another method is the following. 0.2500 g of sodium carbonate required 45.00 cm^3 of acid, hence 1 dm^3 of acid is equivalent to $1000 \times 0.2500/45.00 = 5.556$ g of sodium carbonate. But a dm^3 of N-acid is equivalent to 53.00 g of sodium carbonate, hence the acid is 5.556/53.00 = 0.1048N.

In the method described in Note 1 above, the normality of the sodium carbonate is first computed from the weight of sodium carbonate used. The mode of calculation described in Section **X, 4** is employed. If V_A is the volume in cm^3 of the standard solution of normality n_A required to react completely with V_B cm^3 of the unknown solution of normality n_B, then:

$$V_A \times n_A = V_B \times n_B$$

from which the value of n_B is readily deduced. Thus if 1.3890 g of anhydrous sodium carbonate is dissolved in 250 cm^3 of water, the normality of the sodium carbonate solution is $1.3890 \times 4/53.00 = 0.1048N$. If 25 cm^3 of the sodium carbonate solution exactly neutralise 25.45 cm^3 of the hydrochloric acid, then:

$$25.00 \times 01048 = 25.45 \times n_B$$

or the acid is 0.1030N.

B. Standardisation against sodium tetraborate. The advantages of sodium tetraborate decahydrate (borax) are: (i) it has a large equivalent, 190.72 g (that of anhydrous sodium carbonate is 53.00); (ii) it is easily and economically purified by recrystallisation; (iii) heating to constant weight is not required; (iv) it is practically non-hygroscopic; and (v) a sharp end-point can be obtained with methyl red at room temperatures, since this indicator is not affected by the very weak boric acid.

$$B_4O_7{}^{2-} + 2H^+ + 5H_2O = 4H_3BO_3$$

Pure sodium tetraborate. The A.R. salt is recrystallised from distilled water; 50 cm^3 of water is used for every 15 g of solid. Care must be taken that the crystallisation does not take place above 55 °C; above this temperature there is a possibility of the formation of the pentahydrate, since the transition temperature, decahydrate $\rightleftharpoons$ pentahydrate, is 61 °C. The crystals are filtered at the pump, washed twice with water, then twice with portions of 95 per cent ethanol, followed by two portions of A.R. diethyl ether. Five-cm^3 portions of water, ethanol, and ether are used for 10 g of crystals. Each washing must be followed by suction to remove the wash liquid. After the final washing, the solid is spread in a thin layer on a watch or clock glass and allowed to stand at room temperature for 12–18 hours. The sodium tetraborate is then dry, and may be kept in a well-stoppered tube for three to four weeks without appreciable change. An alternative method of drying is to place the recrystallised product (after having been washed twice with water) in a desiccator over a solution saturated with respect to sugar (sucrose) *and* sodium chloride. The substance is dry after about three days, and may be kept indefinitely in the desiccator without change. The latter method is more time-consuming; the product is identical with that obtained by the ethanol–ether process.

Procedure. Weigh out accurately from a weighing bottle 0.4–0.5 g of pure sodium tetraborate into a 250-cm^3 conical flask (1), dissolve it in about 50 cm^3 of water and add a few drops of methyl red. Titrate with the hydrochloric acid contained in a burette (for details, see under **A**) until the colour changes to pink (2). Repeat the titration with two other portions. Calculate the strength of the hydrochloric acid from the weight of sodium tetraborate and the volume of acid used. The variation of these results should not exceed 1–2 parts per thousand. If it is greater, further titrations must be performed until the variation falls within these limits. The arithmetical mean is used to calculate the concentration of the solution.

Notes. 1. For *elementary students*, an approximately 0.1N solution of sodium tetraborate may be prepared by weighing out accurately 4.7–4.8 g of A.R. material on a watch glass or in a small beaker, transferring it to a 250-cm^3 graduated flask, dissolving it in water (Section **III, 32**), and making up to the mark. The contents of the flask are well mixed by shaking. Twenty-five cm^3 portions are withdrawn with a pipette and titrated with the acid as detailed under *Method* **A**. Individual titrations should not differ by more than 0.1 cm^3.

2. For work of the highest precision a comparison solution or colour standard may be prepared for detecting the equivalence point. For 0.1N solutions, this is made by adding 5 drops of methyl red to a solution containing 1.0 g of sodium chloride and 2.2 g of boric acid in 500 cm^3 of water; the solution must be boiled to remove any carbon dioxide which may be present in the water. It is assumed that 20 cm^3 of wash water are used in the titration.

Calculation of the normality. This is carried out as described in *Method* **A**. The equivalent of sodium tetraborate is 190.72 g.

C. Standardisation by an iodometric method. The experimental details are given in Section **X, 130**.

D. An **argentimetric method** is described in Section **X, 84**.

X, 38. PREPARATION OF STANDARD ALKALI. *Discussion.* The

hydroxides of sodium, potassium, and barium are generally employed for the preparation of solutions of standard alkalis; they are water-soluble strong bases. Solutions made from aqueous ammonia are undesirable, because they tend to lose ammonia, especially if the concentration exceeds 0.5M; moreover, it is a weak base, and difficulties arise in titrations with weak acids (compare Section **X, 15**). Sodium hydroxide is most commonly used because of its cheapness. None of these solid hydroxides can be obtained pure, so that a standard solution cannot be prepared by dissolving a known weight in a definite volume of water. Both sodium and potassium hydroxides are extremely hygroscopic; a certain amount of alkali carbonate and water are always present. Exact results cannot be obtained in the presence of carbonate with some indicators, and it is therefore necessary to discuss methods for the preparation of carbonate-free alkali solutions. For many purposes A.R. sodium hydroxide (which contains 1–2 per cent of sodium carbonate) is sufficiently pure.

To prepare carbonate-free sodium hydroxide solution one of several methods may be used:

1. Rinse sodium hydroxide sticks rapidly with water; this removes the carbonate from the surface. A solution prepared from the washed sticks is satisfactory for most purposes.

2. If a concentrated solution of sodium hydroxide (equal weights of sticks or pellets and water) is prepared, covered, and allowed to stand, the carbonate remains insoluble; the clear supernatant liquid may be poured or siphoned off, and suitably diluted. (Potassium carbonate is too soluble in the concentrated alkali for this method to be applicable.)

3. A method, which yields a product completely free from carbonate ions, consists in the electrolysis of a saturated solution of A.R. sodium chloride with a mercury cathode and a platinum anode in the apparatus shown in Fig. X, 14. About 20–30 cm^3 of *re-distilled* mercury are placed in a 250-cm^3 pear-shaped Pyrex separatory funnel; 100—125 cm^3 of an almost saturated solution of A.R. sodium chloride are then carefully introduced. Two short lengths of platinum wire are sealed into Pyrex glass tubing; one of these dips into the mercury (cathode), and the other into the salt solution (anode). A little mercury is placed

in the glass tubes, and electrical contact is made by means of amalgamated copper wires dipping into the mercury in the tubes. Electrolysis is carried out using 6–8 volts and 0.5–1 amp for several hours; the funnel is shaken at intervals in order to break up the amalgam crystals that form on the surface of the mercury. The weight of the sodium dissolved in the amalgam may be roughly computed from the total current passed; the current efficiency is 75–80 per cent. When sufficient amalgam has formed, the mercury is run into a Pyrex flask containing about 100 cm^3 of boiled-out distilled water and closed with a rubber bung carrying a soda-lime guard tube. Decomposition of the amalgam, to give the sodium hydroxide solution, is complete after several days;

Fig. X, 14 after 12–18 hours about 75 per cent of the amalgam is decomposed.

4. In the **anion exchange method**, which is recommended, carbonate maybe removed from either sodium or potassium hydroxide. The solution is passed through a strong base anion exchange column (e.g., Zerolit FF or Amberlite IRA-400) in the chloride form (see Chapter VII). Initially the alkali hydroxide converts the resin into the hydroxide form; the carbonate ion has a greater affinity for the resin than the hydroxide ion, and hence is retained on the resin: the first portions of the effluent contain chloride ion. If it is desired not to dilute the standard base appreciably and if chloride ion is objectionable, the effluent is discarded until it shows no test for chloride. Thus if a column containing one of the above resins, 35 cm long, is prepared in a 50-cm^3 burette, about 150 cm^3 of 4 per cent sodium hydroxide solution must be passed through the column of resin at a flow rate of 5–6 cm^3 per minute before the effluent is chloride-free; subsequently the effluent may be collected in a 500 cm^3 filter flask with side arm

carrying a soda-lime guard tube. When about 125 cm^3 of liquid have been collected, 105 cm^3 are measured out and diluted with boiled-out-distilled water to 1 litre. The resulting sodium hydroxide solution is carbonate-free and is about 0.1M. When the column becomes saturated with carbonate ion it is readily re-converted to the chloride form by passing dilute hydrochloric acid through it, followed by water to remove the excess acid.

Strong base anion exchangers in the hydroxide form may be used to prepare standard solutions of sodium or potassium hydroxide using weighed amounts of pure sodium chloride or potassium chloride. The resin, after conversion into the hydroxide form by passage of 1M-sodium hydroxide (prepared from 18M-sodium hydroxide so as to be carbonate-free), is washed with freshly boiled distilled water until the effluent contains no chloride ions and is neutral: about 2 litres of 1M-sodium hydroxide are required for 40 g of resin, and washing is with about 2 litres of water. About 2.92 g of A.R. sodium chloride, accurately weighed, are dissolved in 100 cm^3 of water. The solution is passed through the column at the rate of 4 cm^3 per minute; this is followed by about 300 cm^3 of freshly boiled distilled water. The eluate is collected in a 500-cm^3 graduated flask by means of an adapter permitting the use of a soda-lime guard tube. Towards the end the flow rate is decreased to permit careful adjustment to volume. A ca. 0.1M solution of sodium hydroxide results.

A number of firms supplying laboratory chemicals offer solutions of known concentration which can be employed for titrimetric analysis, and amongst these some manufacturers catalogue sodium hydroxide solutions 'free from carbonate', as for example BDH Chemicals Ltd 'AVS' range of solutions, and if only occasional need for carbonate-free sodium hydroxide solutions arises, this is the simplest way of satisfying this need. The merits of barium hydroxide solution (Section **X, 41**) as a carbonate-free alkali should also be borne in mind, but this suffers from the disadvantage that the maximum concentration available is between 0.05N to 0.1N. Whenever carbonate-free alkali is employed, it is essential that all the water used in the analyses should also be carbon dioxide free. With de-ionised water, there will be little cause for worry provided that the water is protected from atmospheric carbon dioxide, and with ordinary distilled water, dissolved carbon dioxide is readily removed by slowly aspirating a current of air which has been passed through a tube containing soda asbestos or soda lime through the water for 5–6 hours.

Attention must be directed to the fact that alkaline solutions, particularly if concentrated, attack glass. They may be preserved, if required, in polythene bottles, which are resistant to alkali. Furthermore, solutions of the strong bases absorb carbon dioxide from the air. If such solutions are exposed to the atmosphere for any appreciable time they become contaminated with carbonate. This may be prevented by the use of a storage vessel such as is shown in Fig. III, 24; the guard tube should be filled with soda-lime or with soda-asbestos. A short exposure of an alkali hydroxide solution to the air will not, however, introduce any serious error. If such solutions are quickly transferred to a burette and the latter fitted with a soda-lime guard tube, the error due to contamination by carbon dioxide may be neglected.

The solution of alkali hydroxide prepared by any of the above methods must be standardised. Alkaline solutions that are subsequently to be used in the presence of carbon dioxide or with strong acids are best standardised against solutions prepared from constant boiling-point hydrochloric acid or potassium hydrogeniodate or sulphamic acid, or against hydrochloric acid which has been standardised by means of sodium tetraborate or sodium carbonate. If the alkali solution is to be used in the titration of weak acids, it is best standardised against organic acids or against acid salt of organic diprotic acids, such as benzoic acid or potassium hydrogenphthalate, respectively. These substances are commercially available in a purity exceeding 99.9 per cent; potassium hydrogenphthalate is preferable, since it is more soluble in water and has a greater equivalent.

Procedure A.   Weigh out rapidly about 4.2 g of A.R. sodium hydroxide on a watch glass or into a small beaker, dissolve it in water, make up to 1 litre with boiled-out distilled water, mix thoroughly by shaking, and pour the resultant solution into the stock bottle, which should be closed by a rubber stopper.

Procedure B (**carbonate-free sodium hydroxide**). Dissolve 50 g of sodium hydroxide in 50 cm³ of distilled water in a Pyrex flask, transfer to a 75-cm³ test-tube of Pyrex glass, and insert a well-fitting stopper covered with tinfoil. Allow it to stand in a vertical position until the supernatant liquid is clear. For a 0.1M-sodium hydroxide solution *carefully* withdraw, using a pipette fitted with a filling device, 6.5 cm³ of the concentrated clear solution into a litre bottle or flask, and dilute quickly with 1 litre of recently boiled-out water.

A clear solution can be obtained more quickly, and incidentally the transfer can be made more satisfactorily, by rapidly filtering the solution through a sintered glass funnel with exclusion of carbon dioxide with the aid of the apparatus shown in Fig. X, 15. It is advisable to calibrate the test-tube in approximately 5-cm³ intervals and to put the graduations on a thin slip of paper gummed to the outside of the tube.

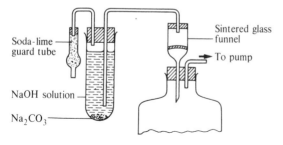

Fig. X, 15

X, 39. STANDARDISATION OF THE APPROXIMATELY 0.1M-SODIUM HYDROXIDE.

If the solution contains carbonate (*Procedure A*), methyl orange, methyl orange–indigo carmine, or bromophenol blue must be used in standardisation against hydrochloric acid of known normality. Phenolphthalein or indicators with a similar pH range, which are affected by carbon dioxide, cannot be used at the ordinary temperature (compare Section **X, 7**). With carbonate-free sodium hydroxide (*Procedure B*) phenolphthalein or thymol blue (Section **X, 13**) may be employed, and standardisation may be effected against hydrochloric acid, potassium hydrogeniodate, potassium hydrogenphthalate, benzoic acid, or other organic acids (Section **X, 40**).

Procedure A.   **With standard hydrochloric acid.**   Place the standardised (approx. 0.1M) hydrochloric acid in the burette. Transfer 25 cm³ of the sodium hydroxide solution into a 250-cm³ conical flask with the aid of a pipette, dilute with a little water, add 1–2 drops of methyl orange or 3–4 drops of methyl orange–indigo carmine indicator, and titrate with the previously standardised hydrochloric acid. Repeat the titrations until duplicate determinations agree within 0.05 cm³ of each other.

Calculation of the normality.   The normality is readily computed from the simple relationship:

$$V_A \times n_A = V_B \times n_B$$

where V_A and n_A refer to the volume and known normality of the acid respectively, V_B is the volume of alkali solution required for the neutralisation, and n_B is its (unknown) normality.

Procedure   B.   **With   potassium   hydrogenphthalate.**   A.R.   potassium hydrogenphthalate has a purity of at least 99.9 per cent; it is almost non-hygroscopic, but, unless a product of guaranteed purity is purchased, it is advisable to dry it at 120 °C for 2 hours, and allow it to cool in a covered vessel in a desiccator. Weigh out three 0.6–0.7 g portions of the salt into 250 cm³ Pyrex conical flasks (1), add 75 cm³ of boiled-out water to each portion, stopper each flask and shake gently until the solid has dissolved. Titrate each solution with the sodium hydroxide solution contained in a burette, using phenolphthalein or thymol blue as indicator.

Calculation of normality.   This is similar to that described in Section **X, 37**. The equivalent of potassium hydrogenphthalate is 204.22 g. The variation in the results should not exceed 0.1–0.2 per cent.

$$HK(C_8H_4O_4) + NaOH = NaK(C_8H_4O_4) + H_2O$$

Note.   1. For *elementary students*, an approximately 0.1M solution is prepared by weighing out accurately about 5.1 g of the ordinary A.R. product, dissolving it in water, and making it up to 250 cm³ in a graduated flask. Twenty-five-cm³ portions are employed in the titrations with the sodium hydroxide solution. Individual titrations should not differ by more than 0.1 cm³.

X, 40.   OTHER STANDARD SUBSTANCES FOR ACIDIMETRY AND ALKALIMETRY.

In addition to the standard substances already detailed for use in standardising acids and alkalis, numerous others have been proposed. A number of these will be briefly described.

A. Benzoic acid $(C_6H_5COOH$; equivalent = 122.12 g). The A.R. product has a purity of at least 99.9 per cent. For work demanding the highest accuracy, the acid should be dried before use by careful fusion in a platinum crucible placed in an oven at about 130 °C, and then powdered in an agate mortar. Benzoic acid is sparingly soluble in water (which is a disadvantage) and must therefore be dissolved in 95 per cent ethanol. The mode of use is similar to that already described for potassium hydrogenphthalate (Section **X, 39,** *B*). For a 0.1M solution, of, say, sodium hydroxide, weigh out accurately 0.4 g portions of the acid into a 250-cm³ conical flask, add 10–20 cm³ of ethanol, shake until dissolved, and then titrate the solution with the strong alkali using phenolphthalein as indicator. A blank test should be made with the same volume of ethanol and the indicator: deduct, if necessary, the volume of the alkali solution consumed in the blank test.

B. Succinic   acid $\{(CH_2COOH)_2$;   equivalent =   59.045 g$\}$.   The A.R. product or a pure commercial product should be recrystallised from pure acetone and dried in a vacuum desiccator. The purity is checked by means of a melting-point determination (185–185.5 °C). The acid is fairly soluble in water; phenolphthalein is a suitable indicator.

C. Potassium hydrogeniodate $\{KH(IO_3)_2$; equivalent $= 389.95$ g$\}$.   Unlike the other solid standards already described, this is a strong acid and thus permits the use of any indicator having a pH range between 4.5 and 9.5 for titration with strong bases. It may be employed for the standardisation of weak bases which are subsequently to be used with strong acids; an indicator, such as methyl red, must then be used. The salt is moderately soluble in water (1.33 g/100 cm³ at 15°), is anhydrous and non-hygroscopic, and its aqueous solution is stable for long periods; the equivalent is high, and a 0.01N solution contains 3.8995 g per dm³.

Preparation of pure potassium hydrogeniodate.   Dissolve 27 g of A.R. potassium iodate in 125 cm³ of boiling water, and add a solution of 22 g of A.R. iodic acid in 45 cm³ of warm water acidified with 6 drops of concentrated hydrochloric acid. Potassium hydrogeniodate separates on cooling. Filter on a sintered-glass funnel, and wash with cold water. Recrystallise three times from hot water: use 3 parts of water for 1 part of the salt and stir continuously during each cooling. Dry the crystals at 100 °C for several hours. The purity exceeds 99.95 per cent.

D. Sulphamic acid (NH_2SO_2OH; equivalent $= 97.09$ g).   A product of high purity (>99.9 per cent) is available commercially. It is a colourless, crystalline, non-hygroscopic solid, melting with decomposition at 205 °C. The acid is moderately soluble in water (21.3 g and 47.1 g in 100 g of water at 20° and 80 °C respectively). Sulphamic acid acts as a strong acid, so that any indicator with a colour change in the pH range 4–9 may be employed; bromothymol blue is particularly suitable for use with strong bases. It undergoes hydrolysis in aqueous solution:

$$NH_2SO_2OH + H_2O = NH_4HSO_4$$

Aqueous solutions should, preferably, not be stored; the titre does not alter on keeping if an indicator which changes in the acid range is used.

X, 41.   STANDARD BARIUM HYDROXIDE (BARYTA) SOLUTION.   This solution is widely employed, particularly for the titration of organic acids. Barium carbonate is insoluble, so that a clear solution is a carbonate-free strong alkali. The equivalent of $Ba(OH)_2,8H_2O$ is 157.75 g, but a standard solution cannot be prepared by direct weighing owing to the uncertainty of the hydration and the possible presence of carbonate. To prepare an approximately 0.1N solution, dissolve 18 g of A.R. crystallised baryta (or 20 g of the commercial substance) in about 1 litre of hot water in a large flask. Stopper the flask and allow the solution to stand for 2 days or until all the barium carbonate has completely settled out. Decant or siphon off the clear solution into a storage bottle of the type depicted in Fig. III, 24. A soda-lime guard tube must be provided to prevent ingress of carbon dioxide. The solution may be standardised against standard 0.1M-hydrochloric acid, succinic acid or potassium hydrogenphthalate; phenolphthalein or thymol blue is employed as indicator.

X, 42.   DETERMINATION OF THE Na_2CO_3 CONTENT OF WASHING SODA.   *Procedure.*   Weigh out accurately about 3.6 g of the washing-soda crystals, dissolve in water, and make up to 250 cm^3 in a graduated flask. Mix thoroughly. Titrate 25 cm^3 of the solution with standard hydrochloric acid of approximately 0.1M concentration using methyl orange, or, better, methyl orange–indigo carmine or bromo-cresol green as indicator. Two consecutive titrations should agree within 0.05 cm^3.

Calculation. The weight of anhydrous sodium carbonate Na$_2$CO$_3$ which has reacted with the standard hydrochloric acid can be readily computed from the equation:

$$Na_2CO_3 + 2HCl = 2NaCl + H_2O + CO_2$$
$$106.01 \quad 2 \times 36.46$$

The percentage of Na$_2$CO$_3$ can then be calculated from the known weight of washing soda employed.

A simpler and more general procedure is to employ the normality method. An actual example will make this clear.

Weight of weighing bottle + substance = 16.7910 g

Weight of weighing bottle + residual substance = 13.0110 g

∴ Weight of sample used = 3.7800 g

This was dissolved in water and made up to 250 cm^3.

Titration of 25.00 cm^3 of the carbonate solution with 0.1060N-HCl, using methyl orange–indigo carmine as indicator.

Experiment	Reading 1	Reading 2	Difference
1	0.50 cm^3	26.60 cm^3	26.10 cm^3 (preliminary)
2	0.55 cm^3	26.45 cm^3	25.90 cm^3
3	0.50 cm^3	26.45 cm^3	25.95 cm^3
			Mean 25.93 cm^3

1 cm^3 M-HCl ≡ 0.05300 g Na$_2$CO$_3$

25.93 × 0.1060 ≡ 2.749 cm^3 M-HCl

2.749 × 0.05300 = 0.1457 g Na$_2$CO$_3$ in portion titrated.

Weight of washing soda in portion titrated
= 3.7800 × 25.0/250 = 0.3780 g

∴ Percentage of Na$_2$CO$_3$ = 0.1457 × 100/0.3780 = 38.54 *per cent.*

Alternative method of calculation. 25.0 cm^3 of the carbonate solution required 25.93 cm^3 of 0.1060M-HCl:

∴ 25.0 × normality of carbonate solution = 25.93 × 0.1060, whence the carbonate solution is 25.93 × 0.1060/25.0 = 0.1099N. But N-Na$_2$CO$_3$ contains 106.00/2 = 53.00 g Na$_2$CO$_3$ per dm^3.

∴ the given solution contains 0.1099 × 53.00 = 5.8271 g Na$_2$CO$_3$ per dm^3, and 250 cm^3 would contain $5.8271 \times 250/1000 = 1.4568$ g.

Thus percentage of $Na_2CO_3 = 1.4568 \times 100/3.7800 = 38.54$ *per cent.*

X, 43. DETERMINATION OF THE STRENGTH OF CONCENTRATED ACIDS. (a) Glacial acetic acid.   Weigh a dry, stoppered 50 cm^3 conical flask, introduce about 2 g of glacial acetic acid and weigh again. Add about 20 cm^3 of water and transfer the solution quantitatively to a 250-cm^3 graduated flask. Wash the small flask several times with water and add the washings to the graduated flask. Make up to the mark with distilled, preferably boiled-out water. Shake the flask well to ensure thorough mixing. Titrate 25-cm^3 portions of the acid with 0.1M standard sodium hydroxide solution, using phenolphthalein or thymol blue as indicator.

$$NaOH + CH_3COOH = CH_3COONa + H_2O$$

$$1 \text{ cm}^3 \text{ } M\text{-NaOH} \equiv 0.06005 \text{ g } CH_3COOH$$

Calculate the percentage of CH_3COOH in the sample of glacial acetic acid.

Note on the determination of the acetic acid content of vinegar.   Vinegar usually contains 4–5 per cent acetic acid. Weigh out about 20 g vinegar as described above, and make up to 100 cm^3 in a graduated flask. Remove 25 cm^3 with a pipette, dilute with an equal volume of water, add a few drops of phenolphthalein, and titrate with standard 0.1M-sodium hydroxide solution. As a result of the dilution of the vinegar, its natural colour will be so reduced that it will not interfere with the colour change of the indicator. Calculate the acetic acid content of the vinegar, and express the result in g of acetic acid per 100 grams.

(b) Concentrated sulphuric acid.   Place about 100 cm^3 of water in a 250 cm^3 graduated flask, and insert a short-stemmed funnel in the neck of the flask. Charge a weight pipette with a few grams of the acid to be evaluated, and weigh. Add about 1.3–1.5 g of the acid to the flask and reweigh the pipette. Alternatively, the acid may be weighed out in a stoppered weighing bottle, and after adding the acid to the flask, the weighing bottle is reweighed. Rinse the funnel thoroughly, remove the flask, and allow the flask to stand for 1–2 hours to regain the temperature of the laboratory when the solution can be made up to the mark. Shake and mix thoroughly, and then titrate 25 cm^3 portions with standard 0.1M sodium hydroxide, using methyl orange or methyl orange–indigo carmine as indicator.

$$1 \text{ cm}^3 \text{ } M\text{-NaOH} \equiv 0.4904 \text{ g } H_2SO_4.$$

Fuming sulphuric acid (oleum) should be weighed in a Lunge–Rey pipette (Fig. III, 11, b).

(c) Syrupy phosphoric acid.   In this case we are dealing with a triprotic acid and theoretically three equivalence points are possible, but in practice the pH changes in the neighbourhood of the equivalence points are not very marked (see Fig. X, 6). For the first stage neutralisation (pH 4.6) we may employ methyl orange, methyl orange–indigo carmine or bromocresol green as indicator, but it is advisable to use a comparator solution of sodium dihydrogenphosphate (0.03M) containing the same amount of indicator as in the solution being titrated. For the second stage (pH 9.7), phenolphthalein is not altogether satisfactory (it changes colour on the early side), thymolphthalein is better; but the best indicator is a mixture of phenolphthalein (2 parts) with 1-naphtholphthalein

(1 part) which changes from pale rose through green to violet at pH 9.6. For the third stage (pH 12.6) there is no satisfactory indicator.

Procedure. Weigh an empty stoppered weighing bottle, add about 2 g of syrupy phosphoric acid and reweigh. Transfer the acid quantitatively to a 250 cm^3 graduated flask, and then proceed as detailed for sulphuric acid, but using the phenolphthalein-1-naphtholphthalein mixed indicator.

$$H_3PO_4 + 2OH^- = HPO_4^{2-} + 2H_2O$$

$$1 \text{ cm}^3 \text{ } M\text{-NaOH} \equiv 0.04902 \text{ g } H_3PO_4$$

X, 44. DETERMINATION OF A MIXTURE OF CARBONATE AND HYDROXIDE. Analysis of commercial caustic soda. *Discussion.* Two methods may be used for this analysis. In the first method the total alkali (carbonate + hydroxide) is determined by titration with standard acid, using methyl orange, methyl orange-indigo carmine, or bromo-phenol blue as indicator. In a second portion of solution the carbonate is precipitated with a slight excess of barium chloride solution, and, without filtering, the solution is titrated with standard acid using thymol blue or phenolphthalein as indicator. The latter titration gives the hydroxide content, and by subtracting this from the first titration, the volume of acid required for the carbonate is obtained.

$$Na_2CO_2 + BaCl_2 = BaCO_3 \text{ (insoluble)} + 2NaCl$$

The second method utilises two indicators. It has been stated in Section **X, 17** that the pH of half-neutralised sodium carbonate, i.e., at the sodium hydrogencarbonate stage, is about 8.3, but the pH changes comparatively slowly in the neighbourhood of the equivalence point; consequently the indicator colour-change with phenolphthalein (pH range 8.3–10.0) or thymol blue (pH range (base) 8.0–9.6) is not too sharp. This difficulty may be surmounted by using a comparison solution containing sodium hydrogencarbonate of approximately the same concentration as the unknown and the same volume of indicator. A simpler method is to employ a mixed indicator (Section **X, 9**) composed of 6 parts of thymol blue and 1 part of cresol red; this mixture is violet at pH 8.4, blue at pH 8.3, and rose at pH 8.2. With this mixed indicator the mixture has a violet colour in alkaline solution and changes to blue in the vicinity of the equivalence point; in making the titration the acid is added slowly until the solution assumes a rose colour. At this stage all the hydroxide has been neutralised and the carbonate converted into hydrogencarbonate. Let the volume of standard acid consumed be v cm^3.

$$OH^- + H^+ = H_2O$$
$$CO_3^{2-} + H^+ = HCO_3^-$$

Another titration is performed with methyl orange, methyl orange–indigo carmine or bromophenol blue as indicator. Let the volume of acid be V cm^3.

$$OH^- + H^+ = H_2O$$
$$CO_3^{2-} + 2H^+ = H_2CO_3$$
$$H_2CO_3 \rightleftharpoons H_2O + CO_2$$

Then $V - 2(V-v)$ corresponds to the hydroxide, $2(V-v)$ to the carbonate, and V to the total alkali. To obtain satisfactory results by this method the solution titrated must be cold (as near 0 °C as is practicable), and loss of carbon dioxide must be prevented as far as possible by keeping the tip of the burette immersed in the liquid.

Procedure A. Weigh out accurately in a glass-stoppered weighing bottle about 2.5 g of commercial sodium hydroxide (e.g., in flake form). Transfer quantitatively to a 500-cm^3 graduated flask and make up to the mark. Shake the flask well. Titrate 25 or 50 cm^3 of this solution with standard 0.1M-hydrochloric acid, using methyl orange or methyl orange–indigo carmine as indicator. Carry out two or three titrations: these should not differ by more than 0.1 cm^3. This gives the total alkalinity (hydroxide + carbonate). Warm another 25 or 50 cm^3 portion of the solution to 70 °C and add 1 per cent barium chloride solution slowly from a burette or pipette in *slight* excess, i.e., until no further precipitate is produced. Cool to room temperature, add a few drops of phenolphthalein to the solution, and titrate very slowly and with constant stirring with standard 0.1M-hydrochloric acid; the end-point is reached when the colour just changes from pink to colourless. If thymol blue is used as indicator, the colour change is from blue to yellow. The amount of acid used corresponds to the hydroxide present.

This method yields only approximate results because of the precipitation of basic barium carbonate in the presence of hydroxide. More accurate results are obtained by considering the above titration as a preliminary one in order to ascertain the approximate hydroxide content, and then carrying out another titration as follows. Treat 25–50 cm^3 of the solution with sufficient standard hydrochloric acid to neutralise most of the hydroxide, then heat and precipitate as before. Under these conditions, practically pure barium carbonate is precipitated.

$$1 \text{ cm}^3 \, M\text{-HCl} \equiv 0.0401 \text{ g NaOH}$$

$$1 \text{ cm}^3 \, M\text{-HCl} \equiv 0.05300 \text{ g Na}_2\text{CO}_3$$

Procedure B. The experimental details for the preparation of the initial solution are similar to those given under *Procedure A*. Titrate 25 or 50 cm^3 of the cold solution with standard 0.1M-hydrochloric acid and methyl orange, methyl orange–indigo carmine, or bromo-phenol blue as indicator. Titrate another 25 or 50 cm^3 of the cold solution, diluted with an equal volume of water, slowly with the standard acid using phenolphthalein or, better, the thymol blue–cresol red mixed indicator; in the latter case, the colour at the end-point is rose.

Calculate the result as described in the *Discussion* above.

X, 45. DETERMINATION OF A MIXTURE OF CARBONATE AND HYDROGENCARBONATE. The two methods available for this determination are modifications of those described in the previous Section for hydroxide–carbonate mixtures. In the first procedure, which is particularly valuable when the sample contains relatively large amounts of carbonate and small amounts of hydrogencarbonate, the total alkali is first determined in one portion of the solution by titration with standard 0.1M-hydrochloric acid using methyl orange, methyl orange–indigo carmine, or bromophenol blue as indicator:

$$CO_3^{2-} + 2H^+ = H_2CO_3$$

$$HCO_3^- + H^+ = H_2CO_3$$

$$H_2CO_3 \rightleftharpoons H_2O + CO_2$$

Let this volume correspond to V cm^3 M-HCl. To another sample, a measured excess of standard $0.1M$-sodium hydroxide (free from carbonate) over that required to transform the hydrogencarbonate to carbonate is added:

$$HCO_3^- + OH^- = CO_3^{2-} + H_2O$$

A slight excess of 10 per cent barium chloride solution is added to the hot solution to precipitate the carbonate as barium carbonate, and the excess of sodium hydroxide solution immediately determined without filtering off the precipitate by titration with the same standard acid; phenolphthalein or thymol blue is used as indicator. If the volume of excess of sodium hydroxide solution added be equivalent to v cm^3 of M-sodium hydroxide and v' cm^3 M-acid corresponds to the excess of the latter, then $v-v' = $ hydrogencarbonate, and $V-(v-v')$ = carbonate.

In the second procedure a portion of the cold solution is slowly titrated with standard $0.1M$-hydrochloric acid, using phenolphthalein or, better, the thymol blue–cresol red mixed indicator. This (say, Y cm^3) corresponds to half the carbonate (compare Section **X, 44**):

$$CO_3^{2-} + H^+ \rightleftharpoons HCO_3^-$$

Another sample of equal volume is then titrated with the same standard acid using methyl orange, methyl orange–indigo carmine or bromophenol blue as indicator. The volume of acid used (say, y cm^3) corresponds to carbonate + hydrogencarbonate. Hence $2Y$ = carbonate, and $y - 2Y$ = hydrogencarbonate.

X, 46. DETERMINATION OF BORIC ACID. *Discussion.* Boric acid acts as

a weak monoprotic acid ($K_a = 6.4 \times 10^{-10}$); it cannot therefore be titrated accurately with $0.1N$ standard alkali (compare Section **X, 13**). However, by the addition of certain organic polyhydroxy compounds, such as mannitol, glucose, sorbitol, or glycerol, it acts as a much stronger acid (for mannitol $K_a \simeq 1.5 \times 10^{-4}$) and can be titrated to a phenolphthalein end-point.

The effect of polyhydroxy compounds has been explained on the basis of the formation of 1,1- and 1,2-mole ratio complexes between the hydrated borate ion and 1,2 or 1,3 diols:

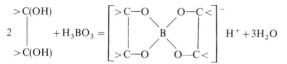

Glycerol has been widely employed for this purpose but mannitol and sorbitol are more effective, and have the advantage that being solids they do not materially increase the volume of the solution being titrated: 0.5–0.7 g of mannitol or sorbitol in 10 cm^3 of solution is a convenient quantity.

The method may be applied to commercial boric acid, but as this material may contain ammonium salts it is necessary to add a slight excess of sodium carbonate solution and then to boil down to half bulk to expel ammonia. Any precipitate which separates is filtered off and washed thoroughly, then the filtrate is neutralised to methyl red, and after boiling, mannitol is added, and the solution titrated with standard $0.1 M$-sodium hydroxide solution:

$$H[boric\ acid\ complex] + NaOH = Na[boric\ acid\ complex] + H_2O$$

$$1\ cm^3\ M\text{-NaOH} = 0.06184\ g\ H_3BO_3$$

A mixture of boric acid and a strong acid can be analysed by first titrating the strong acid using methyl red indicator, and then after adding mannitol or sorbitol, the titration is continued using phenolphthalein as indicator. Mixtures of sodium tetraborate and boric acid can be similarly analysed by titrating the salt with standard hydrochloric acid (Section **X, 37, B**), and then adding mannitol and continuing the titration with standard sodium hydroxide solution: it must of course be borne in mind that in this second titration the boric acid liberated in the first titration will also react.

Procedure. To determine the purity of a sample of boric acid, weigh accurately about 0.8 g of the acid, transfer quantitatively to a 250 cm³ graduated flask and make up to the mark. Pipette 25 cm³ of the solution into a 250 cm³ conical flask, add an equal volume of distilled water, 2.5–3 g of mannitol or sorbitol, and titrate with standard $0.1M$-sodium hydroxide solution using phenolphthalein as indicator. It is advisable to check whether any blank correction must be made: dissolve a similar weight of mannitol (sorbitol) in 50 cm³ distilled water, add phenolphthalein, and ascertain how much sodium hydroxide solution must be added to produce the characteristic end-point colour.

X, 47. DETERMINATION OF AMMONIA IN AN AMMONIUM SALT.

Discussion. Two methods, the direct and indirect, may be used for this determination. In the **direct method**, a solution of the ammonium salt is treated with a solution of a strong base (e.g., sodium hydroxide) and the mixture distilled. Ammonia is quantitatively expelled, and is absorbed in an excess of standard acid. The excess of acid is back-titrated in the presence of methyl red (or methyl orange, methyl orange–indigo carmine, bromo-phenol blue, or bromo-cresol green). Each cm³ of N acid consumed in the reaction is equivalent to 0.017032 g NH_3:

$$NH_4^+ + OH^- \longrightarrow NH_3\uparrow + H_2O$$

In the **indirect method**, the ammonium salt (other than the carbonate or bicarbonate) is boiled with a known excess of standard sodium hydroxide solution. The boiling is continued until no more ammonia escapes with the steam. The excess of sodium hydroxide is titrated with standard acid, using methyl red (or methyl orange–indigo carmine) as indicator.

Procedure (**direct method**). Fit up the apparatus shown in Fig. X, 16, note that in order to provide some flexibility, the spray trap is joined to the condenser by a hemispherical ground joint and this makes it easier to clamp both the flask and the condenser without introducing any strain into the assembly. The flask may be of round bottom form (capacity 500–1000 cm³), or (as shown in the diagram), a Kjeldahl flask. The latter is particularly suitable when nitrogen in organic compounds is determined by the Kjeldahl method: upon completion of the digestion with concentrated sulphuric acid, cooling, and dilution of the contents, the digestion (Kjeldahl) flask is attached to the apparatus as shown in Fig. X, 16. The purpose of the spray trap is to prevent droplets of sodium

Fig. X, 16

hydroxide solution being driven over during the distillation process. The lower end of the condenser is allowed to dip into a known volume of standard acid contained in a suitable receiver, e.g. a conical flask. A commercial distillation assembly may be purchased (Quickfit and Quartz Ltd), in which the tap funnel shown is replaced by a special liquid addition unit: this is similar in form to the tap funnel, but the tap and barrel are replaced by a small vertical ground joint which can be closed with a tapered glass rod. This modification is especially useful when numerous determinations have to be made as it obviates the tendency of glass taps to 'stick' after prolonged contact with concentrated solutions of sodium hydroxide.

For practice, weigh out accurately about 1.5 g of A.R. ammonium chloride, dissolve it in water, and make up to 250 cm^3 in a graduated flask. Shake thoroughly. Transfer 50.0 cm^3 of the solution into the distillation flask and dilute with 200 cm^3 of water: add a few anti-bumping granules (fused alumina) to promote regular ebullition in the subsequent distillation. Place 100.0 cm^3 of standard 0.1 M-hydrochloric acid in the receiver and adjust the flask so that the end of the condenser just dips into the acid. Make sure that all the joints are fitting tightly. Place 100 cm^3 of 10 per cent sodium hydroxide solution in the funnel. Run the sodium hydroxide solution into the flask by opening the tap; close the tap as soon as the alkali has entered. Heat the flask so that the contents boil gently. Continue the distillation for 30–40 minutes, by which time all the ammonia should have passed over into the receiver; open the tap before removing the flame. Disconnect the trap from the top of the condenser. Lower the receiver and rinse the condenser with a little water. Add a few drops of methyl red* and titrate the excess of acid in the solution with standard 0.1 M-sodium hydroxide. Repeat the determination.

Calculate the percentage of NH$_3$ in the solid ammonium salt employed.

$$1 \text{ cm}^3 \ 0.1M\text{-HCl} \equiv 1.703 \text{ mg NH}_3$$

Procedure (**indirect method**). Weigh out accurately 0.1–0.2 g of the ammonium salt into a 500-cm^3 Pyrex conical flask, and add 100 cm^3 of standard 0.1M-sodium hydroxide. Place a small funnel in the neck of the flask in order to

---

* A sharper colour change is obtained with the mixed indicator methyl red–bromo-cresol green (prepared from 1 part of 0.2 per cent methyl red in ethanol and 3 parts of 0.1 per cent bromo-cresol green in ethanol).

prevent mechanical loss, and boil the mixture until a piece of filter paper moistened with mercury(I) nitrate solution and held in the escaping steam is no longer turned black. Cool the solution, add a few drops of methyl red, and titrate with standard 0.1M-hydrochloric acid. Repeat the determination.

X, 48. DETERMINATION OF NITRATES. *Discussion.* Nitrates are reduced to ammonia by means of aluminium, zinc or, most conveniently, by Devarda's alloy (50%Cu, 45%Al, 5%Zn) in strongly alkaline solution:

$$3NO_3^- + 8Al + 5OH^- + 2H_2O = 8AlO_2^- + 3NH_3$$

The ammonia is distilled into excess of standard acid as in the previous Section.

Nitrites are similarly reduced, and must be allowed for if nitrate alone is to be determined.

Procedure. Weigh out accurately about 1.0 g of the nitrate. Dissolve it in water and transfer the solution quantitatively to the distillation flask of Fig. X, 16. Dilute to about 240 cm^3. Add 3 g of pure, finely divided Devarda's alloy (it should all pass a 20-mesh sieve). Fit up the apparatus completely and place 75–100 cm^3 standard 0.2N-hydrochloric acid in the receiver (500-cm^3 Pyrex conical flask). Introduce 10 cm^3 of 20 per cent sodium hydroxide solution through the funnel, and immediately close the tap. Warm *gently* to start the reaction, and allow the apparatus to stand for an hour, by which time the evolution of hydrogen should have practically ceased and the reduction of nitrate to ammonia be complete. Then boil the liquid gently and continue the distillation until 40–50 cm^3 of liquid remain in the distillation flask. Open the tap before removing the flame. Wash the condenser with a little distilled water, and titrate the contents of the receiver plus the washings with standard 0.2M-sodium hydroxide, using methyl red as indicator. Repeat the determination. For very accurate work, it is recommended that a blank test be carried out with distilled water.

$$1 \text{ cm}^3 \text{ } M\text{-HCl} \equiv 0.06201 \text{ g NO}_3^-$$

X, 49. DETERMINATION OF PHOSPHATE (PRECIPITATION AS QUINOLINE MOLYBDOPHOSPHATE). *Discussion.* When a solution of an orthophosphate is treated with a large excess of ammonium molybdate solution in the presence of nitric acid at a temperature of 20–45 °C, a precipitate is obtained, which after washing is converted into ammonium molybdophosphate with the composition $(NH_4)_3[PO_4,12MoO_3]$. This may be titrated with standard sodium hydroxide solution using phenolphthalein as indicator, but the end-point is rather poor due to the liberation of ammonia. If, however, the ammonium molybdate is replaced by a reagent containing sodium molybdate and quinoline, then quinoline molybdophosphate is precipitated which can be isolated and titrated with standard sodium hydroxide:

$$(C_9H_7N)_3[PO_4,12MoO_3] + 26NaOH$$
$$= Na_2HPO_4 + 12Na_2MoO_4 + 3C_9H_7N + 14H_2O$$

The main advantages over the ammonium molybdophosphate method are: (i) quinoline molybdophosphate is less soluble and has a constant composition, and (ii) quinoline is a sufficiently weak base not to interfere in the titration.

Calcium, iron, magnesium, alkali metals, and citrates do not affect the analysis. Ammonium salts interfere and must be eliminated by means of sodium nitrite or sodium hypobromite. The hydrochloric acid normally used in the analysis may be replaced by an equivalent amount of nitric acid without any influence on the course of the reaction. Sulphuric acid leads to high and erratic results. If hydrochloric acid is present in amount slightly in excess of the sulphuric acid the interference is prevented, but the total acidity should not be greatly in excess of $2N$.

The method may be standardised, if desired, with pure potassium dihydrogen phosphate (see below): sufficient 1:1-hydrochloric acid must be present to prevent precipitation of quinoline molybdate; the molybdophosphate complex is readily formed at a concentration of 20 cm³ of concentrated hydrochloric acid per 100 cm³ of solution, especially when warm, and precipitation of the quinoline salt should take place *slowly* from boiling solution. A 'blank' determination should always be made; it is mostly due to silica.

Solutions required. *Sodium molybdate solution.* Prepare a 15 per cent solution of A.R. sodium molybdate, $Na_2MoO_4,2H_2O$. Store in a polythene bottle.

Quinoline hydrochloride solution. Add 20 cm³ of redistilled quinoline to 800 cm³ of hot water containing 25 cm³ of pure concentrated hydrochloric acid, and stir well. Cool to room temperature, add a little filter paper pulp ('accelerator'), and again stir well. Filter with suction through a paper-pulp pad, but do not wash. Dilute to 1 litre with water.

Mixed indicator solution. Mix two volumes of 0.1 per cent phenolphthalein solution and three volumes of 0.1 per cent thymol blue solution (both in ethanol).

To standardise the procedure, A.R. potassium dihydrogenphosphate which has been dried at 105 °C is usually suitable; if necessary it may be further purified by dissolving 100 g in 200 cm³ of boiling distilled water, keeping on a boiling water bath for several hours, filtering through paper pulp from any turbidity which may appear, and cooling rapidly with constant stirring. The crystals are filtered with suction on hardened filter paper, washed twice with ice-cold water and once with 50 per cent ethanol, and dried at 105 °C.

Procedure. This will be described by reference to the standardisation with potassium dihydrogenphosphate. Weigh accurately 0.08–0.09 g of the pure salt into a 250 cm³ conical flask and dissolve in about 50 cm³ of distilled water. Add 20 cm³ of concentrated hydrochloric acid, then 30 cm³ of the sodium molybdate solution. Heat to boiling, and add a few drops of the quinoline reagent from a burette while swirling the solution in the flask. Again heat to boiling and add the quinoline reagent drop by drop with constant swirling until 1 or 2 cm³ have been added. Boil again, and to the gently boiling solution add the reagent a few cm³ at a time, with swirling, until 60 cm³ in all have been introduced. A coarsely crystalline precipitate is thus produced. Allow the suspension to stand in a boiling water bath for 15 minutes, and then cool to room temperature. Prepare a paper-pulp filter in a funnel fitted with a porcelain cone, and tamp well down. Decant the clear solution through the filter and wash the precipitate twice by decantation with about 20 cm³ of hydrochloric acid (1:9); this removes most of the excess of quinoline and of molybdate. Transfer the precipitate to the pad with cold water, washing the flask well; wash the filter and precipitate with 30-cm³ portions of water, letting each washing run through before applying the next, until the washings are acid free (test for acidity with pH test paper; about six washings are usually required). Transfer the filter pad and precipitate back to the original flask: insert the funnel into the flask and wash with about 50 cm³ of water to ensure the transfer of all traces of precipitate. Shake the flask well so that filter paper and precipitate are completely broken up. Run in 50.0 cm³ of standard (carbonate-free) 0.5M-sodium hydroxide, swirling during the addition. Shake until the precipitate is *completely* dissolved. Add a few drops of the mixed indicator solution and titrate with standard 0.5M-hydrochloric acid to an end-point which changes sharply from pale green to pale yellow.

Run a blank on the reagents, but use 0.1N-acid and alkali solutions for the titrations; calculate the blank to 0.5M-sodium hydroxide. Subtract the blank (which should not exceed 0.5 cm³) from the volume neutralised by the original precipitate.

$$1 \text{ cm}^3 \, 0.5M\text{-NaOH} \equiv 1.830 \text{ mg } PO_4^{3-}$$

Wilson (Ref. 12) has recommended that the hydrochloric acid added before precipitation be replaced by citric acid, and the subsequent washing of the precipitate is then carried out solely with distilled water.

The method can be applied to the determination of phosphorus in a wide variety of materials, e.g., phosphate rock, phosphatic fertilisers and metals, and is suitable for use in conjunction with the oxygen flask procedure (Section **III, 46**). In all cases it is essential to ensure that the material is so treated that the phosphorus is converted to orthophosphate; this may usually be done by dissolution in an oxidising medium such as concentrated nitric acid or in 60 per cent perchloric acid.

B. 2   COMPLEXATION TITRATIONS

The simple complexation titration described in Section **X, 22**, i.e., the determination of cyanide by titration with standard silver nitrate solution and which involves formation of the complex cyanoargentate ion $[Ag(CN)_2]^-$, is most conveniently included with the use of standard silver nitrate solutions in Precipitation Titrations (Section **X, 87**). The following Sections will therefore be devoted to applications of 1,2-diaminoethanetetra-acetic acid (ethylenediamine-tetra-acetic acid, EDTA) and its congeners. These reagents possess great versatility arising from their inherent potency as complexing agents and from the availability of numerous metal-ion indicators (Section **X, 28**), each effective over a limited range of pH, but together covering a wide range of pH values: to these factors must be added the additional refinements offered by 'masking' and 'demasking' techniques (Section **X, 27**).

It is clearly impossible, within the scope of the present volume, to give details for all the cations (and anions) which can be determined by EDTA or similar types of titration. Accordingly, details of a few typical determinations are given which serve to illustrate the general procedures to be followed and the use of various buffering agents and of some different indicators. A conspectus of some selected procedures for the commoner cations is then given, followed by some examples of the uses of EDTA for the determination of the components of mixtures; finally, some examples of the determination of anions are given. The relevant theoretical Sections (**X, 19–28**) should be consulted before commencing the determinations.

X, 50.   STANDARD EDTA SOLUTIONS.   Disodium dihydrogenethylene-diaminetetra-acetate of analytical reagent quality is available commercially but this may contain a trace of moisture. After drying the Analar material at 80 °C its composition agrees exactly with the formula $Na_2H_2C_{10}H_{12}O_8N_2,2H_2O$ (molecular weight 372.24), but it should not be used as a primary standard. If necessary, the commercial material may be purified by preparing a saturated solution at room temperature: this requires about 20 g of the salt per 200 cm^3 of water. Add ethanol slowly until a permanent precipitate appears; filter. Dilute the filtrate with an equal volume of ethanol, filter the resulting precipitate through a sintered glass funnel, wash with acetone and then with diethyl ether. Air-dry at room temperature overnight and then dry in an oven at 80 °C for at least 24 hours. Consult Chapter XXIII for a discussion of the thermal behaviour of EDTA.

Solutions of EDTA of the following concentrations are suitable for most experimental work; $0.1M$, $0.05M$, and $0.01M$ and contain respectively 37.224 g, 18.612 g, and 3.7224 g of the dihydrate per dm^3 of solution. As already indicated, the dry Analar salt cannot be regarded as a primary standard and the solution must be standardised; this can be done by titration of nearly neutralised zinc chloride or zinc sulphate solution prepared from a known weight of A.R. zinc pellets; nearly neutralised magnesium chloride (or sulphate) solution prepared from a known weight of pure magnesium; or a manganese chloride solution prepared from spectroscopically pure manganese.

The water employed in making up solutions, particularly dilute solutions, of EDTA should contain no traces of polyvalent ions. The distilled water normally used in the laboratory may require distillation in an all-Pyrex glass apparatus or, better, passage through a column of cation exchange resin in the sodium form— the latter procedure will remove all traces of heavy metals. De-ionised water is also satisfactory; it should be prepared from distilled water since tap water sometimes contains non-ionic impurities not removed by an ion exchange column. The solution may be kept in Pyrex (or similar borosilicate glass) vessels, which have been thoroughly steamed out before use. For prolonged storage in borosilicate vessels, the latter should be boiled with a strongly alkaline, 2 per cent EDTA solution for several hours and then repeatedly rinsed with de-ionised water. Polythene bottles are the most satisfactory, and should always be employed for the storage of very dilute (e.g., $0.001M$) solutions of EDTA. Vessels of ordinary (soda) glass should not be used; in the course of time such soft glass containers will yield appreciable amounts of cations (including calcium and magnesium) and anions to solutions of EDTA.

Water purified or prepared as described above should be used for the preparation of *all* solutions required for EDTA or similar titrations.

X, 51.   SOME PRACTICAL CONSIDERATIONS.   The following points should be borne in mind when carrying out complexometric titrations.

A. Adjustment of pH.   For many EDTA titrations the pH of the solution is extremely critical; often limits of ± 1 unit of pH, and frequently limits of ± 0.5 unit of pH must be achieved for a successful titration to be carried out. To achieve such narrow limits of control it is necessary to make use of a pH meter whilst adjusting the pH value of the solution, and even for those cases where the latitude is such that a pH test paper can be used to control the adjustment of pH, only a paper of the narrow range variety should be used.

In some of the details which follow, the addition of a recommended volume of buffer solution is referred to, and in all such cases, to ensure that the requisite buffering action is in fact achieved, it is necessary to make certain that the original solution has been made almost neutral by the cautious addition of sodium or ammonium hydroxide, or of dilute acid, before adding the buffer solution. When an acid solution containing a metallic ion is neutralised by the addition of alkali care must be taken to ensure that the metal hydroxide is not precipitated.

B. Concentration of the metal ion to be titrated. Most titrations are successful with 0.25 millimole of the metal ion concerned in a volume of 50–150 cm³ of solution. If the metal ion concentration is too high, then the end-point may be very difficult to discern, and if difficulty is experienced with an end-point then it is advisable to start with a smaller portion of the test solution, and to dilute this to 100–150 cm³ before adding the buffering medium and the indicator, and then repeating the titration.

C. Amount of indicator. The addition of too much indicator is a fault which must be guarded against. In many cases the colour due to the indicator intensifies considerably during the course of the titration, and further, many indicators exhibit dichroism, i.e., there is an intermediate colour change one to two drops before the real end-point. Thus, for example, in the titration of lead using xylenol orange as indicator at pH = 6, the initial reddish-purple colour becomes orange red, and then with the addition of one or two further drops of reagent, the solution acquires the final lemon yellow colour. This 'end-point anticipation', which is of great practical value, may be virtually lost if too much of the indicator is added so that the colour is too intense. In general, a satisfactory colour is achieved by the use of 30–50 mg of a solid mixture of the indicator with A.R. potassium nitrate which contains about 1 per cent of the indicator.

D. Attainment of the end-point. In many EDTA titrations the colour change in the neighbourhood of the end-point may be slow. In such cases, cautious addition of the titrant coupled with continuous stirring of the solution is advisable; the use of a magnetic stirrer is recommended. Frequently, a sharper end-point may be achieved if the solution is warmed to about 40 °C. Titrations with CDTA (see Section **X, 23**) are always slower in the region of the end-point than the corresponding EDTA titrations.

E. Detection of the colour change. With all of the metal-ion indicators used in complexometric titrations, detection of the end-point of the titration is dependent upon the recognition of a specified change in colour; for many observers this can be a difficult task, and for those affected by colour blindness may be virtually impossible. These difficulties may be overcome by replacing the eye by a photocell which is much more sensitive, and eliminates the human element. To carry out the requisite operations it is necessary to have available a colorimeter or a spectrophotometer in which the cell compartment is large enough to accommodate the titration vessel (a conical flask or a tall form beaker); the Unicam SP 500 spectrophotometer is one example of an instrument suitable for this purpose, and a number of phototitrators are available commercially. A simple apparatus may be readily constructed in which light passing through the solution is first allowed to strike a suitable filter and then a photocell; the current generated in the latter is measured with a galvanometer. Whatever form of instrument is used, the wavelength of the incident light is selected (either by an optical filter or by the controls on the instrument) so that the titration solution (including the indicator) shows a maximum transmittance. The titration is then carried out stepwise, taking readings of the transmittance after each addition of EDTA; these readings are then plotted against volume of EDTA solution added, and at the end-point (where the indicator changes colour), there will be an abrupt alteration in the transmittance, i.e., a break in the curve, from which the end-point may be assessed accurately.

F. Alternative methods of detecting the end-point. In addition to the visual and spectrophotometric detection of end-points in EDTA titrations with the aid of metal-ion indicators, the following methods are also available for end-point detection.

1. Potentiometric titration using a mercury electrode (see Section **XIV, 30**).
2. Potentiometric titration using a selective ion electrode (Section **XIV, 9–12**) responsive to the ion being titrated.
3. Potentiometric titration using a bright platinum-saturated calomel electrode system; this can be used when the reaction involves two different oxidation states of a given metal (see Section **XIV, 31**).
4. By conductometric titration (Chapter XV).
5. By amperometric titration (Chapter XVII).
6. By enthalpimetric titration (Chapter XXV).

A method of coulometric analysis (Chapter XIII) has also been devised; see Reference 13.

Determination of Cations

X, 52. DETERMINATION OF ALUMINIUM: BACK TITRATION USING SOLOCHROME BLACK INDICATOR. Pipette 25 cm^3 of an aluminium ion solution (approximately $0.01M$) into a conical flask and run in from a burette a slight excess of $0.01M$-EDTA solution; adjust the pH to between 7 and 8 by the addition of ammonia solution (test drops on phenol red paper or use a pH meter). Boil the solution for a few minutes to ensure complete complexation of the aluminium; cool to room temperature and adjust the pH to 7–8. Add 50 mg of Solochrome Black/KNO_3 mixture (see Section **X, 51**, *C*) and titrate rapidly with standard $0.01M$-zinc sulphate solution until the colour changes from blue to wine red.

After standing for a few minutes the fully titrated solution acquires a reddish-violet colour due to the transformation of the zinc dye complex into the aluminium–Solochrome Black complex; this change is irreversible, so that over-titrated solutions are lost.

Every cm^3 difference between the volume of $0.01M$-EDTA added and the $0.01M$-zinc sulphate solution used in the back-titration corresponds to 0.2698 mg Al.

The standard zinc sulphate solution required is best prepared by dissolving about 1.63 g (accurately weighed) of A.R. granulated zinc in dilute sulphuric acid, nearly neutralising with sodium hydroxide solution, and then making up to 250 cm^3 in a graduated flask; alternatively, the requisite quantity of A.R. zinc sulphate may be used. In either case, de-ionised water must be used.

X, 53. DETERMINATION OF BARIUM: DIRECT TITRATION WITH METHYL THYMOL BLUE INDICATOR. Pipette 25 cm^3 barium ion solution (*ca.* $0.01M$) into a 250-cm^3 conical flask and dilute to about 100 cm^3 with de-ionised water. Adjust the pH of the solution to 12 by the addition of 3–6 cm^3 of

$1M$ sodium hydroxide solution; the pH must be checked with a pH meter as it must lie between 11.5 and 12.7. Add 50 mg Methyl Thymol Blue/KNO_3 (see Section **X, 51,** *C*) mixture and titrate with standard ($0.01M$) EDTA solution until the colour changes from blue to grey.

1 mole EDTA $\equiv$ 1 mole Ba^{2+}

X, 54.   DETERMINATION OF BISMUTH: DIRECT TITRATION USING XYLENOL ORANGE INDICATOR.

Pipette 25 cm^3 of the bismuth solution (*approx.* $0.01M$) into a 500 cm^3 conical flask and dilute with de-ionised water to about 150 cm^3. If necessary, adjust the pH to about 1 by the cautious addition of dilute aqueous ammonia or of dilute nitric acid; use a pH meter. Add 30 mg of the Xylenol Orange/KNO_3 (see Section **X, 51,** *C*) mixture and then titrate with standard $0.01M$-EDTA solution until the red colour starts to fade. From this point add the titrant slowly until the end-point is reached and the indicator changes to yellow.

1 mole EDTA $\equiv$ 1 mole Bi^{3+}

X, 55.   DETERMINATION OF CALCIUM: SUBSTITUTION TITRATION USING SOLOCHROME BLACK (ERIOCHROME BLACK T) INDICATOR.

Discussion. When calcium ions are titrated with EDTA a relatively stable calcium complex is formed:

$$Ca^{2+} + H_2Y^{2-} \rightleftharpoons CaY^{2-} + 2H^+$$

With calcium ions alone, no sharp end-point can be obtained with Solochrome Black (Eriochrome Black T) indicator and the transition from red to pure blue is not observed. With magnesium ions, a somewhat less stable complexonate is formed:

$$Mg^{2+} + H_2Y^{2-} \rightleftharpoons MgY^{2-} + 2H^+$$

and the magnesium indicator complex is more stable than the calcium–indicator complex but less stable than the magnesium–EDTA complex. Consequently, during the titration of a solution containing magnesium and calcium ions with EDTA in the presence of Solochrome Black (Eriochrome Black T) the EDTA reacts first with the free calcium ions, then with the free magnesium ions, and finally with the magnesium–indicator complex. Since the magnesium–indicator complex is wine red in colour and the free indicator is blue between pH 7 and 11, the colour of the solution changes from wine red to blue at the end-point:

$$MgD^- \text{ (red)} + H_2Y^{2-} = MgY^{2-} + HD^{2-} \text{ (blue)} + H^+$$

If magnesium ions are not present in the solution containing calcium ions they must be added, since they are required for the colour change of the indicator. A common procedure is to add a small amount of magnesium chloride to the EDTA solution before it is standardised. Another procedure, which permits the EDTA solution to be used for other titrations, is to incorporate a little magnesium–EDTA (MgY^{2-}) (1–10 per cent) in the buffer solution or to add a little $0.1M$-magnesium–EDTA (Na_2MgY) to the calcium-ion solution:

$$MgY^{2-} + Ca^{2+} = CaY^{2-} + Mg^{2+}$$

Traces of many metals interfere in the determination of calcium and magnesium using Solochrome Black (Eriochrome Black T) indicator, e.g., Co, Ni, Cu, Zn, Hg, and Mn. Their interference can be overcome by the addition of a little hydroxylammonium chloride (which reduces some of the metals to their lower valency states) and also of sodium or potassium cyanide, which forms very stable cyanide complexes ('masking'). Iron may be rendered harmless by the addition of a little sodium sulphide.

The titration with EDTA, using Solochrome Black (Eriochrome Black T) as indicator, will yield the calcium content of the sample (if no magnesium is present) or the total calcium and magnesium content if both metals are present. To determine the individual elements, calcium may be evaluated by titration using a suitable indicator, e.g., Patton and Reeder's indicator or Calcon—see Sections **X, 62** and **X, 28**, or by titration with EGTA using Zincon as indicator—see Section **X, 63**. The difference between the two titrations is a measure of the magnesium content.

Procedure.   Prepare an *ammonia–ammonium chloride buffer solution* (pH 10), by adding 142 cm^3 concentrated ammonia solution (sp. gr. 0.88–0.90) to 17.5 g A.R. ammonium chloride and diluting to 250 cm^3 with de-ionised water.

Prepare the *magnesium complex of EDTA*, Na_2MgY, by mixing equal volumes of 0.2M solutions of EDTA and of magnesium sulphate. Neutralise with sodium hydroxide solution to a pH between 8 and 9 (phenolphthalein just reddened). Take a portion of the solution, add a few drops of the buffer solution (pH 10), and a few milligrams of the Solochrome Black (Eriochrome Black T)/ KNO_3 (see Section **X, 51**, C) indicator mixture. A violet colour should be produced which turns blue on the addition of a drop of 0.01M-EDTA solution and red on the addition of a single drop of 0.01M-$MgSO_4$ solution; this confirms the equimolarity of magnesium and EDTA. If the solution does not pass this test, it may be treated with more EDTA or with more magnesium sulphate solution until the required condition of equimolarity is attained; this gives an approximately 0.1M solution. Alternatively solid Mg-EDTA complex may be used, which is available from Hopkin and Williams Ltd.

Pipette 25.0 cm^3 of the calcium-ion solution 0.01M into a 250-cm^3 conical flask, dilute it with about 25 cm^3 distilled water, add 2 cm^3 buffer solution, 1 cm^3 0.1M-Mg–EDTA, and 30–40 mg Solochrome Black (Eriochrome Black T)/ KNO_3 mixture. Titrate with the EDTA solution until the colour changes from wine red to clear blue. No tinge of reddish hue should remain at the equivalence point. Titrate slowly near the end-point.

1 mole EDTA $\equiv$ 1 mole Ca^{2+}

X, 56. DETERMINATION OF COPPER: DIRECT TITRATION USING FAST SULPHON BLACK F INDICATOR.

This indicator is virtually specific in its colour reaction with copper in ammoniacal solution; it forms coloured (red) complexes with only copper and nickel, but the indicator action with nickel is poor.

Procedure.   Prepare the *indicator solution* by dissolving 0.5 g of the solid in 100 cm^3 of de-ionised water.

Pipette 25 cm^3 of the copper solution (0.01M) into a conical flask, add 100 cm^3 de-ionised water, 5 cm^3 concentrated ammonia solution and 5 drops of the indicator solution. Titrate with standard EDTA solution (0.01M) until the colour changes from purple to dark green.

1 mole EDTA = 1 mole Cu^{2+}

It should be noted that this method is only applicable to solutions containing up to 25 mg copper ions in 100 cm^3 of water; if the concentration of Cu^{2+} ions is too high, the intense blue colour of the copper(II) ammine complex masks the colour change at the end-point. The indicator solution must be freshly prepared.

X, 57.   DETERMINATION OF IRON(III): DIRECT TITRATION USING VARIAMINE BLUE INDICATOR. *Procedure.* Prepare the *indicator solution* by dissolving 1 g Variamine Blue in 100 cm^3 de-ionised water: as already pointed out (Section **X, 28**), Variamine Blue acts as a redox indicator.

Pipette 25 cm^3 iron(III) solution (0.05M) into a conical flask and dilute to 100 cm^3 with de-ionised water. Adjust the pH to 2–3; Congo Red paper may be used—to the first perceptible colour change. Add 5 drops of the indicator solution, warm the contents of the flask to 40 °C, and titrate with standard (0.05M) EDTA solution until the initial blue colour of the solution turns grey just before the end-point, and with the final drop of reagent changes to yellow.

This particular titration is well adapted to be carried out potentiometrically (Section **XIV, 31**).

1 mole EDTA = 1 mole Fe^{3+}

X, 58.   DETERMINATION OF NICKEL: DIRECT TITRATIONS USING (a) MUREXIDE AND (b) BROMOPYROGALLOL RED AS INDICATORS. *Procedure (a).*   Prepare the *indicator* by grinding 0.1 g murexide with 10 g A.R. potassium nitrate; use about 50 mg of the mixture for each titration.

Also prepare a 1M *solution of ammonium chloride* by dissolving 26.75 g of the A.R. solid in de-ionised water and making up to 500 cm^3 in a graduated flask.

Pipette 25 cm^3 nickel solution (0.01M) into a conical flask and dilute to 100 cm^3 with de-ionised water. Add the solid indicator mixture (50 mg) and 10 cm^3 of the 1M ammonium chloride solution, and then add concentrated ammonia solution dropwise until the pH is about 7 as shown by the yellow colour of the solution. Titrate with standard (0.01M) EDTA solution until the end-point is approached, then render the solution strongly alkaline by the addition of 10 cm^3 of concentrated ammonia solution, and continue the titration until the colour changes from yellow to violet. The pH of the final solution must be 10; at lower pH values an orange-yellow colour develops and more ammonia solution must be added until the colour is clear yellow. Nickel complexes rather slowly with EDTA, and consequently the EDTA solution must be added dropwise near the end-point.

Procedure (b).   Prepare the *indicator* by dissolving 0.05 g Bromopyrogallol Red in 100 cm^3 of 50 per cent ethanol, and a *buffer solution* by mixing 100 cm^3 of 1M ammonium chloride solution with 100 cm^3 of 1M aqueous ammonia solution.

Pipette 25 cm^3 nickel solution (0.01M) into a conical flask and dilute to 150 cm^3 with de-ionised water. Add about 15 drops of the indicator solution, 10 cm^3 of the buffer solution and titrate with standard EDTA solution (0.01M) until the colour changes from blue to claret red.

1 mole EDTA $\equiv$ 1 mole Ni^{2+}

X, 59.   DETERMINATION OF SILVER: INDIRECT METHOD USING POTASSIUM TETRACYANONICKELATE(II) AND NICKEL ION–MUREXIDE INDICATOR.   Silver halides can be dissolved in a solution of potassium tetracyanonickelate(II) in the presence of an ammonia–ammonium chloride buffer, and the nickel ion set free may be titrated with standard EDTA using murexide as indicator.

$$2Ag^+ + [Ni(CN)_4]^{2-} \rightleftharpoons 2[Ag(CN)_2]^- + Ni^{2+}$$

It can be shown from a consideration of the overall stability constants of the ions $[Ni(CN)_4]^{2-}$ (10^{27}) and $[Ag(CN)_2]^-$ (10^{21}) that the equilibrium constant for the above ionic reaction is 10^{15}, i.e., the reaction proceeds practically completely to the right. An interesting exercise is the analysis of a solid silver halide, e.g., silver chloride.

Procedure.   Prepare the *murexide indicator* as detailed in Section **X, 58** (*a*), and an *ammonium chloride solution* (1*M*) by dissolving 26.75 g A.R. ammonium chloride in de-ionised water in a 500 cm^3 graduated flask.

The *potassium tetracyanonickelate(II)* which is required is prepared as follows. Dissolve 25 g of A.R. NiSO$_4$,7H$_2$O in 50 cm^3 distilled water and add portionwise, with agitation, 25 g A.R. KCN. (**Caution.** Use a fume cupboard.) A yellow solution forms and a white precipitate of K$_2$SO$_4$ separates. Gradually add, with stirring, 100 cm^3 of 95 per cent ethanol, filter off the precipitated K$_2$SO$_4$ with suction, and wash twice with 2 cm^3 ethanol. Concentrate the filtrate at about 70 °C —an infrared heater is convenient for this purpose. When crystals commence to separate, stir frequently. When the crystalline mass becomes thick (without evaporating completely to dryness), allow to cool and mix the crystals with 50 cm^3 ethanol. Separate the crystals by suction filtration and wash twice with 5-cm^3 portions ethanol. Spread the fine yellow crystals in thin layers upon absorbent paper, and allow to stand for 2–3 days in the air, adequately protected from dust. During this period the excess of KCN is converted into K$_2$CO$_3$. The preparation is then ready for use; it should be kept in a stoppered bottle.

Treat an aqueous suspension of about 0.072 g (accurately weighed) silver chloride with a mixture of 10 cm^3 of concentrated ammonia solution and 10 cm^3 of 1*M* ammonium chloride solution, then add about 0.2 g of potassium cyanonickelate and warm gently. Dilute to 100 cm^3 with de-ionised water, add 50 mg of the indicator mixture and titrate with standard (0.01*M*) EDTA solution, adding the reagent dropwise in the neighbourhood of the end-point, until the colour changes from yellow to violet.

1 mole EDTA ≡ 2 moles Ag$^+$

Palladium(II) compounds can be determined by a similar procedure, but in this case, after addition of the cyanonickelate, excess of standard (0.01*M*) EDTA solution is added, and the excess is back titrated with standard (0.01*M*) manganese(II) sulphate solution using Solochrome Black indicator.

Gold may be titrated similarly.

X, 60.   DETERMINATION OF SODIUM: INDIRECT TITRATION USING SOLOCHROME BLACK (ERIOCHROME BLACK T) INDICATOR.   The sodium is precipitated as sodium zinc uranyl acetate NaZn (UO$_2$)$_3$(CH$_3$COO)$_9$,6H$_2$O (Section **XI, 53, B**), and the zinc is then determined by titration with EDTA.

Procedure.   Prepare the *indicator* by grinding 0.1 g of Solochrome Black (Eriochrome Black T) with 10 g A.R. potassium nitrate, a *buffer solution* (pH 10) as in Section **X, 55**, an ammonium carbonate solution (1M) by dissolving 78.6 g of the A.R. solid in 500 cm^3 of de-ionised water, and the *precipitation reagent* as detailed in Section **XI, 53, B**: hydrochloric acid (1M) will also be required.

The solution which contains not more than 5 mg of sodium must be concentrated to a small bulk (1–2 cm^3), and precipitation carried out as detailed in Section **XI, 53, B**. After filtering through a porcelain filter crucible the precipitate of zinc uranyl acetate is washed four times with 2 cm^3 portions of the precipitating reagent, and finally with ten 2 cm^3 portions of 95 per cent ethanol which has been saturated with sodium zinc uranyl acetate. The filter crucible is then stood inside a 400 cm^3 Pyrex beaker and 5 cm^3 of 1M-hydrochloric acid added to the crucible. After a few minutes add 50 cm^3 of de-ionised water and boil. Allow the solution to cool somewhat, remove the crucible with crucible tongs and wash carefully into the beaker. Then set up the crucible again for filtration and wash the sinter thoroughly. Neutralise the combined filtrate and washings (total volume about 100 cm^3) with M-ammonium carbonate and add 2 cm^3 in excess to hold the uranium in solution as the carbonato-complex. Add 2 cm^3 buffer mixture (pH = 10) and 30 mg of Solochrome Black (Eriochrome Black T/KNO$_3$) indicator. Titrate with standard 0.001M-EDTA. The colour change at the end-point is from yellowish-red to greenish-blue.

1 mole EDTA ≡ 1 mole Na$^+$

Sodium, free from all other cations, may be determined by passage through a cation exchange column in the magnesium form and titration of the liberated magnesium ion with standard EDTA solution.

X, 61.   DETAILS FOR THE DETERMINATION OF A SELECTION OF METAL IONS BY EDTA TITRATION.   With the detailed instructions given in Sections **X, 52–60** it should be possible to carry out any of the following determinations in Table X, 13 without serious problems arising. In all cases it is recommended that the requisite pH value for the titration should be established by use of a pH meter, but in the light of experience the colour of the indicator at the required pH may, in some cases, be a satisfactory guide. Where no actual buffering agent is specified, the solution should be brought to the required pH value by the cautious addition of dilute acid or of dilute sodium hydroxide solution or aqueous ammonia solution as required.

Table X, 13.   Summarised procedures for EDTA titrations of some selected cations

Metal	Titration type	pH	Buffer	Indicator (1)	Colour change (2)		Notes
Aluminium*	Back	7–8	Aq. NH$_3$	SB	B	R	
Barium*	Direct	12		MTB	B	Gr	(3)
Bismuth	Direct*	1		XO	R	Y	
	Direct	0–1		MTB	B	Y	
Cadmium	Direct	5	Hexamine	XO	R	Y	
Calcium	Direct	12		MTB	B	Gr	
	Substn.*	7–11	Aq. NH$_3$/NH$_4$Cl	SB	R	B	
Cobalt	Direct	6	Hexamine	XO	R	Y	(4)

Table X, 13.

Metal	Titration type	pH	Buffer	Indicator (1)	Colour change (2)		Notes
Copper*	Direct			FSB	P	G	(6)
Gold	See silver						
Iron(III)*	Direct	2–3		VB	B	Y	(4)
Lead	Direct	6	Hexamine	XO	R	Y	
Magnesium	Direct	10	Aq. NH_3/NH_4Cl	SB	R	B	(4*)
Manganese	Direct	10	Aq. NH_3/NH_4Cl	SB	R	B	(5)
	Direct	10	Aq. NH_3	TPX	B	PP	(5)
Mercury	Direct	6	Hexamine	XO	R	Y	
	Direct	6	Hexamine	MTB	B	Y	
Nickel	Direct*	7–10	Aq. NH_3/NH_4Cl	M	Y	V	
	Direct*	7–10	Aq. NH_3/NH_4Cl	BPR	B	R	
	Back	10	Aq. NH_3/NH_4Cl	SB	B	R	
Palladium	See silver						
Silver*	Indirect			M	Y	V	
Sodium*	Indirect	10	Aq. NH_3/NH_4Cl	SB	R	B	
Strontium	Direct	12		MTB	B	Gr	
	Direct	10–11		TPX	B	PP	
Thorium	Direct	2–3		XO	R	Y	
	Direct	2–3		MTB	B	Y	
	Direct	3–3.5		CV	B	Y	(4)
Tin(II)	Direct	6	Hexamine	XO	R	Y	
Zinc	Direct	10	Aq. NH_3/NH_4Cl	SB	R	B	
	Direct	6	Hexamine	XO	R	Y	
	Direct	6	Hexamine	MTB	B	Y	

Notes to Table X, 13.
*/Details in Sections **X, 52–60**.
(1) BPR = Bromopyrogallol Red; CV = Catechol Violet; FSB = Fast Sulphon Black F;
    M = Murexide; MTB = Methylthymol Blue; SB = Solochrome Black (Eriochrome Black T);
    TPX = Thymolphthalexone; VB = Variamine Blue; XO = Xylenol Orange.
(2) B = Blue; G = Green; Gr. = Grey; O = Orange; P = Purple; PP = Pale pink; R = Red; V
    = Violet; Y = Yellow.
(3) Can also be determined by precipitation as $BaSO_4$ and dissolution in excess EDTA (Section **X,
    75**).
(4) Temperature 40 C; (4*) Warming optional.
(5) Add 0.5 g hydroxylammonium chloride (to prevent oxidation), and 3 cm³ triethanolamine (to
    prevent precipitation in alkaline solution); use boiled-out (air-free) water.
(6) In presence of concentrated aqueous ammonia.

Analysis of mixtures of cations

X, 62. DETERMINATION OF CALCIUM AND MAGNESIUM USING PATTON AND REEDER'S INDICATOR. *Discussion.* Patton and Reeder's indicator (HHSNNA), see Section **X, 28**, permits the determination of calcium in the presence of magnesium, and finds application in the determination of the hardness of water and in the analysis of limestone and dolomite. Titration using Solochrome Black (Eriochrome Black T) gives calcium and magnesium together, and the difference between the two titrations gives the magnesium content of the mixture (1).

Calcon may also be used for the titration of Ca in the presence of Mg (compare Section **X, 28**). The neutral solution (say, 50 cm³) is treated with 5 cm³ of diethylamine (giving a pH of about 12.5, which is sufficiently high to precipitate the magnesium quantitatively as the hydroxide) and 4 drops of Calcon indicator are added. The solution is stirred magnetically and titrated with standard EDTA solution until the colour changes from pink to a pure blue.

A sharper end-point may be obtained by adding 2–3 drops of 1 per cent aqueous polyvinyl alcohol to the sample solution, then adjusting the pH to 12.5 with sodium hydroxide, adding 2–3 drops of 10 per cent aqueous potassium cyanide solution, warming to 60 °C (**Caution:** use a fume cupboard), and treating the warm solution with 3–4 drops of Calcon indicator. The solution is titrated with $0.01M$-EDTA to a red-blue end-point. The polyvinyl alcohol reduces the adsorption of the dye on the surface of the precipitate. The solution is prepared by mixing 1.0 g of medium-viscosity polyvinyl alcohol with 100 cm^3 of boiling water in a mechanical homogenizer.

Procedure. Prepare the *indicators* by grinding (*a*) 0.5 g HHSNNA with 50 g A.R. potassium chloride, and (*b*) 0.2 g Solochrome Black (Eriochrome Black T) with 50 g A.R. potassium chloride. The following solutions will also be required:

Magnesium chloride solution $(0.01M)$. Dissolve 0.608 g pure magnesium turnings in dilute hydrochloric acid, nearly neutralise with sodium hydroxide solution $(1M)$ and make up to 250 cm^3 in a graduated flask with de-ionised water. Pipette 25 cm^3 of the resulting $0.1M$ solution into a 250 cm^3 graduated flask and make up to the mark with de-ionised water.

Potassium hydroxide solution ca. 8M. Dissolve 112 g A.R. potassium hydroxide pellets in 250 cm^3 of de-ionised water.

Buffer solution. Add 55 cm^3 concentrated hydrochloric acid to 400 cm^3 de-ionised water and mix thoroughly. Slowly pour 310 cm^3 redistilled mono-ethanolamine with stirring into the mixture and cool to room temperature (2). Titrate 50.0 cm^3 of the standard MgCl$_2$ solution with standard $(0.01M)$ EDTA solution using 1 cm^3 of the monoethanolamine–hydrochloric acid solution as the buffer and Solochrome Black (Eriochrome Black T) as the indicator. Add 50.0 cm^3 of the MgCl$_2$ solution to the volume of EDTA solution required to complex the magnesium exactly (as determined in the last titration), pour the mixture into the monoethanolamine–hydrochloric acid solution, and mix well. Dilute to 1 litre (3).

Determination of calcium. Pipette two 25.0 cm^3 portions of the mixed calcium- and magnesium-ion solution (not more than $0.01M$ with respect to either ion) into two separate 250-cm^3 conical flasks and dilute each with about 25 cm^3 of distilled water. To the first flask add 4 cm^3 $8M$-potassium hydroxide solution (a precipitate of magnesium hydroxide may be noted here), and allow to stand for 3–5 minutes with occasional swirling. Add about 30 mg each of A.R. potassium cyanide (**Caution:** poison) and A.R. hydroxylammonium chloride and swirl the contents of the flask until the solids dissolve. Add about 50 mg of the HHSNNA indicator mixture and titrate with $0.01M$-EDTA until the colour changes from red to blue. Run into the second flask from a burette a volume of EDTA solution equal to that required to reach the end-point less 1 cm^3. Now add 4 cm^3 of the potassium hydroxide solution, mix well and complete the titration as with the first sample; record the exact volume of EDTA solution used. Perform a blank titration, replacing the sample with distilled water.

Determination of total calcium and magnesium. Pipette 25.0 cm^3 of the mixed calcium- and magnesium-ion solution into a 250-cm^3 conical flask, dilute to about 50 cm^3 with distilled water, add 5 cm^3 of the buffer solution, and mix by swirling. Add about 30 mg A.R. potassium cyanide and A.R. hydroxylammonium chloride, shake gently until the solids dissolve, and then add about 50 mg of the Solochrome Black (Eriochrome Black T) indicator mixture. Titrate with the EDTA solution to a pure blue end-point. Perform a blank titration, replacing the 25-cm^3 sample solution with de-ionised water.

Calculate the volume of standard EDTA solution equivalent to the magnesium by subtracting the total volume required for the calcium from the volume required for the total calcium and magnesium for equal amounts of the test sample.

Notes. 1. The usefulness of the HHSNNA indicator for the titration of calcium depends upon the fact that the pH of the solution is sufficiently high to ensure the quantitative precipitation of the magnesium as magnesium hydroxide and that calcium forms a more stable complex with EDTA than does magnesium. The EDTA does not react with magnesium (present as $Mg(OH)_2$) until all the free calcium and the calcium–indicator complex have been complexed by the EDTA. If the indicator is added before the potassium hydroxide, a satisfactory end-point is not obtained because magnesium salts form a lake with the indicator as the pH increases and the magnesium indicator-lake is coprecipitated with the magnesium hydroxide.

2. The monoethanolamine–hydrochloric acid buffer has a buffering capacity equal to the ammonia–ammonium chloride buffer commonly employed for the titration of calcium and magnesium with EDTA and Solochrome Black (Eriochrome Black T) (compare Section **X, 55**). The buffer has excellent keeping qualities, sharp end-points are obtainable, and the strong ammonia solution is completely eliminated.

3. When relatively pure samples of calcium are titrated using Solochrome Black (Eriochrome Black T) as indicator, magnesium must be added to obtain a sharp end-point, hence magnesium is usually added to the buffer solution (compare Section **X, 55**). The addition of magnesium to the EDTA solution prevents a sharp end-point when calcium is titrated using HHSNNA as indicator. The introduction of complexed magnesium into the buffer eliminates the need for two EDTA solutions and ensures an adequate amount of magnesium, even when small amounts of this element are titrated.

X, 63. DETERMINATION OF CALCIUM IN THE PRESENCE OF MAGNESIUM USING EGTA AS TITRANT. *Discussion.* Calcium may be determined in the presence of magnesium by using EGTA as titrant, because whereas the stability constant for the calcium–EGTA complex is about 1×10^{11}, that of the magnesium–EGTA complex is only about 1×10^5, and thus magnesium does not interfere with the reagent. The method described in the preceding section, which involves precipitation of magnesium hydroxide, is not satisfactory if the magnesium content of the mixture is much greater than about 10 per cent of the calcium content, since co-precipitation of calcium hydroxide may occur. Titration with EGTA is therefore to be recommended for the determination of small amounts of calcium in the presence of larger amounts of magnesium.

The indicator used in the titration is Zincon (Section **X, 28**) which gives rise to an indirect end-point with calcium. Detection of the end-point is dependent upon the reaction

$$ZnEGTA^{2-} + Ca^{2+} = Zn^{2+} + CaEGTA^{2-}$$

327

and the zinc ions liberated form a blue complex with the indicator. At the end-point, the zinc–indicator complex is decomposed:

$$ZnIn^{2-} + EGTA = ZnEGTA^{2-} + HIn^{3-}$$

and the solution acquires the orange-red colour of the indicator.

Procedure.   Prepare an *EGTA solution* (0.05M) by dissolving 19.01 g in 100 cm^3 sodium hydroxide solution (1M) and diluting to 1 dm^3 in a graduated flask with de-ionised water. Prepare the *indicator* by dissolving 0.065 g Zincon in 2 cm^3 sodium hydroxide solution (0.1M) and diluting to 100 cm^3 with de-ionised water, and a *buffer solution* (*pH* 10) by dissolving 25 g sodium tetraborate, 3.5 g ammonium chloride, and 5.7 g sodium hydroxide in 1 litre of de-ionised water.

Prepare 100 cm^3 of *Zn–EGTA complex* solution by taking 50 cm^3 of 0.05M zinc sulphate solution and adding an equivalent volume of 0.05M EGTA solution; exact equality of zinc and EGTA is best achieved by titrating a 10 cm^3 portion of the zinc sulphate solution with the EGTA solution using zincon indicator, and from this result the exact volume of EGTA solution required for the 50 cm^3 portion of zinc sulphate solution may be calculated.

The EGTA solution may be standardised by titration of a standard (0.05M) calcium solution, prepared by dissolving 5.00 g A.R. calcium carbonate in dilute hydrochloric acid contained in a dm^3 graduated flask, and then after neutralising with sodium hydroxide solution diluting to the mark with de-ionised water: use zincon indicator in the presence of Zn–EGTA solution (see below).

To determine the calcium in the calcium–magnesium mixture, pipette 25 cm^3 of the solution into a 250-cm^3 conical flask, add 25 cm^3 of the buffer solution and check that the resulting solution has a pH of 9.5–10.0. Add 2 cm^3 of the Zn–EGTA solution and 2–3 drops of the indicator solution. Titrate slowly with the standard EGTA solution until the blue colour changes to orange-red.

X, 64.   DETERMINATION OF THE TOTAL HARDNESS (PERMANENT AND TEMPORARY) OF WATER USING SOLOCHROME BLACK (ERIOCHROME BLACK T) INDICATOR.

The hardness of water is generally due to dissolved calcium and magnesium salts and may be determined by complexometric titration.

Procedure.   To a 50 cm^3 sample of the water to be tested add 1 cm^3 buffer solution (aq.NH$_3$/NH$_4$Cl, pH 10, Section **X, 55**) and 30–40 mg Solochrome Black (Eriochrome Black T)/KNO$_3$ indicator mixture. Titrate with standard EDTA solution (0.01M) until the colour changes from red to pure blue. Should there be no magnesium present in the sample of water it is necessary to add 0.1 cm^3 magnesium–EDTA solution (0.1M) before adding the indicator (see Section **X, 55**). The total hardness is expressed in parts of CaCO$_3$ per million of water.

If the water contains traces of interfering ions, then 4 cm^3 of buffer solution should be added, followed by 30 mg hydroxylammonium chloride and then 50 mg A.R. potassium cyanide (**Caution**) before adding the indicator.

Notes.   1. Somewhat sharper end-points may be obtained if the sample of water is first acidified with dilute hydrochloric acid, boiled for about a minute to drive off carbon dioxide, cooled, neutralised with sodium hydroxide solution, buffer and indicator solution added, and then titrated with EDTA as above.

2. The **permanent hardness** of a sample of water may be determined as follows. Place 250 cm^3 of the sample of water in a 600-cm^3 beaker and boil gently for 20–30 minutes. Cool and filter it directly into a 250-cm^3 graduated flask: do not wash

the filter paper, but dilute the filtrate to volume with distilled water and mix well. Titrate 50.0 cm^3 of the filtrate by the same procedure as was used for the total hardness. This titration measures the permanent hardness of the water. Calculate this hardness as parts per million of $CaCO_2$.

Calculate the **temporary hardness** of the water by subtracting the permanent hardness from the total hardness.

3. If it is desired to determine both the calcium and the magnesium in a sample of water, determine first the total calcium and magnesium content as above, and calculate the result as parts per million of $CaCO_3$.

The calcium content may then be determined by titration with EDTA using either Patton and Reeder's indicator or Calcon (Section **X, 62**), or alternatively by titration with EGTA (see previous Section).

X, 65. DETERMINATION OF CALCIUM IN THE PRESENCE OF BARIUM USING CDTA AS TITRANT.

There is an appreciable difference between the stability constants of the CDTA complexes of barium ($\log K = 7.99$) and calcium ($\log K = 12.50$), with the result that calcium may be titrated with CDTA in the presence of barium; the stability constants of the EDTA complexes of these two metals are too close together to permit independent titration of calcium in the presence of barium.

The indicator Calcichrome (see Section **X, 28**) is specific for calcium at a pH 11–12 in the presence of barium.

Procedure. Prepare the *CDTA solution* (0.02M) by dissolving 6.880 g of the solid reagent in 50 cm^3 of sodium hydroxide solution (1M) and making up to 1 dm^3 with de-ionised water; the solution may be standardised against a standard calcium solution prepared from 2.00 g A.R. calcium carbonate (see Section **X, 63**). The *indicator is prepared* by dissolving 0.5 g of the solid in 100 cm^3 of water.

Pipette 25 cm^3 of the solution to be analysed into a 250 cm^3 conical flask and dilute to 100 cm^3 with de-ionised water: the original solution should be about 0.02M with respect to calcium and may contain barium to a concentration of up to 0.2M. Add 10 cm^3 sodium hydroxide solution (1M) and check that the pH of the solution lies between 11–12; then add 3 drops of the indicator solution. Titrate slowly with the standard CDTA solution until the pink colour changes to blue.

X, 66. DETERMINATION OF CALCIUM AND LEAD IN ADMIXTURE USING METHYLTHYMOL BLUE INDICATOR.

With Methylthymol Blue lead may be titrated at a pH of 6 without interference by calcium; the calcium is subsequently titrated at pH 12.

Procedure. Pipette 25 cm^3 of the test solution (which may contain both calcium and lead at concentrations of up to 0.01M) into a 250 cm^3 conical flask and dilute to 100 cm^3 with de-ionised water. Add about 50 mg of the MTB/KNO_3 mixture followed by dilute nitric acid until the solution is yellow, and then add powdered hexamine until the solution has an intense blue colour (pH *ca.* 6). Titrate with standard (0.01M) EDTA solution until the colour turns to yellow; this gives the titration value for lead.

Now carefully add sodium hydroxide solution (1M) until the pH of the solution has risen to 12 (pH meter); 3–6 cm^3 of the sodium hydroxide solution will be required. Continue the titration of the bright blue solution with the EDTA solution until the colour changes to grey; this gives the titration value for calcium.

X, 67. DETERMINATION OF MAGNESIUM, MANGANESE, AND ZINC IN ADMIXTURE: USE OF FLUORIDE ION AS A DEMASKING AGENT.

Discussion. In mixtures of magnesium and manganese the sum of both ion concentrations may be determined by direct EDTA titration. Fluoride ion will demask magnesium selectively from its EDTA complex, and if excess of a standard solution of manganese ion is also added, the following reaction occurs at room temperature:

$$MgY^{2-} + 2F^- + Mn^{2+} = MgF_2 + MnY^{2-}$$

The excess of manganese ion is evaluated by back-titration with EDTA. The amount of standard manganese ion solution consumed is equivalent to the EDTA 'liberated' by the fluoride ion, which is in turn equivalent to the magnesium in the sample.

Mixtures of manganese, magnesium, and zinc can be similarly analysed. The first EDTA end-point gives the sum of the three ions. Fluoride ion is added and the EDTA liberated from the magnesium–EDTA complex is titrated with manganese ion as detailed above. Following the second end-point, cyanide ion is added to displace zinc from its EDTA chelate and to form the stable cyanozincate complex $[Zn(CN)_4]^{2-}$; the liberated EDTA (equivalent to the zinc) is titrated with standard manganese-ion solution.

Details for the analysis of Mn–Mg–Zn mixtures will be given.

Procedure. Prepare a *manganese(II) sulphate solution* (approx. 0.05M) by dissolving 11.15 g of the A.R. solid in 1 litre of de-ionised water; standardise the solution by titration with 0.05M-EDTA solution using Solochrome Black indicator after the addition of 0.25 g hydroxylammonium chloride—see below.

Prepare a *buffer solution* (*pH* 10) by dissolving 8.0 g A.R. ammonium nitrate in 65 cm^3 of de-ionised water and adding 35 cm^3 of concentrated ammonia solution (sp. gr. 0.88).

Pipette 25 cm^3 of the solution containing magnesium, manganese and zinc ions (each approx. 0.02M), into a 250 cm^3 conical flask and dilute to 100 cm^3 with de-ionised water. Add 0.25 g hydroxylammonium chloride (this is to prevent oxidation of Mn(II) ions), followed by 10 cm^3 of the buffer solution and 30–40 mg of the indicator/KNO$_3$ mixture. Warm to 40 °C and titrate (preferably stirring magnetically) with the standard EDTA solution to a pure blue colour.

After the end-point, add 2.5 g of sodium fluoride and stir (or agitate) for 1 minute. Now introduce the standard manganese(II) sulphate solution from a burette in 1-cm^3 portions until a *permanent* red colour is obtained; note the exact volume added. Stir for 1 minute. Titrate the excess of manganese ion with EDTA until the colour changes to pure blue.

After the second end-point, add 4–5 cm^3 of 15 per cent aqueous potassium cyanide solution, and run in the standard manganese-ion solution from a burette until the colour changes sharply from blue to red. Record the exact volume of manganese(II) sulphate solution added.

Calculate the weights of magnesium, zinc, and manganese in the sample solution.

Example of calculation. In the standardisation of the Mn(II) solution, 25.0 cm^3 of the solution required 30.30 cm^3 of 0.0459M-EDTA solution.

∴ Molarity of Mn(II) solution $= (30.30 \times 0.0459)/25.0 = 0.0556M$

First titration of mixture with EDTA (Mg + Mn + Zn) = 33.05 cm

Second titration (after adding NaF; gives Mg) = 9.85 cm^3 of Mn(II) solution and the excess Mn(II) required 1.26 cm^3 of standard EDTA solution.

∴ Millimoles of EDTA liberated by NaF

$$= (9.85 \times 0.0556) - (1.26 \times 0.0459)$$
$$= 0.4899$$

and weight of magnesium per cm^3 $\quad = (0.4899 \times 24.31)/1000$ g
$$= 11.91 \text{ mg cm}^{-3}.$$

Third titration (after adding KCN; gives Zn) = 8.46 cm^3 of Mn(II) solution.

∴ Millimoles of EDTA liberated by KCN

$$= (8.46 \times 0.0556)$$
$$= 0.4703$$

and weight of zinc per cm^3 $\quad = (0.4703 \times 65.38)/1000$ g
$$= 30.75 \text{ mg cm}^{-3}$$

In the first titration, (33.05×0.0459) millimoles of EDTA were used
$$= 1.5170 \text{ millimoles,}$$

which represents the total amount of metal ion titrated (Mn + Mg + Zn).

Hence amount of Mn = $1.5170 - (0.4899 + 0.4703) = 0.5568$ millimoles

and weight of manganese per cm^3 $\quad = (0.5568 \times 54.94)/1000$ g
$$= 30.60 \text{ mg cm}^{-3}.$$

X, 68. DETERMINATION OF CHROMIUM(III) AND IRON(III) IN ADMIXTURE: AN EXAMPLE OF KINETIC MASKING.

Iron (and nickel, if present) can be determined by adding an excess of standard EDTA to the cold solution, and then back-titrating the solution with lead nitrate solution using Xylenol Orange as indicator; provided the solution is kept cold, chromium does not react. The solution from the back titration is then acidified, excess of standard EDTA solution added and the solution boiled for 15 minutes when the red-violet Cr(III)–EDTA complex is produced. After cooling and buffering to pH 6, the excess EDTA is then titrated with the lead nitrate solution.

Procedure. Place 10 cm^3 of the solution containing the two metals (the concentration of neither of which should exceed 0.01 M) in a 600 cm^3 beaker fitted with a magnetic stirrer, and dilute to 100 cm^3 with de-ionised water. Add 20 cm^3 of standard (*approx.* 0.01 M) EDTA solution and add hexamine to adjust the pH to 5–6. Then add a few drops of the indicator solution (0.5 g Xylenol Orange dissolved in 100 cm^3 of water) and titrate the excess EDTA with a standard lead nitrate solution (0.01 M), i.e., to the formation of a red-violet colour.

To the resulting solution now add a further 20 cm^3 portion of the standard EDTA solution, add nitric acid (1 M) to adjust the pH to 1–2, and then boil the solution for 15 minutes. Cool, dilute to 400 cm^3 by the addition of de-ionised water, add hexamine to bring the pH to 5–6, add more of the indicator solution, and titrate the excess EDTA with the standard lead nitrate solution.

The first titration determines the amount of EDTA used by the iron, and the second, the amount of EDTA used by the chromium.

X, 69. DETERMINATION OF MANGANESE IN PRESENCE OF IRON: ANALYSIS OF FERRO-MANGANESE.

After dissolution of the alloy in a mixture of concentrated nitric and hydrochloric acids the iron is masked with triethanolamine in an alkaline medium, and the manganese titrated with standard EDTA solution using Thymolphthalexone as indicator. The amount of iron(III) present must not exceed 25 mg per 100 cm^3 of solution otherwise the colour of the iron(III)–triethanolamine complex is so intense that the colour change of the indicator is obscured. Consequently, the procedure can only be used for samples of ferro-manganese containing more than about 40 per cent manganese.

Procedure. Dissolve a weighed amount of ferro-manganese (about 0.40 g) in concentrated nitric acid and then add concentrated hydrochloric acid (or use a mixture of the two concentrated acids); prolonged boiling may be necessary. Evaporate to a small volume on a water bath. Dilute with water and filter directly into a 100-cm^3 graduated flask, wash with distilled water and finally dilute to the mark. Pipette 25.0 cm^3 of the solution into a 350-cm^3 conical flask, add 5 cm^3 of 10 per cent aqueous hydroxylammonium chloride solution, 10 cm^3 of 20 per cent aqueous triethanolamine solution, 30–35 cm^3 of concentrated ammonia solution, about 100 cm^3 of water, and 6 drops of Thymolphthalexone indicator solution. Titrate with standard 0.05M-EDTA until the colour changes from blue to colourless (or a very pale pink).

X, 70. DETERMINATION OF NICKEL IN PRESENCE OF IRON: ANALYSIS OF NICKEL STEEL.
Nickel may be determined in the presence of a large excess of iron(III) in weakly acidic solution by adding EDTA and triethanolamine; the intense brown precipitate dissolves upon the addition of aqueous sodium hydroxide to yield a colourless solution. The iron(III) is present as the triethanolamine complex and only the nickel is complexed by the EDTA. The excess of EDTA is back-titrated with standard calcium chloride solution in the presence of Thymolphthalexone indicator. The colour change is from colourless or very pale blue to an intense blue. The nickel–EDTA complex has a faint blue colour; the solution should contain less than 35 mg of nickel per 100 cm^3.

In the back-titration small amounts of copper and zinc and trace amounts of manganese are quantitatively displaced from the EDTA and are complexed by the triethanolamine: small quantities of cobalt are converted into a triethanolamine complex during the titration. Relatively high concentrations of copper can be masked in the alkaline medium by the addition of thioglycollic acid until colourless. Manganese, if present in quantities of more than 1 mg may be oxidised by air and forms a manganese(III)–triethanolamine complex, which is intensely green in colour; this does not occur if a little hydroxylammonium chloride solution is added.

Procedure. Prepare a *standard calcium chloride solution (0.01M)* by dissolving 1.000 g of A.R. calcium carbonate in the minimum volume of dilute hydrochloric acid and diluting to 1 dm^3 with de-ionised water in a graduated flask. Also prepare a 20 *per cent aqueous solution of triethanolamine.*

Weigh out accurately a 1.0 g sample of the nickel steel and dissolve it in the minimum volume of concentrated hydrochloric acid (about 15 cm^3) to which a little concentrated nitric acid (*ca.* 1 cm^3) has been added. Dilute to 250 cm^3 in a graduated flask. Pipette 25.0 cm^3 of this solution into a conical flask, add 25.0 cm^3 of 0.01M-EDTA and 10 cm^3 of triethanolamine solution. Introduce 1M-sodium hydroxide solution, with stirring, until the pH of the solution is 11.6 (use a pH meter). Dilute to about 250 cm^3. Add about 0.05 g of the indicator–KNO$_3$ mixture; the solution acquires a very pale blue colour. Titrate with 0.01M-calcium chloride solution until the colour changes to an intense blue. If it is felt that the end-point colour change is not sufficiently distinct, add a further small amount of the indicator, a known volume of 0.01M-EDTA and titrate again with 0.01M-calcium chloride.

X, 71. DETERMINATION OF LEAD AND TIN IN ADMIXTURE: ANALYSIS OF SOLDER. A mixture of tin(IV) and lead(II) ions may be complexed by adding an excess of standard EDTA solution, the excess EDTA being determined by titration with a standard solution of lead nitrate; the total lead plus tin content of the solution is thus determined. Sodium fluoride is now added and this displaces the EDTA from the tin(IV)–EDTA complex; the liberated EDTA is determined by titration with a standard lead solution.

Procedure. Prepare a *standard EDTA solution (0.2M)*, a *standard lead solution (0.01M)*, a *30 per cent aqueous solution of hexamine*, and a *0.2 per cent aqueous solution* of Xylenol Orange.

Dissolve a weighed amount (about 0.4 g) of solder in 10 cm³ concentrated hydrochloric acid and 2 cm³ concentrated nitric acid; gentle warming is necessary. Boil the solution gently for about 5 minutes to expel nitrous fumes and chlorine, and allow to cool slightly, whereupon some lead chloride may separate. Add 25.0 cm³ of standard 0.2M-EDTA and boil for 1 minute; the lead chloride dissolves and a clear solution is obtained. Dilute with 100 cm³ of de-ionised water, cool and dilute to 250 cm³ in a graduated flask. Without delay, pipette two or three 25.0 cm³ portions into separate conical flasks. To each flask add 15 cm³ hexamine solution, 110 cm³ de-ionised water, and a few drops of Xylenol Orange indicator. Titrate with the standard lead nitrate solution until the colour changes from yellow to red. Now add 2.0 g A.R. sodium fluoride; the solution acquires a yellow colour owing to the liberation of EDTA from its tin complex. Titrate again with the standard lead nitrate solution until a permanent (i.e., stable for 1 minute) red colour is obtained. Add the titrant dropwise near the end-point; a temporary pink or red colour gradually reverting to yellow signals the approach of the end-point.

X, 72. DETERMINATION OF BISMUTH, CADMIUM AND LEAD IN ADMIXTURE: ANALYSIS OF A LOW-MELTING ALLOY. The analysis of low-melting alloys such as **Wood's metal** is greatly simplified by complexometric titration, and tedious gravimetric separations are avoided. The alloy is treated with concentrated nitric acid, evaporated to a small volume, and after dilution the precipitated tin(IV) oxide is filtered off; heavy metals adsorbed by the precipitate are removed by washing with a known volume of standard EDTA solution previously made slightly alkaline with aqueous ammonia. The hydrated tin(IV) oxide is ignited and weighed. The Bi, Pb, and Cd are determined in the combined filtrate and washings from the tin separation: these are diluted to a known volume and aliquots used in the subsequent titrations. The Bi content is determined by titration with standard EDTA at pH 1–2 using Xylenol Orange as indicator; then, after adjustment of the pH to 5–6 with hexamine, the combined Pb + Cd can be titrated with EDTA. 1,10-Phenanthroline may now be added to mask the cadmium, and the liberated EDTA is titrated with standard lead nitrate solution; this gives the cadmium content and thence the Pb content is obtained by difference.

Procedure. Prepare a *standard solution of lead nitrate (0.05M)*, a 0.05 per cent *aqueous solution* of Xylenol Orange indicator, and a *1,10-Phenanthroline solution (0.05M)* by dissolving 0.90 g of pure 1,10-phenanthroline in 1.5 cm³ of concentrated nitric acid and 100 cm³ of water.

Weigh out accurately 2.0–2.5 g of Wood's metal and dissolve it in hot concentrated nitric acid (*ca.* 50 cm³). Evaporate the resulting solution to a small volume, dilute to about $150 \, cm^3$ with water and boil for 1–2 minutes. Filter off the precipitate of hydrated tin(IV) oxide through a quantitative filter paper (Whatman No. 542) and keep the filtrate. Render a known volume (say $50.0 \, cm^3$) of a $0.05M$-EDTA solution slightly basic with aqueous ammonia. Wash the precipitate on the filter with this solution and then with $50 \, cm^3$ of water. The final wash liquid should give no precipitate with 5 per cent sodium sulphide solution. Transfer the filtrate and washings (containing the other metals as well as the excess of EDTA) quantitatively to a 500-cm^3 graduated flask, and dilute to the mark with de-ionised water. Char the filter paper in the usual way and, after ignition, weigh the tin(IV) oxide.

Into a conical flask, pipette a 50.0 or $100.0 \, cm^3$ aliquot of the solution and adjust the pH to 1–2 with aqueous ammonia solution (use pH test paper). Add 5 drops of Xylenol Orange indicator and titrate with additional $0.05M$-EDTA until the colour changes sharply from red to yellow. This gives the Bi content. Record the total (combined) volume of EDTA solution used. Now add small amounts of hexamine (*ca.* 5 g) until an intense red-violet coloration persists, and titrate with the standard EDTA to a yellow end-point; the further consumption of EDTA corresponds to the Pb + Cd content.

To determine the Cd content, add $20–25 \, cm^3$ of the 1,10-phenanthroline solution and titrate the liberated EDTA with the $0.05M$-lead nitrate solution until the colour change from yellow to red-violet occurs—a little practice is required to discern the end-point precisely. Introduce further 2–5-cm^3 portions of the 1,10-phenanthroline solution and note whether the indicator colour changes: if so, continue the titration with the lead nitrate solution. The consumption of lead nitrate solution corresponds to the Cd content.

Determination of anions

Anions do not complex directly with EDTA, but methods can be devised for the determination of appropriate anions which involve either adding an excess of a solution containing a cation which reacts with the anion to be determined, and then using EDTA to measure the excess of cation added; alternatively, the anion is precipitated with a suitable cation, the precipitate is collected, dissolved in excess EDTA solution and then the excess EDTA is titrated with a standard solution of an appropriate cation. The procedure involved in the first method will be self-evident but some details are given for determinations carried out by the second method.

X, 73. DETERMINATION OF HALIDES (EXCLUDING FLUORIDE) AND THIOCYANATES. The procedure involved in the determination of these anions is virtually that discussed in Section **X, 59** for the indirect determination of silver. The anion to be determined is precipitated as the silver salt, the precipitate is collected and dissolved in a solution of potassium tetracyanonickelate(II) in the presence of an ammonia/ammonium chloride buffer. Nickel ions are liberated and titrated with standard EDTA solution using murexide as indicator:

$$2Ag^+ + [Ni(CN)_4]^{2-} = Ni^{2+} + 2[Ag(CN)_2]^-$$

The method may be illustrated by the determination of bromide; details for the preparation of the potassium tetracyanonickelate are given in Section **X, 59**.

Pipette 25.0 cm^3 of the bromide ion solution (0.01–0.02M) into a 400-cm^3 beaker, add excess of dilute silver nitrate solution, filter off the precipitated silver bromide on a sintered glass filtering crucible, and wash it with cold water. Dissolve the precipitate in a warm solution prepared from 15 cm^3 of concentrated ammonia solution, 15 cm^3 of 1M-ammonium chloride, and 0.3 g of potassium tetracyanonickelate. Dilute to 100–200-cm^3, add 3 drops of murexide indicator, and titrate with standard EDTA (.01M) (slowly near the end-point) until the colour changes from yellow to violet.

1 mole EDTA $\equiv$ 2 moles Br$^-$

X, 74. DETERMINATION OF PHOSPHATES. The phosphate is precipitated as $Mg(NH_4)PO_4,6H_2O$, the precipitate is filtered off, washed, dissolved in dilute hydrochloric acid, an excess of standard EDTA solution added, the pH adjusted to 10, and the excess of EDTA titrated with standard magnesium chloride or magnesium sulphate solution using Solochrome Black (Eriochrome Black T) as indicator. The initial precipitation may be carried out in the presence of a variety of metals by first adding sufficient EDTA solution (1M) to form complexes with all the polyvalent metal cations, then adding excess of magnesium sulphate solution, followed by ammonia solution: alternatively, the cations may be removed by passing the solution through a cation exchange resin in the hydrogen form.

Procedure. Prepare a *standard* (0.05M) *solution of magnesium sulphate or chloride* from pure magnesium (Section **X, 62**), an ammonia–ammonium chloride *buffer solution* (*pH* 10) (Section **X, 55**), and a *standard* (0.05M) *solution of EDTA.*

Pipette 25.0 cm^3 of the phosphate solution (*approx.* 0.05M) into a 250 cm^3 beaker and dilute to 50 cm^3 with de-ionised water; add 1 cm^3 of concentrated hydrochloric acid and a few drops of methyl red indicator. Treat with an excess of 1M-magnesium sulphate solution (*ca.* 2 cm^3), heat the solution to boiling, and add concentrated ammonia solution dropwise and with vigorous stirring until the indicator turns yellow, followed by a further 2 cm^3. Allow to stand for several hours or overnight. Filter the precipitate through a sintered-glass crucible (porosity G4) and wash thoroughly with 1M-ammonia solution (about 100 cm^3). Rinse the beaker (in which the precipitation was made) with 25 cm^3 of hot 1M-hydrochloric acid and allow the liquid to percolate through the filter crucible, thus dissolving the precipitate. Wash the beaker and crucible with a further 10 cm^3 of 1M-hydrochloric acid and then with about 75 cm^3 of water. To the filtrate and washings in the filter flask add 35.0 cm^3 of 0.05M-EDTA, neutralise the solution with 1M-sodium hydroxide, add 4 cm^3 of buffer solution and a few drops of Solochrome Black (Eriochrome Black T) indicator. Back-titrate with standard 0.05M-magnesium chloride until the colour changes from blue to wine red.

X, 75. DETERMINATION OF SULPHATES. The sulphate is precipitated as barium sulphate from acid solution, the precipitate is filtered off and dissolved in a measured excess of standard EDTA solution in the presence of aqueous ammonia. The excess of EDTA is then titrated with standard magnesium chloride solution using Solochrome Black (Eriochrome Black T) as indicator.

Procedure. Prepare a *standard magnesium chloride solution* (0.05M) and a *buffer solution* (*pH* 10); see previous Section. *Standard EDTA* (0.05M) will also be required.

Pipette 25.0 cm³ of the sulphate solution (0.02–0.03*M*) into a 250-cm³ beaker, dilute to 50 cm³, and adjust the pH to 1 with 2*M*-hydrochloric acid; heat nearly to boiling. Add 15 cm³ of a nearly boiling barium chloride solution (*ca.* 0.05*M*) fairly rapidly and with vigorous stirring: heat on a steam bath for 1 hour. Filter with suction through a filter-paper disc (Whatman filter paper No. 42) supported upon a porcelain filter disc or a Gooch crucible, wash the precipitate thoroughly with cold water, and drain. Transfer the filter-paper disc and precipitate quantitatively to the original beaker, add 35.0 cm³ standard 0.05*M*-EDTA solution, 5 cm³ concentrated ammonia solution and boil gently for 15–20 minutes; add a further 2 cm³ concentrated ammonia solution after 10–15 minutes to facilitate the dissolution of the precipitate. Cool the resulting clear solution, add 10 cm³ of the buffer solution (pH = 10), a few drops of Solochrome Black (Eriochrome Black T) indicator, and titrate the excess of EDTA with the standard magnesium chloride solution to a clear red colour.

Sulphate can also be determined by an exactly similar procedure by precipitation as lead sulphate from a solution containing 50 per cent (by volume) of propan-2-ol (to reduce the solubility of the lead sulphate), separation of the precipitate, dissolution in excess of standard EDTA solution, and back-titration of the excess EDTA with a standard zinc solution using Solochrome Black (Eriochrome Black T) as indicator.

B. 3 PRECIPITATION TITRATIONS

In the following Sections we are concerned with the use of standard solutions of reagents such as silver nitrate, sodium chloride, potassium (or ammonium) thiocyanate, and potassium cyanide. As already pointed out, some of the determinations which will be considered strictly involve complex formation rather than precipitation reactions, but it is convenient to group them here as reactions involving the use of standard silver nitrate solutions. Before commencing the experimental work, the theoretical Sections **X, 29** and **X, 30** should be studied.

X, 76. PREPARATION OF 0.1*M*-SILVER NITRATE. *Discussion.* Very pure silver can be obtained commercially, and a standard solution can be prepared by dissolving a known weight (say, 10.787 g) in pure dilute nitric acid in a conical flask having a funnel in the neck to prevent mechanical loss, and making up to a known volume (say, 1 dm³ for a 0.1*M* solution). The presence of acid must, however, be avoided in determinations with potassium chromate as indicator or in determinations employing adsorption indicators. It is therefore preferable to employ a neutral solution prepared by dissolving silver nitrate (molecular weight, 169.87) in water.

A.R. Silver nitrate has a purity of at least 99.9 per cent, so that a standard solution can be prepared by direct weighing. If, however, commercial re-crystallised silver nitrate be employed, or if an additional check of the molarity of the silver nitrate solution is required, standardisation may be effected with pure sodium chloride. A.R. Sodium chloride has a purity of 99.9–100.0 per cent; the substance is therefore an excellent primary standard. Sodium chloride is very slightly hygroscopic, and for accurate work it is best to dry the finely powdered solid in an electric oven at 250–350 °C for 1–2 hours, and allow it to cool in a desiccator.

Procedure. **From A.R. silver nitrate.** Dry some finely powdered A.R. silver nitrate at 120 °C for 2 hours and allow it to cool in a covered vessel in a desiccator. Weigh out accurately 8.494 g, dissolve it in water and make up to 500 cm^3 in a graduated flask. This gives a 0.1000M solution. Alternatively, about 8.5 g of pure, dry silver nitrate may be weighed out accurately, dissolved in 500 cm^3 of water in a graduated flask, and the molar concentration calculated from the weight of silver nitrate employed.

In many cases the A.R. material may be replaced by 'pure recrystallised' silver nitrate, but in that case it is advisable to standardise the solution against sodium chloride. Solutions of silver nitrate should be protected from light and are best stored in amber-coloured glass bottles.

X, 77.   STANDARDISATION OF THE SILVER NITRATE SOLUTION.

Sodium chloride has a molecular weight of 58.44. A 0.1000M solution is prepared by weighing out 2.922 g of the pure dry A.R. salt (see Section **X, 76**) and dissolving it in 500 cm^3 of water in a graduated flask. Alternatively about 2.9 g of the pure salt is accurately weighed out, dissolved in 500 cm^3 of water in a graduated flask, and the molar concentration calculated from the weight of sodium chloride employed.

A. With potassium chromate as indicator. The Mohr titration. The reader is referred to Section **X, 30** for the detailed theory of the titration. Prepare the *indicator solution* by dissolving 5 g A.R. potassium chromate in 100 cm^3 of water. The final volume of the solution in the titration is 50–100 cm^3, and 1 cm^3 of the indicator solution is used, so that the indicator concentration in the actual titration is 0.005–0.0025M.

Alternatively, and preferably, dissolve 4.2 g A.R. potassium chromate and 0.7 g A.R. potassium dichromate in 100 cm^3 of water; use 1 cm^3 of indicator solution for each 50 cm^3 of the final volume of the test solution.

Pipette 25 cm^3 of the standard 0.1M-sodium chloride into a 250-cm^3 conical flask resting upon a white tile (1), and add 1 cm^3 of the indicator solution (preferably with a 1-cm^3 pipette). Add the silver nitrate solution slowly from a burette, swirling the liquid constantly, until the red colour formed by the addition of each drop begins to disappear more slowly: this is an indication that most of the chloride has been precipitated. Continue the addition dropwise until a faint but distinct change in colour occurs. This *faint* reddish-brown colour should persist after brisk shaking. If the end-point is overstepped (production of a deep reddish-brown colour), add more of the chloride solution and titrate again. Determine the indicator blank correction by adding 1 cm^3 of the indicator to a volume of water equal to the final volume in the titration (2), and then 0.01M-silver nitrate solution until the colour of the blank matches that of the solution titrated. The indicator blank correction, which should not amount to more than 0.03–0.10 cm^3 of silver nitrate, is deducted from the volume of silver nitrate used in the titration. Repeat the titration with two further 25-cm^3 portions of the sodium chloride solution. The various titrations should agree within 0.1 cm^3.

Notes.   1. The end-point is very readily detected in a large porcelain basin. The solution is stirred with a short glass stirring rod.

2. A better blank is obtained by adding about 0.5 g of A.R. calcium carbonate before determining the correction. This gives an inert white precipitate similar to

that obtained in the titration of chlorides and materially assists in matching the colour tints.

B. With an adsorption indicator. *Discussion.* The detailed theory of the process is given in Section **X, 30**. Both fluorescein and dichlorofluorescein are suitable for the titration of chlorides. In both cases the end-point is reached when the white precipitate in the greenish-yellow solution suddenly assumes a pronounced reddish tint. The change is reversible upon the addition of chloride. With fluorescein the solution must be neutral or only faintly acidic with acetic acid; acid solutions should be treated with a slight excess of sodium acetate. The chloride solution should be diluted to about $0.01–0.05M$, for if it is more concentrated the precipitate coagulates too soon and interferes. Fluorescein cannot be used in solutions more dilute than $0.005M$. With more dilute solutions resort must be made to dichlorofluorescein, which possesses other advantages over fluorescein. Dichlorofluorescein gives good results in very dilute solutions (e.g., for drinking water) and is applicable in the presence of acetic acid and in weakly acid solutions. For this reason the chlorides of copper, nickel, manganese, zinc, aluminium, and magnesium, which cannot be titrated according to the method of Mohr, can be determined by a direct titration when dichlorofluorescein is used as indicator.

For the reverse titration (chloride into silver nitrate), tartrazine (4 drops of a 0.2 per cent solution per 100 cm^3) is a good indicator. At the end-point, the almost colourless liquid assumes a blue colour.

The indicator solutions are prepared as follows:

Fluorescein. Dissolve 0.2 g fluorescein in 100 cm^3 of 70 per cent ethanol, or dissolve 0.2 g sodium fluoresceinate in 100 cm^3 of water.

Dichlorofluorescein. Dissolve 0.1 g dichlorofluorescein in 100 cm^3 of 60–70 per cent ethanol, or dissolve 0.1 g sodium dichlorofluoresceinate in 100 cm^3 of water.

Procedure. Pipette 25 cm^3 of the standard $0.1M$-sodium chloride into a 250-cm^3 conical flask. Add 10 drops of either fluorescein or dichlorofluorescein indicator, and titrate with the silver nitrate solution in a diffuse light, while rotating the flask constantly. As the end point is approached, the silver chloride coagulates appreciably, and the local development of a pink colour upon the addition of a drop of the silver nitrate solution becomes more and more pronounced. Continue the addition of the silver nitrate solution until the precipitate suddenly assumes a pronounced pink or red colour. Repeat the titration with two other 25-cm^3 portions of the chloride solution. Individual titrations should agree within 0.1 cm^3.

Calculate the molar concentration of the silver nitrate solution.

X, 78. DETERMINATION OF CHLORIDES. Either the Mohr titration or the adsorption-indicator method may be used for the determination of chlorides in neutral solution by titration with standard $0.1M$-silver nitrate. If the solution is acid, neutralisation may be effected with chloride-free calcium carbonate, sodium tetraborate, or sodium hydrogen carbonate; the A.R. substances are suitable. Mineral acid may also be removed by neutralising most of the acid with ammonia solution and then adding an excess of A.R. ammonium acetate. Titration of the neutral solution, prepared with calcium carbonate, by the adsorption indicator method is rendered easier by the addition of 5 cm^3 of 2 per cent dextrin solution; this offsets the coagulating effect of the calcium ion. If the solution is basic, it may be neutralised with chloride-free nitric acid, using phenolphthalein as indicator.

Similar remarks apply to the **determination of bromides**; the Mohr titration can be used, and the most suitable adsorption indicator is **eosin** which can be used in dilute solutions and even in the presence of $0.1M$ nitric acid, but in general, acetic acid solutions are preferred. Fluorescein may be used but is subject to the same limitations as experienced with chlorides (Section **X, 77, B**). With eosin indicator, the silver bromide flocculates approximately 1 per cent before the equivalence point and the local development of a red colour becomes more and more pronounced with the addition of silver nitrate solution: at the end-point the precipitate assumes a magenta colour.

The **indicator** is prepared by dissolving 0.1 g eosin in 100 cm^3 of 70 per cent ethanol, or by dissolving 0.1 g of the sodium salt in 100 cm^3 of water.

For the reverse titration (bromide into silver nitrate), rhodamine 6G (10 drops of a 0.05 per cent aqueous solution) is an excellent indicator. The solution is best adjusted to $0.05M$ with respect to silver ion. The precipitate acquires a violet colour at the end-point.

Thiocyanates may also be determined using adsorption indicators in exactly similar manner to chlorides and bromides, but an iron(III) salt indicator is usually preferred (Section **X, 82**).

X, 79.   DETERMINATION OF IODIDES. *Discussion.* The Mohr method cannot be applied to the titration of iodides (or of thiocyanates), because of adsorption phenomena and the difficulty of distinguishing the colour change of the potassium chromate. Eosin is a suitable adsorption indicator, but di-iododimethylfluorescein is better. Eosin is employed as described under bromides (Section **X, 78**).

The **di-iododimethylfluorescein indicator** is prepared by dissolving 1.0 g in 100 cm^3 of 70 per cent ethanol. The colour change is from an orange-red to a blue-red on the precipitate.

X, 80.   DETERMINATION OF MIXTURES OF HALIDES WITH ADSORPTION INDICATORS.   A. Chloride and iodide in admixture. These two ions differ considerably in the ease with which they are adsorbed on the corresponding silver halide. This makes it possible to select adsorption indicators which will permit the determination of chloride and iodide in the presence of one another. Thus the iodide may be determined by titration with standard $0.1M$-silver nitrate using di-iododimethylfluorescein and the iodide + chloride by a similar titration using fluorescein. Chloride is obtained by difference. If a large excess of chloride is present the result for iodide may be as much as 1 per cent high. If, however, Rose Bengal (dichlorotetraiodofluorescein) is used as indicator (colour change, carmine red to blue-red) in the presence of ammonium carbonate, the iodide titration is exact.

B. Bromide and iodide in admixture. The total halide (bromide + iodide) is determined by titration with standard $0.1M$-silver nitrate using eosin or fluorescein as indicator. The iodide is determined by titration with $0.01–0.2M$-silver nitrate, using di-iododimethylfluorescein as indicator. Bromide is obtained by difference.

Numerous adsorption indicators have been suggested for various purposes, but a full treatment is outside the scope of this work.

X, 81. DETERMINATION OF MIXTURES OF HALIDES BY AN INDIRECT METHOD. *Discussion.* The method is applicable to the determination of a mixture of two salts having the same anion (e.g., sodium and potassium chlorides) or the same cation (e.g., potassium chloride and potassium bromide). Let us first suppose that it is desired to determine the amount of sodium and potassium chlorides in a mixture of the two salts. A known weight (w_1 g) of the solid mixture is taken, and the total chloride is determined with standard $0.1M$-silver nitrate, using Mohr's method or an adsorption indicator. Let w_2 g of silver nitrate be required for the complete precipitation of w_1 g of the mixture, which contains x g of NaCl and y g of KCl. Then:

$$x + y = w_1$$

$$\frac{169.87x}{58.44} + \frac{169.87y}{74.55} = w_2$$

Upon solving these two simultaneous equations, the values for x and y are deduced.

Let us now suppose that the determination of potassium chloride and potassium bromide in admixture is desired. The total halide is determined by Mohr's method or with an adsorption indicator. Let the weight of the mixture be w_3 g, w_4 g be the weight of silver nitrate required for complete precipitation, p g be the weight of the potassium chloride, and q g be the weight of the potassium bromide. Then:

$$p + q = w_3$$

$$\frac{169.87p}{74.55} + \frac{169.87q}{119.00} = w_4$$

The values of p and q can be obtained by solving the simultaneous equations.

It can be shown that the method depends upon the difference between the molecular weights of the two components of the mixture and that *inter alia* it is most satisfactory when the two constituents are present in approximately equal proportions.

X, 82. PREPARATION AND USE OF 0.1M-AMMONIUM OR POTASSIUM THIOCYANATE. **Titrations according to Volhard's method.** *Discussion.* Volhard's original method for the determination of silver in *dilute nitric acid* solution by titration with standard thiocyanate solution in the presence of an iron(III) salt as indicator has proved of great value not only for silver determinations, but also in numerous indirect analyses. The theory of the Volhard process has been given in Section **X, 30**. In this connection it must be pointed out that the concentration of the nitric acid should be from 0.5–1.5M (strong nitric acid retards the formation of the thiocyanatoiron(III) complex $[FeSCN]^{2+}$) and at a temperature not exceeding 25 °C (higher temperatures tend to bleach the colour of the indicator). The solutions must be free from nitrous acid, which gives a red colour with thiocyanic acid, and may be mistaken for 'iron(III) thiocyanate'. Pure nitric acid is prepared by diluting the usual pure (e.g., A.R.) acid with about one-fourth of its volume of water and boiling until perfectly colourless; this eliminates any lower oxides of nitrogen which may be present.

The method may be applied to those anions (e.g., chloride, bromide, and iodide) which are completely precipitated by silver and are sparingly soluble in dilute nitric acid. Excess of standard silver nitrate solution is added to the solution containing free nitric acid, and the residual silver nitrate solution is titrated with standard thiocyanate solution. This is sometimes termed the **residual process**. Anions whose silver salts are slightly soluble in water, but which are soluble in nitric acid, such as phosphate, arsenate, chromate, sulphide, and oxalate, may be precipitated in neutral solution with an excess of standard silver nitrate solution. The precipitate is filtered off, thoroughly washed, dissolved in dilute nitric acid, and the silver titrated with thiocyanate solution. Alternatively, the residual silver nitrate in the filtrate from the precipitation may be determined with thiocyanate solution after acidification with dilute nitric acid.

Both ammonium and potassium thiocyanates are usually available as deliquescent solids; the A.R. products are, however, free from chlorides and other interfering substances. An approximately $0.1M$ solution is therefore first prepared, and this is standardised by titration against standard $0.1M$-silver nitrate.

Procedure. Weigh out about 8.5 g A.R. ammonium thiocyanate, or 10.5 g A.R. potassium thiocyanate, and dissolve it in 1 litre of water in a graduated flask. Shake well.

Standardisation. Use $0.1M$-silver nitrate, which has been prepared and standardised as described in Section **X, 77**.

The **iron(III) indicator solution** consists of a cold, saturated solution of A.R. ammonium iron(III) sulphate in water (about 40 per cent) to which a few drops of $6M$-nitric acid has been added. One cm^3 of this solution is employed for each titration.

Pipette 25 cm^3 of the standard $0.1M$-silver nitrate into a 250 cm^3 conical flask, add 5 cm^3 of $6M$-nitric acid and 1 cm^3 of the iron(III) indicator solution. Run in the potassium or ammonium thiocyanate solution from a burette. At first a white precipitate is produced, rendering the liquid of a milky appearance, and as each drop of thiocyanate falls in, it produces a reddish-brown cloud, which quickly disappears on shaking. As the end-point approaches, the precipitate becomes flocculent and settles easily; finally one drop of the thiocyanate solution produces a faint brown colour, which no longer disappears upon shaking. This is the end-point. The indicator blank amounts to 0.01 cm^3 of $0.1M$-silver nitrate. It is essential to shake vigorously during the titration in order to obtain correct results.*

The standard solution thus prepared is stable for a very long period if evaporation is prevented.

Use of tartrazine as indicator. Satisfactory results may be obtained by the use of tartrazine as indicator. Proceed as above, but add 4 drops of tartrazine (0.5 per cent aqueous solution) in lieu of the iron(III) indicator. The precipitate will appear pale yellow during the titration, but the supernatant liquid (best viewed by placing the eye at the level of the liquid and looking through it) is colourless. At the end-point, the supernatant liquid assumes a bright lemon-yellow colour. The titration is sharp to one drop of $0.1M$-thiocyanate solution.

---

* The freshly precipitated silver thiocyanate adsorbs silver ions, thereby causing a false end-point which, however, disappears with vigorous shaking.

X, 83. DETERMINATION OF SILVER IN A SILVER ALLOY. A commercial silver alloy in the form of wire or foil is suitable for this determination. Clean the alloy with emery cloth and weigh it accurately. Place it in a 250-cm^3 conical flask, add 5 cm^3 water and 10 cm^3 concentrated nitric acid; place a funnel in the mouth of the flask to avoid mechanical loss. Warm the flask gently until the alloy has dissolved. Add a little water and boil for 5 minutes in order to expel oxides of nitrogen. Transfer the cold solution quantitatively to a 100-cm^3 standard flask and make up to the mark with distilled water. Titrate 25-cm^3 portions of the solution with standard 0.1M-thiocyanate.

$$1 \text{ mole KSCN} \equiv 1 \text{ mole Ag}^+$$

Note. The presence of metals whose salts are colourless does not influence the accuracy of the determination, except that mercury and palladium must be absent since their thiocyanates are insoluble. Salts of metals (e.g., nickel and cobalt) which are coloured must not be present to any considerable extent. Copper does not interfere, provided it does not form more than about 40 per cent of the alloy.

X, 84. DETERMINATION OF CHLORIDES (VOLHARD'S METHOD). *Discussion.* The chloride solution is treated with excess of standard silver nitrate solution, and the residual silver nitrate determined by titration with standard thiocyanate solution. Now silver chloride is more soluble than silver thiocyanate, and would react with the thiocyanate thus:

$$AgCl\,(\text{solid}) + SCN^- \rightleftharpoons AgSCN\,(\text{solid}) + Cl^-$$

It is therefore necessary to remove the silver chloride by filtration. The filtration may be avoided by the addition of a little nitrobenzene (about 1 cm^3 for each 0.05 g of chloride); the silver chloride particles are probably surrounded by a film of nitrobenzene. Another method, applicable to chlorides, in which filtration of the silver chloride is unnecessary, is to employ tartrazine as indicator (Section **X, 82**).

Procedure A **(HCl content of concentrated hydrochloric acid).** Ordinary concentrated hydrochloric acid is usually 10–11M, and must be diluted first. Measure out accurately 10 cm^3 of the concentrated acid from a burette into a 1-dm^3 graduated flask and make up to the mark with distilled water. Shake well. Pipette 25 cm^3 into a 250-cm^3 conical flask, add 5 cm^3 6M-nitric acid and then add 30 cm^3 standard 0.1M-silver nitrate (or sufficient to give 2–5 cm^3 excess). Shake to coagulate the precipitate,* filter through a quantitative filter paper (or through a porous porcelain or sintered-glass crucible), and wash thoroughly with very dilute nitric acid (1:100). Add 1 cm^3 of the iron(III) indicator solution to the combined filtrate and washings, and titrate the residual silver nitrate with standard 0.1M-thiocyanate.

Calculate the volume of standard 0.1M-silver nitrate that has reacted with the hydrochloric acid, and therefrom the percentage of HCl in the sample employed.

Procedure B. Pipette 25 cm^3 of the diluted solution into a 250-cm^3 conical flask containing 5 cm^3 6M-nitric acid. Add a slight excess of standard 0.1M-silver

---

* It is better to boil the suspension for a few minutes to coagulate the silver chloride and thus remove most of the adsorbed silver ions from its surface before filtration.

nitrate (about 30 cm^3 in all) from a burette. Then add 2–3 cm^3 pure (e.g., A.R.) nitrobenzene and 1 cm^3 of the iron(III) indicator, and shake vigorously to coagulate the precipitate. Titrate the residual silver nitrate with standard 0.1M-thiocyanate until a permanent faint reddish-brown coloration appears.

From the volume of silver nitrate solution added, subtract the volume of silver nitrate solution that is equivalent to the volume of standard thiocyanate required. Then calculate the percentage of HCl in the sample.

Procedure C. Pipette 25 cm^3 of the diluted solution into a 250-cm^3 conical flask containing 5 cm^3 of 6M-nitric acid, add a slight excess of 0.1M-silver nitrate (30–35 cm^3) from a burette, and 4 drops of tartrazine indicator (0.5 per cent aqueous solution). Shake the suspension for about a minute in order to ensure that the indicator is adsorbed on the precipitate as far as possible. Titrate the residual silver nitrate with standard 0.1M-ammonium or potassium thiocyanate with swirling of the suspension until the very pale yellow supernatant liquid (viewed with the eye at the level of the liquid) assumes a rich lemon-yellow colour.

Bromides can likewise be determined by the Volhard method, but as silver bromide is less soluble than silver thiocyanate it is not necessary to filter off the silver bromide (compare chloride). The bromide solution is acidified with dilute nitric acid, an excess of standard 0.1M-silver nitrate added, the mixture thoroughly shaken, and the residual silver nitrate determined with standard 0.1M-ammonium or potassium thiocyanate, using iron(III) alum as indicator.

Iodides can also be determined by this method, and in this case too there is no need to filter off the silver halide, since silver iodide is very much less soluble than silver thiocyanate. In this determination the iodide solution must be very dilute in order to reduce adsorption effects. The dilute iodide solution (*ca.* 300 cm^3), acidified with dilute nitric acid, is treated very slowly and with vigorous stirring or shaking with standard 0.1M-silver nitrate until the yellow precipitate coagulates and the supernatant liquid appears colourless. Silver nitrate is then present in excess. One cm^3 of iron(III) alum solution is added, and the residual silver nitrate is titrated with standard 0.1M-ammonium or potassium thiocyanate.

X, 85. DETERMINATION OF FLUORIDE; PRECIPITATION AS LEAD CHLOROFLUORIDE COUPLED WITH VOLHARD TITRATION. *Discussion.* This method is based upon the precipitation of lead chlorofluoride, in which the chlorine is determined by Volhard's method, and from this result the fluorine content can be calculated. The advantages of the method are: the precipitate is granular, settles readily, and is easily filtered; the factor for conversion to fluorine is low; the procedure is carried out at pH 3.6–5.6, so that substances which might be co-precipitated, such as phosphates, sulphates, chromates, and carbonates, do not interfere. Aluminium must be entirely absent, since even very small quantities cause low results; a similar effect is produced by boron (>0.05 g), ammonium (>0.5 g), and sodium or potassium (>10 g) in the presence of about 0.1 g of fluoride. Iron must be removed, but zinc is without effect. Silica does not vitiate the method, but causes difficulties in filtration.

Procedure. Pipette 25.0 cm^3 of the solution containing between 0.01–0.1 g fluoride into a 400-cm^3 beaker, add 2 drops of bromo-phenol blue indicator, 3 cm^3 of 10 per cent sodium chloride, and dilute the mixture to 250 cm^3. Add dilute nitric acid until the colour just changes to yellow, and then add dilute sodium hydroxide solution until the colour just changes to blue. Treat with 1 cm^3 of concentrated hydrochloric acid, then with 5.0 g of A.R. lead nitrate, and heat on the steam bath. Stir gently until the lead nitrate has dissolved, and then immediately add 5.0 g crystallised sodium acetate and stir vigorously. Digest on the steam bath for 30 minutes, with occasional stirring, and allow to stand overnight.

Meanwhile, a washing solution of lead chlorofluoride is prepared as follows. Add a solution of 10 g of lead nitrate in 200 cm^3 of water to 100 cm^3 of a solution containing 1.0 g of sodium fluoride and 2 cm^3 of concentrated hydrochloric acid, mix it thoroughly, and allow the precipitate to settle. Decant the supernatant liquid, wash the precipitate by decantation with 5 portions of water, each of about 200 cm^3. Finally add 1 litre of water to the precipitate, shake the mixture at intervals during an hour, allow the precipitate to settle, and filter the liquid. Further quantities of wash liquid may be prepared as needed by treating the precipitate with fresh portions of water. The solubility of lead chlorofluoride in water is 0.325 g dm^{-3} at 25 °C.

Separate the original precipitate by decantation through a Whatman No. 542 or No. 42 paper. Transfer the precipitate to the filter, wash once with cold water, four or five times with the saturated solution of lead chlorofluoride, and finally once more with cold water. Transfer the precipitate and paper to the beaker in which precipitation was made, stir the paper to a pulp in 100 cm^3 of 5 per cent nitric acid, and heat on the steam bath until the precipitate has dissolved (5 minutes). Add a slight excess of standard 0.1M-silver nitrate, digest on the steam bath for a further 30 minutes, and allow to cool to room temperature while protected from the light. Filter the precipitate of silver chloride through a sintered glass crucible, wash with a little cold water, and titrate the residual silver nitrate in the filtrate and washings with standard 0.1M-thiocyanate. Subtract the amount of silver found in the filtrate from that originally added. The difference represents the amount of silver that was required to combine with the chlorine in the lead chlorofluoride precipitate.

1 Mole AgNO$_3$ ≡ 1 Mole F$^-$

X, 86. DETERMINATION OF ARSENATES. *Discussion.* Arsenates in solution are precipitated as silver arsenate, Ag$_3$AsO$_4$, by the addition of neutral silver nitrate solution: the solution must be neutral, or if slightly acid, an excess of sodium acetate must be present to reduce the acidity; if strongly acid, most of the acid should be neutralised by aqueous sodium hydroxide. The silver arsenate is dissolved in dilute nitric acid, and the silver titrated with standard thiocyanate solution. The silver arsenate has nearly six times the weight of the arsenic, hence quite small amounts of arsenic may be determined by this procedure.

Arsenites may also be determined by this procedure but must first be oxidised by treatment with nitric acid. Small amounts of antimony and tin do not interfere, but chromates, phosphates, molybdates, tungstates, and vanadates, which precipitate as the silver salts, should be absent. An excessive amount of ammonium salts has a solvent action on the silver arsenate.

Procedure. Place 25 cm^3 of the arsenate solution in a 250-cm^3 beaker, add an equal volume of distilled water and a few drops of phenolphthalein solution. Add sufficient sodium hydroxide solution to give an alkaline reaction, and then discharge the red colour from the solution by just acidifying with acetic acid. Add a slight excess of silver nitrate solution with vigorous stirring, and allow the precipitate to settle in the dark. Pour off the supernatant liquid through a sintered glass crucible, wash the precipitate by decantation with cold distilled water, transfer the precipitate to the crucible, and wash it free from silver nitrate solution. Wash out the receiver thoroughly. Dissolve the silver arsenate in dilute nitric acid (*ca. M*) (which leaves any silver chloride undissolved), wash with very dilute nitric acid, and make up the filtrate and washings to 250 cm^3 in a graduated flask. Titrate a convenient aliquot portion with standard ammonium (or potassium) thiocyanate solution in the presence of iron(III) alum as indicator.

3 moles KSCN ≡ 1 mole AsO_4^{3-}

X, 87. DETERMINATION OF CYANIDES. *Discussion.* The theory of the titration of cyanides with silver nitrate solution has been given in Section **X, 22**. All silver salts except the sulphide are readily soluble in excess of a solution of an alkali cyanide, hence chloride, bromide, and iodide do not interfere. The only difficulty in obtaining a sharp end-point lies in the fact that silver cyanide is often precipitated in a curdy form which does not readily re-dissolve, and, moreover, the end-point is not easy to detect with accuracy.

There are two methods for overcoming these disadvantages. In the first the precipitation of silver cyanoargentate at the end-point can be avoided by the addition of ammonia solution, in which it is readily soluble and if a little potassium iodide solution is added before the titration is commenced, sparingly soluble silver iodide, which is insoluble in ammonia solution, will be precipitated at the end point. The precipitation is best seen by viewing against a black background.

In the second method diphenylcarbazide is employed as an adsorption indicator. The end-point is marked by the pink colour becoming pale violet (amost colourless) on the colloidal precipitate in dilute solution (*ca.* 0.01*M*) before the opalescence is visible. In 0.1*M* solutions, the colour change is observed on the precipitated particles of silver cyanoargentate.

Procedure. **NOTE. Potassium cyanide and all other cyanides are deadly poisons, and extreme care must be taken in their use. Details for the disposal of cyanides and other dangerous and toxic chemicals may be found in Refs 22 and 23.**

For practice in the method, the cyanide content of potassium cyanide (laboratory reagent grade) may be determined.

Method A. Weigh out accurately about 3.5 g potassium cyanide from a glass-stoppered weighing bottle, dissolve it in water and make up to 250 cm^3 in a graduated flask. Shake well. Transfer 25.0 cm^3 of this solution **by means of a burette and NOT a pipette** to a 250-cm^3 conical flask, add 75 cm^3 water, 5–6 cm^3 6*M*-ammonia solution, and 2 cm^3 10 per cent potassium iodide solution. Place the flask on a sheet of black paper, and titrate with standard 0.1*M*-silver nitrate. Add the silver nitrate solution dropwise as soon as the yellow colour of silver iodide shows any signs of persisting. When 1 drop produces a permanent turbidity, the end-point has been reached.

Method B. Prepare the solution and transfer 25 cm^3 of it to a 250-cm^3 conical flask as detailed under *Method A*. Add 2 to 3 drops of diphenylcarbazide indicator and titrate with standard 0.1*M*-silver nitrate solution until a permanent violet colour is just produced.

The diphenylcarbazide indicator is prepared by dissolving 0.1 g of the solid in 100 cm^3 of ethanol.

1 mole AgNO$_3$ = 2 moles CN$^-$.

X, 88. DETERMINATION OF CHLORIDES BY TITRATION WITH MERCURY(II) NITRATE SOLUTION. *Discussion.* When chloride ions are titrated with mercury(II) ions (a solution of mercury(II) nitrate acidified with nitric acid), the reaction

$$Hg^{2+} + 2Cl^- \rightleftharpoons HgCl_2$$

is essentially stoichiometric. The reaction is not strictly a precipitation reaction, but it is convenient to include this alternative method of determining chlorides with the argentimetric method of Section **X, 84**. The end-point may be detected with diphenylcarbazone, which forms a blue-violet complex with mercury(II) ions. Alternatively, a mixture of diphenylcarbazone and bromophenol blue may be used; the bromophenol blue changes from blue (alkaline) to yellow (acid) at *ca.* pH 3.6, which is the acidity sometimes recommended for the titration in dilute solutions. At the equivalence point the yellow colour of the solution becomes blue-violet owing to the reaction of the excess of mercury(II) ions with diphenylcarbazone. Satisfactory results can, however, be obtained in the pH range 3–8.

The mercurimetric method may be applied to the titration of chlorides in very dilute solutions—down to 0–100 p.p.m. range. Bromides, thiocyanates, and cyanides may be determined by similar methods, but there is no particular advantage over the usual silver procedures. Sodium, potassium, calcium, magnesium, aluminium, manganese, zinc, fluoride, sulphate, nitrate, and acetate individually in concentrations at least equal to that of the chloride do not interfere: chromate and iron(III) ions react with diphenylcarbazone and must be removed if present.

The main advantage of the mercurimetric method of determining chloride, i.e., its applicability to very dilute solutions of chloride, is only realised to the maximum by working in an 80 per cent ethanolic medium; in purely aqueous solution the end-point is not very sharp. As the reaction is not strictly stoichiometric, it is necessary to standardise the mercury(II) nitrate reagent against sodium (or potassium) chloride.

Procedure. Prepare a *mercury(II) nitrate solution* (0.02N) by dissolving 3.4 g recrystallised mercury(II) nitrate Hg(NO$_3$)$_2$,H$_2$O, in 800 cm^3 distilled water containing 20 cm^3 2M nitric acid. Dilute to 1 litre in a graduated flask and then standardise with A.R. sodium chloride as described below. Prepare the *indicator* by dissolving 0.1 g diphenylcarbazone in 100 cm^3 of ethanol, and the *mixed indicator* by dissolving 0.5 g diphenylcarbazone and 0.5 g bromophenol blue in 100 cm^3 of 95 per cent ethanol.

To standardise the mercury(II) nitrate solution, weigh accurately about 1.5 g A.R. sodium chloride and dissolve in 1 dm^3 distilled water in a graduated flask. To 25.0 cm^3 of this solution add 1 cm^3 of the diphenylcarbazone indicator, and titrate with the 0.02N mercury(II) nitrate solution until the first permanent blue-purple coloration appears. Repeat the titration with a fresh portion of the solution but using the mixed indicator, and decide which gives the most easily recognisable change at the end-point. At this concentration of chloride the end-point should be quite sharp in purely aqueous solution, but at lower concentrations about 80 cm³ of 95 per cent ethanol should be added before starting the titration.

The determination of the chloride content of a given solution will be apparent from the above standardisation details.

X, 89. DETERMINATION OF POTASSIUM. *Discussion.* Potassium may be precipitated with excess of sodium tetraphenylborate solution as potassium tetraphenylborate. The excess of reagent is determined by titration with mercury(II) nitrate solution. The indicator consists of a mixture of iron(III) nitrate and dilute sodium thiocyanate solution. The end-point is revealed by the decolorisation of the iron(III)–thiocyanate complex due to the formation of the colourless mercury(II) thiocyanate. The reaction between mercury(II) nitrate and sodium tetraphenylborate under the experimental conditions used is not quite stoichiometric, hence it is necessary to determine a factor (cm³ of $Hg(NO_3)_2$ solution equivalent to 1 cm³ of $NaB(C_6H_5)_4$ solution). Halides must be absent.

Procedure. Prepare the *sodium tetraphenylborate solution* by dissolving 6.0 g of the solid in about 200 cm³ of distilled water in a glass-stoppered bottle. Add about 1 g of moist aluminium hydroxide gel, and shake well at five minute intervals for about twenty minutes. Filter through a Whatman No. 40 filter paper, pouring the first runnings back through the filter if necessary, to ensure a clear filtrate. Add 15 cm³ of 0.1M-sodium hydroxide to the solution to give a pH of about 9, then make up to one litre and store the solution in a polythene bottle. Prepare the *mercury(II) nitrate solution* as in the last Section, but using a weight of approximately 10 g in a litre of solution; this solution may be standardised by titrating with a standard thiocyanate solution using iron(III) alum as indicator. Prepare *the indicator solutions* for the main titration by dissolving separately 5 g hydrated iron(III) nitrate in 100 cm³ of distilled water and filtering, and 0.08 g sodium thiocyanate in 100 cm³ distilled water.

Standardisation. Pipette 10.0 cm³ of the sodium tetraphenylborate solution into a 250 cm³ beaker and add 90 cm³ water, 2.5 cm³ 0.1M-nitric acid, 1.0 cm³ iron(III) nitrate solution, and 1.0 cm³ sodium thiocyanate solution. Without delay stir the solution mechanically, then slowly add from a burette 10 drops of mercury(II) nitrate solution. Continue the titration by adding the mercury(II) nitrate solution at a rate of 1–2 drops per second until the colour of the indicator is temporarily discharged. Continue the titration more slowly, but maintain the rapid rate of stirring. The end-point is arbitrarily defined as the point when the indicator colour is discharged and fails to reappear for 1 minute. Perform at least three titrations, and calculate the mean volume of mercury(II) nitrate solution equivalent to 10.0 cm³ of the sodium tetraphenylborate solution.

Pipette 25.0 cm³ of the potassium ion solution (about 10 mg K⁺) into a 50-cm³ graduated flask, add 0.5 cm³ M-nitric acid and mix. Introduce 20.0 cm³ of the sodium tetraphenylborate solution, dilute to the mark, mix, then pour the mixture into a 150-cm³ flask provided with a ground stopper. Shake the stoppered flask for 5 minutes on a mechanical shaker to coagulate the precipitate, then filter most of the solution through a dry Whatman No. 40 filter paper into a dry beaker. Transfer 25.0 cm³ of the filtrate into a 250-cm³ conical flask and add 75 cm³ of water, 1.0 cm³ of iron(III) nitrate solution, and 1.0 cm³ of sodium thiocyanate solution. Titrate with the mercury(II) nitrate solution as described above.

B. 4 OXIDATION – REDUCTION TITRATIONS

In the following Sections we are concerned with the titration of reducing agents with oxidising agents such as potassium permanganate, potassium dichromate, cerium(IV) sulphate, iodine, potassium iodate and potassium bromate, and with the titration of oxidising agents by reducing reagents such as arsenic(III) oxide and sodium thiosulphate.

The relevant theoretical Sections (**X, 31–33**) should be studied, and it should also be noted that in many cases, before titration with an oxidising reagent is carried out, it is necessary to ensure that the substance to be titrated is in a suitable lower oxidation state, i.e., it may be necessary to reduce the test solution before titration can be carried out. A selection of methods for carrying out such reductions is given at the end of this Chapter (Sections **X, 142–6**).

Oxidations with potassium permanganate

X, 90. DISCUSSION. This valuable and powerful oxidising agent was first introduced into titrimetric analysis by F. Margueritte for the titration of iron(II). In *acid solutions*, the reduction can be represented by the following equation:

$$MnO_4^- + 8H^+ + 5e \rightleftharpoons Mn^{2+} + 4H_2O$$

from which it follows that the equivalent is one-fifth of the mole, i.e. 158.03/5, or 31.606. The standard potential in acid solution, $E^\ominus$, has been calculated to be 1.51 volts, hence the permanganate ion in acid solution is a strong oxidising agent.

Sulphuric acid is the most suitable acid, as it has no action upon permanganate in dilute solution. With hydrochloric acid, there is the likelihood of the reaction:

$$2MnO_4^- + 10Cl^- + 16H^+ = 2Mn^{2+} + 5Cl_2 + 8H_2O$$

taking place, and some permanganate may be consumed in the formation of chlorine. This reaction is particularly liable to occur with iron salts unless special precautions are adopted (see below). With a small excess of free acid, a very dilute solution, low temperature and slow titration with constant shaking, the danger from this cause is minimised. There are, however, some titrations, such as those with arsenic(III) oxide, trivalent antimony, and hydrogen peroxide, which can be carried out in the presence of hydrochloric acid.

In the analysis of iron ores, solution is frequently effected in concentrated hydrochloric acid; the iron(III) is reduced and the iron(II) is then determined in the resultant solution. To do this, it is best to add about 25 cm³ of *Zimmermann and Reinhardt's solution* (this is sometimes termed *preventive solution*), which is prepared by dissolving 50 g crystallised manganese(II) sulphate $MnSO_4,4H_2O$ in 250 cm³ water, adding a cooled mixture of 100 cm³ concentrated sulphuric acid and 300 cm³ water, followed by 100 cm³ syrupy phosphoric acid. The manganese(II) sulphate lowers the reduction potential of the MnO_4^-–Mn(II) couple (compare Sections **II, 23–24**) and thereby makes it a weaker oxidising agent; the tendency of the permanganate ion to oxidise chloride ion is thus reduced. It has been stated that a further function of the manganese(II) sulphate is to supply an adequate concentration of Mn^{2+} ions to react with any local excess of permanganate ion. Mn(III) is probably formed in the reduction of permanganate ion to manganese(II); the Mn(II), and also the phosphoric acid, exert a depressant effect upon the potential of the Mn(III)–Mn(II) couple, so that

Mn(III) is reduced by Fe^{2+} ion rather than by chloride ion. The phosphoric acid combines with the yellow Fe^{3+} ion to form the complex ion $[Fe(HPO_4)]^+$, thus rendering the end-point more clearly visible. The phosphoric acid lowers the reduction potential of the Fe(III)–Fe(II) system by complexation, and thus tends to increase the reducing power of the Fe^{2+} ion. Under these conditions permanganate ion oxidises iron(II) rapidly and reacts only slowly with chloride ion.

For the titration of colourless or slightly coloured solutions, the use of an indicator is unnecessary, since as little as 0.01 cm^3 of 0.01N-potassium permanganate imparts a pale-pink colour to 100 cm^3 of water. The intensity of the colour in dilute solutions may be enhanced, if desired, by the addition of a redox indicator (such as sodium diphenylamine sulphonate, N-phenylanthranilic acid, or ferroin) just before the end-point of the reaction; this is usually not required, but is advantageous if more dilute solutions of permanganate are used.

Potassium permanganate also finds some application in *strongly* alkaline solutions. Here two consecutive partial reactions take place:
(i) the relatively rapid reaction:

$$MnO_4^- + e \rightleftharpoons MnO_4^{2-}$$

and (ii) the relatively slow reaction:

$$MnO_4^{2-} + 2H_2O + 2e \rightleftharpoons MnO_2 + 4OH^-$$

The standard potential $E^\ominus$ of reaction (i) is 0.56 volt and of reaction (ii) 0.60 volt. By suitably controlling the experimental conditions (e.g., by the addition of barium ions, which form the sparingly soluble barium manganate as a fine, granular precipitate), reaction (i) occurs almost exclusively; the equivalent is then 1 mole. In moderately alkaline solutions permanganate is reduced quantitatively to manganese dioxide. The half-cell reaction is:

$$MnO_4^- + 2H_2O + 3e \rightleftharpoons MnO_2 + 4OH^-$$

and the standard potential $E^\ominus$ is 0.59 volt.

Potassium permanganate is not a primary standard. It is difficult to obtain the substance perfectly pure and completely free from manganese dioxide. Moreover, ordinary distilled water is likely to contain reducing substances (traces of organic matter, etc.) which will react with the potassium permanganate to form manganese dioxide. The presence of the latter is very objectionable because it catalyses the auto-decomposition of the permanganate solution on standing. The decomposition:

$$4MnO_4^- + 2H_2O = 4MnO_2 + 3O_2 + 4OH^-$$

is catalysed by solid manganese dioxide. Permanganate is inherently unstable in the presence of manganese(II) ions:

$$2MnO_4^- + 3Mn^{2+} + 2H_2O = 5MnO_2 + 4H^+;$$

this reaction is slow in acid solution, but is very rapid in neutral solution. For these reasons, potassium permanganate solution is rarely made up by dissolving weighed amounts of the highly purified (e.g., A.R.) solid in water; it is more usual to heat a freshly prepared solution to boiling and keep it on the steam bath for an hour or so, and then filter the solution through a non-reducing filtering medium, such as purified glass wool or a sintered glass filtering crucible (porosity No. 4).

Alternatively, the solution may be allowed to stand for 2–3 days at room temperature before filtration. The glass-stoppered bottle or flask should be carefully freed from grease and prior deposits of manganese dioxide: this may be done by rinsing with dichromate–sulphuric acid cleaning mixture and then thoroughly with distilled water. Acidic and alkaline solutions are less stable than neutral ones. Solutions of permanganate should be protected from unnecessary exposure to light; a dark-coloured bottle is recommended. Diffuse daylight causes no appreciable decomposition, but bright sunlight slowly decomposes even pure solutions.

Potassium permanganate solutions may be standardised using arsenic(III) oxide or sodium oxalate as primary standards: secondary standards include metallic iron, and iron(II) ethylenediammonium sulphate (or ethylenediamine iron(II) sulphate), $FeSO_4,C_2H_4(NH_3)_2SO_4,4H_2O$.

Of these substances sodium oxalate was formerly regarded as the most trustworthy, since it is readily obtained pure and anhydrous, and the ordinary A.R. substance has a purity of at least 99.9 per cent. The experimental procedure hitherto employed was due to R. S. McBride. A solution of the oxalate, acidified with dilute sulphuric acid and warmed to 80–90 °C, was titrated with the permanganate solution slowly (10–15 cm^3 per minute) and with constant stirring until the first permanent faint pink colour was obtained; the temperature near the end-point was not allowed to fall below 60 °C. R. M. Fowler and H. A. Bright have, however, shown that with McBride's procedure the results may be 0.1–0.45 per cent high; the titre depends upon the acidity, the temperature, the rate of addition of the permanganate solution, and upon the speed of stirring. These authors recommend *a more rapid* addition of 90–95 per cent of the permanganate solution (about 25–35 cm^3 per minute) to a solution of sodium oxalate in M-sulphuric acid at 25–30 °C, the solution is then warmed to 55–60 °C and the titration completed, the last 0.5–1-cm^3 portion being added dropwise. The method is accurate to 0.06 per cent. Full experimental details are given in *Procedure B* below.

$$2Na^+ + C_2O_4{}^{2-} + 2H^+ \rightleftharpoons H_2C_2O_4 + 2Na^+$$

$$2MnO_4{}^- + 5H_2C_2O_4 + 6H^+ = 2Mn^{2+} + 10CO_2 + 8H_2O$$

It should be mentioned that if oxalate is to be determined it is often not convenient to use the room-temperature technique for unknown amounts of oxalate. The permanganate solution may then be standardised against sodium oxalate at about 80 °C using the same procedure in the standardisation as in the analysis.

The procedure of H. A. Bright, which utilises arsenic(III) oxide as a primary standard and potassium iodide or potassium iodate as a catalyst for the reaction, is more convenient in practice and is a trustworthy method for the standardisation of permanganate solutions. A.R. arsenic(III) oxide has a purity of at least 99.8 per cent, and the results by this method agree to within 1 part in 3000 with the sodium oxalate procedure of Fowler and Bright. Full experimental details are given in *Procedure A* (Section **X, 92**).

$$As_2O_3 + 4OH^- = 2HAsO_3{}^{2-} + H_2O$$

$$5H_3AsO_3 + 2MnO_4{}^- + 6H^+ = 5H_3AsO_4 + 2Mn^{2+} + 3H_2O$$

Potassium iodide, if specially purified, may be used as a primary standard. For many practical purposes, the dry A.R. reagent is sufficiently pure. The potentiometric method (see Chapter XIV) should be employed: a bright platinum indicator electrode and a saturated calomel electrode are required. The concentration of the sulphuric acid should be about $0.4M$.

$$10I^- + 2MnO_4^- + 16H^+ = 5I_2 + 2Mn^{2+} + 8H_2O$$

Iron wire of 99.9 per cent purity is available commercially and the A.R. reagent is a suitable standard, particularly if the potassium permanganate solution is subsequently to be employed in the determination of iron. If the wire exhibits any sign of rust, it should be drawn between two pieces of fine emery cloth, and then wiped with a clean, dry cloth before use. The reaction which occurs is:

$$MnO_4^- + 5Fe^{2+} + 8H^+ = Mn^{2+} + 5Fe^{3+} + 4H_2O$$

Ethylenediammonium iron(II) sulphate, $FeSO_4,C_2H_4(NH_3)_2SO_4,4H_2O$ is relatively stable and has a high molecular weight (382.16). The preparation is as follows.

To 10.0 g of a 99 per cent solution of ethylenediamine, add 60 cm³ 6N-sulphuric acid and 46.3 g A.R. iron(II) sulphate heptahydrate. Dilute to 300 cm³ with distilled water, and to the resulting solution introduce 300 cm³ of ethanol slowly and with constant stirring. Filter through a sintered glass funnel, wash the precipitate with 50 per cent ethanol, and redissolve it in slightly acidulated water. Add two-thirds the volume of ethanol. Filter again as before, and wash the solid successively with 65 per cent ethanol and 95 per cent ethanol. Dry in the air or at 50 °C for about 12 hours. The yield is about 50 g.

X, 91. PREPARATION OF 0.1N-POTASSIUM PERMANGANATE.
Weigh out about 3.2–3.25 g A.R. potassium permanganate on a watch glass, transfer it to a 1500-cm³ beaker, add 1 litre water, cover the beaker with a clock glass, heat the solution to boiling, boil gently for 15–30 minutes and allow the solution to cool to the laboratory temperature. Filter the solution through a funnel containing a plug of purified glass wool, or through a Gooch crucible provided with a pad of purified asbestos, or, most simply, through a sintered glass or porcelain filtering crucible or funnel. Collect the filtrate in a vessel which has been cleaned with chromic acid mixture and then thoroughly washed with distilled water. The filtered solution should be stored in a clean, glass-stoppered bottle, and kept in the dark or in diffuse light except when in use: alternatively, it may be kept in a bottle of dark-brown-coloured glass.

X, 92. STANDARDISATION OF PERMANGANATE SOLUTIONS.
Procedure A. **With arsenic(III) oxide.** Dry some A.R. arsenic(III) oxide at 105–110 °C for 1–2 hours, cover the container, and allow to cool in a desiccator. Accurately weigh approximately 0.25 g of the dry oxide, and transfer it to a 400-cm³ beaker. Add 10 cm³ of a cool solution of sodium hydroxide, prepared from 20 g sodium hydroxide and 100 cm³ water (1). Allow to stand for 8–10 minutes, stirring occasionally. When solution is complete, add 100 cm³ water, 10 cm³ pure concentrated hydrochloric acid, and 1 drop $0.0025M$-potassium iodide or potassium iodate (2). Add the permanganate solution from a burette until a faint pink colour persists for 30 seconds. Add the last 1–1.5 cm³ dropwise, allowing each drop to become decolorised before the next drop is introduced. For the most accurate work it is necessary to determine the volume of

permanganate solution required to duplicate the pink colour at the end-point. This is done by adding permanganate solution to a solution containing the same amounts of alkali, acid, and catalyst as were used in the test. The correction should not be more than 0.03 cm³. Repeat the determination with two other similar quantities of oxide. Calculate the normality of the potassium permanganate solution. Duplicate determinations should agree within 0.1 per cent.

Notes.   1. For *elementary students*, it is sufficient to weigh out accurately about 1.25 g of A.R. arsenic(III) oxide, dissolve this in 50 cm³ of a cool 20 per cent solution of sodium hydroxide, and make up to 250 cm³ in a graduated flask. Shake well. Measure 25.0 cm³ of this solution by means of a burette and **not** with a pipette (**caution—the solution is highly poisonous**) into a 250–350-cm³ conical flask, add 100 cm³ water, 10 cm³ pure concentrated hydrochloric acid, 1 drop potassium iodide solution, and titrate with the permanganate solution to the first permanent pink colour as detailed above. Repeat with two other 25-cm³ portions of the solution. Successive titrations should agree within 0.1 cm³.

2. $0.0025M$-potassium iodide = 0.41 g KI dm⁻³, $0.0025M$-potassium iodate = 0.54 g KIO_3 dm⁻³.

Calculation.   It is evident from the equation given in Section **X, 90** and also from the equation:

$$As_2O_3 + 2O = As_2O_5$$

that the equivalent of arsenic(III) oxide is one quarter of a mole, 197.84/4 or 49.460 g. One cm³ of a normal solution contains the milli-equivalent, or 0.04946 g. If the weight of arsenic(III) oxide be divided by the number of cm³ of potassium permanganate solution to which it is equivalent as found by titration, we have the weight of primary standard equivalent to 1 cm³ of the permanganate solution. If this last value be divided by the milli-equivalent of arsenic(III) oxide, the normality of the permanganate solution is obtained.

Procedure B.   **With sodium oxalate.**   Dry some A.R. sodium oxalate at 105–110 °C for 2 hours, and allow it to cool in a covered vessel in a desiccator. Weigh out accurately from a weighing bottle about 0.3 g of the dry sodium oxalate into a 600-cm³ beaker, add 240 cm³ of recently prepared distilled water, and 12.5 cm³ of concentrated sulphuric acid (*caution*) or 250 cm³ of $2N$-sulphuric acid. Cool to 25–30 °C and stir until the oxalate has dissolved (1). Add 90–95 per cent of the required quantity of permanganate solution from a burette at a rate of 25–35 cm³ per minute while stirring slowly (2). Heat to 55–60 °C (use a thermometer as stirring rod), and complete the titration by adding permanganate solution until a faint pink colour persists for 30 seconds. Add the last 0.5–1 cm³ dropwise, with particular care to allow each drop to become decolorised before the next is introduced. For the most exact work, it is necessary to determine the excess of permanganate solution required to impart a pink colour to the solution. This is done by matching the colour produced by adding permanganate solution to the same volume of boiled and cooled diluted sulphuric acid at 55–60 °C. This correction usually amounts to 0.03–0.05 cm³. Repeat the determination with two other similar quantities of sodium oxalate.

Notes.   1. For *elementary students*, it is sufficient to weigh out accurately about 1.7 g of A.R. sodium oxalate, transfer it to a 250-cm³ graduated flask, and make up to the mark. Shake well. Use 25 cm³ of this solution per titration and add 150 cm³ of *ca. M*-sulphuric acid. Carry out the titration rapidly at the ordinary temperature until the *first* pink colour appears throughout the solution, and allow to stand until the solution is colourless. Warm the solution to 50–60 °C and continue the titration to a permanent faint pink colour. It must be remembered that oxalate solutions attack glass, so that the solution should not be stored more than a few days.

2. An approximate value of the volume of permanganate solution required can be computed from the weight of sodium oxalate employed. In the first titration about 75 per cent of this volume is added, and the determination is completed at 55–60 °C. Thereafter, about 90–95 per cent of the volume of permanganate solution is added at the laboratory temperature.

Calculation. This is similar to that described under *Procedure A*. The equivalent of sodium oxalate is $\frac{1}{2}$ mole or 67.00 g.

Procedure C. **With metallic iron.** Use A.R. iron wire of 99.9 per cent assay value. Insert a well-fitting rubber stopper provided with a bent delivery tube into a 350-cm³ conical flask and clamp the flask in a retort stand in an inclined

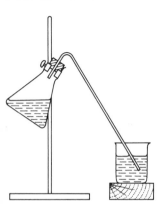

position, the tube being so bent as to dip into a small beaker containing saturated sodium hydrogencarbonate solution or 20 per cent potassium hydrogencarbonate solution (prepared from the A.R. solids) (Fig. X, 17). Place 100 cm³ 3N-sulphuric acid (from 92 cm³ water and 8 cm³ concentrated sulphuric acid) in the flask, and add 0.5–1 g A.R. sodium hydrogencarbonate in two portions; the carbon dioxide produced will drive out the air. Meanwhile, weigh out accurately about 0.15 g of iron wire, place it quickly into the flask, replace the stopper and bent tube, and warm gently until the iron has dissolved completely. Cool the flask rapidly under a stream of cold water,* and then run the permanganate solution cautiously from a

Fig. X, 17

burette, with constant shaking, until the faint pink colour is permanent. The addition of about 5 cm³ of pure syrupy phosphoric acid facilitates the detection of the end-point. Repeat the determination with two other samples of the iron wire.

The reaction is:

$$MnO_4^- + 5Fe^{2+} + 8H^+ = Mn^{2+} + 5Fe^{3+} + 4H_2O$$

1 equivalent $MnO_4^- \equiv 1$ mole Fe

Procedure D. **With ethylenediammonium iron(II) sulphate** (see Section **X, 90**). Weigh out accurately from a weighing bottle about 1.3–1.5 g of the salt into a 350-cm³ conical flask, add 60 cm³ M-sulphuric acid, and swirl the contents of the flask until the solid has dissolved. Titrate immediately with permanganate solution to the first permanent faint pink tinge. (The addition of about 5 cm³ syrupy phosphoric acid sp. gr. 1.75, facilitates the detection of the end-point).

Provided that it is stored with due regard to the precautions referred to in

---

* As the flask cools, the hydrogencarbonate solution is automatically drawn in until the pressure of the carbon dioxide inside the flask is equal to the atmospheric pressure.

Section **X, 90**, the standardised permanganate solution will keep for a long time, but it is advisable to re-standardise the solution frequently to confirm that no decomposition has set in.

X, 93.   DETERMINATION OF IRON(II).   The detailed experimental method has already been given under *Procedures C* and *D* of Section **X, 92**; the solution is acidified with dilute sulphuric acid. If chloride ion is present high results are obtained, because the reaction between iron(II) and permanganate induces the oxidation of hydrochloric acid. The chloride ion is rendered almost harmless by the addition of a manganese(II) salt, preferably in the form of the so-called Zimmermann–Reinhardt or preventive solution (Section **X, 90**), and by slow titration.

The test solution should be approximately $0.1M$ with respect to iron(II), and should contain about 10 per cent (by volume) of dilute sulphuric acid to reduce the tendency for atmospheric oxidation of the iron solution. Pipette 25 cm³ of the solution into a 250-cm³ conical flask, add 25 cm³ sulphuric acid ($0.5M$), and titrate with the standard ($0.1N$) potassium permanganate solution until a faint permanent pink coloration is produced.

A solution containing **iron in the trivalent condition** may also be analysed by titration with standard potassium permanganate after reduction of the iron to the divalent condition; this may be done, preferably by use of a Jones reductor, or by one of the other methods discussed in Sections **X, 142–6**.

The **iron content of an iron ore** may be similarly determined by dissolving a known weight of the ore in dilute hydrochloric acid and making the resulting solution up to the mark in a graduated flask. An aliquot portion of the solution is subjected to a suitable reduction procedure and is then titrated with standard permanganate solution in the presence of Zimmermann–Reinhardt solution.

X, 94.   DETERMINATION OF CALCIUM.   The calcium is precipitated as oxalate, the washed precipitate is dissolved in dilute sulphuric acid, and the oxalic acid liberated is titrated with standard permanganate solution. Precipitation of the calcium oxalate is best carried out by **homogeneous precipitation** using the urea hydrolysis method, in which acid is added to the solution to produce a pH of about 1.0; this is followed by ammonium oxalate and urea. Upon boiling the solution, the urea gradually undergoes hydrolysis and the pH rises to the point of calcium oxalate precipitation; this may take 10–15 minutes. The crystals precipitated from the hot solution are relatively large, and may be filtered off shortly after formation; this eliminates the digestion period which is otherwise required. The solution must remain clear until boiling is commenced to hydrolyse the urea.

When sulphate is present, both this and the normal method of precipitation yield high results in a single precipitation. With a double precipitation the error by the urea method is considerably smaller and any magnesium present is almost completely eliminated.

Procedure.   Weigh out accurately 0.15–0.20 g calcium carbonate, preferably of A.R. grade, into a 400-cm³ beaker. Add 20 cm³ water and cover the beaker with a clock glass. Introduce 10 cm³ dilute hydrochloric acid (1:1) and warm, if necessary, until the solid has dissolved. Dilute to 200 cm³, and add a few drops of methyl red indicator; sufficient acid must be present in the solution to prevent the precipitation of calcium oxalate when ammonium oxalate solution is added. Now introduce 15 cm^3 saturated ammonium oxalate solution and 15 g urea. Boil the solution gently until the methyl red changes colour to yellow (at pH 5). Filter through a coarse filter paper, or, with suction, on a small filter paper supported in a Gooch crucible. Wash the precipitate with small volumes of cold water until free from chloride. Transfer the filter paper and precipitate (or the Gooch crucible and precipitate) to the original beaker, dissolve the precipitate in hot dilute sulphuric acid, and titrate immediately with standard 0.1N-potassium permanganate solution: follow Procedure B, Section **X, 92**.

X, 95. ANALYSIS OF HYDROGEN PEROXIDE. Hydrogen peroxide is usually encountered in the form of an aqueous solution containing about 6 per cent, 12 per cent or 30 per cent hydrogen peroxide, and frequently referred to as '20 volume', '40 volume', and '100 volume' hydrogen peroxide respectively; this terminology is based upon the volume of oxygen liberated when the solution is decomposed by boiling. Thus 1 cm^3 of '100 volume' hydrogen peroxide will yield 100 cm^3 of oxygen measured at s.t.p.

The following reaction occurs when potassium permanganate solution is added to hydrogen peroxide solution acidified with dilute sulphuric acid:

$$2MnO_4^- + 5H_2O_2 + 6H^+ = 2Mn^{2+} + 5O_2 + 8H_2O$$

This forms the basis of the method of analysis given below.

It is good practice to use a fairly high concentration of acid and a reasonably low rate of addition in order to reduce the danger of forming manganese dioxide, which is an active catalyst for the decomposition of hydrogen peroxide. For slightly coloured solutions or for titrations with dilute permanganate, the use of ferroin as indicator is recommended. Organic substances may interfere. A fading end-point indicates the presence of organic matter or other reducing agents, in which case the iodimetric method is better (Section **X, 124**).

Procedure. Transfer 25.0 cm^3 of the '20-volume' solution by means of a burette to a 500-cm^3 graduated flask, and dilute with water to the mark. Shake thoroughly. Transfer 25.0 cm^3 of this solution to a conical flask, dilute with 200 cm^3 water, add 20 cm^3 dilute sulphuric acid (1:5), and titrate with standard 0.1N-potassium permanganate to the first permanent, faint pink colour. Repeat the titration; two consecutive determinations should agree within 0.1 cm^3.

Calculate: (i) the weight of hydrogen peroxide per dm^3 of the original solution and (ii) the 'volume strength', i.e., the number of cm^3 of oxygen at s.t.p. that can be obtained from 1 cm^3 of the original solution.

A **metallic peroxide**, such as **sodium peroxide**, can be analysed in similar manner, provided that care is taken to avoid loss of oxygen during the dissolution of the peroxide. This may be done by working in a medium containing boric acid which is converted to the relatively stable 'perboric acid' upon the addition of the peroxide.

Procedure. To 100 cm^3 of distilled water, add 5 cm^3 concentrated sulphuric acid, cool and then add 5 g pure boric acid; when this has dissolved cool the mixture in ice. Transfer gradually from a weighing bottle about 0.5 g (accurately weighed), of the sodium peroxide sample to the well-stirred, ice-cold solution. When the addition is complete, transfer the solution to a 250-cm^3 graduated flask, make up to the mark, and then titrate 50-cm^3 portions of the solution with standard (0.1N) permanganate solution.

X, 96.   DETERMINATION OF MANGANESE DIOXIDE IN PYROLU-SITE. *Discussion.* Manganese dioxide occurs in nature as the mineral pyrolusite. For many purposes, a knowledge of the percentage of MnO_2 is required. This may be determined by treatment with an excess of an acidified solution of a reducing agent, such as sodium oxalate, or arsenic(III) oxide.

$$MnO_2 + H_2C_2O_4 + 2H^+ = Mn^{2+} + 2CO_2 + 2H_2O$$

$$2MnO_2 + 2H_3AsO_3 + 4H^+ = 2Mn^{2+} + 2H_3AsO_4 + 2H_2O$$

The excess of reducing agent is determined by titration with standard permanganate solution. Arsenic(III) oxide is somewhat more trustworthy in this determination than is sodium oxalate, because oxalic acid decomposes very slowly at high temperatures into carbon monoxide and carbon dioxide, the decomposition being catalysed by manganese(II) salts; the extent of decomposition under ordinary circumstances is, however, very small. Both procedures will be described.

Procedure A (**arsenic(III) oxide method**).   Dry the finely powdered sample of pyrolusite at 120 °C to constant weight. Weigh out accurately from a weighing bottle about 0.2 g of the sample into a 250-cm³ conical flask, add 50 cm³ of standard 0.1N-arsenic(III) oxide (Section **X, 92, A, Note** 1) and 10 cm³ of concentrated sulphuric acid. Place a short funnel in the mouth of the flask, and boil until the pyrolusite has decomposed completely; no brown or black particles should then be present. Cool the solution, add 1 drop of 0.0025M-potassium iodide solution, and titrate the excess arsenic(III) oxide with standard 0.1N-potassium permanganate. Repeat the determination with two other samples of the solid.

Calculate the percentage of MnO_2 in the pyrolusite from the amount of arsenic(III) oxide consumed in the reaction.

Procedure B (**sodium oxalate method**).   Weigh out accurately about 0.2 g of the finely powdered, dry pyrolusite into a conical flask, add 50 cm³ of standard 0.1N-sodium oxalate (Section **X, 92, B, Note** 1), add 50 cm³ of 2M-sulphuric acid (*ca.* 10 per cent), and place a short funnel in the mouth of the flask. Boil the mixture gently until no black particles remain. Allow to cool, and titrate the excess of oxalate with standard 0.1N-potassium permanganate as detailed in Section **X, 92,** *Procedure B.* Repeat the determination with two other samples of similar weight.

Calculate the amount of sodium oxalate consumed in the reaction, and from this the percentage of MnO_2 in the pyrolusite.

X, 97.   DETERMINATION OF NITRITES. *Discussion.*   Nitrites react in warm acid solution (*ca.* 40 °C) with permanganate solution in accordance with the equation:

$$2MnO_4^- + 5NO_2^- + 6H^+ = 2Mn^{2+} + 5NO_3^- + 3H_2O$$

If a solution of a nitrite is titrated in the ordinary way with potassium permanganate, poor results are obtained, because the nitrite solution has first to be acidified with dilute sulphuric acid. Nitrous acid is liberated, which being volatile and unstable, is partially lost. If, however, a measured volume of standard potassium permanganate solution, acidified with dilute sulphuric acid, is treated with the nitrite solution, added from a burette, until the permanganate is just decolorised, results accurate to 0.5–1 per cent may be obtained. This is due to the fact that nitrous acid does not react instantaneously with the permanganate. This method may be used to determine the purity of commercial potassium nitrite.

Procedure. Weigh out accurately about 1.1 g of commercial potassium nitrite, dissolve it in cold water, and dilute to 250 cm^3 in a graduated flask. Shake well. Measure out 25.0 cm^3 of standard 0.1N-potassium permanganate into a 500-cm^3 flask, add 300 cm^3 of 0.75N-sulphuric acid, and heat to 40 °C. Place the nitrite solution in the burette, and add it slowly and with constant stirring until the permanganate solution is just decolorised. Better results are obtained by allowing the tip of the burette to dip under the surface of the diluted permanganate solution. Towards the end the reaction is sluggish, so that the nitrite solution must be added very slowly.

More accurate results may be secured by adding the nitrite to an acidified solution in which permanganate is present in excess (the tip of the pipette containing the nitrite solution should be below the surface of the liquid during the addition), and back-titrating the potassium permanganate with a solution of ammonium iron(II) sulphate which has recently been compared with the permanganate solution.

X, 98. DETERMINATION OF PERSULPHATES. *Discussion.* Alkali persulphates (peroxydisulphates) can readily be evaluated by adding to their solutions a known excess of an acidified iron(II) salt solution, and determining the excess of iron(II) by titration with standard potassium permanganate solution.

$$S_2O_8{}^{2-} + 2Fe^{2+} + 2H^+ = 2Fe^{3+} + 2HSO_4{}^-$$

By adding phosphoric acid or hydrofluoric acid, the reduction is complete in a few minutes at room temperature. Many organic compounds interfere.

Another procedure utilises standard oxalic acid solution. When a sulphuric acid solution of a persulphate is treated with excess of standard oxalic acid solution in the presence of a little silver sulphate as catalyst, the following reaction occurs:

$$H_2S_2O_8 + H_2C_2O_4 = 2H_2SO_4 + 2CO_2$$

The excess of oxalic acid is titrated with standard potassium permanganate solution.

Procedure A. Prepare an approximately 0.1N-solution of ammonium iron(II) sulphate by dissolving about 9.8 g of the A.R. solid in 200 cm^3 of sulphuric acid (0.5M) in a 250-cm^3 graduated flask, and then making up to the mark with freshly boiled and cooled distilled water. Standardise the solution by titrating 25-cm^3 portions with standard potassium permanganate solution (0.1N) after the addition of 25 cm^3 sulphuric acid (0.5M).

Weigh out accurately about 0.3 g potassium persulphate into a conical flask and dissolve it in 50 cm^3 of water. Add 5 cm^3 syrupy phosphoric acid or 2.5 cm^3 35–40 per cent hydrofluoric acid, 10 cm^3 5N-sulphuric acid, and 50.0 cm^3 of the *ca.* 0.1N-iron(II) solution. After 5 minutes, titrate the excess of Fe^{2+} ion with standard 0.1N-potassium permanganate.

From the difference between the volume of 0.1N-permanganate required to oxidise 50 cm^3 of the iron(II) solution and that required to oxidise the iron(II) salt remaining after the addition of the persulphate, calculate the percentage purity of the sample.

Procedure B.   Prepare an approximately $0.1N$ solution oxalic acid by dissolving about 1.6 g of the A.R. material and making up to 250 cm³ in a graduated flask. Standardise the solution with standard $(0.1N)$ potassium permanganate solution using the procedure described in Section **X, 92**, *B*.

Weigh out accurately 0.3–0.4 g potassium persulphate into a 350-cm³ conical flask, add 50 cm³ $0.1N$-oxalic acid, followed by 0.2 g of silver sulphate dissolved in 20 cm³ 10 per cent sulphuric acid. Heat the mixture in a water bath until no more carbon dioxide is evolved (15–20 minutes), dilute the solution to about 100 cm³ with water at about 40 °C, and titrate the excess of oxalic acid with standard $0.1N$-potassium permanganate.

X, 99.   DETERMINATION OF MANGANESE IN STEEL.   A. Bismuthate method.   *Discussion.*   The steel is dissolved in nitric acid and the resulting cooled solution is treated with sodium bismuthate when permanganic acid is formed:

$$2Mn^{2+} + 5NaBiO_3 + 14H^+ = 2MnO_4^- + 5Bi^{3+} + 7H_2O + 5Na^+$$

Excess bismuthate is removed by filtration, a measured volume (excess) of a standardised ammonium iron(II) sulphate solution is added to reduce the permanganic acid, and the excess iron(II) is then determined by titration with standard potassium permanganate. The solution should be free from cobalt, chromium and chloride, and since many steels contain chromium in addition to manganese, *Procedure B* is of more general application.

B.   Persulphate–arsenite method.   *Discussion.*   Manganese salts are oxidised to permanganic acid by persulphate in the presence of silver nitrate solution as catalyst:

$$2Mn^{2+} + 5S_2O_8^{2-} + 8H_2O = 2MnO_4^- + 10SO_4^{2-} + 16H^+$$

If the oxidation with persulphate is carried out in the presence of phosphoric acid, it is possible to oxidise as much as 50 mg of manganese to permanganic acid without the separation of oxides of manganese. No satisfactory method is known for removing the excess of persulphate; boiling will destroy it, but some permanganic acid will be decomposed at the same time. Use is made of the fact that an arsenite solution reacts rapidly with permanganic acid in the cold, but no reaction occurs with the persulphate. A little chloride is added to precipitate the silver catalyst and thus prevent the re-oxidation of the manganese(II) salt formed by reduction. The reduction of the permanganic acid by the arsenite does not proceed completely to bivalent manganese, and it is therefore advisable to standardise the arsenite solution against a steel of known manganese content.

Chromium is oxidised to chromate, but the yellow colour has little effect if the chromium content does not exceed 10 mg per 100 cm³.

Procedure.   Weigh out accurately about 1.0 g of the steel into a 350-cm³ conical flask and add successively 15 cm³ water, 3 cm³ concentrated sulphuric acid, 4 cm³ 85 per cent phosphoric acid, and 8 cm³ concentrated nitric acid. Heat until solution is complete, and boil to expel oxides of nitrogen. Add 50 cm³ water, 5 cm³ $0.1M$-silver nitrate solution, and 2.5 g pure ammonium persulphate dissolved in a little water. Heat to boiling and boil briskly for $\frac{1}{2}$ minute. Cool rapidly to 25 °C or lower, add 75 cm³ cold water, and 5 cm³ $0.2M$-sodium chloride solution. Titrate immediately with 0.025N-sodium arsenite solution (1) to a clear yellow end-point which does not change upon the addition of more arsenite solution.

Standardise the arsenite solution against a similar steel of known manganese content.

Note. 1. Prepare the 0.025N-sodium arsenite solution by dissolving 1.230 g A.R. arsenic(III) oxide in a solution of 10 g A.R. sodium hydroxide in 30 cm^3 water, warming if necessary. Dilute to about 500 cm^3, neutralise by the addition of 29–30 cm^3 1M-hydrochloric acid, then add 10 g A.R. sodium hydrogen-carbonate, and dilute to 1 dm^3 in a graduated flask.

Once the oxidation to permanganic acid has been effected, the determination may be completed more rapidly spectrophotometrically (see Section **XVIII, 23**), and an alternative procedure is to carry out a potentiometric titration in which the solution containing Mn(II) is titrated with a standard solution of potassium permanganate (Section **XIV, 28**).

Oxidations with potassium dichromate

X, 100. DISCUSSION. Potassium dichromate is not such a powerful oxidising agent as potassium permanganate (compare reduction potentials in Table II, 4 in Section **II, 23**), but it has several advantages over the latter substance. It can be obtained pure, is stable up to its fusion point, and is therefore an excellent primary standard. Standard solutions of exactly known strength can be prepared by weighing out the pure dry salt and dissolving it in the proper volume of water. Furthermore, the aqueous solutions are stable indefinitely if adequately protected from evaporation. Potassium dichromate is used only in acid solution, and is reduced rapidly at the ordinary temperature to a green chromium(III) salt. It is not reduced by cold hydrochloric acid, provided the acid concentration does not exceed 1 or 2M. Dichromate solutions are also less easily reduced by organic matter than are those of permanganate and are also stable towards light. Potassium dichromate is therefore of particular value in the determination of iron in iron ores: the ore is usually dissolved in hydrochloric acid, the iron(III) reduced to iron(II), and the solution then titrated with standard dichromate solution:

$$Cr_2O_7^{2-} + 6Fe^{2+} + 14H^+ = 2Cr^{3+} + 6Fe^{3+} + 7H_2O$$

In acid solution, the reduction of potassium dichromate may be represented as:

$$Cr_2O_7^{2-} + 14H^+ + 6e \rightleftharpoons 2Cr^{3+} + 7H_2O$$

from which it follows that the equivalent is one-sixth of the mole, i.e., 294.18/6 or 49.030 g. A 0.1N-solution therefore contains 4.9030 g dm^{-3}.

The green colour due to the Cr^{3+} ions formed by the reduction of potassium dichromate makes it impossible to ascertain the end-point of a dichromate titration by simple visual inspection of the solution and so a redox indicator must be employed which gives a strong and unmistakable colour change; this procedure has rendered obsolete the external indicator method which was formerly widely used. Suitable indicators for use with dichromate titrations include N-phenylanthranilic acid (0.1 per cent solution in 0.005M-NaOH) and sodium diphenylamine sulphonate (0.2 per cent aqueous solution); the latter must be used in presence of phosphoric acid.

X, 101.   PREPARATION OF 0.1*N*-POTASSIUM DICHROMATE.   A.R. potassium dichromate has a purity of not less than 99.9 per cent and is satisfactory for most purposes.* Powder finely about 6 g of the A.R. material in a glass or agate mortar, and heat for 30–60 minutes in an air oven at 140–150 °C. Allow to cool in a closed vessel in a desiccator. Weigh out accurately about 4.9 of the dry potassium dichromate into a weighing bottle and transfer the salt quantitatively to a 1-dm^3 graduated flask, using a small funnel to avoid loss. Dissolve the salt in the flask in water and make up to the mark; shake well. Alternatively, place a little over 4.9 g of potassium dichromate in a weighing bottle, and weigh accurately. Empty the salt into a 1-dm^3 graduated flask, and weigh the bottle again. Dissolve the salt in water, and make up to the mark.

The normality of the solution can be calculated directly from the weight of salt taken, but if the salt has only been weighed out approximately, then the solution must be standardised as in the following Section.

X, 102.   STANDARDISATION OF POTASSIUM DICHROMATE SOLUTION AGAINST IRON.   Use the method described in Section **X, 92**, *Procedure C*, with 0.2 g accurately weighed, of A.R. iron wire. Titrate the cooled solution immediately with the dichromate solution, using either sodium diphenylamine sulphonate or *N*-phenylanthranilic acid as indicator. If the former is selected, add 6–8 drops of the indicator, followed by 5 cm^3 of syrupy phosphoric acid: titrate slowly with the dichromate solution, stirring well, until the pure green colour changes to a grey-green. Then add the dichromate solution dropwise until the first tinge of blue-violet, which remains permanent on shaking, appears. If the latter indicator is selected, add 200cm^3 of *M*-sulphuric acid, then 0.5 cm^3 of the indicator; add the dichromate solution, with shaking until the colour changes from green to violet-red.

$$1 \text{ mole K}_2\text{Cr}_2\text{O}_7 \equiv 6 \text{ moles Fe}$$

The standardisation may also be effected with ethylenediammonium iron(II) sulphate as described in Section **X, 92**, *Procedure D*.

X, 103.   DETERMINATION OF IRON(II).   The conditions are very similar to those outlined in Section **X, 93** with the exception that the presence of moderate amounts of chloride ion have no effect on the determination.

The test solution should be approximately 0.1*M* with respect to iron(II), and should contain dilute sulphuric acid to reduce the tendency for atmospheric oxidation. Titrate 25.0-cm^3 portions of this solution with the standard (0.1*N*) potassium dichromate solution using either sodium diphenylamine sulphonate (I) or N-phenylanthranilic acid (II) as internal indicator.

Use 8 drops (say 0.4 cm^3) of the indicator I, add 200 cm^3 of 2.5 per cent sulphuric acid, followed by 5 cm^3 of 85 per cent phosphoric acid, and titrate slowly, whilst stirring constantly, with the standard dichromate until the solution

---

* If only a 'pure' grade (as distinct from A.R.) of commercial salt is available, or if there is some doubt as to the purity of the salt, the following method of purification should be used. A concentrated solution of the salt in hot water is prepared and filtered. The crystals which separate on cooling are filtered on a sintered glass filter funnel and sucked dry. The resultant crystals are recrystallised again. The purified crystals are then dried at 180–200 °C, ground to a fine powder in a glass or agate mortar, and again dried at 140–150 °C to constant weight.

assumes a bluish-green or greyish-blue tint near the end-point. Continue the titration, adding the dichromate solution dropwise and maintaining an interval of a few seconds between each drop, until the addition of 1 drop causes the formation of an intense purple or violet-blue coloration, which remains permanent after shaking and is unaffected on further addition of the dichromate.

Use 0.5 cm^3 of indicator II. Add about 200 cm^3 of M-sulphuric acid and then titrate with the 0.1N-potassium dichromate until the colour changes from green to violet-red. This titration is sharp to within 1 drop.

A solution containing **iron in the trivalent condition** may be analysed after reduction of the iron to the divalent condition with a Jones reductor, or by one of the other methods described in Sections **X, 143–5**.

The iron content of an iron ore may be similarly determined by weighing out about 2 g, dissolving in dilute hydrochloric acid, and making up to the mark in a 250-cm^3 graduated flask. Portions of the solution (25.0 cm^3) are then subjected to a suitable reduction procedure, and the solutions titrated with standard potassium dichromate.

The iron ore will usually contain both iron(II) and iron(III) compounds, and the procedure just described measures the total iron content of the ore. The **proportion of iron(II)** can be determined by the following procedure.

Fit a 350-cm^3 conical flask with a rubber stopper carrying two glass tubes bent at right angles, one passing to the bottom of the flask and the other ending just inside the stopper. Join the longer tube to a gas wash-bottle which is connected to a source of carbon dioxide (Kipp's apparatus or a cylinder of the compressed gas), and attach a gas bubbler to the shorter tube; the wash bottle and the bubbler contain distilled water. Weigh out accurately about 0.4 g of the finely powdered ore into the flask, and then pass a stream of carbon dioxide through the flask to displace the air; this ensures that the iron(II) chloride formed in the subsequent dissolution of the ore does not undergo atmospheric oxidation. Open the stopper of the flask momentarily and introduce 30 cm^3 of 1:1 hydrochloric acid, then warm the flask gently and pass a slow stream of carbon dioxide until the ore has been completely attacked; in most cases a small white residue of silica will remain. Allow the flask to cool whilst still maintaining the current of carbon dioxide, then wash down the tubes and neck of the flask with a little cold, air-free distilled water, add 200 cm^3 of 2.5 per cent sulphuric acid (prepared with air-free water), and then titrate with standard potassium dichromate solution using an internal indicator. The iron(II) content of the ore thus determined, subtracted from the total iron, gives the iron(III) content of the ore.

X, 104. DETERMINATION OF CHROMIUM IN A CHROMIUM(III) SALT. *Discussion.* Chromium(III) salts are oxidised to dichromate by boiling with excess of a persulphate solution in the presence of a little silver nitrate (catalyst). The excess of persulphate remaining after the oxidation is complete is destroyed by boiling the solution for a short time. The dichromate content of the resultant solution is determined by the addition of excess of a standard iron(II) solution and titration of the excess of the latter with standard 0.1N-potassium dichromate.

$$2Cr^{3+} + 3S_2O_8^{2-} + 7H_2O \xrightarrow{\text{(AgNO}_3)} Cr_2O_7^{2-} + 6HSO_4^- + 8H^+$$

$$2S_2O_8^{2-} + 2H_2O = O_2\uparrow + 4HSO_4^-$$

Procedure.   Weigh out accurately an amount of the salt which will contain about 0.25 g of chromium, and dissolve it in 50 cm³ distilled water. Add 20 cm³ of *ca.* 0.1*M*-silver nitrate solution, followed by 50 cm³ of a 10 per cent solution of ammonium or potassium persulphate. Boil the liquid gently for 20 minutes. Cool, and dilute to 250 cm³ in a graduated flask. Remove 50 cm³ of the solution with a pipette, add 50 cm³ of 0.1*N*-ammonium iron(II) sulphate solution (Section **X, 98**, *Procedure A*), 200 cm³ of 2*N*-sulphuric acid, and 0.5 cm³ of *N*-phenylanthranilic acid indicator. Titrate the excess of the iron(II) salt with standard 0.1*N*-potassium dichromate until the colour changes from green to violet-red.

Standardise the ammonium iron(II) sulphate solution against the 0.1*N*-potassium dichromate, using *N*-phenylanthranilic acid as indicator. Calculate the volume of the iron(II) solution which was oxidised by the dichromate originating from the chromium salt, and from this the percentage of chromium in the sample.

Note.   Lead or barium can be determined by precipitating the sparingly soluble chromate, dissolving the washed precipitate in dilute sulphuric acid, adding a known excess of ammonium iron(II) sulphate solution, and titrating the excess of Fe^{2+} ion with 0.1*N*-potassium dichromate in the usual way.

$$2PbCrO_4 + 2H^+ = 2Pb^{2+} + Cr_2O_7^{2-} + H_2O$$

X, 105.   DETERMINATION OF CHROMIUM IN CHROMITE.   *Discussion.*   The highly refractory mineral chromite is brought into solution by fusion with excess of sodium peroxide:

$$\text{?}Fe(CrO_2)_2 + 7Na_2O_2 = 2NaFeO_2 + 4Na_2CrO_4 + 2Na_2O$$

or $$2Fe(CrO_2)_2 + 7O_2^{2-} = 2FeO_2^- + 4CrO_4^{2-} + 2O^{2-}$$

Upon leaching the melt with water, the sodium chromate dissolves and the iron is precipitated as iron(III) hydroxide:

$$NaFeO_2 + 2H_2O = NaOH + Fe(OH)_3$$

$$2Na_2O + 2H_2O = 4NaOH$$

The excess of peroxide is decomposed by boiling the alkaline solution. The precipitate is filtered off after diluting the solution; the filtrate is acidified with hydrochloric acid, a known volume of excess of ammonium iron(II) sulphate solution is added, and the excess of iron(II) is titrated with standard potassium dichromate solution.

$$2CrO_4^{2-} + 2H^+ \rightleftharpoons Cr_2O_7^{2-} + H_2O$$

$$Cr_2O_7^{2-} + 6Fe^{2+} + 14H^+ = 2Cr^{3+} + 6Fe^{3+} + 7H_2O$$

Procedure.   Weigh out accurately about 0.5 g of the very finely powdered ore into a 30–35-cm³ nickel, or heavy-walled porcelain, crucible, add 4 g of sodium peroxide, and mix thoroughly by means of a thin glass rod. Remove any powder adhering to the rod by stirring about 1 g of sodium peroxide with it; cover the mixture in the crucible with this peroxide. Place the lid on the crucible, and gently heat the covered crucible in the fume cupboard over a small flame until the mass is quite liquid (about 10 minutes); keep fused for a further 10 minutes at a dull red heat. Allow to cool, and when a solid crust has formed, add 4 g more of the sodium peroxide, and fuse the mixture again at a cherry-red heat for 10 minutes. Allow the crucible to cool and place it in a 600-cm^3 Pyrex beaker containing a little distilled water. Cover the beaker with a clock glass, add a little warm water, and, after the violent action has subsided, remove the crucible and wash it thoroughly, collecting the washings in the same beaker. Boil the liquid for 30 minutes, keeping the beaker covered (this decomposes the hydrogen peroxide), add 250 cm^3 boiling water, and allow the precipitate to settle. Filter through a hardened 15-cm filter paper or, better, through a sintered glass filtering crucible, and wash the residue thoroughly with boiling water until free from chromate. (The residue should be completely soluble in concentrated hydrochloric acid; no black gritty particles should remain. If this is not the case, decomposition is not complete, and the determination must be started afresh.) Evaporate the filtrate to about 200 cm^3, cool, and add 4.5M-sulphuric acid cautiously until acid. Cool, transfer to a 250-cm^3 graduated flask, and make up to the mark with distilled water. Shake well. Remove 50 cm^3 of this solution with a pipette, add 50 cm^3 of 0.1N-ammonium iron(II) sulphate and proceed as in the previous Section.

X, 106. **DETERMINATION OF CHLORATE.** *Discussion.* Chlorate ion is reduced by warming with excess of iron(II) in the presence of a relatively high concentration of sulphuric acid:

$$ClO_3^- + 6Fe^{2+} + 6H^+ = Cl^- + 6Fe^{3+} + 3H_2O$$

The excess Fe^{2+} ion is determined by titration with standard dichromate solution in the usual way.

Procedure. To obtain experience in the method, the purity of A.R. potassium chlorate may be determined. Prepare a 0.02M-potassium chlorate solution using the A.R. solid. Into a 250-cm^3 or 350-cm^3 conical flask, place 25.0 cm^3 of the potassium chlorate solution, 25.0 cm^3 of 0.2N-ammonium iron(II) sulphate solution in 4N-sulphuric acid and add cautiously 12 cm^3 concentrated sulphuric acid. Heat the mixture to boiling (in order to ensure completion of the reduction), and cool to room temperature by placing the flask in running tap water. Add 20 cm^3 1:1-phosphoric acid, followed by 0.5 cm^3 sodium diphenylaminesulphonate indicator. Titrate the excess Fe^{2+} ion with standard 0.1N-potassium dichromate to a first tinge of purple coloration which remains on stirring.

Standardise the ammonium iron(II) sulphate solution by repeating the procedure but using 25 cm^3 distilled water in place of the chlorate solution. The difference in titres is equivalent to the amount of potassium chlorate added.

Oxidations with cerium(IV) sulphate solutions

X, 107. **GENERAL DISCUSSION.** Cerium(IV) sulphate is a powerful oxidising agent; its reduction potential in 1–8N-sulphuric acid at 25 °C is 1.43 ±0.05 volts. It can be used only in acid solution, best in 0.5N or higher concentrations: as the solution is neutralised, cerium(IV) hydroxide (hydrated cerium(IV) oxide) or basic salts precipitate. The solution has an intense yellow colour, and in hot solutions which are not too dilute the end-point may be detected without an indicator; this procedure, however, necessitates the application of a blank correction, and it is therefore preferable to add a suitable indicator.

The advantages of cerium(IV) sulphate as a standard oxidising agent are:

363

1. Cerium(IV) sulphate solutions are remarkably stable over prolonged periods. They need not be protected from light, and may even be boiled for a short time without appreciable change in concentration. The stability of sulphuric acid solutions covers the wide range of $10-40$ cm^3 of concentrated sulphuric acid per litre. It is evident, therefore, that an acid solution of cerium(IV) sulphate surpasses a permanganate solution in stability.

2. Cerium(IV) sulphate may be employed in the determination of reducing agents in the presence of a high concentration of hydrochloric acid (contrast potassium permanganate, Section **X, 90**).

3. Cerium(IV) solutions in $0.1N$ solution are not too highly coloured to obstruct vision when reading the meniscus in burettes and other titrimetric apparatus.

4. In the reaction of cerium(IV) salts in acid solution with reducing agents, the simple valency change

$$Ce^{4+} + e \rightleftharpoons Ce^{3+}$$

is assumed to take place; the equivalent weight is therefore the mole. With permanganate, of course, a number of reduction products are produced according to the experimental conditions.

5. The cerium(III) ion is colourless (compare colourless manganese(II) ion from potassium permanganate, and green chromium(III) ion from potassium dichromate).

6. Cerium(IV) sulphate is a very versatile oxidising agent. It may be employed in most titrations in which permanganate has been used, and also for other determinations.

7. Cerium(IV) sulphate solutions are best standardised with arsenic(III) oxide or with sodium oxalate.

Solutions of cerium(IV) sulphate in dilute sulphuric acid are stable even at boiling temperatures. Hydrochloric acid solutions of the salt are unstable because of reduction to cerium(III) by the acid with the simultaneous liberation of chlorine:

$$2Ce^{4+} + 2Cl^- = 2Ce^{3+} + Cl_2$$

This reaction takes place quite rapidly on boiling, and hence hydrochloric acid cannot be used in oxidations which necessitate boiling with excess of cerium(IV) sulphate in acid solution: sulphuric acid must be used in such oxidations. However, direct titration with cerium(IV) sulphate in a dilute hydrochloric acid medium (e.g., for iron(II) may be accurately performed at room temperature, and in this respect cerium(IV) sulphate is superior to potassium permanganate (cf. 2 above). The presence of hydrofluoric acid is harmful, since fluoride ion forms a stable complex with Ce(IV) and decolorises the yellow solution.

Formal potential measurements show that the redox potential of the Ce(IV)–Ce(III) system is greatly dependent upon the nature and the concentration of the acid present; thus the following values are recorded for the acids named in molar solution: H_2SO_4 1.44 V, HNO_3 1.61 V, $HClO_4$ 1.70 V, and in $8M$ perchloric acid solution the value is 1.87 V.

It has been postulated on the basis of the formal potential measurements that Ce(IV) exists as anionic complexes $[Ce(SO_4)_4]^{4-}$ or $[Ce(SO_4)_3]^{2-}$, $[Ce(NO_3)_6]^{2-}$, and $[Ce(ClO_4)_6]^{2-}$; in consequence, solid salts such as ammonium cerium(IV) sulphate $2(NH_4)_2SO_4.Ce(SO_4)_2,2H_2O$ and ammonium cerium(IV) nitrate

$2NH_4NO_3.Ce(NO_3)_4,4H_2O$ have been formulated as ammonium tetrasulph-atocerate(IV) $(NH_4)_4[Ce(SO_4)_4]2H_2O$ and ammonium hexanitratocerate(IV) $(NH_4)_2[Ce(NO_3)_6]4H_2O$ respectively. For convenience, the term cerium(IV) sulphate will be retained.

Solutions of cerium(IV) sulphate may be prepared by dissolving cerium(IV) sulphate or the more soluble ammonium cerium(IV) sulphate in dilute $(1-2N)$ sulphuric acid. Ammonium cerium(IV) nitrate may be purchased of A.R. quality, and a solution of this in M-sulphuric acid may be used for many of the purposes for which cerium(IV) solutions are employed, but in some cases the presence of nitrate ion is undesirable. The nitrate ion may be removed by evaporating the solid reagent with concentrated sulphuric acid, or alternatively a solution of the nitrate may be precipitated with aqueous ammonia and the resulting cerium(IV) hydroxide filtered off and dissolved in sulphuric acid.

Internal indicators suitable for use with cerium(IV) sulphate solutions include N-phenylanthranilic acid, ferroin, and 5,6-dimethylferroin.

X, 108. PREPARATION OF 0.1N CERIUM(IV) SULPHATE. *Method A.* Dissolve about 28 g A.R. ammonium cerium(IV) nitrate (equivalent $= 548.23$) in 100 cm^3 water in a 600-cm^3 beaker, add dilute ammonia solution slowly and with stirring until a slight excess is present (about 60 cm^3 *ca.* 2.5N-ammonia solution are required). Filter the precipitated cerium(IV) hydroxide with suction through a 7-cm sintered glass funnel, and wash with five 50-cm^3 portions of water to remove ammonium nitrate; leave the precipitate 'on the water pump' for about 30 minutes to remove as much water as possible. Transfer the precipitate back to the original beaker as far as possible and remove the residual hydroxide on the sintered glass filter by washing with four 50-cm^3 portions of 2M-sulphuric acid previously warmed to about 60 °C. Add the washings to the precipitate in the beaker, and warm until the precipitate dissolves completely. Allow to cool, transfer the solution to a 500-cm^3 graduated flask, and make up to the mark with distilled water. The resulting solution of cerium(IV) sulphate is about 0.1N, and requires standardisation before use.

Method B. Evaporate 55.0 g of A.R. ammonium cerium(IV) nitrate almost to dryness with excess (48 cm^3) of concentrated sulphuric acid in a Pyrex evaporating-dish. Dissolve the resulting cerium(IV) sulphate in M-sulphuric acid (28 cm^3 concentrated sulphuric acid to 500 cm^3 water), transfer to a 1-dm^3 graduated flask, add M-sulphuric acid until near the graduation mark, and make up to the mark with distilled water. Shake well.

Method C. The molecular weight and also the equivalent of cerium(IV) sulphate $Ce(SO_4)_2$ and ammonium cerium(IV) sulphate $(NH_4)_4[Ce(SO_4)_4],2H_2O$ are 333.25 and 632.56 respectively.

Weigh out 35–36 g of pure cerium(IV) sulphate into a 600-cm^3 beaker, add 56 cm^3 of 1:1-sulphuric acid and stir, with frequent additions of water and gentle warming, until the salt is dissolved. Transfer to a 1-dm^3 glass-stoppered graduated flask and, when cold, dilute to the mark with distilled water. Shake well.

Alternatively, weigh out 64–66 g of ammonium cerium(IV) sulphate into a solution prepared by adding 28 cm^3 of concentrated sulphuric acid to 500 cm^3 of water: stir the mixture until the solid has dissolved. Transfer to a 1-dm^3 graduated flask, and make up to the mark with distilled water.

Method D. Place about 21 g of cerium(IV) hydroxide in a 1500-cm^3 beaker, and add, with stirring, 100 cm^3 of concentrated sulphuric acid. Continue the stirring and introduce 300 cm³ of distilled water slowly and cautiously. Allow to stand overnight, and if any residue remains, filter the solution into a 1-dm³ graduated flask and dilute to the mark.

X, 109.   STANDARDISATION OF CERIUM(IV) SULPHATE SOLUTIONS.

Discussion. The most trustworthy method for standardising cerium(IV) sulphate solutions is with pure arsenic(III) oxide. The reaction between cerium(IV) sulphate solution and arsenic(III) oxide is very slow at the ordinary temperature; it is necessary to add a trace of osmium tetroxide as catalyst. The arsenic(III) oxide is dissolved in sodium hydroxide solution, the solution acidified with dilute sulphuric acid, and after adding 2 drops of an 'osmic acid' solution prepared by dissolving 0.1 g osmium tetroxide in 40 cm³ $0.1N$-sulphuric acid, and the indicator (1–2 drops ferroin or 0.5 cm³ N-phenylanthranilic acid), it is titrated with the cerium(IV) sulphate solution to the first sharp colour change: orange-red to very pale blue or yellowish-green to purple respectively.

$$2Ce^{4+} + H_3AsO_3 + H_2O = 2Ce^{3+} + H_3AsO_4 + 2H^+$$

Standardisation may also be carried out using pure iron, and also with A.R. sodium oxalate; in this last case, an indirect procedure must be used as the redox indicators are themselves oxidised at the elevated temperatures which are necessary. The procedure, therefore, is to add an excess of the cerium(IV) solution, and then, after cooling, the excess is determined by back titration with an iron(II) solution. It is possible to carry out a direct titration of the sodium oxalate if a potentiometric procedure is used (Chapter XIV).

Procedure A. **Standardisation with arsenic(III) oxide.** Weigh out accurately about 0.2 g of A.R. arsenic(III) oxide, previously dried at 105–110 °C for 1–2 hours, and transfer to a 400-cm³ beaker or to a 350-cm³ conical flask. Add 20 cm³ of about $2M$-sodium hydroxide solution, and warm the mixture gently until the arsenic(III) oxide has *completely* dissolved. Cool to room temperature, and add 100 cm³ water, followed by 25 cm³ $2.5M$-sulphuric acid. Then add 3 drops $0.01M$-osmium tetroxide solution (0.25 g osmium tetroxide dissolved in 100 cm³ $0.05M$-sulphuric acid) and 0.5 cm³ N-phenylanthranilic acid indicator (or 1–2 drops of ferroin). Titrate with the $0.1N$-cerium(IV) sulphate solution until the first sharp colour change occurs (see *Discussion* above). Repeat with two other samples of approximately equal weight of arsenic(III) oxide.

Procedure B. **Standardisation with pure iron.** Weigh out accurately 0.15–0.20 g A.R. iron wire, and then proceed exactly as described in Section **X, 92**, *Procedure C*. Titrate the resulting solution with the cerium(IV) sulphate solution, using any of the indicators referred to in the *Discussion*.

Procedure C. **Standardisation with sodium oxalate.** Prepare an approximately $0.1N$ solution of ammonium iron(II) sulphate in dilute sulphuric acid and titrate with the cerium(IV) sulphate solution using ferroin indicator.

Weigh out accurately about 0.2 g A.R. sodium oxalate into a 250-cm³ conical flask and add 25–30 cm³ M-sulphuric acid. Heat the solution to about 60 °C and then add about 30 cm³ of the cerium(IV) solution to be standardised dropwise, adding the solution as rapidly as possible consistent with drop formation. Reheat the solution to 60 °C, and then add a further 10 cm³ of the cerium(IV) solution. Allow to stand for three minutes, then cool and back-titrate the excess cerium(IV) with the iron(II) solution using ferroin as indicator.

Practically all the determinations described under potassium permanganate and potassium dichromate may be carried out with cerium(IV) sulphate. Use is made of the various indicators already detailed and also, in some cases where great accuracy is not required, of the pale yellow colour produced by the cerium(IV) sulphate itself. Only a few determinations will therefore be considered in some detail.

X, 110. DETERMINATION OF COPPER. *Discussion.* Divalent copper is quantitatively reduced in 2N-hydrochloric acid solution by means of the silver reductor (Section **X, 145**) to the copper(I) state. The solution of the copper(I) salt is collected in a solution of ammonium iron(III) sulphate, and the Fe^{2+} ion formed is titrated with standard cerium(IV) sulphate solution using ferroin or N-phenylanthranilic acid as indicator.

Comparatively large amounts of nitric acid, and also zinc, cadmium, bismuth, tin, and arsenate have no effect upon the determination; the method may therefore be applied to determine copper in brass.

Procedure (**copper in crystallised copper sulphate**). Weigh out accurately about 3.1 g A.R. copper sulphate crystals, dissolve in water, and make up to 250 cm^3 in a graduated flask. Shake well. Pipette 50 cm^3 of this solution into a small beaker, add an equal volume of *ca.* 4M-hydrochloric acid. Pass this solution through a silver reductor at the rate of 25 cm^3 per minute, and collect the filtrate in a 350-cm^3 conical flask charged with 20 cm^3 0.5M-iron(III) alum solution (prepared by dissolving the appropriate quantity of A.R. iron(III) alum in 0.5M-sulphuric acid). Wash the reductor column with six 25-cm^3 portions of 2M-hydrochloric acid. Add 1 drop of ferroin indicator or 0.5 cm^3 N-phenylanthranilic acid, and titrate with 0.1N-cerium(IV) sulphate solution. The end-point is sharp, and the colour imparted by the Cu^{2+} ions does not interfere with the detection of the equivalence point.

Procedure (**copper in copper(I) chloride**). Prepare an ammonium iron(III) sulphate solution by dissolving 10.0 g of the A.R. salt in about 80 cm^3 of 6N-sulphuric acid and dilute to 100 cm^3 with acid of the same strength. Weigh out accurately about 0.3 g of the sample of copper(I) chloride into a dry 250-cm^3 conical flask and add 25.0 cm^3 of the iron(III) solution. Swirl the contents of the flask until the copper(I) chloride dissolves, add a drop of two of ferroin indicator, and titrate with standard 0.1N-cerium(IV) sulphate.

Repeat the titration with 25.0 cm^3 of the iron solution, omitting the addition of the copper(I) chloride. The difference in the two titrations gives the volume of 0.1N-cerium(IV) sulphate which has reacted with the known weight of copper(I) chloride.

X, 111. DETERMINATION OF MOLYBDATE. *Discussion.* Molybdates [Mo(VI)] are quantitatively reduced in 2M-hydrochloric acid solution at 60–80 °C by the silver reductor to quinquevalent molybdenum [Mo(V)]. The reduced molybdenum solution is sufficiently stable over short periods of time in air to be titrated with standard cerium(IV) sulphate solution using ferroin or N-phenylanthranilic acid as indicator. Nitric acid must be completely absent; the presence of a little phosphoric acid during the reduction of the molybdenum(VI) is not harmful and, indeed, appears to increase the rapidity of the subsequent oxidation with cerium(IV) sulphate. Elements such as iron, copper, and vanadium interfere; nitrate interferes, since its reduction is catalysed by the presence of molybdates.

Procedure. Weigh out accurately about 2.5 g A.R. ammonium molybdate $(NH_4)_6Mo_7O_{24},4H_2O$, dissolve in water and make up to 250 cm^3 in a graduated flask. Pipette 50 cm^3 of this solution into a small beaker, add an equal volume of 4M-hydrochloric acid, then 3 cm^3 of 85 per cent phosphoric acid, and heat the solution to 60–80 °C. Pour hot 2M-hydrochloric acid through a silver reductor, and then pass the molybdate solution through the hot reductor at the rate of about 10 cm^3 per minute. Collect the reduced solution in a 400-cm^3 beaker or 350-cm^3 conical flask, and wash the reductor with six 25-cm^3 portions of 2M-hydrochloric acid; the first two washings should be made with the hot acid (rate: 10 cm^3 per minute) and the last four washings with the cold acid (rate: 20–25 cm^3 per minute). Cool the solution, add one drop of ferroin or 0.5 cm^3 N-phenylanthranilic acid, and titrate with standard 0.1N-cerium(IV) sulphate. The precipitate of cerium(IV) phosphate, which is initially formed, dissolves on shaking. Add the last 0.5 cm^3 of the reagent dropwise and with vigorous stirring or shaking.

X, 112. DETERMINATION OF TELLURITE. A measured excess of standard 0.1N-cerium(IV) sulphate is added to the tellurium(IV) solution (200 cm^3) containing 10 cm^3 of concentrated hydrochloric acid and about 0.05 g of chromium(III) sulphate as catalyst.* The solution is boiled for 10 minutes, then cooled and back-titrated with standard 0.1N-ammonium iron(II) sulphate, using N-phenylanthranilic acid or ferroin as indicator. The tellurium is oxidised from the tetra- to the hexavalent stage. Selenium does not interfere.

X, 113. DETERMINATION OF CERIUM(III). *Method A.* The cerium(III) salt in the form of sulphate in 100 cm^3 of 1:4-sulphuric acid is treated with 2 g of ammonium sulphate, 1 g of A.R. sodium bismuthate is added, and the solution heated to boiling. The mixture is cooled somewhat, 50 cm^3 of 2 per cent sulphuric acid added, filtered through a Gooch, porcelain, or sintered glass filtering crucible, and the crucible washed with 100–150 cm^3 of 2 per cent sulphuric acid.

$$2Ce^{3+} + BiO_3^- + 6H^+ = 2Ce^{4+} + Bi^{3+} + 3H_2O$$

Excess of 0.025N-ammonium iron(II) sulphate is added (as shown by the change from yellow to colourless, and the consequent complete reduction of the cerium(IV) to the cerium(III) salt), and the excess of iron(II) salt titrated with 0.1N-potassium permanganate to the first appearance of a pink colour.

Method B. 100–300 cm^3 of the solution, containing 0.1–0.3 g of cerium and 2.5–7.5 cm^3 of concentrated sulphuric acid, are treated with 1–1.5 g of A.R. ammonium persulphate and 10 drops of 0.1M-silver nitrate solution (catalyst), and then boiled for 10 minutes. The solution is cooled to room temperature, and is ready for titration. Two procedures may be used.

---

* Cerium(IV) sulphate alone does not oxidise selenite or tellurite, but it oxidises chromium to the hexavalent state and this, in turn, oxidises tellurite (but not selenite) to the hexavalent condition; any chromium(VI) at the end is reduced by the iron(II) sulphate. The cerium(IV) ion acts as a potential mediator (compare Section **X, 33,** *A*).

(i) Add 10–20 cm^3 of 10 per cent potassium iodide solution, and titrate the liberated iodine with 0.1N- or 0.025N-sodium thiosulphate. To avoid the oxidation of the hydriodic acid by the air, the titration should be performed in an inert atmosphere (N$_2$ or CO$_2$).

(ii) Titrate the solution with 0.1N- or 0.025N-ammonium iron(II) sulphate, using N-phenylanthranilic acid or ferroin as indicator.

X, 114. DETERMINATION OF NITRITES. *Discussion.* Satisfactory results are obtained by adding the nitrite solution to excess of standard 0.1N-cerium(IV) sulphate, and determining the excess of cerium(IV) sulphate with a standard iron(II) solution (compare Section **X, 97**).

$$2Ce^{4+} + NO_2^- + H_2O = 2Ce^{3+} + NO_3^- + 2H^+$$

For practice, determine the percentage of NO$_2$ in potassium nitrite, or the purity of sodium nitrite, preferably of A.R. quality.

Procedure. Weigh out accurately about 1.5 g of sodium nitrite and dissolve it in 500 cm^3 of boiled-out water in a graduated flask. Shake thoroughly. Place 50 cm^3 of standard 0.1N-cerium(IV) sulphate in a conical flask, and add 10 cm^3 of 2M-sulphuric acid. Transfer 25 cm^3 of the nitrite solution to this flask by means of a pipette, and keep the tip of the pipette below the surface of the liquid during the addition. Allow to stand for 5 minutes, and titrate the excess of cerium(IV) sulphate with standard 0.1N-ammonium iron(II) sulphate, using ferroin or N-phenylanthranilic acid as indicator. Repeat the titration with two further portions of the nitrite solution. Standardise the iron solution by titrating 25 cm^3 of it with the cerium(IV) solution in the presence of dilute sulphuric acid.

Determine the volume of the standard cerium(IV) sulphate solution which has reacted with the nitrite solution, and therefrom calculate the purity of the sodium nitrite employed.

Note. Cerium(IV) sulphate may also be used for the following analyses.

1. **Hydrogen peroxide.** The diluted solution, which may contain nitric, sulphuric, or hydrochloric acid in any concentration between 0.5 and 3N, is titrated directly with standard cerium(IV) sulphate solution, using ferroin or N-phenylanthranilic acid as indicator. The reaction is:

$$2Ce^{4+} + H_2O_2 = 2Ce^{3+} + O_2 + 2H^+$$

2. **Persulphate** (peroxydisulphate). Persulphate cannot be determined directly by reduction with iron(II) because the reaction is too slow:

$$S_2O_8^{2-} + 2Fe^{2+} = 2SO_4^{2-} + 2Fe^{3+}$$

An excess of a standard solution of iron(II) must therefore be added and the excess back-titrated with standard cerium(IV) sulphate solution. Erratic results are obtained, depending upon the exact experimental conditions, because of induced reactions leading to oxidation by air of iron(II) ion or to decomposition of the persulphate; these induced reactions are inhibited by bromide ion in concentrations not exceeding 1M and, under these conditions, the determination may be carried out in the presence of organic matter.

To 25.0 cm^3 of 0.01–0.015M-persulphate solution in a 150-cm^3 conical flask, add 7 cm^3 of 5M-sodium bromide solution and 2 cm^3 of 3M-sulphuric acid. Stopper the flask. Swirl the contents, then add excess of 0.05N-ammonium iron(II) sulphate (15.0 cm^3), and allow to stand for 20 minutes. Add 1 cm^3 of

0.001M-ferroin indicator, and titrate the excess of Fe^{2+} ion with 0.02N-cerium(IV) sulphate in 0.5M-sulphuric acid to the *first* colour change from orange to yellow.

3. **Uranium.** Uranium, as uranyl sulphate in solutions 4M in hydrochloric acid, is reduced quantitatively to tetravalent uranium on passage through a silver reductor at 60–90 °C: the uranium(IV) can be titrated with standard cerium(IV) sulphate solution.

Dissolve the uranium salt, containing 0.1–0.3 g of uranium, in 50 cm³ of 4M-hydrochloric acid and heat to 60–90 °C. Pre-treat a silver reductor (Section **X, 145**) with hot 4M-hydrochloric acid and pass the uranium(VI) solution through it at a rate of 20 cm³ per minute. Wash with hot 4M-hydrochloric acid. Cool the reduced solution, add 3 cm³ of 85 per cent phosphoric acid and one drop of ferroin indicator. Titrate with standard 0.1N-cerium(IV) sulphate to the disappearance of the pink colour. A little silver chloride may precipitate, but this does not affect the analysis. Run a blank determination and subtract the value found from the titre found in the uranium titration.

Calculate the percentage of uranium in the sample.

$$U^{4+} + 2Ce^{4+} + 2H_2O = UO_2^{2+} + 2Ce^{3+} + 4H^+$$

4. **Iron.** The determination of iron (e.g., in an iron ore) can be carried out by following the procedure given in Section **X, 109**, *Procedure B*.

5. **Oxalates.** Oxalates can be determined by means of the indirect method described in Section **X, 109**, *Procedure C*.

6. Hexacyanoferrate(II) can be determined by titration in M-H_2SO_4 using N-phenylanthranilic acid.

Oxidation and Reduction processes involving iodine Iodometric titrations

X, 115. GENERAL DISCUSSION. The **direct iodometric titration method** (sometimes termed *iodimetry*) refers to titrations *with* a standard solution of iodine. The **indirect iodometric titration method** (sometimes termed *iodometry*) deals with the titration *of* iodine liberated in chemical reactions. The normal reduction potential of the reversible system:

$$I_2 \text{ (solid)} + 2e \rightleftharpoons 2I^-$$

is 0.5345 volt. The above equation refers to a saturated aqueous solution in the presence of solid iodine; this half-cell reaction will occur, for example, towards the end of a titration of iodide with an oxidising agent such as potassium permanganate, when the iodide ion concentration becomes relatively low. Near the beginning, or in most iodometric titrations, when an excess of iodide ion is present, the tri-iodide ion is formed

$$I_2 \text{ (aq.)} + I^- \rightleftharpoons I_3^-$$

since iodine is readily soluble in a solution of iodide. The half-cell reaction is better written:

$$I_3^- + 2e \rightleftharpoons 3I^-$$

and the standard reduction potential is 0.5355 volt. Iodine or the tri-iodide ion is therefore a much weaker oxidising agent than potassium permanganate, potassium dichromate, and cerium(IV) sulphate.

In most direct titrations with iodine (iodimetry) a solution of iodine in potassium iodide is employed, and the reactive species is therefore the tri-iodide ion I_3^-. Strictly speaking, all equations involving reactions of iodine should be written with I_3^- rather than with I_2, e.g.,

$$I_3^- + 2S_2O_3^{2-} = 3I^- + S_4O_6^{2-}$$

is more accurate than

$$I_2 + 2S_2O_3^{2-} = 2I^- + S_4O_6^{2-}$$

For the sake of simplicity, however, the equations in this book will usually be written in terms of molecular iodine rather than the tri-iodide ion.

Strong reducing agents (substances with a much lower reduction potential), such as tin(II) chloride, sulphurous acid, hydrogen sulphide, and sodium thiosulphate, react completely and rapidly with iodine even in acid solution. With somewhat weaker reducing agents, e.g., trivalent arsenic, or trivalent antimony, complete reaction occurs only when the solution is kept neutral or very faintly acid; under these conditions the reduction potential of the reducing agent is a minimum, or its reducing power is a maximum.

If a strong oxidising agent is treated in neutral or (more usually) acid solution with a large excess of iodide ion, the latter reacts as a reducing agent and the oxidant will be quantitatively reduced. In such cases, an equivalent amount of iodine is liberated, and is then titrated with a standard solution of a reducing agent, which is usually sodium thiosulphate.

The normal reduction potential of the iodine–iodide system is independent of the pH of the solution so long as the latter is less than about 8; at higher values iodine reacts with hydroxide ions to form iodide and the extremely unstable hypoiodite, the latter being transformed rapidly into iodate and iodide by self-oxidation and reduction:

$$I_2 + 2OH^- = I^- + IO^- + H_2O$$
$$3IO^- = 2I^- + IO_3^-$$

The reduction potentials of certain substances increase considerably with increasing hydrogen-ion concentration of the solution. This is the case with systems containing permanganate, dichromate, arsenate, antimonate, bromate, etc., i.e., with anions which contain oxygen and therefore require hydrogen for complete reduction. Many weak oxidising anions are completely reduced by iodide ions if their reduction potentials are raised considerably by the presence in solution of a large amount of acid.

By suitable control of the pH of the solution, it is sometimes possible to titrate the reduced form of a substance with iodine, and the oxidised form, after the addition of iodide, with sodium thiosulphate. Thus with the arsenite–arsenate system:

$$H_3AsO_3 + I_2 + H_2O \rightleftharpoons H_3AsO_4 + 2H^+ + 2I^-$$

the reaction is completely reversible. At pH values between 4 and 9, arsenite can be titrated with iodine solution. In strongly acid solutions, however, arsenate is reduced to arsenite and iodine is liberated. Upon titration with sodium thiosulphate solution, the iodine is removed and the reaction proceeds from right to left.

Two important **sources of error in titrations involving iodine** are: (*a*) loss of iodine owing to its appreciable volatility; and (*b*) acid solutions of iodide are oxidised by oxygen from the air:

$$4I^- + O_2 + 4H^+ = 2I_2 + 2H_2O$$

In the presence of excess of iodide, the volatility is decreased markedly through the formation of the tri-iodide ion; at room temperature the loss of iodine by volatilisation from a solution containing at least 4 per cent of potassium iodide is negligible provided the titration is not prolonged unduly. Titrations should be performed in cold solutions in conical flasks and not in open beakers. If a solution is to stand it should be kept in a glass-stoppered vessel. The atmospheric oxidation of iodide is negligible in neutral solution in the absence of catalysts, but the rate of oxidation increases rapidly with decreasing pH. The reaction is catalysed by certain metal ions of variable valency (particularly copper), by nitrite ion, and also by strong light. For this reason titrations should not be performed in direct sunlight, and solutions containing iodide should be stored in amber glass bottles. Furthermore, the air oxidation of iodide ion may be induced by the reaction between iodide and the oxidising agent, especially when the main reaction is slow. Solutions containing an excess of iodide and acid must therefore not be allowed to stand longer than necessary before titration of the iodine. If prolonged standing is necessary (as in the titration of vanadate or Fe^{3+} ions) the solution should be free from air before the addition of iodide and the air displaced from the titration vessel by carbon dioxide (e.g., by adding small portions (0.2–0.5 g) of pure sodium hydrogencarbonate to the acid solution, or a little Dry Ice); potassium iodide is then introduced and the glass stopper replaced immediately.

It seems appropriate to refer at this point to the uses of a **standard solution containing potassium iodide and potassium iodate.** This solution is quite stable and yields iodine when treated with acid:

$$IO_3^- + 5I^- + 6H^+ = 3I_2 + 3H_2O$$

The standard solution is prepared by dissolving a weighed amount of pure potassium iodate in a solution containing a slight excess of pure potassium iodide, and diluting to a definite volume. This solution has two important uses. The first is as a source of a known quantity of iodine in titrations (compare Section **X, 118A**); it must be added to a solution containing strong acid; it cannot be employed in a medium which is neutral or possesses a low acidity.

The second use is in the determination of the acid content of solutions iodometrically or in **the standardisation of solutions of strong acids.** It is evident from the above equation that the amount of iodine liberated is equivalent to the acid content of the solution. Thus if, say, 25 cm^3 of an approximately $0.1N$ solution of a strong acid is treated with a slight excess of potassium iodate (say, 30 cm^3 of $0.1N$-potassium iodate solution, Section **X, 132**) and a slight excess of potassium iodide solution (say, 10 cm^3 of a 10 per cent solution), and the liberated iodine titrated with standard $0.1N$-sodium thiosulphate with the aid of starch as an indicator, the normality of the acid may be readily evaluated.

X, 116.   DETECTION OF THE END-POINT. A solution of iodine in aqueous iodide has an intense yellow to brown colour. One drop of $0.1N$-iodine solution imparts a perceptible pale yellow colour to 100 cm^3 of water, so that in otherwise colourless solutions iodine can serve as its own indicator. The test is made much more sensitive by the use of a solution of starch as indicator. Starch reacts with iodine in the presence of iodide to form an intensely blue-coloured complex, which is visible at very low concentrations of iodine. The sensitivity of the colour reaction is such that a blue colour is visible when the iodine concentration is $2 \times 10^{-5} M$ and the iodide concentration is greater than 4 $\times 10^{-4} M$ at 20 °C. The colour sensitivity decreases with increasing temperature of the solution; thus at 50 °C it is about ten times less sensitive than at 25 °C. The sensitivity decreases upon the addition of solvents, such as ethanol: no colour is obtained in solutions containing 50 per cent ethanol or more. It cannot be used in a strongly acid medium because hydrolysis of the starch occurs.

Starches can be separated into two major components, amylose and amylopectin, which exist in different proportions in various plants. Amylose, which is a straight-chain compound and is abundant in potato starch, gives a blue colour with iodine and the chain assumes a spiral form. Amylopectin, which has a branched-chain structure, forms a red-purple product, probably by adsorption.

The great merit of starch is that it is inexpensive. It possesses the following disadvantages: (i) insolubility in cold water; (ii) instability of suspensions in water; (iii) it gives a water-insoluble complex with iodine, the formation of which precludes the addition of the indicator early in the titration (for this reason, in titrations of iodine, the starch solution should not be added until just prior to the end-point when the colour begins to fade); and (iv) there is sometimes a 'drift' end-point, which is marked when the solutions are dilute.

Most of the shortcomings of starch as an indicator are absent in **sodium starch glycollate**. This is a white, non-hygroscopic powder, readily soluble in hot water to give a faintly opalescent solution, which is stable for many months; it does not form a water-insoluble complex with iodine, and hence the indicator may be added at any stage of the reaction. With excess of iodine (e.g., at the beginning of a titration with sodium thiosulphate) the colour of the solution containing 1 cm^3 of the indicator (0.1 per cent aqueous solution) is green; as the iodine concentration diminishes the colour changes to blue, which becomes intense just before the end-point is reached. The end-point is very sharp and reproducible and there is no 'drift' in dilute solution.

Carbon tetrachloride has been used in certain reactions instead of starch solution. One dm^3 of water at 25 °C will dissolve 0.335 g of iodine, but the same volume of carbon tetrachloride will dissolve about 28.5 g. Iodine is therefore about eighty-five times as soluble in carbon tetrachloride as it is in water, and the carbon tetrachloride solution is highly coloured. When a little carbon tetrachloride is added to an aqueous solution containing iodine and the solution well shaken, the great part of the iodine will dissolve in the carbon tetrachloride; the latter will fall to the bottom since it is immiscible with water, and the colour of the organic layer will be much deeper than that of the original aqueous solution. The reddish-violet colour of iodine in carbon tetrachloride is visible in very low concentrations of iodine; thus on shaking 10 cm^3 of carbon tetrachloride with 50 cm^3 of 2×10^{-5} N-iodine, a distinct violet coloration is produced in the organic layer. This enables many iodometric determinations to be carried out with comparative ease. The titrations are performed in 250-cm^3 glass-stoppered bottles or flasks with accurately ground stoppers. After adding the excess of potassium iodide solution and 5–10 cm^3 of carbon tetrachloride to the reaction mixture, the titration with sodium thiosulphate is commenced. At first the presence of iodine in the aqueous solution will be apparent, and gentle rotation of the liquid causes sufficient mixing. Towards the end of the titration the bottle or flask is stoppered and shaken after each addition of sodium thiosulphate solution; the end-point is reached when the carbon tetrachloride just becomes colourless. Equally satisfactory results can be obtained with chloroform.

Preparation and use of starch solution.   Make a paste of 1.0 g of soluble starch with a little water, and pour the paste, with constant stirring, into 100 cm^3 of boiling water, and boil for 1 minute. Allow the solution to cool and add 2–3 g of potassium iodide. Keep the solution in a stoppered bottle.

Only freshly prepared starch solution should be used. Two cm^3 of a 1 per cent solution per 100 cm^3 of the solution to be titrated is a satisfactory amount; the same volume of starch solution should always be added in a titration. In the titration of iodine, starch must not be added until just before the end-point is reached. Apart from the fact that the fading of the iodine colour is a good indication of the approach of the end-point, if the starch solution is added when the iodine concentration is high, some iodine may remain adsorbed even at the end-point. The indicator blank is negligibly small in iodimetric and iodometric titrations of 0.1N-solutions; with more dilute solutions, it must be determined in a liquid having the same composition as the solution titrated has at the end-point.

A solid solution of starch in urea may also be employed. Reflux 1 g soluble starch and 19 g urea with xylene. At the boiling point of the organic solvent the urea melts with little decomposition, and the starch dissolves in the molten urea. Allow to cool, then remove the solid mass and powder it; store the product in a stoppered bottle. A few milligrams of this solid added to an aqueous solution containing iodine then behaves like the usual starch indicator.

Preparation and use of sodium starch glycollate indicator.   Sodium starch glycollate, prepared as described below, dissolves slowly in cold but rapidly in hot water. It is best dissolved by mixing, say, 5.0 g of the finely powdered solid with 1–2 cm^3 ethanol, adding 100 cm^3 cold water, and boiling for a few minutes with vigorous stirring: a faintly opalescent solution results. This 5 per cent stock solution is diluted to 1 per cent strength as required. The most convenient concentration for use as an indicator is 0.1 mg/cm^{-3}, i.e., 1 cm^3 of the 1 per cent aqueous solution is added to 100 cm^3 of the solution being titrated.

X, 117.   PREPARATION OF 0.1N-SODIUM THIOSULPHATE.   *Discussion.*   Sodium thiosulphate $Na_2S_2O_3,5H_2O$ is readily obtainable in a state of high purity, but there is always some uncertainty as to the exact water content because of the efflorescent nature of the salt and for other reasons. The substance is therefore unsuitable as a primary standard. It is a reducing agent by virtue of the half-cell reaction:

$$2S_2O_3{}^{2-} \rightleftharpoons S_4O_6{}^{2-} + 2e$$

the equivalent of sodium thiosulphate pentahydrate is the mole, or 248.18. An approximately 0.1N solution is prepared by dissolving about 25 g A.R. crystallised sodium thiosulphate in 1 litre of water in a graduated flask. The solution is standardised by any of the methods described below.

Before dealing with these, it is necessary to refer briefly to the stability of thiosulphate solutions. Solutions prepared with conductivity (equilibrium) water are perfectly stable. However, ordinary distilled water usually contains an excess of carbon dioxide; this may cause a slow decomposition to take place with the formation of sulphur:

$$S_2O_3{}^{2-} + H^+ = HSO_3{}^- + S$$

Moreover, decomposition may also be caused by bacterial action (e.g., *thiobacillus thioparus*), particularly if the solution has been standing for some time. For these reasons, the following recommendations are made:
1. Prepare the solution with recently boiled distilled water.
2. Add 3 drops of chloroform or 10 mg of mercury(II) iodide per litre; these compounds improve the keeping qualities of the solution.
(Bacterial activity is least when the pH lies between 9 and 10. The addition of a *small* amount, 0.1 g per litre, of sodium carbonate is advantageous to ensure the correct pH. In general, alkali hydroxides, sodium carbonate (>0.1 g/l), and sodium tetraborate should not be added, since they tend to accelerate the decomposition:

$$S_2O_3{}^{2-} + 2O_2 + H_2O \rightleftharpoons 2SO_4{}^{2-} + 2H^+)$$

3. Avoid exposure to light, as this tends to hasten the decomposition.
The standardisation of thiosulphate solutions may be effected with potassium iodate, potassium dichromate, copper and iodine as primary standards, or with potassium permanganate or cerium(IV) sulphate as secondary standards. Owing to the volatility of iodine and the difficulty of preparation of perfectly pure iodine, this method is not a suitable one for beginners. If, however, a standard solution of iodine (see Sections **X, 119, 120**) is available, this may be used for the standardisation of thiosulphate solutions.
Procedure. Weigh out 25 g A.R. sodium thiosulphate crystals, $Na_2S_2O_3$, $5H_2O$, dissolve in boiled-out distilled water, and make up to 1 litre in a graduated flask with boiled-out water. If the solution is to be kept for more than a few days, add 0.1 g sodium carbonate or 3 drops of chloroform.

X, 118. STANDARDISATION OF SODIUM THIOSULPHATE SOLU-TIONS. A. With potassium iodate.

A.R. potassium iodate has a purity of at least 99.9 per cent: it can be dried at 120 °C. This reacts with potassium iodide in acid solution to liberate iodine:

$$IO_3{}^- + 5I^- + 6H^+ = 3I_2 + 3H_2O$$

Its equivalent as an oxidising agent is $\frac{1}{6}$ mole or 214.00/6; a $0.1N$ solution therefore contains 3.567 g of potassium iodate per dm^3.
Weigh out accurately 0.14–0.15 g of pure dry potassium iodate, dissolve it in 25 cm^3 of cold, boiled-out distilled water, add 2 g of iodate-free potassium iodide* and 5 cm^3 of M-sulphuric acid (1). Titrate the liberated iodine with the thiosulphate solution with constant shaking. When the colour of the liquid has become a pale yellow, dilute to *ca.* 200 cm^3 with distilled water, add 2 cm^3 of starch solution, and continue the titration until the colour changes from blue to colourless. Repeat with two other similar portions of potassium iodate.
Note. 1. Potassium iodate has a small equivalent (35.67) so that the error in weighing 0.14–0.15 g may be appreciable. In this case it is better to weigh out accurately 3.567 g of the A.R. salt (if a slightly different weight is used, the exact normality is calculated), dissolve it in water, and make up to 1 dm^3 in a graduated flask. Twenty-five cm^3 of this solution are treated with excess of pure potassium

---

* The absence of iodate is indicated by adding dilute sulphuric acid when no immediate yellow coloration should be obtained. If starch is added, no immediate blue coloration should be produced.

iodide (1 g of the solid or 10 cm^3 of 10 per cent solution), followed by 3 cm^3 of M-sulphuric acid, and the liberated iodine is titrated as detailed above.

B. With potassium dichromate.   Potassium dichromate is reduced by an acid solution of potassium iodide, and iodine is set free:

$$Cr_2O_7{}^{2-} + 6I^- + 14H^+ = 2Cr^{3+} + 3I_2 + 7H_2O$$

This reaction is subject to a number of errors: (1) the hydriodic acid (from excess of iodide and acid) is readily oxidised by air, especially in the presence of chromium(III) salts, and (2) it is not instantaneous. It is accordingly best to pass a current of carbon dioxide through the reaction flask before and during the titration (a more convenient but less efficient method is to add some solid sodium hydrogencarbonate to the acid solution, and to keep the flask covered as much as possible), and to allow 5 minutes for its completion.

Place 100 cm^3 of cold, recently boiled distilled water in a 500-cm^3 conical, preferably glass-stoppered, flask, add 3 g of iodate-free potassium iodide and 2 g of pure sodium hydrogencarbonate, and shake until the salts dissolve. Add 6 cm^3 of concentrated hydrochloric acid slowly whilst gently rotating the flask in order to mix the liquids; run in 25.0 cm^3 of standard 0.1N-potassium dichromate (1), mix the solutions well, and wash the sides of the flask with a little boiled-out water from the wash bottle. Stopper the flask (or cover it with a small watch glass), and allow to stand in the dark for 5 minutes in order to complete the reaction. Rinse the stopper or watch glass, and dilute the solution with 300 cm^3 of cold, boiled-out water. Titrate the liberated iodine with the sodium thiosulphate solution contained in a burette, whilst constantly rotating the liquid so as to thoroughly mix the solutions. When most of the iodine has reacted as indicated by the solution acquiring a yellowish-green colour, add 2 cm^3 of starch solution and rinse down the sides of the flask; the colour should change to blue. Continue the addition of the thiosulphate solution dropwise, and swirling the liquid constantly, until 1 drop changes the colour from greenish-blue to light green. The end-point is sharp, and is readily observed in a good light against a white background. Carry out a blank determination, substituting distilled water for the potassium dichromate solution; if the potassium iodide is iodate-free, this should be negligible.

Note.   1. If preferred, about 0.20 g of A.R. potassium dichromate may be accurately weighed out, dissolved in 50 cm^3 of cold, boiled-out water, and the titration carried out as detailed above.

The following *alternative procedure* utilises a trace of copper sulphate as a catalyst to increase the speed of the reaction; in consequence, a weaker acid (acetic acid) may be employed and the extent of atmospheric oxidation of hydriodic acid reduced. Place 25.0 cm^3 of 0.1N-potassium dichromate in a 250-cm^3 conical flask, add 5.0 cm^3 of glacial acetic acid, 5 cm^3 of 0.001M-copper sulphate, and wash the sides of the flask with distilled water. Add 30 cm^3 of 10 per cent potassium iodide solution, and titrate the iodine as liberated with the approximately 0.1N-thiosulphate solution, introducing a little starch indicator towards the end. The titration may be completed in 3–4 minutes after the addition of the potassium iodide solution. Subtract 0.05 cm^3 to allow for the iodine liberated by the copper sulphate catalyst.

A **standardised solution of potassium permanganate** may be used in place of the potassium dichromate solution, adding 2 cm^3 of concentrated hydrochloric acid to each 25-cm^3 portion of potassium permanganate solution; in this case the alternative procedure of weighing out a portion of the salt cannot be used.

C. **With a standard solution of iodine.**   If a standard solution of iodine is available (see Section **X, 119**), this may be used to standardise the thiosulphate solution. Measure a 25.0-cm^3 portion of the standard iodine solution into a 250-cm^3 conical flask, add about 150 cm^3 distilled water and titrate with the thiosulphate solution, adding 2 cm^3 of starch solution when the liquid is pale yellow in colour.

When thiosulphate solution is added to a solution containing iodine the overall reaction, which occurs rapidly and stoichiometrically under the usual experimental conditions (pH $<$ 5), is:

$$2S_2O_3{}^{2-} + I_2 = S_4O_6{}^{2-} + 2I^-$$

or $$2S_2O_3{}^{2-} + I_3{}^- = S_4O_6{}^{2-} + 3I^-$$

It has been shown that the colourless intermediate $S_2O_3I^-$ is formed by a rapid reversible reaction:

$$S_2O_3{}^{2-} + I_2 \rightleftharpoons S_2O_3I^- + I^-$$

The intermediate reacts with thiosulphate ion to provide the main course of the overall reaction:

$$S_2O_3I^- + S_2O_3{}^{2-} = S_4O_6{}^{2-} + I^-$$

The intermediate also reacts with iodide ion:

$$2S_2O_3I^- + I^- = S_4O_6{}^{2-} + I_3{}^- ;$$

this explains the reappearance of iodine after the end-point in the titration of very dilute iodine solutions by thiosulphate.

D. **With cerium(IV) sulphate.**   This method for standardising sodium thiosulphate solutions makes use of a secondary standard, but gives satisfactory results provided the experimental conditions given below are rigidly adhered to; this is due to the fact that cerium(IV) sulphate solution contains free acid, which may otherwise lead to appreciable errors.

For 0.1N-cerium(IV) sulphate (Sections **X, 108–109**), use 25.0 cm^3 of the *ca.* 0.1N-sodium thiosulphate solution, 0.3–0.4 g of pure potassium iodide, 2 cm^3 of 0.2 per cent starch solution, dilute to 250 cm^3, and titrate with the cerium(IV) sulphate solution to the starch–iodine end-point, i.e., to a first permanent blue colour.

The reaction is:

$$2Ce^{4+} + 2I^- = 2Ce^{3+} + I_2$$

X, 119.   PREPARATION OF 0.1N-IODINE SOLUTION.   *Discussion.*

0.335 gram of iodine dissolves in 1 dm^3 of water at 25 °C. In addition to this small solubility, aqueous solutions of iodine have an appreciable vapour pressure of iodine, and therefore decrease slightly in concentration on account of volatilisation when handled. Both difficulties are overcome by dissolving the iodine in an aqueous solution of potassium iodide. Iodine dissolves readily in aqueous potassium iodide, the more concentrated the solution, the greater is the solubility of the iodine. The increased solubility is due to the formation of a tri-iodide ion:

$$I_2 + I^- \rightleftharpoons I_3{}^-$$

The resulting solution has a much lower vapour pressure than a solution of iodine in pure water, consequently the loss by volatilisation is considerably diminished. Nevertheless, the vapour pressure is still appreciable so that *precautions should always be taken to keep vessels containing iodine closed except during the actual titrations.* When an iodide solution of iodine is titrated with a reductant, the free iodine reacts with the reducing agent, this displaces the equilibrium to the left, and eventually all the tri-iodide is decomposed; the solution therefore behaves as though it were a solution of free iodine.

For the preparation of standard iodine solutions, A.R. or resublimed iodine and iodate-free (e.g., A.R.) potassium iodide should be employed. The solution may be standardised against pure arsenic(III) oxide or with a sodium thiosulphate solution which has been recently standardised against potassium iodate.

The equation:

$$I_2 + 2e \rightleftharpoons 2I^-$$

indicates that the equivalent is equal to the atomic weight, or 126.905 g.

Procedure. **Preparation of 0.1N-iodine.** Dissolve 20 g of iodate-free potassium iodide (e.g., A.R.) in 30–40 cm^3 of water in a glass-stoppered 1 dm^3 graduated flask. Weigh out about 12.7 g of A.R. or resublimed iodine on a watch glass on a rough balance (never on an analytical balance on account of the iodine vapour), and transfer it by means of a small dry funnel into the concentrated potassium iodide solution. Insert the glass stopper into the flask, and shake in the cold until all the iodine has dissolved. Allow the solution to acquire room temperature, and make up to the mark with distilled water.

The iodine solution is best preserved in small glass-stoppered bottles. These should be filled completely and kept in a cool, dark place.

X, 120. STANDARDISATION OF IODINE SOLUTIONS. A. With arsenic(III) oxide. *Discussion.* As already indicated (Section **X, 92**), A.R. arsenic(III) oxide which has been dried at 105–110 °C for two hours is an excellent primary standard. The reaction between this substance and iodine is a reversible one:

$$H_3AsO_3 + I_2 + H_2O \rightleftharpoons H_3AsO_4 + 2H^+ + 2I^-$$

and only proceeds quantitatively from left to right if the hydrogen iodide is removed from the solution as fast as it is formed. This may be done by the addition of sodium hydrogencarbonate: sodium carbonate and sodium hydroxide cannot be used, since they react with the iodine, forming iodide, hypoiodite, and iodate. Actually it has been shown that complete oxidation of the arsenite occurs when the pH of the solution lies between 4 and 9, the best value being 6.5, which is very close to the neutral point. Buffer solutions are employed to maintain the correct pH. A 0.12N solution of sodium hydrogencarbonate saturated with carbon dioxide has a pH of 7; a solution saturated with both sodium tetraborate and boric acid has a pH of about 6.2, whilst a Na_2HPO_4–NaH_2PO_4 solution is almost neutral. Any of these three buffer solutions is suitable, but as already stated the first-named is generally employed.

Procedure. Weigh out accurately about 2.5 g of finely powdered A.R. arsenic(III) oxide, transfer to a 400-cm^3 beaker, and dissolve it in a concentrated solution of sodium hydroxide, prepared from 2 g of iron-free sodium hydroxide (e.g., A.R.) and 20 cm^3 of water. Dilute to about 200 cm^3, and neutralise the solution with M-hydrochloric acid, using a strip of litmus paper as indicator. When the solution is faintly acid, remove the litmus paper by means of a stirring rod and carefully rinse both the rod and the paper. Transfer the contents of the beaker quantitatively to a 500-cm^3 graduated flask, add 2 g of pure sodium hydrogencarbonate, and, when all the salt has dissolved, dilute to the mark and shake well.

Measure out from a **burette** (this is necessary owing to the poisonous properties of the solution) 25.0 cm^3 of the arsenite solution into a 250-cm^3 conical flask, add 25–50 cm^3 of water, 5 g of sodium hydrogencarbonate, and 2 cm^3 of starch solution. Swirl the solution carefully until the hydrogencarbonate has dissolved. Then titrate slowly with the iodine solution, contained in a burette, to the first blue colour.

Alternatively, the arsenite solution may be placed in the burette, and titrated against 25.0 cm^3 of the iodine solution contained in a conical flask. When the solution has a pale yellow colour, add 2 cm^3 of starch solution, and continue the titration slowly until the blue colour is just destroyed.

If it is desired to base the standardisation directly upon arsenic(III) oxide, proceed as follows. Weigh out accurately about 0.20 g of pure arsenic(III) oxide into a conical flask, dissolve it in 10 cm^3 of M-sodium hydroxide, and add a small excess of dilute sulphuric acid (say, 12–15 cm^3 of N acid). Mix thoroughly and cautiously. Then add carefully a solution of 2 g of sodium hydrogencarbonate in 50 cm^3 of water, followed by 2 cm^3 of starch solution. Titrate slowly with the iodine solution to the first blue colour. Repeat with two other similar quantities of the oxide.

B. With standard sodium thiosulphate solution. Sodium thiosulphate solution, which has been recently standardised, preferably against pure potassium iodate, is employed. Transfer 25 cm^3 of the iodine solution to a 250-cm^3 conical flask, dilute to 100 cm^3 and add the standard thiosulphate solution from a burette until the solution has a pale-yellow colour. Add 2 cm^3 of starch solution, and continue the addition of the thiosulphate solution slowly until the solution is just colourless.

X, 121. DETERMINATION OF COPPER IN CRYSTALLISED COPPER SULPHATE.

Procedure. Weigh out accurately about 3.0 g of the salt, dissolve it in water, and make up to 250 cm^3 in a graduated flask. Shake well. Pipette 50.0 cm^3 of this solution into a 250-cm^3 conical flask, add 1 g potassium iodide (or 10 cm^3 of a 10 per cent solution) (1), and titrate the liberated iodine with standard 0.1N-sodium thiosulphate (2). Repeat the titration with two other 50-cm^3 portions of the copper sulphate solution.

The reaction, written in molecular form, is:

$$2CuSO_4 + 4KI = 2CuI + I_2 + 2K_2SO_4$$

from which it follows that:

$$2CuSO_4 \equiv I_2 \equiv 2Na_2S_2O_3$$

Notes. 1. If in a similar determination, free mineral acid is present, a few drops of dilute sodium carbonate solution must be added until a *faint* permanent precipitate remains, and this is removed by means of a drop or two of acetic acid. The potassium iodide is then added and the titration continued. For accurate results, the solution should have a pH of 4–5.5.

2. After the addition of the potassium iodide solution, run in standard 0.1N-sodium thiosulphate until the brown colour of the iodine fades, then add 2 cm^3 of starch solution, and continue the addition of the thiosulphate solution until the blue colour commences to fade. Then add about 1 g of A.R. potassium or ammonium thiocyanate, preferably as a 10 per cent aqueous solution: the blue colour will instantly become more intense. Complete the titration as quickly as possible. The precipitate possesses a pale flesh colour, and a distinct permanent end-point is readily obtained.

X, 122.   DETERMINATION OF COPPER IN AN ORE.   *Discussion.*   Of the common elements which are usually associated with copper ores, those that interfere with the iodometric determination are iron, arsenic, and antimony. Trivalent iron is reduced by iodide:

$$2Fe^{3+} + 2I^- \rightleftharpoons 2Fe^{2+} + I_2$$

but by the addition of excess of fluoride, the iron(III) is converted into the complex $[FeF_6]^{3-}$, which yields so small a concentration of Fe^{3+} ions that it has no oxidising action upon the iodide. Arsenic and antimony in the trivalent form react with iodine, but in consequence of the oxidising medium usually employed to bring the sample into solution they will be present in the quinquevalent form. Arsenic(V) and antimony(V) compounds will not oxidise iodide in a solution having a pH greater than about 3.2. By the use of excess of ammonium hydrogen-fluoride, NH_4HF_2, which acts as a buffer, the pH of the solution can be maintained above 3.2; under these conditions the reduction of the Cu^{2+} ion proceeds to completion. The concentration of the fluoride should be 1.0–1.6M.

Procedure.   The ore (copper pyrites) may be dissolved in concentrated nitric acid but it is then necessary to evaporate down with concentrated sulphuric acid to remove the nitric acid which would liberate iodine from potassium iodide. It is therefore preferable to employ perchloric acid which does not give rise to this problem.

Weigh out accurately about 0.6 g of the dry, finely ground ore into a dry, narrow-mouth Pyrex flask. Add about 15 cm^3 of 72 per cent perchloric acid (**Caution**), two small glass beads to promote regular ebullition, and insert a short-necked glass funnel into the neck of the flask. Heat the flask *gently* in the fume cupboard: the acid should reflux down the sides of the flask, but no pronounced white fumes of perchloric acid should leave the flask. The sample should dissolve in about 5 minutes, by which time the condensation ring of acid will have reached half-way up the walls of the flask. Remove the burner beneath the flask and allow to cool for 3–4 minutes. Add 50 cm^3 of water carefully through the funnel, mix well, and boil the solution for 5 minutes. The boiling will remove the free chlorine formed in the oxidation of the mineral. Cool to room temperature. Add dilute aqueous ammonia solution (1:1) dropwise until the solution smells slightly of ammonia. This will precipitate iron hydroxide; an excess of ammonia solution should be avoided. Add 2.0 g ammonium hydrogenfluoride NH_4HF_2 and shake until all the iron hydroxide has dissolved. Now add 3 g A.R. potassium iodide dissolved in 5–10 cm^3 water, and titrate at once with standard 0.1N-sodium thiosulphate, adding 2 cm^3 starch solution when the brown colour of the iodine decreases in intensity. Continue the addition of the thiosulphate solution until the blue colour becomes faint. Then add 20 cm^3 of 10 per cent aqueous ammonium of potassium thiocyanate solution, and complete the titration without delay.

X, 123. DETERMINATION OF CHLORATES. *Discussion.* One procedure is based upon the reaction between chlorate and iodide in the presence of concentrated hydrochloric acid:

$$ClO_3^- + 6I^- + 6H^+ = Cl^- + 3I_2 + 3H_2O$$

The liberated iodine is titrated with standard sodium thiosulphate solution.

In another method the chlorate is reduced with bromide in the presence of *ca.* 8*M*-hydrochloric acid, and the bromine liberated is determined iodimetrically:

$$ClO_3^- + 6Br^- + 6H^+ = Cl^- + 3Br_2 + 3H_2O$$

Procedure. A. Place 25 cm^3 of the chlorate solution (0.1*N*) in a glass-stoppered conical flask and add 3 cm^3 of concentrated hydrochloric acid followed by two portions of about 0.3 g each of pure sodium hydrogencarbonate to remove air. Add immediately about 1.0 g of iodate-free potassium iodide and 22 cm^3 of concentrated hydrochloric acid. Stopper the flask, shake the contents, and allow to stand for 5–10 minutes. Titrate the solution with standard 0.1*N*-sodium thiosulphate in the usual manner.

B. Place 10.0 cm^3 of the chlorate solution in a glass-stoppered flask, add *ca.* 1.0 g A.R. potassium bromide and 20 cm^3 concentrated hydrochloric acid (the final concentration of acid should be about 8*M*). Stopper the flask, shake well, and allow to stand for 5–10 minutes. Add 100 cm^3 of 1 per cent potassium iodide solution, and titrate the liberated iodine with standard 0.1*N*-sodium thiosulphate.

X, 124. ANALYSIS OF HYDROGEN PEROXIDE. *Discussion.* Hydrogen peroxide reacts with iodide in acid solution in accordance with the equation:

$$H_2O_2 + 2H^+ + 2I^- = I_2 + 2H_2O$$

The reaction velocity is comparatively slow, but increases with increasing concentration of acid. The addition of 3 drops of a neutral 20 per cent ammonium molybdate solution renders the reaction almost instantaneous, but as it also accelerates the atmospheric oxidation of the hydriodic acid, the titration is best conducted in an inert atmosphere (N_2 or CO_2).

The iodometric method has the advantage over the permanganate method (Section **X, 95**) that it is less affected by stabilisers which are sometimes added to commercial hydrogen peroxide solutions. These preservatives are often boric acid, salicylic acid, and glycerol, and render the results obtained by the permanganate procedure less accurate.

Procedure. Dilute the hydrogen peroxide solution to *ca.* 0.3 per cent H_2O_2. Thus, if a '20-volume' hydrogen peroxide is used, transfer 10.0 cm^3 by means of a burette or pipette to a 250-cm^3 graduated flask, and make up to the mark. Shake well. Remove 25.0 cm^3 of this diluted solution, and add it gradually and with constant stirring to a solution of 1 g of pure potassium iodide in 100 cm^3 of *M*-sulphuric acid (1:20) contained in a stoppered bottle. Allow the mixture to stand for 15 minutes, and titrate the liberated iodine with standard 0.1*N*-sodium thiosulphate, adding 2 cm^3 starch solution when the colour of the iodine has been nearly discharged. Run a blank determination at the same time.

Better results are obtained by transferring 25.0 cm^3 of the diluted hydrogen peroxide solution to a conical flask, and adding 100 cm^3 M (1:20) sulphuric acid. Pass a slow stream of carbon dioxide or nitrogen through the flask, add 10 cm^3 of 10 per cent potassium iodide solution, followed by 3 drops of 3 per cent ammonium molybdate solution. Titrate the liberated iodine immediately with standard 0.1N-sodium thiosulphate in the usual way.

Note.   The above method may also be used for all per-salts.

X, 125.   DETERMINATION OF THE AVAILABLE CHLORINE IN BLEACHING POWDER.   *Discussion.*   Bleaching powder consists essentially of a mixture of calcium hypochlorite $Ca(OCl)_2$ and the basic chloride $CaCl_2$, $Ca(OH)_2,H_2O$; some free slaked lime is usually present. The active constituent is the hypochlorite, which is responsible for the bleaching action. Upon treating bleaching powder with hydrochloric acid, chlorine is liberated.

$$OCl^- + Cl^- + 2H^+ = Cl_2 + H_2O$$

The **available chlorine** refers to the chlorine liberated by the action of dilute acids, and is expressed as the percentage by weight of the bleaching powder. The bleaching powder of commerce contains 36–38 per cent of available chlorine.

Two methods are in common use for the determination of the available chlorine. In the first, the bleaching powder solution or suspension is treated with an excess of a solution of potassium iodide, and strongly acidified with acetic acid:

$$OCl^- + 2I^- + 2H^+ \rightleftharpoons Cl^- + I_2 + H_2O$$

The liberated iodine is titrated with standard sodium thiosulphate solution. The solution should not be strongly acidified with hydrochloric acid, for the little calcium chlorate which is usually present, by virtue of the decomposition of the hypochlorite, will react slowly with the potassium iodide and liberate iodine:

$$ClO_3^- + 6I^- + 6H^+ = Cl^- + 3I_2 + 3H_2O$$

In the second method, the bleaching powder solution or suspension is titrated against standard 0.1N-sodium arsenite solution; this is best done by adding an excess of the arsenite solution and then back-titrating with standard iodine solution.

*Procedure A (**iodometric method**).*   Weigh out accurately about 5.0 g of the bleaching powder into a clean glass mortar. Add a little water, and rub the mixture to a smooth paste. Add a little more water, triturate with the pestle, allow the mixture to settle, and pour off the milky liquid into a 500-cm^3 graduated flask. Grind the residue with a little more water, and repeat the operation until the whole of the sample has been transferred to the flask either in solution or in a state of very fine suspension, and the mortar washed quite clean. The flask is then filled to the mark with distilled water, well shaken, and 50.0 cm^3 of the turbid liquid immediately withdrawn with a pipette. This is transferred to a 250-cm^3 conical flask, 25 cm^3 of water added, followed by 2 g of iodate-free potassium iodide (or 20 cm^3 of a 10 per cent solution) and 10 cm^3 of glacial acetic acid. Titrate the liberated iodine with standard 0.1N-sodium thiosulphate.

*Procedure B (**arsenite method**).*   Prepare a bleaching powder solution (suspension) as above and transfer 50 cm^3 to a 350-cm^3 conical flask. Add from a burette 75 cm^3 of standard (approximately 0.1N) sodium arsenite solution (Section **X, 120**), then titrate the excess arsenite with standard (approx. 0.1N) iodine solution.

The **concentration of any hypochlorite solution** can be determined by either of the procedures detailed above.

X, 126. DETERMINATION OF ARSENIC(V). The reaction is the reverse of that employed in the standardisation of iodine with sodium arsenite solution (Section **X, 120**):

$$As_2O_5 + 4H^+ + 4I^- \rightleftharpoons As_2O_3 + 2I_2 + 2H_2O$$

$$\text{or} \quad H_3AsO_4 + 2H^+ + 2I^- \rightleftharpoons H_3AsO_3 + I_2 + H_2O$$

For good results, the following experimental conditions must be observed: (i) the hydrochloric acid concentration in the final solution should be at least 4M; (ii) air should be displaced from the titration mixture by adding a little solid sodium hydrogencarbonate; (iii) the solution must be allowed to stand for at least 5 minutes before the liberated iodine is titrated; and (iv) constant stirring is essential during the titration to prevent decomposition of the thiosulphate in the strongly acid solution.

Treat the arsenate solution (say, 20.0 cm^3 of $ca.$ 0.1N) in a glass-stoppered conical flask with concentrated hydrochloric acid to give an $ca.$ 4M solution in hydrochloric acid. Displace the air by introducing two 0.4 g portions of pure sodium hydrogencarbonate into the flask. Add 1.0 g of pure potassium iodide, replace the stopper, mix the solution, and allow to stand for at least 5 minutes. Titrate the solution, whilst stirring vigorously, with standard 0.1N-sodium thiosulphate.

A similar procedure may also be used for the determination of **pentavalent antimony**, whilst **trivalent antimony** may be determined like arsenic(III) by direct titration with standard iodine solution (Section **X, 120 A**), but in the antimony titration it is necessary to include some tartaric acid in the solution; this acts as complexing agent and prevents precipitation of antimony as hydroxide or as basic salt in alkaline solution. On the whole, however, the most satisfactory method for determining antimony is by titration with potassium bromate (Section **X, 139**).

X, 127. DETERMINATION OF SULPHUROUS ACID AND OF SUL-PHITES. *Discussion.* The iodimetric determination is based upon the equations:

$$SO_3^{2-} + I_2 + H_2O = SO_4^{2-} + 2H^+ + 2I^-$$

$$HSO_3^- + I_2 + H_2O = SO_4^{2-} + 3H^+ + 2I^-$$

For accurate results, the following experimental conditions must be observed:
(a) the solutions should be very dilute;
(b) the sulphite must be added slowly and with constant stirring to the iodine solution, and not conversely; and
(c) exposure of the sulphite to the air should be minimised.

In determinations of sulphurous acid and sulphites, excess of standard 0.1N-iodine is diluted with several volumes of water, acidified with hydrochloric or sulphuric acid, and a known volume of the sulphite or sulphurous acid solution is added slowly and with constant stirring from a burette, with the jet close to the surface of the liquid. The excess of iodine is then titrated with standard $0.1N$-sodium thiosulphate. Solid soluble sulphites are finely powdered and added directly to the iodine solution. Insoluble sulphites (e.g., calcium sulphite) react very slowly, and must be in a very fine state of division.

Procedure. Pipette 25.0 cm³ standard $(0.1N)$ iodine solution into a 350-cm³ conical flask and add 5 cm³ $2M$-hydrochloric acid and 150 cm³ distilled water. Weigh accurately sufficient solid sulphite to react with about 20 cm³ $0.1N$-iodine solution and add this to the contents of the flask; swirl the liquid until all the solid has dissolved and then titrate the excess iodine with standard $(0.1N)$ sodium thiosulphate using starch indicator. If the sulphite is in solution, then a volume of this equivalent to about 20 cm³ of $0.1N$-iodine should be pipetted into the contents of the flask in place of the weighed amount of solid.

X, 128. DETERMINATION OF HYDROGEN SULPHIDE AND SULPHIDES. *Discussion.* The iodimetric method utilises the reversible reaction

$$H_2S + I_2 \rightleftharpoons 2H^+ + 2I^- + S$$

For reasonably satisfactory results, the sulphide solution must be dilute (concentration not greater than 0.04 per cent or $0.02N$), and the sulphide solution added to excess of acidified $0.01N$- or $0.1N$-iodine and not conversely. Loss of hydrogen sulphide is thus avoided, and side reactions are almost entirely eliminated. (With solutions more concentrated than about $0.02N$, the precipitated sulphur encloses a portion of the iodine, and this escapes the subsequent titration with the standard sodium thiosulphate solution.) The excess of iodine is then titrated with standard thiosulphate solution, using starch as indicator.

Excellent results are obtained by the following method, which is of wider applicability. When excess of standard sodium arsenite solution is treated with hydrogen sulphide solution and then acidified with hydrochloric acid, arsenic(III) sulphide is precipitated:

$$As_2O_3 + 3H_2S = As_2S_3 + 3H_2O$$

The excess of arsenic(III) oxide is determined with $0.1N$-iodine and starch.

The procedure is illustrated by determination of the strength of hydrogen sulphide water.

Procedure. Prepare a saturated solution of hydrogen sulphide by bubbling the gas through distilled water. Place 50.0 cm³ standard $0.1N$-sodium arsenite in a 250-cm³ graduated flask, add 20 cm³ of the hydrogen sulphide water, mix well, and add sufficient hydrochloric acid to render the solution distinctly acid. A yellow precipitate of arsenic(III) sulphide is formed, but the liquid itself is colourless. Make up to the mark with distilled water, and shake thoroughly. Filter the mixture through a dry filter paper into a dry vessel. Remove 100 cm³ of the filtrate, neutralise it with sodium hydrogencarbonate, and titrate with standard $0.1N$-iodine to the first blue colour with starch. The quantity of residual arsenic(III) oxide is thus determined, and is deducted from the original 50 cm³ employed.

Note. If certain sulphides are treated with hydrochloric acid, hydrogen sulphide is evolved and can be absorbed in an ammoniacal cadmium chloride solution: upon acidification hydrogen sulphide is released.

Hydrogen sulphide and soluble sulphides can also be determined by **oxidation with potassium iodate in an alkaline medium**. Mix 10.0 cm^3 of the sulphide solution containing about 2.5 mg sulphide with 15.0 cm^3 0.1N-potassium iodate (Section **X, 132**) and 10 cm^3 of 10M-sodium hydroxide. Boil gently for 10 minutes, cool, add 5 cm^3 of 5 per cent potassium iodide solution and 20 cm^3 of 4M-sulphuric acid. Titrate the liberated iodine, which is equivalent to the unused iodate, with standard 0.1N-sodium thiosulphate in the usual manner.

X, 129.   DETERMINATION OF HEXACYANOFERRATES(III).   *Discussion.* The reaction between hexacyanoferrates(III) (ferricyanides) and soluble iodides is a reversible one:

$$2[Fe(CN)_6]^{3-} + 2I^- \rightleftharpoons 2[Fe(CN)_6]^{4-} + I_2$$

In strongly acid solution the reaction proceeds from left to right, but is reversed in almost neutral solution. Oxidation also proceeds quantitatively in a slightly acid medium in the presence of a zinc salt. The very sparingly soluble potassium zinc hexacyanoferrate(II) is formed, and the hexacyanoferrate(II) ions are removed from the sphere of action:

$$2[Fe(CN)_6]^{4-} + 2K^+ + 3Zn^{2+} = K_2Zn_3[Fe(CN)_6]_2$$

The procedure may be used to determine the purity of potassium hexacyanoferrate(III).

Procedure.   Weigh out accurately about 10 g of the salt and dissolve it in 250 cm^3 of water in a graduated flask. Pipette 25 cm^3 of this solution into a 250-cm^3 conical flask, add about 20 cm^3 of 10 per cent potassium iodide solution, 2 cm^3 of M-sulphuric acid, and 15 cm^3 of a solution containing 2.0 g crystallised zinc sulphate. Titrate the liberated iodine immediately with standard 0.1N-sodium thiosulphate and starch; add the starch solution (2 cm^3) after the colour has faded to a pale yellow. The titration is complete when the blue colour has just disappeared. When great accuracy is required, the process should be conducted in an atmosphere of carbon dioxide.

X, 130.   STANDARDISATION OF AN ACID.   When an iodate is allowed to react with iodide ions in solutions of moderate acidity, free iodine is liberated and it is apparent from the equation for the reaction (Section **X, 115**) that the amount of iodine liberated is equivalent to the acid content of the solution provided that excess of iodate and iodide are present.

Procedure.   Pipette 25 cm^3 of the acid solution to be standardised into a 250-cm^3 conical flask, add 1.0–1.5 g potassium iodide crystals or 10 cm^3 of a 10 per cent solution of potassium iodide, followed by 25 cm^3 of 0.5 per cent potassium iodate solution. Titrate the liberated iodine with standard sodium thiosulphate solution.

When dealing with a solution containing a weak acid, the rate of reaction is rather slow, and it is then preferable to add 50 cm^3 of standard sodium thiosulphate solution after adding the potassium iodate solution. The thiosulphate removes the iodine as it is liberated, and the speed of reaction is increased. Allow the solution to stand for ten minutes after adding the thiosulphate and then back titrate the excess thiosulphate with a standard iodine solution.

Oxidations with potassium iodate

X, 131.   GENERAL DISCUSSION.   Potassium iodate is a powerful oxidising agent, but the course of the reaction is governed by the conditions under which it is employed. The reaction between potassium iodate and reducing agents such as iodide ion or arsenic(III) oxide in solutions of moderate acidity (0.1–2.0M hydrochloric acid) stops at the stage when the iodate is reduced to iodine:

$$IO_3^- + 5I^- + 6H^+ = 3I_2 + 3H_2O$$

$$2IO_3^- + 5H_3AsO_3 + 2H^+ = I_2 + 5H_3AsO_4 + H_2O.$$

As already indicated (Section **X, 115**), the first of these reactions is very useful for the generation of known amounts of iodine, and it also serves as the basis of a method for standardising solutions of acids (Section **X, 130**).

With a more powerful reductant, e.g., titanium(III) chloride, the iodate is reduced to iodide:

$$IO_3^- + 6Ti^{3+} + 6H^+ = I^- + 6Ti^{4+} + 3H_2O$$

In more strongly acid solutions (3–6M-hydrochloric acid) reduction occurs to iodine monochloride, and it is under these conditions (due to Andrews) that it is most widely used (Refs 14, 15).

$$IO_3^- + 6H^+ + Cl^- + 4e \rightleftharpoons ICl + 3H_2O$$

In hydrochloric acid solution, iodine monochloride forms a stable complex ion with chloride ion:

$$ICl + Cl^- \rightleftharpoons ICl_2^-$$

The overall half-cell reaction may therefore be written as:

$$IO_3^- + 6H^+ + 2Cl^- + 4e \rightleftharpoons ICl_2^- + 3H_2O;$$

the reduction potential is 1.23 volts, hence under these conditions potassium iodate acts as a very powerful oxidising agent. Furthermore, *under these particular conditions* the equivalent of potassium iodate is one-fourth of a mole $KIO_3/4$, and a 0.1N solution will contain $KIO_3/(4 \times 10)$, or $214.00/40 = 5.3500$ g dm^{-3}. This is by contrast with the situation where reduction to iodine occurs (i.e., conditions of mild acidity), when the equivalent is one-sixth of a mole and a 0.1N solution contains $214.00/6 = 3.5667$ g dm^{-3}.

Oxidation by iodate ion in a strong hydrochloric acid medium proceeds through several stages:

$$IO_3^- + 6H^+ + 6e \rightleftharpoons I^- + 3H_2O$$

$$IO_3^- + 5I^- + 6H^+ = 3I_2 + 3H_2O$$

$$IO_3^- + 2I_2 + 6H^+ = 5I^+ + 3H_2O$$

In the initial stages of the reaction free iodine is liberated: as more titrant is added, oxidation proceeds to iodine monochloride, and the dark colour of the solution gradually disappears. The overall reaction may be written as:

$$IO_3^- + 6H^+ + 4e \rightleftharpoons I^+ + 3H_2O$$

The reaction has been used for the determination of many reducing agents: the optimum acidity for reasonably rapid reaction varies from one reductant to another within the range 2.5–9M-hydrochloric acid; in many cases the concentration of acid is not critical, but for Sb(III) it is 2.5–3.5M.

Under these conditions starch cannot be used as indicator because the characteristic blue colour of the starch–iodine complex is not formed at high concentrations of acid. In the original procedure, a few cm^3 of an immiscible solvent (carbon tetrachloride or chloroform) were added to the solution being titrated contained in a glass-stoppered bottle or conical flask. The end-point is marked by the disappearance of the last trace of violet colour, due to iodine, from the solvent: iodine monochloride is not extracted and imparts a pale yellowish colour to the aqueous phase. The extraction end-point is very sharp. The main disadvantage is the inconvenience of vigorous shaking with the extraction solvent in a stoppered vessel after each addition of the reagent near the end-point.

The immiscible solvent may be replaced by certain dyes, e.g., Amaranth (C.I. 16185), colour change red to colourless; Xylidine Ponceau (C.I. 16255), colour change orange to colourless; Naphthalene Black 12B (C.I. 20470), colour change green to faint pink; the first two of these are generally preferred. The indicators are used as 0.2 per cent aqueous solutions and about 0.5 cm^3 per titration is added near the end-point. The dyes are destroyed by the first excess of iodate, and hence the indicator action is irreversible. The indicator blank is equivalent to 0.05 cm^3 of 0.1N-potassium iodate per 1.0 cm^3 of indicator solution, and is therefore virtually negligible.

p-Ethoxychrysoidine is a moderately satisfactory reversible indicator. It is used as a 0.1 per cent solution in ethanol (about 12 drops per titration), and the colour change is from red to orange; the colour is red-purple just before the end-point. The indicator is added after the colour of the iodine commences to fade. A blank determination should be made for each new batch of indicator.

X, 132. PREPARATION OF 0.025M-POTASSIUM IODATE. Dry some A.R. potassium iodate at 120 °C for 1 hour and allow it to cool in a covered vessel in a desiccator. Weigh out exactly 5.350 g of the finely powdered potassium iodate on a watch glass, and transfer it by means of a clean camel-hair brush directly into a dry 1-dm^3 graduated flask. Add about 400–500 cm^3 of water, and gently rotate the flask until the salt is completely dissolved. Make up to the mark with distilled water. Shake well. The solution will keep indefinitely.

It must be emphasised again that the solution is 0.1N only for the reaction:

$$IO_3^- + 6H^+ + Cl^- + 4e \rightleftharpoons ICl + 3H_2O$$

and when used in solutions of moderate acidity leading to the liberation of free iodine, the 0.1N solution requires 3.5667 g KIO_3 per litre; the method of preparation will be as described above with suitable adjustment of the weight of salt taken.

X, 133. DETERMINATION OF ARSENIC OR OF ANTIMONY. *Discussion.* The determination of arsenic in arsenic(III) compounds is based upon the following reaction:

$$IO_3^- + 2H_3AsO_3 + 2H^+ + Cl^- = ICl + 2H_3AsO_4 + H_2O$$

A similar reaction occurs with antimony(III) compounds. The determination of antimony(III) in the presence of tartrate is not very satisfactory with an immiscible solvent to assist in indicating the end-point; Amaranth, however, gives excellent results.

$$IO_3^- + 2[SbCl_4]^- + 6H^+ + 5Cl^- = ICl + 2[SbCl_6]^- + 3H_2O$$

To assay a sample of arsenic(III) oxide the following procedure may be used.

Procedure. Weigh out accurately about 1.1 g of the oxide sample, dissolve in a small quantity of warm 10 per cent sodium hydroxide solution, and make up to 250 cm³ in a graduated flask. Use a burette to measure 25.0 cm³ of this solution into a stoppered reagent bottle of about 250 cm³ capacity,* add 25 cm³ water, 60 cm³ concentrated hydrochloric acid and about 5 cm³ carbon tetrachloride or chloroform. Cool to room temperature. Run in the standard 0.1N-potassium iodate from a burette until the solution, which at first is strongly coloured with iodine, becomes pale brown. The bottle is then stoppered and vigorously shaken, and the organic solvent layer acquires the purple colour due to iodine. Continue to add small volumes of the iodate solution, shaking vigorously after each addition, until the organic layer is only very faintly violet. Continue the addition dropwise, with shaking after each drop, until the solvent loses the last trace of violet and has only a very pale-yellow colour (due to iodine chloride). The end-point is very sharp and, after a little experience, is rarely overshot. If this should occur, a small volume of the oxide solution is added from a graduated pipette, and the end-point re-determined. Allow to stand for ten minutes and observe whether the organic layer shows any purple colour; the absence of colour confirms that the titration is complete.

The acidity of the mixture at the end of the titration should be not less than $3M$ and not more than $5M$; if the acidity is too high the reaction takes place slowly.

X, 134. DETERMINATION OF MERCURY. *Discussion.* The mercury is precipitated as mercury(I) chloride and the latter is reacted with standard potassium iodate solution:

$$IO_3^- + 2Hg_2Cl_2 + 6H^+ + 13Cl^- = ICl + 4[HgCl_4]^{2-} + 3H_2O.$$

Thus $KIO_3 \equiv 4Hg \equiv 2Hg_2Cl_2$.

To determine the purity of a sample of a mercury(II) salt the following procedure in which the compound is reduced with phosphorous acid may be used; to assay a sample of a mercury(I) salt, the reduction with phosphorous acid is omitted.

Procedure. Weigh out accurately about 2.5 g of finely powdered mercury(II) chloride, and dissolve it in 100 cm³ of water in a graduated flask. Shake well. Transfer 25.0 cm³ of the solution to a conical flask, add 25 cm³ water, 2 cm³ N-hydrochloric acid, and excess of 50 per cent phosphorous acid solution. Stir thoroughly and allow to stand for 12 hours or more. Filter the precipitated mercury(I) chloride through a quantitative filter paper or through a Gooch crucible with asbestos, and wash the precipitate moderately with cold water.

---

* A 250-cm³ graduated flask with a short neck and a well-fitting ground glass stopper may also be used. The colour of the organic layer is readily seen by inverting the flask so that the layer of solvent indicator collects in the neck.

Alternatively, in this and all subsequent titrations with 0.1N-potassium iodate, a 250- or 350-cm³ conical flask may be used and the carbon tetrachloride or chloroform indicator replaced by 0.5 cm³ Amaranth or Xylidine Ponceau indicator, which is added after most of the iodine colour has disappeared from the reaction mixture (see Section X, 131).

Transfer the precipitate with the filter paper or asbestos quantitatively to a 250-cm^3 reagent bottle, add 30 cm^3 concentrated hydrochloric acid, 20 cm^3 water, and 5 cm^3 carbon tetrachloride or chloroform. Titrate the mixture with standard $0.1N$-potassium iodate in the usual manner (Section **X, 133**).

$$2HgCl_2 + H_3PO_3 + H_2O = Hg_2Cl_2 + 2HCl + H_3PO_4$$

Many other metallic ions which are capable of undergoing oxidation by potassium iodate can also be determined. Thus for example **copper(II) compounds** can be analysed by precipitation of copper(I) thiocyanate which is titrated with potassium iodate:

$$7IO_3^- + 4CuSCN + 18H^+ + 7Cl^- = 7ICl + 4Cu^{2+} + 4HSO_4^- + 4HCN + 5H_2O.$$

As a typical example, 0.8 g of copper(II) sulphate $CuSO_4,5H_2O$ is dissolved in water, 5 cm^3 of $0.5M$-sulphuric acid added, and the solution made up to 250 cm^3 in a graduated flask. 25.0 cm^3 of the resulting solution are pipetted into a 250-cm^3 conical flask, 10–15 cm^3 of freshly prepared sulphurous acid solution added, and then after heating to boiling, 10 per cent ammonium thiocyanate solution is added slowly from a burette with constant stirring until there is no further change in colour, and then 4 cm^3 of reagent is added in excess. After allowing the precipitate to settle for 10–15 minutes, it is filtered through a Gooch crucible containing asbestos and then washed with cold 1 per cent ammonium sulphate solution until free from thiocyanate. It is then transferred quantitatively into the vessel in which the titration is to be performed, and after adding 30 cm^3 of concentrated hydrochloric acid, followed by 20 cm^3 of water, the titration is carried out in the usual manner with either an organic solvent present, or an internal indicator is added as the end-point is approached.

Thallium(I) salts are oxidised in accordance with the equation:

$$IO_3^- + 2Tl^+ + 6H^+ + Cl^- = ICl + 2Tl^{3+} + 3H_2O$$

so that $$KIO_3 \equiv 2Tl.$$

The solution should contain 0.25–0.30 g Tl^+ in 20 cm^3 plus 60 cm^3 of concentrated hydrochloric acid and is titrated as usual with $0.025M$ KIO_3 solution.

Tin(II) salts are likewise oxidised in accordance with the equation

$$IO_3^- + 2Sn^{2+} + 6H^+ + Cl^- = ICl + 2Sn^{4+} + 3H_2O$$

so that $$KIO_3 = 2Sn.$$

If the bulk of the iodate solution is added rapidly, atmospheric oxidation does not present a serious problem, but the method cannot be used in the presence of salts of antimony(III), copper(I) or iron(II). The solution which should contain for example 0.15 g $SnCl_2,2H_2O$ in 25 cm^3 is treated with 30 cm^3 of concentrated hydrochloric acid and 20 cm^3 of water and is then titrated in the usual manner with standard potassium iodate solution.

X, 135. DETERMINATION OF HYDRAZINE. *Discussion.* Hydrazine

reacts with potassium iodate under the usual Andrews conditions thus:

$$IO_3^- + N_2H_4 + 2H^+ + Cl^- = ICl + N_2 + 3H_2O$$

Thus $$KIO_3 \equiv N_2H_4$$

To determine the N_2H_4, H_2SO_4 content of hydrazinium sulphate, use the following method.

Procedure.   Weigh out accurately 0.08–0.1 g of hydrazinium sulphate into a 250-cm^3 reagent bottle, add a mixture of 30 cm^3 of concentrated hydrochloric acid, 20 cm^3 of water, and 5 cm^3 of chloroform or carbon tetrachloride. Run in the standard 0.025M-potassium iodate slowly from a burette, with shaking the stoppered bottle between the additions, until the organic layer is just decolorised.

X, 136. DETERMINATION OF VANADATES. *Discussion.* Vanadates are reduced by iodides in strongly acid (hydrochloric) solution in an atmosphere of carbon dioxide to the quadrivalent condition:

$$2VO_4{}^{3-} + 2I^- + 12H^+ = 2VO^{2+} + I_2 + 6H_2O$$

The liberated iodine and the excess of iodide is determined by titration with standard potassium iodate solution; the hydrochloric acid concentration must not be allowed to fall below 7M in order to prevent re-oxidation of the vanadium compound by iodine chloride.

$$2I_2 + IO_3{}^- + 6H^+ + 5Cl^- = 5ICl + 3H_2O$$
$$2I^- + IO_3{}^- + 6H^+ + 3Cl^- = 3ICl + 3H_2O$$

The total result of the reaction is:

$$4VO_4{}^{3-} + 4I^- + IO_3{}^- + 5Cl^- + 30H^+ = 4VO^{2+} + 5ICl + 15H_2O$$

This method is applicable in the presence of arsenate, phosphate, or iron(III), and also in the presence of tungstic acid, which may be held in solution by adding phosphoric acid.

Procedure.   Place 25.0 cm^3 of the solution containing 0.05–0.10 g of vanadium (as vanadate) in a 250-cm^3 glass-stoppered reagent bottle, and pass a rapid current of carbon dioxide for 2–3 minutes into the bottle, but not through the solution. Then add sufficient concentrated hydrochloric acid through a funnel to make the solution 6–8M during the titration. Introduce a known volume (excess) of approximately 0.05M-potassium iodide, which has been titrated against the standard iodate solution. Mix the contents of the bottle, allow to stand for 1–2 minutes, add 5 cm^3 of carbon tetrachloride, and then titrate as rapidly as possible with standard 0.025M-potassium iodate until no more iodine colour can be detected in the organic layer. Add concentrated hydrochloric acid as needed during the titration so that the concentration does not fall below 7M.

Oxidations with potassium bromate

X, 137. GENERAL DISCUSSION.   Potassium bromate is a powerful oxidising agent which is reduced smoothly to bromide:

$$BrO_3{}^- + 6H^+ + 6e \rightleftharpoons Br^- + 3H_2O$$

The equivalent is therefore $\frac{1}{6}$ mole ($KBrO_3/6$), or 167.00/6, or 27.833, and a 0.1N solution contains 2.7833 g potassium bromate per dm^3. At the end of the titration free bromine appears:

$$BrO_3{}^- + 5Br^- + 6H^+ = 3Br_2 + 3H_2O$$

The presence of free bromine, and consequently the end-point, can be detected by its yellow colour, but it is better to use indicators such as methyl orange, methyl red, Naphthalene Black 12B, Xylidine Ponceau, and Fuchsine. These indicators have their usual colour in acid solution, but are destroyed by the first excess of bromine. With all irreversible oxidation indicators the destruction of the indicator is often premature to a slight extent: a little additional indicator is usually required near the end-point. The quantity of bromate solution consumed by the indicator is exceedingly small, and the 'blank' can be neglected for $0.1N$ solutions. Direct titrations with bromate solution in the presence of irreversible dyestuff indicators are usually made in hydrochloric acid solution, the concentration of which should be at least $1.5-2M$. At the end of the titration some chlorine may appear by virtue of the reaction:

$$10Cl^- + 2BrO_3^- + 12H^+ = 5Cl_2 + Br_2 + 6H_2O$$

this immediately bleaches the indicator.

The titrations should be carried out slowly so that the indicator change, which is a time reaction, may be readily detected. If the determinations are to be executed rapidly, the volume of the bromate solution to be used must be known approximately, since ordinarily with irreversible dyestuff indicators there is no simple way of ascertaining when the end-point is close at hand. With the highly coloured indicators (Xylidine Ponceau, Fuchsine, or Naphthalene Black 12B), the colour fades as the end-point is approached (owing to local excess of bromate) and another drop of indicator can be added. At the end-point the indicator is irreversibly destroyed and the solution becomes colourless or almost so. If the fading of the indicator is confused with the equivalence point, another drop of the indicator may be added. If the indicator has faded, the additional drop will colour the solution; if the end-point has been reached, the additional drop of indicator will be destroyed by the slight excess of bromate present in the solution.

The introduction of reversible redox indicators for the determination of trivalent arsenic and trivalent antimony has considerably simplified the procedure; those at present available include 1-naphthoflavone, and p-ethoxychrysoidine. The addition of a little tartaric acid or potassium sodium tartrate is recommended when antimony(III) is titrated with bromate in the presence of the reversible indicators; this will prevent hydrolysis at the lower acid concentrations. The end-point may be determined with high precision by potentiometric titration (see Chapter XIV).

Examples of determinations utilising direct titration with bromate solutions are expressed in the following equations:

$$BrO_3^- + 3H_3AsO_3 \xrightarrow{\text{(HCl)}} Br^- + 3H_3AsO_4$$

$$2BrO_3^- + 3N_2H_4 \xrightarrow{\text{(HCl)}} 2Br^- + 3N_2 + 6H_2O$$

$$BrO_3^- + NH_2OH \xrightarrow{\text{(HCl)}} Br^- + NO_3^- + H^+ + H_2O$$

$$BrO_3^- + 6[Fe(CN)_6]^{4-} + 6H^+ \longrightarrow Br^- + 6[Fe(CN)_6]^{3-} + 3H_2O$$

Various substances cannot be oxidised directly with potassium bromate, but react quantitatively with an excess of bromine. Acid solutions of bromine of exactly known concentration are readily obtainable from a standard potassium bromate solution by adding acid and an excess of bromide:

$$BrO_3^- + 5Br^- + 6H^+ = 3Br_2 + 3H_2O$$

In this reaction 1 mole of bromate yields six atoms of bromine, hence the equivalent is $KBrO_3/6$, identical with that of potassium bromate alone. Bromine is very volatile, and hence such operations should be conducted at as low a temperature as possible and in conical flasks fitted with ground-glass stoppers. The excess of bromine may be determined iodometrically by the addition of excess of potassium iodide and titration of the liberated iodine with standard thiosulphate solution:

$$2I^- + Br_2 = I_2 + 2Br^-$$

Potassium bromate is readily available in a high state of purity; the A.R. product has an assay value of at least 99.9 per cent. The substance can be dried at 120–150 °C, is anhydrous, and the aqueous solution keeps indefinitely. It can therefore be employed as a primary standard. Its only disadvantage is that the equivalent is comparatively small.

X, 138. PREPARATION OF 0.1N-POTASSIUM BROMATE. Dry some finely powdered A.R. potassium bromate for 1–2 hours at 120 °C, and allow to cool in a closed vessel in a desiccator. Weigh out accurately 2.783 g of the pure potassium bromate, and dissolve it in 1 dm³ of water in a graduated flask.

X, 139. DETERMINATION OF ANTIMONY OR OF ARSENIC. *Discussion.* The antimony or the arsenic must be present in the trivalent condition. The reaction of trivalent arsenic or antimony with potassium bromate may be written:

$$2KBrO_3 + 3M_2O_3 + 2HCl = 2KCl + 3M_2O_5 + 2HBr \text{ (M = As or Sb)}$$

The presence of tin and of considerable quantities of iron and copper interfere with the determinations.

To determine the purity of a sample of arsenic(III) oxide follow the general procedure outlined in Section **X, 133** but when the 25-cm³ sample of solution is being prepared for titration, add 25 cm³ water, 15 cm³ of concentrated hydrochloric acid and then two drops of indicator solution (Xylidine Ponceau or Naphthalene Black 12B; see Section **X, 131**). Titrate slowly with the standard 0.1N-potassium bromate with constant swirling of the solution. As the end-point approaches, add the bromate solution dropwise with intervals of 2–3 seconds between the drops until the solution is colourless or very pale yellow. If the colour of the indicator fades, add another drop of indicator solution. (The immediate discharge of the colour indicates that the equivalence point has been passed and the titration is of little value.)

As an alternative, a reversible indicator may be employed, either (*a*) 1-naphthoflavone (0.5% solution in ethanol, which gives an orange-coloured solution at the end-point), or (*b*) *p*-ethoxychrysoidine (0.1% aqueous solution, colour change pink to pale yellow). Under these conditions, the measured 25-cm³ portion of the arsenic solution is treated with 10 cm³ of 10 per cent potassium bromide solution, 6 cm³ of concentrated hydrochloric acid, 10 cm³ of water and either 0.5 cm³ of indicator (*a*) or two drops of indicator (*b*).

X, 140. DETERMINATION OF METALS BY MEANS OF 8-HYDROXYQUINOLINE ('OXINE'). *Discussion.* Various metals (e.g., aluminium, iron, copper, zinc, cadmium, nickel, cobalt, manganese, and magnesium) under specified conditions of pH yield well-defined crystalline precipitates with 8-hydroxyquinoline. These precipitates have the general formula $M(C_9H_6ON)_n$, where n is the valency of the metal M (see, however, Section **XI, 11C**). Upon treatment of the oxinates with dilute hydrochloric acid, the oxine is liberated. Oxine reacts with 4 equivalents of bromine to give 5,7-dibromo-8-hydroxyquinoline:

$$C_9H_7ON + 2Br_2 = C_9H_5ONBr_2 + 2H^+ + 2Br^-$$

Hence 1 mole of the oxinate of a divalent metal requires 8 equivalents of bromine, whilst that of a trivalent metal requires 12 equivalents. The bromine is derived by the addition of standard 0.1N-potassium bromate and excess of potassium bromide to the acid solution.

$$BrO_3^- + 5Br^- + 6H^+ = 3Br_2 + 3H_2O$$

Full details are given for the determination of aluminium by this method. Many other metals may be determined by this same procedure, but in many cases complexometric titration offers a simpler method of determination. In cases where the oxine method offers advantages, the experimental procedure may be readily adapted from the details given for aluminium.

Determination of aluminium. Prepare a 2 per cent solution of A.R. 8-hydroxyquinoline (see Section **XI, 11C**) in 2M acetic acid; add ammonia solution until a *slight* precipitate persists, then redissolve it by warming the solution.

Transfer 25 cm³ of the solution to be analysed, containing about 0.02 g of aluminium, to a conical flask, add 125 cm³ of water and warm to 50–60 °C. Then add a 20 per cent excess of the oxine solution (1 cm³ will precipitate 0.001 g of Al), when the complex $Al(C_9H_6ON)_3$ will be formed. Complete the precipitation by the addition of a solution of 4.0 g of ammonium acetate in the minimum quantity of water, stir the mixture, and allow to cool. Filter the granular precipitate through a sintered glass crucible of porosity No. 4 (or through a porcelain filtering crucible), and wash with warm water (1). Dissolve the complex in warm concentrated hydrochloric acid, collect the solution in a 250-cm³ reagent bottle, add a few drops of indicator (0.1 per cent solution of the sodium salt of methyl red or 0.1 per cent methyl orange solution), and 0.5–1 g of pure potassium bromide. Titrate slowly with standard 0.1N (i.e., M/60) potassium bromate until the colour becomes pure yellow (with either indicator). The exact end-point is not easy to detect, and the best procedure is to add an excess of potassium bromate solution, i.e., a further 2 cm³ beyond the estimated end-point, so that the solution now contains free bromine. Dilute the solution considerably with 2M-hydrochloric acid (to prevent the precipitation of 5,7-dibromo-8-hydroxyquinoline during the titration), then add (after 5 minutes) 10 cm³ of 10 per cent potassium iodide solution, and titrate the liberated iodine with standard 0.1N-sodium thiosulphate, using starch as indicator (2).

From the above discussion, it is evident that Al ≡ 12Br, i.e., to 12 dm³ of 1N-bromate (or 1N-thiosulphate).

Notes. 1. This will remove the excess of oxine. Complications due to adsorption of iodine will thus be avoided.

2. A brown additive compound of iodine with the dibromo compound may

separate during the titration; this compound usually dissolves during the subsequent titration with thiosulphate, yielding a yellow solution so that the end-point with starch may be found in the usual manner. Occasionally, the dark-coloured compound, which contains adsorbed iodine, may not dissolve readily and thus introduces an uncertainty in the end-point: this difficulty may be avoided by adding 10 cm^3 of carbon disulphide before introducing the potassium iodide solution.

X, 141. DETERMINATION OF HYDROXYLAMINE. The method based upon the reduction of iron(III) solutions in the presence of sulphuric acid, boiling, and subsequent titration in the cold with standard 0.1N-potassium permanganate frequently yields high results unless the experimental conditions are closely controlled:

$$2NH_2OH + 4Fe^{3+} = N_2O + 4Fe^{2+} + 4H^+ + H_2O$$

Better results are obtained by oxidation with potassium bromate in the presence of hydrochloric acid:

$$NH_2OH + BrO_3^- = NO_3^- + Br^- + H^+ + H_2O$$

The hydroxylamine solution is treated with a measured volume of 0.1N (i.e., $M/60$) potassium bromate so as to give 10–30 cm^3 excess, followed by 40 cm^3 of 5M-hydrochloric acid. After 15 minutes the excess of bromate is determined by the addition of potassium iodide solution and titration with standard 0.1N-sodium thiosulphate (compare Section **X, 140**).

The Reduction of Higher Oxidation States

X, 142. GENERAL DISCUSSION. It has already been indicated that before titration with an oxidising agent can be carried out, it may in some cases be necessary to reduce the compound supplied to a lower state of oxidation. Such a situation is frequently encountered with the determination of iron; iron(III) compounds must be reduced to iron(II) before titration with potassium permanganate or potassium dichromate can be performed. It is possible to carry out such determinations directly as a *reductimetric titration* by the use of solutions of powerful reducing agents such as chromium(II) chloride, titanium(III) chloride or vanadium(II) sulphate, but the problems associated with the preparation, storage and handling of these reagents have militated against their widespread use. Titanium(III) sulphate has found application in the analysis of certain types of organic compounds (Ref. 3), but is of limited application in the inorganic field. An apparatus suitable for the preparation, storage and manipulation of chromium(II) and vanadium(II) solutions is described in Reference 16; with both these reagents it is necessary (and is also advisable with Ti(III) solutions), to carry out titrations in an atmosphere of hydrogen, nitrogen or carbon dioxide, and in view of the instability of most indicators in the presence of these powerful reducing agents, it is frequently necessary to determine the end-point potentiometrically.

The most important method for reduction of compounds to an oxidation state suitable for titration with one of the common oxidising titrants is based upon the use of metal amalgams, but there are various other methods which can be used, and these will be discussed in the following Sections.

X, 143. REDUCTION WITH AMALGAMATED ZINC; THE JONES REDUCTOR. Amalgamated zinc is an excellent reducing agent for many metallic ions. Zinc reacts rather slowly with acids, but upon treatment with a dilute solution of a mercury(II) salt, the metal is covered with a thin layer of mercury; the amalgamated metal reacts quite readily. Reduction with amalgamated zinc is usually carried out in the 'reductor', due to C. Jones. This consists of a column of amalgamated zinc contained in a long glass tube provided with a stopcock, through which the solution to be reduced may be drawn. A large surface is exposed, and consequently such a zinc column is much more efficient than pieces of zinc placed in the solution.

A suitable form of the Jones reductor, with approximate dimensions, is shown in Fig. X, 18. A perforated porcelain plate, covered with purified asbestos or glass wool, supports the zinc column. The tube below the tap passes through a tightly fitting one-holed rubber stopper into a 750-cm³ filter flask. It is advisable to connect another filter flask in series with the water-pump, so that if any water 'sucks back' it will not spoil the determination. The amalgamated zinc is prepared as follows. About 300 g of A.R. granulated zinc (or zinc shavings, or pure 20–30-mesh zinc) are covered with 2 per cent mercury(II) chloride solution in a beaker. The mixture is stirred for 5–10 minutes, then the solution is decanted from the zinc, which is washed three times with water by decantation. The resulting amalgamated zinc should have a bright silvery lustre. The porcelain plate is placed in position, covered with a layer of purified asbestos or glass wool and then the amalgamated zinc added: the latter should reach to the shoulder of the tube. The zinc is washed with distilled water (500 cm³), using gentle suction. If the reductor is not to be used immediately, it must be left full of water in order to prevent the formation of basic salts by atmospheric oxidation, which impair the reducing surface. If the moist amalgam is exposed to the moisture of the atmosphere, hydrogen peroxide may be generated:

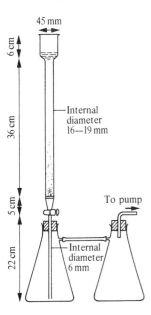

45 mm

6 cm

36 cm

Internal diameter 16–19 mm

5 cm

To pump

22 cm

Internal diameter 6 mm

Fig. X, 18

$$Zn + O_2 + 2H_2O = Zn(OH)_2 + H_2O_2$$

but no hydrogen peroxide is formed if acid is present.

To use the reductor for the reduction of iron(III), proceed as follows. The zinc is activated by filling the cup (which holds about 50 cm³) with M (ca. 5 per cent) sulphuric acid, the tap being closed. The flask is connected to a filter pump, the tap opened, and the acid *slowly* drawn through the column until it has fallen to *just above the level of the zinc*; the tap is then closed and the process repeated twice. The tap is shut, the flask detached, cleaned, and replaced. The reductor is now ready for use. It is important to note that during use the level of the liquid should always be just above the top of the zinc column. The solution to be reduced should have a volume of 100–150 cm³, contain not more than 0.25 g of iron, and be about M in sulphuric acid. The cold iron solution is passed through the reductor, using gentle suction, at a rate not exceeding 75–100 cm³ per minute.

As soon as the reservoir is nearly emptied of the solution, 100 cm^3 of 2.5 per cent (or *ca.* 0.5*M*) sulphuric acid is passed through in two portions, followed by 100–150 cm^3 of water. The last washing is necessary in order to wash out all of the reduced compound and also the acid, which would otherwise cause unnecessary consumption of the zinc. Disconnect the flask from the reductor, wash the end of the delivery tube, and titrate immediately with standard 0.1*N*-potassium permanganate.

Carry out a blank determination, preferably before passing the iron solution through the reductor, by running the same volumes of acid and water through the apparatus as are used in the actual determination. This should not amount to more than about 0.1 cm^3 of 0.1*N*-permanganate, and should be deducted from the volume of permanganate solution used in the subsequent titration.

It must be emphasised that if hydrochloric acid has been employed in the original solution of the iron-bearing material, the volume should be reduced to *ca.* 25 cm^3 and then diluted to *ca.* 150 cm^3 with 5 per cent sulphuric acid. The determination is carried out as detailed above, but 25 cm^3 of Zimmermann–Reinhardt or 'preventive solution' must be added before titration with standard potassium permanganate solution. For the determination of iron in hydrochloric acid solution, it is more convenient to reduce the solution in a silver reductor (Section **X, 145**) and to titrate the reduced solution with either standard potassium dichromate or standard cerium(IV) sulphate solution.

Applications and limitations of the Jones reductor. 1. Solutions containing 1–10 per cent by volume of sulphuric acid or 3–15 per cent by volume of concentrated hydrochloric acid can be used in the reductor. Sulphuric acid is, however, generally used, as hydrochloric acid may interfere in the subsequent titration, e.g., with potassium permanganate.

2. Nitric acid must be absent, for this is reduced to hydroxylamine and other compounds which react with permanganate. If nitric acid is present, evaporate the solution just to dryness, wash the sides of the vessel with about 3 cm^3 of water, carefully add 3–4 cm^3 of concentrated sulphuric acid, and evaporate until fumes of the latter are evolved. Repeat this operation twice to ensure complete removal of the nitric acid, dilute to 100 cm^3 with water, add 5 cm^3 of concentrated sulphuric acid, and proceed with the reduction.

3. Organic matter (acetates, etc.) must be absent. It is removed by heating to fumes of sulphuric acid in a covered beaker, then carefully adding drops of a saturated solution of potassium permanganate until a permanent colour is obtained, and finally continuing the fuming for a few minutes.

4. Solutions containing compounds of copper, tin, arsenic, antimony, and other reducible metals must never be used. These must be removed before the reduction by treatment with hydrogen sulphide.

5. Other ions which are reduced in the reductor to a definite valency stage are those of titanium to Ti^{3+}, chromium to Cr^{2+}, molybdenum to Mo^{3+}, niobium to Nb^{3+}, and vanadium to V^{2+}. Uranium is reduced to a mixture of U^{3+} and U^{4+}, but by bubbling a stream of air through the solution in the filter flask for a few minutes, the dirty dark green colour changes to the bright apple-green colour characteristic of pure uranium(IV) salts. Tungsten is reduced, but not to any definite valency state.

With the exception of iron(II) and quadrivalent uranium, the reduced solutions are extremely unstable and readily re-oxidise upon exposure to air. They are best stabilised in a five-fold excess of a solution of iron(III) alum (150 g of A.R. iron(III) alum and 150 cm^3 of concentrated sulphuric acid per litre: approximately $0.3N$ with respect to iron) contained in the filter flask. The iron(II) formed is then titrated with a standard solution of a suitable oxidising agent. Titanium and chromium are completely oxidised and produce an equivalent amount of iron(II) sulphate; molybdenum is reoxidised to the quinquevalent (red) stage, which is fairly stable in air, and complete oxidation is effected by the permanganate, but the net result is the same, viz., Mo(III) → Mo(VI); vanadium is re-oxidised to the quadrivalent, condition, which is stable in air, and the final oxidation is completed by slow titration with potassium permanganate solution or with cerium(IV) sulphate solution.

X, 144. REDUCTION WITH LIQUID AMALGAMS.

Makazono introduced liquid zinc amalgam as a reducing agent and subsequent Japanese workers have used liquid amalgams of cadmium, bismuth, and lead. The advantages claimed for liquid amalgam reductions are: (*a*) complete reduction is achieved in a few minutes; (*b*) the amalgam can be used repeatedly; and (*c*) no blank correction is required as in the Jones reductor. The reduction potentials of the saturated metal amalgams are as follows:

$$Zn^{2+} + 2e \rightleftharpoons Zn; \qquad -0.76 \text{ volt}$$

$$Cd^{2+} + 2e \rightleftharpoons Cd; \qquad -0.40 \text{ volt}$$

$$Pb^{2+} + 2e \rightleftharpoons Pb; \qquad -0.13 \text{ volt}$$

$$BiO^{+} + 2H^{+} + 3e \rightleftharpoons Bi + H_2O; \quad +0.32 \text{ volt}$$

The most powerful reductant is therefore zinc amalgam, while bismuth amalgam is the least reducing. The final reduction products obtained with these amalgams for a few elements are collected in the table.

Liquid amalgam	Iron	Titanium	Molybdenum	Vanadium	Uranium	Tungsten
Zinc	Fe^{2+}	Ti^{3+}	Mo^{3+}	V^{2+}	U^{4+}†	W^{3+}
Cadmium	Fe^{2+}	Ti^{3+}	Mo^{3+}	V^{2+}	U^{4+}	—
Lead	Fe^{2+}	Ti^{3+}	Mo^{3+}	V^{2+}	U^{4+}	W^{3+}
Bismuth	Fe^{2+}	Ti^{3+}	Mo^{3+} or Mo^{5+}*	VO^{2+}	U^{4+}	W^{5+}

* The exact product depends upon the pH of the solution.
† Some U^{3+} is also formed.

The **zinc amalgam** is **prepared** by washing 15 g of pure, fine-mesh zinc shot (e.g., A.R.) with dilute sulphuric acid, and then heating for 1 hour on the water bath with 300 g of mercury plus 5 cm^3 of 1:4 sulphuric acid. (CAUTION. Mercury vapour is highly poisonous; the operation must therefore be performed in a fume cupboard with a good draught.) The whole is allowed to cool, the amalgam washed several times with dilute sulphuric acid, and the liquid portion separated from the solid by means of a separating-funnel. The solid is reserved for another preparation of the amalgam. The liquid amalgam is preserved under dilute sulphuric acid; reaction with the latter is very slow, and the same sample of amalgam may be employed for several reductions. The other amalgams are prepared in similar manner, except that for bismuth and lead, hydrochloric acid is used in place of sulphuric acid. All the amalgams can also be prepared electrolytically using a mercury cathode.

Reductions with liquid amalgams are usually carried out in a separating funnel in an atmosphere of carbon dioxide, so that when the reduction is complete the remaining amalgam can be easily removed before the titration of the reduced solution is attempted. In a simplified procedure, 30–50 cm^3 of carbon tetrachloride are added to the flask in which reduction has been carried out, at the stage when the amalgam is normally separated: this produces three layers in the flask, the amalgam, surmounted by the carbon tetrachloride, which thus separates the aqueous solution from the amalgam. A mechanically operated stirrer is then inserted so that it will agitate the aqueous layer without disturbing the layers below, and the aqueous solution can then be titrated with the appropriate reagent.

X, 145.   THE SILVER REDUCTOR.   The silver reductor has a relatively low reduction potential (the Ag/AgCl electrode potential in M-hydrochloric acid is 0.2245 volt), and consequently it is not able to effect many of the reductions which can be made with amalgamated zinc. The silver reductor is preferably used with hydrochloric acid solutions, and this is frequently an advantage. The various reductions which can be effected with the silver and the amalgamated zinc reductors are summarised in the following table:*

Silver reductor Hydrochloric acid solution	Amalgamated zinc (Jones) reductor Sulphuric acid solution
$Fe^{3+} \longrightarrow Fe^{2+}$	$Fe^{3+} \longrightarrow Fe^{2+}$
Ti^{4+} not reduced	$Ti^{4+} \longrightarrow Ti^{3+}$
$Mo^{6+} \longrightarrow Mo^{5+}$ (2M-HCl; 60–80 °C)	$Mo^{6+} \longrightarrow Mo^{3+}$
Cr^{3+} not reduced	$Cr^{3+} \longrightarrow Cr^{2+}$
$UO_2^{2+} \longrightarrow U^{4+}$ (4M-HCl; 60–90 °C)	$UO_2^{2+} \longrightarrow U^{3+} + U^{4+}$
$V^{5+} \longrightarrow V^{4+}$	$V^{5+} \longrightarrow V^{2+}$
$Cu^{2+} \longrightarrow Cu^+$ (2M-HCl)	$Cu^{2+} \longrightarrow Cu^0$

* Strictly speaking the higher valency states 6 + and 5 + should be presented as oxidation states (VI) and (V).

The silver reductor (shaped like a short, squat Jones reductor tube) may be constructed from a tube 12 cm long and 2 cm internal diameter fused to a reservoir bulb of 50–75 cm^3 capacity. It is not always necessary to use suction. The silver is conveniently prepared as follows on a large scale; for preparations on a smaller scale the procedure must be appropriately adapted. A solution of 500 g of silver nitrate in 2500 cm^3 of water, slightly acidified with dilute nitric acid, is placed in a 4-litre beaker. Cathodes consisting of two heavy-gauge platinum plates, each 10 cm square, are suspended in the electrolyte by the use of a heavy copper bus-bar connection to a source of current. The anode consists of either a silver rod 200 mm long and 10–25 mm in diameter or a similar weight of silver as a heavy-gauge rectangular sheet; it is suspended in the centre of the electrolyte with the platinum cathodes placed at the outer edges of the deposition cell. Silver is deposited as granular crystals with high ratio of surface to mass by a current of 60–70 amp at 5–6 volts. These crystals, obtained in excellent yield, are deposited on the four outside edges of the cathodes; they should be dislodged by gentle tapping, and washed by decantation with dilute sulphuric acid. About 30 g of silver in this form occupy a volume of 40–50 cm^3—sufficient to fill one reductor tube.

The necessary quantity of silver is introduced into the reductor above a small plug of glass wool: by means of a glass rod flattened at one end, it is compressed to as great an extent as necessary without restricting the free flow of solution through the column. The reductor is rinsed with 100 cm^3 of M-hydrochloric acid, added in five equal portions, each consecutive portion being allowed to pass through the reductor to just above the level of the silver.

The dark silver chloride coating which covers the silver of the upper part of the reductor when hydrochloric acid solutions are employed moves farther down the column in use, and when it extends to about three-quarters of the length of the column, the reductor must be regenerated by the following method. The reductor is rinsed with water and filled completely with 1:3-ammonia solution. The silver chloride dissolves; after 10 minutes, the solution is rinsed out of the reductor tube with water, followed by M-hydrochloric acid and is then ready for re-use. As a precautionary measure, the ammoniacal solution of silver chloride should be immediately acidified. The wastage of silver associated with this method of regeneration may be avoided by filling the tube with sulphuric acid (0.1M) and then inserting a rod of zinc with its lower end well buried in the silver; when the reduction is complete (as evidenced by loss of the dark colour), the column is well washed with water and is then ready for use.

Examples of the use of the silver reductor are given in Sections **X, 110** and **X, 111**.

X, 146. OTHER METHODS OF REDUCTION. Although as already stated the use of metal amalgams, and in particular use of the Jones reductor or of the related silver reductor, is the best method of reducing solutions in preparation for titration with an oxidant, it may happen that for occasional use there is no Jones reductor available, and a simpler procedure will commend itself. In practical terms, the need is most likely to arise in connection with the determination of iron, and the following reagents are amongst those most commonly employed for the reduction of iron(III) to iron(II).

A. Tin(II) chloride solution.   Many iron ores are brought into solution with concentrated hydrochloric acid and the resulting solution may be readily reduced with tin(II) chloride:

$$2Fe^{3+} + Sn^{2+} = 2Fe^{2+} + Sn^{4+}$$

The hot solution (70–90 °C) from about 0.3 g of iron ore which should occupy a volume of 25–30 cm^3 and be 5–6M with respect to hydrochloric acid, is reduced by adding concentrated tin(II) chloride solution dropwise from a separating-funnel or a burette, with stirring, until the yellow colour of the solution has *nearly* disappeared. The reduction is then completed by diluting the concentrated solution of tin(II) chloride with 2 volumes of dilute hydrochloric acid, and adding the dilute solution dropwise, with agitation after each addition, until the liquid has a faint green colour, quite free from any tinge of yellow. The solution is then rapidly cooled under the tap to about 20 °C, with protection from the air, and the slight excess of tin(II) chloride present removed by adding 10 cm^3 of a saturated solution (*ca.* 5 per cent) of mercury(II) chloride rapidly in one portion and with thorough mixing; a *slight* silky white precipitate of mercury(I) chloride should be obtained.

The small amount of mercury(I) chloride in suspension has no appreciable effect upon the oxidising agent used in the subsequent titration, but if a heavy precipitate forms, or a grey or black precipitate is obtained, too much tin(II) solution has been used; the results are inaccurate and the reduction must be repeated. Finely divided mercury reduces permanganate or dichromate ions and also slowly reduces Fe^{3+} ions in the presence of chloride ion.

After the addition of the mercury(II) chloride solution, the whole is allowed to stand for five minutes, then diluted to about 400 cm^3 and titrated with standard potassium dichromate solution (Section **X, 103**) or with standard permanganate solution in the presence of 'preventive solution' (Section **X, 93**).

Blank runs on the reagents should be carried through all the operations, and corrections made, if necessary.

The **concentrated solution of tin(II) chloride** is prepared by dissolving 12 g of pure tin or 30 g of A.R. crystallised tin(II) chloride ($SnCl_2,2H_2O$) in 100 cm^3 of concentrated hydrochloric acid and diluting to 200 cm^3 with water.

B. Reduction with sulphurous acid. The solution must be feebly acid ($<1N$) and fairly dilute, say, 500 cm^3 for 0.5 g of iron. If the concentration of the acid exceeds $5N$, sulphurous acid will oxidise iron(II) solutions. Hydrochloric acid–chloride solutions are reduced more rapidly than sulphuric acid–sulphate solutions. Either sulphur dioxide from a siphon of the liquid gas or *freshly prepared* sulphurous acid solution or ammonium hydrogensulphite solution may be used. The operation is best carried out in a special all-glass wash bottle or, if this is not available, in a flask fitted with a rubber stopper carrying two 'wash-bottle' tubes.

Treat the hydrochloric acid or sulphuric acid solution of the iron slowly and with constant shaking with dilute ammonia solution until a faint permanent precipitate is obtained. Dilute to about 100 cm^3, pass sulphur dioxide through the solution for 2–3 minutes, and then gradually heat to boiling, still continuing the passage of the gas. When the solution is colourless (15–30 minutes), replace the sulphur dioxide by a stream of washed carbon dioxide (from a Kipp's apparatus or cylinder), and boil vigorously until all the sulphur dioxide is expelled (20–30 minutes) as shown by passing the escaping gas for 30 seconds through dilute sulphuric acid containing 2 drops of $0.1N$-permanganate. Allow the solution to cool in a stream of carbon dioxide, add more acid, and titrate with a standard solution of a suitable oxidising agent.

A simpler method is to place the acidified iron solution in a conical flask, add dilute ammonia solution slowly until a faint permanent precipitate is obtained, and then add either 25 cm^3 of a freshly prepared saturated solution of sulphur dioxide or excess of freshly prepared ammonium hydrogensulphite solution; in the latter case a little dilute sulphuric acid is added. A small funnel is placed in the mouth of the flask, and the mixture boiled for 30 minutes. All the sulphur dioxide will then have been expelled. Cool the solution in an atmosphere of carbon dioxide, add 10 cm^3 of dilute sulphuric acid (1:6), and titrate at once with a standard solution of the oxidising agent.

It has been found that a higher acidity can be tolerated ($<2N$ in H_2SO_4), and the reaction is accelerated, in the presence of thiocyanate ion. The procedure for the thiocyanate-accelerated reduction is as follows. Add 10 cm^3 of $0.1N$-potassium thiocyanate to the solution of iron(III), which should be less than M in sulphuric acid. Saturate the solution in the cold with sulphur dioxide, or add 50 cm^3 of freshly prepared sulphurous acid solution. Heat slowly to the boiling point. The solution rapidly becomes colourless or pale yellow. Displace the sulphur dioxide with carbon dioxide or nitrogen, cool, and add 10 cm³ of 0.1M-mercury(II) nitrate solution (to complex the thiocyanate). Titrate with standard permanganate (or dichromate) solution in the usual way.

Members of the hydrogen sulphide group of metals must be absent. If present, they must be removed first. Titanium and chromium are unaffected by the treatment; vanadium(V) is reduced to vanadium(IV).

C. Reduction with hydrogen sulphide. The method is not frequently employed. A typical procedure is as follows. The solution for reduction (*ca.* 200 cm³) should be about 0.5M in sulphuric acid. Heat to boiling, and pass a stream of washed hydrogen sulphide until the solution is saturated. Remove from the source of heat, and continue to pass the gas for a further 15 minutes. Boil the solution down to about 50 cm³ during 30–60 minutes while a stream of oxygen-free carbon dioxide is passed through. The solution is allowed to cool in a stream of the gas, diluted to 200 cm³ with distilled water, and titrated with standard permanganate solution. The precipitated sulphur is coagulated during the concentration, and usually need not be removed before titration.

Hydrogen sulphide is sometimes used for the reduction of Fe(III) to Fe(II) because of its selectivity. Copper is precipitated as sulphide and is filtered off; vanadium(V) is reduced to vanadium(IV), which does not interfere in the subsequent titration provided dichromate is used. Molybdenum is largely precipitated and is filtered off; the hydrogen sulphide is boiled out of the filtrate, a few drops of permanganate solution are added to re-oxidise the reduced molybdenum, hydrogen sulphide passed again, and the remaining molybdenum sulphide separated by filtration.

The reagents listed above can be applied to the reduction of many other ions in addition to Fe³⁺, and there are also a number of other substances which can be employed as reducing agents; thus for example hydroxylammonium salts are frequently added to solutions to ensure that reagents do not undergo atmospheric oxidation, and as an example of an unusual reducing agent, phosphorous acid may be used to reduce mercury(II) to mercury(I); see Section **X, 134**.

X, 147. References

1. International Union of Pure and Applied Chemistry (Aug. 1974). *Information Bulletin*. No. 36.
2. M. L. McGlashan (1971). *Physico-Chemical Quantities and Units*. 2nd edn., p. 45. London; Royal Institute of Chemistry.
3. A. I. Vogel (1958). *Elementary Practical Organic Chemistry*. Pt. III. *Quantitative Organic Analysis*. London; Longmans Green and Co.
4. F. A. Cotton and G. Wilkinson (1972). *Advanced Inorganic Chemistry*. 3rd edn. London; Interscience Publishers.
5. S. F. A. Kettle (1969). *Co-ordination Compounds*. London; T. Nelson and Sons Ltd.
6. G. Schwarzenbach and H. Flaschka (1969). *Complexometric Titrations*. 2nd edn. London; Methuen and Co.
7. R. Pribil and V. Veselý (1961). *Chemist-Analyst*, **50**, 100.
8. T. S. West. *Complexometry* (1969). 3rd edn. Poole; BDH Chemicals Ltd.
9. Society of Dyers and Colourists (1956). *Colour Index*. 2nd edn. Bradford.
10. R. A. Close and T. S. West (1960). *Talanta*, **5**, 221.
11. C. Woodward and H. N. Redman (1973). *High-precision Titrimetry*. London; Society for Analytical Chemistry.

12. H. N. Wilson (1951). *Analyst*, **76**, 65.
13. C. C. Reilly and W. W. Porterfield (1956). *Analytical Chemistry*, **28**, 443.
14. L. W. Andrews (1903). *J. Am. Chem. Soc.*, **25**, 756.
15. G. S. Jamieson (1926). *Volumetric Iodate Methods*. New York; Reinhold.
16. C. M. Ellis and A. I. Vogel (1956). *Analyst*, **81**, 693.
17. B. W. Smith and M. L. Parsons (1973). *J. Chem. Ed.*, **50**, 679.
18. A. R. Morrison (1972). *Lab. Practice*, **21**, 726.
19. International Union of Pure and Applied Chemistry (Sept. 1975). *Information Bulletin*. No. 45.
20. A. Ringbom (1963). *Complexation in Analytical Chemistry*. New York; Interscience.
21. *Stability Constants of Metal-Ion Complexes*. Chemical Society Special Publications Nos. 17 and 25, London.
22. *Laboratory Waste Disposal Manual* (1969). 2nd ed. Washington D.C.; Manufacturing Chemists Association. (rev. edn. 1974).
23. P. J. Gaston (1964). *The Care, Handling and Disposal of Dangerous Chemicals*. Aberdeen; Northern Publishers Ltd.

X, 148. Selected bibliography

1. D. Betteridge and H. E. Hallam (1972). *Modern Analytical Methods*. London; The Chemical Society.
2. E. Bishop (1972). *Indicators*. Oxford; Pergamon Press Ltd.
3. N. H. Furman (1962). *Standard Methods of Chemical Analysis*. 6th edn. Princeton, N.J.; Van Nostrand.
4. G. Jander (1956). *Neuere massanalytischen Methoden*. Stuttgart; Ferdinand Enke Verlag.
5. I. M. Kolthoff and V. A. Stenger. *Volumetric Analysis*. Vol. I (1942), Vol. II (1947), I. M. Kolthoff and R. Belcher. Vol. III (1957). New York; Interscience Publishers.
6. I. M. Kolthoff and P. J. Elving (1961). *Treatise on Analytical Chemistry*. New York; Interscience Publishers.
7. L. Meites (1963). *Handbook of Analytical Chemistry*. New York; McGraw-Hill.
8. R. Pribil (1972). *Analytical Applications of EDTA and Related Compounds*. Oxford; Pergamon Press Ltd.
9. G. Schwarzenbach and H. Flaschka (1969). *Complexometric Titrations*. 2nd edn. London; Methuen and Co.
10. W. Wagner and C. J. Hull (1971). *Inorganic Titrimetric Analysis*. New York; Marcel Dekker Inc.
11. C. L. Wilson and D. W. Wilson (1962). *Comprehensive Analytical Chemistry*. Amsterdam; Elsevier.
12. L. F. Hamilton, S. G. Simpson and D. W. Ellis (1969). *Calculations of Analytical Chemistry*. 7th edn. New York; McGraw-Hill.

Many of the general textbooks listed in Chapter I will also be relevant.

CHAPTER XI GRAVIMETRY

XI, 1. INTRODUCTION TO GRAVIMETRIC ANALYSIS. Gravimetric analysis or quantitative analysis by weight is the process of isolating and weighing an element or a definite compound of the element in as pure a form as possible. The element or compound is separated from a weighed portion of the substance being examined. A large proportion of the determinations in gravimetric analysis is concerned with the transformation of the element or radical to be determined into a pure stable compound which can be readily converted into a form suitable for weighing. The weight of the element or radical may then be readily calculated from a knowledge of the formula of the compound and the atomic weights of the constituent elements.

The separation of the element or of the compound containing it may be effected in a number of ways, the most important of which are: (a) precipitation methods, (b) volatilisation or evolution methods, (c) electroanalytical methods, and (d) extraction and chromatographic methods. Only (a) and (b) will be discussed in this chapter: (c) is considered in Part E, and (d) in Part C.

It may be mentioned at this stage that the great advantage of gravimetric over titrimetric analysis is that the constituent is isolated and may be examined for the presence of impurities and a correction applied, if necessary; the disadvantage of gravimetric methods is that they are generally more time-consuming.

XI, 2. PRECIPITATION METHODS. These are perhaps the most important with which we are concerned in gravimetric analysis. The constituent being determined is precipitated from solution in a form which is so slightly soluble that no appreciable loss occurs when the precipitate is separated by filtration and weighed. Thus in the determination of silver, a solution of the substance is treated with an excess of sodium or potassium chloride solution, the precipitate is filtered off, well washed to remove soluble salts, dried at 130–150 °C, and weighed as silver chloride. Frequently the constituent being estimated is weighed in a form other than that in which it was precipitated. Thus magnesium is precipitated, as ammonium magnesium phosphate $Mg(NH_4)PO_4,6H_2O$, but is weighed, after ignition, as the pyrophosphate $Mg_2P_2O_7$. The factors which determine a successful analysis by precipitation are:

1. The precipitate must be so insoluble that no appreciable loss occurs when it is collected by filtration. In practice this usually means that the quantity remaining in solution does not exceed the minimum detectable by the ordinary analytical balance, viz., 0.1 mg.

2. The physical nature of the precipitate must be such that it can be readily separated from the solution by filtration, and can be washed free of soluble impurities. These conditions require that the particles are of such size that they do not pass through the filtering medium, and that the particle size is unaffected (or, at least, not diminished) by the washing process.

3. The precipitate must be convertible into a pure substance of definite chemical composition; this may be effected either by ignition or by a simple chemical operation, such as evaporation, with a suitable liquid.

Factor 1, which is concerned with the completeness of precipitation, has already been dealt with in connection with the solubility product principle (Sections **II, 8** and **9**), and the influence upon the solubility of the precipitate of (i) a salt with a common ion, (ii) salts with no common ion, (iii) acids and bases, and (iv) temperature.

It was assumed throughout that the compound which separated out from the solution was chemically pure, but this is not always the case. The purity of the precipitate depends *inter alia* upon the substances present in solution both before and after the addition of the reagent, and also upon the exact experimental conditions of precipitation. In order to understand the influence of these and other factors, it will be necessary to give a short account of the properties of colloids.

Problems which arise with certain precipitates include the coagulation or flocculation of a colloidal dispersion of a finely divided solid to permit its filtration and to prevent its repeptisation upon washing the precipitate. It is therefore desirable to understand the basic principles of the colloid chemistry of precipitates.

XI, 3.   THE COLLOIDAL STATE.   The colloidal state of matter is distinguished by a certain range of particle size, as a consequence of which certain characteristic properties become apparent. Before discussing these, mention must be made of the various units which are employed in expressing small dimensions. The most important of these are:

$$1 \mu m = 10^{-3} \, mm \qquad\qquad 1 \, nm = 10^{-6} \, mm$$
$$1 \, \text{Ångström unit} = \text{Å} = 10^{-10} \, metre = 10^{-7} \, mm = 0.1 \, nm$$

Colloidal properties are, in general, exhibited by substances of particle size ranging between 0.1 μm and 1 nm. Ordinary quantitative filter paper will retain particles up to a diameter of about 10^{-2} mm or 10 μm, so that colloidal solutions in this respect behave like true solutions (size of molecules is of the order of 0.1 nm or 10^{-8} cm). The limit of vision under the microscope is about 0.2 μm. If a powerful beam of light is passed through a colloidal solution and the solution viewed at right angles to the incident light, a scattering of light is observed. This is the so-called **Tyndall effect**. True solutions, i.e., those with particles of molecular dimensions, do not exhibit a Tyndall effect, and are said to be 'optically empty'. Use is made of the Tyndall effect in the ultra-microscope; here the Tyndall cone or beam is observed in a microscope which is situated at right angles to the path of the incident light. The diffraction images are thus seen, and it is possible to observe the light scattered by each particle separately. The limit of visibility under the ultra-microscope is about 10 nm.

By the use of X-rays the physical structure of the smallest unit of colloidal substances may be ascertained. It has been found that most substances consist of minute crystalline particles; a few, such as silica and tin(IV) oxide, are amorphous. An intermediate stage is also possible: a gradual development of crystalline particles may occur with some amorphous substances upon 'ageing' or with suitable treatment, such as digestion with hot water or solutions of electrolytes (Section XI, 5).

An important consequence of the smallness of the size of the particles is that the ratio of surface area to weight is extremely large. Phenomena, such as adsorption, which depend upon the size of the surface will therefore play an important part with substances in the colloidal state. Table XI, 1 clearly shows the influence of particle size in connection with a 1-cm cube decimally divided.

Table XI, 1   Increase in number and total surface of particles as a one centimetre cube is decimally divided

Number of particles	Length of edge in cm	Total surface in cm^2
1	1	6
10^6	10^{-2}	6×10^2
10^{12}	$10^{-4} (= 1 \mu m)$	6×10^4
10^{15}	10^{-5}	6×10^5
10^{21}	$10^{-7} (= 1 nm)$	6×10^7
10^{24}	$10^{-8} (= 1 Å)$	6×10^8

The characteristic properties of most types of colloidal particles encountered in inorganic analysis are:

(a) they exhibit a Tyndall effect when viewed with proper illumination (see, however, the table below);

(b) they may be separated from true solutions of substances by means of a collodion or parchment membrane, i.e., by the process of dialysis;

(c) they possess electrical charges since they migrate under the influence of a suitable potential gradient;

(d) they possess a very large surface area.

For convenience, we may divide colloids into two main groups, designated as **lyophobic** and **lyophilic colloids**. The chief properties of each class are summarised in the following table, although it must be emphasised that the distinction is not an absolute one, since some gelatinous precipitates (e.g., aluminium and other metallic hydroxides) have properties intermediate between those of lyophobic and lyophilic colloids.

Lyophobic colloids	Lyophilic colloids
1. The dispersion (or *sols*) are only slightly viscous. Examples: sols of metals, silver halides, metallic sulphides, etc.	1. The dispersions are very viscous; they set to jelly-like masses known as *gels*. Examples: sols of silicic acid, tin(IV) oxide, gelatin.
2. A comparatively minute concentration of an electrolyte results in flocculation. The change is, in general, irreversible; water has no effect upon the flocculated solid.	2. Comparatively large concentrations of electrolytes are required to cause precipitation ('salting out'). The change is, in general, reversible, and reversal is effected by the addition of a solvent (water).
3. Lyophobic colloids, ordinarily, have an electric charge of definite sign, which can be changed only by special methods.	3. Most lyophilic colloids change their charge readily, e.g., they are positively charged in acid medium and negatively charged in an alkaline medium.
4. The ultra-microscope reveals bright particles in vigorous motion (Brownian movement).	4. Only a diffuse light cone is exhibited under the ultra-microscope.

The process of dispersing a gel or a flocculated solid to form a sol is called **peptisation**.

The stability of lyophobic colloids is intimately associated with the electrical charge on the particles.* Thus in the formation of an arsenic(III) sulphide sol by precipitation with hydrogen sulphide in acid solution, sulphide ions are primarily adsorbed (since every precipitate has a tendency to adsorb its own ions), and

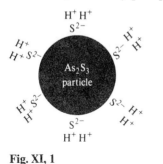

some hydrogen ions are secondarily adsorbed. The hydrogen ions or other ions which are secondarily adsorbed have been termed **counter ions**. Thus the so-called electrical double layer is set up between the particles and the solution. An arsenic(III) sulphide particle is represented diagrammatically in Fig XI, 1. The colloidal particle of arsenic(III) sulphide has a negatively charged surface, with positively charged counter ions which impart a positive charge to the liquid immediately surrounding it. If an electric current is passed

Fig. XI, 1

through the solution, the negative particles will move towards the anode; the speed is comparable with that of electrolytic ions. The electrical conductivity of a sol is, however, quite low because the number of current-carrying particles is small compared with that in a solution of an electrolyte at an appreciable concentration; the large charge carried by the colloidal particles is not sufficient to compensate for their smaller number.

If the electrical double layer is destroyed, the sol is no longer stable, and the particles will flocculate, thereby reducing the large surface area. Thus if barium chloride solution is added, barium ions are preferentially adsorbed by the particles; the charge distribution on the surface is disturbed and the particles flocculate. After flocculation, it is found that the dispersion medium is acid owing to the liberation of the hydrogen counter ions. It appears that ions of opposite charge to those primarily adsorbed on the surface are necessary for cogulation. The minimum amount of electrolyte necessary to cause flocculation of the colloid is called the **flocculation** or **coagulation value**. It has been found that the latter depends primarily upon the valency of the ions of the opposite charge to that on the colloidal particles: the nature of the ions has some influence also. This is clearly shown by the results collected in Table XI, 2.

Table XI, 2   Coagulation values in milli-mols of coagulating ion per litre

Negative arsenic(III) sulphide sol.		Positive hydrated iron(III) oxide sol.	
Salt	Coag. value	Salt	Coag. value
$AlCl_3$	0.062	$K_4Fe(CN)_6$	0.06
$Al_2(SO_4)_3$	0.074	$K_3Fe(CN)_6$	0.09
$FeCl_3$	0.136	K_2SO_4	0.22
$CaCl_2$	0.649	$K_2Cr_2O_7$	0.19
$BaCl_2$	0.691	K_2CrO_4	0.33
$MgCl_2$	0.717	$K_2C_2O_4$	0.24
$Ba(NO_3)_2$	0.687	$KBrO_3$	31
KCl	49.5	KSCN	47
NaCl	51.0	KCl	103
LiCl	58.4	KNO_3	131
KNO_3	50.0	KBr	138
HCl	30.8	KI	154

* Lyophilic colloids are mainly stabilised by solvation.

If two sols of opposite sign are mixed, mutual coagulation usually occurs owing to the neutralisation of charges. The above remarks apply largely to lyophobic colloids. Lyophilic colloids are generally much more difficult to coagulate than lyophobic colloids. If a lyophilic colloid, e.g., of gelatin, is added to a lyophobic colloid, e.g., of gold, then the lyophobic colloid appears to be strongly protected against the flocculating action of electrolytes. It is probable that the particles of the lyophilic colloid are adsorbed by the lyophobic colloid and impart their own properties to the latter. The lyophilic colloid is known as a **protective colloid**. This explains the relative stability produced by the addition of a little gelatin to the otherwise unstable gold sols. For this reason also, organic matter, which might form a protective colloid, is generally destroyed before proceeding with an inorganic analysis.

During the flocculation of a colloid by an electrolyte, the ions of opposite sign to that of the colloid are adsorbed to a varying degree on the surface; the higher the valency of the ion, the more strongly is it adsorbed. In all cases, the precipitate will be contaminated by surface adsorption. Upon washing the precipitate with water, part of the adsorbed electrolyte is removed, and a new difficulty may arise. The electrolyte concentration in the supernatant liquid may fall below the coagulation value, and the precipitate may pass into colloidal solution again. This phenomenon, which is known as **peptisation**, is of great importance in quantitative analysis. By way of illustration, let us consider the precipitation of silver by excess of chloride ions in acid solution and the subsequent washing of the coagulated silver chloride with water; the adsorbed hydrogen ions will be removed by the washing process and a portion of the precipitate may pass through the filter. If, however, washing is carried out with dilute nitric acid, no peptisation occurs. For this reason, precipitates are always washed with a suitable solution of an electrolyte which does not interfere with the subsequent steps in the determination.

The adsorptive properties of colloids find a number of applications in analysis, e.g., in the removal of phosphates by hydrated tin(IV) oxide in the presence of nitric acid, in the use of adsorption indicators Section **X, 30, C**, in the qualitative detection and colorimetric determination of elements and radicals with many organic reagents (for example, magnesium with Titan yellow, Section **XVIII, 22, A**).

XI, 4. SUPERSATURATION AND PRECIPITATE FORMATION. The solubility of a substance at any given temperature in a given solvent is the amount of the substance dissolved by a known weight of that solvent when the substance is in equilibrium with the solvent. The solubility depends upon the particle size, when these are smaller than about 0.01 mm in diameter; the solubility increases greatly the smaller the particles, owing to the increasing rôle played by surface effects (compare Table XI, 1). (The definition of solubility given above refers to particles larger than 0.01 mm.) A supersaturated solution is one that contains a greater concentration of solute than corresponds to the equilibrium solubility at the temperature under consideration. Supersaturation is therefore an unstable state which may be brought to a state of stable equilibrium by the addition of a crystal of the solute ('seeding' the solution) or of some other substance, or by mechanical means such as shaking or stirring. The difficulty of precipitation of ammonium magnesium phosphate will at once come to mind as an example of supersaturation.

According to von Weimarn supersaturation plays an important part in determining the particle size of a precipitate. He deduced that the initial velocity of precipitation is proportional to $(Q - S)/S$, where Q is the total concentration of the substance that is to precipitate, and S is the equilibrium solubility; $(Q - S)$ will denote the supersaturation at the moment precipitation commences. The expression applies approximately only when Q is large as compared with S. The influence of the degree of supersaturation is well illustrated by von Weimarn's results for the formation of barium sulphate from solutions of barium thiocyanate and manganese sulphate respectively. These are collected in Table XI, 3. The results clearly show that the particle size of a precipitate decreases with

Table XI, 3 Separation of BaSO$_4$ at various degrees of supersaturation (von Weimarn)

Concentration of reagents	$(Q-S)/S$	Type of precipitate
$7N$	175 000	A gelatinous precipitate is formed, and practically the whole of the water is immobilised; the containing vessel can be inverted without the contents running out. The gel is unstable, and growth of the large crystals at the expense of small ones is very rapid; after a few hours the precipitate becomes opaque.
$3N$	75 000	Gelatinous films formed; become turbid after one minute.
N	25 000	Primary precipitate is curdy and of colloidal dimensions. Particles appear as points at a magnification of 1500 ×.
$0.05N$	1300	Primary precipitate consists of feathery and star-shaped crystal skeletons.
$0.005N$	125	Precipitate consists of compact crystal skeletons.
$ca.\ 0.001N$	25	Solution becomes opalescent during first 5 minutes, and precipitation continues for 2–3 hours. After that time crystals have a length of 0.005 mm.
$ca.\ 0.0002N$	5	Precipitate appears after about a month. At the end of six months, the length of the largest crystals is about 0.03 mm and their breadth 0.015 mm.

increasing concentration of the reactants. For the production of a crystalline precipitate, for which the adsorption errors will be least and filtration will be easiest, $(Q - S)/S$ should be as small as possible. There is obviously a practical limit to reducing $(Q - S)/S$ by making Q very small, since for a precipitation to be of value in analysis, it must be complete in a comparatively short time and the volumes of solutions involved must not be too large. There is, however, another method which may be used, viz., that of increasing S. For example, barium sulphate is about fifty times more soluble in $2M$-hydrochloric acid than in water: if $0.05M$ solutions of barium chloride and sulphuric acid are prepared in $2M$ boiling hydrochloric acid and the solutions mixed, a typical crystalline precipitate of barium sulphate is slowly formed (Refs. 1 and 2).

Applications of the above conceptions are to be found in the following recognised procedures in gravimetric analysis:

1. Precipitation is usually carried out in hot solutions, since the solubility generally increases with rise in temperature.
2. Precipitation is effected in dilute solution and the reagent is added slowly and with thorough stirring. The slow addition results in the first particles precipitated acting as nuclei which grow as further material precipitates.
3. A suitable reagent is often added to increase the solubility of the precipitate and thus lead to larger primary particles.

4. A procedure which is commonly employed to prevent supersaturation from occurring is that of precipitation from homogeneous solution. This is achieved by forming the precipitating agent within the solution by means of an homogeneous reaction at a similar rate to that required for precipitation of the species.

XI, 5   THE PURITY OF THE PRECIPITATE. CO-PRECIPITATION.

When a precipitate separates from a solution, it is not always perfectly pure: it may contain varying amounts of impurities dependent upon the nature of the precipitate and the conditions of precipitation. The contamination of the precipitate by substances which are normally soluble in the mother liquor is termed **co-precipitation**. We must distinguish between two important types of co-precipitation. The first is concerned with adsorption at the *surface* of the particles exposed to the solution, and the second relates to the occlusion of foreign substances during the process of crystal growth from the primary particles.

With regard to surface adsorption, this will, in general, be greatest for gelatinous precipitates and least for those of pronounced macro-crystalline character. Precipitates with ionic lattices appear to conform to the Paneth–Fajans–Hahn adsorption rule, which states that the ion that is most strongly adsorbed by an ionic substance (crystal lattice) is that ion which forms the least soluble salt. Thus on sparingly soluble sulphates calcium ions are adsorbed preferentially over magnesium ions because calcium sulphate is less soluble than magnesium sulphate. Also silver iodide adsorbs silver acetate much more strongly than silver nitrate under comparable conditions, since the former is the less soluble. The deformability of the adsorbed ions and the electrolytic dissociation of the adsorbed compound also have a considerable influence; the smaller the dissociation of the compound, the greater is the adsorption. Thus hydrogen sulphide, a weak electrolyte, is strongly adsorbed by metallic sulphides.

The second type of co-precipitation may be visualised as occurring during the building up of the precipitate from the primary particles. The latter will be subject to a certain amount of surface adsorption, and during their coalescence the impurities will either be partially eliminated if large single crystals are formed and the process takes place slowly, or, if coalescence is rapid, large crystals composed of loosely bound small crysals may be produced and some of the impurities may be entrained within the walls of the large crystals. If the impurity is isomorphous or forms a solid solution with the precipitate, the amount of co-precipitation may be very large, since there will be no tendency for elimination during the 'ageing' process. The latter actually occurs during the precipitation of barium sulphate in the presence of alkali nitrates; in this particular case X-ray studies have shown that the abnormally large co-precipitation (which may be as high as 3.5 per cent if precipitation occurs in the presence of high concentrations of nitrate) is due to the formation of solid solutions. Fortunately, however, such cases are comparatively rare in analysis.

Appreciable errors may also be introduced by **post-precipitation**. This is the precipitation which occurs on the surface of the first precipitate *after* its formation. It occurs with sparingly soluble substances which form super-saturated solutions; they usually have an ion in common with the primary precipitate. Thus in the precipitation of calcium as oxalate in the presence of magnesium, magnesium oxalate separates out gradually upon the calcium oxalate; the longer the precipitate is allowed to stand in contact with the solution, the greater is the error due to this cause. A similar effect is observed in the precipitation of copper or mercury(II) sulphide in $0.3M$-hydrochloric acid in the presence of zinc ions; zinc sulphide is slowly post-precipitated.

Post-precipitation differs from co-precipitation in several respects:

(*a*) The contamination increases with the time that the precipitate is left in contact with the mother liquor in post-precipitation, but usually decreases in co-precipitation.

(*b*) With post-precipitation, contamination increases the faster the solution is agitated by either mechanical or thermal means. The reverse is usually true with co-precipitation.

(*c*) The magnitude of contamination by post-precipitation may be much greater than in co-precipitation.

It is convenient to consider now the influence of **digestion**. This is usually carried out by allowing the precipitate to stand for 12–24 hours at room temperature, or sometimes by warming the precipitate for some time, in contact with the liquid from which it was formed: the object is, of course, to obtain complete precipitation in a form which can be readily filtered. During the process of digestion or of the ageing of precipitates, at least two changes occur. The very small particles, which have a greater solubility than the larger ones, will, after precipitation has occurred, tend to pass into solution, and will ultimately re-deposit upon the larger particles; co-precipitation on the minute particles is thus eliminated and the total co-precipitation on the ultimate precipitate reduced. The rapidly formed crystals are probably of irregular shape and possess a compara-tively large surface; upon digestion these tend to become more regular in character and also more dense, thus resulting in a decrease in the area of the surface and a consequent reduction of adsorption. The net result of digestion is usually to reduce the extent of co-precipitation and to increase the size of the particles, rendering filtration easier.

XI, 6.   CONDITIONS OF PRECIPITATION.   No universal rules can be given which are applicable to all cases of precipitation, but, with the aid of an intelligent application of the facts enumerated in the foregoing paragraphs, a number of fairly general rules may be stated:

1. Precipitation should be carried out in dilute solution, due regard being paid to the solubility of the precipitate, the time required for filtration, and the subsequent operations to be carried out with the filtrate. This will minimise the errors due to co-precipitation.

2. The reagents should be mixed slowly and with constant stirring. This will keep the degree of supersaturation small and will assist the growth of large crystals. A slight excess of the reagent is all that is generally required; in exceptional cases a large excess may be necessary. In some instances the order of mixing the reagents may be important. Precipitation may be effected under conditions which increase the solubility of the precipitate, thus further reducing the degree of supersaturation (compare Section **XI, 5**).

3. Precipitation is effected in hot solutions, provided the solubility and the stability of the precipitate permit. Either one or both of the solutions should be heated to just below the boiling point or other more favourable temperature. At the higher temperature: (*a*) the solubility is increased with a consequent reduction in the degree of supersaturation, (*b*) coagulation is assisted and sol

formation decreased, and (*c*) the velocity of crystallisation is increased, thus leading to better-formed crystals.

4. Crystalline precipitates should be digested for as long as practical, preferably overnight, except in those cases where post-precipitation may occur. As a rule, digestion on the steam bath is desirable. This process decreases the effect of co-precipitation and gives more readily filterable precipitates. Digestion has little effect upon amorphous or gelatinous precipitates.

5. The precipitate should be washed with the appropriate dilute solution of an electrolyte. Pure water may tend to cause peptisation. (For theory of washing, see Section **XI, 8** below.)

6. If the precipitate is still appreciably contaminated as a result of co-precipitation or other causes, the error may often be reduced by dissolving it in a suitable solvent and then reprecipitating it. The amount of foreign substance present in the second precipitation will be small, and consequently the amount of the entrainment by the precipitate will also be small.

XI, 7. PRECIPITATION FROM HOMOGENEOUS SOLUTION. The
major objective of a precipitation reaction is the separation of a pure solid phase in a compact and dense form which can be filtered easily. The importance of a small degree of supersaturation has long been appreciated, and it is for this reason that a dilute solution of a precipitating agent is added slowly and with stirring. In the technique known as precipitation from homogeneous solution the precipitant is not added as such, but is slowly generated by a homogeneous chemical reaction within the solution. The precipitate is thus formed under conditions which eliminate the undesirable concentration effects which are inevitably associated with the conventional precipitation process. The precipitate is dense and readily filterable; co-precipitation is reduced to a minimum. Moreover, by varying the rate of the chemical reaction producing the precipitant in homogeneous solution, it is possible to alter further the physical appearance of the precipitate—the slower the reaction, the larger (in general) are the crystals formed.

Many different anions can be generated at a slow rate; the nature of the anion is important in the formation of compact precipitates. It is convenient to deal with the subject under separate headings.

(*a*) *Hydroxides and basic salts.*   The necessity for careful control of the pH has long been recognised. This is accomplished by making use of the hydrolysis of urea, which decomposes into ammonia and carbon dioxide as follows:

$$CO(NH_2)_2 + H_2O = 2NH_3 + CO_2$$

Urea possesses negligible basic properties ($K_b = 1.5 \times 10^{-14}$), is soluble in water and its hydrolysis rate can be easily controlled. It hydrolyses rapidly at 90–100 °C, and hydrolysis can be quickly terminated at a desired pH by cooling the reaction mixture to room temperature. The use of a hydrolytic reagent *alone* does not result in the formation of a compact precipitate; the physical character of the precipitate will be very much affected by the presence of certain anions. Thus in the precipitation of aluminium by the urea process, a dense precipitate is obtained in the presence of succinate, sulphate, formate, oxalate, and benzoate, but not in the presence of chloride, chlorate, perchlorate, nitrate, sulphate, chromate, and acetate. The preferred anion for the precipitation of aluminium is succinate. It would appear that the main function of the 'suitable anion' is the formation of a basic salt which seems responsible for the production of a compact precipitate. The pH of the initial solution must be appropriately adjusted.

The following are suitable anions for urea precipitations of some metals: sulphate for gallium, tin, and titanium; formate for iron, thorium, and bismuth; succinate for aluminium and zirconium.

The urea method generally results in the deposition on the surface of the beaker of a thin, tenacious, and somewhat transparent film of the basic salt. This film cannot be removed by scraping with a 'policeman'. It is dissolved by adding a few cm^3 of hydrochloric acid, covering the beaker with a clock glass, and refluxing for 5–10 minutes; the small amount of metallic ion is precipitated by ammonia solution and filters readily through the same filter containing the previously precipitated basic salt.

The urea hydrolysis method may be applied also to:
(i) the precipitation of barium as barium chromate in the presence of ammonium acetate;
(ii) the precipitation of large amounts of nickel as the dimethylglyoximate; and
(iii) the precipitation of aluminium as the oxinate.

(*b*) *Phosphates.* Insoluble phosphates may be precipitated with phosphate ion derived from trimethyl or triethyl phosphate by stepwise hydrolysis. Thus 1.8M-sulphuric acid containing zirconyl ions and trimethyl phosphate on heating gives a dense precipitate of variable composition, which is ignited to and weighed as the pyrophosphate ZrP_2O_7.

Metaphosphoric acid may also be used; it hydrolyses in warm acid solution forming orthophosphoric acid. Thus bismuth may be precipitated as bismuth phosphate in a dense, crystalline form.

(*c*) *Oxalates.* Urea may be employed to raise the pH of an acid solution containing hydrogenoxalate ion $HC_2O_4{}^-$, thus affording a method for the slow generation of oxalate ion. Calcium oxalate may thus be precipitated in a dense form:

$$CO(NH_2)_2 + 2HC_2O_4{}^- + H_2O = 2NH_4{}^+ + CO_2 + 2C_2O_4{}^{2-}$$

Dimethyl and diethyl oxalate can be hydrolysed to serve as reagents for oxalate ion:

$$(C_2H_5)_2C_2O_4 + 2H_2O = 2C_2H_5OH + 2H^+ + C_2O_4{}^{2-}$$

Diethyl oxalate is usually preferred because of its slower rate of hydrolysis. Satisfactory results are obtained in the precipitation of calcium, magnesium, and zinc: thorium is precipitated using dimethyl oxalate.

Calcium can be determined as the oxalate by precipitation from homogeneous solution by cation release from the EDTA complex in the presence of oxalate ion (Ref. 3).

(*d*) *Sulphates.* Sulphate ion may be generated by the hydrolysis of sulphamic acid:

$$NH_2SO_3H + H_2O = NH_4{}^+ + H^+ + SO_4{}^{2-}$$

The reaction has been used to produce barium sulphate in a coarsely crystalline form.

The hydrolysis of dimethyl sulphate also provides a source of sulphate ion, and the reaction has been used for the precipitation of barium, strontium, and calcium as well as lead:

$$(CH_3)_2SO_4 + 2H_2O = 2CH_3OH + 2H^+ + SO_4^{2-}$$

XI, 8. WASHING OF THE PRECIPITATE.   The experimental aspect of this important subject is dealt with in Section **III, 42**. Only some general theoretical considerations will be given here. Most precipitates are produced in the presence of one or more soluble compounds, and it is the object of the washing process to remove these as completely as possible. It is evident that only surface impurities will be removed in this way. The composition of the wash solution will depend upon the solubility and chemical properties of the precipitate and upon its tendency to undergo peptisation, the impurities to be removed, and the influence of traces of the wash liquid upon the subsequent treatment of the precipitate before weighing. Pure water cannot, in general, be employed owing to the possibility of producing partial peptisation of the precipitate and, in many cases, the occurrence of small losses as a consequence of the slight solubility of the precipitate: a solution of some electrolyte is employed. This should possess a common ion with the precipitate in order to reduce solubility errors, and should easily be volatilised in the preparation of the precipitate for weighing. For these reasons, ammonium salts, ammonia solution, and dilute acids are commonly employed. If the filtrate is required in a subsequent determination, the selection is limited to substances which will not interfere in the sequel. Also hydrolysable substances will necessitate the use of solutions containing an electrolyte which will depress the hydrolysis (compare Section **II, 18**). Whether the wash liquid is employed hot or at some other temperature will depend primarily upon the solubility of the precipitate; if permissible, hot solutions are to be preferred because of the greater solubility of the foreign substances and the increased speed of filtration.

It is convenient to divide wash solutions into three classes:

1. *Solutions which prevent the precipitate from becoming colloidal and passing through the filter*. This tendency is frequently observed with gelatinous or flocculated precipitates but rarely with well-defined crystalline precipitates. The wash solution should contain an electrolyte. The nature of the electrolyte is immaterial, provided it is without action upon the precipitate either during washing or ignition. Ammonium salts are therefore widely used. Thus dilute ammonium nitrate solution is employed for washing iron(III) hydroxide (hydrated iron(III) oxide), and 1 per cent nitric acid for washing silver chloride.

2. *Solutions which reduce the solubility of the precipitate*.   The wash solution may contain a moderate concentration of a compound with one ion in common with the precipitate, use being made of the fact that substances tend to be less soluble in the presence of a slight excess of a common ion. Most salts are insoluble in ethanol and similar solvents, so that organic solvents can sometimes be used for washing precipitates. Sometimes a mixture of an organic solvent (e.g., ethanol) and water or a dilute electrolyte is effective in reducing the solubility to negligible proportions. Thus $100\,cm^3$ of water at $25\,°C$ will dissolve 0.7 mg of calcium oxalate, but the same volume of dilute ammonium oxalate solution dissolves only a negligible weight of the salt. Also $100\,cm^3$ of water at room temperature will dissolve 4.2 mg of lead sulphate, but dilute sulphuric acid or 50 per cent aqueous ethanol has practically no solvent action on the compound.

3. *Solutions which prevent the hydrolysis of salts of weak acids and bases*. If the

413

precipitate is a salt of a weak acid and is slightly soluble it may exhibit a tendency to hydrolyse, and the soluble product of hydrolysis will be a base; the wash liquid must therefore be basic. Thus $Mg(NH_4)PO_4$ may hydrolyse appreciably to give acid phosphate ion HPO_4^{2-} and hydroxide ion, and should accordingly be washed with dilute aqueous ammonia. If salts of weak bases, such as hydrated iron(III), chromium or aluminium ion, are to be separated from a precipitate, e.g., silica, by washing with water, the salts may be hydrolysed and their insoluble basic salts or hydroxides may be produced together with an acid:

$$[Fe(H_2O_6)]^{3+} \rightleftharpoons [Fe(OH)(H_2O)_5]^{2+} + H^+$$

The addition of an acid to the wash solution will prevent the hydrolysis of iron(III) or similar salts: thus dilute hydrochloric acid will serve to remove iron(III) and aluminium salts from precipitates that are insoluble in this acid.

Solubility losses are reduced by employing the minimum quantity of wash solution consistent with the removal of impurities. It can be readily shown that washing is more efficiently carried out by the use of many small portions of liquid than with a few large portions, the volume being the same in both instances. Under ideal conditions, where the foreign body is simply mechanically associated with the particles of the precipitate, the following expression may be shown to hold:

$$x_n = x_0 \left(\frac{u}{u+v} \right)^n$$

where x_0 is the concentration of impurity before washing, x_n is the concentration of impurity after n washings, u is the volume in cm^3 of the liquid remaining with the precipitate after draining, and v is the volume in cm^3 of the solution used in each washing. It follows from this expression that it is best: (a) to allow the liquid to drain as far as possible in order to maintain u at a minimum, and (b) to use a relatively small volume of liquid and to increase the number of washings. Thus if $u = 1 cm^3$ and $v = 9 cm^3$, five washings would reduce the surface impurity to 10^{-6} of its original value; one washing with the same volume of liquid, viz., $45 cm^3$, would only reduce the concentration to $1/46$ or 2.2×10^{-2} of its initial concentration.

In practice, the washing process is not quite so efficient as the above simple theory would indicate, since the impurities are not merely mechanically associated with the surface. Furthermore, solubility losses are not so great as one would expect from the solubility data because the wash solution passing through the filter is not saturated with respect to the precipitate. Frequent qualitative tests must be made upon portions of the filtrate for some foreign ion which is known to be present in the original solution; as soon as these tests are negative, the washing is discontinued.

XI, 9.   IGNITION OF THE PRECIPITATE. THERMOGRAVIMETRIC METHOD OF ANALYSIS.

In addition to superficially adherent water, precipitates may contain:
(a) adsorbed water, present on all solid surfaces in amount dependent on the humidity of the atmosphere;
(b) occluded water, present in solid solution or in cavities within crystals;
(c) sorbed water, associated with substances having a large internal surface development, e.g., hydrous oxides; and

(*d*) essential water, present as water of hydration or crystallisation (e.g., CaC_2O_4,H_2O or $Mg(NH_4)PO_4,6H_2O$) or as water of constitution (the water is not present as such but is formed on heating, e.g., $Ca(OH)_2 \rightarrow CaO + H_2O$).

In addition to the evolution of water, the ignition of precipitates often results in thermal decomposition reactions involving the dissociation of salts into acidic and basic components, e.g., the decomposition of carbonates and sulphates; the decomposition temperatures will obviously be related to the thermal stabilities.

The temperatures at which precipitates may be dried or ignited are determined from a knowledge of the thermogravimetric curves for the individual substances (the nature of these curves is explained in detail in Chapter XXIII). Thus calcium oxalate remains as the anhydrous salt between about 200° and 350 °C, above which it progressively decomposes. By using these curves it is possible to select appropriate drying and ignition temperatures to convert precipitates into pre-determined chemical forms. Thermogravimetric curves must be interpreted with due regard to the fact that in obtaining them the temperature is changing (usually at a regular rate), whereas in routine gravimetric analysis a precipitate is brought to a specified temperature and maintained at that temperature for a definite time.

Quantitative separations based upon precipitation methods

XI, 10. FRACTIONAL PRECIPITATION. The simple theory of fractional precipitation has been given in Section **II, 10**. It was shown that when the solubility products of two sparingly soluble salts having an ion in common differ sufficiently, then one salt will precipitate almost completely before the other commences to separate. This separation is actually possible for a mixture of chloride and iodide, but in other cases the theoretical predictions must be verified experimentally because of the danger of co-precipitation (Section **XI, 5**) affecting the results. Some separations based upon fractional precipitation, which are of practical importance, will now be considered.

A. Precipitation of sulphides. In order to understand fully the separations dependent upon the sulphide ion, we shall consider first the quantitative relationships involved in a saturated solution of hydrogen sulphide. The following equilibria are present:

$$H_2S \rightleftharpoons H^+ + HS^-$$
$$HS^- \rightleftharpoons H^+ + S^{2-}$$
$$[H^+] \times [HS^-]/[H_2S] = K_1 = 1.0 \times 10^{-7} \tag{1}$$
$$[H^+] \times [S^{2-}]/[HS^-] = K_2 = 1.0 \times 10^{-14} \tag{2}$$

The very small value of K_2 indicates that the secondary dissociation and consequently $[S^{2-}]$ is exceedingly small. It follows therefore that only the primary ionisation is of importance, and $[H^+]$ and $[HS^-]$ are practically equal in value. A saturated aqueous solution of hydrogen sulphide at 25 °C, at atmospheric pressure, is approximately $0.1\,M$, and calculation shows (see Section **II, 6**) that in this solution

$$[H^+] = [HS^-] = 1 \times 10^{-4}\,mol\,dm^{-3},$$
$$[S^{2-}] = 1 \times 10^{-14}\,mol\,dm^{-3},$$

and $[S^{2-}]$ is inversely proportional to the square of the hydrogen ion concentration. Clearly, by varying the pH of the solution the sulphide ion concentration may be controlled, and in this way, separations of metallic sulphides may be effected.

As shown in Section **II, 7**, in a solution of $0.25\,M$ hydrochloric acid saturated with hydrogen sulphide (this is the solution employed for the precipitation of the sulphides of the Group II metals in qualitative analysis),

$$[HS^-] = 4 \times 10^{-8}\,mol\,dm^{-3}$$

and

$$[S^{2-}] = 1.6 \times 10^{-21}\,mol\,dm^{-3}.$$

Thus by changing the acidity from $9.5 \times 10^{-5}\,M$ (that present in saturated hydrogen sulphide water) to $0.25\,M$, the sulphide ion concentration is reduced from 1×10^{-14} to 1.6×10^{-21}.

With the aid of a table of solubility products of metallic sulphides (see Appendix), we can calculate whether certain sulphides will precipitate under any given conditions of acidity and also the concentration of the metallic ions remaining in solution. Precipitation of a metallic sulphide MS will occur when $[M^{2+}] \times [S^{2-}]$ exceeds the solubility product, and the concentration of metallic ions remaining in the solution may be calculated from the equation:

$$[M^{2+}] = \frac{S_{MS}}{[S^{2-}]} = \frac{S_{MS} \times [H^+]^2}{1.0 \times 10^{-21} \times [H_2S]} \tag{3}$$

As an example we may consider the precipitation of copper(II) sulphide ($K_{S.CuS} = 8.5 \times 10^{-45}$) and iron(II) sulphide ($K_{S.FeS} = 1.5 \times 10^{-19}$) from $0.01\,M$ solutions of the metallic ions in the presence of $0.25\,M$-hydrochloric acid. For copper(II) sulphide, the solubility product is readily exceeded;

$$[S^{2-}] = 1.6 \times 10^{-21},\ [Cu^{2+}] = 0.01\ \text{and precipitation will occur until}$$

$$[Cu^{2+}] = 8.5 \times 10^{-45} \times [H^+]^2/1.0 \times 10^{-21} \times [H_2S]$$
$$= 8.5 \times 10^{-45} \times (0.25)^2/1.0 \times 1^{-21} \times 0.1$$
$$= 5 \times 10^{-24}$$

i.e., precipitation is virtually complete. With iron(II) sulphide, the solubility product cannot be exceeded and precipitation will not occur under these conditions. If, however, the acidity is sufficiently decreased, and consequently $[S^{2-}]$ increased, iron(II) sulphide will be precipitated.

The case of zinc sulphide is of special interest. Various values are given in the literature for its solubility product: the most trustworthy figures vary between 1×10^{-24} and 8×10^{-26}. If we accept the latter figure, then we should expect precipitation to occur in a, say, $0.01\,M$ solution of zinc ions in the presence of $0.25\,M$-hydrochloric acid, since the S.P. should be exceeded; furthermore, the residual zinc ion concentration should be 4.7×10^{-4} when calculated as described above. In practice, precipitation does not occur at this acidity. This may be partly due to the great tendency that zinc sulphide possesses to remain in supersaturated solution, but is perhaps best explained as follows. The above figure for the solubility product refers to a solution in equilibrium with relatively large particles, whereas for precipitation to occur it is necessary that the S.P. of the particles *actually formed* should be exceeded. It may well be that under the above experimental conditions these are extremely small, thus possessing a greater solubility (Section **XI, 3**) and a greater solubility product; precipitation will therefore not take place. This view is supported by the fact that post-precipitation of zinc sulphide (compare Section **XI, 5**) will occur upon the surface of other metallic sulphides, such as those of copper and mercury. It is possible to precipitate zinc in acid solution provided the experimental conditions are very carefully controlled, e.g., when the pH of the solution lies between 2 and 3 and ammonium salts are present as coagulants: this is attained by the use of a buffer mixture of formic acid and ammonium formate, sulphuric acid and ammonium sulphate or of chloroacetic acid and sodium chloroacetate. It is probable that large particles of zinc sulphide are initially formed under these conditions.

It must be pointed out that the above calculations are approximate only, and may be regarded merely as illustrations of the calculations involved in considering the precipitation of sulphides under various experimental conditions; the solubility products of most metallic sulphides are not known with any great accuracy. It is by no means certain that the sulphide ion S^{2-} is the most important reactant in acidified solutions; it may well be that in many cases the active precipitant is the hydrogensulphide ion HS^-, the concentration of which is considerable, and that intermediate products are formed. Also much co-precipitation and post-precipitation occur in sulphide precipitations unless the experimental conditions are rigorously controlled.

B. Precipitation and separation of hydroxides at controlled hydrogen-ion concentration or pH.   The underlying theory is very similar to that just given for sulphides. Precipitation will depend largely upon the solubility product of the metallic hydroxide and the hydroxide-ion concentration, or since $pH + pOH = pK_w$ (Section **II, 16**), upon the hydrogen-ion concentration of the solution.

We have seen that the sulphide ion concentration of a saturated aqueous solution of hydrogen sulphide may be controlled within wide limits by suitably changing the concentration of hydrogen ions—a common ion—of the solution. In a like manner the hydroxide ion concentration of a solution of a weak base, such as aqueous ammonia ($K_b = 1.8 \times 10^{-5}$), may be regulated by the addition of a common ion, e.g., ammonium ions in the form of the completely dissociated ammonium chloride. The magnitude of the effect is best illustrated by means of an example. In a $0.1M$-ammonia solution, the degree of dissociation is given (Section **II, 5**) approximately by:

$$\alpha = \sqrt{KV} = \sqrt{1.8 \times 10^{-5} \times 10} = 0.0013$$

Hence $[OH^-] = 0.0013$, $[NH_4^+] = 0.0013$, and $[NH_3] = 0.0987$. As shown in Section **II, 7**, by the addition of 0.5 mole of ammonium chloride to 1 dm^3 of this solution, $[OH^-]$ is reduced to 3.6×10^{-6} mol dm^{-3}.

Thus the addition of half a mole of ammonium chloride to a $0.1M$ solution of aqueous ammonia has decreased the hydroxide-ion concentration from 0.0013 to

0.000 003 6, or has changed pOH from 2.9 to 5.4, i.e., the pH has changed from 11.1 to 8.6.

An immediate application of the use of the aqueous ammonia–ammonium chloride mixture may be made to the familiar example of the prevention of precipitation of magnesium hydroxide (S.P. 1.5×10^{-11}). We can first compute the minimum hydroxide ion concentration necessary to prevent precipitation in, say, $0.1M$-magnesium solution.

$$[OH^-] = \sqrt{\frac{K_{\text{S. Mg(OH)}_2}}{[Mg^{2+}]}} = \sqrt{\frac{1.5 \times 10^{-11}}{0.1}} = 1.22 \times 10^{-5}M$$

or $pOH = 4.9$ and $pH = 14.0 - 4.9 = 9.1$

If we employ an aqueous ammonia solution which is $0.1M$, the concentration of $[NH_4^+]$ ion as ammonium chloride or other ammonium salt necessary to prevent the precipitation of magnesium hydroxide can be readily calculated as follows. Substituting in the mass-action equation:

$$\frac{[NH_4^+] \times [OH^-]}{[NH_3]} = 1.8 \times 10^{-5}$$

$$\frac{[NH_4^+] \times 1.22 \times 10^{-5}}{0.1} = 1.8 \times 10^{-5}$$

or $[NH_4^+] = 1.48 \times 10^{-1}M$

This corresponds to an ammonium chloride concentration of $1.48 \times 10^{-1} \times 53.5 = 7.9 \text{ g dm}^{-3}$.

We will now consider the conditions necessary for the practically complete precipitation of magnesium hydroxide from a $0.1M$ solution of, say, magnesium chloride. A pOH slightly in excess of 4.9 (i.e., $pH = 9.1$) might fail to precipitate the hydroxide owing to supersaturation. Let us suppose the hydroxide ion concentration is increased ten times, i.e., to pOH 3.9 or pH 10.1, then, provided no supersaturation is present:

$$[Mg^{2+}] = \frac{K_{\text{S. Mg(OH)}_2}}{[OH^-]^2} = \frac{1.5 \times 10^{-11}}{(1.22 \times 10^{-4})^2} = 0.001M$$

i.e.,the concentration of the magnesium ions remaining in solution is $0.001M$, or 1 per cent of the magnesium ions would remain unprecipitated. If pOH is changed to 2.9 or pH to 11.1, it can be shown in a similar way that the concentration of the magnesium ions left in solution is *ca.* $1 \times 10^{-5}M$, so that the precipitation error is 0.1 per cent, a negligible quantity. We may therefore say that magnesium is precipitated quantitatively at a pH of 11.1.

Our knowledge of the solubility products of metallic hydroxides is, however, not very precise, so that it is not always possible to make exact theoretical calculations. The approximate pH values at which various hydroxides begin to precipitate from dilute solution are collected in Table XI,4.

The precipitated metallic hydroxides or hydrated oxides are gelatinous in character, and they tend to be contaminated with anions by adsorption and occlusion, and sometimes with basic salts. The values presented inTable XI, 4 suggest that many separations should be possible by fractional precipitation of

Table XI, 4 pH values at which various hydroxides are precipitated

pH	Metal ion	pH	Metal ion
	Sn^{2+}, Fe^{3+}, Zr^{4+}	7	Fe^{2+}
	Th^{4+}	8	Co^{2+}, Ni^{2+}, Cd^{2+}
	Al^{3+}	9	Ag^{+}, Mn^{2+}, Hg^{2+}
	Zn^{2+}, Cu^{2+}, Cr^{3+}	11	Mg^{2+}

the hydroxides. These separations are not always practical owing to high local concentrations of base when the solution is treated with alkali. Such unequal concentrations of base result in regions of high local pH and lead to the precipitation of more soluble hydroxides, which may be occluded in the desired precipitate. Slow, or preferably homogeneous, neutralisation overcomes this difficulty, and much sharper separations may be achieved.

The common tripositive cations may be separated from many dipositive cations by the *basic acetate* or *basic benzoate method*. These separations are based upon the fact that the equilibria for the first dissociation of the typical ions are:

$$[M(H_2O)_x]^{3+} + H_2O \rightleftharpoons$$
$$[M(H_2O)_{x-1}(OH)]^{2+} + H_3O^{+} \quad (K = 5 \times 10^{-3} - 1 \times 10^{-5})$$
$$[M(H_2O)_y]^{2+} + H_2O \rightleftharpoons$$
$$[M(H_2O)_{y-1}(OH)]^{+} + H_3O^{+} \quad (K = 10^{-7} - 10^{-12})$$

Any strong acid that may be present is first neutralised. Then, by selecting an appropriate base, whose conjugate acid has a K_a of about 10^{-5}, the equilibrium for the trivalent cations will be forced to the right; the base is too weak, however, to remove the hydroxonium ions from the equilibrium of the divalent cations. Since a large excess of the basic ion is added, a basic salt of the trivalent metal usually precipitates instead of the normal hydroxide. Acetate or benzoate ions (in the form of the sodium salts) are the most common bases that are employed for this procedure. The precipitation of basic salts may be combined with homogeneous precipitation, and thus very satisfactory separations may be obtained.

XI, 11. ORGANIC PRECIPITANTS.   Separation of one or more inorganic ions from mixtures may be made with the aid of organic reagents, with which they yield sparingly soluble and often coloured compounds. These compounds usually have high molecular weights, so that a small amount of the ions will yield a relatively large amount of the precipitate. The ideal organic precipitant should be *specific* in character, i.e., it should give a precipitate with only one particular ion. In few cases, however, has this ideal been attained; it is more usual to find that the organic reagent will react with a group of ions, but frequently by a rigorous control of the experimental conditions it is possible to precipitate only one of the ions of the group. Sometimes the precipitated organic compound may be weighed after drying at a suitable temperature; in other cases the composition is not quite definite and the substance is converted by ignition to the oxide of the metal; in a few instances, a titrimetric method is employed which utilises the quantitatively precipitated organic complex.

The original work in this field was largely empirical in character and was directed towards a search for specific, or at least highly selective, reagents for particular metal ions. A more fundamental approach is now possible, attention being directed to theoretical factors which lead to selectivity and also to a quantitative consideration of the equilibria involved. Frequently sufficient selectivity can be achieved for a particular purpose by controlling such variables as the concentration of the reagent and the pH, and also by taking advantage of secondary complexing agents (masking agents—see Section **X, 27**).

It is difficult to give a rigid classification of the numerous organic reagents. The most important are those which form chelate complexes, which involve the formation of one or more (usually five- or six-membered) rings incorporating the metal ion; ring formation leads to a relatively great stability. One classification of organic reagents is concerned with the number of hydrogen ions displaced from a neutral molecule in forming one chelate ring. A guide to qualitative predictions about the applicability of organic reagents for analytical purposes may be obtained from a study of the formation constant of the coordination compound (which is a measure of its stability), the effect of the nature of the metallic ion and of the ligand on the stability of complexes, and of the precipitation equilibria involved, particularly in the production of uncharged chelates. For further details, the reader is referred to Sections **X, 19–21** and to the books on chelate compounds listed in the Bibliography at the end of this chapter. Selected examples of precipitation reagents follow.

A. Dimethylglyoxime. This reagent (I) was discovered by L. Tschugaeff and was applied by O. Brunck for the determination of nickel in steel. It gives a bright red precipitate (II) of $Ni(C_4H_7O_2N_2)_2$ with nickel salt solutions; precipitation is usually carried out in ammoniacal solution or in a buffer solution containing ammonium acetate and acetic acid. The complex is weighed after drying at 110–120 °C. A slight excess of the reagent exerts no action on the precipitate, but a large excess should be avoided because: (a) of the possible precipitation of the dimethylglyoxime itself due to its low solubility in water (it is used in ethanolic solution),* and (b) the increased solubility of the precipitate in water–ethanol mixtures.* The interference of iron(III), aluminium, or bismuth is prevented by the addition of a soluble tartrate or citrate; when much cobalt, zinc, or manganese is present, precipitation should take place in a sodium acetate, rather than an ammonium acetate, buffer.

Solutions of palladium(II) salts give a characteristic yellow precipitate in dilute hydrochloric or sulphuric acid solution; the composition is similar to that of nickel, viz., $Pd(C_4H_7O_2N_2)_2$, and the precipitate can be dried at 110–120 °C and weighed. The precipitate is almost insoluble in hot water, but dissolves readily in ammonia and cyanide solutions. Gold is reduced to the metal by the reagent, and platinum (if present in appreciable quantity) is partially precipitated either as a greenish complex compound or as the metal, upon boiling the solution. The precipitation of palladium is not complete in the presence of nitrates.

Solutions of bismuth salts in the presence of EDTA give a yellow precipitate with dimethylglyoxime: precipitation is quantitative at pH 11.0–11.5, the precipitate is believed to be a polymer (Ref. 4) with an apparent composition $Bi_2O_2(C_4H_6N_2O_2)$, and may be dried at 105–125 °C. In the presence of EDTA and of cyanide ion, Al, As, Ba, Cd, Ca, Co, Cu, Pb, Mg, Hg, Ni, Pd, Pt, Ag, Sr, W, and Zn do not interfere.

---

* These possible errors may be avoided by employing disodium dimethylglyoxime, which is soluble in water—see below.

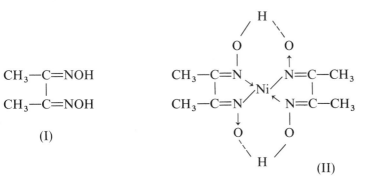

$$CH_3—C=NOH$$
$$CH_3—C=NOH$$

(I)

(II)

Dimethylglyoxime is only slightly soluble in water (0.40 g dm^{-3}), consequently it is employed as a 1 per cent solution in ethanol. The sodium salt of dimethylglyoxime $Na_2C_4H_6O_2N_2,8H_2O$ is available commercially: this is soluble in water and maybe employed as 2–3 per cent aqueous solution.

Furil-α-dioxime (III) has also been proposed for the determination of nickel. It gives a red precipitate with nickel salts in ammoniacal solution. The complex is less soluble than nickel dimethylglyoxime, and has a smaller nickel content, thus

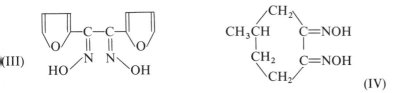

(III)

(IV)

giving a larger weight of precipitate for a given weight of nickel. The great advantage of furil-α-dioxime is its solubility in water, which precludes the possibility of contaminating the precipitate of the nickel derivative with the free reagent. A 2 per cent aqueous solution is normally used. The reagent is, however, expensive.

Cyclohexane-1,2-dione dioxime (nioxime) is more soluble in water (8.2 g dm^{-3} at $21\,°C$) than dimethylglyoxime: it is an excellent reagent for the gravimetric determination of palladium, but an empirical factor is required for the determination of nickel owing to co-precipitation of the reagnet. **4-Methylcyclohexane-1,2-dione dioxime (4-methyl-nioxime)** (IV) is fairly soluble in water (3.4 g dm^{-3} at $25\,°C$) and precipitates nickel quantitatively as a scarlet-coloured complex down to pH 3; the precipitate is uncontaminated by excess of reagent and filters easily: it is equally useful for the gravimetric determination of palladium.

B. Cupferron (ammonium salt of N-nitroso-N-phenylhydroxylamine), (V).*
This reagent, the ammonium salt of nitrosophenylhydroxylamine, forms
V).* This reagent, the ammonium salt of nitrosophenylhydroxylamine, forms
insoluble compounds with a number of metals in both weakly acid and strongly acid solutions. It is most useful when employed in strongly acid solutions (5–10 per cent by volume of hydrochloric or sulphuric acid) and then precipitates iron(III), vanadium(V), titanium(IV), zirconium(IV), cerium(IV), niobium(V),

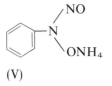

(V)

tantalum(V), tungsten(VI), gallium(III) and tin(IV) separating these elements from aluminium, beryllium chromium, manganese, nickel cobalt, zinc, uranium(VI) calcium, strontium and barium. The presence of tartrate and oxalate has no effect upon the precipitation of metals by cupferron.

The cupferron method is very satisfactory for the separation of iron, titanium, zirconium, vanadium and, in special cases, tin, tantalum, uranium, and gallium.

The reagent is usually employed as a 6 per cent aqueous solution; this should be freshly prepared, since it does not keep satisfactorily for more than a few days. The solid reagent should be stored in amber bottles containing a few lumps of ammonium carbonate. Precipitation is always carried out in the cold, since cupferron is decomposed into nitrosobenzene on heating. Sufficient reagent is added to form the curdy precipitate of the metallic derivative of cupferron and to give a white flocculent precipitate of free nitrosophenylhydroxylamine (needles). Precipitates should be filtered as soon after their formation as possible, since excess of cupferron is not very stable in acid solution. Nitric acid solutions cannot be used for the precipitation, since oxidising agents destroy the reagent. The addition of macerated filter paper assists the filtration of the precipitate and also the subsequent gradual ignition. The precipitates cannot be weighed after drying, but must be ignited to the corresponding oxide and weighed in this form. The ignition must be done cautiously in a *large* crucible with a gradual increase in temperature to avoid mechanical loss.

Neo-cupferron (ammonium salt of N-nitroso-N-2-naphthyl-hydroxylamine), (VI) forms less soluble and more bulky precipitates than cupferron. It may be employed for the direct separation of iron and copper in mineral and sea-waters without preliminary concentration.

N-Benzoyl-N-phenylhydroxylamine, $C_6H_5CO(C_6H_5)NOH$, has been proposed as a reagent similar to cupferron in its reactions, but is more stable. The reagent is moderately soluble in hot water but easily soluble in ethanol and other organic solvents. The Cu(II), Fe(III) and Al complexes can be weighed as such {e.g., as $Cu(C_{13}H_{10}O_2N)_2$} but the Ti compound must be ignited to the oxide.

C. 8-Hydroxyquinoline (oxine), (VII). Oxine (C_9H_7ON) forms sparingly soluble derivatives with metallic ions, which have the composition $M(C_9H_6ON)_2$ if the co-ordination number of the metal is four (e.g., magnesium, zinc, copper, cadmium, lead, and indium), $M(C_9H_6ON)_3$ if the co-ordination number is six (e.g., aluminium, iron, bismuth, and gallium), and $M(C_9H_6ON)_4$ if the co-ordination number is eight (e.g., thorium and zirconium). There are, however, some exceptions, for example, $TiO(C_9H_6ON)_2$, $MnO_2(C_9H_6ON)_2$, $WO_2(C_9H_6ON)_2$, and $UO_2(C_9H_6ON)_2$. By proper control of the pH of the solution, by the use of complex-forming reagents and by other methods, numerous separations may be carried out: thus aluminium may be separated

---

* The name *cupferron* was assigned to the compound by O. Baudisch, and is derived from the fact that the reagent precipitates both copper and iron. Cupferron precipitates iron completely in strong mineral acid solution, and copper is only quantitatively precipitated in faintly acid solution. The selectivity of the reagent is greatest in strongly acid solution.

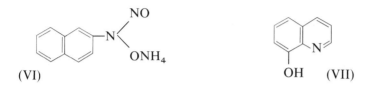

(VI)   (VII)

from beryllium in an ammonium acetate–acetic acid buffer, and magnesium from
the alkaline-earth metals in ammoniacal buffers. The pH values, extracted from
the literature, for the quantitative precipitation of metal oxinates are collected in
Table XI, 5.

Table XI, 5   pH range for precipitation of metal oxinates

Metal	pH	
	initial precipitation	complete precipitation
Aluminium	2.9	4.7– 9.8
Bismuth	3.7	5.2– 9.4
Cadmium	4.5	5.5–13.2
Calcium	6.8	9.2–12.7
Cobalt	3.6	4.9–11.6
Copper	3.0	3.3+
Iron(III)	2.5	4.1–11.2
Lead	4.8	8.4–12.3
Magnesium	7.0	8.7+
Manganese	4.3	5.9– 9.5
Molybdenum	2.0	3.6– 7.3
Nickel	3.5	4.6–10.0
Thorium	3.9	4.4– 8.8
Titanium	3.6	4.8– 8.6
Tungsten	3.5	5.0– 5.7
Uranium	3.7	4.9– 9.3
Vanadium	1.4	2.7– 6.1
Zinc	3.3	4.4+

8-Hydroxyquinoline is an almost colourless, crystalline solid, m.p. 75–76°; it is
almost insoluble in water. The reagent is prepared for use in either of the
following ways:

(*a*) Two grams of A.R. oxine are dissolved in 100 cm^3 of 2*M*-acetic acid, and
ammonia solution is added dropwise until a turbidity begins to form; the
solution is clarified by the addition of a little acetic acid. This solution is stable for
long periods, particularly if it is kept in an amber bottle.

(*b*) Two grams of A.R. oxine are dissolved in 100 cm^3 of methanol or ethanol
(this reagent cannot be used for the determination of aluminium) or in acetone.
The solution is stable for about ten days if protected from light. It is stated that
the alcoholic solution may be employed in cases where precipitation occurs at a
high pH, and the acetic acid solution for precipitations at low pH.

The following general conditions for conducting precipitations with 8-
hydroxyquinoline may be given:

1. The reagent is added to the cold solution (or frequently at 50–60 °C) until the
yellow or orange-yellow colour of the supernatant liquid indicates that a small
but definite excess is present.

2. The precipitate is coagulated by a short period of heating at a temperature not exceeding 70°C.

3. The precipitate may be filtered through paper or any variety of filtering crucible.

4. The filtrate should possess a yellow or orange colour, indicating the presence of excess of precipitant. If a turbidity appears, a portion should be heated; if the turbidity disappears, it may be assumed to be due to excess of reagent crystallising out, and is harmless. Otherwise, more reagent should be added, and the solution filtered again.

5. Washing of the precipitate may often be effected with hot or cold water (according to the solubility of the metal 'oxinate') and is continued until the filtrates become colourless. The use of ethanol is permissible if it is known to have no effect upon the precipitate.

6. The washed precipitate may be dried at 105–110 °C (usually hydrated 'oxinate') or at 130–140 °C (anhydrous 'oxinate'). In cases where prolonged heating at 130–140 °C is required, slight decomposition may occur. Frequently ignition to the oxide yields a more suitable form for weighing, but care must be exercised to prevent loss, since many 'oxinates' are appreciably volatile; it is usually best to cover the complex with oxalic acid (1–3 g) and heat gradually. The determination may also be completed titrimetrically by dissolving the precipitate in dilute hydrochloric acid and titrating with a standard solution of potassium bromate as detailed in Section **X, 140**.

D. Benzoin-α-oxime (cupron), (VIII).   This compound yields a green precipitate, $CuC_{14}H_{11}O_2N$, with copper in dilute ammoniacal solution, which may be dried to constant weight at 110 °C. Ions which are precipitated by aqueous ammonia are kept in solution by the addition of tartrate; the reagent is then specific for copper. Copper may thus be separated from cadmium, lead, nickel, cobalt, zinc, aluminium, and small amounts of iron.

From strongly acidic solutions benzoin-α-oxime precipitates molybdate and tungstate ions quantitatively; chromate, vanadate, niobate, tantalate, and palladium(II) are partially precipitated. The molybdate complex is best ignited at 500–525 °C to MoO_3 before weighing; alternatively, the precipitate may be dissolved in ammonia solution and the molybdenum precipitated as lead molybdate, in which form it is conveniently weighed.

Benzoin-α-oxime is a white, crystalline solid, m.p. 152 °C, which is sparingly soluble in water but fairly soluble in ethanol. The reagent is employed as a 2 per cent solution in ethanol.

C₆H₅—CH—OH
|
C₆H₅—C=NOH

(VIII)

CH=NOH
OH

(IX)

E. Salicylaldehyde oxime (IX).   This compound is chiefly employed for the determination of copper: a greenish-yellow precipitate $Cu(C_7H_6O_2N)_2$ is obtained in the presence of acetic acid (the precipitation is complete at pH 2.6), which is weighed after drying at 100–105 °C. Iron(III) is carried down with the copper complex in acetic acid solution and interferes seriously, but silver, cadmium, mercury, arsenic, and zinc have no effect. Salicylaldehyde oxime reacts with many other ions, and has found application in the determination of lead, bismuth, zinc, nickel, and palladium. As with similar non-selective reagents (e.g., oxine) the pH of the solution is an important factor, particularly if it is desired to separate one divalent metal from another. Thus copper is completely precipitated at a pH of 2.6, nickel commences to precipitate at a pH of 3.3, and hence for the separation of copper from nickel the pH must be maintained between 2.6 and 3.3.

A bright yellow, insoluble basic salt is formed with bismuth ions in almost neutral solution, which must be ignited to the oxide Bi_2O_3 for weighing. Lead is quantitatively precipitated as a yellow complex $PbC_7H_5O_2N$ at pH 8.9 or higher; the use of a strongly ammoniacal solution permits the separation of lead from silver, cadmium, and zinc. Palladium(II) is precipitated quantitatively as yellow $Pd(C_7H_6O_2N)_2$ from acid solution, and can thus be separated from platinum. Nickel may also be satisfactorily determined as the green complex $Ni(C_7H_6O_2N)_2$.

Salicylaldehyde oxime is a white, crystalline solid, m.p. 57 °C, which is sparingly soluble in water. The reagent is prepared by dissolving 1.0 g of salicylaldehyde oxime in 5 cm³ of 95 per cent ethanol, and pouring the solution slowly into 95 cm³ of water at a temperature not exceeding 80 °C; the mixture is shaken until clear and filtered, if necessary. Another procedure is to add 2.22 g of pure salicylaldehyde dissolved in 8 cm³ of 90% ethanol to 1.27 g of A.R. hydroxylammonium chloride dissolved in 2 cm³ of water: the resulting solution is diluted with 15 cm³ of 90% ethanol and poured slowly and with stirring into 225 cm³ of water at 80 °C; when cold, the solution is filtered if necessary and stored in an amber bottle. The reagent decomposes in solution, and should not be kept for more than about three days.

F. 1-Nitroso-2-napthol (X). This organic reagent precipitates quantitatively cobalt, iron(III), palladium, and zirconium from slightly acid solutions; it precipitates partially tin, silver, bismuth, chromium(III), titanium, tungsten(VI), uranium(VI), and vanadium(V). The following elements are not precipitated: lead, cadmium, mercury, arsenic, antimony, beryllium, aluminium, nickel, manganese, zinc, calcium, and magnesium. The principal use of 1-nitroso-2-naphthol is in the separation of cobalt from large amounts of nickel after any

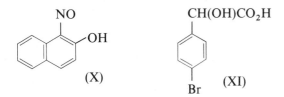

iron(III) present has been removed. The red-brown, bulky precipitate obtained in dilute hydrochloric acid solution is reported to have the composition $Co(C_{10}H_6O_2N)_3$, but it is doubtful whether the complex is pure; careful ignition in the presence of oxalic acid gives a cobalt oxide to which the formula Co_3O_4 has been assigned, but this also is not perfectly pure, and should not be used except when dealing with minute amounts of cobalt. For larger amounts the cobalt oxide may (*a*) be reduced in hydrogen in a Rose crucible and weighed as the metal, or (*b*) be treated with a few drops of concentrated nitric acid to convert it into the nitrate, the excess of nitric acid expelled by evaporating cautiously, and then converted into the sulphate by at least two evaporations with concentrated sulphuric acid, followed by a few drops of water, and weighed as $CoSO_4$ after heating for a short time at 450–500 °C (very dull red heat); the cobalt sulphate solution may also be electrolysed and the resulting metal weighed.

1-Nitroso-2-naphthol is a brown powder, m.p. 109 °C; it is insoluble in water. The reagent is prepared by dissolving 4 g of 1-nitroso-2-naphthol in 100 cm³ of glacial acetic acid and then adding 100 cm³ of hot distilled water. The cold, filtered solution should be used immediately.

G. 4-Bromomandelic acid (XI). Mandelic acid, $C_6H_5CH(OH)COOH$, is a highly selective and sensitive reagent for zirconium: precipitation is usually effected in a hydrochloric acid medium. Owing to the high concentration of mandelic acid that must be used and the difficulty of removing the excess by washing (unless a special wash solution is used—a hot solution containing 2 per cent HCl and 5 per cent mandelic acid), because of the appreciable solubility of zirconium mandelate, 4-bromomandelic acid is preferred as a reagent for the gravimetric determination of zirconium: the precipitate is insoluble in water and can be washed with water without loss. The precipitate is ignited to and weighed as the oxide.

Precipitation is best effected in a hydrochloric acid medium (up to 3–5M) and from hot solution. Cu(II), Cd(II), Hg(I), Hg(II), Sn(II), Th(IV), Sb(III) and Fe(III) interfere, as do chromate and vanadate. Reduction of chromate, vanadate, and iron(III) eliminates the interference. The following elements do not interfere: Be, Mg, Ca, Ba, Zn, Al, Ti(IV), V(IV), Cr(III), Mn(II), Fe(II), Co and Ni.

4-Bromomandelic acid is a white crystalline solid, m.p. 118°, and is slightly soluble in water. The reagent for precipitation is a 0.1M-solution; it remains stable indefinitely.

H. Nitron (XII). The strong organic base 4,5-dihydro-1,4-diphenyl-3,5-phenylimino-1,2,4-triazole, which is named nitron, yields a sparingly soluble crystalline nitrate $C_{20}H_{16}N_4HNO_3$ in solutions acidified with acetic or sulphuric acid. Perchlorate, perrhenate, tetrafluoborate, and tungstate also form insoluble salts, and can be determined in a similar manner. Numerous other anions, including bromide, iodide, chlorate, thiocyanate, nitrite, and chromate, interfere, but may easily be removed by preliminary treatment. The results in the presence of chloride are generally high, possibly because of co-precipitation.

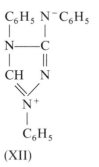
(XII)

Nitron is a yellow, crystalline solid, m.p. 189 °C, which is insoluble in water. The reagent consists of a 10 per cent solution in 5 per cent acetic acid; it should be filtered, if necessary, and the clear solution protected from light.

I. Tannic acid. This reagent is essentially a colloidal suspension of negatively charged particles capable of flocculating positively charged hydrated oxide sols, such as those of WO_3, Nb_2O_5, and Ta_2O_5. Thus if a tungstate solution is treated with tannic acid and acidified most of the tungsten is precipitated: a small amount remains colloidally dispersed and may be flocculated with a tannic acid precipitant such as cinchonine. The separation of various elements depends to a large extent upon the proper adjustment of the pH of the solution.

This reagent in the form of a freshly prepared 3 or 10 per cent aqueous solution is useful for the separation of some of the so-called rarer elements. It may be employed *inter alia* for the quantitative determination of titanium and tungstates, for the separation of aluminium, chromium, iron, etc., from beryllium, and of niobium from tantalum (the latter is precipitated selectively from a slightly acidic oxalate solution). In most cases the flocculent precipitate of the tannic acid complex of the element is ultimately ignited and weighed as the oxide.

J. The arsonic acids, R-AsO(OH)$_2$. Alkyl and arylarsonic acids are of particular value for the precipitation of the quadrivalent metals (tin, thorium, titanium, and zirconium). Phenyl- (R = C_6H_5), propyl- (R = C_3H_7), and 4-hydroxyphenyl- (R = HO·C_6H_4) arsonic acids are available commercially; the last-named is the least expensive.

Phenylarsonic acid, employed as a 10 per cent aqueous solution, will precipitate tin from fairly concentrated acid solutions, and separates it from all the common elements except titanium and zirconium. Thorium is precipitated quantitatively from acetic acid–ammonium acetate solution; rare earths and aluminium do not interfere, but titanium, zirconium, hafnium, and other quadrivalent ions are precipitated. The reagent provides a good separation of thorium from the rare-earth elements. The thorium phenylarsonate is dissolved in dilute hydrochloric acid, precipitated as the oxalate, ignited, and weighed as ThO_2.

Propylarsonic acid, employed as a 2.5 per cent aqueous solution, precipitates zirconium but not titanium in strongly acid solution. The following elements do not interfere: aluminium, chromium, cobalt, nickel, copper, vanadium, uranium, thorium, and molybdenum; a possible exception is tin, which can be removed by heating the ignited oxide with ammonium iodide. The precipitate may be ignited to the oxide; heating is carried out first with a Bunsen and then with a Meker or Fisher burner.

4-Hydroxyphenylarsonic acid, employed as a 4 per cent aqueous solution, gives precipitates with titanium and with zirconium in acid solution, and permits the separation of these elements from iron and all the common elements except tin and cerium(IV). Hydrogen peroxide prevents the precipitation of titanium, but does not affect the quantitative separation of zirconium.

K. Pyridine. Pyridine forms insoluble complexes with the thiocyanates of cadmium, copper, nickel, cobalt, zinc, and manganese; these have the general formula $M(SCN)_2.(C_5H_5N)_n$ ($n = 4$ for Co, Ni, and Mn; $n = 2$ for Cu, Cd, and Zn). In practice, alkali thiocyanate and a few cm^3 of pure pyridine are added to the neutral or very faintly acid solution of the metal ions. The complexes are readily filtered. They are washed first with water, then with dilute ethanol (both containing a little alkali thiocyanate and pyridine), followed successively by absolute ethanol and diethyl ether, each containing a little pyridine. The precipitates are weighed after drying in a vacuum desiccator for 5–30 minutes at the laboratory temperature. The method is rapid, but, as is evident, many ions interfere. The results for manganese are not satisfactory because of the slight solubility of the complex in the wash solutions. The method is applicable in the presence of alkali and alkaline-earth metals and of magnesium; considerable quantities of ammonium salts must be absent.

L. Anthranilic acid (XIII). The sodium salt of anthranilic acid precipitates in neutral or weakly acid solution zinc, cadmium, cobalt, nickel, copper, lead, silver, and mercury. Several of these salts, including the anthranilates of

427

cadmium, zinc, nickel, cobalt, and copper, are suitable for the quantitative precipitation and gravimetric determination of these elements; the salts have the

(XIII)   (XIV)

general formula $M(C_7H_6O_2N)_2$, and may be dried at 105–110 °C. The precipitations must be carried out at a controlled pH range: in too strongly acidic solutions the precipitates will not form, while in too strongly basic solutions the organo-metallic complexes undergo decomposition. At present, sodium anthranilate is limited in use to the precipitation of a single listed cation from a relatively pure solution in which small amounts of ammonium, alkaline earths, and alkali-metal salts may be present.

The reagent consists of a 3 per cent aqueous solution of pure sodium anthranilate.

M. Quinaldic acid (XIV). This organic reagent gives insoluble complexes with copper, cadmium, zinc, manganese, silver, cobalt, nickel, lead, mercury, iron(II), palladium(II), and platinum(II), and insoluble basic salts with iron(III) aluminium, chromium, beryllium, and titanium. The formation of insoluble quinaldates is influenced by the pH of the solution. Thus copper quinaldate $Cu(C_{10}H_6NO_2)_2,H_2O$ (after drying at 110–115 °C) may be precipitated from relatively acidic solutions, whilst under the same conditions the more soluble cadmium and zinc quinaldates remain in solution. Complexing reagents may also assist in rendering the reagent more selective. The quinaldates of copper, cadmium, and zinc are well-defined crystalline salts, which are readily filtered, washed, and dried.

The reagent consists of a 2 per cent aqueous solution of the acid or its sodium salt.

N. Pyrogallol (XV). This compound yields insoluble complex salts with bismuth and with antimony, and may be employed for the quantitative determination of these elements either alone or in the presence of arsenic, lead, cadmium, or zinc.

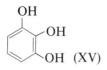

Pyrogallol is a white solid, m.p. 133–134 °C, and is freely soluble in water. The reagent consists of a 3 per cent solution in deoxygenated water; alternatively, solid A.R. pyrogallol may be added to the solution for analysis.

O. Ethylenediamine (1,2-diaminoethane) $(NH_2 \cdot CH_2 \cdot CH_2 \cdot NH_2)$. Ethylenediamine yields a complex cation with copper(II) ions:

$$Cu^{2+} + 2NH_2 \cdot CH_2 \cdot CH_2 \cdot NH_2 = [Cu(NH_2 \cdot CH_2 \cdot CH_2 \cdot NH_2)_2]^{2+}$$
$$\equiv [Cu\,en_2]^{2+}$$

This reacts with the complex ions $[HgI_4]^{2-}$ or $[CdI_4]^{2-}$ to yield the insoluble complex salts $[Cu\,en_2][HgI_4]$ and $[Cu\,en_2][CdI_4]$ respectively:

$$HgCl_2 + 4KI = K_2[HgI_4] + 2KCl$$
$$K_2[HgI_4] + [Cu\,en_2](NO_3)_2 = [Cu\,en_2][HgI_4] + 2KNO_3$$

The complex salts are insoluble in water, 95 per cent ethanol, and diethyl ether, and hence may be employed in the rapid determination of mercury, cadmium, and copper respectively. The mercury complex is stable in air and in a vacuum, and its precipitation is unaffected by the presence of ammonium salts; a valuable rapid method is thus available for the determination of mercury. The cadmium complex has similar properties, but is slightly soluble in the presence of ammonium salts or in strongly ammoniacal solution.

The reagent may be prepared by either of the following methods:

(a) Heat an aqueous solution containing 1 part of copper(II) nitrate and 2 parts of ethylenediamine on a water bath until a crust forms on the surface of the violet-blue solution. Allow to cool, filter off the separated crystals of $[Cu\,en_2](NO_3)_2,2H_2O$ at the pump, and wash them several times with ethanol, followed by diethyl ether. A concentrated solution of this salt is used for precipitations.

(b) Treat a solution of copper(II) sulphate with an aqueous solution of ethylenediamine (five to six times the theoretical quantity) until the dark blue-violet coloration, due to the $[Cu\,en_2]^{2+}$ ion, appears and does not increase in intensity upon further addition of ethylenediamine. The presence of excess of the latter in the reagent has no harmful influence. Here the reagent consists of a solution of $[Cu\,en_2]SO_4$, and is as satisfactory as (a) for the determination of mercury.

P. 8-Hydroxyquinaldine (XVI). The reactions of 8-hydroxyquinaldine are, in general, similar to 8-hydroxyquinoline (**C**), but unlike the latter it does not produce an insoluble complex with aluminium. In acetic acid–acetate solution precipitates are formed with bismuth, cadmium, copper, iron(II) and iron(III), chromium, manganese, nickel, silver, zinc, titanium (TiO^{2+}), molybdate, tungstate, and vanadate. The same ions are precipi-

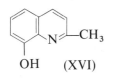

tated in ammoniacal solution with the exception of molybdate, tungstate, and vanadate, but with the addition of lead, calcium, strontium, and magnesium; aluminium is not precipitated, but tartrate must be added to prevent the separation of aluminium hydroxide.

8-Hydroxyquinaldine (2-methyl-oxine) is a pale yellow, crystalline solid, m.p. 72 °C; it is insoluble in water, but readily soluble in hot ethanol, benzene, and diethyl ether. The reagent is prepared by dissolving 5 g of 8-hydroxyquinaldine in 12 g of glacial acetic acid and diluting to 100 cm³ with water: the solution is stable for about a week.

Q. Tetraphenylarsonium chloride, $[(C_6H_5)_4As]^+Cl^-$. This reagent has been proposed as a precipitant for thallium(III) as $[(C_6H_5)_4As]^+TlCl_4^-$, which is weighed in this form. Precipitation is effected in ca. $1M$-HCl solution, the precipitate is washed with $1M$-HCl, and then dried at 110 °C. Cations that form insoluble chlorides interfere as do various anions (Br^-, I^-, F^-, NO_3^-, SCN^-, etc.) other than chloride.

R. Sodium tetraphenylboron, $Na^+[B(C_6H_5)_4]^-$. This is an excellent reagent for potassium: the solubility product of the potassium salt is 2.25×10^{-8}. Precipitation is usually effected at pH 2 or at pH 6.5 in the presence of EDTA. Rubidium and cesium interfere; ammonium ion forms a slightly soluble salt and can be removed by ignition; mercury(II) interferes in acid solution but does not do so at pH 6.5 in the presence of EDTA.

XI, 12. VOLATILISATION OR EVOLUTION METHODS. Evolution or volatilisation methods depend essentially upon the removal of volatile constituents. This may be effected in several ways: (i) by simple ignition in air or in a current of an indifferent gas; (ii) by treatment with some chemical reagent whereby the desired constituent is rendered volatile; and (iii) by treatment with a chemical reagent whereby the desired constituent is rendered non-volatile. The volatilised substance may be absorbed in a weighed quantity of a suitable medium when the estimation is a *direct* one, or the weight of the residue remaining after the volatilisation of a component is determined, and the proportion of the constituent calculated from the loss in weight; the latter is the *indirect method*. Examples of each of these procedures are given in the following paragraphs; full experimental details will be found later in this Chapter.

The determination of superficially bound moisture or of water of crystallisation in hydrated compounds may be carried out simply by heating the substance to a suitable temperature and weighing the residue (see Section **XVII, 17** for a method involving the use of the Karl Fischer reagent). Substances that decompose upon heating can be studied more fully by thermal analysis (Chapter XXIII). The water may also be absorbed in a weighed quantity of an appropriate drying agent, such as anhydrous calcium chloride or magnesium perchlorate.

The determination of carbon dioxide in carbonate-containing materials may be effected by treating the sample with excess of acid and absorbing the carbon dioxide in an alkaline absorbent, such as soda lime, soda lime–asbestos, or sodium hydroxide-asbestos ('ascarite'). The gas is completely expelled by heating the solution and by passing a current of purified air through the apparatus; it is, of course, led through a drying agent to remove water vapour before passing to the carbon-dioxide-absorption apparatus. The gain in weight of the latter is due to carbon dioxide.

In the determination of carbon in steels and alloys, the substance is burnt in pure oxygen in the presence of catalysts and the carbon dioxide absorbed as in the previous example. Precautions are taken to remove other volatile constituents such as sulphur dioxide. This method is employed in the determination of carbon and hydrogen in organic compounds; the sample is burnt in a controlled stream of oxygen, and the water and carbon dioxide are absorbed separately in an appropriate absorbent, e.g., in calcium chloride saturated with carbon dioxide and in soda lime (or soda asbestos).

Some elements, such as sodium and potassium, in combination with radicals of volatile acids or organic acids, may be determined by evaporating to dryness with sulphuric acid; the residual sulphate is then weighed:

$$2NaX + H_2SO_4 = Na_2SO_4 + 2HX$$

Interfering metals must, of course, be removed first.

An example of a related kind is the determination of pure silica in an impure ignited silica residue. The latter is treated in a platinum crucible with a mixture of sulphuric and hydrofluoric acids; the silica is converted into the volatile silicon tetrafluoride:

$$SiO_2 + 4HF \rightleftharpoons SiF_4\uparrow + 2H_2O$$

The residue consists of the impurities, and the loss in weight of the crucible gives the amount of pure silica present, provided that the contaminants are in the same form before and after the hydrofluoric acid treatment and are not volatilised in the operation. Although silicon is not the only element that forms a volatile fluoride, it is by far the most abundant and most often encountered element; consequently the volatilisation method of separation is generally satisfactory.

The separation of chromium as chromyl chloride CrO_2Cl_2 is a convenient method for removing chromium where aluminium and other trivalent elements are to be determined.

Distillation may be used to separate certain inorganic chlorides and bromides. Only molecular compounds can be distilled, and so only those elements that can be converted into volatile halides under certain experimental conditions can be separated by distillation. Thus arsenic(III), antimony(III), and tin(IV) form volatile chlorides; only arsenic(III) can be distilled quantitatively from concentrated hydrochloric acid. By increasing the b.p. by the addition of phosphoric acid, antimony(III) can be separated. Tin can be removed by distillation from a mixture of concentrated hydrochloric and hydrobromic acids.

Boron in the form of borates or boric acid can be separated from complex mixtures by distilling in acid solution with methanol. The boron volatilises as methyl borate, is collected in water or other suitable reagent, and is determined by titration, after the addition of mannitol, with standard alkali.

Practical gravimetric analysis

XI, 13. GENERAL DISCUSSION. Before commencing experimental work in gravimetric analysis, the student should be familiar with the general theory underlying the chief experimental processes outlined in Sections **XI, 1–12**. He should also read the account of the technique of gravimetric analysis given in Sections **III, 33–45**, this will assume a greater significance when the various processes have actually been employed in practice. It is proposed, in the first place, to give an account of a number of typical gravimetric determinations. These determinations may be performed with substances which are readily obtainable in a state of purity (e.g., of analytical reagent quality), and the experimental error can therefore be checked by calculation. Many may, however, prefer to carry out the analyses with solutions or solids of 'unknown' composition. These determinations should be carried out before those described under the heading of 'Systematic gravimetric analysis' are attempted. In general, the experimental procedures will not be given in such detail in this latter section which is arranged in alphabetical order with cations (metals) first, followed by anions.

For all gravimetric determinations described in this chapter, the phrase 'Allow to cool in a desiccator' should be interpreted as cooling the crucible, etc., *provided with a well-fitting cover* in a desiccator. The crucible, etc., may be weighed as soon as it has acquired the laboratory temperature (for a detailed discussion, see Section **III, 26**).

XI, 14. CALCULATIONS OF GRAVIMETRIC ANALYSIS. The calculation of the weight of a constituent in a given precipitate follows directly from the proportion:

$$M_nA_p : nM :: w : x$$

where M_nA_p is the molecular weight of the precipitate, M the atomic (or molecular) weight of the element (or radical) sought, n the number of atomic (or molecular) weights of M in the molecular weight M_nA_p, w is the weight of precipitate, and x is the weight of the constituent desired. Furthermore, if W is the weight of the sample used, the percentage of the constituent sought y is given by:

$$x:W::y:100$$

or $y = x \times 100/W$

Example.   1.000 gram of an iron compound, after suitable treatment, yielded 0.1565 g of iron(III) oxide. Calculate the percentage of iron in the compound.

$$Fe_2O_3:2Fe \qquad ::0.1565:x$$
$$159.68:2 \times 55.84::0.1565:x$$

(Mol. wt. of Fe_2O_3) ($2 \times$ At. wt. of Fe)

$$x = \frac{111.68}{159.68} \times 0.1565 = 0.6994 \times 0.1565$$
$$= 0.1095 \text{ g of Fe}$$

Now $0.1095:1.0000::y:100$

or $y = 10.95$ *per cent of Fe*

Simple gravimetric determinations

XI, 15. DETERMINATION OF WATER OF HYDRATION IN CRYS-TALLISED BARIUM CHLORIDE. *Discussion.* Barium chloride dihydrate loses all its water of crystallisation above 100 °C. Much higher temperatures (up to 800–900 °C) can be used in this dehydration, for anhydrous barium chloride is non-volatile and stable even at fairly high temperatures.

$$BaCl_2,2H_2O = BaCl_2 + 2H_2O$$

With some hydrated salts, special temperature limits must be observed.

Procedure.   Heat a crucible and lid to dull redness for several minutes, allow to cool in a desiccator, and weigh after 20 minutes. Introduce into the crucible 1–1.5 g of A.R. barium chloride, and weigh again. Place the covered crucible, resting upon a pipe-clay or silica triangle, about 15 cm above a small flame (not more than 5–6 cm high). At intervals of a few minutes increase the flame gradually until the bottom of the crucible is heated to dull redness. Maintain the crucible at this temperature for about 10 minutes, allow it to cool in a desiccator for 20 minutes, and weigh. Repeat the process until constant weight (two consecutive weighings agreeing within 0.0002 g) is obtained.

From the loss in weight, calculate the percentage of water in barium chloride dihydrate.

Similar determinations may be carried out with magnesium sulphate heptaphydrate ($MgSO_4,7H_2O = MgSO_4 + 7H_2O$), sodium tetraborate decahydrate ($Na_2B_4O_7,10H_2O = Na_2B_4O_7 + 10H_2O$), and with disodium hydrogen phosphate dodecahydrate ($2Na_2HPO_4,12H_2O = Na_4P_2O_7 + 25H_2O$).

XI, 16. OTHER DETERMINATIONS BY IGNITION. In addition to the removal of water of hydration a number of other determinations may be carried

out by simple ignition. These include iron in ammonium iron(III) sulphate ('ferric alum'),

$$(NH_4)_2SO_4.Fe_2(SO_4)_3,24H_2O \rightarrow Fe_2O_3$$

aluminium in aluminium ammonium sulphate ('ammonium alum'),

$$(NH_4)_2SO_4.Al_2(SO_4)_3,24H_2O \rightarrow Al_2O_3$$

bismuth in bismuth oxycarbonate (the residue obtained being the oxide Bi_2O_3). More detailed thermal analysis is carried out by the methods described in Chapter XXIII.

XI, 17. DETERMINATION OF CHLORIDE AS SILVER CHLORIDE.

Discussion. The aqueous solution of the chloride is acidified with dilute nitric acid in order to prevent the precipitation of other silver salts, such as the phosphate and carbonate, which might form in neutral solution, and also to produce a more readily filterable precipitate. A slight excess of silver nitrate solution is added, whereupon silver chloride is precipitated:

$$Cl^- + Ag^+ = AgCl$$

The precipitate, which is initially colloidal, is coagulated into curds by heating the solution and stirring the suspension vigorously; the supernatant liquid becomes almost clear. The precipitate is collected in a filtering crucible, washed with very dilute nitric acid, in order to prevent it from becoming colloidal (Section **XI, 8**), dried at 130–150 °C, and finally weighed as AgCl.

Silver chloride has a solubility in water of $1.4\,mg\,dm^{-3}$ at 20 °C, and $21.7\,mg\,dm^{-3}$ at 100 °C. The solubility is less in the presence of very dilute nitric acid (up to 1 per cent), and is very much less in the presence of moderate concentrations of silver nitrate (see Section **II, 8**; the optimum concentration of silver nitrate is $0.05\,g\,dm^{-3}$, but the solubility is negligibly small up to about $1.7\,g\,dm^{-3}$). The solubility is increased by the presence of ammonium and of alkali metal salts, and by large concentrations of acids. Under the conditions of the precipitation, very little occlusion occurs. If silver chloride is washed with pure water, it may become colloidal and run through the filter. For this reason the wash solution should contain an electrolyte (compare Sections **XI, 3** and **XI, 8**). Nitric acid is generally employed because it is without action on the precipitate and is readily volatile; its concentration need not be greater than $0.01M$. Completeness of washing of the precipitate is tested for by determining whether the excess of the precipitating agent, silver nitrate, has been removed. This may be done by adding 1 or 2 drops of $0.1M$-hydrochloric acid to $3–5\,cm^3$ of the washings collected after the washing process has been continued for some time; if the solution remains clear or exhibits only a very slight opalescence, all the silver nitrate has been removed.

Silver chloride is light sensitive; decomposition occurs into silver and chlorine, the silver remains colloidally dispersed in the silver chloride and thereby imparts a purple colour to it. The decomposition by light is only superficial, and is negligible unless the precipitate is exposed to direct sunlight and is stirred frequently. Hence the determination must be carried out in as subdued a light as possible, and when the solution containing the precipitate is set aside, it should be placed in the dark (e.g., in a locker), or the vessel containing it should be covered with thick brown paper.

It has been found that in a solution containing silver chloride and 1–2 per cent excess of 0.2M-silver nitrate, exposure to direct sunlight for 5 hours with occasional stirring leads to a positive error of about 2.1 per cent while exposure in a bright laboratory, with no direct or reflected sunlight and occasional stirring, gives a positive error of about 0.2 per cent. This positive error is due to the liberation of chlorine during exposure to light: the chlorine is largely changed back to chloride ions, which cause further precipitation of silver chloride. A possible reaction is:

$$3Cl_2 + 5Ag^+ + 3H_2O = 5AgCl + ClO_3^- + 6H^+$$

On the other hand, in the determination of silver by precipitation with a slight excess of 0.2M-hydrochloric acid (Section **XI, 52**), the error is negative, e.g., 0.4 per cent after 2 hours exposure in direct sunlight with no stirring, and 0.1 per cent after 2 hours exposure in a bright laboratory, with no direct or reflected sunlight and occasional stirring. This arises from the loss of chlorine which escapes from the precipitate. The weight of the precipitate may be brought to the correct value by treatment with nitric acid, followed by hydrochloric acid.

Procedure. Weigh out accurately about 0.2 g of the solid chloride (or an amount containing approximately 0.1 g of chlorine)* into a 250–350 cm³ beaker provided with a stirring rod and covered with a clock glass. Add about 150 cm³ of water, stir until the solid has dissolved, and add 0.5 cm³ of concentrated nitric acid. To the cold solution add 0.1M-silver nitrate slowly and with constant stirring. Only a slight excess should be added; this is readily detected by allowing the precipitate to settle and adding a few drops of silver nitrate solution, when no further precipitate should be obtained. *Carry out the determination in subdued light.* Heat the suspension nearly to boiling, while stirring constantly, and maintain it at this temperature until the precipitate coagulates and the supernatant liquid is clear (2–3 minutes). Make certain that precipitation is complete by adding a few drops of silver nitrate solution to the supernatant liquid. If no further precipitate appears, set the beaker aside in the dark, and allow the solution to stand for about 1 hour before filtration. In the meantime prepare a filtering crucible (Gooch, porcelain or sintered glass—the last-named is most convenient); the crucible must be dried at the same temperature as is employed in heating the precipitate (130–150 °C) and allowed to cool in a desiccator (see Section **III, 40, 41** for details). Collect the precipitate in the weighed filtering crucible (Section **III, 43**). Wash the precipitate two or three times by decantation with about 10 cm³ of cold very dilute nitric acid (say, 0.5 cm³ of the concentrated acid added to 200 cm³ of water) before transferring the precipitate to the crucible. Remove the last small particles of silver chloride adhering to the beaker with a 'policeman' (Section **III, 27**). Wash the precipitate in the crucible with very dilute nitric acid added in small portions (see Sections **XI, 8** and **III, 42**) until 3–5 cm³ of the washings, collected in a test-tube, give no

---

* A.R. potassium or sodium chloride, dried at 110–120 °C, is suitable.

turbidity with 1 or 2 drops of $0.1M$-hydrochloric acid.* Place the crucible and contents in an oven at 130–150 °C for 1 hour, allow to cool in a desiccator, and weigh. Repeat the heating and cooling until constant weight is attained.

Calculate the percentage of chlorine in the sample.

In this and all other gravimetric determinations, duplicate estimations are recommended. Both determinations may be carried out simultaneously, or if this is not convenient, the second should be commenced as soon as possible after the first is in progress.

Note on the gravimetric standardisation of hydrochloric acid. The gravimetric standardisation of hydrochloric acid by precipitation as silver chloride is a convenient and accurate method, which has the additional advantage of being independent of the purity of any primary standard (compare Section **X, 37**). Measure out from a burette 30–40 cm³ of the, say, $0.1M$-hydrochloric acid which is to be standardised. Dilute to 150 cm³, precipitate (but omit the addition of nitric acid), filter, and weigh the silver chloride. From the weight of the precipitate, calculate the chloride concentration of the solution, and thence the concentration of the hydrochloric acid.

XI, 18. DETERMINATION OF ALUMINIUM AS ALUMINIUM OXIDE.

Discussion. The aluminium is precipitated as the hydrated oxide by means of ammonia solution in the presence of ammonium chloride. The gelatinous precipitate is washed, converted into the oxide by ignition, and weighed as Al_2O_3.

This determination is subject to several sources of error, most of which will now be discussed. Aluminium hydroxide is amphoteric in character:

$$Al(OH)_3 + 3H^+ = Al^{3+} + 3H_2O$$
$$\text{or} \quad Al(OH)_3 + OH^- + 2H_2O = [Al(OH)_4(H_2O)_2]^-$$

Precipitation commences at approximately pH 4, and is complete when the pH lies between 6.5 and 7.5. The latter pH range can be adjusted with the aid of methyl red as indicator. The pH employed for precipitation must clearly be controlled. This is achieved by the addition of ammonium chloride, which exerts a buffering effect (Section **II, 19**) and also assists the coagulation of the initially colloidal precipitate. The presence of ammonium salts reduces to a minimum the co-precipitation of the divalent metals, such as calcium and magnesium (see Section **XI, 21**) and other cations. A readily filterable precipitate is obtained by precipitation in hot solution. The precipitate cannot be washed with hot water, for the aluminium hydroxide is readily peptised (Section **XI, 3**), and will run through the filter. A 2 per cent solution of either ammonium chloride or ammonium nitrate is satisfactory; the presence of ammonium chloride in the precipitate causes no appreciable volatilisation of aluminium during the subsequent ignition (contrast iron(III) oxide).

---

* A rapid method for drying the silver chloride, collected in a porcelain or sintered glass filtering crucible, is as follows. (This method should *not* be used by elementary students or beginners in the study of quantitative analysis.) After washing the precipitate with very dilute nitric acid, wash the walls of the crucible five or six times with small volumes of ethanol (a small pipette or a drawn-out glass tube is useful for this purpose), and then several times with small volumes of anhydrous diethyl ether. Suck the precipitate dry at the pump for 10 minutes, wipe the outside of the crucible with a clean linen cloth, leave in a vacuum desiccator for 10 minutes, and weigh as AgCl. The procedure may be employed for silver bromide, iodide, and thiocyanate. The results are usually slightly high.

The aluminium oxide obtained by igniting aluminium hydroxide is hygroscopic unless the temperature has been raised to at least 1200 °C, when apparently a non-hygroscopic form of the oxide is formed. For this reason the precipitate is ignited in a silica crucible (porcelain is slightly hygroscopic when heated to a high temperature) over a Meker or Fisher burner, or with a blast lamp. The best procedure is to finally heat for 10–15 minutes in an electric muffle furnace at 1200 °C.

   Procedure.   Weigh out accurately about 1.8 g of A.R. aluminium ammonium sulphate $(NH_4)_2SO_4.Al_2(SO_4)_3,24H_2O$ (or a weight of a sample containing about 0.1 g of aluminium) into a 400- or 600-cm³ beaker, provided with a clock-glass cover and a stirring rod. Dissolve it in 200 cm³ of water, add 5 g of pure ammonium chloride, a few drops of methyl red indicator (0.2 per cent alcoholic solution) (1), and heat just to boiling. Add pure dilute ammonia solution (1:1) dropwise from a burette until the colour of the solution changes to a distinct yellow. Boil the solution for 1 or 2 minutes, and filter at once through a suitable quantitative filter paper (Section **III, 38**) (2). Wash the precipitate thoroughly with hot 2 per cent ammonium nitrate or chloride solution made neutral with ammonia solution to methyl red (or to phenol red). Place the paper with the precipitate in a previously ignited silica or platinum crucible, dry, char, and ignite for 10–15 minutes with a Meker or Fisher high-temperature burner. Allow the crucible, covered with a well-fitting lid, to cool in a desiccator containing a good desiccant, and weigh as soon as cold. Ignite to constant weight.

   Calculate the percentage of aluminium in the sample.

   Notes.   1. Phenol red [pH range: 6.4 (yellow) to 8.0 (red)] has also been recommended. 0.5 cm³ of a 0.1 per cent solution of the indicator is added; the colour change upon the addition of ammonia solution is from yellow to orange. Bromocresol purple has also been used; the purple end point (pH = 6.8) is taken.

   2. Ashless paper pulp (Section **III, 39**) may be added to assist the subsequent filtration.

XI, 19.   DETERMINATION OF ALUMINIUM AS THE 8-HYDROXY-QUINOLATE, $Al(C_9H_6ON)_3$, WITH PRECIPITATION FROM HOMOGENEOUS SOLUTION.   *Discussion.*   Some of the details of this method have already been given in Section **XI, 11, C**. This procedure separates aluminium from beryllium (see, however, Section **XI, 31, A**), the alkaline earths, magnesium, and phosphate. For the gravimetric determination a 2 per cent or 5 per cent solution of oxine in 2M-acetic acid may be used: 1 cm³ of the latter solution is sufficient to precipitate 3 mg of aluminium. For practice in this determination, use about 0.40 g, accurately weighed, of A.R. aluminium ammonium sulphate. Dissolve it in 100 cm³ of water, heat to 70–80 °C, add the appropriate volume of the oxine reagent, and (if a precipitate has not already formed) slowly introduce 2M-ammonium acetate solution until a precipitate just appears, heat to boiling, and then add 25 cm³ of 2M-ammonium acetate solution dropwise and with constant stirring (to ensure complete precipitation). If the supernatant liquid is yellow, enough oxine reagent has been added. Allow to cool, and collect the precipitated aluminium 'oxinate' on a weighed sintered glass (porosity No. 4) or porous porcelain filtering crucible, and wash well with cold water. Dry to constant weight at 130–140 °C. Weigh as $Al(C_9H_6ON)_3$.

   Precipitation may also be effected from **homogeneous solution**. The experimental conditions must be carefully controlled. The solution containing 25–50 mg of aluminium should also contain $1.25-2.0\,cm^3$ of concentrated hydrochloric acid in a total volume of $150-200\,cm^3$. After addition of excess of the oxine reagent, 5 g of urea is added for each 25 mg of aluminium present, and the solution is heated to boiling. The beaker is covered with a clock glass and heated for 2–3 hours at 95 °C. Precipitation is complete when the supernatant liquid, originally greenish-yellow, acquires an orange-yellow colour. The cold solution is filtered through a sintered glass filtering crucible (porosity No. 3 or 4), washed well with cold water, and dried to constant weight at 130 °C.

Procedure. The solution should contain 25–50 mg of aluminium and $1.0-2.0\,cm^3$ of concentrated hydrochloric acid in a volume of $150-200\,cm^3$. For practice in this determination, weigh out accurately about 0.45 g of A.R. aluminium ammonium sulphate, dissolve it in water containing about $1.0\,cm^3$ of concentrated hydrochloric acid, and dilute to about $200\,cm^3$. Add $5-6\,cm^3$ of oxine reagent (a 10 per cent solution in 20 per cent acetic acid) and 5 g of urea. Cover the beaker with a clock glass and heat on an electric hot plate at 95 °C for 2.5 hours. Precipitation is complete when the supernatant liquid, originally greenish-yellow, acquires a pale orange-yellow colour. The precipitate is compact and filters easily. Allow to cool and collect the precipitate in a sintered glass filtering crucible (porosity No. 3 or No. 4), wash with a little hot water and finally with cold water. Dry at 130 °C. Weigh as $Al(C_9H_6ON)_3$.

XI, 20. DETERMINATION OF CALCIUM AS OXALATE. *Discussion.*

The calcium is precipitated as calcium oxalate CaC_2O_4,H_2O by treating a hot hydrochloric acid solution with ammonium oxalate, and slowly neutralising with aqueous ammonia solution:

$$Ca^{2+} + C_2O_4^{2-} + H_2O = CaC_2O_4,H_2O$$

The precipitate is washed with dilute ammonium oxalate solution and then weighed in one of the following forms:

(i) as CaC_2O_4,H_2O by drying at 100–105 °C for 1–2 hours. This method is not recommended for accurate work, because, *inter alia*, of the hygroscopic nature of the oxalate and the difficulty of removing the co-precipitated ammonium oxalate at this low temperature. The results are usually 0.5–1 per cent high.

(ii) As $CaCO_3$ by heating at 475–525 °C in an electric muffle furnace. This is the most satisfactory method, since calcium carbonate is non-hygroscopic.

$$CaC_2O_4 = CaCO_3 + CO$$

(iii) As CaO by igniting at 1200 °C. This method is widely used, but the resulting calcium oxide has a comparatively small molecular weight and is hygroscopic; precautions must therefore be taken to prevent absorption of moisture (and of carbon dioxide).

$$CaCO_3 = CaO + CO_2$$

Calcium oxalate monohydrate has a solubility of 0.0067 g and $0.0140\,g\,dm^{-3}$ at 25° and 95 °C respectively. The solubility is less in neutral solutions containing moderate concentrations of ammonium oxalate owing to the common ion effect (Section **II, 9**); hence a dilute solution of ammonium oxalate is employed as the wash liquid in the gravimetric determination. Calcium oxalate being the salt of a weak acid, its solubility increases with increasing hydrogen ion concentration of the solution because of the removal of the oxalate ions (compare Section **II, 12**) to form hydrogenoxalate ions and oxalic acid:

$$CaC_2O_4 \text{ (solid)} \rightleftharpoons Ca^{2+} + C_2O_4^{2-}$$
$$C_2O_4^{2+} + H^+ \rightleftharpoons HC_2O_4^-$$
$$HC_2O_4^- + H^+ \rightleftharpoons H_2C_2O_4$$

Calculation shows that precipitation is quantitative at a pH of 4 or higher.

Precipitation from cold neutral or ammoniacal solutions yields a very finely divided precipitate, which is difficult to filter. Satisfactory results are obtained by adding ammonium oxalate to a hot *acid* solution of the calcium salt (more or less calcium oxalate may precipitate, depending upon the pH of the solution), and finally neutralising with aqueous ammonia solution. The precipitate formed, after digesting for about an hour, consists of relatively coarse crystals which are readily filtered. Better results are given by precipitation from homogeneous solution according to the urea-hydrolysis method, details of which will be found in Section **X, 94** (see also Section **XI, 34**).

In this determination all those metals (e.g., copper, lead, zinc) which form slightly soluble oxalates must be absent. The problem which frequently arises in practice is the precipitation of calcium in the presence of magnesium and the alkali metals. The amount of the alkali metals which is precipitated is usually small; in the presence of large amounts of sodium, re-precipitation may be desirable. Magnesium may be co-precipitated (Section **XI, 5**) to a considerable extent, but the amount of this may be considerably reduced by not boiling the solution and not allowing the precipitate to stand in contact with the solution too long before filtration (post-precipitation, Section **XI, 5**, is thus minimised). By using a very large excess of ammonium oxalate, magnesium is held in solution in the form of a complex salt with the oxalate ion; furthermore, magnesium oxalate readily forms quite stable supersaturated solutions. If the concentration ratio of magnesium to calcium is extremely large, a second precipitation is usually necessary.

As already pointed out, calcium, when precipitated as oxalate, is best weighed as the carbonate or oxide. The theory of the decomposition of calcium oxalate is of some interest in this connection. Decomposition of the oxalate into the carbonate is rapid at about 475 °C. At higher temperatures, the dissociation of calcium carbonate ($CaCO_3 \rightleftharpoons CaO + CO_2$) comes into play. At any given temperature, a mixture of $CaCO_3$, CaO, and CO_2 in equilibrium with one another exerts a certain definite pressure of carbon dioxide. If the partial pressure of the carbon dioxide in the surrounding atmosphere is greater than the equilibrium pressure for that temperature, the above reaction will proceed from right to left, and eventually the oxide will be completely converted into the carbonate. Otherwise expressed, calcium carbonate cannot be decomposed into the oxide so long as the pressure of carbon dioxide in the surrounding atmosphere is greater than the equilibrium pressure of the system $CaCO_3$–CaO–CO_2 at the temperature of heating. The atmosphere contains about 0.03 per cent of carbon dioxide by volume; when the pressure is 760 mm, this corresponds to $760 \times 0.0003 = 0.228$ mm of mercury. Calcium carbonate will therefore be perfectly stable in the atmosphere so long as the decomposition pressure does not exceed 0.23 mm of mercury. The dissociation pressures of calcium carbonate, expressed in mm of mercury, at various temperatures are collected in the following table:

Temp., °C	Dissociation pressure	Temp., °C	Dissociation pressure
200	7.8×10^{-9}	700	31.2
400	0.3×10^{-3}	800	208
500	0.15	882	760
600	2.98	900	984

Thus calcium carbonate will not commence to dissociate appreciably in air until a temperature of slightly above 500 °C is reached. Actual experiment has shown that complete decomposition of calcium oxalate into the carbonate occurs at a temperature between 475° and 525 °C; the rate of the decomposition $CaC_2O_4 \rightarrow CaCO_3 + CO$ is slow at 450 °C but becomes reasonably rapid at 475 °C; above 530 °C the calcium carbonate commences to lose carbon dioxide. For the weighing of calcium oxalate as calcium carbonate, fine temperature control is necessary; this can be achieved only by the use of an electrically heated muffle furnace, provided with a pyrometer or suitable thermometer. If such equipment is available, the method should be used in preference to all others; the oxalate must be filtered through a Gooch (preferably of silica) or porcelain filtering crucible and not through filter paper.

Above 882 °C calcium carbonate is completely decomposed into the oxide, but unless the carbon dioxide is removed by diffusion, convection, etc., by conducting the ignition in a loosely covered crucible, there will be a re-combination of calcium oxide and carbon dioxide on cooling, with the formation of some calcium carbonate. In practice, it is found that the rate of decomposition at about 900 °C is very slow, and it is best to use a temperature of 1100–1200 °C. This temperature is not easily attained in a porcelain or silica crucible unless an electrically heated muffle furnace is employed. However, quantities up to 1 g can be completely decomposed in a platinum crucible by the use of a Meker or Fisher high-temperature burner. The residue of calcium oxide is hygroscopic (unless heated for a considerable time above 1200 °C). The crucible should be kept well-covered in a desiccator, containing pure concentrated sulphuric acid, freshly ignited quicklime, or phosphorus pentoxide, and weighed immediately it has acquired the laboratory temperature. Although anhydrous calcium oxalate appears to be stable between 226 °C and 398 °C, this is not used as a weighing form because of its hygroscopicity.

Procedure. Weigh out accurately sufficient of the sample to contain 0.2 g of calcium* into a 400- or 600-cm³ beaker covered with a clock glass and provided with a stirring rod. Add 10 cm³ of water, followed by about 15 cm³ of dilute hydrochloric acid (1:1). Heat the mixture until the solid has dissolved, and boil gently for several minutes in order to expel carbon dioxide. Rinse down the sides of the beaker and the clock glass, and dilute to 200 cm³: add 2 drops of methyl red indicator. Heat the solution to boiling, and add very slowly a warm solution of 2.0 g of ammonium oxalate in 50 cm³ of water. Add to the resultant hot solution (about 80 °C) filtered dilute ammonia solution (1:1) dropwise and

---

* 0.5 gram of A.R. calcium carbonate, or of calcite, which has been finely powdered in an agate mortar and dried at 110–130 °C for 1 hour, is suitable.

with stirring until the mixture is neutral or faintly alkaline (colour change from red to yellow). Allow the solution to stand without further heating for at least an hour. After the precipitate has settled, test the solution for complete precipitation with a few drops of ammonium oxalate solution. The subsequent procedure will depend on whether the calcium oxalate is to be weighed as the carbonate or as the oxide.

Weighing as calcium carbonate.   Decant the clear supernatant liquid through a weighed silica Gooch crucible or a porcelain filtering crucible. Transfer the precipitate to the crucible with a jet of water from the wash bottle; any precipitate adhering to the beaker or to the stirring rod is transferred with the aid of a rubber-tipped rod ('policeman'). Wash the precipitate with a cold, very dilute ammonium oxalate solution (0.1–0.2 per cent) at least five times, or until the washings give no test for chloride ion (add dilute nitric acid and a few drops of silver nitrate solution to 5 cm^3 of the washings). Dry the precipitate in the steam oven or at 100–120 °C for 1 hour, and then transfer to an electrically heated muffle furnace, maintained at 500 ± 25 °C for 2 hours. Cool the crucible and contents in a desiccator, and weigh. Further heating at 500 °C should not affect the weight. As a final precaution, moisten the precipitate with a few drops of saturated ammonium carbonate solution, dry at 110 °C, and weigh again. A gain in weight indicates that some oxide was present; this should not occur.

$$CaO + (NH_4)_2CO_3 = CaCO_3 + 2NH_3 + H_2O$$

Calculate the percentage of calcium in the sample.

Weighing as calcium oxide.   Decant the clear, supernatant liquid through a Whatman No. 40 or 540 filter paper, transfer the precipitate to the filter (Section **III, 38**), and wash with a cold 0.1–0.2 per cent ammonium oxalate solution until free from chloride. Transfer the moist precipitate to a previously ignited and weighed platinum crucible, and ignite gently at first over a Bunsen flame and finally for 10–15 minutes with a Meker or Fisher high-temperature burner until two successive weighings do not differ by more than 0.0003 g. The covered crucible and contents are placed in a desiccator containing pure concentrated sulphuric acid or phosphorus pentoxide (but not calcium chloride), and weighed as soon as cold.

Calculate the percentage of calcium in the sample.

For other methods for the determination of calcium, including precipitation from homogeneous solution, see Section **XI, 34**.

XI, 21.   DETERMINATION OF IRON AS IRON(III) OXIDE.   *Discussion.* The solution containing the iron(III) salt* is treated with a slight excess of aqueous ammonia solution to precipitate the hydrated oxide Fe_2O_3,xH_2O. The precipitate has no definite stoichiometric composition, but contains a variable amount of water, partly bound chemically and partly adsorbed. It is convenient, however, in writing equations for reactions involving hydrated oxides and also for calculating solubility product constants, to assume the hydroxide formula, although, in most cases, the composition of the precipitate does not correspond to this formula. The equation for the

---

* Iron(II) is only partially precipitated by ammonia solution in the presence of ammonium salts.

precipitation of hydrated iron(III) oxide may be written as:

$$[Fe(H_2O)_6]^{3+} + 3NH_3 = Fe(H_2O)_3(OH)_3 + 3NH_4^+$$
$$\text{or as} \quad Fe^{3+} + 3NH_3 + 3H_2O = Fe(OH)_3 + 3NH_4^+$$

Other elements that are precipitated by ammonia solution must of course, be absent. These include aluminium, trivalent chromium, titanium, and zirconium. In the presence of an oxidising agent (even atmospheric oxygen) manganese may be precipitated as the hydrated dioxide. Anions, such as arsenate, phosphate, vanadate and silicate, which yield insoluble compounds of iron in weakly basic media, must be absent. The presence of salts of organic hydroxy-acids (e.g., citric, tartaric, and salicylic acids), hydroxy compounds (e.g., glycerol and sugars), alkali pyrophosphate and fluorides must be guarded against, because of the formation of complex salts and the consequent non-precipitation of iron(III) hydroxide.

The solubility product of iron(III) hydroxide is of the order of 10^{-38}, so that quantitative precipitation occurs even in weakly acid solution, and errors due to washing will be negligibly small. The precipitate first forms as a dispersed phase, but on heating in the presence of electrolytes it coagulates to a gelatinous mass, which settles out of suspension; prolonged heating tends to break up the aggregates and causes the precipitate to become slimy. The hydrated iron(III) oxide is a typical example of a flocculated colloid. The coagulation of a colloidal precipitate, and especially the agglomeration of the primary particles, is aided considerably by raising the temperature of the solution. Hence precipitation is carried out at or near the boiling point, and the liquid is maintained at this temperature for a *short time* after precipitation.

As might be expected from its colloidal character, hydrated iron(III) oxide has a great tendency to adsorb other ions present. If precipitation is made from basic solution, the primary adsorbed ion is the hydroxide ion (Section **XI, 3**), and this readily holds by secondary adsorption positive ions which may be present. If there is a large excess of ammonium ions in the precipitating and wash solutions, the adsorption of other cations can be kept at a minimum; since ammonium salts are volatilised upon ignition of the precipitate, little harm is caused by their adsorption. Divalent ions are more strongly adsorbed than monovalent ions (Section **XI, 5**). If the extent of co-precipitation is large, purification may be effected by re-precipitation, since the precipitate is soluble in dilute acids.

The gelatinous precipitate of iron(III) hydroxide is always filtered through filter paper. Application of suction, in order to hasten filtration, should not be attempted, since the effect of the suction is merely to force the small particles of the precipitate into the pores of the filtering medium. It may often happen that with suction the liquid will pass through more rapidly; this does not mean that the washing process is accelerated, since the liquid runs through small channels and does not permeate the main body of the precipitate. For this reason iron(III) hydroxide is best washed by decantation; the precipitate may then be thoroughly stirred with the wash liquid. To prevent peptisation and the production of slimy material, an electrolyte is used in the wash liquid. The most satisfactory is ammonium nitrate; this volatilises upon ignition and assists somewhat in the subsequent ignition of the precipitate. Ammonium chloride is unsuitable, because iron(III) chloride, which is volatile, is formed during the ignition:

$$Fe_2O_3 + 6NH_4Cl = 2FeCl_3 + 6NH_3 + 3H_2O$$

It is advisable, therefore, to wash out nearly all the ammonium chloride present in the hydrated iron(III) oxide: very small amounts, however, will not lead to any significant error. To assist filtration, a hot wash solution should be employed. The filtration and washing of any gelatinous precipitate is hastened by the use of ashless filter pulp (Section **III, 39**). Under no circumstances should the precipitate be allowed to stand in the filter paper before washing is complete, because it shrinks rapidly as it partially dries, and channels, which permit the wash liquid to run through, are formed in the precipitate.

Hydrated iron(III) oxide upon ignition at $1000\,°C$ yields iron(III) oxide; at higher temperatures tri-iron tetroxide is slowly formed. The ignition should be carried out under good oxidising conditions, especially during the burning of the filter paper, for otherwise partial reduction to the magnetic oxide Fe_3O_4, or even to the metal, may occur. These reduction products are only slowly converted into iron(III) oxide upon continued heating with free access of air. Such reduction is avoided by burning off the carbon at a low heat, by maintaining at all times free access of air, and by excluding the reducing gases from the flame.

Procedure.   For practice in this determination, the student may employ either A.R. ammonium iron(II) sulphate $(NH_4)_2SO_4 \cdot FeSO_4, 6H_2O$ or A.R. ammonium iron(III) sulphate $(NH_4)_2SO_4 \cdot Fe_2(SO_4)_3, 24H_2O$. The former is to be preferred, as this determination involves oxidation of the iron(II) salt to the iron(III) state. Weigh out accurately about 0.8 g of ammonium iron(II) sulphate into a 400-cm^3 beaker provided with a clock glass and stirring rod. Dissolve it in $50\,cm^3$ of water and $10\,cm^3$ of dilute hydrochloric acid (1:1). Add 1–2 cm^3 of concentrated nitric acid* to the solution, and boil gently until the colour is clear yellow (3–5 minutes is usually necessary) (1). Dilute the solution to $200\,cm^3$,† heat to boiling, and slowly add pure 1:1 ammonia solution (2) in a slow stream from a small beaker until a slight excess is present, as is shown by the odour of the steam above the liquid (3). Boil the liquid gently for 1 minute, and allow the precipitate to settle. The supernatant liquid should be colourless. As soon as most of the precipitate has settled, decant the supernatant liquid through an ashless filter paper, but leave as much of the precipitate as possible in the beaker. It is essential that the filter paper fits the funnel properly (Section **III, 38**), so that the stem of the funnel is always filled with liquid, otherwise filtration will be very slow. Add about $100\,cm^3$ of boiling 1 per cent ammonium nitrate solution to the precipitate, stir the mixture thoroughly, and allow to settle. Decant as much liquid as possible through the filter. Wash the precipitate three to four times by

---

* The reaction is:

$$3Fe^{2+} + NO_3{}^- + 4H^+ = 3Fe^{3+} + NO + 2H_2O$$

The disadvantage of this procedure is that the presence of nitrates is undesirable if sulphate is subsequently to be determined in the filtrate as $BaSO_4$, necessitating one or more evaporations to dryness with hydrochloric acid to remove the nitric acid. This difficulty may be avoided by employing either bromine water ($2Fe^{2+} + Br_2 = 2Fe^{3+} + 2Br^-$) or hydrogen peroxide for the oxidation. Add 10–15 cm^3 of saturated bromine water to the hot solution (i.e., in moderate excess as indicated by the colour of the solution and the persistent odour of bromine—caution!) and boil to complete the oxidation and to remove most of the excess of bromine. Hydrogen peroxide is conveniently employed as the '100-volume' solution. Add 1 cm^3 of the latter and destroy the excess of reagent by boiling.

† If A.R. (ammonium iron(III) sulphate) is used, dissolve 1.3 g (or a sufficient amount of an iron(III) salt containing about 0.15 g of iron) in 200 cm^3 of water, add 10 cm^3 of 1:1 hydrochloric acid, and proceed as described.

decantation with 75–100-cm^3 portions of hot 1 per cent ammonium nitrate solution. Transfer the precipitate (and ashless filter pulp, if employed) to the filter (Section **III, 39**); any small particles adhering to the sides of the vessel or to the glass rod are dislodged with the aid of a 'policeman', and subsequently transferred to the main precipitate with the assistance of hot water from a wash bottle. Wash the iron(III) hydroxide several times with hot ammonium nitrate solution (4) until no test (or, at most, a very slight test) for chloride is obtained from the washings. Allow each portion of the wash liquid to run through before adding the next portion; do not fill the filter more than three-quarters full of the precipitate. While the filtration is in progress, ignite a clean crucible (porcelain, silica, or platinum) at a red heat, cool in a desiccator for 20 minutes, and weigh. When the filter paper has drained thoroughly, fold over the edges, and transfer to the weighed crucible. Proceed as described in Section **XI, 18**. Heat gradually until dry, char the paper without inflaming, and burn off the carbon at as low a temperature as possible under good oxidising conditions, i.e., with free access of air in order to avoid reduction of the iron(III) oxide. Finally, ignite the precipitate at a red heat for 15 minutes and take care to exclude the flame gases from the interior of the crucible, cool in a desiccator for 15 minutes, and weigh. Alternatively, heat in an electric muffle furnace at 500–550 °C. Repeat the ignition (10–15 minues) until constant weight is obtained (to within 0.0002 g).

From the weight of iron(III) oxide obtained, calculate the percentage of iron in the salt used.

Notes. 1. At this stage test the solution for the complete oxidation of the iron. Transfer a drop of the solution to a test tube by means of a stirring rod, and dilute with about 1 cm^3 of water. Add a few drops of a freshly prepared potassium hexacyanoferrate(III) solution (0.1%). If a blue colour appears iron(II) is still present in the solution, and more nitric acid must be added. Alternatively, use a drop or two of 0.1 per cent aqueous 1,10-phenanthroline: iron(II) gives a red colour.

2. Filtered ammonia solution should be used in order to prevent the introduction of silica, which is often present in suspension in alkaline solutions.

3. At this point it is advantageous to add a little ashless filter pulp, best in the form of a Whatman 'accelerator' or one-fourth of an 'ashless tablet'. For further details, see Section **III, 39**.

4. If desired, hot water from a wash bottle may be substituted at this stage; peptisation is negligible.

It is interesting to note that if a few drops of hydrazine hydrate are added immediately after the ammonia solution and the suspension boiled for 30–60 seconds, the precipitated iron(III) hydroxide is in a relatively compact form and filters fairly easily. The precipitate may be filtered through a Whatman No. 541 filter paper, washed with 1 per cent aqueous ammonium nitrate solution, and finally three times with warm water. After charring the filter paper as described above, the precipitate is heated at 450 °C, cooled, and weighed as Fe_2O_3. This is an improvement on the conventional method of precipitation using ammonia solution alone. The reader may wish to repeat the determination using this modified procedure and compare the results obtained.

Iron(III) can be **precipitated from homogeneous solution** as a dense basic formate by the urea hydrolysis method (compare Section **XI, 7**). The basic iron(III) formate is easily filtered and readily washed, and adsorbs fewer impurities than the precipitate obtained by the ammonia and other methods. Ignition of the basic formate yields iron(III) oxide. For experimental details of this and other methods, see Section **XI, 40**.

XI, 22. DETERMINATION OF LEAD AS CHROMATE. *Discussion.*

Although this method is limited in its applicability because of the general insolubility of chromates it is a useful procedure for gaining experience in gravimetric analysis. The best results are obtained by precipitation from homogeneous solution utilising the homogeneous generation of chromate ion produced by slow oxidation of chromium(III) by bromate at 90–95 °C in the presence of an acetate buffer. For further details see Section **XI, 36B**.

Procedure. Use a sample solution containing 0.1–0.2 g lead. Neutralise the solution by adding sodium hydroxide until a precipitate just begins to form. Add 10 cm³ acetate buffer solution (6M in acetic acid and 0.6M in sodium acetate), 10 cm³ chromium nitrate solution (2.4 g per 100 cm³), and 10 cm³ potassium bromate solution (2.0 g per 100 cm³). Heat to 90–95 °C. After generation (of chromate) and precipitation are complete (about 45 minutes) as shown by a clear supernatant liquid, cool, filter through a weighed sintered glass or porcelain filtering crucible, wash with a little 1 per cent nitric acid, and dry at 120 °C. Weigh as $PbCrO_4$.

XI, 23. DETERMINATION OF MAGNESIUM AS THE AMMONIUM PHOSPHATE HEXAHYDRATE AND AS THE PYROPHOSPHATE.

Discussion. A cold acid solution of the magnesium salt is treated with an excess of diammonium hydrogenphosphate, and then excess of ammonia solution is added to precipitate ammonium magnesium phosphate hexahydrate, $MgNH_4PO_4,6H_2O$, at room temperature:*

$$Mg^{2+} + HPO_4{}^{2-} + NH_4{}^+ + OH^- = MgNH_4PO_4 + H_2O$$

This precipitate possesses a relatively high solubility (about 65 mg dm^{-3} at 10 °C in pure water, but less in the presence of dilute aqueous ammonia), and it also has a tendency to form supersaturated solutions: the solution should therefore stand for several hours before filtration. The precipitate is washed with 0.8M aqueous ammonia solution† (say, 1:19) and then weighed either as the hexahydrate $MgNH_4PO_4,6H_2O$ or as the pyrophosphate $Mg_2P_2O_7$.

For the former, the precipitate is washed with ethanol, followed by anhydrous diethyl ether, and weighed after standing at room temperature (preferably in a desiccator) for about 20 minutes. This method is of moderate accuracy, and is recommended here because of difficulties which attend the ignition of the precipitate and the time saving achieved.

---

* The precipitation should be carried out at 15–30 °C in order to ensure the absence of the monohydrate, $MgNH_4PO_4,H_2O$. The latter salt forms and is stable in solutions above 62 °C; when once formed, it takes about 24 hours standing at room temperature before it is converted into the hexahydrate.

† The approximate solubilities, expressed as mg of $MgNH_4PO_4,6H_2O$ dm^{-3}, at room temperature in aqueous ammonia solutions of various concentrations are: 0.12M, 12; 0.3M, 6; 0.6M, 3; 1.2M, 1; 1.7M, 0.8.

For the latter, the precipitate is ignited at a high temperature (>1000 °C for 1 hour) to magnesium pyrophosphate and weighed as such:

$$2MgNH_4PO_4 = Mg_2P_2O_7 + 2NH_3 + H_2O$$

To obtain a precipitate of the correct composition ($MgNH_4PO_4$) at the first precipitation is a difficult matter owing to the co-precipitation of ammonium phosphate and magnesium phosphates; however, if the experimental conditions are carefully chosen and a pure magnesium salt is used, a precipitate of normal composition is formed. If the precipitation takes place in the presence of much ammonium salts, the precipitate may contain $Mg(NH_4)_4(PO_4)_2$; the latter gives magnesium metaphosphate $Mg(PO_3)_2$ upon ignition. If the precipitation is made in the presence of much potassium or sodium salts, the precipitate is contaminated with magnesium potassium (or sodium) phosphate. Hence if much ammonium, potassium, or sodium salts are present, re-precipitation is essential. In any case, re-precipitation is desirable to procure the best results. The double-precipitation process will accordingly be described. It is an experimental fact that the precipitate is practically insoluble in 5 per cent ammonia solution: this is accordingly used as the wash liquid.

Great care must be taken in the conversion of ammonium magnesium phosphate into the pyrophosphate. The carbon must be burnt off at as low a temperature as possible, because of the danger of the reduction of the phosphate precipitate if the heating is strong while carbon remains; if a platinum crucible is used, the resulting phosphorus may lead to serious damage to the crucible. Furthermore, if the heating is rapid, a dark-coloured product is obtained. For these reasons, the charring of the paper and the burning off of the carbon are conducted at as low a temperature as possible; the temperature must be raised very gradually. Some authors recommend, particularly for elementary students, that the filter paper be ignited apart from the precipitate (Section **III, 44**) in order to minimise this danger. It is preferable, however, to collect the precipitate in a porcelain filtering crucible: this is then heated in an electric muffle furnace at 1000–1100 °C.

Procedure. To a neutral or slightly acid (hydrochloric) solution of a magnesium compound, containing not more than 0.1 g of magnesium,* add 5 cm^3 concentrated hydrochloric acid, and dilute to 150 cm^3. Add a few drops of methyl red indicator to the cold solution, and then 10 cm^3 of the freshly prepared ammonium phosphate reagent (25 g of A.R. $(NH_4)_2HPO_4$ dissolved in 100 cm^3 of water). Now add pure concentrated ammonia solution slowly while stirring the solution vigorously until the indicator turns yellow. Avoid scratching the sides of the beaker with the stirring rod, for wherever there is contact, an adhering crystalline deposit forms quickly. Continue to stir the solution for 5 minutes, adding ammonia solution dropwise to keep the solution yellow, and finally add 5 cm^3 concentrated ammonia solution in excess. Allow the solution to stand in a cool place for at least 4 hours or preferably overnight. The precipitate may be weighed either as $MgNH_4PO_4,6H_2O$ or as $Mg_2P_2O_7$.

---

* About 0.6 g of A.R. magnesium sulphate, accurately weighed, is a convenient quantity of magnesium salt to employ for this estimation.

Weighing as MgNH$_4$PO$_4$,6H$_2$O. Filter through a sintered glass or porcelain filtering crucible which has been washed with ethanol and diethyl ether and weighed. Wash the precipitate with small portions of dilute ammonia solution (1:19; *ca.* 0.8*M*) until a few cm^3 of the filtrate, when acidified with dilute nitric acid and tested with silver nitrate solution, gives no test for chloride. Now wash with three 10-cm^3 portions rectified spirit (95 per cent ethanol), draining well after each washing; this serves to remove most of the adhering water. Finally, wash with five 5 cm^3 portions of anhydrous diethyl ether, draining after each washing. Then draw air through the crucible for 10 minutes, wipe the outside of the cold crucible with a clean linen cloth, and allow to stand in the air or in a desiccator for 20 minutes. Weigh as MgNH$_4$PO$_4$,6H$_2$O.

Weighing as Mg$_2$P$_2$O$_7$. Filter through a porcelain filtering crucible, taking great care to remove all the precipitate from the beaker and stirring rod, wash with small portions of cold 0.8*M*-aqueous ammonia solution until the washings give no turbidity with dilute nitric acid and silver nitrate solution. Dry the filtering crucible in an air oven at 100–150 °C for an hour, and then heat it gradually in an electric muffle furnace to 1000–1100 °C, and maintain it at this temperature until constant weight is attained. If an electric furnace is not available place the porcelain filter crucible inside a nickel crucible (or use the ignition dish supplied with the crucible), and then heat gradually to the full heat of a Meker, Fisher, or equivalent burner. Heat for 25–30-minute periods until constant weight is attained. Weigh as Mg$_2$P$_2$O$_7$.

Alternatively, but less satisfactorily, the precipitate may be filtered through a quantitative filter paper. Wash the precipitate on the paper with cold 0.8*M*-aqueous ammonia solution until the washings give no turbidity with dilute nitric acid and silver nitrate solution. Dry the precipitate at 100 °C and place it in a previously ignited and weighed platinum crucible (1). Char the paper slowly without allowing it to ignite, and burn off the carbon at as low a temperature as possible with free access of air (gradually increase the flame, but do not heat the crucible to more than the faintest red), and then ignite to constant weight in an electric muffle furnace at 1000–1100 °C or, less desirably, over a Meker or Fisher high-temperature burner.

Calculate the percentage of magnesium in the compound.

If there is time for a second precipitation, or if a pure magnesium salt is not used and consequently the purity of the precipitate may be suspect, it is advisable to carry out a second precipitation. In this case the initial precipitate is conveniently collected on a quantitative filter paper (Whatman No. 40 or No. 540): a little 0.8*M*-aqueous ammonia solution should be used to assist the transfer of most of the precipitate to the filter paper. Dissolve the precipitate on the filter paper in approximately 50 cm^3 of warm dilute hydrochloric acid (1:10), and wash the paper thoroughly with hot very dilute hydrochloric acid (1:100) into the beaker used for the initial precipitation. Dilute to 125–150 cm^3, add a few drops of methyl red indicator, 0.3 g of A.R. diammonium hydrogen phosphate, and again precipitate the ammonium magnesium phosphate by the addition of concentrated ammonia solution dropwise (preferably from a burette) and with constant stirring until the solution is yellow, followed by a further 5 cm^3 of the ammonia solution. Allow the solution to stand for at least 4 hours or, better, overnight. Weigh the precipitate as MgNH$_4$PO$_4$,6H$_2$O or as Mg$_2$P$_2$O$_7$ as detailed above. Magnesium may also be determined as the 8-hydroxyquinaldinate; see Section **XI, 43**.

Note. 1. For elementary students it is sufficient to dry the filter with the precipitate in the steam-oven (or at $100\,°C$), and to incinerate the filter paper apart from the precipitate (Section **III, 44**) at as low a temperature as possible; the paper should not be allowed to take fire. After the volatile carbonaceous matter has been burnt off, the residue may be ignited strongly with the lid of the crucible displaced, to allow circulation of the air, until the residue is as white as possible. The main precipitate is added, and the whole ignited to constant weight as described above.

XI, 24. DETERMINATION OF NICKEL AS THE DIMETHYL-GLYOXIMATE.

Discussion. The nickel is precipitated by the addition of an ethanolic solution of dimethylglyoxime $\{CH_3·C(:NOH)·C(:NOH)·CH_3,$ referred to in what follows as $H_2DMG\}$ to a hot, faintly acid solution of the nickel salt, and then adding a slight excess of aqueous ammonia solution (free from carbonate). The precipitate is washed with cold water and then weighed as nickel dimethylglyoximate after drying at $110–120\,°C$. With large precipitates, or in work of high accuracy, a temperature of $150\,°C$ should be used: any reagent that may have been carried down by the precipitate is volatilised.

$$Ni^{2+} + 2H_2DMG = Ni(HDMG)_2 + 2H^+$$

(For the structure of the complex and further details about the reagent, see Section **XI, 11A**).

The precipitate is soluble in free mineral acids (even as little as is liberated by reaction in neutral solution), in solutions containing more than 50 per cent of ethanol by volume, in hot water ($0.6\,mg\ 100\,cm^{-3}$), and in concentrated ammoniacal solutions of cobalt salts, but is insoluble in dilute ammonia solution, in solutions of ammonium salts, and in dilute acetic acid–sodium acetate solutions. Large amounts of aqueous ammonia and of cobalt, zinc, or copper retard the precipitation; extra reagent must be added, for these elements consume dimethylglyoxime to form various soluble compounds. Better results are obtained in the presence of cobalt, manganese, or zinc by adding sodium or ammonium acetate to precipitate the complex; iron(III), aluminium, and chromium(III) must, however, be absent.

Dimethylglyoxime forms sparingly soluble compounds with palladium, platinum, and bismuth. Palladium and gold are partially precipitated in weakly ammoniacal solution; in weakly acid solution palladium is quantitatively precipitated and gold partially. Bismuth is precipitated in strongly basic solution. These elements, and indeed all the elements of the hydrogen sulphide group, should be absent. Iron(II) yields a red-coloured soluble complex in ammoniacal solution and leads to high results if much of it is present. Silicon and tungsten interfere only when present in amounts of more than a few milligrams. Iron(III), aluminium, and chromium(III) are rendered inactive by the addition of a soluble tartrate or citrate, with which these elements form complex ions.

Dimethylglyoxime is almost insoluble in water, and is added in the form of a 1 per cent solution in rectified spirit or absolute ethanol; $1\,cm^3$ of this solution is sufficient for the precipitation of $0.0025\,g$ of nickel. As already pointed out, the reagent is added to a hot feebly acid solution of a nickel salt, and the solution is then rendered faintly ammoniacal. This procedure gives a more easily filterable precipitate than does direct precipitation from cold or from ammoniacal solutions. Only a slight excess of the reagent should be used, since dimethylglyoxime is not very soluble in water or in very dilute ethanol and may precipitate; if a very large excess is added (such that the alcohol content of the solution exceeds 50 per cent), some of the precipitate may dissolve.

Procedure. **A. Nickel in a nickel salt.** Weigh out accurately 0.3–0.4 g of pure (preferably A.R.*) ammonium nickel sulphate $(NH_4)_2SO_4 \cdot NiSO_4, 6H_2O$ into a 400-cm^3 beaker provided with a clock-glass cover and stirring rod. Dissolve it in water, add 5 cm^3 of dilute hydrochloric acid (1:1) and dilute to 200 cm^3. Heat to 70–80 °C, add a slight excess of the dimethylglyoxime reagent (at least 5 cm^3 for every 10 mg of Ni present), and immediately add dilute ammonia solution dropwise, directly to the solution and not down the beaker wall, and with constant stirring until precipitation takes place, and then in slight excess. Allow to stand on the steam bath for 20–30 minutes, and test the solution for complete precipitation when the red precipitate has settled out. Allow the precipitate to stand for 1 hour, cooling at the same time. Filter the cold solution through a Gooch, sintered glass or porcelain filtering crucible, previously heated to 110–120 °C and weighed after cooling in a desiccator. Wash the precipitate with cold water until free from chloride, and dry it at 110–120 °C for 45–50 minutes. Allow to cool in a desiccator and weigh. Repeat the drying until constant weight is attained. Weigh as $Ni(C_4H_7O_2N_2)_2$, which contains 20.32 per cent of Ni.

Calculate the percentage of nickel in the salt.

B. Nickel in nickel steel. Weigh out accurately about 1 g of the drillings or borings of the nickel steel† (or sufficient of the sample to contain 0.03–0.04 g of nickel) into a 100–150-cm^3 beaker or porcelain basin, dissolve it in the minimum volume of concentrated hydrochloric acid (about 20 cm^3 should suffice), and boil with successive additions of concentrated nitric acid (*ca.* 5 cm^3) to ensure complete oxidation of the iron to the iron(III) state. Dilute somewhat, filter, if necessary, from any solid material, and wash the paper with hot water; dilute the filtrate (or solution) to 250 cm^3 in a 400-cm^3 beaker. Add 5 g of citric or tartaric acid, neutralise the solution with dilute aqueous ammonia solution,‡ and then barely acidify (litmus) with dilute hydrochloric acid. Warm the solution to 60–80 °C, add a slight excess of a 1 per cent ethanolic solution of dimethyl-glyoxime (20–25 cm^3), immediately followed by dilute ammonia solution dropwise until the liquid is slightly ammoniacal, stir well, and allow to stand on the steam bath for 20–30 minutes. Allow the solution to stand at least 1 hour and cool to room temperature during this time. Filter off the precipitate through a weighed filtering crucible; test the filtrate for complete precipitation with a little dimethylglyoxime solution, and wash the precipitate with cold water until free from chloride. Dry the precipitate at 100–120 °C for 45–60 minutes, and weigh as $Ni(C_4H_7O_2N_2)_2$.

Calculate the percentage of nickel in the steel.

For other methods for the determination of nickel, see Section **XI, 47**.

---

* Alternatively, sufficient of a nickel salt to contain about 0.03–0.05 g of nickel may be used.

† Bureau of Analysed Samples 'Nickel Steel, No. 222' (a British Chemical Standard) is suitable. This steel contains about 3.5 per cent of nickel.

‡ If a precipitate appears or if the solution is not clear when it is rendered ammoniacal, more tartaric or citric acid must be added until a perfectly clear solution is obtained upon adding dilute ammonia solution. Any insoluble matter should be filtered off and washed with hot water containing a little ammonia solution.

Systematic gravimetric analysis

XI, 25. GENERAL DISCUSSION. In the succeeding sections a brief account will be given of a number of selected methods for the gravimetric determination of the various elements and radicals. It is believed that these will suffice to meet the needs of the student during the whole period of his training; for a more detailed study, particularly of the limitations of some of the various methods, reference must be made to other treatises (see Selected Bibliography at end of this chapter).

As no arrangement of ions or metals is ideal the following procedures have been arranged in alphabetical order with the metals (cations) listed first and the anions similarly arranged afterwards.

Cations

XI, 26. ALUMINIUM. Methods for the determination of aluminium as the oxide and as the 8-hydroxyquinolate have already been given (Sections **XI, 18** and **19**); the following procedure, which also involves ignition to aluminium oxide, is also widely used.

Determination of aluminium as basic aluminium succinate and subsequent ignition to the oxide, Al_2O_3 (precipitation from homogeneous solution).
Discussion. Aluminium can be precipitated from homogeneous solution as the dense basic succinate by boiling an acid solution containing succinic acid with urea (starting pH = 3.1–3.5). The hydrolysis of urea gradually produces ammonia, resulting in a pH of 4.2–4.6:

$$CO(NH_2)_2 + H_2O = CO_2 + 2NH_3$$

The dense precipitate is easily filtered and washed, and exhibits much less tendency to adsorption of other salts than does the precipitate obtained by precipitation as the hydroxide. Upon ignition, the basic succinate is readily converted into aluminium(III) oxide.

The method permits the separation of aluminium from large amounts of calcium, barium, magnesium, manganese, or cadmium, or from an equal amount of nickel or cobalt; for large amounts of nickel and cobalt, a double precipitation is necessary. Owing to the relatively low solubility of copper(II) succinate, the copper(II) must first be reduced by hydroxylammonium chloride solution or ammonium hydrogensulphite solution at the boiling point. A double precipitation is essential if zinc is present. Iron(III) must be reduced to iron(II): the hot hydrochloric acid solution containing the aluminium sample is first reduced with fresh ammonium sulphite solution, precipitation is then effected in the presence of $2\,cm^3$ of phenylhydrazine, and the precipitate is washed with 1 per cent succinic acid solution (containing some phenylhydrazine) rendered neutral to methyl red with aqueous ammonia.

Procedure.   The solution should contain about 0.1 g of Al and be acid with hydrochloric acid. Add dilute ammonia solution until the solution becomes slightly turbid, remove the turbidity with dilute hydrochloric acid, and add 1–2 drops in excess. Add a solution of 5 g of A.R. succinic acid in $100\,cm^3$ of water, followed by 10 g of ammonium chloride and 4 g of urea; dilute to $500\,cm^3$ with distilled water. Heat the solution to gentle boiling and continue the boiling for 2 hours after the solution has become turbid (*ca.* 45 minutes).* The insertion of a boiling rod into the solution is recommended as this reduces the tendency to 'bump' during the heating. Allow the precipitate to settle for a few minutes, filter through a Whatman No. 40 or No. 540 filter paper, and wash several times with a 1 per cent succinic acid solution made neutral to methyl red with aqueous ammonia solution. If any precipitate adheres to the sides of the beaker, dissolve it in a little dilute hydrochloric acid, add a drop of methyl red or phenol red indicator, and then dilute ammonia solution until just alkaline; filter off the precipitate of aluminium hydroxide on a separate small filter paper, and wash it with a 2 per cent solution of ammonium nitrate. Place both papers and precipitates in a silica or, preferably, a platinum crucible, and ignite to constant weight at 1200 °C (compare Section **XI, 18**). Weigh as Al_2O_3.

XI, 27. AMMONIUM. *Discussion.* For the determination of ammonium by a gravimetric procedure, it must be present as the chloride; all other cations must be absent. A little hydrochloric acid is added, followed by excess of chloroplatinic acid reagent (see Section **XI, 50C**). The mixture is evaporated almost to dryness on the water bath; the residue is triturated with absolute ethanol to remove excess of chloroplatinic acid, and then transferred to a weighed filtering crucible (Gooch, sintered glass, or porcelain). The crucible is dried at 130 °C, and the residual **ammonium chloroplatinate, $(NH_4)_2PtCl_6$,** weighed. (For details of experimental procedure, see Section **XI, 50C**.)

Ammonium may also be determined by precipitation with sodium tetraphenylborate as the sparingly soluble **ammonium tetraphenylboron** $NH_4[B(C_6H_5)_4]$, using a similar procedure to that described for potassium; it is dried at 100 °C. For further details of the reagent, including interferences, notably potassium, rubidium, and caesium, see Section **XI, 50C**.)

If the ammonium salt is present with other cations and anions, a titrimetric procedure (see Chapter X) is usually employed.

XI, 28. ANTIMONY. Antimony may be determined in the following forms:

A. Antimony(III) sulphide, Sb_2S_3. *Discussion.* This method is of limited application, since no other elements that are precipitable by hydrogen sulphide in acid solution can be present, and the sulphide must be dried and finally heated in an atmosphere of carbon dioxide at 280–300 °C. Arsenic can be separated by removal by distillation as arsenic trichloride; tin can be removed by precipitation in the presence of oxalic and tartaric acids or of phosphoric acid.

Procedure. Quickly heat the solution of the antimony compound in 1:4-hydrochloric acid ($100\,cm^3$) (1) contained in a conical flask to boiling and immediately pass a rapid stream of washed hydrogen sulphide; maintain the solution at 90–100 °C. Shake the flask gently at intervals after the sulphide has turned red, and keep the precipitate, as far as possible, below the surface of the solution. As the precipitate darkens in colour, reduce the gas stream. Continue the passage of gas until the precipitate is crystalline and black in colour (total time required for precipitation is 30–35 minutes). Dilute the solution with an equal volume of water, mix, and heat again whilst the gas is slowly passed into the

---

* The boiling period (after the appearance of a turbidity) may be reduced to 1 hour by first partially neutralising the hot solution to bromo-phenol blue or to methyl orange by the drop-wise addition of dilute ammonia solution: a very faint opalescence will appear.

suspension for some minutes. When the solution is clear, cool, and filter through a filtering crucible (Gooch, sintered glass, or porcelain) that has been heated at 280–300 °C and weighed. Wash the precipitate a few times with water to remove acid, and then with ethanol, draw air through the crucible to dry the precipitate as far as possible. Place the crucible and contents in a wide glass tube passing through an electrically heated tube furnace. Heat for 2 hours at 100–130 °C in a current of carbon dioxide (this will completely dry the precipitate), and then heat for a further 2 hours at 280–300 °C (this process will convert any Sb_2S_5 present into Sb_2S_3 and will volatilise the sulphur). Cool in a slow stream of carbon dioxide, then place in a desiccator for 20–30 minutes, and weigh as Sb_2S_3.

Note. 1. A solution, suitable for practice in this determination, may be prepared by dissolving 0.5 g, accurately weighed, of A.R. anhydrous antimony potassium tartrate in 150 cm³ of 1 : 4-hydrochloric acid.

B. Antimony pyrogallate, $Sb(C_6H_5O_3)$. Antimony(III) salts in the presence of tartrate ions may be quantitatively precipitated with a large excess of aqueous pyrogallol as the dense antimony pyrogallate. The method allows of a simple separation from arsenic; the latter element may be determined in the filtrate from the precipitation of antimony by direct treatment with hydrogen sulphide.

Procedure. The solution should contain the antimony (0.1–0.2 g) in the trivalent condition. Add a slight excess over the calculated quantity of potassium sodium tartrate to avoid the formation of basic salts upon dilution. Dissolve approximately five times the theoretical quantity of pure pyrogallol (Section **XI, 11N**) in 100 cm³ of air-free water, add this all at once to the antimony solution, and dilute to 250 cm³. After 30–60 seconds the clear mixture becomes turbid, and then a dense, cloudy precipitate forms which separates out rapidly. Allow to stand for 2 hours, filter through a weighed sintered glass or porcelain filtering crucible, wash several times with cold water to remove the excess of pyrogallol (50 cm³ is usually sufficient), dry at 100–105 °C to steady weight. Wash again with cold water, dry at 100–105 °C, and weigh; repeat the operation until the weight is constant. Weigh as $Sb(C_6H_5O_3)$.

It should be pointed out that the titrimetric methods described for the determination of antimony (Chapter X) are to be preferred to the gravimetric methods as they are simpler, more rapid, and quite as accurate.

XI, 29. ARSENIC. Arsenic may be determined in the following forms:

A. Arsenic(III) sulphide, As_2S_3. *Discussion.* The arsenic must be present in the trivalent state. Arsenic in the trivalent state (ensured by the addition of, for example, iron(II) sulphate, copper(I) chloride, pyrogallol, or phosphorous acid) may be separated from other elements by distillation from a hydrochloric acid solution, the temperature of the vapour being held below 108 °C; arsenic trichloride (also germanium chloride, if present) volatilises and is collected in water or in hydrochloric acid.

Procedure. Pass a rapid stream of washed hydrogen sulphide through a solution of the arsenic(III) (1) in 9M-hydrochloric acid at 15–20 °C. Allow to stand for an hour or two, and filter through a weighed filtering crucible (Gooch, sintered glass, or porcelain) (2). Wash the precipitate with 8M-hydrochloric acid saturated with hydrogen sulphide, then successively with ethanol, carbon disulphide (to remove any free sulphur which may be present), and ethanol. Dry at 105 °C to constant weight, and weigh as As_2S_3.

Notes. 1. A suitable solution for practice in this determination is prepared by dissolving about 0.3 g of A.R. arsenic(III) oxide, accurately weighed, in $9M$-hydrochloric acid.

2. Sometimes a film of the sulphide adheres to the glass vessel in which precipitation was carried out; this can be dissolved in a little ammonia solution and the sulphide re-precipitated with the acid washing liquor.

B. Ammonium uranyl arsenate, $NH_4UO_2AsO_4, xH_2O$, and subsequent weighing as the oxide, U_3O_8. The addition of a uranyl salt solution to an *arsenate* solution containing excess of ammonium ions results in the precipitation of ammonium uranyl arsenate, which is soluble in mineral acids but insoluble in acetic acid. Upon igniting the precipitate, the arsenic is completely volatilised, leaving a moss-green residue which consists mainly of U_3O_8: this residue is dissolved in concentrated nitric acid, and the resultant uranyl nitrate upon cautious evaporation and subsequent ignition yields pure black U_3O_8, and is weighed in this form.

If the solution contains arsenite, the latter must first be oxidised to arsenate with $0.1N$-potassium bromate in hydrochloric acid solution at 70 °C in the usual way (Section **X, 139**). A method is thus available for the **determination of arsenite and arsenate in admixture**. The arsenite is first determined with standard potassium bromate solution, and the total arsenate in the resulting liquid is then determined by precipitation as the uranium salt.

Procedure. The solution (150 cm^3) should contain about 0.06 g of As as arsenate. Add 30 cm^3 of $4M$-ammonia solution, acidify with acetic acid, heat to boiling, and add 50 cm^3 (excess) of approximately $0.1N$-uranyl acetate solution. Allow to stand for several hours, but preferably overnight: during this period the pale-yellow granular precipitate will become coarser. Filter through a fine quantitative filter paper, wash free from soluble salts, and transfer the filter and precipitate to a weighed silica crucible. *Heat in a fume chamber provided with a good draught* until all the carbon has burnt off—the arsenic is simultaneously volatilised. Moisten the residue with a few drops of concentrated nitric acid, and ignite to constant weight over an ordinary Bunsen burner. Weigh as U_3O_8.

XI, 30. BARIUM. The various methods available for the determination of barium all suffer from the disadvantage that they are affected by a number of interfering metals. Two main procedures are commonly used:

A. Determination of barium as sulphate. *Discussion.* This method is most widely employed. The effect of various interfering elements and radicals (e.g., calcium, strontium, lead, nitrate, etc., which contaminate the precipitate) is fully dealt with in Section **XI, 84**. The solubility of barium sulphate is *ca.* 1 part in 400 000 of cold water or about 2.5 mg dm^{-3}. The solubility is greater in hot water or in dilute hydrochloric or nitric acid, and less in solutions containing a common ion.

The barium sulphate may be precipitated either by the use of sulphuric acid, or from homogeneous solution by the use of sulphamic acid solution which produces sulphate ions on boiling:

$$NH_2SO_3H + H_2O = NH_4^+ + SO_4^{2-} + H^+$$

Procedure. **Precipitation with sulphuric acid.** The solution (100 cm^3) should contain not more than 0.15 g of barium (1), and not more than 1 per cent by volume of concentrated hydrochloric acid. Heat to boiling, add a slight excess of hot 0.5M-sulphuric acid slowly and with constant stirring. Digest on the steam bath until the precipitate has settled, filter, wash with hot water containing two drops of sulphuric acid per litre, and then with a little water until the acid is removed. Full experimental details of the filtration, washing, and ignition processes (900–1000 °C) are given in Section **XI, 84**. Weigh as BaSO$_4$.

Note. 1. A suitable solution for practice may be prepared by dissolving about 0.3 g, accurately weighed, A.R. barium chloride in 100 cm^3 water and adding 1 cm^3 of concentrated hydrochloric acid.

Procedure. **Precipitation from homogeneous solution; sulphamic acid method.** The sample solution may contain up to 100 mg of barium, preferably present as the chloride. A solution prepared from about 0.18 g, accurately weighed, A.R. barium chloride may be used to obtain experience in the determination. Dilute the solution to about 100 cm^3; add 1.0 g sulphamic acid. Heat the covered beaker on an electric hot plate at 97–98 °C; continue the heating for 30 minutes after the first turbidity appears. Filter through a weighed porcelain filtering crucible and wash with warm distilled water. Ignite to constant weight at 900 °C (preferably in an electric muffle furnace). Weigh as BaSO$_4$.

B. Determination of barium as chromate. This method is of limited application because of the influence of numerous interfering elements. It is useful, however, in the separation of barium from both calcium and strontium. Thermogravimetric analysis suggests that a drying temperature of 120–180 °C is to be preferred to the much higher temperatures usually recommended; the loss of oxygen amounts to about 1 per cent at 1000 °C.

Precipitation of barium chromate is usually carried out in a dilute acetic acid solution which is buffered with ammonium acetate; a double precipitation is desirable in the presence of much strontium and/or calcium.

Barium chromate is soluble in an acid solution of pH about 2. If the pH of such a solution is slowly raised by the use of urea the barium chromate will precipitate in large, readily filterable crystals; ammonium acetate is added to prevent the pH increasing too rapidly. For quantitative precipitation the final pH should be near 5.7. Above pH 5.7, strontium will precipitate if present in large quantities. Barium (100 mg) may be separated satisfactorily from calcium (100 mg) and strontium (40 mg) in a single precipitation; for larger quantities of strontium (50–100 mg) a double precipitation is necessary. If strontium is absent the final pH may be 6.8–7.0.

Procedure. The solution (200 cm^3) (1) should contain not more than 0.4 g of Ba, and be neutral in reaction. Add 1.0 cm^3 6M-acetic acid and 10 cm^3 neutral 3M-ammonium acetate to the solution, heat to boiling, and treat with a slight excess of a hot dilute solution of ammonium chromate (2) which is added dropwise from a burette with constant stirring. Place the beaker on a water bath until the precipitate settles; test for completeness of precipitation by adding a little more of the reagent. Allow to cool, filter through a weighed porcelain or sintered glass filtering crucible, wash with hot water until 1 cm^3 of the washings gives scarcely any reddish-brown coloration with neutral silver nitrate solution. Dry to constant weight at 120 °C. Weigh as BaCrO$_4$.

Notes. 1. A suitable solution for practice in this determination may be prepared by dissolving about 0.3 g, accurately weighed, A.R. barium chloride in 200 cm^3 of water.

2. The ammonium chromate solution is prepared by dissolving 10 g pure

ammonium dichromate (free from sulphate) in 100 cm³ water, and adding dilute ammonia solution until the colour of the solution is clear yellow.

XI, 31. BERYLLIUM. Beryllium may be determined in the following forms:*

A. Determination of beryllium by precipitation with ammonia solution and subsequent ignition to beryllium oxide. *Discussion.* Beryllium may be determined by precipitation with aqueous ammonia solution in the presence of ammonium chloride or nitrate, and subsequently igniting and weighing as the oxide BeO. The method is not entirely satisfactory owing to the gelatinous nature of the precipitate, its tendency to adhere to the sides of the vessel, and the possibility of adsorption effects.

Beryllium is sometimes precipitated together with aluminium hydroxide, which it resembles in many respects. Separation from aluminium (and also from iron) may be effected by means of oxine. An acetic acid solution containing ammonium acetate is used; the aluminium and iron are precipitated as oxinates, and the beryllium in the filtrate is then precipitated with ammonia solution. Phosphate must be absent in the initial precipitation of beryllium and aluminium hydroxides.

The precipitation by ammonia solution of such elements as Al, Bi, Cd, Cr, Ca, Cu, Fe, Pb, Mn, Ni, and Zn may be prevented by complexation with EDTA: upon boiling the ammoniacal solution, beryllium hydroxide is precipitated quantitatively.

In all the above methods the element is weighed as the oxide, BeO, which is somewhat hygroscopic (compare aluminium(III) oxide). The ignited residue, contained in a covered crucible, must be cooled in a desiccator containing concentrated sulphuric acid or phosphorus pentoxide, and weighed immediately it has acquired the laboratory temperature.

Procedure. The beryllium solution (200 cm³), prepared with nitric acid or hydrochloric acid and containing about 0.1 g of Be, must be almost neutral and contain no other substance precipitable by ammonia solution. Heat to boiling, and add dilute ammonia solution slowly and with constant stirring until present in *very slight* excess. Add a Whatman accelerator or one-half of a Whatman ashless tablet, boil for 1 or 2 minutes, and filter on a Whatman No. 41 or 541 filter paper. Transfer as much of the precipitate as possible by rinsing with hot 2 per cent ammonium nitrate solution. Remove any precipitate adhering to the walls of the beaker by dissolving in the minimum volume of hot very dilute nitric acid, heating to boiling, and precipitating as before. Filter through the same paper, and wash thoroughly with the ammonium nitrate solution. Place the paper and precipitate in a weighed silica or platinum crucible, dry, and then slowly decompose the hydroxide by raising the temperature gradually to 700 °C, and then ignite at about 1000 °C for at least 1 hour. Cool in a covered crucible in a desiccator charged with concentrated sulphuric acid, phosphorus pentoxide, or anhydrous magnesium perchlorate, and weigh immediately when cold as BeO.

In the **presence of interfering elements**, proceed as follows. Neutralise 80–120 cm³ of the solution containing 15–25 mg of beryllium with ammonia

---

* Beryllium and its compounds are toxic and care should be taken to avoid inhalation of dusts or contact with eyes and skin.

solution until the hydroxides commence to precipitate. Redissolve the precipitate by the addition of a few drops of dilute hydrochloric acid. Add 0.5 g of ammonium chloride and sufficient 0.5M-EDTA solution to complex all the heavy elements present. Add a slight excess of dilute ammonia solution, with stirring, boil for 2–3 minutes, add a little ashless filter pulp, filter, and complete the determination as above.

B. Ammonium beryllium phosphate and subsequent ignition to beryllium pyrophosphate. *Discussion.* Beryllium may be precipitated as ammonium beryllium phosphate and subsequently ignited to the pyrophosphate, $Be_2P_2O_7$. If the conditions of precipitation are not carefully controlled, slight departures from the expected theoretical $Be_2P_2O_7$ composition may occur, resulting in loss of accuracy. In the presence of a slight excess of EDTA, metal ions such as those of aluminium, iron, copper, nickel, calcium, and magnesium need not be removed.

Procedure. Precipitate the hydroxides, containing from 1 to 10 mg of beryllium, with ammonia solution at pH 8–9. Filter off the precipitate, transfer back to the original beaker, and dissolve it in dilute mineral acid. Dilute the solution to 100 cm^3 and adjust the pH to 2. Add 5 cm^3 15 per cent diammonium hydrogen phosphate solution and a slight excess of 15 per cent EDTA solution (according to the amount of interfering elements present): both reagent solutions should be previously adjusted to a pH of 5.5. Now add 0.5M-ammonium acetate to the resulting solution until the pH is 5.5 (use a pH meter). Digest the solution just below the boiling point for 5–10 minutes, cool, filter the granular precipitate, redissolve in the minimum volume of hot 6M-hydrochloric acid, and reprecipitate, using only 1 cm^3 of each of the reagents. Filter again, wash with a 0.5M-acetate buffer (3.5 g ammonium acetate and 3.0 cm^3 glacial acetic acid per 100 cm^3 of water), ignite at 1000 °C, and weigh as $Be_2P_2O_7$ in the usual manner.

XI, 32.   BISMUTH.   Bismuth may be satisfactorily determined in the following forms:

A. Determination of bismuth as oxyiodide. *Discussion.* The cold bismuth solution, weakly acid with nitric acid, is treated with an excess of potassium iodide when BiI_3 and some $K[BiI_4]$ are formed:

$$Bi(NO_3)_3 + 3KI = BiI_3 + 3KNO_3$$
$$BiI_3 + KI = K[BiI_4]$$

Upon dilution and boiling, bismuth oxyiodide is formed, and is weighed as such after suitable drying.

$$\underset{\text{(black)}}{BiI_3} + H_2O = BiOI + 2HI$$
$$\underset{\text{(yellow)}}{K[BiI_4]} + H_2O = BiOI + KI + 2HI$$

A large excess of potassium iodide should be avoided, since the complex salt is not so readily hydrolysed as the tri-iodide. This is an excellent method, because the oxyiodide is precipitated in a form which is very convenient for filtration and weighing.

Procedure. The cold bismuth nitrate solution, containing 0.1–0.15 g of Bi (1), must be slightly acid with nitric acid (2), and occupy a volume of about 20 cm^3. Add finely powdered solid potassium iodide, slowly and with stirring, until the supernatant liquid above the black precipitate of bismuth tri-iodide is just coloured yellow (due to $K[BiI_4]$). Dilute to $200\,cm^3$ with boiling water, and boil for a few minutes. The black tri-iodide is converted into the copper-coloured precipitate of the oxyiodide. The supernatant liquid should be colourless; if this is yellow, a further $100\,cm^3$ of water should be added, and the boiling continued until colourless. Add a few drops of methyl orange indicator, and then sodium acetate solution ($25\,g\,dm^{-3}$) from a burette until the solution is neutral. Filter off the precipitate through a weighed Gooch, sintered glass, or porcelain filtering crucible, wash with hot water, and dry at $105-110\,°C$ to constant weight. Weigh as BiOI.

Notes. 1. A suitable solution for practice can be obtained by dissolving about $0.15\,g$ of pure bismuth, accurately weighed, in the minimum volume of $1:4$ nitric acid. Alternatively, A.R. bismuth nitrate may be used.

2. Chloride and bromide should be absent. If the solution is strongly acid with nitric acid, it should be evaporated to dryness on the water bath, and the residue dissolved in a little dilute nitric acid.

B. Determination of bismuth as pyrogallate. *Discussion.* The precipitation of bismuth with pyrogallol is quantitative only if the acidity (hydrochloric, sulphuric, or nitric acid) does not exceed $0.1N$. The method is an excellent one for the determination of bismuth in the presence of lead, cadmium, and zinc. Antimony, which forms a similar complex, must, of course, be absent.

Procedure. The solution ($150\,cm^3$) should be weakly acid with nitric acid and contain $0.1-0.2\,g$ of Bi. Treat the solution with dilute ammonia solution until a permanent turbidity is obtained; render the solution clear by the cautious addition of a little dilute nitric acid. Heat to boiling, and add a slight excess of a solution of pure pyrogallol (Section **XI, 11N**) in air-free water. A yellow, finely crystalline precipitate is immediately formed. Boil for a short time, test for completeness of precipitation with a little of the reagent, dilute slightly, and filter through a weighed sintered glass or porcelain filtering crucible after the precipitate settles out. Wash with $0.05M$-nitric acid, and finally with water. Dry to constant weight at $105\,°C$. Weigh as $Bi(C_6H_3O_3)$.

The following alternative method has been recommended. Add $1.0\,g$ of pure pyrogallol to the solution ($150\,cm^3$) containing about $0.1\,g$ of Bi and heated to $70\,°C$. Then add $0.5M$-aqueous ammonia solution dropwise until a distinct turbidity forms. Heat the resulting solution to boiling, add 2 drops of thymol blue indicator, and then more of the ammonia solution until the solution is basic. Heat on a water bath for 10 minutes, filter on a sintered glass or porcelain filtering crucible, wash, and dry to constant weight at $105\,°C$. Weigh as $Bi(C_6H_3O_3)$.

XI, 33. CADMIUM. Cadmium may be determined in the following forms:

A. Determination of cadmium as the 2-naphthaquinoline complex. *Discussion.* This method permits of the separation of cadmium from relatively large quantities of zinc, iron, chromium, aluminium, cobalt, nickel, manganese, and magnesium, and also from antimony and tin if ammonium oxalate or large amounts of sodium tartrate are used.

Procedure. The cadmium salt solution, containing about $0.15\,g$ of Cd, should occupy a volume of about $50\,cm^3$ and be M with respect to sulphuric acid. Add $50\,cm^3$ of 10 per cent sodium tartrate solution, followed successively by a 2.5 per cent solution of 2-naphthaquinoline in 0.25M-sulphuric acid, a few drops of dilute sulphuric acid, and then 0.2M-potassium iodide in excess. After 20 minutes, filter the precipitate of the cadmium complex through a weighed Gooch, sintered glass, or porcelain filtering crucible, wash with a solution containing 10 cm^3 of 0.2M-potassium iodide, 10 cm^3 of 2.5 per cent 2-naphthaquinoline in 0.25M-sulphuric acid, 80 cm^3 of water, and 1–2 drops of dilute sulphur dioxide solution, and finally suck as free as possible from the wash liquor. Dry the precipitate to constant weight at 130 °C. Weigh as $[(C_{13}H_9N)_2H_2](CdI_4)$.

B. Determination of cadmium as quinaldate. *Discussion.* Quinaldic acid or its sodium salt precipitates cadmium quantitatively from acetic acid or neutral solutions. The precipitate is collected on a Gooch type of crucible, and dried at 125 °C. A determination may be completed in about 90 minutes. For the limitations of the method, see Section **XI, 11M**.

Procedure. The solution (150 cm^3) should be neutral or weakly acid with acetic acid, and should contain 0.1–0.15 g of Cd. Heat the solution to boiling, and remove the source of heat. Add the reagent (a 3.3 per cent solution of quinaldic acid or of the sodium salt in water) dropwise with vigorous stirring until present in slight excess. Then neutralise carefully with dilute ammonia solution, and allow the white curdy precipitate to settle. When cold, wash with cold water by decantation, filter through a sintered glass or porcelain filtering crucible, wash thoroughly with cold water, and dry at 125 °C to constant weight. Weigh as $Cd(C_{10}H_6O_2N)_2$.

C. Determination of cadmium by the pyridine method. *Discussion.* If a hot neutral or faintly acid solution of a cadmium salt is treated with ammonium thiocyanate and pyridine, dipyridinecadmium thiocyanate is quantitatively precipitated. This precipitate is collected, and washed, *inter alia*, with ethanol and diethyl ether containing a little pyridine; it may be dried simply by leaving in a vacuum desiccator for 15–20 minutes. A determination can thus be completed in less than an hour. If the solution is weakly acid, ammonium thiocyanate may be added, followed by pyridine, until a precipitate just forms, the latter dissolved by warming, and a further 1 cm^3 of pyridine added.

Procedure. The solution (75–100 cm^3) should contain about 0.1 g of Cd (1) and be neutral or very feebly acid. Add 0.5–1.0 g A.R. ammonium thiocyanate, stir, heat to boiling, and treat the solution with 1 cm^3 pure pyridine dropwise and with stirring. The complex slowly separates as the solution cools. Filter the *cold* solution through a weighed sintered glass or porcelain filtering crucible, transfer the precipitate to the crucible with the aid of *Solution 1*. Wash four to five times with *Solution 2*, then twice with 1-cm^3 portions of *Solution 3*, and finally five to six times with small volumes (*ca.* 1 cm^3) of *Solution 4*. (For further experimental details, see under Zinc, Section **XI, 62**.) Dry the precipitate in a vacuum desiccator (Fig. III, 15) for 10–15 minutes and weigh. Repeat the drying until constant weight is attained. Weigh as $[Cd(C_5H_5N)_2](SCN)_2$.

Solution 1. 100 cm^3 water containing 0.3 g NH$_4$SCN and 0.5 cm^3 pyridine.

Solution 2. 73 cm^3 water, 25 cm^3 95 per cent ethanol, 0.1 g NH$_4$SCN, and 2 cm^3 of pyridine.

Solution 3. 10 cm^3 absolute ethanol and 1 cm^3 pyridine.

Solution 4. 15 cm^3 diethyl ether (sodium dried) and 2 drops pyridine.

Note. 1. For practice in this determination use about 0.3 g, accurately weighed, A.R. cadmium sulphate or A.R. cadmium iodide.

XI, 34. CALCIUM. Of the methods available for the determination of calcium, that via calcium carbonate by initial precipitation of the oxalate is the best and most widely used.

A. Determination of calcium as calcium carbonate, by precipitation from homogeneous solution as the oxalate. *Discussion.* This method has been fully described in Section **XI, 20**. Precipitation is first effected as calcium oxalate which is subsequently converted into calcium carbonate or calcium oxide.

Precipitation of calcium oxalate may also be made **from homogeneous solution** either by use of urea as reagent (see Section **X, 94**) or by the use of dimethyl oxalate. Both procedures lead to satisfactory separations from magnesium.

Procedure. To obtain experience in this method, weigh out accurately about 0.25 g A.R. calcium carbonate, dissolve it in 5 cm³ dilute hydrochloric acid (1:5), and dilute to 150 cm³. Adjust the pH to 4.7 with dilute ammonia solution (use a pH meter). Add 100 cm³ ammonium acetate–acetic acid buffer (2.5M with respect to each) and 5.0 g pure dimethyl oxalate (1). Cover the beaker and heat on a temperature controlled hot plate at 90 °C for 2.5 hours: stir occasionally. Precipitation usually commences after 10 minutes. As a precautionary measure add, 10 minutes before filtration, 5 cm³ of a solution containing 0.25 g ammonium oxalate. Cool the solution rapidly to room temperature, filter through a weighed porcelain filtering crucible of medium porosity, and wash with 1 per cent ammonium oxalate solution. Dry the precipitate for 1 hour at 120 °C and then ignite in an electric muffle furnace at 500 °C for 2 hours. Weigh as $CaCO_3$.

Note. 1. Unless pure dimethyl oxalate is used, immediate precipitation of some fine calcium oxalate will occur. Impure dimethyl oxalate should be recrystallised from ethanol and stored in a desiccator.

B. Determination of calcium as tungstate. *Discussion.* The calcium is precipitated in neutral solution (pH 7–8) with a solution of sodium tungstate as calcium tungstate $CaWO_4$. This procedure is applicable in the presence of considerable amounts of magnesium.

Procedure. The solution (100 cm²) should contain about 0.04 g Ca and possess a pH of 7–8 (1). Add either dilute sodium hydroxide solution or dilute acetic acid to attain the correct pH with the aid of cresol red indicator or a pH meter. Large quantities of ammonium salts hinder precipitation. Heat the solution to about 80 °C and introduce, with stirring, 2.0 cm³ of the sodium tungstate reagent (2). Calcium tungstate is precipitated immediately. Cool in ice for 30 minutes, filter through a sintered glass (porosity No. 3) or porcelain filtering crucible, wash with 20 cm³ warm water, and dry at 110 °C for 1 hour or to constant weight. Weigh as $CaWO_4$.

Notes. 1. For practice in this determination a solution may be prepared by dissolving 1.0 g (accurately weighed) A.R. calcium carbonate in a little dilute hydrochloric acid, and diluting to 250 cm³ in a graduated flask. A 25-cm³ portion of this solution, diluted to 100 cm³, may be used.

2. The reagent is prepared by dissolving 19.0 g A.R. sodium tungstate, $Na_2WO_4.2H_2O$, in 100 cm³ of water.

XI, 35. CERIUM. Determination of cerium as cerium(IV) iodate and subsequent ignition to cerium(IV) oxide. *Discussion.* Cerium may be determined as **cerium(IV) iodate, $Ce(IO_3)_4$**, which is ignited to, and weighed as, the **oxide, CeO_2**. Thorium (also titanium and zirconium) must, however, be first removed (see Section **XI, 56**); the method is then applicable in the presence of relatively large quantities of rare earths. Titrimetric methods (see Section **X, 113**) are generally preferred.

Procedure. The solution should not exceed 50 cm³ in volume, all metallic elements should be present as nitrates, and the cerium content should not exceed 0.10 g. Treat the solution with half its volume of concentrated nitric acid, and add 0.5 g potassium bromate (to oxidise the cerium). When the latter has dissolved, add ten to fifteen times the theoretical quantity of potassium iodate in nitric acid solution (1) slowly and with constant stirring, and allow the precipitated cerium(IV) iodate to settle. When cold, filter the precipitate through a fine filter paper (e.g., Whatman No. 42 or 542), allow to drain, rinse once, and then wash back into the beaker in which precipitation took place by means of a solution containing 0.8 g potassium iodate and 5 cm³ concentrated nitric acid in 100 cm³. Mix thoroughly, collect the precipitate on the same paper, drain, wash back into the beaker with hot water, boil, and treat at once with concentrated nitric acid dropwise until the precipitate just dissolves (20–25 cm³ of acid are required per 0.1 g of cerium). Add 0.25 g potassium bromate and as much potassium iodate–nitric acid solution as before. When cold, collect the cerium(IV) iodate upon the same filter paper, wash once with the washing solution, return to the beaker, stir with the washing solution, filter again, and wash thrice with the same solution. Place the filter paper and precipitate in the same beaker, add 5–8 g oxalic acid and 50 cm³ water, and heat to boiling. After all the iodine has been expelled, set aside for several hours, filter, wash with cold water, dry, and ignite (at 500–600 °C) to constant weight in a platinum crucible. Weigh as CeO_2.

Note. 1. This is prepared by dissolving 50 g of potassium iodate in 167 cm³ of concentrated nitric acid, and diluting to 500 cm³.

XI, 36. CHROMIUM. Chromium may be determined in one of the following forms:

A. Determination of chromium as barium chromate. *Discussion.* The chromium must be present as chromate. The method is of limited application because of the general insolubility of chromates. Chlorides do not interfere, but sulphates must, of course, be absent. For further properties of barium chromate, see Section **XI, 30B**.

To convert a **chromium(III) salt into a chromate**, treat the chromium solution contained in a porcelain dish with several cm³ of bromine water, followed by freshly prepared potassium hydroxide solution until alkaline. Warm until the odour of bromine disappears.

Procedure. The solution should contain about 0.1 g of Cr as chromate, be neutral or weakly acid with acetic acid, and occupy a volume of 200–300 cm³. Add a 10 per cent solution of barium acetate dropwise from a burette and with constant stirring to the boiling solution (1) until present in slight excess. Place the beaker on a water bath until the precipitate settles; test for completeness of precipitation by adding a little more of the reagent. Allow to cool, filter with gentle suction through a weighed Gooch, sintered glass, or porcelain filtering crucible, wash with hot water until 1 cm³ of the washings gives no precipitate with a little dilute sulphuric acid. Complete the determination as described in Section **XI, 30B**. Weigh as $BaCrO_4$.

Note. 1. A solution for practice may be prepared by dissolving about 0.5 g A.R. potassium dichromate, accurately weighed, in 300 cm³ of water, adding ammonia solution until neutral, and then 1 cm³ 6*M*-acetic acid.

B. Determination of chromium as lead chromate (precipitation from homogeneous solution).   *Discussion.*   Use is made of the homogeneous generation of chromate ion produced by the slow oxidation of chromium(III) by bromate at 90–95 °C in the presence of excess of lead nitrate solution and an acetate buffer. The crystals of lead chromate produced are relatively large and easily filtered; the volume of the precipitate is about half that produced by the standard method of precipitation.

$$2Cr^{3+} + BrO_3^- + 5H_2O = 2CrO_4^{2-} + Br^- + 10H^+$$
$$BrO_3^- + 5Br^- + 6H^+ = 3Br_2 + 3H_2O$$
$$Pb^{2+} + CrO_4^{2-} = PbCrO_4$$

Cations forming insoluble chromates, such as those of silver, barium, mercury(I), mercury(II), and bismuth, do not interfere because the acidity is sufficiently high to prevent their precipitation. Bromide ion from the generation may be expected to form insoluble silver bromide, and so it is preferable to separate silver prior to the precipitation. Ammonium salts interfere, owing to competitive oxidation by bromate, and should be removed by treatment with sodium hydroxide.

$$BrO_3^- + 2NH_4^+ = Br^- + N_2 + 2H^+ + 3H_2O$$

Procedure.   Use a sample solution containing about 50 mg of chromium(III). Neutralise the solution by the addition of sodium hydroxide solution until a precipitate just begins to form. Add 10 cm³ acetate buffer solution ($6M$ in acetic acid and $0.6M$ in sodium acetate), 10 cm³ lead nitrate solution (3.5 g per 100 cm³), and 10 cm³ potassium bromate solution (2.0 g per 100 cm³). Heat to 90–95 °C: after generation (of chromate) and precipitation are complete (about 45 minutes), as shown by the clear supernatant liquid, cool, filter on a weighed sintered glass or porcelain filtering crucible, wash with a little 0.1 per cent nitric acid, and dry to constant weight at 120 °C. Weigh as $PbCrO_4$.

XI, 37.   COBALT.   Cobalt may be separated in one of the following forms:

A. Determination of cobalt with 1-nitroso-2-naphthol.   *Discussion.*   1-Nitroso-2-naphthol gives a red precipitate with solutions of cobalt salts. There is some doubt as to the exact composition of the precipitate: the formula $Co(C_{10}H_6O_2N)_3$ has been assigned to it, but it is probably not pure. The complex is best converted into cobalt sulphate, $CoSO_4$, or into metallic cobalt, and weighed in either of these forms. For the limitations of the method, see Section **XI, 11F**.

It has been stated that a precipitate of definite composition $Co(C_{10}H_6ONO)_3$ is obtained if the cobalt(II) (1–30 mg) is first oxidised to cobalt(III) with a little 30 per cent hydrogen peroxide in faintly acid solution. Sodium hydroxide solution ($2M$) is added until black cobalt(III) hydroxide commences to precipitate; the latter is dissolved in warm acetic acid and the solution diluted to 200 cm³. The 1-nitroso-2-naphthol reagent (10–20 cm³) is added with stirring to the warm solution, which is then heated with vigorous stirring until the precipitate coagulates. The coloured precipitate is filtered through a weighed porcelain or sintered glass filtering crucible, washed with a little dilute acetic acid (1:2) and thrice with hot water. It is dried to constant weight at 130 °C and weighed as $Co(C_{10}H_6ONO)_3$.

An important application is to the separation of nickel and cobalt: a double precipitation is desirable when nickel is present in large amount.

Procedure. Dilute the solution containing not more than 0.1 g of Co as chloride or as sulphate (1) to 200 cm³, add sufficient concentrated hydrochloric acid to give a total of 5 cm³ of the concentrated acid in the solution, and warm to about 80 °C. Add the freshly prepared 1-nitroso-2-naphthol reagent (for preparation, see Section **XI, 11F**) until precipitation is considered complete: about 0.25 g of 1-nitroso-2-naphthol is required for each 0.01 g of Co. Heat to gentle boiling with stirring until the precipitate coagulates or settles out. The supernatant liquid should be clear and yellow. Test whether precipitation is complete by adding a little more of the reagent to the clear solution. Allow to stand for 2–3 hours and decant the clear solution through a quantitative filter paper (e.g., Whatman, No. 541). Wash the precipitate by decantation with a little hot (*ca.* 80 °C) dilute hydrochloric acid (1:2), finally transfer the precipitate to the filter paper, and wash with hot water until free from acid. Dry the bulky precipitate at 100–110 °C for 1 hour; it will shrink considerably. Place the dried filter paper and precipitate in a silica crucible, just cover the precipitate with A.R. oxalic acid (this will prevent sudden decomposition of the complex and consequent mechanical loss during the ignition process), and heat gently until all the organic matter has burned off. Ignite the precipitate for a few minutes, allow to cool, treat with a few drops of concentrated nitric acid to oxidise any residual carbon and to convert the oxide into nitrate: heat carefully until the excess of nitric acid has been expelled. Finally, add enough sulphuric acid to convert the nitrate into sulphate, heat cautiously until the excess of acid has been expelled, and then for a few moments to incipient redness (450–500 °C). Allow to cool, moisten with a drop or two of water, and again heat cautiously as before to expel any free sulphuric acid. Allow to cool in a desiccator and weigh as the sulphate, $CoSO_4$.

In an alternative procedure, the cobalt is weighed as the metal. Transfer the dried precipitate and filter paper to a crucible. Ignite the precipitate and filter paper in the presence of oxalic acid as before: it is important to heat very slowly at first and subsequently to oxidise any residual carbon with a little concentrated nitric acid. Fit a crucible-cover on to the crucible, and continue the ignition in a stream of pure hydrogen for at least 30 minutes. Withdraw the burner beneath the crucible and, after the crucible is almost at room temperature, momentarily stop the stream of hydrogen so as to extinguish the flame burning at the cover. Continue the passage of the stream of hydrogen until the crucible is at room temperature, and weigh as metallic cobalt. Repeat the treatment with hydrogen (heating period of 30 minutes in hydrogen) until constant weight is attained.

Note. 1. A suitable solution for practice in this determination may be prepared from 0.1 g, accurately weighed, of A.R. cobalt(II) sulphate (clear, uneffloresced crystals) or pure ammonium cobalt sulphate.

B. Determination of cobalt as cobalt tetrathiocyanatomercurate(II) (mercurithiocyanate). *Discussion.* This method is based upon the fact that cobalt(II) in almost neutral solution forms a blue complex salt $Co[Hg(SCN)_4]$ with a reagent prepared by dissolving 1 mol of mercury(II) chloride and 4 mols of ammonium thiocyanate in water. The precipitate is sparingly soluble in water, soluble in acids and in a large excess of the reagent, soluble in diethyl ether, chloroform, and carbon tetrachloride, and sparingly soluble in absolute ethanol. It may be dried at 100–110 °C. The following elements interfere: copper, cadmium, zinc, iron(II), iron(III), nickel, manganese(II), bismuth, silver and mercury(II); iron(III) may be rendered innocuous by the addition of phosphate.

Procedure. The almost neutral sample solution may conveniently contain 35–40 mg of cobalt in 25 cm^3 (1) and be free from the interfering elements mentioned above. Add, with constant stirring, 4.8 cm^3 of the mercury(II) chloride solution (2) followed by 5.2 cm^3 of the ammonium thiocyanate reagent (2). Do not scratch the sides of the beaker with the stirring rod. A dark blue precipitate forms after stirring for 1–3 minutes; continue the stirring for a further 2–3 minutes and allow to stand for 2 hours at room temperature. Collect the precipitate in a weighed sintered glass (porosity No. 4) or porcelain filtering crucible; use the filtrate to assist the transfer of any residual precipitate in the beaker. Wash the precipitate with 2–3 cm^3 of a dilute solution of the precipitating reagent (3) and finally with 5 cm^3 ice-cold water. Dry at 100 °C. Weigh as Co[Hg(SCN)$_4$].

Notes. 1. A suitable solution for practice analysis may be prepared by dissolving about 5.0 g, accurately weighed, pure ammonium cobalt sulphate in water and diluting to 500 cm^3 in a graduated flask. Use 25.0 cm^3 for each determination.

The cobalt content may be rapidly checked by titration with standard EDTA solution in the presence of Xylenol Orange as indicator (see Section **X, 61**).

2. *Solution* (i): dissolve 5.4 g finely powdered A.R. mercury(II) chloride in 100 cm^3 of distilled water; slight warming may be necessary.

Solution (ii): dissolve 6.0 g A.R. ammonium thiocyanate in 100 cm^3 of distilled water.

It is preferable to add the solutions separately to the cobalt solution: a slight excess (up to 10 per cent) of Solution (ii) is not harmful. About 1.0 cm^3 of each solution is required for the precipitation of 10 mg of cobalt. The excess of the reagent should not be more than 10–15 per cent owing to the solubility of the precipitate in the ammonium mercurithiocyanate solution.

3. The washing solution is prepared by adding 1.0 cm^3 each of Solutions (i) and (ii) to 100 cm^3 of water.

XI, 38. COPPER. Copper may be determined in the following forms:

A. Determination of copper as copper(I) thiocyanate. *Discussion.* This is an excellent method, since most thiocyanates of other metals are soluble. Separation may thus be effected from bismuth, cadmium, arsenic, antimony, tin, iron, nickel, cobalt, manganese, and zinc. The addition of 2–3 g of tartaric acid is desirable for the prevention of hydrolysis when bismuth, antimony, or tin are present. Excessive amounts of ammonium salts or of the thiocyanate precipitant should be absent, as should also oxidising agents; the solution should only be slightly acidic, since the solubility of the precipitate increases with decreasing pH. Lead, mercury, the precious metals, selenium, and tellurium interfere and contaminate the precipitate.

The essential experimental conditions are:

1. Slight acidity of the solution with respect to hydrochloric acid or sulphuric acid, since the solubility of the precipitate increases appreciably with decreasing pH.
2. The presence of a reducing agent, such as sulphurous acid or ammonium hydrogensulphite, to reduce copper(II) to copper(I).
3. A slight excess of ammonium thiocyanate, since a large excess increases the

solubility of the copper(I) thiocyanate due to the formation of a complex thiocyanate ion.

4. The absence of oxidising agents.

The reaction may be represented as:

$$2Cu^{2+} + HSO_3^- + H_2O = 2Cu^+ + HSO_4^- + 2H^+$$
$$Cu^+ + SCN^- = CuSCN$$

The precipitate is curdy (compare silver chloride) and is readily coagulated by boiling. It is washed with dilute ammonium thiocyanate solution: a little sulphurous acid or ammonium hydrogensulphite is added to the wash solution to prevent any oxidation of the copper(I) salt.

Procedure. Weigh out accurately about 0.4 g of the copper salt (1) into a 250-cm³ beaker, and dissolve it in 50 cm³ of water. Add a few drops of dilute hydrochloric acid, and then a slight excess (about 20–30 cm³ are required) of freshly prepared saturated sulphurous acid solution. Alternatively, add 25 cm³ ammonium hydrogensulphite solution: the latter is prepared by diluting to ten times its volume the commercial concentrated solution, which has a specific gravity of 1.33 and contains about 54 per cent sulphur dioxide. Dilute the cold liquid to 150–200 cm³, heat nearly to boiling, and add freshly prepared 10 per cent ammonium thiocyanate solution, slowly and with constant stirring, from a burette until present in *slight* excess. The precipitate of copper(I) thiocyanate should be white; the mother liquor should be colourless and smell of sulphur dioxide. Allow to stand for two hours, but preferably overnight. Filter through a weighed filtering crucible (Gooch, sintered glass, or porcelain), and wash the precipitate ten to fifteen times with a cold solution prepared by adding to every 100 cm³ of water 1 cm³ of a 10 per cent solution of ammonium thiocyanate and 5–6 drops of saturated sulphurous acid solution, and finally several times with 20 per cent ethanol to remove ammonium thiocyanate (2). Dry the precipitate to constant weight at 110–120 °C (3). Weigh as CuSCN.

Notes. 1. A.R. copper sulphate pentahydrate is suitable for practice in this determination. 0.4 gram of this contains about 0.1 g of Cu.

2. Alternatively, but less desirably, the precipitate may be washed with cold water until the filtrate gives only a slight reddish coloration with iron(III) chloride, and finally with 20 per cent ethanol.

3. The precipitate, collected in a sintered glass (porosity No. 4) or porcelain filtering crucible, may be weighed more rapidly as follows. Wash the copper(I) thiocyanate five or six times with ethanol, followed by a similar treatment with small volumes of anhydrous diethyl ether, then suck the precipitate dry at the pump for 10 minutes, wipe the outside of the crucible with a clean linen cloth and leave it in a vacuum desiccator for 10 minutes. Weigh as CuSCN.

B. Determination of copper with benzoin-α-oxime. *Discussion.* Benzoin-α-oxime (cupron) is a specific reagent for the determination of copper in ammoniacal solutions (compare Section **XI, 11D**). A green, heavy, and readily filterable precipitate is obtained: this is insoluble in water, dilute ammonia solution, acetic acid, tartaric acid, and ethanol, is slightly soluble in concentrated ammonia solution, and readily soluble in mineral acids. Precipitation is quantitative in ammoniacal tartrate solutions: separation can thus be effected from iron and other metals whose hydroxides are not precipitated in tartrate solutions. Separation can also be made from cadmium, zinc, cobalt, and nickel, which are not precipitated in ammoniacal solutions.

Procedure. Treat the neutral solution, which should be free from ammonium salts and contain not more than 0.05 g copper, with dilute ammonia solution until a clear blue solution is obtained. Heat to boiling and precipitate the copper by the addition, dropwise, of a 2 per cent ethanolic solution of the reagent. Precipitation is complete when the blue colour of the solution disappears. Filter the heavy green precipitate on a weighed sintered glass or porcelain filtering crucible, wash with hot dilute ammonia solution (1 : 100), then with hot water, and finally with warm ethanol. Dry to constant weight at 105–115 °C. Weigh as $Cu(C_{14}H_{11}O_2N)$. It is recommended that completeness of washing be tested for by washing the dry precipitate again with warm ethanol followed by hot water. Dry again to constant weight at 105–110 °C.

XI, 39.   GOLD.   Determination of gold as the metal.   *Discussion.*   Gold is nearly always determined as the metal. The reducing agents generally employed are sulphur dioxide, oxalic acid, and iron(II) sulphate. If nitric acid is present it must be removed by repeated evaporation with concentrated hydrochloric acid, and the solution diluted with water. With sulphurous acid, small amounts of the platinum metals (particularly platinum) may be carried down with the precipitate. It is therefore usually necessary to re-dissolve the solid in dilute aqua regia and to re-precipitate the gold; oxalic acid gives a better separation from the platinum metals in the second precipitation, although the precipitate is somewhat finely divided. Iron(II) sulphate gives satisfactory results for gold alone, but difficulties are introduced if the platinum metals are subsequently to be determined. Oxalic acid is slow in its action, and yields a precipitate which is difficult to filter.

The best results are obtained with quinol as the reducing agent. Precipitation in hot 1.2*M*-hydrochloric acid solution is rapid, the gold is readily filtered, and occlusion of the platinum metals is negligible. Precipitation in the cold is complete in 2 hours. Palladium in the filtrate can be precipitated directly with dimethylglyoxime, whilst platinum in the filtrate may be determined either by evaporating to dryness in order to destroy organic matter and then digesting with a little aqua regia or by reduction with sodium formate and formic acid.

Gold may also be separated from hydrochloric acid solutions of the platinum metals by extraction with diethyl ether or with ethyl acetate (compare Chapter VI); except in special cases these methods do not offer any special advantages over the reduction to the metal.

Procedure A.   The solution should contain not more than 5 cm³ concentrated hydrochloric acid per 100 cm³ of solution, not more than 0.5–1 g Au, and be free from lead, selenium, tellurium, and the alkaline earths. Add 25 cm³ of a freshly prepared saturated sulphur dioxide solution, and digest for 1 hour on the steam bath in order to coagulate the precipitate. Add 5–10 cm³ more of the sulphur dioxide solution, and allow to cool. If the cold solution smells strongly of sulphur dioxide, the precipitation of gold is complete. Some of the metal is finely divided, and it is therefore advisable to make use of a Whatman accelerator or ashless tablet. Pour the supernatant liquid through a Whatman No. 42 or 542 filter paper, preferably containing some filter-paper pulp, and transfer as little as possible of the precipitate to the paper unless one precipitation is thought sufficient; this will only be the case if very small amounts of platinum or palladium are present. Wash well by decantation with hot dilute hydrochloric acid (1 : 99). Transfer the filter to the beaker, and re-dissolve the gold in dilute aqua regia; use $8\,cm^3$ of concentrated hydrochloric acid, $2\,cm^3$ of concentrated nitric acid, and $10\,cm^3$ of water for each gram or less of gold. Filter from the paper pulp, and wash thoroughly with hot dilute hydrochloric acid (1:99). Evaporate the filtrate to dryness on the water bath, add 2–$3\,cm^3$ of concentrated hydrochloric acid, and evaporate to dryness again; repeat this operation twice in order to eliminate all the nitric acid. Treat the residue with $3\,cm^3$ of concentrated hydrochloric acid, 5 drops of concentrated sulphuric acid, and $75\,cm^3$ of water, disregard the small amount of gold which may separate, add $25\,cm^3$ of a saturated solution of oxalic acid, and boil for a minute or two. If no further visible precipitation of gold occurs, digest the solution on the water bath for at least 4 hours. Filter off the gold through a filter paper (as described above), and wipe the inside of the beaker with small pieces of quantitative filter paper to ensure that all the metal is transferred from the beaker; wash well with 1:99 hydrochloric acid. Transfer the filter to a weighed porcelain or silica crucible, burn off the paper carefully, and ignite to constant weight. Weigh as Au.

Procedure B. The solution must be free from nitric acid, be about $1.2M$ with respect to hydrochloric acid (*ca.* $5\,cm^3$ of concentrated hydrochloric acid in $50\,cm^3$ of water), and contain up to $0.2\,g$ of Au in $50\,cm^3$. Heat the solution to boiling, add excess of 5 per cent aqueous quinol solution ($3\,cm^3$ for every 25 mg of Au), and boil for 20 minutes. Allow to cool, and filter either through a weighed porcelain filtering crucible or through a Whatman No. 42 or 542 filter paper; wash thoroughly with hot water. The small particles of gold remaining in the bottom of the beaker (easily seen with a small flash lamp) are best removed with pieces of ashless filter paper. Ignite the porcelain filtering crucible to constant weight. If filter paper is used, transfer to a weighed porcelain or silica crucible, and complete the determination as described in *Procedure A*.

XI, 40. IRON. Iron may be determined in the following forms:

A. Determination of iron as iron(III) oxide by initial formation of basic iron(III) formate. *Discussion.* The precipitation of iron as iron(III) hydroxide by ammonia solution, etc., and its conversion into iron(III) oxide, in which form it is weighed, is fully described in Section **XI, 21**. Precipitation with ammonia solution yields a gelatinous precipitate which is somewhat difficult to wash and to filter; this difficulty is largely overcome by the addition of a few drops of hydrazine hydrate after the ammonia solution.

There are, however, three methods of precipitation which yield iron(III) hydroxide in a relatively dense and granular form, which is easily washed and filtered. These include the precipitation of iron(III) as basic iron(III) formate from homogeneous solution by hydrolysis of urea in a hydrochloric acid solution containing formic acid. The precipitate thus obtained is denser, more readily filtered and washed, and adsorbs fewer impurities than the precipitate obtained by other hydrolytic procedures. Ignition yields iron(III) oxide. The pH at which basic iron(III) formate begins to precipitate depends upon several factors, which include the initial iron and chloride concentration: a high concentration of ammonium chloride is essential to prevent colloid formation. It is important to use an optimum initial pH to avoid a large excess of free acid, which would have to be neutralised by urea hydrolysis, and yet there must be present sufficient acid to prevent the formation of a gelatinous precipitate prior to boiling the solution: ideally, a turbidity should appear about 5–10 minutes after the solution has begun to boil. For iron contents of 5 mg to 55 mg per $100\,cm^3$, the optimum initial pH is between 2.00 to 1.70. Some reduction occurs during the precipitation (due to the presence of both formate and chloride): re-oxidation of iron(II) to iron(III) is easily effected by the addition of hydrogen peroxide towards the end of the procedure. Precipitation as basic formate enables iron to be separated from manganese(II), cobalt, nickel, copper, zinc, cadmium, magnesium, calcium, and barium. When copper is present the solution must be cooled before hydrogen peroxide is added, otherwise the vigorous decomposition of the hydrogen peroxide may result in loss of some of the solution.

Attention is directed to the fact that if ignition is carried out in a platinum crucible at a temperature above 1100 °C some reduction to the oxide Fe_3O_4 may occur, and at temperatures above 1200 °C some of the oxide may be reduced to the metal and alloy with the platinum. This accounts in part for the contamination of the platinum crucible by iron which sometimes occurs in analytical work. No magnetic oxide of iron is produced if silica crucibles are employed for the ignitions.

Procedure.   For practice in this determination A.R. ammonium iron(III) sulphate (iron alum) may be used. Weigh out accurately about 1.0 g A.R. iron alum, dissolve it in dilute hydrochloric acid, add 2.0 cm^3 formic acid (sp. gr. 1.20; *ca.* 90 per cent), 10 g ammonium chloride, and 4.5 g urea (as a 10 per cent aqueous solution); mix well. Dilute to about 350 cm^3. Adjust the pH to 1.80 (use a pH meter) by the addition of hydrochloric acid.* Dilute to 400 cm^3, insert a boiling rod, and boil gently for about 90 minutes or until a pH of about 4.0 is reached. Add 5 cm^3 of 3 per cent hydrogen peroxide solution and boil for a further 5 minutes. Then add 10 cm^3 0.02 per cent gelatin solution: the latter will improve the filtering and washing properties of the precipitate. Filter on a Whatman No. 40 or No. 540 filter paper and wash the precipitate fifteen times with hot 1 per cent ammonium nitrate solution adjusted to pH 4. Remove as much as possible of the precipitate adhering to the walls of the beaker with the aid of a rubber-tipped stirring rod.

Dissolve any film adhering tenaciously to the walls of the beaker by adding 4–5 cm^3 of concentrated hydrochloric acid, cover with a clock glass, and reflux gently for a few minutes. Then wash the beaker, clock glass, and stirring rod with 25 cm^3 distilled water, add a few drops of methyl red indicator and then dilute ammonia solution dropwise until the colour of the solution is a distinct yellow. Boil for 2–3 minutes to coagulate the precipitate, filter, and wash on a separate filter paper. Place both filter papers in a weighed porcelain or silica crucible, char the filter papers over a small flame, and then ignite at a red heat or in an electric muffle furnace at 850 °C. Heating for 1 hour is usually sufficient. Weigh as Fe_2O_3.

B. Determination of iron with cupferron and subsequent weighing as iron(III) oxide. *Discussion.*   Cupferron, the ammonium salt of nitrosophenylhydroxylamine, $C_6H_5N(NO)\cdot ONH_4$, precipitates iron, tin, uranium(IV), vanadium, titanium, and zirconium from strongly acid solutions, thus affording a separation from aluminium, chromium, beryllium, phosphorus, boron, manganese, zinc, nickel, cobalt, and hexavalent uranium (compare Section **XI, 11B**). Copper

---

* Alternatively, add pure aqueous ammonia solution (1:1) by means of a dropper pipette until a definite precipitate of iron(III) hydroxide just begins to form. Now add concentrated hydrochloric acid dropwise, stirring and allowing to stand for a minute or two after each 2–3 drops, until a clear solution is obtained: then add 1.5 cm^3 of concentrated hydrochloric acid and mix well.

and thorium must be precipitated from weakly acid solutions. Several metals (e.g., lead, silver, mercury, bismuth, tungsten, and cerium) interfere, but most of these may usually be removed by other methods—for example, by hydrogen sulphide in acid solution. The precipitate cannot be weighed as such, but must be ignited to the oxide. The ignition must be very carefully carried out in the early stages in order to avoid mechanical losses, for wet precipitates tend to liquefy and effervesce, whilst dry precipitates give off considerable volatile matter. The precipitate is rather bulky, and the amount of material taken should therefore be such as to yield 0.1–0.2 g of oxide.

Procedure. The solution (1) (150–200 cm^3) should contain about 0.1 g of Fe in the iron(III) state and be strongly acid with hydrochloric acid or sulphuric acid. To the cooled solution (*ca.* 10 °C) add a freshly prepared, filtered 5 per cent aqueous cupferron solution (2) slowly and with constant and vigorous stirring until no further formation of a brown precipitate takes place. The formation of a white precipitate of nitrosophenylhydroxylamine indicates when the reagent is present in excess. Do not warm, since the reagent is rapidly decomposed in hot acid solution. Add a Whatman accelerator (or a third of a Whatman ashless tablet), stir for 2–3 minutes, and without further delay filter through a Whatman No. 41 or 541 filter paper, preferably supported upon a Whatman filter cone. Wash several times with 10 per cent by volume of hydrochloric acid containing 1.5 g of cupferron dm^{-3}, then twice with 5M-ammonia solution to remove excess of cupferron, and finally once with water. Ignite the precipitate with the paper in a weighed porcelain, silica, or platinum crucible, very gently at first until all the organic matter is destroyed, and then strongly to constant weight. Weigh as Fe$_2$O$_3$.

Notes. 1. A suitable solution for practice may be prepared by weighing out accurately 0.8–0.9 g A.R. iron alum, dissolving it in 150 cm^3 water, and adding 25 cm^3 concentrated hydrochloric acid.

2. Only freshly prepared solutions should be employed, because the solution only keeps for a day or two. The dry reagent should be kept in a cool, dark place, and preferably with a bag of ammonium carbonate suspended in the bottle.

XI, 41. LEAD. Lead may be determined in a number of forms in addition to that of lead chromate already described (Section **XI, 22**). Two useful procedures follow:

A. Determination of lead as molybdate. *Discussion.* This is an excellent method, since the substance has a high molecular weight, is less soluble than the sulphate, and suffers no change upon ignition. Substances which form insoluble molybdates (e.g., the alkaline-earth metals, copper, and cadmium), which are easily hydrolysed (e.g., tin or titanium), and which form insoluble compounds with lead (e.g., chromates, arsenates, or phosphates) must be absent.

Procedure. Weigh out accurately about 0.30 g of the lead salt, dissolve it in 200 cm^3 water, and add 4 drops concentrated nitric acid. Heat to boiling, and slowly add from a burette or pipette, with stirring, a 2.5 per cent aqueous solution of ammonium molybdate. When precipitation appears to be complete, boil for 1 minute, allow the precipitate to settle, and add a few drops of the precipitant to the supernatant liquid. If a precipitate forms, repeat the process until the ammonium molybdate is present in *slight* excess. When precipitation is complete, add dilute ammonia solution (1:2) dropwise until the solution is neutral or slightly alkaline to litmus or methyl red. Acidify with a few drops of acetic acid, and allow to stand for a few minutes. Decant the supernatant liquid through a weighed porcelain or silica filtering crucible, and wash the precipitate three or four times by decantation with 75-cm^3 portions of 2 per cent ammonium nitrate solution. Transfer the precipitate to the filter, and wash until the washings give no test for molybdenum (e.g., no brown precipitate with potassium hexacyanoferrate(II) solution). Place the filtering crucible inside a nickel crucible or upon a crucible-ignition dish, and gradually heat to dull redness. Maintain the crucible at dull redness for 10 minutes, cool in a desiccator, and weigh. Repeat the heating, etc., until constant weight is attained. Alternatively, heat in an electric muffle furnace at 500–600 °C to constant weight. Weigh as PbMoO$_4$.

B. Determination of lead as the salicylaldoximate. *Discussion.* Lead may be precipitated in strongly ammoniacal solution (pH 9.3 or higher) with salicylaldehydeoxime as the lead complex Pb(C$_7$H$_5$O$_2$N); it should therefore be possible to separate lead from silver, cadmium and zinc, the salicylaldoximate of which are soluble in ammoniacal solution (see Section **XI, 11E**).

Procedure. To a solution (*ca.* 25 cm^3) of lead nitrate or lead acetate containing about 0.1 g of Pb, add 10 cm^3 freshly prepared 1 per cent salicylaldehydeoxime solution (for preparation, see Section **XI, 11E**), dilute to 50 cm^3 and add 12.5 cm^3 concentrated ammonia solution. Stir the resulting precipitate for 1 hour and allow to settle. Decant the supernatant liquid through a sintered glass crucible (porosity No. 4), wash the precipitate by decantation until free from salicylaldehydeoxime (as shown by the absence of a colour with iron(III) chloride solution), dry at 105 °C for 1 hour, and weigh as Pb(C$_7$H$_5$O$_2$N).

XI, 42. LITHIUM. It is usually necessary to determine lithium in the presence of sodium and/or potassium. The following procedures are suitable for this:

A. Determination of lithium in the presence of sodium and potassium by extraction with organic solvents. *Discussion.* This procedure is dependent upon the fact that lithium chloride is very soluble in many organic solvents, whilst the chlorides of sodium and potassium are only very slightly soluble. A number of solvents have been suggested as being suitable for this purpose, including dioxan, hexanol, 3-methyl-1-butanol and 2-ethyl hexanol. The following table gives the solubilities expressed as grams dissolved by 100 cm^3 of the anhydrous solvent at 25 °C.

	3-methyl-1-butanol	Hexanol	2-Ethylhexanol
LiCl	7.3	5.8	3.0
NaCl	0.0016	0.0008	0.0001
KCl	0.0006	0.00004	<0.00001

The preferred solvent is 2-ethylhexanol.

Procedure. Treat a concentrated solution prepared from 0.3–0.4 g or less of the mixed chlorides, accurately weighed, with a suitable volume of 2-ethylhexanol, introduce a little platinum foil or a few fragments of porous porcelain to prevent bumping, and distil until the water has passed over and the boiling point becomes constant (175–180 °C) for some time. Sodium and potassium chlorides are deposited, and lithium chloride is dehydrated and held in solution. Allow to cool, filter through a sintered glass filtering crucible, and wash thoroughly with successive small volumes of the anhydrous alcohol. Dry the crucible at 200–210 °C to volatilise the residual solvent, and weigh. The loss in weight is due to the lithium chloride.

If the weight of the lithium chloride exceeds 20 mg, a second extraction is necessary in order to remove the small quantity of lithium hydroxide present in the residual solid (formed by hydrolysis at the boiling point of the 2-ethylhexanol): the solid must be dissolved in a little water containing a few drops of hydrochloric acid.

B. Determination of lithium as lithium aluminate. *Discussion.* Lithium may be determined as **lithium aluminate** by precipitation with excess of sodium aluminate solution in the cold, the final pH of the solution being adjusted to 12.6–13.0. The precipitate is washed with water until free from alkali and weighed as $2Li_2O \cdot 5Al_2O_3$ after heating at 500–550 °C. The solubility in water is 0.008 g dm^{-3} at room temperature; it is 0.09 g dm^{-3} at pH 12.6.

Procedure. The sample solution (20 cm^3) may contain up to 10 mg of lithium, and the pH should be about 3.0. Add 40 cm^3 of the cold reagent (1) for each 10 mg of lithium. Adjust the pH to 12.6 by the addition of 1M-sodium hydroxide solution: use a pH meter. Allow to stand for 30 minutes, and collect the voluminous precipitate in a porcelain filtering crucible. Wash with small volumes of ice-cold water until the washings are no longer alkaline to phenolphthalein. Ignite at 500–550 °C in an electric muffle furnace. Weigh as $2Li_2O \cdot 5Al_2O_3$.

Note. 1. Prepare the precipitating reagent by dissolving 5.0 g A.R. aluminium potassium sulphate (potash alum) in 90 cm^3 warm water. Cool and add dropwise with stirring, while cooling in ice, a solution of 2.0 g sodium hydroxide in 5.0 cm^3 water until the initially formed precipitate redissolves. After standing for 12 hours, filter, adjust the pH to 12.6, and dilute to 100 cm^3 with water.

XI, 43. MAGNESIUM. Methods for the determination of magnesium as ammonium magnesium phosphate hexahydrate and as magnesium pyrophosphate have already been given in Section **XI, 23**. The following method is also useful:

Determination of magnesium as the 8-hydroxyquinaldinate. *Discussion.* Magnesium may be precipitated by 8-hydroxyquinaldine (2-methyloxine) in ammoniacal solution (pH at least 9.3) as the complex $Mg(C_{10}H_8ON)_2$, and weighed as such after drying at 130–140 °C. Numerous ions interfere (see Section **XI, 11P**).

Procedure. The solution (150–200 cm^3) may contain up to 0.05 g of Mg. Add 3 cm^3 of the 8-hydroxyquinaldine (2-methyl-oxine) reagent (1) for every 10 mg of magnesium present, and then add concentrated ammonia solution until the pH is at least 9.3 or until no further precipitate forms. Digest the solution at 60–80 °C for 20 minutes and filter through a sintered glass or porcelain filtering crucible. Wash the precipitate with hot water and dry to constant weight at 130–140 °C. Weigh as $Mg(C_{10}H_8ON)_2$.

Note. 1. The reagent is prepared by dissolving 5 g of 8-hydroxyquinaldine in 12 g of glacial acetic acid and diluting to 100 cm^3 with water.

XI, 44. MANGANESE. Determination of manganese as the ammonium phosphate or as the pyrophosphate. *Discussion.* The only method which is at all widely used for the gravimetric estimation of manganese is the precipitation as **ammonium manganese phosphate, $MnNH_4PO_4,H_2O$,** in slightly ammoniacal solution containing excess of ammonium salts. The precipitate may be weighed in this form after drying at 100–105 °C, or it may be ignited and subsequently weighed as **manganese pyrophosphate, $Mn_2P_2O_7$.** The latter procedure is by far the better one. The method is, however, of limited application because of the interfering influence of numerous other elements. Titrimetric methods are generally preferred (see Chapter X); the potentiometric determination of manganese (see Section **XIV, 28**) may also be recommended.

Procedure. The solution (200 cm^3) should be slightly acid, contain not more than 0.2 g of Mn in 200 cm^3, and no other cations except those of the alkali metals (1). Almost neutralise the solution with dilute ammonia solution, add 20 g of ammonium chloride and a considerable excess of diammonium hydrogen phosphate $(NH_4)_2HPO_4$ (say, 2 g of the solid). If a precipitate forms at this point, dissolve it by the addition of a few drops of 1:3 hydrochloric acid. Heat the solution almost to boiling (90–95 °C), and add dilute ammonia solution (1:3) dropwise and with constant stirring until a precipitate $(Mn_3(PO_4)_2)$ begins to form; immediately suspend the addition of the alkali. Continue the heating and stirring until the precipitate becomes crystalline ($MnNH_4PO_4,H_2O$). Then add another drop or two of ammonia solution, stir as before, etc., and so continue until no more precipitate is produced and its silky appearance remains unchanged. The precipitate must be maintained at 90–95 °C throughout; a large excess of ammonia solution must be avoided. Allow the solution to stand at room temperature (or, better, at 0 °C) for 2 hours. Filter through a quantitative filter paper or through a weighed porcelain filtering crucible, and wash the precipitate with cold, 1 per cent ammonium nitrate solution until free from chloride. Dry at a gentle heat, ignite at as low a temperature as possible until the carbon is oxidised (2), and then heat at 700–800 °C (in an electric crucible furnace or within a larger nickel crucible) to constant weight. Weigh as $Mn_2P_2O_7$. Alternatively, but less desirably, the precipitate in the porcelain filtering crucible may be dried at 100–105 °C to constant weight and weighed as $MnNH_4PO_4,H_2O$; in this case, a Gooch or a sintered glass filtering crucible may also be used.

Notes. 1. A suitable solution for practice may be prepared by one of the following methods:

(*a*) Dissolve 0.7 g, accurately weighed, A.R. manganese(II) sulphate $MnSO_4,4H_2O$ in 200 cm^3 of water.

(*b*) Dissolve 0.5 g, accurately weighed, A.R. potassium permanganate in very dilute sulphuric acid, and reduce the solution with sulphur dioxide or with ethanol. Remove the excess of sulphur dioxide or of acetaldehyde (and ethanol) by boiling. Dilute to 200 cm^3.

2. These remarks apply, of course, when filter paper is used.

XI, 45. MERCURY. Mercury may be determined in the following forms:

A. Determination of mercury as sulphide. *Discussion.* The precipitation of mercury as mercury(II) sulphide by hydrogen sulphide in hydrochloric acid solution is an accurate procedure in the absence of copper, cadmium, tin, zinc, and thallium; the latter metals complicate reactions which are based upon the behaviour of pure mercury(II) sulphide. Unless the experimental conditions detailed below are strictly followed, the precipitate is liable to be contaminated with a little sulphur, which must be removed by extraction with carbon disulphide. Oxidising agents (nitric acid, chlorine, iron(III) chloride, etc.) must be absent.

Procedure. Weigh out accurately about 0.15 g of the mercury(II) salt (1), dissolve it in 100 cm^3 of water, and add a few cm^3 of dilute hydrochloric acid. Saturate the cold solution with washed hydrogen sulphide (2), allow the precipitate to settle, and filter through a weighed Gooch, sintered glass or porcelain filtering crucible. Wash the precipitate with cold water (3), and weigh it, as HgS, after drying at 105–110 °C.

Notes. 1. A.R. mercury(II) chloride is suitable. Alternatively, the solution should contain not more than 0.1 g of mercury(II) per 100 cm^3, and should be free from oxidising agents.

2. The colour of the mercury(II) sulphide precipitate will become perfectly black as soon as the liquid is saturated with the gas.

3. If the presence of sulphur is suspected, the precipitate is washed with hot water, ethanol, carbon disulphide, or ethanol + diethyl ether, and then dried at 105–110 °C.

B. Determination of mercury as mercury(II) thionalide. *Discussion.* Thionalide $C_{10}H_7 \cdot NH \cdot CO \cdot CH_2 \cdot SH$ may be used for the quantitative precipitation of mercury(II) as $Hg(C_{12}H_{10}ONS)_2$. Sulphate does not interfere. Attention is drawn to the following experimental points:

1. The chloride ion concentration of the solution should not exceed 0.1M; the results are high if the chloride ion concentration is excessive.

2. If nitric acid solutions of mercury(II) nitrate are used, the latter must be converted into mercury(II) chloride by the addition of at least an equivalent amount of chloride ion.

3. A three-fold excess of reagent should be employed.

Procedure. The sample solution may contain 5 to 75 mg of mercury(II). A solution, prepared from A.R. mercury(II) chloride and containing, say, 20 mg mercury in 150 cm^3 of water, may be used for practice in this determination. Heat the solution to 80–85 °C and add, with constant stirring, a three-fold excess of a 1 per cent solution of thionalide in acetic acid. The precipitate coagulates upon stirring. Filter the hot solution through a sintered glass filtering crucible (porosity No. 3) which has been preheated by pouring hot water through it. (The use of a warm filtering crucible is essential; the separation of thionalide in the pores of the sintered plate of the crucible, which would render filtration difficult, is thus avoided.) Wash with hot water until free from acid, and dry to constant weight at 105 °C. Weigh as $Hg(C_{12}H_{10}ONS)_2$.

XI, 46. MOLYBDENUM. Molybdenum may be determined in the following forms:

A. Determination of molybdenum as lead molybdate. *Discussion.* Precipitation as lead molybdate is usually made by the slow addition of a solution of lead acetate to a hot, dilute acetic acid–ammonium acetate solution containing the molybdenum. The method is applicable in the presence of copper, mercury, cobalt, nickel, manganese, zinc, and magnesium. Alkali salts are not objectionable except sulphates, which must be absent when the alkaline earths are present. A great excess of the precipitant should be avoided if chlorides are present. Free mineral acids prevent complete precipitation, and elements such as iron, chromium, vanadium, tungsten, silicon, phosphorus, arsenic, antimony, tin, and titanium interfere.

Procedure. Weigh out accurately about 0.4 g ferro-molybdenum (1), dissolve it in 10 cm^3 concentrated hydrochloric acid and 2 cm^3 concentrated nitric acid, evaporate to 2–3 cm^3, dilute to 50 cm^3, and transfer to a separatory funnel. Dissolve 5 g sodium hydroxide in 200 cm^3 water in the original beaker, heat to boiling, and run in the solution from the separatory funnel dropwise and with constant stirring. Rinse out the funnel twice with boiling water, and add the washings to the main solution. Filter off the precipitated iron(III) hydroxide and wash with hot water. Dissolve the precipitate in the minimum volume of dilute hydrochloric acid, and re-precipitate by slowly pouring into a solution of about 4 g sodium hydroxide in 100 cm^3 of water. Filter off the precipitate, wash with hot water, and add the filtrate and washings to the main solution. Acidify with acetic acid, add 50 cm^3 of a 50 per cent solution of ammonium acetate, and make up to 500 cm^3 in a graduated flask. Remove 250 cm^3 of the solution (2), heat to boiling, and maintain near the boiling point with a small flame; add from a burette a solution of lead acetate (containing 4 g of the salt and 1 cm^3 of glacial acetic acid per 100 cm^3) dropwise and with constant stirring. When a slight excess of the precipitant has been added, the milky solution clears appreciably. When this occurs, boil for 2–3 minutes whilst the solution is stirred, allow to settle, and add a few drops of the reagent to see if precipitation is complete. A large excess of precipitant should be avoided. Digest on the steam bath for 15–30 minutes. Decant the clear solution through a weighed Gooch or porcelain filtering crucible, wash by decantation three or four times with 75-cm^3 portions of hot 2 per cent ammonium nitrate solution, transfer the precipitate to the filter, and wash until the soluble salts have been removed. Dry and ignite the precipitate at a *dull* red heat (*ca.* 600 °C) as described in Section **XI, 41A**. Weigh as PbMoO$_4$.

Notes.   1. The Bureau of Analysed Samples 'Ferro-Molybdenum, No. 231' (a British Chemical Standard) is suitable.

2. If molybdenum is being determined in a simple salt, e.g., in A.R. molybdic acid or molybdic anhydride, commence at this point. The solution should contain about 0.1 g Mo in 200 cm^3 and may be prepared as follows. Dissolve 0.15 g, accurately weighed, A.R. molybdic acid or anhydride, in 50 cm^3 dilute ammonia solution, acidify with acetic acid, add 25 cm^3 of a 50 per cent solution of ammonium acetate, and dilute to 200 cm^3.

B. Determination of molybdenum with oxine. *Discussion.* Molybdates yield sparingly soluble orange-yellow molybdyl 'oxinate' with oxine solution; the pH of the solution should be between the limits 3.3–7.6. The complex differs from other 'oxinates' in being insoluble in organic solvents and in many concentrated inorganic acids. The freshly precipitated compound dissolves only in concentrated sulphuric acid and in hot solutions of caustic alkalis. This determination is of particular interest, as it allows a complete separation of molybdenum and rhenium.

Procedure. Neutralise the solution of alkali molybdate, containing up to 0.1 g of Mo, to methyl red, and then acidify with a few drops of *M*-sulphuric acid. Add 5 cm^3 2*M*-ammonium acetate, dilute to 50–100 cm^3, and heat to boiling. Precipitate the molybdenum by the addition of 3 per cent solution of oxine in dilute acetic acid (**Note 1**), until the supernatant liquid becomes perceptibly yellow. Boil gently and stir for 3 minutes, filter through a filtering crucible (sintered glass or porcelain), wash with hot water until free from the reagent, and dry to constant weight at 130–140 °C. Weigh as $MoO_2(C_9H_6ON)_2$.

Notes. 1. The oxine reagent may be **prepared** by dissolving 4 g A.R. oxine in 8.5 cm³ of warm A.R. glacial acetic acid, pouring into 80 cm³ water, and diluting to 100 cm³.

XI, 47. NICKEL. The determination of nickel as nickel dimethylglyoximate has already been described in Section **XI, 24**; the following methods are also useful:

A. Determination of nickel in the presence of copper with salicylaldehyde-oxime. *Discussion.* The complex is precipitated in neutral or very faintly acid solutions (best at pH = 7) in contrast to that of copper, which is formed in the presence of acetic acid. The experimental details are similar to those for copper, except that the solution must be neutral or very faintly acid. Iron(III) interferes, and should therefore be absent.

Procedure. Treat the solution, free from mineral acid and containing both nickel (not more than 0.03 g) and copper (about 0.03 g), with 1 g sodium acetate and 10 cm³ glacial acetic acid per 100 cm³ of solution.* Add excess of salicylaldehydeoxime reagent over the quantity required to precipitate both metals, and stir the solution vigorously during the addition. Filter off the precipitated copper complex on a weighed filtering crucible (Gooch, sintered glass, or porcelain), wash well with cold water, and dry at 100–105 °C to constant weight. Weigh as $Cu(C_7H_6O_2N)_2$. Add dilute ammonia solution to the filtrate and washings (diluted to 300–350 cm³) until the solution remains very faintly acidic. Stir thoroughly to coagulate the precipitate of nickel salicylaldoximate. Filter through a weighed sintered glass or porcelain filtering crucible, wash with cold water until the washings give no coloration with iron(III) chloride solution, and dry at 100 °C to constant weight. Weigh as $Ni(C_7H_6O_2N)_2$.

B. Determination of nickel by the pyridine method. *Discussion.* This is a rapid method with similar advantages and limitations to those described for zinc (Section **XI, 62**). A determination may be completed in about 30 minutes.

Procedure. The solution (100 cm³) should contain about 0.1 g nickel (1) and be neutral in reaction. Stir in 0.5–1.0 g A.R. ammonium thiocyanate, heat to boiling, add 1–2 cm³ pure pyridine, and immediately remove the flame. Stir for 2–5 seconds until the precipitate commences to separate in sky-blue prisms. (The precipitate separates immediately or after standing a short time, according to the quantity of nickel present.) When cold, filter through a weighed sintered glass or porcelain filtering crucible, and use *Solution 1* to assist in the transfer of the precipitate to the crucible. Wash four to five times with *Solution 2*, then twice with 1-cm³ portions of *Solution 3*, and finally five or six times with 1-cm³ portions of *Solution 4*. Dry in a vacuum desiccator at room temperature for 10 minutes and weigh. Repeat the drying until the weight is constant. Weigh as $[Ni(C_5H_5N)_4](SCN)_2$ (2).

Solution 1. 100 cm³ water containing 0.4 g NH_4SCN and 0.6 cm³ pyridine.

Solution 2. 61.5 cm³ water, 37.0 cm³ of 95 per cent ethanol, 0.1 g NH_4SCN, and 1.5 cm³ pyridine.

Solution 3. 10 cm³ absolute alcohol and 0.5 cm³ pyridine.

Solution 4. 20 cm³ diethyl ether (sodium dried) and 2 drops pyridine.

---

* The best pH range for the precipitation of copper in the presence of nickel is 2.6–3.1.

Notes. 1. For practice in this estimation, employ 0.3 g, accurately weighed, A.R. nickel sulphate or pure ammonium nickel sulphate.

2. For further experimental details, see under Zinc.

XI, 48. PALLADIUM. Palladium may be determined in one of the following forms:

A. Determination of palladium with dimethylglyoxime. *Discussion.* This is one of the best methods for the determination of the element. Gold must be absent, for it precipitates as the metal even from cold solutions. The platinum metals do not, in general, interfere. Moderate amounts of platinum cause little contamination of the precipitate, but with large amounts a second precipitation is desirable. The precipitate is decomposed by digestion on the water bath with a little aqua regia, and diluted with an equal volume of water; the resulting solution is largely diluted with water, and the palladium re-precipitated with dimethylglyoxime.

An objection to the precipitation of palladium with dimethylglyoxime is the voluminous character of the precipitate. Hence if much palladium is present, an aliquot part of the solution should be used.

Procedure. The solution should contain not more than 0.1 g Pd in 250 cm^3, be 0.25M with respect to hydrochloric or nitric acid, and be free from nickel and gold. Add, at room temperature, a 1 per cent solution of dimethylglyoxime in 95 per cent ethanol. Use 2–5 cm^3 of the reagent for every 10 mg of palladium. Allow the solution to stand for 1 hour, and then filter through a weighed filtering crucible (Gooch, sintered glass, or porcelain). Test the filtrate with a little of the reagent to make sure that precipitation is complete. Wash the orange-yellow precipitate of palladium dimethylglyoximate thoroughly, first with cold water and then with hot water. Dry at 110 °C to constant weight. Weigh as $Pd(C_4H_7O_2N_2)_2$.

B. Determination of palladium with cyclohexane-1,2-dionedioxime. *Discussion.* Cyclohexane-1,2-dionedioxime (nioxime) yields a highly insoluble yellow compound, $Pd(C_6H_9O_2N_2)_2$, with palladium salts at pH values between 1 and 5 (see Section **XI, 11A**); it can be filtered from the hot solution after a brief digestion period. The reagent, unlike dimethylglyoxime, is soluble in water, and hence the palladium precipitate is unlikely to be contaminated with excess of reagent. The precipitate is rather bulky, so that determinations are conducted with quantities not exceeding 20–30 mg of palladium. Common anions do not interfere, nor do beryllium, aluminium, lanthanum, uranium(VI), and the alkaline-earth ions. Amounts of platinum (up to that of the palladium) do not interfere; gold(I) at 60 °C is partially reduced to metallic gold.

Procedure. The solution (volume about 200 cm^3) may contain 5–30 mg palladium: the pH may vary from 1 to 5. Heat the solution to 60 °C, add slowly from a graduated pipette with stirring 0.50 cm^3 of a 0.8 per cent aqueous solution of nioxime for each milligram of Pd present. Digest the solution with occasional stirring for 30 minutes at 60 °C, filter through a sintered glass or porcelain filtering crucible, and wash well with hot water. Dry at 110 °C to constant weight, and weigh as $Pd(C_6H_9O_2N_2)_2$.

XI, 49. PLATINUM. Platinum is determined preferentially as the metal.

Determination as metallic platinum. *Discussion.* The platinum solution is treated with formic acid, best at pH 6, and the precipitated platinum weighed. A

2 per cent solution of hypophosphorous acid may also be used as the reducing agent.

Procedure.   In this determination any excess of nitric and/or hydrochloric acid present must be removed. Evaporate the solution of platinum, containing no other platinum metals (ruthenium, rhodium, palladium, osmium, and iridium) or gold, to a syrup on the steam bath so as to remove as much hydrochloric acid as possible. If nitric acid was present, dissolve the residue in 5 cm³ of water, heat on the water bath for a few minutes, add 5 cm³ of concentrated hydrochloric acid, and again evaporate to a syrupy consistency. Dissolve the residue in water, and dilute so that the solution does not contain more than 0.5 g of Pt in 100 cm³. For each 100 cm³ of solution, add 3 g of anhydrous sodium acetate and 1 cm³ of formic acid. Heat on the boiling water bath for several hours. Filter through a quantitative filter paper. Add a little more sodium acetate and formic acid to the filtrate and digest in order to ensure complete precipitation. Wash the precipitate with water until free from chloride, dry and ignite the filter paper in contact with the precipitate to constant weight. Weigh as metallic Pt.

XI, 50.   POTASSIUM.   Potassium may be determined in one of the following forms:

A. Determination of potassium as dipotassium sodium hexanitritoco-baltate(III) (cobaltinitrite).   *Discussion.*   By precipitation of potassium solutions with sodium hexanitritocobaltate(III) reagent under the experimental conditions given below, a quantitative yield of dipotassium sodium hexanitritocobaltate(III) is obtained. This method is applicable in the presence of sulphate.

Formerly precipitation was made in an acetic acid solution with a reagent prepared by mixing a solution of cobalt nitrate or acetate in dilute acetic acid and one of sodium nitrite in water. There is some evidence that the composition of the precipitate may, upon occasion, vary slightly from the formula given below. If precipitation is made from a nitric acid solution by a solution of sodium hexanitritocobaltate(III), $Na_3[Co(NO_2)_6]$, the heavy, crystalline precipitate invariably has the the composition $K_2Na[Co(NO_2)_6]H_2O$ for amounts of potassium from 2 to 15 mg in 10 cm³ of solution. Acidification with dilute nitric acid tends to prevent decomposition of the nitrite. The precipitate may be weighed as such after drying at 100–110 °C for 2 hours, or it may be converted into potassium perchlorate.

Procedure.   Weigh out accurately 0.03–0.04 g A.R. potassium sulphate and dissolve it in 10 cm³ water. Add 1 cm³ of M-nitric acid and a freshly prepared solution of 1 g A.R. sodium cobaltinitrite in 5 cm³ water, mix, and allow to stand for 2 hours. Filter through a weighed sintered glass or porcelain filtering crucible and transfer the precipitate completely with the aid of 0.01 M-nitric acid. Wash ten times with 2-cm³ portions of 0.01 M-nitric acid and five times with 2-cm³ portions of 95 per cent ethanol. Suck the precipitate as dry as possible on the pump, dry for 1 hour at 110 °C, cool in a desiccator, and weigh as $K_2Na[Co(NO_2)_6],H_2O$.

B. Determination of potassium as potassium tetraphenylboron.*
Discussion.   A   solution   of   sodium   tetraphenylboron $Na^+[B(C_6H_5)_4]$

---

* This method is only useful where a high degree of accuracy is not required.

is probably the best precipitant for potassium, but is expensive. Precipitation may be effected at a temperature below 20 °C in dilute mineral acid solution (pH 2), in which interference from most foreign ions is negligible. The precipitate is granular and settles readily; it is washed with a saturated aqueous solution of the precipitate (prepared independently), and the potassium tetraphenylboron is dried at 120 °C and weighed. The compound decomposes at temperatures above 265 °C. The precipitate is of constant composition $K[B(C_6H_5)_4]$, and is sparingly soluble in water ($\equiv 5.1$ mg dm^{-3} of potassium at 20 °C). Very few elements interfere with the determination: these include the ions of silver, mercury(II), thallium(I), rubidium, and cesium; ammonium ion, which forms a slightly soluble salt, can be removed by ignition prior to the addition of the reagent.

Procedure. For practice in this determination, weigh out accurately about 0.10 g A.R. potassium chloride and dissolve it in 50 cm^3 distilled water. Add 10 cm^3 of 0.1M-hydrochloric acid. Then introduce from a burette 40 cm^3 of the sodium tetraphenylboron reagent (1) slowly (5–10 minutes) and stir continuously. The temperature throughout must be below 20 °C. Allow the precipitate to settle during 1 hour. Collect the precipitate on a sintered glass filtering crucible (porosity No. 4), wash the precipitate with a small volume (5–10 cm^3 in small portions) of saturated potassium tetraphenylboron solution (2), and finally with 1–2 cm^3 ice-cold distilled water (3). Dry at 120 °C and cool the covered crucible in a desiccator. Weigh as $K[B(C_6H_5)_4]$.

Notes. 1. Prepare the sodium tetraphenylboron reagent by dissolving 3.0 g of the solid reagent in 500 cm^3 distilled water in a glass-stoppered bottle. Add about 1 g moist aluminium hydroxide gel, break up the gel if necessary, and shake the suspension for 15 minutes. Filter through a Whatman No. 40 filter paper. Refilter the first part of the filtrate, if necessary, to ensure a clear filtrate.

2. Precipitate about 0.1 g potassium (present as potassium chloride in 50 cm^3 water) with 40 cm^3 of the sodium tetraphenylboron solution added slowly and with constant stirring. Allow to stand for 30 minutes, filter through a sintered glass filtering crucible, wash with distilled water, and dry for 1 hour at 120 °C. Shake 20–25 mg of the dry precipitate with 200 cm^3 distilled water in a stoppered bottle at 5-minute intervals during 1 hour. Filter through a Whatman No. 40 filter paper, and use the filtrate as the wash liquid.

3. The reagent is expensive; it is therefore desirable to recover it from potassium TPB precipitates remaining from gravimetric determinations or obtained by adding potassium chloride to filtrates, wash liquids, etc. The potassium tetraphenylboron is dissolved in acetone and the acetone solution is passed through a strongly acidic cation exchange resin (sodium form); the effluent contains sodium TPB, and is evaporated to dryness on a water bath. The resulting sodium tetraphenylboron is recrystallised from acetone.

C. Determination of potassium as chloroplatinate and subsequent weighing as metallic platinum. *Discussion.* This method is applicable only to those potassium compounds which can be completely converted into potassium chloride by evaporation with hydrochloric acid (as by the technique of Section **XI, 53A**), because it is only from a solution containing chloride that potassium can be completely precipitated as $K_2[PtCl_6]$ by chloroplatinic acid solution. Ammonium salts and all metals other than sodium and potassium must be removed, as must also sulphate, phosphate, and similar radicals. Sodium chloroplatinate is soluble in 80 per cent ethanol, hence this method provides a means of separation of sodium from potassium. As the composition of the precipitate may vary slightly from that expressed by the formula $K_2[PtCl_6]$, it is preferable to convert the potassium chloroplatinate to the metal by reduction with magnesium ribbon in acid solution, and weigh the platinum. This modified procedure admits of determining potassium in the presence of sulphates, phosphates, chlorides, nitrates, borates, sodium, the alkaline-earth metals, iron, aluminium, and magnesium. Rubidium, cesium, and ammonium salts must be absent.

Procedure. Weigh out accurately about 0.25 g of the mixed sodium and potassium chlorides (1) into a small porcelain dish and dissolve it in 5–10 cm³ of water. Add 5 cm³ of hydrochloric acid. Treat with a slight excess of chloroplatinic acid reagent (2) over that required by the potassium; the presence of sodium and other salts causes no interference. Crush the precipitate with a glass rod flattened at one end and collect it in a sintered glass or porcelain filtering crucible. Wash with ethanol (80–90 per cent by volume) (3), dissolve the precipitate by pouring hot water over it, and transfer the filtrate and washings quantitatively to a small beaker. Add 2 cm³ concentrated hydrochloric acid, followed by about 0.5 g magnesium ribbon (previously washed in water) for every 0.20 g of potassium present. Stir the solution and hold the ribbon at the bottom of the beaker by means of a glass rod with a flattened end. When the magnesium has nearly disappeared, add a few cm³ dilute hydrochloric acid, and allow the platinum to settle. If the reduction is complete, the liquid is clear and colourless. To make sure add a little more magnesium, and note whether the solution darkens. Add dilute hydrochloric acid, boil to dissolve any basic salts which may be present, collect the platinum on a small filter paper, wash with water until free from chlorides, and ignite in a weighed porcelain or, preferably, platinum crucible to constant weight. Weigh the platinum, and calculate the potassium equivalent by the proportion $Pt \equiv 2K$.

Notes. 1. For practice in this procedure, employ either A.R. potassium chloride or an artificial mixture of, say, equal weights of A.R. sodium and potassium chlorides.

2. The **chloroplatinic acid reagent** is **prepared** by dissolving 1 g chloroplatinic acid in 10 cm³ water.

3. All washings must be kept, and the platinum contained in them subsequently recovered.

XI, 51. SELENIUM AND TELLURIUM. *Discussion.*

The gravimetric determination depends upon the separation and weighing as elementary selenium or tellurium (or as tellurium dioxide). Alkali selenites and selenious acid are reduced in hydrochloric acid solution with sulphur dioxide, hydroxylammonium chloride, hydrazinium sulphate or hydrazine hydrate. Alkali selenates and selenic acid are not reduced by sulphur dioxide alone, but are readily reduced by a saturated solution of sulphur dioxide in concentrated hydrochloric acid. In working with selenium it must be remembered that appreciable amounts of the element may be lost on warming strong hydrochloric acid solutions of its compounds: if dilute acid solutions (concentration $< 6M$) are heated at temperatures below 100 °C the loss is negligible.

With tellurium, precipitation of the element with sulphur dioxide is slow in dilute hydrochloric acid solution and does not take place at all in the presence of excess of acid; moreover, the precipitated element is so finely divided that it oxidises readily in the subsequent washing process. Satisfactory results are obtained by the use of a mixture of sulphur dioxide and hydrazinium chloride as the reducing agent, and the method is applicable to both tellurites and tellurates. Another method utilises excess of sodium hypophosphite in the presence of dilute sulphuric acid as the reducing agent.

A process for the gravimetric determination of mixtures of selenium and tellurium is also described. Selenium and tellurium* occur in practice either as the impure elements or as selenides or tellurides. They may be brought into solution by mixing intimately with 2 parts of sodium carbonate and 1 part of potassium nitrate in a nickel crucible, covering with a layer of the mixture, and then heating gradually to fusion. The cold melt is extracted with water, and filtered. The elements are then determined in the filtrate.

A. Determination of selenium. *Procedure.* The selenium must be present in the quadrivalent state, and the selenium content of the solution must not exceed 0.25 g per 150 cm^3. Take an amount of the oxide, selenite, etc., that will contain not more than 0.25 g selenium, and dissolve it in 100 cm^3 concentrated hydrochloric acid. Add, with constant stirring and at not over 25 °C, 50 cm^3 cold concentrated hydrochloric acid that has been saturated with sulphur dioxide at room temperature, allow the solution to stand until the red selenium settles out, filter through a weighed filtering crucible (Gooch, sintered glass, or porcelain), wash well successively with cold concentrated hydrochloric acid, cold water until free from chloride, ethanol, and diethyl ether. Dry the precipitate for 3–4 hours at 30–40 °C to remove ether, and then to constant weight at 100–110 °C. Weigh as Se.

B. Determination of tellurium. *Procedure.* The solution should contain not more than 0.2 g Te in 50 cm^3 of 3M-hydrochloric acid (*ca.* 25 per cent by volume of hydrochloric acid). Heat to boiling, add 15 cm^3 of a freshly prepared, saturated solution of sulphur dioxide, then 10 cm^3 of a 15 per cent aqueous solution of hydrazinium chloride, and finally 25 cm^3 more of the saturated solution of sulphur dioxide. Boil until the precipitate settles in an easily filterable form; this should require not more than 5 minutes. Allow to settle, filter through a weighed filtering crucible (Gooch, sintered glass, or porcelain), and immediately wash with hot water until free from chloride. Finally wash with ethanol (to remove all water and prevent oxidation), and dry to constant weight at 105 °C. Weigh as Te.

In the alternative method of reduction, which is particularly valuable for the determination of small amounts of tellurium, the procedure is as follows. Treat the solution containing, say, up to about 0.01 g Te in 90 cm^3 with 10 cm^3 of 1 : 3-sulphuric acid, then add 10 g sodium hypophosphite, and heat on a steam bath for 3 hours. Collect and weigh the precipitated tellurium as above.

C. Determination of mixtures of selenium and tellurium.

Procedure. Dissolve the mixed oxides (not exceeding 0.25 g of each) in 100 cm^3 of concentrated hydrochloric acid, and add with constant stirring 50 cm^3 cool concentrated hydrochloric acid which has been saturated with sulphur dioxide at the ordinary temperature. Allow the solution to stand until the *red* selenium has settled, filter through a weighed filtering crucible (Gooch,

---

* Tellurium and its compounds are toxic and cause irritation to eyes and skin: contact and inhalation should be avoided.

sintered glass, or porcelain), and complete the determination as described in **A**. Preserve the filtrate, hydrochloric acid, and water washings. Concentrate the latter on a water bath below 100 °C (above 100 °C tellurium is lost as chloride) to 50 cm³, and determine the tellurium as described under **B**.

XI, 52. SILVER. Determination of silver as chloride. *Discussion.* The theory of the process has been given under Chloride (Section **XI, 17**). Lead, copper(I), palladium(II), mercury(I), and thallium(I) ions interfere, as do cyanides and thiosulphates. If a mercury(I) (or copper(I) or thallium(I)) salt is present, it must be oxidised with concentrated nitric acid before the precipitation of silver; this process also destroys cyanides and thiosulphates. If lead is present, the solution must be diluted so that it contains not more than 0.25 g of the substance in 200 cm³, and the hydrochloric acid must be added very slowly. Compounds of bismuth and antimony that hydrolyse in the dilute acid medium used for the complete precipitation of silver must be absent. For possible errors in the weight of silver chloride due to the action of light, see Section **XI, 17**.

Procedure. The solution (200 cm³) should contain about 0.1 g of silver (1) and about 1 per cent by volume of nitric acid. Heat to about 70 °C, and add approximately 0.2M pure hydrochloric acid slowly and with constant stirring until no further precipitation occurs; avoid a large excess of the acid. Do not expose the precipitate to too much bright light. Warm until the precipitate settles, allow to cool to about 25 °C, and test the supernatant liquid with a few drops of the acid to be sure that precipitation is complete. Allow the precipitate to settle in a dark place for several hours or, preferably, overnight. Pour the supernatant liquid through a weighed Gooch, sintered glass or porcelain filtering crucible, wash the precipitate by decantation with 0.1M-nitric acid, transfer the precipitate to the crucible, and wash again with 0.01M-nitric acid until free from chloride. Dry the precipitate first at 100 °C and then at 130–150 °C, allow to cool in a desiccator and weigh. Repeat the heating, etc., until constant weight is obtained (2). Weigh as AgCl.

Notes. 1. For example, from 0.2 g of A.R. silver nitrate.
2. See last footnote in Section **XI, 17**.

XI, 53. SODIUM. Sodium may be determined in one of the following forms:

A. Determination of sodium as sulphate. *Discussion.* Any sodium compound of a volatile acid may be converted into sodium sulphate by repeated evaporation with sulphuric acid. Some sodium hydrogensulphate is formed in the process, and this is converted (via the pyrosulphate, $Na_2S_2O_7$) into the normal salt with some difficulty. The latter change is facilitated by the addition of a little powdered ammonium carbonate; this is because ammonium sulphate, which is completely volatilised on heating, is formed:

$$Na_2S_2O_7 + (NH_4)_2CO_3 = Na_2SO_4 + (NH_4)_2SO_4 + CO_2$$

Procedure. This determination is carried out in a silica or platinum crucible. Evaporate the solution (1) to dryness in a weighed crucible on the water bath. Transfer to a triangle (or to a hot plate, the temperature of which can be controlled), add a few cm³ of concentrated sulphuric acid dropwise, and evaporate gently to dryness *in the fume cupboard* until fuming ceases. Repeat this operation twice. Allow to cool, add a few small pieces (about the size of a pea) of solid ammonium carbonate to decompose any pyrosulphate present, and heat to dull redness (or at 400–700 °C) (2) for 15 minutes. Allow to cool in a desiccator and weigh the covered crucible immediately it has acquired the laboratory temperature. Repeat the treatment with ammonium carbonate until constant weight is attained. Weigh as Na_2SO_4.

Notes. 1. A suitable solution for practice may be prepared by weighing out accurately about 0.3 g of sodium chloride, and dissolving it in a little water.

2. The temperature should not exceed 850 °C; any temperature between 400 °C and 700 °C is satisfactory. Anhydrous sodium sulphate is slightly hygroscopic.

B. Determination of sodium as sodium zinc uranyl acetate. *

Discussion. Treatment of a concentrated solution of a sodium salt with a large excess of zinc uranyl acetate reagent results in the precipitation of sodium zinc uranyl acetate. This substance is moderately soluble in water (58.8 g per 1000 g of water at 20 °C) so that a special washing technique must be used. The solubility in a solution containing excess of the reagent is less. About 10 volumes of the reagent is added for each volume of the sample solution, which should not contain more than 8 mg of sodium per cm^3; precipitation of the triple acetate is usually complete in 1 hour. One mg of sodium yields 66.88 mg of the triple salt; the latter is relatively bulky, so that the amount of sodium that can be handled in a single determination is limited.

Lithium interferes, since it forms a sparingly soluble triple acetate. Potassium has no effect provided not more than 50 mg cm^{-3} are present. Sulphate must be absent when potassium is present, for potassium sulphate is sparingly soluble in the reagent. Moderate amounts of ammonium salts, calcium, barium, and magnesium may be tolerated; for larger amounts, a double precipitation is necessary. Phosphates, arsenates, molybdates, oxalates, tartrates, sulphates (in the presence of potassium), and strontium interfere.

Procedure. The neutral or feebly acid sample solution, free from the interfering substances mentioned above, should contain not more than 8 mg of sodium per cm^3, preferably as chloride. Treat the sample solution (say, 1.5 cm^3) (1) with 15 cm^3 of zinc uranyl acetate reagent (2), and stir vigorously, preferably mechanically for at least 30 minutes. Allow to stand for 1 hour, and filter through a weighed sintered glass or porcelain filtering crucible (porosity No. 4). Wash the precipitate four times with 2-cm^3 portions of the precipitating reagent (allow the wash liquid to drain completely before adding the next portion), then ten times with 95 per cent ethanol saturated with sodium zinc uranyl acetate at room temperature (2-cm^3 portions), and finally with a little dry diethyl ether or acetone. Dry for 30 minutes only at 55–60 °C (3). Weigh as $NaZn(UO_2)_3(C_2H_3O_2)_9,6H_2O$.

Notes. 1. A suitable solution for practice may be prepared by evaporating 20.0 cm^3 of 0.02M-sodium chloride, prepared from the A.R. salt, to 1.5 cm^3 on a water bath.

2. The reagent is prepared by mixing equal volumes of solutions A and B and filtering after standing overnight.

Solution A: dissolve 20 g crystallised uranyl acetate $UO_2(C_2H_3O_2)_2 2H_2O$ in 4 cm^3 glacial acetic acid and 100 cm^3 water (warming may be necessary).

---

* This method is only of use where a high degree of accuracy is not required.

Solution B: dissolve 60 g crystallised zinc acetate $Zn(C_2H_3O_2)_2,3H_2O$ in 3 cm³ glacial acetic acid and 100 cm³ of water.

3. Alternatively, draw air through the crucible for 5 minutes to volatilise the solvent, wipe off any condensed moisture on the outside with a clean linen cloth, allow to stand in the air or in a desiccator for 10–15 minutes, and weigh.

XI, 54. STRONTIUM. Strontium may be determined in one of the following forms:

A. Determination of strontium as sulphate. *Discussion.* In this determination (probably the most accurate for strontium) calcium, barium, and lead must be absent, and the solution (preferably of the chloride) should be nearly neutral. If considerable quantities of acid are present, this must be removed by evaporation. Strontium sulphate dissolves appreciably in an acid medium because of the reaction:

$$SrSO_4 + H^+ \rightleftharpoons HSO_4^- + Sr^{2+}$$

Strontium sulphate has a solubility of about 0.14 g dm⁻³ at the laboratory temperature; the solubility is decreased by the addition of a *slight* excess of sulphuric acid, and of ethanol (50 per cent).

Procedure. The solution (100 cm³) should contain about 0.2 g of strontium and be very slightly acid with hydrochloric acid (1). Add slowly a ten-fold excess of dilute sulphuric acid, followed by a volume of ethanol equal to that of the solution. Stir well, and allow to stand for at least 12 hours. Transfer the precipitate to a weighed Vitreosil or porcelain filtering crucible, wash with 75 per cent ethanol to which a few drops of sulphuric acid have been added, and finally with pure ethanol until the washings are free from sulphate. Dry and ignite (crucible ignition-dish or in a large nickel crucible) at dull redness (or in an electric muffle furnace at 500–600 °C) to constant weight. Alternatively, a filter paper may be used: here the paper should be burnt apart from the precipitate (to prevent possible reduction of the latter to the sulphide), and the residue then ignited together with the main precipitate in a weighed porcelain, silica, or platinum crucible. Weigh as $SrSO_4$.

Note. 1. A solution for practice in this determination may be prepared by dissolving 0.3–0.4 g, accurately weighed, of pure strontium carbonate in a little dilute hydrochloric acid (see Section **XI, 20**), and diluting to 100 cm³.

B. Determination of strontium as strontium hydrogen phosphate, SrHPO₄. *Discussion.* Strontium (30–200 mg) may be precipitated as $SrHPO_4$ using potassium dihydrogen phosphate; precipitation commences at pH 4 and is quantitative at pH 5.7–6. The flocculent precipitate soon becomes crystalline. It may be weighed as $SrHPO_4$ after drying at 120 °C; alternatively, it may be ignited and weighed as $Sr_2P_2O_7$. Ions which yield insoluble phosphates should be absent; the sodium, potassium, or ammonium ion concentration should not exceed $0.2M$.

Procedure. Treat the sample solution (60 cm³; say, prepared by weighing out accurately about 0.15 g pure $SrCl_2,6H_2O$ and dissolving in water) with 10–20 cm³ $0.5M$-potassium dihydrogen phosphate and heat to the boiling point. Add $1M$-potassium hydroxide from a dropper pipette until an appreciable precipitate is formed. The final pH should be about 6; this can be detected by adding the base until bromocresol purple indicator in the solution just turns purple, or a pH meter may be used. Boil until the initial flocculent precipitate becomes crystalline (30–60 minutes). Allow to stand for 1 hour. Collect the precipitate in a sintered glass (porosity No. 4) or porcelain filtering crucible; remove any precipitate adhering to the walls of the beaker with a rubber-tipped stirring rod, and wash the precipitate with a little cold water. Dry at 120 °C. Weigh as $SrHPO_4$.

XI, 55. THALLIUM. Thallium may be determined in either of the following forms:

A. Determination of thallium as chromate. *Discussion.* The thallium* must be present in the thallium(I) state. If present as a thallium(III) salt, reduction must be effected (before precipitation) with sulphur dioxide; the excess of sulphur dioxide is boiled off.

Procedure. The solution (100 cm³) should contain about 0.1 g of Tl, no excessive amounts of ammonium salts, and no substances that form precipitates with ammonia solution, or reduce potassium chromate, or react with potassium or thallium(I) chromate in ammoniacal solution. Neutralise the thallium solution with dilute ammonia solution (2:1), and add 3 cm³ in excess. Heat to about 80 °C, and add 2 g of potassium chromate in the form of a 10 per cent solution slowly and with constant stirring. Allow to stand at the laboratory temperature for at least 12 hours. Filter through a weighed filtering crucible (Gooch, sintered glass, or porcelain), wash with 1 per cent potassium chromate solution, then sparingly with 50 per cent ethanol, and dry at 120 °C to constant weight. Weigh as Tl_2CrO_4.

B. Determination of thallium with tetraphenylarsonium chloride.

Discussion. Thallium(III) in excess of hydrochloric acid reacts with tetra-phenylarsonium chloride $[(C_6H_5)_4As]Cl$ to give an insoluble tetraphenylar-sonium chlorothallate:

$$[C_6H_5)_4As]^+ + [TlCl_4]^- = [(C_6H_5)_4As][TlCl_4]$$

The precipitate can be dried at 110 °C. Thallium(I) is readily oxidised to thallium(III) by hydrogen peroxide in alkaline solution. The principal interferences are cations which form insoluble chlorides and also various anions (e.g., fluoride, iodide, bromide, thiocyanate, nitrate, perchlorate, periodate, permanganate, per-rhenate, molybdate, chromate, and tungstate). The precipitate must be washed with dilute hydrochloric acid, otherwise hydrolysis occurs and low results are obtained.

Procedure. Oxidise thallium(I) in the sample solution (75 cm³ containing up to 90 mg of thallium) (1) to thallium(III) by adding 2.0 cm³ '100-volume' hydrogen peroxide in the presence of sodium hydroxide solution. Acidify with hydrochloric acid and add a few cm³ concentrated hydrochloric acid in excess: a white precipitate, probably a Tl(I)–Tl(III) complex, forms, but this will dissolve upon the addition of a further 1 cm³ of 30 per cent hydrogen peroxide to the acid solution. Dilute the solution to render it 0.5–2.0M in hydrochloric acid, and add excess of the reagent solution (2). Heat to boiling to coagulate the white precipitate and keep overnight. Collect the precipitate on a weighed sintered glass filtering crucible (porosity No. 4), wash with 20–40 cm³ 1M-hydrochloric acid, and dry at 110 °C. Weigh as $[(C_6H_5)_4As][TlCl_4]$.

---

* Thallium and its compounds are toxic and cause irritation to eyes and the skin: contact and inhalation should be avoided.

Notes. 1. A solution of 0.05–0.10 g, accurately weighed, A.R. thallium(I) sulphate in 50–75 cm³ water may be used for practice in this determination.

2. The reagent solution is prepared by dissolving 6.7 g $[(C_6H_5)_4As]Cl$ in 100 cm³ water. Ten cm³ suffice for the precipitation of 90 mg of thallium. Tetraphenylarsonium chloride is available from the G. Frederick Smith Chemical Co., Columbus, Ohio, and from Fluka A.G., Buchs, Switzerland.

XI, 56. THORIUM. Thorium may be determined in either of the following forms:

A. Determination of thorium as sebacate and subsequent ignition to the oxide, ThO₂. *Discussion.* This procedure permits of the separation by a single precipitation of thorium from relatively large amounts of the lanthanoids (Ce, La, Pr, Nd, Sm, Gd) and also from quadrivalent cerium.

Procedure. The solution (100 cm³) should be neutral or faintly acid, and contain not more than 0.1 g Th. Heat the solution to boiling and add slowly and with constant stirring a hot almost saturated solution of pure sebacic acid in slight excess. The precipitate is voluminous, but granular, and therefore easily manipulated. Filter off immediately, wash thoroughly with hot water, dry, and ignite (use either a Meker or Fisher burner or an electric muffle furnace at 700–800 °C) in a weighed platinum, porcelain, or silica crucible to constant weight. Weigh as ThO₂.

B. Determination of thorium as iodate, and subsequent ignition to the oxide, ThO₂, via the oxalate. *Discussion.* Thorium iodate is precipitated quantitatively by potassium iodate from nitric acid solution: a separation from the lanthanoids, trivalent cerium, iron, aluminium, and phosphoric acid is thus achieved. Titanium, zirconium, and cerium(IV) accompany thorium, and must therefore be absent. The thorium iodate is dissolved in hydrochloric acid, precipitated as the oxalate, and ignited to and weighed as ThO₂.

Procedure. The solution (100 cm³) should be chloride-free (1) and contain not more than 0.2 g Th. Add 50 cm³ concentrated nitric acid and cool in ice water. Add a cold solution of 15 g A.R. potassium iodate in 30 cm³ water and 50 cm³ concentrated nitric acid; stir occasionally during 30 minutes. Allow to settle, break up any lumps of precipitate with a glass rod flattened at one end, filter through a hardened quantitative filter paper, wash with 250 cm³ of a cold solution containing 8 g potassium iodate and 200 cm³ dilute nitric acid (1 : 1) per dm³, and allow to drain. Transfer the precipitate back into the original beaker with the aid of 100 cm³ of the wash solution, stir thoroughly and filter through the same filter paper. Allow to drain and again transfer the precipitate back into the beaker, but this time with a little hot water. Heat nearly to boiling, and dissolve the precipitate by adding 30 cm³ of concentrated nitric acid slowly and with stirring. Dilute to 60–100 cm³, and re-precipitate the thorium as iodate by adding a solution of 4 g A.R. potassium iodate dissolved in a little hot water acidified with nitric acid: allow to cool. Filter, wash by decantation as before with 100 cm³ of the wash solution, and transfer the precipitate to the paper.

To remove any titanium, zirconium, or cerium(IV) which may be present, place the filter and precipitate in the original beaker and dissolve the precipitate by boiling with hot dilute hydrochloric acid and a little sulphurous acid. Dilute, precipitate with ammonia solution, filter, and wash the precipitate with hot water until free from iodides. Dissolve the precipitate again in hydrochloric acid, and precipitate the thorium as oxalate by adding slowly and with constant stirring sufficient of a boiling 10 per cent solution of oxalic acid to combine with all the thorium and leave an excess of $20 \, \text{cm}^3$. Allow the solution to cool and stand overnight. Filter through a quantitative filter paper, wash with a solution containing $3.5 \, \text{cm}^3$ concentrated hydrochloric acid and $2.5 \, \text{g}$ oxalic acid per $100 \, \text{cm}^3$. Ignite the precipitate as in method **A** above. Weigh as ThO_2.

Note. 1. The solution may contain sulphuric acid; for example, that obtained by dissolving monazite sand in sulphuric acid.

XI, 57. TIN. Tin may be determined in any of the following ways.

A. Determination of tin with cupferron and weighing as tin(IV) oxide, SnO_2. *Discussion.* This process permits of the precipitation of tin in the presence of aluminium, chromium, cobalt, nickel, and manganese. In the presence of $5 \, \text{cm}^3$ of 48 per cent hydrofluoric acid per $300 \, \text{cm}^3$ of solution which is *ca.* $0.2M$ in hydrochloric acid, tin(IV) (about $0.15 \, \text{g}$) is not precipitated by hydrogen sulphide, whereas copper, lead, arsenic(III) and antimony(III) are precipitated; after adding about 3 g boric acid, and boiling to expel hydrogen sulphide, the tin is precipitated in the filtrate by cupferron.

Procedure. Remove metals such as copper, lead, trivalent arsenic, and antimony, if present, by precipitation with hydrogen sulphide in the presence of hydrofluoric acid (see *Discussion*). The solution should contain about $0.15 \, \text{g}$ tin and occupy a volume of $250–300 \, \text{cm}^3$. Add 3 g boric acid, boil off the excess of hydrogen sulphide, introduce $2.5 \, \text{cm}^3$ concentrated sulphuric acid cautiously, followed by a liberal excess of a filtered 10 per cent aqueous solution of cupferron. Stir vigorously (*ca.* 30–40 minutes) until the precipitate becomes compact and brittle; it may then be crushed to a fine powder with a glass rod. Filter upon a Whatman No. 41 or 541 filter paper, wash with cold water, dry in a weighed crucible, expel the organic matter by gentle ignition, and then ignite to constant weight. Weigh as SnO_2.

B. Determination of tin with N-benzoyl-N-phenylhydroxylamine.
Discussion. N-Benzoyl-N-phenylhydroxylamine $C_6H_5CON(OH)C_6H_5$, as a 1 per cent solution in ethanol, precipitates a complex $(C_{13}H_{11}O_2N)_2SnCl_2$, m.p. $171 \, °C$, from tin(IV) solutions containing 1–8 per cent concentrated hydrochloric acid: the complex can be dried at $110 \, °C$. Apparently the reagent reduces tin(IV) to tin(II) and then forms the addition compound. Copper can be quantitatively precipitated by the reagent at pH 3.6–6.0; no interference is encountered from copper, lead, or zinc in precipitating tin from, for example, brass solutions containing 7 per cent by volume of concentrated hydrochloric acid.

N-Benzoyl-N-phenylhydroxylamine is a white crystalline solid, m.p. $121 \, °C$: its solubility in water is $0.04 \, \text{g}$ per $100 \, \text{cm}^3$ at $25 \, °C$ and $0.5 \, \text{g}$ per $100 \, \text{cm}^3$ at about $80 \, °C$.

The reagent has been used for the determination of copper, iron, and aluminium. The pH ranges for quantitative precipitation are: Cu, 3.6–6.0; Fe, 3.0–5.5; and Al, 3.6–6.4. Incomplete precipitation occurs at a lower pH, and high results are obtained at higher pH values. Titanium must be precipitated below $25 \, °C$ and ignited to, and weighed as, the dioxide. Zirconium is also precipitated. Iron and aluminium cannot be precipitated in the presence of phosphate; chromium(III) interferes with the precipitation of iron(III). The following elements do not give precipitates with the reagent at pH 4: bismuth, cadmium, cobalt, manganese, nickel, uranium(IV), and zinc.

Procedure. To the sample solution of tin(IV) chloride (containing 5–20 mg of tin), add 10 cm³ concentrated hydrochloric acid and dilute to *ca.* 600 cm³ with distilled water. Add from a separatory funnel, dropwise and with constant stirring, 5 cm³ of a 1 per cent solution of the reagent in ethanol for each 10 mg of tin present plus 8 cm³ in excess. Cool in an ice bath for 4 hours, filter on a filtering crucible (sintered glass or porcelain), wash with a few cm³ of ice water, and dry at 110 °C. Weigh as $(C_{13}H_{11}O_2N)_2SnCl_2$.

XI, 58. TITANIUM.

Titanium may be determined in one of the following forms.

A. Determination of titanium with tannic acid and phenazone.
Discussion. This method affords a separation from iron, aluminium, chromium, manganese, nickel, cobalt, and zinc, and is applicable in the presence of phosphates and silicates. Small quantities of titanium (2–50 mg) may be readily determined.

Procedure. The Ti content of the solution should not exceed 0.1 g of TiO_2, and the titanium should be present as the sulphate or chloride. Add dilute ammonia to the solution until the odour persists, then (cautiously) 10 cm³ concentrated sulphuric acid and 40 cm³ of 10 per cent tannic acid solution. Dilute to 400 cm³, stir thoroughly, and cool. Introduce a 20 per cent aqueous solution of 'phenazone' (antipyrine) 2,3-dimethyl-1-phenyl-5-pyrazatone with constant stirring until an orange-red flocculent precipitate is obtained. Stop the stirring, and continue the addition of the phenazone solution until a white, cheese-like precipitate (produced by the interaction of tannic acid and phenazone) is formed in addition to the red precipitate. Boil the mixture, remove the flame, add 40 g ammonium sulphate, and allow to cool with occasional stirring. Filter the bulky precipitate through a Whatman No. 41 or 541 filter paper, supported on a Whatman filter cone (hardened, No. 51), with slight suction, and wash with a solution of 100 cm³ water, 3 cm³ concentrated sulphuric acid, 10 g ammonium sulphate, and 1 g phenazone. Dry the precipitate at 100 °C, transfer to a weighed crucible, heat gently at first, and then ignite at 700–800 °C to constant weight. Weigh as TiO_2.

Note. If the wet precipitate is heated directly, caking occurs which renders the complete oxidation of the carbonaceous matter very slow. If alkali metals were originally present, the ignited oxide must be washed with hot water, filtered, and re-ignited to constant weight.

B. Determination of titanium with 4-hydroxyphenylarsonic acid.
Discussion. This procedure will separate titanium from most other commonly occurring ions by a single precipitation. Zirconium, tin, cerium(IV), and hydrogen peroxide must be absent.

Procedure. Dissolve the sample (1) containing not more than about 0.06 g TiO_2 in sulphuric or hydrochloric acid, and dilute to 200 cm³. The amount of the acid present should be such that the solution will be approximately, but not more than, $0.6M$ in hydrochloric acid or $0.9M$ in sulphuric acid after the reagents have been added and the precipitation is complete. Heat the solution to boiling; if iron is present, add 2–3 g A.R. ammonium thiocyanate: add 100 cm³ of a 4 per cent aqueous solution of 4-hydroxyphenylarsonic acid, $HO \cdot C_6H_4 \cdot AsO_3H_2$. Boil gently for 15 minutes to coagulate the precipitate. Allow to cool to room temperature, and filter with suction on a Whatman No. 542 or 42 filter paper supported on on a filter cone (Whatman, No. 51, hardened). Wash the precipitate five or six times with a wash liquid of 0.25N-hydrochloric or sulphuric acid containing about 0.5 g of the solid reagent per 100 cm³ (if iron is present, 1–2 g ammonium thiocyanate should be also added to each 100 cm³ of wash liquor). Finally, wash the precipitate two or three times with 2 per cent aqueous ammonium nitrate solution. Transfer the filter to a silica crucible, ignite gently at first until all the carbon is burnt off (*this operation must be carried out in a fume chamber (hood) provided with an efficient draught*) and then with a Fisher burner or in an electric muffle furnace at 700–800 °C until constant weight is attained. Weigh as TiO_2.

Note. 1. For practice in this determination the Bureau of Analysed Samples 'Iron Ore, No. 175' may be used. Dissolve 4 g of this in 100 cm³ dilute hydrochloric acid and filter. Fuse the undissolved residue with sodium carbonate, wash the melt into the main filtrate, remove the silica in the usual manner, add 4 g A.R. ammonium thiocyanate, dilute to 200–250 cm³, and continue the determination as above.

XI, 59. TUNGSTEN.   Tungsten, as tungstate, may be determined in one of the following forms.

A. Determination of tungsten as the trioxide (tannic acid–phenazone method).   *Discussion.* Tungstic acid is incompletely precipitated from solutions of tungstates by tannic acid. If, however, phenazone (2,3-dimethyl-1-phenyl-5-pyrazalone) is added to the cold solution after treatment with excess of tannic acid, precipitation is quantitative. This process effects a separation from aluminium, and also from iron, chromium, manganese, zinc, cobalt, and nickel if a double precipitation is used.

Procedure.   The solution of tungstate (200–250 cm³) should contain not more than 0.15 g of WO_3, and be faintly ammoniacal. Add 6–7 cm³ of concentrated sulphuric acid and 7–8 g of ammonium sulphate, and heat to boiling. Treat with 6 cm³ of 10 per cent aqueous tannic acid solution, keep the mixture on the water bath for a few minutes, and allow to cool to room temperature. A flocculent dark-brown precipitate separates. When cold, stir in 10 cm³ of a 10 per cent aqueous solution of phenazone. Filter the precipitate through a weighed silica, Gooch, or porcelain filtering crucible (1), wash with the special wash liquid (2), and ignite to constant weight at 800–900 °C. Weigh as WO_3.

Notes. 1. The filtrate must be colourless. If it is yellow, insufficient phenazone has been added.

2. The special wash liquid contains 1 cm³ concentrated sulphuric acid, 10 g ammonium sulphate, and 0.4 g phenazone in 200 cm³ of water.

B. Determination of tungsten as barium tungstate.   *Discussion.* A dilute neutral solution of a tungstate (pH about 7.7) is precipitated by barium chloride solution as barium tungstate. It is important that the solution be dilute: in concentrated solutions high results are obtained, due to co-precipitation of barium chloride. In extremely dilute solutions and at low temperatures, a fine precipitate is slowly formed, which tends to pass through the filter and adhere to the walls of the beaker. The solubility of barium tungstate is 4 mg dm⁻³ at 22 °C, and 0.02 mg dm⁻³ in the presence of a 50 per cent excess of barium chloride; it increases rapidly with decreasing pH.

Procedure.   The solution of tungstate (250 cm³) may contain about 0.15 g of W (1) and be almost neutral (pH 7–8). Adjust the pH of the solution, if necessary, by the addition of dilute acetic acid or of dilute sodium hydroxide solution. Heat to boiling and add a saturated solution of crystallised barium chloride in $10 cm^3$ of water dropwise and with constant stirring. Allow the suspension to stand to acquire the laboratory temperature, filter through a porcelain filtering crucible, wash with cold water until the washings are free from chloride, dry at about $750 °C$ (in an electric crucible furnace) to constant weight. Weigh as $BaWO_4$.

Note. 1. Use a solution prepared from about 0.25 g A.R. sodium tungstate, $Na_2WO_4,2H_2O$ (accurately weighed), for practice in this determination.

XI, 60. URANIUM. Uranium, as uranyl salts, may be determined in either of the following forms.

A. Determination of uranium with oxine. *Discussion.* The formula of the compound is noteworthy, for it differs from all other metallic 'oxinates' (compare Section **XI, 11C**). This method may also be employed for the titrimetric determination of uranium with standard potassium bromate solution (compare Section **X, 140**).

Procedure. The uranium should be present as uranyl nitrate or chloride in 1–2 per cent acetic acid solution (1); up to 0.3 g of U may be present in $200 cm^3$ of solution. Add 5 g A.R. ammonium acetate, heat to boiling, and add 4 per cent oxine solution (2) dropwise and with stirring: use $0.5 cm^3$ of the reagent for every 10 mg of U present and a further $4–5 cm^3$. Heat on a boiling water bath for 5–10 minutes. Allow to cool, and filter through a sintered glass or porcelain filtering crucible; wash several times with hot water and then with cold water. Dry to constant weight at $105–110 °C$, and weigh as $UO_2(C_9H_6ON)_2 \cdot C_9H_7ON$. Alternatively, the precipitate may be ignited to and weighed as U_3O_8.

Notes. 1. If the solution contains mineral acid, almost neutralise with ammonia solution (or add dilute ammonia solution until a faint turbidity persists and render the solution just clear with a few drops of dilute hydrochloric acid), add 5 g A.R. ammonium acetate and then sufficient acetic acid to give a 1–2 per cent solution.

2. Details of the oxine solution are given under Section **XI, 46B**.

B. Determination of uranium with cupferron. *Discussion.* Cupferron does not react with hexavalent uranium, but tetravalent uranium is quantitatively precipitated. These facts are utilised in the separation of iron, vanadium, titanium, and zirconium from uranium(VI). After precipitation of these elements in acid solution with cupferron, the uranium in the filtrate is reduced to the tetravalent state by means of a Jones reductor and then precipitated with cupferron (thus separating it from aluminium, chromium, manganese, zinc, and phosphate). Ignition of the uranium(IV) cupferron complex affords U_3O_8.

Procedure. If uranium is to be determined in the filtrate from the precipitation of the iron group by cupferron, concentrate the solution to $50 cm^3$, add $20 cm^3$ of concentrated nitric acid and $10 cm^3$ of concentrated sulphuric acid (if not already present) and evaporate until fumes of sulphur trioxide appear. If organic matter still remains (as shown by the appearance of a dark colour upon evaporation), repeat the treatment with nitric acid. Finally, expel the nitric acid by evaporating to strong fuming, after the addition of a little water. Dilute the solution so that it contains about $6 cm^3$ of concentrated sulphuric acid per $100 cm^3$. Cool to room temperature and pass the solution through a Jones reductor (Section **X, 143**); wash the reductor with 5 per cent sulphuric acid, cool the combined reduced solution and washings to 5–10 °C, and add excess of a freshly prepared 6 per cent solution of cupferron. The precipitate does not usually form until about 5 cm³ cupferron solution has been added. Introduce a Whatman 'accelerator' or one-quarter of an 'ashless tablet', allow to settle for a few minutes, and filter through a quantitative filter paper. Wash with cold 4 per cent sulphuric acid containing 1.5 g of cupferron per dm³. Dry the precipitate at 100 °C, ignite cautiously in a platinum crucible, first at a low temperature and then at 1000 °C, to constant weight. Weigh as U_3O_8.

XI, 61. VANADIUM.   This element, as vanadate, may be determined in the following form.

Determination of vanadium as silver vanadate.   *Discussion.*   Vanadates are precipitated by excess of silver nitrate solution in the presence of sodium acetate; after boiling, the precipitate consists of silver orthovanadate. The following reactions occur with a solution of a metavanadate:

$$2NaVO_3 + 2CH_3COONa + H_2O \rightleftharpoons Na_4V_2O_7 + 2CH_3COOH$$
$$Na_4V_2O_7 + 4AgNO_3 \rightleftharpoons Ag_4V_2O_7 + 4NaNO_3$$
$$Ag_4V_2O_7 + 2AgNO_3 + 2CH_3COONa + H_2O \rightleftharpoons$$
$$2Ag_3VO_4 + 2CH_3COOH + 2NaNO_3$$

Titrimetric methods (see Chapter X) are, however, more convenient, less influenced by interfering elements, and are generally preferred.

Procedure.   Neutralise the solution (200 cm³), containing not more than 0.2 g of alkali vanadate, if acid, by aqueous sodium hydroxide, or, if alkaline, by the addition of nitric acid to the boiling solution until it becomes yellow, followed by decolorisation with dilute ammonia solution. Add 3 g of ammonium acetate, 0.5 cm³ of concentrated ammonia solution, and then excess of silver nitrate solution, heat to boiling and then keep on a steam bath for 30 minutes. Test for complete precipitation with more silver nitrate solution; if a turbidity is produced, boil the liquid until it becomes clear. Allow the dense brown precipitate of silver vanadate to settle, and collect it on a weighed filtering crucible (Gooch, sintered glass, or porcelain), wash with hot water, and dry at 110 °C. Weigh as Ag_3VO_4.

It has been stated that the results obtained by precipitation of vanadate as silver ortho-vanadate Ag_3VO_4 are not altogether satisfactory. Better results are obtained by precipitation at pH 4.5 as silver meta-vanadate $AgVO_3$; the precipitate is weighed after drying at 100–105 °C.

XI, 62. ZINC.   Zinc may be determined in any of the following forms.

A. Determination of zinc as quinaldate.   *Discussion.*   Quinaldic acid or its sodium salt precipitates zinc quantitatively from dilute acetic acid or slightly ammoniacal solutions. Iron, aluminium, chromium, beryllium, titanium, and uranium interfere in acid solution, but in the presence of alkali tartrate in alkaline solution only zinc precipitates; copper and cadmium must be absent. The reagent is described in Section **XI, 11M.**

Procedure.   The solution may contain not more than 0.1 g Zn, and should be acidified with 2–5 cm³ acetic acid (to pH 3–4). Heat to boiling and add 3 per cent sodium quinaldate solution with stirring until precipitation is complete; an excess of 25 per cent should be used. Allow to cool to room temperature. Wash the precipitate by decantation with cold water, collect it on a sintered glass or porcelain filtering crucible, wash with a little ethanol, and dry at 105–110 °C to constant weight. Weigh as $Zn(C_{10}H_6O_2N)_2,H_2O$.

B. Determination of zinc by the pyridine method. *Discussion.* This method is a very rapid one, but unless the various wash solutions are carefully prepared, low results will be obtained. The complex may be kept unchanged in a vacuum desiccator for 2–3 hours (see Section **XI, 11K**). Large quantities of ammonium salts must not be present, as these exert a slight solvent action upon the precipitate. If the solution is strongly acid, it must be evaporated to dryness and the residue dissolved in water.

Procedure.   The solution (75 cm³) should contain about 0.05 g zinc (1) and be neutral or very faintly acid. To the cold solution add 1 g solid A.R. ammonium thiocyanate, followed by 1 cm³ pure pyridine. Shake vigorously, when a white crystalline precipitate will separate. (Precipitation may also be carried out in hot solution; the complex separates in comparatively large crystals on cooling.) Allow to stand for 15 minutes, and stir frequently. Filter through a weighed sintered glass or porcelain filtering crucible, and transfer the precipitate to the crucible with the aid of *Solution 1*. Wash the precipitate four times with *Solution 2*, then wash the walls of the crucible with 1-cm³ portions of *Solution 3* (use a 1- or 2-cm³ pipette for this process), and finally five to six times with 1–2 cm³ volumes of *Solution 4*. It is important to suck well on the pump between each washing; it is also advantageous to stir the precipitate with a thin glass rod when washing with *Solutions 3* and *4*. Dry the crucible and precipitate in a vacuum desiccator for 15 minutes, and weigh. Repeat the drying process until the weight is constant. Weigh as $[Zn(C_5H_5N)_2](SCN)_2$.

Solution 1. 100 cm³ water containing 0.3 g NH₄SCN and 0.5 cm³ pyridine.

Solution 2. 85.5 cm³ water, 13 cm³ of 95 per cent ethanol, 0.1 g NH₄SCN, and 1.5 cm³ pyridine.

Solution 3. 10 cm³ absolute ethanol + 1 cm³ pyridine.

Solution 4. 15 cm³ diethyl ether (sodium dried) + 2 drops pyridine.

If the wash solutions have been prepared, the determination should be completed within an hour.

Note. 1. For practice in this determination, employ about 0.25 g, accurately weighed, A.R. zinc sulphate, or about 0.6 g pure ammonium zinc sulphate $(NH_4)_2SO_4·ZnSO_4,6H_2O$; prepared by mixing equimolecular amounts of A.R. zinc sulphate and A.R. ammonium sulphate dissolved in boiling water, and re-crystallising the product twice from hot water. The crystals are air dried, and 0.6 g, accurately weighed, is dissolved in 75 cm³ water.

C. Determination of zinc as 8-hydroxyquinaldinate. *Discussion.* Zinc may be precipitated by 8-hydroxyquinaldine (2-methyloxine) in acetic acid–acetate solution: it can thus be separated from aluminium and magnesium (see Section **XI, 11P**). It can be weighed as $Zn(C_{10}H_8ON)_2$ after drying at 130–140 °C. The co-precipitated reagent is volatile at 130 °C.

Procedure.   The solution may contain up to 0.05 g Zn in 200 cm³. Add dilute aqueous ammonia solution until a white precipitate of zinc hydroxide just appears. Re-dissolve the zinc hydroxide with a drop of acetic acid. Add a slight excess of the reagent (1) (2 cm³ for each 10 mg of Zn present) and then 2–3 drops of concentrated ammonia solution; the pH should be at least 5.5. Digest the precipitate at 60–80 °C for 15 minutes, allow to stand for 10–20 minutes, and filter through a sintered glass or porcelain filtering crucible. Dry to constant weight at 130–140 °C. Weigh as $Zn(C_{10}H_8ON)_2$.

If aluminium is present, add 1 g of ammonium tartrate to the clear, slightly acid solution. Introduce the reagent (2 cm³ for each 10 mg of Zn present), dilute the solution to 200 cm³, and heat to 60–80 °C. Neutralise the excess of acid by adding dilute ammonia solution (1:5) dropwise until the complex salt which forms on the addition of each drop just re-dissolves on stirring. Add, with stirring, 45 cm³ of 2M-ammonium acetate solution. The pH should·be at least 5.5. Allow the solution to stand for 10–20 minutes, and complete the determination as above.

Note.   1. The reagent is prepared by dissolving 5 g 8-hydroxyquinaldine in 12 g glacial acetic acid and diluting to 100 cm³ with water.

XI, 63.   ZIRCONIUM.   Zirconium may be determined in one of the following forms.

A. Determination of zirconium with selenious acid, and subsequent ignition to the dioxide, ZrO_2.   *Discussion.*   The zirconium is precipitated as the basic selenite with selenious acid in hot dilute hydrochloric acid solution, the precipitate washed with dilute hydrochloric acid, and then ignited to, and weighed as, ZrO_2. No other acids should be present, and the hydrochloric acid content should preferably be 5 per cent and not over 7 per cent by volume. This method enables a separation to be effected by a single precipitation from the rare earths (cerium being in the trivalent condition) and from aluminium. Iron, if present up to 10 per cent of the weight of the zirconium, does not interfere if precipitation is made in dilute solution (100–200 cm³ of solution containing 0.05 g ZrO_2) and double the quantity of selenious acid solution is used for precipitation.

In an alternative method the zirconium is converted into the normal selenite, $Zr(SeO_3)_2$, after digestion at 80–100 °C and weighed as such after drying at 110–150 °C.

Procedure.   The solution (200 cm³) should contain about 5 per cent by volume of hydrochloric acid (sulphuric acid is undesirable) and not more than 0.2 g zirconium (as ZrO_2). Treat with 20 cm³ of 12.5 per cent aqueous selenious acid solution, and boil for a few minutes. Allow the precipitate of basic selenite to settle, filter through a quantitative filter paper, wash with hot 3 per cent hydrochloric acid containing a little selenious acid, dry in a weighed porcelain, or platinum crucible, and ignite (a temperature of 900–1000 °C is satisfactory) to constant weight. Weigh as ZrO_2.

Alternatively, precipitate the zirconium as the normal selenite, filter through a sintered glass filtering crucible (porosity No. 4), wash with hot dilute hydrochloric acid, followed by cold water until the washings are free from selenious acid, and dry at 110–150 °C to constant weight. The precipitate contains 26.43 per cent Zr. Some reduction may occur with large quantities of precipitate, leading to slightly high results.

B. Determination of zirconium with mandelic acid and subsequent ignition to the dioxide, ZrO_2.   *Discussion.*   Zirconium may be precipitated from a hydrochloric acid solution with mandelic acid ($C_6H_5 \cdot CH(OH) \cdot COOH$) as zirconium mandelate, $Zr(C_8H_7O_3)_4$, which is ignited to and weighed as the dioxide (see Section **XI, 11G**). Quantitative separation is thus made from titanium, iron, vanadium, aluminium, chromium, thorium, cerium, tin, barium, calcium, copper, bismuth, antimony, and cadmium. If sulphuric acid is employed, the concentration should not exceed 5 per cent: higher concentrations give low results.

Procedure. The solution (20–30 cm^3) may contain 0.05–0.2 g Zr, and should possess a hydrochloric acid content of about 20 per cent by volume. Add 50 cm^3 of 16 per cent aqueous mandelic acid solution and dilute to 100 cm^3. Raise the temperature slowly to 85 °C and maintain this temperature for 20 minutes. Filter off the resulting precipitate through a quantitative filter paper, wash it with a hot solution containing 2 per cent hydrochloric acid and 5 per cent mandelic acid. Ignite the filter and precipitate to the oxide in the usual manner; a temperature of 900–1000 °C is satisfactory. Weigh as ZrO$_2$.

Note. 1. Bromomandelic acid is a superior reagent for this determination, but is more expensive. A similar procedure to that above is employed.

Anions

XI, 64. BORATE. Determination of borate as nitron tetrafluoroborate.
Discussion. Boric acid (100–250 mg) in aqueous solution may be determined by conversion into tetrafluoroboric acid and precipitation of the latter with a large excess of nitron (see Section **XI, 11H**) as **nitron tetrafluoroborate**, which is weighed after drying at 110 °C. The accuracy is about 1 per cent.

$$H_3BO_3 + 4HF \rightleftharpoons HBF_4 + 3H_2O$$
$$HBF_4 + C_{20}H_{16}N_4 \rightleftharpoons C_{20}H_{16}N_4 \cdot HBF_4$$

Fluoride ion, and weak acids and bases do not interfere, but nitrate, nitrite, perchlorate, thiocyanate, chromate, chlorate, iodide, and bromide do. Since analysis of almost all boron-containing compounds requires a preliminary treatment which ultimately results in an aqueous boric acid sample, this procedure may be regarded as a gravimetric determination of boron.

Procedure. Place the aqueous sample solution of boric acid (containing 100–250 mg of H$_3$BO$_3$) in a 250-cm^3 polythene beaker and dilute to about 60 cm^3 with distilled water. Add 15.0 cm^3 of the nitron solution (1) and 1.0–1.3 g of A.R. 48 per cent hydrofluoric acid (**CARE!**). Allow the solution to stand for 10–20 hours, and cool in an ice bath for 2 hours. Collect the precipitate in a porcelain filtering crucible, and wash it with five 10-cm^3 portions of saturated nitron tetrafluoroborate solution (2); drain the precipitate after each washing. Dry at 105–110 °C for 2 hours, and weigh as C$_{20}$H$_{16}$N$_4$·HBF$_4$.

Notes. 1. Prepare the **nitron reagent** by dissolving 3.75 g nitron in 25 cm^3 5 per cent acetic acid (by volume). Store in a dark bottle.

2. Prepare the wash solution by adding an excess of solid nitron tetrafluoroborate to 100 cm^3 of water and shaking mechanically for 2 hours.

XI, 65. BROMATE AND BROMIDE. *Discussion.* These anions are both determined as **silver bromide, AgBr**, by precipitation with silver nitrate solution in the presence of dilute nitric acid. With the bromate, initial reduction to the bromide is achieved by the procedures described for the chlorate (Section **XI, 67**) and the iodate (Section **XI, 75**). Silver bromide is less soluble in water than is the chloride. The solubility of the former is 0.11 mg dm^{-3} at 21 °C as compared with 1.54 mg dm^{-3} for the latter, hence the procedure for the determination of bromide is practically the same as that for chloride. Protection from light is even more essential with the bromide than with the chloride because of its greater sensitivity.

XI, 66. CARBONATE. Determination of carbonate by the evolution of carbon dioxide. *Discussion.* The carbonate is decomposed by dilute acid, and either the loss in weight due to the escape of carbon dioxide determined (**indirect method**) or the carbon dioxide evolved is absorbed in a suitable medium and the increase in weight of the absorbent determined (**direct method**). The direct method gives more satisfactory results, and will therefore be described. The indirect method is often employed, however, for samples containing relatively large amounts of carbonate.

The decomposition of the carbonate may be effected with dilute hydrochloric acid, dilute perchloric acid, or syrupy phosphoric acid. The last-named acid is perhaps the most convenient because of its comparative non-volatility and the fact that the reaction can be more easily controlled than with the other acids. If dilute hydrochloric acid is employed, a short, water-cooled condenser should be inserted between the decomposition flask and the absorption train (see below).

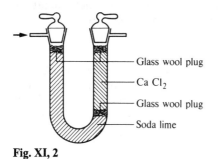

Glass wool plug
Ca Cl₂
Glass wool plug
Soda lime

Fig. XI, 2

Two absorbents are required, one for water vapour the other for carbon dioxide. The absorbents for water vapour which are generally employed are: (*a*) anhydrous calcium chloride (14–20 mesh), (*b*) anhydrous calcium sulphate ('Drierite' or 'Anhydrocel'), and (*c*) anhydrous magnesium perchlorate ('Anhydrone'). Both (*b*) and (*c*) are preferable to (*a*); (*c*) absorbs about 50 per cent of its weight of water, but is expensive. Anhydrous calcium chloride usually contains a little free lime, which will absorb carbon dioxide also; it is essential to saturate the U-tube containing calcium chloride with dry carbon dioxide for several hours and then to displace the carbon dioxide by a current of pure dry air before use.

The absorbents for carbon dioxide in general use are: (*d*) soda lime (this is available also in the form of self-indicating granules, 'Carbosorb', which indicate when the absorbent is exhausted), (*e*) soda lime–asbestos (the 'Carbosorb' variety gives a marked colour change and therefore indicates the degree of exhaustion), and (*f*) sodium hydroxide–asbestos ('Ascarite'). In all cases the carbon dioxide is absorbed in accordance with the following equation:

$$2NaOH + CO_2 = Na_2CO_3 + H_2O$$

Water is formed in the reaction, hence it is essential to fill one-quarter or one-third of the tube with any of the desiccants referred to above (Fig. XI, 2).

Procedure. Fit up the apparatus shown in Fig. XI, 3. A is a flask of about 100 cm³ capacity, B is a dropping funnel containing 20–25 cm³ of A.R. syrupy phosphoric acid, C is a soda-lime guard tube, D is a bubbler containing syrupy phosphoric acid, E is a U-tube containing calcium chloride which has been saturated with carbon dioxide and the residual carbon dioxide displaced by air (this may be replaced by anhydrous calcium sulphate, or by anhydrous magnesium perchlorate, if available).* F and G are U-tubes containing soda-lime

---

* The first third of this tube may be filled with anhydrous copper sulphate to remove any hydrogen sulphide or hydrogen chloride present from sulphides or chlorides in the limestone.

(this absorbent may be replaced by 'Carbosorb', soda-lime–asbestos, or 'Ascarite'), and H is a guard U-tube containing the same desiccant as in E. The U-tubes may be suspended by silver wires attached to hooks on the glass or metal rod I, or by some other means. All joints are made with short lengths of stout-walled rubber tubing, and the two ends of the glass tubing should be in contact. Rubber bungs are employed in A, B, and C. Before proceeding with the actual determination, make sure that the apparatus is gas-tight.

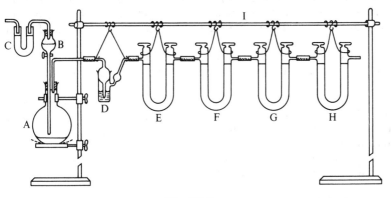

Fig. XI, 3

Weigh out accurately 0.5–0.6 g of the carbonate (1) into the flask A, which should be clean and dry. Remove the two soda-lime or 'Ascarite' tubes F and G, wipe them with a clean linen handkerchief or cloth, and leave them in the balance case for 45 minutes. Open the taps of the U-tubes momentarily to the air in the balance case, and weigh them separately. Replace them on the drying train; place 25 cm³ A.R. syrupy phosphoric acid in B, and see that the apparatus is connected up as in Fig. XI, 3. Open the taps of the U-tubes. Run in sufficient phosphoric acid from the tap funnel to cover the solid in the flask (the 25 cm³ will more than suffice). Close the tap of the funnel and heat the flask carefully; regulate the temperature so that not more than 2 bubbles of gas per second pass through the bubbler D. After about 30–40 minutes, the contents of the flask should be boiling; boil for 2–3 minutes. Remove the flame, and immediately attach a filter pump and a large bubbler (similar to D, and containing syrupy phosphoric acid) to the end of the tube H. Open the tap funnel, and draw air through the apparatus at the rate of about 2 bubbles per second for 20 minutes. Remove the tubes F and G, close the taps, treat them as before, and weigh them. From the increase in weight, calculate the percentage of CO_2 in the sample (2).

Notes. 1. For practice in this determination, the student may employ A.R. calcium carbonate or 'Limestone, 15e' (Analysed Samples for Students) from the Bureau of Analysed Samples.

2. For the most accurate work, and particularly when the amount of carbon dioxide is small, a 'blank' experiment must be run with the reagents alone before the determination proper is carried out.

XI, 67. CHLORATE. Determination of chlorate as silver chloride.

Discussion. The chlorate is reduced to chloride, and the latter is determined as **silver chloride, AgCl.** The reduction may be performed with iron(II) sulphate solution, sulphur dioxide, or by zinc powder and acetic acid. Alkali chlorates may be quantitatively converted into chlorides by three evaporations with concentrated hydrochloric acid, or by evaporation with three times the weight of ammonium chloride.

Procedure. The chlorate solution should have a volume of about $100 \, cm^3$, and contain about $0.2 \, g \, ClO_3$. Add $50 \, cm^3$ of a 10 per cent solution of A.R. crystallised iron(II) sulphate, heat with constant stirring to the boiling point, and boil for 15 minutes. Allow to cool, add nitric acid until the precipitated basic iron(III) salt is dissolved, precipitate the chloride by means of silver nitrate solution, and collect and weigh as AgCl after the usual treatment (Section **XI, 17**).

Alternatively, treat the chlorate solution with excess of sulphur dioxide, boil the solution to remove the excess of the gas, render slightly acid with nitric acid, and precipitate the silver chloride as above.

For the reduction with zinc, render the chlorate solution strongly acid with acetic acid, add excess of zinc, and boil the mixture for 1 hour. Dissolve the excess of unused zinc with nitric acid, filter, and treat the filtrate with silver nitrate in the usual manner.

Note. Hypochlorites and **chlorites** may be reduced to chlorides with sulphur dioxide, and determined in the same way.

XI, 68. CHLORIDE. *Discussion.* This anion is determined as **silver chloride, AgCl**; full details are given in Section **XI, 17**. Anions which give silver salts which are insoluble in dilute nitric acid must be absent; these include bromide, iodide, thiocyanate, sulphide, thiosulphate, hexacyanoferrate(II), and hexacyanoferrate(III). Heavy metals interfere, and must be removed by precipitation.

If the chloride is insoluble, it is necessary to boil it, with a large excess of saturated sodium carbonate solution or, better, to fuse it with sodium carbonate and extract the melt with water. In either case the chloride passes into solution, and is determined in the usual way after acidification with nitric acid.

XI, 69. CYANIDE. *Discussion.* This anion is determined as **silver cyanide, AgCN**; the experimental details are similar to those given for Chloride, except that, owing to the volatility of hydrocyanic acid, the solution must not be heated. The cold solution of alkali cyanide is treated with a slight excess of silver nitrate solution, *faintly* acidified with nitric acid, the precipitate allowed to settle, collected on a weighed filtering crucible, and weighed as AgCN after drying at $100 \, °C$.

XI, 70. FLUORIDE. This anion may be determined in one of the following forms:

Determination of fluoride as triphenyltin fluoride.*

Discussion. Triphenyltin chloride reagent precipitates fluorides quantitatively as the corresponding fluoride. The precipitate is crystalline, easily filtered, and washed, and is quite stable. Owing to the insolubility of the reagent in water, precipitation is carried out in 60–70 per cent ethanol solution, and washing is

---

* This method is only of use where a high degree of accuracy is not required.

effected with an ethanolic solution of the reagent saturated with triphenyltin fluoride.

The method is well adapted for the determination of small quantities of fluorides; the maximum amount that can be conveniently handled is 0.04 g of F. The solution should have a pH of 5–7; if acid some fluorine will be lost on heating to boiling, and if basic, triphenyltin hydroxide will be precipitated along with the fluoride. Metals other than the alkali metals should preferably be absent; the latter may be removed by washing the precipitate several times with the ethanol wash solution, followed by cold water. Small quantities of nitrates, chlorides, bromides, iodides, and sulphates do not interfere, but silicates (with ammoniacal zinc hydroxide), phosphates (with silver nitrate), and carbonates must be removed before precipitation. Carbonate is best removed by neutralising with dilute nitric acid (to phenolphthalein), and boiling off the carbon dioxide. A disadvantage of the method is that the reagent is expensive.

Procedure. The solution (say, 25 cm^3) should contain not more than 0.04 g of F, and be almost neutral (1). Add 95 per cent ethanol to the aqueous solution of the fluoride so that it comprises about 60–70 per cent of the final volume. Heat to boiling and treat with twice the calculated quantity of the reagent (2) diluted with an equal volume of 95 per cent ethanol and also heated to boiling. The latter is run slowly into the hot fluoride solution with vigorous stirring, and the whole again heated to the boiling point. Remove the source of heat, and continue the stirring until the solution has cooled somewhat (3). Allow to stand overnight, and cool for 1 hour in ice (4). Filter through a weighed, sintered glass or porcelain filtering crucible, wash with 95 per cent ethanol which has been saturated with triphenyltin fluoride (about 50 cm^3). Dry for 40 minutes at 110 °C, cool in a desiccator, and weigh as $(C_6H_5)_3SnF$.

Notes. 1. For practice in this estimation, A.R. sodium fluoride may be used. If desired, a sample of *pure sodium fluoride* may be prepared as follows. Treat A.R. anhydrous sodium carbonate with an excess of A.R. hydrofluoric acid **(CARE!)** in a platinum dish, and allow to stand for a few hours. Remove the excess of acid by heating (fume cupboard; hood), allow to cool, and add more acid. Mix thoroughly with a platinum spatula, heat the dish gently at first, and then strongly until the sodium fluoride is entirely fused. Pulverise in an agate mortar, dry the powder in platinum at 110 °C, and store in a desiccator over calcium chloride.

2. The reagent is prepared by shaking vigorously 4.0 g triphenyltin chloride with 200 cm^3 95 per cent ethanol; filter from the small undissolved residue. This is practically a saturated solution.

About 55 cm^3 of this reagent are required for 0.04 g F.

3. If the quantity of fluoride is large, precipitation as white crystals commences in about a minute after the addition of the reagent, but with small quantities it does not take place until the solution has cooled to room temperature.

4. This is unnecessary if the amount of fluoride is large and the total volume of the solution small.

XI, 71. FLUOROSILICATE. *Discussion.* The determination of this anion is of little practical importance. The methods available for its determination will, however, be outlined. Alkali fluorosilicates are decomposed by heating with sodium carbonate solution into a fluoride and silicic acid:

$$Na_2[SiF_6] + 2Na_2CO_3 + H_2O = 6NaF + H_2SiO_3 + 2CO_2$$

Insoluble fluorosilicates are brought into solution by fusion with four times the bulk of fusion mixture, and extracting the melt with water. In either case, the solution is treated with a considerable excess of ammonium carbonate, warmed to 40 °C, and, after standing for 12 hours, the precipitated silicic acid is filtered off, and washed with 2 per cent ammonium carbonate solution. The filtrate contains a little silicic acid, which may be removed by shaking with a little freshly precipitated cadmium oxide. The fluoride in the filtrate is determined as described in Section **XI, 70**.

If an acid solution of a fluorosilicate is rendered faintly alkaline with aqueous sodium hydroxide and then shaken with freshly precipitated cadmium oxide, all the silicic acid is adsorbed by the suspension. The alkali fluoride is then determined in the filtrate.

XI, 72.   HEXACYANOFERRATE(III).   No satisfactory gravimetric method is available. For titrimetric methods, see Chapter X.

XI, 73.   HEXACYANOFERRATE(II).   No satisfactory gravimetric procedure is available. Titrimetric methods are described in Chapter X.

XI, 74.   HYPOPHOSPHITE.   This anion is determined similarly to phosphite (Section **XI, 82**) either indirectly as **mercury(I) chloride, Hg_2Cl_2,** or as **ammonium magnesium phosphate hexahydrate, $MgNH_4PO_4,6H_2O$,** or as **magnesium pyrophosphate, $Mg_2P_2O_7$.** In this case the reaction with mercury(II) chloride solution is:

$$4HgCl_2 + H_3PO_2 + 2H_2O = 2Hg_2Cl_2 + H_3PO_4 + 4HCl$$

so that $\qquad\qquad 2Hg_2Cl_2 \equiv H_3PO_2$

XI, 75.   IODATE.   Determination of iodate as silver iodide.
Discussion. Iodates are readily reduced by sulphurous acid to iodides; the latter are determined by precipitation with silver nitrate solution as **silver iodide, AgI**. Iodates cannot be converted quantitatively into iodides by ignition, for the decomposition takes place at a temperature at which the iodide is appreciably volatile.

Periodates are also reduced by sulphurous acid, and may therefore be similarly determined. Similar remarks apply to **bromates**; these are ultimately weighed as silver bromide, AgBr.

Procedure. Acidify the iodate solution (100 cm^3 containing *ca.* 0.3 g of IO_3) (1) with sulphuric acid, and pass in sulphur dioxide (or add a freshly prepared saturated solution of sulphurous acid) until the solution which at first becomes yellow, on account of the separation of iodine, is again colourless. Boil off the excess of sulphur dioxide, and precipitate the iodide with dilute silver nitrate solution as described in Section **XI, 76**. Weigh as AgI.

Note. 1. For practice in this determination, A.R. potassium iodate may be employed.

XI, 76.   IODIDE. Two procedures are commonly employed for the determination of iodides.
   A. Determination of iodide as silver iodide. *Discussion.* This anion is

usually determined by precipitation as **silver iodide, AgI**. Silver iodide is the least soluble of the silver halides; $1 \, dm^3$ of water dissolves $0.0035 \, mg$ at $21 \, °C$. Co-precipitation and similar errors are more likely to occur with iodide than with the other halides.

Procedure. Precipitation is therefore made by adding a very dilute solution, say $0.05 M$, of silver nitrate slowly and with constant stirring to a dilute ammoniacal solution of the iodide until precipitation is complete, and then adding excess of nitric acid (1 per cent by volume). The precipitate is collected in the usual manner, washed with 1 per cent nitric acid, and finally with a *little* water to remove nitric acid. Peptisation tends to occur with excess of water. Other details of the determination will be found in Section **XI, 17**.

B. Determination of iodide as palladium(II) iodide. *Discussion.* Iodide may also be determined by precipitation as **palladium(II) iodide, PdI_2**. Substances, such as ethanol, which cause reduction to metallic palladium must be absent; bromides and chlorides are not precipitated and therefore do not interfere. The precipitate is insoluble in water and in dilute hydrochloric acid (1:99). The reagent, palladium(II) chloride, is expensive, and the method is therefore rarely employed except for gravimetric separation from other halides.

Procedure. The iodide solution should contain 1 per cent by volume of hydrochloric acid, and not more than $0.1 \, g$ iodide. Warm to $70 \, °C$, and add palladium(II) chloride solution, dropwise and with stirring, until no more precipitate is formed. Allow the solution to stand for 24–48 hours at $20-30 \, °C$, filter the brownish-black precipitate on a weighed filtering crucible (Gooch, sintered glass, or porcelain), and wash four times with warm water. Dry at $100 \, °C$, for 1 hour, and weigh as PdI_2.

XI, 77.   NITRATE.   Determination of nitrate as nitron nitrate.
Discussion. The mono-acid base nitron, $C_{20}H_{16}N_4$, forms a fairly in-soluble crystalline nitrate, $C_{20}H_{16}N_4, HNO_3$ (solubility is $0.099 \, g/dm^{-3}$ at about $20 \, °C$), which can be used for the quantitative determination of nitrates (see Section **XI, 11H**). The sulphate and acetate are soluble so that precipitation may be made in sulphuric or acetic acid solution. Perchlorates $(0.08 \, g)$, iodides $(0.17 \, g)$, thiocyanates $(0.4 \, g)$, chromates $(0.6 \, g)$, chlorates $(1.2 \, g)$, nitrites $(1.9 \, g)$, bromides $(6.1 \, g)$, hexacyanoferrate(II), hexacyanoferrate(III), oxalates, and considerable quantities of chlorides interfere, and should be absent. The figures in parentheses are the approximate solubilities of the nitron salts in $g \, dm^{-3}$ at about $20 \, °C$.

Procedure. The solution $(75-100 \, cm^3)$ should be neutral and contain about $0.1 \, g \, NO_3$. Add $1 \, cm^3$ glacial acetic acid or $0.5 \, cm^3$ M-sulphuric acid and heat the solution nearly to the boiling point. Then introduce in one portion $10-12 \, cm^3$ of the nitron reagent (1), stir, and cool in ice-water for 2 hours. Filter through a weighed filtering crucible (Gooch, sintered glass, or porcelain). Wash with $10-15 \, cm^3$ of a cold saturated solution of nitron nitrate, added in several portions, and drain the precipitate well after each washing. Finally, wash twice with $3-cm^3$ portions of ice-cold water. Dry at $105 \, °C$ (1 hour is usually required), and weigh as $C_{20}H_{16}N_4, HNO_3$.

Note.   1.   Prepare the reagent by dissolving $5 \, g$ of nitron in $50 \, cm^3$ of 5 per cent acetic acid. Store in an amber bottle.

XI, 78.   NITRITE.   No satisfactory gravimetric procedure is available. Titrimetric methods are described in Chapter X.

XI, 79.   OXALATE.   Determination of oxalate as calcium oxalate and as calcium carbonate or calcium oxide.   *Discussion.*   The neutral solution of alkali oxalate is acidified with acetic acid, heated to boiling, and precipitated with boiling calcium chloride solution. After standing for 12 hours, the precipitate is filtered off, washed with hot water, and weighed either as calcium oxalate, or after heating, as **calcium carbonate, $CaCO_3$**, or as **calcium oxide, CaO**.

Procedure.   The following rapid method yields results of moderate accuracy. Precipitation of the oxalate is effected in boiling solution containing a little ammonium chloride by a hot solution of calcium chloride. The solution is allowed to cool, treated with one-third of its volume of 90 per cent ethanol, and allowed to stand for 30 minutes. The precipitate is washed by decantation through a weighed porcelain or sintered glass filtering crucible with warm water (50–60 °C) until the chloride reaction is negative. The calcium oxalate is then transferred to the filtering crucible, washed once with cold water, five times with ethanol, and several times with small volumes of anhydrous diethyl ether. The precipitate is sucked dry at the pump for 10 minutes, the outside of the crucible wiped dry with a clean linen cloth, and then left in a vacuum desiccator for 10 minutes. It is weighed as CaC_2O_4,H_2O, or may be converted to the other two forms (see Section **XI, 20**).

XI, 80.   PERCHLORATE.   Determination of perchlorate as silver chloride.   *Discussion.*   Perchlorates are not reduced by iron(II) sulphate solution, sulphurous acid, or by repeated evaporation with concentrated hydrochloric acid; reduction occurs, however, with titanium(III) sulphate solution. Ignition of perchlorates with ammonium chloride in a platinum crucible or in a porcelain crucible in the presence of a little platinum powder results in reduction to the chlorides (the platinum acts as a catalyst), which may be determined in the usual manner. Losses occur when perchlorates are ignited alone.

Procedure.   The perchlorate, if supplied as a solution, is evaporated to dryness on the water bath; otherwise the solid perchlorate is used directly. Intimately mix about 0.4 g of the perchlorate (1) with 1.5 g of A.R. ammonium chloride in a *platinum* crucible covered with a watch glass or lid, ignite gently until fuming ceases and continue the heating for 1 hour. Do not fuse the resulting chloride, as the crucible may be attacked. Repeat the ignition with another 1.5 g of ammonium chloride. Dissolve the residue in a little water, filter through a small quantitative filter paper to remove any platinum powder which may be present, and determine the chloride in the filtrate as silver chloride (Section **XI, 17**).

Note.   1. For practice in this determination, employ A.R. potassium perchlorate.

XI, 81.   PHOSPHATE.   Phosphates may be determined by either of the following methods.

A. Determination of phosphate as ammonium magnesium phosphate hexahydrate or as magnesium pyrophosphate.   *Discussion.*   Orthophosphates may be precipitated as **ammonium magnesium phosphate, $MgNH_4PO_4,6H_2O$**, by magnesium chloride and ammonium chloride in ammoniacal solution ('magnesia' reagent). Most elements, other than those of the alkalis, interfere, however, by giving precipitates with 'magnesia mixture'.* It is therefore necessary in the majority of cases to separate the phosphate first from interfering substances. This may be readily effected by precipitation as **ammonium molybdophosphate** with excess of ammonium molybdate in warm nitric acid solution: arsenic, vanadium, titanium, zirconium, silica, and excessive amounts of ammonium salts interfere. When first precipitated (in the presence of a large excess of nitric acid and of ammonium nitrate), the yellow precipitate has the composition $(NH_4)_2H[PMo_{12}O_{40}],H_2O$. Upon washing with a dilute solution of ammonium nitrate, the diammonium salt passes easily into the triammonium salt $(NH_4)_3[PMo_{12}O_{40}]$. The precipitate thus obtained is dissolved in dilute ammonia solution, and the phosphate is then precipitated as ammonium magnesium phosphate. A double precipitation of the latter is usually necessary in order to obtain a precipitate entirely free from molybdate.

Procedure. To a neutral or weakly acid solution (50–100 cm³) of the phosphate, containing not more than 0.10 g of P_2O_5 and free from interfering elements (1), add 3 cm³ of concentrated hydrochloric acid and a few drops of methyl red indicator. Introduce 25 cm³ of magnesia mixture (2), followed by pure concentrated ammonia solution slowly, whilst stirring the solution vigorously until the indicator turns yellow. The procedure from this stage is the same as described for the determination of magnesium in Section **XI, 23**, except that when carrying out the re-precipitation from the hydrochloric acid solution 2 cm³ of the magnesia mixture are added instead of the 1 cm³ of ammonium phosphate solution. Weigh as $MgNH_4PO_4,6H_2O$ or as $Mg_2P_2O_7$.

Note. 1. A suitable solution for practice may be prepared by dissolving about 0.4 g, accurately weighed, A.R. anhydrous Na_2HPO_4 in 100 cm³ water. The appropriate weight of A.R. KH_2PO_4 may also be used and is perhaps to be preferred.

2. The magnesia mixture is prepared as follows. Dissolve 25 g magnesium chloride $MgCl_2,6H_2O$ and 50 g ammonium chloride in 250 cm³ of water. Add a slight excess of ammonia solution, allow to stand overnight, and filter if a precipitate is present. Acidify with dilute hydrochloric acid, add 1cm³ concentrated hydrochloric acid, and dilute to 500 cm³.

B. Determination of phosphate as ammonium molybdophosphate.
Discussion. If interfering elements are absent, the original yellow precipitate obtained in **A.** above may be weighed either as **ammonium molybdophosphate, $(NH_4)_3[PMo_{12}O_{40}]$**, after drying at 200–400 °C (280 °C is recommended) or as **$P_2O_5,24MoO_3$**, after heating at 800–825 °C for about 30 minutes. For practice in this determination the student may determine the percentage of P_2O_5 (or P) in anhydrous A.R. disodium hydrogen phosphate Na_2HPO_4 or, preferably, in A.R. potassium dihydrogen phosphate KH_2PO_4. Some experimental details will be found in Section **X, 49**, a slightly modified procedure is described below.

Procedure. Prepare a solution of anhydrous A.R. disodium hydrogen phosphate or of A.R. potassium dihydrogen phosphate containing about 125 mg of P_2O_5 in 150 cm³. Warm to 60 °C, and run in 100 cm³ of ammonium molybdate

---

* Phosphate may be precipitated directly as ammonium magnesium phosphate in the presence of elements such as iron, aluminium, titanium, zirconium, tin, and calcium by adding excess of citric acid and using an excess of magnesia mixture.

reagent (1) also warmed to 60 °C: use a fast-flowing pipette for the addition and stir well. Heat to 60 °C for about 1 hour with frequent stirring. Collect the precipitate in a weighed porcelain filtering crucible using two 20-cm^3 portions of 2 per cent ammonium nitrate solution to transfer it from the beaker (remove any precipitate adhering to the walls of the beaker with a rubber-tipped glass rod); wash the precipitate in the crucible with five 10-cm^3 portions of 2 per cent ammonium nitrate solution. Dry the precipitate at 280 °C, and weigh as $(NH_4)_3[PMo_{12}O_{40}]$. As an additional check, ignite the precipitate at 800–825 °C in an electric muffle furnace; weigh as $P_2O_5,24MoO_3$. Both solids are appreciably hygroscopic; the covered crucible, after cooling in a desiccator, should be weighed as soon as it has acquired the laboratory temperature.

Note. 1. Prepare the **ammonium molybdate reagent** as follows. Dissolve 125 g ammonium nitrate in 125 cm^3 water in a flask and add 175 cm^3 nitric acid, sp. gr. 1.42. Dissolve 12.5 g A.R. ammonium molybdate* in 75 cm^3 of water and add this slowly and with constant shaking to the nitrate solution. Dilute to 500 cm^3 with water, heat the flask in a water bath at 60 °C for 6 hours, and allow the solution to stand for 24 hours. If a precipitate forms, filter through a Whatman No. 42 filter paper. This reagent has good keeping qualities; it is said that no precipitate is formed for at least 3 months.

XI, 82. PHOSPHITE. This anion may be determined in either of the following forms:

A. Determination of phosphite as mercury(I) chloride. *Discussion.* The acid solution of phosphite reduces mercury(II) chloride solution to mercury(I) chloride which is weighed. The reaction is:

$$2HgCl_2 + H_3PO_3 + H_2O = Hg_2Cl_2 + H_3PO_4 + 2HCl$$
whence $\qquad\qquad Hg_2Cl_2 \equiv H_3PO_3$

Procedure. The phosphite solution (30 cm^3) should contain about 0.1 g HPO_3^{2-}. Place 50 cm^3 of 3 per cent mercury(II) chloride solution, 20 cm^3 of 10 per cent sodium acetate, and 5 cm^3 of glacial acetic acid in a 250-cm^3 beaker, and add the phosphite solution dropwise, and with stirring, in the cold. Allow to stand on a water bath at 40–50 °C for 2 hours. When cold, filter through a weighed filtering crucible (Gooch, sintered glass, or procelain), wash two or three times with 1 per cent hydrochloric acid, and then four times with warm water. Dry at 105–110 °C, and weigh as Hg_2Cl_2.

B. Determination of phosphite as ammonium magnesium phosphate hexahydrate or as the pyrophosphate. *Discussion.* The phosphite is oxidised by nitric acid to phosphate, and the latter is determined as ammonium magnesium phosphate hexahydrate or as the pyrophosphate.

Procedure. Treat the aqueous solution of the phosphite (100 cm^3) with 5 cm^3 concentrated nitric acid, evaporate to a small volume on the water bath, add 1 cm^3 fuming nitric acid, and heat again. Dilute the solution, and precipitate the phosphoric acid by magnesia mixture and ammonia solution, and weigh as $MgNH_4PO_4,6H_2O$ or as $Mg_2P_2O_7$ (Section **XI, 81A**).

---

* This is actually the hepta-molybdate $(NH_4)_6Mo_7O_{24},4H_2O$.

XI, 83.   SILICATE.   For analytical purposes silicates may be conveniently divided into the following two classes: (*a*) those ('soluble' silicates) which are decomposed by acids, such as hydrochloric acid, to form silicic acid and the salts (e.g., chlorides) of the metals present, and (*b*) those ('insoluble' silicates) which are not decomposed by any acid, except hydrofluoric acid. There are also many silicates which are partially decomposed by acids; for our purpose these will be included in class (*b*).

A.   Determination of silica in a 'soluble' silicate.   *Discussion.*   Most of the silicates which come within the classification of 'soluble' silicates are the orthosilicates formed from $SiO_4{}^{4-}$ units in combination with just one or two cations. More highly condensed silicate structures give rise to the 'insoluble' silicates.

Procedure.   Weigh out accurately about 0.4 g of the finely powdered silicate (1) into a platinum or porcelain dish, add 10–15 cm^3 water, and stir until the silicate is thoroughly wet. Place the dish, covered with a clock glass, on the water bath, and add gradually 25 cm^3 1:1 hydrochloric acid. The contents of the dish must be continuously stirred with a glass rod; when no gritty particles remain, the powder will have been completely decomposed. Evaporate the liquid to dryness: stir the residue continuously and break up any lumps with the glass rod. When the powder appears to be dry, place the basin in an air oven at 100–110 °C for 1 hour in order to dehydrate the silica. Moisten the residue with 5 cm^3 of concentrated hydrochloric acid, and bring the acid into contact with the solid with the aid of a stirring rod. Add 75 cm^3 of water, rinse down the sides of the dish, and heat on a steam bath for 10–20 minutes to assist in the solution of the soluble salts. Filter off the separated silica on a Whatman No. 41 or 541 filter paper. Wash the precipitate first with warm, dilute hydrochloric acid (approx. 0.5*M*), and then with hot water until free from chlorides. Pour the filtrate and washings into the original dish, evaporate to dryness on the steam bath, and heat in an air oven at 100–110 °C for 1 hour. Moisten the residue with 5 cm^3 concentrated hydrochloric acid, add 75 cm^3 water, warm to extract soluble salts, and filter through a fresh, but smaller, filter paper. Wash with warm dilute hydrochloric acid (approx. 0.1*M*), and finally with a little hot water. Fold up the moist filters, and place them in a weighed platinum crucible. Dry the paper with a small flame, char the paper, and burn off the carbon over a low flame; take care that none of the fine powder is blown away. When all the carbon has been oxidised, cover the crucible, and heat for an hour at the full temperature of a Meker type burner in order to complete the dehydration. Allow to cool in a desiccator, and weigh. Repeat the ignition, etc., until the weight is constant.

To determine the exact SiO_2 content of the residue, moisten it with 1 cm^3 water, add 2 or 3 drops concentrated sulphuric acid and about 5 cm^3 of the purest available (A.R.) hydrofluoric acid. **(CARE!)** Place the crucible in an air bath (Section III, 25) and evaporate the hydrofluoric acid in a fume cupboard (hood) with a small flame until the acid is completely expelled; the liquid should not be boiled. (The crucible may also be directly heated with a small non-luminous flame.) Then increase the heat to volatilise the sulphuric acid, and finally heat with a Meker-type burner for 15 minutes. Allow to cool in a desiccator and weigh. Re-heat to constant weight. The loss in weight represents the weight of the silica (2).

Notes.   1. For practice in this determination, powdered, fused sodium silicate may be used.

2. It is advisable to carry out a blank determination with the hydrofluoric acid, and to allow for any non-volatile substances, if necessary.

B. Determination of silica in an 'insoluble' silicate, and ultimate weighing as silica, SiO$_2$. *Discussion.* Insoluble silicates are generally fused with sodium carbonate, and the melt, which contains the silicate in acid-decomposable form, is then treated with hydrochloric acid. The acid solution of the decomposed silicate is evaporated to dryness on the water bath to separate the gelatinous silicic acid SiO$_2$,xH$_2$O as insoluble silica SiO$_2$,yH$_2$O; the residue is heated at 110–120 °C to partially dehydrate the silica and render it as insoluble as possible. The residue is extracted with hot dilute hydrochloric acid to remove salts of iron, aluminium, and other metals which may be present. The greater portion of the silica remains undissolved, and is filtered off. The filtrate is evaporated to dryness, and the residue heated at 110–120 °C as before in order to render insoluble the small amount of silicic acid that has escaped dehydration. The residue is treated with dilute hydrochloric acid as before, and the second portion of silica is filtered off on a fresh filter. The two washed precipitates are combined, and ignited in a platinum crucible at about 1050 °C to **silicon dioxide, SiO$_2$**, and the latter is weighed. The ignited residue is not usually pure silicon dioxide; it will generally contain small amounts of the oxides of iron, aluminium, titanium, etc. The amount of impurity may be determined, if desired, by treating the weighed residue in the platinum crucible with an excess of hydrofluoric acid and a little concentrated sulphuric acid. The silica is expelled as the volatile silicon tetrafluoride; the impurities (e.g., Al$_2$O$_3$ and Fe$_2$O$_3$) are first converted into the fluorides, which pass into the sulphates in contact with the less-volatile sulphuric acid, whilst the subsequent brief ignition (at 1050–1100 °C for a few minutes) converts the sulphates back into oxides. Thus, for example:

$$SiO_2 + 6HF = H_2[SiF_6] + 2H_2O$$
$$H_2[SiF_6] = SiF_4 + 2HF$$
$$Al_2O_3 + 6HF = 2AlF_3 + 3H_2O$$
$$2AlF_3 + 3H_2SO_4 = Al_2(SO_4)_3 + 6HF$$
$$Al_2(SO_4)_3 = Al_2O_3 + 3SO_3$$

The loss in weight therefore represents the amount of pure silicon dioxide present.

Procedure. Weigh out accurately into a platinum crucible about 1.0 g of the finely powdered dry silicate (1), add six times the weight of anhydrous A.R. sodium carbonate (or, better, of A.R. fusion mixture), and mix the solids thoroughly by stirring with a thin, rounded glass rod. Cover the mixture with a little more of the carbonate, and then cover the crucible. Heat the mixture gradually until after about 20 minutes a tranquil melt is obtained; the cover of the crucible is lifted occasionally to examine the contents. Maintain the temperature of a quiet liquid fusion for about 30 minutes. Allow to cool. Place the crucible and lid in a covered deep porcelain or platinum basin (or in a large casserole), cover it with water, and leave overnight, or warm on the water bath until the contents are well disintegrated. Introduce very slowly by means of a pipette or a bent funnel about 25 cm^3 of concentrated hydrochloric acid into the covered vessel. Warm on the steam bath until the evolution of carbon dioxide has ceased. Remove and rinse the cover glass, crucible, and lid, and evaporate the contents of the dish to *complete* dryness on the steam bath, crushing all lumps with a glass rod. Heat the residue for an hour at 100–110 °C to dehydrate the silica. Complete the determination as described in **A**.

Note. 1. 'Feldspar (Potash), No. 29dG' (one of the Analysed Samples for Students) available from the Bureau of Analysed Samples is suitable.

C. Determination of silica in an 'insoluble' silicate as quinoline molybdosilicate. *Discussion.* Silica may also be determined gravimetrically as **quinoline molybdosilicate**. The solution of silicic acid is treated with ammonium molybdate to form molybdosilicic acid $H_4[SiO_4,12MoO_3]$, which is then precipitated as quinoline molybdosilicate, $(C_9H_7)_4H_4[SiO_4,12MoO_3]$. The latter is weighed after drying at 150 °C. The experimental conditions lead to quinoline molybdosilicate in a pure form suitable for weighing.

Phosphate, arsenate, and vanadate interfere. Borate, fluoride, and large amounts of aluminium, calcium, magnesium, and the alkali metals have no effect in the determination, but large amounts of iron (> 5 per cent) appear to produce slightly low results.

Procedure. The method to be described is especially suitable for ceramic materials such as fireclay, firebrick, or silica brick. The finely ground sample should be dried at 110 °C. The weight of sample to be employed depends largely upon the silica content of the material, since not more than 35–40 mg of silica should be present in the aliquot employed for the determination. For samples containing more than 65 per cent SiO_2 (1), use 0.25 g; for samples containing less than 65 per cent SiO_2, use 0.50 g (2).

Place 7 g of A.R. sodium hydroxide pellets in a nickel crucible (4.5 × 4.5 cm) and fuse gently until the water is expelled and a clear melt results. Allow to cool, introduce the weighed sample evenly on to the solidified melt, moisten with a little ethanol and gently evaporate the ethanol on a hot plate; this reduces the tendency to spirting in the subsequent fusion. Heat gently over a Bunsen burner, with occasional rotation of the crucible, until the sodium hydroxide is just molten, after which raise the temperature to a dull red heat for 2–5 minutes; the sample should then have dissolved completely. Carefully cool the crucible by partial immersion in cold water; when the melt has just solidified transfer the hot crucible to a 400-cm³ nickel beaker and cover with a clock glass. Raise the clock glass slightly, fill the crucible with boiling water and replace the cover; this should suffice to dissolve the fused mass, otherwise add a little more boiling water. When the vigorous reaction has subsided, wash the clock glass and sides of the beaker with hot water; remove the crucible with clean tongs, carefully rinsing it inside and out with hot water. Dilute the suspension to 175 cm³; do not exceed this volume. Place 20 cm³ of concentrated hydrochloric acid in a 500-cm³ conical flask; pour the fusion extract, with swirling, into the acid, rinse the beaker with a little hot water, and add the rinsings to the flask. Cool rapidly to room temperature and dilute to 250 cm³ in a graduated flask.

Withdraw an aliquot part containing about 35 mg of silica, dilute it to about 250 cm³ in a 800-cm³ beaker, add 3 g of sodium hydroxide pellets and swirl until dissolved. Add 10 drops of thymol blue indicator (0.04 per cent solution in dilute ethanol, 1:4) followed by concentrated hydrochloric acid dropwise, swirling constantly, until the colour of the indicator changes from blue, through yellow, just to red; do not allow the solution to become too hot (3). Now add 10 cm³ of dilute hydrochloric acid (1:9) and dilute to 400 cm³. Add 50 cm³ of 10 per cent ammonium molybdate solution (4) from a burette: stir vigorously during the addition and for 1 minute afterwards. Allow to stand for 10 minutes, add 50 cm³ of concentrated hydrochloric acid, and immediately precipitate the yellow molybdosilicate by introducing 50 cm³ of the quinoline reagent (5) from a burette, stirring constantly. A cream-coloured, finely-divided precipitate of quinoline molybdosilicate forms. Warm the suspension to about 80 °C during about 10 minutes and maintain this temperature for 5 minutes in order to coagulate the precipitate. Cool in running water below 20 °C and collect the precipitate in a sintered glass filtering crucible (porosity No. 4); wash the precipitate six times with the special wash solution (6), taking care not to allow the precipitate to run dry during the filtration and washing. Dry at 150 °C for 2 hours and cool the covered crucible in a desiccator. Weigh as $(C_9H_7)_4H_4[SiO_4,12MoO_3]$.

    Notes. 1. 'Silica brick, No. 267' (a British Chemical Standard) may be used.

    2. 'Firebrick, No. 269' (a British Chemical Standard) may be used.

    3. This process is to ensure that the silica is in the correct form for reaction with ammonium molybdate. If the solution is too hot, the red colour may not develop.

    4. Prepare the **10 per cent ammonium molybdate solution** by dissolving 25 g ammonium molybdate in water and diluting to 250 cm³ in a polythene bottle. It keeps for about 4 weeks.

    5. Prepare the **2 per cent quinoline hydrochloride solution** by adding 10 cm³ pure quinoline to about 400 cm³ hot water containing 12.5 cm³ of concentrated hydrochloric acid, and stirring constantly. Cool the solution, add a little ashless filter pulp, and leave to settle. Filter the solution through a paper pulp pad, but do not wash. Dilute the filtrate with water to 500 cm³.

    6. Prepare the wash solution by diluting 5 cm³ of the 2 per cent quinoline hydrochloride solution with water to 200 cm³.

XI, 84. SULPHATE. Determination of sulphate as barium sulphate.

Discussion. The method consists in slowly adding a dilute solution of barium chloride to a hot solution of the sulphate slightly acidified with hydrochloric acid:

$$Ba^{2+} + SO_4^{2-} \rightarrow BaSO_4$$

The precipitate is filtered off, washed with water, carefully ignited at a red heat, and weighed as barium sulphate.

    The reaction upon which the determination depends appears to be a simple one, but is in reality subject to numerous possible errors; satisfactory results can be obtained only if the experimental conditions are carefully controlled. Before some of these are discussed, the student is recommended to read Sections **XI, 3–6**.

    Barium sulphate has a solubility in water of about 3 mg dm⁻³ at the ordinary temperature. The solubility is increased in the presence of mineral acids, because of the formation of the hydrogensulphate ion $(SO_4^{2-} + H^+ \rightleftharpoons HSO_4^-)$; thus the solubilities at room temperature in the presence of 0.1, 0.5, 1.0, and 2.0M-hydrochloric acid are 10, 47, 87, and 101 mg dm⁻³ respectively, but the solubility is less in the presence of a moderate excess of barium ions. Nevertheless, it is customary to carry out the precipitation in weakly acid solution in order to prevent the possible formation of the barium salts of such anions as chromate, carbonate, and phosphate, which are insoluble in neutral solutions; moreover, the precipitate thus obtained consists of large crystals, and is therefore more easily filtered (compare Section **XI, 4**). It is also of great importance to carry out the precipitation at boiling temperature, for the relative supersaturation is less at higher temperatures (compare Section **XI, 4**). The concentration of hydrochloric acid is, of course, limited by the solubility of the barium sulphate, but it has been found that a concentration of $0.05M$ is suitable; the solubility of the precipitate in the presence of barium chloride at this acidity is negligible. The precipitate may be washed with cold water, and losses, owing to solubility influences, may be neglected except for the most accurate work.

Barium sulphate exhibits a remarkable tendency to carry down other salts (see co-precipitation, Section **XI, 5**). Whether the results will be low or high will depend upon the nature of the co-precipitated salt. Thus barium chloride and barium nitrate are readily co-precipitated. These salts will be an addition to the true weight of the barium sulphate, hence the results will be high, since the chloride is unchanged upon ignition and the nitrate will yield barium oxide. The error due to the chloride will be considerably reduced by the very slow addition of hot dilute barium chloride solution to the hot sulphate solution, which is constantly stirred; that due to the nitrate cannot be avoided, and hence nitrate ion must always be removed by evaporation with a large excess of hydrochloric acid before precipitation. Chlorate has a similar effect to nitrate, and is similarly removed.

In the presence of certain cations (sodium, potassium, lithium, calcium, aluminium, chromium, and iron(III)), co-precipitation of the sulphates of these metals occurs, and the results will accordingly be low. This error cannot be entirely avoided except by the removal of the interfering ions. Aluminium, chromium, and iron may be removed by precipitation, and the influence of the other ions, if present, is reduced by considerably diluting the solution and by digesting the precipitate (Section **XI, 5**). It must be pointed out that the general method of re-precipitation, in order to obtain a purer precipitate, cannot be employed, because no simple solvent (other than concentrated sulphuric acid) is available in which the precipitate may be easily dissolved.

Positively charged barium sulphate, which is obtained when sulphate is precipitated by excess of barium ions, can be coagulated by the addition of a trace of agar-agar. About 1 mg of agar-agar as a 1 per cent aqueous solution will cause the flocculation of about 0.1 g of barium sulphate, but in practice somewhat larger quantities are generally used. The resulting precipitate does not creep up the sides of the vessel.

Negatively charged barium sulphate, obtained in the determination of barium is not appreciably improved by agar-agar; this precipitate as a rule, presents little difficulty in filtration.

Pure barium sulphate is not decomposed when heated in dry air until a temperature of about 1400 °C is reached:

$$BaSO_4 = BaO + SO_3$$

The precipitate is, however, easily reduced to sulphide at temperatures above 600 °C by the carbon of the filter paper:

$$BaSO_4 + 4C = BaS + 4CO$$

The reduction is avoided by first charring the paper without inflaming, and then burning off the carbon slowly at a low temperature with free access of air. If a reduced precipitate is obtained, it may be re-oxidised by treatment with sulphuric acid, followed by volatilisation of the acid and re-heating. The final ignition of the barium sulphate need not be made at a higher temperature than 600–800 °C (dull red heat). A Vitreosil or porcelain filtering crucible may be used, and the difficulty of reduction by carbon is entirely avoided.

Procedure. Weigh out accurately about 0.3 g of the solid* (or a sufficient amount to contain 0.05–0.06 g of sulphur) into a 400-cm³ beaker, provided with a stirring rod and clock-glass cover. Dissolve the solid† in about 25 cm³ of water, add 0.3–0.6 cm³ of concentrated hydrochloric acid, and dilute to 200–225 cm³. Heat the solution to boiling, add dropwise from a burette or pipette 10–12 cm³ of warm 5 per cent barium chloride solution (5 g BaCl₂,2H₂O in 100 cm³ of water — *ca.* 0.2*M*). Stir the solution constantly during the addition. Allow the precipitate to settle for a minute or two. Then test the supernatant liquid for complete precipitation by adding a few drops of barium chloride solution. If a precipitate is formed, add slowly a further 3 cm³ of the reagent, allow the precipitate to settle as before, and test again; repeat this operation until an excess of barium chloride is present. When an excess of the precipitating agent has been added, keep the covered solution hot, but not boiling, for an hour (steam bath, low-temperature hot plate, or small flame) in order to allow time for complete precipitation.‡ The volume of the solution should not be allowed to fall below 150 cm³; if the clock glass covering the beaker is removed, the under side must be rinsed off into the beaker by means of a stream of water from a wash bottle. The precipitate should settle readily, and a clear supernatant liquid should be obtained. Test the latter with a few drops of barium chloride solution for complete precipitation. If no precipitate is obtained, the barium sulphate is ready for filtration. The determination may be completed by either of the following processes.

(i) Filter paper method. Decant the clear solution through an ashless filter paper (Whatman, No. 40 or 540), and collect the filtrate in a clean beaker. Test the filtrate with a few drops of barium chloride: if a precipitate forms, the entire sample must be discarded and a new determination commenced. If no precipitate forms discard the liquid, rinse out the beaker, and place it under the funnel; this is in order to avoid the necessity of re-filtering the whole solution if any precipitate should pass through the filter. Transfer the precipitate to the filter with the aid of a jet of hot water from the wash bottle. Use a rubber-tipped rod ('policeman') to remove any precipitate adhering to the walls of the beaker or to the stirring rod, and transfer the precipitate to the filter paper. Wash the precipitate with small portions of hot water. Direct the jet as near the top of the filter paper as possible, and let each portion of the wash solution run through before adding the next. Continue the washing until about 5 cm³ of the wash solution gives no

---

* A.R. Potassium sulphate may be employed.

† For sulphates which are insoluble in water and acids, it is best to mix the finely powdered solid with six to twelve times its bulk of anhydrous sodium carbonate in a platinum crucible (Section **III, 45**), heat the covered crucible slowly to fusion, and maintain in the fused state for 15 minutes. The melt is extracted with water, the solution filtered, the residue washed with hot 1 per cent sodium carbonate solution, and the cold filtrate carefully acidified with hydrochloric acid (to methyl orange). The sulphate is determined as above.

‡ An equivalent result is obtained by allowing the solution to stand at the laboratory temperature for about 18 hours.

opalescence with a drop or two of silver nitrate solution. Eight or ten washings are usually necessary.

Fold the moist paper around the precipitate and place it in a porcelain or silica (Vitreosil) crucible, previously ignited to redness, cooled in a desiccator and weighed. Dry the paper by placing the loosely covered crucible upon a triangle several centimetres above a small flame. Then gradually increase the heat until the paper chars and volatile matter is expelled. Do not allow the paper to burst into flame, as mechanical loss may thus ensue. When the charring is complete, raise the temperature of the crucible to dull redness and burn off the carbon with free access of air* (crucible slightly inclined with cover displaced, Fig. III, 29). When the precipitate is white,† ignite the crucible at a red heat for 10–15 minutes. Then allow the crucible to cool somewhat in the air, transfer it to a desiccator, and, when cold, weigh the crucible and contents. Repeat the ignition with 10-minute periods of heating, subsequent cooling in a desiccator, etc., until constant weight (± 0.0002 g) is attained.

Calculate the percentage of SO_4 in the sample.

(ii) Filtering crucible method. Clean, ignite, and weigh either a porcelain filtering crucible or a Vitreosil filtering crucible (porosity, No. 4). Carry out the ignition either upon a crucible ignition-dish or by placing the crucible inside a nickel crucible at a red heat (or, if available, in an electric muffle furnace at 600–800 °C), allow to cool in a desiccator and weigh. Filter the supernatant liquid, after digestion of the precipitate, through the weighed crucible, using gentle suction. Reject the filtrate, after testing for complete precipitation with a little barium chloride solution. Transfer the precipitate to the crucible and wash with warm water until 3–5 cm^3 of the filtrate give no precipitate with a few drops of silver nitrate solution. Dry the crucible and precipitate in the oven or at 100–110 °C, and then ignite in a manner similar to that used for the empty crucible for periods of 15 minutes until constant weight is attained (1).

Note. 1. A rapid method for weighing the precipitate is as follows. (This procedure should *not* be employed by elementary students or beginners in the study of quantitative analysis.) Filter off the precipitated barium sulphate through a weighed filtering crucible (Gooch, sintered glass, or porcelain) and wash it with hot water until the chloride reaction of the washings is negative. Then wash five or six times with small volumes of ethanol, followed by five or six small volumes of anhydrous diethyl ether. Suck the precipitate dry on the pump for 10 minutes, wipe the outside of the crucible dry with a clean linen cloth, leave in a vacuum desiccator for 10 minutes (or until constant in weight), and weigh as $BaSO_4$. The result is of a moderate order of accuracy.

XI, 85. SULPHIDE. Determination of sulphur in mineral sulphides.

Introduction. The methods to be described apply to most insoluble sulphides. In these the sulphur is oxidised to sulphuric acid, and determined as barium sulphate. Two procedures are available for effecting the oxidation.

---

* Any dark matter on the crucible cover may be removed by placing it, clean side down, on a triangle, and heating it for some time.

† If the precipitate is slightly discoloured, add a drop or two of dilute sulphuric acid, evaporate gently, etc.

A. Dry process. *Discussion.* The oxidation is carried out by fusion with sodium peroxide, or, less efficiently, with sodium carbonate and potassium nitrate:

$$2FeS_2 + 15Na_2O_2 = Fe_2O_3 + 4Na_2SO_4 + 11Na_2O$$

The sulphide is fused with the sodium peroxide in an iron or nickel crucible (platinum is strongly attacked—Sections **III, 24** and **III, 35**), the fused mass treated with water, filtered, and acidified. The excess peroxide is removed by boiling, and the sulphate ion precipitated with barium chloride. The decomposition of the sulphide is rapid, but the method has several disadvantages. Amongst these may be mentioned: the slight attack on the metal crucible, thus preventing the subsequent determination of the metal content of the sample; the introduction of appreciable quantities of sodium salts, thus increasing the error due to co-precipitation (Sections **XI, 5** and **XI, 84**); and the possible contamination by sulphur from the flame gases, since sulphur dioxide is rapidly absorbed by the alkaline melt. The last error may be minimised by fitting the crucible into a hole in a sheet of asbestos or 'uralite', and keeping the crucible covered during the ignition (see Section **III, 45**).

Procedure. Dry some finely powdered pyrites* at 100 °C for 1 hour. Fit an iron or nickel crucible into a hole in an asbestos or 'uralite' board sufficiently large to allow two-thirds of the crucible to project below the board. Place about 1 g of A.R. anhydrous sodium carbonate into the crucible, and weigh accurately into it 0.4–0.5 g of the pyrites. Add 5–6 g of sodium peroxide, and mix well with a stout copper or nickel wire or with a thin glass rod. Wipe the wire or rod, if necessary, with a small piece of quantitative filter paper, and add the latter to the crucible; cover the mixture with a thin layer of peroxide. Place the crucible in the hole in the asbestos or 'uralite' sheet, and heat it with a very small flame. Increase the temperature gradually until after 10–15 minutes the crucible is at a *dull* red heat (the lower the temperature, the less is the crucible attacked) and just sufficiently hot to keep the mass completely fused. Remove the cover occasionally and examine the contents; be sure that the whole mass is fluid. Maintain the mass fluid for 15 minutes to complete the oxidation. Allow to cool, extract the crucible with water in a covered 600 cm³ beaker, rinse off the crucible-cover into the beaker, remove the crucible with a glass rod and wash it well; dilute to 300 cm³. Boil the solution for 15 minutes in order to destroy the excess of peroxide ($Na_2O_2 + 2H_2O = 2NaOH + H_2O_2$), neutralise part of the alkali by adding 5–6 cm³ of concentrated hydrochloric acid with stirring, add a Whatman 'accelerator' or a quarter of an 'ashless tablet', and filter through a Whatman No. 541 filter paper. Wash the residue at least ten times with hot 1 per cent sodium carbonate solution (10–20-cm³ portions). Acidify the combined filtrate and washings contained in a 800–1000 cm³ beaker with concentrated hydrochloric acid, using methyl red or methyl orange as indicator, and add 2 cm³ of acid in excess. Dilute, if necessary, to 600 cm³, and heat to boiling. Precipitate the sulphate by the slow addition with stirring of a boiling 5 per cent solution of barium chloride; the latter is added in slight excess of the calculated amount

---

* 'Burnt Pyrites, No. 45aG' (one of the Analysed Samples for Students) available from the Bureau of Analysed Samples, is suitable.

required, assuming the pyrites to be pure FeS_2. Complete the determination as in Section **XI, 84**.

Calculate the percentage of sulphur in the sample.

B. Wet process. *Discussion.* The sulphide is oxidised (i) by bromine in carbon tetrachloride solution, followed by nitric acid, (ii) by sodium chlorate and hydrochloric acid, or (iii) by a mixture of nitric and hydrochloric acids and a little bromine. The use of the first-named oxidising agent will be described; the reaction may be represented by:

$$2FeS_2 + 6HNO_3 + 15Br_2 + 16H_2O = 2Fe(NO_3)_3 + 4H_2SO_4 + 30HBr$$

The method has the advantage of not introducing any metallic ions, but it is essential to remove the excess of nitric acid (see *Discussion* in Section **XI, 84**). The action is slower than by the fusion method.

Procedure. Dry some finely powdered iron pyrites (1) at 100 °C for 1 hour. Weigh out accurately 0.4–0.5 g of the pyrites into a dry 400-cm^3 beaker, add 6 cm^3 of a mixture of 2 volumes of pure liquid bromine and 3 volumes of pure carbon tetrachloride (fume cupboard!), and cover with a clock glass. Allow to stand in the fume cupboard for 15–20 minutes and swirl the contents of the beaker occsionally during this period. Then add 10 cm^3 of concentrated nitric acid down the side of the beaker, and allow to stand for another 15–20 minutes, swirling occasionally as before. Heat the covered beaker below 100 °C by placing it on a hot plate or thermostatically controlled water bath until all action has ceased and most of the bromine has been expelled (about 1 hour). Raise the clock glass cover by glass hooks resting on the rim of the beaker, or displace it to one side, and evaporate the liquid to dryness on the steam bath. Add 10 cm^3 concentrated hydrochloric acid, mix well, and again evaporate to dryness to eliminate most of the nitric acid. Place the beaker in an oven or in an air bath at 95–100 °C for 30–60 minutes in order to dehydrate any silica which may be present (2). If the dry residue is heated at a temperature above 100 °C, loss of sulphuric acid may occur and the determination will be rendered useless. Moisten the cold, dry residue with 1–2 cm^3 of concentrated hydrochloric acid and, after an interval of 3–5 minutes, dilute with 50 cm^3 of hot water, and rinse the sides of the beaker and the cover glass with water. Digest the contents of the beaker at 100 °C for 10 minutes in order to dissolve all soluble salts. Allow the solution to cool for 5 minutes, and add 0.2–0.3 g of aluminium powder to reduce the iron(III). Gently swirl (or stir) until the solution becomes colourless. Allow to cool, add a Whatman 'accelerator', stir, and rinse down the cover glass and the sides of the beaker. Filter through a Whatman No. 540 paper, and collect the filtrate in an 800-cm^3 beaker; wash the filter thoroughly with hot water. Dilute the combined filtrate and washings to 600 cm^3 and add 2 cm^3 of concentrated hydrochloric acid. Precipitate the sulphate in the cold (3) by running in from a burette, *without stirring*, a 5 per cent solution of barium chloride at a rate not exceeding 5 cm^3 per minute until an excess of 5–10 cm^3 is present (4). When all the precipitant has been added, stir gently and allow the precipitate to settle for 2 hours, but preferably overnight. Filter through a No. 540 filter paper or, preferably, through a porcelain filtering crucible, wash with warm water until free from chloride, and ignite to constant weight as described under **A**.

Calculate the percentage of sulphur in the sample.

Notes. 1. The procedure is applicable to most **mineral sulphides**; many of these contain silica, and provision is made for the removal of this impurity in the experimental details.

2. If the iron pyrites or the sample of sulphide contains no appreciable proportion of silica, the heating at 95–100 °C may be omitted.

3. If a drop or two of tin(II) chloride solution is added to prevent reoxidation of the Fe(II) salt by air, precipitation of the barium sulphate may be made in boiling solution according to the usual procedure (Section **XI, 84**).

4. Calculate the volume of 5 per cent barium chloride solution which must be added from the approximate sulphur content of the iron pyrites FeS_2 or of the mineral sulphide.

XI, 86. SULPHITE. Determination of sulphite by oxidation to sulphate and precipitation as barium sulphate. *Discussion.* Sulphites may be readily converted into sulphates by boiling with excess of bromine water, sodium hypochlorite, sodium hypobromite, or ammoniacal hydrogen peroxide (equal volumes of 20-volume hydrogen peroxide and 1:1 ammonia solution). The excess of the reagent is decomposed by boiling, the solution acidified with hydrochloric acid, precipitated with barium chloride solution, and the **barium sulphate** collected and weighed in the usual manner (Section **XI, 84**).

XI, 87. THIOCYANATE. This anion may be determined in one of the following forms.

A. Copper(I) thiocyanate, CuSCN. The solution (100 cm³) should be neutral or slightly acid (hydrochloric or sulphuric acid), and contain not more than 0.1 g SCN. It is saturated with sulphur dioxide in the cold (or 50 cm³ of freshly prepared saturated sulphurous acid solution added), and then treated dropwise and with constant stirring with about 60 cm³ of 0.1 M-copper sulphate solution. The mixture is again saturated with sulphur dioxide (or 10 cm³ of saturated sulphurous acid solution added), allowed to stand for a few hours, collected on a weighed filtering crucible (Gooch, sintered glass, or porcelain), washed several times with cold water containing sulphurous acid until the copper is removed (potassium hexacyanoferrate(II) test), and finally once with ethanol. The precipitate is dried at 110–120 °C to constant weight, and weighed as CuSCN.

B. Barium sulphate, BaSO₄. The thiocyanate is oxidised with bromine water to sulphate, and the latter is determined by precipitation as barium sulphate. All other compounds containing sulphur must be absent. The alkali thiocyanate solution is treated with excess of bromine water, heated for 1 hour on the water bath, the solution acidified with hydrochloric acid, and the sulphuric acid precipitated and weighed as $BaSO_4$ (see Section **XI, 84**).

$$SCN^- + 4Br_2 + 4H_2O = SO_4^{2-} + 7Br^- + 8H^+ + BrCN$$

XI, 88. THIOSULPHATE. Two methods are commonly used for the determination of thiosulphates.

A. Conversion of thiosulphate to sulphate and determination as barium sulphate. *Discussion.* Thiosulphates are oxidised to sulphates by methods similar to those described for sulphites (Section **XI, 86**), e.g., by heating on a water bath with an ammoniacal solution of hydrogen peroxide, followed by boiling to expel the excess of the reagent. The sulphate is then determined as

barium sulphate, BaSO₄. One molecule of thiosulphate corresponds to two molecules of barium sulphate.

    B. Determination of thiosulphate as silver sulphide. *Discussion.* Thiosulphate may also be determined in almost neutral solution by the addition of a slight excess of $0.1M$-silver nitrate solution in the cold and, after 2–3 minutes, heating at 60 °C in a covered vessel. After cooling, the precipitate of **silver sulphide Ag₂S** is collected, washed with ammonium nitrate solution, water, and finally with ethanol. The precipitate is dried at 110 °C to constant weight, and weighed as Ag₂S.

XI, 89. References

1. T. B. Smith (1940). *Analytical Processes.* 2nd edn. London; Arnold.
2. H. A. Laitinen and W. E. Harris (1975). *Chemical Analysis.* 2nd edn. New York; McGraw-Hill.
3. R. Grzeskowiak and T. A. Turner (1973). *Talanta,* **20,** 351.
4. J. Bassett, G. B. Leton and A. I. Vogel (1967). 'Dioximes of Large Ring 1,2-Diketones and their Applications to the Determination of Bismuth, Nickel and Palladium', *Analyst,* **92,** 279.

XI, 90. Selected bibliography

1. C. J. Rodden (1950). *Analytical Chemistry of the Manhatten Project.* New York; McGraw-Hill.
2. R. Fresenius and G. Jander (1940–58). *Handbuch der Analytischen Chemie. Dritte Teil. Bestimmungs und Trennungsmethoden.* Berlin; Springer-Verlag.
3. C. Duval (1954–57). *Traité de Micro-Analyse Minérale.* Vol. I–IV. Paris; Presses Scientifiques.
4. G. Charlot and D. Bézier (1957). *Quantitative Inorganic Analysis* (trans. R. C. Murray). London; Methuen and Co.
5. L. Gordon, M. L. Salutsky, and H. H. Willard. (1959). *Precipitation from Homogeneous Solution.* New York; John Wiley.
6. C. N. Reilley (ed) (1960). *Advances in Analytical Chemistry and Instrumentation.* Vol. I. H. Flashka and A. J. Barnard, Jr. *Tetraphenylboron (TPB) as an Analytical Reagent.* New York; Interscience Publishers.
7. F. E. Beamish and J. A. Page. 'Inorganic Gravimetric and Volumetric Analysis', *Analytical Chemistry,* 1956, **28,** 694; 1958, **30,** 805.
8. F. E. Beamish and A. D. Westland (1960). 'Volumetric and Gravimetric Analytical Methods for Inorganic Compounds', *Analytical Chemistry,* **32,** 249R.
9. C. L. Wilson and D. W. Wilson (ed.) (1960). *Comprehensive Analytical Chemistry.* Vol. 1A, *Classical Analysis.* Amsterdam and London; Elsevier.
10. L. Erdey. *Gravimetric Analysis.* Part 1 (1963) and Parts 2 and 3 (1965). Oxford; Pergamon Press.
11. N. H. Furman (ed.) (1962). *Standard Methods of Chemical Analysis.* Vol. 1, *The Elements,* 6th edn. Princeton, New Jersey; Van Nostrand.
12. Hopkins and Williams (1964). *Organic Reagents for Metals and for Certain Radicals.* Vol. II, London.

PART E <u>ELECTROANALYTICAL METHODS</u>

CHAPTER XII   ELECTRO-GRAVIMETRY

XII, 1. THEORY OF ELECTRO-GRAVIMETRIC ANALYSIS. In electro-gravimetric analysis the element to be determined is deposited electrolytically upon a suitable electrode. Filtration is thus avoided, and co-deposition, if the experimental conditions are carefully controlled, is very rare. The method, when applicable, has many advantages, and we shall therefore study the theory of the process in order to understand how and when it may be applied.

Electro-deposition is governed by Ohm's law and by Faraday's two laws of electrolysis (1833–34). The latter state:

1. *The amounts of substances liberated at the electrodes of a cell are directly proportional to the quantity of electricity which passes through the solution.*

2. *The amounts of different substances which are deposited, or liberated, by the same quantity of electricity are proportional to their chemical equivalents.*

It follows from the second law that when a given current is passed in series through solutions containing, say, copper sulphate and silver nitrate respectively, then the weights of copper and silver deposited will be in the ratio of their equivalents, viz., 63.54/2: 107.868.

Ohm's law expresses the relation between the three fundamental quantities, current, electromotive force, and resistance.

The current I is directly proportional to the electromotive force E *and inversely proportional to the resistance* R, i.e.,

$$I = E/R$$

Electrical units. The fundamental SI unit is the unit of current which is called the **ampere** and which is defined as the constant current which, if maintained in two parallel rectilinear conductors of negligible cross section and of infinite length and placed one metre apart in a vacuum, would produce between these conductors a force equal to 2×10^{-7} newton per metre length.

The unit of electrical potential is the **volt** which is the difference of potential between two points of a conducting wire which carries a constant current of one ampere, when the power dissipated between these two points is one watt, or one joule per second.

The unit of electrical resistance is the **ohm**, which is the resistance between two points of a conductor when a constant difference of potential of one volt applied between these two points produces a current of one ampere.

Prior to the introduction of the above absolute units in 1948, 'international' units were in use; the relationships between the two sets of units are:

1 international ohm $= 1.00049$ absolute ohms;
1 international ampere $= 0.99985$ absolute ampere;
1 international volt $= 1.00034$ absolute volts;

for electro-gravimetric analysis, the differences are insignificant.

The unit quantity of electricity is the **coulomb**, and is defined as the quantity of electricity passing when 1 ampere flows for 1 second. Each coulomb will deposit $1.11800\,mg$ of silver.

The weight of an element liberated by the passage of 1 coulomb of electricity (or 1 ampere for 1 second) is called the **electrochemical equivalent** of the element. The equivalent of silver is 107.868, hence $107.868/0.00111800$, i.e., 96 483 coulombs will be required to liberate 1 equivalent of silver. The value generally employed is 96 500 coulombs and this is termed the **Faraday constant** (F); this is the charge associated with one mole of electrons and has the accurate value $96\,487\,C\,mol^{-1}$.

Some terms used in electro-gravimetric analysis.   *Voltaic (galvanic) and electrolytic cells.*   A cell consists of two electrodes and one or more solutions in an appropriate container. If the cell can furnish electrical energy to an external system it is called a *voltaic* (or *galvanic*) cell. The chemical energy is converted more or less completely into electrical energy, but some of the energy may be dissipated as heat. If the electrical energy is supplied from an external source the cell through which it flows is termed an *electrolytic cell* and Faraday's laws account for the material changes at the electrodes. A given cell may function at one time as a galvanic cell and at another as an electrolytic cell: a typical example is the lead accumulator or storage cell. During an electro-gravimetric operation a galvanic cell is built up as the products form on the electrodes. If the current is switched off the products tend to produce a current in a direction opposite to the direction in which the electrolysis current was passed. The voltage applied to the electrolysis cell $(E_{appl.})$ must exceed that of the galvanic cell which is produced (this is always in opposition to the applied e.m.f. and can be written E_{back}) and must also overcome the resistance of the solution to the passage of current (i.e., the IR drop). The amount of current that flows is given by Ohm's law:

$$I = (E_{appl.} - E_{back})/R$$

Cathode.   The cathode is the electrode at which reduction occurs. In an electrolytic cell it is the electrode attached to the negative terminal of the source, since electrons leave the source and enter the electrolysis cell at that terminal. The cathode is the positive terminal of a galvanic cell, because such a cell accepts electrons at this terminal.

Anode.   The anode is the electrode at which oxidation occurs. It is the positive terminal of an electrolysis cell or the negative terminal of a voltaic cell.

Polarised electrode.   An electrode is polarised if its potential deviates from the reversible or equilibrium value. An electrode is said to be *depolarised* by a substance if that substance lowers the amount of polarisation.

Current density.   The current density is defined as the current per unit area of electrode surface. It is generally expressed in amperes per square cm (or per square dm) of the electrode surface.

Current efficiency.   By measuring the amount of a particular substance that is deposited and comparing this with the theoretical quantity (calculated by Faraday's laws), the actual current efficiency may be obtained. In general,

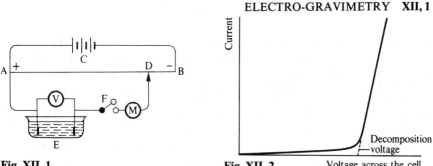

Fig. XII, 1

Fig. XII, 2   Voltage across the cell

analytical depositions show low current efficiencies owing to other reactions which occur during the electrolysis; for example, liberation of hydrogen during the later stages of the deposition of a metal at a cathode.

Decomposition potential.   If a small voltage of, say, 0.5 volt is applied to two smooth platinum electrodes immersed in a solution of M-sulphuric acid, then an ammeter placed in the circuit will at first show that an appreciable current is flowing, but its strength decreases rapidly, and after a short time it becomes virtually equal to zero. If the applied voltage is gradually increased, there is a slight increase in the current until, when the applied voltage reaches a certain value, the current suddenly increases rapidly with increase in the e.m.f. It will be observed, in general, that at the point at which there is a sudden increase in current, bubbles of gas commence to be freely evolved at the electrodes. The experiment may be carried out by means of the apparatus shown diagrammatically in Fig. XII, 1. A storage battery C is connected across a uniform resistance wire AB, along which a contact maker D can be moved; the fall of potential between A and D can thus be varied gradually. Two smooth platinum electrodes are immersed in M-sulphuric acid in the cell E. V is a suitable voltmeter placed between the two electrodes of the cell; M is a milliammeter and F is a switch. When the sliding contact is near to A only a small potential is applied to the electrodes of the cell; the fall of potential across the cell and the current flowing through it are read off on the instruments V and M respectively. The applied voltage is slowly increased by moving D towards B, and the readings of the voltmeter and ammeter noted after allowing a short time for the values to become steady. Upon plotting the current against the applied voltage, a curve similar to that shown in Fig. XII, 2 is obtained; the point at which the current suddenly increases is evident, and in the instance under consideration is about 1.7 volts. The voltage at this point is termed the **decomposition potential**, and it is at this point that the evolution of both hydrogen and oxygen in the form of bubbles is first observed. We may define the **decomposition potential** of an electrolyte as the minimum external voltage that must be applied in order to bring about continuous electrolysis.

If the circuit is broken after the e.m.f. has been applied, it will be observed that the reading on the voltmeter V is at first fairly steady, and then decreases, more or less rapidly, to zero. The cell E is clearly behaving as a source of current, and is said to exert a **back** or **counter** or **polarisation e.m.f.**, since the latter acts in a direction opposite to that of the applied e.m.f. This back e.m.f. arises from the accumulation of oxygen and hydrogen at the anode and cathode respectively; two gas electrodes are consequently formed, and the potential difference between them opposes the applied e.m.f. When the primary current from the battery is shut off, the cell produces a moderately steady current until the gases at the electrodes are either used up or have diffused away; the voltage then falls to zero. This back e.m.f. is present even when the current from the battery passes through the cell and accounts for the shape of the curve in Fig. XII, 2. It is evident that the minimum value of the counter e.m.f. may be computed, for it is equal to the algebraic difference of the electrode potentials which exist at the anode and cathode respectively. This calculation will be referred to again in the succeeding paragraphs.

The back e.m.f. is usually regarded as being made up of three components:

(*a*) The reversible back e.m.f. This is the reversible e.m.f. of the voltaic cell set up by the passage of the electrolytic current, and is based upon concentrations of solutes in the bulk of the solution.

(*b*) A concentration polarisation e.m.f. or concentration overvoltage. This is the effect of changes in concentration at an electrode surface with reference to the concentration of the bulk of the solution. Thus in the electrolysis of an acidic solution of copper sulphate between platinum electrodes, concentration changes occur both at the cathode and the anode. At the cathode depletion of copper ions occurs near the surface; the reversible potential of the copper electrode therefore shifts in the negative direction. At the anode accumulation of hydrogen ions $(2H_2O \rightarrow O_2 + 4H^+ + 4e)$ and perhaps of oxygen (if the solution is not already saturated with it) causes the reversible potential of the oxygen electrode to shift in the positive direction. Both effects tend to increase the back e.m.f. The concentration overvoltage is increased by increased current density and decreased by stirring.

(*c*) An activation overvoltage. This is the departure of the potential of an electrode from its reversible value due to the passage of the electrolytic current.

It is observed that whereas the decomposition potentials of salt solutions vary considerably, those for acids and alkalis, with the exception of the halogen acids, are all approximately 1.7 volts. It is therefore concluded that the same electrolytic process occurs with these acids and bases; this can only be the evolution of hydrogen at the cathode and of oxygen at the anode:

$2H^+ + 2e \rightleftharpoons H_2$ (acidic medium)
$2H_2O + 2e \rightleftharpoons H_2 + 2OH^-$ (basic medium)
$\quad 2H_2O \rightleftharpoons O_2 + 4H^+ + 4e$ (acidic medium)
$\quad 4OH^- \rightleftharpoons O_2 + 2H_2O + 4e$ (basic medium)

The net cell reaction is the decomposition of water:

$$2H_2O = 2H_2 + O_2$$

With the halogen acids in M solution, the halogen and not oxygen is liberated at the anode, since the discharge of the halogen ion can occur more readily than that of hydroxide ion; the discharge potential varies with the halogen.

For a similar electrolysis of M-zinc sulphate solution, the reactions at the cathode and anode are respectively:

$Zn^{2+} + 2e \rightleftharpoons Zn$
$2H_2O \rightleftharpoons O_2 + 4H^+ + 4e$

an oxygen electrode being produced at the anode.

XII, 2. ELECTRODE REACTIONS. In electro-gravimetric analysis we are largely concerned with the electrolysis of salt solutions, and it is therefore proposed to study in some detail the reactions which take place at the electrodes. Let us consider the electrolysis of a molar solution* of zinc bromide between smooth platinum electrodes. The application of a voltage will result in the deposition of zinc on the cathode (thus producing a zinc electrode) and of bromine at the anode (thus producing a bromine electrode). The reaction at the cathode is:

$$Zn^{2+} + 2e \rightleftharpoons Zn$$

i.e., a reduction (Section **II, 23**), and that at the anode is:

$$2Br^- \rightleftharpoons Br_2 + 2e$$

i.e., an oxidation. Thus reduction occurs at the cathode and oxidation at the anode. We may calculate the potential at the cathode at 25 °C from the formula (Section **II, 20**):

$$E_{cathode} = E^{\ominus}{}_{Zn} + \frac{0.0591}{2}\log [Zn^{2+}] = E^{\ominus}{}_{Zn}$$

since $[Zn^{2+}] = 1$ mole dm^{-3}.

At the anode:

$$E_{anode} = E^{\ominus}{}_{Br_2} - \frac{0.0591}{1}\log [Br^-] = E^{\ominus}{}_{Br_2} - \frac{0.0591}{1}\log 2$$

since $[Br^-] = 2$ mole dm^{-3}.

The e.m.f. of the resulting cell will therefore be:

$$E^{\ominus}{}_{Zn} - \{E^{\ominus}{}_{Br_2} - \frac{0.0591}{1}\log 2\} = 0.76 - (-1.07 - 0.02) = 1.85 \text{ volts}$$

In general, it may be stated that the theoretical back or polarisation e.m.f. E_{back} is given by:

$$E_{back} = E_{cathode} - E_{anode}$$

where $E_{cathode}$ and E_{anode} are calculated as already described (Section **II, 20**).

XII, 3. OVERPOTENTIAL.

It has been found by experiment that the decomposition voltage of an electrolyte varies with the nature of the electrodes employed for the electrolysis and is, in many instances, higher than that computed from the difference of the *reversible* electrode potentials. The excess voltage over the calculated back e.m.f. is termed the **overvoltage**.† Overpotential may occur at the anode as well as at the cathode. The decomposition voltage E_D is therefore:

$$E_D = E_{cathode} + E_{o.c.} - (E_{anode} + E_{o.a.})$$

where $E_{o.c.}$ and $E_{o.a.}$ are the overpotentials at the cathode and anode respectively.

The overpotential at the anode or cathode is a function of the following variables:

1. The nature and the physical state of the metal employed for the electrodes.

---

* i.e. molar with respect to Zn^{2+}. Ideally, concentration should be replaced by activity; the former is, however, sufficiently accurate for our purpose.

† The term **overvoltage** should strictly be applied to a cell, and **overpotential** to a single electrode.

The fact that reactions involving gas evolution usually require less overvoltage at platinised than at polished platinum electrodes is due to the much larger area of the platinised electrode and the smaller current density at a given electrolysis current.

2. The physical state of the substance deposited. If it is a metal, the overpotential is usually small; if it is a gas, such as oxygen or hydrogen, the overpotential is relatively great.

3. The current density employed. For current densities up to 0.01 ampere cm^{-2}, the increase in overpotential is very rapid; above this figure the increase in overpotential continues, but less rapidly.

4. The change in concentration, or the concentration gradient, existing in the immediate vicinity of the electrodes; as this increases, the overpotential rises. The concentration gradient depends upon the current density, the temperature, and the rate of stirring of the solution.

5. The overpotential decreases, often very considerably, with increasing temperature.

The overpotential of hydrogen is of great importance in electrolytic determinations and separations. Some values are collected in Table XII, 1.

Table XII, 1   Hydrogen overpotential on various cathodes (in volts)

Cathode	Solution	First visible gas bubbles	Current density	
			0.01 amp. cm^{-2}	0.1 amp. cm^{-2}
Pt (bright)	$1M$-H_2SO_4	0.00	0.09	0.16
Au	$1M$-H_2SO_4	0.02	0.4	1.0
Ag	$1M$-H_2SO_4	0.10	0.3	0.9
Co	$0.005M$-H_2SO_4	0.07	0.2	—
Ni	$1M$-H_2SO_4	0.14	0.3	0.7
Cu	$1M$-H_2SO_4	0.19	0.4	0.8
Bi	$1M$-H_2SO_4	0.39	0.4	—
Sn	$0.05M$-H_2SO_4	0.40	0.5	1.2
Cd	$0.005M$-H_2SO_4	0.4	—	1.2
Pb	$1M$-H_2SO_4	0.04	0.4	1.2
Zn	$1M$-H_2SO_4	0.5	0.7	1.1
Hg	$1M$-H_2SO_4	0.8	1.2	1.3

The hydrogen overpotential is greatest with the relatively soft metals, such as bismuth, cadmium, tin, lead, and zinc, and especially mercury (in the last case the value is about 1.0 volt, but is dependent on the current density). The existence of hydrogen overpotential renders possible the electro-gravimetric determination of metals, such as cadmium and zinc, which otherwise could not be deposited before the reduction of hydrogen ion. In alkaline solution, the hydrogen overpotential is slightly higher (0.05–0.3 volt) than in acid solution.

The oxygen overpotential is about 0.4–0.5 volt at a polished platinum anode in acid solution, and is of the order of 1 volt in alkaline solution with current densities of 0.02–0.03 amp. cm^{-2}. As a rule the overpotential associated with the deposition of some metals (Ag, Hg, Cu, Pb, Cd, and Zn) on the cathode is quite small (about 0.1–0.3 volt) because the depositions proceed nearly reversibly. When the depositions do not proceed reversibly the overpotential may attain the same order of magnitude as for hydrogen evolution. Thus nickel ions show an overpotential of about 0.6 volt at a mercury cathode.

XII, 4. COMPLETENESS OF DEPOSITION. The voltage $E_{appl.}$ applied to an electrolytic cell must overcome the decomposition potential E_D or back e.m.f., as well as the ohmic resistance of the solution, i.e., $E_{appl.}$ must be equal to or greater than $E_D + IR$. It has been shown (Section **XII, 3**) that:

$$E_D = E_{cathode} + E_{o.c.} - (E_{anode} + E_{o.a.})$$

where $E_{cathode}$ and E_{anode} are the reversible cathode and anode potentials respectively, $E_{o.c.}$ is the overpotential effect at the cathode and $E_{o.a.}$ is the overpotential effect at the anode. Overpotential at the cathode makes the effective cathode potential more negative than the equilibrium value, and at the anode causes the effective anode potential to be more positive. Let us consider the variations in e.m.f. at the cathode during the deposition of a metal in an electrolytic determination. Let the ionic concentration at the commencement of the estimation be c_i. For a bivalent metal, e.g., copper, the cathode potential at 25 °C will be:

$$E^{\ominus}_{M^{II}} + \frac{0.0591}{2} \log c_i = E^{\ominus}_{M^{II}} + 0.0296 \log c_i \text{ volts}$$

If the ionic concentration is reduced by deposition to one-ten-thousandth of its original value (i.e., to secure an accuracy of 0.01 per cent in the determination), the new cathode potential will be:

$$E^{\ominus}_{M^{II}} + 0.0296 \log (c_i \times 10^{-4})$$
$$= E^{\ominus}_{M^{II}} + 0.0296 \log c_i + 0.0296 \log 10^{-4}$$
$$= (E^{\ominus}_{M^{II}} + 0.0296 \log c_i) - 4 \times 0.0296$$
$$= \text{Potential at commencement of deposition} - 0.118 \text{ volt}$$

This reduction in potential is independent of the value of c_i, and hence whenever the ionic concentration is reduced to one-ten-thousandth of its initial value (this may be regarded as the ultimate limit of an electro-gravimetric determination, although for most purposes an accuracy of 0.1 per cent is regarded as sufficient), the potential is altered by $4 \times 0.0591/2 = 0.118$ volt for a bivalent ion. For a univalent ion, the change is $4 \times 0.0591/1 = 0.236$ volt, and for a trivalent ion it is $4 \times 0.0591/3 = 0.079$ volt. Since the back e.m.f. is produced by a metal cathode acting as the negative pole of a cell, the positive pole being, say, oxygen, it follows that the back e.m.f. will become greater as the cathode becomes more negative during the course of the analysis. Otherwise expressed, the decomposition potential increases as the deposition of the metal proceeds. For *quantitative deposition* the applied e.m.f. must equal or exceed the decomposition voltage when the concentration of the given cation is negligibly small (say 10^{-4} of the initial value).

It is important to know the conditions for the deposition of the metal in preference to hydrogen in an electrolysis. The condition for the deposition of the metal is evidently that the potential difference between the electrolyte and the cathode, $E_{solution, metal}$, must be less than the reversible deposition potential of hydrogen plus the overpotential of hydrogen (o_{H_2}) for the metal under consideration. The relationship may be expressed in several ways:

$$E_{solution, metal} < (E_{H^+, H_2} + o_{H_2}),$$
$$< (1 - E_{H_2, H^+} + o_{H_2}),$$
$$< (0.059 \, pH + o_{H_2}) \quad \text{(at 25 °C)}$$

XII, 5. ELECTROLYTIC SEPARATION OF METALS. When a constant current is passed through a solution containing two or more electrolytes the electrochemical process with the most positive reduction potential will occur first at the cathode, followed by the next most positive electrochemical process, etc. Thus if a current is passed through a solution containing copper, hydrogen and cadmium ions, copper will be deposited first at the cathode. As the copper deposits, the electrode potential decreases, and when the potential is equal to that of the hydrogen ions hydrogen gas will form at the cathode. The potential at the cathode will remain virtually constant as long as hydrogen is evolved, which is usually until all the water is electrolysed: the potential of the cathode cannot therefore become sufficiently negative to cause the deposition of cadmium ions (see Section **II, 20**). Thus metal ions with positive reduction potentials may be separated, without external control of the cathode potential, from metal ions having negative reduction potentials. In practice, the hydrogen overpotential on the cathode plus the reversible reduction potential of the hydrogen ions must be less than the negative reduction potential of any metal ions that remain in solution. For example, copper ions in a solution containing $1M$-hydrogen ions may be separated from all metallic ions whose reduction potentials are more negative than about -0.8 volt (the hydrogen overpotential on a copper electrode); the reversible reduction potential of hydrogen ions is 0.0 volt in this medium. The principle of separating, by electro-deposition at constant current, metallic ions whose reduction potentials are on different sides of the potential of the hydrogen electrode finds application for analytical separations and determinations. It will be clear from what has been stated in the preceding Section that the initial deposition potentials of two metals must differ by at least 0.25 volt for a virtually quantitative separation to be theoretically possible. This minimum value would require a very precise control of the potential drop at the cathode; for most practical purposes, the difference in potential should be at least 0.4 volt. (The procedure for controlled cathode potential is discussed in Sections **XII, 7–14**.) Certain metals can be separated electrolytically with great ease, for example, copper from zinc, nickel, and cobalt, silver from copper, etc; when, however, the standard potentials of the two metals differ only slightly, the electro-separation is more difficult. The obvious method is to alter the electrode potential of one of the metals in some way. This is most simply achieved by decreasing the ionic concentration of the ion being discharged by incorporating it in a complex ion of large stability constant (Section **II, 11**). The deposition potential of the metal forming a complex ion is thus raised. Furthermore, the overpotential at the small ionic concentration is also usually increased. A possible consequence of changing the electrode potential in this way is that a metal which in simple ionic solution is liberated at a lower voltage than another metal, may exhibit the reverse behaviour in a complex-forming medium.

Some results for the deposition potentials $(= -E_{cathode})$ of some metals in simple and alkali cyanide solutions are given in Table XII, 2.

An interesting application of these results is to the direct quantitative separation of copper and cadmium. The copper is first deposited in acid solution; the solution is then neutralised with pure aqueous sodium hydroxide, potassium cyanide is added until the initial precipitate just re-dissolves, and the cadmium is deposited electrolytically. Another application is to the separation of copper and bismuth: these two metals cannot be separated electrolytically from solutions of their simple salts. If cyanide is added, the copper ions form a cyano complex, and

Table XII, 2 Deposition potentials E_D of some metals in simple and in alkali cyanide solutions

Metal	E_D for 0.1M solutions of M^{2+} ions (volt)	Concentration of excess of KCN per 0.1 mol. of simple metallic cyanide		
		0.2M	0.4M	M
Zn	+0.79	+1.03	+1.18	+1.23
Cd	+0.44	+0.71	+0.87	+0.90
Cu	−0.31	+0.61	+0.96	+1.17

the deposition potential is much more negative than before; the bismuth-ion concentration and the electrode potential are hardly affected and a separation from copper becomes possible, the bismuth depositing first. The separation is improved if tartrate is also added.

XII, 6. CHARACTER OF THE DEPOSIT. The ideal deposit for analytical purposes is adherent, dense, and smooth; in this form it is readily washed without loss. Flaky, spongy, powdery, or granular deposits adhere only loosely to the electrode, and for this and other reasons should be avoided.

As a rule, more satisfactory deposits are obtained when the metal is deposited from a solution in which it is present as complex ions rather than as simple ions. Thus silver is obtained in a more adherent form from a solution containing the $[Ag(CN)_2]^-$ ion than from silver nitrate solution. Nickel when deposited from solutions containing the complex ion $[Ni(NH_3)_6]^{2+}$ is in a very satisfactory state for drying and weighing. Mechanical stirring often improves the character of the deposit, since large changes of concentration at the electrode are reduced, i.e., concentration polarisation is brought to a minimum.

Increased current density up to a certain critical value leads to a diminution of grain size of the deposit. Beyond this value, which depends *inter alia* upon the nature of the electrolyte, the rate of stirring, and the temperature, the deposits tend to become unsatisfactory. At sufficiently high values of the current density, evolution of hydrogen may occur owing to the depletion of metal ions near the cathode. If appreciable evolution of hydrogen occurs, the deposit will usually become broken up and irregular; spongy and poorly adherent deposits are generally obtained under such conditions. For this reason the addition of nitric acid or ammonium nitrate is often recommended in the determination of certain metals, such as copper; bubble formation is thus considerably reduced. The action of the nitrate ion at the copper cathode can be represented by:

$$NO_3^- + 10H^+ + 8e = NH_4^+ + 3H_2O$$

The nitrate ion is reduced to ammonium ion at a lower (i.e., less negative) cathode potential than that at which hydrogen ion is discharged, and therefore acts to decrease hydrogen evolution. The nitrate ion acts as a **cathodic depolariser**.

Raising the temperature, say, to between 70 and 80 °C often improves the physical properties of the deposit. This is due to several factors, which include the decrease in resistance of the solution, increased rate of stirring and of diffusion, and changed over potential effects. Stirring by heat or by mechanical means makes possible the use of a high current density and therefore a rapid deposition.

In practice, two methods of electrolysis are utilised. In the first method stationary electrodes are used and the solution is not stirred; small current densities must of necessity be applied in order to secure a coherent deposit, and the procedure is a slow one (**slow electrolysis**). In the second method, which has largely superseded the first, the solution is rapidly stirred (**rapid electrolysis**). Various devices are employed for stirring. An independent mechanical stirrer may be used, but it is more usual to have a rotating anode, which may consist, after Sand (Ref. 7), of a platinum gauze cylinder surrounded by a similar (but stationary) cylinder, which constitutes the cathode, the intervening space being small (3–5 mm). A very much higher current density may then be applied without seriously affecting the purity or the physical character of the deposit. The stirring results in a liberal supply of metal ions always being present near the cathode, and consequently the current is principally used in the deposition of the metal. A considerable saving of time is thus effected, and this accounts for the popularity of the method. It must be emphasised that when the electrolysis is complete, the current must not be switched off as long as the electrodes are in the solution. If the circuit were broken, the counter e.m.f. would come into play, and this would cause part of the metallic deposit to pass back into solution.

XII, 7. ELECTROLYTIC SEPARATION OF METALS WITH CONTROLLED CATHODE POTENTIAL.
For electrolysis to proceed the e.m.f. which must be applied to an aqueous solution of an electrolyte is given by the expression (compare Section **XII, 4**):

$$E_{appl.} = E_{cathode} + E_{o.c.} - (E_{anode} + E_{o.a.}) + IR$$

In the common method of electro-gravimetric analysis, a voltage slightly greater than E_D is applied and the electrolysis is allowed to proceed without further attention, except perhaps to occasionally increase the applied voltage to maintain the current at approximately the same value. This process, termed **constant current electrolysis**, when applied to the separation of metals is limited to the separation of those metals below hydrogen in the electrochemical series from those above hydrogen. Following the deposition of the first metal (the one lower in the series) hydrogen is evolved at the cathode, and as long as the solution remains acid the second metal is not deposited. This is exemplified by the separation of copper from nickel, and from zinc in a sulphuric acid solution. If the second metal lies only slightly above the other in the electrochemical series, separation is no longer possible unless the decomposition potentials can be displaced either through the formation of an appropriate complex ion or by other means (compare Section **XII, 5**). The separation of such a mixture may be effected by the application of **controlled cathode potential electrolysis**. An auxiliary standard electrode (which may be a saturated calomel electrode with the tip of the salt bridge very close to the cathode or working electrode) is inserted in the solution, and thus the voltage between the cathode and the reference half-cell may be measured. It is thus possible to isolate the effect at the cathode and to limit the potential at this electrode during the electrolysis to a definite value by decreasing the overall voltage applied to the cathode and anode. We have already seen (Section **XII, 4**) that every ten-fold decrease in metal-ion concentration makes the cathode potential $0.0591/n$ volt more negative at $25\,^{\circ}\text{C}$ (n is the valency of the ion). For an accuracy of 0.1 per cent, the concentration of the ion is reduced to 10^{-3} of the original value, consequently the potential will decrease by

$3 \times 0.0591/n$ volt, i.e., 0.177 volt for a univalent ion, 0.088 volt for a divalent ion, etc. Thus by controlling the cathode potential with the aid of an auxiliary electrode, the separation of one metal from another lying somewhat higher in the electrochemical series becomes possible. Manual control of the potential may become tedious except for occasional determinations, but the time may be materially reduced by the use of high current densities; however methods for the automatic control of cathode potential have been developed (see Refs. 1–4).

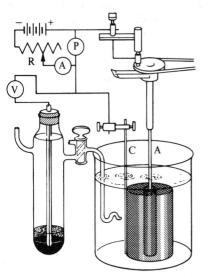

Fig. XII, 3

A simple circuit and apparatus for controlled cathode potential electro-analysis is shown in Fig. XII, 3; this will serve to illustrate the principles of the technique involved. The various components of the apparatus are: a source of current, which may be a large storage battery; a saturated calomel electrode; a voltmeter P to indicate the e.m.f. applied to the cell for the electrolysis; an adjustable resistance R capable of carrying current up to 10–15 amps.; a platinum-gauze cathode C; and a platinum-gauze anode A which can be rotated. The potential between the saturated calomel electrode and the cathode must be measured with an instrument which draws little or negligible current from the reference cell: a digital voltmeter V is satisfactory. The total potential measured E is equal to the difference between the potentials of the calomel electrode and the cathode:

$$E = E_{cal.\,sat.} - (E_{cathode} + E_{o.c.})$$

Since $E_{cal.\,sat.}$ is known, the electrode potential of the cathode can be easily referred to the hydrogen scale. In order to prevent the cathode potential from exceeding a fixed value, it is simply necessary to decrease the potential applied to the cathode and anode by increasing the value of the resistance R.

It may be noted that in evaluating the limiting cathode potential to effect the separation of one metal from another, a simple computation of the equilibrium potential from the Nernst equation is insufficient; the equilibrium potential must be increased by the overpotential. The latter depends, *inter alia*, upon the rate of stirring, the current density, the temperature, as well as upon the nature of the metal surface; in consequence, the limiting potential must be established by other methods.

One procedure utilises current–working electrode potential curves for the two substances to be separated. The current–electrode potential curve is determined for each reaction under exactly the same conditions that will prevail in the actual analysis. The potential of the working electrode is increased in regular increments by increasing the total voltage applied to the cell and is measured against a standard reference electrode (usually a saturated calomel electrode). The current is observed at each value of the electrode potential: to minimise change in the composition of the solution, especially when the current is large, the cell circuit should be closed only long enough to make the current measurement at each value of the electrode potential. Schematic current–cathode potential curves for the reduction of two substances X and Y are shown in Fig. XII, 4. To initiate the deposition of the substance X, the cathode potential E_c must be at least as large as the value of the 'decomposition' potential indicated by x, but the potential should not exceed y, for if it does Y will deposit also. Consequently, for the complete deposition of X, the cathode potential should be limited to a value slightly less than that which corresponds to the potential y. The initial current should not exceed the value indicated by z. As the deposition of the substance X proceeds, the cathode potential tends to become more negative but is prevented from exceeding the value y by decreasing the voltage applied to the cell.

The course of the current and of the total applied voltage during a typical controlled potential electro-deposition of copper is shown in Fig. XII, 5 (due to Lingane). It concerns the deposition of 0.2 g of copper at a platinum cathode from an acid tartrate solution (pH = 4.5), in the presence of hydrazine as anodic depolariser, with the cathode potential maintained at -0.36 ± 0.02 volt *vs.* S.C.E.

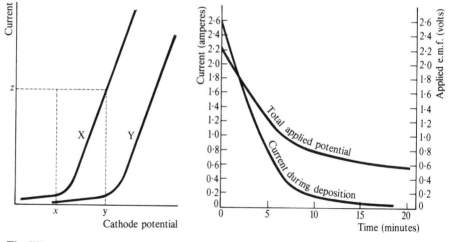

Fig. XII, 4   **Fig. XII, 5**

The initial value of the current at 2.6 amp decreased to 0.03 amp after 17 minutes and finally fell to less than 0.01 amp after 30 minutes. The potentiostat continuously decreased the total applied voltage from 2.2 volts to 0.48 volt to maintain the cathode potential constant.

In general, it may be stated that the control potential (constant potential of the working electrode) for a two-electron reaction need be no more than 0.15–0.20 volt greater than the decomposition potential to obtain rapid electrolysis. For a reversible one-electron reaction, the difference between the decomposition voltage and the control potential must be twice as great as for a two-electron reaction for equally complete deposition: if 99.9 per cent complete reaction is acceptable a difference of 0.18 volt would suffice.

Ordinary polarograms obtained with the dropping mercury electrode (see Chapter XVI) may be used to define the optimum control potential for controlled potential electro-deposition with a large mercury working electrode. The potential corresponds to that at the top of the corresponding polarographic wave where the diffusion plateau begins: usually it is about 0.1–0.15 volt beyond the polarographic half-wave potential (compare Section **XVI, 3**).

Anodic re-oxidation of the metal if it can exist in more than one valency state, or any reaction between the plated metal and the anodic oxidation products, must be reduced to a minimum for trustworthy results, and also to minimise the time required for the deposition. This may be achieved by such methods as (i) the use of a reducing agent which will be oxidised in preference to the intermediate valency state, i.e., an **anodic depolariser** (e.g., a hydrazinium salt, $N_2H_5^+ = N_2 + 5H^+ + 4e$, or hydroxylamine, largely $2NH_2OH = N_2O + 4H^+ + H_2O + 4e$), (ii) isolation of the anode by means of a membrane, porous cup, or its equivalent, and (iii) reduction of the anode potential to a value which will not oxidise the ion in the intermediate valency state.

Electrolytic determinations at constant current

XII, 8. APPARATUS. A. Electrolysis unit. The actual set-up employed will vary from one laboratory to another; a simple circuit, employing the d.c. mains (200–240 volts or 110 volts), is shown in Fig. XII, 6. M is the d.c. mains, R_1 is a fixed resistance (which may consist of a bank of lamps), R_2 is a small high wattage, variable resistance, A is an ammeter reading up to 10 amps, V is a voltmeter reading up to 10–15 volts, E is an electrolysis vessel, and S is a switch,

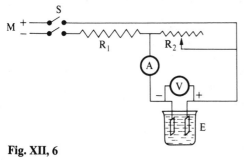

Fig. XII, 6

Alternatively, the source of d.c. may be a large-capacity 6-volt or 12-volt car battery, or a number of accumulators connected in series. If a d.c. mains supply is not at hand, any of the commercial d.c. power supply units operating from the a.c. mains may be used. In these a transformer steps the voltage down to 3–15 volts, the current is then passed through a rectifier, and finally through a smoothing filter circuit.

If the polarities of the terminals are not known, they may be determined by touching the two wires from the terminals on to paper moistened with potassium iodide solution: a brown stain of iodine will form at the positive pole.

Many types of commercial apparatus for electrolytic analysis are available; Fig. XII, 7 shows the B.T.L. apparatus (Baird and Tatlock Ltd). This is designed for use on 200–250 volt a.c. mains and incorporates silicon rectifiers providing low tension direct current for the electrolysis. Stirring is accomplished by means of a magnetic stirrer mounted beneath the stainless steel drip tray on which, the beaker containing the solution to be electrolysed, is placed. The electrode holders are mounted in an arm which can be adjusted for height.

B. Electrodes. These are generally made of platinum or of platinum-iridium, although platinum-coated titanium electrodes are available (Baird and Tatlock): gauze electrodes are preferred as they assist the circulation of the solution and thus help to reduce any tendency to local depletion of the electrolyte. Typical electrodes are shown in Fig. XII, 8: (*a*) and (*b*) represent a

Fig. XII, 7

pair of electrodes of which the inner is rotated whilst the outer is kept fixed; (c) represents another pair of electrodes (Fischer type), in which both electrodes are stationary: a glass tube is slid into the loops on the lead wire of the outer electrode and the lead to the inner electrode passes through this tube. With the Fischer type electrodes an independent glass paddle stirrer or a magnetic stirring bar must be used.

It is frequently necessary to calculate the current density. For a smooth

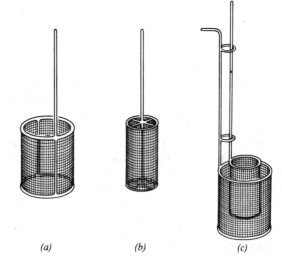

Fig. XII, 8   (a)   (b)   (c)

platinum surface, the estimation of the surface area presents no particular difficulty. As a rough approximation the usual gauze electrode may be regarded as having an effective area twice that of a plain foil electrode of the same dimensions. If an accurate value is required, the actual surface area of the electrode material must be calculated. The total length of wire can be calculated from the number of meshes and the dimensions of the electrode. The effective area will then be the total length of the wire multiplied by πd, where d is the diameter of the wire.

A **mercury cathode** finds widespread application for separations by constant current electrolysis. The most important use is the separation of the alkali and alkaline-earth metals, Al, Be, Mg, Ta, V, Zr, W, U, and the Lanthanoids from such elements as Fe, Cr, Ni, Co, Zn, Mo, Cd, Cu, Sn, Bi, Ag, Ge, Pd, Pt, Au, Rh, Ir, and Tl, which can, under suitable conditions, be deposited on a mercury cathode. The method is therefore of particular value for the determination of Al, etc, in steels and alloys: it is also applied in the separation of iron from such elements as titanium, vanadium, and uranium. In an uncontrolled constant-current electrolysis in an acid medium the cathode potential is limited by the potential at which hydrogen ion is reduced: the overvoltage of hydrogen on mercury is high (about 0.8 volt), and consequently more metals are deposited from an acid solution at a mercury cathode than with a platinum cathode. Four types of mercury cell are shown in Fig. XII, 9. In (a) the platinum wire is sealed into the side of a lipped Pyrex beaker (250 cm³), whilst in (b) the platinum wire is sealed into the side tube; the latter type permits the almost complete separation of the aqueous and mercury layers. Apparatus (c) is perhaps the most useful form; the diagram is almost self-explanatory. The electrolysis vessel contains the platinum anode (preferably of the rotating type) immersed in the electrolyte.

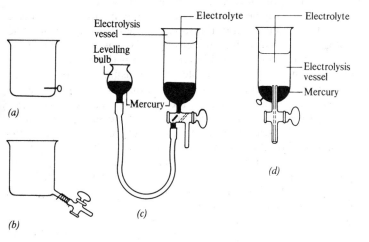

Fig. XII, 9

Electrical contact to the mercury is made through a platinum wire sealed into the side of the vessel (alternatively, a piece of amalgamated copper wire dipping into mercury contained in a glass tube, into the lower end of which a short platinum wire is sealed, may be used for electrical contact); the mercury acts as cathode: the stirring should agitate both the mercury and the solution. When electrolysis is complete, the levelling-bulb is lowered until the mercury reaches the upper end of the stopcock bore, keeping the circuit closed at all times; the stopcock is then turned through 180° and the electrolyte is collected in a suitable vessel. Apparatus (d) utilises a slightly modified separatory funnel which permits easy removal of the electrolyte and washings. Mercury is placed in the cell to within 1 mm of the top of a small inwardly projecting tube of capillary bore sealed to the stopcock entrance.

Controlled potential separations of many metals can be effected with the aid of the mercury cathode. This is because the optimum control potential and the most favourable solution conditions for a given separation can be deduced from polarograms recorded with the dropping mercury electrode: the optimum control potential corresponds to the beginning of the polarographic diffusion plateau or about 0.15 volt greater than the half-wave potential; see Ref. 5 and 6. The combination of controlled potential separation with polarographic analysis has been employed to analyse mixtures, such as copper, bismuth, lead, and cadmium; also in the separation of large amounts of copper from small amounts of nickel and zinc, of large amounts of cadmium from minute amounts of zinc, and the determination of trace impurities in refined copper.

C. Electrode vessel. Tall-form beakers, without lip, are generally employed to hold the solution to be electrolysed. These should be of such size that there is the smallest practicable volume of liquid between the cathode and the glass. The

Fig. XII, 10

apparatus should be so arranged that the beaker can be easily removed without touching any other part of the apparatus, and also so that the electrolyte can be warmed, if necessary. To prevent loss by spraying, split clock glasses with a central hole (Fig XII, 10) are used for covering the electrolytic vessel. An additional hole may be drilled for the lead from the stationary electrode.

D. Stirring. It has already been pointed out that stirring of the electrolyte during electrolysis reduces considerably the time required for deposition. The method often employed is to rotate the inner electrode; this is usually made the anode, except in lead determinations (as PbO_2), when it is the cathode. Speeds up to 500–1000 rev min^{-1} are permissible; those usually employed are 300–600 rev min^{-1}. An independent glass stirrer, operated by means of a motor mounted above the electrolysis vessel and incorporated in the electrolysis apparatus, is also employed, but this is generally not quite so satisfactory as rotating one of the electrodes. The inner electrode (gauze cylinder or, occasionally, wire spiral) should be placed concentrically with the cathode so that the electrical field is as uniform as possible. When a glass stirrer is used the propeller blades should preferably be located just below the cathode, and their pitch should be such that the solution is drawn downwards rather than forced upwards. Magnetic stirring may also be used

Some commercially available forms of electrolysis apparatus do not provide for direct rotation of the inner electrode; an independent unit suitable for this purpose is shown in Fig. XII, 11. The inner (rotating) electrode is attached to the chuck. The metal bar (shown enlarged), clamped at a suitable height, incorporates an ebonite plug screwed into the metal ring attached to the clamping device on the vertical rod of the stand; there is also an ebonite washer, and both the ebonite screw and washer serve to insulate the two binding posts C and D. D is connected electrically through the metal stand and by means of a small mercury pool A in the pulley wheel to the chuck. The outer electrode is

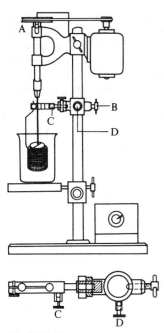

Fig. XII, 11

fitted into two jaws at the end of the metal bar and is fixed in position by the clamping screw B. The speed of the motor is controlled by the adjustable resistance on the base of the stand. The direct current may be obtained from the laboratory d.c. supply, or from the terminals of a commercial electro-analyser designed for stirring of the electrolyte with a glass or magnetic stirrer.

E. Use and care of electrodes. Electrodes must be free from grease, otherwise an adherent deposit may not be obtained. For this reason an electrode should never be touched on the deposition surface with the fingers; it should always be handled by the platinum wire or rod attached to the main body of the electrode. Platinum electrodes are easily rendered grease free by heating them to redness in a flame.

Before use electrodes must be carefully cleaned to remove any previous deposits. Deposits of copper, silver, cadmium, mercury, and many other metals can be removed by immersion in dilute nitric acid (1:1), rinsing with water, then boiling with fresh 1:1 nitric acid for 5–10 minutes, followed by a final washing with water. Copper may also be removed by means of a solution composed of 20 g trichloroacetic acid, 100 cm³ concentrated ammonia solution, and 100 cm³ water. Deposits of lead dioxide are best removed by means of 1:1 nitric acid containing a little hydrogen peroxide to reduce the lead to the bivalent form; ethanol or oxalic acid may replace the hydrogen peroxide. In all cases it is recommended that the electrodes be heated to bright redness over the colourless flame of a Bunsen burner before use.

When the electrolysis is complete, stirring is stopped, and the electrolysis beaker (where possible) is lowered away from the electrodes *before breaking the circuit*; the latter is necessary, since otherwise the electrolyte in contact with the electrode may dissolve some of the deposit. The electrodes are washed *immediately* with a fine stream of distilled water directed uniformly around the upper rims of both electrodes from a wash bottle: the first 10–15 cm³ will then contain virtually all the electrolyte adhering to the electrodes. It is unnecessary to save subsequent washings, thus avoiding excessive dilution of the residual solution. The electrode is then disconnected and rinsed with pure (e.g., A.R.) acetone (about 15 cm³ delivered from a small, all-glass wash bottle), and then dried at 100–110 °C for 3–4 minutes. The electrode with its deposit is weighed after cooling for about 5 minutes at the laboratory temperature. Cooling in a desiccator is usually unnecessary and, in any case, requires a much longer time. In a few cases there is evidence that a small transfer of platinum takes place during electrolysis. If this is suspected, the electrode plus deposit should be weighed, the deposit dissolved, the electrode weighed again, and the difference taken. The occurrence of anodic corrosion is most simply detected by weighing the anode before and after electrolysis.

Electrolysis may be carried out in chloride solutions provided a sufficient amount (1–5 g) of either a hydrazinium or a hydroxylammonium salt (usually the chloride) is added as an anodic depolariser. If no depolariser is added to an acidic chloride solution, corrosion of the anode occurs and the dissolved platinum is deposited on the cathode, leading to erroneous results and to destruction of the anode. A number of metals (for example, zinc and bismuth) should not be deposited on a platinum surface. These metals, particularly zinc, appear to react with the platinum in some way, for when they are dissolved off with nitric acid the platinum surface is dulled or blackened. Injury to the platinum can be prevented in these cases by first plating it with copper, and then depositing the metal on this surface; alternatively, a silver gauze electrode may be used. It must be emphasised that a platinum cathode should have a surface as smooth and bright as possible, because any surface unevenness tends to increase during metal deposition, and may lead to rough deposits (Ref. 7).

The determinations described in the following Sections include:

copper (Section **XII, 9**): deposition from acid solution;

lead (Section **XII, 10**): an example of anodic deposition;

cadmium (Section **XII, 11**) and silver (Section **XII, 12**): deposition from cyanide complexes;

copper and nickel in a cupro-nickel alloy (Section **XII, 13**): an example of electrolytic separation and of deposition from an ammoniacal solution;

copper, lead, antimony and tin in a bearing metal (Section **XII, 14**): an example of the use of controlled cathode potential.

XII, 9 COPPER. *Discussion.* Copper may be deposited from either sulphuric or nitric acid solution, but, usually, a mixture of the two acids is employed. If such a solution is electrolysed with an e.m.f. of 2–3 volts the following reactions occur:

Cathode: $Cu^{2+} + 2e \rightleftharpoons Cu$
$$2H^+ + 2e \rightleftharpoons H_2$$
Anode: $4OH^- \rightleftharpoons O_2 + 2H_2O + 4e$

The acid concentration of the solution must not be too great, otherwise the deposition of the copper may be incomplete or the deposit will not adhere satisfactorily to the cathode. The beneficial effect of nitrate ion is due to its depolarising action at the cathode:

$$NO_3^- + 10H^+ + 8e \rightleftharpoons NH_4^+ + 3H_2O$$

The reduction potential of the nitrate ion is lower than the discharge potential of hydrogen, and therefore hydrogen is not liberated in the free state. The nitric acid must be free from nitrous acid, as the nitrite ion hinders complete deposition and introduces other complications. The nitrous acid may be removed by boiling the nitric acid before adding it, or by the addition of urea to the solution:

$$2H^+ + 2NO_2^- + CO(NH_2)_2 = 2N_2 + CO_2 + 3H_2O$$

Nitrous acid is most efficiently removed by the addition of a little sulphamic acid:

$$H^+ + NO_2^- + {}^-O \cdot SO_2 \cdot NH_2 = N_2 + HSO_4^- + H_2O$$

the action is rapid, and the acidity of the electrolyte is unaffected. The error due to nitrous acid is increased by the presence of a large amount of iron; iron is reduced by the current to the iron(II) state, whereupon the nitric acid is reduced. This error may be minimised by the proper regulation of the pH and by the addition of ammonium nitrate instead of nitric acid, or, best, by the removal of the iron prior to the electrolysis, or by complexation with phosphate or fluoride.

The solution should be free from the following, which either interfere or lead to an unsatisfactory deposit: silver, mercury, bismuth, selenium, tellurium, arsenic, antimony, tin, molybdenum, gold and the platinum metals, thiocyanate, chloride, oxidising agents such as oxides of nitrogen, or excessive amounts of iron(III) nitrate or nitric acid. Chloride ion is avoided for two reasons:

1. Chlorine, if set free at the anode, may attack the platinum and some of the latter may plate out at the cathode: the use of an anodic depolariser, such as a hydrazinium or hydroxylammonium salt, prevents this.
2. Cu(I) is stabilised as a chloro-complex and remains in solution to be re-oxidised at the anode.

The electrolytic deposit should be salmon-pink in colour, silky in texture, and adherent. If it is dark, the presence of foreign elements and/or oxidation is indicated. Spongy or coarsely crystalline deposits are likely to yield high results; they arise from the use of too high current densities or improper acidity and absence of nitrate ion.

Procedure. The solution ($100 \, cm^3$) may contain 0.2–0.3 g of Cu (1). Add cautiously $2 \, cm^3$ of concentrated sulphuric acid, $1 \, cm^3$ of concentrated nitric acid (free from nitrous acid by boiling or by the addition of a little urea, or, better, 0.5 g of sulphamic acid), and transfer to, unless already present in, the electrolysis vessel. For simplicity of description, it will be assumed that the cathode is of the gauze form (Fig. XII, 8, *a*) and the anode is a gauze cylinder (Fig. XII, 8, *b*). Clean the platinum gauze cathode by heating it in 1:1 nitric acid, and washing it thoroughly with distilled water, followed by pure acetone. Dry the electrode at 100–110 °C for 3–4 minutes, cool in the air for about 5 minutes, and weigh. Handle the electrode by the stem and not by the gauze, since a trace of grease may cause a non-adherent deposit of copper. Arrange the circuit as shown in Fig. XII, 6, or suitably adapted according to the source of current (if accumulators or storage batteries supplying not more than 12 volts are employed, one small resistance R_2 will suffice), but do not connect the source of current. Be sure that the cathode is connected to the negative terminal and the anode to the positive terminal. If a rotating anode is to be used subsequently, make certain that it can rotate without coming into contact with the cathode at any point. Place the electrolysis vessel in position (e.g., a beaker resting on a wooden block or upon a stand; a 'lab-jack' is very convenient), and adjust the height so that the electrodes extend nearly to the bottom of the beaker and the cathode is 80–90 per cent immersed in the solution. Cover the electrolysis vessel (beaker) with a split clock glass (Fig. XII, 10), and with all the resistance in the circuit, so that only a small current will flow, close the circuit, and proceed as **A** or **B**, depending upon whether or not stirring is employed.

A. Slow electrolysis, without stirring. With an applied potential of 2–2.5 volts, adjust the resistance until a current of approximately 0.3 amp flows, as indicated by the ammeter in the circuit; electrolyse, preferably overnight (2). Rinse off the split clock glass and test for complete deposition after the blue colour of the solution has disappeared. This is best done by adding more water to raise the level of the electrolyte (say, by 0.5 cm) and continuing the electrolysis for

30 minutes. If at the end of this time no copper deposit has appeared on the freshly immersed surface of the cathode, it may be assumed that the deposition is complete. If a deposit does form, continue the passage of the current as long as may be judged necessary, and again test as before. If the solution is not to be used for a further determination, a drop or two may be removed and tested with sodium acetate and hydrogen sulphide water or with potassium hexacyanoferrate(II) solution.

B. Rapid electrolysis, with stirring. Start the motor driving the anode (or, less satisfactorily, the glass propeller) and adjust its speed so that the solution is vigorously stirred but with no danger of mechanical loss of liquid. Use a voltage across the terminals of the cell of 3–4 volts, and adjust the resistance so that the current is 2–4 amps. Continue the electrolysis until the blue colour of the solution has entirely disappeared (usually somewhat less than 1 hour), reduce the current to 0.5–1 amp, and test for completeness of deposition by rinsing the split clock glass, raising the level of the liquid by about 0.5 cm by the addition of distilled water, and continuing the electrolysis for 15–20 minutes. If no copper plates out on the fresh surface of the cathode, electrolysis may be regarded as complete.

When electrolysis has been shown to be complete, the subsequent procedure is the same whether slow or rapid electrolysis has been employed. Two methods may be employed: (i) is used only when the residual electrolyte is not required for further determination, and (ii) is of universal application.

(i) Siphon off the liquid (for example, with a glass siphon provided with a stopcock). Whilst siphoning the heavier liquid from the bottom of the beaker, add water to the top so as to keep the level of the liquid in the electrolysis vessel nearly constant. Continue this process until the ammeter needle drops practically to zero. Then rinse the cathode with A.R. acetone (3), and dry it for 3–4 minutes at 100–110 °C. Weigh after cooling in air for about 5 minutes.

(ii) Lower the beaker very slowly, or raise the electrodes, and at the same time direct a continuous stream of distilled water from a wash bottle against the upper edge of the cathode. This washing must be done immediately the cathode is removed out of the solution, and the circuit must not be broken during the process. When the cathode has been thoroughly washed, break the circuit, dip the cathode into a beaker of distilled water, and then rinse it with A.R. acetone (3). Dry at 100–110 °C for 3–4 minutes, and weigh after cooling in air for about 5 minutes.

From the increase in weight of the cathode, calculate the copper content of the solution. After the cathode has been weighed, it should be cleaned with nitric acid as described in Section **XII, 8**, and re-weighed; the loss in weight will serve as a check.

Notes. 1. Larger quantities of copper may be present, particularly if rapid electrolysis is employed; the quantity given is, however, convenient for instructional purposes. For practice in the determination, prepare the solution *either* by weighing out accurately about 1.0 g of A.R. copper sulphate pentahydrate *or* by dissolving about 0.25 g, accurately weighed, of A.R. copper in 1:1 nitric acid, boiling to remove nitrous fumes, just neutralising with ammonia solution, and then just acidifying with dilute sulphuric acid and diluting to 100 cm^3.

2. If conditions do not allow this practice, use a current of 1.5–3.0 amps: deposition is usually complete in 2–4 hours. This procedure is, however, less satisfactory if stirring is not employed.

3. This is best effected by directing a stream of A.R. acetone from a small all-glass wash bottle on to the electrodes: about 10–$15\,cm^3$ are required.

XII, 10. LEAD. *Discussion.* Lead is deposited quantitatively as the dioxide at the anode in the presence of a high concentration of nitric acid (10–$15\,cm^3$ of concentrated acid per $100\,cm^3$ of the electrolyte). The addition of 3–4 drops of concentrated sulphuric acid is said to make the deposit more adherent. It is probable that the lead is oxidised to the Pb(IV) state in the nitric acid medium, and the Pb(IV) ion is converted into hydrated lead dioxide, PbO_2,xH_2O, at the anode. Alternatively, the Pb(II) ion is oxidised quantitatively to hydrated lead dioxide at the anode. The action of nitric acid is an example of a cathodic depolariser. Nitrate ion is reduced more easily than Pb(II) ion, and thus functions as a cathodic depolariser to maintain the cathode potential below the value required for the reduction of the lead ion. With a gauze electrode of the usual size about $0.1\,g$ of lead is the maximum that can be firmly deposited.

It is difficult to remove all the water from the electrode by drying at low temperatures. For a temperature of $120\,°C$, a conversion factor of 0.864, instead of the theoretical conversion factor of 0.8662, is employed. A useful method is to dissolve the dioxide, without drying, in standard oxalic acid solution, and to titrate the excess acid with standard potassium permanganate solution.

The following interfere in this determination: mercury, arsenic, antimony, tin, selenium, tellurium, phosphorus, chromium, chloride, iodide, silver, bismuth, and manganese (the last three metals tend to form peroxides at the anode).

Procedure. For a platinum gauze electrode, the solution ($100\,cm^3$) should contain not more than $0.1\,g$ of Pb as lead nitrate, $15\,cm^3$ of concentrated nitric acid (free from nitrous acid), and none of the interfering elements mentioned above.

Heat the anode to $120\,°C$ in an electric oven for 20–30 minutes, allow to cool in air for about 5 minutes, and weigh. Connect the positive terminal of the source of current to the gauze anode, and the negative terminal to the wire or gauze cathode (1). Adjust the current to 0.05–0.1 amp at 2 volts with the aid of the rheostat, and allow the electrolysis to proceed overnight. Test for completeness of deposition by adding about $20\,cm^3$ of distilled water and continuing the electrolysis for 15 minutes; if no darkening of the freshly covered anode surface occurs, deposition is complete. When the electrolysis is complete, either lower the beaker from the electrodes or slowly raise the electrodes out of the solution without interrupting the current, and at the same time rinse the electrodes very thoroughly with a jet of water from a wash bottle. Then disconnect the source of current, wash the anode with A.R. acetone, and dry at $120\,°C$ for 20–30 minutes; cool in air for about 5 minutes and weigh.

The determination may be carried out more rapidly by one of the following methods. In these the outer electrode is the anode.

(*a*) With a current of 1.5–2.0 amps at 2 volts at the ordinary temperature. Electrolysis is complete in about 1.5 hours.

(*b*) With 2 volts and an initial current of 0.5 amp, which is subsequently raised to 5 amps. 0.1 gram of lead dioxide may be deposited in 6–8 minutes at room temperature: this time is still further reduced by working at about $60\,°C$.

(*c*) Up to $0.3\,g$ of lead may be deposited in about 10 minutes from a solution (total volume 85–$100\,cm^3$) containing $10\,cm^3$ concentrated nitric acid by electrolysis at 90–$95\,°C$, with a current of 5 amps. Under these conditions electro-osmosis expels most of the water from the deposit, and it may be dried by washing with A.R. acetone in the ordinary way.

The deposited dioxide is removed from the electrode, after weighing, by immersion in warm 1:1 nitric acid to which a little pure hydrogen peroxide has been added.

Note. 1. Full experimental details of the general technique are given under Copper (Section **XII, 9**), and will therefore not be repeated in the description of this and the succeeding determinations.

XII, 11. CADMIUM. *Discussion.* Cadmium is best determined from a faintly alkaline solution containing only enough potassium cyanide to keep the cadmium in solution, i.e., containing the complex $K_2[Cd(CN)_4]$ (*Procedure A*). Elements, such as zinc and silver, interfere. Less accurate results, but sufficiently precise for most routine analyses, may be obtained in very dilute sulphuric acid solution (*Procedure B*). The elements that interfere are essentially those mentioned under Copper (Section **XII, 9**). Deposition may also be made from a hydrochloric acid solution in the presence of hydroxylammonium chloride or hydrazinium chloride, which acts as an anodic depolariser (*Procedure C*).

Procedure A. The solution should contain about 0.4 g of Cd as the sulphate, acetate, or, less desirably, the nitrate. Add a drop of phenolphthalein, followed by *ca.* 0.1M-sodium or potassium hydroxide until a permanent pink colour is just obtained. Then add a solution of pure (e.g., A.R.) potassium or sodium cyanide dropwise and with constant stirring until the precipitated cadmium hydroxide just dissolves. A large excess of alkali cyanide should be avoided. Dilute to $100-150 \text{ cm}^3$, and electrolyse the cold solution, preferably with a platinum gauze cathode, and a current of 0.5–0.7 amp at 4.8–5.0 volts (1). At the end of 6 hours, increase the current to 1.0–1.2 amps, and continue the electrolysis for another hour. Wash the split clock glass and the sides of the beaker with about 20 cm^3 of water, and continue the electrolysis for 15 minutes. If the newly exposed surface of the cathode remains bright, thus indicating that the deposition of the cadmium is complete (2), remove the electrolyte from the electrodes, rinse immediately with water, stop the current, and rinse the cathode with ethanol or A.R. acetone. Dry at 100 °C, cool, and weigh. Test the residual electrolyte for cadmium by any of the recognised tests.

The determination may be carried out more rapidly by using a rotating cathode with a current of 1.5–2.0 amps at 2.7–3.0 volts. 0.2 gram of Cd may thus be deposited in 30 minutes.

Procedure B. The cadmium should be present in the solution (100 cm^3) as sulphate; nitrates and chlorides must be absent. The maximum concentration of free sulphuric acid is 0.5N, and 5 g of potassium hydrogensulphate is added. Electrolyse at room temperature with 0.1–0.2 amp at 2.4–2.8 volts; after 3 hours increase to 0.5 amp until electrolysis is complete. 0.3 gram of Cd is thus deposited in 3–4 hours.

Alternatively, use a rotating electrode with 1.5–1.7 amps at 2.7 volts. 0.2 gram of Cd may thus be deposited in 20 minutes.

Procedure C. The cadmium (up to 0.3 g) should be present in the solution (200 cm^3) as chloride. Add 2 g of hydroxylammonium chloride or hydrazinium chloride, acidify slightly with hydrochloric acid, and electrolyse at the laboratory temperature, using a current of about 1 amp.

Notes. 1. If left overnight, use a current of 0.2–0.3 amp at 2.8–3.2 volts.

2. It is sometimes difficult to detect the deposition of the bright cadmium on the platinum surface. This difficulty is readily overcome by heavily plating the platinum electrode with copper or silver first and then proceeding with the electrolysis in the usual manner. An added advantage is that the removal of the cadmium after the electrolysis is easier; also, if the temperature of drying should accidentlly exceed 100 °C, there is little danger of harming the platinum electrode.

XII, 12. SILVER. *Discussion.* Silver may be determined by electrolysis in nitrate, ammoniacal, or cyanide solutions. In cyanide solution the silver is present largely as the complex ion:

$$[Ag(CN)_2]^- \rightleftharpoons Ag^+ + 2CN^-$$

an excellent plate is obtained, and separation from other elements (e.g., copper and lead) may be effected. The cyanide method will be described.

The disadvantage of the electrolytic method is that so many other elements are also deposited, either wholly or in part, that a number of preliminary separations are usually required before it can be applied. For this reason, it is not widely employed.

Procedure. The silver (ca. 0.2 g) should be present in neutral or *faintly* acidic solution as the nitrate. Add pure (e.g., A.R.) potassium cyanide until the precipitate of silver cyanide is dissolved, and then add an excess such that about 2 g of potassium cyanide is present in the solution. Dilute to 100–120 cm³. Electrolyse with 0.2–0.5 amp at 3.7–4.8 volts at 20–30 °C; about 0.1 g of Ag is deposited in 3 hours. Alternatively, electrolyse with a rotating electrode with 0.5–1.0 amp. at 2.5–3.2 volts; 0.2 g of Ag is deposited in 20–25 minutes. Completeness of deposition is tested for by transferring a few drops of the electrolyte to a test-tube, acidifying with a little nitric acid, boiling off the hydrocyanic acid (**caution: poisonous**), rendering ammoniacal, and adding a few drops of ammonium sulphide solution; no brown precipitate should be obtained. The determination is completed as under Copper (Section **XII, 9**).

Note. If insoluble silver salts are to be analysed, e.g., chloride, bromide, iodide, and oxalate, these may be dissolved directly in the potassium cyanide solution.

XII, 13. ELECTROLYTIC SEPARATION AND DETERMINATION OF COPPER AND NICKEL. *Discussion.* This determination has been included to indicate the use of constant current electrolysis in the separation and determination of metals in simple alloys. More complex alloys require the application of methods utilising controlled potential at the cathode; *see* Section **XII, 14**. The theory of simple separations is discussed in Section **XII, 5**. There are a number of alloys, which include Monel metal, certain coinage alloys, and 'cupro-nickel,' which are composed principally of copper and nickel, together with small amounts of iron and manganese and not more than traces of other elements. These are suitable for electrolytic separation.

The copper is determined in strongly acid solution at a potential not exceeding 4 volts (above this potential nickel may plate out). The solution is evaporated to fuming in order to remove excess of nitric acid, the iron present is precipitated with ammonia solution, and the nickel deposited from the filtrate after the addition of a large excess of ammonia solution.

Procedure (**analysis of a copper–nickel alloy**).   Weigh out accurately about 0.5 g of the clean alloy into a 150-cm^3 tall form beaker, which should be suitable as an electrolytic vessel. Add a mixture of 10 cm^3 water, 1 cm^3 concentrated sulphuric acid, and 2 cm^3 concentrated nitric acid to dissolve the alloy. When solution is complete, boil off the oxides of nitrogen, and dilute to 100 cm^3. The solution is now ready for the deposition of copper.

Copper.   Proceed as directed in Section **XII, 9**, employing either the slow or rapid method of electrolysis. Wash the copper deposit thoroughly with water, and keep the solution for the determination of iron and nickel.

Iron.   Evaporate the solution and washings from which the copper has been removed on a low-temperature hot plate as far as possible, and then heat at a higher temperature until fumes of sulphur trioxide appear. Cool the residue, and carefully add water until the volume is about 25 cm^3. Precipitate the small quantity of iron that is now present in the Fe(III) state by adding to the warm solution about 10 cm^3 of 1:1 ammonia solution in excess. Filter through a small quantitative filter paper, and collect the filtrate in a 150-cm^3 electrolysis beaker (*A*). Wash the precipitate three times with water. Place the original beaker under the filter; dissolve the precipitate in a little hot 1:5 sulphuric acid and wash the paper with water. Precipitate the iron again with the same large excess of 1:1 ammonia solution, and filter through the same paper. Wash the precipitate, and collect the filtrate and washings in beaker (*A*) containing the filtrate and washings from the first precipitation. Ignite and weigh as Fe$_2$O$_3$ (Section **XI, 21**).

Alternatively, since the iron content is small, the washed precipitate may be dissolved in dilute hydrochloric acid and the iron determined colorimetrically (Section **XVIII, 21**).

Nickel.   Add 15 cm^3 concentrated ammonia solution to the ammoniacal nickel solution, and dilute to 100–120 cm^3. Carry out electrolysis using a rotating electrode with a current of 4 amps at 3–4 volts; 0.1 g of nickel is deposited in about 10 minutes. Test for completeness of precipitation in the usual way by adding about 20 cm^3 of water, and continuing the electrolysis for 15–20 minutes; no nickel should be deposited on the freshly immersed surface. Alternatively, the dimethylglyoxime test may be applied after neutralisation of the ammoniacal solution with hydrochloric acid. Wash, dry, and weigh the cathode as in the determination of copper.

The nickel is removed from the electrode by means of dilute nitric or sulphuric acid; concentrated nitric acid should not be employed because of the danger of inducing passivity. If difficulty is experienced in stripping the nickel from the platinum cathode, anodic solution of the metal in warm dilute nitric acid may be employed.

Electrolytic determinations with controlled cathode potential

The principles of electrolysis using controlled cathode potentials have been discussed in Section **XII, 7**, and the details for determination of antimony, copper, lead, and tin in a bearing metal, which are given below, serve to illustrate the practical details of this procedure. As indicated previously, the cathode potential may be controlled manually, but it is preferable to make use of commercially available equipment, a **potentiostat**, which will automatically maintain the potential of the working electrode constant.

XII, 14. ANTIMONY, COPPER, LEAD, AND TIN IN AN ALLOY (e.g., BEARING METAL).

Weigh accurately 0.2–0.4 g of the alloy (as drillings or fine filings) into a small beaker. Dissolve the alloy by warming with a mixture of 10 cm^3 concentrated hydrochloric acid, 10 cm^3 water, and 1 g ammonium chloride (the last-named to minimise the loss of tin as tetrachloride). Solution may be hastened by the addition, drop by drop, of a saturated solution of potassium chlorate or of concentrated nitric acid. When all the alloy has dissolved, boil off the excess of chlorine or of nitrous fumes, add 5 cm^3 concentrated hydrochloric acid, dilute to 150 cm^3, and then add 1 g hydrazinium chloride. Stir the solution efficiently and electrolyse, limiting the cathode potential to -0.36 volt *vs.* S.C.E.; copper and antimony are deposited together. After 30–45 minutes the current becomes constant (usually at about 20 milliamps): remove the saturated calomel electrode, stop the stirrer, lower the electrolysis beaker, and at the same time wash the electrodes with a fine stream of water from a wash bottle directed at the upper rims. Now break the circuit, remove the cathode, rinse it with A.R. acetone, dry for 3–4 minutes at 105 °C, and weigh after cooling in air for 5 minutes.

Separate the copper and antimony by dissolving the deposit in a mixture of 5 cm^3 concentrated nitric acid, 5 cm^3 40 per cent hydrofluoric acid (**CARE**), and 10 cm^3 water: boil off the oxides of nitrogen, dilute to 150 cm^3, and add dropwise a solution of potassium dichromate until the liquid is distinctly yellow. Deposit the copper by electrolysing the solution at room temperature and limiting the cathode–S.C.E. potential to -0.36 volt. Evaluate the weight of antimony by difference.

To the solution from which the copper and antimony have been separated as above, add 5 cm^3 concentrated hydrochloric acid and 1 g hydrazinium chloride. Electrolyse, using a copper-plated cathode;* the purpose of this is to prevent alloy formation of the platinum with the lead. Add water to the solution until the cathode is completely immersed, and then electrolyse with the cathode maintained at -0.70 volt *vs.* S.C.E. Continue the electrolysis for 45 minutes: the final value of the current is, in this instance, often an unreliable indication of the completeness of the deposition. Neutralise the electrolyte by adding dilute ammonia solution (1 : 1)—otherwise the deposited metals will partially dissolve during the washing process—immediately lower the electrolysis beaker, wash the electrodes with water, rinse the cathode with A.R. acetone, dry, and weigh in the usual manner. The increase in weight gives the weight of lead and tin in the sample.

Dissolve the deposit from the cathode in 15 cm^3 nitric acid, sp. gr. 1.20, in a 400-cm^3 beaker, and finally wash the cathode with water. Evaporate the resulting solution almost to dryness, cool, and add a further 15 cm^3 nitric acid, sp. gr. 1.2. Digest hot for a time and then filter the hydrated tin(IV) oxide on a paper-pulp

---

* Prepare this electrode by plating about 50 mg of copper—use a measured volume, e.g., 25.00 cm^3 of a standard copper sulphate solution, say, 0.00500M—from a H_2SO_4–HNO_3 solution, washing with water, followed by A.R. acetone, drying at 110 °C for 3–4 minutes, and weighing after cooling for 5 minutes in air.

pack, and wash it four times with hot water. Dilute the resulting filtrate and washings to $100\,cm^3$, and heat to boiling. Electrolyse the hot solution with a small platinum gauze anode at 4–5 amps until the deposition of PbO_2 is complete (about 5 minutes). Remove the anode, dry, and weigh as before. Calculate the percentage of lead from the weight of PbO_2 using the empirical factor of 0.864. Evaluate the tin content by subtraction from the combined weight of tin and lead.

Calculate the percentages of antimony, copper, lead, and tin in the alloy.

XII, 15. INTERNAL ELECTROLYSIS. The term internal electrolysis was applied by H. J. S. Sand (1930) to electro-analysis in which an attackable anode is used and there is an external wire connection between the cathode and anode so that electrolysis proceeds spontaneously without the application of an external e.m.f. The arrangement is, in effect, a short-circuited voltaic cell. Internal electrolysis has also been described as spontaneous electro-gravimetric analysis. The method is a special case of controlled potential electro-analysis using a platinum gauze cathode: potential control is achieved by appropriate choice of anode, and no external voltage source is required. The driving voltage is, of course, small and, in consequence, the cell resistance is a critical factor in determining the rate of metal deposition. The applications of the procedure are, in general, restricted to the determination of small amounts ($\not> 25\,mg$) if the time of electrolysis is not to be excessively long. See Refs. 8 and 9.

Applications of internal electrolysis are mainly confined to the determination of small amounts of relatively noble metal impurities in relatively base metals or alloys. These include:

Silver in lead, galena, and pyrites.

Mercury in copper and brass.

Copper in lead and in steel.

Bismuth and copper in lead, in lead–tin alloys, and in galena.

Lead and also cadmium in zinc.

XII, 16. References

1. H. Diehl (1948). *Electrochemical Analysis with Graded Cathode Potential Control.* Columbus, Ohio; G. F. Smith Chemical Co.
2. C. W. C. Milner and R. N. Whittem (1952). *Controlled Potential in the Analysis of Copper-base Alloys.* Analyst, **77**, 11.
3. B. Alfonsi (1958). 'Determination of Copper, Lead, Tin and Antimony by Controlled Potential Electrolysis'. Parts I–III. *An. Chimica Acta*, **19**, 276; 389; 569.
4. J. F. Heringshaw and P. F. Halfhide (1960). 'A Potentiostat for Electrogravimetric Analysis'. *Analyst*, **85**, 69.
5. J. A. Maxwell and R. P. Graham (1950). 'The Mercury Cathode and its Applications'. *Chem. Rev.*, **46**, 471.
6. J. A. Page, J. A. Maxwell and R. P. Graham (1962). 'Analytical Applications of the Mercury Electrode'. *Analyst*, **87**, 245.
7. H. J. S. Sand (1940). *Electrochemistry and Electrochemical Analysis.* Vols. I and II. Blackie; London.
8. B. L. Clarke, L. A. Wooten and C. L. Luke (1936). 'Analysis by "Internal" Electrolysis'. *Indl. Eng. Chem., Anal. Edn.*, **8**, 411.
9. B. L. Clarke and L. A. Wooten (1939). 'Internal Electrolysis as a Method of Analysis'. *Trans. Electrochem. Soc.*, **76**, 63.

XII, 17. Selected Bibliography

1. D. R. Browning (1969). *Electrometric Methods*. London; McGraw Hill.
2. C. W. Davis (1967). *Electrochemistry*. London; Geo. Newnes Ltd.
3. P. Delahay (1957). *Instrumental Analysis*. New York; Macmillan Co.
4. G. A. Ewing (1975). *Instrumental Methods of Chemical Analysis*. 4th edn. New York; McGraw Hill.
5. I. M. Kolthoff and P. J. Elving (1959). *Treatise on Analytical Chemistry*. Part I. Vol. 4. New York; Wiley.
6. H. A. Laitinen (1960). *Chemical Analysis. An Advanced Text and Reference*. New York; McGraw Hill.
7. J. J. Lingane (1958). *Electroanalytical Chemistry*. 2nd edn. New York; Interscience.
8. G. W. C. Milner (1957). *The Principles and Applications of Polarography and other Electroanalytical Processes*. London; Longmans Green and Co.
9. W. F. Pickering (1971). *Modern Analytical Chemistry*. New York; Marcel Dekker Inc.
10. H. A. Strobel (1960). *Chemical Instrumentation. A Systematic Approach to Instrumental Analysis*. Reading, Mass; Addison-Wesley.
11. C. R. N. Strouts, J. H. Gilfillan, and H. N. Wilson (1962). *Analytical Chemistry. The Working Tools*. Vol. 1. 2nd edn. London; Oxford University Press.
12. F. J. Welcher (1966). *Standard Methods of Chemical Analysis*. Vol. 3-A. 6th edn. Princeton; Van Nostrand.
13. H. H. Willard, L. L. Merritt, and J. A. Dean (1974). *Instrumental Methods of Analysis*. 5th edn. New York; Van Nostrand.
14. C. L. Wilson and D. W. Wilson (1964). *Comprehensive Analytical Chemistry*. Part A. Vol. 2. Amsterdam; Elsevier.

CHAPTER XIII   COULOMETRY

XIII, 1. GENERAL DISCUSSION. Coulometric analysis is an application of Faraday's first law of electrolysis which may be expressed in the form that the extent of chemical reaction at an electrode is directly proportional to the quantity of electricity passing through the electrode. For each equivalent of chemical change at an electrode 96 487 coulombs of electricity (the Faraday constant) are required; a coulomb is that quantity of electricity represented by the flow of one ampere for one second.

The fundamental requirement of a coulometric analysis is that the electrode reaction used for the determination proceeds with 100 per cent efficiency so that the quantity of substance reacted can be expressed by means of Faraday's law from the measured quantity of electricity (coulombs) passed. The substance being determined may directly undergo reaction at one of the electrodes (*primary coulometric analysis*), or it may react in solution with another substance generated by an electrode reaction (*secondary coulometric analysis*).

The weight corresponding to one equivalent of substance being electrolysed is its atomic weight or its molecular weight divided by the number of electrons involved in the electrode reaction. The weight W of substance produced or consumed in an electrolysis involving Q coulombs is therefore given by the expression

$$W = \frac{W_m Q}{96487n}$$

where W_m is the atomic weight or the molecular weight of the substance being electrolysed, and n is the number of electrons involved in the electrode reaction. Analytical methods based upon the measurement of a quantity of electricity and the application of the above equation are termed **coulometric methods**—a term derived from 'coulomb'.

Two distinctly different coulometric techniques are available: (i) coulometric analysis with controlled potential of the working electrode, and (ii) coulometric analysis with constant current. In the former method the substance being determined reacts with 100 per cent current efficiency at a working electrode the potential of which is controlled. The completion of the reaction is indicated by the current decreasing to practically zero, and the quantity of the substance reacted is computed from the reading of a coulometer in series with the cell or by means of a current–time integrating device. In method (ii) a solution of the substance to be determined is electrolysed with constant current until the reaction is completed (as detected by a visual indicator in the solution or by amperometric, potentiometric, or spectrophotometric methods) and the circuit is then opened. The total quantity of electricity passed is derived from the product current (amperes) × time (seconds): an accurate electric stop-clock may be used or, more conveniently, a low inertia integrating motor–counter unit.

XIII, 2. COULOMETRY AT CONTROLLED POTENTIAL. In a controlled potential coulometric analysis, the current generally decreases exponentially with time according to the equation

$$I_t = I_0 e^{-k't} \quad \text{or} \quad I_t = I_0 10^{-kt}$$

where I_0 is the initial current, I_t the current at time t, and k (k') is a constant. It can be shown (Lingane, Ref. 7) that $k = 25.8 DA/\delta V$, where D is the diffusion coefficient of the reducible substance ($cm^2 s^{-1}$), A is the electrode area (cm^2), δ is the thickness of the diffusion layer (cm), and V is the total volume (cm^3) of the solution of concentration C. A typical time–current curve is shown in Fig. XIII, 1, the current decreases more or less exponentially to almost zero. In many cases an appreciable 'background current' is observed with the supporting electrolyte alone, and in such instances the current finally decays to the background current rather than to zero; a correction can be applied by assuming that the background current is constant during the electrolysis. The reaction, strictly speaking, is never complete; nevertheless, when the ratio I_t/I_0 reaches a sufficiently low value (e.g., 0.001), the analysis may be terminated.

In electrolysis at controlled potential, the quantity of electricity Q (coulombs) passed from the beginning of the determination to time t is given by

$$Q = \int_0^t I_t \, dt$$

where I_t is the current at time t.

The integration may be performed graphically by measuring the area under the current–time curve or automatically by means of a mechanical current–time integrator. Alternatively, I_t may be measured at a series of suitable time intervals, and then $\log I_t$ plotted against t; a straight line of slope equal to $k'/2.303$ is obtained. Then

$$Q = \int_0^t I_t \, dt = \int_0^t I_0 \, e^{-k't} \, dt = I_0/k' \qquad \text{for large values of } t.$$

Clearly, this process is time-consuming and it is better to use a coulometer or an integrating device.

The apparatus employed in controlled potential coulometry may be considered under three headings:
1. the coulometer or other method for determining the quantity of electricity;
2. the controlled source of current;
3. the electrolysis vessel.

Coulometers suitable for measuring the total quantity of electricity passed include (i) the silver coulometer, (ii) the iodine coulometer, (iii) the hydrogen–oxygen coulometer, and (iv) the hydrogen–nitrogen coulometer; the coulometer is connected in series with the electrolysis vessel. Of those listed, the silver coulometer is the most accurate, and consists of a platinum basin which serves as cathode for the electrolysis of 10 per cent silver nitrate solution, together

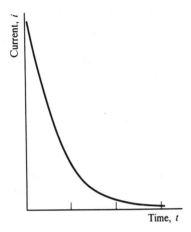

Time, t

Fig. XIII, 1

with a silver rod anode. A small filter crucible, also containing silver nitrate solution is placed inside the basin, and the silver anode is situated within the crucible: in this way, any small particles which may break off from the anode are prevented from adhering to the basin, the increase in weight of which measures the quantity of electricity passed.

The iodine coulometer contains a pair of platinum electrodes immersed in potassium iodide solution: at the end of the determination, the liberated iodine is titrated with standard thiosulphate solution and the number of coulombs passed can be calculated. The hydrogen–oxygen coulometer consists of a tube about 40 cm in length and 2 cm internal diameter, terminating in a tap at its upper end and with the lower end joined by flexible tubing to a levelling tube. Two platinum electrodes are sealed into the lower end of the tube, and the upper end, which is graduated, is surrounded by a water jacket so that gas within the tube can be maintained at constant temperature. The electrolyte is $0.5M$ potassium sulphate solution which must be saturated with hydrogen and oxygen just prior to the experiment; this is done by passing a current of 50–100 mA through the solution for about 5 minutes with the tap open and the liquid close to the top of the tube. The stop-cock is then closed, the liquid levels in the two tubes equalised, and then the coulometer is attached to the main electrolysis circuit and the current switched on: the temperature and the barometric pressure must be recorded. When the volume of gas no longer increases, its volume is read and corrected to s.t.p. {vapour pressure of the potassium sulphate solution 17.2 mm (20°); 23.2 mm (25°); 31.2 mm (30°)}. Theoretically, the volume of the hydrogen–oxygen mixture at s.t.p. should be 0.1741 cm³ per coulomb, but the observed value is 0.1739 cm³. On account of this discrepancy, the hydrogen–nitrogen coulometer is often preferred: this is set up and used as described above, but the electrolyte is a solution of hydrazinium sulphate.

A method sometimes used to measure the quantity of electricity passed is to include a standard resistor in the circuit and to connect a potentiometric recorder across the resistor. Upon completion of the electrolysis, the chart below the recorder trace is cut out and weighed on an analytical balance, thus permitting evaluation of the time–current integral. For accurate work however this method is unsuitable owing to variations in the density of the paper, and Bishop (Ref. 1) has described an accurate procedure based upon resistance–capacity integration.

The current source for the electrolysis may be a large storage battery or a mains operated power-supply unit together with a large series resistor. A simple circuit showing how the electrode potential may be controlled manually is shown in Fig. XIII, 2 (compare Fig. XII, 3), but the coulometer or other device for measuring the quantity of electricity is not included. The ammeter A indicates the electrolysis current; the voltmeter V records the e.m.f. being applied between the anode and the cathode (total applied voltage). The potential of the cathode with respect to the reference electrode S.C.E. (usually a saturated calomel electrode) is directly indicated by a high-resistance voltmeter G previously calibrated against

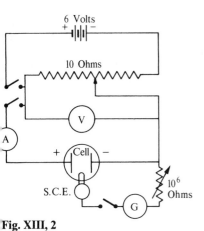

Fig. XIII, 2

a potentiometer* or, better, a pH meter provided with a millivolt scale. The experimental procedure consists in adjusting the resistance manually until the potential difference between the cathode and the reference electrode attains the desired value. As the electrolysis proceeds, the cathode tends to become more negative with respect to the reference electrode, and it is necessary to adjust the rheostat so as to restore the cathode to the desired potential. The ammeter reading decreases throughout the electrolysis and generally attains a low constant value signalling the completion of the determination. Frequent adjustment and constant attention are demanded by this procedure if the cathode potential is to be kept constant to ± 0.05 volt or better. The development of instruments, known as 'potentiostats', which automatically maintain the potential of an electrode constant to ± 5–10 millivolts at any predetermined value, has led to considerable development of the method. A number of potentiostats have been described in the literature (Refs. 2, 3), and many instruments are commercially available: some of these (e.g., Solartron, Beckman 'Electroscan', McKee-Pedersen), can function as both potentiostats and amperostats (producing constant current).

The electrode whose potential is being controlled (which may be either the cathode or the anode) is generally called the **'working electrode'** of the cell. The non-controlled electrolysis electrode is termed the **'auxiliary electrode'**, and the third electrode is a **'reference electrode'**; this does not conduct any of the electrolysis current and merely serves to permit observation of the potential of the working electrode.

Two types of **electrolytic cell**, due to Lingane, suitable for coulometric analysis at controlled potential will be described; both use a mercury cathode. In the first (Fig. XIII, 3) the cell has a capacity of about $100 \, \text{cm}^3$ and is fitted with a two-way stopcock for the introduction of the cathode mercury from the reservoir and also for the withdrawal of the solution after the completion of the electrolysis. It is closed with a Bakelite cover (fitted on to the top of the glass cell, which is ground flat) and a gas delivery tube is provided for removing dissolved air from the solution with nitrogen or other inert gas; the excess nitrogen escapes through the loosely fitted glass sleeve through which the shaft of the glass stirrer passes. Removal of the air is necessary, because oxygen is reduced at the mercury cathode at about -0.05 volt vs. S.C.E., and this would interfere with the determination of most substances. The area of the mercury cathode is about $20 \, \text{cm}^2$. Two kinds of anode, immersed directly in the test solution, may be used, viz., a large helical silver wire (*ca.* 2.6 mm diameter, helix 5 cm long, and 3 cm diameter; area about $100 \, \text{cm}^2$; as shown in the figure) or a platinum gauze

---

* This may be improvised, as shown, from a portable galvanometer (for which the deflection is directly proportional to the current) by inserting a large resistance in series with it.

cylinder (area $75\,cm^2$) mounted vertically and co-axially with the stirrer shaft. The silver anode is employed, *inter alia*, when the solution contains metals, such as bismuth, which tend to be oxidised to insoluble higher oxides at a platinum anode: chloride ion, at least equivalent to the quantity of the cathode reaction (but preferably in 50–100 per cent excess) is added. The reaction at the silver anode is

$$Ag + Cl^- = AgCl + e$$

Hydrazine is used as the depolariser at the platinum anode for metals which are not reduced by this compound:

$$N_2H_5^+ \rightleftharpoons N_2 + 5H^+ + 4e;$$

the evolution of nitrogen aids in removing dissolved air. The salt bridge (4-mm tube) from the saturated calomel electrode is filled with 3 per cent agar gel saturated with potassium chloride and its tip is placed within 1 mm of the mercury cathode when the mercury is not being stirred; this ensures that the tip

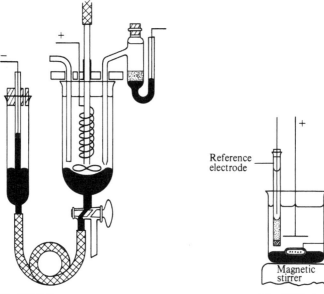

Reference electrode

Stirring bar

Magnetic stirrer

Fig. XIII, 3   **Fig. XIII, 4**

trails in the mercury surface when the latter is stirred. It is essential that the mercury–solution interface (not merely the solution) be vigorously stirred, and for this purpose the propeller blades of the glass stirrer are partially immersed in the mercury.

The second type of mercury cathode cell, shown in Fig. XIII, 4, utilises magnetic stirring. A 250-cm³ Pyrex beaker serves as the electrolysis vessel; electrical contact with the cathode mercury is made by a short length of platinum wire sealed into the side at the bottom or by means of a platinum wire sealed into the bottom of a glass tube and immersed in the mercury cathode. The stirring bar floats on the mercury and results in smooth and efficient stirring of the mercury–solution interface. The anode is a stout platinum wire coiled into a flat spiral. The reference electrode is a silver–silver chloride electrode with its end just barely brushing the surface of the mercury cathode; the top is held in a burette clamp. It consists of a glass tube of about 10 mm internal diameter, the bottom of which is closed by a sintered glass disc. The lower half of the tube is filled with a 3 per cent agar gel in saturated potassium chloride, and the upper half contains saturated potassium chloride solution to which a drop of molar silver nitrate solution has been added to saturate it with silver chloride. The electrode proper is a length of pure silver wire, about 2 mm in diameter, dipping into the solution and held in the rubber stopper. The potential of this electrode is $+0.197$ volt (hydrogen electrode scale) or -0.045 volt vs. S.C.E. When not in use the electrode is stored with its lower end immersed in saturated potassium chloride solution in a test tube.

One of the outstanding advantages of the mercury cathode is that the optimum control potential for a given separation is easily determinable from polarograms recorded with the dropping mercury electrode. This potential corresponds to the beginning of the polarographic diffusion current plateau; there is usually no advantage in employing a control potential more than about 0.15 volt greater than the half-wave potential. Some values for the half-wave potential ($E_{1/2}$) and suitable values for the cathode potential are collected in Table XIII, 1.

Table XIII, 1   Deposition of metals at controlled potential of the mercury cathode

Element	Supporting electrolyte	Volts vs. S.C.E.	
		$E_{1/2}$	$E_{cathode}$
Cu	0.5M-acid sodium tartrate, pH 4.5	-0.09	-0.16
Bi	0.5M-acid sodium tartrate, pH 4.5	-0.23	-0.40
Pb	0.5M-acid sodium tartrate, pH 4.5	-0.48	-0.56
Cd	1M-$NH_4Cl + 1M$-aq. NH_3	-0.81	-0.85
Zn	1M-$NH_4Cl + 1M$-aq. NH_3	-1.33	-1.45
Ni	1M-pyridine + HCl, pH 7.0	-0.78	-0.95
Co	1M-pyridine + HCl, pH 7.0	-1.06	-1.20

By means of the controlled cathode potential technique it is possible to effect such difficult separations as Cu and Bi, Cd and Zn, and Ni and Co. The electrolysis is best conducted by using a potentiostat to automatically control the potential of the mercury cathode at the desired value against a saturated calomel or a silver–silver chloride reference electrode.

The general technique for performing a coulometric determination at controlled potential of the mercury cathode is as follows. The supporting electrolyte (50–60 cm^3) is first placed in the cell and the air is removed by passing a rapid stream of nitrogen through the solution for about 5 minutes. The cathode mercury is then introduced through the stopcock at the bottom of the cell (Fig. XIII, 3) by raising the mercury reservoir. The stirrer is started and the tip of the bridge from the reference electrode is adjusted so that it just touches, or trails slightly in, the stirred mercury cathode. The potentiostat is adjusted to maintain the desired control potential and the solution is electrolysed, with nitrogen passing continuously, until the current decreases to a very small constant value (the 'background current'). This preliminary electrolysis removes traces of reducible impurities; the current usually decreases to 1 milliamp or less after about 10 minutes. A known volume (say, 10–40 cm^3) of the sample solution is then pipetted into the cell, and the electrolysis is allowed to proceed until the current decreases to the same small value observed with the supporting electrolyte alone. Electrolysis is usually complete within an hour. The hydrogen–oxygen coulometer is then read, and the weight W of metal deposited is calculated from the expression

$$W = \frac{M(Q - I_b t)}{nF}$$

where M is the atomic weight of the metal, Q is the total quantity of electricity (coulombs) deduced from the reading of the coulometer (or current–time integrator), I_b is the final background current (amperes), t the electrolysis time (seconds), n the number of electrons required for the reduction, and F is the faraday constant. In many cases the correction for the background current is negligible and the factor $I_b t$ may be neglected.

Experimental details (in outline) for the separation of nickel and cobalt follow.

XIII, 3.   SEPARATION OF NICKEL AND COBALT BY COULOMETRIC ANALYSIS AT CONTROLLED POTENTIAL. Reagents. *Standard nickel- and cobalt-ion solutions.* Prepare standard solutions of nickel and cobalt ion (*ca.* 10 mg per cm³) from pure ammonium nickel sulphate and pure ammonium cobalt sulphate respectively.

Pyridine. Redistil A.R. pyridine and collect the middle fraction boiling within a 2° range.

Supporting electrolyte. Prepare a supporting electrolyte composed of 1.00*M*-pyridine and 0.50*M*-chloride ion, adjusted to a pH of 7.0 ± 0.2 for use with a silver anode, or 1.00*M*-pyridine, 0.30*M*-chloride ion and 0.20*M*-hydrazinium sulphate, adjusted to a pH of 7.0 ± 0.2, for use with a platinum cathode. A small background current is obtained with the latter.

Procedure. Place 90 cm³ of the supporting electrolyte in the cell (Fig. XIII, 3), remove dissolved air with pure nitrogen, and subject the solution to a preliminary electrolysis with the potential of the mercury cathode -1.20 volt *vs.* S.C.E. to remove traces of reducible impurities; stop the electrolysis when the background current (*ca.* 2 milliamp) has decreased to a constant value (30–60 minutes). Prepare the coulometer, adjust the potentiostat to maintain the potential of the cathode at the value to be used in the determination (-0.95 volt *vs.* S.C.E. for nickel and -1.20 volt *vs.* S.C.E. for cobalt), and add 20.0 cm³ of the sample solution. Continue the electrolysis under automatic control until the current decreases to a constant minimal value (2–3 hours for each metal). Record the total quantity of electricity passed (evaluated from the coulometer readings), the electrolysis time, and the final current.

Calculate the weight of metal deposited at each potential using the relation given in Section **XIII, 2.**

Coulometry at constant current: coulometric titrations

XIII, 4.   GENERAL DISCUSSION. Coulometry at controlled potential is applicable only to the limited number of substances which undergo quantitative reaction at an electrode during electrolysis. By using coulometry at controlled or constant current, the range of substances that can be determined may be extended considerably, and includes many which do not react quantitatively at an electrode. Constant-current electrolysis is employed to generate a reagent which reacts stoichiometrically with the substance to be determined. The quantity of substance reacted is calculated with the aid of Faraday's law, and the quantity of electricity passed can be evaluated simply by timing the electrolysis at constant current. Since the current can be varied from, say, 0.1 to 100 milliamp, amounts of material corresponding to 1×10^{-9} to 1×10^{-6} equivalent per second of electrolysis time can be determined. In titrimetric analysis the reagent is added from a burette; in **coulometric titrations** the reagent is generated electrically and its amount is evaluated from a knowledge of the current and the generating time. The electron becomes the standard reagent. In many respects, e.g., detection of end-points, the procedure differs only slightly from ordinary titrations.

The fundamental requirements of a coulometric titration are: (i) that the reagent-generating electrode reaction proceeds with 100 per cent efficiency, and (ii) that the generated reagent reacts stoichiometrically and, preferably rapidly, with the substance being determined. The reagent may be generated directly within the test solution or, less frequently, it may be generated in an external solution which is allowed to run continuously into the test solution.

Since a small quantity of electricity can be readily measured with a high degree of accuracy, the method has high sensitivity. Coulometric titrimetry has several important advantages:

1. Standard solutions are not required and in their place the coulomb becomes the primary standard.
2. Unstable reagents, such as bromine, chlorine, silver(II) ion (Ag^{2+}), and titanium(III) ion, can be utilised, since they are generated and consumed immediately; there is no loss on storage or change in titre.
3. When necessary very small amounts of titrants may be generated: this dispenses with the difficulties involved in the standardisation and storage of dilute solutions. The procedure is ideally adapted for use on a micro- or semimicro-scale (e.g., Ref. 8).
4. The sample solution is not diluted in the internal generation procedure.
5. By pre-titration of the generating solution before the addition of the sample, more accurate results can be obtained. The end-point indicator corrections are thus automatically cancelled and the effect of impurities in the generating solution is minimised.
6. The method (which is largely electrical in nature) is readily adapted to remote control: this is significant in the titration of radioactive or dangerous materials. It may also be adapted to automatic control because of the relative ease of the automatic control of current.

Several methods are available for the detection of end points in coulometric titrations. These are:

(*a*) Use of chemical indicators: these must not be electro-active. Examples include methyl orange for bromine, starch for iodine, dichlorofluorescein for chloride, and eosin for bromide and iodide.

(*b*) By potentiometric observations. Electrolytic generation is continued until the e.m.f. of a reference electrode-indicating electrode assembly placed in the test solution attains a pre-determined value corresponding to the equivalence point.

(*c*) By amperometric procedures.

These are based upon the establishment of conditions such that either the substance being determined or, more usually, the titrant undergoes reaction at an indicator electrode to produce a current which is proportional to the concentration of the electro-active substance. With the potential of the indicator electrode maintained constant, or nearly so, the end point can be established from the course of the current change during the titration. The voltage impressed upon the indicator electrode is well below the 'decomposition voltage' of the pure supporting electrolyte but close to or above the 'decomposition voltage' of the supporting electrolyte plus free titrant; consequently, as long as any of the substance being determined remains to react with the titrant, the indicator current remains very small but increases as soon as the end point is passed and free titrant is present. There is a relatively inexhaustible supply of titrant ion (e.g., bromide ion in coulometric titrations with bromine), and the indicator current beyond the equivalence point is therefore governed largely by the rate of diffusion of the free titrant (e.g., bromine) to the surface of the indicator electrode. The indicator current is consequently proportional to the concentration of the free titrant (e.g., bromine) in the bulk of the solution and to the area of the indicator electrode (cathode for bromine). The indicator current will increase with increasing rate of stirring, since this decreases the thickness of the diffusion layer at the electrode; it is also somewhat temperature-dependent. The generation time at which the equivalence point is reached may be determined by calibrating the indicator electrode system with the supporting electrolyte alone by generating the titrant (e.g., bromine) for various times (say, 10–50 seconds) to evaluate the constant in the relation $I_i = Kt$, where I_i is the indicator current and t is the time. The generating time to the equivalence point may then be obtained from the observed final value of the indicator current in the actual titration, calculating the excess generating time and subtracting this from the total generating time in the titration. Alternatively, and more simply, the equivalence point time may be located by measuring three values of the indicator current at three measured times beyond the equivalence point and extrapolating to zero current.

(*d*) By application of the biamperometric (dead-stop) method (compare Section **XVII, 14**).

(*e*) By spectrophotometric observations (compare Sections **XVIII, 40–45**).

The titration cell consists of a spectrophotometer cuvette (2 cm light path). The motor-driven glass propeller stirrer and the working platinum electrode are placed in the cell in such a way as to be out of the light path: a platinum electrode in dilute sulphuric acid in an adjacent cuvette also placed in the cell holder serves as an auxiliary electrode and is connected with the titration cell by an inverted U-tube salt bridge. The appropriate wavelength is set on the instrument. Before the end point the absorbance changes only very slowly, but a rapid and linear response occurs beyond the equivalence point. Examples are: the titration of Fe(II) in dilute sulphuric acid with electro-generated Ce(IV) at 400 nm, and the titration of arsenic(III) with electro-generated iodine at 342 nm.

The principle of coulometric titration, involving the generation of a titrant by electrolysis, may be illustrated by reference to the titration of iron(II) with electro-generated cerium(IV). A large excess of Ce(III) is added to the solution containing the Fe(II) ion in the presence of, say $1M$-sulphuric acid. Let us consider what happens at a platinum anode when a solution containing Fe(II) ions alone is electrolysed at constant current. Initially the reaction $Fe^{2+} \rightleftharpoons Fe^{3+} + e$ will proceed with 100 per cent current efficiency. At the anode surface the concentration of Fe(III) ions formed is relatively large, while that of the Fe(II) ions, which is governed by the rate of transfer from the bulk of the solution, is very small: the potential of the anode gradually acquires a value which is much more positive (more oxidising) than the standard potential of the Fe(III)/Fe(II) couple (0.77 volt). As electrolysis proceeds, the anode potential becomes more and more positive (oxidising) at a rate that depends on the current density, and ultimately it becomes so positive (*ca.* 1.23 volt) that oxygen evolution from the oxidation of water begins ($2H_2O = O_2 + 4H^+ + 4e$), and this occurs before all the Fe(II) ions in the bulk of the solution is oxidised. As soon as oxygen evolution commences, the current efficiency for the oxidation of Fe(II) falls below 100 per cent and the quantity of Fe(II) initially present cannot be computed from Faraday's law. If the electrolysis is conducted in the presence of a relatively large concentration of Ce(III) ions the following reactions will take place at the anode. At a certain potential of the anode, which is considerably less than that required for oxygen evolution, oxidation of Ce(III) to Ce(IV) sets in, and the Ce(IV) thus produced is transferred to the bulk of the solution, where it oxidises Fe(II). The potential of the working electrode is thus stabilised by the reagent-generating reaction, and hence is prevented from drifting to a value such that an interfering reaction may result. The resulting Ce(IV) ions readily react with the remaining Fe(II) ions, according to the reaction:*

$$Ce^{4+} + Fe^{2+} = Ce^{3+} + Fe^{3+}$$

Stoichiometrically, the total quantity of electricity passed is exactly the same as it would have been if the Fe(II) ions had been directly oxidised at the anode and the oxidation of Fe(II) proceeds with 100 per cent efficiency. The equivalence point is marked by the first persistence of excess Ce(IV) in the solution, and may be detected by any of the methods described above.

Side reactions are avoided at the generating electrode provided there is not complete depletion (at the electrode surface) of the substance involved in the generation of the titrant. The concentration of the titrant depends upon the current through the cell, the area of the generating electrode, and the rate of stirring; the concentration of the generating substance is usually between $0.01 M$ and $0.1 M$.

XIII, 5. INSTRUMENTATION. Constant current sources. The currents used in coulometric titrations are usually in the range 1–50 milliamp. Fairly constant currents are conveniently obtained from batteries with a series regulating resistance; seven 6-volt car (or storage) batteries in series yielding the equivalent of a 42-volt battery will be found satisfactory. Periodic adjustment of the series resistance may be required to maintain constant current.

---

* The oxidation of Ce(III) ions at the surface of the electrode probably proceeds, at least in part, by the reaction:

$$Ce^{3+} + 3SO_4^{2-} \rightleftharpoons [Ce(SO_4)_3]^{2-} + e$$

the cerium complex thus produced reacts with Fe(II) ions in the solution:

$$[Ce(SO_4)_3]^{2-} + Fe^{2+} = Ce^{3+} + 3SO_4^{2-} + Fe^{3+}.$$

The net reaction is

$$Fe^{2+} \rightleftharpoons Fe^{3+} + e.$$

Regardless of the actual path of the reaction, one mole of Fe(II) is oxidised by each faraday constant.

If V_B is the voltage provided by the batteries (*ca.* 42 volts), V_C is the voltage across the cell (1–2 volts), R the series regulating resistance (maximum value, say, 10000 ohms), R_b the internal resistance of the battery itself, and R_c the resistance of the cell (probably not greater than about 20 ohms), the current I flowing through the circuit is given by:

$$I = \frac{V_B - V_C}{R + R_b + R_c}.$$

The maximum variation of the cell voltage V_C is of the order of 0.5–0.7 volt. For $V_B = 42$ volts, the maximum variation is about 1 per cent. To maintain the batteries in good condition and the series resistance in thermal equilibrium, it is advisable to employ a switching arrangement whereby the titration cell is replaced by a dummy resistance during the intervals between titrations. The dummy resistance is selected so that the current will have about the same value as during titrations.

More rigorous control of the current may be achieved by the use of commercially available stabilised power units, but better, by the use of a purpose-designed constant current source (e.g., as in Ref. 4), or a commercial 'amperostat' (Section **XIII, 2**).

Current-measuring devices. The obvious means for measuring the current is a carefully calibrated milliammeter; a more precise method is to determine with a good potentiometer the voltage drop E_1 across a precision resistance (50–100 ohms, R_1) in series with the electrolysis cell. The electrolysis current I_e is calculated from the equation $I_e = E_1/R_1$.

Time measurement. An ordinary electric stop-clock operated by closing and opening the electrolysis circuit is not very satisfactory because of the appreciable lag and 'coast' of the motor; the error may amount to several tenths of a second. The best type of electric stop-clock is one fitted with magnetic brakes: this starts and stops simultaneously with the starting and stopping of the current. Such electric stop-clocks (capacity 1000 seconds or more) are available commercially, and readings may be made precise to 0.01 second per operation. Short-period variations in the frequency of the a.c. mains supply can cause significant errors in the measurement of time intervals of several minutes duration; the error may be eliminated by the use of a frequency-regulated power supply for the clock. The electric timer should be controlled by the same switch which starts and stops the electrolysis current.

A method which is convenient, but which does not yield results of the highest precision, is to include an integrating electric motor in the circuit, and driven by the voltage drop across a precision fixed resistance in series with the electrolysis cell: it is fitted with a counter which gives the product, current × time. An integrating milliammeter may also be employed (Ref. 8).

For the highest precision, a quartz crystal clock may be used (Ref. 4), or a dekatron timer (Ref. 5).

XIII, 6. CIRCUIT AND CELL FOR COULOMETRY AT CONTROLLED CURRENT. Fig. XIII, 5 is a schematic diagram of the circuit for coulometric titration with internal generation of the titrant using the dead stop or amperometric end-point technique. The d.c. supply may be obtained from a bank of storage batteries (accumulators) delivering 42 volts; the two variable

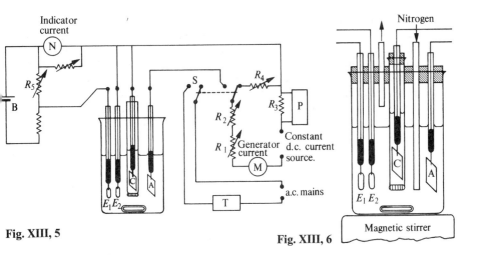

Fig. XIII, 5

Fig. XIII, 6

high-wattage resistances R_1 (large) and R_2 (small) permit the current to be varied. Alternatively, and more conveniently, an electronically controlled current supply unit may be used (see Section **XIII, 5**). The calibrated milliammeter M records the generating current; a more accurate value of the current is obtained by measuring the voltage drop across a standard resistance R_3 (say, 100 ohms) with a potentiometer P. The variable resistance R_4 (high-wattage type) is so connected that the electrolysis current flows through it whenever the electrolysis cell is disconnected from the circuit; its value (*ca.* 20 ohms) should not differ greatly from that of the cell, and is easily adjusted by arranging that the current will have nearly the same magnitude as in the titration. This arrangement ensures that the resistances R_1 and R_2 are at constant temperature, and so minimises the variations in their resistance which would occur if the current through them were interrupted periodically.

The electrolysis cell contains the working or generator electrode A, at which the reagent is electro-generated, and the auxiliary electrode C. Electrode A may be of platinum, silver, or mercury (the last-named will, of course, be a pool at the bottom of the cell); electrode C is usually of platinum. The auxiliary electrode C is generally placed in a separate glass tube closed at its lower end by a fine-porosity glass disc: the level of the solution in this compartment must be maintained at a higher level and greater ionic strength than the solution in the titration cell, so as to prevent diffusion of the latter into the isolated compartment. The electrolyte in which C is immersed may be either the same as the supporting electrolyte in the test solution or some other innocuous electrolyte appropriate to the particular case.

E_1 and E_2 are the indicator electrodes. These may consist of a tungsten pair for a biamperometric end-point; for an amperometric end-point the indicator electrodes may both be of platinum foil or one can be platinum and the other a saturated calomel reference electrode. The voltage impressed upon the indicator electrodes is supplied by battery B (*ca.* 1.5 volts) via a variable resistance R_5; N records the indicator current. For a potentiometric end-point E_1 and E_2 may consist of either platinum–tungsten bimetallic electrodes, or E_1 may be an S.C.E. and E_2 a glass electrode. These are connected directly to a pH meter with a subsidiary scale calibrated in millivolts.

S is a double-pole toggle switch. This permits the simultaneous operation of the electrolysis current and the electric timer T; when the current is not passing through the cell, it passes through the equivalent resistance R_4.

A more detailed drawing of the titration cell is shown in Fig. XIII, 6. It consists of a tall-form beaker (without lip) of about 150 or 200 cm³ capacity. Provision is made for magnetic stirring and for passing a stream of inert gas (e.g., nitrogen) through the solution. The main generator electrode A may consist of platinum foil (1 × 1 cm or 4 × 2.5 cm) and the auxiliary electrode C may consist of platinum foil (1 × 1 cm or 4 × 2.5 cm) bent into a half cylinder so as to fit into a wide glass tube (*ca.* 1 cm diameter). The isolation of the auxiliary generator electrode C within the glass cylinder (closed by a sintered-glass disc) from the bulk of the solution avoids any effects arising from undesirable reactions at this electrode. The nature of the indicator electrodes E_1 and E_2 will depend upon the procedure adopted for the detection of the end-point—biamperometric, amperometric, or potentiometric—as described above. In general, the indicator electrodes should be positioned outside the electric field (current path) between the generator electrodes, otherwise spurious indicator currents may be produced, particularly in the amperometric detection of the equivalence point.

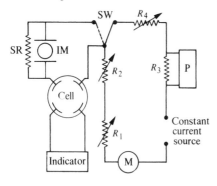

If an integrating motor is used, the circuit shown in Fig. XIII, 5 is modified as shown in Fig XIII, 7. The right-hand side of the diagram is similar to that of Fig. XIII, 5, with the obvious omission of the timing device; the milliammeter need not be calibrated. The integrating motor-counter unit IM is connected across the high stability resistance SR. The electrolysis cell contains two generating electrodes and two indicator electrodes; the latter are connected to a device (denoted by *Indicator*) for the detection of the equivalence point by biamperometric, amperometric, or potentiometric methods. SW is a switch, usually of the double-pole type.

Before use, the integrating motor must be calibrated by passing an accurately measured constant current through a high-stability resistance for an accurately measured time and reading the corresponding count on the integrating motor. The calibration factor may be expressed as:

$$F_c = \text{Coulombs per count} = \frac{\text{Current (amperes)} \times \text{time (seconds)}}{\text{Number of counts}}$$

The general procedure for making a determination is as follows: The electrolysis cell is set up with both generator and indicator electrodes in position and provision is made, if necessary, for passing an inert gas (e.g., nitrogen) through the solution. The titration cell is charged with the solution from which the titrant will be generated electrolytically, together with the solution to be titrated. The auxiliary electrode compartment is filled with a solution of the appropriate electrolyte at a higher level than the solution in the titration cell. The indicator electrodes are connected to a suitable apparatus for the detection of the end-point, e.g., a pH meter with additional millivolt scale, or a galvanometer.

Stirring is effected with a magnetic stirrer. The reading of the digital indicator instrument is taken. The current, previously adjusted to a suitable value, is then switched on and reaction between the internally generated titrant and the test solution allowed to proceed. Readings are taken periodically (more frequently as the end point is approached) of the integrating motor counter and of the indicating instrument (e.g., pH meter); it is usually necessary to switch off the electrolysing current while the readings of the indicating instrument are recorded. The end-point of the titration is readily evaluated from the plot of the reading of the indicating instrument (e.g., millivolts) against the counter reading; the first or second derivative curve is drawn to locate the equivalence point accurately. It is possible to repeat the titration with a fresh volume of the test solution. If the end point is determined potentiometrically subsequent determinations may be stopped at the potential found for the equivalence point in the initial titration.

XIII, 7. EXTERNAL GENERATION OF TITRANT. The limitations of coulometric titration with internal generation of the titrant include:
1. No substance may be present which undergoes reaction at the generator electrodes; for example, in acidimetric titrations the test solutions must not contain substances which are reduced at the generator cathode.
2. When applied on a macro scale—samples of 1–5 milli-equivalents—generation rates of 100–500-milliamp are required: parasitic currents are induced in the indicator electrodes at currents in excess of about 10–20 milliamp, consequently precise location of the equivalence point by amperometric methods is not trustworthy.

To overcome these limitations, the reagent can be generated at constant current with 100 per cent efficiency in an external generator cell and subsequently delivered to the titration cell. This technique is identical with an ordinary titration except that the reagent is generated electrolytically. A double-arm electrolytic cell for external generation of the titrant is shown in Fig. XIII, 8. The generator electrodes consist of two small platinum spirals near the centre of the inverted U-tube. The space between the electrodes is packed with glass wool to prevent turbulent mixing; the downward legs of the generator tube are constructed of 1-mm capillary tubing to reduce the inconvenience due to hold-up. The solution of the electrolyte, which upon electrolysis will yield the desired titrant, is fed continuously into the top of the generator cell. The solution is divided at the T joint so that about equal quantities flow through each of the arms of the cell. As these portions of the solution flow past the electrodes, electrolysis occurs: the products of electrolysis are swept along by the flow of the solution through the arms and emerge from the delivery tips. A beaker containing the substance to be titrated is placed beneath the appropriate delivery tip, and the solution from the other tip is run to waste. Thus for the titration of acids, electrode a functions as a cathode in sodium sulphate generator electrolyte and the hydroxide ion generated by the reaction $2H_2O + 2e = 2OH^- + H_2$ flows into the test solution. The hydrogen ion and oxygen generated at the other electrode by the reaction $2H_2O = 4H^+ + O_2 + 4e$ are swept out of the other arm into the drain. For the titration of bases, the generator electrode which delivers to the titration cell is employed as the anode. For titrations with electrically generated iodine, the generator electrolyte consists of potassium iodide solution, and the iodine solution formed at the anode flows into the titration vessel.

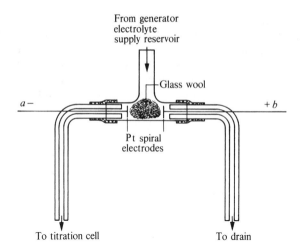

From generator
electrolyte
supply reservoir

Glass wool

$a-$ $+b$

Pt spiral
electrodes

To titration cell   To drain

Fig. XIII, 8

A minor disadvantage of external generation of titrant is the dilution of the contents of the titration cell; care is therefore necessary in suitably adjusting the rate of flow and the concentration of the generator solution. The procedure is, however, admirably suited for automatic control.

XIII, 8. EXPERIMENTAL DETAILS FOR TYPICAL COULOMETRIC TITRATIONS AT CONSTANT CURRENT. In the following pages experimental details will be given for some typical coulometric titrations at constant current. Either of two procedures for determining the total quantity of electricity passed may be used:

1. Maintenance of constant current and exact measurement of the time during which current is passed. The current may be read on a calibrated milliammeter or can be evaluated with the aid of a potentiometer and standard resistance. The time can be measured with an electric timer provided with solenoid brakes and operated from the a.c. supply mains. Results of moderate accuracy can usually be obtained by the use of a good stop-clock or stop-watch.

2. Maintenance of a reasonably constant current coupled with the use of a low-inertia integrating motor; this technique is generally more convenient but is somewhat less accurate.

Alternatively, a commercial coulometric titrator may be employed;* for details of precision measurements, Refs. (1)–(6) should be consulted. It is considered that, with appropriate attention to detail, coulometric analysis is probably capable of better precision than any other technique, and the suggestion has been made that the Faraday constant should be regarded as the prime chemical standard.

By virtue of its inherent accuracy, coulometric titration is very suitable for the

---

* Suppliers include, *inter alia*, Dohrmann-Envirotech; McKee Pedersen Instruments; Metrohm, Herisau; Princeton Allied Research; Radiometer Ltd.

determination of substances present in small amount, and quantities of the order of 10^{-7} to 10^{-5} of a mole are typical. Larger amounts of material require very long electrolysis times unless an amperostat capable of delivering relatively large currents (up to 2 A) is available. In such cases, a common procedure is to start the electrolysis with the heavy duty apparatus, and then to switch to one with a much lower output as the end point is approached.

In the determinations described below, use of an integrating motor is assumed; the modifications necessary if other equipment is used should be evident.

XIII, 9. ANTIMONY(III). *Discussion.* Iodine (or tri-iodide ion $I_3^- \rightleftharpoons I_2 + I^-$) is readily generated with 100 per cent efficiency by the oxidation of iodide ion at a platinum anode, and can be used for the coulometric titration of antimony(III). The optimum pH is between 7.5 and 8.5, and a complexing agent (e.g., tartrate ion) must be present to prevent hydrolysis and precipitation of the antimony. In solutions more alkaline than pH of about 8.5, disproportionation of iodine to iodide and hypoiodite (or iodate) occurs. The reversible character of the iodine–iodide complex renders equivalence-point detection easy by both potentiometric and amperometric techniques; for macro titrations, the usual visual detection of the end point with starch is possible.

Apparatus. Use the apparatus shown in Figs. XIII, 5, 6 and 7. The generator cathode (isolated auxiliary electrode) consists of platinum foil (4×2.5 cm, bent into a half cylinder) and the generator anode (working electrode) is a rectangular platinum foil (4×2.5 cm). For potentiometric end-point detection, use a platinum-foil electrode 1.25×1.25 cm (or a silver-rod electrode) in combination with a saturated calomel reference electrode connected to the cell by a potassium chloride- or potassium nitrate–agar bridge.

Reagents. *Supporting electrolyte.* Prepare a $0.1M$-phosphate buffer of pH = 8 containing $0.1M$-potassium iodide and $0.025M$-potassium tartrate (say, 0.17 g of $Na_2HPO_4,12H_2O$, 3.38 g $NaH_2PO_4,2H_2O$, 4.15 g KI and 1.5 g potassium tartrate, all A.R. salts, in 250 cm³ of water).

Antimony potassium tartrate, 0.01M; use the dried A.R. salt.

Procedure. Place 45 cm³ of the supporting electrolyte in the cell and fill the isolated cathode compartment with the same solution to a level well above that in the cell. Pipette 5.00, 10.00, or 15.00 cm³ of the $0.01M$-antimony solution into the cell and titrate coulometrically with a current of 40 milliamp. Stir the solution continuously by means of the magnetic stirrer and take e.m.f. readings of the Pt–S.C.E. electrode combination at suitable time intervals: the readings may be somewhat erratic initially, but become steady and reproducible after about three minutes. Evaluate the end point of the titration from the graph of e.m.f. *vs.* counter reading: this will be similar in shape to the curve shown in Fig. XIII, 9 (Section **XIII, 12**), but in the reverse order. If it proves difficult to locate the end-point precisely, recourse may be made to the first and second differential plots.

If it is desired to use the biamperometric method for detecting the end-point, then the calomel electrode and also the silver rod (if used) must be removed and replaced by two platinum plates 1.25 cm × 1.25 cm. The potentiometer (or pH meter) used to measure the e.m.f. must also be removed, and one of the indicator electrodes is then joined to a sensitive galvanometer fitted with a variable shunt. The indicator circuit is completed through a potential divider placed across a 1.5 V dry battery (see Fig. XVII, 5; Section **XVII, 14**). Charge the electrolysis cell as described above, adjust the potential across the indicator electrodes to about

150 mV, and set the galvanometer shunt to give maximum deflection on the galvanometer. Switch on the electrolysis current and read the indicator current from time to time. Plot indicator current against counter reading and extrapolate to zero current to locate the end point.

For determination of the end-point by a visual method, add $1-2 \, cm^3$ of 1 per cent starch solution, and stop the titration immediately the solution has acquired a uniform blue colour.

XIII, 10. THIOSULPHATE. *Discussion.* Thiosulphate may be titrated coulometrically with electro-generated iodine, using starch for visual end-point detection. As the end point is approached, deep blue streaks appear which spread into the solution upon stirring; at the end point the anolyte suddenly acquires a uniform blue colour. Care must be taken that the titration is stopped at the very first darkening in colour of the test solution, otherwise high results are obtained.

Apparatus. See Section **XIII, 9**.

Reagents. *Supporting electrolyte.* Dissolve 0.5 g pure potassium iodide in $40 \, cm^3$ water.

Sodium thiosulphate, 0.01M. Prepare from the A.R. salt using boiled-out water.

Catholyte. A 1–2 per cent potassium chloride solution acidified faintly with dilute hydrochloric acid.

Procedure. Place $40 \, cm^3$ of the supporting electrolyte in the cell, together with 5.00, 10.00, or $15.00 \, cm^3$ of the $0.01M$-thiosulphate solution and $1-2 \, cm^3$ of starch solution. Fill the isolated cathode compartment with the acidified potassium chloride solution. Stir the solution magnetically. Pass a current of 30 milliamp until the anolyte first acquires a uniform dark colour.

XIII, 11. OXINE (8-HYDROXYQUINOLINE). *Discussion.* Bromine may be electro-generated with 100 per cent current efficiency by the oxidation of bromide ion at a platinum anode. Bromination of oxine proceeds according to the equation:

$$C_9H_7ON + 2Br_2 = C_9H_5ONBr_2 + 2H^+ + 2Br^-$$

and thus four faraday constants are required per mole of oxine. The end-point is detected amperometrically.

Apparatus. Set up the apparatus as in Section **XIII, 9** with two small platinum plates connected to apparatus for the amperometric detection of the end-point.

Reagents. *Supporting electrolyte.* Prepare $0.2M$-potassium bromide from the A.R. salt.

Oxine solution. $0.003M$-oxine (use the A.R. material) in $0.0025M$-hydrochloric acid.

Procedure. Place $40 \, cm^3$ of the supporting electrolyte in the coulometric cell and pipette $10.00 \, cm^3$ of the oxine solution into it. Charge the cathode compartment with the $0.2M$-potassium bromide. Pass a current of 30 milliamp while stirring the solution magnetically. Adjust the sensitivity of the indicating apparatus to a suitable value. Near the end point transient deflections occur and serve to give warning of its approach. The end point is at the first permanent deflection and the reading of the counter is taken.

XIII, 12. POTASSIUM DICHROMATE (DICHROMATE ION)

Discussion. Iron(II) ions are generated electrolytically by reduction of iron(III) ions at a smooth platinum cathode. They reduce dichromate ions present in the same solution; the equivalence point is determined potentiometrically using a platinum–tungsten electrode pair.

$$Fe^{3+} + e \rightleftharpoons Fe^{2+}$$
$$6Fe^{2+} + Cr_2O_7^{2-} + 14H^+ = 6Fe^{3+} + 2Cr^{3+} + 7H_2O$$

Apparatus. Set up the circuit and the electrolytic cell (as in Section **XIII, 9**). Fill the isolated anode compartment with *ca.* 0.2*M*-sodium sulphate (65 g of A.R. $Na_2SO_4,10H_2O$ per litre) and maintain its level above that of the solution in the cell. Use a platinum sheet (1.2 cm^2) and a tungsten helix as indicating electrodes. Measure the potential change by means of a digital voltmeter or a pH meter provided with an additional millivolt scale.

Reagents. *Ammonium iron(III) sulphate solution,* ca. 0.3M. Dissolve 145 g A.R. ammonium iron(III) sulphate in 125 cm^3 water containing 10 cm^3 concentrated sulphuric acid in a 600-cm^3 beaker, add 45 cm^3 of concentrated sulphuric acid cautiously, followed by 15 cm^3 100-volume hydrogen peroxide. Maintain the solution at 50–70 °C until evolution of oxygen ceases (*ca.* 30 minutes). When cold, filter through a fine-porosity sintered-glass funnel and dilute to 500 cm^3. The above treatment removes any iron(II) which may be present.

Sulphuric acid, ca. 9M.

Potassium dichromate solution, 0.1N. Prepare a 0.1*N* solution from the dried powdered A.R. salt. Prepare also a more dilute solution, *e.g.,* *0.01N*, by dilution.

Procedure. Pipette the dichromate solution (say, 5.00 cm^3) into the titration cell, add 2 cm^3 9*M*-sulphuric acid, and then the ammonium iron(III) sulphate solution (25 cm^3 for 0.1*N*- or 10 cm^3 for 0.01*N*-$K_2Cr_2O_7$). Dilute the solution in the cell sufficiently to cover the electrodes. Pass pure nitrogen through the solution for about 15 minutes in order to remove dissolved oxygen and continue the passage of gas during the electrolysis.

Adjust the current before the titration to the desired value (15–20 or 40–50 milliamp). Commence the electrolysis and record the counter reading. Follow the potential change across the indicator electrodes on a digital voltmeter or on a pH meter (millivolt scale) as the titration proceeds. Interrupt the electrolysis current and the gas stream momentarily while taking voltmeter readings; this procedure is essential near the end-point to allow the system to reach equilibrium. The voltage change (*ca.* 200 millivolts) is abrupt at the stoichiometric end point. The latter can be evaluated precisely by plotting a voltage–count curve and the first or second differential derived therefrom. A typical titration curve (for 5.00 cm^3 of 0.01.*N*-$K_2Cr_2O_7$)

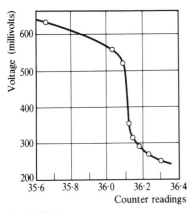

Fig. XIII, 9

is shown in Fig. XIII, 9. Having once located the end-point potential, subsequent titrations may be stopped at this value; it is essential, however, to de-oxygenate the solution before each titration and to maintain the level of the solution in the isolated anode compartment above that in the cell.

XIII, 13.   IRON(II).   *Discussion.*   Cerium(IV) ions are generated at a bright platinum anode from a supporting electrolyte containing a high concentration of sulphuric acid and of Ce(III): the equivalence point is determined potentiometrically. The oxidation potential of the Ce(III)–Ce(IV) couple in sulphuric acid is $+1.43$ volt and is relatively close to the potential at which water is oxidised at a platinum anode $(2H_2O \rightleftharpoons O_2 + 4H^+ + 4e)$. The Ce(IV) is generated in a solution containing a high concentration of Ce(III), and consequently the generator anode operates at a potential well below the standard potential.

   Apparatus.   Use the apparatus described in Section **XIII, 9**, except that the generating electrodes are reversed, i.e., the auxiliary cathode is in the isolated cathode compartment containing 15 per cent ammonium sulphate solution or $1.5M$-sulphuric acid. The end-point is determined potentiometrically using a platinum indicator electrode and a saturated calomel reference electrode.

   Reagents.   *Sulphuric acid, 9M and 1.5M.*

   Cerium(III) sulphate solution, ca. $0.1M$. Dissolve cerium(III) sulphate or, better, ammonium cerium(III) sulphate ('low in rare earths') in boiled-out distilled water.

   Iron(II) solution, 0.01M. Prepare a $0.1M$ ammonium iron(II) sulphate solution from the A.R. salt, a little M-sulphuric acid, and boiled-out distilled water. Dilute with boiled-out distilled water to $0.01M$ concentration.

   Procedure.   Place $40\,\text{cm}^3$ of the cerium(III) solution and $10\,\text{cm}^3$ of $9M$-sulphuric acid in the titration cell, and $1.5M$-sulphuric acid in the isolated cathode compartment (the level must be above that of the ultimate level in the main cell). Pass nitrogen for 10 minutes to remove dissolved air and maintain the stream of gas during the titration. Pipette $5.00\,\text{cm}^3$ of the iron(II) solution into the coulometric cell, and adjust the level of the liquid in the cathode compartment by adding $1.5M$-sulphuric acid with a dropper pipette. Titrate coulometrically at about 50 milliamp and follow the potential continuously with a digital voltmeter or a pH meter provided with a millivolt scale: allow about 10 seconds near the end-point before taking potential readings, since equilibrium does not appear to be established immediately.

XIII, 14.   CHLORIDE, BROMIDE, AND IODIDE.   *Discussion.* Mercury(I) ions can be generated at 100 per cent efficiency from mercury-coated gold or from mercury pool anodes, and employed for the coulometric titration of halides. The end-point is conveniently determined potentiometrically. In titrations of chloride ion, the addition of methanol (up to 70–80 per cent) is desirable in order to reduce the solubility of the mercury(I) chloride.

   The standard potentials (*vs.* N.H.E.) of the fundamental couples involving uncomplexed mercury(I) and mercury(II) ions are:

$$Hg_2^{2+} + 2e = 2Hg; \qquad E^{\ominus} = +0.80 \text{ volt}$$

$$Hg^{2+} + 2e = Hg; \qquad E^{\ominus} = +0.88 \text{ volt}$$

$$2Hg^{2+} + 2e = Hg_2^{2+}; \qquad E^{\ominus} = +0.91 \text{ volt}$$

The oxidation of Hg to Hg_2^{2+} requires a smaller (less oxidising) potential than to Hg^{2+}; mercury(I) ions are the main product when a mercury electrode is subjected to anodic polarisation in a non-complexing medium. From a stoichiometric standpoint it matters not whether oxidation of a mercury anode produces the mercury(I) or mercury(II) salt of a given anion, because the same quantity of electricity per mol of the anion is involved in either case: thus the same number of coulombs per mol of the anion are required to form either Hg_2Cl_2 or $HgCl_2$.

Apparatus. The apparatus is similar to that described in Section **XIII, 9**. The generator anode A now consists of a mercury pool, 0.5–1 cm deep, at the bottom of the cell; electrical connection is made by means of a platinum wire sealed through glass tubing and dipping into the mercury. For titrations of chloride and bromide, the mercury pool generator anode serves also as the indicator electrode and is used in conjunction with a saturated calomel reference electrode; the latter is connected to the cell through a saturated potassium nitrate salt bridge. For titrations of iodide, the indicator electrode consists of a silver rod fitted through glass tubing and held by the cover of the cell. During the titration the contents of the electrolysis cell are stirred vigorously with a magnetic stirrer; the stirrer bar floats on the surface of the mercury pool anode.

Reagents. *Supporting electrolyte.* For chloride and bromide, use 0.5*M*-perchloric acid. For iodide, use 0.1*M*-perchloric acid plus 0.4*M*-potassium nitrate. It is recommended that a stock solution of about five times the above concentrations be prepared {2.5*M*-perchloric acid for chloride and bromide; 0.5*M*-perchloric acid + 2.0*M*-potassium nitrate for iodide}, and dilution be effected in the cell according to the volume of test solution used. The reagents must be chloride-free.

Catholyte. The electrolyte in the isolated cathode compartment may be either the same supporting electrolyte as in the cell or 0.1*M*-sulphuric acid: the formation of mercury(I) sulphate causes no difficulty.

Chloride. Experience in this determination may be obtained by the titration of, say, carefully standardised *ca.* 0.005*M*-hydrochloric acid.

Pipette 5.00 or 10.00 cm³ of the hydrochloric acid into the cell, add 35–40-cm³ of methanol and 10 cm³ of the stock solution of the supporting electrolyte. Fill the isolated cathode compartment with supporting electrolyte of the same concentration as that in the main body of the solution or with 0.1*M*-sulphuric acid; the level of the liquid must be kept above that in the titration cell. Note the counter reading, stir magnetically, and commence the electrolysis at about 50 milliamp. Stop the generating current periodically, record the counter reading, and observe the potential between the mercury pool and the S.C.E. Plot a potential–counter reading curve and evaluate the equivalence point from the first or second differential graph. The approach of the equivalence point is readily detected in practice: successive small increments of 0.05 or 0.1 counter unit result in a relatively large change of potential (*ca.* 30 millivolts per 0.1 counter unit).

Bromide. A 0.01*M* solution of potassium bromide, prepared from the A.R. salt previously dried at 110 °C, is suitable for practice in this determination. The experimental details are similar to those given above for *Chloride* except that no methanol need be added. The titration cell may contain 10.00 cm³ of the bromide solution, 30 cm³ of water, and 10 cm³ of the stock solution of supporting electrolyte.

Iodide. A 0.01*M* solution of potassium iodide, prepared from the dry A.R.

salt with boiled-out water, is suitable for practice in this determination. The experimental details are similar to those given for *Bromide*, except that the indicator electrode consists of a silver rod immersed in the solution. The titration cell may be charged with 10.00 cm³ of the iodide solution, 30 cm³ of water, and 10 cm³ of the stock solution of perchloric acid + potassium nitrate. In the neighbourhood of the equivalence point it is necessary to allow at least 30–60 seconds to elapse before steady potentials are established.

XIII, 15. BROMIDE AND IODIDE. *Discussion.* Silver ion can be electrogenerated with 100 per cent efficiency at a silver anode and can be applied to precipitation titrations. The end-points can be determined potentiometrically or, less accurately, visually with adsorption indicators (in halide determinations: eosin for bromide and iodide; dichlorofluorescein for chloride). The insoluble silver salt deposits on the silver anode and the solution remains clear until the residual concentration of halide ion becomes so small that its rate of transfer to the anode is smaller than the rate at which silver ion is generated; thenceforth the silver ion produced at the anode diffuses into the solution and the precipitation occurs in the solution.

The supporting electrolyte may be 0.5M-potassium nitrate for bromide and iodide; for chloride, 0.5M-potassium nitrate in 25–50 per cent ethanol must be used because of the appreciable solubility of silver chloride in water.

Apparatus. Use the apparatus of Section **XIII, 9**. The generator anode is of *pure* silver foil (3 × 3 cm); the cathode in the isolated compartment is a platinum foil (3 × 3 cm) bent into a half-cylinder. For the potentiometric end-point detection, use a short length of platinum or silver wire as the indicator electrode; the electrical connection to the saturated calomel reference electrode is made by means of an agar–potassium nitrate bridge.

Reagents. *Supporting electrolyte.* Prepare 0.5M-potassium nitrate from the A.R. salt.

Potassium bromide solution, ca. *0.025*M. Weigh accurately the appropriate amount of dry A.R. potassium bromide.

Procedure. **Bromide.** Place 40 cm³ of the supporting electrolyte in the cell and add 5.00, 10.00, or 15.00 cm³ of the potassium bromide solution. Charge the isolated cathode compartment with 0.5M-potassium nitrate. Pass a current of 30 milliamp, stir vigorously, and measure the potential of the indicator electrode: the potential change at the end-point is about 200 millivolts.

Iodide. Proceed as for *Bromide*.

XIII, 16. TITRATION OF ACIDS. *General discussion.* The limiting reactions in aqueous solution at platinum electrodes are:

$$2H_2O \rightleftharpoons O_2 + 4H^+ + 4e \text{ (anode)}$$
$$2H_2O + 2e \rightleftharpoons H_2 + 2OH^- \text{ (cathode)}$$

consequently anodic electro-generation of hydrogen ion for the titration of bases and cathodic electro-generation of hydroxide ion* for the titration of acids is

---

* Direct reduction of hydrogen ion ($2H^+ + 2e \rightleftharpoons H_2$) may occur with small current densities at a platinum cathode. It is immaterial in a stoichiometric sense whether the titration proceeds indirectly by the electro-generated hydroxide ion or directly by the reduction of hydrogen ion; the current efficiency remains at 100 per cent up to very large current densities.

readily accomplished. One of the many advantages of coulometric titration of acids is that difficulties associated with the presence of carbon dioxide in the test solution or of carbonate in the standard titrant base are easily avoided: carbon dioxide can be removed completely by passing nitrogen or carbon dioxide-free air through the original acid solution before the titration is commenced. The presence of any substance that is reduced more easily than hydrogen ion or water at a platinum cathode, or which is oxidised more easily than water at a platinum anode, will, of course, interfere.

When internal generation is used in association with a platinum auxiliary electrode the latter must be placed in a separate compartment (see Fig. XIII, 10); contact between the auxiliary electrode compartment and the sample solution is made through some sort of a diaphragm, e.g., a tube with a sintered glass disc or an agar–salt bridge. For the titration of acids a silver anode may be used in combination with a platinum cathode in presence of bromide ions; the silver electrode is placed inside a straight tube closed by a sintered disc at its lower end and this can be inserted directly into the test solution. A bromide ion concentration of about $0.05M$ is satisfactory.

A. With isolated platinum auxiliary generating electrode. Apparatus. Use the cell (*ca.* 150 cm³ capacity) shown in Fig. XIII, 6

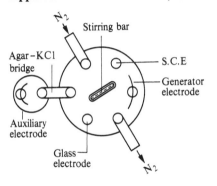

Fig. XIII, 10

Section **XIII, 6**), but equipped as presented diagrammatically in Fig. XIII, 10. The working electrode consists of a platinum foil (4 × 2 cm); the auxiliary electrode is a small platinum sheet (1 × 1 cm) immersed in a small beaker and connected to the cell by means of an inverted U-tube salt bridge containing 3 per cent agar gel in saturated potassium chloride. The glass electrode and the saturated calomel reference electrode are those supplied with commercial pH meters. Efficient stirring is provided by a magnetic stirrer.

Reagents. Supporting electrolyte. $0.1M$-sodium chloride solution.

Catholyte. This consists of $0.1M$-sodium chloride solution to which a little dilute sodium hydroxide solution is added.

Hydrochloric acid, 0.01M and 0.001M. Prepare with boiled-out water using A.R. hydrochloric acid, and standardise.

Procedure. Place 50 cm³ of the supporting electrolyte in the coulometric cell, and pass nitrogen through the solution until a pH of 7.0 is attained: thenceforth pass nitrogen over the surface of the solution. Pipette 10.00 cm³ of the acid into the cell. Adjust the current to a suitable value (40 or 20 milliamp). Turn on the current and read the counter of the integrating motor simultaneously: stop the titration when the equivalence point pH (7.00) is reached.

B. With silver auxiliary electrode. Apparatus. The titration cell is shown diagrammatically in Fig. XIII, 11, but note that the silver anode is placed inside a glass tube (not shown) with a sintered disc at the lower end—see Fig. XIII, 5. It consists of a 100–150-cm³ rimless beaker. The cork or plastic cover has holes for (i) inlet and outlet tubes for nitrogen, (ii) platinum cathode and silver anode, and

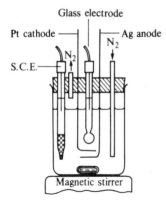

Glass electrode

Pt cathode — Ag anode

N_2

S.C.E.— N_2

Magnetic stirrer

Fig. XIII, 11

(iii) a glass electrode and a saturated calomel reference electrode such as are supplied with commercial pH meters. If the S.C.E. cannot be accommodated conveniently in the cell it may be placed in a small beaker of saturated potassium chloride solution and connected to the test solution by a U-tube salt bridge containing saturated potassium chloride solution in 3 per cent agar. Both the platinum cathode and the silver anode consist of stout wires coiled into helices. The silver anode may be used repeatedly before the silver bromide coating becomes so thick that it must be removed — about thirty successive titrations of 0.1 meq. samples at 20 milliamp. When finally necessary, the silver bromide coating may be removed by dissolution in potassium cyanide solution.

Reagent. *Supporting electrolyte.* Prepare a 0.05M-sodium bromide solution using the A.R. salt.

Procedure. Place 50 cm^3 of the supporting electrolyte in the beaker and add some of the same solution to the tube carrying the silver electrode so that the liquid level in this tube is just above the beaker. Pass nitrogen into the solution until the pH is 7.0. Pipette 10.00 cm^3 of either 0.01M- or 0.001M-hydrochloric acid into the cell. Continue the passage of nitrogen. Proceed with the titration as described under **A** above.

Several successive samples may be titrated without renewing the supporting electrolyte.

Note. The above techniques are generally applicable to many other acids, both strong and weak. The only limitation is that the anion must not be reducible at the platinum cathode and must not react in any way with the silver anode or with silver bromide (e.g., by complexation).

The coulometric determination of acids has been extensively studied by Bishop and co-workers (Ref. 4 and subsequent papers).

XIII, 17. TITRATION OF BASES. *Discussion.* When a base is titrated with electro-generated hydrogen ion at a platinum anode ($2H_2O \rightleftharpoons O_2 + 4H^+ + 4e$) a platinum auxiliary cathode is used and must be separated from the test solution by placing it in a separate compartment. The apparatus described under *Electrolytically Generated Hydroxide Ion* (Section **XIII, 16, A**) may be employed; the electrodes are, of course, reversed.

Reagent. *Supporting electrolyte.* Prepare a 0.2M-sodium sulphate solution using the A.R. salt.

Procedure. Experience in this titration may be acquired by titration of, say, 5.00 cm^3 of accurately standardised 0.01N-sodium hydroxide solution. Use 50 cm^3 of supporting electrolyte and a current of 30 milliamp.

XIII, 18. References

1. E. Bishop and P. H. Hitchcock (1973). 'Potentiostatic Coulometric Determination of Vanadium, Vanadium–Manganese and Vanadium–Iron Mixtures'. *Analyst*, **98**, 574.

2. I. R. Juniper (1974). 'A Solid-state Potentiostat for Controlled Cathode-potential Electrolysis'. *Analyst*, **99**, 58.
3. E. Bishop and P. H. Hitchcock (1973). 'Mass and Charge Transfer Kinetics and Coulometric Current Efficiencies'. *Analyst*, **98**, 470.
4. E. Bishop and M. Riley (1973). 'Precise Coulometric Determination of Acids in Cells without Liquid Junctions'. Part I. *Analyst*, **98**, 305.
5. J. A. Pike and G. C. Goode (1967). 'Precise Constant-current Coulometer'. *Anal. Chimica Acta*, **39**, 1.
6. G. Marinenko and J. K. Taylor (1968). 'Electrochemical Equivalents of Benzoic and Oxalic Acid'. *Anal. Chem.*, **40**, 1645.
7. J. J. Lingane (1958). *Electroanalytical Chemistry*. 2nd edn. New York; Interscience.
8. V. J. Jennings, A. Dodson and A. Harrison (1974). 'Coulometric Micro-titration of Arsenic(III) and Isoniazid Using a Vitreous Carbon Generating Electrode'. *Analyst*, **99**, 145.

XIII, 19.   Selected bibliography

1. D. R. Browning (1969). *Electrometric Methods*. London; McGraw-Hill.
2. D. G. Davis (1972). 'Electroanalysis and Coulometric Analysis' (review article). *Anal. Chemistry*, **44**, 79R.
3. P. Delahay (1957). *Instrumental Analysis*. New York; The Macmillan Co.
4. J. G. Dick (1973). *Analytical Chemistry*. New York; McGraw-Hill Inc.
5. G. W. Ewing (1975). *Instrumental Methods of Chemical Analysis*. 4th edn. New York; McGraw-Hill Book Co.
6. B. Fleet and R. D. Jee (1973). *Electrochemistry*. Vol. 3. Specialist Periodical Report. London; The Chemical Society.
7. G. G. Guilbault and L. G. Hargis (1970). *Instrumental Analysis Manual*. New York; Marcel Dekker Inc.
8. I. M. Kolthoff and P. J. Elving (1963). *Treatise on Analytical Chemistry*. Part I. Vol. 4. New York; Wiley-Interscience.
9. J. J. Lingane (1958). *Electroanalytical Chemistry*. 2nd edn. New York; Interscience.
10. L. Meites and H. C. Thomas (1958). *Advanced Analytical Chemistry*. New York; McGraw-Hill Book Co.
11. W. F. Pickering (1971). *Modern Analytical Chemistry*. New York; Marcel Dekker Inc.
12. H. A. Strobel (1973). *Chemical Instrumentation. A Systematic Approach to Instrumental Analysis*. 2nd edn. Reading, Mass., Addison-Wesley Publishing Co.
13. T. S. West (1973). *Analytical Chemistry. Part 2*. (MTP Series). London; Butterworth and Co.
14. H. H. Willard, L. L. Merritt and J. A. Dean (1974). *Instrumental Methods of Analysis*. 5th edn. New York; Van Nostrand.
15. C. Woodward and H. N. Redman (1973). *High-precision Titrimetry*. Analytical Sciences Monograph. No. 1. London; Society for Analytical Chemistry.
16. D. A. Skoog and D. M. West (1971). *Principles of Instrumental Analysis*. New York; Holt, Rinehart and Winston Inc.
17. C. L. Wilson and D. W. Wilson (1971). *Comprehensive Analytical Chemistry*. Vol. 2B. London; Elsevier.

CHAPTER XIV   **POTENTIOMETRY**

XIV, 1.   INTRODUCTION.   As shown in Section **II, 20**, when a metal M is immersed in a solution containing its own ions M^{n+}, then an electrode potential is established, the value of which is given by the **Nernst equation**

$$E = E^{\ominus} + (RT/nF) \ln a_{M^{n+}}$$

where $E^{\ominus}$ is a constant, the standard electrode potential of the metal. E can be measured by combining the electrode with a reference electrode (commonly a saturated calomel electrode: see Section **XIV, 3**), and measuring the e.m.f. of the resultant cell. It follows that knowing the potential E_r of the reference electrode, we can deduce the value of the electrode potential E, and provided the standard electrode potential $E^{\ominus}$ of the given metal is known, we can then proceed to calculate the metal ion activity $a_{M^{n+}}$ in the solution. For a dilute solution the measured ionic activity will be virtually the same as the ionic concentration, and for stronger solutions, given the value of the activity coefficient, we can convert the measured ionic activity into the corresponding concentration.

This procedure of using a single measurement of electrode potential to determine the concentration of an ionic species in solution is referred to as **direct potentiometry**. The electrode whose potential is dependent upon the concentration of the ion to be determined is termed the **indicator electrode**, and when, as in the case above, the ion to be determined is directly involved in the electrode reaction, we are said to be dealing with an *electrode of the first kind*.

It is also possible in appropriate cases to measure by direct potentiometry the concentration of an ion which is not directly concerned in the electrode reaction. This involves the use of an *electrode of the second kind*, an example of which is the silver–silver chloride electrode which is formed by coating a silver wire with silver chloride; this electrode can be used to measure the concentration of chloride ions in solution.

The silver wire can be regarded as a silver electrode with a potential given by the Nernst equation as

$$E = E^{\ominus}_{Ag} + (RT/nF) \ln a_{Ag^+}$$

The silver ions involved are derived from the silver chloride, and by the solubility product principle (Section **II, 8**), the activity of these ions will be governed by the chloride ion activity

$$a_{Ag^+} = K_{s(AgCl)}/a_{Cl^-}.$$

566

Hence the electrode potential can be expressed as

$$E = E_{Ag}^{\ominus} + (RT/nF)\ln K_s - (RT/nF)\ln a_{Cl^-}$$

and is clearly governed by the activity of the chloride ions, so that the value of the latter can be deduced from the measured electrode potential.

In the Nernst equation the term RT/nF involves known constants, and introducing the factor for converting natural logarithms to logarithms to base 10, the term has a value at a temperature of 25 °C of 0.0591 V when n is equal to 1. Hence, for a univalent metal, a tenfold change in ionic activity will alter the electrode potential by about 60 millivolts, whilst if the metal is bivalent, a similar change in activity will alter the electrode potential by approximately 30 millivolts, and it follows that to achieve an accuracy of 1 per cent in the value determined for the ionic concentration by direct potentiometry, the electrode potential must be capable of measurement to within 0.26 mV for the univalent metal, and to within 0.13 mV for the bivalent metal.

An element of uncertainty is introduced into the e.m.f. measurement by the **liquid junction potential** which is established at the interface between the two solutions, one pertaining to the reference electrode and the other to the indicator electrode. This liquid junction potential can be largely eliminated however if one solution contains a high concentration of potassium chloride or of ammonium nitrate; electrolytes in which the ionic conductivities of the cation and the anion have very similar values.

One way of overcoming the liquid junction potential problem is to replace the reference electrode by an electrode composed of a solution containing the same cation as in the solution under test, but at a known concentration, together with a rod of the same metal as that used in the indicator electrode: in other words we set up a **concentration cell** (Section **II, 21**). The activity of the metal ion in the solution under test is given by

$$E_{cell} = (RT/nF)\ln \frac{(activity)_{known}}{(activity)_{unknown}}$$

As a further refinement of this procedure, provided that we start with a solution containing a known ionic concentration which is greater than that in the solution under measurement, then by a process of accurate dilution of the standard solution, we can adjust its concentration to be the same as that in the solution under test. This process will be accompanied by a gradual fall in the e.m.f. of the concentration cell, and when the two solutions have the same concentration the cell e.m.f. will be zero; this procedure is termed **null point potentiometry**.

In view of the problems referred to above in connection with direct potentiometry, much attention has been directed to the procedure of **potentiometric titration** as an analytical method. As the name implies, it is a titration procedure in which potentiometric measurements are carried out in order to fix the end point. In this procedure we are concerned with *changes* in electrode potential rather than in an accurate value for the electrode potential with a given solution, and under these circumstances the effect of the liquid junction potential may be ignored. In such a titration, the change in cell e.m.f. occurs most rapidly in the neighbourhood of the end point, and as will be explained later (Section **XIV, 23**), various methods can be used to ascertain the point at which the rate of potential change is at a maximum: this is the end point of the titration.

In the present Chapter consideration will be given to various types of indicator and reference electrodes, to the procedures and instrumentation (potentiometers) for measuring cell e.m.f., to some selected examples of determinations carried out by direct potentiometry, and to some typical examples of potentiometric titrations.

Reference electrodes

XIV, 2. THE HYDROGEN ELECTRODE. All electrode potentials are quoted with reference to the standard hydrogen electrode (Section **II, 20**), and hence this must be regarded as the primary reference electrode. A typical hydrogen electrode has already been described (Section **II, 20**), and the electrode shown in Fig. II, 2 is the **Hildebrand bell-type electrode**. The platinum electrode is surrounded by an outer tube into which hydrogen enters through a side inlet, escaping at the bottom through the test solution. There are several small holes near the bottom of the bell; when the speed of the gas is suitably adjusted, the hydrogen escapes through the small openings only. Because of the periodic formation of bubbles, the level of the liquid inside the tube fluctuates, and a part of the foil is alternately exposed to the solution and to hydrogen. The lower end of the foil is continuously immersed in the solution to avoid interruption of the electric current. It should be noted that although in Fig. II, 2 an open vessel is shown, in practice the electrode will be used in a stoppered flask with a suitable exit for the hydrogen, so that an oxygen-free atmosphere can be maintained in the flask.

The **Lindsey hydrogen electrode**, illustrated in Fig. XIV, 1 has many valuable features and utilises $5-7 \, \text{cm}^3$ of the test solution. The introduction and the removal of the test solution is simple, and rapid saturation of the platinum is

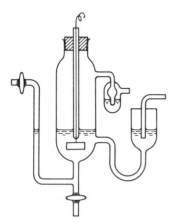

readily attained. The hydrogen-outlet trap is at right angles to the plane of the paper; and not in the same plane as indicated in the figure. The funnel limb serves for filling and washing out the vessel, and also supplies the connection to the reference electrode. The hydrogen stream is admitted through the left-hand tube and is adjusted to produce a pulsating movement up and down the platinum electrode.

The following notes on **the preparation and the use of the hydrogen electrode** may be useful. The hydrogen ions of the solution are brought into equilibrium with the gaseous hydrogen by means of platinum black; the latter adsorbs the hydrogen and acts catalytically. The platinum black may be supported

Fig. XIV, 1

on platinum foil of about $1 \, \text{cm}^2$ total area but a platinum wire, $1 \, \text{cm}$ long and $0.3 \, \text{mm}$ diameter, is often satisfactory. The platinum electrode is first carefully cleaned with hot chromic acid mixture and thoroughly washed with distilled water. Then it is plated from a solution containing $3.0 \, \text{g}$ of chloroplatinic acid and $25 \, \text{mg}$ of lead acetate per $100 \, \text{cm}^3$ with platinum foil as an anode. The current may be obtained from two accumulators connected to a suitable sliding resistance; the current is adjusted to produce a moderate evolution of hydrogen, and the process is complete in about 2 minutes. It is important that only a *thin*, jet-black deposit be made; thick deposits lead to unsatisfactory hydrogen electrodes. After platinising, the electrode must be freed from traces of chlorine: it is washed thoroughly with water, electrolysed in *ca.* 0.25M-sulphuric acid as cathode for about 30 minutes, and again well washed with water. Hydrogen electrodes should be stored in distilled water; they should never be touched with the fingers. It is advisable to have two hydrogen electrodes so that the readings obtained with one can be periodically checked against the other.

The most convenient source of hydrogen is the compressed gas, sold in cylinders; a steady stream of hydrogen gas may be readily obtained by means of a reducing valve. The gas may be passed through all-glass wash bottles containing respectively 0.2M-potassium permanganate solution, alkaline pyrogallol solution (1–2 g of pyrogallol in *ca.* 35 cm^3 of 4M-sodium hydroxide solution), dilute sulphuric acid (*ca.* 0.05M; to neutralise alkali which might splash over), and distilled water, before reaching the electrode.

The alkaline pyrogallol serves to remove any traces of oxygen from the gas: this is most important, for otherwise it is difficult to establish a steady electrode potential owing to interaction between hydrogen and oxygen on the platinised surface of the electrode. An alternative procedure for removing oxygen is to pass the gas over heated platinised asbestos, but if this method is used, particular care must be taken to ensure that the gas is properly cooled and saturated with water vapour before admission to the electrode. Connections in the gas supply line should preferably be made with polythene tubing: rubber tubing should not be used unless it has been treated with hot concentrated sodium hydroxide solution and then thoroughly washed in order to remove traces of sulphur compounds which might 'poison' the electrode.

Of the two electrodes shown, the Lindsey pattern is particularly suited for use as a reference electrode, whilst the alternative Hildebrand electrode has advantages if the electrode is to function as an indicator electrode (Section **XIV, 6**), and especially for potentiometric titrations.

Although the hydrogen electrode is the primary reference electrode, in practise, subsidiary standard electrodes which can be kept permanently set up and which are therefore available for immediate use are preferred for most purposes, thus obviating the careful setting up (including gas purification) which is required in order to establish a satisfactory hydrogen electrode. When used as a standard electrode, the hydrogen electrode operates in a solution containing hydrogen ions at constant (unit) activity based usually on hydrochloric acid, and the hydrogen gas must be at one atmosphere pressure; the effect of change in gas pressure is discussed in Ref. 1.

XIV, 3. THE CALOMEL ELECTRODE. The most widely used reference electrode, due to its ease of preparation and constancy of potential, is the **calomel electrode**. A calomel half-cell is one in which mercury and calomel (mercury(I) chloride) are covered with potassium chloride solution of definite concentration; this may be 0.1N, 1.0N, 3.5N, or saturated. The potassium chloride solution must be saturated with the calomel. The potentials of the 0.1N, 1.0N, and saturated calomel electrodes at 25 °C relative to the normal hydrogen electrode are 0.3371, 0.2846, and 0.2458 volt respectively.

Various forms of the calomel electrode are illustrated in Fig. XIV, 2; 0.1N, N, or saturated potassium chloride may be used, but the last-named is generally preferred for outine work. One of these (a), will be described in detail; the others will then be self-evident. It consists of a glass vessel provided with a bent side tube A and another side tube B, over the end of which a piece of rubber tubing is placed which can be closed by a spring or screw clip. Electrical connection with the electrode is made by means of a platinum wire, sealed through a glass tube C; the latter contains a little pure mercury into which an amalgamated copper wire

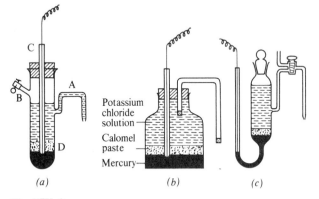

(a)   (b)   (c)

Fig. XIV, 2

dips. To set up the electrode, a saturated solution of analytically pure potassium chloride containing some of the solid salt is first prepared. Pure mercury to a depth of 0.5–1 cm is placed in the bottom of the dry electrode vessel; the mercury is then covered with a layer of calomel paste D. The latter is prepared by rubbing pure calomel, mercury, and saturated potassium chloride solution together in a glass mortar; the supernatant liquid is poured off and the rubbing process repeated twice with fresh quantities of saturated potassium chloride solution. The rubber bung carrying the glass tube and platinum wire is then inserted, care being taken that the platinum wire dips into the mercury. The vessel is then filled with a saturated solution of potassium chloride (previously saturated with calomel by shaking with the solid salt) by drawing in the solution through the bent tube A, and then closing the rubber tube B with a clip. The electrode is then ready for use. In electrode (b), the siphon tube or salt bridge may be filled with a jelly of 3 per cent agar in saturated potassium chloride solution. The electrode (c) is suitable for precision work; it has a three-way stopcock for flushing away the contaminated potassium chloride after it has been employed in a titration. Compact calomel electrodes are available commercially (see Fig. XIV, 3).

For special purposes, modifications of the calomel electrode may be preferred.

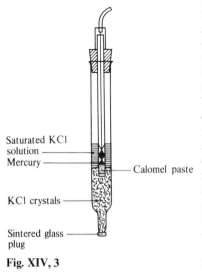

Saturated KCl solution

Mercury

Calomel paste

KCl crystals

Sintered glass plug

Fig. XIV, 3

Thus, if it is necessary to avoid the presence of potassium ions (see for example the determination of potassium by amperometric titration—Section **XVII, 9**), the electrode may be prepared with sodium chloride solution replacing the potassium chloride. In some cases the presence of chloride ions may be inimical and a mercury(I) sulphate electrode may then be used: this is prepared in similar manner to a calomel electrode using mercury(I) sulphate and potassium or sodium sulphate solution.

XIV, 4.   THE SILVER–SILVER CHLORIDE ELECTRODE. This electrode is perhaps next in importance to the calomel electrode as a reference electrode. It consists of a silver wire or a silver-plated platinum wire, coated electrolytically with a thin layer of silver chloride, dipping into a potassium chloride solution of known concentration. The potentials of the $0.1M$ and saturated silver–silver chloride electrodes at 25 °C with respect to the normal (or standard) hydrogen electrode are 0.290 and 0.199 volt respectively. In certain circumstances electrodes other than those described above may serve as reference electrodes, but generally these are of limited application.

Indicator electrodes

XIV, 5.   GENERAL DISCUSSION. As already stated, the indicator electrode of a cell is one whose potential is dependent upon the activity (and therefore the concentration) of a particular ionic species whose concentration is to be determined. In direct potentiometry or the potentiometric titration of a metal ion, a simple indicator electrode will usually consist of a carefully cleaned rod or wire of the appropriate metal: it is most important that the surface of the metal to be dipped into the solution is free from oxide films or any corrosion products. In some cases a more satisfactory electrode can be prepared by using a platinum wire which has been coated with a thin film of the appropriate metal by electro-deposition.

When hydrogen ions are involved, a hydrogen electrode can obviously be used as indicator electrode, but its function can also be performed by other electrodes, foremost amongst which is the glass electrode. This is an example of a **membrane electrode** in which the potential developed between the surface of a glass membrane and a solution is a linear function of the pH of the solution, and so can be used to measure the hydrogen ion concentration of the solution. Since the glass membrane contains alkali metal ions, it is also possible to develop glass electrodes which can be used to determine the concentration of these ions in solution, and from this development (which is based upon an ion exchange mechanism), a whole range of membrane electrodes have evolved based upon both solid state and liquid membrane ion exchange materials: these electrodes constitute the important series of **ion sensitive electrodes** which are now available for many different ions (Sections **XIV, 9–12**).

Indicator electrodes for anions may take the form of a gas electrode (e.g., oxygen electrode for OH^-; chlorine electrode for Cl^-), but in many instances consist of an appropriate electrode of the second kind: thus as shown in Section **XIV, 1**, the potential of a silver–silver choride electrode is governed by the chloride ion activity of the solution. Ion sensitive electrodes are also available for many anions.

The indicator electrode employed in a potentiometric titration will of course be dependent upon the type of reaction which is under investigation. Thus, for an acid-base titration, the indicator electrode may be a hydrogen electrode or some other hydrogen-ion responsive electrode (Sections **XIV, 7–8**); for a precipitation titration (halide with silver nitrate, or silver with chloride) a silver electrode will be used, and for a redox titration (e.g., iron(II) with dichromate) a plain platinum wire is used as the redox electrode.

XIV, 6. THE HYDROGEN ELECTRODE. In addition to its function as a standard electrode, the hydrogen electrode can be used as indicator electrode to measure the hydrogen ion concentration or the pH of solutions and can also be employed for potentiometric acid-base titrations.

The construction and operation of such electrodes have already been described (Section **XIV, 2**), but it must be noted that the hydrogen electrode cannot be used in solutions containing oxidising agents, e.g., permanganate, nitrate, cerium(IV) and iron(III) ions, or of other substances capable of reduction, such as unsaturated organic compounds, or in the presence of sulphides, compounds of arsenic, etc. (catalytic poisons) which destroy the catalytic property of platinum black. It is also unsatisfactory in the presence of salts of the noble metals, e.g., copper, silver, and gold, and also in solutions containing lead, cadmium, and thallium(I) salts. There are many other electrodes which are more convenient to use in the range in which they are applicable. Some of these will be described below.

Mention must, however, be made of the advantages of the hydrogen electrode: (1) it is a fundamental electrode to which all measurements of pH are ultimately referred: (2) it can be applied over the entire pH range; and (3) it exhibits no salt error.

Just as for reference electrode purposes the hydrogen electrode is more conveniently replaced by alternatives such as the calomel or the silver–silver chloride electrode, so too alternative electrodes are preferred to the hydrogen electrode as an indicator electrode. Amongst these alternative hydrogen-ion responsive electrodes may be mentioned (i) the quinhydrone electrode (of historic interest and now rarely used for analytical purposes), (ii) the antimony electrode (of limited application in the analytical field, but of some industrial importance on account of its simple nature and robust character), and (iii) the glass electrode: the latter has virtually superseded all other electrodes for the measurement of hydrogen ion concentration.

XIV, 7. THE ANTIMONY ELECTRODE. The so-called 'antimony electrode' is really an antimony–antimony trioxide electrode. The electrode reaction is:

$$Sb_2O_3\,(s) + 6H^+ + 6e \rightleftharpoons 2Sb\,(s) + 3H_2O$$

and the potential at 25 °C is theoretically given by:

$$E = E^{\ominus}_{Sb_2O_3,Sb} - \frac{0.0591}{6} \log \frac{1}{a^6_{H^+}} = E^{\ominus}_{Sb_2O_3,Sb} - 0.0591\ pH$$

the activities of the solid antimony and antimony trioxide, and of the water, being taken as unity. In practice, it is found that the pH response of the Sb,Sb_2O_3 electrode is roughly given by the above equation, but the exact limits of the validity of this relation are uncertain.

The electrode is generally prepared by casting a stick of antimony in the presence of air: sufficient oxidation occurs in this way to render further addition of oxide unnecessary. A wire is attached to one end of the antimony rod, while the other end is inserted into the experimental solution: the potential is then measured against a convenient reference electrode. As the potentials differ from one electrode to another, it is necessary to standardise each antimony electrode by means of solutions of known pH (buffer solutions) and also under the same experimental conditions to which it will be subjected in use; for example, in the presence or absence of oxygen, etc. The most useful pH range is 2–8.

The antimony electrode cannot be applied: (a) in the presence of strong oxidising agents or of complexing reagents (such as tartrates and organic hydroxy acids); (b) in solutions with a pH lower than 3, since the oxide then beomes appreciably soluble; and (c) in the presence of metals more noble than antimony. The electrode is not readily poisoned, is simple to use (no reagents are usually required), and is rugged; it has therefore found application for the continuous recording or control of pH in conditions where it is applicable.

XIV, 8.   THE GLASS ELECTRODE.   The glass electrode is the most widely used hydrogen-ion responsive electrode, and its use is dependent upon the fact that when a glass membrane is immersed in a solution, a potential is developed which is a linear function of the hydrogen ion concentration of the solution. The basic arrangement of a glass electrode is shown in Fig. XIV, 4 (a); the bulb B is immersed in the solution of which it is required to measure the hydrogen ion concentration, and the electrical circuit is completed by filling the bulb with a solution of hydrochloric acid (usually 0.1M), and inserting a silver–silver chloride electrode. Provided that the internal hydrochloric acid solution is maintained at constant concentration, the potential of the silver–silver chloride electrode inserted into it will be constant, and so too will the potential between the hydrochloric acid solution and the inner surface of the glass bulb. Hence the only potential which can vary is that existing between the outer surface of the glass bulb and the test solution into which it is immersed, and so the overall potential of the electrode is governed by the hydrogen ion concentration of the test solution. To ensure that the concentration of the inner hydrochloric acid solution remains constant, the upper end of the electrode must be sealed, thus giving the typical glass electrode depicted in Fig. XIV, 4(b).

The nature of the glass used for construction of the glass electrode is very important. Hard glasses of the Pyrex type are not suitable, and for many years a

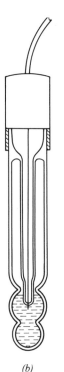

(a)   *(b)*

Fig. XIV, 4

A

B

lime-soda glass (Corning 015) of the approximate composition SiO_2 72 per cent. Na_2O 22 per cent, CaO 6 per cent was universally used for the manufacture of glass electrodes. Such electrodes were extremely satisfactory over the pH range 1–9, but in solutions of higher alkalinity the electrode was subject to an 'alkaline error' and tended to give low values for the pH. The error increased with the concentration of alkali metal ions in solution, and for example at pH 12 in the presence of sodium ions, the error varied from $-1.0\,pH\,([Na^+] = 1M)$ to $-0.4\,pH\,([Na^+] = 0.1M)$; the errors were smaller in solutions containing lithium, potassium, barium or calcium ions. Attempts were therefore made to discover glasses which would give electrodes free from this alkaline error, and it was found that the required result could be achieved by replacing most or all of the sodium content of the glass by lithium, and an electrode constructed of a glass having the composition SiO_2 63 per cent, Li_2O 28 per cent, Cs_2O 2 per cent, BaO 4 per cent, La_2O_3 3 per cent has an error of only $-0.12\,pH$ at pH 12.8 in the presence of sodium ions at a concentration of $2M$. Lithium-based glasses are now exclusively used for hydrogen-ion responsive glass electrodes.

To measure the hydrogen ion concentration of a solution the glass electrode must be combined with a reference electrode, for which purpose the saturated calomel electrode is most commonly used, thus giving the cell:

Ag,AgCl(s) | HCl(0.1M) | Glass | Test solution ‖ KCl(sat'd),Hg_2Cl_2(s) | Hg.

Owing to the high resistance of the glass membrane, a simple potentiometer (Section **XIV, 13**) cannot be employed for measuring the cell e.m.f. and specialised instrumentation (Section **XIV, 14**) must be used. The e.m.f. of the cell may be expressed by the equation:

$$E = K + (RT/F)\ln a_{H^+}$$

or at a temperature of 25 °C by the expression:

$$E = K + 0.0591\,pH.$$

In these equations K is a constant partly dependent upon the nature of the glass used in the construction of the membrane, and partly upon the individual character of each electrode; its value may vary slightly with time. This variation of K with time is related to the existence of an *asymmetry potential* in a glass electrode which is determined by the differing responses of the inner and outer surfaces of the glass bulb to changes in hydrogen ion activity; this may originate as a result of differing conditions of strain in the two glass surfaces. Owing to the asymmetry potential, if a glass electrode is inserted into a test solution which is in fact identical with the internal hydrochloric acid solution, then the electrode has a small potential which is found to vary with time. On account of the existence of this asymmetry potential of time-dependent magnitude, a constant value cannot be assigned to K, and every glass electrode must be standardised frequently by placing in a solution of known hydrogen ion activity (a buffer solution).

So-called combination electrodes may be purchased in which the glass electrode and the saturated calomel reference electrode are combined into a single unit, thus giving a more robust piece of equipment, and the convenience of having to insert and support a single probe in the test solution instead of the two separate components.

As will be apparent from the above discussion, the operation of a glass electrode is related to the situations existing at the inner and outer surfaces of the glass membrane. Glass electrodes require soaking in water for some hours before use and it is concluded that a hydrated layer is formed on the glass surface, inside which an ion exchange process can take place. If the glass contains sodium, the exchange process can be represented by the equilibrium

$$H^+_{soln} + Na^+_{glass} \rightleftharpoons H^+_{glass} + Na^+_{soln}$$

The concentration of the solution within the glass bulb is fixed, and hence on the inner side of the bulb an equilibrium condition leading to a constant potential is established. On the outside of the bulb, the potential developed will be dependent upon the hydrogen ion concentration of the solution in which the bulb is immersed. Within the layer of 'dry' glass which exists between the inner and outer hydrated layers, the conductivity is due to the interstitial migration of sodium ions within the silicate lattice. For a detailed account of the theory of the glass electrode a text book of electrochemistry should be consulted.

In view of the equilibrium shown in the equation above it is not surprising that if the solution to be measured contains a high concentration of sodium ions, say a sodium hydroxide solution, the pH determined is too low. Under these conditions sodium ions from a solution pass into the hydrated layer in preference to hydrogen ions, and consequently the measured e.m.f. (and hence the pH) are too low. This is the reason for the 'alkaline error' encountered with the glass electrode constructed from a lime-soda glass. Likewise in strongly acid solutions (hydrogen ion concentration in excess of $1M$), errors also arise but to a much smaller degree; this effect is related to the fact that in the relatively concentrated solutions involved, the activity of the water in the solution is reduced and this can affect the hydrated layer of the electrode which is involved in the ion exchange reaction.

The glass electrode can be used in the presence of strong oxidants and reductants, in viscous media, and in the presence of proteins and similar substances which seriously interfere with other electrodes. It can also be adapted for measurements with small volumes of solutions. It may give erroneous results when used with very poorly buffered solutions which are nearly neutral.

The glass electrode should be thoroughly washed with distilled water after each measurement and then rinsed with several portions of the next test solution before making the following measurement. The glass electrode should not be allowed to become dry, except during long periods of storage: it will return to its responsive condition when immersed in distilled water for at least twelve hours prior to use.

Ion-sensitive electrodes
XIV, 9.   ALKALI METAL ION-RESPONSIVE GLASS ELECTRODES.
As mentioned in Section **XIV, 8**, a glass electrode used for pH measurements, will if it is constructed of a lime-soda glass, be subject to an 'alkaline error' which stems from the ion exchange equilibrium between hydrogen ions in solution and sodium ions in the layer of hydrated glass. If the composition of the glass is altered, then so too is the position of equilibrium, and indeed, as already stated, if the sodium in the glass is replaced by lithium, then the 'alkaline error' virtually disappears.

If the preference for hydrogen ion exchange shown by lime-soda glasses can be reduced, then other cations will become involved in the ion exchange process and we can see the possibility of an electrode responsive to metallic ions such as sodium and potassium. The required effect can be achieved by the introduction of aluminium, and as shown in Table XIV, 1, this approach has led to new glass electrodes of great importance to the analyst.

In all cases some sensitivity to hydrogen ions remains, and in any potentiometric determination with these modified glass electrodes the hydrogen ion concentration of the solution must be reduced so as to be not more than 1 per cent of the concentration of the ion being determined, and in a solution containing more than one kind of alkali metal cation, some interference will be encountered.

Table XIV, 1   Composition of glasses for cation-sensitive glass electrodes

Composition	For determination of
Na_2O 22%, CaO 6%, SiO_2 72%	H^+ (Subject to alkaline error)
Li_2O 28%, Cs_2O 2%, BaO 4%, La_2O_3 3%, SiO_2 63%	H^+ (Alkaline error reduced)
Li_2O 15%, Al_2O_3 25%, SiO_2 60%	Li^+
Na_2O 11%, Al_2O_3 18%, SiO_2 71%	Na^+, Ag^+
Na_2O 27%, Al_2O_3 5%, SiO_2 68%	K^+

The construction of these electrodes is exactly similar to that already described for the pH-responsive glass electrode. They must of course be used in conjunction with a reference electrode and for this purpose a silver–silver chloride electrode is usually preferred. A 'double junction' reference electrode is often used in which an inner tube containing the silver–silver chloride electrode immersed in potassium chloride solution is surrounded by an outer tube which can be filled with any appropriate solution (potassium chloride, potassium, sodium or ammonium nitrate, etc.), the choice being governed by the nature of the solution under test. The inner tube is closed at its lower end by a porous diaphragm, and the base of the outer tube is closed by a ground-on glass cap; electrical connection is achieved through the film of liquid trapped in the ground joint. This arrangement makes it easy to change the junction liquid in the outer tube, and it is claimed that this glass seal type of junction gives a particularly reproducible liquid junction potential and minimum diffusion of junction electrolyte into the test solution; this last feature is important owing to the possibility of interference from the ions thus introduced.

The electrode response to the activity of the appropriate cation is given by the usual Nernst equation:

$$E = K + (RT/nF)\log a_{M^{n+}}$$

and for a monovalent cation, since $-\log a_{M^+} = pM$ (cf. pH)

$$E = K - 0.0591\, pM \text{ (at } 25\,°C).$$

Such an electrode may however also show a response to certain other cations, and when an interfering cation B^{x+} is present, then the expression for electrode potential becomes:

$$E = K + \frac{2.303RT}{nF}\log a_{M^{n+}} + \frac{2.303RT}{nF}\log k_{M,B}(a_{B^{x+}})^{n/x}$$

where $a_{M^{n+}}$ is the activity of the ion to be determined, $a_{B^{x+}}$ is the activity of the interfering ion, and $k_{M,B}$ is the **Selectivity Coefficient** of the electrode,

$$k_{M,B} = \frac{a_{M^{n+}}}{(a_{B^{x+}})^{n/x}}$$

The value of this coefficient will be determined by the nature of the interfering species present: if the interfering ion B^{x+} is replaced by another interfering ion C^{z+}, then the Selectivity Coefficient will acquire a different value $k_{M,C}$. Hence Selectivity Coefficients should always be denoted in the manner shown, which indicates what particular interference is involved, and the activity of each species must also be specified when quoting a value for the coefficient.

The selectivity coefficient can be evaluated by measuring the e.m.f. response of the ion-sensitive electrode in solutions containing a constant activity of the interfering ion B and varying activities of the principal ion M; the smaller the value of $k_{M,B}$ the greater the preference of the electrode for the principal ion.

Values for the selectivity coefficients of a given electrode with respect to common interferences are frequently quoted by the manufacturer, but it should be noted that they are not always quoted in the recommended form (see Ref. 11).

XIV, 10. OTHER SOLID MEMBRANE ELECTRODES. The glass membrane of the electrodes discussed above may be replaced by other materials such as a single crystal or a solid ion exchange material; it may be advantageous to incorporate the ion exchange material into an inert carrier such as paraffin wax or a suitable polymer.

Pungor (Ref. 2) developed an iodide-ion-sensitive electrode by incorporating finely dispersed silver iodide into a silicone rubber monomer and then carrying out polymerisation. A circular portion of the resultant silver iodide-impregnated polymer was used to seal the lower end of a glass tube which was then partly filled with potassium iodide solution $(0.1M)$, and then a silver wire was inserted to dip into the potassium iodide solution. When the membrane end of the assembly is inserted into a solution containing iodide ions, we have a situation exactly similar to that encountered with glass membrane electrodes. The silver iodide particles in the membrane set up an exchange equilibrium with the solutions on either side of the membrane. Inside the electrode, the iodide ion concentration is fixed and a stable situation results. Outside the electrode, the position of equilibrium will be governed by the iodide ion concentration of the external solution, and a potential will therefore be established across the membrane and this potential will vary according to the iodide ion concentration of the test solution.

This original Pungor or **heterogeneous membrane** type of electrode has been extended to give electrodes capable of measuring the concentration of Cl^-, Br^-, CN^-, S^{2-}, and many other anions, and electrodes suitable for measuring the concentration of Cl^-, Br^-, and I^- can be obtained by using a membrane cast from the appropriate pure silver halide; that is to say, the inert matrix is dispensed with, and we are dealing with a **solid state electrode**. A particularly useful application of this last technique is the single crystal lanthanum fluoride electrode developed by Orion Research Inc, which can be used to measure the concentration of fluoride ions in solution.

XIV, 11. LIQUID MEMBRANE ELECTRODES. Another type of selective ion electrode is based upon the use of liquid ion exchange materials, usually consisting of an ion exchange material dissolved in an organic solvent which is not miscible with water to any great extent and thus obviating undue mixing of the electrode material with the solution to be analysed. Two different types of electrode are used: (*a*) those in which the liquid exchanger contains the ion to which the electrode is responsive, and (*b*) those in which the liquid exchanger is electrically neutral and does not contain any ions.

Important electrodes of the first type are (i) the calcium-responsive electrode based upon the calcium salt of didecyl hydrogen phosphate dissolved in di-n-octylphenylphosphonate (Ref. 3), and (ii) the anion-responsive electrodes based upon the methyl tri-octanoyl-ammonium cation (Ref. 4): these are suitable, *inter alia*, for the determination of ClO_4^-, SO_4^{2-}, and many organic anions. An example of the second kind of electrode is the Philips potassium electrode in which the ion exchange material is an antibiotic (valinomycin) dissolved in diphenyl ether. Valinomycin forms an association complex with alkali metal ions with the important feature that the selectivity coefficient for K^+ as compared with Na^+ is about 4000, and for K^+ as compared with H^+ is about 18 000, so that the electrode can be used to determine potassium in the presence of large amounts of sodium, and in relatively strongly acid solutions.

In these liquid membrane electrodes, the solution of the ion exchange material is placed in a tube closed by a porous diaphragm at its lower end, and the internal silver–silver chloride electrode in potassium (or sodium) chloride solution is placed in a narrow tube which is mounted inside the wider one.

In a recent development it has been shown (Ref. 8) that if the active components of a liquid membrane electrode (exchange medium plus solvent) are added to a solution of polyvinyl chloride in tetrahydrofuran, and the resulting mixture allowed to stand for some days for the tetrahydrofuran to evaporate, then a solid residue is left, from which a circle may be cut and cemented to the end of a PVC tube. This arrangement then functions as a heterogeneous membrane type of electrode, responding to the same ion(s) as the original liquid membrane electrode.

XIV, 12. COMMERCIALLY AVAILABLE ION-SENSITIVE ELEC-TRODES. At the present time a number of ion selective electrodes are available from laboratory supply houses and new ones are frequently being added: whilst not intended to be an exhaustive list, Table XIV, 2 serves to indicate the range of determinations for which electrodes are now available; see Refs. 5, 6, 7.

A range of gas-sensing electrodes are also available which can be used to determine soluble gases such as hydrogen chloride, ammonia, sulphur dioxide, and carbon dioxide. In these electrodes the gas stream is passed through a tube containing a semi-permeable membrane separating it from a solution of carefully selected pH. The soluble gas will pass through the membrane, dissolve in the solution and thus affect the pH. The actual measuring electrode is a pH-responsive glass electrode, and the measured change in pH can be related to the concentration of the gas under investigation.

As already explained, some care must be exercised in using an ion-sensitive electrode to ensure that interferences do not arise from other ions, and it is of course also necessary to ensure that the ion which is to be measured has not undergone complex formation with any of the reagents which have been added to the solution; conversely, it may be possible to reduce the interference due to a

Table XIV, 2.   A selection of commercially available ion sensitive electrodes.

Type of membrane	Ion	Lower limit of detection (mol dm^{-3})
Glass	H^+	10^{-14}
	Na^+	10^{-6}
	K^+	10^{-6}
Liquid	K^+	10^{-6}
	Ca^{2+}	10^{-5}
	NO_3^-	10^{-5}
	ClO_4^-	10^{-5}
Solid	Ag^+	10^{-17}
	Pb^{2+}	10^{-7}
	Cd^{2+}	10^{-7}
	Cu^{2+}	10^{-8}
	F^-	10^{-6}
	Cl^-	5×10^{-5}
	Br^-	5×10^{-6}
	I^-	5×10^{-8}
	CN^-	10^{-6}
	SCN^-	10^{-5}
	S^{2-}	10^{-17}

given ion by adding a reagent which will complex the interfering ion. As an example, if it is required to measure the fluoride ion concentration of a solution with a fluoride-responsive electrode, it is important to ascertain whether the solution contains any aluminium which causes formation of the ion AlF_6^{3-}; the fluoride ion which is thus bound will not affect the fluoride ion electrode. If the solution does contain aluminium, it is treated with a complexing agent (e.g., cyclohexane-diamine-tetra-acetic acid) which complexes the aluminium and releases the fluoride ion from the AlF_6^{3-} ion.

Instrumentation and measurement of cell e.m.f.

XIV, 13.   POTENTIOMETERS. The most satisfactory method for the measurement of the e.m.f. of a cell is that known as **Poggendorff's compensation method**, an outline of which is given below. The principle of the method is to balance the unknown e.m.f. against a known e.m.f., which can easily be varied. When these two e.m.f.s are exactly equal, no current will flow through a galvanometer placed in the circuit: the galvanometer is therefore employed as a null instrument. The essential details are shown in Fig. XIV, 5.

A 2- or 4-volt accumulator furnishes the opposing e.m.f.; this is connected in series with a rheostat and with the terminals of a slide wire AB. The latter is a thin wire of uniform cross-section, and is often termed the 'potentiometer wire'. The cell, the e.m.f. of which is to be determined, is connected to one end A of the slide wire, and though a galvanometer G and a key S_2 to a sliding contact C, which can

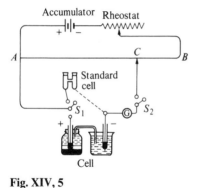

Accumulator Rheostat

Fig. XIV, 5

be moved along *AB*. A special double-throw switch S_1 may be provided to permit the standard cell to be placed in the circuit. In connecting the accumulator and the cell to the bridge, it is essential that the positive poles should be connected to the same end of the bridge wire; the unknown cell will then send a current through the circuit in a direction opposite to that furnished by the accumulator.

If we assume that the potentiometer wire has uniform cross-section and resistance, then the fall of potential along the slide wire will be uniform. The difference of potential between *A* and any point *C* will be proportional to the length *AC*, and will be equal to the fraction *AC/AB* of the total fall of potential along the wire. If the standard cell is now placed in circuit and the position of *C* adjusted to say *C'* so that when the switch S_2 is depressed, no current passes through the galvanometer *G*, then the e.m.f. of the cell is equal to that of the accumulator multiplied by *AC'/AB*. For much potentiometric work, only *changes* of potential are required, so that for example variations of the length *AC* are all that are required during a titration. In general, however, the e.m.f. of the accumulator is not quite constant, and as indicated above a standard cell is therefore employed to calibrate the slide wire. This is usually a Weston cell which has an e.m.f. of 1.0183 volts at 20 °C or

$$1.0183 - 0.0000406\,(t - 20\,°)$$

at any other temperature $t\,°C$. It should be noted however that there are often small differences in the e.m.f. of Weston cells supplied by different makers, and it is essential to use the exact value quoted for the cell in use, or preferably the value obtained by calibration of the cell in use against a Weston cell of known e.m.f. which is reserved for calibration purposes.

If the standard cell is placed in circuit by means of the switch S_1 and the point of balance *C'* on the bridge is determined, then the unknown e.m.f. may be calculated from the expression:

$$\frac{AC}{AC'} = \frac{\text{e.m.f. of unknown cell}}{\text{e.m.f. of standard cell}}$$

For approximate work, the slide wire *AB* may consist of a simple meter bridge, and the indicating instrument may be a milliammeter. It is preferable, however, to employ a commercial type of potentiometer which utilises the more compact spiral type of bridge wire.

The most convenient type of indicating instrument is the direct-vision type of mirror galvanometer; the galvanometer, lamp, and scale are incorporated in a blackened wooden (or plastic) box or compartment, and the 'spot' is clearly visible in daylight; a switch on the front of the instrument allows a choice of sensitivity. Alternatively, a solid state d.c. Null Detector (H. Tinsley and Co, or Croydon Precision Instrument Co) may be used; these instruments have the advantage of being less susceptible to vibration than the conventional mirror galvanometer.

In commercial potentiometers, a rheostat is provided in series with the 2-volt accumulator which can be adjusted so that the effective e.m.f. applied across the potentiometer is such that scale readings are directly in volts (or millivolts). If the bridge is divided into, say, 2000 equal parts, then the rheostat may be adjusted with the standard cell in circuit and with the sliding contact C at a position corresponding to 1018.3 divisions so that no current flows through the galvanometer. The position of the sliding contact will then give the e.m.f. of any unknown cell directly in milli-volts. It is usual for the rheostat to contain both coarse and fine adjustments: the fine adjustment may be used to compensate for the slight variations of the accumulator during the measurements.

The **Tinsley general-utility potentiometer** (type 3387B) shown in Fig. XIV, 6 is an excellent commercial potentiometer. Balancing is effected upon a main dial having eighteen steps of 0.1 volt and a calibrated circular slide wire range -0.005 to $+0.105$ volt, which can be read to 0.0001 volt by estimation, the smallest division being 0.0005 volt. The instrument has three range multipliers of $\times 1$, $\times 0.1$, and $\times 0.01$, giving the following ranges of direct calibration: 1.9 volts to 1 millivolt, 0.19 volt to 100 microvolts, and 0.019 volt to 10 microvolts respectively.

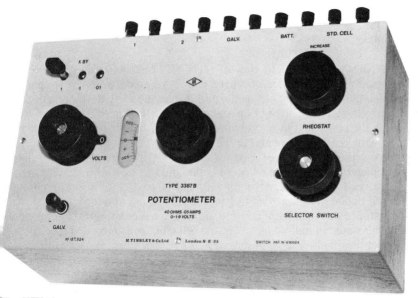

Fig. XIV, 6

There is an independent standardising circuit, adjusted to 20 °C, so that standardisation may be effected with a Weston cell independently of the dial setting. The selector switch has three positions, one for the standardising circuit, and two for external test circuits.

For work of the highest precision, one of the Tinsley 5590 series of Precision Vernier Potentiometers may be employed; Model 5590C for example is a five-dial potentiometer readable to seven decimal places, and with the smallest subdivision equal to one microvolt.

Whatever kind of potentiometer is used, the working cell (the accumulator) must be joined to the instrument which is then left for twenty to thirty minutes before attempting any readings; during this period the resistance coils of the instrument may warm up slightly and sufficient time must be allowed for thermal equilibrium to be established. The standard cell is then switched into circuit and the balance point on the potentiometer slide wire determined, or with an instrument such as the Tinsley potentiometer which reads directly in volts, the Selector switch is turned to the 'Calibrate' position, and the balancing rheostat is adjusted until the galvanometer shows no deflection. At the commencement of the standardising operation, the galvanometer sensitivity switch must be set to the lowest sensitivity, and then the sensitivity can be gradually increased as the point of balance is approached. Finally, the standard cell is replaced by the test cell (on the Tinsley potentiometer this is achieved by turning the Selector switch to position '1' or '2', depending upon which pair of terminals the cell has been joined to), and with the galvanometer at its lowest sensitivity, first the step-wise control (steps of 0.1 V) is adjusted, and then the slide wire control until the balance point is reached; once again as the balance point is approached, the galvanometer sensivity is increased until maximum sensitivity is achieved.

In much analytical work, the measurement of cell e.m.f. may be simplified by use of a solid state millivoltmeter (see pH meters and pIon meters in the following Sections); the e.m.f. is thus obtained immediately from a single reading without the need for the time-consuming balancing process needed with a conventional potentiometer.

XIV, 14. pH METERS.   In view of the high resistance of a glass electrode (1–100 megohms) a simple potentiometer cannot be used to measure the e.m.f. of a cell which includes a glass electrode, and in fact the early development of the glass electrode was dependent upon advances in the design of thermionic valves permitting the construction of 'valve voltmeters'. Since these instruments were designed with the requirements of the glass electrode in mind, and the glass electrode was used to measure the pH of solutions, the instruments were referred to as **pH meters**. Early pH meters were classified as (*a*) direct reading, or (*b*) potentiometric type meters. In meters of type (*a*) the e.m.f. of the cell containing the glass electrode was impressed upon a high resistance and the current flowing in the resistance was then amplified and applied to a sensitive moving coil meter; this was calibrated in millivolts so that the cell e.m.f. was recorded directly, and since in fact the quantity to be measured was pH, the scale was also calibrated in pH units, a selector switch being provided to allow choice of scale reading. In meters of type (*b*) a potentiometric circuit was employed in conjunction with an electronic amplifier and a milliammeter as balance point detector. The potentiometer was balanced against a standard cell contained within the instrument, and then the e.m.f. of the cell containing the glass electrode was applied to the potentiometer and balance achieved in the usual manner by adjustment first of a 'coarse' (stepwise) control, and then of a 'fine' (slide wire) control; these controls were calibrated in millivolts and also in pH units.

With the introduction of solid state circuitry which has simplified the problem of measuring small d.c. potentials in circuits of high impedance, the direct reading type of pH meter is now standard; and in the most modern type of meter a digital voltmeter is used, scaled to read pH directly. Such instruments are supplied by many makers; a typical example is the Electronic Instruments Ltd Model 7060 pH meter shown in Fig. XIV, 7.

As already explained, a glass electrode has an 'asymmetry potential' which makes it impossible to relate a measured electrode potential directly to the pH of

Fig. XIV, 7

the solution, and makes it necessary to calibrate the electrode. A pH meter therefore always includes a control ('Set Buffer', 'Standardise' or 'Calibrate') so that with the electrode assembly (glass plus reference electrode or a combination electrode) placed in a buffer solution of known pH, the scale reading of the instrument can be adjusted to the correct value.

The Nernst equation shows that the glass electrode potential for a given pH value will be dependent upon the temperature of the solution. A pH meter therefore includes a biasing control so that the scale of the meter can be adjusted to correspond to the temperature of the solution under test. This may take the form of a manual control, calibrated in °C, and which is set to the temperature of the solution as determined with an ordinary mercury thermometer. In some instruments, arrangements are made for automatic temperature compensation by inserting a temperature probe (a resistance thermometer) into the solution, and the output from this is fed into the pH meter circuit.

Some instruments also include what is known as a 'Slope control'. This is to allow for the fact that in some cases, if a meter is calibrated at a certain pH (say pH 4.00), then when the electrode assembly is placed in a new buffer solution of different pH (say 9.20), the meter reading may not agree exactly with the known pH of the solution. In this event, the slope control is adjusted so that the meter reading in the second solution agrees with the known pH value. The meter is again checked in the first buffer solution, and provided the scale reading is correct (4.00), it is assumed that the meter will give accurate readings for all pH values falling within the limits of the two buffer solutions.

Using a given glass electrode–reference electrode assembly, if we measure the cell e.m.f. over a range of pH, all measurements being at the same temperature, and if the readings are then repeated for a series of different temperatures, then on plotting the results as a series of isothermal curves, we find that at some pH value (pH_i), the cell e.m.f. is independent of temperature; pH_i was referred to by Jackson (Ref. 9) as the 'isopotential pH'. If the composition of the solution surrounding the inner silver–silver chloride is altered, or if an entirely different external reference electrode is used, then the value of pH_i changes, and some pH meters include an 'Isopotential' control which can be used to take account of such changes in the electrode system.

Mode of operation

Before use, it is obviously necessary to become familiar with the instruction manual issued with the pH meter it is proposed to employ, but the general procedure for making a pH measurement is similar for all instruments, and will follow a pattern such as that detailed below.

1. Switch on and allow the instrument to warm up; the time for this will be quite short if the circuit is of the solid-state type. Whilst this is taking place, make certain that the requisite buffer solutions for calibration of the meter are available, and if necessary prepare any required solutions: this is most conveniently done by dissolving an appropriate 'buffer tablet' (these are obtainable from many suppliers of pH meters and from laboratory supply houses) in the specified volume of distilled water.

2. If the instrument is equipped with a manual temperature control, take the temperature of the solutions and set the control to this value; if automatic control is available, then place the temperature probe into some of the first standard buffer solution contained in a small beaker which has been previously rinsed with a little of the solution.

3. Insert the electrode assembly into the same beaker, and if available, set the selector switch of the instrument to read pH.

4. Adjust the 'Set Buffer' control until the meter reading agrees with the known pH of the buffer solution.

5. Remove the electrode assembly (and the thermometer probe if used), rinse in distilled water, and place into a small beaker containing a little of the second buffer solution. If the meter reading does not agree exactly with the known pH, adjust the 'Slope' control until the required reading is obtained.

6. Remove the electrode assembly, rinse in distilled water, place in the first buffer solution and confirm that the correct pH reading is shown on the meter: if not, repeat the calibration procedure.

7. If the calibration is satisfactory, rinse the electrodes, etc., with distilled water, and introduce into the test solution contained in a small beaker. Read off the pH of the solution.

8. Remove the electrodes, etc., rinse in distilled water, and leave standing in distilled water.

XIV, 15. SELECTIVE ION METERS. Direct reading meters suitable for use with specific ion electrodes are available from a number of manufacturers; they are sometimes referred to as **ion activity meters**. They are very similar in construction to pH meters, and most can in fact be used as a pH meter, but by virtue of the extended range of measurements for which they must be used (anions as well as cations, and divalent as well as monovalent ions), the circuitry is necessarily more complex and scale expansion facilities are included. A typical meter of this type is the Electronic Instruments Ltd Model 7050 specific ion meter shown in Fig. XIV, 8.

As with a pH meter, the electrode appropriate to the measurement to be

Fig. XIV, 8

undertaken must be calibrated in solutions of known concentration of the chosen ion; at least two reference solutions should be used, differing in concentration by 2–5 units of pM according to the particular determination to be made. The general procedure for carrying out a determination with one of these instruments is outlined in Section **XIV, 17**.

Direct potentiometry

The use of a pH meter or a specific ion meter to measure the concentration of hydrogen ions or of some other ion in a solution is clearly an example of direct potentiometry. In view of the discussion in the preceding Sections the procedure involved will be evident, and two examples will suffice to illustrate the experimental method.

XIV, 16.   DETERMINATION OF pH.   At this stage it should be pointed out that the original definition of pH $= -\log c_H$ (due to Sørensen, 1909; this may be written as pcH) is not exact, and cannot be determined exactly by electrometric methods. It is realised that the activity rather than the concentration of an ion determines the e.m.f. of a galvanic cell of the type commonly used to measure pH, and hence pH may be defined as

$$pH = -\log a_{H^+}$$

where a_{H^+} is the activity of the hydrogen ion. This quantity as defined is also not capable of precise measurement, since any cell of the type

$$H_2, Pt \mid H^+ \text{ (unknown)} \mid \text{Salt Bridge} \mid \text{Reference Electrode}$$

used for the measurement inevitably involves a liquid junction potential of more or less uncertain magnitude. Nevertheless the measurement of pH by the e.m.f. method gives values corresponding more closely to the activity than the concentration of hydrogen ion. It can be shown that the pcH value is nearly equal to $-\log 1.1\, a_{H^+}$, hence:

$$pH = pcH + 0.04$$

This equation is a useful practical formula for converting tables of pH based on the Sørensen scale to an approximate activity basis, in line with the practical definition of pH given below.

The modern definition of pH is an operational one and is based on the work of standardisation and the recommendations of the US National Bureau of Standards (NBS). The NBS definition and the British definition are consistent in nearly all respects: the British definition is confined to one standard solution whereas the NBS definition extends to a number of standard solutions. The *difference* in pH between two solutions S (a standard) and X (an unknown) at the same temperature with the same reference electrode and with hydrogen electrodes at the same hydrogen pressure is given by:

$$pH\,(X) - pH\,(S) = \frac{E_X - E_S}{2.3026\, RT/F}$$

where E_X is the e.m.f. of the cell

H_2,Pt | Solution X | 3.5M KCl | Reference electrode,

E_S is the e.m.f. of the cell

H_2,Pt | Solution S | 3.5M KCl | Reference electrode.

The pH difference is thus a pure number. The scale is anchored by defining the nature of the standard solution and assigning a pH value to it.

In the British standard, S is a 0.05M solution of pure potassium hydrogen phthalate, the pH of which is 4.000 at 15 °C. At any other temperature, t, between 0 and 55 °C the pH is given by:

$$pH = 4.000 + \tfrac{1}{2}\left|\frac{t-15}{100}\right|^2$$

In the NBS recommendations equal importance is given to a small number of solutions for which pH values have been determined with great care over a range of temperatures: one of these solutions is the potassium hydrogenphthalate solution on which the British standard is based, and the pH values are almost identical. Data for four of the NBS standards now accepted by the IUPAC are collected in Table XIV, 3. The tartrate, phthalate, phosphate, and sodium tetraborate solutions are regarded as primary standards: at least two reference solutions should be used for the standardisation of cells with the glass electrode. Below pH 2 and above pH 12 the liquid–liquid junction potential is suspect; for this reason the tetroxalate solution and the calcium hydroxide solution are designated as secondary standards, and the pH values obtained by their use may be 0.02–0.04 unit lower than those obtained with an instrument standardised with the primary standards.

In passing it may be noted that the British Standard when applied to dilute solutions ($< 0.1M$) at pH between 2 and 12 conforms approximately to the equation

$$pH = -\log\{c_{H^+} y_{1:1}\} \pm 0.02$$

where $y_{1:1}$ is the mean activity coefficient which a typical 1:1-electrolyte would have in that solution.

Table XIV, 3.　pH of NBS standards from 0 to 95 °C

Temperature, °C	Secondary standard 0.05m-K tetroxalate $KH_3(C_2O_4)_2,2H_2O$	Primary standards					Secondary standard
		K H tartrate (satd. at 25°) $KHC_4H_4O_6$	0.05m-KH phthalate $KHC_8H_5O_4$	0.025m-KH_2PO_4, 0.025m-Na_2HPO_4	0.01m-borax, $Na_2B_4O_7,10H_2O$		$Ca(OH)_2$ (satd. at 25°)
0	1.67	—	4.00	6.98	9.46		13.43
5	1.67	—	4.00	6.95	9.40		13.21
10	1.67	—	4.00	6.92	9.33		13.00
15	1.67	—	4.00	6.90	9.28		12.81
20	1.68	—	4.00	6.88	9.23		12.63
25	1.68	3.56	4.01	6.86	9.18		12.45
30	1.69	3.55	4.02	6.85	9.14		12.30
35	1.69	3.55	4.02	6.84	9.10		12.14
40	1.70	3.55	4.03	6.84	9.07		11.99
45	1.70	3.55	4.05	6.83	9.04		11.84
50	1.71	3.55	4.06	6.83	9.01		11.70
55	1.72	3.55	4.08	6.83	8.99		11.58
60	1.72	3.56	4.09	6.84	8.96		11.45
70	1.74	3.58	4.13	6.85	8.92		
80	1.77	3.61	4.16	6.86	8.89		
90	1.80	3.65	4.21	6.88	8.85		
95	1.81	3.67	4.23	6.89	8.83		

The fifth NBS primary standard (KH_2PO_4, $0.00870m$; Na_2HPO_4, $0.0304m$), which only covers a very limited pH range (7.37–7.53), has not been included in Table XIV, 3, and it may be noted that the five primary standard solutions, with concentrations expressed on a molal basis (i.e., moles of solute per kilogram of solution), have been adopted as international standards by the IUPAC. The pH values for these international standards are quoted to the third decimal place (Ref. 10), and the values given in Table XIV, 3 agree with these when rounded off to the second place of decimals.

Details for the preparation of the solutions referred to in the table are as follows (*note that concentrations are expressed in molalities*): All reagents must be of the highest purity, e.g., A.R. products. Freshly distilled water protected from carbon dioxide during cooling, having a pH of 6.7–7.3, should be used, and is essential for basic standards. De-ionised water is also suitable. Standard buffer solutions may be stored in well-closed Pyrex or polythene bottles. If the formation of mould or sediment is visible the solution must be discarded.

0.05m-Potassium tetroxalate.　Dissolve 12.70 g of the dihydrate in water and dilute to 1 kg. The salt $KHC_2O_4,H_2C_2O_4,2H_2O$ must not be dried above

50 °C. The solution is stable and the buffer capacity is relatively high.

Saturated potassium hydrogen tartrate solution.   The pH is insensitive to changes of concentration and the temperature of saturation may vary from 22 to 28 °C: the excess of solid must be removed. The solution does not keep for more than a few days unless a preservative (crystal of thymol) is added.

0.05m-Potassium hydrogenphthalate.   Dissolve 10.21 g of the solid (dried below 130 °C) in water and dilute to 1 kg. The pH is not affected by atmospheric carbon dioxide: the buffer capacity is rather low. The solution should be replaced after 5–6 weeks, or earlier if mould-growth is apparent.

0.025m-Phosphate buffer.   Dissolve 3.40 g of KH_2PO_4 and 3.55 g of Na_2HPO_4 (dried for 2 hours at 110–130 °C) in carbon dioxide-free water and dilute to 1 kg. The solution is stable when protected from undue exposure to the atmosphere.

0.01m Borax.   Dissolve 3.81 g of sodium tetraborate $Na_2B_4O_7,10H_2O$ in carbon dioxide-free water and dilute to 1 kg. The solution should be protected from exposure to atmospheric carbon dioxide, and replaced about a month after preparation.

Saturated calcium hydroxide solution.   Shake a large excess of finely divided calcium hydroxide vigorously with water at 25 °C, filter through a sintered glass filter (porosity 3) and store in a polythene bottle. Entrance of carbon dioxide into the solution should be avoided. The solution should be replaced if a turbidity develops. The solution is $0.0203M$ at 25 °C (pH 12.45), $0.0211M$ at 20 °C (pH 12.47), and $0.0195M$ at 30 °C (pH 12.44).

To measure the pH of a given solution the normal procedure is to use a glass electrode together with a saturated calomel reference electrode and to measure the e.m.f. of the cell with a pH meter; the scale of the meter is calibrated to read pH directly. The procedure for use of a pH meter has already been described in Section **XIV, 14**; the instruction manual of the instrument available for use should however be consulted for details of minor variations in the controls supplied. The glass electrode supplied with the instrument should be standing in distilled water: if for any reason it is necessary to make use of a new electrode, then this must be left soaking in distilled water for at least twelve hours before measurements are attempted. Never handle the bulb of the electrode, and remember that the assembly is necessarily somewhat fragile and treat it with great care: in particular the electrode must always be supported within the measuring vessel (special electrode stands are usually supplied with pH meters) and not allowed to stand on the base of the vessel.

Prepare the buffer solutions for calibration of the pH meter if these are not already available; the potassium hydrogenphthalate buffer (pH 4), and the sodium tetraborate buffer (pH 9.2) are the most commonly used for calibration purposes. The solutions can be prepared in accordance with the details given above, but the simplest procedure is to make use of the buffer tablets which can be purchased.

Check whether the instrument supplied is equipped for automatic temperature compensation, and, if so, that the temperature probe (resistance thermometer) is available. If it is not so equipped, then the temperature of the solutions to be used must be measured, and the appropriate setting made on the manual temperature control of the instrument.

Proceed to measure the pH of the given solution, following the steps outlined in Section **XIV, 14**. On completion of the determination, remember to wash down the electrodes with distilled water, and to leave them standing in distilled water.

XIV, 17. DETERMINATION OF FLUORIDE. This determination involves the use of a ion sensitive electrode (Sections **XIV, 9–11**) in conjunction with an ion activity meter. The electrode must (as with the glass electrode used for pH measurements), be calibrated, using solutions of the appropriate ion at known concentrations. In view of the influence of ionic strength on activity coefficients, it is important that the test and the standard solutions should be of comparable ionic strength. When dealing with a test solution containing a single electrolyte little difficulty will be encountered in arranging for the test and standard solutions to be of similar ionic strength, but this may not be the case when the test solution has arisen from an involved analytical procedure. In such a case, an **ionic strength adjuster buffer** is added to both test and standard solutions so as to achieve comparable values for the ionic strength in all solutions, the value being governed by the ionic strength adjuster buffer rather than by the sample itself; a number of electrolytes may be used in this fashion, due attention being paid to ensure that errors do not arise owing to complexation, or to poor selectivity of the electrode in use with respect to the added ion in relation to the ion whose concentration is to be determined. Whereas for pH measurements it suffices to calibrate the glass electrode at two pH values, for ion-sensitive electrodes it is advisable to plot a calibration curve by making measurements with a number (usually five to six) of standard solutions of varying concentration.

As an alternative to plotting a calibration curve, the method of standard addition may be used. We first set up the appropriate ion-sensitive electrode together with a suitable reference electrode in a known volume of the test solution, and then measure the resultant e.m.f. (E_t). Applying the usual Nernst equation we can say

$$E_t = K + k \log y_t C_t$$

where K is the electrode constant, k is theoretically $2.303RT/nF$ but in practice is the experimentally determined slope of the E vs log C plot for the given electrode, f_t and C_t are the activity coefficient and the concentration respectively of the ion to be determined in the test solution. A known volume V_1 of a standard solution (concentration C_s) of the ion to be determined is added to the test solution, and the new e.m.f. E_1 is measured; C_s should be 50–100 times greater than the value of C_t. For the new e.m.f. E_1 we can write:

$$E_1 = K + k \log y_1 (V_t C_t + V_1 C_s)/(V_t + V_1)$$

where V_t is the original volume of the test solution.

Provided that the first and second solutions are of similar ionic strength, the activity coefficients will be the same in each solution, and the difference between the two e.m.f. values can be expressed as

$$\Delta E = (E_1 - E_t) = k \log (V_t C_t + V_1 C_s)/C_t(V_t + V_1)$$

from whence $$C_t = \frac{C_s}{10^{\Delta E/k}(1 + V_t/V_1) - V_t/V_1}$$

Hence provided the value of the slope constant k is known, the unknown concentration C_t can be calculated.

Procedure. Set up the ion activity meter (a digital pH/millivoltmeter, e.g., Corning-EEL Model 109, used in the millivolt mode is equally satisfactory) in accordance with the manual supplied with the instrument.

The electrodes required are a fluoride ion-sensitive electrode (e.g., Corning No. 476042) and a calomel reference electrode of the type supplied for use with pH meters.

Prepare the following solutions.

Sodium fluoride standards. Using A.R. sodium fluoride and de-ionised water, prepare a standard solution which is approximately $0.05M$ ($2.1\,g\,dm^{-3}$), and of accurately known concentration (Solution A). Take $10\,cm^3$ of solution A and dilute to $1\,dm^3$ in a graduated flask to obtain solution B which contains approximately 10 p.p.m. fluoride ion. $20\,cm^3$ of solution B further diluted (graduated flask) to $100\,cm^3$ gives a standard (solution C), containing approximately 2 p.p.m. fluoride ion, and by diluting $10\,cm^3$ and $5\,cm^3$ portions of solution B to $100\,cm^3$, we obtain standards D and E, containing respectively 1 and 0.5 p.p.m. F^-.

Total Ionic Strength Adjustment Buffer (TISAB). Dissolve $57\,cm^3$ A.R. acetic acid, 58 g A.R. sodium chloride and 4 g cyclohexane diamino-tetra-acetic acid (CDTA) in $500\,cm^3$ of de-ionised water contained in a large beaker. Stand the beaker inside a water bath fitted with a constant level device, and place a rubber tube connected to the cold water tap *inside* the bath. Allow water to flow slowly into the bath and discharge through the constant level: this will ensure that in the subsequent treatment the solution in the beaker will remain at constant temperature.

Insert into the beaker a calibrated glass electrode-calomel electrode assembly which is joined to a pH meter, then with constant stirring and continuous monitoring of the pH, add slowly sodium hydroxide solution ($5M$), until the solution acquires a pH of 5.0–5.5. Pour into a dm^3 graduated flask and make up to mark with de-ionised water.

The resulting solution will exert a buffering action in the region pH 5–6, the CDTA will complex any polyvalent ions which may interact with fluoride, and by virtue of its relatively high concentration the solution will furnish a medium of high total ionic strength, thus obviating the possibility of variation of e.m.f. owing to varying ionic strength of the test solutions.

Pipette $25\,cm^3$ of solution B into a $100\,cm^3$ beaker mounted on a magnetic stirrer and add an equal volume of TISAB from a pipette. Stir the solution to ensure thorough mixing, stop the stirrer, insert the fluoride ion–calomel electrode system and measure the e.m.f. The electrode rapidly comes to equilibrium, and a stable e.m.f. reading is obtained immediately. Wash down the electrodes and then insert into a second beaker containing a solution prepared from $25\,cm^3$ each of standard solution C and TISAB; read the e.m.f. Carry out further determinations using the standards D and E.

Plot the observed e.m.f. values against the concentrations of the standard solutions, using a semi-log graph paper which covers four cycles (i.e., spans four decades on the log scale): use the log axis for the concentrations which should be in terms of fluoride ion concentration. A straight line plot (calibration curve) will be obtained. With increasing dilution of the solutions there tends to be a departure from the straight line: with the electrode combination and measuring system referred to above, this becomes apparent when the fluoride ion concentration is reduced to *ca.* 0.2 p.p.m.

Now take $25\,\text{cm}^3$ of the test solution, add $25\,\text{cm}^3$ TISAB and proceed to measure the e.m.f. as above. Using the calibration curve, the fluoride ion concentration of the test solution may be deduced. The procedure described is suitable for measuring the fluoride ion concentration of tap water in areas where fluoridation of the supply is undertaken.

Potentiometric titrations

XIV, 18. POTENTIOMETRIC TITRATIONS; CLASSICAL METHOD.
In the two previous sections dealing with direct potentiometry the procedure involved measurement of the e.m.f. between two electrodes; an indicator electrode, the potential of which is a function of the concentration of the ion to be determined, and a reference electrode of constant potential: accurate determination of the e.m.f. is crucial. In potentiometric titrations absolute potentials or potentials with respect to a standard half-cell are not usually required, and measurements are made whilst the titration is in progress. The equivalence point of the reaction will be revealed by a sudden change in potential in the plot of e.m.f. readings against the volume of the titrating solution; any method which will detect this abrupt change of potential may be used. One electrode must maintain at a constant, but not necessarily known, potential; the other electrode must serve as an indicator of the changes in ion concentration, and must respond rapidly. The solution must, of course, be stirred during the titration. Simple arrangements for potentiometric titration are given in Fig. XIV, 9, and in Fig. XIV, 10. In the former diagram, A is a reference

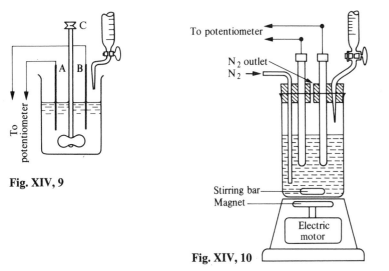

Fig. XIV, 9

Fig. XIV, 10

electrode (e.g., a saturated calomel half-cell), B is the indicator electrode, and C is a mechanical stirrer (it may be replaced, with advantage, by a magnetic stirrer); the solution to be titrated is contained in the beaker. When basic or other solutions requiring the exclusion of atmospheric carbon dioxide or of air are titrated it is advisable to use either a three- or four-necked flask or a tall lipless beaker equipped as shown in Fig. XIV, 10. It is convenient to use as reference electrode a compact calomel half-cell as supplied with pH meters (see Fig. XIV, 3); nitrogen may be bubbled through the solution before and, if necessary, during the titration.

The e.m.f. of the cell containing the initial solution is determined, and relatively large increments (1–5 cm³) of the titrant solution are added until the equivalence point* is approached; the e.m.f. is determined after each addition. The approach of the e.p. is indicated by a somewhat more rapid change of the e.m.f. In the vicinity of the equivalence point, equal increments (e.g., 0.1 or 0.05 cm³) should be added; the equal additions in the region of the e.p. are particularly important when the equivalence point is to be determined by the analytical method described below. Sufficient time should be allowed after each addition for the indicator electrode to reach a reasonably constant potential (to *ca.* ±1–2 millivolts) before the next increment is introduced. Several points should be obtained well beyond the e.p.

To measure the e.m.f. the electrode system must be connected to a potentiometer; the simple slide wire system shown in Fig. XIV, 5 will often suffice, but it will usually be more convenient to employ a commercial potentiometer, as for example that shown in Fig. XIV, 6. If the indicator electrode is a membrane electrode (e.g., a glass electrode), then a simple potentiometer is unsuitable and either a pH meter or a selective ion meter must be employed: the meter readings may give directly the varying pH (or pM) values as titration proceeds, or the meter may be used in the millivoltmeter mode, so that e.m.f. values are recorded. Used as a millivoltmeter, such meters can be used with almost any electrode assembly to record the results of many different types of potentiometric titrations, and in many cases the instruments have provision for connection to a recorder so that a continuous record of the titration results can be obtained, i.e., a titration curve is produced.

A number of commercial 'potentiometric titration units' are available which comprise an electrode system (frequently a selection of electrodes is offered) together with a special stand providing supports for the electrodes, and one or two burette holders. The base of the stand incorporates a magnetic stirrer, and in some cases a hot plate, so that if necessary the solution to be titrated may be heated. A separate unit embodies a potentiometer and a compact galvanometer; or if the potentiometer is designed for operation from the a.c. mains, the balance point indicator may be a 'magic eye' electronic indicator similar to that used in mains-operated conductivity bridges (Section **XV, 4**).

XIV, 19. USE OF BIMETALLIC ELECTRODE SYSTEMS. A tungsten electrode does not respond readily to changes in potential in certain oxidation–reduction systems (e.g., $Cr_2O_7^{2-}$, Fe^{2+}) whereas a platinum electrode does. Hence a platinum–tungsten couple can be used instead of the usual combination of a platinum electrode and reference (e.g., calomel) electrode to indicate the end point. With the Pt–W couple the tungsten appears to undergo anodic oxidation and acts as a kind of 'attackable reference electrode'. The use of such bimetallic systems is empirical, and the optimum conditions should be established by trial. With the Pt–W pair the potential is small at first, remains at

---

* The abbreviation e.p. will be used for equivalence point.

this value until very near the equivalence point, when it usually increases slightly, and then there is an abrupt change at the equivalence point. It is unsuitable for the titration of very dilute solutions ($<0.001M$).

XIV, 20. POLARISED INDICATOR ELECTRODES. Some redox couples (e.g., $Cr_2O_7^{2-}$, Cr^{3+}; MnO_4^-, Mn^{2+}; and $S_2O_3^{2-}$, $S_4O_6^{2-}$) encountered in titrimetric analysis are somewhat slow in establishing steady potentials at a platinum electrode when the measurement is made in the ordinary way (i.e., with zero current). To eliminate long waiting periods for the attainment of steady potentials in cases of this kind polarised indicator electrodes, at which electrolysis is forced to occur at a slow rate, may be used; polarised mono-metallic and polarised bimetallic systems have been employed. The former consists of a polarised metallic electrode and an unpolarised electrode (e.g., a calomel electrode). A bimetallic system consists of two identical pure platinum wires, one polarised anodically and the other cathodically with a polarising current of the order of a few microamperes; these appear to behave as two dissimilar metals, and their single electrode potentials respond in a different manner. At all events a distinct change in behaviour is apparent at the equivalence point, and if the potential difference between the electrodes is plotted against the volume of reagent added the usual differential type of curve is obtained. The potential difference developed at the end point may be of the order of 100–200 millivolts. When one or both redox couples involved in the titration reaction behave irreversibly a polarised electrode may show a considerably different change in potential at the equivalence point than when the measurement is made with an unpolarised electrode at zero current. If both couples behave reversibly the potential change of the indicator electrode will be about the same with or without electrolysis.

In potentiometric titrations with polarised electrodes the measured quantity is the change in e.m.f. at constant current (compare Amperometric Titrations, Chapter XVII, in which the change in current at constant applied e.m.f. is measured). In some cases titration curves obtained with one or two polarised electrodes exhibit larger changes in e.m.f. at the end point than curves obtained with an unpolarised indicator electrode, e.g., in the titration of iodine with thiosulphate ion using either two identical polarised platinum electrodes or a single platinum electrode polarised cathodically. The use of polarised electrodes can entail an error corresponding to the amount of electrolysis that occurs at the electrode; this is always present with a single polarised electrode, and also occurs with two identical polarised electrodes whenever the titration couple behaves irreversibly. The error can be reduced to negligible proportions by using small electrodes and a small electrolysis current. The main value of polarised electrodes is for titrations involving irreversible couples where an unpolarised electrode is often very slow in acquiring a constant potential whereas a polarised indicator electrode reaches a steady potential quickly at constant current, and large variations of potential are observed at the end point.

XIV, 21. DIFFERENTIAL POTENTIOMETRIC TITRATION. As indicated in Section **XIV, 18**, as the end point of a titration is approached the e.m.f. of the system changes more rapidly; and as shown by curve (b) in Fig. XIV, 14 (Section **XIV, 23**), a plot of $\Delta E/\Delta V$ against V (the volume of titrant added) is a maximum at the end point. It is possible to measure directly $\Delta E/\Delta V$ as a function of V and this procedure is referred to as **differential potentiometric titration**. The desired result is accomplished by placing two identical indicator electrodes (e.g. platinum wires) in the solution to be titrated, but one of these (the 'isolated' electrode) is in a small portion of the liquid that is separated from the main body of the solution, and hence isolated from immediate reaction with the titrant. A simple device for this purpose is depicted in Fig. XIV, 11. A small volume of the solution is withdrawn into a dropper provided with platinum-wire electrodes as shown in the figure: the latter are connected to a high-resistance galvanometer, serving as a voltmeter. In practice, the titration is conducted by adding small uniform increments of titrant. The potential difference, corresponding to ΔE, is read after each addition, then the liquid within the bulb is expelled and the bulb is refilled with a portion of the main solution. Thus at each stage the isolated electrode is kept one increment, ΔV, behind the electrode in the main solution, but

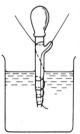

Fig. XIV, 11

is allowed to catch up before the next increment is added by completely expelling the liquid within the bulb. The value of ΔE begins to increase rapidly near the equivalence point: the latter is indicated by the maximum value of Δ. The main advantage of the differential method is that it does not require a reference electrode; it is slower and less convenient than the technique of titrating to the equivalence point potential. Differential methods are not suited for titrations where the electrodes in the solution reach equilibrium very slowly.

XIV, 22.   AUTOMATIC POTENTIOMETRIC TITRATIONS.   As already mentioned (Section **XIV, 18**), by joining a recorder to a mains-operated potentiometer, it is possible to produce directly the titration curve relating to the potentiometric titration under investigation. If the delivery of titrant from the burette is linked to the movement of the recorder chart, then the process becomes automatic, and a number of firms market titration units which fulfil this function. A typical example is the Metrohm **'Potentiograph'** (Model E536) shown in Fig. XIV, 12, which includes the control unit/chart recorder linked to a motor-driven piston burette, and to the electrode assembly of the titration vessel on the right-hand side of the photograph; the paper feed of the recorder is coupled to the motor drive of the piston burette, and the pen of the recorder follows the change in e.m.f. of the electrode assembly. The instrument will also provide a plot of $\Delta E/\Delta V$.

Automation has also been extended to stop delivery of the titrant when the potential of the indicator electrode attains the value corresponding to the equivalence point of the particular titration involved; this feature is clearly of great value when a number of repetitive titrations have to be performed. It is necessary to carry out a preliminary experiment to determine the equivalence point potential of the indicator electrode (or more precisely, the equivalence point e.m.f. for the indicator electrode-standard electrode combination in use), and to prevent over-shooting the end point provision must be made for reducing the rate of addition of titrant as the end-point is approached. Various control units can be purchased (e.g., the Metrohm **End-Point Titrator** E526) which carry the requisite instrumentation. A potentiometer is included, and the final equivalence point e.m.f. to be attained is set on this. At the start of the titration, the e.m.f. set up by the electrode assembly in the titration vessel will be far removed from the equivalence point value, and so a 'difference potential' will

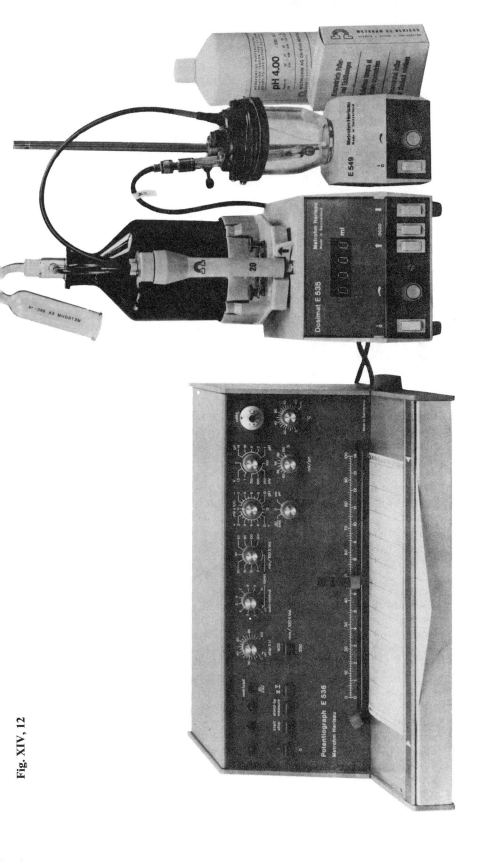

Fig. XIV, 12

exist between the electrodes and the pre-set reading of the potentiometer; it is this 'difference potential' which controls the subsequent operations. The delivery of titrant from the burette is controlled in some systems by a solenoid-controlled valve, or with a piston burette, by the rate at which the motor drives the burette. It is usually necessary to select another e.m.f., somewhat lower than the equivalence point e.m.f., from which point the titrant will be added slowly, and this value must also be set on the control unit. The dials on the control unit are frequently scaled in pH units as well as in millivolts, so that for an acid-base titration the controls may be set in terms of pH rather than of e.m.f.

With the control unit set up with the readings appropriate to the titration to be carried out, a measured volume of the solution to be titrated is introduced into the titration vessel and diluted to a suitable volume (usually $50-100\,cm^3$), and the burette is charged with the titrant. The magnetic stirrer is set into operation, and the titration started. Initially, with a large difference potential, titrant will be added rapidly to the solution and this will continue until the e.m.f. is equal to the pre-selected 'change-over potential'. At this point the addition of titrant is slowed down; the piston burette will be driven more slowly, or in the case of the solenoid-controlled valve burette, the valve is largely closed down, or, in some systems, two burettes are provided, one with a coarse jet, and the other with a fine jet; at the change-over point, the burette with the coarse jet is shut off completely and the titration is completed through the fine-jet burette. Addition of titrant will finally cease when the difference potential between electrodes and potentiometer disappears. As a further aid to avoid over-shooting the end-point, it is usually recommended that the burette be provided with a drawn-out tip which is inserted directly into the solution to be titrated, and in such a position, that as the solution is stirred, the liquid in the neighbourhood of the burette tip is directed towards the indicator electrode, which thus tends to cut off delivery of titrant a little on the early side. With continued stirring and dispersal of the added titrant, if the end point has in fact not been attained, a difference potential will reappear and the burette will be actuated again: step-wise addition will thus continue until the true end-point is reached. Auto-titrators are not suitable for use in cases where the indicator electrode response is slow, or when the chemical reaction involved in the titration is slow.

XIV, 23.   LOCATION OF END POINTS.   When a titration curve has been obtained (i.e., a plot of e.m.f. readings against volume of titrant added) either by manual plotting of the experimental readings, or with suitable equipment, plotted automatically during the course of the titration; it will in general be of the same form as the neutralisation curve for an acid, i.e., an S-shaped curve as shown in Fig. X, 2 (Section **X, 12**). The central portion of such a curve is shown in Fig. XIV, 13 and also in Fig. XIV, 14(*a*), and clearly the end-point will be located on the steeply rising portion of the curve, and it will in fact occur at the point of inflection. Although when the curve shows a very clearly marked steep portion, one can give an approximate value of the end-point as being mid-way along the steep part of the curve, it is generally necessary to carry out some geometrical construction in order to fix the end point exactly. Three procedures may be adopted for this purpose:

(*a*) the method of bisection;
(*b*) the method of parallel tangents;
(*c*) the method of circle fitting.

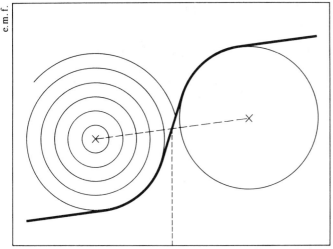

Fig. XIV, 13

Volume of titrant

(*a*) *The method of bisection.* This can be applied when the curve shows reasonably good straight lines before and after the steep part of the curve. Each of these straight lines is extended (the lower portion to the right, the upper portion to the left), and then at suitable points vertical lines are erected, one to the right of the steep part of the titration curve and one to the left. These vertical lines are then bisected, and the midpoints joined; where the line joining the midpoints cuts the titration curve is the end point of the titration.

(*b*) *The method of parallel tangents.* For this method a thin rigid plastic sheet large enough to cover the titration curve is required, and on this is marked a central horizontal line, together with a number of pairs of parallel lines drawn on either side of the central line, i.e., line 1 (the first line above the central marker) will be paired with the corresponding line 1' equidistant below the marker. The scale of the titration curves normally encountered will determine the dimensions of the markings on the plastic sheet, but up to about ten pairs of lines will usually be adequate, and it will be found convenient for identification purposes to mark different pairs of lines in different coloured inks. When the sheet has been prepared, a thin slot is then cut along the central marker so that the point of a pencil can be introduced.

The method is used when the portions of the curve on either side of the steep portion shows a marked curvature, and the procedure is to lay the plastic sheet on top of the titration curve in such a position that a given pair of the parallel lines (easily identified if they are similarly coloured), are tangential to the upper and lower parts of the titration curve. A pencil mark made through the central slot at the point where it cuts the steep part of the titration curve then identifies the end-point.

(*c*) *The method of circle fitting.* For this method also a thin rigid plastic sheet will be needed on which is marked a series of circles of varying sizes; the circles may be drawn independently or they may be concentric, but in either case a small hole must be drilled in the plastic at the centre of each circle (or at the common centre) so that a pencil point may be inserted. The circles should

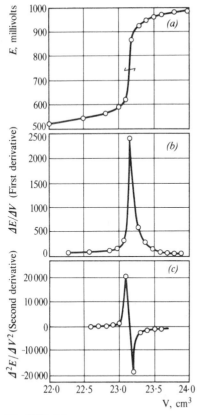

Fig. XIV, 14

increase in diameter in steps of about 1 cm and a maximum diameter of about 30 cm usually suffices; if larger diameters are found to be necessary then it is not essential to draw the complete circle thus obviating the need for an excessively large plastic sheet. The method of procedure is indicated in Fig. XIV, 13; the plastic sheet is laid on the titration curve and its position altered until one of the circles fits the lower bend in the curve and the position of the centre of the circle is marked on the titration curve. The sheet is then moved to the upper bend of the curve and when a circle which fits the bend is found, the position of the centre is marked. The marks indicating the two centres are then joined by a straight line (XX in the diagram), and where this line cuts the steep part of the titration curve is the end point.

Unless the curve has been plotted automatically, the accuracy of the results obtained by any of the above procedures will be dependent upon the skill with which the titration curve has been drawn through the points plotted on the graph from the experimental observations. It is therefore usually considered preferable to employ **analytical (or derivative) methods** of locating the end point; these consist in plotting the first derivative curve ($\Delta E/\Delta V$ against V), or the second derivative curve ($\Delta^2 E/\Delta V^2$) against V). The first derivative curve gives a maximum at the point of inflection of the titration curve, i.e., at the end point, whilst the second derivative ($\Delta^2 E/\Delta V^2$) is zero at the point where the slope of the $\Delta E/\Delta V$ curve is maximum.

The procedure may be illustrated by the actual results obtained for the potentiometric titration of 25.0 cm³ of *ca.* 0.1*M*-ammonium iron(II) sulphate with standard (0.1095*M*)-cerium(IV) sulphate solution using platinum and saturated calomel electrodes:

$$Fe^{2+} + Ce^{4+} = Fe^{3+} + Ce^{3+}$$

The results are collected in Table XIV, 4, as are also the calculated values for the first derivative $\Delta E/\Delta V$ (millivolt/cm³) and the second derivative $\Delta^2 E/\Delta V^2$. It is clear that for locating the end-point, only the experimental figures in the vicinity of the equivalence point are required: all the observed results for the potentiometric titration are given for the sake of completeness. It is convenient, and simplifies the calculations, if small equal volumes of titrant are added in the neighbourhood of the end-point, but this is not essential.

Table XIV, 4   Potentiometric titration of Fe^{2+} solution with 0.1095M-Ce^{4+} solution, using platinum and calomel electrodes

Ce^{4+} solution added, cm^3 (V)	$E(mV)$	$\Delta E/\Delta V$ (mV/cm^3)	$\Delta^2 E/\Delta V^2$
1.00	373		
		10.5	
5.00	415		
		4.6	
10.00	438		
		4.2	
15.00	459		
		6.4	
20.00	491		
		12	
21.00	503		
		20	
22.00	523		
		40	
22.50	543		
		70	
22.60	550		0
		70	
22.70	557		100
		80	
22.80	565		200
		100	
22.90	575		500
		150	
23.00	590		1500
		300	
23.10	620		21000
		2400	
23.20	860		−18500
		550	
23.30	915		−2600
		290	
23.40	944		−1500
		140	
23.50	958		−840
		56	
24.00	986		−300
		40.5	
26.00	1067		−90
		14.5	
30.00	1125		

In Fig. XIV, 14, are presented: (*a*) the part of the experimental titration curve in the vicinity of the equivalence point; (*b*) the first derivative curve, i.e., the slope of the titration curve as a function of V (the equivalence point is indicated by the maximum, which corresponds to the inflexion in the titration curve); and (*c*) the second derivative curve, i.e., the slope of curve (*b*) as a function of V (the second derivative becomes zero at the inflexion point and provides a more exact measurement of the equivalence point).

The optimum volume increment ΔV depends upon the magnitude of the slope of the titration curve at the equivalence point and this can easily be estimated from a preliminary titraton. In general, the greater the slope at the e.p., the smaller should ΔV be, but it should also be large enough so that the successive values of ΔE exhibit a significant difference.

When the titration curve is symmetrical about the equivalence point the end-point defined by the maximum value of $\Delta E/\Delta V$ is identical with the true stoichiometrical equivalence point. A symmetrical titration curve is obtained when the indicator electrode is reversible and when in the titration reaction one mol. or ion of the titrant reagent reacts with one mol. or ion of the substance titrated. Asymmetrical titration curves result when the number of molecules or ions of the reagent and the substance titrated are unequal in the titration reaction, e.g., in the reaction

$$5Fe^{2+} + MnO_4^- + 8H^+ = 5Fe^{3+} + Mn^{2+} + 4H_2O$$

In such reactions, even though the indicator electrode functions reversibly, the maximum value of $\Delta E/\Delta V$ will not occur exactly at the stoichiometric equivalence point. The resulting *titration error* (difference between end point and equivalence point) can be computed or can be determined by experiment and a correction applied. The titration error is small when the potential change at the equivalence point is large. With most of the reactions used in potentiometric analysis, the titration error is usually small enough to be neglected. It is assumed that sufficient time is allowed for the electrodes to reach equilibrium before a reading is recorded.

As has been indicated, if suitable automatic titrators are used, then the derivative curve may be plotted directly and there is no need to undertake the calculations described above. Likewise if a differential titration is carried out (Section **XIV, 21**), then data is available which can be plotted directly to give the first derivative curve.

When the potential of the indicator electrode at the equivalence point is known, either from a previous experiment or from calculations, the end-point can be determined simply by adding the titrant solution until this equivalence-point potential is reached. This technique is analogous to ordinary titrations with indicators and is very convenient and rapid. The potentiometer is set to this potential, and the titrant solution is added (dropwise near the end point) until the galvanometer shows no deflection, or reverses the direction of deflection, when the tapping key is closed momentarily. The accuracy of this technique will depend upon the reproducibility of the equivalence point potential: it need only be known approximately when $\Delta E/\Delta V$ is large.

XIV, 24. SOME GENERAL CONSIDERATIONS. In this and succeeding Sections experimental details are given for some typical potentiometric titrations; with this information it should be possible to deduce the appropriate procedure to be followed in other cases. The majority of potentiometric titrations involve chemical reactions which can be classified as (*a*) neutralisation reactions, (*b*) oxidation–reduction reactions, (*c*) precipitation reactions or (*d*) complexation reactions, and for each of these different types of reaction, certain general principles can be enunciated.

(*a*) **Neutralisation reactions.** The indicator electrode may be a hydrogen, glass, or antimony electrode; a calomel electrode is generally employed as the reference electrode.

The accuracy with which the end point can be found potentiometrically depends upon the magnitude of the change in e.m.f. in the neighbourhood of the equivalence point, and this depends upon the concentration and the strength of the acid and alkali (compare Sections **X, 13–16**). Satisfactory results are obtained in all cases except: (*a*) those in which either the acid or the base is very weak ($K < 10^{-8}$) and the solutions are dilute, and (*b*) those in which both the acid and the base are weak. In the latter case an accuracy of about 1 per cent may be obtained in $0.1M$ solution.

The method may be used to titrate a mixture of acids which differ greatly in their strengths, e.g., acetic and hydrochloric acids; the first break in the titration curve occurs when the stronger of the two acids is neutralised, and the second when neutralisation is complete. For this method to be successful, the two acids or bases should differ in strength by at least 10^5 to 1.

(b) Oxidation–reduction reactions. The theory of oxidation–reduction reactions is given in Sections **II, 25**. The determining factor is the ratio of the concentrations of the oxidised and reduced forms of certain ion species. For the reaction:

*Ox*idised form + n electrons $\rightleftharpoons$ *Red*uced form

the potential E acquired by the indicator electrode at 25 °C is given by:

$$E = E^\ominus + \frac{0.0591}{n} \log \frac{[Ox]}{[Red]}$$

where $E^\ominus$ is the standard potential of the system. The potential of the immersed electrode is thus controlled by the *ratio* of these concentrations. During the oxidation of a reducing agent or the reduction of an oxidising agent the ratio, and therefore the potential, changes more rapidly in the vicinity of the end point of the reaction. Thus titrations involving such reactions (e.g., iron(II) with potassium permanganate or potassium dichromate or cerium(IV) sulphate) may be followed potentiometrically and afford titration curves characterised by a sudden change of potential at the equivalence point. The indicator electrode is usually a bright platinum wire or foil, and the oxidising agent is generally placed in the burette.

(c) Precipitation reactions. The theory of precipitation reactions is given in Sections **X, 29–30**. The ion concentration at the equivalence point is determined by the solubility product of the sparingly soluble material formed during the titration. In the precipitation of an ion I from solution by the addition of a suitable reagent, the concentration of I in the solution will clearly change most rapidly in the region of the end-point. The potential of an indicator electrode responsive to the concentration of I will undergo a like change, and hence the change can be followed potentiometrically. Here one electrode may be a saturated calomel or silver–silver chloride electrode, and the other must be an electrode which will readily come into equilibrium with one of the ions of the precipitate. For example, in the titration of silver ions with a halide (chloride, bromide, or iodide) this must be a silver electrode. It may consist of a silver wire, or of a platinum wire or gauze plated with silver and sealed into a glass tube. Since a halide is to be determined, the salt bridge must be a saturated solution of potassium nitrate. Excellent results are obtained by titrating, for example, silver nitrate solution with thiocyanate ions. Mechanical stirring is desirable to accelerate the attainment of solubility equilibrium.

(d) Complexation reactions. In many cases of this type of titration, complex formation results from the interaction of a sparingly soluble precipitate with an excess of reagent; this occurs for example when we titrate a solution of potassium cyanide with silver nitrate, where silver cyanide initially produced dissolves in excess potassium cyanide to give the complex ion $[Ag(CN)_2]^-$ and consequently only a very small concentration of silver ions. This situation continues up to the point where all the cyanide ion has been converted to the complex ion, the increasing concentration of which also means a gradually increasing concentration of free silver ions and consequently a gradual rise in the potential of a silver electrode in the solution. At the end point, there is a marked rise in potential which enables the end point to be determined, but if the addition of silver nitrate is continued past this point, the e.m.f. changes only very gradually and silver cyanide is precipitated. Finally a second rapid change in potential is observed at the point where all the cyanide ion has been precipitated as silver cyanide. For this particular titration a silver electrode is the obvious indicator electrode, and as reference electrode either a mercury–mercury(I) sulphate electrode, or a calomel electrode which is isolated from the solution to be titrated by means of a potassium nitrate or potassium sulphate salt bridge.

For complexation titrations involving the use of EDTA, an indicator electrode can be set up by using a mercury electrode in the presence of mercury(II)-EDTA complex (see Section **XIV, 29**).

XIV, 25. SOME EXPERIMENTAL DETAILS FOR POTENTIOMETRIC TITRATIONS. A few simple experiments will be briefly described, the performance of which will enable the reader to obtain experience of the technique. Experiment 1 will require the use of a pH meter (or specific ion meter) which should be employed in the millivolt mode, and it is suggested that experiment 2 be carried out using a simple potentiometer to measure the e.m.f. According to the availability of apparatus, the other experiments may be carried out using a commercial potentiometric titration apparatus (manual measurement of e.m.f.), and with a commercial apparatus which plots the titration curve automatically. In this way a wide range of experience will be acquired, but of course if need be, all the experiments can be carried out using a simple potentiometer, apart from the experiment involving use of a glass electrode for which a pH meter is essential.

Experiment 1. **Neutralisation reactions.** Prepare solutions of acetic acid and of sodium hydroxide, each approximately 0.1 M and set up a pH meter as described in Section **XIV, 14**.

The following general instructions are applicable to most potentiometric titrations and are given in detail here to avoid subsequent repetition.

(*a*) Fit up the apparatus shown in Fig. XIV, 9 with the electrode assembly (or combination electrode) supplied with the pH meter supported inside the beaker. The beaker has a capacity of about $400\,cm^3$ and contains $50\,cm^3$ of the solution to be titrated (the acetic acid).

(*b*) Select a burette, and by means of a piece of polythene tubing attach to the jet a piece of glass capillary tubing about 8–10 cm in length. Charge the burette with the sodium hydroxide solution taking care to remove all air bubbles from the capillary extension, and then clamp the burette so that the end of the capillary is immersed in the solution to be titrated. This procedure ensures that all additions recorded on the burette have in fact been added to the solution, and no drops have been left adhering to the tip of the burette; a factor which can be of some significance for e.m.f. readings made near the end point of the titration.

(*c*) Stir the solution in the beaker gently. Read the potential difference between the electrodes with the aid of the meter. Record the reading and also the volume of alkali in the burette.

(*d*) Add 2–3 cm^3 of solution from the burette, stir for about 30 seconds, and, after waiting for a further half minute, measure the e.m.f. of the cell.

(*e*) Repeat the addition of 1-cm^3 portions of the base, stirring and measuring the e.m.f. after each addition until a point is reached within about 1 cm^3 of the expected end-point. Henceforth, add the solution in portions of 0.1 cm^3 or less, and record the potentiometer readings after each addition. Continue the additions until the equivalence point has been passed by 0.5–1.0 cm^3.

(*f*) Plot potentials as ordinates and volumes of reagent added as abscissæ; draw a smooth curve through the points. The equivalence point is the volume corresponding to the steepest portion of the curve. In some cases the curve is practically vertical, one drop of solution causing a change of 100–200 millivolts in the e.m.f. of the cell; in other cases the slope is more gradual.

(*g*) Locate the end-point of the titration by plotting $\Delta E / \Delta V$ for small increments of the titrant in the vicinity of the equivalence point ($V = 0.1$ cm^3 or 0.05 cm^3) against V. There is a maximum in the plot at the end point (compare Fig. XIV, 14(*b*)).

(*h*) Plot the second derivative curve, $\Delta^2 E / \Delta V^2$, against V: the second derivative becomes zero at the end-point (compare Fig. XIV, 14(*c*)). This method, although laborious, gives the most exact evaluation of the end-point.

Other suggested experiments include titration of 0.05M-Na$_2$CO$_3$ with 0.1M-HCl, and of 0.1M-boric acid in the presence of 4 g of mannitol with 0.1M-NaOH.

Experiment 2. **Oxidation–reduction reaction.** Experience in this kind of titration may be obtained by determining the iron(II) content of a solution by titration with a standard potassium dichromate solution.

Prepare 250 cm^3 of 0.1N-potassium dichromate solution (using the dry A.R. solid) and an equal volume of *ca.* 0.1M ammonium iron(II) sulphate solution; the latter must contain sufficient dilute sulphuric acid to produce a clear solution, and the exact weight of A.R. ammonium iron(II) sulphate employed should be noted. Place 25 cm^3 of the ammonium iron(II) sulphate solution in the beaker, add 25 cm^3 of *ca.* 2.5M-sulphuric acid and 50 cm^3 of water. Charge the burette with the 0.1N-potassium dichromate solution, and add a capillary extension tube. Use a bright platinum electrode as indicator electrode and an S.C.E. reference electrode. Set the stirrer in motion. Proceed with the titration as directed in Experiment 1. After each addition of the dichromate solution measure the e.m.f. of the cell. Determine the end-point: (i) from the potential–volume curve and (ii) by the differential method. Calculate the molarity of the ammonium iron(II) sulphate solution, and compare this with the value computed from the actual weight of solid employed in preparing the solution.

Repeat the experiment using another 25 cm^3 of the ammonium iron(II) sulphate solution but with a pair of **polarised platinum electrodes**. Set up two small platinum plate electrodes (0.5 cm square) in the titration beaker and remove the two electrodes previously in use. Connect the platinum plates to a polarising circuit consisting of a 50 volt dry battery joined to a 20 megohm resistor so that a minute current will flow between the electrodes when they are placed in solution. Also join the electrodes to the circuit used for measuring the cell e.m.f.; a simple potentiometer which is excellent for the first part of the experiment cannot be used with the polarised electrodes, and the most satisfactory procedure is to use the millivolt scale of a pH meter. Some commercial potentiometric titration units make provision for titration with polarised electrodes. The end-point of the titration is indicated by the large jump in e.m.f.

The experiment may also be repeated using a platinum (indicator) electrode and a tungsten wire reference electrode. If the tungsten electrode has been left idle for more than a few days, the surface must be cleaned by dipping into just molten sodium nitrite (**CARE!**). The salt should be only just at the melting point or the tungsten will be rapidly attacked; it should remain in the melt for a few seconds only and is then thoroughly washed with distilled water.

Experiment 3. **Precipitation reactions.** The indicator electrode must be reversible to one or the other of the ions which is being precipitated. Thus in the titration of a potassium iodide solution with standard silver nitrate solution, the electrode must be either a silver electrode or a platinum electrode in the presence of a little iodine (best introduced by adding a little of a freshly prepared alcoholic solution of iodine), i.e., an iodine electrode (reversible to I^-). The exercise recommended is the standardisation of silver nitrate solution with pure sodium chloride.

Prepare an approximately $0.1M$-silver nitrate solution. Place 0.1169 g of dry A.R. sodium chloride in the beaker, add 100 cm^3 of water, and stir until dissolved. Use a silver-wire electrode (or a silver-plated platinum wire), and a silver–silver chloride or a saturated calomel reference electrode separated from the solution by a potassium nitrate–agar bridge (see below). Titrate the sodium chloride solution with the silver nitrate solution following the general procedure described in Experiment 1; it is important to have efficient stirring and to wait long enough after each addition of titrant for the e.m.f. to become steady. Continue the titration to 5 cm^3 beyond the end-point. Determine the end-point and thence the molarity of the silver nitrate solution.

The **salt bridge** which is required in this experiment is prepared from a piece of narrow glass tubing which is first bent at right angles giving a limb long enough to reach to near the bottom of the titration vessel. The tube is then given a second right angle bend in such a position that the horizontal limb will extend from the titration vessel to a suitable position in which a small beaker can be supported; the two vertical limbs of the bridge should be of equal length. Clean the tube thoroughly and then clamp with the two vertical limbs extending upwards. Dissolve 3 g of agar in 100 cm^3 of hot (almost boiling) distilled water, and then add 40 g of A.R. potassium nitrate. As soon as the salt has dissolved, allow to cool for a few minutes, and then carefully pour the hot liquid into the inverted bridge tube so that it is filled *completely* and with no air bubbles entrained in the liquid; a drawn-out thistle funnel will be found useful for this operation. Allow the tube to cool completely still in the inverted position, and when cold it may be found that at the ends of the tube the gel has contracted somewhat, so that when the tube is placed in a liquid, an air bubble is trapped at the bottom of the tube; if this has happened, the extreme ends of the tube should be carefully cut off.

For this particular titration, a three- or four-necked flat bottom flask is conveniently used as titration vessel; the salt bridge can then be inserted into one of the necks of the flask and held in position by means of a cork. The free end of the bridge is allowed to dip into a small beaker containing potassium nitrate solution $(3M)$, and the side arm of the reference electrode is then inserted into the beaker. When not in use, the salt bridge should be stored with the two ends immersed in potassium nitrate solution contained in two test tubes. A **potassium chloride–agar bridge** is obtained by replacing the potassium nitrate by 40 g of A.R. potassium chloride.

An interesting extension of the above experiment is the **titration of a mixture of halides** (chloride/iodide) with silver nitrate solution. Prepare a solution (100 cm³) containing both potassium chloride and potassium iodide; weigh each substance accurately and arrange for the solution to be about 0.025M with respect to each salt. A silver nitrate solution of known concentration (about 0.05M) will also be required.

Pipette 10 cm³ of the halide solution into the titration vessel and dilute to about 100 cm³ with distilled water. Insert a silver electrode, an agar–potassium nitrate salt bridge and complete the cell with a saturated calomel electrode. Fit a 10 cm³ micro burette with a capillary tube extension, and fill with the silver nitrate solution. Add 1 cm³ of the silver nitrate solution to the contents of the titration vessel and read the cell e.m.f. after allowing adequate time for the value to become stable; complete the titration in accordance with the details previously given, but remember that there will be two end points, one in the neighbourhood of 5 cm³ of silver nitrate (I⁻), and the other in the neighbourhood of 10 cm³ (Cl⁻).

A comment on the polarity of the electrodes of the silver–calomel electrode cell may be helpful at this point. With the respective values of electrode potentials (calomel 0.245 V, $E_{Ag}^{\ominus}$ 0.799 V) one would normally expect the silver electrode to be the positive electrode of the cell, but at the start of the above titration, the concentration of silver ions in solution is so minute that the log term in the Nernst equation in fact has a large negative value, and the potential of the silver electrode actually becomes smaller than that of the calomel electrode. With continued addition of silver nitrate, the concentration of silver ions in solution gradually rises and the potential of the silver electrode increases, and at a point which occurs near the first end-point of the titration it becomes equal to, and subsequently greater than, the potential of the calomel electrode. When this point is reached, it is necessary to reverse the connections of the leads from the cell to the potentiometer, and in order to plot a satisfactory titration curve the subsequent readings must be regarded as negative: conversely, of course, the initial readings may be regarded as negative, and those after the change over point as positive.

XIV, 26. DETERMINATION OF COPPER. Prepare a solution of the sample, containing about 0.1 g copper and no interfering elements, by any of the usual methods; any large excess of nitric acid and all traces of nitrous acid must be removed. Boil the solution to expel most of the acid, add about 0.5 g urea (to destroy the nitrous acid) and boil again. Treat the cooled solution with concentrated ammonia solution dropwise until the deep-blue cuprammonium compound is formed, and then add a further two drops. Decompose the cuprammonium complex with glacial acetic acid and add 0.2 cm³ in excess. Too great a dilution of the final solution should be avoided, otherwise the reaction between the copper(II) acetate and the potassium iodide may not be complete.

Place the prepared copper acetate solution in the beaker and add 10 cm³ of 20 per cent potassium iodide solution. Set the stirrer in motion and add distilled water, if necessary, until the platinum plate electrode is fully immersed. Use a saturated calomel reference electrode, and carry out the normal potentiometric titration procedure using a standard sodium thiosulphate solution as titrant.

XIV, 27. DETERMINATION OF CHROMIUM. The chromium in the substance is converted into chromate or dichromate by any of the usual methods. A platinum indicator electrode and a saturated calomel electrode are used. Place a known volume of the dichromate solution in the titration beaker, add 10 cm^3 of 10 per cent sulphuric acid or hydrochloric acid per 100 cm^3 of the final volume of the solution and also 2.5 cm^3 of 10 per cent phosphoric acid. Insert the electrodes, stir, and after adding 1cm^3 of a standard ammonium iron(II) sulphate solution, the e.m.f. is measured. Continue to add the iron solution, reading the e.m.f. after each addition, then plot the titration curve and determine the end point.

XIV, 28. DETERMINATION OF MANGANESE. The method is based upon the titration of manganese(II) ions with permanganate in neutral pyrophosphate solution:

$$4Mn^{2+} + MnO_4^- + 8H^+ + 15H_2P_2O_7^{2-} = 5Mn(H_2P_2O_7)_3^{3-} + 4H_2O$$

The manganese(III) pyrophosphate complex has an intense reddish-violet colour, consequently the titration must be performed potentiometrically. A bright platinum indicator electrode and a saturated calomel reference electrode may be used. The change in potential at the equivalence point at a pH between 6 and 7 is large (about 300 millivolts); the potential of the platinum electrode becomes constant rapidly after each addition of the potassium permanganate solution, thus permitting direct titration to almost the equivalence point and reducing the time required for a determination to less than 10 minutes. With relatively pure manganese solutions, a sodium pyrophosphate concentration of 0.2–0.3M, a pH between 6 and 7, the equivalence point potential is $+0.47 \pm 0.02$ volt *vs.* the saturated calomel electrode. At a pH above 8 the pyrophosphate complex is unstable and the method cannot be used.

The method is at least as accurate as the bismuthate procedure (Section **X, 99**) and is even less subject to interferences. Large amounts of chloride, cobalt(II), and chromium(III) do not interfere; iron(III), nickel, molybdenum(VI), tungsten(VI), and uranium(VI) are innocuous; nitrate, sulphate, and perchlorate ions are harmless. Large quantities of magnesium, cadmium, and aluminium yield precipitates which may co-precipitate manganese and should therefore be absent. Vanadium causes difficulties only when the amount is equal to or larger than the amount of manganese; when it is present originally in the $+4$ state, it is oxidised slowly in the titration to the $+5$ state along with the manganese. Small amounts of vanadium (up to about one-fifth of the amount of manganese) cause little error. The interference of large amounts of vanadium(V) can be circumvented by performing the titration at a pH of 3–3.5. Oxides of nitrogen interfere because of their reaction with potassium permanganate: hence when nitric acid is used to dissolve the sample, the resulting solution must be boiled thoroughly and a small amount of urea or sulphamic acid must be added to the acid solution to remove the last traces of oxides of nitrogen before introducing the sodium pyrophosphate solution.

For initial practice in the method determine the manganese content of anhydrous A.R. manganese(II) sulphate. Heat A.R. manganese(II) sulphate crystals to 280 °C, allow to cool, grind to a fine powder, reheat at 280 °C for 30 minutes, and allow to cool in a desiccator. Weigh accurately about 2.2 g of the anhydrous manganese(II) sulphate, dissolve it in water and make up to 250 cm^3 in a graduated flask.

Prepare a 0.02M solution of potassium permanganate and standardise it against A.R. arsenic(III) oxide.

Prepare 5M-sodium hydroxide solution using the A.R. solid: test a 10-cm^3 sample for reducing agents by adding a drop of the permanganate solution; no green coloration should develop.

Prepare also a saturated solution of the purest available sodium pyrophosphate (do not heat above 25 °C, otherwise appreciable hydrolysis may occur); 12 g of the hydrated solid $Na_4P_2O_7,10H_2O$ will dissolve in 100–150 cm^3 of water according to the purity of the compound. It is essential to employ freshly made sodium pyrophosphate solution in the determination.

Place 150 cm^3 of the sodium pyrophosphate solution in a 250–400-cm^3 beaker, adjust the pH to 6–7 by the addition of concentrated sulphuric acid from a 1-cm^3 graduated pipette (use the appropriate indicator test-paper or a pH meter). Add 25 cm^3 of the manganese(II) sulphate solution and adjust the pH again to 6–7 by the addition of 5M-sodium hydroxide solution. Introduce a bright platinum electrode into the solution, and connect the latter through a saturated potassium chloride bridge to a saturated calomel electrode; complete the assembly for potentiometric titrations as in Fig. XIV, 9. Stir the mixture, add the potassium permanganate solution in 2-cm^3 portions at first, reduce this to 0.1-cm^3 portions in the vicinity of the end-point: determine the potential after each addition. Plot the e.m.f. values (ordinates) against the volume of potassium permanganate solution added (abscissæ), and determine the equivalence point. From your curve read off the potential at the equivalence point; this should be +0.47 volt. Calculate the percentage of Mn in the sample.

$$1 \text{ cm}^3 \ 0.02M\text{-KMnO}_4 \equiv 0.00439 \text{ g Mn}$$

Further practice may be obtained by determining manganese in a manganese ore and in a steel.

Pyrolusite. Dissolve 1.5–2 g, accurately weighed, pyrolusite in a mixture of 25 cm^3 of 1:1 hydrochloric acid and 6 cm^3 concentrated sulphuric acid, and dilute to 250 cm^3. Filtration is unnecessary. Titrate an aliquot part containing 80–100 mg manganese: add 200 cm^3 freshly prepared, saturated sodium pyrophosphate solution, adjust the pH to a value between 6 and 7, and perform the potentiometric titration as described above.

Steel. Dissolve 5 g, accurately weighed, of a steel in 1:1 nitric acid with the aid of the minimum volume of hydrochloric acid in a Kjeldahl flask. Boil the solution down to a small volume with excess of concentrated nitric acid to re-oxidise any vanadium present reduced by the hydrochloric acid: this step is unnecessary if vanadium is known to be absent. Dilute, boil to remove gaseous oxidation products, allow to cool, add 1 g of urea or sulphamic acid, and dilute to 250 cm^3. Titrate 50-cm^3 portions as above.

XIV, 29. POTENTIOMETRIC   EDTA   TITRATIONS   WITH   THE MERCURY ELECTRODE. *Discussion.* The indicator electrode employed is a mercury | mercury(II)-EDTA-complex electrode. A mercury electrode in contact with a solution containing metal ions M^{n+} (to be titrated) and a small added quantity of a mercury(II)–EDTA complex HgY^{2-} (EDTA = Na_2H_2Y)

exhibits a potential corresponding to the half-cell:

$$\text{Hg} \mid \text{Hg}^{2+}, \text{HgY}^{2-}, \text{MY}^{(n-4)+}, \text{M}^{n+}$$

It can be shown that the potential at equilibrium is given by:

$$E = E^{\ominus}_{\text{Hg}^{2+},\text{Hg}} + \frac{RT}{2F} \ln \frac{[\text{HgY}^{2-}]}{[\text{MY}^{(n-4)+}]} \cdot \frac{K_{\text{MY}}}{K_{\text{HgY}}} + \frac{RT}{2F} \ln [\text{M}^{n+}]$$

where K_{MY} and K_{HgY} are the stability (or formation) constants of the metal–EDTA and mercury–EDTA complexes respectively. The first two terms on the right-hand side of this equation are essentially constant during a potentiometric titration, especially in the region of the end point, hence the measured potential of the electrode becomes a linear function of pM. The mercury | mercury(II)-EDTA complex electrode will, for convenience, be subsequently described as a mercury indicator electrode; it is clearly a pM indicator electrode.

The potential of the mercury indicator electrode depends upon the total mercury concentration in the solution. In practice it is found that the addition of 1 drop of a $0.001–0.01M$ solution of the mercury–EDTA complex HgY^{2-} is sufficient to establish a reasonably constant value for the mercury content so that trace additions of this metal do not seriously alter the shape of the titration curve. In complexometric titrations of metal ions with the mercury electrode, the experimental conditions, such as pH, the kind and nature of the buffer solution, must be carefully controlled. The buffer should be present in an amount sufficient to prevent pH changes during the titration: a large excess of buffer should be avoided, as this may decrease the extent of the potential break at the end-point. Halide ions must not be present in appreciable concentrations because they may interfere with the electrode reaction, especially for titrations performed under acid conditions (e.g., chloride interferes at a pH less than about 6.5). At a pH lower than 2, the mercury(II)–EDTA complex dissociates to such an extent that a poorly defined titration curve results. At a pH above about 11, oxygen reacts with mercury, leading to a distorted titration curve; this may be often avoided by bubbling nitrogen through the solution before and during the titration. Direct and also back-titration procedures have been used for the determination of numerous metal ions and a selection of these is given below.

Apparatus. *Mercury electrode.* The electrode, together with the essential dimensions, is shown in Fig. XIV, 15; it is easily constructed from Pyrex tubing.

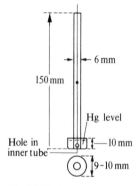

The platinum wire dipping into the mercury may be welded to a copper wire; but it is preferable to use a platinum wire sufficiently long to protrude at the top of the electrode tube. The mercury must be pure and clean; in case of doubt, the mercury should be washed with dilute nitric acid and then thoroughly rinsed with distilled water. The electrode is filled with mercury from above, and it is allowed to pass into the annular space through the hole until the outside compartment is almost filled. It is most important that no mercury is spilled into the titration vessel during the titration. After each titration the electrode is repeatedly washed with distilled water.

Fig. XIV, 15

Alternatively, an amalgamated gold electrode may

be employed. This may be prepared by dipping a commercial gold plate electrode for about a minute into pure mercury; after rinsing with water, it is ready for use. The electrode can be used only a few times, and then must be re-amalgamated.

Titration assembly. The electrode system consists of a mercury electrode and a saturated calomel (or, in some cases, a mercury–mercury(I) sulphate) reference electrode, both supported in a 250-cm³ Pyrex beaker. Provision is made for magnetic stirring and the potential is followed by means of a precision slide wire potentiometer and a sensitive 'spot' galvanometer or better with a potentiograph.

Reagents required. *Standard EDTA solution, 0.05*M. See Section **X, 50**.

Mercury–EDTA solution. Mix small equal volumes of 0.05*M*-mercury(II) nitrate (prepared from the A.R. solid) and 0.05*M*-EDTA; neutralise the liberated acid by the addition of a few drops of 3*M*-ammonia solution. (In acid solution an insoluble precipitate, probably HgH_2Y, forms after a few days). Dilute 10.0 cm³ of this solution to 100 cm³ with distilled water. The resulting *ca.* 0.0025*M*-mercury-EDTA solution is used for most titrations.

Ammonia buffer solution. Mix 20 g ammonium nitrate and 35 cm³ concentrated ammonia solution, and make up to 100 cm³ with distilled water. Dilute 80 cm³ to 1 dm³ with distilled water. The pH is about 10.1.

Acetate buffer solution. Mix equal volumes of 0.5*M*-sodium acetate solution and 0.5*M*-acetic acid solution. The resulting solution has a pH of about 4.7.

Triethanolamine buffer solution. Prepare a *ca.* 0.5*M* aqueous solution of triethanolamine and add 2.5*M*-nitric acid until the pH is 8.5 (use a pH meter).

Procedure. The general procedure is as follows. Place 25.0 cm³ of the metal-ion solution (approximately 0.05*M*) in a 250-cm³ Pyrex beaker, add 25 cm³ of the appropriate buffer solution and 1 drop of 0.0025*M*-mercury–EDTA solution (1 drop of 0.025*M*-mercury–EDTA solution for calcium and magnesium). Use the titration assembly described above. Stir magnetically. Titrate potentiometrically with standard 0.05*M*-EDTA solution added from a burette supported over the beaker. Reduce the volume of EDTA solution added to 0.1 cm³ or less as soon as the potential begins to rise; wait for a steady potential to be established after each addition. Soon after the end-point the change of potential with each addition of EDTA becomes smaller and only a few large additions need be made. Care must be taken that mercury is not spilled into the solution either during the insertion of the mercury electrode or during the titration.

Plot the titration curve (potential in millivolts *vs.* S.C.E. against volume of standard EDTA solution) and evaluate the end-point. In general, results accurate to better than 0.1 per cent are obtained. Brief notes on determinations with various metal-ion solutions follow.

Calcium. 25.0 cm³ calcium-ion solution + 25 cm³ ammonia or tri-ethanolamine buffer.

Magnesium. 25.0 cm³ magnesium-ion solution + 25 cm³ ammonia buffer.

Nickel. 25.0 cm³ nickel-ion solution + 25 cm³ ammonia buffer.

Cobalt. 20.0 cm³ cobalt-ion solution + 25 cm³ ammonia buffer.*

Copper. 25.0 cm³ of copper(II)-ion solution + 25 cm³ of acetate buffer.

---

* A larger excess of ammonia buffer is required to ensure the formation of the cobalt-ammine.

Mercury. $25.0 \, cm^3$ mercury(II)-ion solution $+ 25 \, cm^3$ acetate buffer.

Zinc. $25.0 \, cm^3$ zinc-ion solution $+ 25 \, cm^3$ acetate buffer.

Bismuth. $25.0 \, cm^3$ bismuth-ion solution $+$ solid hexamine to pH about 4.6; the precipitate of basic bismuth salt dissolves as the EDTA solution is added but the titration is slow.

Lead. $25.0 \, cm^3$ lead-ion solution $+$ solid hexamine to pH about 4.6.

Thorium. (i) $25.0 \, cm^3$ thorium-ion solution $+ 90 \, cm^3$ $0.001M$-nitric acid, $3M$-ammonia solution added until pH about 3.2; mercury–mercury(I) sulphate reference electrode.

(ii) $25.0 \, cm^3$ thorium-ion solution $+$ solid hexamine to pH about 4.6; $40.0 \, cm^3$ $0.05M$-EDTA added, and back-titrate excess with standard lead nitrate solution.

Chromium. $25.0 \, cm^3$ chromium(III)-ion solution $(0.02M$, prepared by dilution of stock solution) $+ 50.0 \, cm^3$ $0.02M$-EDTA $+ 50 \, cm^3$ acetate buffer, boiled for 10 minutes, solution cooled, pH adjusted to 4.6 with hexamine, 1 drop of mercury–EDTA solution added, and then back-titrated with standard zinc-ion solution.

Aluminium. $25.0 \, cm^3$ aluminium-ion solution, acidified with a few drops of $2.5M$-nitric acid (to pH 1–2), boiled for 1 minute, $50.0 \, cm^3$ $0.05M$-EDTA added to hot solution, solution cooled, $50 \, cm^3$ acetate buffer and 1 drop of $0.0025M$-mercury–EDTA added, excess of EDTA back-titrated with standard zinc-ion solution.

XIV, 30. DETERMINATION OF IRON(III) WITH EDTA. *Discussion.* This is an example of a potentiometric titration involving two different oxidation states of the same metal. Thus in the titration of iron(III) with EDTA, the potential measured is that of the Fe(III)–Fe(II) couple (compare Section **X, 31**, and Variamine Blue in Section **X, 28**). The titration is conducted at pH *ca.* 3 using a bright platinum indicator electrode: the iron(II) remains uncomplexed with EDTA during the titration and a large potential change accompanies the abrupt change of iron(III) concentration in the vicinity of the end-point. Metals which form complexes with EDTA interfere.

Procedure. For practice in this determination, prepare a *ca.* $0.05M$-ammonium iron(III) sulphate solution by dissolving about 6.0 g of the A.R. salt, accurately weighed, in $250 \, cm^3$ of water in a graduated flask. Dilute $25.0 \, cm^3$ of this solution to $100 \, cm^3$ with distilled water, and add dilute ammonia solution dropwise until the pH is about 3.0. Titrate potentiometrically with standard $0.05M$-EDTA (Section **X, 50**): use a small platinum foil as the indicator electrode and a saturated calomel half-cell as the reference electrode.

XIV, 31. STANDARDISATION OF POTASSIUM PERMANGANATE SOLUTION WITH POTASSIUM IODIDE. *Discussion.* Potassium permanganate solution may be standardised very accurately by potentiometric titration with A.R. potassium iodide. The latter is only slightly hygroscopic; it may be dried, if necessary, by heating at $200 \, ^\circ C$. The titration apparatus consists of a 250-cm^3 Pyrex beaker (or, better, a four-necked flask), a bright platinum foil electrode ($1.2 \times 1.0 \, cm$) and a mercury–mercury(I) sulphate reference electrode; a saturated calomel electrode is also satisfactory.

It is claimed that the results by the potassium iodide method are reproducible to ± 0.01 per cent and agree with the arsenic(III) oxide procedure (Section **X, 92**) to within 0.03 per cent; the standardisation with arsenic(III) oxide may also be performed potentiometrically.

Procedure. Weigh out accurately about 0.35 g dry A.R. potassium iodide, dissolve it in 50 cm³ distilled water in the titration vessel, and add 1.0 cm³ concentrated sulphuric acid from a dropper pipette; dilute further to about 100 cm³. Stir magnetically and also bubble a slow stream of nitrogen through the solution. Assemble the titration apparatus described above. Titrate potentiometrically with the potassium permanganate solution (*ca.* 0.1N) to be standardised.

XIV, 32. DETERMINATION OF NICKEL AND OF COBALT BY COMPLEXATION WITH CYANIDE. *Discussion.*

The concentration of the cyanide solution is first determined by potentiometric titration with standard silver nitrate solution using a silver indicator electrode and a mercury–mercury(I) sulphate reference electrode. Two points of inflection (indicated by a rapid fall in potential) will be found in the titration curve corresponding to the reactions:

$$Ag^+ + 2CN^- = [Ag(CN)_2]^-$$
$$[Ag(CN)_2]^- + Ag^+ = 2AgCN(\downarrow)$$

The silver-ion concentration at the second point of inflection is almost exactly twice that at the first inflection point; the latter is employed for the calculation of the cyanide concentration.

For the determination of nickel, a nickel(II) salt in ammoniacal solution is treated with excess of potassium cyanide solution, and the excess of the latter is titrated potentiometrically with standard silver nitrate solution:

$$[Ni(NH_3)_4]^{2+} + 4CN^- = [Ni(CN)_4]^{2-} + 4NH_3$$

For the determination of cobalt, a cobalt(II) salt in almost neutral solution is treated with excess of potassium cyanide solution, and the excess of the latter is titrated potentiometrically with standard silver nitrate solution. The cobalt(II) cyanide complex is pentacovalent $[Co(CN)_5]^{3-}$, and this fact must be borne in mind when calculating the cobalt-ion concentration. Somewhat less satisfactory results are obtained in slightly ammoniacal solution.

Reagents. *Silver nitrate solution,* ca. *0.1M.* Weigh out accurately about 8.5 g A.R. silver nitrate, dissolve it in water, and dilute to 500 cm³ in a graduated flask.

Potassium cyanide solution, ca. *0.1M.* Weigh out about 6.5 g A.R. potassium cyanide (**CAUTION!**) and dissolve it in 1 dm³ of water in a graduated flask.

Nickel-ion solution, ca. *0.05M.* Weigh out accurately about 2.9 g pure nickel pellets, dissolve in the minimum volume of concentrated nitric acid, boil gently to remove nitrous fumes, cool, and dilute to 1 dm³ with distilled water in a graduated flask.

Cobalt-ion solution, ca. *0.05M.* Weigh out about 14.0 g A.R. cobalt(II) sulphate and dissolve it in 1 dm³ of water in a graduated flask. Determine the exact cobalt content by titration with standard 0.05M-EDTA using Xylenol Orange as indicator.

Apparatus. The titration vessel may consist of a 250-cm³ Pyrex beaker; provision is made for magnetic stirring. A silver rod (3 mm diameter) is used as an indicator electrode, and a mercury–mercury(I) sulphate half-cell as a reference electrode; the latter may be replaced by a saturated calomel electrode connected to the titration vessel by means of an agar–potassium sulphate bridge.

The potassium cyanide solution is conveniently measured out with a pipette attached to a 'Pumpett'; if the latter is not available, a burette may be used. Attention is directed to the *highly poisonous character* of the potassium cyanide solution; the hands should always be thoroughly washed immediately after handling the reagent.

Procedure. **Standardisation of potassium cyanide solution.** Place 25.0 cm³ of the potassium cyanide solution in the titration vessel and dilute to 100 cm³ with distilled water. Stir magnetically. Titrate potentiometrically in the usual manner with standard 0.1M-silver nitrate solution. Plot the titration curve (potential in millivolts against volume of silver nitrate solution) and evaluate the end-point (at the first sharp change in potential) using the first or second differential curve.

Calculate the concentration of the potassium cyanide solution.

Determination of nickel. Place 10.00 cm³ of the nickel-ion solution in the titration vessel, dilute to about 100 cm³ with distilled water, and add 6M-ammonia solution until the pH is about 10. Add 50.0 cm³ of the potassium cyanide solution, and titrate the excess of potassium cyanide potentiometrically with standard 0.1M-silver nitrate. Evaluate the end-point of the latter titration. Calculate the volume of potassium cyanide solution which has reacted with the nickel, and thence the nickel content of the solution ($Ni^{2+} \equiv 4CN^-$).

Determination of cobalt. Place 10.00 cm³ of the cobalt-ion solution in the titration vessel, and dilute to 100 cm³ with distilled water. Add 50.0 cm³ of the potassium cyanide solution, and titrate potentiometrically the excess of potassium cyanide with standard 0.1M-silver nitrate. Deduce the volume of potassium cyanide solution which has reacted with the cobalt and then calculate the cobalt content of the solution ($Co^{2+} \equiv 5CN^-$).

XIV, 33. DETERMINATION OF FLUORIDE BY A NULL-POINT METHOD.

Discussion. The method is based upon the complexing by fluoride of one of the oxidation states of a redox couple; the Ce(IV)–Ce(III) couple is generally used. The potential measured is that between two half-cells, each initially containing the same volume of a Ce(IV)–Ce(III) solution. The sample solution is added to one half-cell and an equal volume of water to the other. The magnitude of the e.m.f. of the cell gives some measure of the fluoride concentration. Standard fluoride solution is added to the second cell until the e.m.f. of the cell is zero. During the addition of the titrant, water is added at the same rate to the first half-cell so that concentrations, etc., remain similar in each half-cell. For the analysis of 0.05M-fluoride solutions, the optimum working conditions are with a fluoride: cerium(IV): cerium(III) ratio of 2:1:1. The approximate fluoride concentration of the sample solution must be known, and this can be evaluated very approximately by a preliminary titration.

Sulphate and nitrate do not interfere, nor does chloride in amount about equal to the fluoride concentration. Bromide interferes owing to the reduction of cerium(IV) to cerium(III); this can be overcome by using an iron(III)–iron(II) redox couple. Acetate gives slightly high results; oxalate, molybdate, and phosphate interfere and must be absent. Alkali metals and ammonium have little influence on the titrations. Other cations should be removed by passage through an ion exchange column or by precipitation: the latter method is applicable in those cases (e.g., silver chloride from silver fluoride) where co-precipitation of fluoride is negligible.

Reagents. *Cerium(IV)–cerium(III) solution.* Dissolve 6.35 g pure ammonium cerium(IV) sulphate dihydrate in 200 cm^3 water and 14 cm^3 18M-sulphuric acid. Then add 2.8 g pure cerium(III) sulphate, stir until dissolved, and dilute the resulting solution to 1 dm^3 with distilled water. The solution is 0.01M in cerium(IV) and in cerium(III) and 0.25M in sulphuric acid.

Standard sodium fluoride solution, 0.05M. Prepare from dry A.R. sodium fluoride.

Apparatus. The two half-cells consist of two 250-cm^3 Pyrex or polythene beakers, standing on two small but similar magnetic stirrers, and connected by an agar–potassium chloride bridge. A clean platinum wire electrode is supported in each beaker; the electrodes are connected to a precision slide-wire potentiometer.

Procedure. For practice in this determination, the fluoride content of A.R. potassium fluoride may be determined: prepare a *ca.* 0.05M solution from an accurately weighed amount of the dry solid, and regard this as the sample solution.

Into each of the two dry and clean 250-cm^3 beakers, place 50.0 cm^3 of the cerium(IV)–cerium(III) solution. Connect the two half-cells and check that the e.m.f. of the cell is zero. Pipette, say, 20.0 cm^3 of the sample solution into one half-cell, and add the same volume of distilled water to the other. The magnitude of the e.m.f. at this stage gives an approximate measure of the fluoride concentration of the sample. Add the standard fluoride solution portion-wise from a burette to the second half-cell. During the titration add distilled water from another burette to the first half-cell at the same rate as the titrant is introduced. Stir the contents of each beaker magnetically at about the same rate throughout the titration. After each addition, measure the potential difference between the electrodes. When the end-point is approached (i.e., when the e.m.f. is about 10 millivolts, and for about 1 cm^3 on either side of the equivalence point), run in the titrant in 0.2-cm^3 portions.

Plot the values of the e.m.f. against the volume of the standard sodium fluoride solution, and read from the graph the exact volume required to produce zero e.m.f. Calculate the fluoride content of the sample of potassium fluoride.

XIV, 34. References

1. G. Mattock (1961). *pH Measurement and Titration.* London; Heywood and Co. Ltd.
2. E. Pungor, J. Havas and K. Toth (1965). *Z. Chem.*, **5**, 9.
3. J. W. Ross (1967). *Science.*, **156**, 1378.
4. J. Ruzicka and J. C. Tjell (1970). *Anal. Chim. Acta.*, **49**, 346; **51**, 1.
5. A. K. Covington (1970–3). *Electrochemistry.* Vols. I/III. Specialist Periodical Reports. London; The Chemical Society.
6. N. K. Lakshminarayanaiah (1972). *Electrochemistry.* Vol. II. Specialist Periodical Report. London; The Chemical Society.
7. G. J. Moody and J. D. R. Thomas (1973). *Selected Annual Reviews of the Analytical Sciences.* Vol. 3. London; Society for Analytical Chemistry.
8. G. J. Moody and J. D. R. Thomas (1974). *Chemistry and Industry.*, 644.
9. J. Jackson (1948). *Chemistry and Industry*, 7.
10. International Union of Pure and Applied Chemistry (1969). *Manual of Symbols and Terminology for Physicochemical Quantities and Units.* London; Butterworths.
11. G. J. Moody and J. D. R. Thomas (1971). *Talanta*, **18**, 1251.

XIV, 35. Selected bibliography

1. I. M. Kolthoff and N. H. Furman (1931). *Potentiometric Titrations*, 2nd edn. New York; John Wiley.
2. H. H. Willard, L. L. Merritt and J. A. Dean (1974). *Instrumental Methods of Analysis*, 5th edn. New York; Van Nostrand.
3. A. L. Beilby (1970). *Modern Classics in Analytical Chemistry*. Washington DC; American Chemical Society.
4. L. Meites (1963). *Handbook of Analytical Chemistry*. New York; McGraw-Hill.
5. T. S. West (1972). *Analytical Chemistry*. MTP Series I, Vols. 12/13. London; Butterworths.
6. H. A. Strobel (1973). *Chemical Instrumentation—A Systematic Approach to Instrumental Analysis*. 2nd edn. Reading, Mass.; Addison-Wesley Pub. Co.
7. H. F. Walton and J. Reyes (1973). *Modern Chemical Analysis and Instrumentation*. New York; Marcel Dekker Inc.
8. D. J. G. Ives and G. J. Janz (1961). *Reference Electrodes*. London; Academic Press.
9. R. G. Bates (1973). *Determination of pH*. 2nd edn. New York; Wiley.
10. G. Eisenman (1967). *Glass Electrodes for Hydrogen and Other Cations*. New York; Marcel Dekker Inc.
11. R. A. Durst (1969). *Ion Selective Electrodes*. Special Publication No. 314. Washington DC; National Bureau of Standards.
12. A. J. Bard (from 1966). *Electroanalytical Chemistry*. A Series of Advances. Various volumes. New York; Marcel Dekker Inc.
13. R. P. Buck (1974). 'Ion Selective Electrodes, Potentiometry, Potentiometric Titrations'. Review article. *Analytical Chemistry*, **46**, 28R.
14. J. W. Robinson (1970). *Undergraduate Instrumental Analysis*. New York; Marcel Dekker Inc.
15. K. Cammann (1973). *Das Arbeiten mit Ionenselektiven Elektroden*. Anleitungen für die chemische Laboratoriumspraxis. Vol. XIII. Berlin; Springer Verlag.
16. T. Anfält and D. Jagner (1973). 'Computation of Intrinsic End-point Errors in Titrations with Ion Selective Electrodes'. *Anal. Chem.*, **45**, 2412.
17. IUPAC (1975). *Recommendations for Nomenclature of Ion-selective Electrodes*. Appendices on Provisional Nomenclature, Symbols, Units and Standards, No. 43. Oxford; IUPAC Secretariat.
18. G. J. Moody and J. D. R. Thomas (1971). *Selective Ion Sensitive Electrodes*. Watford; Merrow Publishing Co. Ltd.

CHAPTER XV   **CONDUCTOMETRIC TITRATIONS**

XV, I.   GENERAL CONSIDERATIONS. Ohm's law states that the current I (amperes) flowing in a conductor is directly proportional to the applied electromotive force E (volts) and inversely proportional to the resistance R (ohms) of the conductor:

$$I = E/R$$

(compare Section **XII, 1**). The reciprocal of the resistance is termed the **conductance** (G): this is measured in reciprocal ohms (ohm^{-1}), for which the name siemens (S) has been proposed (Ref. 1). The resistance of a sample of homogeneous material, length l, and cross-sectional area a, is given by:

$$R = \rho . l/a$$

where ρ is a characteristic property of the material termed the **resistivity** (formerly called specific resistance). In SI units, l and a will be measured respectively in metres and square metres, so that ρ refers to a metre cube of the material, and

$$\rho = R . a/l$$

is measured in ohm metres. Hitherto, resistivity measurements have been made in terms of a centimetre cube of substance, giving ρ the units ohm cm. The reciprocal of resistivity is the **conductivity**, κ (formerly specific conductance), which in SI units is the conductance of a one metre cube of substance and has the units $ohm^{-1} m^{-1}$ (or $S\,m^{-1}$), but if ρ is measured in ohm cm, then κ will be measured in $ohm^{-1} cm^{-1}$ (or $S\,cm^{-1}$). Virtually all the data at present recorded in the literature is expressed in terms of κ measured in $S\,cm^{-1}$ units, and these values will therefore be adopted in this book. Furthermore, as pointed out in Section **XII, 1**, most of the existing data is expressed in terms of the 'international ohm' and not the SI 'absolute' unit introduced in 1948.

The conductivity of an electrolytic solution at any temperature depends only on the ions present, and their concentration. When a solution of an electrolyte is diluted, the conductivity will decrease, since fewer ions are present per cm^3 of solution to carry the current. If all the solution be placed between two electrodes 1 cm apart and large enough to contain the whole of the solution, the conductance will increase as the solution is diluted. This is due largely to a decrease in inter-ionic effects for strong electrolytes and to an increase in the degree of dissociation for weak electrolytes.

The **molar conductivity** (Λ) of an electrolyte is defined as the conductivity due to one mole and is given by:

$$\Lambda = 1000\,\kappa/C = \kappa \,.\, 1000\,V,$$

where C is the concentration of the solution in moles per dm^3, and V is the dilution in dm^3 (i.e., the number of dm^3 containing one mole). Clearly, since κ has the dimensions $S\,cm^{-1}$, the units of Λ are $S\,cm^2\,mol^{-1}$, or in SI units, $S\,m^2\,mol^{-1}$.

For strong electrolytes the molar conductivity increases as the dilution is increased, but it appears to approach a limiting value known as the **molar conductivity at infinite dilution** Λ_∞; this quantity is written as Λ_0 when concentration, rather than dilution, is considered. The quantity Λ_0 can be determined by extrapolation for dilute solutions of strong electrolytes. For weak electrolytes the extrapolation method cannot be used for the determination of Λ_0 but it may be computed from the molar conductivities at infinite dilution of the respective ions, use being made of the 'law of independent migration of ions'. At infinite dilution the ions are independent of each other, and each contributes its part to the total conductivity, thus:

$$\Lambda_0 = \Lambda_0\,(\text{cat}) + \Lambda_0\,(\text{an})$$

where Λ_0 (cat) and Λ_0 (an) are the **ionic molar conductivities** at infinite dilution of the cation and anion respectively. The values for the limiting ionic molar conductivities for some ions in water at 25 °C are collected in Table XV, 1.

Table XV, 1   Limiting ionic molar conductivities at 25 °C

Cation	Λ_0 (cat)	Anion	Λ_0 (an)
H^+	349.8	OH^-	198.3
Na^+	50.1	F^-	55.4
K^+	73.5	Cl^-	76.3
Li^+	38.7	Br^-	78.1
$NH_4{}^+$	73.5	I^-	76.8
Ag^+	61.9	$NO_3{}^-$	71.5
Tl^+	74.7	$ClO_3{}^-$	64.6
$\frac{1}{2}Ca^{2+}$	59.5	$ClO_4{}^-$	67.4
$\frac{1}{2}Sr^{2+}$	59.5	$BrO_3{}^-$	55.7
$\frac{1}{2}Ba^{2+}$	63.6	$IO_3{}^-$	40.5
$\frac{1}{2}Mg^{2+}$	53.1	$IO_4{}^-$	54.6
$\frac{1}{2}Zn^{2+}$	52.8	$HCO_3{}^-$	44.5
$\frac{1}{2}Pb^{2+}$	69.5	$\frac{1}{2}CO_3{}^{2-}$	69.3
$\frac{1}{2}Cu^{2+}$	53.6	$\frac{1}{2}SO_3{}^{2-}$	80.0
$\frac{1}{2}Ni^{2+}$	53	$\frac{1}{3}PO_4{}^{3-}$	80
$\frac{1}{2}Co^{2+}$	55	$\frac{1}{2}C_2O_4{}^{2-}$	74.2
$\frac{1}{2}Fe^{2+}$	54	$HCOO^-$	54.6
$\frac{1}{3}Fe^{3+}$	68.4	CH_3COO^-	40.9
$\frac{1}{3}La^{3+}$	69.7	$CH_3CH_2COO^-$	35.8
$NMe_4{}^+$	44.9	$\frac{1}{3}Fe(CN)_6{}^{3-}$	100.9
$NEt_4{}^+$	32.7	$\frac{1}{4}Fe(CN)_6{}^{4-}$	110.5

XV, 2.   THE MEASUREMENT OF CONDUCTIVITY. To measure the conductivity of a solution it is placed in a cell of which the cell constant has been determined by calibration with a solution of accurately known conductivity, e.g., a standard potassium chloride solution. The cell is placed in one arm of a

Wheatstone bridge circuit as in Fig. XV, 1, and the resistance measured. For details of the procedure a textbook of practical physical chemistry should be consulted; see Ref. 2.

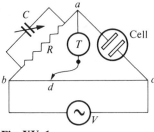

Fig. XV, 1

The passage of a current through a solution of an electrolyte may produce changes in the composition of the solution in the vicinity of the electrodes; potentials may thus arise at the electrodes, with the consequent introduction of serious errors in the conductivity measurements, unless such polarisation effects can be reduced to negligible proportions. These difficulties are generally overcome by the use of alternating currents for the measurements so that the extent of electrolysis and the polarisation effects are greatly reduced. The source of alternating current (V) may be either the electric mains with a frequency of 50–60 hertz, or a mains-operated oscillator giving current with a frequency of up to 3000 Hz. Since alternating current is being used, the cell will have a capacitance which will not be counter-balanced in the standard resistance box R, and it is therefore necessary to include a variable condenser in parallel with the resistance box so that the capacitance in a–c can be matched in a–b.

If the frequency of the current is greatly increased to 10^6–10^8 Hz, then the capacitance and inductive effects become highly important, and the apparatus must be modified to take account of these effects. It is therefore necessary to consider separately (*a*) *conductometric titrations* carried out with current of low frequency (up to 3000 Hz), and (*b*) titrations carried out using current at high frequencies: in these cases we measure changes in capacitance or in inductance rather than in conductance, and such titrations are therefore usually referred to as *high frequency titrations*.

Conductometric (low frequency) titrations

XV, 3. THE BASIS OF CONDUCTOMETRIC TITRATIONS. The

addition of an electrolyte to a solution of another electrolyte under conditions producing no appreciable change in volume will affect the conductance of the solution according to whether or not ionic reactions occur. If no ionic reaction takes place, such as in the addition of one simple salt to another (e.g., potassium chloride to sodium nitrate), the conductance will simply rise. If ionic reaction occurs, the conductance may either increase or decrease; thus in the addition of a base to a strong acid, the conductance decreases owing to the replacement of the hydrogen ion of high conductivity by another cation of lower conductivity. This is the principle underlying conductometric titrations, i.e., the substitution of ions of one conductivity by ions of another conductivity.

Let us consider how the conductance of a solution of a strong electrolyte A^+B^- will change upon the addition of a reagent C^+D^-, assuming that the cation A^+ (which is the ion to be determined) reacts with the ion D^- of the reagent. If the product of the reaction AD is relatively insoluble or only slightly ionised, the reaction may be written:

$$A^+B^- + C^+D^- = AD + C^+B^-$$

Thus in the reaction between A^+ ions and D^- ions, the A^+ ions are replaced by C^+ ions during the titration. As the titration proceeds the conductance increases or decreases, depending upon whether the conductivity of the C^+ ions is greater or less than that of the A^+ ion.

During the progress of neutralisations, precipitations, etc., changes in conductivity may, in general, be expected, and these may therefore be employed in determining the end points as well as the progress of the reactions. The conductivity is measured after each addition of a small volume of the reagent, and the points thus obtained are plotted to give a graph which consists of two straight lines intersecting at the equivalence point. The accuracy of the method is greater the more acute the angle of intersection and the more nearly the points of the graph lie on a straight line. The volume of the solution should not change appreciably; this may be achieved by employing a titrating reagent which is 20 to 100 times more concentrated than the solution being titrated, and the latter should be as dilute as practicable. Thus if the conductivity cell contains about $100\,cm^3$ of the solution at the beginning of the titration, the reagent (say, of fifty times the concentration of the solution being analysed) may be placed in a 5-cm³ microburette, graduated in 0.01 or 0.02 cm³. A correction for the dilution effect may, however, be made by multiplying the values of the conductivity by the factor $(V+v)/V$, in which V is the original volume of the solution and v is the volume of reagent added.

In contrast to potentiometric titration methods (see Chapter XIV), but similar to amperometric titration methods (see Chapter XVII), measurements near the equivalence point have no special significance. Indeed, owing to hydrolysis, dissociation, or solubility of the reaction product, the values of the conductivity measured in the vicinity of the equivalence point are usually worthless in the construction of the graph, since one or both curves will give a rounded portion at this point. Even if the conductivity of the reaction product at the equivalence point is appreciable, the reaction may frequently be employed for conductometric titration if the conductivity of the reaction product AD is practically completely suppressed by a reasonable excess of A^+ or D^-. Thus conductometric methods may be applied where visual or potentiometric methods fail to give results owing to considerable solubility or hydrolysis at the equivalence point, for example, in many precipitation reactions producing moderately soluble substances, in the direct titration of weak acids by weak bases, and in the displacement titration of salts of moderately weak acids or bases by strong acids or bases. A further important advantage is that the method is as accurate in dilute as in more concentrated solutions; it can also be employed with coloured solutions.

It may be noted that very weak acids, such as boric acid and phenol, which cannot be titrated potentiometrically in aqueous solution, can be titrated conductometrically with relative ease. Mixtures of certain acids can be titrated more accurately by conductometric than by potentiometric (pH) methods. Thus mixtures of hydrochloric acid (or any other strong acid) and acetic acid (or any other weak acid of comparable strength) can be titrated with a weak base (e.g., aqueous ammonia) or with a strong base (e.g., sodium hydroxide): reasonably satisfactory end-points are obtained.

Attention is directed to the importance of temperature control in conductance measurements. While the use of a thermostat is not essential in conductometric titrations, constancy of temperature is required but it is usually only necessary to place the conductivity cell in a large vessel of water at the laboratory temperature.

The relative change of conductivity of the solution during the reaction and upon the addition of an excess of reagent largely determines the accuracy of the titration; under optimum conditions this is about 0.5 per cent. Large amounts of foreign electrolytes, which do not take part in the reaction, must be absent, since these have a considerable effect upon the accuracy. In consequence, the conductometric method has much more limited application than visual, potentiometric, or amperometric procedures.

XV, 4. APPARATUS AND MEASUREMENTS. A conductivity cell for conductometric titrations may be of any kind that lends itself to thorough stirring of the contents (preferably by mechanical means), and permits the periodical addition of reagents. As explained above, it may be necessary to place the cell in a large vessel of water in order to maintain constancy of temperature, but in most circumstances the cell may be used at the ambient temperature of the laboratory. The cell should be constructed of Pyrex or other resistance glass and fitted with platinised platinum electrodes; the platinising helps to minimise polarisation effects. The size and separation of the electrodes will be governed by the change of conductance during the titration: for low-conductance solutions (e.g., when extremely dilute), the electrodes should be large and close together; for precipitation reactions the electrodes must be vertical.

The following procedure may be used for **platinising the electrodes**. The conductivity vessel and electrodes are thoroughly cleaned by immersion in a warm solution of potassium dichromate in concentrated sulphuric acid. After washing with distilled water until free from acid, the electrodes are plated from a solution containing 3 g chloroplatinic acid and 0.025 g lead acetate per 100 cm^3. The current may be obtained from two accumulators (4 volts), the poles of which are connected to the ends of a suitable sliding resistance. The current is adjusted so as to produce a moderate evolution of hydrogen. Each electrode should be used alternately as anode and cathode (i.e., the current should be reversed every half minute) and electrolysis should be continued until both electrodes are covered with a jet-black deposit. The time may vary from about two to about five minutes. After platinising, the electrodes must be freed from traces of chlorine; dilute sulphuric acid is electrolysed during 15 minutes using the two platinised electrodes (connected together) as cathode and another platinum electrode as anode. The electrodes are then washed with distilled water and afterwards kept immersed in distilled water until required for use.

A cell suitable for conductometric titration is depicted in Fig. XV, 2(a); the electrodes are firmly fixed in the perspex lid which is provided with openings for the stirrer and the jet of the burette. The stirrer shown may be replaced by a magnetic stirrer. Alternatively, a three- or four-necked flat bottom flask may be used, with the tubes carrying the electrodes fitting into

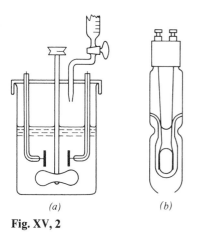

(a) *(b)*

Fig. XV, 2

ground glass joints and thus being accurately located: with the three-necked flask a magnetic stirrer must be used.

For most purposes a special cell is not required and good results are obtained by clamping a commercially available dip cell (shown diagramatically in Fig. XV, 2(b)) inside a beaker which is placed on a magnetic stirrer. With this arrangement, the dipping cell should be lifted clear of the solution after each addition from the burette to ensure that the liquid between the electrodes becomes thoroughly mixed. Since absolute conductivity values are not required it is not necessary to know the cell constant.

The conductance measurements are made using a Wheatstone bridge circuit as explained in Section XV, 2, and the **MEL Conductivity Bridge** type 7566/3 (Pye Unicam Ltd) is a very suitable instrument for this purpose. It is basically a mains-operated Wheatstone bridge working at the mains frequency (50 Hz) for the measurement of very low conductivities ($0.1–10\,\mu\mathrm{S\,cm^{-1}}$), but from a built-in oscillator of frequency 2.9 kHz for higher conductivities; the bridge incorporates an electronic 'magic eye' balance point indicator. A range switch selects any one of a series of standard resistances, and these provide standard conductance values of 1, 100, and 10 000 μS and of 1 S, whilst the main dial of the bridge moves over a scale which is directly calibrated to give the ratio of the two arms $b–d$ and $d–c$ (see Fig. XV, 1). Hence, for any setting of the main dial, the observed conductance is given by (standard conductance × scale reading).

To check the correct operation of the bridge, the instrument is set at the 'calibrate' (CAL) position in which two equal resistances are connected across the bridge arms so that the main dial balances in the central 'unity' position; it will be found that in the unbalanced condition the whole area of the magic-eye indicator will fluoresce brightly, but on rotating the dial a position will be found (exactly over the centre division marked '1') where the area of fluorescence contracts to a minimum.

To measure the conductance of a solution, the latter is placed in a suitable conductivity cell, or a dip cell (Fig. XV, 2(b)) is supported in the solution, and then connected to the TEST terminals of the conductivity bridge. The selector switch is set to the appropriate conductance range, and the dial is rotated until a balance is indicated on the magic eye. The conductivity may be calculated by multiplying the observed conductance by the cell constant.

It is also possible to use the bridge with an external standard resistance box: this is connected across the terminals marked 'STD' and the selector switch is set to an intermediate position, marked by a red spot, in which the internal standards are disconnected. In this mode of operation, a resistance R selected in the resistance box corresponds to a standard conductance of $1/R$, and the unknown conductance is then given by ($1/R$ × reading on main scale).

For conductometric titrations a convenient procedure is to use an external resistance box as described above, to set the dial of the bridge to the central position of the scale, and then to adjust the resistance box until balance is attained. This process is repeated after each addition of titrant and the recorded resistance values are then plotted against the volume of titrant. This produces a curve which is a mirror-image inversion of the usual conductance vs. volume of titrant curve, but is equally satisfactory for determining the end point.

A number of other conductance bridges are commercially available also operating on the Wheatstone bridge principle, but the Wayne-Kerr bridge (Wayne-Kerr Laboratories Ltd) is a transformer ratio-arm bridge which is less affected by stray capacitance than is a Wheatstone bridge (Ref. 3).

XV, 5. APPLICATIONS OF CONDUCTOMETRIC TITRATIONS. Some typical conductometric titration curves are collected in Fig. XV, 3, *a–h*.

Strong acid with a strong base. The conductance first falls, due to the replacement of the hydrogen ion (conductivity 350, Table XV, 1) by the added cation (conductivity 40–80) and then, after the equivalence point has been reached, rapidly rises with further additions of strong alkali due to the large conductivity of the hydroxyl ion (198). The two branches of the curve are straight lines provided the volume of the reagent added is negligible, and their intersection gives the end-point (curve (*a*)). This titration is of practical interest when the solutions are dark or deeply coloured or if they are very dilute (10^{-3}–10^{-4} M); in the latter case carbon dioxide must be excluded.

Strong acid with a weak base. The titration of a strong acid with a moderately weak base (K *ca.* 10^{-5}) may be illustrated by the neutralisation of dilute sulphuric acid by dilute ammonia solution (curve (*b*)). The first branch of the graph reflects the disappearance of the hydrogen ions during the neutralisation, but after the end-point has been reached the graph becomes almost horizontal, since the excess aqueous ammonia is not appreciably ionised in the presence of ammonium sulphate.

Weak acid with a strong base. In the titration of a weak acid with a strong base, the shape of the curve will depend upon the concentration and the dissociation constant K of the acid. Thus in the neutralisation of acetic acid ($K_a = 1.8 \times 10^{-5}$) with sodium hydroxide solution, the salt (sodium acetate) which is formed during the first part of the titration tends to repress the ionisation of the acetic acid still present so that its conductance decreases. The rising salt concentration will, however, tend to produce an increase in conductance. In consequence of these opposing influences the titration curves may have minima, the position of which will depend upon the concentration and upon the strength of the weak acid. As the titration proceeds, a somewhat indefinite break will occur at the end point, and the graph will become linear after all the acid has been neutralised. Some curves for acetic acid–sodium hydroxide titrations are shown in diagram (*c*).

For moderately strong acids (K *ca.* 10^{-3}) the influence of the rising salt concentration is less pronounced, but, nevertheless, difficulty is also experienced in locating the end-point accurately. Thus curve 1 in diagram (*d*) is obtained upon titrating $0.005M$ *o*-nitrobenzoic acid with $0.130M$-potassium hydroxide; the neutralisation line is slightly curved in the neighbourhood of the end point. There are two procedures for determining the end-point in the titration of weak acids with bases. The acid is first titrated with aqueous ammonia solution; if the end-point cannot be obtained from this curve with the desired accuracy, a second titration is carried out using potassium hydroxide solution of the same concentration. The two curves are practically identical up to the neutralisation point, and beyond this straight lines are obtained in both titrations, the intersection of which gives the end-point. In diagram (*d*), curve 2 is obtained with $0.130M$-aqueous ammonia solution. If the end-point is required with great accuracy, a correction should be applied for the fact that the conductance of the ammonium salt is approximately 0.6 per cent lower than that of the potassium salt, and the point of intersection should therefore be found when the final section of curve 2 is raised by this amount.

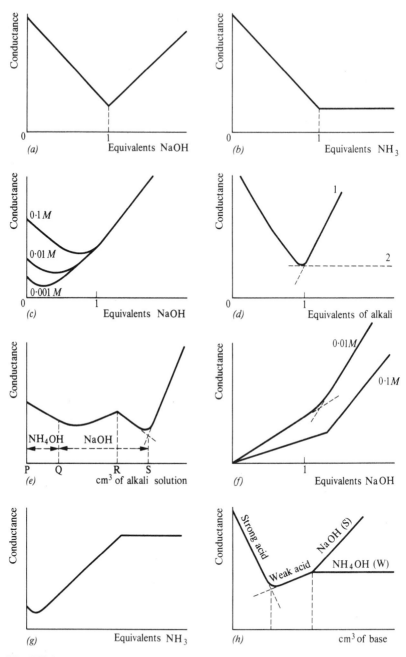

Fig. XV, 3

In the second procedure, the titration is commenced with a small amount of aqueous ammonia solution (sufficient, say, to neutralise one-third of the acid), and is then completed with sodium hydroxide solution of the same concentration. A typical curve for $0.005M$-mandelic acid is shown in diagram (e).

When all the acid ($\overline{M}$) has been neutralised the conductance of the mixture falls owing to the replacement of the ammonium ion by the less conducting sodium ion $\{(NH_4^+ + \overline{M}^-) + (Na^+ + OH^-) = NH_3 + H_2O + Na^+ + \overline{M}^-\}$; when the displacement of the ammonia is complete, the conductance rises abruptly. At this end-point (S), the total amount of sodium hydroxide solution added (QS) is equivalent to the acid originally present (PQ represents the first stage of the titration performed with aqueous ammonia solution). Alternatively, the acid present is also measured by the total alkali added to the point (R) at which the conductivity falls, i.e., by (PR). A double check is thus obtained in the titration. The method may be employed to improve the end-point of any titration if the acid is sufficiently strong to form an ammonium salt.

Very weak acid with a strong base. The initial conductance is very small, but increases as the neutralisation proceeds owing to the salt formed. The conductance values near the equivalence point are high because of hydrolysis; beyond the equivalence point the hydrolysis is considerably reduced by the excess alkali. To determine the end-point, values of the conductance considerably removed from the equivalence point must therefore be used for extrapolation. Some titration curves for boric acid and sodium hydroxide solution are given in diagram (f).

Weak acids with weak bases. The titration of a weak acid and a weak base can be readily carried out, and frequently it is preferable to employ this procedure rather than use a strong base. Curve (g) is the titration curve of 0.003M-acetic acid with 0.0973M-aqueous ammonia solution. The neutralisation curve up to the equivalence point is similar to that obtained with sodium hydroxide solution, since both sodium and ammonium acetates are strong electrolytes; after the equivalence point an excess of aqueous ammonia solution has little effect upon the conductivity, as its dissociation is depressed by the ammonium salt present in the solution. The advantages over the use of strong alkali are that the end-point is more easy to detect, and in dilute solution the influence of carbon dioxide may be neglected.

Mixture of a strong acid and a weak acid with a strong base. Upon adding a strong base to a mixture of a strong acid and a weak acid (e.g., hydrochloric and acetic acids), the conductance falls until the strong acid is neutralised, then rises as the weak acid is converted into its salt, and finally rises more steeply as excess of alkali is introduced. Such a titration curve is shown as S in diagram (h). The three branches of the curve will be straight lines except in so far as: (a) increasing dissociation of the weak acid results in a rounding off at the first end point, and (b) hydrolysis of the salt of the weak acid causes a rounding off at the second end point. Usually, extrapolation of the straight portions of the three branches leads to definite location of the end-points. Here also titration with a weak base, such as aqueous ammonia solution, is frequently preferable to strong alkali for reasons already mentioned in discussing weak acids: curve W in diagram (h) is obtained by substituting aqueous ammonia solution for the strong alkali. The procedure may be applied to the determination of mineral acid in vinegar or other weak organic acids ($K \nleq 10^{-5}$).

Displacement (or replacement) titrations. When a salt of a weak acid is titrated with a strong acid, the anion of the weak acid is replaced by that of the stronger one and the weak acid itself is liberated in the undissociated form. Similarly, in the addition of a strong base to the salt of a weak base, the cation of the weak base is replaced by that of the stronger one and the weak base itself is

623

liberated in the undissociated form. If, for example, M-hydrochloric acid is added to a $0.1M$-solution of sodium acetate, the curve shown in Fig. XV, 4 is obtained; the acetate ion is replaced by the chloride ion. The initial increase in conductivity is due to the fact that the conductivity of the chloride ion is slightly greater than that of the acetate ion. Until the replacement is nearly complete, the solution contains sufficient sodium acetate to suppress the ionisation of the liberated acetic acid and thereby render negligible its contribution to the conductivity of the solution. Near the equivalence point the acetic acid is sufficiently ionised to affect the conductivity, thus leading to higher values of the conductivity and the rounded portion of the curve. Beyond the equivalence point when excess of hydrochloric acid is present, the ionisation of the acetic acid is again suppressed and the conductivity rises rapidly. It can easily be calculated that to titrate a $0.1M$-salt solution the dissociation constant must not be greater than 5×10^{-4}, for a $0.01M$-salt solution $K \not> 5 \times 10^{-5}$, and for a $0.001M$-salt solution $K \not> 5 \times 10^{-6}$, i.e., the ionisation constant of the displaced acid or base divided by the original concentration of the salt must not exceed about 5×10^{-3}. Fig. XV, 4 also includes the titration curve of $0.01M$-ammonium chloride solution with $0.1M$-sodium hydroxide solution. The decrease in conductivity during the displacement is caused by the substitution of the ammonium ion by the sodium ion (see Table XV, 1 in Section **XV, 1**).

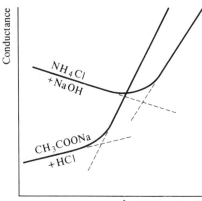

cm³ of HCl or NaOH

Fig. XV, 4

Precipitation and complex formation reactions. A reaction may be made the basis of a conductometric titration if the reaction product is sparingly soluble or is a stable complex. The following factors must be considered in connection with the usefulness and accuracy of the titration:

1. In order to reduce the influence of errors in the conductometric titration to a minimum the angle between the two branches of the curve should be as small as possible. If the angle is very obtuse, a small error in the conductance data can cause a large deviation. The following approximate rules will be found useful:

(a) The smaller the conductivity of the ion which replaces the reacting ion, the more accurate will be the result. (Thus it is preferable to titrate a silver salt with lithium chloride rather than with hydrochloric acid; cations may be titrated with lithium salts, and anions with acetates.)

(b) The larger the conductivity of the anion of the reagent which reacts with the cation to be determined, or vice versa, the more acute is the angle.

(c) The titration of a slightly ionised salt does not give good results, since the conductivity increases continuously from the commencement. Hence the salt present in the cell should be virtually completely dissociated; for a similar reason, the added reagent should also be a strong electrolyte.

2. The solubility of the precipitate (or the dissociation of the complex) should be less than 5 per cent. The addition of ethanol is sometimes recommended to reduce the solubility, but its influence on the factors detailed in 1 must be borne in

mind. An experimental curve is given in Fig. XV, 5 (ammonium sulphate in aqueous-ethanol solution with barium acetate). If the solubility of the precipitate were negligibly small, the conductance at the equivalence point would be given by AB and not the observed AC. The addition of excess of reagent depresses the solubility of the precipitate and, if the solubility is not too large, the position of the point B can determined by continuing the straight portions of the two arms of the curve until they intersect.

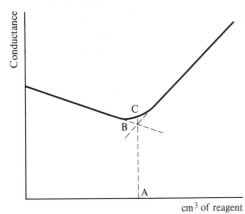

Fig. XV, 5

3. A slow rate of precipitation, particularly with micro-crystalline precipitates, prolongs the time of titration. Seeding or the addition of ethanol (concentration up to 30–40 per cent) may have a favourable effect.

4. If the precipitate has pronounced adsorptive properties, the composition of the precipitate will not be constant, and appreciable errors may result. Occlusion may take place with micro-crystalline precipitates.

In spite of the obvious limitations of the method, quite a large number of precipitation titrations have been carried out; thus silver nitrate, lead nitrate, barium acetate or barium chloride, uranyl acetate, lithium sulphate, and lithium oxalate have been employed in precipitation reactions (for details, see Selected bibliography—Section **XV, 12**).

Oxidation–reduction (redox) titrations. The conductometric method is not well suited to the study of oxidation–reduction titrations. Almost all such reactions must be carried out in the presence of a large excess of acid or base, which more or less completely masks the change in conductance due to the redox reaction. A typical example is the titration of iron(II) with permanganate in which, say, a $0.01M$ solution of iron(II) in $0.5M$-sulphuric acid is titrated with $0.02M$-potassium permanganate. Although the reaction

$$5Fe^{2+} + MnO_4^- + 8H^+ = 5Fe^{3+} + Mn^{2+} + 4H_2O$$

does consume hydrogen ions, thus decreasing the conductance of the solution up to the equivalence point, the fraction of hydrogen ion thus removed is relatively small. The entire change in conductance is not great and cannot be detected with accuracy by the usual equipment employed for conductometric titrations.

XV, 6. SOME EXPERIMENTAL DETAILS FOR CONDUCTOMETRIC TITRATIONS. One of the cells described in Section **XV, 4** is used in conjunction with a Wheatstone bridge operating with alternating current. The dip cell (Fig. XV, 2(b)) clamped inside a beaker which is set up for magnetic stirring is recommended, and the conductance measurements are conveniently made with the MEL Conductivity bridge. As explained in Section **XV, 4**, it is convenient to use the bridge in conjunction with an external resistance box, and to adjust this, with the ratio arms of the bridge set to unity, to find the balance point.

After measuring the solution to be titrated (up to 25 cm³) into the beaker, it is diluted with distilled water to at least 100 cm³, and the stirrer set in motion. The titrating agent (concentration at least 10 times that of the solution being titrated) should be placed in a 5 or 10-cm³ micro-burette; the reagent is added in small portions, and the solution is stirred or shaken after each addition. The conductivity is measured after the well-mixed solution has been allowed to stand for a minute or two. The addition of the titrating reagent is continued until at least five readings beyond the equivalence point have been made. It is often advisable to carry out a preliminary titration; this will provide information as to the increments of the reagent best suited for the particular titration, e.g., increments of 0.5 cm³, etc. The conductance (or resistance) is plotted as ordinates against the volume of the titrating reagent as abscissae; the two straight portions of the curve are extrapolated until they intersect and the point of intersection is taken as the equivalence point of the reaction.

In order to obtain satisfactory results by extrapolation procedures from the titration curves, the following points must be borne in mind. Firstly, extrapolation can only be performed satisfactorily with *straight lines*. Curvature in the immediate neighbourhood of the end point as shown in curves XV, 3, (e), (f) and (h) may be ignored, provided there are adequate numbers of readings on *both* sides of the end-point to permit the drawing of unequivocal straight lines: this requires that sufficient readings be taken after the equivalence point has been reached to give at least five points lying on a straight line.

Secondly, it is most important that the dilution correction factor $(V + v)/V$ (see Section XV, 3) be applied to the readings before plotting the curves.

High frequency titrations

XV, 7.   GENERAL CONSIDERATIONS.   In the high-frequency method of titration a suitable cell containing the chemical system is made part of, or is coupled to, an oscillator circuit resonating at a frequency of several megahertz. As the composition of the chemical system changes the resistance (impedance) and/or capacitance of the circuit are altered, and changes are produced in oscillator characteristics, such as frequency, grid current and voltage, and plate current and voltage. Any of these quantities may be measured and taken as an indication of the change in composition of the chemical system, e.g., as a solution is titrated with an appropriate reagent: curves may be generally obtained which show inflexion or breaks at the equivalence point. The fundamental properties of the chemical system which affect the oscillator characteristics are the dielectric constant and the conductivity. An important advantage of the high frequency method is that the electrodes may be placed on the outside of the cell, and are therefore not in direct contact with the test solution. Measurements can accordingly be made without danger of electrolysis or electrode polarisation; furthermore, errors resulting from the coating of the electrodes by a precipitate and other surface phenomena are eliminated. One disadvantage is that the response of a high-frequency titrimeter is non-specific, being dependent only on the conductivity and dielectric constant of the system and is independent of the chemical identity of the components of the system.

It is interesting to consider what happens to the individual ions of an electrolyte and to polar molecules when exposed to a rapidly alternating field.

Each ion or dipolar molecule tends to move or align itself in the direction of the electrode of opposite polarity. The electrode polarity changes once every cycle, and the ion or dipole must reverse its motion or orientation. The conductance of the solution is the result of the movement of negative and positive ions relative to their neighbours and the solvent molecules. Each ion tends to move ahead of its ionic atmosphere, and in consequence an unsymmetrical charge distribution forms around each central ion and exerts a retarding force on the ion in a direction opposite to its motion. At alternating frequencies greater than one megahertz the central ion changes its direction of motion so rapidly with every cycle of the applied field that there is little chance for dissymmetry of the ionic atmosphere to arise, hence the conductance increases. Also at high frequencies the ions undergo such small oscillations that the oppositely charged ionic atmosphere exerts a relatively much smaller drag than at low frequencies. Since the apparent dielectric constant and also the time of formation and decay of the ionic atmosphere (relaxation time) are both concentration dependent, the response curves of high-frequency instruments are greatly influenced by ionic concentration. Indeed, with oscillators which operate at frequencies of the order of about 2 megahertz, the maximum concentration of electrolyte that can be employed is of the order of $0.01M$. The high-frequency technique is most sensitive in titrations where the total concentration of dissolved ions changes, e.g., in precipitation and complex formation reactions. It is applicable also to cases where a fast-moving ion is replaced by a slow-moving ion, as in acid–base titrations.

In ordinary conductance measurements at frequencies of the order of 1000 hertz the influence of cell capacitance is small, but at megahertz frequencies it plays an important part. It has been shown that high-frequency titration graphs can be interpreted on the assumption that the solution in the cell behaves like a capacitor and resistor in parallel in the oscillator circuit. A simple cell for high frequency use may consist of two metallic plates fixed to the walls of a rectangular glass container (Fig. XV, 6(a)). The cell may be represented by the equivalent

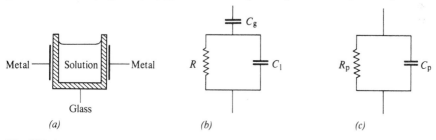

Fig. XV, 6

circuit, (b), where C_g is the capacitance through the glass container walls, C_l is the capacitance of the solution (liquid), and R is the resistance of the solution. The resistance of the glass container walls is assumed to be so high that it may be neglected. Circuit (b) may be represented by the simpler equivalent parallel circuit, (c), and it can be demonstrated by conventional methods of the theory of alternating currents that:

$$\frac{1}{R_p} = \frac{\kappa\omega^2 C_g^{\,2}}{\kappa^2 + \omega^2(C_g + C_l)^2} \qquad (1)$$

$$C_p = \frac{C_g\kappa^2 + \omega^2(C_g C_l^2 + C_g^2 C_l)}{\kappa^2 + \omega^2(C_g + C_l)^2} \tag{2}$$

where $\omega = 2\pi f$, f being the frequency (hertz), κ is the low-frequency conductance of the solution $(= 1/R)$; and $1/R_p$ the high-frequency conductance. These expressions give the relationship between the low-frequency conductance, the frequency, and either the high-frequency conductance $(1/R_p)$ or the capacitance (C_p).

It can be deduced from equation (1) that for a given electrolyte, the optimum frequency is related to concentration by the expression

$$f = \frac{1.8 \times 10^{12}\kappa}{D}$$

where D is the dielectric constant and concentration is expressed in terms of the normal (low frequency) conductivity of the solution (κ). For an electrolyte such as sodium chloride, this equation shows that the appropriate concentrations for frequencies of 5, 30 and 100 megahertz are respectively 0.0025 M, 0.014 M, and 0.05 M, whilst for an electrolyte of appreciably higher conductivity (i.e., acids and bases), the appropriate concentrations are significantly reduced: thus for hydrochloric acid the concentrations corresponding to the same three frequencies are respectively 0.0006M, 0.003M, and 0.01M. Since 30 megahertz represents a practical upper limit for the construction of apparatus which is reasonably easy to operate, it follows that there are limitations to the concentration range over which the method may be used.

In practice, the cell portrayed in Fig. XV, 6(a) will consist either of a glass vessel with two conducting metal (aluminium, copper, gold or silver) bands encircling it as in Fig. XV, 7 (Section **XV, 8**), or the glass container will simply fit snugly inside a coil of wire which covers an appreciable length of the cell. The first type of cell is termed a **capacitative cell** and equations (1) and (2) are applicable to it: the circuit responds to changes in both the conductance and the dielectric constant of the solution in the cell. The second type of cell is termed an **inductive cell** and the associated circuit responds only to changes in the conductance of the cell contents. The capacitative type of cell is usually preferred for titrimetry.

XV, 8.   APPARATUS.   Since its introduction as an analytical technique in 1946 (Ref. 4) many instruments have been designed for high frequency titrimetry: typical examples will be found in the papers listed in Ref. 5. In addition, a number of commercial instruments have also been produced: the Fisher High Frequency Titrimeter (Fisher Scientific Co), the Sargent Oscillometer (Sargent and Co), the PCL-Lee Titrimeter (Polymer Consultants Ltd). In each case, changes in the composition of the solution contained in the titration cell affects the high frequency conductance and the capacitance of the cell with a consequent effect upon the circuit of which the cell forms a component: the various instruments differ in the response (voltage, current, capacitance or inductance) which is observed, and in the way in which the response is measured. The instruction manual supplied with the instrument in use must therefore be consulted for the precise operational details.

A satisfactory **high frequency conductance cell** is shown in Fig. XV, 7. It consists of a Pyrex tube, 150 mm × 25 mm, provided with two 25-mm wide bands of aluminium sheet (1.5 mm thick) fitted tightly around the tube and held in

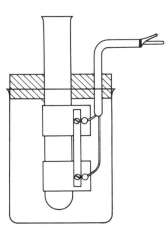

Fig. XV, 7

position (about 2.5 cm apart) by a Bakelite strip. The cell is mounted inside a glass vessel, which *inter alia* prevents moisture depositing on the metal strips, etc.; the glass vessel may be filled, if desired, with paraffin wax, Connection to the titrimeter is made with a screened cable. The level of the liquid in the cell should be about 1 cm above the top of the upper metal strip before measurements are made: the initial volume is about 35 cm³. The liquid in the cell is stirred mechanically with an efficient glass stirrer; alternatively, the bottom of the cell may be flattened and magnetic stirring employed. For titrations with somewhat larger volumes of liquid (> 60 cm³), a Pyrex cell with dimensions 200 mm × 38 mm is suitable.

Unless the information is given in the instruction manual, it is advisable to ascertain the **response of the instrument in relation to electrolyte concentration**: each instrument with a given cell shows maximum response to changes in the compositon of the test solution over a somewhat limited range of ionic strength. It is also advantageous if the operating frequency can be varied, as the shape of the titration curve may well alter with frequency, and for example, an M-shaped curve which is difficult to interpret, may, by alteration of frequency be converted to a readily interpreted V-shaped curve.

XV, 9. ADVANTAGES OF THE TECHNIQUE. As indicated in Section **XV, 8**, a suitable conductivity cell for high frequency measurements can be easily constructed from glass tubing by contrast with the special cells with expensive platinum electrodes which are required for normal conductance measurements. A special advantage is the fact that the electrodes of the high frequency cell are not in direct contact with the solution under investigation, and thus no electrolysis occurs and there can therefore be no polarisation. Furthermore, the electrodes are protected from contamination, e.g., by precipitated solid during precipitation titrations, and there is no danger of chemical reactions arising due to catalysis by the metal surfaces. A further advantage is the applicability of the method to solutions of low concentration, 10^{-3}–$10^{-4} M$.

Drawbacks to the method are (i) the need for adequate electrical screening, (ii) the fact that because the electrodes are separated from the solution by the non-conducting walls of the cell the sensitivity is necessarily less than that associated with normal conductometric measurements, (iii) the limited concentration range over which the method can be safely used, and (iv) unless the correct frequency is chosen, the titration plot may produce curves which are not readily interpreted.

XV, 10. SOME EXAMPLES OF HIGH FREQUENCY TITRATIONS. The solution to be titrated must, if necessary, be appropriately diluted, so that when using the cell shown in Fig. XV, 7 (Section **XV, 8**), a 5 or 10 cm³ aliquot of the final test solution, further diluted in the cell to a total volume of about 35 cm³ (so that the liquid level is just over 1 cm above the upper electrode of the cell), gives a concentration which lies within the optimum operating range of the titrimeter to be employed: see Section **XV, 8**. The stirrer is inserted, and after allowing an adequate warm-up period for the instrument, the initial reading is made in accordance with the operating instructions. The titrant should have a concentration five to ten times that of the test solution, and small portions are added to the cell from a microburette. After each addition of reagent the titrimeter is readjusted; it may be found advisable to stop the stirrer whilst the adjustments are made: finally, the instrument readings are plotted against the volume of titrant added. Some typical results are shown by the curves in Fig. XV, 8.

Curves (A) and (B) are typical curves for neutralisation reactions: it is noteworthy that curve (B) shows two breaks corresponding to the reactions $CO_3^{2-} + H^+ = HCO_3^-$ and $HCO_3^- + H^+ = H_2O + CO_2$.

Curve (C) is typical of a complexation titration, and curves (D), (E) and (F) are examples of those obtained during precipitation titrations. Hara and West (Ref. 7) recommend that EDTA titrations should be performed in buffered solutions, but the Ni–EDTA titration curve shown (C), was obtained by working in an unbuffered medium. The Th-oxalate titration (D) was carried out using test and titrant solutions both $0.01M$ in nitric acid, whilst the $0.01M$ lanthanum acetate reagent for the fluoride titration (Curve E), contained a trace of acetic acid ($0.5\,cm^3$ per litre). These curves were obtained using the apparatus described in Ref. 5b.

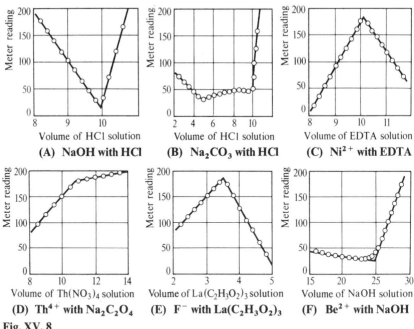

(A) NaOH with HCl

(B) Na₂CO₃ with HCl

(C) Ni²⁺ with EDTA

(D) Th⁴⁺ with Na₂C₂O₄

(E) F⁻ with La(C₂H₃O₂)₃

(F) Be²⁺ with NaOH

Fig. XV, 8

XV, 11. References

1. M. L. McGlashan (1971). *Physico-chemical Quantities and Units*. 2nd edn. London; Royal Institute of Chemistry.
2. B. P. Levitt (1973). *Findlay's Practical Physical Chemistry*. 9th edn. Harlow; Longman.
3. D. R. Browning (1969). *Electrometric Methods*. London; McGraw-Hill.

4. F. W. Jensen and A. L. Parrack (1946). 'Use of High-frequency Oscillators in Titrations and Analyses.' *Indl. and Engrg. Chem., Analytical Edition*, **18**, 595.
5. (a) J. L. Hall, J. A. Gibson, H. O. Phillips and F. E. Critchfield (1954). 'The Use of a Radiofrequency Oscillator in Student Analysis.' *J. Chem. Ed.*, **31**, 54.
    (b) V. Kyte and A. I. Vogel (1959). 'An Inexpensive High-frequency Titration Apparatus for General Laboratory Use.' *Analyst*, **84**, 1004.
    (c) F. Kovacs, O. Klug, M. Gombos and F. Farkas (1971). 'Apparatus for the Oscillometric Determination of Concentrations.' *Chem. Anal. (Warsaw)*, **16**, 251.
6. C. L. Wilson and D. W. Wilson (1964). *Comprehensive Analytical Chemistry*. Vol. IIa. London; Elsevier.
7. R. Hara and P. W. West (1954). 'High-frequency Titrations Involving Chelation with EDTA.' *Anal. Chim. Acta.*, **11**, 264.

XV, 12.   Selected bibliography

1. W. G. Berl (1956). *Physical Methods in Chemical Analysis*. Vols. II/III. New York; Academic Press.
2. H. T. S. Britton (1934). *Conductometric Analysis*. London; Chapman and Hall.
3. K. Cruse and R. Huber (1957). *Hochfrequenztitration*. Weinheim; Verlag Chemie.
4. P. Delahay (1957). *Instrumental Analysis*. New York; The Macmillan Co.
5. G. W. Ewing (1975). *Instrumental Methods of Chemical Analysis*. 4th edn. New York; McGraw-Hill Book Co.
6. I. M. Kolthoff and P. J. Elving (1959). *Treatise on Analytical Chemistry*. Part 1, Vol. 4. New York; Wiley (Interscience).
7. I. M. Kolthoff and H. A. Laitinen (1941). *pH and Electro Titrations*. New York; John Wiley.
8. J. J. Lingane (1958). *Electroanalytical Chemistry*. 2nd edn. New York; Interscience.
9. H. J. S. Sand (1941). *Electrochemistry and Electrochemical Analysis*. Vol. III. London; Blackie.
10. P. H. Sherrick, G. A. Dawe, R. Karr, and E. F. Ewen (1954). *Manual of Chemical Oscillometry*. Chicago; E. H. Sargent and Co.
11. C. R. N. Strouts, J. H. Gilfillan and H. N. Wilson (1955). *Analytical Chemistry. The Working Tools*. Vol. II. London; Oxford University Press.
12. H. A. Strobel (1973). *Chemical Instrumentation*. 2nd edn. Reading, Mass.; Addison-Wesley Publishing Co.
13. A. Weissberger (1960). *Physical Methods of Organic Chemistry*. Vol. I, Part 4. 3rd edn. New York; Interscience.
14. F. J. Welcher (1966). *Standard Methods of Chemical Analysis*. 6th edn. Vol. 3-A. New York; Van Nostrand.
15. H. H. Willard, L. L. Merritt, and J. A. Dean (1974). *Instrumental Methods of Analysis*. 5th edn. New York; Van Nostrand.

CHAPTER XVI **VOLTAMMETRY**

XVI, 1. INTRODUCTION. Voltammetry is concerned with the study of voltage–current–time relationships during electrolysis carried out in a cell where one electrode is of relatively large surface area, and the other (the working electrode) has a very small surface area and is often referred to as a micro-electrode: the technique commonly involves studying the influence of voltage changes on the current flowing in the cell. The micro-electrode is usually constructed of some inert, conducting material such as gold, platinum or carbon, and in some circumstances a dropping mercury electrode (D.M.E) may be used; for this special case the technique is referred to as **polarography**.

In view of the relative surface areas of the two electrodes, it follows that at the large auxiliary or counter electrode the current density will be very small, whilst at the working electrode it may be high. In consequence, the counter electrode is not readily polarised, and when small currents flow through the cell, the concentration of the ions in the electrode layer (i.e., the layer of solution immediately adjacent to the electrode) remains virtually equal to the concentration in the bulk solution, and the potential of the electrode is maintained at a constant value. By contrast, at the micro-electrode, the electrode layer tends to become depleted of the ions being discharged at the electrode, and if the solution is not stirred, then the diffusion of ions across the resultant concentration gradient becomes an important factor in determining the magnitude of the current flowing.

The total current flowing will in fact be equal to the current carried by the ions undergoing normal electrolytic migration, plus the current due to the diffusion of ions

$$I = I_d + I_m$$

where I is the total current, I_d the diffusion current, and I_m the migration current. There is, however, a complicating factor in that, in dilute solution, the depletion of the electrode layer leads to an increase in the resistance of the solution, and thus to a change in the Ohm's law potential drop ($I \times R$) in the cell; consequently the exact potential operative at the electrode is open to doubt. To overcome this, it is usual to add an excess of an indifferent electrolyte to the system (e.g. $0.1M$-KCl), and under these conditions the solution is maintained at a low, constant resistance, whilst the migration current of the species under investigation virtually disappears, i.e., $I = I_d$.

The rate of diffusion of the ion to the electrode surface is given by Fick's law as

$$\frac{\partial c}{\partial t} = \frac{D\partial^2 c}{\partial x^2}$$

where D is the diffusion coefficient, c = concentration, t = time, and x = distance from the electrode surface (see Ref. 1), and the potential of the electrode is controlled by the Nernst equation

$$E = E^{\ominus} + \frac{RT}{nF} \ln \frac{a_{Ox}}{a_{Red}}.$$

Techniques which come under the general heading of voltammetry and which will be treated in this Chapter are:

Polarography (d.c. and a.c.).
Anodic stripping voltammetry.
Chronopotentiometry.

Polarography

XVI, 2. BASIC PRINCIPLES.   If a steadily increasing voltage is applied to a cell incorporating a relatively large quiescent mercury anode and a minute mercury cathode (composed of a succession of small mercury drops falling slowly from a fine capillary tube), it is frequently possible to construct a reproducible current–voltage curve. The electrolyte is a dilute solution of the material under examination (which must be electro-active) in a suitable medium containing an excess of an indifferent electrolyte (*base or ground solution, or supporting electrolyte*) to carry the bulk of the current and raise the conductivity of the solution, thus ensuring that the material to be determined, if charged, does not migrate to the dropping mercury cathode. From an examination of the current–voltage curve, information as to the nature and concentration of the material may be obtained (Heyrovsky, Ref. 2). Heyrovsky and Shikata (Ref. 3) developed an apparatus which increased the applied voltage at a steady rate and simultaneously recorded photographically the current–voltage curve. Since the curves obtained with this instrument are a graphical representation of the polarisation of the dropping electrode, the apparatus was called a **polarograph**, and the records obtained with it, **polarograms**; the photographic recorder is now replaced by a pen recorder, and in some circumstances, by an oscilloscope.

The basic apparatus for polarographic analysis is depicted in Fig. XVI, 1. The dropping mercury electrode is here shown as the cathode (its most common function); it is sometimes referred to as the working or micro-electrode. The anode is a pool of mercury, and its area is correspondingly large, so that it may be regarded as incapable of becoming polarised, i.e., its potential remains almost constant in a medium containing anions capable of forming insoluble salts with mercury (Cl^-, SO_4^{2-}, etc.); it acts as a convenient non-standardised reference electrode, the exact potential of which will depend upon the nature and the concentration of the supporting electrolyte. The polarisation of the cell is therefore governed by the reactions occurring at the dropping mercury cathode. Inlet and outlet tubes are provided to the cell for expelling dissolved oxygen from the solution by the passage of an inert gas (hydrogen or nitrogen) before, but not during, an actual measurement—otherwise the polarogram of the dissolved oxygen will appear in the current–voltage curve. *P* is a potentiometer by which any e.m.f. up to 3 volts may be gradually applied to the cell. *S* is a shunt for adjusting the sensitivity of the galvanometer *G* appropriate to the nature and concentration of the substance being investigated. It may be mentioned that under these conditions the current–voltage curve is really a current–cathode potential curve, but displaced by a constant voltage corresponding to the potential of the anode. For some purposes it is advisable to employ an external anode of known potential (e.g., a saturated calomel electrode): an internal electrode is more convenient for most analytical work, since absolute values of the cathode potential are not usually required.

The initial potential of the dropping mercury cathode is indeterminate, and will assume any potential applied to it from an external source; when it acquires a potential different from that which it had in the absence of electrical connections, the working electrode is said to be *polarised*.

Let us consider what will occur if an external e.m.f. is applied to the cell shown in Fig. XVI, 1, charged with, say, a dilute, oxygen-free solution of cadmium chloride. All the positively charged ions present in the solution will be attracted to the negative working electrode by: (*a*) an electrical force, due to the attraction of oppositely charged bodies to each other, and by (*b*) a diffusive force, arising from the concentration gradient produced at the electrode surface. The total current passing through the cell can be regarded as the sum of these two factors. A typical simple current–voltage curve is shown in Fig. XVI, 2. The working electrode,

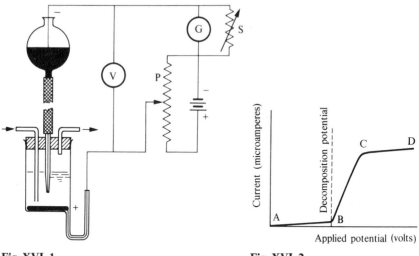

Fig. XVI, 1   **Fig. XVI, 2**

being perfectly polarisable, assumes the correspondingly increasing negative potential applied to it; from A to B practically no current will pass through the cell. At B, where the potential of the micro-electrode is equal to the deposition potential of the cadmium ions with respect to a metallic cadmium electrode, the current suddenly commences to increase and the working electrode becomes depolarised by the cadmium ions, which are then discharged upon the electrode surface to form metallic cadmium, consequently a rapid increase in the current flowing through the cell will be observed. At the point C the current no longer increases linearly with applied potential but approaches a steady limiting value at the point D: no increase in current is observed at higher cathode potentials unless a second compound able to depolarise the working electrode is present in the solution. At any point on the curve between B and C (usually spoken of as the **polarographic wave**) the number of cadmium ions reaching the micro-electrode surface as a result of migration and diffusion from the main bulk of the solution always exceeds the number of cadmium ions which react at and are deposited upon the electrode. At the point C the rate of supply of the cadmium ions from the main bulk of the solution to the working electrode surface has become equal to the rate of their deposition. Hence at potentials more negative than point D, the concentration of undischarged cadmium ions at the micro-electrode surface is negligibly small relative to the cadmium-ion concentration in the bulk of the solution; no further increase in current passing through the electrolytic cell can be expected, since the limiting current is now fixed by the rate at which cadmium ions can reach the electrode surface.

A number of **polarisable micro-electrodes** (e.g., a rotating platinum wire, *ca.* 3 mm long and 0.5 mm diameter, suitably mounted, or stationary noble-metal electrodes) have been used in determining current–voltage curves, but the most satisfactory is a slowly growing drop of mercury issuing, under a head of 40–60 cm of mercury, from a resistance-glass capillary (0.05–0.08 mm in diameter and 5–9 cm long) in small, uniform drops. The dropping mercury electrode has the following advantages:

(a) Its surface is reproducible, smooth, and continuously renewed; this is conducive to good reproducibility of the current–potential curve and eliminates passivity or poisoning effects.

(b) Mercury forms amalgams (solid solutions) with many metals.

(c) The diffusion current assumes a steady value immediately after each change of applied potential, and is reproducible.

(d) The large hydrogen overpotential on mercury renders possible the deposition of substances difficult to reduce, e.g., the alkali metal ions, aluminium ion and manganese(II) ion. (The current–potential curves of these ions are inaccessible with a platinum micro-electrode.)

(e) The surface area can be calculated from the weight of the drops.

The dropping mercury electrode may be applied over the range + 0.4 to about − 2.0 volts with reference to the S.C.E. Above + 0.4 volt mercury dissolves and gives an anodic wave; it begins to oxidise to mercury(I) ion. At potentials more negative than about − 1.8 volts *vs.* S.C.E., visible hydrogen evolution occurs in acid solutions and the usual supporting electrolytes commence to discharge. The range may be extended to about − 2.6 volts *vs.* S.C.E. by using supporting electrolytes having higher reduction potentials than the alkali metals; tetra-alkyl ammonium hydroxides or their salts are satisfactory for this purpose.

Reference has already been made to the convenience for routine analytical work of using a mercury pool covering the bottom of the electrolysis cell as the **non-polarisable reference electrode**; the mercury pool is connected to an external circuit via a platinum wire sealed through the wall of the cell. If the solution covering the mercury pool contains chloride ion, the mercury pool acts as a calomel electrode of the particular chloride-ion concentration. Whilst convenient for routine polarographic determinations, the mercury pool never possesses a definite, known potential and the potential does not attain a constant value in the absence of chloride ions or other depolarising ions: further the internal reference electrode cannot be used in the presence of powerful oxidising or reducing agents. With the mercury pool anode, the *apparent* half-wave potential of a given substance will be in terms of the total e.m.f. applied to the cell and varies with the potential of the mercury anode. If absolute values are required it is best to use a cell which incorporates a reference electrode (usually a saturated calomel electrode); see Section **XVI, 9**. External reference electrodes have potentials which are accurately known and may be used with solutions containing strong oxidising or reducing agents, and the sample solution need not contain a depolarising anion.

Clearly, with the apparatus described above we are dealing with direct current, and the technique as thus carried out is termed **d.c. polarography** to distinguish it from the modifications which become possible when using alternating current (**a.c. polarography**), and from which many of the recent advances in polarographic techniques stem.

Direct current polarography

XVI, 3. THEORETICAL PRINCIPLES.   We will now consider *the factors affecting the limiting current* with a dropping mercury cathode.

Residual (or condenser) current.   Mercury is unique in remaining electrically uncharged when it is dropping freely into a solution containing an indifferent electrolyte, such as potassium chloride or potassium nitrate. If a current–voltage curve is determined for a solution containing ions with a strongly negative reduction potential (e.g., potassium ions), a small current will flow before the decomposition of the solution begins. This current increases almost linearly with the applied voltage, and it is observed even when the purest, air-free solutions are used, so that it cannot be due to the reduction of impurities. It must therefore be considered a non-faradaic or condenser current, made appreciable by the continual charging of new mercury drops to the applied potential. It is known that metals, when submerged in an electrolyte, are covered with an electrical double layer of positively and negatively charged ions. The capacity of the double layer and hence the charging current vary, depending upon the potential which is imposed upon the metal.

In practice, one often finds that the indifferent electrolyte contains traces of impurities so that small, almost imperceptible currents are superimposed upon the condenser current. It is customary to include all these in the **residual current**. As will be shown later, in practical polarographic work the residual current is automatically subtracted from the total observed current by proper extrapolation and placement of tangents to the wave.

Migration current.   Electro-active material reaches the surface of the electrode largely by two processes. One is the migration of charged particles in the electric field caused by the potential difference existing between the electrode surface and the solution; the other is concerned with the diffusion of particles, and will be discussed in a succeeding paragraph. Heyrovsky showed that the migration current can be practically eliminated if an *indifferent electrolyte* is added to the solution in a concentration so large that its ions carry essentially all the current. (An indifferent electrolyte is one which conducts the current but does not react with the material under investigation, nor at the electrodes within the potential range studied.) In practice, this means that the concentration of the added electrolyte (*supporting electrolyte*) must be at least 100-fold that of the electro-active material.

An example will make this conception of supporting electrolyte clear. Let us imagine an electrolytic solution is composed of potassium ions $0.10M$ and copper(II) ions $0.005M$. If we assume that the molar conductivities of K^+ and $\frac{1}{2}Cu^{2+}$ are approximately equal, then it follows that *ca.* 90 per cent of the current will be transported to the cathode by the potassium ions and only 10 per cent by the copper ions. Both ions will tend to diffuse towards any portion of the solution where a concentration gradient exists, but the rate of diffusion will be slow. If the concentration of the potassium ions be increased until it represents 99 per cent of the total cations present, practically all the current passing through the cell will be transported by the potassium ions. Under such conditions the electro-active material can reach the electrode surface only by diffusion. It must be emphasised that the supporting electrolyte must be composed of ions which are discharged at higher potentials than, and which will not interfere or react chemically with, the ions under investigation.

Diffusion current.   When an excess of supporting electrolyte is present in the solution the electrical force on the reducible ions is nullified; this is because the ions of the added salt carry practically all the current and the potential gradient is compressed or shortened to a region so very close to the electrode surface that it is no longer operative to attract electro-reducible ions. Under these conditions the limiting current is almost solely a diffusion current. Ilkovic (Ref. 7) examined the various factors which govern the diffusion current and deduced the following equation:

$$I_d = 607\, n\, D^{1/2}\, C\, m^{2/3}\, t^{1/6}$$

where $I_d = $ the average diffusion current in microamperes during the life of the drop;

$n = $ the number of faradays of electricity required per mol of the electrode reaction (or the number of electrons consumed in the reduction of one mol of the electro-active species);

$D = $ the diffusion coefficient of the reducible or oxidisable substance expressed as $cm^2\ sec^{-1}$;

$C = $ its concentration in millimoles per dm^3;

$m = $ the rate of flow of mercury from the dropping electrode expressed in mg per second; and

$t = $ drop time in seconds.

The constant 607 is a combination of natural constants, including the Faraday constant; it is slightly temperature dependent and the value 607 is for 25 °C. The **Ilkovic equation** is important because it accounts quantitatively for the many factors which influence the diffusion current: in particular, the linear dependence of the diffusion current upon n and C. Thus, with all the other factors remaining constant, the diffusion current is directly proportional to the concentration of the electro-active material—this is of great importance in quantitative polarographic analysis.

The original Ilkovic equation neglects the effect on the diffusion current of the curvature of the mercury surface. This may be allowed for by multiplying the right-hand side of the equation by $(1 + AD^{1/2}\, t^{1/6}\, m^{-1/3})$, where A is a constant and has a value of 39. The correction is not large (the expression in parentheses usually has a value between 1.05 and 1.15) and need only be taken account of in very accurate work.

The diffusion current I_d depends upon several factors, such as temperature, the viscosity of the medium, the composition of the base electrolyte, the molecular or ionic state of the electro-active species, the dimensions of the capillary, and the pressure on the dropping mercury. The temperature coefficient is about 1.5–2 per cent/°C; precise measurements of the diffusion current require temperature control to about 0.2 °C, which is generally achieved by immersing the cell in a water thermostat (preferably at 25 °C). A metal-ion complex usually yields a different diffusion current from the simple (hydrated) metal ion. The drop time t depends largely upon the pressure on the dropping mercury and to a smaller extent upon the interfacial tension at the mercury–solution interface; the latter is dependent upon the potential of the electrode. Fortunately t appears only as the sixth root in the Ilkovic equation, so that variation in this quantity will have a relatively small effect upon the diffusion current. The product $m^{2/3} t^{1/6}$ is important because it permits results with different capillaries under otherwise identical conditions to be compared: the ratio of the diffusion currents is simply the ratio of the $m^{2/3} t^{1/6}$ values.

Unless the individual drops fall under their own weight when they are completely formed, the diffusion currents are not reproducible: stirring of the solution under investigation is therefore not permissible.

Polarographic maxima. Current–voltage curves obtained with the dropping mercury cathode frequently exhibit pronounced maxima, which are reproducible and which can be usually eliminated by the addition of certain appropriate 'maximum suppressors'. These maxima vary in shape from sharp peaks to rounded humps, which gradually decrease to the normal diffusion-current curve as the applied voltage is increased. A typical example is shown in Fig. XVI, 3. Curve A is that for copper ions in $0.1M$-potassium hydrogencitrate solution, and curve B is the same polarogram in the presence of 0.005 per cent acid fuchsine solution.

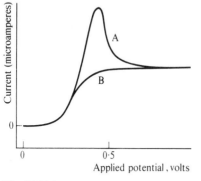

Fig. XVI, 3

To measure the true diffusion current, the maxima must be eliminated or supressed. Fortunately this can be done easily by the addition of a very small quantity of a surface-active substance, such as a dyestuff (as in the example above), gelatin, or other colloids. The function of any such maximum suppressor is probably to form an adsorbed layer on the aqueous side of the mercury–solution interface which resists compression; this prevents the streaming movement of the diffusion layer (which is believed to be responsible for the current maximum) at the interface.

Gelatin is widely used: the amount present in the solution should lie between 0.002 and 0.01 per cent; higher concentrations will usually suppress the diffusion current. Other maximum suppressors which are very effective are Triton X-100 (a non-ionic detergent), which is used at a concentration of 0.002–0.004 per cent, added in the form of a 0.1 per cent stock solution in water, and methyl cellulose (0.005 %).

Half-wave potentials. We are now in a position to appreciate the significance of half-wave potentials. The salient features of a typical current–

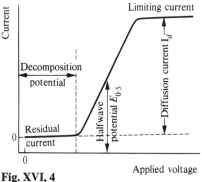

Fig. XVI, 4

applied voltage curve (*polarogram*) are shown in Fig. XVI, 4.

The conventional method of drawing the current–voltage curves is to plot the applied e.m.f. as abscissae reading in increasing negative values on the right: current is plotted as ordinates, cathodic currents (resulting from reduction) being regarded as positive and anodic currents negative. The height of the curve (*wave height*) is the diffusion current, and is a function of the concentration of the reacting material; the potential corresponding to the point of inflection of the curve (*half-wave potential*) is characteristic of the nature of the reacting material. This is the essential basis of quantitative and qualitative polarographic analysis.

The underlying theory may be simplified as follows. Polarography is concerned with electrode reactions at the indicator or micro-electrode, i.e., with reactions involving a transfer of electrons between the electrode and the components of the solution. These components are called *oxidants* when they can accept electrons, and *reductants* when they can lose electrons. The electrode is a *cathode* when a reduction can take place at its surface, and an *anode* when oxidation occurs at its surface. During the reduction of an oxidant at the cathode, electrons leave the electrode with the formation of an equivalent amount of the reductant in solution. Similarly, during the oxidation of a reductant at the anode electrons pass from the solution to the electrode and form an equivalent amount of the oxidant. Free electrons cannot exist in solution, consequently any process of reduction at the cathode is accompanied by a simultaneous oxidation. We may summarise the above discussion by the equation

$$\text{Oxidant} + n \text{ Electrons} \rightleftharpoons \text{Reductant}$$

or
$$\text{Ox} + ne \rightleftharpoons \text{Red} \tag{1}$$

The reductant differs from the oxidant merely by n electrons, and together they form an oxidation–reduction system. We will consider the reversible reduction of an oxidant to a reductant at a dropping mercury cathode. The electrode potential is given by:

$$E = E^\ominus + \frac{RT}{nF} \ln \frac{a_{\text{ox.}}}{a_{\text{red.}}} \tag{2}$$

where $a_{\text{ox.}}$ and $a_{\text{red.}}$ are the activities of the oxidant and reductant respectively *as they exist at the electrode surface* (henceforth called the electrode–solution interface and denoted by the subscript 's'), R is the gas constant, T the absolute temperature, n the number of electrons involved in the reaction, and F the Faraday constant; $E^\ominus$ is the electrode potential of the system when the activities of the oxidant and the reductant are equal. Polarographic measurements are seldom more accurate than ± 1 millivolt, hence substitution of concentrations for activities will not introduce any appreciable error. Equation (2) may therefore be written as:

$$E = E^\ominus + \frac{RT}{nF} \ln \frac{[\text{Ox}]_s}{[\text{Red}]_s} \tag{3}$$

639

Here $E^{\ominus}$ is the standard potential of the reaction against the reference electrode used to measure the potential of the dropping electrode, and the potential E refers to the average value during the life of a mercury drop. Before the commencement of the polarographic wave only a small residual current flows, and the concentration of any electro-active substance must be the same at the electrode interface as in the bulk of the solution. As soon as the decomposition potential is exceeded, some of the reducible substance (oxidant) at the interface is reduced, and must be replenished from the body of the solution by means of diffusion. The reduction product (reductant) does not accumulate at the interface, but diffuses away from it into the solution or into the electrode material. If the applied potential is increased to a value at which all the oxidant reaching the interface is reduced, only the newly formed reductant will be present; the current then flowing will be the diffusion current. The current I at any point on the wave is determined by the rate of diffusion of the oxidant from the bulk of the solution to the electrode surface under a concentration gradient $[Ox]$ to $[Ox]_s$.

$$I = K([Ox] - [Ox]_s) * \tag{4}$$

When $[Ox]_s$ is reduced to almost zero, equation (4) may be written:

$$I = K[Ox] = I_d \tag{5}$$

where I_d is the diffusion current. From equations (4) and (5), it follows that:

$$[Ox]_s = (I_d - I)/K \tag{6}$$

If the reductant (Red) is soluble in water and none was originally present with the oxidant, it will diffuse from the surface of the electrode to the bulk of the solution. The concentration of $[Red]_s$ at the surface at any value of I will be proportional to the rate of diffusion of the reductant from the surface of the electrode to the solution (under a concentration gradient $[Red]_s$) and hence also the current:

$$I = k[Red]_s * \tag{7}$$

If the reductant is insoluble in water but soluble in the mercury phase (amalgam formation), equation (7) still holds. Substituting in equation (3), we have:

$$E = E^{\ominus} - \frac{RT}{nF} \ln \frac{K}{k} + \frac{RT}{nF} \ln \frac{I_d - I}{I}$$

$$= E^{\ominus\prime} + \frac{RT}{nF} \ln \frac{I_d - I}{I} \tag{8}$$

where $E^{\ominus\prime} = (E^{\ominus} - K')$ and $K' = \dfrac{RT}{nF} \ln \dfrac{K}{k}$.

When I is equal to $I_d/2$, equation (8) reduces to:

$$E = E_{1/2} = E^{\ominus\prime} + \frac{RT}{nF} \ln \frac{I_d/2}{I_d/2} = E_0{}' \tag{9}$$

The potential at the point on the polarographic wave where the current is equal to one-half the diffusion current is termed the **half-wave potential** and is designated

---

* The constants K and k may be evaluated from the Ilkovic equation.

by $E_{1/2}$. It is quite clear from equation (9) that $E_{1/2}$ is a characteristic constant for a reversible oxidation–reduction system and that its value is independent of the concentration of the oxidant [Ox] in the bulk of the solution. It follows from equations (8) and (9) that at 25°:

$$E = E_{1/2} + \frac{0.0591}{n} \log \frac{I_d - I}{I} \tag{10}$$

This equation represents the potential as a function of the current at any point on the polarographic wave; it is sometimes termed the **equation of the polarographic wave**.

The half-wave potential is also independent of the electrode characteristics, and can therefore serve for the qualitative identification of an unknown substance. Owing to the proximity of many different half-wave potentials, its use for qualitative analysis is of limited application unless the number of possibilities is strictly limited by the nature of the unknown. The theoretical treatment for anodic waves is similar to the above.

It follows from equation (10) that when $\log (I_d - I)/I$ (where I is the current at any point on the polarographic wave minus the residual current) is plotted against the corresponding potential of the microelectrode (ordinates), a straight line should be obtained with a slope of $0.0591/n$ for a reversible reaction; the intercept of the graph upon the vertical axis gives the half-wave potential of the system. Hence n, the number of electrons taking part in the reversible reaction, may be determined. In applying equation (10), it is necessary to correct both I and I_d for the residual currents at the corresponding values of the applied potential and to correct the applied potential itself for any IR drop in the cell circuit. After these corrections have been made, both $E_{1/2}$ and the slope of the log plot are found to be independent of the concentration of the electro-active ion. Because it is concentration independent, the half-wave potential is generally preferred to the somewhat vague 'decomposition potential'. It also follows from equation (10) that the range of potentials over which the polarographic wave extends decreases with increasing values of n; thus the wave is steeper in the reduction of the trivalent aluminium or lanthanum ion than for the lead or cadmium ion, which in turn is steeper than that of an alkali metal ion or thallium(I) ion.

If the reaction at the indicator electrode involves **complex ions**, satisfactory polarograms can be obtained only if the dissociation of the complex ion is very rapid as compared with the diffusion rate, so that the concentration of the simple ion is maintained constant at the electrode interface. Let us consider the general case of the dissociation of a complex ion:

$$MX_p^{(n-pb)+} \rightleftharpoons M^{n+} + pX^{b-} \tag{11}$$

The instability constant may be written:

$$K_{\text{instab.}} = \frac{[M^{n+}][X^{b-}]^p}{[MX_p^{(n-pb)+}]} \tag{12}$$

(strictly, activities should replace concentrations).

We may imagine the electrode reaction to be (assuming amalgam formation):

$$M^{n+} + ne + Hg \rightleftharpoons M(Hg) \tag{13}$$

Combining (11) and (13), we have:

$$MX_p^{(n-pb)+} + ne + Hg \rightleftharpoons M(Hg) + pX^{b-} \tag{14}$$

It can be shown* that the expression for the electrode potential can be written:

$$E_{1/2} = E^{\ominus} + \frac{0.0591}{n} \log K_{\text{instab.}} - \frac{0.0591}{n} \log [X^{b-}]^{p} \tag{15}$$

Here p is the coordination number of the complex ion formed, X^{b-} is the ligand and n is the number of electrons involved in the electrode reaction. The concentration of the complex ion does not enter into equation (15), so that the observed half-wave potential will be constant and independent of the concentration of the complex metal ion. Furthermore, the half-wave potential is more negative the smaller the value of $K_{\text{instab.}}$, i.e., the more stable the complex ion. The half-wave potential will also shift with a change in the concentration of the ligand, and if the former is determined at two different concentrations of the complex forming agent, we have:

$$\Delta E_{1/2} = -p \cdot \frac{0.0591}{n} \times \Delta \log [X^{b-}] \tag{16}$$

This relationship enables one to determine the coordination number p of the complex ion and thus its formula.

It can also be shown that:

$$(E_{1/2})_c - (E_{1/2})_s = \frac{0.0591}{n} \log K_{\text{instab.}} - p \cdot \frac{0.0591}{n} \log [X^{b-}] \tag{17}$$

where $(E_{1/2})_c$ and $(E_{1/2})_s$ are the half-wave potentials of the complex and simple ions respectively at 25 °C, $K_{\text{instab.}}$ is the instability (or dissociation) constant and $[X^{b-}]$ is the concentration of the complexing agent X^{b-} in the body of the solution. It is assumed that the ligand is present in sufficiently large amount so that its concentration is practically the same at the surface of the dropping electrode as in the bulk of the solution. This formula may be employed to evaluate **the instability constant of the complex ion**: it involves merely the comparison of the half-wave potential at a given concentration of the complexing agent with that of the simple metal ion.

The shift of the half-wave potentials of metal ions by complexation is of value in polarographic analysis to eliminate the interfering effect of one metal upon another, and to promote sufficient separation of the waves of metals in mixtures to make possible their simultaneous determination. Thus in the analysis of copper-base alloys for nickel, lead, etc., the reduction wave of copper(II) ions in most supporting electrolytes precedes that of the other metals and swamps those of the other metals present; by using a cyanide supporting electrolyte, the copper is converted into the difficultly reducible cyanocuprate(I) ion and, in such a medium, nickel, lead, etc., can be determined.

XVI, 4. QUANTITATIVE TECHNIQUE. General considerations. Polarographic analysis is most conveniently carried out if the concentration of the electro-active substance is $10^{-4}–10^{-3}$ molar and the volume of the solution is between 2 and 25 cm^3. It is, however, possible to deal with concentrations as high as 10^{-2} molar or as low as 10^{-5} molar and to employ volumes appreciably less

---

* See any text-book on Polarography, e.g., Ref. 4.

than 1 cm^3. Under normal conditions (in particular, concentrations of 0.0001–0.001M) and with strict adherence to established technique, the reproducibility of duplicate analyses may be as good as ± 2 per cent.

Oxygen dissolved in electrolytic solutions is easily reduced at the dropping mercury electrode, and produces a polarogram consisting of two waves of approximately equal height and extending over a considerable voltage range; their position depends upon the pH of the solution, being displaced to higher voltages by alkali. The concentration of oxygen in aqueous solutions that are saturated with air at room temperature is about $2.5 \times 10^{-4}M$, consequently its polarographic behaviour is of considerable practical importance. A typical polarogram for air-saturated M-potassium chloride solution (in the presence of 0.01 per cent methyl red) is given in Fig. XVI, 5 (curve A). It has been stated that the first wave (starting at about -0.1 volt relative to S.C.E.) is due to the reduction of oxygen to hydrogen peroxide:

$$O_2 + 2H_2O + 2e = H_2O_2 + 2OH^- \text{ (neutral or alkaline solution)}$$

$$O_2 + 2H^+ + 2e = H_2O_2 \text{ (acid solution)}$$

The second wave is ascribed to the reduction of the hydrogen peroxide either to hydroxyl ions or to water:

$$H_2O_2 + 2e = 2OH^- \text{ (alkaline solution)}$$

$$H_2O_2 + 2H^+ + 2e = 2H_2O \text{ (acid solution)}$$

It is therefore necessary to remove any dissolved oxygen from the electrolytic solution whenever cathodic regions are being investigated in which oxygen interferes. This is easily accomplished by bubbling an inert gas (nitrogen or hydrogen) through the solution for about 10–15 minutes before determining the current–voltage curve. Curve B in Fig. XVI, 5 was obtained after the removal of the oxygen by unpurified nitrogen from a cylinder of the compressed gas. The gas stream must be discontinued during the actual measurements to prevent its stirring effect interfering with the normal formation of drops of mercury or with the diffusion process near the microelectrode. Commercial nitrogen or hydrogen derived from a cylinder of compressed gas (usually containing less than 0.05 per cent of oxygen) may be purified by passing through a Pyrex tube charged with copper gauze heated to about 450 °C and then through a wash bottle to saturate it with water vapour before passing through the test solution in the polarographic cell; the latter procedure minimises any change in volume of the test solution due to evaporation. It is more convenient, however, to use pure nitrogen (oxygen free), which can be purchased in cylinders.

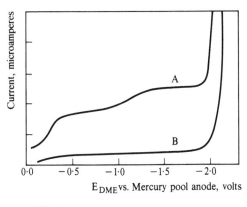

E_{DME} vs. Mercury pool anode, volts

Fig. XVI, 5

The influence of temperature has already been discussed. The electrolytic cell should be immersed in a thermostat bath maintained at ± 0.2 °C; for many purposes a temperature variation of ± 0.5 °C is permissible. A temperature of 25 °C is usually employed.

As a precautionary measure to prevent the appearance of maxima, sufficient gelatin to give a final concentration of 0.005 per cent should be added. The gelatin should preferably be prepared fresh each day; bacterial action usually appears after a few days. Other maximum suppressors (e.g., Triton X-100 and methyl cellulose) are sometimes used.

Two or more electro-active ions may be determined successively if their half-wave potentials differ by at least 0.4 volt for univalent ions and 0.2 volt for bivalent ions provided that the ions are present in approximately equal concentrations. If the concentrations differ considerably, the difference between the half-wave potentials must be correspondingly larger. If the waves of two ions overlap or interfere, various experimental devices may be employed. The half-wave potential of one of the ions may be displaced to more negative potentials by the use of suitable complexing agents which are incorporated in the supporting electrolyte; for example, Cu^{2+} ions may be complexed by the addition of potassium cyanide. Sometimes one ion may be removed by precipitation (e.g., with lead and zinc, the lead can be rendered harmless by precipitation as sulphate; the lead sulphate formed need not be removed by filtration); the possibilities of adsorption or co-precipitation of part of the other ions must, however, be borne in mind. Electrolytic separations are very useful and have the advantage over chemical separations of eliminating the necessity of adding large amounts of reagents whose presence during the subsequent steps of the procedure may be undesirable, or which may contain traces of the substance being determined and so give rise to an excessively high blank. Electrolytic methods often provide cleaner separations and are more rapid. Electrolysis of an acid solution of the mixture between a large stirred mercury cathode and a platinum anode may be used for the removal of many elements, including Fe, Cu, Ni, Mn, and Cr, which are reduced to the metallic state, from others such as Al, Ti, V, W, U, the alkaline-earth and alkali metals. Such electrolyses have been used as procedures in the polarographic determination of vanadium and of aluminium in steels and alloys. It is better to apply electrolysis with controlled potential at the mercury cathode whereby much finer separations can be made. Thus to determine nickel and zinc in a 'pure' copper salt, the latter can be dissolved in an ammoniacal ammonium chloride solution and electrolysed at about -0.7 volt *vs.* S.C.E.; copper is reduced to metal while nickel and zinc are unaffected and can be determined polarographically in the residual solution. As little as 0.00001 per cent nickel or zinc can be determined in this way.

XVI, 5.   EVALUATION OF QUANTITATIVE RESULTS.   Three methods which have been widely used in practice will be described.

Wave height–concentration plots.   Solutions of several different concentrations of the ion under investigation are prepared, the composition of the supporting electrolyte and the amount of maximum suppressor added being the same for the comparison standards and for the unknown. The heights of the waves obtained are measured in any convenient manner and plotted as a function of the concentration. The polarogram of the unknown is produced exactly as the standards, and the concentration is read from the graph. The method is strictly empirical, and no assumptions, except correspondence with the conditions of the calibration, are made. The wave height need not be a linear function of the concentration, although this is frequently the case. For results of the highest precision the unknown should be bracketed by standard solutions run consecutively.

Internal standard (pilot ion) method. The relative diffusion currents of ions in the same supporting electrolyte are independent of the characteristics of the capillary electrode and, to a close approximation, of the temperature. Hence upon determining the relative wave heights with the unknown ion and with some standard or 'pilot' ion added to the solution in known amount, and comparing these with the ratio for known amounts of the same two ions, previously determined, the concentration of the unknown ion may be computed. This procedure has limited application, primarily because only a small number of ions are available to act as pilot or reference ions. The main requirement for such an ion is that its half-wave potential should differ by at least 0.2 volt from the unknown or any other ion in the solution with which it might interfere. When a single unknown is present, this condition can usually be satisfied, but in complex mixtures there is seldom sufficient difference between the half-wave potentials to introduce additional waves.

Method of standard addition. The polarogram of the unknown solution is first recorded, after which a known volume of a standard solution of the same ion is added to the cell and a second polarogram is taken. From the magnitude of the heights of the two waves, the known concentration of ion added, and the volume of the solution after the addition, the concentration of the unknown may be readily calculated as follows. If I_1 is the observed diffusion current ($\equiv$ wave height) of the unknown solution of volume V cm^3 and of concentration C_u, and I_2 is the observed diffusion current after v cm^3 of a standard solution of concentration C_s have been added, then according to the Ilkovic equation we have:

$$I_1 = k\,C_u$$

and $$I_2 = k\,(V C_u + v C_s)(V + v)$$

Thus $$k = I_2(V + v)/(V C_u + v C_s)$$

whence $$C_u = \frac{I_1\, v\, C_s}{(I_2 - I_1)(V + v) + I_1 v}$$

The accuracy of the method depends upon the precision with which the two volumes of solution and the corresponding diffusion currents are measured. The material added should be contained in a medium of the same composition as the supporting electrolyte, so that the latter is not altered by the addition. The assumption is made that the wave height is a linear function of the concentration in the range of concentration employed. The best results would appear to be obtained when the wave height is about doubled by the addition of the known amount of standard solution.

XVI, 6. MEASUREMENT OF WAVE HEIGHTS. With a well-defined polarographic wave where the limiting current plateau is parallel to the residual current curve, the measurement of the diffusion current is relatively simple. In the exact procedure, illustrated in Fig. XVI, 6, the actual residual current curve is determined separately with the supporting electrolyte alone; by subtracting the residual current from the value of the current at the diffusion-current plateau (both measured at the same applied voltage), the diffusion current is obtained. It may be noted that when employing polarograms produced with a pen recorder, a line is drawn through the mid points of the recorder oscillations. For subsequent electroactive substances, the diffusion current would be evaluated by subtracting both the residual current and all preceding diffusion currents.

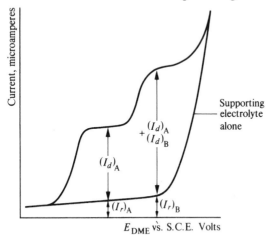

Fig. XVI, 6

It is simpler, though less exact, to apply the extrapolation method. The part of the residual current curve preceding the initial rise of the wave is extrapolated: a line parallel to it is drawn through the diffusion current plateau as shown in Fig. XVI, 7. For succeeding waves, the diffusion current plateau of the preceding wave is used as a pseudo–residual current curve.

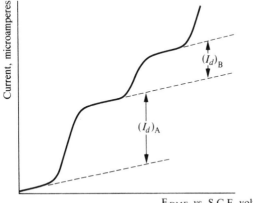

Fig. XVI, 7

When the wave is *slightly* distorted the following graphical method may be used. The distortion of the wave is shown somewhat exaggerated in Fig. XVI, 8. Draw the lines AB and CD perpendicularly to the abscissa axis, and divide these at F and G so that $AF = FB$ and $CG = GD$. The intersection of FG with the wave, i.e., at $E_{1/2}$, gives the position of the half-wave potential. Draw the vertical line HK through the point $E_{1/2}$: HK is the wave height or diffusion current. This method gives wave heights which are obviously too low when the polarogram is very much distorted. In practice, extensive distortion of polarograms can often be averted by application of a 'compensating' or 'counter' current: see Section **XVI, 8**.

XVI, 7. MANUAL NON-RECORDING POLAROGRAPHS. The essential requirements for measuring polarographic current–voltage curves are:
1. a means of applying a variable and known d.c. voltage ranging from 0 to 2 or 3 volts to the cell, and
2. a method for measuring the resulting current, which is usually less than about 50 microamperes.

The applied voltage should be known to about 1 millivolt or better, and the current measuring device should have a sensitivity of at least 0.01 microampere. The simple circuit shown in Fig. XVI, 9, may be used. The voltage applied to the cell is controlled by the potential divider R_1 (50- or 100-ohm radio-type potentiometer), which is powered by two 1.5-volt dry cells. The current is determined by measuring with the potentiometer the IR drop across the 10000-ohm precision fixed resistance R_2. By reversing the double-pole double-throw switch, the voltage applied to the cell may be measured with the same potentiometer: any potentiometer with a precision of ± 0.1 millivolt may be used.

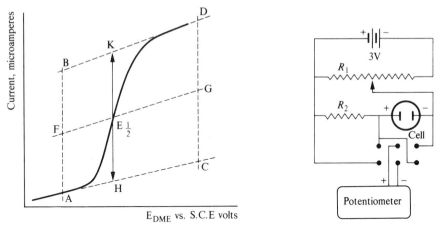

Fig. XVI, 8 | Fig. XVI, 9

The range of the potentiometer should be from 0 to 3 volts; if the range is 1.6 volts it can be extended to 0–3.2 volts by standardising against the Weston standard cell with the potentiometer set at one-half the voltage of the standard cell and then multiplying the observed readings by 2. The galvanometer used as a null-point detector in the potentiometer circuit should have a period three to four times longer than the drop time. The characteristics of the galvanometer should be approximately as follows: sensitivity (critically damped) 0.005–0.01 microampere per mm, critically damped period 3–5 seconds, internal resistance about 1000 ohms, and critical damping resistance about 7000 ohms; a resistance of 500–1000 ohms across the terminals will increase the period to the desirable value.

The current–voltage curve is obtained by increasing the applied voltage stepwise and measuring it with the potentiometer. The current is computed from Ohm's Law $I = E/R_2$. If $R_2 = 10000$ ohms, then each millivolt potential across R_2 corresponds to $10^{-3}/10^4$ amp or 0.1 microamp. To expedite the measurements, the entire voltage range of interest should first be explored using fairly large voltage increments, e.g., 0.2 volt, and measuring the current at each value of the applied voltage. The points should be plotted as the measurements are made so that the general character of the current–voltage curve becomes apparent: appropriate additional points required to define the rapidly changing parts of the curve may then be taken using small voltage increments, say, of 0.01 volt.

Manual polarographs can be purchased, and are useful for mastering the basic techniques and for understanding the principles involved, but they are necessarily somewhat slow to operate and consequently are rarely used for routine analytical procedures.

XVI, 8. COMMERCIAL POLAROGRAPHS.

In commercial polarographs, provision is made for carrying out the voltage scan automatically by continuous, steady adjustment of the potentiometer, and at the same time plotting the e.m.f. values and the corresponding currents on a chart recorder: the polarogram is thus obtained immediately and very rapidly. Many polarographs incorporate a 'counter current' control which applies a small opposing current and can be adjusted to compensate for the residual current; this leads to better defined polarograms. Although there are a number of instruments available which can be used for d.c. polarography, most modern instruments are also equipped to carry out operations which fall in the realm of a.c. polarography and will therefore be referred to in that connection. Recent developments in d.c. polarography are discussed in Ref. 10.

A useful feature of some commercial polarographs is the facility of plotting **derivative polarograms**, i.e., curves obtained by plotting dI/dE against E. Such curves show a peak at the half-wave potential, and by measuring the height of the peak it is possible to obtain quantitative data on the reducible substance; the height of the peak is proportional to the concentration of the ion being discharged.

A typical conventional polarogram for $0.003 M$-cadmium sulphate in $1 M$-potassium chloride in the presence of 0.001 per cent gelatin, and the corresponding derivative curve is shown in Fig. XVI, 10, $(I_G)_{max}$ is the maximum current recorded on the galvanometer in the derivative circuit.

The derivative circuit is used for measuring half-wave potentials closer than 0.15 volts, since with the direct wave, the residual current of the second

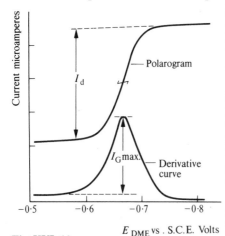

Polarogram

I_d

$I_G$max.

Derivative curve

−0·5   −0·6   −0·7   −0·8

Current microamperes

E_{DME} vs . S.C.E. Volts

Fig. XVI, 10

step is always affected by the ion previously discharged. Also in those cases where the element with the lower half-wave potential is present in much higher concentration, e.g., in the analysis of copper for cadmium content, the analysis is almost impossible without previous chemical separation. When the normal polarographic waves are differentiated and the rate of change of current with voltage is recorded (i.e., derivative polarograms) the series of peaks obtained in positions approximating to the half-wave potentials enable one to identify the individual elements. Fig. XVI, 11, illustrates the polarogram (A) obtained with copper and cadmium ions in the ratio of 40 to 1, and the corresponding derivative polarogram (B): the two peaks are clearly visible.

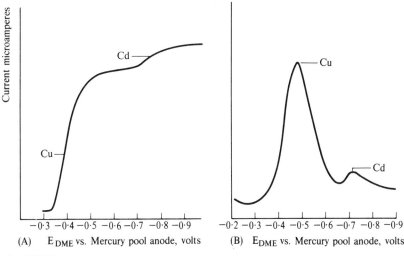

(A) E DME vs. Mercury pool anode, volts   (B) E$_{DME}$ vs. Mercury pool anode, volts

Fig. XVI, 11

Many modern polarographs provide for **potentiostatic control** of the dropping electrode potential: this is particularly valuable when solutions of high resistance are involved, as for example when non-aqueous solvents (or mixtures of water and organic solvents) are used. A high resistance in the polarographic cell leads to a large Ohm's law voltage drop ($I \times R$) across the cell, which not only influences the measured electrode potential but may also distort the polarogram: in extreme cases with some non-aqueous solvents, a straightforward I vs. E plot may appear to be virtually a straight line, and only after correction for the ohmic voltage drop is a normal polarogram obtained.

Potentiostatic control requires the introduction of a third electrode (a counter electrode) into the polarographic cell thus leading to an arrangement such as that depicted in Fig. XIII, 2 (Section **XIII, 2**). The 'cell' in that diagram is now the polarographic cell, the negative electrode is the dropping mercury electrode, S.C.E. is the reference electrode (saturated calomel or other suitable electrode), and the positive electrode is the counter electrode.

The reference electrode must be sited as close as possible to the D.M.E. so that the resistance of the solution between the two electrodes is reduced to a minimum, and the potentiostat then maintains the e.m.f. of the D.M.E.–reference electrode combination at the correct value. This arrangement has the further advantage that no sensible current is passed through the reference electrode and hence there is no possibility of polarisation of the latter with consequent variation in potential.

At least one manufacturer (Metrohm Ltd, Herisau, Switzerland) has developed a system of **rapid polarography** in which the flow of mercury through the dropping electrode is increased mechanically, and the voltage sweep rate increased in proportion, so that the polarogram is recorded in about one minute, i.e., in about one-tenth of the time normally required. Apart from the saving in time, this procedure has the result that much less damping is required in the recording circuit and in consequence, neighbouring waves are more clearly resolved.

Another feature available in some modern polarographs is the facility of **linear sweep polarography**. In the normal polarographic procedure described in the preceding sections, the total voltage sweep of the polarogram is spread out over a succession of several mercury drops, and the rate of change of potential may be of the order of 10 mV per second. In the linear sweep method the rate of potential change is greatly increased, up to say 500 mV per second, with the result that the polarogram is scanned within the lifetime of a single drop, and in fact, to minimise the effect of changing drop area, which is greatest when the drop is small and least near the end of the lifetime of the drop, the potential sweep is accomplished in a short interval just prior to the fall of the drop. This necessitates the use of a fast recording device, and in practice an oscilloscope is usually employed.

Owing to the rapid rise in potential of the D.M.E. the resulting current also increases rapidly, and initially it is not diffusion controlled. However, with the consequent discharge of the reducible ions near to the electrode surface, diffusion from the bulk solution sets in and the current then falls to the diffusion current level: in other words a peaked polarogram is obtained, somewhat similar in character to a derivative polarogram. This leads to better resolution than that achieved in a normal polarogram, and since the peak current is larger than the diffusion current, greater sensitivity results: it is claimed that concentrations down to $5 \times 10^{-7} M$ may be determined. Examples of the use of this technique are discussed in Ref. 16.

XVI, 9. ANCILLARY EQUIPMENT FOR POLAROGRAPHY. Mercury.

Doubly distilled mercury is usually recommended for polarographic work. The re-distilled mercury of commerce is generally satisfactory for most determinations; it should be filtered through a filter-paper cone with a small pinhole in the tip (or through a sintered-glass funnel) before use in order to remove any surface oxides or dust.

Used mercury should be washed with water, thoroughly agitated for about 12 hours in contact with 10 per cent nitric acid (a filter flask, arranged to admit air through the bottom of the mercury and connected to a water pump, is satisfactory), then thoroughly washed with distilled water, dried with filter paper, and re-distilled under reduced pressure.

CAUTION. Mercury vapour is a cumulative poison. All vessels containing mercury should be stoppered. Any spilled mercury should be immediately collected and placed in a flask containing water, and the bench (floor) dusted with powdered sulphur. Employ a tray under all vessels containing mercury and for all operations involving the transfer of mercury.

Dropping mercury electrode assembly. The assembly consists of a mercury reservoir (e.g., a 100-cm³ levelling bulb), a connecting-tube between the reservoir and the capillary tube, and a small glass electrolysis cell in which the unknown

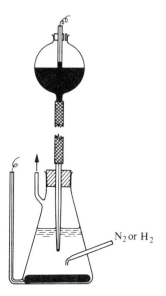

Fig. XVI, 12

N₂ or H₂

solution is placed. A simple arrangement is shown in Fig. XVI, 12. The heavy-walled rubber tubing, 80–100 cm long, should be sulphur-free. Neoprene tubing is generally employed; the inside surface should be steamed out for 30 minutes before use, followed by drying with air filtered through a cotton filter-plug. The electrolysis cell shown in the figure is the original type devised by Heyrovsky. Electrical connection to the mercury in the reservoir is affected by a platinum wire sealed into the end of a softglass tube, which is partly filled with mercury and held in place by the stopper of the reservoir.

The effective capillary tube has a length of 5–10 cm and a bore diameter of about 0.05 mm (range 0.04–0.07 mm); the outside diameter is usually about 6–7 mm: the delivery tip is cut accurately horizontal. Suitable capillary tubes may be purchased from any manufacturer of commercial polarographs. At a given pressure the drop time (which is the time that elapses between the fall of two successive drops) is directly proportional to the length of the capillary, but inversely to the third power of its internal radius; it is also inversely proportional to the pressure on the drop. A capillary suitable for use in polarography should have a length and bore such that the application of a pressure of about 50 cm of mercury will cause a drop weighing 6–10 mg to fall every 3–6 seconds when the top of the capillary is immersed in distilled water. The dropping mercury electrode must be mounted so that it is within $\pm 5°$ of the vertical; deviation from this angle produces erratic dropping.

With careful treatment, a capillary should remain serviceable for many months. It is absolutely essential that no solid matter of any kind should be allowed to reach the inside of the capillary. The electrode must never be allowed to stand in a solution when the mercury is not flowing.

The following procedure is recommended. The sample solution is deaerated, then, with the tip of the capillary in the air, the mercury pressure is raised at least 10 cm above the previously found equilibrium height, the capillary is inserted into the cell, and the mercury level is finally adjusted to the desired value. After the completion of the measurements the capillary is withdrawn from the cell and washed thoroughly with a stream of water from a wash bottle while the mercury is still issuing from the tip and is being collected in a micro beaker. The mercury reservoir is then lowered until the mercury flow *just ceases* (not further) and the electrode is allowed to stand in the air. It is good technique, at the beginning of each period of use, to immerse the capillary for *ca.* 1 min. in 1:1-nitric acid while mercury is flowing, then wash it well with distilled water: a further precaution is to allow the mercury drops to form in distilled water for about 15 minutes.

If the capillary becomes partly or completely blocked it is sometimes possible to clear it by carefully drawing strong nitric acid through it until the foreign matter has been completely dissolved, followed by distilled water to remove all traces of acid: the capillary is finally dried by drawing a stream of warm air (filtered through a cotton-wool plug) through it.

For reproducible results with the same dropping mercury electrode, it is important that the height of the mercury in the reservoir above the capillary tip should be constant, i.e., the same pressure on the dropping mercury tip be maintained; the small quantity of mercury passing through the capillary does not appreciably affect the volume in the reservoir. The stand supporting the electrode should permit the rapid immersion and removal of the capillary from the solution in the polarographic cell, particularly when the latter is in position in a thermostat. Suitable stands are available from the manufacturers of polarographs.

Polarographic cells. Numerous types of polarographic cells have been described and various forms are available commercially; the choice may well be dictated by the electrode stand in use.

The original Heyrovsky cell is depicted in Fig. XVI, 12, and can be readily constructed in the laboratory from a conical flask. The H-type cell devised by Lingane and Laitinen and shown in Fig. XVI, 13, will be found satisfactory for most purposes; a particular feature is the built-in reference electrode. Usually a

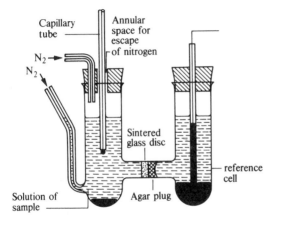

Fig. XVI, 13

saturated calomel electrode is employed, but if the presence of chloride ion is harmful a mercury(I) sulphate electrode (Hg/Hg_2SO_4 in potassium sulphate solution: potential $ca. +0.40$ volt $vs.$ S.C.E.) may be used. It is usually designed to contain $10–50$ cm^3 of the sample solution in the left-hand compartment, but it can be constructed to accommodate a smaller volume down to $1–2$ cm^3. To avoid polarisation of the reference electrode the latter should be made of tubing at least 20 mm in diameter, but the dimensions of the solution compartment can be varied over wide limits. The compartments are separated by a cross-member filled with a 4 per cent agar-saturated potassium chloride gel, which is held in position by a medium-porosity sintered Pyrex glass disc (diameter at least 10 mm) placed as near the solution compartment as possible in order to facilitate deaeration of the test solution. By clamping the cell so that the cross-member is vertical, the molten agar gel is pipetted into the cross-member and the cell is allowed to stand undisturbed until the gel has solidified.

In use, the solution compartment (either dried by aspiration of air through it or rinsed with several portions of the test solution) is charged with at least enough test solution to cover the entire sintered-glass disc. Dissolved air is removed by bubbling pure nitrogen through the solution via the side arm: by means of a two-way tap in the gas stream, the gas is then diverted over the surface of the solution. Measurements should not be attempted while gas is bubbling through the solution, for the stirring causes high and erratic currents. Finally, the dropping electrode is inserted through another hole in the stopper (which should be large enough for ease of insertion and removal of the capillary) and the measurements are made. When the H-cell is not in use the left-hand compartment should be kept filled either with water or with saturated potassium chloride solution (or other electrolyte appropriate to the reference electrode being used) to prevent the agar plug from drying out.

Maximum suppressors.   Gelatin is widely used as a maximum suppressor in spite of the fact that its aqueous solution deteriorates fairly rapidly, and must therefore be prepared afresh every few days as needed. Usually a 0.2 per cent stock solution is prepared as follows. Allow 0.2 g of pure powdered gelatin (the grade sold for bacteriological work is very satisfactory) to stand in 100 cm^3 of boiled-out distilled water for about 30 minutes with occasional swirling: warm the flask containing the mixture to about 70 °C on a water bath for about 15 minutes or until all the solid has dissolved. The solution must not be boiled or heated with a free flame. Stopper the flask firmly. This solution does not usually keep for more than about 48 hours. Its stability may be increased to a few days by adding a few drops of sulphur-free toluene or a small crystal of thymol, but the addition is rarely worth while and is not recommended.

A gelatin concentration of 0.005 per cent, which corresponds to 0.25 cm^3 of the stock 0.2 per cent solution in each 10 cm^3 of the solution being analysed, usually suffices to eliminate maxima. Higher concentrations (certainly not above 0.01 per cent) should not be used, since these will distort the wave form and decrease the diffusion current markedly.

Triton X-100, like gelatin, suppresses both positive and negative maxima, but, unlike gelatin, its aqueous solution is stable. A stock 0.2 per cent solution is prepared by shaking 0.20 g of Triton X-100 thoroughly with 100 cm^3 of water. About 0.1 cm^3 of this solution should be added to each 10 cm^3 of the sample solution to give a Triton X-100 concentration of 0.002 per cent.

XVI, 10.   DETERMINATION OF THE HALF-WAVE POTENTIAL OF THE CADMIUM ION IN M-POTASSIUM CHLORIDE SOLUTION.   The following experiments (Sections XVI, 10–XVI, 12), which can well be performed with a manual polarograph, serve to illustrate the general procedure to be followed in d.c. polarography.

Follow the operating instructions for the particular apparatus in use. Make sure that the reservoir of the dropping electrode contains an adequate supply of mercury, and that mercury drops freely from the capillary when the tip is immersed in distilled water whilst the reservoir is raised to near the maximum height of the stand: allow the mercury to drop for 5–10 minutes. Replace the beaker of water by one containing M-potassium chloride solution and adjust the rate of dropping by varying the height of the mercury reservoir until the dropping rate is 20–24 per minute: then clamp the mercury reservoir in position.

When the measurements have been completed (*vide infra*), rinse the capillary well with a stream of distilled water from a wash bottle and then dry by blotting with filter paper. Insert the capillary through an inverted cone of quantitative filter paper and clamp vertically over a small beaker. Lower the levelling bulb until the mercury drops just cease to flow.

Pipette 10 cm^3 of a cadmium sulphate solution (1.0 g Cd^{2+} dm^{-3}) into a 100-cm^3 measuring flask, add 2.5 cm^3 of 0.2 per cent gelatin solution, 50 cm^3 of $2M$-potassium chloride solution and dilute to the mark. The resulting solution (A) will contain 0.100 g Cd^{2+} dm^{-3} in a base solution (supporting electrolyte) of M-potassium chloride with 0.005 per cent gelatin solution as suppressor.

Measurements. Place 5.0 cm^3 of the solution A in a polarographic cell equipped with an external reference electrode (saturated calomel electrode). Pass pure nitrogen through the solution at a rate of about 2 bubbles per second for 10–15 minutes in order to remove dissolved oxygen. Raise the mercury reservoir to the previously determined height and insert the capillary into the cell so that the capillary tip is immersed in the solution. Connect the S.C.E. to the positive terminal and the mercury in the reservoir to the negative terminal of the polarograph. After about 15 minutes stop the passage of inert gas through the solution; the electrical measurements may now be commenced.

Carry out a preliminary test to ascertain the optimum position of the galvanometer shunt. Slowly turn the applied potential dial and depress the tapping key (or equivalent switch) at intervals. It will be found that at a certain point the current (as indicated by the deflection of the galvanometer spot) will increase rapidly. Decrease the galvanometer sensitivity by means of the shunt switch until the spot remains on the scale even at maximum applied potential; the latter should not, of course, exceed the decomposition potential of the supporting electrolyte.

Now, with the desired sensitivity in circuit and commencing from zero, increase the applied potential in suitable steps (say, in 0.05 volt) and read the galvanometer deflection for each value of the applied voltage. When the deposition potential of the cadmium ion is reached, the galvanometer deflection (i.e., current) increases rapidly, and smaller increments of the applied voltage (say, in 0.01 volt steps) are then advisable until the rate of change decreases considerably. It should be noted that the *maximum* deflection of the spot must be recorded. Plot the applied voltage (abscissae) against current (ordinates) as represented by the galvanometer deflections. (The actual current flowing at each value of the applied voltage may be easily computed, if desired, from the known sensitivity of the galvanometer, as

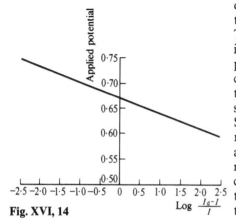

Fig. XVI, 14

determined by the manufacturers, and the position of the sensitivity switch.) The graph should have the form shown in Fig. XVI, 4. Determine the half-wave potential from the current–voltage curve as described in Section **XVI, 5**; the value in M-potassium chloride should be about -0.60 volts *vs*, the S.C.E. Alternatively, measure the maximum height of the diffusion wave after correction has been made for the residual current; this is the diffusion current I_d, and is proportional to the total concentration of cadmium ions in the solution.

Measure the height of the diffusion wave I, after correcting for the residual current at each increment of the applied voltage. Plot the values of $\log (I_d - I)/I$ as abscissae against applied voltage as ordinates (Fig. XVI, 14; strictly speaking, the values of the applied voltage are negative). Determine the slope of the graph, which should be equal to about 0.030, and read off the intercept on the voltage axis. The latter is the desired half-wave potential of the cadmium ion vs. S.C.E.

As an additional exercise, the current–voltage curve of the supporting electrolyte (M-potassium chloride) may be evaluated; this gives the residual current directly and no extrapolation is required for the determination of I and I_d.

XVI, 11.   DETERMINATION OF CADMIUM IN SOLUTION. Two procedures may be employed: (i) that dependent upon wave height–concentration plots, and (ii) the method of standard additions. The theory has been given in Section **XVI, 5**.

(i) **Wave height–concentration plots.** Prepare from the stock solution containing 1.000 g Cd^{2+} dm^{-3} solutions containing respectively 0.1, 0.05, 0.025, and 0.01 g Cd^{2+} dm^{-3} by transferring to 100-cm^3 graduated flasks 10, 5.0, 2.5, and 1.0 cm^3 of the stock solution, adding 50 cm^3 of $2M$-potassium chloride solution and 2.5 cm^3 of 0.2 per cent gelatin solution to each flask and then diluting to the mark with distilled water. Mix 10 cm^3 of the unknown solution (which may contain, say, about 0.04 g of cadmium dm^{-3}) in a 100-cm^3 measuring-flask with 50 cm^3 of $2M$-potassium chloride solution and 2.5 cm^3 of 0.2 per cent gelatin solution and dilute to the mark. Record the polarograms of the four standard solutions and of the unknown solution following the procedure described in Section **XVI, 10**, and determine the wave heights from each polarogram. Draw a calibration curve (wave heights as ordinates, concentrations as abscissae) for the four standard solutions: read off from the curve the concentration corresponding to the wave height of the unknown solution.

(ii) **Method of standard additions.** The polarogram of the unknown solution will have been determined under (i). A new polarogram must now be recorded after the addition of a known volume of a standard solution containing the same ion, care being taken that in the resulting solution the concentrations of the supporting electrolyte and the suppressor are maintained constant.

Place 10 cm^3 of the unknown solution, 5 cm^3 of the stock solution (1.000 g Cd^{2+} dm^{-3}), 50 cm^3 of $2M$-potassium chloride solution, and 2.5 cm^3 of 0.2 per cent gelatin solution in a 100-cm^3 graduated flask, and dilute to the mark with distilled water. Transfer a suitable volume to the polarographic cell in a thermostat, remove the dissolved air with nitrogen, and record the polarogram in the usual way. It is important that the galvanometer sensitivity be kept at the previous value. Determine the new wave height. Calculate the concentration of the unknown solution with the aid of the formula given in Section **XVI, 5**. Compare this value of the concentration with that found by method (i).

XVI, 12.   INVESTIGATION OF THE INFLUENCE OF DISSOLVED OXYGEN. The solubility of oxygen in water at the ordinary laboratory temperature is about 8 mg (or 2.5×10^{-4} mol) per litre. Oxygen gives two polarographic waves ($O_2 \longrightarrow H_2O_2 \longrightarrow H_2O$) which occupy a considerable voltage range, and their positions depend upon the pH of the solution. Unless the test solution contains a substance which yields a large wave or waves compared with which those due to oxygen are negligible, dissolved oxygen will interfere. In general, particularly in dilute solution, dissolved oxygen must be removed by passing pure nitrogen or hydrogen through the solution.

Place some M-potassium chloride solution containing 0.005 per cent gelatin in a polarographic cell immersed in a thermostat. Make the usual preliminary adjustments with regard to sensitivity control of the galvanometer, observe the current (galvanometer deflection) at increasing values of applied voltage, and plot the current–applied voltage curve. Now pass oxygen-free nitrogen through the solution for 10–15 minutes. Plot the polarogram using the same galvanometer sensitivity. It will be observed that the two oxygen waves are absent in the new polarogram (compare Fig. XVI, 5).

XVI, 13. DETERMINATION OF LEAD AND COPPER IN STEEL. In the application of the polarographic method of analysis to steel a serious difficulty arises owing to the reduction of iron(III) ions at or near zero potential in many base electrolytes. One method of surmounting the difficulty is to reduce iron(III) to iron(II) with hydrazinium chloride in a hydrochloric acid medium. The current near zero potential is eliminated, but that due to the reduction of iron(II) ions at about -1.4 volt *vs.* S.C.E. still occurs. Other metals (including copper and lead) which are reduced at potentials less negative than this can then be determined without interference from the iron. Alternatively, the Fe^{3+} to Fe^{2+} reduction step may be shifted to more negative potentials by complex ion formation.

The following procedure may be used for the simultaneous determination of copper and lead in plain carbon steels. Dissolve 5.0 g of the steel, accurately weighed, in a mixture of 25 cm³ of water and 25 cm³ of concentrated hydrochloric acid: heat gently to minimise the loss of acid. Add a few drops of saturated potassium chlorate solution to dissolve carbides, etc., and boil the mixture until the solution is clear. Cool and dilute to 50 cm³ with water in a graduated flask. Pipette 2.00 cm³ of this solution into a polarographic cell and add: 1.0 cm³ of 20 per cent hydrazinium chloride solution to reduce any iron(III) to iron(II) state, 1.0 cm³ of 0.2 per cent methyl cellulose to act as a maximum suppressor, and 5.5 cm³ of 2.0M-sodium formate solution to adjust the pH of the solution to that at which reduction of Fe(III) and Cu(II) ions takes place. Place the cell in a nearly boiling water bath for 10 minutes in order to complete the reduction. Cool. Analyse the solution polarographically: use a saturated calomel reference electrode. The first step in the polarogram is due to the reduction of copper(I) ions to the metal and has a half-wave potential of -0.25 volt *vs.* S.C.E. The second step, which is due to lead, has a half-wave potential of -0.45 volt *vs.* S.C.E. Carry out a calibration by adding known amounts of copper and lead to a solution of steel of low copper and lead content, and determine the increase in wave heights due to the additions.

Calculate the percentage of copper and of lead in the sample of steel.

Alternating current polarography

XVI, 14. THE NATURE OF a.c. METHODS. The use of alternating current in polarographic measurements has developed in two distinct ways: (*a*) by replacement of the direct current used in d.c. polarography by an alternating current; (*b*) by the introduction of an a.c. voltage into a polarographic circuit operating with direct current. Methods in which alternating current only is employed are referred to under the heading of **Oscillographic Polarography**, and the term **a.c. Polarography** is restricted to the combined use of a.c. and d.c.: it is in this field that the greatest advances in polarography have occurred in recent years (Ref. 5).

XVI, 15.   SIMPLE a.c. POLAROGRAPHY.   The fundamental features of a simple a.c. polarograph (Ref. 6) can be indicated by considering the modifications introduced into the basic circuit shown in Fig. XVI, 1 (Section **XVI, 2**). The galvanometer G and voltmeter V were removed, and resistance S was joined to the dropping mercury electrode through the secondary winding of a variable step-down transformer, so that when the transformer was switched on, the 50 Hertz mains voltage produced a resultant a.c. voltage of 1–100 millivolts which was superimposed on the d.c. current flowing in the polarograph. The measuring circuit, which was joined across S, consisted of a condenser (to suppress the d.c. current), an amplifier A, and a valve voltmeter B as in Fig. XVI, 15; the voltage reading on B coupled with the known value of the resistance S, allowed the value of the a.c. current to be calculated.

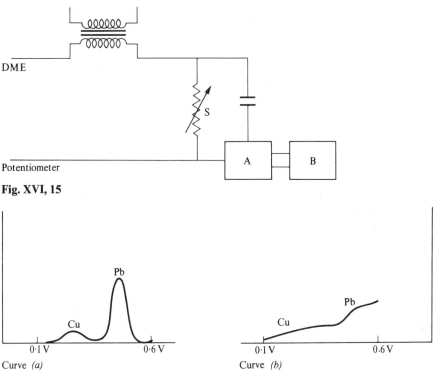

Fig. XVI, 15

Fig. XVI, 16

If the values of the a.c. current are plotted against the potential applied by the potentiometer, a series of peaks are obtained as illustrated in Fig. XVI, 16 (*a*): the normal d.c. polarogram of the same solution is also shown (curve *b*).

The a.c. curve is seen to be similar in character to a derivative polarogram (Figs. XVI, 10; XVI, 11, B: Section **XVI, 8**) but must not be confused with this type of curve. Each peak in the a.c. curve corresponds to a step in the normal

657

polarographic record. The voltage of the peak is the same as that of the midpoint of the step, and the height of a peak above the base line is proportional to the concentration of the depolariser, and thus corresponds to the step height. It will be apparent that with closely separated waves, measurements are much more readily made from the a.c. polarogram than from the d.c. polarogram, and it is considered that peaks separated by 40 mV can be resolved as compared with the separation of 200 mV required in d.c. polarography: the limit of sensitivity $(10^{-5}M)$ is, however, not greatly different from that achieved in d.c. polarography, and this is related to the fact that the residual current is rather large. This arises because the condenser current is relatively large as compared with the diffusion or faradaic current.

XVI, 16. SQUARE-WAVE POLAROGRAPHY.   Barker and Jenkins (Ref. 8) attempted to solve the problem arising from the large condenser current by replacing the normal a.c. sine wave current by a **square wave** current, which means that for a large part of each half cycle, the applied a.c. voltage is constant and the associated charging and faradaic currents rise rapidly to their maximum values. Subsequently each of these currents decays during the half cycle, but whereas the charging current decays virtually to zero, the faradaic current decreases slowly, and hence if the current is measured near the end of the half cycle (over the last 2×10^{-2} sec), the value recorded should be the faradaic current free from the effect of the charging current. This device for minimising the influence of the charging current is termed Tast polarography or strobe polarography. In fact, however, complications arise because the mercury drop is still increasing in area during the measurement period, and it was found that this could be compensated by modifying the completely square wave form into one with a slight downward slope on the upper edge and a corresponding upward slope on the lower edge. It was also found necessary to keep the resistance of the cell low: this could be achieved by increasing the concentration of the supporting electrolyte, which, however, needed to be very pure, otherwise trace impurities gave rise to complications. To avoid possible interaction arising from mains frequency harmonics, the square wave current was generated at a frequency of 225 Hz.

Elaborate electronic techniques were required to achieve the requisite wave form and frequency and the necessary constant periodicity of current measurement, and although concentrations as low as $5 \times 10^{-8}M$ could be determined in certain cases, problems were encountered in connection with the so-called 'capillary response' (Ref. 15). As each mercury drop falls from the capillary a little aqueous solution may be drawn into the tip of the capillary: the amount varies with each drop. The capacity current of this trapped liquid is found to be significant even at the end of the half cycle of the square wave voltage sweep thus upsetting the premise upon which the procedure is based. The procedure was accordingly modified by Barker and Gardner (Ref. 9) to give rise to what is termed pulse polarography.

XVI, 17. PULSE POLAROGRAPHY.   Barker and Gardner argued that given sufficient time, the disturbing effects due to the 'capillary response' would disappear, and if therefore the frequency of the square wave current were reduced, thus giving a longer period between measurements, the required result would be achieved. It was discovered, however, that the requisite reduction in frequency was so great as to be impracticable; it would need to be reduced from the 225 Hz of the square wave polarograph to 10 Hz.

The solution of this problem was found by replacing the square wave current by a series of potential pulses; one pulse of approximately 0.05 second duration being applied during the growth of a mercury drop, and at a fixed point near the end of the life of the drop. Two different procedures may, however, be employed: (*a*) pulses of increasing amplitude may be superimposed upon a constant d.c. potential, or (*b*) pulses of constant amplitude may be applied to a steadily increasing d.c. potential.

In method (*a*) the onset of the pulse is marked by a sudden rise in the total current passing: this is largely due to the condenser (charging) current which, however, soon decays to zero. The faradaic current also decays, but only to the level of the diffusion current, and if the current measurement is made in the last stages of the pulse (in the final 33 ms of its duration), it gives the faradaic current alone. The situation is thus similar to that encountered in square wave polarography, but the time scale is long enough for the disturbing effects of the 'capillary response' to be removed. The resulting polarogram is similar to a conventional d.c. polarogram except that the characteristic saw-tooth pattern of the latter is replaced by a stepped curve.

In method (*b*) the applied potential varies with time as shown in Fig. XVI, 17. The current is read near the end of each pulse, i.e., at points corresponding to C. When the faradaic current is small the pulse current will also be small, but as a normal d.c. polarographic wave sets in and the faradaic current rises, so too does the pulse current, and it will attain a maximum value at the half wave potential for the system under investigation. Consequently, the pulse current–d.c. voltage plot will be a peaked curve similar to a derivative polarogram of conventional d.c. polarography, and will in fact closely resemble a square wave polarogram.

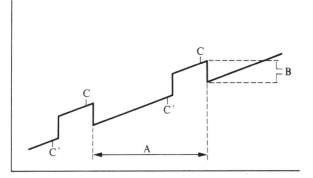

Fig. XVI, 17

For this reason, procedure (*b*), i.e., the application of pulses of constant amplitude to a steadily increasing d.c. voltage, is often referred to as **differential pulse polarography**. The alternative procedure (*a*) is consequently referred to as **normal pulse polarography** or as **integral pulse polarography**.

In apparatus produced by one supplier (Princeton Applied Research, Model 170 Electrochemistry System and also Model 174 Polarographic Analyser), the current is measured *twice* during the lifetime of each mercury drop: once just before the application of the pulse (points corresponding to C' in Fig. XVI, 17) as well as at the usual point C near the end of the pulse. The current at C' is that which would be observed in normal d.c. polarography; its value is stored in the instrument. The onset of the pulse is then marked by a sudden rise in current, which as in method (a) soon settles down as the condenser current decays, and near the end of the pulse (point C), the current is again read. This value is then compared with that stored in the instrument (the value for point C'), and the difference between them is amplified and recorded. Clearly, if the measurements of current are made on the residual current curve, or on a plateau of the d.c. polarogram, the difference in current between points C and C' will be small, but if the measurements are made on a polarographic wave, an appreciable current will be recorded, and it will of course reach a maximum value when the applied d.c. potential is equal to the half wave potential. The difference current plotted against the applied d.c. potential will therefore be a peaked curve with the height of the peak proportional to the concentration of the reducible substance in the solution, just as with a derivative polarogram. Differential pulse polarography is a very satisfactory method for the determination of many substances at the p.p.m. level.

An important feature of both square wave and pulse polarography is the sampling of the current at definite points in the lifetime of the mercury drop, and it is essential to establish an exact timing procedure. Various methods have been adopted to achieve the desired result: these include mechanical tapping of the capillary to dislodge the mercury drop at precise time intervals, or alternatively the natural drop time of the capillary is utilised, with the fall of the mercury drop serving to actuate the timing circuits which control the measurement of the current.

Oscillographic polarography

With the introduction of a.c. methods of polarography, the application of the cathode ray oscilloscope to polarographic investigations was an obvious further step. Developments have followed two main lines: investigations with controlled currents (this was the procedure adopted initially by Heyrovsky—Ref. 11), and investigations at controlled potentials (Ref. 12). Results given by the second procedure most closely resemble those given by other polarographic methods and will therefore be considered first.

XVI, 18. CONTROLLED POTENTIAL METHODS. Some of the problems encountered in simple a.c. and in square wave polarography are also found in this technique; in particular the existence of large charging or condenser currents, and it was Randles (Ref. 12) who first used the technique of a voltage pulse applied near the end of the lifetime of the mercury drop to overcome this difficulty. The voltage pulse covered the complete voltage range required for the particular polarogram, and for approximately two-thirds of its lifetime the mercury drop was allowed to grow with no voltage applied. Then over the remaining one-third of the life span, the complete voltage sweep was applied so that the maximum voltage was reached immediately before the drop fell. When the drop fell, the applied voltage simultaneously dropped back to zero and consequently, the applied voltage–time plot was of a saw tooth character.

The resultant display on the oscilloscope screen was somewhat similar to a conventional polarogram showing a maximum (Fig. XVI, 3, Section **XVI, 3**), and it was established that there is a linear relationship between the peak current and the concentration of the depolariser as shown in the **Randles–Sevcik equation**:

$$I_p = 2344\, n^{3/2}\, D^{1/2}\, m^{2/3}\, t^{2/3}\, \alpha^{1/2}\, C$$

where I_p is the peak current, α the rate of voltage sweep (volts per second), and the other terms have the same significance as in the Ilkovic equation (Section **XVI, 3**).

The replacement of the conventional polarogram by a peaked curve leads to increased sensitivity and better resolution than is achieved in normal d.c. polarography: the peak current is approximately $4 \times \sqrt{n}$ times the diffusion current of normal polarography, where n is the number of electrons involved in the electrode reaction.

Davis and co-workers (Ref. 13) further developed the above observations to produce a polarograph suitable for analytical purposes. A particular feature of this work is the development of a dual dropping electrode system. If the two electrodes, both dropping at the same rate, and with capillaries as near alike as possible, are placed in separate cells, then it is possible to measure the *difference* between the currents flowing in the two cells. If one cell contains the test solution, and the other contains the pure base electrolyte, then only the current due to the component to be determined in the test solution will be measured: this procedure is referred to as **subtractive (or differential) polarography**.

In **comparative polarography** a similar procedure is used but with the test solution in one cell and a reference solution in the second cell: this reference solution has the same base electrolyte as the test solution plus an accurately known amount of the substance to be measured. The measured current is now due to the difference in concentration of the substance to be determined in the test and reference solutions.

Derivative polarograms can be obtained by using a single dropping electrode in conjunction with the appropriate electrical circuit of the instrument, and if this same circuit is employed together with the dual electrode system, then the second derivative curve is obtained.

XVI, 19. CONTROLLED CURRENT METHODS. In this technique, a controlled current is passed through the cell and the variation of potential with time is recorded. In the early investigations of Heyrovsky and Forjet (Ref. 11), a streaming mercury electrode was employed so as to overcome the effect of the changing size of the mercury drop in a conventional dropping electrode, but the latter may be employed if arrangements are made to record the oscilloscope trace only towards the end of the lifetime of the drop in much the same way as measurements are made in pulse polarography.

The trace displayed on the oscilloscope may be either E vs. t, dE/dt vs. t or dE/dt vs. E; for quantitative purposes the last of these is generally found to be the most satisfactory. Using a base solution free from impurities the resultant trace is an ellipse, but in the presence of an electro-active substance, indentations appear on the ellipse giving the result shown in Fig. XVI, 18. The ellipse is symmetrical about the horizontal axis provided that the electrode reaction is reversible; the indentations occur at the half wave potential of the electrode reaction concerned. The depth of the indentations is a function of the concentration of the substance giving rise to them, and is most easily measured by employing a dual beam

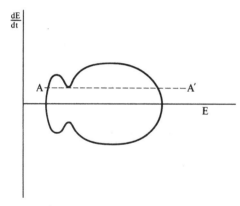

Fig. XVI, 18

oscilloscope with one beam utilised to produce the horizontal trace AA′ (Fig. XVI, 18). The height of this trace can be altered by application of a voltage derived from a potentiometer and a note is made of the voltage (E_1) required to displace AA′ from the axis of the ellipse to the position shown. If the experiment is repeated using a solution containing the depolariser at a different concentration, a different voltage (E_2) will be required to displace AA′ from the axis to the trough of the new indentation, and proceeding in this manner, a calibration curve can be constructed in which the voltage (E) is plotted against the concentration of depolariser: for further details see Ref. 14.

XVI, 20. INSTRUMENTATION.   A number of polarographs are available commercially which enable various aspects of a.c. polarography to be carried out, and which usually also make provision for d.c. polarography. Typical instruments include Bruker Electrospin Ltd, Model E 310 Universal Modular Polarograph (Fig. XVI, 19); the Princeton Applied Research Model 174 Polarographic Analyser and the more sophisticated Model 170 Electrochemistry

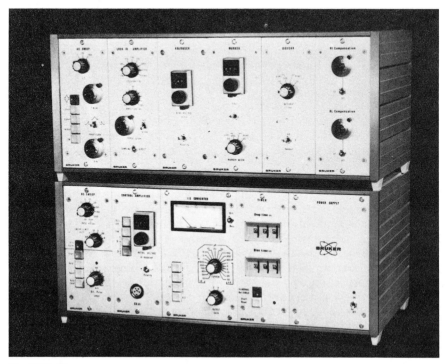

Fig. XVI, 19

System; the Metrohm 'Polarecord' used in conjunction with the a.c. modulator unit E 393. All such instruments incorporate, or they may be connected to, chart recorders so that the polarograms are recorded directly, and in many cases, connection can be made to an oscilloscope. The Davis Differential Cathode Ray Polarograph, Model A1660 (Fig. XVI, 20), supplied by Shandon Southern Instruments Ltd, contains a built-in cathode ray tube: it is designed specifically for controlled potential oscillographic polarography and is equipped with a dual dropping mercury electrode system as described in Section **XVI, 18**.

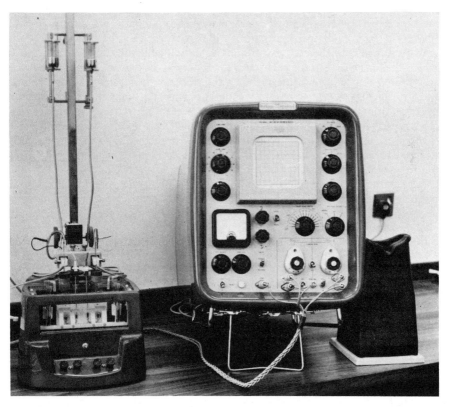

Fig. XVI, 20

XVI, 21. QUANTITATIVE DETERMINATIONS. Any of the determinations described under d.c. polarography (Sections **XVI, 11–13**) may be carried out by a.c. methods: the procedures available will, of course, be governed by the polarographic facilities available to the analyst. The general procedure to be followed will be similar to that described for d.c. polarography, with due observance of the operating instructions for the particular instrument in use. It is instructive to carry out one of the determinations (copper and lead in steel, Section **XVI, 13**, is particularly apt), by a number of the techniques, and to ascertain the concentration limits which can be determined by each one. Specific examples of the applications of differential pulse polarography are given in Ref. 23.

Anodic stripping voltammetry

XVI, 22.   BASIC PRINCIPLES.   If a conventional d.c. polarographic system is set up, with an oxygen-free solution containing one or more ions reducible at the DME, and with the height of the mercury column of the electrode carefully reduced until dropping has ceased, so that a single mercury drop is left attached to the capillary, we have a **Hanging Mercury Drop Electrode** (HMDE) system. If the potentiometer of the polarograph is then set to a *fixed value* which is chosen to be 0.2–0.4 V more negative than the highest reduction potential encountered amongst the reducible ions, then electrolysis will occur, deposition of metals will take place on the HMDE cathode and usually, amalgam formation will take place. The rate of amalgam formation will be governed by the magnitude of the current flowing, by the concentrations of the reducible ions, and by the rate at which the ions travel to the electrode; the latter can be controlled by stirring the solution. Given sufficient time, virtually the whole of the reducible ion content of the solution may be transferred to the mercury cathode, but complete exhaustion of the solution is not really necessary for the present procedure, and in practice, electrolysis is carried out for a carefully controlled time interval so that a fraction (say 10 per cent) of the reducible ions are discharged. This operation is often referred to as a **concentration step**; the metals become concentrated into the relatively small volume of the mercury drop.

If the connections to the cell are now reversed, and the potentiometer of the polarograph set to its lowest range and then allowed to vary in the normal manner (the potentiometer should be motor driven to ensure a steady rate of change of potential), then a gradually increasing positive potential is applied to the HMDE which is now the anode of the cell. If the current is measured and plotted against the anodic voltage (a recorder is used), then initially a gradually increasing current corresponding to the residual current of conventional polarography, and due mainly to the ground solution, is observed. As the anodic potential approaches the oxidation potential of one of the metals dissolved in the mercury, then ions of that metal pass into solution from the amalgam and the current increases rapidly and attains a maximum value when the anodic potential has a value approximating to the appropriate oxidation potential. The metal is said to be **stripped** from the amalgam, and if the potential were held at the value corresponding to the maximum current, all of the metal would eventually be returned to the solution. In actual fact, however, the potential is not held stationary, and as the potential sweep continues, the current declines from its maximum value and settles down to a new approximately steady value; in other words, the curve shows a peak. With continuing rise in the anodic potential, fresh peaks will be produced in the curve as the oxidation potentials of the different metals contained in the amalgam are reached: by analogy with polarogram, the resulting curve is termed a **voltammogram** (or **stripping voltammogram**).

The peaks are characterised by the **peak potential** E_p, by the **peak current** I_p (i.e., the height of the peak) and by the **breadth** b (i.e., the voltage span of the peak at the point where the current is $0.5 I_p$): these parameters are, however, dependent upon characteristics of the electrode and upon the rate of the voltage sweep during the stripping process. The magnitude of the peak current is proportional to the concentration in the amalgam of the metal being stripped, and is therefore proportional to its concentration in the original solution.

From the nature of the process described above it has been referred to as

inverse polarography and also as **stripping polarography**, but the term **anodic stripping voltammetry** is preferred (Ref. 17). It is also possible to reverse the polarity of the two electrodes of the cell thus leading to the technique of cathodic stripping voltammetry.

In just the same way as differential pulse polarography represents a vast improvement over conventional polarography (see Section **XVI, 17**), the application of a pulsed procedure leads to the greatly improved technique of **differential pulsed anodic (cathodic) stripping voltammetry**. A particular feature of this technique is that owing to the much better resolution and greater sensitivity which is achieved, the concentration of metals in the HMDE can be reduced and consequently, the time needed for the concentration step can be cut considerably.

The technique can be used to measure concentrations in the range 10^{-6}–$10^{-9}M$ and as such is eminently suitable for the determination of trace metal impurities; of recent years it has found application in the analysis of semi-conductor materials (Ref. 18) and in the investigation of pollution problems.

XVI, 23. SOME FUNDAMENTAL FEATURES. In view of the limitations referred to above, and particularly the influence of electrode characteristics upon the peaks in the voltammogram, some care must be exercised in setting up an apparatus for stripping voltammetry. The optimum conditions require

(a) in the concentration step: a small mercury volume by comparison with the volume of solution to be electrolysed; efficient stirring of the solution during the electrolysis, otherwise the deposition procedure may be unduly prolonged.

(b) in the stripping operation, as fast a voltage scan as possible consonant with the avoidance of peak tailing.

Electrodes. The **Hanging Mercury Drop Electrode** is traditionally associated with the technique of stripping voltammetry and its capabilities were investigated by Kemula and Kublik (Ref. 19). In view of the importance of drop size it is essential to be able to set up *exactly* reproducible drops, and this can be done by attaching the capillary tube at the end of which the drops are to be produced, to a small reservoir containing mercury and from which mercury can be expelled in controllable quantities by means of a plunger operated by a micrometer screw gauge. Problems which may be encountered with the HDME include: (i) diffusion of the amalgam formed in the concentration process into the capillary, with the result that it is not readily available during the stripping process and the relationship between peak current and concentration breaks down; (ii) possible penetration of aqueous solution into the capillary, which may lead to the drop breaking away; this eventuality may to some extent be guarded against by giving the bore of the capillary a coating of silicone before filling it with mercury. Both of these problems can also be mitigated by using a Sessile Mercury Drop Electrode (SMDE); in this variation, the end of the capillary tube is bent upwards through an angle of $180°$ so that the mercury drop sits on top of the capillary instead of hanging from the end.

In an alternative technique, an HDME is set up on the end of a short platinum wire sealed into a glass tube. The wire is first thoroughly cleaned and is then used as anode for the electrolysis of a solution of pure perchloric acid: this treatment exerts a polishing effect. The current is then reversed, and the electrode used as cathode to ensure that no oxide or adsorbed oxygen films remain on the surface of the electrode. Still using it as cathode, the electrode is now used for the electrolysis of mercury(II) nitrate solution, and it thus becomes plated with mercury. A counted number of mercury drops from a conventional dropping mercury electrode can then be attached to the platinum wire.

If the same procedure is carried out using a longer platinum wire, and the electrolysis conditions (current and time) with the mercury(II) nitrate solution carefully controlled, then a Mercury Film Electrode (MFE) is produced. This is claimed to possess advantages related to its rigidity and also to the greater surface area/volume ratio as compared with a mercury drop. With both the electrodes based on platinum wires, however, should there be any bare platinum areas which have escaped plating in contact with the solution, then complications may arise owing to the smaller hydrogen overpotential on platinum than on mercury, and metals having high positive electrode potentials may fail to deposit when the electrode is used for the concentration step.

Of recent years the use of mercury film electrodes based on substrates other than platinum has been explored, and increased sensitivity is claimed for electrodes based on wax-impregnated graphite, on carbon paste and on vitreous carbon: a technique of simultaneous deposition of mercury and of the metals to be determined has also been developed. For further details the review article (Ref. 10) may be consulted.

Cells. The cell employed can be a suitable polarographic cell, or can be specially constructed to fulfill the following requirements. Efficient *reproducible* stirring of the solution is essential, and for this purpose a magnetic stirrer is usually suitable. Exclusion of oxygen is important, and so the cell must be provided with a cover, and provision made for passing pure (oxygen-free) nitrogen through the solution before commencing the experiment, and over the surface of the liquid during the determination. The cover of the cell must provide a firm seating for the HMDE (or other type of electrode used), and must also have openings for the reference electrode (usually a SCE) and for a platinum counter electrode if it is required to operate under conditions of controlled potential (see Section **XVI, 8**). If the solution under investigation is to be analysed for mercury, then the reference electrode should be isolated from the solution by use of a salt bridge.

Reagents. In view of the sensitivity of the method, the reagents employed for preparing the ground solutions must be very pure, and the water used should be redistilled in an all-glass apparatus: the traces of organic material sometimes encountered in demineralised water (Section **III, 21**) make such water unsuitable for this technique. The common supporting electrolytes include potassium chloride, sodium acetate–acetic acid buffer solutions, ammonia–ammonium chloride buffer solutions, hydrochloric acid and potassium nitrate.

The normal A.R. grade chemicals often contain trace impurities which are quite unimportant for most analytical purposes, but in terms of stripping voltammetry may represent serious contamination: this is especially true if heavy metals are involved. It is therefore necessary to employ reagents of very high purity (e.g., the B.D.H. 'Aristar' reagents or similar grade), or alternatively to subject the purest material available to an electrolytic purification process. The stirred solution is electrolysed with a small current (10 mA) for twenty-four hours, using a pool of mercury at the bottom of a beaker as cathode and a platinum anode; pure nitrogen is passed through the solution before commencing the electrolysis so as to remove dissolved oxygen, and during the purification process, a current of pure nitrogen is maintained over the surface of the solution. It will usually be necessary to use some form of potentiostatic control during the electrolysis process. When electrolytic purification of reagents needs to be undertaken on a routine basis it may be considered advisable to make use of commercially available apparatus such as, for example, The Princeton Applied Research Model 9500 Electrolyte Purification Apparatus.

In view of the foregoing remarks, it is clear that all glassware used in the preliminary treatment of samples to be subjected to stripping voltammetry, as well as the apparatus to be used in the actual determination, must be scrupulously cleaned. It is usually recommended that glassware be soaked for some hours in *pure* nitric acid ($6M$), or in a 10 per cent solution of *pure* 70 per cent perchloric acid.

XVI, 24. INSTRUMENTATION. Many of the more sophisticated polarographs referred to in Section **XVI, 20** are suitable for carrying out stripping voltammetry and some manufacturers supply equipment specifically designed for this technique: amongst these may be mentioned the 'ElectRoCell-ASV' apparatus supplied by McKee-Pedersen Instruments which incorporates a rotated cell and by this means achieves the controlled stirring of the solution which is required, and the Environmental Sciences Associates Inc. Anodic Stripping Voltameter. The latter can be purchased in single cell (Model SA2011) and multi-cell (Model 2014) versions: the standard (20 mm) cell holds a working volume of about 5 cm^3 and stirring is accomplished by a stream of pure nitrogen,

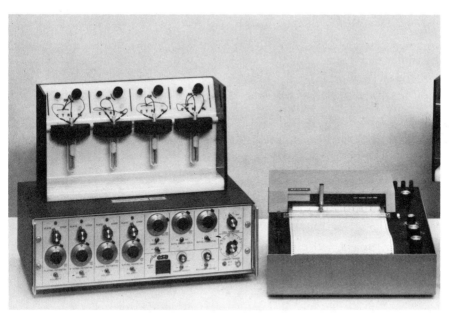

Fig. XVI, 21

whilst the standard electrode assembly consists of a mercury-coated, wax-impregnated graphite rod, a Ag/AgCl reference electrode and a platinum counter electrode. The multiple cell model (2014) together with a four-cell holder is shown in Fig. XVI, 21.

Chronopotentiometry

XVI, 25.   BASIC PRINCIPLES.   If an unstirred ground solution containing a small amount of a depolariser is electrolysed at constant current using a cell provided with a working electrode, a counter electrode, and a reference electrode (cf. Fig. XIII, 2; Section **XIII, 2**), and the potential of the working electrode is plotted against time, then the resultant curve is similar to a conventional polarogram (Fig. XVI, 2; Section **XVI, 2**). The curve is referred to as a **chronopotentiogram** and its shape can be explained as follows.

As shown in equation (3) (Section **XVI, 3**), the potential of a cathode at which a reversible reduction reaction is involved

$$Ox + ne \rightleftharpoons Red$$

is given by

$$E = E^{\ominus} + RT/nF \ln [Ox]/[Red].$$

As soon as a small amount of reductant has been produced, the ratio $[Ox]/[Red]$ only changes relatively slowly, and hence the potential of the electrode changes only gradually with respect to time. As electrolysis proceeds, however, the concentration of the oxidant adjacent to the electrode decreases, and although it is to a certain extent replenished by diffusion, if the magnitude of the electrolysis current is high enough, a situation is reached in which the concentration of the oxidant in the electrode layer is virtually reduced to zero. The conditions for a fixed electrode potential no longer apply, and the potential changes rapidly to a value at which a new electrode reaction is possible. The time from the commencement of electrolysis to the rapid change in potential is termed the **transition time** τ, and it was shown by Sand (Ref. 20) that this is related to the concentration of the electro-active species by the expression

$$\tau^{1/2} = \frac{\pi^{1/2} n F A D^{1/2} C_0}{2I}$$

where n = number of electrons involved in the reduction reaction,

A = surface area of electrode,

D = diffusion coefficient of the electro-active species involved,

C_0 = initial concentration of the depolariser,

I = the constant electrolysis current.

The variation of electrode potential with time can be expressed by the equation

$$E = E^{\ominus} + \frac{RT}{nF} \ln \frac{t^{1/2}}{\tau^{1/2} - t^{1/2}}$$

and when $t = \tau/4$, the potential $E_{\tau/4}$ (the **quarter transition time potential**) $\simeq E^{\ominus} \simeq E_{1/2}$ (the polarographic half wave potential).

It follows from these two equations that, like polarography, chronopotentiometry has potentialities for use in both quantitative and qualitative analysis, but from the quantitative viewpoint it is a less sensitive technique than others already described in this chapter, and the lower limit of concentration which can be measured with reasonable accuracy is about $10^{-4} M$. Complications also arise when the solution under investigation contains more than one reducible species. For the first one to be reduced, the transition time (τ_1) will be given as above by

$$\tau_1^{1/2} = \frac{\pi^{1/2} n_1 FA D_1^{1/2} C_1}{2I}$$

but the transition time (τ_2) for reduction of the second species is given by

$$(\tau_1 + \tau_2)^{1/2} - \tau_1^{1/2} = \frac{\pi^{1/2} n_2 FA D_2^{1/2} C_2}{2I}$$

and if a multivalent ion is reduced in two stages involving respectively n_1 and n_2 electrons, then

$$(\tau_1 + \tau_2)/\tau_1 = (n_1 + n_2)^2/n_1^2.$$

This interdependence of transitions times is clearly disadvantageous as compared with the normal polarographic situation where the height of any wave is proportional to the concentration of the species giving rise to it and is quite independent of the height of the preceding waves.

The fact that there is no periodicity in the measurements such as that associated with the use of a DME is sometimes regarded as a point in favour of chronopotentiometry, but on the whole, although the technique has been applied to quantitative determinations (Ref. 21), its main uses lie in the investigation of the kinetics of electrode reactions, in confirming (or disproving) their reversibility, in ascertaining the number of electrons involved in the reaction, and in establishing the formulae of complexes (Ref. 22).

XVI, 26. EXPERIMENTAL PROCEDURE.

The working electrode is commonly a pool of mercury, but it can be a hanging mercury drop, a wax-impregnated graphite rod, or a platinum disc sealed into a glass tube and situated a few millimetres from the end of the tube: this arrangement helps to prevent convection effects. The reference electrode (most commonly a SCE or a Ag/AgCl electrode) is clamped firmly in position and is provided with a drawn-out tip which is situated close to the working electrode. The auxiliary (counter) electrode (a small platinum plate or a wire helix) is usually separated from the solution to be analysed by placing it inside a glass tube fitted at the end with a sintered glass diaphragm; this tube contains the pure ground solution.

A polarographic cell similar to that depicted in Fig. XVI, 13 (Section **XVI, 9**) may be used but with the counter electrode placed in the right-hand compartment, the DME replaced by the reference electrode, and an electrical connection provided to the mercury pool at the bottom of the left-hand compartment: this is the working electrode. Provision must also be made (as in the cell depicted) for passing pure nitrogen through the solution to remove dissolved oxygen before commencing the electrolysis.

The constant current for the electrolysis can be obtained from a battery in series with a variable high resistance, but is best produced by a commercially available amperostat such as the Solartron, Beckman 'Electroscan', McKee-Pedersen, Princeton Applied Research, etc.

Use of a chart recorder enables the voltage–time plot to be produced directly and it is usually considered that the current should be adjusted so that the transition times lie between 10 and 100 seconds; the procedure described in

Section **XVI, 6** can be employed for precise evaluation of the transition times from the graph.

XVI, 27. References

1. D. R. Crow and J. V. Westwood (1968). *Polarography*. London; Methuen and Co. Ltd.
2. J. Heyrovsky (1922). *Chemicke Listy.*, **16**, 256.
3. J. Heyrovsky and M. Shikata (1925). *Rec. trav. chim.*, **44**, 496.
4. D. R. Crow (1969). *Polarography of Metal Complexes*. London; Academic Press.
5. J. B. Flato (1972). 'The Renaissance in Polarographic and Voltammetric Analysis' (review article). *Anal. Chem.*, **44**, 75A.
6. B. Breyer, F. Gutman and S. Hacobian (1950). *Australian J. Sci. Res.*, **A3**, 558.
7. D. Ilkovic (1934). *Coll. Czech. Chem. Comm.*, **6**, 498.
8. G. C. Barker and I. L. Jenkins (1952). *Analyst*, **77**, 685.
9. G. C. Barker and A. W. Gardner (1960). *Z. anal. Chem.*, **173**, 79.
10. B. Fleet and R. D. Jee (1973). *Specialist Periodical Reports, Electrochemistry*. Vol. 3. London; The Chemical Society.
11. J. Heyrovsky and J. Forjet (1943). *Z. Phys. Chem.*, **193**, 77.
12(a). J. E. B. Randles (1947). *Analyst*, **72**, 301.
12(b). J. E. B. Randles (1948). *Trans. Far. Soc.*, **44**, 327; 334.
13(a). H. M. Davis and J. E. Seaborn (1953). *Electronic Engineering*, **26**, 314.
13(b). H. M. Davis and J. E. Seaborn (1960). *Advances in Polarography*. Vol. 1. Oxford; Pergamon Press.
14(a). R. Kalvoda (1965). *Techniques of Oscillographic Polarography*. Amsterdam; Elsevier.
14(b). R. Kalvoda, W. Anstine and M. Heyrovsky (1970). *Anal. Chim. Acta.*, **50**, 93.
15(a). G. C. Barker, R. L. Faircloth and A. W. Gardner (1958). *Nature*, **181**, 247.
15(b). B. S. Bruk and B. M. Sternberg (1970). *Zavod. Lab.*, **36**, 365.
16(a). P. E. Toren (1968). *Anal. Chem.*, **40**, 1152.
16(b). G. C. Whitnack and R. G. Brophy (1969). *Anal. Chim. Acta.*, **48**, 123.
16(c). C. E. Plock and J. Vasquez (1971). *Anal. Chim. Acta.*, **55**, 278.
17. Information Bulletin (1973). Appendices on Tentative Nomenclature, Symbols, Units and Standards; No. 30. *Classification and Nomenclature of Electroanalytical Techniques*. Oxford; IUPAC.
18. P. F. Kane and G. B. Larrabee (1970). *Characterisation of Semiconductor Materials*. New York; McGraw-Hill Book Co.
19. W. Kemula and Z. Kublik (1958). *Anal. Chim. Acta.*, **18**, 104.
20. H. J. S. Sand (1901). *Phil. Mag.*, **1**, 45.
21. D. G. Davis in A. J. Bard (1966). *Electroanalytical Chemistry*. Vol. 1. New York; M. Dekker Inc.
22. J. B. Headridge (1969). *Electrochemical Techniques for Inorganic Chemists*. London; Academic Press.
23. R. S. Nicholson (1972). 'Polarographic Theory, Instrumentation and Methodology' (review article). *Anal. Chem.*, **44**, 478R.

XVI, 28. Selected bibliography

1. D. E. Burge (1970). 'Pulse Polarography', *J. Chem. Ed.*, **47**, A81.
2. D. E. Smith. *a.c. Polarography*, in A. J. Bard (Ref. 21).
3. A. J. Bard (1967). *Electroanalytical Chemistry*. Vol. 2. (*Stripping Voltammetry; Oscillographic Polarography*). New York; M. Dekker Inc.
4. J. Heyrovsky and I. Kuta (1966). *Principles of Polarography*. New York; Academic Press.

5. G. Charlot, J. Badoz-Lambling and B. Tremillon (1962). *Electrochemical Reactions.* Amsterdam; Elsevier.
6. J. Heyrovsky and P. Zuman (1968). *Practical Polarography.* New York; Academic Press.
7. O. P. Bhargava, W. D. Lord and W. G. Hines (Sept./Oct. 1975). *Anodic stripping voltammetric determination of lead and zinc in iron and steelmaking materials. International Laboratory.* Fairfield, Conn.; International Scientific Communications Inc.

CHAPTER XVII **AMPEROMETRY**

XVII, 1. AMPEROMETRIC TITRATIONS. It has been shown in the previous Chapter (Section **XVI, 3**; Fig. XVI, 4) that the limiting current is independent of the applied voltage impressed upon a dropping mercury electrode (or other indicator micro-electrode). The only factor affecting the limiting current, if the migration current is almost eliminated by the addition of sufficient supporting electrolyte, is the rate of diffusion of electro-active material from the bulk of the solution to the electrode surface. Hence the diffusion current (= limiting current − residual current) is proportional to the concentration of the electro-active material in the solution. If some of the electro-active material is removed by interaction with a reagent, the diffusion current will decrease. This is the fundamental principle of amperometric titrations. The observed diffusion current at a suitable applied voltage is measured as a function of the volume of the titrating solution: the end point is the point of intersection of two lines giving the change of current before and after the equivalence point.

It may be noted that when during a titration the potential is measured between an indicator electrode and a reference electrode, the titration is termed a potentiometric one; here it is important to measure the potential relatively accurately near the end point, the latter being characterised by a maximum of the differential $\Delta E/\Delta v$, the rate of change of potential, say, per $0.1\,\text{cm}^3$. In conductometric titrations the electrical conductivity of the solution is measured during the titration, and the end point is found graphically as the point of intersection of two straight lines giving the change of conductivity before and after the equivalence point (compare high-frequency titration methods). In amperometric titrations (derived from *ampere*, the unit of current) the current which passes through the titration cell between an indicator electrode (e.g., the dropping mercury electrode) and the appropriate depolarised reference electrode (e.g., the saturated calomel electrode) at a suitable applied e.m.f. is measured as a function of the volume of the titrating solution. Such titrations have also been termed polarographic and polarometric; the term **amperometric titration** is now recommended.

Some advantages of amperometric titrations may be mentioned:

1. The titration can usually be carried out rapidly, since the end-point is found graphically; a few current measurements at constant applied voltage before and after the end point suffice.
2. Titrations can be carried out in cases in which the solubility relations are such that potentiometric or visual-indicator methods are unsatisfactory; for

example, when the reaction product is markedly soluble (precipitation titration) or appreciably hydrolysed (acid-base titration). This is because the readings near the equivalence point have no special significance in amperometric titrations. Readings are recorded in regions where there is excess of titrant, or of reagent, at which points the solubility or hydrolysis is suppressed by the mass-action effect; the point of intersection of these lines gives the equivalence point.

3. A number of amperometric titrations can be carried out at dilutions (*ca.* $10^{-4}M$) at which visual or potentiometric titrations no longer yield accurate results. (It must be noted, however, that high frequency titrimetry may also be applied to dilute solutions [see Chapter XV]; the potentiometric method is superior for more concentrated solutions.)

4. 'Foreign' salts may frequently be present without interference and are, indeed, usually added as the supporting electrolyte in order to eliminate the migration current.

5. The results of the titration are independent of the characteristics of the capillary.

6. The temperature need not be known provided it is kept constant during the titration.

7. Although a polarograph is convenient as a means of applying the voltage to the cell, its use is not essential in amperometric titrations. The constant applied voltage may be obtained with a simple potentiometric device (see Fig. XVI, 9; Section **XVI, 7**).

If the current–voltage curve of the reagent and of the substance being titrated are not known, the polarograms must first be determined in the supporting electrolyte in which the titration is to be carried out. The voltage applied at the beginning of the titration must be such that the total diffusion current of the substance to be titrated, or of the reagent, or of both, is obtained. In Fig. XVII, 1, are collected the most common types of curves encountered in amperometric titrations together with the corresponding hypothetical polarograms of each individual substance: S refers to the solute to be titrated and R to the titrating reagent. The slight 'rounding off' in the vicinity of the equivalence point is due to the solubility of the precipitate; this curvature does not usually interfere, since the end-point is located by extending the linear branches to the point of intersection. For each amperometric titration the applied voltage is adjusted to a value between X and Y shown in Fig. XVII, 1, A'–D'. In A only the material being titrated gives a diffusion current (see A'), i.e., the electro-active material is removed from the solution by precipitation with an inactive substance (for example, lead ions titrated with oxalate or sulphate ions). In B the solute gives no diffusion current but the reagent does (see B'), i.e., an electro-active precipitating reagent is added to an inactive substance (for example, sulphate ions titrated with barium or lead ions). In C both the solute and the titrating reagent give diffusion currents (see C') and a sharp V-shaped curve is obtained (for example, lead ion titrated with dichromate ion, nickel ion with dimethylglyoxime, and copper ion with benzoin α-oxime). Finally, in D the solute gives an anodic diffusion current (that is, is oxidised at the dropping mercury cathode) at the same potential as the titrating reagent gives a cathodic diffusion current (see D'); here the current changes from anodic to cathodic or *vice versa* and the end-point of the titration is indicated by a zero current. Examples of D include the titration of iodide ion with mercury(II) (as nitrate), of chloride ion with silver ion, and of titanium(III) in an

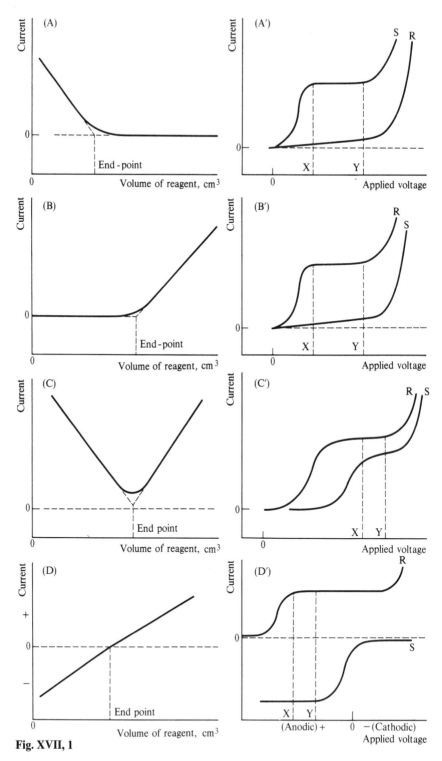

Fig. XVII, 1

acidified tartrate medium with iron(III). Because the diffusion coefficient of the reagent is usually slightly different from the substance being titrated, the slope of the line before the end-point differs slightly from that after the end-point (compare D); in practice, it is easy to add the reagent until the current acquires a zero value or, more accurately, the value of the residual current for the supporting electrolyte.

To take into account the change in volume of the solution during the titration, the observed currents should be multiplied by the factor $(V + v)/V$, where V is the initial volume of the solution and v is the volume of the titrating reagent added. Alternatively, this correction may be avoided (or considerably reduced) by adding the reagent from a semi-micro burette in a concentration ten to twenty times that of the solute. The use of concentrated reagents has the additional advantage that comparatively little dissolved oxygen is introduced into the system, thus rendering unnecessary prolonged bubbling with inert gas after each addition of the reagent. The migration current is eliminated by adding sufficient supporting electrolyte; if necessary, a suitable maximum suppressor is also introduced.

XVII, 2. TECHNIQUE OF AMPEROMETRIC TITRATIONS WITH THE DROPPING MERCURY ELECTRODE.

An excellent and inexpensive titration cell consists of a commercial resistance glass (e.g., Pyrex), 100-cm^3, three-necked, flat or round-bottomed flask to which a fourth neck is sealed. The complete assembly is depicted schematically in Fig. XVII, 2, A. The burette (preferably of the semi-micro type and graduated in 0.01 cm^3), dropping electrode, a two-way gas-inlet tube (thus permitting nitrogen to be passed either through the solution or over its surface), and an agar–potassium salt bridge (*not* shown in the figure) are fitted into the four necks by means of rubber stoppers.

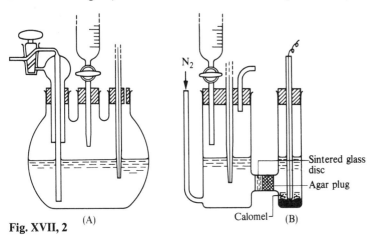

Fig. XVII, 2   (A)   Calomel   (B)

The agar–salt bridge is connected through an intermediate vessel (a weighing bottle may be used) containing saturated potassium chloride solution to a large saturated calomel electrode. The agar–salt bridge is made from a gel which is 3 per cent in agar and contains sufficient potassium chloride to saturate the solution at the room temperature; when chloride ions interfere with the titrations, the connection is made with an agar–potassium nitrate bridge.

Another cell, due to Lingane and Laitinen, is shown in Fig. XVII, 2, B: the special feature of this H-cell is the sintered-glass disc (porosity 3) and the 3 per cent agar–salt plug which separates the saturated calomel electrode from the solution being titrated. A minor disadvantage would appear to be the possibility of breaking the fragile capillary or the burette tip upon removal from the rubber stopper. If desired, the right-hand compartment can be filled with saturated potassium chloride solution and connection with the external reference electrode made with another salt bridge in the usual way.

Thermostatic control is not essential provided the cell is maintained at a fairly constant temperature during the titration. It is advantageous to store the reagent beneath an atmosphere of inert gas: this precaution is not absolutely necessary if the reagent solution has ten to twenty times the concentration of the solution being titrated and is added from a semi-micro burette. If the solute is electro-reducible, sufficient electrolyte should be added to eliminate the migration current; if the reagent is electro-reducible and the solute is not, the addition of a supporting electrolyte is usually not required, since sufficient electrolyte is formed during the titration to eliminate the migration current beyond the end point. It may be necessary to add a suitable maximum suppressor, such as gelatin. If the polarographic characteristics of the solute and the reagent are not known, the current–voltage curve of each must be determined in the medium in which the titration is being carried out. The applied voltage is then adjusted at the beginning of the titration to such a value that the diffusion current of the unknown solute, or of the reagent, or of both, is obtained; frequently the voltage range is comparatively large and, in consequence, great accuracy is not required in adjusting the applied voltage.

The general procedure is as follows. A known volume of the solution under test is placed in the titration cell, which is then assembled as in Fig. XVII, 2, A: the electrical connections are completed (dropping mercury electrode as cathode; saturated calomel half-cell, or mercury pool at bottom of flask, as anode), and dissolved oxygen is removed by passing a slow stream of pure nitrogen for about 15 minutes. The applied voltage is then adjusted to the desired value, and the initial diffusion current is noted. A known volume of the reagent is run in from a semi-micro burette, nitrogen is bubbled through the solution for about 2 minutes to eliminate traces of oxygen from the added liquid and to ensure complete mixing. The flow of gas through the solution is then stopped, but is allowed to pass over the surface of the solution (thus maintaining an inert, oxygen-free atmosphere). The current and burette readings are both noted. This procedure is repeated until sufficient readings have been obtained to permit the end point to be determined as the intersection of the two linear parts of the graph.

XVII, 3. DETERMINATION OF LEAD WITH STANDARD POTASSIUM DICHROMATE SOLUTION.
Both lead ion and dichromate ion yield a diffusion current at an applied potential to a dropping mercury electrode of -1.0 volt against the saturated calomel electrode (SCE). Amperometric titration gives a V-shaped curve (Fig. XVII, 1, C). For convenience in its use by large classes of students, the exercise has been adapted to the determination of lead in A.R. lead nitrate; the application to the determination of lead in dilute aqueous solutions $(10^{-3}–10^{-4}M)$ is self-evident.

Reagents required. Dissolve an accurately weighed amount of A.R. lead nitrate in $250\,cm^3$ water in a graduated flask to give an approximately $0.01M$

solution. For the titration, dilute 10 cm^3 of this solution (use a pipette) to 100 cm^3 in a graduated flask, thus yielding a *ca.* 0.001*M* solution of known strength.

Prepare a *ca.* 0.05*M* solution of potassium dichromate using the appropriate quantity, accurately weighed, of the dry A.R. solid. Dilute this solution to *ca.* 0.005*M*.

Prepare also a *ca.* 0.01*M* solution of potassium nitrate from the A.R. solid for use as the supporting electrolyte.

Procedure. Use any commercial or manual polarograph: see Chapter XVI. Set up the dropping mercury electrode assembly and allow the mercury to drop into distilled water for at least 5 minutes. Meanwhile, place 25.0 cm^3 of the *ca.* 0.001*M*-lead nitrate solution in the titration cell, add 25 cm^3 0.01*M*-potassium nitrate solution, complete the cell assembly, and bubble nitrogen slowly through the solution for 15 minutes. Make the necessary electrical connections. Apply a potential of -1.0 volt *vs.* SCE: at this potential both the lead and the dichromate ions yield diffusion currents. Turn the three-way tap so that the nitrogen now passes over the surface of the solution. Adjust the galvanometer sensitivity so that the spot is on the scale and take the reading. Do not alter the applied voltage during the determination. Add the *ca.* 0.005*M*-dichromate solution in 0.5-cm^3 portions until within 1 cm^3 of the end point, and henceforth in 0.1 cm^3 portions until about 1 cm^3 beyond the end point, and continue with additions of 0.5 cm^3. After each addition pass nitrogen through the solution for 1 minute to ensure thorough mixing and also deoxygenation, turn the tap so that the nitrogen passes over the surface of the solution, and note the deflection of the galvanometer spot, i.e., measure the current. It will be observed that a large initial current will decrease as the titration proceeds to a small value at the equivalence point, and then increase again beyond the equivalence point. Correct the readings of the galvanometer deflection for the change in volume of the solution due to the added reagent using the formula $I_{corr.} = I_{obs.} (V+v)/V$, where V is the initial volume of the solution and v is the volume of the titrating reagent. Plot the values of the corrected current (galvanometer deflections) as ordinates against the volume of reagent added as abscissæ: draw two straight lines through the branches of the 'curve'. The point of intersection is the equivalence point. Calculate the percentage of lead in the sample of lead nitrate.

$$1 \text{ cm}^3 \; 0.01M\text{-K}_2\text{Cr}_2\text{O}_7 \equiv 0.002072 \text{ g Pb}$$

Repeat the titration using 0.05*M*-dichromate solution added from a 5- or 10-cm^3 semi-micro burette.

XVII, 4. DETERMINATION OF SULPHATE WITH STANDARD LEAD NITRATE SOLUTION.

Solutions as dilute as 0.001*M* with respect to sulphate may be titrated with 0.01*M*-lead nitrate solution in a medium containing 30 per cent ethanol with reasonable accuracy. For solutions 0.01*M* or higher in sulphate the best results are obtained in a medium containing about 20 per cent ethanol. The object of the alcohol is to reduce the solubility of the lead sulphate and thus minimise the magnitude of the rounded portion of the titration curve in the vicinity of the equivalence point. The titration is performed in the absence of oxygen at a potential of -1.2 volts (*vs.* SCE) at which potential lead ions yield a diffusion current. A 'reversed L' graph (compare Fig. XVII, 1, B) is obtained: the intersection of the two branches gives the end-point. A supporting electrolyte need not be added, since the current does not increase appreciably until an excess of lead is present in the solution, and the amount of salt formed during the titration suffices to completely suppress the migration current of lead ions.

Reagents required. Prepare an approximately $0.01M$ solution of potassium sulphate in a 100-cm^3 graduated flask using an accurately weighed quantity of the dry A.R. solid. Similarly prepare an approximately $0.1M$-lead nitrate solution in a 100-cm^3 graduated flask from a known weight of the dry A.R. solid.

Procedure. Use the apparatus and technique described in the previous Section. Introduce 25.0 cm^3 of the potassium sulphate solution into the cell, add 2 to 3 drops of thymol blue followed by a few drops of concentrated nitric acid until the colour is just red (pH 1.2); finally, add 25 cm^3 of 95 per cent ethanol. Connect the saturated calomel electrode through an agar–potassium nitrate bridge to the cell. Fill the semi-micro burette with the standard lead nitrate solution. Pass nitrogen through the solution in the cell for 15 minutes and then over the surface of the solution. Meanwhile adjust the applied voltage to -1.2 volt. Set the sensitivity control at the appropriate value and also the galvanometer spot at zero. Introduce the lead nitrate solution from the burette in 0.5-cm^3 portions until within 1 cm^3 of the equivalence point, then in 0.1-cm^3 quantities for the following 2 cm^3, and subsequently in 0.5-cm^3 portions. Pass the gas stream through the solution for about 1 minute after each addition (more dilute solutions will require up to 3 minutes to assist the precipitation of the lead sulphate) and then over the surface before reading the galvanometer deflection (current). Correct the current readings for the change in volume of the solution due to the added reagent as in the previous experiment. Read off the equivalence point from the amperometric titration curve drawn from your results.

Calculate the percentage of SO_4 in the sample of A.R. potassium sulphate.

$$1 \text{ cm}^3 \ 0.1M\text{-Pb(NO}_3)_2 \equiv 0.009606 \text{ g } SO_4$$

XVII, 5. DETERMINATION OF NICKEL WITH DIMETHYLGLY-OXIME. The nickel solution (concentration less than $0.005M$) is introduced into an aqueous ammonia–ammonium chloride supporting medium and, after deoxygenation, titration is carried out at an applied voltage (-1.85 volts *vs.* SCE) at which both nickel and dimethylglyoxime are reducible. A V-shaped titration graph (Fig. XVII, 1, C) is obtained.

Reagents required. (i) Prepare a $0.02M$-dimethylglyoxime solution by dissolving 0.2322 g of A.R. dimethylglyoxime in 95 per cent ethanol (rectified spirit) and make up to 100 cm^3 in a graduated flask with the same solvent.

(ii) Prepare an approximately $0.01M$ solution of ammonium nickel sulphate by weighing out about 0.395 g of the salt (preferably of A.R. quality) and dissolving it in 100 cm^3 of water in a graduated flask. Standardise the solution by an EDTA titration (Section **X, 58**). Dilute 25.0 cm^3 of this solution to 250 cm^3 in a graduated flask, thus giving a *ca.* $0.001M$ solution.

(iii) Prepare the base solution by dissolving 4.0 cm^3 of concentrated ammonia solution (sp. gr. 0.88) and 5.35 g of A.R. ammonium chloride in water and diluting to 1 dm^3 in a graduated flask. The resulting solution is $0.5M$ in aqueous ammonia and $0.1M$ in ammonium chloride.

Procedure. Use the four-necked 100-cm^3 titration flask depicted in Fig. XVII, 2, A including an agar–potassium nitrate bridge and a microburette. Place

25.0 cm^3 of the *ca.* 0.001 *M*-nickel solution, 25 cm^3 of the base solution, and 1 cm^3 of 0.2 per cent gelatin solution in the clean, dry titration vessel; the base solution will now be *ca.* 0.25*M* in aqueous ammonia and *ca.* 0.05*M* in ammonium chloride. Pass oxygen-free nitrogen through the solution for 15 minutes. Raise the dropping mercury electrode reservoir and allow the mercury to drop into distilled water for 5 minutes. Meanwhile connect a saturated calomel electrode through an intermediate saturated potassium chloride solution by means of an agar–salt bridge to the titration vessel. Fill the semi-micro burette with the 0.02*M*-dimethylglyoxime solution and insert the tip inside the titration flask.

Set the applied potential at − 1.85 volts versus the saturated calomel electrode, commence the flow of mercury from the dropping electrode and note the maximum deflection of the galvanometer spot. Add the dimethylglyoxime solution from the semi-micro burette in suitable increments (e.g., of 0.2 cm^3) until within 1 cm^3 of the end-point; then reduce the additions to 0.05–0.1 cm^3 and continue well beyond the equivalence point. After each addition pass nitrogen through the solution for 1 minute to deoxygenate and to mix the solution, and then observe the galvanometer deflection (current). It will be observed that the galvanometer deflection (current) decreases linearly to the end-point and then increases more rapidly. Plot current (ordinates) against volume of dimethylglyoxime solution (abscissæ), making the appropriate correction for the volume of reagent added at each reading. The equivalence point is the point of intersection of the two linear branches of the graph. Calculate the percentage of nickel in the sample of ammonium nickel sulphate.

$\qquad$ 1 cm^3 0.02*M*-dimethylglyoxime ≡ 0.0005869 g Ni

Note. In the above determination, the dimethylglyoxime is assumed to be pure. It is better to check the purity of the dimethylglyoxime with a standard nickel solution and to use the resulting factor for the dimethylglyoxime solution in subsequent calculations. Then determine the nickel content of a solution containing 0.05–1 mg of nickel. Other elements which form complexes with dimethylglyoxime, especially cobalt, copper, and bismuth, must be absent.

Many other metals can be similarly determined by amperometric titration with suitable organic reagents; a full selection is given in Ref. 10.

XVII, 6. DETERMINATION OF FLUORIDE WITH STANDARD THORIUM NITRATE SOLUTION. Neutral solutions of fluoride may be titrated with 0.01*M*-thorium nitrate in a medium of 0.1*M*-potassium chloride at an applied potential of − 1.7 volts *vs.* SCE. Thorium ions are not reducible at the dropping mercury cathode, but they seem to have the property of carrying to the mercury cathode nitrate and nitrite ions which are reduced, producing a step with a half-wave potential of − 1.3 volts. The height of this step is roughly proportional to the concentration of thorium ions in solution, consequently a reversed L-type of titration graph is produced in the titration of fluoride.

Reagents required. (i) Prepare a *ca.* 0.01*M*-thorium nitrate solution by dissolving about 5.8 g A.R. thorium nitrate in 1 litre distilled water. The solution may be standardised against a standard fluoride solution by amperometric titration.

(ii) Prepare a standard 0.01*M* fluoride solution by dissolving 0.1050 g, accurately weighed, dry A.R. sodium fluoride in 250 cm^3 water in a graduated flask. Transfer 25.0 cm^3 of this solution to a 250 cm^3 graduated flask containing

1.86 g A.R. potassium chloride. Shake until dissolved and dilute to the mark with distilled water. The resulting solution is $0.001M$ in fluoride and is also $0.1M$ with respect to potassium chloride.

Procedure. Check the pH of the fluoride solution with a pH meter: it should be in the range 7–8. Place 25.0 cm^3 of the neutral standard fluoride solution in the titration flask (Fig. XVII, 2, A), set the applied voltage at -1.7 volts *vs.* SCE, and titrate with the thorium nitrate solution in the usual manner. Plot the titration curve, and evaluate the exact concentration of the thorium nitrate solution.

Repeat the titration with an 'unknown' neutral fluoride ion solution, say, of *ca.* $0.0005M$ concentration: the base electrolyte should be $0.1M$ in potassium chloride.

$1 \text{ cm}^3 \ 0.01M\text{-Th}(NO_3)_4 \equiv 0.0007600 \text{ g F}^-$

XVII, 7. DETERMINATION OF ZINC WITH EDTA.

Zinc ions may be titrated with standard EDTA solution in a strongly alkaline medium (produced with cyclohexylamine) at an applied potential of -1.4 volts *vs.* SCE. Under these conditions the diffusion current due to zinc ions decreases during the titration and an L-shaped titration graph results.

Reagents required. (i) Prepare a standard $0.02M$ zinc-ion solution by dissolving about 1.31 g, accurately weighed, A.R. zinc in dilute hydrochloric acid and diluting to 1 dm^3 with distilled water in a graduated flask. Dilute 25.0 cm^3 of this solution to 100 cm^3 in a graduated flask, thus giving a *ca.* $0.005M$ zinc-ion solution.

(ii) Prepare a standard $0.01M$-EDTA solution (Section **X, 50**).

Procedure. Place 5.00 cm^3 of the zinc-ion solution in the titration flask, add 1.0 cm^3 pure cyclohexylamine and 19.0 cm^3 distilled water. Set the applied potential at -1.4 volts *vs.* SCE. Deaerate the solution and titrate with the standard EDTA solution in the usual manner. Plot the titration graph, evaluate the concentration of the zinc in the solution, and compare it with the known value.

Repeat the titration using an 'unknown' solution of zinc ions, say, of $0.0005M$ concentration.

$1 \text{ cm}^3 \ 0.01M\text{-EDTA} \equiv 0.0006538 \text{ g Zn}$

XVII, 8. TITRATION OF AN IODIDE SOLUTION WITH MERCURY(II) NITRATE SOLUTION.

This experiment illustrates the titration of a substance yielding an anodic step (iodide ion) with a solution of an oxidant (mercury(II) nitrate) giving a cathodic diffusion current at the same applied voltage. The magnitude of the anodic diffusion current decreases up to the end-point: upon adding an excess of titrant, the diffusion current increases, but in the opposite direction. The type of graph obtained is similar to that in Fig. XVII, 1, D. The end-point of the titration is given by the intersection of the two linear portions of the graph with the volume (of titrant) axis: the diffusion current is then approximately zero. The two linear parts do not usually have the same slope, because the titrant and the substance being titrated have different diffusion currents for equivalent concentrations.

Reagents required. (i) Prepare a *ca.* $0.004M$-potassium iodide solution by dissolving 0.68 g A.R. potassium iodide, accurately weighed, in 1 dm^3 water.

(ii) Prepare a $0.01M$-mercury(II) nitrate solution by dissolving 1.713 g pure mercury(II) nitrate monohydrate in 500 cm^3 of $0.05M$-nitric acid.

(iii) $0.1M$-nitric acid.

Procedure.   Equip a 100-cm^3 four-necked flask (compare Fig. XVII, 2) with a dropping mercury electrode, an agar–KCl bridge connected to a SCE through saturated potassium chloride solution contained in a 10-cm^3 beaker, a nitrogen gas inlet, and a magnetic stirrer. Charge the flask with 25.0 cm^3 of the iodide solution, add 25 cm^3 $0.1M$-nitric acid, and 2.5 cm^3 warm 1 per cent gelatin solution. Connect the dropping mercury electrode to the negative terminal of a polarograph and the positive terminal to the SCE. Set the applied potential at zero and adjust the zero of the galvanometer at the centre of the scale. Pass nitrogen through the solution for at least 5 minutes whilst stirring magnetically. Run in the mercury(II) nitrate solution from a semi-micro burette and take readings of the galvanometer at 0.10-cm^3 intervals. The end point corresponds to zero current, but continue the titration beyond this point to obtain the cathodic current due to excess of mercury(II) nitrate. Plot galvanometer readings against volume of mercury(II) nitrate solution, and evaluate the exact end point from the graph.

The end point may be checked by potentiometric titration. Calculate the concentration of the mercury(II) nitrate solution from the known concentration of the potassium iodide solution; alternatively, assume that the former is $0.01M$ and calculate the molarity of the latter.

Note.   A standard solution of mercury(II) nitrate may be prepared by dissolving a weighed amount of twice-distilled mercury in nitric acid, heating the solution to expel oxides of nitrogen, and then diluting with distilled water to the desired volume. This solution may be used for the determination of iodide.

XVII, 9. DETERMINATION OF POTASSIUM WITH SODIUM TETRA-PHENYLBORON (GRAPHITE INDICATING ELECTRODE).

The tetraphenylboron ion (TPB) gives two anodic voltammetric waves at a graphite electrode in aqueous solution. This electroactivity forms the basis for the direct amperometric titration of potassium via its precipitation as potassium tetraphenylboron. The method is simple and rapid; it is not necessary to filter off the precipitate.

The procedure is relatively free from interferences, tolerating the presence of large amounts of chloride and other commonly encountered anions, such as phosphate, sulphate, and acetate. The tetraphenylboron ion forms insoluble salts with ammonium, Rb, Cs, Tl(I), Ag, and Hg(II) ions; a precipitate is also produced with Hg(I). These constitute the major interferences to the method. Strong oxidising agents should be absent. The test solution should contain at least 0.2 mg of K per cm^3; below this concentration precipitation proceeds very slowly, and the time required for a single measurement becomes excessive.

The procedure may be applied to the direct determination of potassium in silicates and other refractory substances after sulphuric–hydrofluoric acid dissolution and fuming.

Reagents required.   (i) Prepare a 3 per cent sodium tetraphenylboron solution by dissolving about 3.0 g of the pure solid reagent, accurately weighed, in 100 cm^3 of conductivity water. The solution is slightly turbid; satisfactory results are obtained without removal of the turbidity.

(ii) Prepare a $0.5M$ solution of sodium acetate (using the A.R. solid) and add acetic acid until the pH is 5.6 (pH meter).

(iii) Prepare a *ca.* $0.02M$ solution potassium chloride using an accurately weighed amount of the A.R. solid; also a *ca.* $0.01M$-potassium sulphate solution employing the A.R. salt.

Apparatus.   Prepare a saturated sodium chloride–calomel reference electrode using A.R. sodium chloride: allow it to stand for 2–3 days before use. A spectroscopic graphite electrode about 10 cm long and 10 mm in diameter is used as indicator electrode.

Use a potentiometer together with a sensitive galvanometer.

The titration vessel may be a three-necked flask (see Fig. XVII, 2, A) of $100\,cm^3$ capacity. Insert the arm of the saturated sodium chloride–calomel electrode and the graphite electrode into the two side necks and a 5- or 10-cm^3 semi-micro burette into the central neck. Connect the graphite electrode to the potentiometer with the aid of an alligator clip. Stir the solution using a magnetic stirrer, and maintain the same speed of stirring during all the determinations. Alternatively, stir the solution with a glass stirrer at a constant speed of about 600 r.p.m.: this will necessitate the use of a four-necked flask.

Procedure.   Charge the titration flask with $25\,cm^3$ of the acetate buffer solution, introduce $25.0\,cm^3$ of the standard $0.02M$-potassium chloride, and add sufficient water to ensure that at least 1 cm of the graphite electrode is immersed in

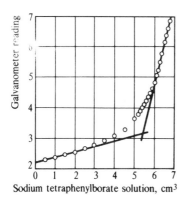

Sodium tetraphenylborate solution, cm^3

Fig. XVII, 3

the solution. Apply a potential of $+0.55$ volt to the graphite electrode. Stir, and add the sodium tetraphenylboron reagent from the semi-micro burette (about 0.5-cm^3 increments before the end point and 0.05–0.10-cm^3 increments after the end-point). After each addition, record the current as soon as it becomes constant (1–3 minutes); after the end point has been reached, the current is usually constant after 30 seconds. Determine the end-point by plotting galvanometer readings against volume of titrant, as in Fig. XVII, 3. Calculate the titre of the reagent, i.e., mg $K \equiv 1.00\,cm^3$ Na TPB reagent.

Determine the potassium content of the 0.01 M-potassium sulphate solution and compare the result obtained with that calculated from the weight of potassium sulphate used. Alternatively, determine the potassium content of 'unknown' potassium chloride solutions (10–25 cm^3) containing between 5 and 20 mg of potassium. A new graphite electrode should be used for each determination.

Titrations with the rotating platinum micro-electrode

XVII, 10.   DISCUSSION AND APPARATUS.   The dropping mercury electrode cannot be used at markedly positive potentials (say, above about 0.4 volt *vs.* SCE) because of the oxidation of the mercury. By replacing the dropping mercury electrode by an inert platinum electrode, it was hoped to extend the range of polarographic work in the positive direction to the voltage approaching that at which oxygen is evolved, namely, 1.1 volts. The attainment of a steady diffusion current is slow with a stationary platinum electrode, but the difficulty may be overcome by rotating the platinum electrode at constant speed: the diffusion layer thickness is considerably reduced, thus increasing the sensitivity and the rate of attainment of equilibrium. Difficulties, however, arise in obtaining reproducible values for the diffusion currents from day to day, and so the applications of the rotating platinum electrode in quantitative polarography are limited. Nevertheless, it is suitable as an indicator electrode in amperometric titrations. The larger currents (about twenty times those at the dropping mercury electrode) attained with the rotating platinum electrode allow correspondingly smaller currents to be measured without loss of accuracy and thus very dilute solutions (up to $10^{-4}M$) may be titrated. In order to obtain a linear relation between current and amount of reagent added, the speed of stirring must be kept constant during the titration: a speed of about 600 revolutions per minute* is generally suitable.

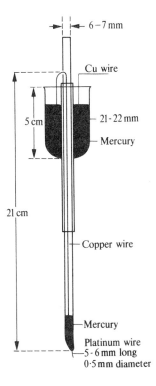

The construction of a simple rotating platinum micro-electrode will be evident from Fig. XVII, 4. The electrode is constructed from a standard 'mercury seal'. About 5 mm of platinum wire (0.5 mm diameter) protrudes from the wall of a length of 6-mm glass tubing; the latter is bent at an angle approaching a right angle a short distance from the lower end. Electrical connection is made to the elecrode by a stout amalgamated copper wire passing through the tubing to the mercury covering the platinum wire seal; the upper end of the copper wire passes through a small hole blown in the stem of the stirrer and dips into mercury contained in the 'mercury seal'. A wire from the latter is connected to the source of applied voltage. The tubing forms the stem of the electrode, which is rotated at a *constant* speed of 600 r.p.m.

Fig. XVII, 4

6–7 mm

Cu wire

5 cm

21–22 mm

Mercury

21 cm

Copper wire

Mercury

Platinum wire
5–6 mm long
0·5 mm diameter

XVII, 11. DETERMINATION OF THIOSULPHATE WITH IODINE.
Dilute solutions of sodium thiosulphate (e.g., $0.001M$) may be titrated with dilute iodine solutions (e.g., $0.005M$) at zero applied voltage. For satisfactory results, the thiosulphate solution should be present in a supporting electrolyte which is $0.1M$ in potassium chloride and $0.004M$ in potassium iodide. Under these conditions no diffusion current is detected until after the equivalence point when excess of iodine is reduced at the electrode; a reversed L-type of titration graph is obtained.

Dilute solutions of iodine, e.g., $0.0001M$, may be titrated similarly with

* The limiting current is proportional to the cube root of the number of revolutions per minute above 200 r.p.m.

standard thiosulphate. The supporting electrolyte consists of $1.0M$-hydrochloric acid and $0.004M$-potassium iodide. No external e.m.f. is required when a SCE is employed as reference electrode.

Reagents required.   (i) Prepare a *ca.* $0.001M$-sodium thiosulphate solution which is $0.1M$ with respect to potassium chloride and $0.004M$ with respect to potassium iodide.

(ii) Prepare a standard $0.005M$-iodine solution in $0.004M$-potassium iodide.

Procedure.   Place $25.0\,\text{cm}^3$ of the thiosulphate solution in the titration cell. Set the applied voltage to zero with respect to the SCE, after connecting the rotating platinum micro-electrode to the manual polarograph. Adjust the sensitivity control of the galvanometer. Titrate with the standard $0.005M$-iodine solution in the usual manner.

Plot the titration graph, evaluate the end point, and calculate the exact concentration of the thiosulphate solution. As a check, repeat the titration using freshly-prepared starch indicator solution.

XVII, 12. DETERMINATION OF ARSENITE WITH STANDARD IODINE SOLUTION.

Dilute solutions of sodium arsenite (e.g., $0.0005M$) may be titrated with standard iodine solution using a rotating platinum micro-electrode and a SCE. The supporting electrolyte consists of $0.1M$-potassium chloride $+ 0.1M$-sodium hydrogencarbonate $+ 0.004M$-potassium iodide. A reversed L-type of titration graph results.

Reagents required.   (i) Prepare a $0.0005M$-sodium arsenite solution which is $0.1M$ in potassium chloride and sodium hydrogencarbonate and $0.004M$ in potassium iodide.

(ii) Prepare standard $0.005M$-iodine solution.

Procedure.   Pipette $25.0\,\text{cm}^3$ of the sodium arsenite solution into the titration flask. Set the applied voltage to zero *vs.* SCE; adjust the sensitivity of the galvanometer. Titrate with the standard $0.005M$-iodine in the usual manner.

Plot the titration graph, evaluate the end point, and calculate the concentration of the arsenite solution. Check the end point with starch indicator.

XVII, 13. DETERMINATION OF ANTIMONY WITH STANDARD POTASSIUM BROMATE SOLUTION.

Dilute solutions of trivalent antimony and arsenic (*ca.* $0.0005M$) may be titrated with standard $0.01N$-potassium bromate in a supporting electrolyte of M-hydrochloric acid containing $0.05M$-potassium bromide. The two electrodes are a rotating platinum micro-electrode and a SCE: the former is polarized to $+0.2$ volt. A reversed L-type of titration graph is obtained.

Reagents required.   (i) Prepare a $0.005M$ solution of A.R. potassium antimonyl tartrate by dissolving $1.625\,\text{g}$ of the A.R. solid in $1\,\text{dm}^3$ of distilled water. Dilute $25.0\,\text{cm}^3$ of this solution to $250\,\text{cm}^3$ with $1M$-hydrochloric acid which is $0.05M$ in potassium bromide.

(ii) Prepare a standard $0.01N$-potassium bromate solution from the A.R. solid.

Procedure.   Pipette $25.0\,\text{cm}^3$ of the antimony solution into the titration cell. Set the applied voltage at 0.2 volt *vs.* SCE, and adjust the sensitivity control of the galvanometer. Titrate in the usual manner, and calculate the concentration of the antimony solution.

Biamperometric titrations

XVII, 14.   GENERAL DISCUSSION.   The titrations so far discussed in this chapter have been concerned with the use of a reference electrode (usually SCE), in conjunction with a polarised electrode (dropping mercury electrode or rotating platinum micro-electrode). Titrations may also be performed in a uniformly stirred solution by using two small but similar platinum electrodes to which a small e.m.f. (1–100 millivolts) is applied: the end point is usually shown by either the disappearance or the appearance of a current flowing between the two electrodes. For the method to be applicable the only requirement is that a reversible oxidation–reduction system be present either before or after the end point.

A simple apparatus suitable for this procedure is shown in Fig. XVII, 5. B is a 3-volt torch battery or 2-volt accumulator, M is a micro-ammeter, R is a 500-ohm, 0.5-watt radio potentiometer, and EE are platinum electrodes. The potentiometer is set so that there is a potential drop of about 80–100 millivolts across the electrodes.

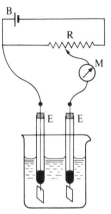

Fig. XVII, 5

In a titration with two indicator electrodes and when the reactant involves a reversible system (e.g., $I_2 + 2e \rightleftharpoons 2I^-$), an appreciable current flows through the cell. The amount of oxidised form reduced at the cathode is equal to that formed by oxidation of the reduced form at the anode. Both electrodes are depolarised until the oxidised component or the reduced component of the system has been consumed by a titrant. After the end point, only one electrode remains depolarised if the titrant (e.g., thiosulphate ion, $2S_2O_3^{2-} \rightarrow S_4O_6^{2-} + 2e$) does not involve a reversible system. Current thus flows until the end point: at or after the end point the current is zero or virtually zero. In the determination of iodine by titration with thiosulphate a rapid decrease in current is observed in the neighbourhood of the end point and this has led to the name **'dead-stop end point'**. The complementary type of end point, which resembles a reversed L-type amperometric graph is probably more desirable in practice, and is obtained in the titration of an irreversible couple (say, thiosulphate) by a reversible couple (say, iodine): the current is very low before the end point, and a very rapid increase in current signals the end point. When both systems are reversible (e.g., iron(II) ions with cerium(IV) or permanganate ions; applied potential 100 millivolts), the current is zero or close to zero at the equivalence point and a V-shaped titration graph results.

XVII, 15.   TITRATION OF THIOSULPHATE WITH IODINE ('DEAD-STOP END POINT').   Reagents required.   Prepare a *ca.* 0.001M-sodium thiosulphate solution and also a standard 0.005M-iodine solution.

Procedure.   Pipette 25.0 cm³ of the thiosulphate solution into the titration cell e.g., a 150-cm³ Pyrex beaker. Insert two similar platinum wire or foil electrodes* into the cell and connect to a manual polarograph or to the apparatus

---

* A length of 6–7 mm of platinum wire of 0.5 mm diameter sealed into a glass tube is satisfactory; electrical connection is made by means of a copper wire dipping into a little mercury in contact with the platinum wire.

of Fig. XVII, 5. Apply 0.10 volt across the electrodes. Adjust the sensitivity of the 'spot' galvanometer to obtain full-scale deflection for a current of 10–25 milliamperes. Stir the solution with a magnetic stirrer. Add the iodine solution from a 5-cm^3 semi-micro burette slowly in the usual manner and read the current (galvanometer deflection) after each addition of the titrant. When the current begins to increase, stop the addition; then add the titrant by small increments of 0.05 or 0.10 cm^3. Plot the titration graph, evaluate the end point, and calculate the concentration of the thiosulphate solution. It will be found that the current is fairly constant until the end point is approached and increases rapidly beyond.

XVII, 16.   DETERMINATION OF NITRATE.   *Discussion.*   'Dead-stop' end point titrimetry may be applied to the determination of nitrate ion by titration with ammonium iron(II) sulphate solution in a strong sulphuric acid medium:

$$4FeSO_4 + 2HNO_3 + 2H_2SO_4 = 2Fe_2(SO_4)_3 + N_2O_3 + 3H_2O$$

Two platinum electrodes are immersed in sulphuric acid of suitable strength containing the nitrate ion to be determined and a potential of about 100 millivolts is applied. Upon titration with 0.4M-ammonium iron(II) sulphate solution there is an initial rise in current followed by a gradual fall, with a marked increase at the end point: the latter is easily determined from a plot of galvanometer reading against volume of iron solution added. The concentration of water should not be allowed to rise above 25 per cent (w/w). The temperature of the solution should not exceed 40 °C.

   Reagents.   *Sulphuric acid*, ca. *25 per cent v/v* (Solution A). Add cautiously 250 cm^3 of concentrated sulphuric acid to 750 cm^3 of water, cool, and dilute to 1 litre. **(Take care with this addition.)**

   Ammonium iron(II) sulphate solution, ca. *0.4*M. Dissolve about 15.6 g, accurately weighed, of A.R. ammonium iron(II) sulphate in 100 cm^3 of Solution A.

   Potassium nitrate solution, ca. *0.3*M. Dissolve about 3.0 g, accurately weighed, of A.R. potassium nitrate in a small volume of Solution A and dilute to 100 cm^3 in a graduated flask with concentrated sulphuric acid.

   Procedure.   Fit up the apparatus as follows. Into a 100-cm^3 four-necked Pyrex flat-bottom flask containing a polythene-covered stirring bar, insert two platinum wire electrodes (0.5 mm diameter; held in position by corks) and a thermometer respectively into the three side necks: insert the tip of a semi-micro burette and a nitrogen inlet tube into the central neck. Place the flask in a beaker charged with an ice–water mixture and clamp the flask in position: mount the beaker on a magnetic stirrer. Pipette 10.0 cm^3 of the potassium nitrate solution into the flask, add 40 cm^3 of concentrated sulphuric acid, and mix well with the aid of the magnetic stirrer. Apply a polarisation voltage of about 100 millivolts: use a galvanometer with adjustable sensitivity control to measure the current. Titrate with the ammonium iron(II) sulphate solution while stirring vigorously: adjust the galvanometer sensitivity to about $\frac{1}{10}$. The galvanometer reading will decrease slightly as the end point is approached (indicated by the fading of the pinkish-brown colour of the solution) and will increase steadily beyond the end-point. The temperature of the solution has some influence upon the galvanometer deflection, and so readings should preferably be taken when the solution temperature is about 20 °C.

Determine the end point from the plot of galvanometer deflection against volume of iron reagent. Calculate the weight of nitrate ion equivalent to $1.0\,cm^3$ of the $0.4M$ iron solution.

When dealing with small amounts of nitrate ion it is advisable to pass a current of pure nitrogen through the solution before commencing the titration, and to maintain an atmosphere of nitrogen in the flask throughout the titration.

If chloride is present, saturated aqueous silver acetate solution should be added in amount slightly more than the calculated quantity prior to the addition of concentrated sulphuric acid. The procedure may be applied to the routine analysis of mixtures of nitric and sulphuric acids, and to the determination of nitrogen in esters such as nitroglycerine and nitrocellulose; the latter are easily hydrolysed by strong sulphuric acid after dispersal in glacial acetic acid.

XVII, 17. DETERMINATION OF WATER WITH THE KARL FISCHER REAGENT.

For the determination of small amounts of water, Karl Fischer (1935) proposed a reagent prepared by the action of sulphur dioxide upon a solution of iodine in a mixture of anhydrous pyridine and anhydrous methanol. Water reacts with this reagent in a two-stage process in which one molecule of iodine disappears for each molecule of water present:

$$3C_5H_5N + I_2 + SO_2 + H_2O = 2C_5H_5NH^+I^- + C_5H_5N\overset{\displaystyle SO_2}{\underset{\displaystyle -O}{\nearrow |}} \qquad (i)$$

$$C_5H_5N\overset{\displaystyle SO_2}{\underset{\displaystyle -O}{\nearrow |}} + CH_3OH = C_5H_5N\overset{\displaystyle OSO_2OCH_3}{\underset{\displaystyle H}{\diagdown}} \qquad (ii)$$

The end point of the reaction is conveniently determined electrometrically using the dead-stop end point procedure. If a small e.m.f. is applied across two platinum electrodes immersed in the reaction mixture a current will flow as long as free iodine is present, to remove hydrogen and depolarise the cathode. When the last trace of iodine has reacted the current will decrease to zero or very close to zero. Conversely, the technique may be combined with a direct titration of the sample with the Karl Fischer reagent: here the current in the electrode circuit suddenly increases at the first appearance of unused iodine in the solution.

The original Karl Fischer reagent prepared with an excess of methanol was somewhat unstable and required frequent standardisation. It was found that the stability was improved by replacing the methanol by 2-methoxyethanol (methyl cellosolve), and a satisfactory reagent may be prepared by dissolving resublimed iodine ($133\,g$) in pure anhydrous pyridine ($425\,cm^3$) contained in a dry, glass-stoppered bottle, and then adding 2-methoxyethanol ($425\,cm^3$). With the bottle cooled in an ice bath, anhydrous liquid sulphur dioxide ($70\,cm^3$) is added in small portions from a graduated cylinder which is kept in an ice-salt bath. Usually, it is not worth the trouble of preparing the reagent which may be purchased from the normal suppliers of laboratory chemicals.

The reagent, whether purchased or prepared in the laboratory must be standardised, and this may be done with pure disodium tartrate dihydrate which contains 15.66 per cent water, or more commonly, by means of a solution of water in methanol. This solution is prepared as follows. Fill a dry 1 dm^3, glass-stoppered graduated flask to within 100 cm^3 of the mark with anhydrous methanol (<0.1 per cent of water), and place it in a thermostatically-controlled water bath at 25 °C, together with a small flask containing about 200 cm^3 of the same methanol. Weigh out accurately about 15 g of distilled water into the dm^3 flask and, after the contents have acquired the temperature of the water bath, adjust the volume to the mark with methanol from the smaller flask.

The Karl Fischer procedure is best carried out with a commercial apparatus, which may be purchased, with slight modifications, from many of the leading laboratory supply houses. Basically, the instrument will carry two burettes, one for the reagent and the other for the standard solution of water in methanol. Each burette is attached to a reservoir which may hold up to 1 litre of liquid, and a series of guard tubes containing dessicant to prevent the ingress of atmospheric moisture are provided. The titration vessel is fitted with an air-tight cover, and is provided with a pair of bright platinum electrodes connected to a micro-ammeter; provision is made for stirring the contents of the vessel by means of a magnetic stirrer. The scale of the micro-ammeter is often marked with 'Excess reagent' and 'Excess water' signs. The usual experimental procedure is to add a slight excess of the reagent so that all the water in the sample under test is reacted, and then the excess Fischer reagent is back-titrated with the standard water-in-methanol solution.

The method is clearly confined to those cases where the test substance does not react with either of the components of the reagent, nor with the hydrogen iodide which is formed during the reaction with water: the following compounds interfere in the Karl Fischer titration.

(i) Oxidising agents, such as chromates, dichromates, copper(II) and iron(III) salts, higher oxides, and peroxides.

$$MnO_2 + 4C_5H_5NH^+ + 2I^- = Mn^{2+} + 4C_5H_5N + I_2 + 2H_2O$$

(ii) Reducing agents, such as thiosulphates, tin(II) salts and sulphides.

(iii) Compounds which can be regarded as forming water with the components of the Karl Fischer reagent, for example:

(a) basic oxides—

$$ZnO + 2C_5H_5NH^+ = Zn^{2+} + 2C_5H_5N + H_2O;$$

(b) salts of weak oxy-acids—

$$NaHCO_3 + C_5H_5NH^+ = Na^+ + H_2O + CO_2 + C_5H_5N$$

XVII, 18.   DETERMINATION OF THE WATER CONTENT OF A SALT HYDRATE. The Karl Fischer procedure may be applied to the determination of water present in hydrated salts or which is absorbed on the surface of solids. The procedure, where applicable, is more rapid and direct than the commonly used drying process. A sample of the finely powdered solid, containing 5–10 millimols (90–180 mg) of water, is dissolved or suspended in 25 cm^3 of dry methanol in a 250-cm^3 glass-stoppered graduated flask. The mixture is titrated with standard Karl Fischer reagent to the usual electrometric end point. An end point stable for 15 seconds usually indicates complete reaction. If the initial titration is incomplete the mixture may be titrated at 10-second intervals until a suitable end point is obtained. The water content of the methanol solution is determined by a separate titration of an equal volume, and the titre of the sample is reduced by this amount. The corrected titre is equivalent to the available water in the sample.

Standardisation of the Karl Fischer reagent. By means of a standard solution of water in methanol. Dry the beaker, stirrer, and electrode system using acetone and a stream of dry air. Rapidly add 2–3 drops water from a weighing bottle (fitted with a stopper and small dropper) to the beaker, and immediately fit the beaker in position on the apparatus. Add the Karl Fischer reagent in 1-cm^3 portions at about two-second intervals. Switch on the stirrer after a few cm^3 have been added, and continue the titration until a permanent iodine colour is obtained. The meter needle will now swing over to 'Excess reagent'. Back-titrate the excess of Fischer reagent with the standard water-in-methanol solution at a rate of about 1 drop per second until the meter needle begins to oscillate; continue the titration until one drop causes a large deflection and the needle reads 'Excess water'. Stirring must not be vigorous, and should be maintained at a steady rate throughout the titration. Record the volumes of reactants added, and also the exact weight of water used.

Calculate the strength of the Fischer reagent in terms of milligrams of water per cm^3 of solution from both results. A useful check on the standard solution of water in methanol is thus available.

Reproducible standardisation figures are sometimes difficult to obtain because of variation in the amount of adsorbed water present in the apparatus. The following modified procedure may be used. Transfer 20.0 cm^3 anhydrous methanol to the titration beaker, stir, and add the Karl Fischer reagent until about 1.0 cm^3 excess is present. Now titrate to the end point with the standard water-in-methanol solution. Introduce 3 drops of water as rapidly as possible through the side arm of the beaker, titrate with the Karl Fischer reagent until a permanent iodine colour is obtained and the needle of the meter is at 'Excess reagent', add 1.0 cm^3 more of the reagent. Titrate the excess of Fischer reagent with the standard water-in-methanol solution. Run in a further 10.0 cm^3 of the water-in-methanol solution, titrate with the Karl Fischer reagent until a known excess is present, and back-titrate the excess with the water-in-methanol solution. Several determinations of the strength of the Karl Fischer reagent can thus be made.

By disodium tartrate dihydrate. Place 25.0 cm^3 absolute methanol in the titration vessel and titrate with the Karl Fischer reagent. Add 0.5–0.6 g pure disodium tartrate dihydrate (15.66 per cent water), accurately weighed, stir, and titrate again with the Karl Fischer reagent. The salt dissolves completely before the titration is completed.

Calculate the mg of water equivalent to 1 cm^3 of the Karl Fischer reagent from the formula:

$$\text{mg of H}_2\text{O per cm}^3 = \frac{\text{mg of sample} \times 0.1566}{\text{cm}^3 \text{ of reagent}}$$

Analysis of the hydrate. To determine the water content of the hydrate (sodium acetate is a satisfactory material for practising the technique), proceed as follows. Place 20 cm^3 of anhydrous methanol in the titration vessel of the Karl Fischer apparatus, add a slight excess of the Karl Fischer reagent and then back titrate with the standard water–methanol mixture; this will remove any water present in the methanol and also water adsorbed on the surface of the vessel.

Immediately add to the methanol about 0.2 g of crystallised sodium acetate which has been previously placed in a weighing bottle, stir the solution, add Karl Fischer reagent until a slight excess is present, and then back titrate with the water in methanol solution. Finally, reweigh the weighing bottle and calculate the water content of the salt.

It may be noted that in some modern Karl Fischer titrators (e.g., the 'Aquatest II' marketed by the Photovolt Corporation) a reagent is used which is deficient in iodine, and then in each determination the requisite amount of iodine is generated electrolytically; i.e., the determination is made coulometrically. This procedure eliminates many of the problems associated with the instability of the normal Karl Fischer reagent, and obviates the necessity for standardisation of the reagent, with the result that determinations can be carried out with great rapidity.

XVII, 19. Selected bibliography

1. P. Delahay (1954). *New Instrumental Methods in Electrochemistry*. New York; Interscience Publishers.
2. G. Charlot and D. Bézier (1954). *Méthodes Electrochimiques d'Analyse*. Paris; Masson et Cie.
3. G. Charlot and D. Bézier (1957). Translated by R. C. Murray. *Quantitative Inorganic Analysis*. Chapter XXV. 'Amperometry'. London; Methuen and Co.
4. K. G. Stone and H. G. Scholten (1952). 'The Dead-Stop End Point', *Analytical Chemistry*, **24**, 671.
5. L. M. Kolthoff (1954). 'Relations between Voltammetry and Potentiometric and Amperometric Titrations', *Analytical Chemistry*, **26**, 1685.
6. P. Delahay (1955). 'Voltammetry at Constant Current', *Analytical Chemistry*, **27**, 478.
7. D. L. Smith, D. R. Jamieson, and P. J. Elving (1960). 'Direct Titration of Potassium with Tetraphenylborate. Amperometric Equivalence-Point Detection', *Analytical Chemistry*, **32**, 1253.
8. H. A. Laitinen. 'Amperometric Titrations', *Analytical Chemistry*, 1956, **28**, 666; 1958, **30**, 657; 1960, **32**, 180R.
9. J. Mitchell and D. M. Smith (1948). *Aquametry: Application of the Karl Fischer Reagent to Quantitative Analysis Involving Water*. New York; Interscience.
10. L. Meites (1963). *Handbook of Analytical Chemistry*. New York; McGraw-Hill.
11. J. T. Stock (1974). Amperometric Titrations (review article). *Anal. Chem.*, **46**, 1R.
12. J. T. Stock (1965). 'Amperometric Titrations'. New York; Interscience.

PART F

SPECTROANALYTICAL METHODS

COLORIMETRY
CHAPTER XVIII **AND SPECTROPHOTOMETRY**

XVIII, 1. GENERAL DISCUSSION. The variation of the colour of a system with change in concentration of some component forms the basis of what the chemist commonly terms **colorimetric analysis**. The colour is usually due to the formation of a coloured compound by the addition of an appropriate reagent, or it may be inherent in the desired constituent itself. The intensity of the colour may then be compared with that obtained by treating a known amount of the substance in the same manner.

Colorimetry is concerned with the determination of the concentration of a substance by measurement of the relative absorption of light with respect to a known concentration of the substance. In **visual colorimetry**, natural or artificial white light is generally used as a light source, and determinations are usually made with a simple instrument termed a **colorimeter** or colour comparator. When the eye is replaced by a photoelectric cell (thus largely eliminating the errors due to the personal characteristics of each observer) the instrument is termed a **photo-electric colorimeter**. The latter is usually employed with light contained within a comparatively narrow range of wavelengths furnished by passing white light through **filters**, i.e., materials in the form of plates of coloured glass, gelatin, etc., transmitting only a limited spectral region: the name **filter photometer** is sometimes applied to such an instrument.

In **spectrophotometric analysis** a source of radiation is used that extends into the ultraviolet region of the spectrum. From this, definite wavelengths of radiation are chosen possessing a bandwidth of less than 1 nm. This process necessitates the use of a more complicated and consequently more expensive instrument. The instrument employed for this purpose is a **spectrophotometer**, and as its name implies, is really two instruments in one cabinet—a **spectrometer** and a **photometer**.

An optical spectrometer is an instrument possessing an optical system which can produce dispersion of incident electromagnetic radiation, and with which measurements can be made of the quantity of transmitted radiation at selected wavelengths of the spectral range. A photometer is a device for measuring the intensity of transmitted radiation or a function of this quantity. When combined in the spectrophotometer the spectrometer and photometer are employed conjointly to produce a signal corresponding to the difference between the transmitted radiation of a reference material and that of a sample at selected wavelengths.

The chief advantage of colorimetric and spectrophotometric methods is that they provide a simple means for determining minute quantities of substances. The upper limit of colorimetric methods is, in general, the determination of constituents which are present in quantities of less than 1 or 2 per cent. The development of inexpensive photoelectric colorimeters has placed this branch of instrumental chemical analysis within the means of even the smallest teaching institution.

In this chapter we are concerned with analytical methods that are based upon the absorption of electromagnetic radiation. Light consists of radiation to which the human eye is sensitive, waves of different wavelengths giving rise to light of different colours, while a mixture of light of these wavelengths constitutes white light. White light covers the entire visible spectrum 400–760 nm. The approximate wavelength ranges of colours are given in Table XVIII, 1.

Table XVIII, 1.   Approximate wavelengths of colours

Ultraviolet	< 400 nm	Yellow	570–590 nm
Violet	400–450 nm	Orange	590–620 nm
Blue	450–500 nm	Red	620–760 nm
Green	500–570 nm	Infrared	> 760 nm

The visual perception of colour arises from the selective absorption of certain wavelengths of incident light by the coloured object. The other wavelengths are either reflected or transmitted, according to the nature of the object, and are perceived by the eye as the colour of the object. If a solid opaque object appears white, all wavelengths are reflected equally; if the object appears black, very little light of any wavelength is reflected; if it appears blue, the wavelengths that give the blue stimulus are reflected, etc.

It must be emphasised that the range of electromagnetic radiation extends considerably beyond the visible region. The approximate limits of wavelength and frequency for the various types of radiation, including the frequency range of sound waves, are shown in Fig. XVIII, 1 (not drawn to scale); this may be regarded as an electromagnetic spectrum. It will be seen that γ-rays and X-rays have very short wavelengths, while ultraviolet, visible, infrared and radio waves have progressively longer wavelengths. For colorimetry and spectrophotometry, the visible region and the adjacent ultraviolet region are of major importance.

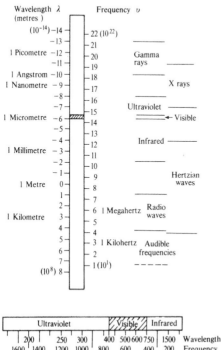

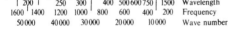

Fig. XVIII, 1

Electromagnetic waves are usually described in terms of (*a*) wavelength λ (distance between the peaks of waves in cm, unless otherwise specified), (*b*) wavenumber $\tilde{\nu}$ (number of waves per cm), and (*c*) the frequency ν (number of waves per second). The three quantities are related as follows:

$$\frac{1}{\text{Wavelength}} = \text{Wavenumber} = \frac{\text{Frequency}}{\text{Velocity of light}}$$

$$\frac{1}{\lambda} = \tilde{\nu} = \frac{\nu}{c}$$

The units in common use are:

1 Ångstrom unit $= 1 \text{ Å} = 10^{-10}$ metre $= 10^{-8}$ cm

1 Nanometre $= 1 \text{ nm} = 10 \text{ Å} = 10^{-7}$ cm

1 Micrometre $= 1 \mu\text{m} = 10^4 \text{ Å} = 10^{-4}$ cm

Velocity of light $= c = 2.99793 \times 10^8 \text{ ms}^{-1}$

Wavenumber $\tilde{\nu} = 1/\lambda$ waves per cm

Frequency $\nu = c/\lambda \approx 3 \times 10^{10}/\lambda$ waves per second.

To comply completely with SI units these functions should be calculated using the metre as the basic unit. It is, however, still common practice to use centimetres for this purpose.

XVIII, 2. THEORY OF SPECTROPHOTOMETRY* AND COLORIMETRY.

When light (monochromatic or heterogeneous) falls upon a homogeneous medium, a portion of the incident light is reflected, a portion is absorbed within the medium, and the remainder is transmitted. If the intensity of the incident light is expressed by I_0, that of the absorbed light by I_a, that of the transmitted light by I_t, and that of the reflected light by I_r, then:

$$I_0 = I_a + I_t + I_r$$

For air–glass interfaces consequent upon the use of glass cells, it may be stated that about 4 per cent of the incident light is reflected. I_r is usually eliminated by the use of a control, such as a comparison cell, hence:

$$I_0 = I_a + I_t \tag{1}$$

Credit for investigating the change of absorption of light with the thickness of the medium is frequently given to Lambert (Ref. 1), although he really extended concepts originally developed by Bouguer (Ref. 2). Beer (Ref. 3) later applied similar experiments to solutions of different concentrations and published his results just prior to those of Bernard (Ref. 4). This very confusing story has been explained by Malinin and Yoe (Ref. 5). The two separate laws governing absorption are usually known as **Lambert's Law** and **Beer's Law**. In the combined form (Ref. 6) they are referred to as the **Beer–Lambert Law**.

Lambert's law. This law states that when monochromatic light passes

---

* Spectrophotometry proper is mainly concerned with the following regions of the spectrum: ultraviolet, 185–400 nm; visible 400–760 nm; and infrared, 0.76–15 μm. Colorimetry is concerned with the visible region of the spectrum. In this chapter our attention will be confined largely to the visible and near ultraviolet region of the spectrum.

through a transparent medium, the rate of decrease in intensity with the thickness of the medium is proportional to the intensity of the light. This is equivalent to stating that the intensity of the emitted light decreases exponentially as the thickness of the absorbing medium increases arithmetically, or that any layer of given thickness of the medium absorbs the same fraction of the light incident upon it. We may express the law by the differential equation:

$$-\frac{dI}{dl} = kI \tag{2}$$

where I is the intensity of the incident light of wavelength λ, l is the thickness of the medium, and k is a proportionality factor. Integrating (2) and putting $I = I_0$ when $l = 0$, we obtain:

$$\ln\frac{I_0}{I_t} = kl$$

or, stated in other terms,

$$I_t = I_0 \cdot e^{-kl} \tag{3}$$

where I_0 is the intensity of the incident light falling upon an absorbing medium of thickness l, I_t is the intensity of the transmitted light, and k is a constant for the wavelength and the absorbing medium used. By changing from natural to common logarithms we obtain:

$$I_t = I_0 \cdot 10^{-0.4343kl} = I_0 \cdot 10^{-Kl} \tag{4}$$

where $K = k/2.3026$ and is usually termed the **absorption coefficient**. The absorption coefficient is generally defined as the reciprocal of the thickness (l cm) required to reduce the light to $\frac{1}{10}$ of its intensity. This follows from equation (4), since:

$$I_t/I_0 = 0.1 = 10^{-Kl} \quad \text{or} \quad Kl = 1 \quad \text{and} \quad K = 1/l$$

The ratio I_t/I_0 is the fraction of the incident light transmitted by a thickness l of the medium and is termed the **transmittance T**. Its reciprocal I_0/I_t is the **opacity**, and the **absorbance A** of the medium (formerly called the **optical density D** or **extinction E**) is given by:

$$A = \log I_0/I_t \tag{5}$$

Thus a medium with absorbance 1 for a given wavelength transmits 10 per cent of the incident light at the wavelength in question.

   Beer's law.   We have thus far considered the light absorption and the light transmission for monochromatic light as a function of the thickness of the absorbing layer only. In quantitative analysis, however, we are mainly concerned with solutions. Beer studied the effect of concentration of the coloured constituent in solution upon the light transmission or absorption. He found the same relation between transmission and concentration as Lambert had discovered between transmission and thickness of the layer {equation (3)}, i.e., the intensity of a beam of monochromatic light decreases exponentially as the concentration of the absorbing substance increases arithmetically. This may be written in the form:

$$I_t = I_0 \cdot e^{-k'c}$$
$$= I_0 \cdot 10^{-0.4343k'c} = I_0 \cdot 10^{-K'c} \tag{6}$$

where c is the concentration, and k' and K' are constants. Combining (4) and (5), we have (Ref. 6):

$$I_t = I_0 \cdot 10^{-acl} \tag{7}$$

or $\quad \log I_0/I_t = acl \tag{8}$

This is the fundamental equation of colorimetry and spectrophotometry, and is often spoken of as the **Beer–Lambert law**. The value of a will clearly depend upon the method of expression of the concentration. If c is expressed in mole dm^{-3} and l in centimetres then a is given the symbol ϵ and is called the **molar absorption coefficient** or molar absorptivity (formerly the **molar extinction coefficient**).

The specific absorption (or extinction) coefficient E_s (sometimes termed absorbancy index) may be defined as the absorption per unit thickness (path length) and unit concentration.

Where the molecular weight of a substance is not definitely known, it is obviously not possible to write down the molecular absorption coefficient, and in such cases it is usual to write the unit of concentration as a superscript, and the unit of length as a subscript.

Thus $\quad E_{1\,\text{cm}}^{1\%}$ 325 nm $= 30$

means that for the substance in question, at a wavelength of 325 nm, a solution of length 1 cm, and concentration 1 per cent (1 per cent by weight of solute *or* 1 g of solid per 100 cm^3 of solution) $\log I_0/I_t$ has a value of 30.

It will be apparent that there is a relationship between the Absorbance A, the Transmittance T, and the molar absorption coefficient, since:

$$A = \epsilon cl = \log \frac{I_0}{I_t} = \log \frac{1}{T} = -\log T \tag{9}$$

The scales of spectrophotometers are often calibrated to read directly in absorbances, and frequently also in percentage transmittance. It may be mentioned that for colorimetric measurements I_0 is usually understood as the intensity of the light transmitted by the pure solvent, or the intensity of the light entering the solution; I_t is the intensity of the light emerging from the solution, or transmitted by the solution. It will be noted that:

the **absorption coefficient** (or extinction coefficient) is the absorbance for unit path length

$$K = A/t \text{ or } I_t = I_0 \cdot 10^{-Kt}$$

the **specific absorption coefficient** (or absorbancy index) is the absorbance per unit path length and unit concentration

$$E_s = A/cl \text{ or } I_t = I_0 \cdot 10^{-E_s cl}$$

the **molar absorption coefficient** (or molar extinction coefficient) is the specific absorption coefficient for a concentration of 1 mole dm^{-3} and a path length of 1 cm.

$$\epsilon = A/cl$$

Application of Beer's law.   Let us consider the case of two solutions of a coloured substance with concentrations c_1 and c_2. These are placed in an instrument in which the thickness of the layers can be altered and measured easily, and which also allows a comparison of the transmitted light (e.g., a

Duboscq colorimeter, Section **XVIII, 6**). When the two layers have the same colour intensity:

$$I_{t_1} = I_0 \cdot 10^{-\epsilon l_2 c_1} = I_{t_2} = I_0 \cdot 10^{-c l_2 c_2} \tag{10}$$

Here l_1 and l_2 are the lengths of the columns of solutions with concentrations c_1 and c_2 respectively when the system is optically balanced. Hence under these conditions and when Beer's law holds:

$$l_1 c_1 = l_2 c_2 \tag{11}$$

A colorimeter can therefore be employed in a dual capacity: (a) to investigate the validity of Beer's law by varying c_1 and c_2 and noting whether equation (11) applies, and (b) for the determination of an unknown concentration c_2 of a coloured solution by comparison with a solution of known concentration c_1. It must be emphasised that equation (11) is valid only if Beer's law is obeyed over the concentration range employed and the instrument has no optical defects.

When a spectrophotometer is used it is unnecessary to make comparison with solutions of known concentration. With such an instrument the intensity of the transmitted light or, better, the ratio I_t/I_0 (the transmittance) is found directly at a known thickness l. By varying l and c the validity of the Lambert–Beer law, equation (6), can be tested and the value of ϵ may be evaluated. When the latter is known, the concentration c_x of an unknown solution can be calculated from the formula:

$$c_x = \frac{\log I_0/I_t}{\epsilon l} \tag{12}$$

Attention is directed to the fact that the extinction coefficient ϵ depends upon the wavelength of the incident light, the temperature, and the solvent employed. In general, it is best to work with light of wavelength approximating to that for which the solution exhibits a maximum selective absorption (or minimum selective transmittance): the maximum sensitivity is thus attained.

For matched cells (i.e., l constant) the Beer–Lambert law may be written:

$$c \propto \log \frac{I_0}{I_t}$$

$$c \propto \log \frac{1}{T}$$

or $c \propto A$ $\tag{13}$

Hence by plotting A $\left(\text{or} \log \dfrac{1}{T}\right)$, as ordinate, against concentration as abscissa, a straight line will be obtained and this will pass through the point $c = 0$, $A = 0$ ($T = 100\%$). This calibration line may then be used to determine unknown concentrations of solutions of the same material after measurement of absorbances.

Deviation from Beer's law. Beer's law will generally hold over a wide range of concentration if the structure of the coloured ion or of the coloured non-electrolyte in the dissolved state does not change with concentration. Small amounts of electrolytes, which do not react chemically with the coloured components, do not usually affect the light absorption; large amounts of electrolytes may result in a shift of the maximum absorption, and may also change the value of the extinction coefficient. Discrepancies are usually found when the coloured solute ionises, dissociates, or associates in solution, since the nature of the species in solution will vary with the concentration. The law does not hold when the coloured solute forms complexes, the composition of which depends upon the concentration. Also discrepancies may occur when monochromatic light is not used. The behaviour of a substance can always be tested by plotting $\log I_0/I_t$, or $\log T$ against the concentration: a straight line passing through the origin indicates conformity to the law.

For solutions which do not follow Beer's law, it is best to prepare a calibration curve using a series of standards of known concentration. Instrumental readings are plotted as ordinates against concentrations in, say, mg per 100 cm^3 or 1000 cm^3 as abscissae. For the most precise work each calibration curve should cover the dilution range likely to be met with in the actual comparison.

XVIII, 3.   CLASSIFICATION OF METHODS OF 'COLOUR' MEASUREMENT OR COMPARISON.

The basic principle of most colorimetric measurements consists in comparing under well-defined conditions the colour produced by the substance in unknown amount with the same colour produced by a known amount of the material being determined. The quantitative comparison of these two solutions may, in general, be carried out by one or more of six methods. It is not essential to prepare a series of standards with the spectrophotometer; the molar absorption coefficient can be calculated from one measurement of the absorbance or transmittance of a standard solution, and the unknown concentration can then be computed with the aid of the molar absorption coefficient and the observed value of the absorbance or transmittance (cf. Section **XVIII, 2**, equations (12) and (13)).

A. Standard series method. (Section **XVIII, 4**) The test solution contained in a Nessler tube is diluted to a definite volume, thoroughly mixed, and its colour compared with a series of standards similarly prepared. The concentration of the unknown is then, of course, equal to that of the known solution whose colour it matches exactly. The accuracy of the method will depend, *inter alia*, upon the concentrations of the standard series; the probable error is of the order of ± 3 per cent, but may be as high as ± 8 per cent.

For convenience, artificial standards, e.g., Lovibond glasses, salt solutions such as iron(III) chloride in aqueous hydrochloric acid (yellow), aqueous cobalt chloride (pink), aqueous copper sulphate (blue), and aqueous potassium dichromate (orange) are sometimes used. It is essential to standardise the artificial standards against known amounts of the substance being determined, the latter always being treated under exactly similar conditions. The disadvantage of this method is that the spectral absorption curves of the test solutions and of the sub-standard glasses or solutions may be far from identical; the error due to this cause is greatly magnified in the case of observers suffering from partial colour blindness.

B. Duplication method. (Section **XVIII, 5**) A standard solution of the component under determination is added to the reagent until the colour produced matches that of the unknown sample in the *same* volume of solution. This method is less accurate than **A**.

C. Dilution method. The sample and standard solution are contained in glass tubes of the same diameter, and are observed *horizontally* through the tubes. The more concentrated solution is diluted until the colours are identical in intensity when observed horizontally through the same thickness of solution. The relative concentrations of the original solutions are then proportional to the heights of the matched solutions in the tubes. This is the least accurate method of all, and will not be discussed further.

D. Balancing method. (Section **XVIII, 6**)   This method forms the basis of all colorimeters of the plunger type, e.g., in the Duboscq colorimeter. The comparison is made in two tubes, and the height of the liquid in one tube is adjusted so that when both tubes are observed vertically the colour intensities in the tubes are equal. The concentration in one of the tubes being known, that in the other may be calculated from the respective lengths of the two columns of liquid and the relation (eqn XVIII, 11):

$$c_1 l_1 = c_2 l_2$$

It must be emphasised again that this simple proportionality holds only if Beer's law is applicable, and that the relation holds with greater exactness if a beam of monochromatic light (obtained with the aid of a suitable colour filter) rather than white light is employed. As a general rule, it is preferable that the solutions under comparison should not differ greatly in concentration, and for the most accurate work an empirically constructed calibration curve should be used. As usually employed with white light, the accuracy obtainable with a Duboscq colorimeter is of the order of ± 7 per cent; the accuracy is increased appreciably if monochromatic light (produced with colour filters) is employed.

E. Photoelectric photometer method. (Section **XVIII, 7**)   In this method the human eye is replaced by a suitable photoelectric cell; the latter is employed to afford a direct measure of the light intensity, and hence of the absorption. Instruments incorporating photoelectric cells measure the light absorption and not the colour of the substance: for this reason the term 'photoelectric colorimeters' is a misnomer; better names are photoelectric comparators, photometers, or, best, absorptiometers.

Essentially most such instruments consist of a light source, a suitable light filter to secure an approximation to monochromatic light (hence the name photoelectric filter photometer), a glass cell for the solution, a photoelectric cell to receive the radiation transmitted by the solution, and a measuring device to determine the response of the photoelectric cell. The comparator is first calibrated in terms of a series of solutions of known concentration, and the results plotted in the form of a curve connecting concentrations and readings of the measuring device employed. The concentration of the unknown solution is then determined by noting the response of the cell and referring to the calibration curve.

These instruments are available in a number of different forms incorporating one or two photocells. With the one-cell type, the absorption of light by the solution is usually measured directly by determining the current output of the photoelectric cell in relation to the value obtained with the pure solvent. It is of the utmost importance to use a light source of constant intensity, and if the photo cells exhibit a 'fatigue effect' it is necessary to allow them to attain its equilibrium current after each change of light intensity. The two-cell type of filter photometer is usually regarded as the more trustworthy (provided the electrical circuit is appropriately designed) in that any fluctuation of the intensity of the light source will affect both cells alike if they are matched for their spectral response. Here the two photocells, illuminated by the same source of light, are balanced against each other through a galvanometer: the test solution is placed before one cell and the pure solvent before the other, and the current output difference is measured.

F. Spectrophotometer method. (Section XVIII, 9) This is undoubtedly the most accurate method for determining *inter alia* the concentration of substances in solution, but the instruments are, of necessity, more expensive. A spectrophotometer may be regarded as a refined filter photoelectric photometer which permits the use of continuously variable and more nearly monochromatic bands of light. The essential parts of a spectrophotometer are: (i) a source of radiant energy, (ii) a monochromator, i.e., a device for isolating monochromatic light or, more accurately expressed, narrow bands of radiant energy from the light source, (iii) glass or silica cells for the solvent and for the solution under test, and (iv) a device to receive or measure the beam or beams of radiant energy passing through the solvent or solution.

In the following Sections it is proposed to discuss the more important of the above methods in somewhat greater detail. For a more complete treatment the reader is referred to the special treatises on the subject (see Selected Bibliography at end of chapter).

XVIII, 4. STANDARD SERIES METHOD. In this method colourless glass tubes of uniform cross-section and with flat bottoms are usually employed. These are termed **Nessler tubes**. The best variety have polished, flat bottoms. They are made in either the 'low' form with a height of 175–200 mm and a diameter of 25–32 mm (Fig. XVIII, 2) or as a 'high' form with a height of 300–375 mm and a diameter of 21–24 mm. The solution of the substance being determined is made

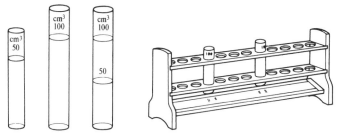

Fig. XVIII, 2   **Fig. XVIII, 3**

up to a definite volume, and the colour is compared with that of a series of standards prepared in the same way from known amounts of the component being determined. Fifty or 100 cm^3 of the unknown and standard solutions are placed in Nessler tubes, and the solutions are viewed *vertically* through the length of the columns of the liquid. The concentration of the unknown is equal to that of the standard having the same colour.* As a general rule, it will be found that the colour intensity of the unknown lies between two successive standards. Another

---

* It is advisable, wherever possible, to make a preliminary determination of the strength of the unknown solution by adding from a burette a solution of the component in known concentration to a Nessler tube containing the reagents diluted with a suitable amount of water until the depth of colour obtained is practically the same as that of an equal volume of the unknown solution also contained in a Nessler cylinder and standing at its side. A series of standards on either side of this concentration is then prepared.

series of standards may then be prepared covering the latter range over smaller concentration intervals. Thus, for example, in the determination of a particular constituent the first series of standards might cover the range 0.1, 0.2, 0.4, 0.6, 0.8, and 1.0 mg dm^{-3}, and it is found that the colour of the unknown lies between 0.4 and 0.6 mg dm^{-3}. The second series of standards may then be prepared containing 0.40, 0.45, 0.50, 0.55, and 0.60 mg dm^{-3}. Further comparison may then show that the value lies between 0.45 and 0.50 mg dm^{-3}, and for many purposes this should be returned as 0.48 mg dm^{-3}. If a more accurate value is required, and provided the colour intensity of the solution and also the apparatus employed will permit of finer comparison, another series of standards covering the range of, say, 0.45, 0.475, and 0.50 mg dm^{-3} may be made up and the unknown compared with these standards.

For the comparison of colours in Nessler tubes, the simplest apparatus consists of a modified test-tube rack (Fig. XVIII, 3). It is constructed of wood, finished dull black, and is provided with an inclined opal glass reflector or mirror, arranged to reflect light up through the tubes. The Nessler tubes rest on a narrow ledge, and do not come into contact with the reflector. The unknown and standards are compared by placing them adjacent to each other and looking vertically down through them.

This procedure serves as the basis for the colorimetric determination of pH by employing a series of buffer solutions and suitable indicators. A series of appropriate buffer solutions is selected, differing successively in pH by about 0.2, covering the pH range of the solutions under investigation; the range of the buffer solutions required will be indicated by the preliminary pH determination. Equal volumes, say 10 cm^3, of the buffer solutions differing successively in pH by about 0.2 are placed in test-tubes of colourless glass and having approximately the same dimensions, and a small equal quantity of a suitable indicator for the particular pH range is added to each tube. A series of different colours corresponding to the different pH values is thus obtained. An equal volume (say 10 cm^3) of the test solution is treated with an equal volume of indicator to that used for the buffer solutions, and the resulting colour is compared with that of the coloured standard buffer solutions. When a complete match is found, the test solution and the corresponding buffer solution have the

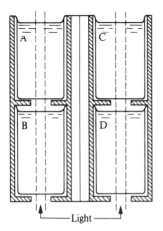

Fig. XVIII, 4

same pH. Sometimes a complete match is not obtained, but the colour of the test solution falls between those of two successive standards, then it is known that the pH value lies between those of the two standards. Further buffer solutions may then be prepared differing by 0.1 pH, if desired, and pH value redetermined. As a general rule, colorimetric methods cannot be relied upon to give values of pH more accurate than to within 0.2 pH unit. For matching the colours, the buffer solutions may be arranged in the holes of a test-tube stand in order of pH: the test solution is then moved from hole to hole until the best colour match is obtained Special stands and standards for making the comparision are available commercially. The commercial standards, prepared from buffer solutions, are not permanent, and must be checked every six months.

For turbid or slightly coloured solutions, the direct-comparison method given above can no longer be applied. The interference due to the coloured substance can be eliminated in a simple way by the following device, suggested by Walpole. In Fig. XVIII, 4, A, B, C, and D are glass cylinders with plane bases standing in a box which is painted dull black on the inside. A contains the coloured solution to be tested (here the test solution + indicator), B contains an equal volume of water, C contains a solution of known strength for comparison (here the standard buffer solution + indicator), while D contains the same volume of the solution to be tested as was originally added to A. The colour of the unknown solution is thus compensated for.

Standard series using glass comparators. A number of devices are manufactured which employ permanent glass standards which can be mounted in special viewers. The **BDH Lovibond Nessleriser Mark 3*** is one of the simplest of these instruments and can be used for a variety of determinations (Fig. XVIII, 5).

Fig. XVIII, 5

It consists essentially of a plastic case for holding two Nessler tubes vertically between a reflector and a detachable rotating disc having nine apertures containing a series of graded, permanent glass standards. Each disc incorporates

---

* Manufactured by The Tintometer Ltd, Salisbury, England, and available from BDH Chemicals Ltd, Poole, BH12 4NN, England.

a series of standards designed for one particular test conducted under specified conditions. Discs are available for many of the common determinations by colorimetric methods and include: ammonia (with Nessler's reagent); dissolved oxygen (with indigo carmine); copper (with dithio-oxamide); nitrate (with phenol-2,4-disulphonic acid) and chlorine (with *p*-amino-N:N-diethyl-aniline sulphate).

A similar device suitable for a wider range of determinations is the **Lovibond '1000' Comparator**,* this also uses series of permanent glass colour standards (Fig. XVIII, 6). The discs containing the nine glass colour standards fit into the *comparator*, which is furnished with four compartments to receive small test-tubes or rectangular cells, and is also provided with an opal glass screen. The disc

Fig. XVIII, 6

can revolve in the comparator, and each colour standard passes in turn in front of an aperture through which the solution in the cell (or cells) can be observed. As the disc revolves, the value of the colour standard visible in the aperture appears in a special recess.

Over 300 standard tests can now be carried out using the Lovibond '1000' Comparator including narrow and broad range pH determinations, and the concentrations of many metal ions, detergents and organic compounds.

XVIII, 5.   DUPLICATION METHOD.   This method finds its chief application in the so-called **colorimetric titration**. A known volume, say 50 or 100 cm^3, of the solution is placed in a Nessler cylinder (Fig. XVIII, 2) and a measured volume of the reagent or reagents is then added. An equal volume of water (50 or 100 cm^3) together with the same volume of the reagent is introduced into another similar Nessler cylinder. For mixing the solutions both cylinders are provided either with a glass tube on which a flattened bulb (*ca.* 1 cm diameter) is blown or with a stirring-rod of which the lower end is flattened to a width of 1 cm and over a length of several centimetres. The tubes should also be provided with black- or brown-paper cylinders to exclude light from the sides. The colour intensities are

---

* Manufactured by The Tintometer Ltd, Salisbury, England, and available from BDH Chemicals Ltd, Poole, BH12 4NN, England.

compared by holding the tubes close together over a white surface, such as a sheet of opal glass or, better, in a Nessler tube-stand (Fig. XVIII, 3). A solution containing a known concentration of the constituent being determined is added to the blank solution from a burette (preferably of the micro type) until the colours of the two solutions viewed by looking down into the tubes match. As a rule, if the volume of the standard solution required to match the colour of the unknown is less than about 2 per cent of the total volume, the volume change due to the addition of the reagent may be neglected. It may, however, be allowed for by a simple calculation, or the determination may be repeated by taking $100 - x$ (or $50 - x$) cm³ of water, where x is the volume of the standard solution employed in the first titration. Several determinations should be carried out, and the positions of the tubes should be interchanged—thus that on the right-hand side should be put to the left of the observer and vice versa.

It must be emphasised that this method can be applied only when the colour is independent of the mode of mixing, for in one tube a very dilute solution of the substance to be determined is mixed with the reagent, whilst in the other tube a comparatively concentrated solution of the substance is mixed with a dilute solution of the reagent. The development of colour should be practically instantaneous and remain permanent during the time required for the measurements; foreign substances present in the unknown should not affect the colour. The method is, at best, only an approximate one, but has the advantage that only the simplest apparatus is required.

XVIII, 6. BALANCING METHOD. Plunger-type colorimeters.

The plunger-type of colorimeter with two halves of the field of view illuminated by the light passing through the unknown and standard solutions respectively was invented by J. Duboscq of Paris in 1854. Various improved modifications of the instrument were subsequently developed by manufacturers of optical apparatus. Before describing the latter, reference must be made to **Hehner cylinders** (Fig. XVIII, 7). These are utilised in pairs, and are the simplest form of apparatus employed in matching colours by the balancing method. Each cylinder has a glass stopcock about 2.5 cm from the bottom through which liquid may be drawn off until the colour in the two cylinders is the same in intensity when viewed vertically. The cylinders are graduated at 1cm³ intervals, and usually have a capacity of 100 cm³; they should have flat, carefully ground and polished bottoms of clear glass and be uniform in bore. It is advisable to place them in a box so arranged that the light is reflected from the bottom of the latter up through the tubes.

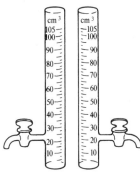

Fig. XVIII, 7

The essential principles of a **Duboscq colorimeter** are illustrated in Fig. XVIII, 8. Light from an even source of illumination concealed in the base of the instrument passes through the windows (matt white screens) in the top of the base through the solutions to be tested and through the plungers. Some of the light is absorbed in passing through the liquids, the amount of absorption being dependent upon the concentration and the depth of the solution. The two beams of light from the plungers are then brought to a common axis by a prism system.

On looking through the eyepiece, a wide, circular field is visible, light from one cup illuminating one half, and light from the second cup illuminating the other half of the field. The depths of the columns of liquids are adjusted by rotating the milled heads on either side of the instrument, which raises and lowers the cups, until the two halves of the field are identical in intensity, i.e., until the dividing line practically disappears. When this condition holds and Beer's law is applicable, the concentrations of the two solutions are inversely proportional to their depths, which are normally read on the scales attached to the cup carriers. The two scales are 60 cm long and are engraved on metal: they are divided into millimetres, and a vernier scale enables readings to be taken to within 0.1 mm.

Use of Duboscq-type colorimeter. The colorimeter must be kept scrupulously clean. The cups and plungers are rinsed with distilled water and either dried with soft lens-polishing material or rinsed with the solution to be measured.

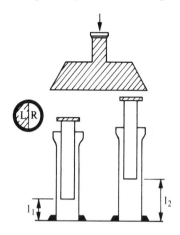

Make sure that the readings are zero when the plungers just touch the bottoms of the cups. Place the standard solution in one cup, and an equal volume of the unknown solution in the other; do not fill the cups above the shoulder. Set the unknown solution at a scale reading of 10.0 mm and adjust the standard until the fields are matched. Carry out at least six adjustments with the cup containing the standard solution, and calculate the mean value.

The plungers should always remain below the surface of the liquid. Since the eye may become fatigued and unable to detect small differences, it is recommended after making adjustment to close the eyes for a moment or to look at something else, and then see if the adjustment still appears satisfactory. It is advisable to approach the match point both from above and below.

Fig. XVIII, 8

If l_1 and l_2 are the average readings for the cups containing the solutions of known and unknown concentration respectively, and c_1 and c_2 are the corresponding concentrations, then if Beer's law holds:

$$c_1 l_1 = c_2 l_2 \quad \text{or} \quad c_2 = c_1 \frac{l_1}{l_2}$$

It will be noted that if $l_2 = 10.0$, the standard scale when multiplied by 10 will give the percentage concentration of the sample in terms of the standard.

Owing to optical and mechanical imperfections of some makes of colorimeters, it is sometimes found that the same reading cannot be obtained in the adjustment for illumination when the cups are filled with the same solution and balanced. In such a case one of the cups (say, the left one) is filled with a reference solution (which may be a solution containing the component to be determined) of the same colour and approximately the same intensity as the unknown and the plunger set at some convenient point (about the middle) of the scale. Fill the other cup with a solution having a colour corresponding to a known concentration of the component to be determined, and adjust this cup to colour balance. Take the reading and repeat the adjustment, say ten times, in such a way that the balancing point is approached five times from the lower and five times from the higher side. Calculate the average reading (l_1). Remove the cup, rinse it thoroughly, and fill it with the unknown solution. Repeat the balancing exactly as for the standard solution, and find the average of, say, 10 readings (l_2). If c_1 is the concentration in the standard solution, then the concentration of the unknown solution is given by:

$$c_2 = c_1 \frac{l_1}{l_2}$$

(This method is comparable in many respects to the method of weighing by substitution.) If Beer's law is not valid for the solution, it is best to arrange matters so that the colour intensity of the standard lies close to that of the unknown.

Immediately the determination has been completed, empty the cups and rinse both the cups and plungers with distilled water. Leave the colorimeter in a scrupulously clean condition.

XVIII, 7. PHOTOELECTRIC PHOTOMETER METHOD. Photoelectric colorimeters (absorptiometers).

One of the greatest advances in the design of colorimeters has been the use of photoelectric cells to measure the intensity of the light, thus eliminating the errors due to the personal characteristics of each observer. Before describing the various types of photoelectric colorimeters and spectrophotometers, a brief account will be given of the construction and properties of the light-sensitive devices employed. Photoemissive and barrier-layer cells are commonly used.

Photoemissive cells. In the simplest form of photoemissive cell (also called phototube) a glass bulb is coated internally with a thin, sensitive layer, such as cesium or potassium oxide and silver oxide (i.e., one which emits electrons when illuminated), a free space being left to permit the entry of the light. This layer is the cathode. A metal ring inserted near the centre of the bulb forms the anode, and is maintained at a high voltage by means of a battery. The interior of the bulb may be either evacuated or, less desirably, filled with an inert gas at low pressure (e.g., argon at about 0.2 mm). When light, penetrating the bulb, falls on the sensitive layer, electrons are emitted, thereby causing a current to flow through an outside circuit; this current may be amplified by electronic means, and is taken as a measure of the amount of light striking the photosensitive surface. Otherwise expressed, the emission of electrons leads to a potential-drop across a high resistance in series with the cell and the battery; the fall in potential may be measured by a suitable potentiometer, and is related to the amount of light falling on the cathode. The action of the photoemissive cell is shown diagrammatically in Fig. XVIII, 9.

The sensitivity of a photoemissive cell (photo-tube) may be considerably increased by means of the so-called **photomultiplier tube**. The latter

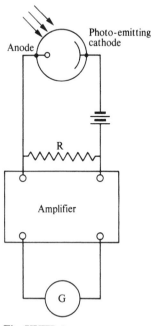

Anode

Photo-emitting cathode

R

Amplifier

G

Fig. XVIII, 9

consists of an electrode covered with a photoemissive material and a series of positively charged plates, each charged at a successively higher potential. The plates are covered with a material which emits several (2–5) electrons for each electron collected on its surface. When the electrons hit the first plate, secondary electrons are emitted in greater number than initially struck the plate, with the net result of a large amplification (up to 10^6) in the current output of the cell. The output of a photomultiplier tube is limited to several milliamperes, and for this reason only low incident radiant energy intensities can be employed. It can measure intensities about 200 times weaker than those measurable with an ordinary photoelectric cell and amplifier.

Barrier-layer cells. A barrier-layer cell (also known as a photovoltaic or photronic cell) is entirely different in design and principle from the photoemissive cell, and it operates without the use af a battery. The commonest form consists essentially of a metal base plate A (usually of iron) upon which is deposited a thin layer of a semiconductor, usually selenium, B (see Fig. XVIII, 10); this is covered, in turn, by a transparent metal layer D, which is lacquered except for a portion mechanically strengthened to form a collecting ring E. The underside of the metal base plate is covered by a non-oxidising metal.

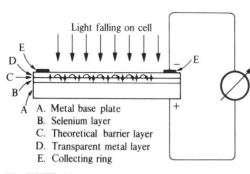

Light falling on cell

E
D
C
B
A

A. Metal base plate
B. Selenium layer
C. Theoretical barrier layer
D. Transparent metal layer
E. Collecting ring

Fig. XVIII, 10

When light passes through the thin metal film of the silver collecting electrode D to the selenium layer electrons are released from the semiconductor surface. These penetrate a hypothetical barrier layer C, passing to the collecting electrode D. Thus, under the action of light, we have a cell of which the negative pole is the metal collecting ring E and the positive pole is the metal base plate. If this cell is connected to a galvanometer, a current will flow which will vary with the intensity of the incident light.

The 'EEL.' selenium photocell* is an excellent example of a highly sensitive barrier-layer cell. Figure XVIII, 11 shows the output obtained from cells of different areas when subjected to an illumination of approx. 54 lux. Figure XVIII, 12 shows how the current output of a typical 'EEL' 45-mm cell varies with the intensity of the light falling upon it and with the resistance of the external circuit.

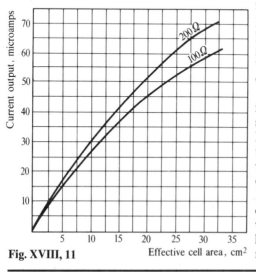

Current output, microamps

70
60
50
40
30
20
10

200 Ω
100 Ω

5   10   15   20   25   30   35

Fig. XVIII, 11   Effective cell area, cm²

---

* Manufactured by Corning-EEL Ltd, Colchester Road, Halstead, Essex, England.

In the single-cell type of photoelectric colorimeter the galvanometer (or micro-ammeter) reading is taken as a measure of the intensity of the light falling upon the cell. It is important therefore that conditions be so chosen that there is a straight-line relationship between the light intensity and the current output; this may be achieved by having a low resistance in the external circuit (< 100–200 ohms) and using a low level of illumination. The linear response is not important for colorimeters of the two-cell type.

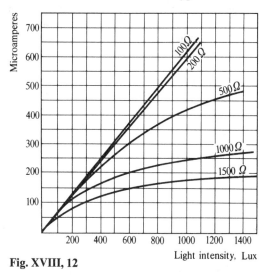

Fig. XVIII, 12

The current output of the barrier-layer cell is dependent also upon the wavelength of the incident light, the variation being similar to that of the human eye, particularly if a corrective filter is employed. The spectral response of the 'EEL' cell compared with that of the eye is shown in Fig. XVIII, 13.

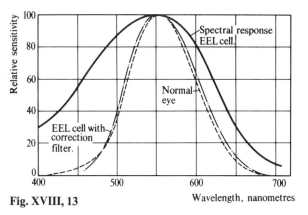

Fig. XVIII, 13

Barrier-layer cells are much more rugged than photoemissive cells and produce higher currents per lumen. The current produced cannot, however, be readily amplified by conventional electronic circuits* because of the low internal

---

* A feedback amplifier can, however, be used.

resistance; barrier-layer cells are largely used where low cost and portability are required. On the other hand, photoemissive cells have very high internal resistances, and their output currents are readily amplified: they are usually employed in the most sensitive devices measuring low intensities of illumination and, indeed, may be ruined by high intensities of incident radiant energy. Barrier-layer cells may exhibit fatigue effects, particularly at sudden exposure to high levels of illumination: the current may rise to a value several per cent higher than the apparent equilibrium value and then fall off gradually. Upon standing in the dark, the original sensitivity is recovered. The fatigue effect can be minimised by careful selection of the optimum level of illumination, resistance of the measuring circuit, etc.

Light filters. Optical filters are used in colorimeters (absorptiometers) for isolating any desired spectral region. They consist of either thin films of gelatin containing different dyes or of coloured glass. The extensive range of Wratten

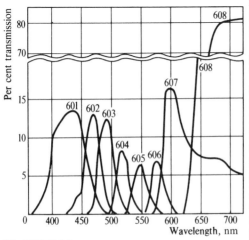

Fig. XVIII, 14

filters supplied by Kodak are of the former type, the gelatin films being about 0.1 mm thick.

Other optical filters are manufactured by Ilford and by Corning. The transmission curves for a series of Ilford Standard Spectrum Filters are depicted in Fig. XVIII, 14. The manufacturers give the following transmission ranges for the various filters: No. 601 Spectrum Violet, 380–470 nm, No. 602, Spectrum Blue, 440–490 nm; No. 603, Spectrum Blue-Green, 470–520 nm. No 604, Spectrum Green, 500–540 nm; No. 605, Spectrum Yellow-Green, 530–570 nm; No. 606, Spectrum Yellow, 560–610 nm; No. 607, Spectrum Orange, 570 nm with absorption increasing from 600 nm onwards; No. 608, Spectrum Red, 620 nm into infrared. In addition to the above Messrs Ilford market a series of Bright Spectrum Filters, Nos. 621–626 (No. 621 Bright Spectrum Violet to No. 626 Bright Spectrum Yellow) which are considerably brighter (i.e., have a higher transmission) than the standard Spectrum Filters but with a slightly wider transmission range.

Interference filters (transmission type) have somewhat narrower transmitted bands than coloured filters and are available commercially.* Interference filters are essentially composed of two highly reflecting but partially transmitting films of metal (usually silver separated by a spacer film of transparent material). The amount of separation of the metal films governs the wavelength position of the pass band, and hence the colour of the light that the filter will transmit. This is the result of an optical interference effect which produces a high transmission of light when the optical separation of the metal films is effectively a half wavelength or a multiple of a half wavelength. Light which is not transmitted is for the most part

* For example, from Bausch and Lomb Inc, 820 Linden Avenue, Rochester, New York, 14625; and from Barr and Stroud Ltd, Caxton Street, Anniesland, Glasgow, G13 1HZ.

reflected. The wavelength region covered is either from 253–390 nm or from 380–1100 nm, peak transmission is between 25–50 per cent and the bandwidth is less than 18 nm for the narrowband filters suitable for colorimetry.

The ideal way of *selecting a filter* for use with a coloured solution is to construct first, by means of a suitable spectrophotometer, the absorption curve for the visible spectrum. Comparison of this curve with the spectral transmission curves of the set of filters supplied by the manufacturers enables a suitable choice to be made. Alternatively, absorbance (or transmittance)/concentration calibration curves may be constructed with a photoelectric colorimeter, using each of the filters in turn. As a general rule, the best filter to use in a particular determination is that which gives the maximum absorption or minimum transmission for a given concentration of the absorbing substance. Less satisfactory methods include the use of a filter that gives the smallest transmission for a given concentration and depth of cell, and the use of a filter whose colour is as close as possible to the complementary colour of the solution. A table of complementary colours is given below.

Complementary colours

Wavelength (nm)	Hue (transmitted)	Complementary hue
400–435	Violet	Yellowish-green
435–480	Blue	Yellow
480–490	Greenish-blue	Orange
490–500	Bluish-green	Red
500–560	Green	Purple
560–580	Yellowish-green	Violet
580–595	Yellow	Blue
595–610	Orange	Greenish-blue
610–750	Red	Bluish-green

Prisms. To obtain improved resolution of spectra in both the visible and ultraviolet regions of the spectrum it is necessary to employ a better optical system than that possible with filters. In many instruments, both manual and automatic, this is achieved by using prisms to disperse the radiation obtained from incandescent tungsten or deuterium sources. The dispersion is dependent upon the fact that the refractive index, n, of the prism material varies with wavelength, λ, the dispersive power being given by $\partial n/\partial \lambda$. The separation achieved between different wavelengths is dependent upon both the dispersive power and the apical angle of the prism.

In instruments in which the radiation is only passed through the prism in a single direction it is common to use a 60° prism. In some cases double dispersion is achieved by reflecting the radiation back through the prism by placing a mirrored surface behind the prism, as

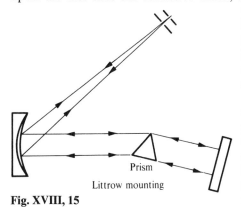

Prism

Littrow mounting

Fig. XVIII, 15

in the Littrow mounting, Fig. XVIII, 15.* Monochromatic radiation of different wavelengths is brought to focus on the instrument slit by rotation of the prism.

Unfortunately no single material is entirely suitable for use over the full range of 200–1000 nm, although fused silica is the favourite compromise material. Glass prisms can be employed between 400 and 1000 nm for the visible region, but are not transparent to ultraviolet radiation. For the region below 400 nm quartz or fused silica prisms are required. If quartz is employed for a 60° single pass prism it is necessary to make the prism in two halves, one half from right-handed quartz and the other from left-handed quartz in order that polarisation effects introduced by one will be reversed by the other.

Prisms have the advantage that, unlike the diffraction gratings described below, they only produce a single order spectrum.

Diffraction gratings. This alternative method of dispersion uses the principle of diffraction of radiation from a series of closely spaced lines marked on a surface. Early diffraction gratings were made of glass through which the radiation passed and became diffracted; these are known as transmission gratings. To achieve the diffraction of ultraviolet radiation, however, modern grating spectrophotometers employ metal reflection gratings with which the radiation is reflected from the surfaces of a series of parallel grooves. These are often known as echelette gratings.

The principle of diffraction is dependent upon the differences in path length experienced by a wavefront incident at an angle to the individual surfaces of the grooves of the grating. If i is the angle of incidence and r the angle of reflection the path difference between rays from adjacent grooves is given by

$$d \sin i - d \sin r$$

where d is the distance between the grooves, Fig. XVIII, 16. Because of the path

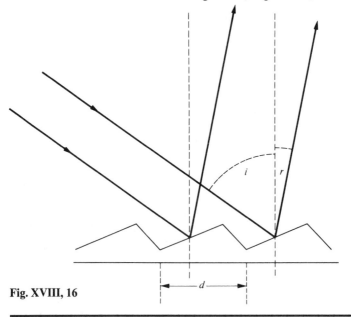

Fig. XVIII, 16

---

* Reproduced by permission from *A Dictionary of Spectroscopy*, by R. C. Denney, Macmillan, 1973.

difference that is created the new wavefronts interfere with each other except when the path difference is an integral number of wavelengths, i.e., when

$$n\lambda = d(\sin i \pm \sin r) \tag{14}$$

When polychromatic radiation is incident upon the diffraction grating this equation can usually only be satisfied for a single wavelength at a time. Rotation of the grating to change the angle of incidence i will bring each wavelength in turn to a position to satisfy the equation, thus serving as a method of monochromation.

Diffraction gratings suffer from the disadvantage that they produce second-order and higher-order spectra which can overlap the desired first-order spectrum. This overlap is most commonly seen between the long wavelength region of the first-order spectrum and the shorter wavelength region of the second-order spectrum. The difficulty is overcome by using carefully positioned filters in the instrument to block the undesired wavelengths.

For ultraviolet/visible spectrophotometers the gratings employed have between 10000 and 30000 lines cm^{-1}. This very fine ruling means that the value of d in equation 14 is small and produces high dispersion between wavelengths in the first-order spectrum. Only a single grating is required to cover the region between 200 and 900 nm. The Unicam SP1700 (Fig. XVIII, 33) is a spectrophotometer in which monochromation is obtained by a diffraction grating.

Instruments

XVIII, 8. PHOTOELECTRIC COLORIMETERS (ABSORPTIO-METERS). Photoelectric colorimeters may be divided into two main classes: one-cell and two-cell instruments. Examples of the former are the 'EEL' portable colorimeter and the long cell absorptiometer,[1] the Unicam SP 300 G.P. photoelectric colorimeter,[2] and the Bausch and Lomb Spectronic 20 colorimeter,[3] while an example of the latter is the Hilger Spekker absorptiometer (type H 760).[4]

The essential parts of a *one-cell filter photoelectric photometer* (Fig. XVIII, 17) are a light source, a light filter, a container for the solution, a barrier-layer photocell to receive the transmitted light, and some means for measuring the response of the photocell. A brief description will now be given of some typical instruments.

Corning colorimeter 252. This particular instrument, which is capable of an accuracy within ± 1 per cent, is illustrated in Fig. XVIII, 18. It employs a series of drop-in gelatine filters (Ilford number 601–608) to cover the wavelength range from 400–710 nm. The transmitted radiation, from a tungsten filament lamp, is detected by means of a phototube which provides a signal to a moving coil analog meter. The instrument can be used with sample volumes as small as 0.8 cm^3.

Manufactured by:
[1] Corning-EEL Ltd.
[2] Pye-Unicam Ltd.
[3] Bausch and Lomb Inc.
[4] Rank Hilger Ltd.

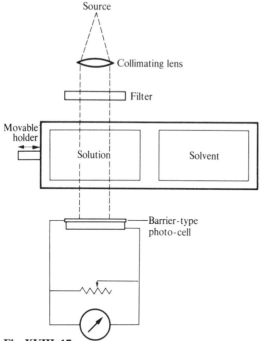

Fig. XVIII, 17

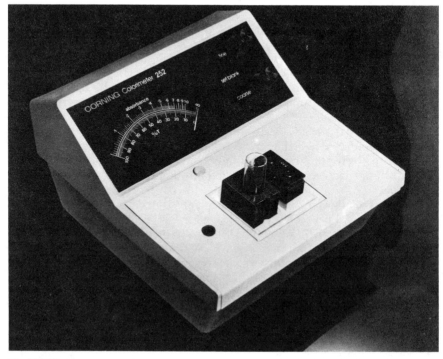

Fig. XVIII, 18

'EEL' absorptiometer (long cell type). This instrument, presented diagrammatically in Fig. XVIII, 19, will accommodate cells of 2.5, 5, 10, 20, 40, and 100 mm optical path, corresponding to volumes of 1.5, 3, 6, 12, 24, and 60 cm^3. The sliding sample-carriage has a simple lever mechanism to bring into position the reference solution and subsequently the sample to be measured.

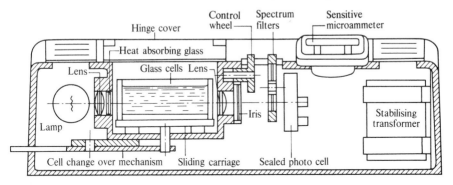

Fig. XVIII, 19

Readings are taken on a sensitive microammeter, calibrated in percentage transmission and absorbance. There is a colour-filter wheel which permits the easy insertion into the light beam of any one of nine spectrum filters.

Unicam SP. 300 G.P. photoelectric colorimeter. This colorimeter (Fig. XVIII, 20, the optical system is shown in Fig. XVIII, 21) operates from a 6-volt battery or from the a.c. mains supply through a constant-voltage transformer.

Fig. XVIII, 20

There is a dual cell holder which brings either cell into the same relative position with respect to the light path. The light is controlled by a shutter mechanism; it passes through an absorption cell with an optical path of 10 mm, which holds the solution under test. The transmitted light passes through the selected filter before

715

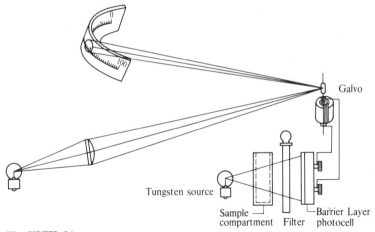

Fig. XVIII, 21

reaching the barrier-layer photocell, the output current of which is indicated on the galvanometer. Ilford spectrum filters can be supplied. The galvanometer is calibrated with a linear scale 0–100, underneath which is a logarithmic scale for calculations of the absorbance.

Bausch and Lomb Spectronic 20 colorimeter. The instrument, shown in Fig. XVIII, 22, consists essentially of a diffraction-grating monochromator and an electronic detection, amplification, and measuring system. It operates from 115-

Fig. XVIII, 22

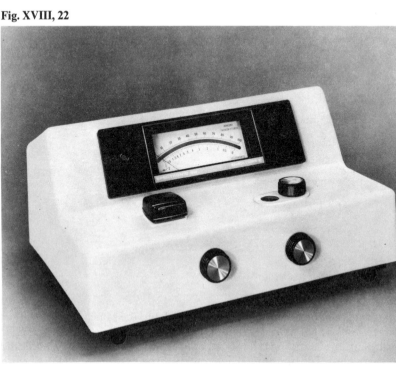

volt, 60-hertz mains or from a battery. The wavelength range is from 375 to 650 nm, and can be extended to 950 nm by the addition of a red filter and exchange of phototubes: the effective band-width is about 20 nm.

The optical system is presented in Fig. XVIII, 23. White light from the tungsten lamp is focused by lens A on the entrance slit; lens B collects the light from the entrance slit and refocuses it on the exit slit after it has been reflected and dispersed by the diffraction grating. To obtain various wavelengths, the grating is rotated by means of an arm which is moved when the cam is rotated; the

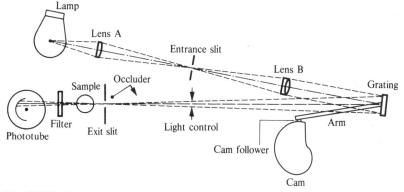

Fig. XVIII, 23

wavelength scale is fastened to the same shaft as the cam. The monochromatic light which passes through the exit slit goes on through the sample to be measured and falls upon the phototube. Whenever the sample is removed from the instrument, an occluder falls into the light beam so that the amplifier control can be adjusted with no further manipulation. A light control is provided for setting the meter to full-scale deflection with a suitable blank in the sample compartment. Cuvettes or special small test-tubes are used as containers for the samples.

The apparatus, although primarily designed as a colorimeter, also serves as an inexpensive spectrophotometer. For colorimetric work, the wavelength control is rotated until the desired wavelength in nm is indicated on the wavelength scale. The amplifier control is adjusted to bring the meter needle to zero on the 'Percent Transmittance Scale' or ∞ on the 'Absorbance Scale'. The test-tube or cuvette containing water or other solvent is then inserted in the sample holder. The light control is then rotated until the meter reads '100' or '0'. The unknown sample is then inserted in place of the blank and the Percent Transmittance or Absorbance read directly from the meter.

Two-cell instruments. In view of possible variations of the operating current of the light source in one-cell colorimeters, two-cell circuits have been proposed based upon the idea that fluctuations would affect the two cells equally and thus be compensated. In addition, the null-point method of balancing the cells against each other, as indicated by a galvanometer, is supposed largely to eliminate errors arising from cell fatigue or temperature changes. The two photocells should be selected on the basis of similarity in spectral response, and should be matched as closely as possible.

Hilger Spekker absorptiometer. The actual instrument (type H 760) is

Fig. XVIII, 24a

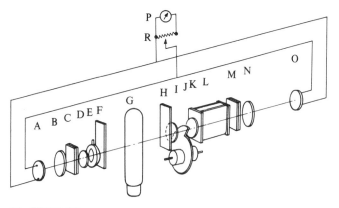

Fig. XVIII, 24b

shown in Fig. XVIII, 24a; the optical arrangements are incorporated in Fig. XVIII, 24b, which includes the photoelectric circuit. A 100-watt projection lamp G, mounted in a central lamphouse and run from the electric-supply mains, is the source of light. To the right of G the light first passes through a heat-absorbing filter H, and then through a lens I to render it parallel before it passes through a cam-shaped diaphragm J, which controls the aperture of the beam and hence its intensity. The light then passes through lens K to a cell containing the absorbing liquid L, a selected colour filter M, and a lens N, which forms an image of the light source on the surface of the photocell O. On the left of G the light passes through a heat-absorbing filter F, symmetrical with H, an iris diaphragm E, a lens system D and B (corresponding with that on the right-hand side of the instrument), a selected colour filter C, and, finally, an image of the light source G is formed on the photocell A.

The cam-shaped disc J is connected with a large, calibrated drum, and enables the intensity of the light falling upon the photocell O to be varied by known amounts. Since there is an image of the filament on the cell there is no change in the photocell area illuminated when the aperture alters; only the quantity of light reaching the cell is controlled by the variable aperture formed by the upper portion of the circumference of the cam and the diaphragm limiting the area of the light beam. The scale associated with the aperture is so calibrated that if R is the reading corresponding to a degree of opening such that the amount of light transmitted is $1/A$ of that admitted when the aperture is fully open, then $R = \log A$. This function, known as absorbance, was chosen because it is approximately linear with the concentration of a solution over small ranges. This function is normally a logarithmic one, but by giving the cam disc J a suitable contour, an evenly divided scale on the drum has been provided; there is also a scale of percentage transmissions.

The cell at A simultaneously receives light from G of an intensity controlled by an iris diaphragm E; the latter is uncalibrated, and is used only for adjusting purposes and also as a fine adjustment for the final setting. The two photocells A and O are connected in opposition across a Cambridge spot galvanometer P, so that when the photoelectric currents given by the cells are equal the galvanometer records zero deflection. A variable resistance R is arranged to provide variable sensitivity.

Use of Spekker photoelectric absorptiometer.   This is best illustrated by describing the procedure for making a determination. Let us suppose that it is desired to compare the depth of colour or, more precisely stated, the amount of light absorbed by two liquids s_1 and s_2, the latter being the more deeply coloured, i.e., the more absorbing.

1. Place s_2, contained in the special cell, into the beam, and open the variable aperture to its full extent by setting the drum at zero.
2. Adjust the *iris diaphragm* in front of the compensation cell until the galvanometer shows zero deflection.
3. Substitute s_1 for s_2, when the galvanometer will be seen to be deflected.
4. Adjust the *calibrated variable aperture* by means of the drum until the galvanometer returns to zero, and take the reading on the drum.

This series of operations takes less than half a minute. The cells containing the liquids are mounted side by side, and are easily interchanged by pushing the slide along an inch or so until it clicks into the correct position.

If we assume that the light intensity remains constant throughout the series of operations, then the current given by the indicating cell at the end of operation 4 is the same as at the end of operation 2 (since in each case it balances the output of the compensating cell). The difference in the illumination condition in the two cases is that in the second case the intensity-reduction produced by closing down the aperture is substituted for the reduction produced by the absorption of the specimen. The ratio of the area of the partly closed aperture to that of the aperture when fully open is thus a measure of the absorption of the liquid. Since both photocells are affected alike by changes in the intensity of the lamp, the reading is unaffected by changes occurring during the series of operations.

The sensitivity of the instrument in detecting small differences between the absorption of two liquids is greatly increased by the use of an appropriate light filter; indeed, as indicated in Fig. XVIII, 24, b, the use of a suitable light filter should be the normal practice (for a discussion as to the choice of filter, see previous paragraphs on *Light Filters*). A set of eight pairs of Ilford spectrum filters, which have a narrow band of transmission and a fairly sharp cut-off (see Fig. XVIII, 14), is normally supplied for use with the instrument. Sliding carriers are provided for two filters which enable them to be interchanged quickly in the absorptiometer or changed at will. The all-glass cells for liquids are available in lengths of 0.25, 0.5, 1, 2, and 4 cm.

It is a well-known fact that photoelectric cells under prolonged illumination tend to behave irregularly. This difficulty is overcome by the use of a gravity-controlled shutter which must be held open while the readings are being taken. In this way the steady burning of the lamp itself is ensured, and at the same time it is impossible to expose the cells for any longer than is required for the reading to be made.

For the routine use of the absorptiometer in colorimetric determinations, it is necessary to prepare a calibration curve by taking readings with a number of coloured solutions of known concentration covering the required range. This calibration curve remains valid so long as appreciable changes do not take place in the spectral sensitivity of the photoelectric cells or in the colour of the filter. The changes in the photocells and in the glass of the filters are generally very gradual, and the calibration curve need be checked only at wide intervals, say, every few months.

The *advantages* of the Spekker absorptiometer include:

(*a*) It runs directly from the electric mains supply; no batteries are required.

(*b*) Owing to the use of the two balanced photocells, the readings are largely independent of the fluctuations of the mains supply.*

(*c*) The scale of the instrument is approximately linear with the concentration of the solution.

(*d*) The instrument readings are not affected by variations in the sensitivity of the photocells or of the galvanometer, since a null method is employed.

(*e*) The galvanometer, which indicates the photoelectric current, is a robust but sensitive instrument of the spot type, and is used as a null indicator.

(*f*) Readings can be taken with as little as 2.5 cm^3 of liquid with the 0.5-cm cells; if a micro-cell is employed the volume can be reduced to 0.5 cm^3. The commonly used cell (1 cm) has a capacity of approximately 8 cm^3.

XVIII, 9. PHOTOELECTRIC SPECTROPHOTOMETERS. Spectrophotometers, from the standpoint of analytical chemistry, are those instruments which enable one to measure transmittance (or absorbance) at various wavelengths. Photoelectric spectrophotometers may be regarded as refined filter photoelectric photometers (absorptiometers) employing continuously variable and more nearly monochromatic bands of light. The less expensive instruments, such as the Bausch and Lomb Spectronic 20 grating colorimeter which give a band-width of 20–30 nm, have already been described: more elaborate spectrophotometers giving a band-width of 5–10 nm (or even less) will now be discussed briefly.

---

* For prolonged experiments, the use of a constant voltage transformer in the lamp supply is recommended.

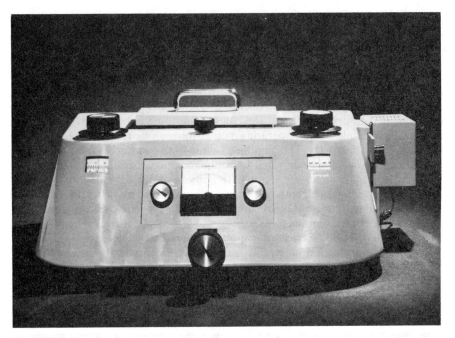

Fig. XVIII, 25

Unicam SP600 UV Spectrophotometer. This precision spectrophotometer (Fig. XVIII, 25) covers the range 220–1000 nm, i.e., the ultraviolet, visible, and near infrared. It operates on either 110–120 V or 200–250 V. The optical system is shown in Fig. XVIII, 26. The main features are the tungsten and deuterium sources, the slit system, including a slit-width indicator fitted to the slit control knob, a Littrow monochromator with a silica prism and two high sensitivity vacuum photocell detectors. An image of the light source is directed through the lower half of the slits to the collimating mirror M5, and then to the silica prism

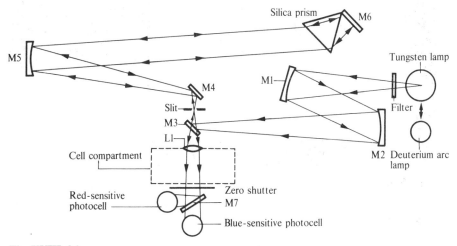

Fig. XVIII, 26

and Littrow mirror M6. On the return path the light passes through the upper half of the slits, and through the absorption cell to the appropriate photocell. The spectral band-width of the instrument is less than 3 nm over most of the wavelength range and not more than 10 nm at the extremes. The cell compartment can accommodate four rectangular cells of light path from 1 to 40 mm. Two vacuum-type photocells are fitted, a red cell for use above *ca.* 620 nm and a blue cell for shorter wavelengths. Both cells are in circuit when the instrument is in operation, and the change from one photocell to the other is effected by a simple control operating a plane mirror, and thus no resetting of the dark current is necessary. The amplified output of the photocells is balanced by a potentiometer, which is calibrated in both percentage transmission (linear) and absorbance (logarithmic); the length of the scale is about 28 cm. The instrument is suitable for ordinary absorption determinations in large numbers (four cells can be accommodated) or for the plotting of absorption spectra over the range 220–1000 nm from direct readings.

Unicam SP500 Series 2 Spectrophotometer. This is a precision photoelectric spectrophotometer with a wide range of applications, including (*a*) plotting the absorption curves of liquids throughout the visible and ultraviolet regions, (*b*) determining absorption (or transmission) at any previously chosen wavelengths, and (*c*) quantitative analysis of mixtures of known components by their visible or ultraviolet absorption. Fig. XVIII, 27 depicts the actual

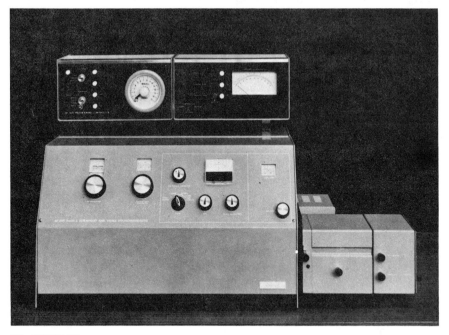

Fig. XVIII, 27

instrument, and Fig. XVIII, 28 is a schematic diagram of the optical system. The two light sources are a deuterium arc lamp for the ultraviolet and a tungsten-filament lamp for the visible range. Light from the lamps is selected by the solenoid-operated mirror M1 either automatically (at 340 nm) or manually. The

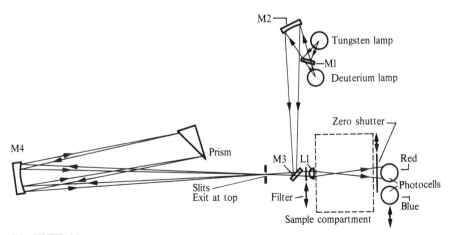

Fig. XVIII, 28

beam is then focused onto the entrance slit of the monochromator and is dispersed by a 30° rear-aluminised silica prism. Monochromatic radiation of the required wavelength is passed through the exit slit and collimated onto the sample by lens L1. Two vacuum photocells are employed as detectors—a red-sensitive cell is used at wavelengths above 625 nm and a blue-sensitive cell used for the shorter wavelengths. The cell compartment has a four-cell holder to accommodate glass or silica cells with a light path of up to 40 mm. Standard cells are available in glass (320–1000 nm), silica (200–1000 nm), and in 'Suprasil' (186–1000 nm), with light paths of 1, 2, 5, 10, 20, 30, and 40 mm. Stoppered cells (10 mm) are also supplied. The power supply for the lamps is provided from an electronically stabilised unit, fed from the mains voltage. This system is claimed to give exceptional baseline stability.

Beckman DU ultraviolet and visible spectrophotometer. This is a precision instrument. Two interchangeable light sources are used: a tungsten-filament lamp and a hydrogen-discharge lamp, the former for measurements down to 320 nm and the latter for measurements in the ultraviolet region below 360 nm. It employs a quartz prism of the Littrow type with a concave mirror of 50 cm focal length for collimation. The slit mechanism is continuously adjustable from 0.01 to 2.0 mm: slit widths are read directly from a calibrated dial and are reproducible to within 0.1 per cent. The wavelength range is from 210 to 1000 nm, and wavelength scale readings are accurate to better than 0.5 nm. The band-spread of the monochromator can, if necessary, be adjusted to less than 1 nm with the

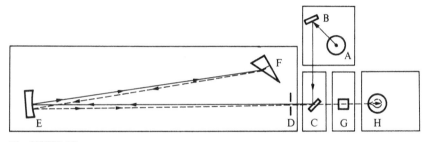

Fig. XVIII, 29

appropriate setting of the slit opening over most of the spectral range of the instrument. Two photocells are employed: a red-sensitive phototube for use above 600 nm and a blue-sensitive phototube for use in the range 320–625 nm (tungsten lamp) and 210–360 nm (hydrogen lamp). The phototube current is measured by a null method utilising a slide-wire potentiometer and an electronic amplifier. The potentiometer is calibrated in per cent transmission from 0 to 100 and in absorbance from 0 to 2.0; a switch is provided which increases the sensitivity by a factor of 10, thus giving greater accuracy in reading transmission values below 10 per cent and at the same time extending the absorbance range from 1.0 to 3.0. Controls are provided *inter alia* for adjusting the dark current to zero and the percentage transmission to 100. Four standard rectangular absorption cells of 10 mm light path are supplied in a four-place cell holder: an

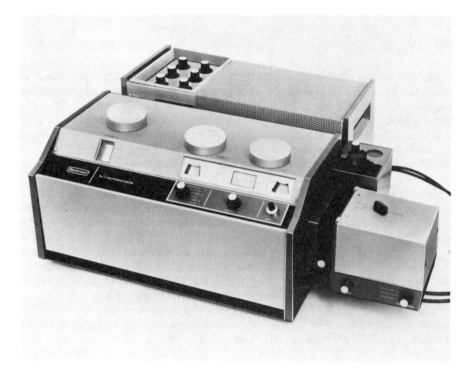

Fig. XVIII, 30

interchangeable cell compartment is available to accommodate either Pyrex or silica cells with path lengths from 2 to 100 mm and with sample volumes from 0.3 to 28.5 cm^3.

A schematic diagram of the optical system is given in Fig. XVIII, 29 and the instrument is shown in Fig. XVIII, 30. An image of the light source A is focused by the condensing mirror B and the diagonal mirror C on the entrance slit at D. The entrance slit is the lower of two slits vertically over each other. Light falling on the collimating mirror E is rendered parallel, and is reflected towards the quartz prism F. The back surface of the prism is aluminised, so that light refracted at the first surface is reflected back through the prism, undergoing further refraction as it emerges from the prism. The collimating mirror focuses the spectrum in the plane of the slits D, and light of the wavelength for which the prism is set passes out of the monochromator through the exit (upper) slit, through the absorption cell G to the photocell H. The photocell response is amplified and is registered on the meter. Both battery operated (6 volt, 120 ah) and a.c. mains operated (115 or 230 volt) models are available.

Double beam spectrophotometers.

Most modern general purpose ultraviolet/visible spectrophotometers are double beam instruments which cover the range between about 200 and 800 nm by a continuous automatic scanning process producing the spectrum as a pen trace on calibrated chart paper.

In these instruments the monochromated beam of radiation, from tungsten and deuterium lamp sources, is divided into two identical beams; one of which passes through the reference cell and the other through the sample cell. The signal for the absorption of the contents of the reference cell is automatically subtracted from that from the sample cell giving a net signal corresponding to the absorption for the components in the sample solution.

Perkin Elmer 402 Spectrophotometer. This is a highly versatile double beam spectrophotometer which covers the range from 190 to 850 nm using a fused silica prism for monochromation and a photomultiplier detector. The instrument (Fig. XVIII, 31) has a continuous flow chart with a linear absorbance scale calibrated from 0 to 1.5. By use of attenuators the absorbance scale can be extended to 3.0 and scan times from 2 to 40 minutes are possible.

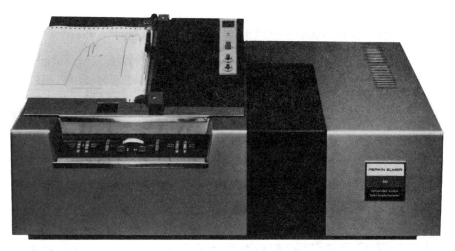

Fig. XVIII, 31

As shown in Fig. XVIII, 32, splitting of the beam into the reference and sample beams does not take place until after monochromation and is achieved by a rotating segmented mirror. Scanning is from the lower wavelength limit to the upper, the value for the absorption being automatically plotted as the chart paper, actuated by a servo-motor, passes under the pen.

This instrument can be used for automatic repeated scans and has scale expansion controls for both the wavelength and absorbance scales. It is suitable

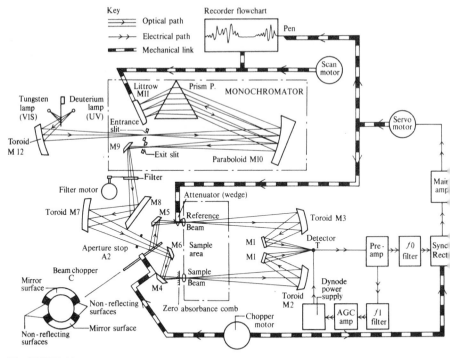

Fig. XVIII, 32

for a wide range of studies including reaction kinetics in addition to standard quantitative determinations.

Unicam SP 1700 Spectrophotometer. This double beam spectrophotometer can be used for the range between 190 and 850 nm, monochromation being

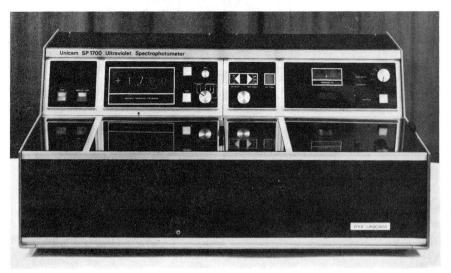

Fig. XVIII, 33

by means of a diffraction grating and producing a resolution of 0.1 nm. As shown in Fig. XVIII, 33, the instrument possesses a digital display panel for the instantaneous reading of the absorbance values as these are measured. In this instrument the signal is fed to a separate chart recorder to produce the complete spectrum as a pen trace. An extra large sample compartment is built into the instrument to enable different types of cells to be fitted easily. The complete spectrum is produced as a continuous chart at any one of eleven different speeds.

Experimental. Colorimetric determinations

XVIII, 10. SOME GENERAL REMARKS UPON COLORIMETRIC DETERMINATIONS. Visual methods have been virtually displaced for most determinations by methods depending upon the use of photoelectric cells (filter photometers or absorptiometers, and spectrophotometers), thus leading to reduction of the experimental errors of colorimetric determinations. The so-called photoelectric colorimeter is a comparatively inexpensive instrument, and should be available in every laboratory. The use of spectrophotometers has enabled determinations to be extended into the ultraviolet region of the spectrum, whilst the use of chart recorders means that the analyst is not limited to working at a single fixed wavelength.

The choice of a colorimetric procedure for the determination of a substance will depend upon such considerations as the following:

(a) A colorimetric method will often give more accurate results at low concentrations than the corresponding titrimetric or gravimetric procedure. It may also be simpler to carry out.

(b) A colorimetric method may frequently be applied under conditions where no satisfactory gravimetric or titrimetric procedure exists, e.g., for certain biological substances.

(c) Colorimetric procedures possess advantages for the routine determination of some of the components of a number of similar samples by virtue of the rapidity with which they may be made: there is often no serious sacrifice of accuracy over the corresponding gravimetric or titrimetric procedures provided the experimental conditions are rigidly controlled.

The criteria for a satisfactory colorimetric analysis are:

1. *Specificity of the colour reaction.* Very few reactions are specific for a particular substance, but many give colours for a small group of related substances only, i.e., are selective. By utilising such devices as the introduction of other complex-forming compounds, by altering the oxidation states, and control of pH, close approximation to specificity may often be obtained. This subject is discussed in detail below.

2. *Proportionality between colour and concentration.* For visual colorimeters it is important that the colour intensity should increase linearly with the concentration of the substance to be determined. This is not essential for photoelectric instruments, since a calibration curve may be constructed relating the instrumental reading of the colour with the concentration of the solution. Otherwise expressed, it is desirable that the system follows Beer's law even when photoelectric colorimeters are used.

3. *Stability of the colour.* The colour produced should be sufficiently stable to permit an accurate reading to be taken. This applies also to those reactions in which colours tend to reach a maximum after a time: the period of

maximum colour must be long enough for precise measurements to be made. In this connection the influence of other substances and of experimental conditions (temperature, pH, stability in air, etc.) must be known.

4. *Reproducibility.*   The colorimetric procedure must give reproducible results under specific experimental conditions. The reaction need not necessarily represent a stoichiometrically quantitative chemical change.

5. *Clarity of the solution.*   The solution must be free from precipitate if comparison is to be made with a clear standard. Turbidity scatters as well as absorbs the light.

6. *High sensitivity.*   It is desirable, particularly when minute amounts of substances are to be determined, that the colour reaction be highly sensitive. It is also desirable that the reaction product absorb strongly in the visible rather than in the ultraviolet; the interfering effect of other substances in the ultraviolet is usually more pronounced.

In view of the selective character of many colorimetric reactions, it is important to control the operational procedure so that the colour is specific for the component being determined. This may be achieved by isolating the substance by the ordinary methods of inorganic analysis; double precipitation is frequently necessary to avoid errors due to occlusion and co-precipitation. Such methods of chemical separation may be tedious and lengthy: if minute quantities are under consideration, they may also lead to appreciable loss owing to solubility, supersaturation, and peptisation effects. Use may be made of any of the following processes in order to render colour reactions specific and/or to separate the individual substances:

(a)   Suppression of the action of interfering substances by the formation of complex ions or of non-reactive complexes.

(b)   Adjustments of the pH; many reactions take place within well-defined limits of pH.

(c)   Removal of the interfering substance by extraction with an organic solvent, sometimes after suitable chemical treatment.

(d)   Isolation of the substance to be determined by the formation of an organic complex, which is then removed by extraction with an organic solvent. This method may be combined with (a) in which an interfering ion is prevented from forming a soluble organic complex by converting it into a complex ion which remains in the aqueous layer.

(e)   Separation by volatilisation. This method is of limited application, but gives good results, e.g., distillation of arsenic as the trichloride in the presence of hydrochloric acid.

(f)   Electrolysis with a mercury cathode or with controlled cathode potential.

(g)   Application of physical methods utilising selective absorption, chromatographic separations, and ion exchange separations.

Some remarks concerning **standard curves** seem appropriate at this point. The usual method of use of a filter photometer or a spectrophotometer requires the construction of a standard curve (also termed the *reference* or *calibration* curve) for the constituent being determined. Suitable quantities of the constituent are taken and treated in the same way as the sample solution for the development of colour and the measurement of the transmission (or absorbance) at the optimum wavelength. The absorbance ($\log I_0/I_t$) is plotted against the concentration: a straight line plot is obtained if Beer's law is obeyed. The curve may then be used for future determinations of the constituent under the same experimental conditions. When the absorbance is directly proportional to the concentration, only a few points are required to establish the line: when the relation is not linear, a greater number of points will generally be necessary. The standard curve should be checked at intervals. When a filter photometer is used, the characteristics of the filter and the light source may change with time.

When plotting the standard curve it is customary to assign a transmission of 100 per cent to the blank solution (reagent solution plus water); this represents zero concentration of the constituent. It may be mentioned that some coloured solutions have an appreciable temperature coefficient of transmission, and the temperature of the determination should not differ appreciably from that at which the calibration curve was prepared.

The following procedures are arranged in alphabetical order, with cations first (Sections **11–30**), followed by the anions (Sections **31–36**).*

Cations

XVIII, 11.   ALUMINIUM.   *Discussion.*   Among the reagents that have been used for the colorimetric determination of aluminium are ammonium aurintricarboxylate (*aluminon*) and Eriochrome Cyanine R. The latter appears to be somewhat superior, and its use will therefore be described. At a pH of 5.9–6.1, zinc, nickel, manganese, and cadmium interfere negligibly, but iron and copper must be absent. One procedure for removing interfering elements, e.g., in the analysis of steels, is to pass the solution through a cellulose column (compare Section **VIII, 7**); iron and other elements are separated by elution with a mixture of concentrated hydrochloric acid and freshly distilled ethyl methyl ketone (8 : 192 v/v). The aluminium, and any nickel present, are recovered by passing dilute hydrochloric acid (1 : 5 v/v) through the column.

   Reagents.   *Eriochrome Cyanine R solution.*   Dissolve 0.1 g of the solid reagent in water, dilute to 100 cm³, and filter through a Whatman No. 541 filter paper if necessary. This solution should be prepared daily.

   Standard aluminium solution.   Dissolve 1.319 g A.R. aluminium potassium sulphate in water and dilute to 1 dm³ in a graduated flask; 1 cm³ ≡ 75 μg Al.

   Buffer solution, concentrated.   Dissolve 27.5 g ammonium acetate and 11.0 g hydrated sodium acetate in 100 cm³ water: add 1.0 cm³ glacial acetic acid and mix well.

   Buffer solution, dilute.   To one volume of concentrated buffer solution, add five volumes water and adjust the pH to 6.1 by adding acetic acid or sodium hydroxide solution.

   Procedure.   Transfer an aliquot of the solution (say, 20.0 cm³), containing 2–70 μg Al and free from interfering elements, to a 250-cm³ beaker, add 5 cm³ of 5-volume hydrogen peroxide and mix well. Adjust the pH of the solution to 6.0 (using either 0.2M-sodium hydroxide or 0.2M-hydrochloric acid), add 5.0 cm³ of Eriochrome Cyanine R solution, and mix. Introduce 50 cm³ of the dilute buffer solution, and dilute without delay to 100 cm³ in a graduated flask. Measure the absorbance after 30 minutes with a spectrophotometer at 535 nm against a reagent blank in a 5 mm cell. For an absorptiometer, use Ilford No. 605 filter and 1 cm cells.

---

* A large number of reagents for metals are available and in some instances a reagent will produce a reaction suitable for colorimetric determination with several metals. The table reproduced in Appendix II gives a clearer idea of the wide range of reagents available.

Construct the calibration curve using 0, 1, 2, 3, 4, and 5 cm^3 of the standard aluminium solution.

XVIII, 12.   DETERMINATION OF AMMONIA. *Discussion.*   J. Nessler in 1856 first proposed an alkaline solution of mercury(II) iodide in potassium iodide as a reagent for the colorimetric determination of ammonia. Various modifications of the reagent have since been made. When Nessler's reagent is added to a dilute ammonium salt solution, the liberated ammonia reacts with the reagent fairly rapidly but not instantaneously to form an orange-brown product, which remains in colloidal solution, but flocculates on long standing. The colorimetric comparison must be made before flocculation occurs.

The reaction with Nessler's reagent (an alkaline solution of potassium tetraiodomercurate(II)) may be represented as:

$$2K_2[HgI_4] + 2NH_3 = NH_2Hg_2I_3 + 4KI + NH_4I$$

The reagent is employed for the determination of ammonia in very dilute ammonia solutions and in water. In the presence of interfering substances, it is best to separate the ammonia first by distillation under suitable conditions. The method is also applicable to the determination of nitrates and nitrites: these are reduced in alkaline solution by Devarda's alloy to ammonia, which is removed by distillation. The procedure is applicable to concentrations of ammonia as low as 0.1 mg dm^{-3}.

Nessler's reagent is prepared as follows.   Dissolve 35 g potassium iodide in 100 cm^3 water, and add 4 per cent mercury(II) chloride solution, with stirring or shaking, until a slight red precipitate remains (about 325 cm^3 are required). Then introduce, with stirring, a solution of 120 g sodium hydroxide in 250 cm^3 water, and make up to 1 dm^3 with distilled water. Add a little more mercury(II) chloride solution until there is a permanent turbidity. Allow the mixture to stand for one day and decant from the sediment. Keep the solution stoppered in a dark-coloured bottle.

The following is an alternative method of preparation. Dissolve 100 g mercury(II) iodide and 70 g potassium iodide in 100 cm^3 ammonia-free water. Add slowly, and with stirring, to a cooled solution of 160 g sodium hydroxide pellets (or 224 g potassium hydroxide) in 700 cm^3 ammonia-free water, and dilute to 1 dm^3 with ammonia-free distilled water. Allow the precipitate to settle, preferably for a few days, before using the pale yellow supernatant liquid.

Ammonia-free water may be **prepared** in a conductivity-water still, or by means of a column charged with a mixed cation and anion exchange resin (e.g., Permutit Bio-Deminrolit or Amberlite MB-1), or as follows. Redistil 500 cm^3 of distilled water in a Pyrex apparatus from a solution containing 1 g potassium permanganate and 1 g anhydrous sodium carbonate; reject the first 100-cm^3 portion of the distillate and then collect about 300 cm^3.

Procedure.   For practice in this determination, employ either a very dilute ammonium chloride solution or ordinary distilled water which usually contains sufficient ammonia for the exercise.

Prepare a **standard ammonium chloride solution** as follows. Dissolve 3.141 g A.R. ammonium chloride, dried at 100 °C, in ammonia-free water and dilute to 1 dm^3 with the same water. This stock solution is too concentrated for most purposes. A standard solution is made by diluting 10 cm^3 of this solution to 1 dm^3 with ammonia-free water: 1 cm^3 contains 0.01 mg of NH$_3$.

If necessary, dilute the sample to give an ammonia concentration of 1 mg dm^{-3} (Hehner cylinders, Fig. XVIII, 7, are useful for this dilution), and fill a 50-cm^3 Nessler tube to the mark. Prepare a series of Nessler tubes containing the following volumes of standard ammonium chloride solution diluted to 50 cm^3: 1.0, 2.0, 3.0, 4.0, 5.0, and 6.0 cm^3. The standards contain 0.01 mg NH$_3$ for each cm^3 of the standard solution. Add 1 cm^3 of Nessler's reagent to each tube, allow to stand for 10 minutes, and compare the unknown with the standards in a Nessler stand (Fig. XVIII, 3) or in a B.D.H. nesslerimeter (Fig. XVIII, 5). This will give an approximate figure which will enable another series of standards to be prepared and more accurate results to be obtained.

A photoelectric colorimeter or a spectrophotometer may, of course, be used. When 1 cm^3 of the Nessler reagent is added to 50 cm^3 of the sample, a blue colour filter in the wavelength region 400–425 nm allows measurements with a 10-mm path in the nitrogen range 20–250 μg. Nitrogen concentrations approaching up to 1 mg can be determined with a green colour filter or in the wavelength range near 525 nm. The calibration curve should be prepared under exactly the same conditions of temperature and reaction time adopted for the sample.

XVIII, 13. ANTIMONY.   *Discussion.*   The procedure is based on the formation of yellow tetraiodoantimonate(III) acid (HSbI$_4$) when antimony(III) in sulphuric acid solution is treated with excess of potassium iodide solution. Spectrophotometric measurements may be made at 425 nm in the visible region or, more precisely, at 330 nm in the ultraviolet region. Appreciable amounts of bismuth, copper, lead, nickel, tin, tungstate, and molybdate interfere.

Reagents.   *Potassium iodide solution.*   Dissolve 14.0 g A.R. potassium iodide and 1.0 g crystallised ascorbic acid in redistilled water and dilute to 100 cm^3.

Standard antimony solution.   Dissolve 0.2668 g A.R. antimonyl potassium tartrate in redistilled water, add 160 cm^3 concentrated sulphuric acid, and dilute to 1 dm^3 with water in a graduated flask.

Procedure.   Use a solution containing 0.15–1.8 mg antimony per 100 cm^3; it should be slightly acidic with sulphuric acid (1.2–1.5M). Transfer a 10-cm^3 aliquot to a 50-cm^3 graduated flask, add 25 cm^3 of the potassium iodide-ascorbic acid reagent, and dilute to the mark with 25 per cent v/v sulphuric acid. Mix thoroughly and measure the absorbance at 425 nm or at 330 nm using a reagent blank as reference solution.

Construct a calibration curve using appropriate volumes of the standard antimony solution treated in the same way as for the sample solution.

XVIII, 14. ARSENIC.   *Discussion.*   Of the numerous procedures available for the determination of minute amounts of arsenic,* only two will be described,

---

* Very small amounts of arsenic (0.001–0.1 mg) may be determined by volatilising the element as arsine AsH$_3$ and comparing the coloration formed upon discs of dry paper impregnated with mercury(II) chloride with that obtained by the use of known amounts of arsenic (**Gutzeit test**). Although the method is still used in practice and suitable apparatus is available from most laboratory supply houses, it is doubtful whether the accuracy exceeds 10 per cent of the true value. Too much dependence is placed upon the rate of evolution of arsine, which is not necessarily the same as the rate of evolution of hydrogen in the reduction apparatus. On the whole, the spectrophotometric procedure, based upon molybdenum blue or the silver diethyldithiocarbamate complex, is far superior. In all evolution methods arsenic must be in the arsenic(III) state.

viz., the molybdenum blue method and the silver diethyldithiocarbamate method. Both possess great sensitivity and precision, and are readily applied colorimetrically or spectrophotometrically.

Molybdenum blue method. When arsenic, as arsenate, is treated with ammonium molybdate solution and the resulting heteropoly molybdioarsenate (arseno-molybdate) is reduced with hydrazinium sulphate or with tin(II) chloride, a blue soluble complex 'molybdenum blue' is formed. The constitution is uncertain, but it is evident that the molybdenum is present in a lower oxidation state. The stable blue colour has a maximum absorption at about 840 nm and shows no appreciable change in 24 hours. Various techniques for carrying out the determination are available, but only one can be given here. Phosphate reacts in the same manner as arsenate (and with about the same sensitivity) and must be absent.

Both macro and micro quantities of arsenic may be isolated by distillation of arsenic(III) chloride from hydrochloric acid solution in an all-glass apparatus in a stream of carbon dioxide or nitrogen: a reducing agent, such as hydrazinium sulphate, is used to reduce arsenic(V) to arsenic(III). The distillate may be collected in cold water. Germanium accompanies arsenic in the distillation; if phosphate is present in large amounts the distillate should be redistilled under the same conditions. Another method of isolation involves volatilisation of arsenic as arsine by the action of zinc in hydrochloric or sulphuric acid solution. Appreciable amounts of certain reducible heavy metals, such as copper, nickel, and cobalt, slow down the evolution of arsine, as do also large amounts of metals that are precipitated by zinc. Copper in more than small quantities prevents complete evolution of arsine; the error amounts to 20 per cent (for 5–10 μg As) with 50 mg of copper. The arsine which is evolved may be absorbed in a sodium hydrogencarbonate solution of iodine. The absorption apparatus should be so designed that the arsine is completely absorbed.

Reagents. * *Potassium iodide solution.* Dissolve 15 g of the A.R. solid in 100 cm^3 water.

Tin(II) chloride solution. Dissolve 40 g A.R. hydrated tin(II) chloride in 100 cm^3 concentrated hydrochloric acid.

Zinc. Use 20–30 mesh or granulated; arsenic-free.

Iodine–potassium iodide solution. Dissolve 0.25 g iodine in a small volume of water containing 0.4 g potassium iodide, and dilute to 100 cm^3.

Sodium disulphite solution. Dissolve 0.5 g of the solid reagent ($Na_2S_2O_5$) in 10 cm^3 water. Prepare fresh daily.

Sodium hydrogencarbonate solution. Dissolve 4.2 g of the solid in 100 cm^3 water.

Ammonium molybdate–hydrazinium sulphate reagent. Solution (a): dissolve 1.0 g A.R. ammonium molybdate in 10 cm^3 water and add 90 cm^3 of 3M-sulphuric acid. Solution (b): dissolve 0.15 g pure hydrazinium sulphate in 100 cm^3 water. Mix 10.0 cm^3 each of solutions (a) and (b) just before use.

Hydrochloric acid. This must be arsenic-free.

Standard arsenic solution. Dissolve 1.320 g A.R. arsenic(III) oxide in the minimum volume of 1M-sodium hydroxide solution, acidify with dilute

---

* Special pure, arsenic-free reagents are available from chemical supply houses (e.g., British Drug Houses) and are symbolised by 'AsT' after the name of the compound; these should be used as far as possible in the determination and for the preparation of the above reagents.

hydrochlorate acid, and make up to 1 dm^3 in a graduated flask: 1 cm^3 contains 1 mg of As. A solution containing 0.001 mg As per cm^3 is prepared by dilution.

Procedure. The arsenic must be in the arsenic(III) state; this may be secured by first distilling in an all-glass apparatus with concentrated hydrochloric acid and hydrazinium sulphate, preferably in a stream of carbon dioxide or nitrogen. Another method consists in reducing the arsenate (obtained by the wet oxidation of a sample) with potassium iodide and tin(II) chloride: the acid concentration of the solution after dilution to 100 cm^3 must not exceed 0.2–0.5M; 1 cm^3 of 50 per cent potassium iodide solution and 1 cm^3 of a 40 per cent solution of tin(II) chloride in concentrated hydrochloric acid are added, and the mixture heated to boiling.

Transfer an aliquot portion of the arsenate solution, having a volume of 25 cm^3 and containing not more than 20 μg of arsenic, to the 50-cm^3 Pyrex evolution vessel A shown in Fig. XVIII, 34, and add sufficient concentrated hydrochloric acid to make the total volume present in the solution 5–6 cm^3, followed by 2 cm^3 of the potassium iodide solution and 0.5 cm^3 of the tin(II) chloride solution. Allow to stand at room temperature for 20–30 minutes to permit the complete reduction of the arsenate.

The tube B is loosely packed with purified glass wool soaked in lead acetate solution (to remove hydrogen sulphide and trap acid spray), and C is a capillary tube (4 mm external and 0.5 mm internal diameter). Place 1.0 cm^3 iodine–potassium iodide solution and 0.2 cm^3 of the sodium hydrogencarbonate solution in the narrow absorption tube D. Mix with the end of the delivery tube.

Rapidly add 2.0 g of zinc to the vessel A, immediately insert the stopper, and allow the gases to bubble through the solution for 30 minutes. At the end of this time the solution in D should still contain some iodine. Disconnect the delivery tube C and leave it in the absorption tube. Add 5.0 cm^3 of the ammonium molybdate–hydrazine reagent and a drop or two of sodium disulphite solution. Heat the resulting colourless solution in a water bath at 95–100°C, cool, transfer to a 10-cm^3 graduated flask, and make up to volume with water.

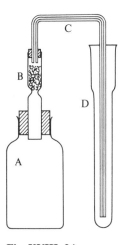

Measure the transmittance of the solution at 840 nm or with a red filter with maximum transmission above 700 nm. Charge the reference cell with a solution obtained by taking the iodine–iodide–hydrogen-carbonate mixture and treating it with molybdate–hydrazinium sulphate–disulphite as in the actual procedure.

Construct the calibration curve by taking, say, 0, 2.5, 5.0, 7.5, and 10.0 μg As (for a final volume of 10 cm^3), mixing with iodine–iodide–hydrogencarbonate solution, adding molybdate–hydrazinium sulphate–disulphite, and heating to 95–100.

Fig. XVIII, 34

The following procedure is recommended by the Analytical Methods Committee of the Society for Analytical Chemistry for the determination of small amounts of arsenic in organic matter (Ref. 7). Organic matter is destroyed by wet oxidation, and the arsenic, after extraction with diethylammonium diethyl-dithiocarbamate in chloroform, is converted into the arsenomolybdate complex: the latter is reduced by means of hydrazinium sulphate to a molybdenum blue complex and determined spectrophotometrically at 840 nm and referred to a calibration graph in the usual manner.

Silver diethyldithiocarbamate method.   Arsine reacts with a solution of silver diethyldithiocarbamate, $AgS \cdot CS \cdot N(C_2H_5)_2$, in pyridine to form a soluble red complex, which has an absorption maximum at 540 nm. This forms the basis of the method; the arsenic must be in the arsenic(III) state. Stibine SbH_3 under similar conditions yields a red colour with maximum absorption at 510 nm, and therefore interferes.

Reagents.   See above under molybdenum blue method for zinc and tin(II) chloride.

Lead acetate solution.   Dissolve 10 g pure lead acetate in 100 cm^3 distilled water.

Silver diethyldithiocarbamate–pyridine solution.   Dissolve 1.0 g pure, dry silver diethyldithiocarbamate in 200 cm^3 pure pyridine. Store in an amber bottle.

Silver diethyldithiocarbamate may be **prepared** as follows. To a solution of 2.25 g A.R. sodium diethyldithiocarbamate $NaS \cdot CS \cdot N(C_2H_5)_2,3H_2O$ in 100 cm^3 of water add, slowly and with constant stirring, a solution of 1.7 g A.R. silver nitrate in 100 cm^3 water. Both solutions should be at 8–10 °C. Collect the lemon yellow precipitate on a sintered glass funnel, wash with about 100 cm^3 cold water, and then dry in a vacuum desiccator at room temperature.

Potassium iodide solution.   Dissolve 15 g A.R. potassium iodide in 100 cm^3 distilled water.

All glassware used should be thoroughly cleaned with either hot concentrated sulphuric acid or boiling concentrated nitric acid, followed by rinsing with distilled water, and then with acetone.

Procedure.   Use the apparatus shown in Fig. XVIII, 35.* The flask has a capacity of 100 or 125 cm^3, and is connected to the scrubber by means of a ground joint; the scrubber is attached to the arsine absorber by means of a ball joint. The arsine absorber has a calibration mark at 4.00 cm^3 to ensure that the same volume of reagent is used in each determination. Impregnate the purified glass wool in the scrubber with lead acetate solution; this will absorb any hydrogen sulphide which may be subsequently evolved. Charge the absorption tube with 4.00 cm^3 of the silver diethyldithiocarbamate reagent.

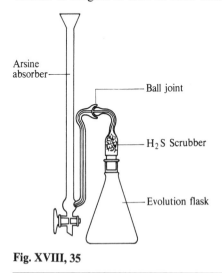

Arsine absorber

Ball joint

H$_2$S Scrubber

Evolution flask

Fig. XVIII, 35

Prepare a calibration curve by pipetting suitable aliquots of the diluted standard arsenic solution into a series of clean evolution flasks fitted with standard taper necks: cover the range 0–10 μg of arsenic. To each of these diluted aliquots, add 5 cm^3 of concentrated hydrochloric acid, 2.0 cm^3 of 15 per cent potassium iodide solution and 8 drops of tin(II) chloride solution. Swirl the contents of the flasks, and allow them to stand for about 15 minutes to ensure

---

* Modified forms are available commercially.

complete reduction to the arsenic(III) state. Add 5.0 g pure granulated zinc to the solution in the flask and insert the hydrogen sulphide scrubber immediately. The evolution of arsine is 99 per cent complete in 30 minutes and virtually complete in about 40 minutes. If necessary, dilute the liquid in the arsine absorber with pure pyridine to the 4.00-cm^3 mark and pass a gentle stream of air through the absorber to mix the solution. Transfer the absorbing solution to a 1-cm cell and measure the absorbance at 540 nm in a spectrophotometer. Repeat the procedure with the remaining flasks. Plot the absorbance of each aliquot (less that of the blank) against its arsenic content in μg.

For the actual determination of arsenic in the sample solution, follow the same procedure as for the calibration, using two flasks, one for the sample solution and the other for the reagent blank. From the absorbance obtained at 540 nm, evaluate the arsenic content of the sample solution by reference to the calibration graph previously prepared.

XVIII, 15. BERYLLIUM. *Discussion.* Minute amounts of beryllium may be readily determined spectrophotometrically by reaction under alkaline conditions with 4-nitrobenzene-azo-orcinol. The reagent is yellow in a basic medium; in the presence of beryllium the colour changes to reddish-brown. The zone of optimum alkalinity is rather critical and narrow; buffering with boric acid increases the reproducibility. Aluminium, up to about 240 mg per 25 cm^3, has little influence provided an excess of 1 mole of sodium hydroxide is added for each mole of aluminium present. Other elements which might interfere are removed by preliminary treatment with sodium hydroxide solution, but the possible co-precipitation of beryllium must be considered. Zinc interferes very slightly but can be removed by precipitation as sulphide. Copper interferes seriously, even in such small amounts as are soluble in sodium hydroxide solution. The interference of small amounts of copper, nickel, iron and calcium can be prevented by complexing with EDTA and triethanolamine.

Procedure. Transfer the almost neutral sample solution of beryllium (containing 5 to 80 μg of the element in a volume of about 10 cm^3) to a 25 cm^3 graduated flask, add 2.8 cm^3 of 2.0M-sodium hydroxide (or more if much aluminium is present), 5.0 cm^3 of 0.64M-boric acid solution, and 6.0 cm^3 of the dye solution (1), dilute to the mark with distilled water, and mix well. Measure the transmittance at 520 nm, preferably using a 2 cm cell.

Construct a calibration curve (for details, see Section **XVIII, 10**) using A.R. beryllium sulphate and the experimental conditions given above: cover the range 5–80 μg of beryllium. Evaluate the concentration of the sample solution of beryllium with the aid of the calibration curve.

Note. 1. Prepare the dye solution by stirring 0.025 g of 4-nitrobenzene-azo-orcinol mechanically for several hours with 0.1M-sodium hydroxide; filter before use.

XVIII, 16. BISMUTH. *Discussion.* When potassium iodide solution is added to a dilute sulphuric acid solution containing a small amount of bismuth a yellow to orange coloration, due to the formation of an iodobismuthate(III) ion, is produced. The colour intensity increases with iodide concentration up to about 1 per cent potassium iodide and then remains practically constant.

The reaction is a sensitive one, but is subject to a number of interferences. The solution must be free from large amounts of lead, thallium(I) copper, tin, arsenic, antimony, gold, silver, platinum, palladium, and from elements in sufficient quantity to colour the solution, e.g., nickel. Metals giving insoluble iodides must be absent, or present in amounts not yielding a precipitate. Substances which liberate iodine from potassium iodide interfere, for example, iron(III), the latter should be reduced with sulphurous acid and the excess of gas boiled off, or by a 30 per cent solution of hypophosphorous acid. Chloride ion reduces the intensity of the bismuth colour. Separation of bismuth from copper can be effected by extraction of the bismuth as dithizonate by treatment in ammoniacal potassium cyanide solution with a 0.1 per cent solution of dithizone in chloroform; if lead is present, shaking of the chloroform solution of lead and bismuth dithizonates with a buffer solution of pH 3.4 results in the lead alone passing into the aqueous phase. The bismuth complex is soluble in a pentanol–ethyl acetate mixture, and this fact can be utilised for the determination in the presence of coloured ions, such as nickel, cobalt, chromium, and uranium.

Procedure.   Prepare a **standard solution of bismuth** by dissolving 0.100 g pure bismuth (Johnson Matthey) in 20 cm^3 concentrated sulphuric acid, and diluting to 1 dm^3 with water: 1 cm^3 contains 0.1 mg Bi. Other standard solutions may be obtained by dilution.

Treat the colourless solution (*ca.* 15 cm^3), free from interfering substances and about M in sulphuric acid, with 1 cm^3 of 30 per cent hypophosphorous acid solution and 1 cm^3 of 10 per cent aqueous potassium iodide solution. Dilute to 25 cm^3 and match the yellow colour produced against standards containing the same concentration of sulphuric acid and hypophosphorous acid. Alternatively, measure the absorbance at or near 460 nm.

In the extraction procedure the yellow solution is allowed to stand for 10 minutes, and then extracted with 3-cm^3 portions of a 3:1 mixture by volume of pentanol and ethyl acetate until the last extract is colourless. Make up the combined extracts to a definite volume (10 cm^3 or 25 cm^3) with the organic solvent, and determine the transmittance (460 nm) at once. Construct the calibration curve by extracting known amounts of bismuth under the same conditions as the sample.

Bismuth in lead.   *Discussion.*   This method is based upon the extraction of bismuth as cupferrate by chloroform from 0.1M-acid solution: as little as 1 μg of bismuth can be separated from 10 g of lead.

Procedure.   Dissolve a suitable weight of the sample of lead in 6M-nitric acid: add a little 50 per cent aqueous tartaric acid to clear the solution if antimony or tin is present. Cool, transfer to a separatory funnel, and dilute to about 25 cm^3. Add concentrated ammonia solution to the point where the slight precipitate will no longer dissolve on shaking, then adjust the pH to 1, using nitric acid or ammonia solution. Add 1 cm^3 freshly prepared 1 per cent cupferron solution, mix, and extract with 5 cm^3 chloroform. Separate the chloroform layer, and repeat the extraction twice with 1-cm^3 portions of cupferron solution + 5 cm^3 of chloroform. Wash the combined chloroform extracts with 5 cm^3 of water. Extract the bismuth from the chloroform by shaking with two 10-cm^3 portions of 1M-sulphuric acid. Run the sulphuric acid solution into a 25-cm^3 graduated flask. Add 3 drops saturated sulphur dioxide solution and 4 cm^3 of 20 per cent aqueous potassium iodide. Dilute to volume and measure the transmission at 460 nm.

XVIII, 17.   BORON.   *Discussion.*   Minute amounts of boron are usually separated by distillation from an acid solution as methyl borate. Borosilicate glass should be avoided, even for the storage of chemicals. The apparatus should be constructed of fused silica;* a platinum dish receiver may also be used. Distillation may be made from a strong acid solution (sulphuric or phosphoric acid). In the simplest apparatus methanol vapour is passed through a flask containing the solution of the sample and is condensed and collected in an excess of either calcium hydroxide or sodium hydroxide solution in a silica or platinum dish. In a more efficient apparatus the methanol is made to cycle between the sample dissolved in the acid medium and a flask containing calcium or sodium hydroxide solution: distillation can thus be continued for several hours with only a small amount of methanol. At the end of the distillation the contents of the receiver in which the methyl borate was collected (which must be strongly alkaline—a minimum of four times the theoretical amount of base) are evaporated to dryness. The residue is used for the colorimetric determination. Most of the reagents, e.g., quinalizarin (1,2,5,8-tetrahydroxyanthraquinone) or 1,1'-dianthrimide (1,1'-iminodianthraquinone) react only in concentrated sulphuric acid solution. With the former the absorption maxima for the reagent and its boron complex lie close together, while with the latter the maximum absorption for the reagent is below 400 nm and for the boron complex is at 620 nm. The use of dianthrimide will accordingly be described. The colour change of 1,1'-dianthrimide from greenish-yellow to blue in the presence of borates in concentrated sulphuric acid is the basis of a trustworthy method for the determination of micro amounts of boron; the effective range of the reagent is 0.5–6 μg and the colour is stable for several hours.

Interferences in the distillation method are fluoride and large amounts of gelatinous silica. Fluoride interference may be overcome by the addition of calcium chloride. Strong oxidising agents, such as chromate and nitrate, interfere, since they destroy the reagent. Boron in natural waters can be determined without separation; the residue obtained after evaporation to dryness with a little calcium hydroxide solution may be used directly in the colour formation. In the analysis of steel by dissolution in sulphuric acid no oxidising compounds are formed which can interfere with the reaction.

Reagents. *Dianthrimide reagent solution.* Dissolve 150 mg of 1,1'-dianthrimide in 1000 cm^3 concentrated sulphuric acid (*ca.* 96 per cent w/w). Keep in the dark and protected from moisture.

Standard boron solution. Dissolve 0.7621 g A.R. boric acid in water and dilute to 1 dm^3. Take 50 cm^3 of this solution and dilute to 1000 cm^3: the resulting solution contains 6.667 μg B per cm^3.

Dilute sulphuric acid. Prepare a 1:3 v/v solution.

Procedure (**boron in steel**).   Dissolve about 3 g of the steel (B content $\not> 0.02$ per cent), accurately weighed, in 40 cm^3 dilute sulphuric acid in a 150-cm^3 Vicor or silica flask fitted with a reflux condenser. Heat until dissolved. Filter through a quantitative filter paper into a 100-cm^3 graduated flask. Wash with hot water, cool to room temperature, and dilute to the mark with water. This flask (A) contains the acid-soluble boron.

Ignite the filter in a platinum crucible, fuse with 2.0 g of A.R. anhydrous sodium carbonate, dissolve the melt in 40 cm^3 of dilute sulphuric acid, and add 1 cm^3 of sulphurous acid solution (about 6 per cent) to reduce any iron(III) salt, etc.,

---

* Corning Vycor glass, containing 96 per cent of silica, is usually suitable.

formed in the fusion, and filter if necessary. Transfer the solution to a 100-cm³ graduated flask, dilute to the mark, and mix. This flask (B) contains the acid-insoluble boron.

Transfer 3.0 cm³ of solutions A and B to two dry, glass-stoppered conical flasks (Vycor or silica). Add 25 cm³ of dianthrimide reagent solution to each with shaking, and insert the glass stoppers loosely. For the blank use 3.0 cm³ of solutions A and B in two similar 50-cm³ conical flasks and add 25 cm³ concentrated sulphuric acid (98 per cent w/w). Heat all four flasks in a boiling water bath for 60 minutes. Cool to room temperature and measure the absorbance of each of the solutions at 620 nm against pure concentrated sulphuric acid in 1-cm or 2-cm cells. Correct for the blanks.

To construct the calibration curve, run 5–50 cm³ of the standard boron solution by means of a burette into 100-cm³ graduated flasks, add 30 cm³ of dilute sulphuric acid, and make up to volume. These solutions contain 1–10 μg of B per 3 cm³. Use 3 cm³ of each solution and of a boron-free comparison solution and proceed as above. Plot a calibration curve relating absorbance and boron content.

Calculate the total boron content of the steel (i.e., acid-soluble plus acid-insoluble boron).

An alternative method for the determination of boron is given under Section **XVI, 8.**

XVIII, 18.   CHROMIUM.   *Discussion.*   Small amounts of chromium (up to 0.5 per cent) may be determined colorimetrically in alkaline solution as chromate; uranium and cerium interfere, but vanadium has little influence. The transmittance of the solution is measured at 365–370 nm or with the aid of a filter having maximum transmission in the violet portion of the spectrum. The standard solution used for the preparation of the reference curve should have the same alkalinity as the sample solution, and should preferably have the same concentration of foreign salts. Standards may be prepared from A.R. potassium chromate.

A more sensitive method is to employ **1,5-diphenylcarbazide** $CO(NH \cdot NHC_6H_5)_2$; in acid solution (*ca.* 0.2*M*) chromates give a soluble violet compound with this reagent.

Molybdenum(VI), vanadium(V), mercury, and iron interfere; permanganates, if present, may be removed by boiling with a little ethanol. If the ratio of vanadium to chromium does not exceed 10:1, nearly correct results may be obtained by allowing the solution to stand for 10–15 minutes after the addition of the reagent, since the vanadium–diphenylcarbazide colour fades fairly rapidly. Vanadate can be separated from chromate by adding oxine to the solution and extracting at a pH of about 4 with chloroform; chromate remains in the aqueous solution. Vanadium as well as iron can be precipitated in acid solution with cupferron and thus separated from chromium(III).

Procedure.   Prepare a 0.25 per cent solution of diphenylcarbazide in 50 per cent acetone as required. The test solution may contain from 0.2 to 0.5 part per million of chromate. To about 15 cm³ of this solution add sufficient 3*M*-sulphuric acid to make the concentration about 0.1*M* when subsequently diluted to 25 cm³, add 1 cm³ of the diphenylcarbazide reagent and make up to 25 cm³ with water. Match the colour produced against standards prepared from 0.001*N*-potassium dichromate solution. A green filter having the transmission maximum at about 540 nm may be used.

Chromium in steel. *Discussion.* The chromium in the steel is oxidised by perchloric acid to the dichromate ion, the colour of which is intensified by iron(III) perchlorate which is itself colourless. The coloured solution is compared with a blank in which the dichromate is reduced with ammonium iron(II) sulphate. The method is not subject to interference by iron or by moderate amounts of alloying elements usually present in steel.

Procedure. Place a 1.000 g sample of the steel (Cr content < 0.1 per cent) in a 100 cm^3 beaker and dissolve it in 10 cm^3 of dilute nitric acid (1 : 1) and 20 cm^3 of A.R. perchloric acid (sp. gr. 1.70; 70–72 per cent). [If the Cr content is 0.1–1 per cent, dissolve a 0.5000 g sample in 10 cm^3 of dilute nitric acid (1 : 1) and 15 cm^3 of perchloric acid (sp. gr. 1.70).] Evaporate to dense fumes of perchloric acid and boil gently for 5 minutes to oxidise the chromium. Cool the beaker and contents rapidly, dissolve soluble salts by adding 20 cm^3 of water, transfer the solution quantitatively to a glass-stoppered 50 cm^3 graduated flask, and dilute to the mark. Remove an aliquot portion to an absorption cell, reduce it with a little (*ca.* 20 mg) A.R. ammonium iron(II) sulphate, and adjust the colorimeter or spectrophotometer so that the reading with this solution is zero; a violet filter having a maximum transmission between 410 and 480 nm may be used in the colorimeter. Discard the solution in the absorption cell, and refill it with an equal volume of the oxidised solution: the reading is a measure of the colour due to the dichromate.

Standardisation may be carried out by the use of solutions prepared from a chromium-free standard steel and standard potassium dichromate solution. After dissolution of the standard steel, the solution is boiled with perchloric acid, potassium dichromate added and the resulting solution is diluted to volume, and measurements are carried out as above. The chromium content of any unknown steel may then be deduced from the colorimeter reading.

XVIII, 19. COBALT. *Discussion.* An excellent method for the colorimetric determination of minute amounts of cobalt is based upon the soluble red complex salt formed when cobalt ions react with an aqueous solution of **nitroso-R-salt** (sodium 1-nitroso-2-hydroxynaphthalene-3,6-disulphonate). Three moles of the reagent combine with 1 mole of cobalt.

The cobalt complex is usually formed in a hot acetate–acetic acid medium. After the formation of the cobalt colour, hydrochloric acid or nitric acid is added to decompose the complexes of most of the other heavy metals present. Iron, copper, cerium(IV), chromium(III and VI), nickel, vanadyl vanadium, and copper interfere when present in appreciable quantities. Excess of the reagent minimises the interference of iron(II), iron(III) can be removed by diethyl ether extraction from a hydrochloric acid solution. Most of the interferences can be eliminated by treatment with potassium bromate, followed by the addition of an alkali fluoride. Cobalt may also be isolated by dithizone extraction from a basic medium after copper has been removed (if necessary) from acidic solution. An alumina column may also be used to adsorb the cobalt nitroso-R-chelate anion in the presence of perchloric acid, the other elements are eluted with warm 1M-nitric acid, and finally the cobalt complex with 1M-sulphuric acid, and the absorbance measured at 500 nm.

Procedure. The test solution should contain between 0.001 and 0.02 mg of cobalt. Evaporate almost to dryness, add 1 cm^3 of concentrated nitric acid, and continue the evaporation just to dryness to oxidise any iron(II) which may be present. Dissolve the residue in 10 cm^3 of water containing 0.5 cm^3 each of 1 : 1 hydrochloric acid and 1 : 10 nitric acid. Boil for a few minutes to dissolve any solid material. Add 2.0 cm^3 of a 0.2 per cent aqueous solution of nitroso-R-salt and also 2.0 g of hydrated sodium acetate. The pH of the solution should be close to 5.5; check with bromocresol green indicator or with a pH meter. Boil for 1 minute, add 1.0 cm^3 of concentrated hydrochloric acid, and boil again for 1 minute. Cool to room temperature, dilute to 25 cm^3 in a graduated flask, and compare the colour with standards or use a spectrophotometer. Determine the absorbance at 425 nm against a reagent blank. In the presence of 2 mg or more of iron it is best to make the measurement at 500 nm to reduce the error resulting from the absorption of light by the yellow solution.

Standard solutions may be conveniently prepared with spectroscopically pure cobalt (Johnson, Matthey).

Cobalt in steel. *Discussion.* An alternative, but less sensitive, method utilises 2-nitroso-1-naphthol, and this can be used for the determination of cobalt in steel. The pink cobalt(III) complex is formed in a citrate medium at pH 2.5–5. Citrate serves as a buffer, prevents the precipitation of metallic hydroxides, and complexes iron(III) so that it does not form an extractable nitrosonaphtholate complex. The cobalt complex forms slowly (*ca.* 30 minutes) and is extracted with chloroform.

Procedure. Prepare the **2-nitroso-1-naphthol reagent** by dissolving 1.0 g of the solid in 100 cm^3 of glacial acetic acid. Add 1 g of activated carbon: shake the solution before use and filter off the required volume.

Dissolve a known weight (*ca.* 0.5 g) of the steel by any suitable procedure. Treat the acidic sample solution ($< 200 \mu g$ Co), containing iron in the iron(II) state, with 10–15 cm^3 of 40 per cent (w/v) sodium citrate solution, dilute to 50–75 cm^3 and adjust the pH to 3–4 (indicator paper) with 2*M*-hydrochloric acid or sodium hydroxide. Cool to room temperature, add 10 cm^3 of 3 per cent (10-volume) hydrogen peroxide and, after 3 minutes, 2 cm^3 of the reagent solution. Allow to stand for at least 30 minutes at room temperature. Extract the solution in a separatory funnel by shaking vigorously for 1 minute with 25 cm^3 of chloroform: repeat the extraction twice with 10-cm^3 portions of chloroform. Dilute the combined extracts to 50 cm^3 with chloroform and transfer to a clean separatory funnel. Add 20 cm^3 of 2*M*-hydrochloric acid, shake for 1 minute, run the chloroform layer into another separatory funnel, and shake for 1 minute with 20 cm^3 of 2*M*-sodium hydroxide. Determine the absorbance of the clear chloroform phase in a 1-cm cell at 530 nm.

For the preparation of standard cobalt solutions, use A.R. cobalt(II) chloride or spectroscopically pure cobalt dissolved in hydrochloric acid; subject solutions containing 0, 5, 10, 25, 50, 100, 150, and 200 μg of Co to the whole procedure.

XVIII, 20. COPPER. *Discussion.* Small quantities of copper may be determined by the diethyldithiocarbamate method (Section **VI, 9**) or by the 'neo-cuproin' method (Section **VI, 10**), an extraction being necessary in both cases. In another somewhat simpler procedure, the copper is complexed with **biscyclohexanone oxalyldihydrazone** and the resulting blue colour is measured by a suitable spectrophotometer within the range 570–600 nm. The solution measured should contain not more than 100 μg of copper.

Reagents. *Bicyclohexanone oxalyldihydrazone solution* (copper reagent). Dissolve 0.1 g of the solid reagent in 10 cm^3 ethanol (or industrial methylated spirit) and 10 cm³ hot water, and dilute to 200 cm³. Filter, if necessary.

Synthetic standard solution (for analysis of steel).   Dissolve an appropriate weight of pure iron (Johnson Matthey) in a mixture of equal volumes of concentrated hydrochloric acid and concentrated nitric acid; with this solution as base, add a suitable amount of copper nitrate solution containing 0.01 g copper per dm³.

Procedure (**copper in steel**).   Weigh out accurately a 0.1-g sample of the steel* into a 150-cm³ conical beaker, add 5 cm³ concentrated hydrochloric acid and 5 cm³ concentrated nitric acid, and warm gently. In the presence of interfering amounts of chromium, add 5 cm³ perchloric acid, sp. gr. 1.70, and evaporate until strong fuming occurs. When the sample has dissolved or after the fuming with perchloric acid, cool, add 50 cm³ cold distilled water, followed by 10 cm³ acid solution (1:HCl/HNO₃). Carefully add 10 cm³ concentrated ammonia solution, sp. gr. 0.88, cool to room temperature, and dilute to 100 cm³ in a graduated flask. Return the solution to the original beaker and transfer a 10-cm³ aliquot to a 100-cm³ graduated flask. Add 20 cm³ of the copper reagent, dilute to 100 cm³ with distilled water, and transfer to a 100-cm³ dry beaker. Allow to stand for 10–15 minutes, and then measure the absorbance with a spectrophotometer.

Construct a calibration curve using the synthetic standard solution: add the standard copper solution immediately before the reagent.

XVIII, 21.   IRON.   Three procedures will be described—the thiocyanate, the 1,10-phenanthroline and the thioglycollic acid methods.

A. Thiocyanate method.   *Discussion.*   Iron(III) reacts with thiocyanate to give a series of intensely red-coloured compounds, which remain in true solution: iron(II) does not react. Depending upon the thiocyanate concentration, a series of complexes can be obtained; these complexes are red and can be formulated as $[Fe(SCN)_n]^{3-n}$, where $n = 1, \ldots 6$. At low thiocyanate concentration the predominant coloured species is $[Fe(SCN)]^{2+}$ ($Fe^{3+} + SCN^- \rightarrow [Fe(SCN)]^{2+}$), at 0.1 M-thiocyanate concentration it is largely $[Fe(SCN)_2]^+$, and at very high thiocyanate concentration it is $[Fe(SCN)_6]^{3-}$. In the colorimetric determination a large excess of thiocyanate should be used, since this increases the intensity and also the stability of the colour. Strong acids (hydrochloric or nitric acid—concentration 0.05–0.5 M) should be present to suppress the hydrolysis:

$$Fe^{3+} + 3H_2O \rightleftharpoons Fe(OH)_3 + 3H^+$$

Sulphuric acid is not recommended, because sulphate ions have a certain tendency to form complexes with iron(III) ions. Silver, copper, nickel, cobalt, titanium, uranium, molybdenum, mercury (> 1 g dm⁻³), zinc, cadmium, and bismuth interfere. Mercury(I) and tin(II) salts, if present, should be converted into the mercury(II) and tin(IV) salts, otherwise the colour is destroyed. Phosphates, arsenates, fluorides, oxalates, and tartrates interfere, since they form fairly stable complexes with iron(III) ions; the influence of phosphates and arsenates is reduced by the presence of a comparatively high concentration of acid.

When large quantities of interfering substances are present, it is usually best to proceed in either of the following ways: (i) remove the iron by precipitation with a

---

* The following British Chemical Standards may be used for practice in this determination: B.C.S. No. 402 (0.23 per cent Cu); No. 410 (0.47 per cent Cu); No. 406 (0.32 per cent Cu).

slight excess of ammonia solution, and dissolve the precipitate in dilute hydrochloric acid; (ii) extract the 'iron(III) thiocyanate' three times either with pure diethyl ether or, better, with a mixture of pentanol and pure diethyl ether (5:2) and employ the organic layer for the colour comparison.

Reagents. Prepare the following solutions:

Standard solution of iron(III). Use method (*a*), (*b*) or (*c*). (*a*) Dissolve 0.7022 g A.R. ammonium iron(II) sulphate in 100 cm^3 water, add 5 cm^3 of 1:5 sulphuric acid, and run in cautiously a dilute solution of potassium permanganate (2 g dm^{-3}) until a slight pink coloration remains after stirring well. Dilute to 1 dm^3 and mix thoroughly. 1 cm^3 ≡ 0.1 mg of Fe. (*b*) Dissolve 0.864 g A.R. ammonium iron(III) sulphate in water, add 10 cm^3 concentrated hydrochloric acid and dilute to 1 dm^3. 1 cm^3 ≡ 0.1 mg of Fe. (*c*) Dissolve 0.1000 g of electrolytic iron or pure iron wire in 50 cm^3 of 1:3 nitric acid, boil to expel oxides of nitrogen, and dilute to 1 dm^3 with de-ionised water.

Potassium thiocyanate solution. Dissolve 20 g A.R. potassium thiocyanate in 100 cm^3 water: the solution is *ca.* 2*M*.

Procedure. Dissolve a weighed portion of the substance in which the amount of iron is to be determined in a suitable acid, and evaporate nearly to dryness to expel excess of acid. Dilute slightly with water, oxidise the iron to the iron(III) state with dilute potassium permanganate solution or with a little bromine water, and make up the liquid to 500 cm^3 or other suitable volume. Take 40 cm^3 of this solution and place in a 50-cm^3 graduated flask, add 5 cm^3 of the thiocyanate solution and 3 cm^3 of 4*M*-nitric acid. Add de-ionised water to dilute to the mark. Prepare a blank using the same quantities of reagents. Measure the absorbance of the sample solution in a spectrophotometer at 480 nm. Determine the concentration of this solution by comparison with values on a reference curve obtained in the same way from different concentrations of the standard iron solution.

B. 1,10-Phenanthroline method. *Discussion.* Iron(II) reacts with 1,10-phenanthroline to form an orange-red complex $[(C_{12}H_8N_2)_3Fe]^{2+}$. The colour intensity is independent of the acidity in the pH range 2–9, and is stable for long periods. Iron(III) may be reduced with hydroxylammonium chloride or with hydroquinone. Silver, bismuth, copper, nickel, and cobalt interfere seriously, as do perchlorate, cyanide, molybdate, and tungstate. The iron–phenanthroline complex (as the perchlorate) may be extracted with nitrobenzene and measured at 515 nm against a reagent blank.

Both iron(II) and iron(III) can be determined spectrophotometrically: the reddish-orange iron(II) complex absorbs at 515 nm, and both the iron(II) and the yellow iron(III) complex have identical absorption at 396 nm, the amount being additive. The solution, slightly acid with sulphuric acid, is treated with 1,10-phenanthroline, and buffered with potassium hydrogenphthalate at a pH of 3.9: the reading at 396 nm gives the total iron and that at 515 nm the iron(II).

Reagents. Prepare the following solutions: 1,10-phenanthroline, 0.25 per cent solution of the monohydrate in water; sodium acetate, 0.2*M* and 2*M*. Hydroxylammonium chloride, 10 per cent aqueous solution, or hydroquinone, 1 per cent solution in an acetic acid buffer of pH, *ca.* 4.5 (mix 65 cm^3 of 0.1*M*-acetic acid and 35 cm^3 of 0.1*M*-sodium acetate solution), is prepared when required.

Procedure. Take an aliquot portion of the unknown slightly acid solution containing 0.1–0.5 mg iron and transfer it to a 50-cm^3 graduated flask. Determine, by the use of a similar aliquot portion containing a few drops of bromophenol blue, the volume of sodium acetate solution required to bring the pH to 3.5 ± 1.0. Add the same volume of acetate solution to the original aliquot part and then 4 cm^3 each of the hydroquinone and 1,10-phenanthroline solutions. Make up to the mark with distilled water, mix well, and allow to stand for 1 hour to complete the reduction of the iron. Compare the intensity of the colour produced with standards, similarly prepared, in any convenient way. If a colorimeter is employed, use a filter showing maximum transmission at 480–520 nm, for a spectrophotometer, use a wavelength of 515 nm.

The iron may also be reduced with hydroxylammonium chloride. Add 5 cm^3 of the 10 per cent hydroxylammonium solution, adjust the pH of the slightly acid solution to 3–6 with sodium acetate, then add 4 cm^3 of the 1,10-phenanthroline solution, dilute to 50 cm^3, mix, and measure the absorbance after 5–10 minutes.

C. Thioglycollic acid method. *Discussion.* The use of thioglycollic acid (mercaptoacetic acid) for the determination of iron (Refs. 8 and 9) is of importance because it is relatively free from interferences giving a red-purple colour with Fe^{3+} which can be measured at 535 nm. Precipitation of Al^{3+} and Cr^{3+} ions is prevented by the addition of ammonium citrate. The reaction with the Fe^{3+} ions is represented as:

$$Fe^{3+} + 2HSCH_2COO^- + 3OH^- = Fe(OH)(SCH_2COO)_2^{2-} + 2H_2O$$

Reagents. Prepare the following solutions:

Thioglycollic acid solution. Dissolve 10 cm^3 of analytical grade thioglycollic acid in water and dilute to 100 cm^3.

Ammonium hydroxide solution. Take 25 cm^3 of 0.880 ammonium hydroxide and dilute to 100 cm^3.

Ammonium citrate solution. Dissolve 20 g analytical grade ammonium citrate (iron free) in water and dilute to 100 cm^3.

Standard iron solution. Dissolve 0.100 g pure iron wire in the minimum quantity of boiling 2*M*-nitric acid and dilute the resulting solution to 1 dm^3, giving a solution containing 100 p.p.m. of iron.

Procedure. Take 10 cm^3 of the standard solution and dilute to 100 cm^3 in a graduated flask, to give a 10 p.p.m. solution. From this solution prepare a series of standards containing 1, 2, 3, 4, and 5 p.p.m. by taking 5, 10, 15, 20 and 25 cm^3 aliquots and placing in 50 cm^3 graduated flasks. Add 5 cm^3 of the thioglycollic acid solution, 2 cm^3 of the 20 per cent ammonium citrate solution and 5 cm^3 of the dilute ammonium hydroxide solution to each flask. Dilute with water and mix to make up to the 50 cm^3 mark. Using a spectrophotometer, measure the absorbances of the solutions at 535 nm by comparison with a blank prepared in the same manner. Plot a concentration/absorbance line from the values obtained.

Take sufficient of the sample under examination to contain approximately 0.1 g iron and dissolve in the minimum amount of dilute nitric acid. If an organic material is being studied or it does not dissolve in dilute nitric acid, heat the sample with a small quantity of concentrated nitric or sulphuric acid and evaporate nearly to dryness. Take the residue into solution with a little dilute nitric acid. Dilute the resulting solution to 1 dm^3. Take 1 cm^3 of this solution and place in a 50-cm^3 graduated flask and add 5 cm^3 thioglycollic acid solution, 2 cm^3 ammonium citrate solution and 5 cm^3 ammonium hydroxide. Dilute with water and make up to the 50 cm^3 mark as with the standard solutions.

Measure the absorbance of this solution at 535 nm and compare the value obtained with the standard line previously plotted. From the value for the concentration of iron in solution calculate the original amount of iron in the starting material.

XVIII, 22. LEAD. *Discussion.* For the determination of small amounts of lead (0.005–0.25 mg) advantage is taken of the fact that when a sulphide is added to a solution containing lead *ions* a brown colour, due to the formation of colloidal lead sulphide, is produced.

Interference is caused by the presence of: (*a*) neutral salts, such as ammonium chloride and particularly tartrates and citrates, and (*b*) other elements, such as copper, bismuth, iron, and aluminium. Errors due to (*a*) may be allowed for by ensuring that the standards for comparison contain amounts of salts approximately equal to those in the solution under test, while those produced by (*b*) are eliminated by the usual analytical procedure or by the use of diphenylthiocarbazone (see below). The disturbing effect due to copper and iron if present in small amount may be overcome by the addition of a few drops of a 10 per cent aqueous solution of potassium cyanide; aluminium may be retained in ammoniacal solution by the addition of ammonium citrate solution, a corresponding amount of the latter being added to the standards.

Reagent. Prepare a **standard lead solution** by dissolving either 0.183 g A.R. lead acetate crystals or 0.160 g A.R. lead nitrate in 100 cm³ water; 10 cm³ of this are diluted to 100 cm³ for a working solution: the latter contains 0.1 mg of Pb per cm³.

Procedure. As an illustration of the simple lead sulphide procedure, the **determination of lead in commercial tartaric acid** will be described. Dissolve 10 g of the tartaric acid in about 40 cm³ of hot water, add 5 to 6 drops of 10 per cent potassium cyanide solution **(danger!)** then 25 cm³ of 1:2 ammonia solution (which should make the solution ammoniacal), and finally, add 0.5 cm³ of 10 per cent (w/v) sodium sulphide solution. Make up to 100 cm³. Prepare a blank solution using 10 g lead-free tartaric acid and treat it in exactly the same way. Measure the absorption of the two solutions with a spectrophotometer at a wavelength of 430 nm.

Construct a calibration curve by adding 0, 0.25, 0.50, 0.75, and 1.00 cm³ of the standard lead solution to the solution prepared as above from 10 g of lead-free tartaric acid.

Minute amounts of lead may also be determined by the dithizone method (see Section **VI, 13**). The procedure may be rendered fairly specific by first removing the lead with sodium diethyldithiocarbamate (compare Section **VI, 9**) at pH 7, and extracting the lead diethyldithiocarbamate with a mixture of equal volumes of pentanol and 'sulphur-free' toluene. The organic layer is treated with dilute hydrochloric acid, whereupon the lead complex passes into the aqueous layer. The latter is mixed with ammoniacal dithizone solution and the lead dithizonate extracted with carbon tetrachloride; the absorbance is measured at 515 nm. The pH of 7 is attained by the use of a buffer solution composed of 25 g sodium citrate, 4 g sodium hydrogencarbonate, and 100 cm³ water; the sodium diethyl-dithiocarbamate is not stable in the presence of this buffer, and so the appropriate amount of the solid reagent is added as required.

XVIII, 23. MAGNESIUM. Two methods are commonly used for the determination of magnesium. Titan Yellow may be used to obtain a coloured colloidal suspension, or Solochrome Black to give a red soluble complex. In most cases the second of these is to be preferred.

A. Titan Yellow method. *Discussion.* When magnesium hydroxide is precipitated with sodium hydroxide solution in the presence of the organic dyestuff **Titan Yellow** (the sodium salt of dehydrothio-*p*-toluidine sulphonic acid; Colour Index No. 19540) a red lake is formed (pH > 12). Fairly stable suspensions of the lake can be obtained if the magnesium concentration is below 4–5 parts per million, and fading of the colour is prevented by the presence of hydroxylammonium chloride. The reagent alone gives a yellow-brown colour in sodium hydroxide solution.

Many metals interfere, particularly those which give insoluble hydroxides in alkali hydroxide solution (e.g. cadmium, nickel, and cobalt) or those of which the hydroxides are soluble in an excess of sodium hydroxide, such as aluminium, zinc, and tin. Appreciable amounts of phosphate destroy the colour, calcium intensifies the colour of the magnesium lake; errors due to calcium can be reduced by adding the same amount of a calcium salt to the sample and standard solutions.

Procedure. Remove all interfering elements, for example, iron, aluminium, and phosphate by double precipitation with aqueous ammonia solution; also calcium (if present in quantity) with ammonium oxalate solution, and other metals by appropriate methods. Evaporate the filtrate to dryness to expel ammonium salts, moisten the residue with a few drops of dilute hydrochloric acid, and dilute to volume in a graduated flask of suitable size.

Transfer into a 50-cm^3 graduated flask a volume of the sample solution (say, 25 cm^3) containing 5 parts per million or less of magnesium. Almost neutralise any acid present with dilute sodium hydroxide solution. Dilute to about 35 cm^3, and add 1.0 cm^3 of 5 per cent aqueous hydroxylammonium chloride solution and 1.00 cm^3 of a 0.15 per cent aqueous solution of Titan Yellow (special quality for analytical use). Then add 5.0 cm^3 1M-sodium hydroxide while swirling the flask. Dilute to the mark with water, mix, and measure the intensity of the colour of the colloidal suspension by suitable means soon after the lake has been formed. Use a green filter (having a maximum transmission at 535 nm) with a filter photometer.

To construct the calibration curve, use standard magnesium solutions (prepared from pure magnesium and dilute hydrochloric acid) containing 0, 1, 2, 3, 4, and 5 μg of magnesium, and treat each solution in the same manner as the sample solution.

B. Solochrome Black method. *Discussion.* The difficulties inherent in the colloidal systems involved in 'lake' methods may be avoided by the use of organic reagents which form soluble coloured complexes with magnesium in basic solution. One such reagent is Solochrome Black, which forms red soluble complexes with magnesium. The colour is not stable: calcium, copper, manganese, iron, aluminium, cobalt, nickel, etc., interfere. By buffering at pH 10.1, a single complex is formed by one magnesium ion and two molecules of the dye. Calcium may be separated from magnesium by precipitation as sulphate in the presence of a large excess of methanol.

Reagents. *Buffer solution* (pH = 10.1). This consists of a 0.75 per cent w/v solution of A.R. ammonium chloride in dilute ammonia solution, prepared by mixing 5 volumes of concentrated ammonia solution (sp. gr. 0.88) and 95 volumes of water.

Solochrome Black solution. Prepare a 0.1 per cent solution in methanol; warm to speed solution and filter.

Procedure. Transfer the neutral sample solution ($< 100 \mu g$ Mg), free from calcium and other metals, to a 100-cm^3 graduated flask with calibrated neck. Add 25 cm^3 of the buffer solution, dilute to just below the 90-cm^3 graduation mark, and shake. Add 10.0 cm^3 of the Solochrome Black solution carefully. Shake to mix and dilute to the 100-cm^3 mark with water. Measure the absorbance immediately at 520 nm against that of a blank solution, similarly prepared but containing no magnesium.

XVIII, 24. MANGANESE. *Discussion.* Small quantities of manganese are usually determined colorimetrically by oxidation to permanganic acid. The two oxidising agents commonly used are ammonium persulphate in a phosphoric acid–nitric acid medium in the presence of a little silver nitrate as catalyst, and potassium periodate. The use of the latter will be described. In hot acid solution (nitric or sulphuric acid), periodate oxidises manganese ion quantitatively to permanganic acid:

$$2Mn^{2+} + 5IO_4^- + 3H_2O = 2MnO_4^- + 5IO_3^- + 6H^+$$

The merits of the periodate method include: (*a*) the concentration of the acid has little influence, and may be varied within wide limits; (*b*) the boiling may be prolonged beyond the time necessary to oxidise the manganese without detriment; and (*c*) the permanganic acid solution will keep for several months unchanged if an excess of periodate is present.

When ready for test the solution should not contain more than 2 mg of manganese per 100 cm^3, otherwise the colour will be too dark and colour matching will be difficult. Frequently iron(III) is added to the standard in amount equal to that found independently to be present in the sample. Phosphoric acid must be present to prevent the precipitation of iron(III) periodate and iodate, and also to decolorise the iron(III) (by complex formation). If chlorides are present it is necessary to evaporate with a mixture of nitric and sulphuric acids until fumes of the latter appear; chlorides react with the periodate. Reducing substances reacting with periodate or permanganate must be destroyed (e.g., by evaporation with nitric acid or a mixture of nitric acid and sulphuric acid) before the periodate oxidation. Chromium(III) and cerium(III) are oxidised by periodate in acid solution.

Procedure. For practice in this determination, the manganese content of a standard steel sample may be evaluated. Weigh out accurately a suitable quantity of the steel (0.1–0.2 g for steels containing up to 1 per cent of Mn) into a conical flask, dissolve it in 20–50 cm^3 of 1:3 nitric acid, and boil for 1 or 2 minutes to expel oxides of nitrogen. Remove from the burner, and add 0.5–1.0 g A.R. ammonium persulphate; boil for 10–15 minutes to oxidise carbon compounds and to destroy the excess of persulphate. If any permanganate colour develops or oxides of manganese separate, add a few drops of sulphurous acid or sodium sulphite solution to reduce the manganese and render the solution clear, and boil for a few minutes to expel the excess of sulphur dioxide. Dilute the solution to *ca.* 100 cm^3, add 5–10 cm^3 A.R. syrupy phosphoric acid and 0.5 g potassium periodate (1): boil for 1 minute and keep hot for 5–10 minutes. Cool the solution and make up to 250 cm^3 in a graduated flask and mix thoroughly. Match the colour against standards.

A wavelength of 545 nm should be used with a spectrophotometer. For steels or other alloys, an aliquot of the sample solution, not oxidised but otherwise treated similarly to the sample, may be used in the reference cell of the instrument. Sometimes the blank consists of a portion of the developed sample to which a reducing agent (nitrite) has been added to remove the permanganate colour: this is not applicable when titanium is present.

Prepare **standard manganese solutions** by any of the following methods:

(a) Use a steel of known manganese content, which has been treated like the unknown sample (2).

(b) Dissolve a known weight of pure electrolytic manganese in dilute nitric acid, boil out oxides of nitrogen, and dilute to give a 0.01 per cent solution. If pure manganese is not available, reduce standard potassium permanganate solution with a little sulphite after the addition of dilute sulphuric acid, and remove the sulphur dioxide by boiling. Dilute the manganese(II) solution and oxidise it with potassium periodate in the same way as the unknown solution.

   Notes. 1. Each 0.1 g of Mn requires 1 g of potassium periodate.

   2. Such standards, containing excess of periodate, are stable for 2–3 months.

XVIII, 25. MOLYBDENUM. *Discussion.* Molybdenum may be determined colorimetrically by the thiocyanate–tin(II) chloride method (for details, see Section **VI, 14**) or by the dithiol method described here.

Toluene-3,4-dithiol, usually called **dithiol**, yields a slightly soluble dark green complex, $(CH_3 \cdot C_6H_3 \cdot S_2)_3 Mo(VI)$, with molybdenum(VI) in a mineral acid medium, which can be extracted by organic solvents. The resulting green solution is used for the colorimetric determination of molybdenum.

Dithiol is a less-selective reagent than thiocyanate for molybdenum. Tungsten interferes most seriously, but does not do so in the presence of tartaric acid or citric acid (see Section **XVIII, 29**). Tin does not interfere if the absorbance is read at 680 nm. Strong oxidants oxidise the reagent; iron(III) salts should be reduced with potassium iodide solution and the liberated iodine removed with thiosulphate.

   Procedure. Prepare the **dithiol reagent** by adding 0.1–0.2 g of dithiol to 100 cm^3 of 0.25M-sodium hydroxide solution, followed by 0.5 cm^3 of thioglycollic acid (to inhibit oxidation of the reagent); keep at 5 °C and prepare fresh daily.

Add to the sample solution (containing 1–25 μg of Mo) 4 cm^3 of 1 : 3 sulphuric acid, 3 drops of 85 per cent phosphoric acid, and 0.5 g of citric acid. Dilute with water to 20 cm^3 and add 2 cm^3 of dithiol solution. Allow to stand at room temperature for 2 hours. Extract the molybdenum complex with 13-cm^3 and 10-cm^3 portions respectively of redistilled butyl acetate, and make up to 25.0 cm^3 with this solvent in a graduated flask; filter through glass wool if not entirely clear. Determine the absorbance of the solution at 670 nm. Prepare a calibration curve as detailed in Section **VI, 14**.

XVIII, 26. NICKEL. *Discussion.* When dimethylglyoxime is added to an alkaline solution of a nickel salt which has been treated with an oxidising agent (such as bromine), a red coloration is obtained. The red soluble complex contains nickel in a higher oxidation state.* The nickel complex formed absorbs at about 445 nm provided absorbance readings are made within 10 minutes of mixing. The

---

* This has been regarded as nickel(III) and also as nickel(IV) dimethylglyoximate.

dimethylglyoxime-oxidising agent method must be distinguished from the divalent nickel–dimethylglyoxime procedure which yields a nickel(II) dimethylglyoximate soluble in chloroform: full details of this solvent extraction method are given in Section **VI, 15**.

Cobalt(II), gold(III), and dichromate ions interfere under the conditions of the test. Metals which precipitate in ammoniacal solution can be removed by double precipitation, or by taking advantage of the solubility of nickel(II) dimethylglyoxime in chloroform (the nickel(III) complex is insoluble, as is also the brown cobalt dimethylglyoxime). Copper may accompany the nickel in the extraction; most of the copper is removed from the chloroform extract when it is shaken with dilute ammonia solution, whereas the nickel remains in the organic solvent. The nickel(II) dimethylglyoxime in the chloroform layer may be decomposed by shaking with dilute hydrochloric acid; most of the dimethylglyoxime remains in the chloroform, the nickel is transferred to the aqueous phase and may be determined colorimetrically. Citrate or tartrate may be added to prevent the precipitation of iron, aluminium, etc. Much manganese may interfere, but this is prevented by adding hydroxylammonium chloride which maintains the element in the divalent state.

Procedure (**nickel in steel**). Dissolve 0.50 g, accurately weighed,* of the steel in 10 cm^3 of warm 1 : 1 nitric acid, boil to expel oxides of nitrogen, cool, and make up to 250 cm^3 with water in a graduated flask. Mix well, and transfer 5 cm^3 of the solution to a 50-cm^3 graduated flask. To 5 cm^3 of this solution add 5 cm^3 of 10 per cent citric acid solution, neutralise with concentrated ammonia solution and add a few drops in excess (pH > 7.5). Add 2 cm^3 of a 1 per cent dimethylglyoxime solution in ethanol (or more if copper or cobalt is present). Extract with three 3-cm^3 portions of chloroform, shaking for 30 seconds each time. Shake the combined chloroform extracts with 6 cm^3 of 0.5M-ammonia solution (1 : 30); shake the ammonia washing with 2 cm^3 of chloroform and add the latter to the main chloroform extract. Return the nickel to the ionic state by shaking the chloroform extract vigorously for 1 minute with two 5-cm^3 portions of 0.5M-hydrochloric acid. Transfer the hydrochloric acid solution to a 25-cm^3 graduated flask, dilute to about 20 cm^3, add 1 cm^3 of saturated bromine water, followed by 2 cm^3 of concentrated ammonia solution. Cool below 30 °C if necessary, add 1 cm^3 of 1 per cent dimethylglyoxime solution, and dilute to volume. Measure the absorbance at 445 nm after 5 minutes. The standard solutions for the construction of the calibration curve should contain approximately the same concentration of iron (nickel-free) as the sample solution.

Prepare the **standard nickel solution** by dissolving 0.673 g pure ammonium nickel sulphate in water and diluting to 1 dm^3: 1 cm^3 contains 0.1 mg of Ni. The solution may be further diluted to a basis of 0.01 mg of Ni per cm^3, if necessary. Pure nickel metal may also be employed for the preparation of the standard solution.

XVIII, 27. TIN. *Discussion.* In acid solution, toluene-3,4-dithiol (**dithiol**) forms a red compound when warmed with tin(II) salts (compare Molybdenum,

---

* The weight of steel to be taken will naturally depend upon the nickel content. The final nickel concentration should not exceed 0.6 mg per 100 cm^3 because a precipitate may form above this concentration.

Section **XVIII, 25**). Tin(IV) also reacts, but more slowly than tin(II); thioglycollic acid may be employed to reduce tin(IV) to tin(II). The reagent is not stable, being easily reduced, and hence should be prepared as required. A dispersant is generally added to the solution under test.

Many heavy metals react with dithiol to give coloured precipitates, e.g., bismuth, iron(III), copper, nickel, cobalt, silver, mercury, lead, cadmium, arsenic, etc.; molybdate and tungstate also react. Of the various interfering elements, only arsenic distils over with the tin when a mixture is distilled from a medium of concentrated sulphuric acid and concentrated hydrobromic acid in a current of carbon dioxide. If arsenic is present in quantities larger than that of the tin it should be removed.

Reagents. *Dithiol reagent.* Dissolve 0.1 g dithiol in 2.5 cm^3 5M-sodium hydroxide solution. Add 0.5 cm^3 thioglycollic acid, and dilute to 50 cm^3 with water. Prepare fresh daily.

Dispersant solution. Prepare a 1 per cent aqueous solution of sodium lauryl sulphate.

Standard tin solution. Dissolve 1.000 g A.R. tin in 100 cm^3 of 1:1 hydrochloric acid and dilute with the same concentration of acid to 1 dm^3: 1 cm^3 contains 1 mg Sn. Prepare more dilute solutions as required (e.g., 0.01 mg Sn per cm^3) by dilution with 1:1 hydrochloric acid.

Procedure. Transfer a 10-cm^3 aliquot of the sample solution, which should be 0.5M in hydrochloric acid and contain not more than 0.25 mg of tin, to a 25-cm^3 graduated flask, and add in the order given 1 drop thioglycollic acid, 2.0 cm^3 concentrated hydrochloric acid, 0.5 cm^3 of the dispersant solution, and 1.0 cm^3 of the dithiol reagent with thorough mixing after each addition. Place the flask in a water bath at 60 °C for 10 minutes, cool, and dilute the contents to the mark. Measure the absorbance at 545 nm against a reagent blank.

Construct a concentration–absorbance curve with the aid of the standard tin solution.

Procedure (**tin in canned foods**). The procedure provides for the removal of interfering copper by the addition of diethylammonium diethyldithiocarbamate in chloroform reagent.*

Weigh 5 or 10 g of the sample, depending on the expected tin content, into a small porcelain crucible. Dry and char the sample on a hot plate; heat to ash in a muffle furnace at 600 °C. Add 1 g of fusion mixture (3 parts Na_2CO_3 + 1 part KCN by weight) and fuse this with the ash by holding the crucible with nickel tongs over a Bunsen or Meker burner. Cool the crucible, place it in a small beaker, and cover the latter with a watch glass. Add 10 cm^3 water, and run 10 cm^3 dilute hydrochloric acid (1:1) cautiously into the crucible **(Fume Cupboard!)**. Boil the contents of the beaker gently for 30 minutes. Cool and filter: wash the beaker, crucible, and filter with water.

If copper is known to be absent or present only in negligible proportions, dilute the solution with water to 50 cm^3 in a graduated flask, and continue as detailed below. Otherwise, transfer the solution to a small separatory funnel and add 5 cm^3 of the diethylammonium diethyldithiocarbamate in chloroform reagent (diluted (1 + 20) with chloroform when required). Shake and run off the chloroform layer, extract the aqueous layer with successive 1-cm^3 portions of the

---

* This reagent is prepared from 3.0 cm^3 of diethylamine in chloroform and 1 cm^3 of carbon disulphide in 9 cm^3 of chloroform. Mix carefully and store in a dark bottle in a refrigerator.

reagent until the chloroform layer is colourless; finally, wash the aqueous layer with a few cm³ of chloroform. Dilute the aqueous solution with water to 50 cm³ in a graduated flask.

To 10.0 cm³ of the solution thus prepared add 0.5 cm³ of dilute hydrochloric acid (1 : 1) and proceed as above. Measure the absorbance at 545 nm, or use an Ilford No. 604 green filter with an absorptiometer.

XVIII, 28.   TITANIUM.   *Discussion.*   With an acidic titanium(IV) solution hydrogen peroxide produces a yellow colour:* with small amounts of titanium (up to 0.5 mg of TiO_2 per cm³), the intensity of the colour is proportional to the amount of the element present. Comparison is usually made with standard titanium(IV) sulphate solutions; a method for their preparation from potassium titanyl oxalate is described below. The hydrogen peroxide solution should be about 3 per cent strength (10-volume) and the final solution should contain sulphuric acid having a concentration from about 0.75 to 1.75M in order to prevent hydrolysis to a basic sulphate and to prevent condensation to metatitanic acid. The colour intensity increases slightly with rise of temperature, hence the solutions to be compared should have the same temperature, preferably 20–25 °C.

Elements which interfere are: (*a*) iron, nickel, chromium, etc., because of the colour of their solutions; (*b*) vanadium, molybdenum, and, under some conditions, chromium, because they form coloured compounds with hydrogen peroxide; (*c*) fluorine (even in minute amount) and large quantities of phosphates, sulphates, and alkali salts (the influence of the last two is largely reduced the greater the concentration of sulphuric acid present—up to 10 per cent). The influence of elements of class (*a*) is overcome, if present in small amount, by matching the colour by the addition of like quantities of the coloured elements to the standard before hydrogen peroxide is added. When large amounts of iron are present, as in the analysis of cast irons and steels, two methods may be adopted: (i) phosphoric acid can be added in like amount to both unknown and standard, after the addition of hydrogen peroxide; (ii) the iron content of the unknown solution is determined, and a quantity of standard iron(III) alum solution, containing the same amount of iron, is added to the standard solution. Large quantities of nickel, chromium, etc., must be removed. Elements of class (*b*) must also be removed; vanadium and molybdenum are most easily separated by precipitation of the titanium with sodium hydroxide solution in the presence of a little iron. Fluoride has the most powerful effect in bleaching the colour; it must be removed by repeated evaporation with concentrated sulphuric acid. The bleaching effect of phosphoric acid is overcome by adding a like amount to the standard, or by adding 1 cm³ of 0.1 per cent uranyl acetate solution for each 0.1 mg of Ti present.

Preparation of standard titanium solution.   Weigh out 3.68 g A.R. potassium titanyl oxalate $K_2TiO(C_2O_4)_2,2H_2O$ into a Kjeldahl flask, add 8 g ammonium sulphate and 100 cm³ concentrated sulphuric acid. Gradually heat the mixture to boiling and boil for 10 minutes. Cool, pour the solution into 750 cm³ of water, and dilute to 1 dm³ in a graduated flask; 1 cm³ ≡ 0.50 mg of Ti.

---

* The coloured species formed has been stated to be $[TiO(SO_4)_2]^{2-}$ or a similar ion; and it has also been formulated as $[Ti(H_2O_2)]^{4+}$ or an analogous complex.

If there is any doubt concerning the purity of the A.R. salt, standardise the solution by precipitating the titanium with ammonia solution or with cupferron solution, and ignite the precipitate to TiO_2.

Procedure. The sample solution should preferably contain titanium as sulphate in sulphuric acid solution, and be free from the interfering constituents mentioned in the *Discussion* above. The final acidity may vary from 0.75 to 1.75M. If iron is present in appreciable amounts, add dilute phosphoric acid from a burette until the yellow colour of the iron(III) is eliminated: the same amount of phosphoric acid must be added to the standards. If alkali sulphates are present in the test solution in appreciable quantity, add a like amount to the standards. Add 10 cm^3 of 3 per cent hydrogen peroxide solution and dilute the solution to 100 cm^3 in a graduated flask: the final concentration of Ti may conveniently be 2–25 parts per million. Compare the colour produced by the unknown solution with that of standards of similar composition by any of the usual methods.

For a filter colorimeter use a blue filter (maximum transmission 400–420 nm); a wavelength of 410 nm is employed for a spectrophotometer. In the latter case, the effect of iron, nickel, chromium(III), and other coloured ions not reacting with hydrogen peroxide may be compensated by using a solution of the sample, not treated with hydrogen peroxide, in the reference cell.

XVIII, 29. TUNGSTEN. *Discussion.* Toluene-3,4-dithiol (**dithiol**) may be used for the colorimetric determination of tungsten; it forms a slightly soluble coloured complex with tungsten(VI) which can be extracted with butyl or pentyl acetate and other organic solvents. Molybdenum reacts similarly (see Section **XVIII, 25**) and must be removed before tungsten can be determined. The molybdenum complex can be preferentially developed in cold weak acid solution and selectively extracted with pentyl acetate before developing the tungsten colour in a hot solution of increased acidity. The procedure will be illustrated by describing the determination of tungsten in steel.

Reagents. *Dithiol reagent solution.* Dissolve 1 g toluene-3,4-dithiol in 100 cm^3 pentyl acetate. This should be prepared immediately before use.

Standard tungsten solution. Dissolve 0.1794 g A.R. sodium tungstate $Na_2WO_4,2H_2O$ in water and dilute to 1 dm^3: 1 $cm^3 \equiv 0.1$ mg W. For use, dilute 100 cm^3 of this solution to 1 dm^3: 1 $cm^3 \equiv 0.01$ mg W.

'Mixed acid'. Mix 15.0 cm^3 concentrated sulphuric acid and 15.0 cm^3 orthophosphoric acid (sp. gr. 1.75), and dilute to 100 cm^3 with distilled water.

Procedure (**tungsten in steel**). Dissolve 0.5 g of the steel, accurately weighed, in 30 cm^3 of the 'mixed acid' by heating, oxidise with concentrated nitric acid, and evaporate to fuming. Extract with 100 cm^3 water, boil, transfer to a 500-cm^3 graduated flask, cool, dilute to the mark with water, and mix. Pipette a 15-cm^3 aliquot into a 50-cm^3 flask, evaporate to fuming, cool, add 5 cm^3 dilute hydrochloric acid (sp. gr. 1.06), warm until the salts dissolve, and cool to room temperature. Add 5 drops of 10 per cent aqueous hydroxylammonium sulphate solution and 10 cm^3 of the dithiol reagent, and allow to stand in a bath at 20–25 °C for 15 minutes with periodic shaking. Transfer the contents quantitatively to a 25-cm^3 separatory funnel, using 3–4 cm^3 portions of pentyl acetate for washing. Shake and allow the layers to separate. Run off the lower acid layer containing the tungsten and reserve it in the original 50-cm^3 flask. Wash the pentyl acetate layer twice consecutively with 5-cm^3 portions of hydrochloric acid (sp. gr. 1.06) and combine the acid washings with the original acid layer. Discard the molybdenum-containing pentyl acetate layer. Evaporate the acid tungsten solution carefully to fuming (to expel dissolved pentyl acetate), then add a few drops of concentrated nitric acid during fuming to clear up any charred organic matter. Add 5 cm^3 of 10 per cent tin(II) chloride solution (in concentrated hydrochloric acid) and heat to 100 °C for 4 minutes: add 10 cm^3 of the dithiol reagent and heat at 100 °C for 10 minutes longer with periodic shaking. Transfer to a 25-cm^3 stoppered separatory funnel, and rinse thrice with 2-cm^3 portions of pentyl acetate. Shake, separate, and draw off the lower acid layer and discard it. Add 5 cm^3 concentrated hydrochloric acid to the organic layer, repeat the extraction and again discard the lower layer. Draw off the pentyl acetate layer containing the tungsten complex into a 50-cm^3 graduated flask and dilute to volume with pentyl acetate. Measure the absorbance with a spectrophotometer at 630 nm in 4-cm cells, or use an absorptiometer and an Ilford Spectrum Red No. 608 filter.

Refer the readings to a calibration curve prepared from spectrographically pure iron to which suitable amounts of standard sodium tungstate solution have been added.

XVIII, 30.   VANADIUM.   Of the two methods commonly used for the determination of vanadium the second, in which phosphotungstovanadic acid is formed, is employed most frequently.

A. Vanadyl Sulphate method.   *Discussion.*   When hydrogen peroxide is added to a solution containing small quantities of vanadium(V) (up to 0.1 mg of V per cm^3) in sulphuric acid solution, a reddish-brown coloration is produced: this is thought to be due to the formation of a compound of the type $(VO)_2(SO_4)_3$. A large excess of hydrogen peroxide tends to reduce the colour intensity and to change the colour from red-brown to yellow. With a hydrogen peroxide concentration of 0.03 per cent, the sulphuric acid concentration can vary between 0.3 and 3M without any appreciable effect on the colour: with higher concentrations of hydrogen peroxide, the acidity must be increased to permit development of the maximum colour intensity.

The colour is unaffected by the presence of phosphate or fluoride. Titanium and molybdenum(VI) (which give colours with hydrogen peroxide) and tungsten interfere. Titanium may be removed by adding fluoride or hydrofluoric acid, which simultaneously remove the yellow colour due to iron(III). If titanium is absent, phosphate may be used to decolorise any iron(III) salt present. Oxalic acid eliminates the interference due to tungsten. In the presence of elements which yield coloured solutions, such as chromium or nickel, it is best to add equal amounts of these elements to the standard solution. If steel is being analysed, the most convenient procedure is to use a like steel as standard.

Prepare a **standard vanadium solution** by dissolving 1.146 g A.R. ammonium vanadate in water and make up to 1 dm^3: 1 cm$^3 \equiv 0.5$ mg of V. The above solution may be diluted further to give a solution containing 0.01 mg V per cm^3.

Procedure.   Make the solution 0.5–1M in sulphuric acid and add 0.25 cm^3 of 3 per cent hydrogen peroxide for each 10 cm^3 of test solution. Compare colorimetrically against a standard having the same acidity and containing the same volume of hydrogen peroxide solution. If titanium is present, add hydrofluoric acid (say, 5–10 per cent of the volume); this will also decolorise the iron(III). If titanium is absent, use phosphoric acid for the decolorisation of the iron.

The absorbance due to the vanadyl sulphate may be measured at 450 nm (or at 290 nm in the ultraviolet) against a reagent blank or compensating blank.

B. Phosphotungstovanadic acid method. *Discussion.* Vanadium may also be determined by making use of the yellow, soluble **phosphotungstovanadic acid** formed upon adding phosphoric acid and sodium tungstate to an acid vanadate solution. The most intense colour is obtained when the molecular ratio of phosphoric acid to sodium tungstate is in the range $3 : 1$ to $20 : 1$, and the tungstate concentration in the test solution is 0.01 to $0.1 M$; the preferred concentrations are $0.5 M$ in phosphoric acid and $0.025 M$ in sodium tungstate.

The following interfere: (*a*) coloured ions, such as chromate, copper, and cobalt; (*b*) titanium, zirconium, bismuth, antimony, and tin yield slightly soluble phosphates or basic salts except in very low concentrations; (*c*) potassium and ammonium ions give sparingly soluble phosphotungstates; (*d*) molybdenum(VI) in relatively high concentration (>0.5 mg cm^{-3}); (*e*) iodide, thiocyanate, etc., reduce phosphotungstic acid; and (*f*) iron in concentration greater than 1 mg cm^{-3} (slight interference even in the presence of phosphoric acid).

Procedure. Render the solution *ca.* $0.5 M$ in mineral acid, and add 1.0 cm^3 of $1 : 2$ phosphoric acid and 0.5 cm^3 of $0.5 M$-sodium tungstate solution (prepared by dissolving 16.5 g A.R. sodium tungstate $Na_2WO_4,2H_2O$ in 100 cm^3 water) for each 10 cm^3 of test solution. Heat to boiling, cool, dilute to volume, and determine the absorbance of the resulting solution* at 400 nm. If small amounts of coloured ions (nickel, cobalt, dichromate, etc.) are present, these should be incorporated in the comparison solution, preferably by employing an aliquot portion of the original sample solution.

Vanadium in steel. Dissolve 1.0 g, accurately weighed, of the steel in 50 cm^3 of $1 : 4$ sulphuric acid. When solution is complete, introduce 10 cm^3 of concentrated nitric acid, and boil until nitrous fumes are no longer evolved. Dilute the solution to 100 cm^3 with hot water, heat to boiling, and add saturated potassium permanganate solution until a pink colour persists or a precipitate is formed. Boil for 5 minutes. Filter off any tungsten(VI) oxide or manganese oxide which may be precipitated. Add a slight excess of freshly prepared sulphurous acid, and boil off the excess. Cool, add 5 cm^3 syrupy phosphoric acid and 5 cm^3 of 10-volume hydrogen peroxide.

Simultaneously with the main determination prepare in an analogous manner a comparison solution from a standard steel which contains no vanadium but is otherwise similar; add a standard solution of vanadium to the control, followed by hydrogen peroxide, etc., and compare this colorimetrically or spectrophotometrically with the solution obtained from the unknown steel.

Anions

XVIII, 31. CHLORIDE. Two procedures are commonly employed for the colorimetric determination of chloride.

A. Mercury(II) chloranilate method. *Discussion.* The mercury(II) salt of chloranilic acid (2,5-dichloro-3,6-dihydroxy-*p*-benzoquinone) may be used for

---

* The yellow colour may also be extracted with 2-methylpropanol, and read at 400 nm against a reagent blank.

the determination of small amounts of chloride ion. The reaction is:

$$HgC_6Cl_2O_4 + 2Cl^- + H^+ = HgCl_2 + HC_6Cl_2O_4^-$$

The amount of reddish-purple acid-chloranilate ion liberated is proportional to the chloride-ion concentration. Methyl cellosolve (2-methoxyethanol) is added to lower the solubility of **mercury(II) chloranilate** and to suppress the dissociation of the mercury(II) chloride; nitric acid is added (concentration $0.05M$) to give the maximum absorption. Measurements are made at 530 nm in the visible or 305 nm in the ultraviolet region. Bromide, iodide, iodate, thiocyanate, fluoride, and phosphate interfere, but sulphate, acetate, oxalate, and citrate have little effect at the 25-p.p.m. level. The limit of detection is 0.2 p.p.m. of chloride ion; the upper limit is about 120 p.p.m. Most cations, but not ammonium ion, interfere and must be removed.

Silver chloranilate cannot be used in the determination because it produces colloidal silver chloride.

Procedure. Remove interfering cations by passing the aqueous solution containing the chloride ion through a strongly acidic ion exchange resin in the hydrogen form (e.g., Zerolit 225 or Amberlite 120) contained in a tube 15 cm long and 1.5 cm in diameter. Adjust the pH of the effluent to 7 with dilute nitric acid or aqueous ammonia and pH paper. To an aliquot containing not more than 1 mg of chloride ion in less than 45 cm³ of water in a 100-cm³ graduated flask, add 5 cm³ $1M$-nitric acid and 50 cm³ methyl cellosolve. Dilute the mixture to volume with distilled water, add 0.2 g mercury(II) chloranilate, and shake the flask intermittently for 15 minutes. Separate the excess of mercury(II) chloranilate by filtration through a fine ashless filter paper or by centrifugation. Measure the absorbance of the clear solution with a spectrophotometer at 530 nm against a blank prepared in the same manner.

Construct a calibration curve using standard ammonium chloride solution (1–100 p.p.m. Cl^-) and deduce the chloride-ion concentration of the test solution with its aid.

Mercury(II) chloranilate may be prepared by adding dropwise a 5 per cent solution of A.R. mercury(II) nitrate in 2 per cent nitric acid to a stirred solution of chloranilic acid at 50 °C until no further precipitate forms. Decant the supernatant liquid, wash the precipitate thrice by decantation with ethanol, once with diethyl ether, and dry in a vacuum oven at 60 °C. The compound is available commercially.

B. Mercury(II) thiocyanate method. This second procedure for the determination of trace amounts of chloride ion depends upon the displacement of thiocyanate ion from mercury(II) thiocyanate by chloride ion; in the presence of iron(III) ion with a highly coloured iron(III) thiocyanate complex is formed, and the intensity of its colour is proportional to the original chloride-ion concentration:

$$2Cl^- + Hg(SCN)_2 + 2Fe^{3+} = HgCl_2 + 2[Fe(SCN)]^{2+}$$

The method is applicable to the range 0.5–100 μg of chloride ion.

Procedure. Place a 20-cm³ aliquot of the chloride solution in a 25-cm³ graduated flask, add 2.0 cm³ of $0.25M$-ammonium iron(III) sulphate {$Fe(NH_4)$ $(SO_4)_2,12H_2O$} in $9M$-nitric acid, followed by 2.0 cm³ of a saturated solution of mercury(II) thiocyanate in ethanol. After 10 minutes measure the absorbance of the sample solution and also of the blank in 5-cm cells in a spectrophotometer at 460 nm against water in the reference cell. The amount of chloride ion in the sample corresponds to the difference between the two absorbances and is obtained from a calibration curve.

Construct a calibration curve using a standard sodium chloride solution containing $10 \mu g$ Cl^- per cm^3: cover the range 0–50 μg as above. Plot absorbance against micrograms of chloride ion.

XVIII, 32.   FLUORIDE.   Fluoride, in the absence of interfering anions (including phosphate, molybdate, citrate, and tartrate) and interfering cations (including cadmium, tin, strontium, iron, and particularly zirconium, cobalt, lead, nickel, zinc, copper, and aluminium), may be determined with **thorium chloranilate** in aqueous 2-methoxyethanol at pH 4.5; the absorbance is measured at 540 nm or, for small concentrations 0–2.0 p.p.m., at 330 nm.

In water as solvent, the reaction is:

$$Th(C_6Cl_2O_4)_2 + 6F^- + 2H^+ \rightleftharpoons ThF_6^{2-} + 2HC_6Cl_2O_4^-$$

In aqueous 2-methoxyethanol, the main reaction is stated to be:

$$Th(C_6Cl_2O_4)_2 + 2F^- + H^+ \rightleftharpoons ThF_2C_6Cl_2O_4 + HC_6Cl_2O_4^-$$

Interfering cations, except aluminium and zirconium, can be removed by passage through an ion exchange column. In the presence of interfering anions and also aluminium and zirconium, fluoride may be separated as hydrofluosilicic acid by distilling with dilute perchloric acid at 135 °C (temperature maintained by the addition of water) in the presence of a few glass beads.

A calibration curve for the range 0.2–10 mg fluoride ion per 100 cm^3 is constructed as follows. Add the appropriate amount of standard sodium fluoride solution, 25 cm^3 of 2-methoxyethanol, and 10 cm^3 of a buffer (0.1M in both sodium acetate and acetic acid) to a 100-cm^3 graduated flask. Dilute to volume with distilled water and add about 0.05 g of thorium chloranilate. Shake the flask intermittently for 30 minutes (the reaction in the presence of 2-methoxyethanol is about 90 per cent complete after 30 minutes and almost complete after 1 hour) and filter about 10 cm^3 of the solution through a dry Whatman No. 42 filter paper. Measure the absorbance of the filtrate in a 1-cm cell at 540 nm against a blank, prepared in the same manner, using a suitable spectrophotometer. Prepare a calibration curve for the concentration range 0.0–0.2 mg fluoride ion per 100 cm^3 in the same way, but add only 10.0 cm^3 of 2-methoxyethanol; measure the absorbance of the filtrate in a 1-cm silica cell at 330 nm.

Treat the fluoride sample solution in the same manner as described for the calibration curve after removing interfering ions and adjusting the pH to about 5 with dilute nitric acid or sodium hydroxide solution. Read off the fluoride concentration from the calibration curve and the observed value of the absorbance.

XVIII, 33.   NITRITE.   *Discussion.*   General procedures for the determination of nitrites are usually based upon some form of diazotisation reaction, often involving carcinogenic materials such as the naphthylamines. In the following method these compounds are avoided.

In this case the nitrite ion, under acidic conditions, causes diazotisation of sulphanilamide (4-aminobenzenesulphonamide) to occur, and the product is coupled with N-(1-naphthyl)-ethylenediamine dihydrochloride.

Reagents.   *Sulphanilamide solution (A).*   Dissolve 0.5 g sulphanilamide in 100 cm^3 of 20 per cent v/v hydrochloric acid.

N-(1-*naphthyl*)-*ethylenediamine dihydrochloride solution* (*B*). Dissolve 0.3 g of the solid reagent in 100 cm³ of 1 per cent v/v hydrochloric acid.

Procedure. To 100 cm³ of the neutral sample solution (containing not more than 0.4 mg nitrite) add 2.0 cm³ of solution *A* and, after 5 minutes, 2.0 cm³ of solution *B*. The pH at this point should be about 1.5. Measure the absorbance after 10 minutes in the wavelength region of 550 nm in a spectrophotometer against a blank solution prepared in the same manner. Calculate the concentration of the nitrite from a calibration plot prepared from a series of standard nitrite solutions.

XVIII, 34. PHOSPHATE. Two methods are commonly used for the determination of phosphate.

A. Molybdenum Blue method. *Discussion.* Orthophosphate and molybdate ions condense in acidic solution to give molybdophosphoric acid (phosphomolybdic acid), which upon selective reduction (say, with hydrazinium sulphate) produces a blue colour, due to *molybdenum blue* of uncertain composition. The intensity of the blue colour is proportional to the amount of phosphate initially incorporated in the heteropoly acid. If the acidity at the time of reduction is $0.5M$ in sulphuric acid and hydrazinium sulphate is the reductant, the resulting blue complex exhibits maximum absorption at 820–830 nm.

Ions which form heteropoly acids, such as silicate (Section **XVIII, 35**), arsenate (Section **XVIII, 14**), germanate, and tungstate, should be absent. Silicate may be separated by fuming with perchloric acid to dehydrate the silicic acid and render it insoluble. Arsenate can be volatilised as arsenic(III) bromide from a hydrobromic acid–sulphuric acid solution; tin and germanium are volatilised simultaneously. Lead, antimony, and copper interfere; these and other metals may be removed by passage of the solution through a cation-exchange column. Oxidising and reducing agents must be absent.

Reagents. *Molybdate solution.* Dissolve 12.5 g of A.R. sodium molybdate ($Na_2MoO_4,2H_2O$) in $5M$-sulphuric acid and dilute to 500 cm³ with $5M$-sulphuric acid.

Hydrazinium sulphate solution. Dissolve 1.5 g of A.R. hydrazinium sulphate in de-ionised water and dilute to 1 dm³.

Standard phosphate solution. Dissolve 0.2197 g of A.R. potassium dihydrogen phosphate in de-ionised water and dilute to 1 dm³ in a graduated flask. 1 cm³ $\equiv$ 0.05 mg P. Dilute as appropriate.

Procedure. The sample solution should contain not more than 0.1 mg of phosphorus as the orthophosphate in 25 cm³ and should be neutral. Transfer 25 cm³ to a 50-cm³ Pyrex graduated flask. Add 5.0 cm³ of the molybdate solution, followed by 2.0 cm³ of the hydrazinium sulphate solution, dilute to the mark with distilled water, and mix well. Immerse the flask in a boiling water bath for 10 minutes, remove, and cool rapidly. Shake the flask, adjust the volume, and measure the absorbance at 830 nm against either de-ionised water or a reagent blank.

Construct the calibration curve, using the standard phosphate solution, in the usual manner.

B. Phosphovanadomolybdate method. *Discussion.* This second method (Ref. 10) is considered to be slightly less sensitive than the previous Molybdenum Blue method, but it has been particularly useful for phosphorus determinations carried out by means of the Oxygen (Schoniger) Flask (Section **III, 46**). The phosphovanadomolybdate complex formed between the phosphate, ammonium vanadate, and ammonium molybdate is bright yellow in colour and its absorbance can be measured between 460 and 480 nm.

Reagents. Ammonium vanadate solution. Dissolve 2.5 g ammonium vanadate (NH_4VO_3) in 500 cm^3 hot water, add 20 cm^3 concentrated nitric acid and dilute with water to 1 dm^3 in a graduated flask.

Ammonium molybdate solution. Dissolve 50 g ammonium molybdate, $(NH_4)_6Mo_7O_{24}\cdot4H_2O$, in warm water and dilute to 1 dm^3 in a graduated flask. Filter the solution before use.

Procedure. Dissolve 0.4 g of the phosphate sample in 2.5M-nitric acid to give 1 dm^3 in a graduated flask. Place a 10-cm^3 aliquot of this solution in a 100-cm^3 graduated flask, add 50 cm^3 water, 10 cm^3 of the ammonium vanadate solution, 10 cm^3 of the ammonium molybdate solution and dilute to the mark. Determine the absorbance of this solution at 465 nm against a blank prepared in the same manner, using 1 cm cells.

Prepare a series of standards from potassium dihydrogenphosphate covering the range 0–2 mg phosphorus per 100 cm^3 and containing the same concentration of acid, ammonium vanadate, and ammonium molybdate as the previous solution. Construct a calibration curve and use it to calculate the concentration of phosphorus in the sample.

XVIII, 35. SILICATE. *Discussion.* Small quantities of dissolved silicic acid react with a solution of a molybdate in an acid medium to give an intense yellow coloration, due probably to the complex molybdosilicic acid $H_4[SiMo_{12}O_{40}]$. The latter may be employed as a basis for the colorimetric determination of silicate (absorbance measurements at 400 nm). It is usually better to reduce the complex acid to **molybdenum blue** (the composition is uncertain); a solution of a mixture of 1-amino-2-naphthol-4-sulphonic acid and sodium hydrogen sulphite solution is satisfactory reducing agents.

Phosphates, arsenates, and germanates give similar colorations and must either be removed or their interferences eliminated by the addition of suitable reagents: arsenic and germanium can be removed by evaporation with hydrochloric acid, and phosphate by precipitation as ammonium magnesium phosphate in acetic acid solution, or may be rendered innocuous by the addition of ammonium citrate. Elements such as barium, bismuth, lead, and antimony give precipitates or turbidities, and must be absent. Water used for dilution should be freshly distilled in an all-Pyrex apparatus or passed through a mixed-bed ion exchange column, and stored in polythene containers. Water tends to dissolve significant traces of silica on standing in glass, particularly soda glass, vessels.

Reagents. *Ammonium molybdate solution.* Dissolve 8.0 g A.R. ammonium molybdate crystals in water, add 9 cm^3 concentrated sulphuric acid, and dilute to 100 cm^3.

Reducing agent. Solution A: dissolve 10 g sodium hydrogensulphite in 70 cm^3 water. Solution B: dissolve 0.8 g anhydrous sodium hydrogensulphite in 20 cm^3 water, and add 0.16 g 1-amino-2-naphthol-4-sulphonic acid. Mix solution A with solution B, and dilute to 100 cm^3

Tartaric acid solution. Prepare a 10 per cent aqueous solution.

Standard solution of silica. Fuse 0.107 g of pure, dry precipitated silica with 1.0 g of A.R. anhydrous sodium carbonate in a platinum crucible. Cool the melt, dissolve it in de-ionised water, dilute to 500 cm³, and store in a polythene bottle. 1 cm³ ≡ 0.1 mg Si. Dilute as appropriate, say, to 1 cm³ ≡ 0.01 mg Si.

Procedure. The sample solution, free from interfering elements and radicals, may conveniently occupy a volume of about 50 cm³ and contain between 0.01 and 0.1 mg of silica; the pH should be 4.5–5.0. Add 1 cm³ of the ammonium molybdate solution and, after 5 minutes, add 5 cm³ of the tartaric acid solution and mix. Introduce 1.0 cm³ of the reducing agent and dilute to 100 cm³ in a graduated flask. Measure the absorbance at *ca.* 815 nm after 20 minutes against de-ionised water.

Construct a calibration curve using 0, 1.0, 2.5, 5.0, 7.5, and 10.0 cm³ of the standard silica solution (1 cm³ ≡ 0.01 mg Si) which have been treated similarly.

XVIII, 36. SULPHATE. *Discussion.* The barium salt of chloranilic acid (2,5-dichloro-3,6-dihydroxy-*p*-benzoquinone) illustrates the principle of a method which may find wide application in the colorimetric determination of various anions. In the reaction

$$Y^- + MA \text{ (solid)} = A^- + MY \text{ (solid)}$$

where Y^- is the anion to be determined and A^- is the coloured anion of an organic acid, MY must be so much less soluble than MA that the reaction is quantitative. MA must be only sparingly soluble so that the blanks will not be too high. Sulphate ion in the range 2–400 p.p.m. may be readily determined by utilising the reaction between **barium chloranilate** with sulphate ion in acid solution to give barium sulphate and the acid-chloranilate ion:

$$SO_4^{2-} + BaC_6Cl_2O_4 + H^+ = BaSO_4 + HC_6Cl_2O_4^-$$

The amount of acid chloranilate ion liberated is proportional to the sulphate-ion concentration. The reaction is carried out in 50 per cent aqueous ethanol, buffered at an apparent pH of 4. Most cations must be removed because they form insoluble chloranilates: this is simply effected by passage of the solution through a strongly acidic ion-exchange resin in the hydrogen form (see Section **VII, 2**). Chloride, nitrate, hydrogencarbonate, phosphate, and oxalate do not interfere at the 100-p.p.m. level. The pH of the solution governs the absorbance of chloranilic acid solutions at a particular wavelength; chloranilic acid is yellow, acid-chloranilate ion is dark purple, and chloranilate ion is light purple. At pH 4 the acid-chloranilate ion gives a broad peak at 530 nm, and this wavelength is employed for measurements in the visible region. A much more intense absorption occurs in the ultraviolet: a sharp band at 332 nm enables the limit of detection of sulphate ion to be extended to 0.06 p.p.m.

Procedure. Pass the aqueous solution containing sulphate ion (2–400 p.p.m.) through a column 1.5 cm in diameter and 15 cm long of Zerolit 225 or equivalent resin in the hydrogen form. Adjust the effluent to pH 4 with dilute hydrochloric acid or ammonia solution. Make up to volume in a graduated flask. To an aliquot containing up to 40 mg of sulphate ion in less than 40 cm³ in a 100-cm³ graduated flask, add 10 cm³ of a buffer (pH = 4; a 0.05M solution of A.R. potassium hydrogenphthalate) and 50 cm³ of 95 per cent ethanol. Dilute to the mark with distilled water, add 0.3 g of barium chloranilate and shake the flask for 10 minutes. Remove the precipitated barium sulphate and the excess of barium chloranilate by filtering or centrifuging. Measure the absorbance of the filtrate with a filter colorimeter or a spectrophotometer at 530 nm against a blank prepared in the same manner. Construct a calibration curve using standard potassium sulphate solutions prepared from the A.R. salt.

Experimental. Determinations with ultraviolet/visible spectrophotometers

XVIII, 37. DETERMINATION OF THE ABSORPTION CURVE AND CONCENTRATION OF A SUBSTANCE (POTASSIUM NITRATE). *Discussion.* Potassium nitrate is an example of an inorganic compound which absorbs mainly in the ultraviolet, and can be employed to obtain experience in the use of a manually operated ultraviolet/visible spectrophotometer. Some of the exercise can also be carried out employing an automatic recording spectrophotometer.

The absorbance and the percentage transmission of an approximately $0.1M$-potassium nitrate solution is measured over the wavelength range 240–360 nm at 5-nm intervals and at smaller intervals in the vicinity of the maxima or minima. Manual spectrophotometers are calibrated to read both absorbance and percentage transmission on the dial settings, while the automatic recording double beam spectrophotometers usually use chart paper printed with both scales. The linear conversion chart, Fig. XVIII, 36, is useful for visualising the relationship between these two quantities.

The three normal means of presenting the spectrophotometric data are described below: by far the most common procedure is to plot absorbance against wavelength (measured in nanometres). The wavelength corresponding to the absorbance maximum (or minimum transmission) is read from the plot and is used for the preparation of the calibration curve. This point is chosen for two reasons: (i) it is the region in which the greatest difference in absorbance between any two different concentrations will be obtained, thus giving the maximum sensitivity for concentration studies, and (ii) as it is a turning point on the curve it gives the least alteration in absorbance value for any slight variation in wavelength.

No general rule can be given concerning the strength of the solution to be prepared, as this will depend upon the spectrophotometer used for the study. Usually a $0.01–0.001M$ solution is sufficiently concentrated for the highest absorbances, and other concentrations are prepared by dilution. The concentrations should be selected such that the absorbance lies between about 0.3 and 1.5.

For the determination of the concentration of a substance, select the wavelength of maximum absorption for the compound (e.g., 302.5–305 nm for potassium nitrate) and construct a calibration curve by measuring the absorbances of four or five concentrations of the substance (e.g., 2, 4, 6, 8 and 10 g KNO_3 dm^{-3}) at the selected wavelength. Plot absorbance (ordinates) against concentration (abiscissae). If the compound obeys Beer's law a linear calibration curve, passing through the origin, will be obtained. If the absorbance of the unknown

Fig. XVIII, 36

solution is measured the concentration can be obtained from the calibration curve.

If it is known that the compound obeys Beer's law the molar absorption coefficient ϵ can be computed from one measurement of the absorbance of a standard solution. The unknown concentration is then calculated using the value of the constant ϵ and the measured value of the absorbance under the same conditions.

Procedure.   Dry some A.R. potassium nitrate at 110 °C for 2–3 hours and cool in a desiccator. Prepare an aqueous solution containing 10.000 g dm^{-3}. With the aid of a precision spectrophotometer* and matched 10-mm rectangular cells, measure the absorbance and the percentage transmission over a series of wavelengths covering the range 240–350 nm. Plot the data in three different ways: (i) absorbance against wavelength; (ii) percentage transmission against wavelength; and (iii) log ϵ (molecular decadic absorption coefficient) against wavelength. The curves obtained for potassium nitrate are shown in Figs. XVIII, 37 to XVIII, 39. From the curves, evaluate the wavelength of maximum absorption (or minimum transmission). Use this value of the wavelength to determine the absorbance of solutions of potassium nitrate containing 2.000, 4.000, 6.000, and 8.000 g of potassium nitrate dm^{-3}. Run a blank on the two cells, filling both the blank cell and the sample cell with distilled water; if the cells are correctly matched no difference in absorbance should be discernible. Plot the absorbances (ordinates) against concentration.

Determine the absorbance of an unknown solution of potassium nitrate and read the concentration from the calibration curve.

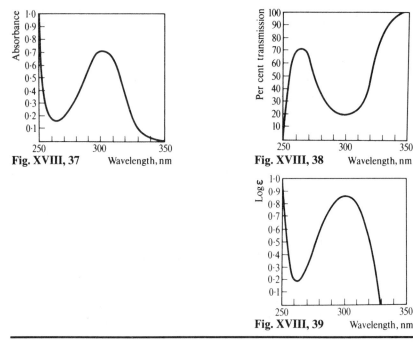

Fig. XVIII, 37   Wavelength, nm

Fig. XVIII, 38   Wavelength, nm

Fig. XVIII, 39   Wavelength, nm

---

* When reporting spectrophotometric measurements, details should be given of the concentration used, the solvent employed, the make and model of the instrument, as well as the slit widths employed, together with any other pertinent information.

XVIII, 38. SPECTROPHOTOMETRIC DETERMINATION OF THE pK VALUE OF AN INDICATOR (THE ACID DISSOCIATION CONSTANT OF METHYL RED). *Discussion.*

The dissociation of an acid–base indicator is well suited to spectrophotometric study; the procedure involved will be illustrated by the determination of the acid dissociation constant of methyl red (MR). The acidic (HMR) and basic (MR$^-$) forms of methyl red are shown below.

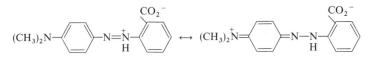

ACID FORM (HMR) RED

$$HO^- \updownarrow H^+$$

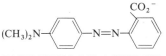

BASIC FORM (MR$^-$) YELLOW

The acid dissociation constant K is given by the equation:

$$K = \frac{[H^+][MR^-]}{[HMR]} \tag{1}$$

$$pK = pH - \log\frac{[MR^-]}{[HMR]} \tag{2}$$

Both HMR and MR$^-$ have strong absorption peaks in the visible portion of the spectrum; the colour change interval from pH 4 to pH 6 can be conveniently obtained with a sodium acetate–acetic acid buffer system.

The determination of pK involves three steps:

(a) Evaluation of the wavelengths at which HMR (λ_A) and MR$^-$ (λ_B) exhibit maximum absorption.

(b) Verification of Beer's law for both HMR and MR$^-$ at wavelengths λ_A and λ_B.

(c) Determination of the relative amounts of HMR and MR$^-$ present in solution as a function of pH.

By using the same concentration of indicator in each of the measurements at different values of pH and measuring the absorbance for each solution at λ_A and at λ_B, the relative amounts of HMR and MR$^-$ in solution can be calculated from the two equations:

$$A_A = d_{A.HMR}[HMR] + d_{A.MR^-}[MR^-] \tag{3}$$

$$A_B = d_{B.HMR}[HMR] + d_{B.MR^-}[MR^-] \tag{4}$$

where $d_{A.HMR}$, $d_{A.MR^-}$, $d_{B.HMR}$ and $d_{B.MR^-}$ are derived from the graphs plotted in (b). By solving these two simultaneous equations, the ratio [MR$^-$]/[HMR] can be obtained and thence pK with the aid of equation (2). Equations (3) and (4) imply that the observed absorbances (A) at λ_A and λ_B are the simple additive sums of the absorbances (d) due to HMR and MR$^-$.

Reagents. *Methyl red solution.* Dissolve 0.10 g pure crystalline methyl red in 30 cm^3 95 per cent ethanol and dilute to 50 cm^3 with water. The solution

761

required in the experiment (standard solution) is prepared by transferring 5.0 cm³ of the above stock solution to 50 cm³ of 95 per cent ethanol contained in a 100-cm³ graduated flask and diluting to 100 cm³ with water.

Sodium acetate, 0.04M *and* 0.01M.

Acetic acid, 0.02M.

Hydrochloric acid, 0.1M *and* 0.01M. The exact concentrations of these two solutions are not critical.

Procedure. The study can be carried out using either a manually operated single beam spectrophotometer, or an automatic recording double beam spectrophotometer. In both cases the wavelengths at which HMR and MR⁻ exhibit absorption maxima are readily obtained from the spectra.

(*a*) Prepare solution A by diluting a mixture of 10.0 cm³ of the standard solution of the indicator (MR) and 10.0 cm³ of 0.1M-hydrochloric acid to 100 cm³; the pH of this solution is about 2, so that the indicator MR is present entirely as HMR. Using 1-cm cells, determine the absorption spectrum of this solution over the range 350–600 nm against a blank of distilled water. For manual plotting cover the range in increments of 25 nm except for the portion between 500 and 550 nm which should be covered in 10 nm increments. From the spectrum of absorbance against wavelength determine the wavelength λ_A at which the maximum absorbance occurs: this is about 520 nm.

Prepare solution B by diluting a mixture of 10.0 cm³ of the standard solution of the indicator and 25.0 cm³ of 0.04M-sodium acetate to 100 cm³. The pH of this solution is about 8, so that the indicator MR is present entirely as MR⁻. Measure the absorbance of solution B over the range 350–600 nm as detailed for solution A: with a manual spectrophotometer use 25-nm steps except for 400–450 nm, where 10-nm steps are recommended. Determine the wavelength λ_B of maximum absorbance as above: this is about 430 nm. The type of plots obtained for solutions A and B is shown in Fig. XVIII, 40. The absorption peaks are not completely separated, but cross at a wavelength of about 460 nm. This point is known as the *isobestic point*. If the absorbance of a solution containing both HMR and MR⁻ is measured at this wavelength, the observed absorbance is independent of the relative amounts of HMR and MR⁻ present and depends only on the total amount of the indicator MR in the solution.

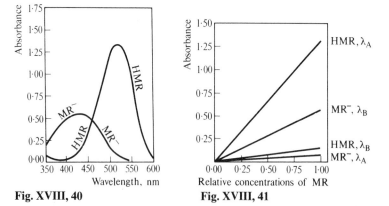

Fig. XVIII, 40   **Fig. XVIII, 41**

(*b*) Using solution A, measure out 40.0 cm³, 25.0 cm³, and 10.0 cm³ into separate 50-cm³ graduated flasks, and dilute in each case to the mark with 0.01M-hydrochloric acid. The resulting solutions will contain 0.8, 0.5, and 0.2 times respectively the initial concentration of HMR. Similarly, using solution B and diluting with 0.01M-sodium acetate, prepare three solutions containing respectively 0.8, 0.5, and 0.2 times the initial concentration of MR^-. Measure the absorbance of each of the six solutions versus water at wavelengths of λ_A and λ_B. It is important in obtaining the experimental absorbance to be sure that all the measurements are made at constant temperature, say, at the temperature of the room housing the spectrophotometer. Plot absorbance against relative concentration of the indicator MR: in each case straight-line plots should be obtained, as in Fig. XVIII, 41.

   (c) Prepare the following solutions in four 100-cm^3 graduated flasks.

Flask number	1	2	3	4
Standard indicator solution MR (cm^3)	10.0	10.0	10.0	10.0
0.04M-Sodium acetate (cm^3)	25.0	25.0	25.0	25.0
0.02M-Acetic acid (cm^3)	50.0	25.0	10.0	5.0
Water (to mark)	15.0	40.0	55.0	60.0
pH	4.84	5.15	5.53	5.81

Determine the pH values of each of the solutions (typical values are incorporated in the table) and measure the absorbance of each solution at wavelengths λ_A and λ_B. All these solutions contain the same concentration of indicator as solutions A and B used in (a). For each prepared solution, obtain the values of the absorbances $d_{A.HMR}$, $d_{A.MR^-}$, $d_{B.HMR}$, and $d_{B.MR^-}$ from the plots in Fig. XVIII, 41, at relative concentrations of 1.0, and solve the simultaneous equations (3) and (4) in order to evaluate the relative amounts of HMR and MR^- in solution. From the relative amounts of HMR and MR^- present as a function of pH, calculate the value of pK for methyl red using equation (2). Some typical results are collected in the following table:

Solution number	Observed pH	Absorbance at λ_A	Absorbance at λ_B	$\dfrac{[MR^-]}{[HMR]}$	pK
1	4.84	0.605	0.204	0.679	5.01
2	5.15	0.442	0.263	1.403	5.00
3	5.53	0.254	0.317	3.436	4.99
4	5.81	0.168	0.348	6.740	4.98
				Mean	5.00

XVIII, 39.   SIMULTANEOUS SPECTROPHOTOMETRIC DETERMINATION (CHROMIUM AND MANGANESE).

Discussion. This section is concerned with the simultaneous spectrophotometric determination of two solutes in a solution. The absorbances are additive, provided there is no reaction between the two solutes. We may write:

$$A_{\lambda_1} = {}_{\lambda_1}A_1 + {}_{\lambda_1}A_2 \tag{1}$$

$$A_{\lambda_2} = {}_{\lambda_2}A_1 + {}_{\lambda_2}A_2 \tag{2}$$

where A_{λ_1} and A_{λ_2} are the *measured* absorbances at the two wavelengths λ_1 and λ_2, and the subscripts 1 and 2 refer to the two different substances, and the subscripts λ_1 and λ_2 refer to the different wavelengths. The wavelengths are selected to coincide with the absorption maxima of the two solutes: the absorption spectra of the two solutes should not overlap appreciably (compare Fig. XVIII, 40), so that substance 1 absorbs strongly at wavelength λ_1 and weakly at wavelength λ_2, and substance 2 absorbs strongly at λ_2 and weakly at λ_1. Now $A = \epsilon c l$, where ϵ is the molar absorption coefficient at any particular wavelength, c is the concentration expressed in mols dm^{-3}, and l is the thickness (length) of the absorbing solution expressed in cm. If l is 1 cm:

$$A_{\lambda_1} = {}_{\lambda_1}\epsilon_1 . c_1 + {}_{\lambda_1}\epsilon_2 . c_2 \tag{3}$$

$$A_{\lambda_2} = {}_{\lambda_2}\epsilon_1 . c_1 + {}_{\lambda_2}\epsilon_2 . c_2 \tag{4}$$

Solution of these simultaneous equations gives:

$$c_1 = \frac{{}_{\lambda_2}\epsilon_2 . A_{\lambda_1} - {}_{\lambda_1}\epsilon_2 . A_{\lambda_2}}{{}_{\lambda_1}\epsilon_1 . {}_{\lambda_2}\epsilon_2 - {}_{\lambda_1}\epsilon_2 . {}_{\lambda_2}\epsilon_1} \tag{5}$$

$$c_2 = \frac{{}_{\lambda_1}\epsilon_1 . A_{\lambda_2} - {}_{\lambda_2}\epsilon_1 . A_{\lambda_1}}{{}_{\lambda_1}\epsilon_1 . {}_{\lambda_2}\epsilon_2 - {}_{\lambda_1}\epsilon_2 . {}_{\lambda_2}\epsilon_1} \tag{6}$$

The values of the molar absorption coefficients ϵ_1 and ϵ_2 can be deduced from measurements of the absorbances of pure solutions of substances 1 and 2. By measuring the absorbance of the mixture at wavelengths λ_1 and λ_2, the concentrations of the two components can be calculated.

The above considerations will be illustrated by the simultaneous determination of manganese and chromium in steel and other ferro-alloys. The absorption spectra of 0.001M-permanganate- and dichromate ions in 1M-sulphuric acid, determined with a spectrophotometer and against 1M-sulphuric acid in the reference cell are shown in Fig. XVIII, 42. The peak at 350 nm for dichromate solutions cannot be used because iron(III) ion absorbs strongly below 425 nm: at a wavelength of 440 nm near the weaker band maximum the correction for iron(III) ion absorption is small. For permanganate, the absorption maximum is at 545 nm, and a small correction must be applied for dichromate absorption. Absorbances for these two ions, individually and in mixtures, obey Beer's law provided the concentration of sulphuric acid is at least 0.5M. Iron(III), nickel, cobalt, and vanadium absorb at 425 nm and 545 nm, and corrections must be made.

Reagents. *Potassium dichromate*, 0.002M, 0.001M, and 0.0005M in 1M-sulphuric acid and 0.7M-orthophosphoric acid, prepared from the A.R. reagents.

Potassium permanganate, 0.002M, 0.001M, and 0.0005M in 1M-sulphuric acid and 0.7M-orthophosphoric acid, prepared from the A.R. reagents. All flasks must be scrupulously clean.

Procedure. (a) *Determination of molar absorption coefficients and verification of additivity of absorbances.*

The molar absorption coefficients must be determined for the particular set of cells and the spectrophotometer employed. For the present purpose we may write:

$$A = \epsilon c l$$

where ϵ is the molar absorption coefficient, c is the concentration (mols dm^{-3}), and l is the cell thickness or length (cm).

Measure the absorbance A of the above three solutions of potassium dichromate and of potassium permanganate, each solution separately, at both 440 nm and 545 nm in 1-cm cells. Calculate ϵ in each case and record the mean values for $Cr_2O_7^{2-}$ and MnO_4^{-} at the two wavelengths.

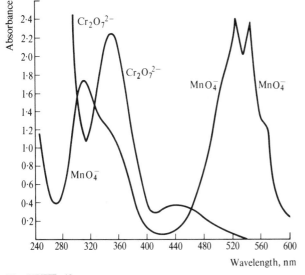

Fig. XVIII, 42

Mix 0.001M-potassium dichromate and 0.0005M-potassium permanganate in the following amounts (plus 1.0 cm^3 of concentrated sulphuric acid), and complete the following table (some typical results are included for guidance only). Measure the absorbance of each of the mixtures at 440 nm. Calculate the absorbance of the mixtures from:

$$A_{440} = {}_{440}\epsilon_{Cr} \cdot c_{Cr} + {}_{440}\epsilon_{Mn} \cdot c_{Mn}$$

Test of additivity principle with $Cr_2O_7^{2-}$ and MnO_4^{-} mixtures at 440 nm

$K_2Cr_2O_7$ Solution, cm^3	$KMnO_4$ Solution, cm^3	A Observed	A Calculated
50	0	0.371	—
45	5	0.338	0.340
40	10	0.307	0.308
35	15	0.277	0.277
25	25	0.211	0.214
15	35	0.147	0.151
5	45	0.086	0.088
0	50	0.057	—

(b) *Determination of chromium and manganese in an alloy steel.** Weigh out accurately about 1.0 g of the alloy steel in a 300-cm³ Kjeldahl flask, add 30 cm³ of water and 10 cm³ of concentrated sulphuric acid (also 10 cm³ of 85 per cent phosphoric acid if tungsten is present). Boil gently until decomposition is complete or the reaction subsides. Then add 5 cm³ of concentrated nitric acid in several small portions. If much carbonaceous residue persists, add 5 cm³ more of concentrated nitric acid, and boil down to copious fumes of sulphuric acid. Dilute to about 100 cm³ and boil until all salts have dissolved. Cool, transfer to a 250-cm³ graduated flask, and dilute to the mark.

Pipette a 25-cm³ or 50-cm³ aliquot of the clear sample solution into a 250-cm³ conical flask, add 5 cm³ concentrated sulphuric acid, 5 cm³ 85 per cent phosphoric acid, and 1–2 cm³ of 0.1*M*-silver nitrate solution, and dilute to about 80 cm³. Add 5 g A.R. potassium persulphate, swirl the contents of the flask until most of the salt has dissolved, and heat to boiling. Keep at the boiling point for 5–7 minutes. Cool slightly, and add 0.5 g pure potassium periodate. Again heat to boiling and maintain at the boiling point for about 5 minutes. Cool, transfer to a 100-cm³ graduated flask, and measure the absorbances at 440 nm and 545 nm in 1-cm cells.

Calculate the percentage of chromium and manganese in the sample. Use equations (5) and (6) and values of the molar absorption coefficients ϵ determined above: these will give concentrations expressed in mols dm⁻³, from which values the percentages can readily be calculated. Each value will require correction for the amounts of vanadium, cobalt, nickel, and iron which may be present, using the following table. The values listed are the equivalent percentages of the respective constituent to be subtracted from the apparent Cr and Mn percentages for each 1 per cent of the element in question.

Substance	Cr correction at 440 nm, %	Mn correction at 545 nm, %
$Cr_2O_7{}^{2-}$	—	0.0025
$MnO_4{}^-$	0.490	—
$VO_2{}^+$	0.0266	—
Co^{2+}	0.0072	0.0011
Ni^{2+}	0.0039	0.0001
Fe^{3+}	0.0005	—

It can be shown that utilising the known (or determined) molar absorption coefficients ($_{545}\epsilon_{Mn}$ 2.35; $_{545}\epsilon_{Cr}$ 0.011; $_{440}\epsilon_{Cr}$ 0.369; $_{440}\epsilon_{Mn}$ 0.095):

$$\text{Mn, per cent} = \frac{0.00549V}{W}(0.426A_{545} - 0.013A_{440})$$

$$\text{Cr, per cent} = \frac{0.01040V}{W}(2.71A_{440} - 0.110A_{545})$$

for a sample of *W* grams in a volume of *V* cm³.

---

* British Chemical Standard No. 225/2 Ni–Cr–Mo steel is suitable for practice in this determination.

Some typical results for Ridsdale's Alloy Steel, No. 60b, which contained 0.64 per cent Mn, 0.75 per cent Cr, 2.59 per cent Ni, and 0.43 per cent Mo, were:

Per cent Mn = 0.63; per cent Cr = 1.10 − (0.31 + 0.01 + 0.05) = 0.73.

Experimental.
Determinations by spectrophotometric titrations

XVIII, 40. SPECTROPHOTOMETRIC TITRATIONS. *General discussion.* In a spectrophotometric titration the end point is evaluated from data on the absorbance of the solution. For monochromatic light passing through a solution, Beer's law may be written as:

Absorbance $= \log I_0/I_t = \epsilon cl$

where I_0 is the intensity of the incident light, I_t that of the transmitted light, ϵ is the molar absorption coefficient, c is the concentration of the absorbing species, and l is the thickness or length of the light path through the absorbing medium. Since spectrophotometric titrations are carried out in a vessel for which the light path is constant, the absorbance is proportional to the concentration. Thus in a titration in which the titrant, the reactant, or a reactive product absorbs radiation, the plot

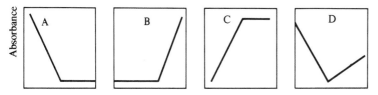

Volume of titrant, cm^3

Fig. XVIII, 43

of absorbance versus volume of titrant added will consist, if the reaction is complete and the volume change is small, of two straight lines intersecting at the end point.

The shape of a photometric titration curve will be dependent upon the optical properties of the reactant, titrant, and products of the reaction at the wavelength used. Some typical titration plots are given in Fig. XVIII, 43.

XVIII, 43, A, is characteristic of systems where the substance titrated is converted into a non-absorbing product.

XVIII, 43, B, is typical of the titration where the titrant alone absorbs.

XVIII, 43, C, corresponds to systems where the substance titrated and the titrant are colourless and the product alone absorbs.

XVIII, 43, D, is obtained when a coloured reactant is converted into a colourless product by a coloured titrant.

Owing to the linear response of absorbance to concentration, an appreciable break will often be obtained in a photometric titration, even though the changes in concentration are insufficient to give a clearly defined inflection point in a potentiometric titration. Photometric titrations have several advantages over direct colorimetric determinations. The presence of other substances absorbing at the same wavelength does not necessarily cause interference, since only the change in absorbance is significant. The precision of locating the titration line (required for the evaluation of the equivalence point) by pooling the information derived from several points is greater than the precision of any single point; furthermore, the procedure may be useful for reactions which tend to be appreciably incomplete near the equivalence point. An accuracy and precision of a few tenths per cent are attainable with comparative ease by spectrophotometric titration. The optimum concentration of the solution to be analysed depends upon the molar absorption coefficient of the absorbing species involved, and is usually of the order of 10^{-4}–10^{-5} molar. The effect of dilution can be made negligible by the use of a sufficiently concentrated titrant. If relatively large volumes of titrant are added the effect of dilution may be corrected by multiplying the observed absorbances by the factor $(V+v)/V$, where V is the initial volume and v is the volume added; if the dilution is of the order of only a few per cent the lines in the titration plots appear straight. The operating wavelength is selected on the basis of two considerations: avoidance of interference by other absorbing substances and need for an absorption coefficient which will cause the change in absorbance to fall within a convenient range. The latter is particularly important, because serious photometric error is possible in high-absorbance regions. Light leakage must, of course, be avoided.

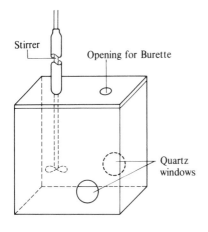

Fig. XVIII, 44

The experimental technique is simple. The cell containing the solution to be titrated is placed in the light path of a spectrophotometer, a wavelength appropriate to the particular titration is selected, and the absorption is adjusted to some convenient value by means of the sensitivity and slit-width controls. A measured volume of the titrant is added to the stirred solution, and the absorbance is read again. This is repeated at several points before the end point and several more points after the end point. The latter is found graphically.

XVIII, 41.   APPARATUS FOR SPECTROPHOTOMETRIC TITRATIONS.   A special titration cell is necessary which completely fills the cell compartment of the spectrophotometer. One for the Unicam SP.500, shown in Fig. XVIII, 44, can be made from 5-mm Perspex sheet, cemented together with special Perspex cement, and has dimensions of $9.0 \times 9.0 \times 4.7$ mm. Since Perspex is opaque to ultraviolet light, two openings are made in the cell to accommodate circular quartz windows* 23 mm in diameter and 1.5 mm thick: the windows are inserted in such a way that the beam of monochromatic light passes through their centres to the photoelectric cell. The Perspex cover of the cell has two small openings for the tip of a 5-cm³ micro-burette and for a micro-stirrer, respectively held by means of rubber bungs: the stirrer is 'sleeved'. The whole of the cell, with the exception of the quartz windows, is covered with black paper and, as a further precaution, the top of the cell is covered with a black cloth: it is most important to exclude all extraneous light.

---

* These were actually fused silica polarimeter end plates and were supplied by Hilger and Watts Ltd.

XVIII, 42.   SIMULTANEOUS DETERMINATION OF ARSENIC(III) AND ANTIMONY(III) IN A MIXTURE. *Discussion.* In acid solution arsenic(III) can be oxidised to arsenic(V) and antimony(III) to antimony(V) by the well-established titration with a solution of potassium bromate and potassium bromide (Section **X, 139**). The end-point for such determinations is usually observed indirectly, and very good results have been obtained by the spectrophotometric method of Sweetser and Bricker (Ref. 11). No change in absorbance at 326 nm is obtained until all the arsenic(III) has been oxidised, the absorbance then decreases to a minimum at the antimony(III) end-point at which it rises again as excess titrant is added.

Reagents. *Bromate/bromide solution.* Prepare a standard bromate/bromide solution by dissolving 2.78 g potassium bromate and 9.9 g potassium bromide in water and diluting to 1 dm³ in a graduated flask. This solution is 0.017M-potassium bromate (0.1N) with a slight excess of the theoretical amount of potassium bromide. Analytical grade reagents should be employed.

Arsenic/antimony solution. Prepare a mixed solution containing approximately 115 mg arsenic and 160 mg antimony in 1 dm³ by dissolving about 150 mg arsenic(III) oxide and 300 mg antimony(III) chloride in 6M-hydrochloric acid.

Procedure. Place 80 cm³ of the arsenic/antimony solution in the titration cell of the spectrophotometer. Titrate with standard bromate/bromide solution at 326 nm taking an absorbance reading at least every 0.2 cm³. From the curve obtained calculate the concentration of arsenic and antimony in the solution.

XVIII, 43.   DETERMINATION OF COPPER(II) WITH EDTA.
Discussion. The titration of a copper-ion solution with EDTA may be carried out photometrically at a wavelength of 745 nm. At this wavelength the copper–EDTA complex has a considerably greater molar absorption coefficient than the copper solution alone. The pH of the solution should be about 2.4.

The effect of different ions upon the titration is similar to that given under iron(III) (Section **XVIII, 44**). Iron(III) interferes (small amounts may be precipitated with sodium fluoride solution): tin(IV) should be masked with 20 per cent aqueous tartaric acid solution. The procedure may be employed for the determination of copper in brass, bronze, and bell metal without any previous separations except the removal of insoluble lead sulphate when present.

Reagents. *Copper-ion solution,* 0.04M. Wash A.R. copper with A.R. petroleum ether (b.p. 40–60°) to remove any surface grease and dry at 100°. Weigh out accurately about 1.25 g of the pure copper, dissolve it in 5 cm³ of concentrated nitric acid, and dilute to 1 dm³ in a graduated flask. Titrate this standard copper solution with the EDTA solution using Fast Sulphon Black as indicator (Section **X, 56**), and thus obtain a further check on the molarity of the EDTA.

EDTA solution, 0.10M, *and Buffer solution.* See Section **XVIII, 44**.

Procedure. Charge the titration cell (Fig. XVIII, 44) with 10.00 cm³ of the copper-ion solution, 20 cm³ of the acetate buffer (pH = 2.2), and about 120 cm³ of water. Position the cell in the spectrophotometer and set the wavelength scale at 745 nm. Adjust the slit width so that the reading on the absorbance scale is zero. Stir the solution and titrate with the standard EDTA: record the absorbance every 0.50 cm³ until the value is about 0.20 and subsequently every 0.20 cm³. Continue the titration until about 1.0 cm³ after the end point; the latter occurs when the absorbance readings become fairly constant. Plot absorbance against

769

cm^3 of titrant added; the intersection of the two straight lines (see Fig. XVIII, 43, C) is the end point.

Calculate the concentration of copper ion (mg cm^{-3}) in the solution and compare this with the true value.

XVIII, 44.   DETERMINATION OF IRON(III) WITH EDTA.   *Discussion.*
Salicylic aid and iron(III) ions form a deep-coloured complex with a maximum absorption at about 525 nm: this complex is used as the basis for the photometric titration of iron(III) ion with standard EDTA solution. At a pH of *ca.* 2.4 the EDTA–iron complex is much more stable (higher stability constant) than the iron–salicylic acid complex. In the titration of an iron–salicylic acid solution with EDTA the iron–salicylic acid colour will therefore gradually disappear as the end point is approached. The spectrophotometric end point at 525 nm is very sharp.

Considerable amounts of zinc, cadmium, tin(IV), manganese(II), chromium(III), and smaller amounts of aluminium cause little or no interference at pH 2.4: the main interferences are lead(II), bismuth, cobalt(II), nickel, and copper(II).

Reagents.   *EDTA solution, 0.10*M.   See Section **X, 50**. Standardise accurately (Section **X, 50**).

Iron(III) solution, 0.05M.   Dissolve about 12.0 g, accurately weighed, of A.R. ammonium iron(III) sulphate in water to which a little dilute sulphuric acid is added, and dilute the resulting solution to 500 cm^3 in a graduated flask. Standardise the solution with standard EDTA using Variamine Blue B as indicator (Section **X, 57**).

Sodium acetate–acetic acid buffer.   Prepare a solution which is 0.2M in sodium acetate and 0.8M in acetic acid. The pH is 4.0.

Sodium acetate–hydrochloric acid buffer.   Add 1M-hydrochloric acid to 350 cm^3 of 1M-sodium acetate until the pH of the mixture is 2.2 (pH meter).

Salicylic acid solution.   Prepare a 6 per cent solution of A.R. salicylic acid in A.R. acetone.

Procedure.   Transfer 10.00 cm^3 of the iron(III) solution to the titration cell (Fig. XVIII, 44), add about 10 cm^3 of the buffer solution of pH = 4 and about 120 cm^3 of water: the pH of the resulting solution should be 1.7–2.3. Insert the titration cell into the spectrophotometer; immerse the stirrer and the tip of the 5-cm^3 micro-burette (graduated in 0.02 cm^3) in the solution. Switch on the tungsten lamp and allow the spectrophotometer to 'warm up' for about 20 minutes. Stir the solution. Add about 4.0 cm^3 of the standard EDTA (note the volume accurately). Set the wavelength at 525 nm, and adjust the slit width of the instrument so that the reading on the absorbance scale is 0.2–0.3. Now add 1.0 cm^3 of the salicylic acid solution; the absorbance immediately increases to a very large value (> 2). Continue the stirring. Add the EDTA solution slowly from the micro-burette until the absorbance approaches 1.8; record the volume of titrant. Introduce the EDTA solution in 0.05-cm^3 aliquots and record the absorbance after each addition. Continue the titration until at least four readings are taken beyond the end point (fairly constant absorbance). Plot absorbance against cm^3 of titrant added: the intersection of the two straight lines (see Fig. XVIII, 43, *A*) gives the true end point.

Calculate the concentration of iron(III) (mg cm^{-3}) in the solution and compare this with the true value.

Determination of iron(III) in the presence of aluminium. Iron(III) (concentration *ca.* 50 mg per 100 cm^3) can be determined in the presence of up to twice the amount of aluminium by photometric titration with EDTA in the presence of 5-sulphosalicylic acid (2 per cent aqueous solution) as indicator at pH 1.0 at a wavelength of 510 nm. The pH of a strongly acidic solution may be adjusted to the desired value with a concentrated solution of sodium acetate: about 8–10 drops of the indicator solution are required. The spectrophotometric titration curve is of the form shown in Fig. XVIII, 43, A.

XVIII, 45. DETERMINATION OF NICKEL ION WITH EDTA.
Discussion. The titration of a nickel-ion solution with EDTA may be performed photometrically at a pH of about 4.0 and at a wavelength of 1000 nm, where the nickel–EDTA complex exhibits characteristic absorption. The titration curve is similar to that obtained for copper ion (see Fig. XVIII, 43, C).

Reagents. *Nickel-ion solution, 0.040M.* Dissolve about 2.45 g, accurately weighed, of pure nickel (Johnson and Matthey) in a mixture of 5 cm^3 of concentrated nitric acid and 5 cm^3 of concentrated sulphuric acid, and dilute to 1 dm^3 in a graduated flask. Check the concentration of the nickel ion by titration with standard EDTA solution using Bromopyrogallol Red (Section **X, 58**) as indicator.

EDTA solution, 0.10M, and Buffer solutions. See Section **XVIII, 44**.

Procedure. Proceed as described for Copper (Section **XVIII, 43**), except that the buffer solution of pH = 4.0 is employed and the wavelength is adjusted to 1000 nm. The pH of the resulting solution should be about 4.0. Evaluate the endpoint from the titration plots.

Calculate the concentration of the nickel ion (mg cm^{-3}) in the solution and compare this with the true value.

XVIII, 46. References

1. J. H. Lambert (1760). *Photometria sive de Mensura et Gradibus Luminus, Colorum et Umbrae.* Augsburg; reprinted in Ostwald (1892), *Klassiker der Exakten Wissenschaften*, No. 32, 64.
2. M. Bouguer (1729). *Essai d'optique sur la Gradation de la Lumiere.* Paris; see also Ostwald (1892), *Klassiker der Exakten Wissenschaften*, No. 33, 58. M. Bouguer (1760). *Traite d'optique sur la Gradation de la Lumiere, Ouvrage Posthume.* Pub. de Lacaille.
3. A. Beer (1852). *Ann. Physik Chem.* (J. C. Poggendorff), **86**, 78. See also H. G. Pfeiffer and H. A. Liebhafsky (1951). *J. Chem. Educ.*, **28**, 123.
4. F. Bernard (1852). *Ann. Chim. Phys.*, **35**, No. 3, 385.
5. D. R. Malinin and J. H. Yoe (1961). *J. Chem. Educ.*, **38**, 129.
6. F. H. Lohman (1955). *J. Chem. Educ.*, **32**, 155.
7. *Determination of Arsenic in Organic Materials* (1960). Analytical Methods Committee, Society for Analytical Chemistry.
8. J. W. McCoy (1969). *Chemical Analysis of Industrial Water.* MacDonald.
9. H. W. Swank and M. G. Mellon (1938). 'The Determination of Iron with Mercaptoacetic Acid', *Ind. Eng. Chem. Anal. Edn.*, **10**.
10. J. E. Barney, J. G. Bergmann and W. G. Tuskan (1959). *Anal. Chem.*, **31**, 1394.
11. P. B. Sweetser and C. E. Bricker (1952). 'Direct Spectrophotometric Titrations with Bromate-Bromide Solutions'. *Anal. Chem.*, **24**, 1107.

XVIII, 47. Selected bibliography

1. E. A. Braude and F. C. Nachod (1955). *Determination of Organic Structures by Physical Methods*. Ch. 4. Ultraviolet and Visible Light Absorption. Ch. 5. Infrared Light Absorption. New York; Academic Press.
2. A. Weissberger (1956). *Technique of Organic Chemistry*. Vol. 9. *Chemical Applications of Spectroscopy*. New York; Interscience.
3. W. R. Brode and M. E. Corning (1960). *Spectrophotometry and Absorptiometry* (in W. G. Berl. *Physical Methods of Analysis*. Vol. I. 2nd edn.) New York and London; Academic Press.
4. R. C. Hirt. 'Ultraviolet Spectrophotometry', *Analytical Chemistry*, 1956, **28**, 579; 1958, **30**, 589; 1960, **32**, 225R.
5. R. F. Godden and D. N. Hume (1954). 'Photometric Titrations', *Analytical Chemistry*, **26**, 1740.
6. J. B. Headridge (1958). 'Photometric Titrations', *Talanta*, **1**, 293.
7. J. B. Headridge (1961). *Photometric Titrations*. Oxford; Pergamon Press.
8. A. E. Gillam, E. S. Stern and C. J. Timmons (1970). *An Introduction to Electronic Absorption Spectroscopy*. 3rd edn. London; Edward Arnold.
9. R. L. Pecsok and L. D. Sheilds (1977). *Modern Methods of Chemical Analysis*. Ch. 8, 9, 10, 11. New York; Wiley.
10. F. J. Welcher (ed.) (1966). *Standard Methods of Chemical Analysis*, Vol. IIIB, 6th edition. New York; Van Nostrand.
11. R. B. Fischer and D. G. Peters (1968). *Quantitative Chemical Analysis*. Ch. 15 and 16. Philadelphia; W. B. Saunders.
12. R. C. Denney (1973). *A Dictionary of Spectroscopy*. London; Macmillan.
13. E. B. Sandell (1959). *Colorimetric Determination of Trace Metals*. 3rd edn. New York; and London; Interscience.

CHAPTER XIX **FLUORIMETRY***

XIX, 1. GENERAL DISCUSSION. Fluorescence is caused by the absorption of radiant energy and the re-emission of some of this energy in the form of visible light. The light emitted is almost always of higher wavelength than that absorbed. In **true fluorescence** the absorption and emission takes place in a short but measurable time—of the order of 10^{-12}–10^{-9} second. If the light is emitted with a time delay ($> 10^{-8}$ second) the phenomenon is known as **phosphorescence**; this time delay may range from a fraction of a second to several weeks, so that the difference between the two phenomena may be regarded as one of degree only. Both fluorescence and phosphorescence are designated by the term **photoluminescence**; the latter is therefore the general term applied to the process of absorption and re-emission of light energy.

Relationship between intensity of fluorescence and concentration. The Beer–Lambert law applies to the intensity of radiation transmitted by a substance or solution; since fluorescent radiation is that emitted by a substance, the law cannot be applied directly. The following relationship has been developed:

$$F = K(I_0 - I) \tag{1}$$

where I_0 = intensity of incident radiant energy;
I = intensity of transmitted radiant energy;
F = intensity of fluorescent radiant energy; and
K = a proportionality constant.

F is assumed proportional to the intensity of the radiant energy absorbed ($I_0 - I$). Applying the Beer–Lambert law (Section **XVIII, 2**),

$$I = I_0 . 10^{-\epsilon l c} \tag{2}$$

and

$$I_0 - I = I_0(1 - 10^{-\epsilon l c}) \tag{3}$$

we have

$$F = KI_0(1 - 10^{-\epsilon l c}) \tag{4}$$

Writing

$$KI_0 = F_0 \tag{5}$$

$$F = F_0 - F_0 . 10^{-\epsilon l c} \tag{6}$$

---

* The term *fluorometry* is usually used in the USA.

$$\text{or} \qquad \log \frac{F_0}{F_0 - F} = \epsilon l c \qquad (7)$$

In these equations K is the fraction of the incident radiation that is absorbed (this is determined by such factors as the dimensions of the light beam, the area of the solution irradiated, the transmission band of the filter before the photocell, and the spectral response of the photocell), ε is the molar absorption coefficient and is dependent upon the substance, l is the thickness of the solution in the cell, and c is the molar concentration of the fluorescent substance.

When $\epsilon c l$ becomes small and approaches a value of 0.01 or less, equation (4) reduces to:*

$$F = 2.303 K I_0 . \epsilon c l \qquad (8)$$

$$\text{or} \quad F = K'c \qquad (9)$$

i.e., the fluorescent intensity is practically proportional to the concentration of the fluorescent substance provided $\epsilon c l \leqslant 0.01$. $\epsilon K'$ is an overall constant for one particular substance in a given instrument. In practice, equation (8) holds up to a few parts per million: at higher concentrations the fluorescence–concentration curve will bend towards the concentration axis.

Factors, such as dissociation, association, or solvation, which would vitiate the Beer–Lambert law, would be expected to have a similar effect in fluorescence. Any material that causes the intensity of fluorescence to be less than the expected value given by equation (8) is known as a quencher, and the effect is termed **quenching**; it is normally caused by the presence of foreign ions or molecules. Fluorescence is affected by the pH of the solution, by the nature of the solvent, the concentration of the reagent which is added in the determination of inorganic ions, and, in some cases, by temperature. The time taken to reach the maximum intensity of fluorescence varies considerably with the reaction.

XIX, 2. INSTRUMENTS FOR FLUORIMETRIC ANALYSIS. Instruments for the measurement of fluorescence are known as **fluorimeters or fluorophotometers**. The essential parts of a fluorimeter are shown in Fig. XIX, 1. The light from a mercury-vapour lamp (a source of ultraviolet light)

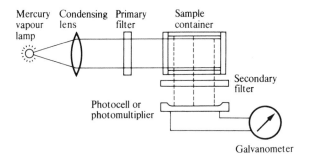

Mercury vapour lamp — Condensing lens — Primary filter — Sample container — Secondary filter — Photocell or photomultiplier — Galvanometer

Fig. XIX, 1

---

* It may be noted that $10^{-\epsilon c l} = e^{-2.303 \epsilon c l}$ and that $e^{-x} = 1 - x + x^2/2! \ldots$;
  $\therefore 1 - 10^{\epsilon c l} \approx \epsilon c l$ for $e^{\epsilon c l} \leqslant 0.01$.

is passed through a condensing lens, a primary filter (to permit the light band required for excitation to pass), a sample container, a secondary filter (selected to absorb the primary radiant energy but transmit the fluorescent radiation), a receiving photocell placed in a position at right angles to the incident beam (in order that it may not be affected by the primary radiation), and a sensitive galvanometer or other device for measuring the output of the photocell. Since fluorescence intensity is proportional to the intensity of irradiation, the light source must be very stable if fluctuations in its intensity are not compensated for. It is usual, therefore, to employ a two-cell instrument; the galvanometer is used as a null instrument, and readings are taken on a potentiometer used in balancing the photocells against each other. Since the two photocells are selected so as to be similar in spectral response, it is assumed that fluctuations in the intensity of the light source are minimised.

The simpler fluorimeters, such as the Locarte,* the EEL 244† and the Coleman fluorimeters, are manual instruments operating only at a single selected wavelength at any one time. Despite this they are perfectly suitable for quantitative measurements, as these are almost always carried out at a fixed wavelength. The experiments listed at the end of this chapter have all been carried out at single fixed wavelengths.

The more advanced fluorescence spectrophotometers are capable of automatically scanning fluorescent spectra between about 200–900 nm and produce a chart record of the spectrum obtained. These can also operate at a fixed wavelength and are equally suitable for carrying out quantitative work, although their main application tends to be for the detection and determination of small concentrations of organic substances.

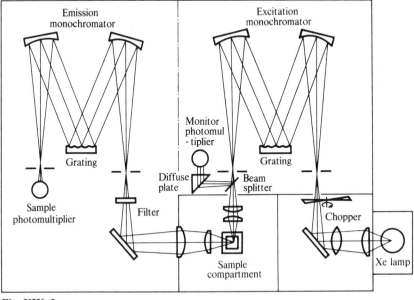

Fig. XIX, 2

---

* The Locarte Company, 8 Wendell Road, London, W12, England.
† Corning-EEL, Halstead, Essex, England.

The optical layout of a typical instrument, the Perkin–Elmer MPF-4, is illustrated in Fig. XIX, 2. This employs a 150 watt xenon lamp power supply and diffraction grating monochromators. An important difference compared with spectrophotometers is that in this case the fluorescent radiation is detected by a photomultiplier, whereas for absorption spectroscopy the detector is usually a photocell.

Spectrofluorimeters can usually scan at several rates between 10 and 500 nm min^{-1} and give a resolution of the order of 0.5 nm.

XIX, 3. SOME APPLICATIONS OF FLUORIMETRY. Fluorimetry is generally used if there is no colorimetric method sufficiently sensitive or selective for the substance to be determined. In inorganic chemistry the most frequent applications are for the determination of metal ions as fluorescent organic complexes, although uranium compounds fluoresce with a brilliant yellow colour. Uranium may be determined by measuring the fluorescence of a bead produced by fusing the substance with a mixture of sodium carbonate and sodium fluoride. Many of the complexes of oxine fluoresce strongly: aluminium, zinc, magnesium, and gallium are sometimes determined at low concentrations by this method. Aluminium forms fluorescent complexes with the dyestuff Eriochrome Blue Black RC (Pontachrome Blue Black R), while beryllium forms a fluorescent complex with quinizarin.

Important applications are to the determination of quinine and the vitamins riboflavin (vitamin B$_2$) and thiamine (vitamin B$_1$). Riboflavin fluoresces in aqueous solution; thiamine must first be oxidised with alkaline hexacyano-ferrate(III) solution to thiochrome, which gives a blue fluorescence in butanol solution.

The intensity and colour of the fluorescence of many substances depend upon the pH of the solution; indeed, some substances are so sensitive to pH that they can be used as pH indicators. These are termed **fluorescent** or **luminescent indicators**. Those substances which fluoresce in ultraviolet light and change in colour or have their fluorescence quenched with change in pH can be used as **fluorescent indicators** in acid–base titrations. The merit of such indicators is that they can be employed in the titration of coloured (and sometimes of intensely

Table XIX, 1   Some fluorescent indicators

Name of indicator	Approx. pH range	Colour change
Acridine	5.2– 6.6	Green to violet-blue
Chromotropic acid	3.0– 4.5	Colourless to blue
2-Hydroxycinnamic acid	7.2– 9.0	Colourless to green
3,6-Dihydroxyphthalimide	0.0– 2.5	Colourless to yellowish-green
	6.0– 8.0	Yellowish-green to green
Eosin	3.0– 4.0	Colourless to green
Erythrosin-B	2.5– 4.0	Colourless to green
Fluorescein	4.0– 6.0	Colourless to green
4-Methyl-aesculetin	4.0– 6.2	Colourless to blue
	9.0–10.0	Blue to light green
2-Naphthoquinoline	4.4– 6.3	Blue to colourless
Quinine sulphate	3.0– 5.0	Blue to violet
	9.5–10.0	Violet to colourless
Quininic acid	4.0– 5.0	Yellow to blue
Umbelliferone	6.5– 8.0	Faint blue to bright blue

coloured) solutions in which the colour changes of the usual indicators would be masked. Titrations are best performed in a silica flask. Examples of fluorescent indicators are given in Table XIX, 1.

It should be noted that a number of these indicators are also used for other purposes, e.g., eosin and fluorescein are frequently employed as adsorption indicators (Section **X, 30, C**).

Experimental

XIX, 4. QUININE. *Discussion.*   Although quinine is an organic compound this determination has been included in this book as it is an ideal experiment with which to gain experience in quantitative fluorimetry. It can be employed particularly for the determination of the amount of quinine in samples of tonic water.

Reagents. *Dilute sulphuric acid,* ca. *0.05M.* Add 3.0 cm^3 concentrated sulphuric acid to 100 cm^3 water, and dilute to 1 dm^3 with distilled water.

Standard solution of quinine. Weigh out accurately 0.100 g quinine and dissolve it in 1 dm^3 0.05M-sulphuric acid in a graduated flask. Dilute 10.0 cm^3 of the above solution to 1 dm^3 with 0.05M-sulphuric acid. The resulting solution contains 0.00100 mg quinine per cm^3.

With the aid of a calibrated burette, run 10.0, 17.0, 24.0, 31.0, 38.0, 45.0, 52.0, and 62.0 cm^3 of the above dilute standard solution into separate 100-cm^3 graduated flasks and dilute each to the mark with 0.05M-sulphuric acid.

Procedure.   Measure the fluorescence of each of the above solutions at 671 nm, using that containing 62.0 cm^3 of the dilute quinine solution as standard for the fluorimeter. Use LF2 as the primary filter and gelatine as the secondary filter if using the Locarte fluorimeter.

Now prepare test solutions containing, say, 0.00025 and 0.00045 mg quinine per cm^3. Determine their concentrations by measuring the fluorescence on the instrument and using the calibration curve.

To determine the quinine content of tonic water it is first necessary to de-gas the sample either by leaving the bottle open to the atmosphere for a prolonged period or by stirring it vigorously in a beaker for several minutes. Take 12.5 cm^3 of the de-gassed tonic water and make up to 25 cm^3 in a graduated flask with 0.1M sulphuric acid. From this solution prepare other dilutions with 0.05M sulphuric acid until a fluorimeter reading is obtained that falls on the calibration line previously prepared. From the value obtained calculate the concentration of quinine in the original tonic water.

XIX, 5. ALUMINIUM.   The procedure utilises Eriochrome Blue Black RC (also called Pontachrome Blue Black R; Colour Index No. 15705) at a pH of 4.8 in a buffered solution. Beryllium gives no fluorescence and does not interfere; iron, chromium, copper, nickel, and cobalt mask the fluorescence; fluoride must be removed if present. The method may be adapted for the determination of aluminium in steel.

Reagents. *Standard solution of aluminium.* Dissolve 1.760 g A.R. aluminium potassium sulphate crystals in distilled water, add 3 cm^3 concentrated sulphuric acid, and dilute to 1 dm^3 in a graduated flask. Pipette 10.0 cm^3 of this solution into a little water, add 2.0 cm^3 concentrated sulphuric acid, and dilute to

1 dm³ with distilled water. This solution contains 0.00100 mg aluminium per dm³.

Ammonium acetate solution, 10 per cent. Dissolve 25 g of the pure salt in water and dilute to 250 cm³.

Dilute sulphuric acid. Add 25 cm³ concentrated sulphuric acid to 200 cm³ water, cool, and dilute to 500 cm³ in a graduated flask.

Eriochrome Blue Black RC, 0.1 per cent. Prepare a 0.1 per cent solution in 90 per cent ethanol.

Procedure. Into 100-cm³ graduated flasks, each containing 10 cm³ of the ammonium acetate solution, 1 cm³ of the dilute sulphuric acid, and 3 cm³ of the Eriochrome Blue Black RC solution, run in from a burette 15.0, 20.0, 25.0, 30.0, 35.0, 40.0, 45.0, and 50.0 cm³ of the standard aluminium solution. Dilute each of the above solutions with distilled water, adjust to a pH of 4.6 ± 0.2 if necessary before making to the 100 cm³ mark. Allow the solutions to stand for at least 1 hour.

Measure the fluorescence of each of the above solutions at 590 nm, using that containing 0.0005 mg cm⁻³ Al as standard. The use of a primary filter (Corning 5543 or 5874 has been recommended) will depend upon the quality of the Eriochrome Blue Black RC: it can often be dispensed with: the secondary filter may be a Chance OR2 or Corning 2408, or LF7 for the Locarte. Draw a calibration curve, plotting instrument readings against concentration of aluminium. Determine the number of mg of Al per dm³ in an unknown solution (say, *ca.* 0.25 mg/dm⁻³), utilising the calibration curve.

XIX, 6. CADMIUM. *Discussion.* Cadmium may be precipitated quantitatively in alkaline solution in the presence of tartrate by 2-(2-hydroxyphenyl)-benzoxazole. The complex dissolves readily in glacial acetic acid, giving a solution with an orange tint and a bright blue fluorescence in ultraviolet light. The acetic acid solution is used as a basis for the fluorimetric determination of cadmium (Ref. 1).

Reagents. *2-(2-Hydroxyphenyl)-benzoxazole solution.* Dissolve 1.0 g of the solid reagent in 1 dm³ of 95 per cent ethanol.

Standard cadmium-ion solution. Prepare a standard cadmium-ion solution containing *ca.* 0.04 mg cm⁻³ Cd using A.R. hydrated cadmium sulphate.

Solutions for calibration curve with fluorimeter. Prepare the cadmium complex of the reagent by precipitating it from a solution of a pure cadmium salt as follows. Introduce a large excess of sodium tartrate, warm to 60 °C, adjust the pH to 9–10 by the addition of 0.5N-sodium hydroxide, add a slight excess of the reagent, and digest at 60 °C for 15 minutes. Filter on a sintered glass crucible (medium porosity), wash with 50 per cent ethanol (rendered faintly ammoniacal) to remove excess of the reagent, and dry at 130–140 °C for 1–2 hours. Weigh out 0.2371 g of the complex (≡ 0.0500 g Cd) and dissolve it in 1 dm³ of glacial acetic acid. Remove volumes of the acetic acid solution equivalent to 2.5, 2.0, 1.0, 0.5, and 0.10 mg Cd, and dilute each to exactly 50 cm³ with glacial acetic acid. Measure the fluorescence of each of the above solutions, using the appropriate filters (e.g., a yellow filter such as Corning 3-74 between the sample and the photocell). Plot fluorimeter readings against concentration of Cd per 50 cm³.

Procedure. Use an aqueous solution of the sample (25–50 cm³) containing from 0.1–2.0 mg of Cd and about 0.1 g of ammonium tartrate. Add an equal volume of 95 per cent ethanol, warm to 60 °C, treat with a slight excess of the reagent solution (4 cm³ ≡ *ca.* 1 mg Cd), adjust the pH to 9–11, digest at 60 °C for 15 minutes, filter on a medium-porosity glass crucible, wash with 20–25 cm³ of 95 per cent ethanol containing a trace of ammonia, and dry the precipitate at 130 °C for 30–45 minutes. Dissolve the precipitate in 50.0 cm³ of glacial acetic acid, and measure the fluorescence of the solution as in the calibration procedure. Evaluate the cadmium content from the calibration curve.

XIX, 7. CALCIUM. *Discussion.* This method is based upon the formation of a fluorescent chelate between calcium ions and Calcein (fluorescein iminodiacetic acid) in alkaline solution (Ref. 2). The procedure described below (Ref. 3) has been employed for the determination of calcium in biological materials (Ref. 4).*

Reagents. *Standard calcium solution.* Prepare a standard solution containing 40.0 mg dm⁻³ calcium by dissolving the calculated quantity of calcium carbonate in the minimum amount of hydrochloric acid and diluting to 1 dm³ in a graduated flask.

Calcein solution. Dissolve sufficient Calcein (fluorescein iminodiacetic acid), or its disodium salt, in the minimum amount of 0.40*M* potassium hydroxide solution and dilute with water to give a concentration of 60 mg dm⁻³ in a graduated flask. A small amount of EDTA solution (about 1.0 cm³ of 0.03*M* for every 100 cm³ Calcein solution) may be needed in the Calcein solution to achieve balancing of the blank on the fluorimeter. This is only necessary in those cases in which the potassium hydroxide used is found to contain a small amount of calcium impurity.

Aqueous solutions of Calcein are not stable for longer than 24 hours and should be kept in the dark as much as possible.

Potassium hydroxide solution. Prepare a 0.4*M* potassium hydroxide solution by dissolving solid potassium hydroxide (preferably calcium free) in de-ionised water and make to 1 dm³ in a graduated flask.

Procedure. Prepare a series of calcium ion solutions covering the concentration range 0–4 *μ*g per 25 cm³ by adding sufficient of the 40 mg dm⁻³ calcium standard to 25 cm³ graduated flasks each containing 5.0 cm³ of 0.4*M* potassium hydroxide solution and 1 cm³ of Calcein solution. Dilute each to 25 cm³ using de-ionised water. Determine the fluorescence for each solution at 540 nm with excitation at either 330 nm or 480 nm, and plot a calibration curve.

Prepare the sample solution in a similar manner to give a fluorescence value falling within the range of the calibration curve, and hence obtain the original calcium concentration in the sample.

XIX, 8. ZINC. The zinc complex of oxine fluoresces in ultraviolet light, and this forms the basis of the following method.

Reagents. *Standard zinc solution.* Dissolve about 4.0 g, accurately weighed, A.R. zinc shot in 35 cm³ concentrated hydrochloric acid, and dilute with distilled water to 1 dm³ in a graduated flask. Pipette 10.0 cm³ of this solution into a dm³ graduated flask and dilute to the mark with distilled water.

8-Hydroxyquinoline (oxine) solution, 5 per cent. Dissolve 5.0 g A.R. oxine in

---

* Calcium in the 10–500 ng range can be determined by using the selective spectrofluorimetric reagent 1,5-bis(dicarboxymethyl-aminomethyl)-2,6-dihydroxynaphthalene at pH 11.7 (Ref. 5).

12 g A.R. glacial acetic acid and dilute to $100\,cm^3$ with distilled water.

Standard dichlorofluorescein solution. Add a 0.1 per cent ethanolic solution of dichlorofluorescein dropwise to $1\,dm^3$ of distilled water until the resulting solution has a fluorescence slightly greater than that produced by the most concentrated zinc solution to be investigated (see below). About $0.8–1.0\,cm^3$ of dichlorofluorescein solution is required.

Gum arabic solution, 2 per cent. Grind finely 2.0 g gum arabic in a glass mortar, dissolve it in water, and dilute to $100\,cm^3$; filter, if necessary.

Ammonium acetate solution, ca. $2M$. Dissolve 15.5 g A.R. crystallised ammonium acetate in water and dilute to $100\,cm^3$.

Procedure. By means of a calibrated burette, run 5.0, 10.0, 15.0, 20.0, and $25.0\,cm^3$ of the standard zinc solution into separate $100\text{-}cm^3$ graduated flasks. To each flask add $10\,cm^3$ of the ammonium acetate solution, $4\,cm^3$ of the gum arabic solution, dilute to about $45\,cm^3$ with distilled water, and mix by swirling. Now add exactly $0.40\,cm^3$ of the oxine solution (use, e.g., a micrometer syringe or a micro-pipette), dilute to the mark with distilled water, shake gently, and transfer immediately to the cell of a fluorimeter for measurement. Employ the dichlorofluorescein solution as standard. Use a Chance OB2 as primary filter and OY2 as the secondary filter. Commence measurements with the most concentrated zinc solution. It is important that the fluorescence of the zinc–oxine mixtures be determined immediately after they are prepared, since the fine suspension of zinc oxinate slowly settles to the bottom of the cell. Plot instrument readings against zinc content $(mg\ cm^{-3})$. Use the calibration curve for determining the zinc content of test solutions containing, say, 4.5 and 6.5 mg of zinc per dm^3.

XIX, 9. References

1. N. Evcim and L. A. Reber (1954). *Anal. Chem.*, **26**, 936.
2. D. F. H. Wallach, D. M. Surgenor, J. Soderberg and E. Delano (1959). *Anal. Chem.*, **31**, 456.
3. B. L. Kepner and D. M. Hercules (1963). *Anal. Chem.*, **35**, 1238.
4. H. M. von Hattingberg, W. Klaus, H. Lüllmann and S. Zepf (1966). *Experientia*, **2**, 553.
5. B. Budĕšínsky and T. S. West (1969). *Talanta*, **16**, 399.

XIX, 10. Selected bibliography

1. P. Delahay (1957). 'Instrumental Analysis'. Ch. 10. *Fluorometry, Turbidimetry, and Nephelometry.* New York; The Macmillan Company.
2. H. H. Willard, L. L. Merritt and J. R. Dean (1974). 'Instrumental Methods of Analysis', Ch. 5. *Molecular Fluorescence and Phosphorescence Methods.* 5th edn. New York; Van Nostrand Reinhold.
3. J. G. Calvert and J. N. Pitts (1966). *Photochemistry.* New York; John Wiley.
4. R. B. Cundall and A. Gilbert (1970). *Photochemistry.* London; Nelson.
5. C. E. White and A. Weissler (1972). 'Fluorometric Analysis', Annual Reviews, in *Anal. Chem.*, **44**, No. 5, 182R.
6. J. S. Fritz and G. H. Schenk (1974). *Quantitative Analytical Chemistry.* Ch. 22, 3rd edn., p. 438. Boston; Allyn and Bacon.
7. M. Pinta (1974). *Detection and Determination of Trace Elements.* Chichester; Wiley.
8. G. G. Guilbault (1967). *Fluorescence—Theory, Instrumentation and Practice.* London; Edward Arnold, New York; Marcel Dekker.
9. C. E. White and R. J. Argauer (1970). *Fluorescence Analysis—A Practical Approach.* New York; Marcel Dekker.

CHAPTER XX

NEPHELOMETRY AND TURBIDIMETRY

XX, 1. GENERAL DISCUSSION. Small amounts of some insoluble compounds may be prepared in a state of aggregation such that moderately stable suspensions are obtained. The optical properties of each suspension will vary with the concentration of the dispersed phase. When light is passed through the suspension, part of the incident radiant energy is dissipated by absorption, reflection, and refraction, while the remainder is transmitted. Measurement of the intensity of the transmitted light as a function of the concentration of the dispersed phase is the basis of **turbidimetric analysis**. When the suspension is viewed at right angles to the direction of the incident light the system appears opalescent due to the reflection of light from the particles of the suspension (Tyndall effect). The light is reflected irregularly and diffusely, and consequently the term scattered light is used to account for this opalescence or cloudiness. The measurement of the intensity of the scattered light (at right angles to the direction of the incident light) as a function of the concentration of the dispersed phase is the basis of **nephelometric analysis** (Gr. *nephele* = a cloud). Nephelometric analysis is most sensitive for very dilute suspensions ($\not> 100$ mg per litre). Techniques for turbidimetric analysis and nephelometric analysis resemble those of filter photometry and fluorimetry respectively.

The construction of calibration curves is recommended in nephelometric and turbidimetric determinations, since the relationship between the optical properties of the suspension and the concentration of the disperse phase is, at best, semi-empirical. If the cloudiness or turbidity is to be reproducible, the utmost care must be taken in its preparation. The precipitate must be very fine, so as not to settle rapidly. The intensity of the scattered light depends upon the number and the size of the particles in suspension and, provided that the average size of particles is fairly reproducible, analytical applications are possible.

The following conditions should be carefully controlled in order to produce suspensions of reasonably uniform character:
1. The concentrations of the two ions which combine to produce the precipitate as well as the ratio of the concentrations in the solutions which are mixed.
2. The manner, the order, and the rate of mixing.
3. The amounts of other salts and substances present, especially protective colloids (gelatin, gum arabic, dextrin, etc.).
4. The temperature.

**XX, 2. INSTRUMENTS FOR NEPHELOMETRY AND TURBIDI-
METRY.** Visual and photoelectric colorimeters may be used as turbidimeters:
a blue filter usually results in greater sensitivity. A calibration curve must be
constructed using several standard solutions, since the light transmitted by a
turbid solution does not generally obey the Beer–Lambert law precisely.

'Visual' nephelometers (comparator type) have been superseded by the
photoelectric type. It is possible to adapt a good Duboscq colorimeter (Section
XVIII, 6) for nephelometric work. Since the instrument is to measure scattered
light, the light path must be so arranged that the light enters the side of the cups at
right angles to the plungers instead of through the bottoms. The usual cups are
therefore replaced by clear glass tubes with opaque bottoms; the glass plungers
are accurately fitted with opaque sleeves. The light, which enters at right angles to
the cups, must be regulated so that equal illumination is obtained on both sides.
A standard suspension is placed in one cup, and the unknown solution is treated

Fig. XX, 1

in an identical manner and placed in the other cup. The dividing line between the
two fields in the eyepiece must be thin and sharp, and seem to disappear when the
fields are matched.

Most fluorimeters (see Chapter XIX) may be adapted for use in
nephelometry.* The **EEL nephelometer†** is a simple, inexpensive, and excellent
instrument, and will be described in detail. The essential part of the instrument is
the nephelometer head; this is illustrated in Figs. XX, 1, and XX, 2, respectively.
A 6-volt, 6-watt lamp is mounted within the base of this unit to shine light
vertically through the orifice of an annular photocell on to the hemispherical base
of a test-tube. A tri-colour filter wheel (containing filters OB2, OR1, and OGR1)
is interposed between the lamp and the photocell, and is also provided with a
position for white-light measurement. The selected filter should be similar in

---

* The Ratio and A4 Fluorimeters manufactured by the Farrand Optical Co. Inc, Valhalla, New
  York, USA are modified for nephelometry and used in clinical analyses.
† Manufactured by Corning-EEL Ltd, Halstead, Essex, England.

colour to that of the sample. If the solution contained in the test-tube is cloudy or turbid to any degree the light is scattered by multiple reflections from the particles. Such scattered light is collected by a reflector, which is mounted above the photocell, and is then directed on to the photocell itself. The current so generated is fed by means of a flexible lead and plug to a sensitive galvanometer (not illustrated). A metal cap is provided to fit over the test-tube when the instrument is in use and will exclude extraneous light. When this cap is removed a micro-switch is operated to disconnect the photocell from the galvanometer and prevent damage to the suspension by large current resulting from the sudden entry of external light. The sensitive galvanometer is of the taut-suspension mirror type: a stabilising transformer is incorporated to supply power to the nephelometer lamp. A large plastic knob protrudes through the top of the

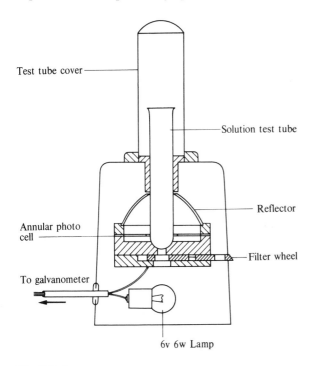

Test tube cover

Solution test tube

Reflector

Annular photo cell

Filter wheel

To galvanometer

6v 6w Lamp

Fig. XX, 2

galvanometer casing to provide a smooth zero setting; a sensitivity control and a clamping device are also incorporated. The standard size of matched test-tube is 1.5 cm diameter and 15 cm long, but adapters are available for test-tubes of 1.25 cm and 1.9 cm diameter. A Perspex standard is supplied, and can be used for standardising the nephelometer; this standard may be immersed in liquids of various refractive indices, if desired.

The **general procedure for operating the EEL nephelometer** is as follows:

1. Adjust the zero control knob of the galvanometer to bring the hair-line to the zero of the scale. Connect the nephelometer head and galvanometer unit by means of the six-pin plug and flexible lead.

2. Remove the cap cover, place the standard in position, and replace the cap.

The standard is generally the matched test-tube containing the most concentrated suspension of the substance being determined: the concentration must, of course, be known.

3. Select the filter required. This should be similar in colour to that of the solution.

4. Adjust the sensitivity control of the galvanometer to obtain a reading of one hundred divisions on the scale.

5. Remove the standard.

6. Fill a test-tube with distilled water or with a 'blank' solution to a depth of not less than 3 cm, and set to zero by means of the galvanometer zero control.

7. Check the reading of the standard (100 divisions).

8. Repeat procedures 6 and 7 until full-scale deflection and zero settings are obtained.

9. Replace the standard suspension with more dilute suspensions, and note the various scale readings. Draw a calibration curve relating galvanometer readings and the concentrations of the substance being determined.

10. Fill a test-tube with the sample to be determined to a depth similar to that used for the standards, and insert into the instrument. Note the galvanometer deflection: evaluate the concentration from the calibration curve.

Some nephelometric determinations

XX, 3. SULPHATE. *Discussion.* The turbidity of a dilute barium sulphate suspension is difficult to reproduce; it is therefore essential to adhere rigidly to the experimental procedure detailed below. The velocity of the precipitation, as well as the concentration of the reactants, must be controlled by adding (after all the other components are present) pure solid barium chloride of definite grain size. The rate of solution of the barium chloride controls the velocity of the reaction. Sodium chloride and hydrochloric acid are added before the precipitation in order to inhibit the growth of microcrystals of barium sulphate; the optimum pH is maintained and minimises the effect of variable amounts of other electrolytes present in the sample upon the size of the suspended barium sulphate particles. A glycerol–ethanol solution helps to stabilise the turbidity. The reaction vessel is shaken gently in order to obtain a uniform particle size: each vessel should be shaken at the same rate and the same number of times. The unknown must be treated exactly like the standard solution. The interval between the time of precipitation and measurement must be kept constant.

Reagents. *Standard sulphate solution.* Dissolve 1.814 g dry A.R. potassium sulphate in distilled water and dilute to 1 dm³ in a graduated flask. This solution contains 1.000 mg of sulphate ion per cm³.

Sodium chloride–hydrochloric acid reagent. Dissolve 60 g A.R. sodium chloride in 200 cm³ distilled water, add 5 cm³ pure concentrated hydrochloric acid, and dilute to 250 cm³.

Barium chloride. Use crystals of A.R. barium chloride that pass through a 20-mesh sieve and are retained by a 30-mesh sieve.

Glycerol–ethanol solution. Dissolve 1 volume of pure glycerol in 2 volumes of absolute ethanol.

Procedure. Run 0.5, 1.0, 1.5, 2.0, 2.5, and 3.0 cm³ of the standard potassium sulphate solution from a calibrated burette into separate 100-cm³ graduated flasks. To each flask add 10 cm³ of the sodium chloride–hydrochloric acid reagent and 20 cm³ of the glycerol–ethanol solution, and dilute to 100 cm³ with distilled water. Add 0.3 g of the sieved barium chloride to each flask, stopper each flask, and shake for 1 minute by inverting each flask once per second: all the barium chloride should dissolve. Allow each flask to stand for 2–3 minutes and measure the turbidity in the EEL nephelometer: take care to avoid small air bubbles adhering to the walls of the matched test-tubes. Use the most concentrated solution as standard and, by means of the sensitivity control, adjust the galvanometer reading to 100 divisions. Prepare a 'blank' solution, repeat the above sequence of operations, but do not add any sulphate solution. Place the 'blank' solution in the nephelometer and adjust to zero reading of the galvanometer scale by means of the zero control above the galvanometer suspension. Check the reading of the most turbid solution, and adjust any deviation from 100 by means of the sensitivity control. Repeat the measurements with the five other standard sulphate solutions. Plot the galvanometer reading against the sulphate-ion content per cm³.

Determine the sulphate-ion content of an unknown solution, say, *ca.* 0.5 mg cm⁻³: use the calibration curve.

XX, 4. PHOSPHATE. *Discussion.* Phosphate ion is determined nephelometrically following the formation of strychnine molybdophosphate. This turbidity is white in colour and consists of extremely fine particles (compare ammonium molybdophosphate, which is yellow and is composed of rather large grains). The precipitate must not be agitated, as it tends to agglomerate easily; it is somewhat sensitive to temperature changes.

Reagents. *Standard phosphate solution.* Dissolve 1.721 g A.R. potassium dihydrogenphosphate (dried at 110 °C) in 1 dm³ of water in a graduated flask. Pipette 10.0 cm³ of this solution into a 1-dm³ graduated flask and dilute to the mark. The resulting dilute solution contains 0.01 mg phosphorus pentoxide per cm³.

*Molybdate–strychnine reagent.** This reagent is prepared in two parts; these are mixed just before use, since the addition of the acid molybdate solution to the strychnine sulphate solution produces a precipitate after 24 hours. Solution A (acid molybdate solution): place 30 g A.R. molybdenum trioxide in a 500 cm³ conical flask, add 10 g A.R. sodium carbonate and 200 cm³ water. Boil the mixture until a clear solution is obtained. Filter the hot solution, if necessary. Add 200 cm³ 5M-sulphuric acid, allow to cool, and dilute to 500 cm³.

Solution B (strychnine sulphate solution): dissolve 1.6 g strychnine sulphate in 100 cm³ warm distilled water, cool and dilute to 500 cm³.

Prepare the reagent by adding Solution B rapidly to an equal volume of Solution A, and shake the resulting mixture thoroughly; filter off the bluish-white precipitate through a Whatman No. 42 filter paper. The resulting clear solution will keep for about 20 hours. Solutions A and B may be kept indefinitely.

Saturated sodium sulphate solution. Prepare a saturated aqueous solution at 50 °C and cool to room temperature. Filter before use.

---

* Strychnine is a toxic alkaloid. It should only be handled with gloves and under no circumstances should it be ingested.

*Sulphuric acid, 1*M. Dilute 27 cm³ of A.R. concentrated sulphuric acid to 500 cm³ in a graduated flask.

Procedure. Run in 1.0, 2.0, 4.0, 6.0, 8.0, and 10.0 cm³ of the standard phosphate solution from a calibrated burette into separate 100-cm³ graduated flasks. To each flask add 18 cm³ 1*M*-sulphuric acid and 16 cm³ saturated sodium sulphate solution, and dilute to approximately 95 cm³ with distilled water. Now add 2.0 cm³ of the molybdate–strychnine reagent and dilute to 100 cm³; mix the contents of the flask by gently inverting several times, but do not shake. Allow the flasks to stand for 20 minutes to permit the turbidities to develop before making the measurements. Prepare a 'blank' solution by repeating the above sequence of operations, but omit the addition of the phosphate solution. Use the most concentrated solution as the initial standard and adjust the galvanometer reading to 100 divisions. Introduce the 'blank' solution into the matched test-tube of the EEL nephelometer and adjust the galvanometer reading to zero. Check the standard solution for a galvanometer reading of 100. Repeat the above with the five other phosphate solutions. Plot galvanometer reading against mg P_2O_5 per cm³.

Determine the phosphate content of an unknown solution, say, containing *ca.* 0.005 mg P_2O_5 per cm³: use the calibration graph.

XX, 5. Selected bibliography

1. R. Barnes and C. R. Stock (1949). 'Apparatus for Transmission Turbidimetry of Slightly Hazy Materials', *Analytical Chemistry*, **21**, 18.
2. C. L. Wilson (1953). 'Nephelometry', *Annual Reports on the Progress of Chemistry*, **50**, 367.
3. P. Delahay (1957). *Instrumental Analysis.* Section on Turbidimetry and Nephelometry. New York; The Macmillan Company.
4. G. W. Ewing (1960). *Instrumental Methods of Chemical Analysis.* Section on Nephelometry and Turbidimetry. 2nd edn. New York; McGraw-Hill Book Co.
5. H. A. Strobel (1960). *Chemical Instrumentation. A Systematic Approach to Instrumental Analysis.* Ch. 8. Light Scattering Photometry. Reading, Mass.; Addison-Wesley Publishing Co.

CHAPTER XXI **EMISSION SPECTROGRAPHY**

XXI, 1. GENERAL DISCUSSION. When certain metals are introduced as salts into the Bunsen flame characteristic colours are produced; this procedure has long been used for detecting elements qualitatively. If the light from such a flame is passed through a spectroscope several lines may be seen, each of which has a characteristic colour: thus calcium gives red, green, and blue radiations, of which the red are largely responsible for the typical colour that this element imparts to the flame. A definite wavelength can be assigned to each radiation, corresponding with its fixed position in the spectrum. Although the flame colours of, for example, calcium, strontium, and lithium are very similar, it is possible to differentiate with certainty between them by observations on their spectra and to detect each in the presence of the others. By extending and amplifying the principles inherent in the qualitative flame test, analytical applications of emission spectrography have been developed. Thus more powerful methods of excitation, such as electric spark or electric arc, are used, and the spectra are recorded photographically by means of a spectrograph: also, since the characteristic spectra of many elements occur in the ultraviolet, the optical system used to disperse the radiation is generally made of quartz.

A detailed discussion of the origin of emission spectra is beyond the scope of this book but a simplified treatment is given in Chapter XXII, Sections 1 and 2.*

It may be stated, however, that there are three kinds of emission spectra: continuous spectra, band spectra, and line spectra. The continuous spectra are emitted by incandescent solids, and sharply defined lines are absent. The band spectra consist of groups of lines that come closer and closer together as they approach a limit, the head of the band: these are caused by excited molecules. Line spectra consist of definite, usually widely and seemingly irregularly spaced, lines; these are characteristic of atoms or atomic ions which have been excited and emit their energy in the form of light of definite wavelengths. The quantum theory predicts that each atom or ion possesses definite energy states in which the various electrons can exist; in the normal or ground state the electrons have the lowest energy. Upon the application of sufficient energy by electrical, thermal, or other means, one or more electrons may be removed to a higher energy state farther from the nucleus; these excited electrons tend to return to the ground

---

* For a more detailed treatment of the theory of atomic spectroscopy the reader is referred to G. Herzberg (1944). *Atomic Spectra and Atomic Structure*. 2nd edn. Dover Publications.

state, and in so doing emit the extra energy as a photon of radiant energy. Since there are definite energy states and since only certain changes are possible according to the quantum theory, there are a limited number of wavelengths possible in the emission spectrum. The greater the energy of the exciting source, the higher the energy of the excited electrons, and therefore the more numerous the lines that may appear. The intensity of a spectral line depends largely upon the probability of the required energy transition or 'jump' taking place. The intensity of some of the stronger lines may occasionally be decreased by self-absorption caused by reabsorption of energy by the cool gaseous atoms in the outer regions of the source. With high-energy sources the atoms may be ionised by the loss of one or more electrons; the spectrum of an ionised atom is different from that of a neutral atom and, indeed, the spectrum of a singly ionised atom resembles that of the neutral atom with an atomic number one less than its own.

The lines in the spectrum from any element always occur in the same positions relative to each other. When sufficient amounts of several elements are present in the source of radiation each emits its characteristic spectrum; this is the basis for qualitative analysis by the spectrochemical method. It is not necessary to examine and identify all the lines in the spectrum, because the strongest lines will be present in definite positions, and they serve to identify unequivocally the presence of the corresponding element. As the quantity of the element in the source is reduced, these lines are the last to disappear from the spectrum: they have therefore been called the **persistent lines** or the **'raies ultimes'** (R.U. lines), and simplify greatly the qualitative examination of spectra.

Lines in an unknown spectrum may be identified by comparing them with those on a spectrum containing a number of lines of known wavelengths. This may be performed either by comparison with charts of spectra of metallic elements such as iron or copper, or by the use of R.U. powder (see Section **XXI, 3**).

The number of lines appearing in the spectrum varies considerably from element to element. The spectra of the transition elements, the lanthanoids and such elements as titanium and molybdenum, produce complex spectra; copper, antimony, tin, and lead are intermediate; while boron, magnesium, aluminium, zinc, and the alkaline-earth metals give relatively simple spectra. The practical result of these differences is that spectrographs with greater dispersion and resolution are required to separate adequately the lines in complex spectra, e.g., iron, nickel, cobalt, or manganese: such spectrographs are necessarily large and expensive.

For quantitative analysis it is necessary to assess the densities of blackening of lines in a spectrogram due to the constituents being determined; this may be done by comparing the spectra from samples of known and unknown composition. Comparisons may be made either visually (best with the aid of a spectrum projector: see Fig. XXI, 6) when no great accuracy is desired, or by photoelectric measurement of line densities with a microphotometer (see Section **XXI, 2**). Details of the procedure are described in Sections **XXI, 4**, and **XXI, 7**.

The applications of emission spectrography include:
1. the examination of a single metal or an alloy for impurities;
2. the analysis of an alloy for its general composition, including a search for minor components and traces of impurities;
3. the analysis of ash of organic substances and other materials (e.g., natural waters) amenable to similar treatment; and

4. the detection of contaminants in food.

The chief advantages of the spectrographic method of analysis are:

(*a*) The procedure is specific for the element being determined, although difficulties occasionally arise when a line of another element overlaps that of the unknown.

(*b*) The method is time-saving; a quantitative determination of traces of the elements in a sample, especially an alloy or a metal, may be made without any preliminary treatment. Most metals and some non-metals (e.g., phosphorus, silicon, arsenic, and boron) may be determined.

(*c*) A permanent record may be obtained on a photographic plate.

(*d*) It may be (and is usually) applied to the determination of small quantities of added constituents or of traces of impurities where conventional methods of analysis are difficult, fail, or give less accurate results. Lengthy and difficult separations by chemical methods, e.g., of zirconium and hafnium and of niobium and tantalum, can be avoided.

The apparent disadvantages are:

(i) Successful use requires wide experience, both in the operation of equipment and in reading and interpreting spectra.

(ii) The spectrograph is essentially a comparator; for quantitative analysis, standards (usually of similar composition to the material under analysis) are necessary. Unknown samples therefore present a relatively difficult problem when quantitative results are required.

(iii) The accuracy and precision are not as high as gravimetric, titrimetric, and some spectrophotometric methods for elements present in quantities greater than 2–5 per cent of the total; indeed, spectrographic methods are not usually applied for elements present to a greater extent than about 3 per cent.

XXI, 2. EQUIPMENT FOR EMISSION SPECTROGRAPHIC ANALYSIS.

This section is concerned with describing the equipment which is necessary for an introduction to spectrographic techniques for the analyst. In this instance the practical work will be described for instruments manufactured by Rank Hilger, Margate, Kent, England, but the comparable products of other manufacturers (e.g., Bausch and Lomb, Rochester, USA) may also be used.

The essential parts of a **spectrograph** are a slit, an optical system, and a camera for recording the spectrum. The light from the source of radiation passes through the slit, which is a narrow vertical aperture, then through the optical system, which includes a prism or grating. An image of the slit is produced by means of lenses at the point where the light is recorded. One such image is produced for each radiation having a specific wavelength, and the result is a series of vertical line images which constitute the spectrum of the element being investigated. The optical system may be either of glass or quartz; the latter transmits in the ultraviolet region, where many useful lines occur, as well as in the visible. The range of wavelengths employed with quartz extends from about 2000 to 10 000 Å.*

Two sizes of prism spectrographs are widely used for analysis, the medium and

---

* Strict compliance with SI units should require that these values are expressed in nanometres; however, the Angstrom is still commonly employed and it is felt desirable to retain it for this particular chapter.

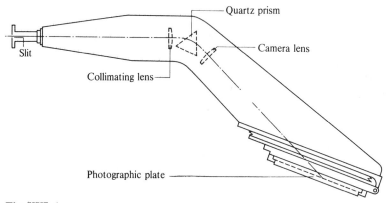

Fig. XXI, 1

the large. The large spectrograph is necessary for the analysis of iron, chromium, cobalt, molybdenum, titanium, tungsten, uranium, and zirconium owing to the numerous lines in the spectra and the need for maximum resolution; it utilises a 25.4 cm × 10.2 cm (10 in. × 4 in.) plate, but adjustments must be made to bring different regions of the spectrum on the plate. The complete spectrum is 76 cm long, so that it is recorded in three separate sections. The **Hilger and Watts medium spectrograph** has ample dispersion for most work with non-ferrous metals and light alloys: it has the advantage that the whole of the spectrum range can be photographed in a single exposure on one plate (25.4 cm × 10.2 cm). The essential features are shown in Fig. XXI, 1, while the actual instrument is depicted in Fig. XXI, 2. The light produced from the sample by either means of excitation enumerated below is received by a narrow slit and passes through a lens system to a prism which deviates each radiation from a direct path by an amount depending upon its wavelength. A second lens system forms an image of the narrow slit upon a photographic plate in the order determined by the prism. The prism employed is known as a Cornu prism;* the 60° prism is composed of two half prisms of quartz in optical contact. The two halves are cut so that they compensate each other, with the result that the combination functions as if it were an equilateral quartz prism.

The slit (20 mm long) is formed of two parallel metal jaws, which are accurately ground flat; the width is controlled to 0.001 mm by means of a micrometer screw which causes both jaws to move symmetrically. For most determinations with spectrographs a slit width of about 0.02 mm is usually satisfactory. The slit is equipped with a **Hartmann diaphragm** (see Fig. XXI, 15) for placing spectra in juxtaposition on the plate; this involves a number (usually three) of square openings arranged in echelon, the square holes being cut with the bottom of one in line with the top of the next. When these holes are placed successively in front of the slit it is possible to record three spectrograms in juxtaposition without moving the plate holder. The shutter is usually placed between the slit jaws and is operated by a lever.

---

* The Littrow prism is also widely used. It is a 30° prism with a mirror back face: the light passes through the prism to the mirror face and is reflected back through the prism, the total path being equal to a 60° prism. The prism mounting results in a fairly compact instrument.

Fig. XXI, 2

The spectrograph is provided with a scale graduated in wavelengths, which can be illuminated and printed directly on the spectrogram plate. It is incorporated in the back of the camera and is controlled by a lever. The camera slide for carrying the plate holder is fitted at the end of the instrument casting. Provision is made for turning the plate-holder carrier mounting through a small angle about a vertical axis. The plate-holder slide is operated by a rack-and-pinion motion that raises or lowers the plate carrier over a range of 75 mm in 1-mm steps. A number of spectra (up to 40, but depending on the length of the slit) can be taken on one plate.

The length of the instrument from the slit to the end of the plate holder is about 1.2 metres, and it is supported on a massive base which raises the optical parts about 30 cm above bench level. An optical bar of steel is attached to the base of the instrument, from which it projects about 90 cm; it is parallel with the optical axis. The bar serves to carry lenses, an arc and spark stand (Gramont stand) for holding samples, and other ancillary equipment.

The holders for arc and spark excitation fit directly on the optical bar attached to the spectrograph. In modern instruments the electrode holders are housed in a metal box fitted with a safety shield. The electrodes are carried in strong screw clamps on horizontal arms which are insulated from the supporting base; the movements include vertical adjustment of the upper electrode with rack and pinion, vertical and horizontal movements of the lower electrode through a collar which may be clamped to the supporting rod, and vertical and rotary movements of both electrodes together. The combinations are such that the discharge can be rapidly located on the optical axis. The holders are primarily designed for cylindrical rods not greater than 1.25 cm diameter.

The slit should be uniformly illuminated along its length. For this purpose a quartz lens, of such focal length that it throws an image of the source on the collimating lens (see Fig. XXI, 1) is placed 2 cm from the entrance slit and located

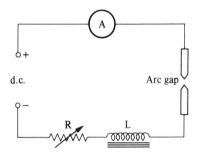

Fig. XXI, 3

by means of the optical bar. The rapid location of the source of light (i.e., positioning of the electrodes) on the optical axis of the spectrograph and the adjustment of the length of the discharge gap is conveniently carried out by projecting an image of the source on to a calibrated screen (gauge plate) provided with the instrument. Both the lens and the screen are mounted on the optical bar at the end remote from the slit (Fig. XXI, 12). After the initial adjustment with the light source, a reading lamp is arranged so that the lamp can be brought quickly into position near the electrodes on the slit side of the stand; it will be found that the electrodes can be seen on the screen. For subsequent work the electrodes can be set in optical alignment with the spectrograph simply by setting the image of the electrodes to agree with the screen gauge without actually passing any current between them. The two outer horizontal lines on the screen gauge are equivalent to an electrode gap of 4 mm and the inner lines are equivalent to a 2-mm gap.

The two most commonly used excitation sources are the low-voltage d.c. arc and the high-voltage a.c. spark. For the **low-voltage d.c. arc** the essential requirements are a source of d.c. at 110–250 volts, a regulating resistance R to control the current (2–12 amperes), an inductance L (this tends to steady the arc and maintain a more constant voltage), a d.c. ammeter A in series with the supply, and an arc gap (see Fig. XXI, 3). The sample under investigation may consist of electrically conducting rods, which then become the electrodes of the arc. In order to strike the arc, the two electrodes must be brought into contact or can be short-circuited by touching them with an insulated third (carbon) electrode. If the sample is in the form of a powder or small pieces it may be supported on a pure carbon or graphite rod which has been hollowed out to hold the specimen. Another similar rod, generally with a pointed or conical tip, serves as the upper electrode. Sometimes pure metal rods, e.g., copper or silver, are used as the supporting electrodes, and the sample may be made either the cathode or the anode. It is advisable to make exposures with the sample respectively positive or negative in order to determine which gives the better results for reproducibility and sensitivity. With two carbon electrodes it is usual to place the sample in the anode (positive electrode), which serves as the lower electrode. The temperature in the arc stream ranges from 2000 °C to about 5000 °C. When an arc is operated between carbon electrodes in air, some cyanogen molecules are formed, and these may emit molecular band spectra in the region 3200–4200 Å.

The d.c. arc is a sensitive source and of wide application, but its reproducibility

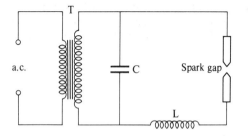

Fig. XXI, 4

is not of the highest order; it is generally used for the identification and determination of elements present in very small concentrations. A comparatively large amount of the substance being analysed passes through the arc, and consequently an average or more representative value of the concentration is shown, provided that the complete sample is burned. This is usually achieved by placing a small amount of the substance on a graphite electrode with a recess drilled to a depth of 4–5 mm,* and continuing the excitation until the sample is completely volatilised.

The **high-voltage a.c. spark** is perhaps the simplest excitation source for many qualitative identifications; its sensitivity is not as high as the d.c. arc source. It is more reproducible and stable than the arc and less material is consumed; this source is well adapted for the analysis of low-melting materials, since the heating effect is less. The basic circuit for spark excitation is shown in Fig. XXI, 4. Alternating current from the mains is fed into a high-tension transformer T that gives up to 15 kilovolts at its secondary terminals. The condenser C across the secondary of the transformer is charged by the current from the latter until the potential is high enough to cause a discharge across the spark gap: with a suitable value of the condenser (about 0.005 microfarad), a brilliant spark is obtained. By placing an inductance L (0.02–1.5 millihenrys) in series with the spark gap, a high-frequency oscillating discharge is produced.

In most spectrographs the intensity of the lines is ultimately registered on a photographic emulsion; the nature of the photographic process is therefore of considerable importance. It is assumed that the recommended plate, developer, developing time, temperature, etc., have been followed and that the spectra are obtained on the plate, together with a wavelength scale. Part of such a spectrum (obtained with the aid of a Hartmann diaphragm) is shown in Fig. XXI, 5; this gives the spectra, from the top downwards, of cadmium, spelter, and zinc. A very satisfactory method of examination of spectra, particularly for qualitative analysis is to employ a Rank Hilger Projection Comparator (Fig. XXI, 6). In this instrument the two plates for comparison are mounted in spring-loaded clips on

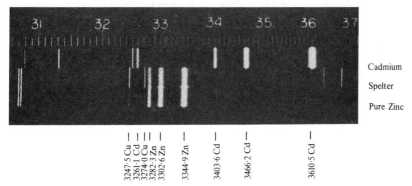

Fig. XXI, 5

adjustable stages. The two plates are illuminated from below by means of 12 V lamps and the transmitted light projected on to a large screen which presents the two images one above the other, each occupying half the screen. Separate

---

* In practice, it is found desirable to have a small 'pip' in the centre of the graphite electrode and to place the powder in a depression around it. The arc tends to pass between the upper pointed electrode and the small projection.

focusing lenses enable clear definitions of the individual spectra to be obtained and a ten-fold magnification is possible. Accurate alignment of the spectra is possible by means of a fine adjustment screw.

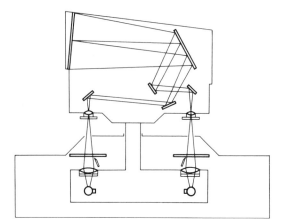

Fig. XXI, 6   The Rank Hilger L150 Projection Comparator

For quantitative analysis it is necessary to compare the relative blackening of lines with one another and with those produced by standard elements. The density of blackening (or simply blackening) B may be defined as:

$$B = \log\frac{i_0}{i}$$

where i_0 is the intensity of the transmitted radiation by a perfectly clear part of the photographic plate and i is the light transmitted by the line in question.

An apparatus for measuring the densities of blackening of very small areas of photographic plates or films is called a microphotometer. The essential features of this type of instrument are shown in Fig. XXI, 7, and an actual instrument, the **Rank Hilger L500 Microphotometer**, is illustrated in Fig. XXI, 8.

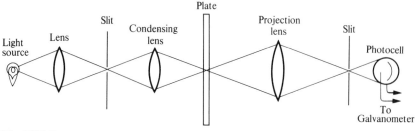

Fig. XXI, 7

Full instructions for use are supplied with the instrument but the main features are as follows. A reduced image of the filament of the lamp is focused on the emulsion side of the photographic plate. An enlargement ($10\times$) of the plate is projected on the whitened jaws of the slit. The light passing through the slit is detected by a photocell whose output can be shown either as a galvanometer

Fig. XXI, 8

deflection or by a photomultiplier from which the output is fed to a chart recorder. The slit width of the microphotometer is controlled by a drum graduated to read to 0.005 mm; the slit height can be reduced by a milled-edged control situated in front of the micrometer drum. An additional lens can be moved into the light path by means of a lever in front of the instrument. The effect of this lens is to flood a small area of the plate with light in order that an image of a short length of the spectrum is seen on the slit-jaw screen and the line to be photometered can be adjusted into the correct position. When the lens is removed to take a reading, a shutter opens before the barrier layer cell or photomultiplier, which is then exposed to light only when the actual readings are taken. The base of the microphotometer supports guide rods on which the plate stage travels. This stage accommodates plates up to 25.4 cm in length and 10.2 cm wide. To bring the spectrum into position the plate holder can be moved at right angles to the traverse by a rack and pinion gearing. A motor-driven leadscrew can move the plate stage to the left or right at a traversing speed of 0.1 mm min^{-1}, but this may be overriden by means of a manual control, even during a motor-driven traverse.

XXI, 3.   QUALITATIVE SPECTROGRAPHIC ANALYSIS.   At least fifty-five elements can be identified under normal conditions of excitation. In a qualitative analysis it is desirable to have a high sensitivity so that the presence of trace elements may be revealed. The d.c. arc usually gives the highest sensitivity. In many cases satisfactory results may be obtained with the a.c. high-voltage spark as the source of excitation; although the sensitivity is lower, the reproducibility is greater. The d.c. arc is preferred for the most difficultly excitable elements and for non-volatile and refractory compounds; a 'complete burn' of the sample may be obtained with comparative ease. A widely used system of qualitative spectrographic analysis is to arc the sample in a depression in a graphite electrode, using a pointed counter electrode. The graphite electrode in air gives a cyanogen band spectrum (headings at 4216, 3883, and 3590 Å) in the near ultraviolet, which may mask some of the lines of interest. A sample of a few milligrams is sufficient; the exposure conditions must be determined by experiment. Non-conducting samples may be more easily excited by being mixed with an equal volume of graphite powder, and these, as well as all refractory materials, are often treated with a carrier to bring them into the arc stream. This carrier, usually a volatile salt such as ammonium chloride or ammonium sulphate, helps to propel the entire sample smoothly up into the arc gap.

The amount of an element that is detectable varies with its concentration, its relative volatility, the energy of excitation, etc.: approximate sensitivity figures with arc excitation for the common elements are collected in Table XXI, 1.

As one dilutes the amount of an element in an arc, the number of lines observable is reduced, and ultimately there remains only a few lines of the element which is diluted. These lines are known as the **'raies ultimes'** or persistent lines. Tables of these persistent lines may be found in the references in the Selected Bibliography at the end of this chapter, in chemical handbooks, and in Appendix XIV. The identification of these lines will permit detection of elements present in low concentration, and all qualitative methods utilise the persistent lines.

Table XXI, 1   Arc sensitivities of some elements

0.1–1 per cent detectable	As, Cs, Nb, P, Ta, W.
0.01–0.1 per cent detectable	B, Bi, Cd, La, Rb, Sb, Si, Tl, Y, Zn, Zr.
0.001–0.01 per cent detectable	Al, Au, Ba, Be, Ca, Fe, Ga, Ge, Hg, Ir, K, Mn, Mo, Pb, Sc, Sn, Sr, Ti, V.
0.0001–0.001 per cent detectable	Ag, Co, Cr, Cu, In, Li, Mg, Na, Ni, Os, Pd, Pt, Rh, Ru.

The simplest and most direct procedure for the qualitative analysis of an unknown sample is the **R.U. powder method**. The R.U. powder is a powder developed in the Research Laboratories of the General Electric Company of Wembley, England, and is marketed by Johnson, Matthey and Co of London. It consists of small quantities of fifty elements incorporated in a base material composed of calcium, magnesium, and zinc oxides. The quantity of each element present has been carefully adjusted so that only the 'raies ultimes' and the most important sensitive lines appear when the spectrum is excited by placing some of the powder (10–20 mg) in the lower (and positive) pole of an arc between graphite electrodes. A current of 5–7 amperes and an arc length of 6 mm is recommended. It is advisable not to expose for a longer time than the powder lasts, or else parts of the spectrum will be unnecessarily masked by the CN bands. A set of seven enlargements of the arc spectra of the R.U. powder is marketed, and these cover a wavelength range of 2284–8000 Å. The spectrum of the iron arc (together with the important wavelengths) is given alongside that of the R.U. powder; this enables the positions of the persistent lines relative to the iron-arc spectra to be seen immediately, and will also permit the position of any sensitive line to be found readily. A qualitative analysis is carried out by producing contiguous arc spectra with the aid of a Hartmann diaphragm of the R.U. powder, the sample, and optionally (if known) of a pure main element present in the sample. The plate is developed, and the lines present in both the R.U. powder and sample spectra are noted; the latter is most simply observed on a Rank Hilger projection comparator (Fig. XXI, 6). The elements are tabulated with the number of lines appearing: three lines,* which are free from interference, are considered proof of the presence of the element. A portion of the spectrum of R.U. powder and a sample of Wood's metal is given in Fig. XXI, 9; both major and minor constituents are readily identified.

Another problem that frequently arises is to decide whether a substance contains a given element or a small number of specified elements. As an example we may take the presence of cadmium in spelter. Contiguous spectra are taken using the Hartmann diaphragm of: (i) the spectroscopically pure metal known to be present in the specimen under test in considerable quantity (e.g., zinc); (ii) the sample under test (e.g., spelter); and (iii) the spectroscopically pure metal whose

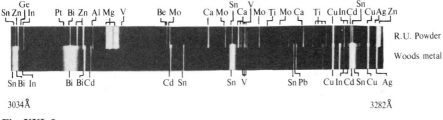

Fig. XXI, 9

presence or absence is to be determined (e.g., cadmium). Part of such a spectrum is shown in Fig. XXI, 5. The graduated scale of wavelengths is also photographed on the plate. Examination of the three spectra with a Rank Hilger projection comparator will reveal at once the presence or absence of the specified element (cadmium in the present example). This instrument, which is described in detail above (Section **XXI, 2**), also enables comparisons to be made between images on two different spectographic plates. Thus if a set of reference plates is available, the need for taking comparison spectra of metals looked for against that of the unknown is obviated and a great deal of tedious measurement is likewise rendered unnecessary. Standard spectrographic samples are available *inter alia* from the Bureau of Standards at Washington.

Mention may be made of the great advantage that the spectrograph offers in the following operations:

(*a*) Rapid qualitative analysis of all the metallic constituents of a substance as a basis for planning a chemical analysis.

(*b*) Approximate analysis of minor components by sight (after some experience has been obtained).

(*c*) Examination of precipitates (after weighing) for freedom from constituents which should have been separated.

(*d*) Detection of traces of metallic impurities or constituents in inorganic residues and powders; in organic substances (foodstuffs, textiles, etc.); in vitreous substances (glasses and slags); and in refractories and clays.

(*e*) Testing the purity of analytical reagents.

---

* With R.U. powder two lines may be acceptable for some elements; these include B, Cu, Au, P, Ag, and Cs.

(*f*) Analysis of substances of which only small quantities are available.

(*g*) Detection of rare or trace metals in minerals.

XXI, 4.   QUANTITATIVE SPECTROGRAPHIC ANALYSIS.   If the excitation conditions are kept constant and the sample composition is varied over a narrow range, the energy emitted for a given spectral line of an element is proportional to the number of atoms that are excited and thus to the concentration of the element in the sample. The energy emitted (i.e., the intensity of the light) is usually measured by the photographic method: the concentration of the unknown is determined from the blackening of the plate for certain lines in the spectrum. The quantitative determination of the blackening of the individual lines is made with a microphotometer (Fig. XXI, 8). Measurements are made of the light transmitted by the line in question (i) and the light transmitted by the clear portion of the plate (i_0), the density D (strictly the density of blackening, also represented by B) may be defined by the expression $D = \log_{10}(i_0/i)$. It is assumed that the galvanometer deflection obtained on the microphotometer is directly proportional to the light falling on the photocell.

The density of the image of the spectral line should ideally be proportional to the concentration of the corresponding element in the sample if the exposure time and conditions of excitation, etc., are held constant. This is often not strictly true; hence, wherever possible, it is desirable to photograph spectra of several samples of varying known composition on the same plate with the unknown. The unknown may then be evaluated by interpolation on a graph of density against concentration. The fact that the unknown and all the standards are on the same plate prevents errors due to differences in sensitivity between plates as well as those due to differences in the time or temperature of photographic processing.

Since the intensity of the lines is ultimately registered on a photographic plate, a brief discussion of the nature of the photographic process is desirable. If one plots the density as a function of the logarithm of the exposure,* a curve such as is shown in Fig. XXI, 10, results. A certain threshold exposure, denoted by A, is necessary before an image is produced. It will be noted that there is a region BC over which the density is proportional to the logarithm of the exposure; this is the useful range of the plate. The slope of this linear portion of the curve is known as the **gamma** (or contrast) of the emulsion: $\gamma = \tan\theta$. Emulsions with a high gamma give images with strong contrast because a small difference in exposure causes a large variation in density: low values of gamma indicate low contrast. The point D (the intercept of BC on the horizontal axis) measures the inertia of the emulsion: its reciprocal is related to the 'speed' of the emulsion. For our purpose the speed is an approximate measure of the minimum amount of light required to produce a useful image. The slope of the characteristic curve of an emulsion varies from emulsion to emulsion, with wavelength, and with the conditions of excitation and of development. In order to determine the curve for any given emulsion, the conditions of excitation (shape of the ends of the electrodes, their distance apart, the electrical circuit, etc.) and also the conditions of development (the type of developer, temperature, and time of development) must be standardised. In selecting a plate on which to photograph the spectrum,

---

* The exposure of the photographic plate is defined as the product of the intensity of the light incident on the undeveloped plate and the time for which it is acting.

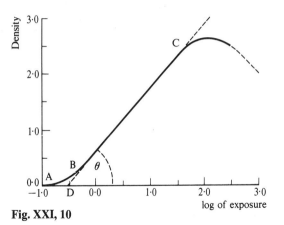

Fig. XXI, 10

one decides first whether a fast emulsion is needed on the basis of the light intensity available and the permissible time of exposure. If very faint spectrum lines are to be detected, high sensitivity at low intensities is needed, which suggests the use of a fast plate. To reproduce both weak and strong spectrum lines on the same spectrogram with correct indication of their relative intensities, medium contrast is needed. For sharp spectrum lines with a clear background, a plate of high contrast is used. Slow contrast plates exhibit high resolving power: such plates are of greatest value for use with spectrographs having low dispersion with high resolving power as is often the case with certain prism instruments of short focus. The most common type of plate used for emission spectrography has a gamma of about 1.

We may now deal with some of the procedures employed in quantitative spectrographic analysis. In the **comparison sample method**, the spectrum of an unknown sample is compared with the spectra of a range of samples of known composition (e.g., those supplied by the US Bureau of Standards) with respect to a particular component or components. The spectra of the unknown and of the various standards are photographed on the same plate under the same conditions. The concentrations of the desired constituent can then be estimated by comparing the blackening of the lines of the particular constituent with the same lines on the standards; visual or photometric comparison of blackening may be used.

In the **internal standard method** the intensity of the unknown line is measured relative to that of an internal standard line. The internal standard line may be a weak line of the main constituent. Alternatively, it may be a strong line of an element known not to be present in the sample and furnished by adding a fixed small amount of a compound of the element in question to the sample. The ratios of the intensities of these lines—the unknown line and the internal standard line—will be unaffected by the exposure and development conditions. This method will provide lines of suitable wavelength and intensity by variations of the added element and the amount added, due regard being paid to the relative volatility of the selected internal standard element. It is important to use as internal standard pairs only those lines of which the relative intensities are insensitive to variations in excitation conditions. The line selected as standard should have a wavelength close to that of the unknown and should, if possible, have roughly the same intensity.

For initial experience in quantitative spectrographic analysis, procedures involving solutions of materials are obviously attractive. The use of solutions has the advantage that the constituents are uniformly distributed but the disadvantage that it is not easy to ensure reproducible conditions for bringing the solution into the light source (arc or spark). One procedure for obtaining spectra from solutions is to add a small amount to spectroscopically pure carbon or graphite electrodes and to arc or spark them after an initial drying period. The solution technique allows a simple preparation of synthetic standards using spectroscopically pure compounds, but it has certain defects: thus elements, such as silicon and tungsten, are difficult to keep in solution. The sensitivity of solution methods is generally lower than that of other techniques.

XXI, 5. DIRECT READING INSTRUMENTS. A detailed discussion of direct reading instruments is beyond the scope of this book as the apparatus is expensive and only available in a few laboratories. However, a brief outline of the principles of direct reading emission spectroscopy will be given.

A diagram of the light path in the **Rank Hilger E950 Polyvac Direct Reading Spectrometer** is shown in Fig. XXI, 11. In this instrument the radiation is dispersed by the holographic grating and the component wavelengths reach a series of exit slits which isolate the selected emission lines for specific elements. The light from each exit slit is directed to fall on the cathode of a photomultiplier tube, one for each spectral line isolated. The light falling on the photomultiplier gives an output which is integrated on a capacitor, thus the resulting voltage is a function of the amount of element present in the sample. A calibration curve of element concentration against capacitor voltage reading can be constructed.

Highly advanced direct reading instruments such as the **Rank Hilger Polyvac E1000** offer a range of gratings, so allowing a wide choice of spectral wavelengths, from a 160 nm to 864 nm (1600–8640 Å). The complete apparatus (with the exception of the excitation source) is enclosed in a vacuum chamber in order to permit adequate transmission at wavelengths below 200 nm (2000 Å) and to avoid interference from the band due to molecular oxygen at 180 nm (1800 Å). Thus elements such as carbon and sulphur, which give rise to emission lines in the vacuum ultraviolet region, may be determined.

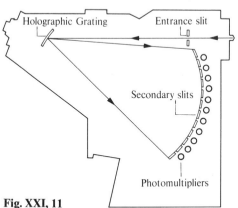

Fig. XXI, 11

The arc or spark excitation discharge is carried out in an argon atmosphere to avoid spectral interference from the components of air. The main advantage of direct reading instruments is that when used in conjunction with an on-line computer they provide a method for the rapid simultaneous analysis of elements with a better precision and accuracy than can be normally obtained from the spectrograph using a photographic plate. Thus results for 25 elements may be obtained within a short time of about 1–2 minutes.

Experimental

XXI, 6. QUALITATIVE SPECTROGRAPHIC ANALYSIS* OF (A) A NON-FERROUS ALLOY AND (B) A COMPLEX INORGANIC MIXTURE. *General discussion.* One procedure for the identification of an alloy is to measure the wavelength values of the observed lines and compare these with the recorded data on known elements (see table of persistent lines in Appendix XIV). A wavelength scale, which has been calibrated with the spectrograph, is photographically reproduced on the plate: this is, of course, only useful as a guide, since the wavelengths cannot be read with sufficient accuracy. A simple method for use with brass, or most other simple non-ferrous alloys, is to determine the spectra with pure samples of the component metals and to compare the spectra by a projection method (see Fig. XXI, 6). At least three persistent lines must be present for positive identification. The spark technique may be used for metals and alloys, but is not altogether satisfactory for powders, including R.U. powder. The d.c. carbon arc is preferred for the qualitative analysis of powders, and it also gives good results for most alloys: low-melting alloys (e.g., Wood's metal) may be dissolved in nitric acid, evaporated to dryness, then evaporated with concentrated sulphuric acid, and the dry sulphated residue employed in the carbon arc.

To gain experience in qualitative analysis, full details will be given for the analysis of brass and an artificial seven radical inorganic mixture.

Adjustment of the optical system. The condensing lens is set between the light source and the slit of the spectrograph so that the beam of light from the d.c. arc source passing through the slit forms a real image at the collimating lens (compare Fig. XXI, 1) of the spectrograph. The adjustments for a Hilger and Watts Medium Spectrograph will be evident from Fig. XXI, 12.

1. Place the Gramont stand so that the electrodes are 38 cm from the jaws of the slit and align them for height; the gap between the electrodes to be 4 mm.

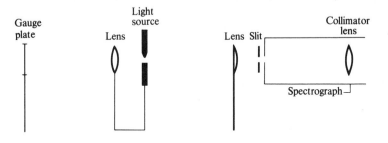

Fig. XXI, 12

2. Set the condensing lens in position at 2 cm from the slit jaws and at the correct height.

3. Set the gauge plate on its stand at the end of the bar at about 23 cm from the

---

* The experiments in qualitative and quantitative analysis described utilise a Hilger and Watts medium spectrograph. They can easily be adapted to other similar spectrographs with the aid of the instruction manuals supplied by the manufacturers of the instruments.

lens on the Gramont stand. An image of the light source will now be seen on the gauge plate, and the height should be set so that the image agrees with the two outer horizontal lines on the gauge. If a reading lamp is so arranged that it can be brought into position near the electrodes on the slit side of the stand it will be found that an image of the electrodes can be seen on the screen of the gauge plate.

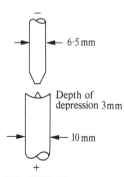

6·5 mm

Depth of depression 3 mm

10 mm

Fig. XXI, 13

Thus for any subsequent work the electrodes can be set in optical alignment with the spectrograph without actually passing any current between them. The two outer horizontal lines of the gauge are equivalent to an elelctrode gap of 4 mm and the inner lines to a 2-mm gap.

d.c. arc.   A 230-volt arc at 4 amperes is suitable for qualitative analysis (see Fig. XXI, 3). The arc gap may be 2 mm and the slit width 0.02 mm.

Electrodes for d.c. arc.   The two electrodes are shown in Fig. XXI, 13. They are conveniently shaped on a lathe from graphite electrodes (Johnson, Matthey; 30 cm long; JM 3B, 10 mm diameter; JM 4B, 6.5 mm diameter). The maximum depression on the lower electrode is 3 mm: the small projection in the centre helps to ensure that the arc passes between it and the upper electrode and does not 'wander' appreciably to the edges of the electrode. A small quantity (about 20 mg) of the alloy or powder is placed on the lower electrode.

Photographic details.   *Dark room.*   A fully equipped dark room is desirable. Ideally the spectrograph should be set up in the dark room. The student should become familiar with its facilities—stainless-steel trays (26 cm × 21 cm), running water, safelights, etc.

Plates.   Ilford R.40 or Kodak V-F. Charge holder in dark room.

Photographic developer.   Ilford ID-2* or equivalent. Dilute 1 volume with 2 volumes of water.

Photographic fixer.   Kodak tropical acid hardening-fixing bath F5† or equivalent.

Procedure.   (a) Load the plate holder with plate, making certain that the sensitised side is placed face down in the open holder. Return the holder to the spectrograph.

Withdraw the safety slide. Expose for several seconds to obtain the wavelength scale in the upper part of the plate, replace slide.

(b) Charge the lower electrode (anode) with about 20 mg of brass or the inorganic mixture.

(c) Withdraw the safety slide which covers the plate. Strike arc between the electrodes. Make an exposure of 6 seconds by opening the shutter of the spectrograph with the Hartmann diaphragm in position. (The best exposure time is evaluated by making a number of consecutive exposures, the plate being lowered to the next position after each exposure.)

---

* This has the following composition: metol, 2 g; sodium sulphite (anhydrous) 75 g; hydroquinone, 8 g; sodium carbonate (anhydrous), 37.5 g; potassium bromide, 2 g; water to make 1 dm³. Dissolve chemicals in order given. (*Ilford Technical Information Book*, Sheet D.20.1, Vol. III, 1971 issue.)

† This has the following composition: sodium thiosulphate, 240.0 g; sodium sulphite (crystals), 30.0 g; acetic acid (glacial), 17.0 cm³; boric acid 7.5 g; potash alum 15.0 g; water to make 1 dm³. Dissolve chemicals in order given. (*Kodak Data Book on Photography*. Data Sheet W.18.)

CAUTION: do not look directly at the arc unless wearing the special dark goggles or dark glasses provided.

(*d*) Replace the safety slide. Turn off the current to the arc and allow to cool. Insert fresh graphite electrodes. Charge the lower electrode with 20 mg of R.U. powder. Move the Hartmann diaphragm to the second position. Strike the arc again and expose for 5 seconds. (The best exposure is determined as above.)

(*e*) If brass was used in the first exposure, change the electrodes, charge the lower electrode with about 20 mg of the inorganic seven radical mixture (say, a mixture of equal weights of magnesium phosphate, zinc borate, copper arsenate, and cadmium oxide), set the Hartmann diaphragm in position, and expose for 5 seconds.

(*f*) Develop the plate in the usual manner in the dark room, say, for 9 minutes with Ilford ID-2 developer at 18 °C, then rinse it rapidly with water, and place it in the fixer (Kodak F5) for *ca.* 20 minutes at 18 °C. Remove the plate from the fixer, wash it with running water for at least 30 minutes, wash with distilled water, and dry in the air.

(*g*) Identify the lines in the spectra of the brass and the inorganic mixture by comparison with those on the spectrum of the R.U. powder with the aid of the Rank Hilger projection comparator (Fig. XXI, 6). Use the series of enlarged photographs of the spectra of R.U. powder for exact identification.

The following are some of the lines which should be identified (wavelengths are given to the nearest Ångstrom):

Brass.
Cu: 3248, 3274.
Zn: 3282, 3303, 3346, 4680, 4722, 4811.
Pb: 2614, 2833, 4058.
Sn: 2707, 2840, 3175.
Fe: 2483, 2488, 2599, 2973, 2984.
Mn: 2795, 2798, and 'triplet' 4031, 4033, 4035.
Ni: 3002, 3415, 3446.
P: 2536, 2555.

Inorganic mixture containing Cd, Zn, Cu, Mg, phosphate, arsenate, borate.
Cd: 2288, 3261, 3404, 3466, 5086.
Zn: 3282, 3303, 3346, 4680, 4722, 4811.
Cu: 3248, 3274.
Mg: 2796, 2803, 2852, and 'quintet' 2777, 2778, 2780, 2781, 2783.
P: 2534, 2536, 2553, 2555.
As: 2350, 2369, 2457.
B: 2497, 2498.

XXI, 7. DETERMINATION OF LEAD IN BRASS. This experiment has been designed with the following objects: (i) to illustrate the use of an internal standard in quantitative spectrographic analysis, (ii) to give the student experience in the use of a non-recording microphotometer, and (iii) to determine an element present in an alloy at as high a proportion as can normally be evaluated by spectrographic analysis and which can also be checked by a purely chemical method. Once this experience has been acquired, the student should have no difficulty in determining minor constituents in metals and non-ferrous alloys provided a suitable internal standard can be found, e.g., lead, cadmium, and copper in zinc using selected zinc lines as internal standards.

Preparation of solutions. Use spectroscopically standardised substances (Johnson, Matthey) throughout, and also analytical reagent acids.

Copper solution. Dissolve about 8.81 g, accurately weighed, of copper* sheet in 1:1-nitric acid (about $60 \, \text{cm}^3$) and dilute to $100 \, \text{cm}^3$ with distilled water in a graduated flask.

Magnesium solution. Dissolve about 0.88 g, accurately weighed, of magnesium in 10 per cent nitric acid (1:9 v/v), and dilute to $500 \, \text{cm}^3$ in a graduated flask with 10 per cent nitric acid.

Lead solution. Dissolve about 1.00 g, accurately weighed, of lead nitrate in the minimum volume of water and dilute to $250 \, \text{cm}^3$ in a graduated flask with 10 per cent nitric acid.

Standard solutions. Prepare the standard solutions by mixing the following volumes of each of the above solutions and diluting each to $25 \, \text{cm}^3$ with distilled water in a graduated flask.

Standard solution	Volumes of solutions, cm^3			Pb concentration,* mg per cm^3
	Cu	Mg	Pb	
1	10.00	5.00	1.50	0.1506
2	10.00	5.00	2.00	0.2008
3	10.00	5.00	3.00	0.3012
4	10.00	5.00	4.00	0.4016
5	10.00	5.00	5.00	0.5020

* These figures apply to 8.8150 g Cu, 0.8861 g Mg, and 1.0088 g $Pb(NO_3)_2$.

Brass solution. Weigh out accurately about 5.0 g of a standard brass containing about 2 per cent Pb, treat with 1:1-nitric acid (about $40 \, \text{cm}^3$) until it is completely attacked, and warm on a water bath for 30 minutes. Dilute with a little water, filter (Whatman paper No. 541), and wash the residue with three $10 \, \text{cm}^3$ portions of hot distilled water. Transfer the combined filtrate and washings to a $100 \, \text{cm}^3$ graduated flask and dilute to the mark with 10 per cent nitric acid.

In order to test the method over a fairly wide concentration range, the brass solution (prepared as above) may be diluted with the standard stock solutions as follows. The final volume is made up with water.

	Brass solution, cm^3	Cu solution, cm^3	Mg solution, cm^3	Total final volume, cm^3
Solution 6	5.00	8.00	5.00	25.0
Solution 7	5.00	6.00	4.00	20.0
Solution 8	5.00	4.00	3.00	15.0

These diluted brass solutions contain $0.2–0.4 \, \text{mg}$ of Pb cm^{-3}; the copper concentration of each solution is about the same ($35.3 \, \text{mg cm}^{-3}$).

---

* The concentration of the main element in the standard and unknown solutions should be approximately equal. The concentration of the internal standard should be constant in all the solutions.

Apparatus. Using the Hilger medium spectrograph, set up the optical system as in Fig. XXI, 14. Place the short focus quartz lens B so that an image of

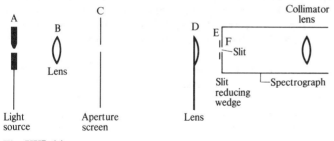

Fig. XXI, 14

the electrode A appears on the aperture screen C; the last-named is a vertical metal screen provided with a horizontal slit about 4 mm wide. Arrange the arc gap (*ca.* 5 mm) so that the image of the tips of the electrodes lies just outside the gap in the screen; if desired, marks may be made on the screen so as to ensure a constant arc gap. By this means radiation from the incandescent electrodes can be excluded from the spectrograph and the entire positioning of the arc image viewed and controlled during the exposure; the 'background' on the spectrum caused by incandescence at the tips of the electrodes is largely eliminated. Place the longer focus lens D close to the slit F: an image of the aperture is focused by the collimator lens on the prism.

The **slit-reducing wedge and Hartmann diaphragm** E is shown in Fig. XXI, 15. This consists of a steel slide which fits into grooves in the face of the slit. Movement of the wedge controls the length of the slit. By means of the Hartmann diaphragm, generally used for qualitative analysis (see Section **XXI, 6**), several

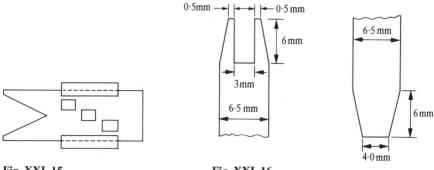

Fig. XXI, 15   **Fig. XXI, 16**

spectrograms can be recorded on the same negative without moving the plate holder. When the slide is moved horizontally any of several apertures allows the light from the source to strike the desired portion of the slit; by appropriate positioning of this device, comparison spectra above and below the unknown can be photographed.

Shape the graphite electrodes (Johnson, Matthey, 4B) as shown in Fig. XXI, 16, with the aid of a lathe. Produce the d.c. arc as described in Section **XXI, 6**.

Use Ilford N30, Kodak III-0 or Agfa 34.B.50 plates. The other photographic details are as in the previous section.

Procedure.   By means of a micro-pipette, transfer $0.05 \, cm^3$ of each solution into the cavity of an electrode: when this volume has been absorbed introduce a further $0.05 \, cm^3$. Use a separate electrode for each solution. When absorption in the walls of the crater is complete, dry the electrodes in an oven at 110° for 30–60 minutes.

Mount a pair of electrodes in the arc stand: the standard solution is absorbed in the lower (positive) electrode. Strike the arc, and photograph the spectrum for an appropriate time immediately the arc is struck. Allow the electrodes to cool, change both the upper and lower electrodes (the latter charged with another standard solution), and repeat the exposure using another portion of the same plate. Repeat with the remaining standard solutions and the various 'unknown' brass solutions.

Develop the plate using ID-2 developer for 3.5 minutes at 16 °C, fix using Kodak tropical acid hardening-fixing bath F5 (or equivalent) for 10–15 minutes, wash for 10–15 minutes, rinse with distilled water, and allow to dry.

Employ a Hilger non-recording microphotometer (Fig. XXI, 8) to determine the blackening of the appropriate lines on the spectrogram with due regard to the following points:

(*a*) The emulsion side of the plate must face the light source.

(*b*) The slit width should be just less than ten times the slit width of the spectrograph (a slit width of 0.12 mm was found to be satisfactory).

(*c*) The deflection of the galvanometer should lie between 5.0 and 25.0.

(*d*) The reference lines should lie close together, e.g., Mg 2776.7 or 2783.0 Å and Pb 2873.3 Å.

(*e*) The clear portion of the plate to be illuminated should be as near as possible to the line to be measured.

Tabulate the results in the form shown below for some typical values obtained for this determination. Plot log ('Density' Pb/'Density' Mg) against log (Concentration of Pb × 10). From the resulting straight-line plot, evaluate the lead concentrations for solutions 6, 7, and 8 (these results are marked with an asterisk in the table). The weights of brass per cm^3 in solutions 6, 7, and 8 are known, and hence the percentage of Pb in the samples of brass can be calculated.

Typical results for the spectrographic analysis for lead (internal standard: magnesium)

Solution number	Densitometer reading		$\log\left(\dfrac{\text{'Density' Pb}}{\text{'Density' Mg}}\right)$	Concn. of Pb, mg cm^{-3}	Log (concn. of Pb × 10)
	Mg	Pb			
1	7.3	17.5	0.380	0.1506	0.178
2	8.7	18.5	0.328	0.2008	0.303
3	7.3	11.0	0.178	0.3012	0.479
4	10.3	12.0	0.066	0.4016	0.604
5	11.6	10.4	− 0.047	0.5020	0.700
6	8.8	15.5	0.246	0.251*	0.400*
7	9.2	12.5	0.133	0.335*	0.525*
8	10.7	12.2	0.057	0.409*	0.612*

The results were: for solution 6, 2.51 per cent; for solution 7, 2.67 per cent; for solution 8, 2.45 per cent, thus giving a mean Pb content of 2.54 per cent with a standard deviation ± 0.12 per cent. The true value for the sample of brass used was 2.52 per cent.

XXI, 8. DETERMINATION OF COPPER AND LEAD IN WHITE METAL.

This experiment has been designed for students with the following primary objects: (i) to illustrate the use of internal standards in the quantitative spectrographic analysis of an alloy for *two* elements, and (ii) to provide experience in the use of a non-recording microphotometer. The alloy selected (white metal) is readily available, and the two elements (copper and lead) can be easily determined by purely chemical methods. It is appreciated that the percentages of the two elements (about 4 per cent each) present in the alloy are very much higher than would normally be determined by spectrographic methods (which are generally confined to percentages less than 1 per cent), nevertheless the experiment will indicate the upper limits possible and the accuracy attainable; the latter may be regarded as secondary objects of the exercise. With the experience so gained, the student should be able to adapt the procedure to the determination of two or more elements present in proportions (< 1 per cent) for which spectrographic techniques are eminently suitable. The approximate proportions of the elements to be determined must be known in order to prepare the standard solutions.

Preparation of solutions. Use spectroscopically standardised substances (Johnson, Matthey) throughout; also analytical reagent acids.

Copper solution. Dissolve about 1.62 g, accurately weighed, of copper in a mixture of about 50 cm^3 concentrated hydrochloric acid and $5-10 \text{ cm}^3$ concentrated nitric acid, boil the solution for a few minutes, cool, and dilute to 500 cm^3 with hydrochloric acid (3.5:1, v/v).

Magnesium solution. Dissolve about 1.05 g, accurately weighed, of magnesium in 50 cm^3 concentrated hydrochloric acid and dilute to 500 cm^3 with hydrochloric acid as above.

Tin solution. Dissolve about 10.5 g, accurately weighed, of tin in 60 cm^3 concentrated hydrochloric acid, and dilute to 100 cm^3 with hydrochloric acid (1:1, v/v).

Lead solution. Dissolve about 2.24 g, accurately weighed, of lead nitrate in 10 cm^3 water and add about 100 cm^3 concentrated hydrochloric acid: boil until the lead chloride dissolves (5–10 minutes), cool, and dilute to 500 cm^3 with concentrated hydrochloric acid.

Standard solutions. Prepare the standard solutions by mixing the following volumes of each of the above solutions and diluting each to 25 cm^3 with concentrated hydrochloric acid in a graduated flask.

Standard solution	Volumes of solutions, cm^3				Cu concentration, mg cm^{-3}*	Pb concentration, mg cm^{-3}*
	Sn	Mg	Pb	Cu		
1	10.00	5.00	1.00	1.00	0.13	0.11
2	10.00	5.00	2.00	2.00	0.26	0.22
3	10.00	5.00	3.00	3.00	0.39	0.34
4	10.00	5.00	4.00	4.00	0.53	0.45
5	10.00	5.00	5.00	5.00	0.66	0.56

* These are approximate values; the exact concentrations will depend upon the weights used in the preparation of the various solutions.

White metal solution. Weigh out accurately about 5.18 g of the white metal alloy, and add about 50 cm³ concentrated hydrochloric acid (vigorous reaction). Treat the suspension dropwise with concentrated nitric acid until a clear green solution results, boil gently to remove nitrous fumes, cool, and make up to 100 cm³ with concentrated hydrochloric acid in a graduated flask. Dilute the alloy solution with known amounts of tin solution so that the resulting solutions contain 0.30–0.60 mg copper cm⁻³ and 0.25–0.55 mg lead cm⁻³: magnesium solution must be added in amounts to ensure that its concentration is almost identical with that in the standard solutions. The final volumes are made up with concentrated hydrochloric acid added from a burette.

	White metal solution, cm³	Sn solution, cm³	Mg solution, cm³	Total final volume, cm³
Solution 6	5.00	6.00	4.00	20.0
Solution 7	5.00	8.00	5.00	25.0
Solution 8	5.00	10.00	6.00	30.0

Procedure. Follow the method given in Section **XXI, 7**. Introduce three separate portions of 0.05 cm³ into the electrodes.

Use the following reference lines: Mg, 2776.7 Å; Cu, 2824.4 Å; and Pb, 2873.3 Å. Measure the densities of the lines with the Hilger non-recording microphotometer.

Plot log ('Density' Pb/'Density' Mg) against log (Concentration of Pb × 10), and log ('Density' Cu/'Density' Mg) against log (Concentration of Cu × 10). Evaluate the Pb and Cu concentrations for Solutions 6, 7, and 8, and thence the corresponding percentages of Pb and Cu in the sample of white metal. Some typical results for Pb were 3.74, 3.60, and 3.84 per cent.

XXI, 9. Selected bibliography

1. W. R. Brode (1943). *Chemical Spectroscopy.* 2nd edn. New York; John Wiley.
2. S. Judd Lewis (1946). *Spectroscopy in Science and Industry.* 2nd edn. London; Blackie.
3. L. N. Ahrens and S. R. Taylor (1960). *Spectrochemical Analysis.* 2nd edn. Reading, Mass.; Addison-Wesley.
4. J. Sherman (1960). *Emission Spectrography,* in W. G. Berl, *Physical Methods of Chemical Analysis.* Vol. 1. 2nd edn. New York; Academic Press.
5. F. Twyman (1951). *Metal Spectroscopy.* London; Charles Griffin.
6. C. R. N. Strouts, H. N. Wilson and T. R. Parry-Jones (1962). *Chemical analysis. The Working Tools.* London; Oxford University Press.
7. G. W. Ewing (1975). *Instrumental Methods of Chemical Analysis.* 4th edn. New York; McGraw-Hill Book Co.
8. H. A. Strobel (1973). *Chemical Instrumentation. A Systematic Approach to Instrumental Analysis.* 2nd edn. Reading, Mass.; Addison-Wesley Publishing Co.
9. A. N. Zaidel *et al.* (1970). *Tables of Spectral Lines.* 3rd edn. New York; Plenum Publishing Co.
10. M. Slavin (1971). *Emission Spectrochemical Analysis* (Chemical Analysis Series). New York; Wiley-Interscience.

11. H. H. Willard, L. L. Merritt and J. A. Dean (1974). *Instrumental Methods of Analysis.* 5th edn. New York; Van Nostrand-Reinhold.
12. I. M. Kolthoff and P. J. Elving (eds.) (1965). *Treatise on Analytical Chemistry*, Part 1, Vol. 6, Ch. 64. B. F. Scribner and M. Margoshes, *Emission Spectroscopy.* New York; Interscience.

CHAPTER XXII FLAME SPECTROMETRY

XXII, 1. GENERAL DISCUSSION. If a solution containing a metallic salt (or some other metallic compound) is aspirated into a flame (e.g., of acetylene burning in air), a vapour which contains atoms of the metal may be formed. Some of these gaseous metal atoms may be raised to an energy level which is sufficiently high to permit the emission of radiation characteristic of that metal; e.g., the characteristic yellow colour imparted to flames by compounds of sodium. This is the basis of **flame emission spectroscopy (FES)** which was formerly referred to as **flame photometry**. However, a much larger number of the gaseous metal atoms will normally remain in an unexcited state or, in other words, in the ground state. These ground state atoms are capable of absorbing radiant energy of their own specific resonance wavelength, which in general is the wavelength of the radiation that the atoms would emit if excited from the ground state. Hence if light of the resonance wavelength is passed through a flame containing the atoms in question, then part of the light will be absorbed, and the extent of absorption will be proportional to the number of ground state atoms present in the flame. This is the underlying principle of **atomic absorption spectroscopy (AAS). Atomic fluorescence spectroscopy (AFS)** is based on the re-emission of absorbed energy by free atoms.

The procedure by which gaseous metal atoms are produced in the flame may be summarised as follows. When a solution containing a suitable compound of the

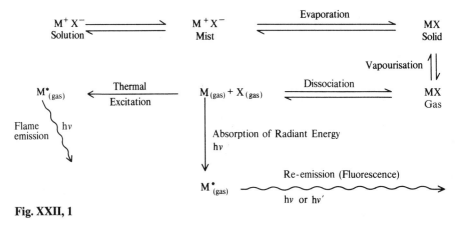

Fig. XXII, 1

metal to be investigated is aspirated into a flame, the following events occur in rapid succession:

1. evaporation of solvent leaving a solid residue;
2. vaporisation of the solid with dissociation into its constituent atoms, which initially, will be in the ground state;
3. some atoms may be excited by the thermal energy of the flame to higher energy levels, and attain a condition in which they radiate energy.

The resulting emission spectrum thus consists of lines originating from excited atoms or ions. These processes are conveniently represented diagrammatically as in Fig. XXII, 1.

XXII, 2. ELEMENTARY THEORY. Consider the simplified energy level diagram shown in Fig. XXII, 2, where E_0 represents the ground state in which the electrons of a given atom are at their lowest energy level and E_1, E_2, E_3, etc., represent higher or excited energy levels.

Transitions between two quantised energy levels, say from E_0 to E_1, correspond to the absorption of radiant energy, and the amount of energy absorbed (ΔE) is determined by Bohr's equation

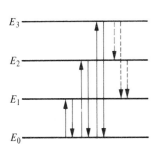

Fig. XXII, 2

$$\Delta E = E_1 - E_0 = h\nu = hc/\lambda$$

where c is the velocity of light, h is Planck's constant, and ν is the frequency and λ the wavelength of the radiation absorbed. Clearly, the transition from E_1 to E_0 corresponds to the *emission* of radiation of frequency ν.

Since an atom of a given element gives rise to a definite, characteristic line spectrum, it follows that there are different excitation states associated with different elements. The consequent emission spectra involve not only transitions from excited states to the ground state, e.g., E_3 to E_0, E_2 to E_0 (indicated by the bold lines in Fig. XXII, 2), but also transitions such as E_3 to E_2, E_3 to E_1, etc. (indicated by the dotted lines). Thus it follows that the emission spectrum of a given element may be quite complex. In theory it is also possible for absorption of radiation by already excited states to occur, e.g., E_1 to E_2, E_2 to E_3, etc., but in practice the ratio of excited to ground state atoms is extremely small, and thus the absorption spectrum of a given element is usually only associated with transitions from the ground state to higher energy states and is consequently much simpler in character than the emission spectrum.

The relationship between the ground state and excited state populations is given by the Boltzmann equation

$$N_1/N_0 = (g_1/g_0)e^{-\Delta E/kT}$$

where
N_1 = number of atoms in the excited state,
N_0 = number of ground state atoms,
g_1/g_0 = ratio of statistical weights for ground and excited states,
ΔE = energy of excitation = $h\nu$,
k = the Boltzmann constant,
T = the temperature in Kelvin.

It can be seen from this equation that the ratio N_1/N_0 is dependent upon both the excitation energy ΔE and the temperature T. An increase in temperature and a decrease in ΔE (i.e., when dealing with transitions which occur at longer wavelengths) will both result in a higher value for the ratio N_1/N_0.

Calculation shows that only a small fraction of the atoms are excited, even under the most favourable conditions, i.e., when the temperature is high and the excitation energy low. This is illustrated by the data in Table XXII, 1 for some typical resonance lines.

Table XXII, 1 **Variation of atomic excitation with wavelength and with temperature**

Element	Wavelength (nm)	N_1/N_0	
		2000 K	4000 K
Na	589.0	9.86×10^{-6}	4.44×10^{-3}
Ca	422.7	1.21×10^{-7}	6.03×10^{-4}
Zn	213.9	7.31×10^{-15}	1.48×10^{-7}

Since as already explained the absorption spectra of most elements are simple in character as compared with the emission spectra, it follows that atomic absorption spectroscopy is less prone to inter-element interferences than is flame emission spectroscopy. Further, in view of the high proportion of ground state to excited atoms, it would appear that atomic absorption spectroscopy should also be more sensitive that flame emission spectroscopy. However, in this respect the wavelength of the resonance line is a critical factor, and elements whose resonance lines are associated with relatively low energy values are more sensitive as far as flame emission spectroscopy is concerned than those whose resonance lines are associated with higher energy values. Thus sodium with an emission line of wavelength 589.0 nm shows great sensitivity in flame emission spectroscopy, whereas zinc (emission line wavelength 213.9 nm) is relatively insensitive.

The integrated absorption is given by the expression

$$K\,dv = fN_0(\pi e^2/mc)$$

where K is the absorption coefficient at frequency v,
        e is the electronic charge,
       m the mass of an electron,
        c the velocity of light,
        f the oscillator strength of the absorbing line (this is inversely proportional to the lifetime of the excited state),
      N_0 is the number of metal atoms per cm^3 capable of absorbing the radiation.

In this expression the only variable is N_0 and it is this which governs the extent of absorption. Thus it follows that the integrated absorption coefficient is directly proportional to the concentration of the absorbing species.

It would appear that measurement of the integrated absorption coefficient should furnish an ideal method of quantitative analysis. In practice, however, the absolute measurement of the absorption coefficients of atomic spectral lines is extremely difficult. The natural line width of an atomic spectral line is about 10^{-5} nm, but owing to the influence of Doppler and pressure effects, the line is broadened to about 0.002 nm at flame temperatures of 2000–3000 K. To measure the absorption coefficient of a line thus broadened would require a spectrometer with a resolving power of 500 000. This difficulty was overcome by Walsh (Ref. 1), who used a source of sharp emission lines with a much smaller half width than the absorption line, and the radiation frequency of which is centred on the absorption frequency. In this way, the absorption coefficient at the centre of the line, K_{max}, may be measured. If the profile of the absorption line is assumed to be due only to Doppler broadening, then there is a relationship between K_{max} and N_0. Thus the only requirement of the spectrometer is that it shall be capable of isolating the required resonance line from all other lines emitted by the source.

It should be noted that in atomic absorption spectroscopy, as with molecular absorption, the absorbance A is given by the logarithmic ratio of the intensity of the incident light signal I_0 to that of the transmitted light I_t, i.e.,

$$A = \log I_0 / I_t = KLN_0$$

where N_0 is the concentration of atoms in the flame (number of atoms per cm^3);

            L is the path length through the flame (cm),
            K is a constant related to the absorption coefficient.

For small values of the absorbance, this is a linear function.

With flame emission spectroscopy, the detector response E is given by the expression

$$E = k\alpha c,$$

where k is related to a variety of factors including the efficiency of atomisation and of self absorption,

            α is the efficiency of atomic excitation,
            c is the concentration of the test solution.

It follows that any electrical method of increasing E, as for example, improved amplification, will make the technique more sensitive.

The basic equation for atomic fluorescence is given by

$$F = QI_0kc$$

where Q is the quantum efficiency of the atomic fluorescence process,

            I_0 is the intensity of the incident radiation,
            k is a constant which is governed by the efficiency of the atomisation process,
            c is the concentration of the element concerned in the test solution.

It follows that the more powerful the radiation source, the greater will be the sensitivity of the technique.

To summarise, in both atomic absorption spectroscopy and in atomic fluorescence spectroscopy, the factors which favour the production of gaseous atoms in the ground state determine the success of the techniques. In flame emission spectroscopy, there is an additional requirement, namely, the production of excited atoms in the vapour state. It should be noted that the conversion of the original solid MX into gaseous metal atoms (M_{gas}) will be governed by a variety of factors including the rate of vapourisation, flame composition and flame temperature, and further, if MX is replaced by a new solid, MY, then the formation of M_{gas} may proceed in a different manner, and with a different efficiency from that observed with MX.

XXII, 3. INSTRUMENTATION. The three flame spectrophotometric procedures require the following essential apparatus.

(*a*) For flame emission spectroscopy a **nebuliser-burner system** which produces gaseous metal atoms by using a suitable combustion flame involving a fuel gas–oxidant gas mixture is needed. Note however that with so-called non-flame cells, the burner is not required.

(*b*) A **spectrophotometer system** which includes a suitable optical train, a photosensitive detector and appropriate display device for the output from the detector.

(*c*) For both atomic absorption spectroscopy and atomic fluorescence spectroscopy, a **resonance line source** is required for each element to be determined: these line sources are usually modulated (see Section **XXII, 9**).

A schematic diagram showing the disposition of these essential components for the different techniques is given in Fig. XXII, 3. The components included

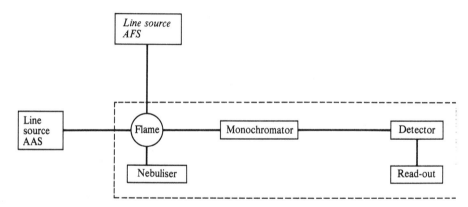

Fig. XXII, 3

within the frame drawn in dashed lines represent the apparatus required for flame emission spectroscopy. For atomic absorption spectroscopy and for atomic fluorescence spectroscopy there is the additional requirement of a resonance line source. In atomic absorption spectroscopy this source is placed in line with the detector, but in atomic fluorescence spectroscopy it is placed in a position at right angles to the detector as shown in the diagram. The essential components of the apparatus required for flame spectrophotometric techniques will be considered in detail in the following sections.

XXII, 4. COMBUSTION FLAMES. For flame spectroscopy an essential requirement is that the flame used shall produce temperatures in excess of 2000 K. In most cases this requirement can only be met by burning the fuel gas in an oxidant gas which is usually air, nitrous oxide, or oxygen diluted with either nitrogen or argon. The flame temperatures attained by the common fuel gases burning in (i) air and (ii) nitrous oxide are given in Table XXII, 2; the value given for town gas/air can only be regarded as approximate since it will depend upon the exact composition of the 'town gas'. The flow rates of both the fuel gas and the oxidant gas should be measured, for some flames are required to be rich in the fuel gas, whilst other flames should be lean in fuel gas: these requirements are discussed in Section **XXII, 20**. The concentration of gaseous atoms within the flame, both in the ground and in the excited states, may be influenced by (a) the flame composition, and by (b) the position considered within the flame.

Table XXII, 2 Flame temperatures with various fuels

Fuel gas	Temperature (T/K)	
	Air	Nitrous oxide
Acetylene	2400	3200
Hydrogen	2300	2900
Propane	2200	3000
Town gas	2100	—

As far as flame composition is concerned, it may be noted that an acetylene–air mixture is suitable for the determination of some thirty metals, but a propane–air flame is to be preferred for metals which are easily converted into an atomic vapour state. For metals such as aluminium and titanium which form refractory oxides, the higher temperature of the acetylene–nitrous oxide flame is essential, and the sensitivity is found to be enhanced if the flame is fuel rich.

With regard to position within the flame, it can be shown that in certain cases the concentration of atoms may vary widely if the flame is moved either vertically or laterally relative to the light path from the resonance line source. Rann and Hambly (Ref. 2) have shown that with certain metals (e.g., calcium and molybdenum), the region of maximum absorption is restricted to specific areas of the flame, whereas the absorption of silver atoms does not alter appreciably within the flame, and is unaffected by the fuel gas/oxidant gas ratio.

For the sake of brevity, the so-called 'cool flame' techniques based upon the use of an oxidant-lean flame such as hydrogen/nitrogen–air, have not been included. Details can however be found in Ref. 13, and it should also be noted that the experiment described in Section **XXII, 26** utilises a 'cool flame'.

XXII, 5. THE NEBULISER–BURNER SYSTEM. The purpose of the nebuliser-burner system is to convert the test solution to gaseous atoms as indicated in Fig. XXII, 1, and the success of flame photometric methods is dependent upon the correct functioning of the nebuliser-burner system. It should however, be noted that some flame photometers have a very simple burner system (see Section **XXII, 12**).

The function of the nebuliser is to produce a mist or aerosol of the test solution. The solution to be nebulised is drawn up a capillary tube by the Venturi action of a jet of air blowing across the top of the capillary; a gas flow at high pressure is necessary in order to produce a fine aerosol.

There are two main types of burner system: (a) the Pre-mix or Laminar-flow burner, and (b) the Total Consumption or Turbulent-flow burner. In the **Pre-mix type of burner**, the aerosol is produced in a vapourising chamber where the larger droplets of liquid fall out from the gas stream and are discharged to waste. The resulting fine mist is mixed with the fuel gas and the carrier (oxidant) gas, and the mixed gases then flow to the burner head. In atomic absorption spectroscopy the burner is a long horizontal tube with a narrow slit along its length. This produces a thin flame of long path length which can be turned into or away from the beam of radiant energy. The flame path of a burner using air–acetylene, air–propane or air–hydrogen mixtures is about 10–12 cm in length, but with a nitrous oxide–acetylene burner it is usually reduced to about 5 cm because of the higher burning velocity of this gas mixture. In addition to a long light path, this type of burner has the advantages of being quiet in action and with little danger of incrustation around the burner head since large droplets of solution have been eliminated from the stream of gas reaching the burner. Its disadvantages are (i) that with solutions made up in mixed solvents, the more volatile solvents are evaporated preferentially; (ii) a potential explosion hazard exists since the burner uses relatively large volumes of gas, but in modern versions of this type of burner this hazard is minimised.

A typical burner of this type is shown in Fig. XXII, 4. In this particular burner (Perkin–Elmer Corporation), the mixing chamber is a steel casting lined with a

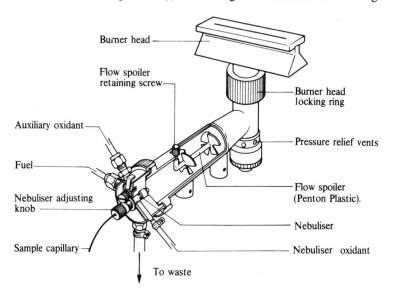

Fig. XXII, 4

plastic ('Penton') which is extremely resistant to corrosion. The burner head is manufactured from titanium, thus avoiding the occasional high readings which are encountered when solutions containing iron and copper in presence of acid are examined with burners having a stainless steel head. The nebuliser is capable of adjustment so that it can handle sample up-take rates of from 1–5 cm³ per minute. The burner can be adjusted in three directions, and horizontal and vertical scales are provided so that its position can be recorded. The head may be turned through an angle of 90° with respect to the light beam, and so the path length of the flame traversed by the resonance line radiation may be varied considerably: by choosing a small path length it becomes possible to analyse solutions of relatively high concentration without the need for prior dilution.

The **Total Consumption type of** burner consists of three concentric tubes as shown in Fig. XXII, 5. The sample solution is carried by a fine capillary tube A directly into the flame. The fuel gas and the oxidant gas are carried along separate tubes so that they only mix at the tip of the burner. Since all the liquid sample which is aspirated up the capillary tube reaches the flame, it would appear that this type of burner should be more efficient than the pre-mix type of burner. However, the total consumption burner gives a flame of relatively short path length, and hence such burners are predominantly used for flame emission studies. This type of burner has the advantages that (i) it is simple to manufacture, (ii) it allows a totally representative sample to reach the flame, and (iii) it is free from explosion hazards arising from unburnt gas mixtures. Its

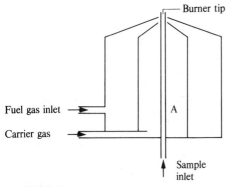

Fig. XXII, 5

disadvantages are that (i) the aspiration rate varies with different solvents, and (ii) there is a tendency for incrustations to form at the tip of the burner which can lead to variations in the signal recorded.

In general terms, Thomerson and Thompson (Ref. 3) have cited the following disadvantages of flame atomisation procedures:
1. Only 5–15 per cent of the nebulised sample reaches the flame (in the case of the pre-mix type of burner) and it is then further diluted by the fuel and oxidant gases so that the concentration of the test material in the flame may be extremely minute.
2. A minimum sample volume of between 0.5 and 1.0 cm^3 is needed to give a reliable reading by aspiration into a flame system.
3. Samples which are viscous (e.g., oils, blood, blood serum) require dilution with a solvent, or alternatively must be 'wet ashed' before the sample can be nebulised.

XXII, 6. NON-FLAME TECHNIQUES. Instead of employing the high temperature of a flame to bring about the production of atoms from the sample, it is possible in some cases to make use of either (*a*) non-flame methods involving the use of electrically heated graphite tubes or rods, or (*b*) vapour techniques. Procedures (*a*) and (*b*) both find applications in atomic absorption spectroscopy and in atomic fluorescence spectroscopy.

   (a) Electrothermal atomisers.   (i) The graphite tube furnace.   A diagram of a graphite tube furnace is shown in Fig. XXII, 6. It consists of a hollow graphite cylinder about 50 mm in length and about 9 mm internal diameter and so situated that the radiation beam passes along the axis of the tube. The graphite

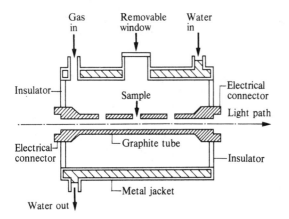

Fig. XXII, 6

tube is surrounded by a metal jacket through which water is circulated and which is separated from the graphite tube by a gas space. An inert gas, usually argon, is circulated in the gas space, and enters the graphite tube through openings in the cylinder wall.

The solution of the sample to be analysed (1–100 μl) is introduced by inserting the tip of a micro-pipette through a port in the outer (water) jacket, and into the gas inlet orifice in the centre of the graphite tube. The graphite cylinder is then heated by the passage of an electric current to a temperature which is high enough to evaporate the solvent from the solution. The current is then increased so that firstly the sample is ashed, and then ultimately it is vapourised so that metal atoms are produced, typically at a temperature of about 3 000 K. For reproducibility, the temperatures and the timing of the drying, ashing and atomisation processes must be carefully selected according to the metal which is to be determined. The absorption signals produced by this method may last for several seconds and can be recorded on a chart recorder. Each graphite tube can be used for 100–200 analyses depending upon the nature of the material to be determined.

   (ii) The graphite rod.   A graphite rod of 2 mm diameter was introduced by West (Ref. 4) as a means of producing atoms from the sample, and a commercial device is now available from Messrs Shandon Southern of Camberley, Surrey. The sample is placed upon the rod which is heated, typically by a current of 100 A from a low voltage (5 V) supply. The rod is placed just below the path of the beam from the radiation source so that vapour from the sample can move upwards into the beam and its absorbance be measured. The whole assembly is contained in a chamber fitted with quartz windows which is purged with argon.

   In some circumstances it is found advantageous to coat graphite rods (or tubes) with a layer of pyrolytic graphite: this leads to improved sensitivity with elements such as vanadium and titanium which are prone to carbide formation.

   The main advantages of flameless techniques is that very small samples (as low as 0.5 μl) can be analysed, and very little or no sample preparation is needed: in fact certain solid samples can be analysed without prior dissolution. It should however be appreciated that greater expertise is required for flameless techniques, and they should be regarded as complementary to the usual flame methods.

   Amongst other devices used to produce the required atoms in the vapour state are the **Delves Cup** which enables the rapid determination of lead in blood samples; the sample is placed in a small nickel cup which is then inserted directly into an acetylene-air flame. The **tantalum boat** is a similar device to the Delves cup; in this case the sample is placed in a small tantalum dish which is then inserted into an acetylene-air flame.

(b) **Vapour technique.** This procedure is strictly confined to the determination of mercury (Ref. 14), which in the elemental state has an appreciable vapour pressure at room temperature so that gaseous atoms exist without the need for any special treatment. As a method for determining mercury compounds, the procedure consists in the reduction of a solution of a mercury(II) compound with tin(II) chloride to form elemental mercury. A diagram of a suitable apparatus (the Rank Hilger H1469 Atomspek accessory) for the determination of mercury is shown in Fig. XXII, 7.

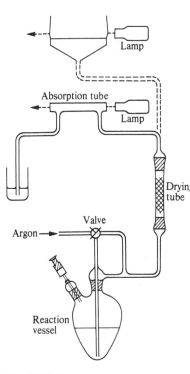

Fig. XXII, 7

This apparatus may also be adapted for the determination of arsenic, antimony, and selenium by conversion to their volatile hydrides by the use of sodium borohydride as reducing agent. In these cases the method differs from that for the determination of mercury since the hydrides thus formed cannot be examined directly in the absorption tube, but they are readily dissociated into atoms in an argon–hydrogen flame. The requisite additional apparatus is indicated by the dashed lines in Fig. XXII, 7.

XXII, 7. RESONANCE LINE SOURCES. As indicated in Fig. XXII, 3, for both atomic absorption spectroscopy and atomic fluorescence spectroscopy a resonance line source is required, and the most important of these is the *hollow cathode lamp* which is shown diagrammatically in Fig. XXII, 8. For any given determination the hollow cathode lamp used has an emitting cathode of the same element as that being studied in the flame. The cathode is in the form of a cylinder, and the electrodes are enclosed in a borosilicate or quartz envelope which contains an inert gas (neon or argon) at a pressure of approximately 5 torr. The application of a high potential across the electrodes causes a discharge which creates ions of the noble gas. These ions are accelerated to the cathode and on collision, excite the cathode element to emission. Multi-element lamps are available in which the cathodes are made from alloys, but in these lamps the resonance line intensities of individual elements are somewhat reduced.

Electrodeless discharge lamps were originally developed as radiation sources for atomic absorption spectroscopy and atomic fluorescence spectroscopy by Dagnall *et al.* (Ref. 5); they give radiation intensities which are much greater than

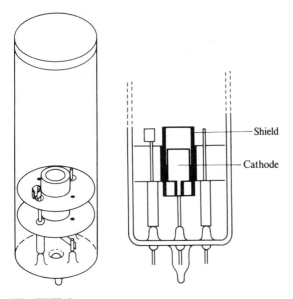

Fig. XXII, 8

those given by hollow cathode lamps. The electrodeless discharge lamp consists of a quartz tube 2–7 cm in length and 8 mm in internal diameter, containing up to 20 mg of the required element or of a volatile salt of the element, commonly the iodide; the tube also contains argon at low pressure (about 2 torr). Under operating conditions the material placed in the tube must have a vapour pressure of about 1 mm at a temperature of 200–400 °C. A microwave frequency of 2000 to 3000 MHz applied through a wave guide cavity provides the energy of excitation.

XXII, 8.   MONOCHROMATOR.   The purpose of the monochromator is to select a given emission line and to isolate it from other lines and occasionally, from molecular band emissions.

In a simple flame (emission) photometer, e.g., the Corning-EEL Model 100 Flame Photometer (see Section **XXII, 12**), an interference filter (Section **XVIII, 7**) is used. In more sophisticated flame emission spectrophotometers which require better isolation of the emitted frequency, a prism or a grating monochromator is employed, and a resolution of 0.1 nm should be achieved.

In atomic absorption spectroscopy the function of the monochromator is to isolate the resonance line from all non-absorbed lines emitted by the radiation source. In most commercial instruments diffraction gratings (Section **XVIII, 7**) are used because the dispersion provided by a grating is more uniform than that given by prisms, and consequently grating instruments can maintain a higher resolution over a longer range of wavelengths.

XXII, 9.   DETECTORS.   For the simple flame emission photometer (Section **XXII, 12**) a barrier layer cell (Section **XVIII, 7**) is a sufficiently good detector because an intense wide band of energy reaches the detector. In atomic absorption spectrophotometers, in view of the improved spectral sensitivity required, photomultipliers (Section **XVIII, 7**) are employed. The output from the detector is fed to a suitable read-out system, and in this connection it must be borne in mind that the radiation received by the detector originates not only from the resonance line which has been selected, but may also arise from emission within the flame. This emission can be due to atomic emission arising from atoms of the element under investigation, and may also arise from molecular band emissions. Hence instead of an absorption signal intensity I_A, the detector may receive a signal of intensity $(I_A + S)$ where S is the intensity of emitted radiation. Since only the measurement arising from the resonance line is required, it is important that this be distinguished from the effects of flame emission. This is achieved by **modulation** of the emission from the resonance line source by either a mechanical chopper device, or electronically, by using an alternating current signal appropriate to the particular frequency of the resonance line, and the detector amplifier is then tuned to this frequency: in this way, the signals arising from the flame, which are essentially d.c. in character, are effectively removed.

The read-out systems available include meters, chart recorders, and digital display; meters have now been virtually superseded by the alternative methods of data presentation.

XXII, 10. INTERFERENCES. Various factors may affect the flame emission of a given element and lead to interference with the determination of the concentration of a given element. These factors may be broadly classified as (a) **spectral interferences** and (b) **chemical interferences**.

Spectral interferences in AAS arise mainly from overlap between the frequencies of a selected resonance line with lines emitted by some other element; this arises because in practice a chosen line has in fact a finite 'band-width'. Since in fact the line width of an absorption line is about 0.005 nm only a few cases of spectral overlap between the emitted lines of a hollow cathode lamp and the absorption lines of metal atoms in flames have been reported: Table XXII, 3 includes some typical examples of spectral interferences which have been observed (see Refs. 6, 7, 8, 9). However most of this data relates to relatively minor resonance lines and the only interferences which occur with preferred resonance lines are with copper where europium at a concentration of about 150 p.p.m. would interfere, and mercury where concentrations of cobalt higher than 200 p.p.m. would cause interference.

Table XXII, 3 Some typical spectral interferences

Resonance source	Wavelength (λ nm)	Analyte	Wavelength (λ nm)
Aluminium	308.216	Vanadium	308.211
Antimony	231.147	Nickel	231.095
Copper	324.754	Europium	324.755
Gallium	403.307	Manganese	403.307
Iron	271.903	Platinum	271.904
Mercury	253.652	Cobalt	253.649

With flame emission spectroscopy, there is greater likelihood of spectral interferences when the line emission of the element to be determined and those due to interfering substances are of similar wavelength, than with atomic absorption spectroscopy. Obviously some of such interferences may be eliminated by improved resolution of the instrument, e.g., by use of a prism rather than a filter, but in certain cases it may be necessary to select other, non-interfering lines for the determination. In some cases it may even be necessary to separate the element to be determined from interfering elements by a separation process such as ion exchange or solvent extraction (see Chapters VI, VII).

Apart from the interferences which may arise from other elements present in the substance to be analysed, some interference may arise from the emission band spectra produced by molecules or molecular fragments present in the flame gases: in particular, band spectra due to hydroxyl and cyanogen radicals arise in many flames. Although in AAS these flame signals are not modulated (Section **XXII, 9**), in practice care should be taken to select an absorption line which does not correspond with the wavelengths due to any molecular bands because of the excessive 'noise' produced by the latter: this leads to decreased sensitivity and to poor precision of analysis.

XXII, 11. CHEMICAL INTERFERENCES. The production of ground state gaseous atoms which is the basis of flame spectroscopy may be inhibited by two main forms of chemical interference: (*a*) by stable compound formation, or (*b*) by ionisation.

(a) **Stable compound formation** leads to incomplete dissociation of the substance to be analysed when placed in the flame, or it may arise from the formation within the flame of refractory compounds which fail to dissociate into the constituent atoms. Examples of these types of behaviour are shown by (i) the determination of calcium in the presence of sulphate or phosphate, and (ii) the formation of stable refractory oxides of titanium, vanadium, and aluminium. Chemical interferences can usually be overcome in one of the following ways.

A. Increase in flame temperature often leads to the formation of free gaseous atoms, and for example aluminium oxide is more readily dissociated in an acetylene–nitrous oxide flame than it is in an acetylene–air flame. A calcium–aluminium interference arising from the formation of calcium aluminate can also be overcome by working at the higher temperature of an acetylene–nitrous oxide flame.

B. By the use of 'Releasing Agents'. If we consider the reaction

$$M - X + R \rightleftharpoons R - X + M$$

then it is clear that an excess of the releasing agent (R) will lead to an enhanced concentration of the required gaseous metal atoms M: this will be especially true if the product $R - X$ is a stable compound. Thus in the determination of calcium in the presence of phosphate, the addition of an excess of lanthanum chloride or of strontium chloride to the test solution will lead to formation of lanthanum (or strontium) phosphate, and the calcium can then be determined in an acetylene–air flame without any interference due to phosphate. The addition of EDTA to a calcium solution before analysis may increase the sensitivity of the subsequent flame spectrophotometric determination: this is possibly due to the formation of an EDTA complex of calcium which is readily dissociated in the flame.

C. Extraction of the analyte or of the interfering element(s) is an obvious method of overcoming the effect of 'interferences'. It is frequently sufficient to perform a simple solvent extraction to remove the major portion of an interfering substance so that, at the concentration at which it then exists in the solution, the interference becomes negligible. If necessary, repeated solvent extraction will reduce the effect of the interference even further, and equally, a quantitative solvent extraction procedure may be carried out so as to isolate the substance to be determined from interfering substances.

(b) Ionisation of the ground state gaseous atoms within a flame

$$M = M^+ + e$$

will reduce the intensity of the emission of the atomic spectral lines in flame emission spectroscopy, or will reduce the extent of absorption in atomic absorption spectroscopy. It is therefore clearly necessary to reduce the possibility of ionisation occurring to a minimum, and an obvious precaution to take is to use a flame operating at the lowest possible temperature which is satisfactory for the element to be determined. Thus the high temperature of an acetylene–air or of an acetylene–nitrous oxide flame may result in the appreciable ionisation of elements such as the alkali metals and of calcium, strontium and barium. The ionisation of the element to be determined may also be reduced by the addition of an excess of an **ionisation suppressant**; this is usually a solution containing a cation having a lower ionisation potential than that of the analyte. Thus, for example, a solution containing potassium ions at a concentration of 2000 p.p.m. added to a solution containing calcium, barium, or strontium ions creates an excess of electrons when the resulting solution is nebulised into the flame, and this has the result that the ionisation of the metal to be determined is virtually completely suppressed.

In addition to the compound formation and ionisation effects which have been considered, it is also necessary to take account of so-called **Matrix effects**. These are predominantly physical factors which will influence the amount of sample reaching the flame, and are related in particular to factors such as the viscosity, the density, the surface tension and the volatility of the solvent used to prepare the test solution. If we wish to compare a series of solutions, e.g., a series of standards to be compared with a test solution, it is clearly essential that the same solvent be used for each, and the solutions should not differ too widely in their bulk composition.

In some circumstances interference may result from **molecular absorptions**. Thus, for example, in an acetylene–air flame a high concentration of sodium chloride will absorb radiation at wavelengths in the neighbourhood of 213.9 nm which is the wavelength of the zinc resonance line: hence sodium chloride would represent an interference in the determination of zinc under these conditions. Such interferences can usually be avoided by choosing a different resonance wavelength for carrying out the determination, or alternatively by using a different flame so that the operating temperature is increased thus leading to dissociation of the interfering molecules.

To summarise, it may be stated that almost all interferences encountered in atomic absorption spectroscopy can be reduced, if not completely eliminated by the following procedures.

1. Ensure if possible that standard and sample solutions are of similar bulk composition to eliminate matrix effects.
2. Alteration of flame composition or of flame temperature can be used to reduce the likelihood of stable compound formation within the flame.
3. Selection of an alternative resonance line will overcome spectral interferences from other atoms or molecules and from molecular fragments.
4. Occasionally, separation, e.g., by solvent extraction or by an ion exchange process, may be necessary to remove an interfering element; such separations

are most frequently necessary when dealing with flame emission spectroscopy.

It may also be noted that the interference referred to as **background absorption**, which arises from the presence in the flame of gaseous molecules, molecular fragments, and in some instances of smoke, is dealt with in many modern instruments by the incorporation of a background correction facility: a typical example is discussed in Section **XXII, 14**.

With regard to the relative merits of the FAAS and FES procedures, it may be stated in general terms that FAAS is a more selective technique than FES, and in terms of sensitivity it is also to be preferred when we are dealing with lines of wavelengths less than about 350 nm. However, for lines of wavelengths appreciably greater than 350 nm, then FES is the more sensitive technique. These general conclusions may be illustrated by the following data relating to some typical metals:

zinc line 213.9 nm; sensitivity 0.009 (FAAS), 80 (FES);
magnesium line 285.2 nm; sensitivity 0.003 (FAAS), 1.0 (FES);
calcium line 422.7 nm; sensitivity 0.02 (FAAS), 0.01 (FES);
sodium line 589.0 nm; sensitivity 0.003 (FAAS), 0.001 (FES);
lithium line 670.8 nm; sensitivity 0.02 (FAAS), 0.007 (FES).

Note however that since the flame conditions differ for each of the above elements, it is not possible to make an absolute comparison of sensitivities; the figures quoted do nevertheless serve as a rough guide for the comparison of sensitivities.

Commercially available instruments

In the following Sections will be found brief descriptions of a selection of commercially available instruments. The general mode of operation will be apparent from the details given later in the Experimental Sections, but for any particular instrument the handbook supplied by the manufacturer must be consulted.

XXII, 12. FLAME PHOTOMETERS. A flame photometer can be compared to a photoelectric absorptiometer and the intensity of the filtered radiation from the flame is measured with a photoelectric detector. The filter, interposed between the flame and the detector, transmits only a strong line of the element. The simplest and least-expensive detector is a barrier-layer cell (Section **XVIII, 7**): if sufficient energy reaches the cell no amplification or external power supply is necessary, and only a sensitive galvanometer is required. The barrier-layer cell has a high temperature coefficient: it must therefore be placed at a cool part of the photometer. In some cases the precision is improved by the use of an internal standard and two filters and, in general, two photocells (one for the standard and one for the unknown) are utilised; the electronic circuit can be devised to give a direct reading of the ratio of line intensities. Flame photometers are intended primarily for analysis of sodium and potassium and also for calcium and lithium, i.e., elements which have an easily excited flame spectrum of sufficient intensity for detection by a photocell. The lay-out of a simple flame photometer is shown in Fig. XXII, 9. Air at a given pressure is passed into an atomiser and the suction this produces draws a solution of the sample into the atomiser, where it joins the air stream as a fine mist and passes into the burner.

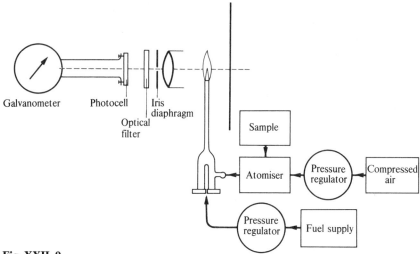

Fig. XXII, 9

Here, in a small mixing chamber, the air meets the fuel gas supplied to the burner at a given pressure and the mixture is burnt. Radiation from the resulting flame passes through a lens, then through an iris diaphragm, and finally through an optical filter which permits only the radiation characteristic of the element under investigation to pass through to the photocell. The output from the photocell is measured on a suitable galvanometer. The flame is surrounded by a chimney to protect it from draughts. The optical path from the chimney to the photocell is enclosed in a light-tight box.

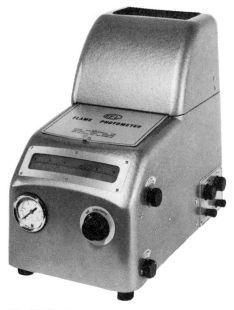

Fig. XXII, 10

An example of an instrument of this type is the **Corning EEL Model 100 Flame Photometer** which is depicted in Fig. XXII, 10; a line diagram of the essential parts is shown in Fig XXII, 11. It is a simple single-cell photometer and will operate satisfactorily with coal gas (or propane or butane or 'Calor' gas) and compressed air. The elements which can be determined are sodium, potassium, calcium, and lithium. The regions of the spectrum appropriate to the elements being determined are isolated by means of optical filters.

The operation of the instrument will be understood by reference to Fig. XXII, 11. Air is introduced to the all-metal atomiser 1 through a control valve 2 at a pressure indicated on a gauge 3 mounted on the front of the instrument.

Liquid from the beaker 4 containing the sample is drawn up the inlet tube 5 by the stream of air which atomises the sample to a fine mist. The atomiser clips into a plug 6 at one end of the spray chamber 7, in which the larger droplets fall from the air stream and flow to waste through the drain tube 8. Gas is introduced into the spray chamber through the inlet tube 9, which is connected by tubing to the automatic gas pressure stabiliser 10 and control valve 11. The gas/air mixture burns in a broad flat flame, and hot gases pass up a well-ventilated chimney 12. The light emitted by the flame is collected by a reflector 13 and focused by a lens 14 through the interchangeable optical filters 15 on to a barrier-layer photocell 16. The current generated by this photocell is taken through a calibrated potentiometer 17 to a suspension galvanometer unit 18. A glass window 19 is interposed between the lens and the filter for cooling purposes. A cover plate 20 having a V-shaped recess is provided to locate the 10-cm^3 beaker holding the sample solution, which may be slid up and held against the stop when the sample is to be sprayed.

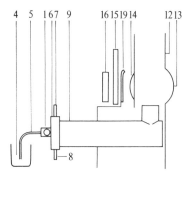

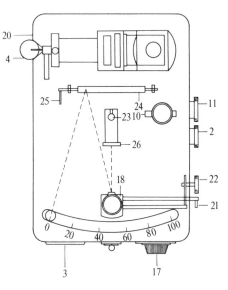

Fig. XXII, 11

For further details of the instrument see Section **XXII, 21**.

XXII. 13. SINGLE BEAM ATOMIC ABSORPTION SPECTROPHOTO-METERS.
METERS. Many commercial instruments are based on the use of a single beam, modulated a.c. system, and a typical example of such an instrument is the **Hilger and Watts Atomspek H1550**, a line diagram of which is shown in Fig. XXII, 12. The important features of this instrument include an easily accessible sample area which can house an automatic sampler capable of handling 40 samples and 8 reference checks in six minutes; the sample volume required is only 0.3 cm^3. A turret holding six hollow cathode lamps is provided with an independent current-stabilised supply to each lamp. The instrument is capable of resolving lines which are less than 0.1 nm apart and is equipped with a photo-multiplier that functions well over the wave-length range of 193 to 853 nm. Integrated measurements can be made with high precision by making use of the crystal-operated digital clock

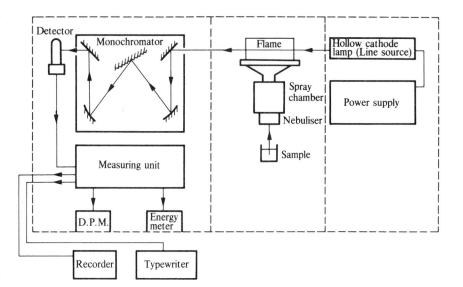

Fig. XXII, 12

which is included; this clock controls the stabilised wave forms for pulsing the hollow cathode lamp supplies as well as the signal demodulation. The analytical data obtained can be recorded on a chart, or printed automatically by digital typewriter.

Particular features of this instrument are that it can also be used as a flame emission spectrophotometer, and that accessories are available for flameless techniques including the vapour technique for mercury, arsenic and tellurium (see Section **XXII, 6**).

The **Varian Techtron Model AA-6** single beam atomic absorption spectrophotometer is shown in Fig. XXII, 13. A feature of this instrument is the incorporation of an 'optical rail' on to which many of the optical and sampling components are fixed so that it is a simple matter to rearrange these to suit specific requirements: a turret holding four hollow cathode lamps is also attached to this rail. The burners are made of titanium and are designed to handle solutions which are of high solids content without clogging of the burners. The monochromator provides uniform dispersion over the wavelength range 185–1000 nm. The presentation of data may be by means of a meter, by a digital display or by means of a digital printer, and facilities are provided for integrating the data over three alternative time periods (3, 10 or 30 seconds).

This instrument can also be used as a flame emission spectrophotometer, and permits the use of graphite rod techniques.

XXII, 14.   DOUBLE BEAM ATOMIC ABSORPTION SPECTRO-PHOTOMETERS. With the introduction of a relatively 'low noise' burner in 1962 having a constant aspiration rate, it was no longer the burner which was the main source of instrument instability: the limiting factor was now the stability of the hollow cathode lamp. The double beam system was designed to overcome the effect of variation in lamp intensity. A typical instrument of this type is the **Perkin Elmer Model 460** atomic absorption spectrophotometer, a simplified optical

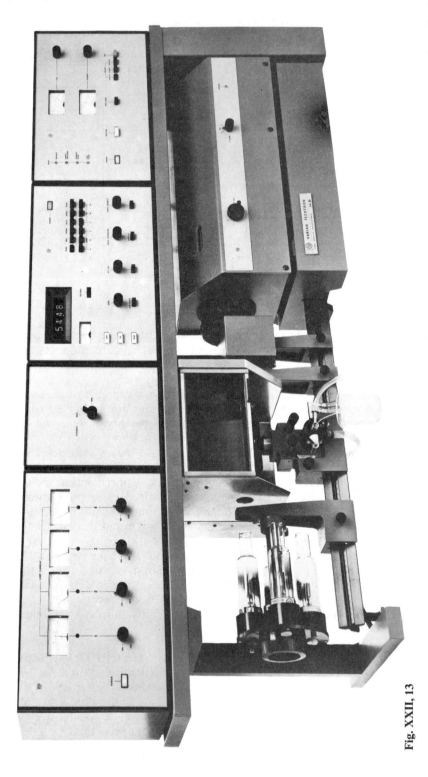

Fig. XXII, 13

diagram of which is shown in Fig. XXII, 14, and the instrument itself in Fig. XXII, 15. In this instrument, a rotating chopper passes the beam from the lamp alternately through the flame (to give the sample beam), and around the flame (to give the reference beam). The sample and reference beams are re-combined by a half silvered mirror, and are then passed together to the photomultiplier detector. The electrical circuits are designed to measure the *ratio* of the two beams, and hence variations in lamp intensity, photomultiplier sensitivity and electronic gain affects both signals similarly, and the ratio of the two signals compensates for variation in the quantities listed.

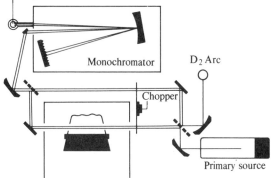

Fig. XXII, 14

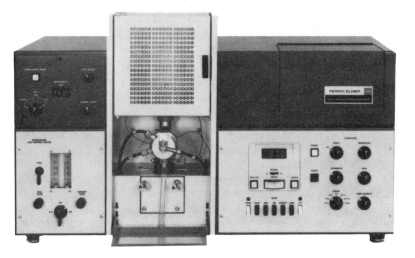

Fig. XXII, 15

The burner chamber of the Model 460 is made of stainless steel lined with a corrosion-resistant plastic; it is provided with a spring-loaded check valve which opens automatically in the rare event of an explosive flash-back and protects the burner from damage. The burner is so designed that only the smallest droplets reach the flame, thus reducing background absorption and providing optimum precision. The uptake to the nebuliser can be varied continuously thus enabling low sample consumption when the quantity available is limited. Provision is made for the connection of one fuel gas and two oxidant gases with rapid switching from one to the other, and an automatic check that the correct burner head is in position before nitrous oxide can be switched on: the burner heads are constructed of titanium.

A background corrector is incorporated which takes the form of a high intensity deuterium arc lamp, producing an emission continuum which travels the same double beam path as does the light from the resonance source (see Fig. XXII, 14). The background absorption affects both the sample and reference beams and so when the ratio of the intensities of the two beams is taken, the background effects are eliminated.

A special feature of the Model 460 is the inclusion of a microcomputer, and this makes it a very simple matter to calibrate the instrument to read directly in concentration units. The concentration values of standard solutions are fed to the computer through the numerical keyboard; normally three standard solutions may be used, but under some conditions a single standard will suffice. A solution blank aspirated through the burner automatically adjusts the zero of the instrument, and on aspiration of the standards, a working curve is automatically computed.

XXII, 15.   ATOMIC FLUORESCENCE SPECTROSCOPY. Within the confines of the present volume it is not possible to provide a detailed discussion of instrumentation for atomic fluorescence spectroscopy. An instrument for simultaneous multi-element determination described by Mitchell and Johansson (Ref. 10) has been developed commercially. Many atomic absorption spectrophotometers can be adapted for fluorescence measurements and details are available from the manufacturers. Detailed descriptions of atomic fluorescence spectroscopy are to be found in many of the volumes listed in the Bibliography (Section **XXII, 28**).

Experimental

XXII, 16.   EVALUATION METHODS.   Before dealing with the experimental details of AAS or FES determinations it is necessary to consider the mode of treatment of the experimental data obtained. To convert the measured absorption values into the concentration of the substance being determined it is necessary either to make use of a calibration curve, or to carry out the 'standard addition' procedure.

(a) *Calibration curve procedure.*   A calibration curve for use in atomic absorption or in flame emission measurements is plotted by aspirating into the flame samples of solutions containing known concentrations of the element to be determined, measuring the absorption (emission) of each solution, and then constructing a graph in which the measured absorption (emission) is plotted against the concentration of the solutions. If we are dealing with a test solution which contains a single component then the standard solutions are prepared by dissolving a weighed quantity of a salt of the element to be determined in a known volume of distilled (de-ionised) water in a graduated flask. If however other substances are present in the test solution, then these should also be incorporated in the standard solutions and at a similar concentration to that existing in the test solution. At least four standard solutions should be used covering the optimum absorbance range of 0.2 to 0.8, and if the calibration curve is found to be non-linear (this often happens at high absorbance values), then measurements with additional standard solutions should be carried out. In common with all absorbance measurements, the readings must be taken after the instrument zero has been adjusted against a 'blank' which may be either distilled water, or a solution of similar composition to the test solution but minus the component to be determined. It is usual to examine the standard solutions in order of increasing concentration, and after making the measurements with one solution, distilled water is aspirated into the flame to remove all traces of solution before proceeding to the next solution. At least two, and preferably three, separate absorption readings should be made with each solution, and an average value taken.

If necessary, the test solution must be suitably diluted using a pipette and a graduated flask, so that it too gives absorbance readings in the range 0.2–0.8. Using the calibration curve it is a simple matter to interpolate from the measured absorbance of the test solution the concentration of the relevant element in the solution. The working graph should be checked occasionally by making measurements with the standard solutions, and if necessary a new calibration curve must be drawn.

(*b*) *The standard addition technique.* When dealing with a test solution which is complex in character, or one whose exact composition is unknown, it may be very difficult and even impossible to prepare standard solutions having a similar composition to the sample. In such a case the method of standard addition can be employed. As described in Section **XVI, 5**, this involves the addition of known amounts of the ion to be determined to a number of aliquots of the sample solution; the solutions thus obtained should all be diluted to the same final volume. Naturally, if the absorbance of the test solution is too high, a quantitative dilution must be carried out, and the measurements made with this diluted solution. The absorbance of the test solution is first measured, and then each of the prepared solutions is examined in turn, leading up to the solution of highest concentration, and remembering to aspirate distilled water into the flame between each solution. The absorbance values are then plotted against the added concentration values; a straight line plot should result and the straight line can be extrapolated to the concentration axis—the point where the axis is cut gives the concentration of the test solution. If the graph is non-linear, then extrapolation cannot be undertaken with any confidence and it is important to realise that an extrapolation procedure is never as reliable as interpolation, and the latter should therefore be chosen if at all possible.

XXII, 17. PREPARATION OF SAMPLE SOLUTIONS. For the application of flame spectroscopic methods the sample must be prepared in the form of a suitable solution unless it is already presented in this form; exceptionally, solid samples can be handled directly in some of the non-flame techniques (Section **XXII, 6**).

Aqueous solutions may sometimes be analysed directly without any pre-treatment, but it is a matter of chance that the given solution should contain the correct amount of material to give a satisfactory absorbance reading. If the existing concentration of the element to be determined is too high then the solution must be diluted quantitatively before commencing the absorption measurements. Conversely, if the concentration of the metal in the test solution is too low, then a concentration procedure must be carried out (see below, under Separatory Methods).

Solutions in organic solvents may, with certain reservations, be used directly provided that the viscosity of the solution is not very different from that of an aqueous solution. The important consideration is that the solvent should not lead to any disturbance of the flame; an extreme example of this is carbon tetrachloride which may extinguish an air–acetylene flame. In many cases, suitable organic solvents {e.g., 4-methylpentan-2-one (methyl isobutyl ketone) and the hydrocarbon mixture sold as 'white spirit'} give enhanced production of ground state gaseous atoms and lead to about three times the sensitivity which is achieved with aqueous solutions. Due regard must of course be paid to the question of safety: see Section **XXII, 19**.

Inorganic solids such as metallic alloys, minerals, cements, etc., must be brought into solution by the usual standard techniques, the aim being to produce a clear solution with no loss of the element to be determined. Generally speaking, the final solution should not contain acid at a greater concentration than about $1M$ since the aspiration of extremely corrosive solutions into the burner of the apparatus should be avoided as far as possible: the instruction manual supplied with the instrument will normally give guidance in this direction.

Organic solids which contain trace elements can sometimes be dissolved in a suitable organic solvent, or alternatively the organic material may be oxidised and the residue treated to give an aqueous solution of the element to be determined.

Separation techniques may have to be applied if the given sample contains substances which act as interferences (Section **XXII, 10**), or, as explained above, if the concentration of the element to be determined in the test solution is too low to give satisfactory absorbance readings. As already indicated (Section **XXII, 10**), the separation methods most commonly used in conjunction with flame spectrophotometric methods are solvent extraction (see Chapter VI) and ion exchange (Chapter VII). When a solvent extraction method is used, it may happen that the element to be determined is extracted into an organic solvent, and as discussed above it may be possible to use this solution directly for the flame photometric measurement.

XXII, 18. PREPARATION OF STANDARD SOLUTIONS.   In flame spectrophotometric measurements we are concerned with solutions having very small concentrations of the element to be determined. It follows that the standard solutions which will be required for the analyses must also contain very small concentrations of the relevant elements, and it is rarely practicable to prepare the standard solutions by weighing out directly the required reference substance. The usual practice therefore is to prepare **stock solutions** which contain about $1000 \, \mu g \, cm^{-3}$ of the required element, and then the working standard solutions are prepared by suitable dilution of the stock solutions. Solutions which contain less than $10 \, \mu g \, cm^{-3}$ are often found to deteriorate on standing owing to adsorption of the solute on to the walls of glass vessels. Consequently, standard solutions in which the solute concentration is of this order should not be stored for more than 1 to 2 days.

The stock solutions are ideally prepared from the pure metal or from the pure metal oxide by dissolution in a suitable acid solution; the solids used must of course be of the highest purity, e.g., the Johnson Matthey 'Specpure' range of reagents. In many cases however it is prepared by dissolution of a suitable metallic salt in de-ionised water provided that the salt satisfies the normal requirements of a primary standard.

XXII, 19.   SAFETY PRACTICES.   Before commencing any experimental work with either a flame (emission) photometer or an atomic absorption spectrophotometer, the following guide lines on safety practices should be studied. These recommendations are a summary of the Code of Practice recommended by the Scientific Apparatus Makers Association (SAMA) of the USA; for full details see Ref. 11.

1. Ensure that the laboratory in which the apparatus is housed is well ventilated and is provided with an adequate exhaust system having air-tight joints on the discharge side; some organic solvents, especially those containing chlorine, give toxic products in a flame.

2. Gas cylinders must be fastened securely in an adequately ventilated room well away from any heat or ignition sources. The cylinders must be clearly marked so that the contents can be immediately identified.

3. When the equipment is turned off, close the fuel gas cylinder valve tightly and bleed the gas line to the atmosphere via the exhaust system.

4. The piping which carries the gases from the cylinders must be securely fixed in such a position that it is unlikely to suffer damage.

5. Make periodic checks for leaks by applying soap solution to joints and seals.

6. The following special precautions should be observed with acetylene.

    (*a*) Never run acetylene at a pressure higher than 15 p.s.i. (103 kN m^{-2}); at higher pressures acetylene can explode spontaneously.

    (*b*) Avoid the use of copper tubing. Use tubing made from brass containing less than 65 per cent copper, from galvanized iron or from any other material that does not react with acetylene.

    (*c*) Avoid contact between gaseous acetylene and silver, mercury or chlorine.

    (*d*) Never run an acetylene cylinder after the pressure has dropped to 50 p.s.i. (3430 kN m^{-2}); at lower pressures the gas will be contaminated with acetone.

7. A nitrous oxide cylinder should not be used after the regulator gauge has dropped to a reading of 100 p.s.i. (6860 kN m^{-2}).

8. A burner which utilises a mixture of fuel and oxidant gases and which is attached to a waste vessel (liquid trap) should be provided with a U-shaped connection between the trap and the burner chamber. The head of liquid in the connecting tube should be greater than the operating pressure of the burner: if this is not achieved, mixtures of fuel and oxidant gas may be vented to the atmosphere and form an explosive mixture. The trap should be made of a material that will not shatter in the event of an explosive flash-back in the burner chamber.

9. Care must be exercised when using volatile inflammable organic solvents for aspiration into the flame. A container fitted with a cover which is provided with a small hole for the sample capillary is recommended.

10. Never view the flame or hollow cathode lamps directly; protective eye wear should always be worn. Safety spectacles will usually provide adequate

protection from ultraviolet light, and will also provide protection for the eyes in the event of the apparatus being shattered by an explosion.

11. Never leave a flame unattended.

Some selected determinations

XXII, 20.   INTRODUCTION.   It is impossible in the present volume for the determination of a wide range of elements by atomic absorption spectroscopy to be discussed in detail. A few detailed examples of the application of atomic absorption and atomic emission methods are given in Sections **XXII, 21–26**; these have been chosen to illustrate the general procedures involved, including the manner in which certain interferences may be overcome and how chemical pre-treatment is often necessary in order to perform a successful analysis by this technique.

In Table XXII, 4 is listed the wavelength of the most widely used resonance line for all the common elements, together with the normal composition of the flame gases. The optimum working range of concentrations is quoted, and although this can vary with the instrument used, the values cited may be regarded as typical. The term **sensitivity** in atomic absorption spectroscopy is defined as the concentration of an aqueous solution of the element which absorbs 1 per cent of the incident resonance radiation; in other words, it is the concentration which gives an absorbance of 0.0044. As a rough guide, the sensitivity may be taken as about one-fiftieth of the lower value of the optimum absorbance range (Section **XXII, 16(a)**). It should be noted that sensitivity is largely dependent upon the reactions occurring in the flame, and is not strictly a characteristic of a given instrument.

The **detection limit** is another value which is often quoted, and this may be defined in a variety of ways. The most widely accepted definition is that the detection limit is the smallest concentration of a solution of an element that can be detected with 95 per cent certainty. This is the quantity of the element that gives a reading equal to twice the standard deviation of a series of at least ten determinations taken with solutions of concentrations which are close to the level of the blank.

Table XXII, 4   FAAS data for the common elements

Element	Wavelength of main resonance line (λ nm)	Flame*	Working range μg cm^{-3}
Ag	328.1	AA(L)	1–5
Al	309.3	NA(R)	40–200
As	193.7	AH(R)[1]	50–200
B	249.8	NA(R)	400–600
Ba	553.6	NA(R)	10–40
Be	234.9	NA(R)	1–5
Bi	223.1	AA(L)	10–40
Ca	422.7	NA(R)	1–4
Cd	228.8	AA(L)	0.5–2
Co	240.7	AA(L)	3–12
Cr	357.9	AA(R)	2–8
Cs	852.1	AP(L)	5–20
Cu	324.7	AA(L)	2–8

Element	Wavelength of main resonance line (λ nm)	Flame*	Working range μg cm^{-3}
Fe	248.3	AA(L)	2.5–10
Ga	294.4	AA(L)	50–200
Ge	265.2	NA(R)	70–280
Hg[2]	253.7	AA(L)	100–400
In	303.9	AA(L)	15–60
Ir	208.9	AA(R)	40–160
K	766.5	AP(L)	0.5–2
Li	670.8	AP(L)	1–4
Mg	285.2	AA(L)	0.1–0.4
Mn	279.5	AA(L)	1–4
Mo	313.3	NA(R)	15–60
Na	589.0	AP(L)	0.15–0.60
Ni	232.0	AA(L)	3–12
Os	290.9	NA(R)	50–200
Pb	217.0	AA(L)	5–20
Pd	244.8	AA(L)	4–16
Pt	265.9	AA(L)	50–200
Rb	780.0	AP(L)	2–10
Rh	343.5	AA(L)	5–25
Ru	349.9	AA(L)	30–120
Sb	217.6	AA(L)	10–40
Sc	391.2	NA(R)	15–60
Se	196.0	AH(R)	20–90
Si	251.6	NA(R)	70–280
Sn	224.6	AH(R)[1]	15–60
Sr	460.7	NA(L)	2–10
Te	214.3	AA(L)	10–40
Ti	364.3	NA(R)	60–240
Tl	276.8	AA(L)	10–50
V	318.5	NA(R)	40–120
W	255.1	NA(R)	250–1000
Y	410.2	NA(R)	200–800
Zn	213.9	AA(L)	0.4–1.6

* *Key.*   L = Fuel lean   R = Fuel rich
         AA = Air/Acetylene   NA = Nitrous oxide/Acetylene
         AP = Air/Propane   AH = Air/Hydrogen
Notes. [1] If there are many interferences then NA is to be preferred.
       [2] The use of the non-flame mercury cell (Section **XXII, 6**) is far more sensitive for the
       determination of mercury.

The data presented in Table XXII, 4, in conjunction with the experimental details given in Sections **XXII, 21**–**26**, will enable the determination of most elements to be carried out successfully. For detailed accounts of the determination of individual elements by atomic absorption spectroscopy, the Bibliography (Section **XXII, 28**) should be consulted. In addition, most instrument manufacturers supply applications handbooks relative to their apparatus in which full experimental details are given.

XXII, 21.   EXPERIMENTS WITH A SIMPLE FLAME PHOTOMETER.
The following account refers to the use of the Corning EEL Model 100 Flame Photometer (see Section **XXII, 12**). Before attempting to use the instrument read

the instruction manual and Sections **XXII, 12** and **XXII, 19**; Fig. XXII, 11 should be consulted in conjunction with the operating instructions which are given.

1. Adjust the sensitivity control (17) to the minimum value.
2. Turn the gas supply on fully and light the gas at the burner with a lighted taper.
3. Adjust the air supply from a cylinder of compressed air (itself fitted with the usual gauges and controls) until the pressure indicated on the pressure gauge mounted on the front of the instrument (3) attains a value of $10\,lb/sq\,in.$ $(690\,kN\,m^{-2})$.
4. Charge the small sample beaker (4) with de-ionised water and place it in position in the instrument. The liquid is drawn up the inlet tube (5) by the stream of air and is atomised to a fine mist.
5. Regulate the gas supply so that the blue cone of the flame just forms ten separate cones, one to each burner hole.
6. Place the appropriate filter in position.
7. Aspirate a standard solution containing the ion to be determined and, by means of the calibrated potentiometer (17), adjust the galvanometer spot to read approximately full-scale deflection.
8. Aspirate de-ionised water and adjust the galvanometer spot to read zero by means of the zero control (22).
9. Aspirate the standard solution again and readjust the sensitivity control (17) for full-scale deflection of the galvanometer.
10. Check the zero by aspirating de-ionised water.
11. Aspirate solutions of known concentration but less than that of the standard solution, and note the galvanometer reading at each concentration. Plot the galvanometer readings (abscissae) against the concentration (ordinates) expressed, say, as p.p.m., and thus prepare a calibration curve for each element. (It is advisable to measure the standard solution periodically in order to check, and if necessary adjust, the full-scale deflection of the galvanometer.)
12. Aspirate the unknown solution in the flame, note the galvanometer deflection, and evaluate the concentration from the calibration curve.

If the metal content of the solution is completely unknown the approximate concentration may be determined with the aid of the potentiometer sensitivity control (17). Aspirate a standard solution of suitable strength into the flame to obtain a reading on the scale at full sensitivity (potentiometer reading of unity). Note the exact scale reading. Turn the sensitivity control right down; aspirate the unknown solution and gently turn the sensitivity control until exactly the same galvanometer reading is obtained. Note the potentiometer reading; this gives approximately the concentration of the unknown solution relative to the standard solution. The unknown solution may then be diluted to give a reading on the scale at full sensitivity; the exact concentration may then be deduced from the calibration curve. The final measurements should always be made by comparing readings obtained on the galvanometer scale with the unknown and standard solutions at the same potentiometer setting.

If it is known that the test solution contains a sufficient concentration of an interfering substance to affect the reading it will be necessary to employ standard solutions which also contain approximately the same concentration of the interfering substance as is present in the sample. The ideal method of removing interferences is to separate the element being determined by chemical means (e.g., calcium as the oxalate), but this procedure is not always practicable.

Preparation of standard solutions for calibration curves. The following concentrations are suitable:

(a) *Sodium.* Dissolve 2.542 g A.R. sodium chloride in 1 dm³ de-ionised water in a graduated flask. This solution contains the equivalent of 1.000 mg Na per cm³. Dilute this stock solution to give four solutions containing 10, 5, 2.5, and 1 p.p.m. of sodium ions.

(b) *Potassium.* Dissolve 1.909 g A.R. potassium chloride in 1 dm³ de-ionised water. This solution contains the equivalent of 1.000 mg K per cm³. Dilute this stock solution to give four solutions containing 20, 10, 5, and 2 p.p.m. of potassium ions.

(c) *Calcium.* Dissolve 2.497 g A.R. calcium carbonate in a little dilute hydrochloric acid, and dilute to 1 dm³ with de-ionised water. This stock solution contains the equivalent of 1.000 mg Ca per cm³. Dilute this solution to give solutions containing 100, 50, 25, and 10 p.p.m. of calcium ions.

(d) *Lithium.* Dissolve 5.324 g pure lithium carbonate in a little dilute hydrochloric acid and dilute to 1 dm³ with de-ionised water. This solution contains 1.000 mg Li per cm³. Dilute the stock solution to give solutions containing 20, 10, 5, and 2 p.p.m. of lithium ions.

Prepare calibration curves for each of the above four elements. With the aid of these calibration curves, carry out the following simple determinations.

(1) **Potassium in A.R. potassium sulphate.** Weigh out accurately about 0.20 g A.R. potassium sulphate and dissolve it in 1 dm³ de-ionised water. Dilute 10.0 cm³ of this solution to 100 cm³, and determine the potassium with the flame photometer using the potassium filter.

(2) **Potassium and sodium in admixture.** Mix suitable volumes of the above stock solutions so that the resulting solution contains, say, 4–10 p.p.m. Na and 10–15 p.p.m. K. Determine the Na and K with the aid of the appropriate filters. Compare the results obtained with the true values.

(3) **Sodium, potassium, and calcium in admixture.** Mix appropriate volumes of the above stock solutions so that the test solution contains, say, 5 p.p.m. Na, 10 p.p.m. K, and 40 p.p.m. Ca. Determine the Na, K, and Ca with the aid of the appopriate filters. Compare the results obtained with the true values.

(4) **Calcium in calcium carbonate.** Determine the calcium in an analysed sample of dolomite. Dissolve about 0.38 g, accurately weighed, in 1:1-hydrochloric acid, warm gently, filter through a quantitative filter paper, wash, dilute the combined filtrate and washings to 1 dm³. Measure the calcium content of the resulting solution: use a calcium filter. Compare the value for Ca thus obtained with the known Ca content.

XXII, 22. DETERMINATION OF MAGNESIUM AND CALCIUM IN TAP WATER (AAS).

The determination of magnesium in potable water is very straightforward; very few interferences are encountered when using an acetylene–air flame. The determination of calcium is however more complicated; many chemical interferences are encountered in the acetylene–air flame and the use of 'releasing agents' such as strontium chloride, lanthanum chloride, or EDTA is necessary. Using the hotter acetylene–nitrous oxide flame the only significant interference arises from the ionisation of calcium, and under these conditions an 'ionisation buffer' such as potassium chloride is added to the test solutions.

(a) **Determination of magnesium.** *Preparation of the standard solutions.* A magnesium stock solution ($1000 \, mg \, dm^{-3}$) is prepared by dissolving $1.000 \, g$ magnesium metal (A.R.) in $50 \, cm^3$ of $5M$ hydrochloric acid. After dissolution of the metal the solution is transferred to a $1 \, dm^3$ graduated flask and made up to the mark with distilled water. An intermediate stock solution containing $50 \, mg \, Mg^{2+} \, dm^{-3}$ is prepared by pipetting $50 \, cm^3$ of the stock solution into a $1 \, dm^3$ graduated flask and diluting to the mark. Dilute accurately four portions of this solution to give four standard solutions of magnesium with known magnesium concentrations lying within the optimum working range of the instrument to be used (typically $0.1–0.4 \, \mu g \, Mg^{2+} \, cm^{-3}$).

Procedure. Although the precise mode of operation may vary according to the particular instrument used, the following procedure may be regarded as typical. Place a magnesium hollow cathode lamp in the operating position, adjust the current to the recommended value (usually $2–3 \, mA$), and select the magnesium line at $285.2 \, nm$ using the appropriate monochromator slit width.

Connect the appropriate gas supplies to the burner following the instructions detailed for the instrument, and adjust the operating conditions to give a fuel-lean acetylene–air flame.

Starting with the least concentrated solution, aspirate in turn the standard magnesium solutions into the flame, and for each take three readings of the absorbance; between each solution, remember to aspirate de-ionised water into the burner. Finally read the absorbance of the sample of tap water; this will usually require considerable dilution in order to give an absorbance reading lying within the range of values recorded for the standard solutions. Plot the calibration curve and use this to determine the magnesium concentration of the tap water.

If the magnesium content of the water is greater than $5 \, \mu g \, cm^{-3}$ it might be considered preferable to work with the less sensitive magnesium line at wavelength $202.5 \, nm$.

(b) **Determination of calcium.** Two procedures are described, (i) involving the use of releasing agents, and (ii) involving the use of an 'ionisation buffer': the latter is the preferred technique provided that an acetylene–nitrous oxide flame is available.

Preparation of the standard solutions. For procedure (i) it is necessary to incorporate a releasing agent in the standard solutions. Three different releasing agents may be used for calcium, (*a*) lanthanum chloride, (*b*) strontium chloride and (*c*) EDTA; of these (*a*) is the preferred reagent, but (*b*) or (*c*) make satisfactory alternatives.

(*a*) Prepare a lanthanum stock solution ($50\,000 \, mg \, dm^{-3}$) by dissolving $67 \, g$ of lanthanum chloride ($LaCl_3 \, 7H_2O$) in $100 \, cm^3$ of $1M$ nitric acid. Warm gently to dissolve the salt, then cool the solution and make up to $500 \, cm^3$ in a graduated flask.

(*b*) A strontium stock solution is prepared by dissolving $76 \, g$ of A.R. strontium chloride ($SrCl_2 \, 6H_2O$) in $250 \, cm^3$ of de-ionised water and then making up to $500 \, cm^3$ in a graduated flask.

(*c*) An EDTA stock solution is prepared by dissolving $75 \, g$ of EDTA disodium salt (A.R. quality) in $800 \, cm^3$ of de-ionised water. Warm gently until the salt is dissolved, then cool and make up to $1 \, dm^3$ in a graduated flask.

For procedure (ii) an ionisation buffer is required and this involves preparing a potassium stock solution ($10\,000 \, mg \, dm^{-3}$). Dissolve $9.6 \, g$ of A.R. potassium chloride in de-ionised water and make up to $500\,\text{cm}^3$ in a graduated flask.

Prepare a calcium stock solution $(1000\,\text{mg}\,\text{dm}^{-3})$ by dissolving $2.497\,\text{g}$ of dried A.R. calcium carbonate in a minimum volume of $1M$ hydrochloric acid: about $50\,\text{cm}^3$ will be required. When dissolution is complete, transfer the solution to a $1\,\text{dm}^3$ graduated flask and make up to the mark with de-ionised water. An intermediate calcium stock solution is prepared by pipetting $50\,\text{cm}^3$ of the stock solution into a $1\,\text{dm}^3$ flask and making up to the mark with de-ionised water.

The working standard solutions for procedure (i) contain between $1\,\mu\text{g}\,\text{Ca}^{2+}\,\text{cm}^{-3}$ to $5\,\mu\text{g}\,\text{Ca}^{2+}\,\text{cm}^{-3}$ and are prepared by mixing appropriate volumes of the intermediate stock solution (measured with a grade A pipette), with suitable volumes of the chosen releasing agent solution, and then making up to $50\,\text{cm}^3$ in a graduated flask; the releasing agent solution is measured in a $25\,\text{cm}^3$ measuring cylinder. Five standard solutions are prepared containing respectively 1.0, 2.0, 3.0, 4.0, and $5.0\,\text{cm}^3$ of the intermediate stock solution and $10\,\text{cm}^3$ of releasing agent (a) or $5\,\text{cm}^3$ of either reagent (b) or (c). A blank solution is similarly prepared but without the addition of any of the intermediate calcium stock solution.

For procedure (ii) the working standard solutions are prepared as detailed for procedure (i) except that the releasing agent solution is replaced by $10\,\text{cm}^3$ of the stock potassium solution.

The unknown calcium solution (the tap water), will normally require to be diluted in order that its absorbance reading shall lie on the calibration curve, and the same amount of releasing agent {procedure (i)}, or of ionisation buffer {procedure (ii)}, must be added as in the standard solutions. So, for example, if the tap water contains about $100\,\mu\text{g}\,\text{cm}^{-3}$ of calcium, $25\,\text{cm}^3$ of it are pipetted into a $100\,\text{cm}^3$ graduated flask and made up to the mark with de-ionised water. Then $5\,\text{cm}^3$ of this solution is pipetted into a $50\,\text{cm}^3$ graduated flask, and if procedure (i) is being followed, $10\,\text{cm}^3$ of reagent (a) is added, or $5\,\text{cm}^3$ of either reagent (b) or (c) and then the solution is made up to the mark. If procedure (ii) is being followed, then $10\,\text{cm}^3$ of the stock potassium solution are used in place of the releasing agent. If any cloudiness should develop during the preparation of the final solution, add $1\,\text{cm}^3$ of $1M$ hydrochloric acid before making up to the mark.

Procedure (i). Set up a calcium hollow cathode lamp selecting the resonance line of wavelength $422.7\,\text{nm}$, and a fuel-lean acetylene–air flame following the details given in the instrument manual. The calibration procedure is similar to that described above for magnesium, but the aspiration of de-ionised water into the burner after taking the readings for each solution is even more important in this case owing to the relatively high concentrations of salts present as releasing agent; remember that de-ionised water should be aspirated into the burner for a few minutes at the conclusion of the series of readings.

Procedure (ii). Make certain that the instrument is fitted with the correct burner for an acetylene–nitrous oxide flame, then set the instrument up with the calcium hollow cathode lamp, select the resonance line of wavelength $422.7\,\text{nm}$, and adjust the gas controls as specified in the instrument manual to give a fuel-rich flame. Take measurements with the blank, the standard solutions, and with the test solution, all of which contain the 'ionisation buffer'; the need, mentioned under procedure (i) for adequate treatment with de-ionised water after each measurement applies with equal force in this case. Plot the calibration graph and ascertain the concentration of the unknown solution.

XXII, 23.   DETERMINATION OF VANADIUM IN LUBRICATING OIL

(AAS). The oil is dissolved in white spirit and the absorption which this solution gives rise to is compared with that produced from standards made up from vanadium naphthenate dissolved in white spirit.

Preparation of the standard solutions. The standard solutions are prepared from a solution of vanadium naphthenate in white spirit (Nuodex Ltd) which contains about 3 per cent of vanadium. Weigh out accurately about 0.6 g of the vanadium naphthenate into a 100 cm^3 graduated flask and make up to the mark with white spirit: this stock solution contains about 180 μg cm^{-3} of vanadium. Dilute portions of this stock solution measured with the aid of a Grade A 50 cm^3 burette to obtain a series of working standards containing from 10–40 μg cm^{-3} of vanadium.

Procedure.   Weigh out accurately about 5 g of the oil sample, dissolve in a small volume of white spirit and transfer to a 50 cm^3 graduated flask; use the same solvent to wash out the weighing bottle and finally to make up the solution to the mark.

A double beam atomic absorption spectrophotometer should be used, e.g., Perkin Elmer Model 306 or Model 460 or equivalent instrument. Set up a vanadium hollow cathode lamp selecting the resonance line of wavelength 318.5 nm, and adjust the gas controls to give a fuel-rich acetylene–nitrous oxide flame in accordance with the instruction manual. Aspirate successively into the flame the solvent blank, the standard solutions, and finally the test solution, in each case recording the absorbance reading. Plot the calibration curve and ascertain the vanadium content of the oil.

XXII, 24.   DETERMINATION OF TRACE LEAD IN A FERROUS ALLOY

(AAS).   The procedure followed entails the removal of gross interferents by solvent extraction, and the selective extraction and concentration of the trace metal by use of a chelating agent. The alloy used should not contain more than 0.1 g of copper in the sample weighed out.

Preparation of solutions. The following solutions are required.

Ammonia solution (concentrated, '0.880', about 35 per cent NH$_3$) either A.R. or preferably the special atomic absorption spectroscopy (A.A.S.) reagent obtainable from laboratory supply houses.

Hydrochloric acid, concentrated, A.R. or preferably A.A.S. reagent; and also a solution prepared by measuring 50 cm^3 (measuring cylinder) of the concentrated acid into a 1 dm^3 graduated flask and making up with de-ionised water.

Nitric acid, concentrated, A.R. or A.A.S. reagent.

Ammonium citrate. Dissolve 50 g A.R. tri-ammonium citrate in 50 cm^3 of concentrated ammonia solution added with care. Cool, and make up to 100 cm^3 with de-ionised water.

Ascorbic acid. Dissolve 20 g of the A.R. solid in 100 cm^3 of de-ionised water. This reagent must be freshly prepared.

Potassium cyanide. **(CAUTION!)** Dissolve 25 g of A.R. salt in 35 cm^3 of de-ionised water to which has been added 5 cm^3 of concentrated ammonia solution. Make up to 50 cm^3 with de-ionised water and filter if necessary.

Sodium diethyldithiocarbamate (NaDDC). Dissolve 1 g of the A.R. solid in 50 cm^3 of de-ionised water and filter if necessary. This reagent must be freshly prepared.

Lead caprate. Prepare a standard stock solution by dissolving 0.1323 g of the solid in 2 cm³ of naphthenic acid with warming. Add 20 cm³ of 4-methylpentan-2-one (methyl isobutyl ketone), cool and then make up to the mark in a 100 cm³ graduated flask with more of the ketone.

Procedure. Weigh accurately 1 g of the alloy and dissolve in 10 cm³ of concentrated hydrochloric acid; warm gently, and if necessary add concentrated nitric acid dropwise (about 3 cm³) to assist the dissolution. When the vigorous reaction is complete, digest the solution with gentle heat for about 15 minutes. Cool, and if necessary filter through a Whatman No. 541 filter paper, washing the beaker and filter paper with small portions of concentrated hydrochloric acid so that a final volume of about 20 cm³ is attained. Transfer the solution to a 250 cm³ separatory funnel using a further 10 cm³ of concentrated hydrochloric acid to effect a quantitative transfer. Add 50 cm³ of butyl acetate, shake for one minute and allow to separate; iron and molybdenum are extracted into the organic layer. Separate the two layers, collecting the acid layer and transferring, with the aid of a further 10 cm³ of concentrated hydrochloric acid, to a clean 250 cm³ separating funnel; extract with a 25 cm³ portion of butyl acetate. Again separate the two layers, collecting the acid layer in a 250 cm³ beaker.

Add cautiously (FUME CUPBOARD), and with constant stirring, 10 cm³ of the ammonium citrate solution; this will prevent the precipitation of metals when, at a later stage, the pH value of the solution is increased. Then add 10 cm³ of the 20 per cent ascorbic acid, and adjust to pH 4 (BDH narrow range indicator paper), by the cautious addition of concentrated ammonia solution down the side of the beaker whilst stirring continuously. Then add 10 cm³ of the 50 per cent potassium cyanide solution (**CAUTION!**) and *immediately* adjust to a pH of 9–10 (BDH indicator paper), by the addition of concentrated ammonia solution.

Transfer the solution to a 250 cm³ separatory funnel, rinsing out the beaker with a little water. Add 5 cm³ of the 2 per cent NaDDC reagent and allow to stand for one minute, and then add a 10 cm³ portion of 4-methylpentan-2-one (methyl isobutyl ketone), shake for one minute and then separate and collect the organic layer. Return the aqueous phase to the funnel, extract with a further 10 cm³ portion of methyl isobutyl ketone, separate and combine the organic layer with that already collected. Finally rinse the funnel with a little fresh ketone and add this rinse liquid to the organic extract. In these operations the lead is converted into a chelate which is extracted into the organic solvent.

In order to concentrate the lead extract, remove the lead from the organic solvent by shaking this with three successive 10 cm³ portions of the dilute hydrochloric acid solution, collecting the aqueous extracts in a 250 cm³ beaker. To the combined extracts add 5 cm³ of 20 per cent ascorbic acid solution and adjust to pH 4 by the addition of concentrated ammonia solution. Place the beaker in a fume cupboard, add 3 cm³ of the 50 per cent potassium cyanide solution and *immediately* adjust the pH to 9–10 with concentrated ammonia solution. Transfer the solution to a 250 cm³ separatory funnel with the aid of a little de-ionised water, add 5 cm³ of the 2 per cent NaDDC reagent, allow to stand for one minute and then add 10 cm³ of methyl isobutyl ketone. Shake for one minute and then separate and collect the organic phase, filtering it through a fluted filter paper. This solution now contains the lead and is ready for the absorption measurement.

Set up a double beam atomic absorption spectrophotometer with a lead hollow cathode lamp and isolate the resonance line at 283.3 nm; adjust the gas controls to give a fuel-lean acetylene–air flame in accordance with the operating manual supplied with the instrument.

Prepare a blank solution by carrying through all the sequences of the separation procedures using a hydrochloric acid solution to which no alloy has been added, and then measure the absorption given by this blank solution, by a series of standard solutions containing from 1–$10\,\mu g\,Pb\,cm^{-3}$ prepared by suitable dilution of the lead caprate stock solution (Note 1), and finally of the extract prepared from the sample of alloy. Plot the calibration curve and determine the lead content of the alloy.

Note 1.   If lead caprate is not available, standard lead solutions can be prepared from aqueous solutions containing known weights of A.R. lead nitrate and following through the extraction procedure as detailed for the final extraction of lead into methyl isobutyl ketone for the alloy. It should also be noted that steps should be taken to avoid excessive inhalation of the vapour of the methyl isobutyl ketone which can cause a headache.

XXII, 25. DETERMINATION OF CHROMIUM IN A NICKEL ALLOY (AAS).
The following details are reproduced by courtesy of Varian Associates Ltd; they refer to the analysis of an alloy containing approximately 15 per cent of chromium.

Sample preparation. Dissolve 1.000 g of the alloy in the form of fine turnings in a mixture of hydrochloric acid ($10\,cm^3$) and nitric acid ($10\,cm^3$), then heat gently until no more fumes of nitrogen dioxide are obtained. Transfer to a PTFE beaker and add dropwise $5\,cm^3$ of hydrofluoric acid (**CAUTION!**); during this operation keep the solution cool and ensure that the temperature does not rise above $30\,°C$. Transfer the solution to a $100\,cm^3$ plastic graduated flask and make up to volume with de-ionised water. With an alloy containing 15 per cent of chromium the chromium concentration of the solution will be approximately $1500\,\mu g\,cm^{-3}$.

Preparation of standard solutions. Prepare a stock chromium solution by dissolving 1.000 g of pure chromium powder in $10\,cm^3$ of hydrochloric acid, and then follow the procedure used with the alloy up to the stage where the solution has been diluted in the $100\,cm^3$ graduated flask. The working standards must contain the same reagents and major matrix elements as does the sample solution, and the concentration of each component should be approximately the same in the standard and test solutions. Thus the standard solutions must all contain nickel ($7000\,\mu g\,cm^{-3}$), iron ($1400\,\mu g\,cm^{-3}$), cobalt ($100\,\mu g\,cm^{-3}$), silicon ($50\,\mu g\,cm^{-3}$), hydrofluoric acid ($5\,cm^3$ in each $100\,cm^3$ of solution) and hydrochloric acid ($5\,cm^3$ in each $100\,cm^3$ of solution). Appropriate dilution of suitable aliquots of the stock chromium solution with a background solution of the above composition can be used to give standards containing 0, 1000, 1500, and $2000\,\mu g\,cm^{-3}$ of chromium.

Procedure. Set up a Varian Techtron Model AA-6 single beam atomic absorption spectrophotometer (or an equivalent instrument) in accordance with the details in the handbook. Set a chromium hollow cathode lamp in the operating position, and select the resonance line at 520.8 nm (note that this is not the main resonance line). Connect up the gas supply in accordance with the handbook to give an acetylene–nitrous oxide flame. Then measure the absorbance of the standard solutions and of the test solution and calculate the chromium content of the alloy.

XXII, 26. DETERMINATION OF SULPHATE ION BY ATOMIC ABSORPTION INHIBITION TITRIMETRY. *Introduction.* The procedure described below serves to demonstrate how atomic absorption spectrophotometry can be adapted to the determination of selected anions such as sulphate, phosphate and silicate by an indirect method. It must be stressed that the method is non-specific for anion determinations, and further, all interfering ions, both anions and cations, must be removed by a preliminary ion exchange treatment. In pure solutions the anions referred to can be determined at very low concentration levels.

The technique, termed **Atomic Absorption Inhibition Titrimetry,** was described by Huber and his co-workers (Ref. 12). In this method, the experimental conditions are deliberately chosen to encourage the occurrence of inhibition and hence a relatively cool argon–hydrogen flame is used. The sample solution containing the anion (SO_4^{2-}) to be determined is titrated with a standard cation solution (Mg^{2+}), with simultaneous aspiration of the mixed solution into the flame. Initially the sulphate ions will cause formation of magnesium sulphate, which at the flame temperature used does not dissociate. Hence the absorption by Mg^{2+} ions in the flame will be inhibited, and only when an excess of Mg^{2+} ions are present in the flame will its absorbance, which is continuously monitored on a chart recorder, become a linear function of the volume of magnesium solution added to the test solution. The delay in attaining this linear behaviour is a measure of the sulphate ion concentration in the test solution.

Preparation of solutions. Prepare a magnesium stock solution by dissolving 0.829 g of dry A.R. magnesium oxide in the minimum quantity (approximately 42 cm^3) of 1M hydrochloric acid, then transfer to a 500 cm^3 graduated flask and make up to the mark with de-ionised water: in preparing this solution it is important to avoid an excess of chloride ions. The working solution is prepared by dilution of the stock solution to give a Mg^{2+} concentration of 200 μg cm^{-3}.

A stock sulphate solution (100 μg cm^{-3}) is conveniently prepared by dilution of a standard sulphuric acid solution.

Apparatus. The reaction vessel consists of a 100 cm^3 tall form beaker fitted with a plastic cover carrying two identical capillary tubes. One capillary is attached to a reservoir burette containing the working solution of magnesium ions, and the other is attached to the nebuliser tube of the atomic absorption spectrophotometer. The beaker is mounted upon a magnetic stirrer so that the solution can be stirred continuously. The uptake rate of the nebuliser must be matched with the flow rate from the burette so that an essentially constant volume is maintained in the reaction vessel; a typical flow rate is 2.4 cm^3 min^{-1}. It is therefore advantageous to use a spectrophotometer fitted with a variable uptake nebulizer such as the Perkin Elmer Models 306 and 460.

The burner is adjusted to take an argon–hydrogen mixture, and a magnesium hollow cathode lamp is placed in position and arranged to give the resonance line of wavelength 285.2 nm. The output from the spectrophotometer is fed to a chart recorder, which typically may be operated at 10 mV full scale deflection and at a chart speed of 30 mm min^{-1}.

Procedure. Pipette 50.0 cm^3 of a standard sulphuric acid solution having a sulphate ion concentration of 5 μg cm^{-3} into the reaction vessel and set the stirrer in motion. Attach one capillary tube to the nebuliser inlet and the other to the burette; set the nebuliser in operation and simultaneously open the tap on the burette and start the recorder chart drive. Continue the addition of the magnesium solution until the linear part of the absorption plot is well established: see Fig. XXII, 16. Record the time taken from the start of the magnesium ion addition to a pre-selected absorption value, t_f in Fig. XXII, 16.

Repeat the procedure using solutions with successive sulphate ion concentrations of 10, 15, and 20 $\mu g\ cm^{-3}$, and also a blank solution which is simply 50 cm^3 of de-ionised water. In each case measure the time taken from the start of the magnesium ion addition to the attainment of the identical pre-selected absorption value, i.e., $(t_f - t_i)$ where t_f is the time of attainment of the pre-selected absorption, and t_i is the time of the start of the magnesium ion addition. Plot a calibration graph of sulphate ion concentration (x-axis) against time $(t_f - t_i)$ in seconds.

Now pipette 50.0 cm^3 of the test solution into the cell and repeat the procedure described above to give the time $(t_f - t_i)$ required to achieve the pre-selected absorption value. Then use the calibration curve to determine the unknown sulphate concentration.

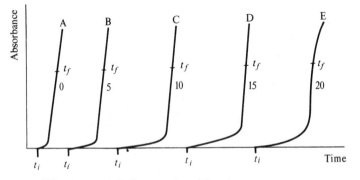

Fig. XXII, 16 t_i = initial start point of titration
t_f = final pre-selected absorbance value
Curve A = blank
Curves B, C, D, E, refer to sulphate solutions containing respectively 5, 10, 15, and 20 $\mu g\ cm^{-3}$ of sulphate ion

XXII, 27. References

1. A. Walsh (1955). *Spectrochimica Acta.*, **7**, 108.
2. C. S. Rann and A. N. Hambly (1965). Anal. Chem., **37**, 879.
3. D. R. Thomerson and K. C. Thompson (1975) *Chemistry in Britain.* **11**, 316.
4. T. S. West and X. K. Williams (1969). *Analytica Chim. Acta.*, **45**, 27.
5. R. M. Dagnall, K. C. Thompson and T. S. West (1967). *Talanta*, **14**, 551.
6. C. W. Frank, W. G. Schrenk and C. E. McLoan (1966). *Anal. Chem.*, **38**, 1005.
7. V. A. Fassel, J. A. Rasmuson and T. G. Cowley (1968). *Spectrochim. Acta.*, **23B**, 579.
8. J. E. Allen (1969). *Spectrochim. Acta.*, **24B**, 13.
9. D. C. Manning and F. Fernandex (1968). *Atom. Absorption Newsletter*, **7**, 24.
10. D. G. Mitchell and A. Johansson (1970). *Spectrochim. Acta.*, **25B**, 175.
11. Safety Practices for Atomic Absorption Spectrophotometers. *International Laboratory*, 1974, May/June, 63. International Scientific Communications Inc, Fairfield, Conn.
12. C. O. Huber and R. W. Looyinga (1971). *Anal. Chem.*, **43**, 498.
13. R. M. Dagnall, K. C. Thompson, and T. S. West (1967). *The Analyst*, **92**, 506.
14. W. R. Hatch and W. L. Ott (1968). *Anal. Chem.*, **40**, 2085.

XXII, 28.   Selected bibliography

1. W. T. Elwell and J. A. F. Gidley (1966). *Atomic Absorption Spectrophotometry.* 2nd edn. Oxford; Pergamon Press.
2. J. W. Robinson (1975). *Atomic Absorption Spectroscopy.* 2nd edn. New York; Marcel Dekker.
3. W. Slavin (1968). *Atomic Absorption Spectroscopy.* New York; Interscience.
4. R. J. Reynolds, K. Aldous and K. C. Thompson (1970). *Atomic Absorption Spectroscopy.* London; Griffin.
5. W. J. Price (1972). *Analytical Atomic Absorption Spectrometry.* London; Heyden.
6. J. A. Dean (1960). *Flame Photometry.* New York; McGraw-Hill.
7. J. A. Dean and T. C. Rains (eds.). *Flame Emission and Atomic Absorption Spectrometry.* Vol. 1: Theory (1969); Vol. 2: Components and Techniques (1971); Vol. 3: Elements and Matrices (1974). New York; Marcel Dekker.
8. B. V. L'vov. (translated from the Russian by J. H. Dixon) (1970). *Atomic Absorption Spectrochemical Analysis.* London; Adam Hilger Ltd.
9. J. Ramirez-Muñoz (1968). *Atomic Absorption Spectroscopy.* Amsterdam; Elsevier.
10. G. F. Kirkbright and M. Sargent (1975). *Atomic Absorption and Fluorescence Spectroscopy.* London; Academic Press.
11. R. Mavrodineau and H. Briteux (1965). *Flame Spectroscopy.* New York; Wiley–Interscience.

PART G THERMAL METHODS

CHAPTER XXIII THERMAL ANALYSIS

XXIII, 1. GENERAL DISCUSSION. Thermal methods of analysis may be defined as those techniques in which changes in physical and/or chemical properties of a substance are measured as a function of temperature. Methods that involve changes in weight or changes in energy come within this definition.

Other thermal analytical techniques such as dilatometry (in which changes in dimensions of a substance are measured as a function of temperature), or evolved gas analysis (where qualitative and quantitative evaluations of volatile products formed during thermal analysis are made), are outside the range of this book.

The thermal analytical techniques discussed in this chapter are:

Thermogravimetry (TG), a technique in which a change in the weight of a substance is recorded as a function of temperature or time.

Differential Thermal Analysis (DTA), which is a method for recording the difference in temperature between a substance and an inert reference material as a function of temperature or time.

Differential Scanning Calorimetry (DSC), a method whereby the energy necessary to establish a zero temperature difference between a substance and a reference material is recorded as a function of temperature or time.

XXIII, 2. THERMOGRAVIMETRY (TG). Introduction. The basic instrumental requirement for thermogravimetry is a precision balance with a furnace programmed for a linear rise of temperature with time. The results may be presented as, (i) a thermogravimetric (TG) curve, in which the weight change is recorded as a function of temperature or time, or (ii) as a derivative thermogravimetric (DTG) curve where the first derivative of the TG curve is plotted with respect to either temperature or time.

A typical thermogravimetric curve, for copper sulphate pentahydrate $CuSO_4.5H_2O$, is given in Fig. XXIII, 1.

The following features of the TG curve should be noted:

(a) the horizontal portions (plateaus) indicate the regions where there is no weight change;

(b) the curved portions are indicative of weight losses;

(c) since the TG curve is quantitative, calculations on compound stoichiometry can be made at any given temperature.

As Fig. XXIII, 1 shows, copper sulphate pentahydrate has four distinct regions of decomposition:

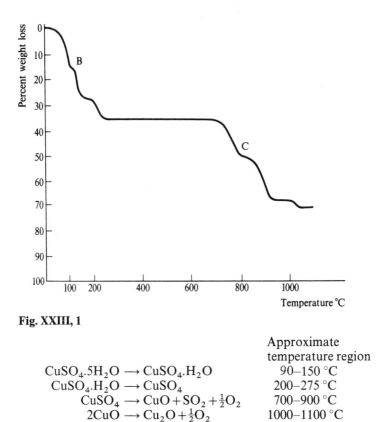

Fig. XXIII, 1

	Approximate temperature region
$CuSO_4.5H_2O \longrightarrow CuSO_4.H_2O$	90–150 °C
$CuSO_4.H_2O \longrightarrow CuSO_4$	200–275 °C
$CuSO_4 \longrightarrow CuO + SO_2 + \frac{1}{2}O_2$	700–900 °C
$2CuO \longrightarrow Cu_2O + \frac{1}{2}O_2$	1000–1100 °C

The precise temperature regions for each of the reactions are dependent upon the experimental conditions (see Section **XXIII, 5**). Although in Fig. XXIII, 1 the ordinate is shown as the percentage weight loss, the scale on this axis may take other forms:

1. as a true weight scale;
2. as a percentage of the total weight;
3. in terms of molecular weight units.

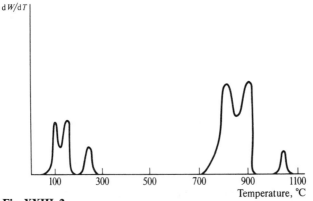

Fig. XXIII, 2

An additional feature of the TG curve (Fig. XXIII, 1) should now be examined, namely the two regions B and C where there are changes in the slope of the weight loss curve. If the rate of change of weight with time dW/dt is plotted against temperature, a derivative thermogravimetric (DTG) curve is obtained (Fig. XXIII, 2). In the DTG curve when there is no weight loss then $dW/dt = 0$. The peak on the derivative curve corresponds to a maximum slope on the TG curve. When dW/dt is a minimum but not zero there is an inflection, i.e., a change of slope on the TG curve. Inflections B and C on Fig. XXIII, 1 may imply the formation of intermediate compounds. In fact the inflection at B arises from the formation of the trihydrate $CuSO_4.3H_2O$, and that at point C is reported by Duval (Ref. 1) to be due to formation of a golden yellow basic sulphate of composition $2CuO.SO_3$. Derivative thermogravimetry is useful for many complicated determinations and any change in the rate of weight loss may be readily identified as a trough indicating consecutive reactions; hence weight changes occurring at close temperatures may be ascertained.

Experimental factors.   In the previous section it was stated that the precise temperature regions for each reaction of the thermal decomposition of copper sulphate pentahydrate is dependent upon experimental conditions. When a variety of commercial thermobalances became available in the early 1960s it was soon realised that a wide range of factors could influence the results obtained. Reviews of these factors have been made by Simons and Newkirk (Ref. 2) and by Coats and Redfern (Ref. 3) as a basis for establishing criteria necessary to obtain meaningful and reproducible results. In addition, several sources of error can arise in thermogravimetry which may lead to both inaccurate temperatures and weight change values. This may necessitate the construction of a **correction curve**. It must be stressed that with some modern instruments (Section **XXIII, 3**) the need for corrections is minimised. However, at the time of writing many laboratories still employ thermobalances of earlier designs and in these cases a correction curve *must* be constructed.

Correction curve.   When an empty crucible is heated from ambient temperature to, say, 1000 °C there is an apparent gain in weight. This weight gain is governed by the heating rate employed and by the crucible weight and volume. A typical correction curve is shown in Fig. XXIII, 3. In this experiment a platinum

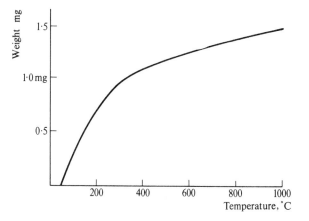

Fig. XXIII, 3

crucible of 1 g weight showed an apparent gain of 1.5 mg when heated at 4 °C min^{-1} from ambient temperature to 1000 °C. Although the error produced is 0.15 per cent of the crucible weight, a 100 mg sample contained in this crucible would suffer an apparent weight change of 1.5 per cent. This apparent weight change is due to a variety of factors including the air buoyancy and convection currents within the furnace. A correction curve must be constructed giving the apparent weight change in order to calculate the actual change occurring in a sample.

The factors which may affect the results can be classified into the two main groups of instrumental effects and the characteristics of the sample:

Instrumental factors

(a) Heating rate. When a substance is heated at a fast heating rate, the temperature of decomposition will be higher than that obtained at a slower rate of heating. The effect is shown for a single-step reaction in Fig. XXIII, 4. The curve

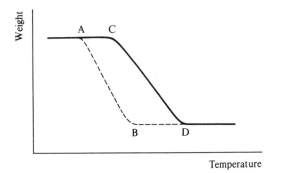

Fig. XXIII, 4

AB represents the decomposition curve at a slow heating rate, whereas the curve CD is that due to the faster heating rate. If T_A and T_C are the decomposition temperatures at the start of the reaction and the final temperatures on completion of the decomposition are T_B and T_D; the following features can be noted:

$$T_A < T_C$$
$$T_B < T_D$$
$$T_B - T_A < T_D - T_C$$

The heating rate has only a small effect when a fast reversible reaction is considered. The points of inflection B and C obtained on the thermogravimetric curve for copper sulphate pentahydrate (Fig. XXIII, 1) may be resolved into a plateau if a slower heating rate is used. Hence the detection of intermediate compounds by thermogravimetry is very dependent upon the heating rate employed.

(b) Furnace atmosphere. The nature of the surrounding atmosphere can have a profound effect upon the temperature of a decomposition stage. For example, the decomposition of calcium carbonate occurs at a much higher temperature if carbon dioxide rather than nitrogen is employed as the surrounding atmosphere. Normally the function of the atmosphere is to remove the gaseous products evolved during thermogravimetry, in order to ensure that the nature of the surrounding gas remains as constant as possible throughout the experiment. This condition is achieved in many modern thermobalances by heating the test sample *in vacuo*.

The most common atmospheres employed in thermogravimetry are:
1. 'static air' (air from the surroundings flows through the furnace);
2. 'dynamic air', where compressed air from a cylinder is passed through the furnace at a measured flow rate;
3. nitrogen gas (oxygen free) which provides an inert environment.

Atmospheres that take part in the reaction—for example, humidified air—have been used in the study of the decomposition of such compounds as hydrated metal salts.

Since thermogravimetry is a dynamic technique, convection currents arising in a furnace will cause a continuous change in the gas atmosphere. The exact nature of this change further depends upon the furnace characteristics so that widely differing thermogravimetric data may be obtained from different designs of thermobalance.

(c) **Crucible geometry.**   The geometry of the crucible can alter the slope of the thermogravimetric curve. Generally, a flat, plate-shaped crucible is preferred to a 'high form' cone shape because the diffusion of any evolved gases is easier with the former type.

Sample characteristics

The weight, particle size, and the mode of preparation (the pre-history) of a sample all govern the thermogravimetric results. A large sample can often create a deviation from linearity in the temperature rise. This is particularly true when a fast exothermic reaction is studied; for example, the evolution of carbon monoxide during the decomposition of calcium oxalate to calcium carbonate. A large volume of sample in a crucible can impede the diffusion of evolved gases through the bulk of the solid large crystals especially those of certain metallic nitrates which may undergo decrepitation ('spitting' or 'spattering') when heated. Other samples may swell, or foam and even bubble. In practice a small sample weight with as small a particle size as practicable is desirable for thermogravimetry.

Diverse thermogravimetric results can be obtained from samples with different pre-histories; for example, TG and DTG curves showed that magnesium hydroxide prepared by precipitation methods has a different temperature of decomposition from that for the naturally occurring material (Ref. 4). It follows that the source and/or the method of formation of the sample should be ascertained.

XXIII, 3.   INSTRUMENTATION FOR THERMOGRAVIMETRY.

Lukaszewski and Redfern (Ref. 5) outlined the following criteria for good thermobalance design:
(a) The thermobalance should be capable of continuously registering the weight change of the sample studied as a function of temperature and time.
(b) The furnace should reach the maximum desired temperature (with commercial thermobalances this can be about 1500 °C).
(c) The rate of heating is linear and reproducible.
(d) The sample holder should be in the hot zone of the furnace and this zone should be of uniform temperature.
(e) The thermobalance should have facilities for the provision of variable

heating rates, to permit heating in a variety of controlled atmospheres and for heating *in vacuo*. The instrument should also be capable of carrying out accurate isothermal studies.

(*f*) The balance mechanism should be protected from the furnace and from the effect of corrosive gases.

(*g*) The temperature of the sample must be measured as accurately as possible.

(*h*) A balance sensitivity suitable for studying small sample weights is necessary.

An additional requirement is a facility for rapid heating and cooling of the furnace to permit several TG analyses to be carried out in a relatively short period of time.

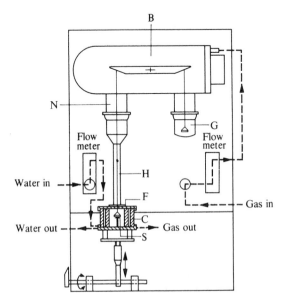

Fig. XXIII, 5

Apparatus. A wide range of commercial instruments is available and these have many similar features. In this text only one instrument, the Stanton Redcroft TG-750 Thermobalance, will be described.

The complete balance and furnace assembly of the TG 750 is shown in Fig. XXIII, 5. The electronic microbalance B is housed in a glass bottle. The balance has a capacity of 1 g with a switched range of sensitivities from 1 to 250 mg full scale deflection. The sample crucible S is suspended in a platinum–rhodium stirrup attached to the beam by an aluminium tube N.

The suspension passes through a narrow bore glass tube H with a glass flange F at one end. The furnace assembly, C, can be raised or lowered mechanically, and seats against F with an O-ring making a complete seal. The gas and water flow paths are also shown in the diagram. The system may be evacuated and then flushed with an inert gas. Access to the reference pan is obtained by removing the glass cap G.

A diagram of the cross-section of the furnace is given in Fig. XXIII, 6. The furnace, F, is approximately 12 mm in diameter and 20 mm long, and the furnace case, C, is water cooled by means of vertical channels. A platinum crucible, S, is used to contain the sample and is heated in the furnace. Measurement of the sample temperature is by means of a platinum *vs* platinum–rhodium thermocouple, T, positioned immediately below the sample crucible.

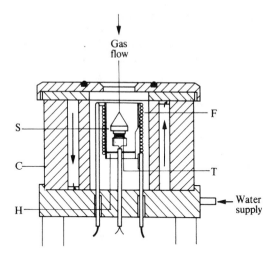

Fig. XXIII, 6

The TG 750 operates over the range ambient temperature to 1000 °C and heating rates from 1–100 °C min^{-1} may be employed. The furnace can be cooled from 1000 °C to 50 °C in four minutes. Incorporation of a sensitive electronic microbalance allows the TG 750 to measure small sample weights (1–10 mg). This is a great advance over earlier thermobalances which used sample weights of between 50 and 200 mg and required slow heating rates to facilitate good resolution. This, coupled with the slow cooling rates of conventional furnaces, meant that thermogravimetry was formerly a lengthy process.

The correction curve, described in Section **XXIII, 2**, due to the so-called 'buoyancy effect', has to be plotted for instruments of earlier designs. Hence the results of TG runs require replotting before graphs of percentage weight loss versus temperature can be obtained on older instruments.

The important features of the TG 750 are that fast heating rates may be employed with good resolution because only small sample weights are used. The cooling time between experiments is a matter of a few minutes. In addition, 'buoyancy effects' are reduced to a minimum so that it is possible to obtain a direct reading of weight changes without any recourse to prior correction. The thermobalance is also suited for isothermal studies, and the furnace can be held at 1000 °C without any balance drift. In addition to TG traces, the TG 750 will also plot DTG curves which are useful for the resolution of weight changes occurring at temperatures close to each other.

XXIII, 4. APPLICATIONS OF THERMOGRAVIMETRY. Some of the applications of thermogravimetry are of particular importance to the analyst. These are:
1. The determination of the purity and thermal stability of both primary and secondary standards.

2. The investigation of correct drying temperatures and the suitability of various weighing forms for gravimetric analysis.
3. Direct application to analytical problems (automatic thermogravimetric analysis).
4. The determination of the composition of complex mixtures.

Thermogravimetry is a valuable technique for the assessment of the purity of materials. Analytical reagents, especially those used in titrimetric analysis as primary standards, e.g., sodium carbonate, sodium tetraborate, and potassium hydrogenphthalate, have been examined. Many primary standards absorb appreciable amounts of water when exposed to moist atmospheres. TG data can show the extent of this absorption and hence the most suitable drying temperature for a given reagent may be determined.

The thermal stability of EDTA as the free acid and also as the more widely used disodium salt, $Na_2EDTA.2H_2O$, has been reported by Wendlandt (Ref. 6). He showed that the dehydration of the disodium salt commences at between 110 and 125 °C, which served to confirm the view of Blaedel and Knight (Ref. 7) that $Na_2EDTA.2H_2O$ could be safely heated to constant weight at 80 °C.

Undoubtedly the most widespread application of thermogravimetry in analytical chemistry has been in the study of the recommended drying temperatures of gravimetric precipitates. Duval studied over a thousand gravimetric precipitates by this method and gave the recommended drying temperatures. He further concluded that only a fraction of these precipitates are suitable weighing forms for the elements. The results recorded by Duval were obtained with materials prepared under specified conditions of precipitation and this must be borne in mind when assessing the value of a given precipitate as a weighing form, since conditions of precipitation can have a profound effect on the pyrolysis curve. It must be stressed that the rejection of a precipitate because it does not give a stable plateau on the pyrolysis curve at one given rate is unjustified. Further, the limits of the plateau should not be taken as indicative of thermal stability within the complete temperature range. The weighing form is not necessarily isothermally stable at all temperatures that lie on the horizontal position of a thermogravimetric curve. A slow rate of heating is to be preferred, especially with a large sample weight, over the temperature ranges in which chemical changes take place. Thermogravimetric curves must be interpreted with due regard to the fact that whilst they are being obtained the temperature is changing at a uniform rate, whereas in routine gravimetric analysis the precipitate is often brought rapidly to a specified temperature and maintained at that temperature for a definite time.

Thermogravimetry may be used to determine the composition of binary mixtures. If each component possesses a characteristic unique pyrolysis curve, then a resultant curve for the mixture will afford a basis for the determination of its composition. In such an automatic gravimetric determination the initial weight of the sample need not be known. A simple example is given by the automatic determination of a mixture of calcium and strontium as their carbonates.

Both carbonates decompose to their oxides with the evolution of carbon dioxide. The decomposition temperature for calcium carbonate is in the temperature range 650–850 °C, whilst strontium carbonate decomposes between 950 and 1150 °C. Hence the amount of calcium and strontium present in a mixture may be calculated from the weight losses due to the evolution of carbon dioxide at the lower and higher temperature ranges respectively. This method could be extended to the analysis of a three-component mixture, as barium carbonate is reported to decompose at an even higher temperature ($\sim 1300\,°C$) than strontium carbonate.

A further example, cited by Duval (Ref. 8), is the automatic determination of a mixture of calcium and magnesium as their oxalates. Calcium oxalate monohydrate has the following three distinct regions of decomposition:

			temperature range °C
(a)	$CaC_2O_4.2H_2O$	$\rightarrow CaC_2O_4 + 2H_2O$	100–250
(b)	CaC_2O_4	$\rightarrow CaCO_3 + CO$	400–500
(c)	$CaCO_3$	$\rightarrow CaO + CO_2$	650–850

In comparison with this, magnesium oxalate dihydrate has only two decomposition stages:

			temperature range °C
(d)	$MgC_2O_4.2H_2O$	$\rightarrow MgC_2O_4 + 2H_2O$	100–250
(e)	MgC_2O_4	$\rightarrow MgO + CO + CO_2$	400–500

A pyrolysis curve for a mixture of these two oxalates would thus show three decomposition steps. The final step would be due entirely to the loss of carbon dioxide from calcium carbonate and hence the amount of calcium present in the mixture may be calculated. The amount of magnesium in the oxalate mixture may be calculated from the second step {at which the stages (b) and (e) occur} because the amount of carbon monoxide due to calcium carbonate may be subtracted from the total observed weight loss, and the remainder is thus due to the loss of carbon dioxide and carbon monoxide from anhydrous magnesium oxalate.

Complex materials (for example, clays and soils) have been the subject of thermogravimetric study by Hoffman et al. (Ref. 9). The pyrolysis curves of most soils examined showed plateaus starting between 150–180 °C and extending to 210–240 °C, indicating that hygroscopic moisture and/or easily volatile organic compounds had been removed. When the clay content of a soil was studied the loss in weight at 500 °C read from a pyrolysis curve gave an estimate of the organic matter which was in reasonable agreement with dry combustion and wet oxidation data. An additional feature of the work suggests that lattice water may be quantitatively determined in pure clays. Because lattice water came off from different clays at different temperatures, these temperatures may possibly be used as a method of identification.

XXIII, 5. EXPERIMENTAL. A limited number of thermogravimetric experiments will be outlined below. For more detailed information on these and other studies the reader is referred to the publications listed in the bibliography at the end of this chapter.

When using a modern thermobalance, e.g., the Stanton Redcroft TG-750, which incorporates an electronic microbalance requiring small sample weights, the following operating precautions should be noted:

(a) The weight of sample selected is dependent upon the actual weight loss anticipated.

(b) The crucible should not be handled, because of the danger of transferring grease or moisture to the crucible. A platinum crucible may be cleaned by placing it in dilute nitric acid. If the platinum crucible is heavily contaminated it may be cleaned by heating with a sodium carbonate/sodium nitrate fusion mixture.

(c) A representative sample should be taken from the original batch. If the material is thought to be inhomogeneous, several samples should be run and different results will confirm the inhomogeneity. The sample particle size should be smaller than 100 mesh ($< 150 \, \mu$m) to ensure that an even layer is distributed in the crucible.

(d) The method of obtaining a sample depends upon the nature of the material; thus a circular disc may be cut from a film of material by the use of an appropriate cork borer or leather punch. Fibrous material, which does not pack easily, may be squeezed between metal foil before being transferred to the crucible. Liquid samples may be transferred to the crucible by means of a hypodermic syringe. Air-sensitive samples should be loaded on to the crucible in a glove box and transferred rapidly to the thermobalance which should be set up all ready for a dry inert gas flow. Materials which creep or froth *should not be used* in a thermobalance. It is always sound practice to heat the test material in a small crucible in an oven or muffle furnace to ascertain whether or not there is any creeping or frothing *before* using the sample for thermogravimetry. Considerable damage can occur to a thermobalance if samples are not monitored in this way prior to analysis in the apparatus.

The following experiments are designed to make the operator familiar with the use of the thermobalance:

A. The thermal decomposition of calcium oxalate monohydrate. This determination may be carried out on any standard thermobalance. In all cases the manufacturer's handbook should be consulted for full detailed instructions for operating the instrument.

Initially, zero the balance on the 10 mg range (in the case of the Stanton Redcroft TG-750) with an empty crucible in position and use an air flow of $10 \, \text{cm}^3$ min^{-1}. Weigh accurately about 2 mg of the calcium oxalate monohydrate directly into the crucible and record the weight on the chart. The recorder variable range may now be used to expand the sample weight to 100 per cent of full scale. Select a suitable heating rate (30 °C min^{-1}) and record the pyrolysis curve of calcium oxalate monohydrate from ambient temperature to 1000 °C in terms of percentage sample weight loss. From the TG curve estimate the purity of the calcium oxalate (see Section **XXIII, 4**).

As additional experiments, investigate the decomposition of calcium oxalate in a static air atmosphere and in a nitrogen atmosphere at a flow rate of $10 \, \text{cm}^3$ min^{-1}. Compare the final stage of the decomposition, i.e., the conversion of calcium carbonate to calcium oxide, using different furnace atmospheres.

B. The thermal decomposition of copper sulphate pentahydrate. Follow the procedure outlined in A above, but in this case weigh out accurately about 6 mg of the copper sulphate. Record the thermal decomposition of copper sulphate from ambient temperature to 1000 °C using a heating rate of 10° C min^{-1} and an air atmosphere with a flow rate of $10 \, \text{cm}^3$ min^{-1}. Examine the effect of varying the heating rate on the dehydration reactions by selecting rates of 2, 20 and 100 °C min^{-1} in addition to 10 °C min^{-1} rate used previously. Further experiments may be designed to study the effect of differing particle size on the dehydration reactions of copper sulphate pentahydrate.

C. Other useful substances to study. The following substances show interesting pyrolysis curves and an assessment of the purity of these materials may be investigated:

cadmium sulphate, $3CdSO_4 \cdot 8H_2O$;

ammonium magnesium phosphate, $MgNH_4PO_4.6H_2O$;
and the disodium salt of ethylenediaminetetraacetic acid, $Na_2EDTA.2H_2O$.

The automatic gravimetric determination of calcium and magnesium as their oxalates, outlined in Section **XXIII, 4**, is an obvious extension of experiment A above and may be readily carried out. The reader should be aware that if an early design of thermobalance requiring a sample weight of several hundred milligrams is used a correction curve (Section **XXIII, 2**) must be applied before determining the calcium and magnesium present in the mixture.

XXIII, 6. DIFFERENTIAL THERMAL ANALYSIS AND DIFFEREN-TIAL SCANNING CALORIMETRY. *Introduction.* In differential thermal analysis (DTA) both the test sample and an inert reference material (usually α alumina) undergo a controlled heating or cooling programme which is usually linear with respect to time. There is a zero temperature difference between the sample and the reference material when the former does not undergo any chemical or physical change. If, however, any reaction takes place, then a temperature difference ΔT will occur between the sample and the reference material. Thus in an endothermic change, e.g., when the sample melts or is dehydrated, the sample temperature is lower than that of the reference material. This condition is only transitory because on completion of the reaction the sample will again show zero temperature difference compared with the reference.

In DTA a plot is made of ΔT against temperature or time, if the heating or cooling programme is linear with respect to time. An idealised DTA curve is shown in Fig. XXIII, 7, in which (1) is an exothermic peak and (2) is an

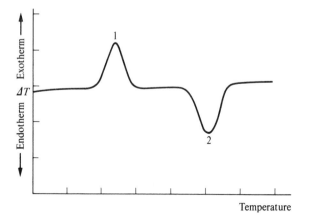

Fig. XXIII, 7

endothermic peak. Both the shape and size of the peaks can give a large amount of information about the nature of the test sample. Thus sharp endothermic peaks often signify changes in crystallinity or fusion processes, whereas broad endotherms arise from dehydration reactions. Physical changes usually result in endothermic curves whilst chemical reactions, particularly those of an oxidative nature, are predominantly exothermic.

Differential scanning calorimetry (DSC) measures the differential energy required to keep both the sample and reference chemicals at the same temperature. Thus, when an endothermic transition occurs, the energy absorbed by the sample is compensated by an increased energy input to the sample in order to maintain a zero temperature difference. Because this energy input is precisely equivalent in magnitude to the energy absorbed in the transition direct calorimetric measurement of the energy of the transition is obtained from this balancing energy. The DSC curve is recorded with the chart abscissa indicating the transition temperature and peak area measures the total energy transfer to or from the sample.

XXIII, 7. INSTRUMENTATION FOR DTA AND DSC. A block diagram of a differential thermal analyser is shown in Fig. XXIII, 8. The basic instrument consists of:

  a sample and reference holder assembly;
  furnace control;
  a reaction chamber allowing analysis in a variety of atmospheric systems;
  a suitable sensor to measure the temperature difference between the sample and reference material;
  an amplifier for ΔT;
  a suitable chart recorder.

Four DTA systems are offered by Stanton Redcroft, operating over a variety of temperature ranges.

The sample platforms and heater assembly of the Stanton Redcroft DTA 671B are shown in Fig. XXIII, 9. Two small metal platforms, (1) and (2), are each welded to a chrome-alumel thermocouple, (3) and (4). The platforms are mounted in the base of a hollow metal cup (5), the walls of which are wound externally with a specially insulated heater (6). The metal disc (7) is above the base of the cylindrical cup through which the thermocouple assemblies protrude. The sample dishes are in good thermal contact with the platforms. The gas inlet (8) to the sample chamber (9) is via a small capillary tube (10).

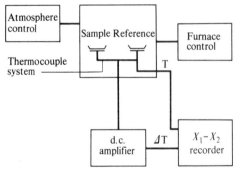

Fig. XXIII, 8

The heater assembly is fitted with a tight fitting metal lid which may incorporate either a Pyrex window or be fitted with a small capillary with one end in position over the sample platform. The total volume of the sample chamber is about 5 cm³.

Sample containers used with this instrument are dish shaped, and are usually made of either aluminium or platinum. The specimen dishes have very flat bases

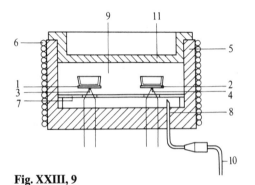

Fig. XXIII, 9

to give good thermal contact between them and the thermocouple platform.

The Perkin Elmer Differential Scanning Calorimeter Model DSC-2 is shown in Fig. XXIII, 10.

In this instrument the sample and reference holders are identical in all respects; both have a built-in heater and temperature sensor. The holders must be thermally and mechanically stable and chemically inert over the entire temperature range of the instrument (-175 °C to 725 °C).

Fig. XXIII, 10

The structure of the holder is in the form of a partially hollowed out cylinder in which the sample or reference pan is placed. Below the base of the cylinder is a platinum resistance thermometer and a platinum wire heating element. The platinum temperature sensors detect the slightest fluctuation of the sample holder temperature compared to the reference holder temperature caused by evolution or absorption of energy by the sample. The electronic system provides differential electrical power to the heaters to compensate exactly for the fluctuation and to maintain a null balance condition. The differential power is read out directly in millicalories per second on the recorder and is equivalent to the rate of energy of sample absorption or evolution. The model DSC-2 operates normally over the range 50° C to 725 °C, but with a sub-ambient accessory the holder enclosure can be cooled to work down to -175 °C.

Eleven different heating rates varying from 0.3125 °C to 320 °C min^{-1} are available from the temperature programmer. The same variety of cooling rates are obtainable from the instrument.

XXIII, 8. EXPERIMENTAL AND INSTRUMENTAL FACTORS. DTA and DSC peaks are also governed by the factors affecting TG curves (Section **XXIII, 2**). Hence, heating rates, atmosphere, and geometry of sample holders can alter the position of DTA and DSC peaks.

However, the most important factor in obtaining reliable results for both techniques is in the preparation of the sample and reference material. Great care should be taken in the preparation of the sample and in the way the crucible or ampoule is loaded. To obtain reproducible results between successive experiments it is essential to ensure that precisely the same packing procedure is carried out each time. The selection and handling of samples for DTA is similar to that outlined in Section **XXIII, 5** (*c*) and (*d*). It is possible, however, to use materials which creep, froth or boil if sealed sample containers are used to ensure no damage occurs to the sample holder assembly. With most modern DTA apparatus a device for encapsulation of the sample is available. In the Perkin Elmer Model DSC-2 it is usual practice to encapsulate the sample in metal pans of high thermal conductivity, to ensure that the sample is in the form of a thin wafer which enables the best thermal contact between sample and temperature sensor.

It is now standard practice to use an empty pan as the reference in DSC (a similar practice is made in DTA when the sample weight is of the order of 1 mg). With higher sample weights it is necessary to use a reference material, for, ideally, the total weight of the sample and its container should be approximately the same as that of the reference and its container. The reference material should be selected so that it possesses similar thermal characteristics to the sample. The most widely used reference material is α alumina which must be of analytical reagent quality. Before use, α alumina should be recalcined and stored over magnesium perchlorate in a desiccator. Kieselguhr is another reference material normally used when the sample is of a fibrous nature. If there is an appreciable difference between the thermal characteristics of the sample and reference materials, or if values of ΔT are large, then dilution of the sample with the reference substance is sensible practice. Dilution may be accomplished by thoroughly mixing suitable proportions of sample and reference material.

XXIII, 9. APPLICATIONS OF DIFFERENTIAL THERMAL ANALYSIS AND DIFFERENTIAL SCANNING CALORIMETRY.

DTA and DSC may both be used in conjunction with TG for certain analytical applications, e.g., the determination of moisture content or the analysis of solid mixtures.

Early applications of DTA were in the qualitative analysis of complex materials. Thus DTA provided a rapid method for the 'fingerprinting' of minerals, clays, and polymeric materials. Indeed, an extremely wide range of materials may be studied by DTA and DSC. The areas of study include thermal stability and decompositions, fusion, phase changes, and purity determinations. A recent important application has been in the measurement of the degree of conversion of high alumina cement, details of which are outlined below.

It must be stressed that all the thermal methods outlined in this chapter are frequently used in conjunction with other techniques. Thus the analysis of evolved gases during a TG, DTA or DSC experiment may be performed by gas chromatography or mass spectrometry. X-ray crystallography may be used to study the structure of reaction intermediates isolated as a result of thermal studies.

XXIII, 10. EXPERIMENTAL.

In this section only one experiment using DTA and one using DSC will be outlined. For detailed information on the considerable number of analyses performed by both of these techniques the reader is referred to the bibliography at the end of this chapter.

(A) DTA studies of copper sulphate pentahydrate. In order to become familiar with the use of the instrument this experiment may be carried out on any standard differential thermal analyser. The manufacturer's handbook should be consulted for detailed instructions on instrument operation.

The dehydration and decomposition peaks of $CuSO_4.5H_2O$ may be compared with those obtained by the TG determination (Section **XXIII, 5, B**).

In the case of the Stanton Redcroft DTA 673 or 674 the following experimental conditions may be used:

Procedure. Weigh accurately the pair of empty crucibles and record each individual weight. Prepare the sample by mixing together equal weights of powdered copper sulphate pentahydrate and α alumina (obtainable from BDH Ltd, Poole, Dorset). Weigh out accurately about 60 mg of the diluted sample into one of the previously weighed crucibles. Load the other crucible with α alumina so that the combined weight of this reference crucible plus α alumina is equal to the weight of the sample crucible and the diluted sample. Insert the crucibles, using tweezers, carefully into the thermocouple wells ensuring that they are correctly located (the reference crucible goes into the front well). Lower the furnace over the sample holder assembly and locate in the furnace mounting. Select the appropriate sample atmosphere, e.g., flowing dry air, at a heating rate of 10 °C min^{-1}. Choose the appropriate amplifier range, chart speed and temperature programme, from ambient temperature to 1000 °C (for full details see the manufacturer's handbook). Record the DTA plot for copper sulphate pentahydrate over the desired temperature range.

To gain further experience with the use of the instrument the dehydration studies of $CuSO_4.5H_2O$ (up to about 550° C) may be performed using, (i) different heating rates, (ii) various gas atmospheres, (iii) samples of different particle size.

Materials that undergo crystalline transition and fusion may be useful for alternative DTA studies. Suitable substances include potassium nitrate, sodium chromate, and potassium sulphate.

(B) The determination of the degree of conversion of high alumina cement. *Introduction.* Both DTA and DSC instruments have been used for the study of conversion of high alumina cements (Ref. 14). In the experiment described below the Perkin Elmer Model DSC-2 Differential Scanning Calorimeter was employed.

Theory. High alumina cements undergo a 'conversion' reaction whereby the metastable compounds $CaO.Al_2O_3.10H_2O$, $2CaO.Al_2O_3.8H_2O$ and alumina gel of the early set cement are converted to more stable materials represented by the following reactions:

(*a*) $3(CaO.Al_2O_3.10H_2O) \longrightarrow 3CaO.Al_2O_3.6H_2O + 2(Al_2O_3.3H_2O)$

(*b*) $3(2CaO.Al_2O_3.8H_2O) \longrightarrow 2(3CaO.Al_2O_3.6H_2O) + Al_2O_3.3H_2O$

(*c*) $\qquad Al_2O_3.xH_2O \longrightarrow Al_2O_3.3H_2O$

Only small quantities of $2CaO.Al_2O_3.8H_2O$ are usually formed during the hydration of the cement, and the alumina gel initially formed disappears after a few months. To determine the degree of conversion the quantities of the materials in the above equations must be estimated. The degree of conversion may be described as follows: half conversion is when the quantity of $3CaO.Al_2O_3.6H_2O$ is equal to the quantity of $CaO.Al_2O_3.10H_2O$. At the stage of approximately 50 per cent conversion the DSC peaks are found to be of equal size. A third peak due to $Al_2O_3 . 3H_2O$ is also present, and it is found to be of an equivalent size.

The degree of conversion may be obtained from the relationship:

$$\frac{\text{amount of } 3CaO . Al_2O_3 . 6H_2O \times 100}{\text{amount of } 3CaO . Al_2O_3 . 6H_2O + \text{amount of } CaO . Al_2O_3 . 10H_2O}$$

When cement is exposed naturally a carbo-aluminate $3CaO . Al_2O_3 . CaCO_3 .$ aq. may be formed due to the carbonation of $3CaO . Al_2O_3 . 6H_2O$ and this could result in a small apparent degree of conversion. Because of this fact and since $Al_2O_3 . 3H_2O$ is not decomposed the progress of the reactions shown by equations (a) and (c) may be determined by the following relationship:

$$\frac{\text{amount of } Al_2O_3 . 3H_2O \times 100}{\text{amount of } Al_2O_3 . 3H_2O + \text{amount of alumina gel} + CaO . Al_2O_3 . 10H_2O}$$

As mentioned above, the alumina gel disappears after a few months and the relationship becomes:

$$\frac{\text{amount of } Al_2O_3 . 3H_2O \times 100}{\text{amount of } Al_2O_3 . 3H_2O + \text{amount of } CaO . Al_2O_3 . 10H_2O}$$

Strong endothermic peaks are obtained for $CaO . Al_2O_3 . 10H_2O$ at 110–120 °C, $Al_2O_3 . 3H_2O$ at 295–310 °C, and for $3CaO . Al_2O_3 . 6H_2O$ at 320–350 °C. The exact value of the transition temperatures depends upon factors such as sample size, sample packing, and heating rate. Using DTA, the percentage of the compounds is proportional to peak height, the relationship does not hold for DSC, and therefore it is necessary to calibrate using standard samples of high alumina cement in which the degree of conversion is known. In this way the following relationship is obtained:

$$\frac{B \times \text{peak height due to } Al_2O_3 . 3H_2O \times 100}{(B \times \text{peak height due to } Al_2O_3 . 3H_2O) + (A \times \text{peak height due to } CaO . Al_2O_3 . 10H_2O)}$$

where 'A' and 'B' are constants obtained from the calibration standards. The peaks chosen are the broad endotherm at 100–120 °C ($CaO . Al_2O_3 . 10H_2O$) and the first of the two peaks (if two peaks are present) at 300–350 °C ($Al_2O_3 . 3H_2O$).

Problems may arise if other materials are present in the sample: e.g., Ettringite, $3CaO . Al_2O_3 . 3CaSO_4 . 32H_2O$, may be formed due to sulphate attack and produces a sharp peak superimposed on the broad endotherm due to $CaO . Al_2O_3 . 10H_2O$.

THERMOMETRIC TITRATIONS

XXIII, 11. INTRODUCTION. Although thermometric titrimetry is not strictly a thermal method of analysis as previously defined (Section **XXIII, 1**), the technique merits inclusion in this part of the book. Thermometric titrations involve the mesurement of the change in temperature of a system as a function of time, or of the volume of titrant. Thus the temperature change of a solution is recorded as titrant is added. The titrations are carried out under as near adiabatic conditions as possible in order to minimise heat losses between the titrated solution and its surroundings. In practice the titrant is added from an automatic burette delivering a constant volume into a thermally insulated vessel containing the titrand (the solution to be titrated). A plot of temperature against volume of titrant (or time) is made. A thermogram of a simple acid-base thermometric titration is shown in Fig. XXIII, 11. AB corresponds to the situation before titrant

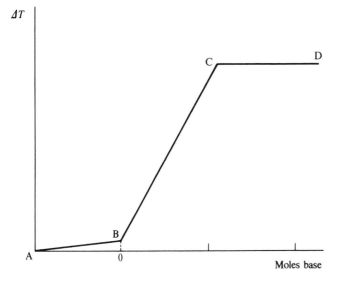

Fig. XXIII, 11

is added and the titration vessel and its contents are gaining or losing heat to the surroundings. At B the titration is started and heat is evolved, mainly from heat of neutralisation. At C the end-point is reached and the heat changes recorded after this, along CD, arise only from heats of dilution and differences in temperature that may exist between the titrant and titrand.

XXIII, 12. THEORY. An essential condition for a successful titration is that there should be a change in free energy, $\Delta G^{\ominus}$, in the end-point region. This condition is based on the equilibrium constants of the reactions involved.

Thus $-\Delta G^{\ominus} = RT \ln K$

A thermometric titration, however, depends upon the heat of the reaction in accordance with the equation

$$\Delta H = \Delta G + T \Delta S$$

Hence a thermometric titration may be feasible when there is a significant change in the entropy term ΔS (providing that free energy changes are favourable), to yield an appreciable overall change in enthalpy ΔH. The potentiometric titration of a weak acid, boric acid, with sodium hydroxide affords a scarcely discernible end point, whereas the thermometric titration of boric acid under the same conditions gives a clearly marked end point (Ref. 10).

The change in temperature ΔT of an acid-base titration is dependent upon the molar heat of neutralisation ΔH_m and is given by the relation:

$$\Delta T = \frac{N\Delta H_m}{Q}$$

where N is the number of moles of water formed by neutralisation and Q is the heat capacity. Since ΔH_m and Q are constant throughout the titration, ΔT is proportional to N.

XXIII, 13. INSTRUMENTATION. The basic instrumentation introduced by Linde, Rogers and Hume (Ref. 11) extended the scope of thermometric titrimetry. The titrant was added continuously by means of a motor-driven syringe burette into a Dewar flask fitted with a mechanical stirrer. A thermistor, which formed part of a bridge circuit, was placed in the titration vessel and acted as the temperature sensor. Continuous recording of the unbalance potential from the bridge circuit enabled rapid automatic titrations to be made for the first time.

A commercial instrument, the Aminco Titra-Thermo-Mat (American Instrument Co, Silver Spring, Maryland, USA) has been developed, in which the titration vessel is located in an insulated enclosure—the 'adiabatic titration tower'. The output from two thermistors, one measuring the titrant temperature and the other the temperature of the titration vessel, is fed into a bridge circuit.

A simple apparatus, employing a single thermistor, which is suitable for elementary studies is described by Williams (Ref. 12), in which a magnetic stirrer mixes the solution contained in an inner beaker which is thermally insulated from the surroundings by an outer beaker. A Perspex lid covers the beakers and supports the thermistor and the capillary tip of a burette. A constant flow is maintained by using a simple constant pressure head device, a separatory funnel fitted with a capillary and screw-clip. The thermistor, Stantel F23 (supplied by Electronic Services, Harlow, Essex) is connected into one arm of the Wheatstone bridge circuit, shown in Fig. XXIII, 12, in which R is a resistance box.

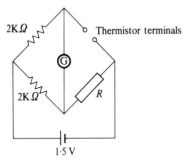

Fig. XXIII, 12

The titration is followed by plotting the thermistor resistance against time as follows. A given resistance value is set on the resistance box and the time noted when the galvanometer G indicates null deflection. The resistance is then changed to another fixed value and the time is noted again for null deflection. A series of readings is obtained and a resistance against time plot is constructed, from which the point of inflection is obtained by drawing lines through the experimental points.

XXIII, 14. APPLICATIONS. Thermometric titrations may be used in aqueous solutions to follow neutralisation, precipitation, complexation, and redox reactions.

Neutralisation titrations. Thermometric titrations of strong acids and bases are widely reported and may be easily performed using the instrumental methods outlined above. However, a more significant application is the titration of weak acids with strong bases, e.g., boric acid and sodium hydroxide. In general, $0.01M$

solutions of all acids with $pK_a < 10$ may be titrated thermometrically with a precision of 1 per cent provided the heat neutralisation is not less than 42 kJ mole^{-1}. Mixtures of a strong acid and a weak acid can be titrated and the resulting thermogram has two inflection points.

Precipitation titrations. When a slightly soluble compound MX is formed heat is either evolved or absorbed according to the equation:

$$M^+_{(aq)} + X^-_{(aq)} \longrightarrow MX_{(s)} \pm \Delta H$$

where ΔH is the heat of reaction. If ΔH is sufficiently large then the heat change may be followed by thermometric titrimetry. Examples include determination of halides with silver and mercury(II) ions, and the estimation of calcium, strontium, or barium with oxalate ion.

Redox titrations. Although relatively little work has been done on redox reactions, the technique appears to be successful for these purposes. Thermometric titrations of permanganate with iron(II) and with oxalic acid compare favourably with the results obtained by conventional titrimetry.

Complexation titrations. Jordan and Alleman (Ref. 13) have studied the thermometric titrations of divalent metal cations with EDTA. An interesting application is to the titration of a binary mixture of calcium and magnesium ions with the tetrasodium salt of EDTA. In the normal titrimetric procedure using Solochrome Black as indicator (Section **X, 55**) only the total calcium and magnesium can be evaluated because the log stability constants of calcium EDTA and magnesium EDTA are 10.7 and 8.7 respectively. Using a thermometric titration the curves obtained (Fig. XXIII, 13) show an exothermic section AB (calcium) and an endotherm BC (magnesium). The endothermic character of the chelation of the magnesium ion and the exothermic character of the chelation of calcium are mainly attributed to a significant difference in the entropies of chelation.

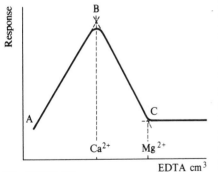

Fig. XXIII, 13

XXIII, 15. EXPERIMENTAL. The following experiment is illustrative of the procedure which has to be followed for thermometric titrations and can be directly adapted for use in other determinations such as those mentioned in Section **XXIII, 14**.

The thermometric titration of chloride

The apparatus described by Williams (Ref. 12) and previously outlined in Section **XXIII, 12** is suitable for this titration. A Pye Scalamp galvanometer and a Cambridge Instrument Decade Resistance Box are used in the bridge circuit together with the electronic components indicated in Fig. XXIII, 12. The constant head device is a 250-cm^3 separatory funnel fitted with a capillary and a screw clip.

Procedure. Prepare the following solutions:*

An approximately $0.1M$-sodium chloride solution by weighing out about 3.0 g of the Analar salt and dissolving it in 500 cm^3 of water.

A 0.05M-silver nitrate solution by weighing out accurately 4.248 g of the salt and dissolving it in 500 cm^3 of water in a graduated flask.

Pipette 50 cm^3 of the 0.05M-silver nitrate solution into a 250-cm^3 beaker which serves as the inner titration vessel and add 10 cm^3 of 2.5M-nitric acid. Place the 250-cm^3 beaker in an outer 600-cm^3 beaker and enclose the whole in a block of expanded polystyrene. Locate the thermistor into the titration cell and connect it to the bridge circuit. Fill the constant head separatory funnel with the sodium chloride solution and adjust the flow rate with the screw-clip to about 2.5 cm^3 min^{-1}. Connect the 'burette' capillary tip to the titration vessel and start the titration. Using the procedure outlined in Section **XXIII, 12**, plot a graph of thermistor resistance against time and determine the point of inflexion. Hence calculate the chloride ion concentration.

XXIII, 16. References

1. C. Duval and M. de Clercq (1951). *Anal. Chim. Acta.*, **5**, 282.
2. E. K. Simons and A. E. Newkirk (1964). *Talanta*, **11**, 549.
3. A. W. Coats and J. P. Redfern (1963). *Analyst*, **88**, 906.
4. R. C. Turner, I. Hoffman and D. Chen (1963). *Can. J. Chem.*, **41**, 243.
5. G. M. Lukaszewski and J. P. Redfern (1961). *Lab. Practice*, **10**, 552.
6. W. W. Wendlandt (1960). *Anal. Chem.*, **32**, 848.
7. W. J. Blaedel and H. T. Knight (1954). *Anal. Chem.*, **26**, 741.
8. C. Duval (1963). *Inorganic Thermogravimetric Analysis*. 2nd edn. p. 93. Amsterdam, London and New York; Elsevier.
9. I. Hoffman, M. Schnitzer and J. R. Wright (1959). *Anal. Chem.*, **31**, 440.
10. J. Jordan (1963). *J. Chem. Educ.*, **40**, A5.
11. H. W. Linde, L. B. Rogers and D. N. Hume (1953). *Anal. Chem.*, **25**, 404.
12. D. R. Williams (1971). *Education in Chemistry*, **8**, 97.
13. J. Jordan and T. G. Alleman (1957). *Anal. Chem.*, **29**, 9.
14. H. G. Midgley and A. Midgley (1975). *Magazine of Concrete Research*, **27**, 55.

XXIII, 17. Selected bibliography

1. C. Duval (1963). *Inorganic Thermogravimetric Analysis*. 2nd edn. Amsterdam, London and New York; Elsevier.
2. P. D. Garn (1966). *Thermoanalytical Methods of Investigation*, New York; Academic Press.
3. C. J. Keatch (1975). *An Introduction to Thermogravimetry*. 2nd edn. London; Heyden.
4. R. C. Mackenzie (ed.). *Differential Thermal Analysis*. Vol. 1 (1970), and Vol. 2 (1972). London; Academic Press.
5. W. W. Wendlandt (1974). *Thermal Methods of Analysis*. 2nd edn. New York; Wiley-Interscience.
6. R. F. Schwenker and P. D. Garn (eds.). *Thermal Analysis*. Vol. 1 (1969), and Vol. 2 (1968). New York; Academic Press.
7. H. J. V. Tyrrell and A. E. Beezer (1968). *Thermometric Titrimetry*. London; Chapman and Hall.
8. L. G. Bark and S. G. Bark (1968). *Thermometric Titrimetry*. Oxford; Pergamon.

---

* It is sound practice to make up the sodium chloride and silver nitrate solutions and leave them in a thermostat for 2–3 hours before proceeding with the thermometric titration. This is to ensure there is no appreciable temperature difference between the titrant and titrand solutions.

APPENDICES

APPENDIX I INTERNATIONAL ATOMIC WEIGHTS, 1973

Element	Symbol	Atomic No.	Atomic Weight	Element	Symbol	Atomic No.	Atomic Weight
Actinium	Ac	89	(227)	Mercury	Hg	80	200.59
Aluminium	Al	13	26.9815	Molybdenum	Mo	42	95.94
Americium	Am	95	(243)	Neodymium	Nd	60	144.24
Antimony	Sb	51	121.75	Neon	Ne	10	20.179
Argon	Ar	18	39.948	Neptunium	Np	93	237.0482
Arsenic	As	33	74.9216	Nickel	Ni	28	58.70
Astatine	At	85	(210)	Niobium	Nb	41	92.9064
Barium	Ba	56	137.34	Nitrogen	N	7	14.0067
Berkelium	Bk	97	(247)	Nobelium	No	102	(255)
Beryllium	Be	4	9.0122	Osmium	Os	76	190.2
Bismuth	Bi	83	208.9804	Oxygen	O	8	15.9994
Boron	B	5	10.81	Palladium	Pd	46	106.4
Bromine	Br	35	79.904	Phosphorus	P	15	30.9738
Cadmium	Cd	48	112.40	Platinum	Pt	78	195.09
Calcium	Ca	20	40.08	Plutonium	Pu	94	(244)
Californium	Cf	98	(251)	Polonium	Po	84	(209)
Carbon	C	6	12.011	Potassium	K	19	39.098
Cerium	Ce	58	140.12	Praseodymium	Pr	59	140.9077
Cesium	Cs	55	132.9054	Promethium	Pm	61	(145)
Chlorine	Cl	17	35.453	Protactinium	Pa	91	231.0359
Chromium	Cr	24	51.996	Radium	Ra	88	226.0254
Cobalt	Co	27	58.9332	Radon	Rn	86	(222)
Copper	Cu	29	63.546	Rhenium	Re	75	186.207
Curium	Cm	96	(247)	Rhodium	Rh	45	102.9055
Dysprosium	Dy	66	162.50	Rubidium	Rb	37	85.4678
Einsteinium	Es	99	(254)	Ruthenium	Ru	44	101.07
Erbium	Er	68	167.26	Samarium	Sm	62	150.4
Europium	Eu	63	151.96	Scandium	Sc	21	44.9559
Fermium	Fm	100	(257)	Selenium	Se	34	78.96
Fluorine	F	9	18.9984	Silicon	Si	14	28.086
Francium	Fr	87	(223)	Silver	Ag	47	107.868
Gadolinium	Gd	64	157.25	Sodium	Na	11	22.9898
Gallium	Ga	31	69.72	Strontium	Sr	38	87.62
Germanium	Ge	32	72.59	Sulphur	S	16	32.06
Gold	Au	79	196.9665	Tantalum	Ta	73	180.9479
Hafnium	Hf	72	178.49	Technetium	Tc	43	(97)
Helium	He	2	4.0026	Tellurium	Te	52	127.60
Holmium	Ho	67	164.9304	Terbium	Tb	65	158.9254
Hydrogen	H	1	1.0079	Thallium	Tl	81	204.37
Indium	In	49	114.82	Thorium	Th	90	232.0381
Iodine	I	53	126.9045	Thulium	Tm	69	168.9342
Iridium	Ir	77	192.22	Tin	Sn	50	118.69
Iron	Fe	26	55.847	Titanium	Ti	22	47.90
Krypton	Kr	36	83.80	Tungsten	W	74	183.85
Lanthanum	La	57	138.9055	Uranium	U	92	238.029
Lawrencium	Lr	103	(260)	Vanadium	V	23	50.9414
Lead	Pb	82	207.2	Xenon	Xe	54	131.30
Lithium	Li	3	6.941	Ytterbium	Yb	70	173.04
Lutetium	Lu	71	174.97	Yttrium	Y	39	88.9059
Magnesium	Mg	12	24.305	Zinc	Zn	30	65.38
Manganese	Mn	25	54.9380	Zirconium	Zr	40	91.22
Mendelevium	Md	101	(258)				

Notes:
1. This table is scaled to the relative atomic mass $A_r(^{12}C) = 12$.
2. Values in parentheses refer to the isotope of longest known half-life for radioactive elements.
3. Information provided here is based upon the Report of the Commission on Atomic Weights, Pure and Applied Chemistry, (1974), **37**, 589.

APPENDIX II INDEX OF ORGANIC CHEMICAL REAGENTS

The following table is included by kind permission of Hopkin and Williams Ltd. Although the authors are in no position to guarantee the claims made for any particular reagent listed, the table is included to enable the reader to find what reagents are available for the detection or determination of the commoner metals, and some of the anions and miscellaneous substances that are included.

The figures in the table refer to the literature items in the list that follows the table. Italic figures indicate that the methods referred to are qualitative only. Some of the reagents and methods are sufficiently well established to appear in standard text books and some of these are cited where they seem appropriate. Apart from those that appear in the list, mention should also be made of six further works that deal particularly with these methods, namely:—

Noel L. Allport and John E. Brocksopp (1963). *Colorimetric Analysis*. Vol. 2. Chapman and Hall.

E. B. Sandell, 3rd edn. (1959). *Colorimetric Determination of Traces of Metals*. Interscience Publishers Inc.

G. Charlot (1964). *Colorimetric Determination of Elements*. Elsevier.

D. F. Boltz (1958). *Colorimetric Determination of Non-metals*. Interscience Publishers Inc.

C. A., C. T. and F. D. Snell. *Colorimetric Methods of Analysis*. 6 vols. 1948–1962. D. van Nostrand.

R. E. Stanton (1966). *Rapid Methods of Trace Analysis*. London; Edward Arnold (Publishers) Ltd.

For the more recently introduced reagents or methods, reference is made to papers in the journal literature. Only one paper is generally given for each method and this has often been selected from many, either because it appears the most important or because it is the most recent and provides, at the same time, a key to previous writings on the subject.

	Aluminium	Ammonium	Antimony	Arsenic	Barium	Beryllium	Bismuth	Boron	Cadmium	Caesium	Calcium	Cerium	Chloride, Chlorine	Chromium	Cobalt	Copper	Cyanide	Fluoride	Gallium	Germanium	Gold	Hafnium	Hydrogen sulphide	Indium	Iridium	Iron	Lanthanoids	Lead	Lithium	Magnesium	Manganese	Mercury	Molybdenum
Alizarin fluorine blue																		2															
Alizarin fluorine blue lanthanum complex preparation																		184															
Alizarin red S	12					13												14															
Aluminon	1			159												31																	
4-Aminophenazone																																	
Ammonium pyrrolidine-dithiocarbamate			190	190			190		190					190	190	190					190					190						190	190
Arsenazo										161																	182						
Arsenazo III																																	
Astrazone pink FG																																	
Barium chloranilate						7																											
Benzoin α-oxime																1																	1
5,6-Benzoquinoline									90										91														
N-Benzoyl-N-phenyl-hydroxylamine	2					2						2			2	2			2						2	2							
Beryllon II				23																													
2,2'-Bipyridyl																										47							43
2,2'-Biquinolyl											2																						
NN'-Bisallylthiocarbamoyl-hydrazine							25									131											24					26	
Biscyclohexanone oxalyldihydrazone																1																	
Bis-(3-methyl-1-phenyl-pyrazol-5-one)		2															2																
Bithionol																										191							
4-Bromomandelic acid																																	
Bromopyrogallol red			193				193													193				193				193					
Cacotheline																																	
Cadion									30																						29		
Calmagite																																	
Carmine						2																											
Chloranilic acid											1																						
2-Chloro-4-nitro-benzenediazonium naphthalene-2-sulphonate																																	
N-Cinnamoyl-N-phenyl-hydroxylamine																																	
Cleve's acid																																	
Cupferron												36							32									33	36				
Copper(II) ethyl acetoacetate															188																		
Curcumin								2																									
Diacetyl dithiol* (see also Toluene-3,4-dithiol)			2	2		2			2							2															2	2	2
Diaminoethane-NN-di-(2-hydroxyphenylacetic acid)																											166						
1,1'-Dianthrimide								1												167													
Dibenzoyl dithiol (see also Toluene-3,4-dithiol)																																	
NN'-Dibenzyldithiooxamide																																	
4-Diethylaminobenzyl-idene-rhodanine																					41												
Diethylammonium diethyldithiocarbamate																1																	
NN-Diethyl-p-phenylene-diamine sulphate													189																				
Di-(2-hydroxy-phenyl-imino)ethane	2								2		2																						

* For the use of diacetyl dithiol as a coagulant, catalyst and precipitant for sulphur and Group II sulphides, see Ref. 2.

	Nickel	Niobium	Nitrate, Nitrite	Osmium	Palladium	Perchlorate	Phenols	Phosphate	Platinum	Potassium	Rhenium	Rhodium	Rubidium	Ruthenium	Scandium	Selenium	Silver	Sodium	Strontium	Sulphate	Sulphides, organic	Sulphur dioxide	Tantalum	Tellurium	Thallium	Thorium	Tin	Titanium	Tungsten	Uranium	Vanadium	Yttrium	Zinc	Zirconium
Alizarin fluorine blue																																		
Alizarin fluorine blue lanthanum complex preparation																																		
Alizarin red S																										15								16
Aluminon																																		
4-Aminophenazone							1																											
Ammonium pyrrolidine-dithiocarbamate	190															190								190	190		190							
Arsenazo																								182						182				
Arsenazo III															165									58						165				113
Astrazone pink FG																						17												
Barium chloranilate																				18														
Benzoin α-oxime																									1									
5,6-Benzoquinoline																																		
N-Benzoyl-N-phenyl-hydroxylamine	2	2														2				2				2	2	2				2	2			2
Beryllon II																																		
2,2'-Bipyridyl																																		
2,2'-Biquinolyl																																		
N,N'-Bisallylthio-carbamoyl-hydrazine	131					28										27																	131	
Biscyclohexanone oxalyldihydrazone																																		
Bis-(3-methyl-1-phenyl-pyrazol-5-one)																																		
Bithionol																																		
4-Bromomandelic acid																																		2
Bromopyrogallol red		193													193		193											193			193			
Cacotheline																											1							
Cadion																																		
Calmagite																											5							
Carmine																																		
Chloranilic acid																			1															1
2-Chloro-4-nitro-benzenediazonium naphthalene-2-sulphonate							170																											
N-Cinnamoyl-N-phenyl-hydroxylamine	2																			2											2			
Cleve's acid			192																															
Cupferron	57																									34	33			61	35			
Copper(II) ethyl acetoacetate																																		
Curcumin																																		
Diacetyl dithiol* (see also Toluene-3,4-dithiol)				2					2							2								2			2	2						
Diaminoethane-NN-di-(2-hydroxyphenylacetic acid)																																		
1,1'-Dianthrimide																173								172										
Dibenzoyl dithiol (see also Toluene-3,4-dithiol)					2																													
N,N'-Dibenzyldithio-oxamide			144				144																											
4-Diethylaminobenzyl-idene-rhodanine																																		
Diethylammonium diethyldithiocarbamate																																	1	
N,N-Diethyl-p-phenylene-diamine sulphate																																		
Di-(2-hydroxy-phenyl-imino)ethane																															4			

	Aluminium	Ammonium	Antimony	Arsenic	Barium	Beryllium	Bismuth	Boron	Cadmium	Cæsium	Calcium	Cerium	Chloride, Chlorine	Chromium	Cobalt	Copper	Cyanide	Fluoride	Gallium	Germanium	Gold	Hafnium	Hydrogen sulphide	Indium	Iridium	Iron	Lanthanoids	Lead	Lithium	Magnesium	Manganese	Mercury	Molybdenum
Dihydroxytartaric acid																																	
Dimercaptothiadiazole							63																										
4-Dimethylaminoazo-benzene-4′-arsonic acid																*43*																	
4-Dimethylaminobenzyl-idene-rhodanine															1				*43*														1
9-(4-Dimethylamino-phenyl)-2,3,7-tri-hydroxy-6-fluorone																				179													
4-Dimethylaminostyryl-β-naphthiazole methiodide																																	
2,9-Dimethyl-4,7-diphenyl-1,10-phenanthroline																2																	
Dimethylglyoxime							1																					1					
2,9-Dimethyl-1,10-phenanthroline																2																	
4,4′-Dinitrodiphenyl-carbazide									*43*																								
1,5-Diphenylcarbazide							1							1																1			
Diphenylcarbazone																																	1
NN-Diphenylhydrazine hydrochloride																																	
4,7-Diphenyl-1.10-phenanthroline																										2							
4,7-Diphenyl-1,10-phenanthrolinedi-sulphonic acid sodium salt																										160							
1,3-Diphenylpropane-1,3-dione																																	
Dithio-oxamide							1								96	1												*1*					*1*
Dithizone									2																			2				2	
Eriochrome cyanine R	2															119		117						117						118			
Formaldoxime hydrochloride																																9	
Furil α-dioxime																											53						
Gallein																																	
Hæmatoxylin	2																																
Hexanitrodiphenylamine										56																							
2-(Hydroxymercury)-benzoic acid																					155												
8-Hydroxyquinaldine				70							65		138	68				62						59			65	59		60		22	
8-Hydroxyquinoline	33								33		65				33			66						67			65			33			
Lanthanum chloranilate																63																	
Magnesium blue																														*185*			
Magnesium uranyl acetate																																	
Magnesons I and II																														1			
Mandelic acid																																	
Mercaptoacetic acid															152													49					
2-Mercaptobenzothiazole						79			78									77			76						73			75			74
Mercury(II) chloranilate													157																				
3-Methoxynitrosophenol															2												2						
α-Methoxyphenylacetic acid																																	
Methylfluorone			1																														
Mordant red 74				19																													
Morin	84			85	88																												
Naphthalhydroxamic acid												89																					
Nickel uranyl acetate																																	
Nioxime																																	
Nitron																																	
4-p-Nitrophenylazo-orcinol						93																											
1-Nitroso-2-naphthol															33	176											176						

	Nickel	Niobium	Nitrate	Osmium	Palladium	Perchlorate	Phenols	Phosphate	Platinum	Potassium	Rhenium	Rhodium	Rubidium	Ruthenium	Scandium	Selenium	Silver	Sodium	Strontium	Sulphate	Sulphides, organic	Sulphur dioxide	Tantalum	Tellurium	Thallium	Thorium	Tin	Titanium	Tungsten	Uranium	Vanadium	Yttrium	Zinc	Zirconium
Dihydroxytartaric acid																	42																	
Dimercaptothiadiazole			64																															
4-Dimethylaminoazo-benzene-4'-arsonic acid																																		44
4-Dimethylaminobenzyl-idenerhodanine																1																		
9-(4-Dimethylamino-phenyl)-2,3,7-tri-hydroxy-6-fluorone																							183											
4-Dimethylaminostyryl-β-naphthiazole methiodide																																	1	
2,9-Dimethyl-4,7-diphenyl-1,10-phenanthroline																																		
Dimethylglyoxime	1		1																															
2,9-Dimethyl-1,10-phenanthroline																																		
4,4'-Dinitrodiphenyl-carbazide																																		
1,5-Diphenylcarbazide																																		
Diphenylcarbazone																																		
NN-Diphenylhydrazine hydrochloride																*43*																		
4,7-Diphenyl-1,10-phenanthroline																																		
4,7-Diphenyl-1,10-phenanthrolinedi-sulphonic acid disodium salt																																		
1,3-Diphenylpropane-1,3-dione																															2			
Dithio-oxamide	96				*1*				*1*	48						*1*																		
Dithizone																	2																2	
Eriochrome cyanine R																																		
Formaldoxime hydrochloride																																		
Furil α-dioxime	2		52		2				2																									
Gallein																																		
Hæmatoxylin																																		
Hexanitrodiphenylamine										55	56																							
2-(Hydroxymercury)-benzoic acid																					180													
8-Hydroxyquinaldine																	59									59	69	59			60			
8-Hydroxyquinoline	33	71																										102	178			67	33	
Lanthanum chloranilate								164																										
Magnesium blue																																		
Magnesium uranyl acetate																		33																
Magnesons I and II																																		
Mandelic acid																																		60
Mercaptoacetic acid																																		168
2-Mercaptobenzothiazole			80										74										79											
Mercury(II) chloranilate																																		
3-Methoxynitrosophenol																																		
α-Methoxyphenylacetic acid																		2																
Methylfluorone																																		140
Mordant red 74																																		
Morin																														87				86
Naphthalhydroxamic acid																																		
Nickel uranyl acetate																		92																
Nioxime	1		124																															
Nitron		1		1						1																			1					
4-p-Nitrophenylazo-orcinol																																		
1-Nitroso-2-naphthol	176			94																										176			176	

875

Reagent	Aluminium	Ammonium	Antimony	Arsenic	Barium	Beryllium	Bismuth	Boron	Cadmium	Caesium	Calcium	Cerium	Chloride, Chlorine	Chromium	Cobalt	Copper	Cyanide	Fluoride	Gallium	Germanium	Gold	Hafnium	Hydrogen sulphide	Indium	Iridium	Iron	Lanthanoids	Lead	Lithium	Magnesium	Manganese	Mercury	Molybdenum
2-Nitroso-1-naphthol															95																		
Nitroso-R salt															2											2							
1,10-Phenanthroline hydrate							169								1											1							21
Pheolphthalein															98	97																	
Phenylarsonic acid				99																													
p-Phenylenediamine dihydrochloride																145																	
Phenylfluorone																				1													116
Phenyl-2-pyridyl ketoxime																												2					
Phenylthiohydantoic acid															50																		
Picrolonic acid										103																							
Potassium dibenzyldithiocarbamate															104																		
1-(2-Pyridylazo)-2-napthol									40										40					175	11	40						40	40
4-(2-Pyridylazo)resorcinol															174												174						
2-(2-Pyridyl)benzimidazole																												105					
2-(2-Pyridyl)imidazoline																												105					
Quinaldic acid									33						33																		
Quinoxaline-2,3-dithiol															107																		
Rhodamine B		108																	110		109												
Rhodamine S		54																															
Salicylaldehyde oxime						111									33												112						
Salicylideneamino-2-thiophenol																																	
Silver diethyldithiocarbamate				142																													
Sodium diethyldithiocarbamate																1																	
Sodium rhodizonate					*43*						*43*																	*43*					
Sodium tetraphenylboron	2									2																						2	
SPADNS																		122															
Stilbazo	123																							154									153
Tannic acid	128					128									129	125																	
2,2':6',2''-Terpyridyl															141																		
1-(2-Thenoyl)-3,3,3-trifluoroacetone															51	6		82										72					
Thioacetamide	For the precipitation of analytical group II and IV metals. Ref. 2																																
Thiourea							148		147																								147
Thiurone																								83									
Thorin								2										2									2						
Tiron																										2							2
Titan yellow																														1			
Toluene-3,4-dithiol (see also Zinc dithiol)			*38*	2	*149*		*149*								38	38			*149*	2	*149*			*149*	*149*	38		2				2	1
2,4,6-Tri(2-pyridyl)-1,3,5-triazine																												81					
Uranyl zinc acetate																																	
Victoria violet					137																												
2,4-Xylenol																																	
2,6-Xylenol																																	
Xylenol orange						135													20			134											
Zinc dibenzyldithiocarbamate																1																	
Zinc OO-di-iso-propyl phosphorodithioate															162																		
Zinc dithiol* (see also Toluene-3,4-dithiol)			*2*	*2*	*2*		*2*								*2*	*2*			*2*	*2*	*2*			*2*	*2*	*2*		*2*			*2*	*2*	*2*
Zincon																																2	

	Nickel	Niobium	Nitrate	Osmium	Palladium	Perchlorate	Phenols	Phosphate	Platinum	Potassium	Rhenium	Rhodium	Rubidium	Ruthenium	Scandium	Selenium	Silver	Sodium	Strontium	Sulphate	Sulphides, organic	Sulphur dioxide	Tantalum	Tellurium	Thallium	Thorium	Tin	Titanium	Tungsten	Uranium	Vanadium	Yttrium	Zinc	Zirconium
2-Nitroso-1-naphthol				94										37																				
Nitroso-R salt																																		
1,10-Phenanthroline hydrate				1																												1		
Phenolphthalein																																		
Phenylarsonic acid																												101						100
p-Phenylenediamine dihydrochloride																																		
Phenylfluorone																																		
Phenyl-2-pyridyl ketoxime																																		
Phenylthiohydantoic acid																																		
Picrolonic acid																																		
Potassium dibenzyldithiocarbamate																																		
1-(2-Pyridylazo)-2-naphthol											11																			158		40		139
4-(2-Pyridylazo)resorcinol		143																												174				
2-(2-Pyridyl)-benzimidazole																																		
2-(2-Pyridyl)-imidazoline																																		
Quinaldic acid				106																													33	
Quinoxaline-2,3-dithiol	107			181																														
Rhodamine B																				3									171					
Rhodamine S																																		
Salicylaldehyde oxime	33																																111	
Salicylideneamino-2-thiophenol																									187									
Silver diethyldithiocarbamate																																		
Sodium diethyldithiocarbamate																																	1	
Sodium rhodizonate																			43	114														
Sodium tetraphenylboron										2			2												2									
SPADNS																											120							121
Stilbazo																																		
Tannic acid		126																					126					126	130					127
2,2′:6′,2″-Terpyridyl																																		
1-(2-Thenoyl)-3,3,3-trifluoroacetone	8																													39				
Thioacetamide	For the precipitation of analytical group II and IV metals. Ref. 2.																																	
Thiourea																																		
Thiurone																																		
Thorin																			177						2									2
Tiron																												2						
Titan yellow																																		
Toluene-3,4-dithiol (see also Zinc dithiol)	2		149	2								2		2 149	2			136						149	38		1		1		38		2	
2,4,6-Tri(2-pyridyl)-1,3,5-triazine																																		
Uranyl zinc acetate																		33																
Victoria violet																																		
2,4-Xylenol			2																															
2,6-Xylenol			186																															
Xylenol orange		10																															133	132
Zinc dibenzyldithiocarbamate																																		
Zinc OO-di-iso-propyl phosphorodithioate																																		
Zinc dithiol* (see also Toluene-3,4-dithiol)	2		2	2					2			2		2	2		2							2	2		2		2	156			2	
Zincon																																	2	

* Zinc dithiol is more stable than toluene-3,4-dithiol. It can be used to prepare a solution of toluene-3,4-dithiol and for many purposes it is added as the solid or as a suspension in ethanol (see Ref. 2).

List of literature references. Key to the table on the preceding pages

1. *Organic Reagents for Metals and other Reagent Monographs*, Vol. 1, 5th Edn., 1955. Hopkin & Williams.
2. *Organic Reagents for Metals and for Certain Radicals*, Vol. II, 1964. Hopkins & Williams.
3. *Anal. Chem.*, **33**, 1128 (1961).
4. *Analyst*, **87**, 703 (1962).
5. *Anal. Chim. Acta*, **26**, 487 (1962).
6. *Anal. Chim. Acta*, **27**, 153 (1962).
7. *Anal. Chem.*, **34**, 209 (1962).
8. *Anal. Chim. Acta*, **27**, 591 (1962).
9. *Anal. Chim. Acta*, **27**, 331 (1962).
10. *Talanta*, **9**, 987 (1962).
11. *Anal. Chem.*, **35**, 149 (1963).
12. *Ind. Eng. Chem. (Anal.)*, **15**, 57 (1943); *J.S.C.I.*, **62**, 187 (1943).
13. *Analyst*, **73**, 395 (1948).
14. *Analyst*, **87**, 197 (1962).
15. *Anal. Chim. Acta*, **13**, 142 (1955).
16. *Analyst*, **87**, 880 (1962).
17. *Analyst*, **80**, 901 (1955).
18. *Anal. Chem.*, **29**, 281 (1957).
19. *Analyst*, **94**, 262 (1969).
20. *Mikrochim. Acta*, 29 (1962).
21. *Anal. Chim. Acta*, **26**, 326 (1962).
22. *Anal. Chem.*, **34**, 571 (1962).
23. *Anal. Abstr.*, **4**, 1448 (1957).
24. *Anal. Chim. Acta*, **15**, 21, 102 (1956).
25. *J. Indian Chem. Soc.*, **28**, 89 (1951); *Chem. Abstr.*, **45**, 8394 (1951).
26. *Anal. Abstr.*, **4**, 3870 (1957).
27. *Anal. Abstr.*, **4**, 3858 (1957).
28. *Anal. Chim. Acta*, **19**, 202 (1958).
29. *J. Proc. Austral. Chem. Inst.*, **3**, 184 (1936); *Brit. Abstr., A. I.* 1082 (1936).
30. *Anal. Chim. Acta*, **19**, 377 (1958).
31. *Anal. Chim. Acta*, **18**, 546 (1958).
32. *Z. anal. Chem.*, **140**, 245 (1953); *Anal. Abstr.*, **1**, 659 (1954).
33. A. I. Vogel (1961). *Text-Book of Quantitative Inorganic Analysis*. 3rd edn. Longmans.
34. *J. Res. Nat. Bur. Stand.*, **33**, 307 (1944); *Brit. Abstr., C*, 91 (1945).
35. *Ann. Chim.*, Roma, **43**, 730 (1953); *Anal. Abstr.*, **1**, 471 (1954).
36. *Anal. Chem.*, **26**, 883 (1954).
37. *Anal. Chem.*, **34**, 94 (1962).
38. *Analyst*, **82**, 177 (1957).
39. *Analyst*, **85**, 376 (1960).
40. *Anal. Chim. Acta*, **25**, 348 (1961).
41. *Anal. Chem.*, **23**, 653 (1951).
42. *J. Inst. Petroleum Tech.*, **19**, 845 (1933); *Brit. Abstr.*, B, 1088 (1933).
43. F. Feigl (1958). *Spot Tests in Inorganic Analysis*. 5th edn. Elsevier.
44. (a) *Ind. Eng. Chem. (Anal.)*, **13**, 603 (1941). (b) *Anal, Abstr.*, **4**, 2935 (1957).
45. *Talanta*, **8**, 579 (1961).
46. *Talanta*, **5**, 231 (1960).
47. *Analyst*, **85**, 823 (1960).
48. *Anal. Chem.*, **22**, 1281 (1950).
49. *A Handbook of Colorimetric Chemical Analytical Methods*. 6th edn. The Tintometer Ltd.
50. A. I. Vogel (1951). *Text-Book of Quantitative Inorganic Analysis*. 2nd edn. Longmans.
51. *Anal. Chem.*, **32**, 1337 (1960).
52. *Anal. Chem.*, **27**, 1932 (1955).
53. 'ANALAR' *Standards for Laboratory Chemicals*. 5th edn., 1957. (Hopkin & Williams and The British Drug Houses Ltd), p. 71.
54. *Analyst*, **95**, 131 (1970).
55. (a) *Anal. Chem.*, **25**, 808 (1953). (b) *Analyst*, **80**, 768 (1955).
56. *Brit. Abstr.*, A. I. 315 (1943).
57. *Anal. Chim. Acta*, **19**, 18 (1958).
58. *Anal. Chim. Acta*, **30**, 176 (1964).
59. (a) *Anal. Chem.*, **24**, 1033 (1952). (b) *Anal. Abstr.*, **4**, 2135 (1957).
60. R. Belcher and C. L. Wilson (1956). *New Methods of Analytical Chemistry*. Chapman & Hall.
61. (a) *Anal. Abstr.*, **6**, 147, 916 (1959). (b) *Anal. Chim. Acta*, **19**, 576 (1958).
62. *Anal. Abstr.*, **6**, 869 (1959).
63. *Analyst (abstr)*, **70**, 189 (1945).
64. *Anal. Chim. Acta*, **19**, 372 (1958).
65. *Science*, **125**, 1042 (1957); *Anal. Abstr.*, **5**, 2560 (1958).
66. *Z. anal. Chem.*, **140**, 252 (1953); *Anal. Abstr.*, **1**, 658 (1954).
67. *Anal. Abstr.*, **4**, 2135 (1957).
68. *Anal. Chim. Acta*, **16**, 121 (1957).
69. *Anal. Abstr.*, **4**, 1494 (1957).
70. *Anal. Abstr.*, **4**, 389 (1957).
71. *Analyst*, **82**, 630 (1957).
72. *Anal. Chim. Acta*, **22**, 223 (1960).
73. *Anal. Chem.*, **23**, 514 (1951).
74. *Anal. Abstr.*, **4**, 386 (1957).
75. *Anal. Chim. Acta*, **12**, 218 (1955).
76. *Anal. Abstr.*, **4**, 1444 (1957).
77. *Z. anal. Chem.*, **102**, 24 (1935).
78. *Z. anal. Chem.*, **102**, 108 (1935).
79. *Z. anal. Chem.*, **104**, 88 (1936).
80. (a) *Z. anal. Chem.*, **162**, 96 (1958); *Anal. Abstr.*, **6**, 944 (1959). (b) *Anal. Chim. Acta*, **20**, 379 (1959).
81. *Anal. Chem.*, **32**, 1117 (1960).
82. *Z. anal. Chem.*, **171**, 241 (1959); *Anal. Abstr.*, **7**, 3135 (1960).
83. *Analyst*, **89**, 707 (1964).
84. *Ind. Eng. Chem. (Anal.)*, **12**, 229 (1940).
85. *Anal. Chem.*, **33**, 1671 (1961).
86. *Talanta*, **9**, 749 (1961).
87. *Anal. Abstr.*, **3**, 1344 (1956).
88. *Anal. Chim. Acta*, **3**, 481 (1949).
89. *Analyst*, **85**, 889 (1960).
90. *Analyst*, **58**, 667 (1933).
91. *Ind. Eng. Chem. (Anal.)*, **16**, 322 (1944).
92. (a) *Analyst*, **56**, 245 (1931). (b) *Anal. Abstr.*, **3**, 3271 (1956).
93. *Anal. Chem.*, **28**, 1728 (1956).

94. *Chem. Abstr.*, **45**, 69 (1951).
95. *Anal. Chim. Acta*, **20**, 340 (1959).
96. *Anal. Chim. Acta*, **20**, 332 (1959).
97. *Bull. Soc. chim. Belg.*, **54**, 186 (1945); *Brit. Abstr.*, C. 2 (1947).
98. *J. Chem. Soc. Japan*, **66**, 37 (1945); *Chem. Abstr.*, **43**, 1682 (1949).
99. (a) *J. Indian Chem. Soc.*, **21**, 119 (1944); *Analyst Abstr.*, **69**, 383 (1944). (b) *J. Indian Chem. Soc.*, **21**, 187, 188 (1944); *Brit. Abstr.*, C, 92 (1945).
100. *Zavod Lab.*, **11**, 254 (1945); *Chem. Abstr.*, **40**, 1418 (1946).
101. *Anal. Chim. Acta*, **21**, 58 (1959).
102. *Analyst*, **76**, 485 (1951).
103. *Analyst*, **76**, 482 (1951).
104. *Analyst*, **79**, 548 (1954).
105. *Anal. Chem.*, **26**, 217 (1954).
106. *Anal. Abstr.*, **6**, 186, 187, 188 (1959).
107. *Anal. Chem.*, **35**, 33 (1963).
108. *Anal. Chem.*, **31**, 1783 (1959).
109. *Anal. Chim. Acta*, **13**, 154 (1955).
110. *Anal. Chim. Acta*, **13**, 159 (1955).
111. *Ind. Eng. Chem. (Anal.)*, **12**, 663 (1940).
112. *Ind. Eng. Chem. (Anal.)*, **14**, 359 (1942).
113. *Talanta*, **8**, 673 (1961).
114. *Brit. Abstr.*, C, 39 (1944).
115. *Talanta*, **8**, 293 (1961).
116. *Anal. Chem.*, **33**, 431 (1961).
117. *Anal. Abstr.*, **6**, 871 (1959).
118. *Metallurgia*, **44**, 207 (1951); *Brit. Abstr.*, C, 134 (1952).
119. *Anal. Chem.*, **22**, 918 (1950).
120. *Anal. Chim. Acta*, **23**, 351 (1960).
121. *Anal. Chim. Acta*, **16**, 62 (1957).
122. *Anal. Chim. Acta*, **13**, 409 (1955).
123. *Anal. Chim. Acta*, **24**, 294 (1961).
124. *Anal. Abstr.*, **5**, 1530 (1958).
125. *Anal. Chim. Acta*, **2**, 254 (1948); see also ref. 60.
126. *Analyst*, **80**, 380 (1955).
127. *Analyst*, **74**, 505 (1949); **75**, 555 (1950).
128. *Ind. Eng. Chem. (Anal.)*, **16**, 598 (1944).
129. *Anal. Chim. Acta*, **3**, 324 (1949).
130. *Analyst*, **70**, 124 (1945).
131. *Mikrochim. Acta*, 571 (1961).
132. *Talanta*, **2**, 266 (1959).
133. *Talanta*, **8**, 203 (1961).
134. *Talanta*, **3**, 81 (1959).
135. *Talanta*, **8**, 753 (1961).
136. *Anal. Chem.*, **33**, 445 (1961).
137. *Anal. Chem.*, **31**, 1102 (1959).
138. *Anal. Chem.*, **33**, 239 (1961).
139. *Anal. Chem.*, **33**, 125 (1961).
140. *Z. anal. Chem.*, **178**, 352 (1961); *Anal. Abstr.*, **8**, 3225 (1961).
141. *Anal. Chem.*, **26**, 1968 (1954).
142. *Analyst*, **88**, 380 (1963).
143. *Talanta*, **10**, 1013 (1963).
144. *Talanta*, **9**, 761 (1962).
145. *Talanta*, **11**, 621(1964).
146. *Anal. Chim. Acta*, **9**, 86 (1953).
147. *Anal. Abstr.*, **4**, 1424 (1957).
148. *Angew. Chem.*, **64**, 608 (1952); *Brit. Abstr.*, C, 246 (1953).
149. *Analyst*, **83**, 396 (1958).
150. *Anal. Chem.*, **25**, 1125 (1953).
151. *Analyst*, **58**, 667 (1933); see also ref. 150.
152. *Anal. Chem.*, **33**, 1933 (1961).
153. *Anal. Abstr.*, **9**, 106 (1962).
154. *Anal. Abstr.*, **9**, 58 (1962).
155. *Analyst*, **83**, 314 (1958).
156. *Analyst*, **84**, 16 (1959).
157. *Anal. Chem.*, **29**, 1187 (1957) (for chloride).
158. *Anal. Chim. Acta*, **22**, 479 (1960).
159. *Anal. Chim. Acta*, **22**, 413 (1960).
160. *Talanta*, **7**, 163 (1961).
161. *Zavod. Lab.*, **27**, 803 (1961); *Anal. Abstr.*, **9**, 558 (1962).
162. *Analyst*, **86**, 407 (1961).
163. *Talanta*, **4**, 126 (1960).
164. *Talanta*, **4**, 244 (1960).
165. *Talanta*, **11**, 1 (1964).
166. *Anal. Chem.*, **30**, 44 (1958).
167. *Anal. Chim. Acta*, **21**, 370 (1959).
168. *Talanta*, **3**, 95 (1959).
169. *Anal. Chim. Acta*, **24**, 167 (1961).
170. (a) *Helv. Chim. Acta*, **31**, 320 (1948). (b) *Analyst*, **74**, 274 (1949).
171. *Analyst*, **83**, 516 (1958).
172. *Anal. Chim. Acta*, **23**, 175 (1960).
173. *Anal. Chim. Acta*, **23**, 565 (1960).
174. *Anal. Chim. Acta*, **20**, 26 (1959).
175. *Anal. Chim. Acta*, **23**, 434 (1960).
176. *Anal. Chem.*, **32**, 1350 (1960).
177. *Anal. Chim. Acta*, **23**, 538 (1960).
178. *Anal. Chem.*, **32**, 1083 (1960).
179. *Talanta*, **8**, 453 (1961).
180. *Analyst*, **86**, 543 (1961).
181. *Anal. Chem.*, **31**, 1985 (1959).
182. *Anal. Chim. Acta*, **26**, 528 (1962).
183. *Zavod. Lab.*, **2B**, 1283 (1957); C. A., **53**, 1387c (1959).
184. *Anal. Chim. Acta*, **45**, 341 (1969).
185. *Mikrochim. Acta*, 512 (1961).
186. *Organic Chemical Reagents.* Monograph No. 70. 1967. Hopkin & Williams.
187. *Organic Chemical Reagents.* Monograph No. 71. 1967. Hopkin & Williams.
188. *Analyst*, **91**, 282 (1966).
189. *J. Amer. Water Wks. Assocn.*, **49**, 873 (1957); C. A., **51**, 15045c (1957) (for chlorine).
190. Monograph No. 74. 1969. Hopkin & Williams.
191. *Anal. Chim. Acta*, **47**, 151 (1969).
192. 'ANALAR' *Standards for Laboratory Chemicals.* 6th edn., 1967 ('ANALAR' Standards Ltd), p. 617.
193. Monograph No. 75. 1969. Hopkin & Williams.

APPENDIX III SPECIFIC GRAVITIES OF ACIDS AT 20 °C

(Specific gravities and percentages by weight are based on weights *in vacuo* and the percentage by weight refers to the formula given)

Per cent by weight	Specific gravity				
	H_2SO_4	HNO_3	CH_3COOH	H_3PO_4	HCl
1	1.0051	1.0036	0.9996	1.0038	1.0032
2	1.0118	1.0091	1.0012	1.0092	1.0082
3	1.0184	1.0146	1.0025	—	—
4	1.0250	1.0201	1.0040	1.0200	1.0181
5	1.0317	1.0256	1.0055	—	—
10	1.0661	1.0543	1.0125	1.0532	1.0474
15	1.1020	1.0842	1.0195	—	—
16	1.1094	1.0903	1.0209	1.0884	1.0776
20	1.1394	1.1150	1.0263	1.1134	1.0980
24	1.1704	1.1404	1.0313	1.1395	1.1187
25	1.1783	1.1469	1.0326	—	—
26	1.1862	1.1534	1.0338	1.1529	1.1290
30	1.2185	1.1800	1.0384	1.1805	1.1493
34	1.2515	1.2071	1.0428	—	1.1691
35	1.2599	1.2140	1.0438	1.216	—
36	1.2684	1.2205	1.0449	—	1.1789
40	1.3028	1.2463	1.0488	1.254	1.1980
45	1.3476	1.2783	1.0534	1.293	—
50	1.3951	1.3100	1.0575	1.335	—
55	1.4453	1.3393	1.0611	1.379	—
60	1.4983	1.3667	1.0642	1.426	—
65	1.5533	1.3913	1.0666	1.475	—
70	1.6105	1.4134	1.0685	1.526	—
75	1.6692	1.4337	1.0696	1.579	—
80	1.7272	1.4521	1.0700	1.633	—
85	1.7786	1.4686	1.0689	1.689	—
90	1.8144	1.4826	1.0661	1.746	—
92	1.8240	1.4873	1.0643	1.770	—
93	1.8279	1.4892	1.0632	—	—
94	1.8312	1.4912	1.0619	1.794	—
95	1.8337	1.4932	1.0605	—	—
96	1.8355	1.4952	1.0588	1.819	—
97	1.8364	1.4974	1.0570	—	—
98	1.8361	1.5008	1.0549	1.844	—
99	1.8342	1.5056	1.0524	—	—
100	1.8305	1.5129	1.0498	1.870	—

APPENDIX IV SPECIFIC GRAVITIES OF ALKALINE SOLUTIONS AT 20°C

(Specific gravities and percentages by weight are based on weights *in vacuo* and the percentage by weight refers to the formula given)

Per cent by weight	Specific gravity			Per cent by weight	Specific gravity		
	KOH	NaOH	NH_3		KOH	NaOH	NH_3
1	1.0083	1.0095	0.9939	26	1.2489	1.2848	0.9040
2	1.0175	1.0207	0.9895	27	1.2592	—	—
3	1.0267	1.0318	—	28	1.2695	1.3064	0.8980
4	1.0359	1.0428	0.9811	29	1.2800	—	—
5	1.0452	1.0538	—	30	1.2905	1.3279	0.8920
6	1.0544	1.0648	0.9730	31	1.3010	—	—
7	1.0637	1.0758	—	32	1.3117	1.3490	—
8	1.0730	1.0869	0.9651	33	1.3224	—	—
9	1.0824	1.0979	—	34	1.3331	1.3696	—
10	1.0918	1.1089	0.9575	35	1.3440	—	—
11	1.1013	—	—	36	1.3549	1.3900	—
12	1.1108	1.1309	0.9501	37	1.3659	—	—
13	1.1203	—	—	38	1.3769	1.4101	
14	1.1299	1.1530	0.9430	39	1.3879	—	—
15	1.1396	—	—	40	1.3991	1.4300	—
16	1.1493	1.1751	0.9362	41	1.4103	—	—
17	1.1590	—	—	42	1.4215	1.4494	—
18	1.1688	1.1972	0.9295	43	1.4329	—	—
19	1.1786	—	—	44	1.4443	1.4685	—
20	1.1884	1.2191	0.9229	45	1.4558	—	—
21	1.1984	—	—	46	1.4673	1.4873	—
22	1.2083	1.2411	0.9164	47	1.4790	—	—
23	1.2184	—	—	48	1.4907	1.5065	—
24	1.2285	1.2629	0.9101	49	1.5025	—	—
25	1.2387	—	—	50	1.5143	1.5253	—

APPENDIX V DATA ON THE STRENGTH OF AQUEOUS SOLUTIONS OF THE COMMON ACIDS AND OF AQUEOUS AMMONIA

Reagent	Approximate			Vol. required to make 1 dm^3 of approx. N solution (cm^3)
	Per cent by weight	Specific gravity	Normality	
Hydrochloric acid	35	1.18	11.3	89
Nitric acid	70	1.42	16.0	63
Sulphuric acid	96	1.84	36.0	28
Perchloric acid	70	1.66	11.6	86
Hydrofluoric acid	46	1.15	26.5	38
Phosphoric acid	85	1.69	41.1	23
Acetic acid	99.5	1.05	17.4	58
Aqueous ammonia	27(NH_3)	0.90	14.3	71

APPENDIX VI   SATURATED SOLUTIONS OF SOME REAGENTS AT 20 °C

Reagent	Formula	Specific gravity	Molarity	Quantities required for 1 dm³ of saturated solution	
				Grams of reagent	cm³ of water
Ammonium chloride	NH_4Cl	1.075	5.44	291	784
Ammonium nitrate	NH_4NO_3	1.312	10.80	863	449
Ammonium oxalate	$(NH_4)_2C_2O_4,H_2O$	1.030	0.295	48	982
Ammonium sulphate	$(NH_4)_2SO_4$	1.243	4.06	535	708
Barium chloride	$BaCl_2,2H_2O$	1.290	1.63	398	892
Barium hydroxide	$Ba(OH)_2$	1.037	0.228	39	998
Barium hydroxide	$Ba(OH)_2,8H_2O$	1.037	0.228	72	965
Calcium hydroxide	$Ca(OH)_2$	1.000	0.022	1.6	1000
Mercury(II) chloride	$HgCl_2$	1.050	0.236	64	986
Potassium chloride	KCl	1.174	4.00	298	876
Potassium chromate	K_2CrO_4	1.396	3.00	583	858
Potassium dichromate	$K_2Cr_2O_7$	1.077	0.39	115	962
Potassium hydroxide	KOH	1.540	14.50	813	727
Sodium carbonate	Na_2CO_3	1.178	1.97	209	869
Sodium carbonate	$Na_2CO_3,10H_2O$	1.178	1.97	563	515
Sodium chloride	$NaCl$	1.197	5.40	316	881
Sodium hydroxide	$NaOH$	1.539	20.07	803	736

APPENDIX VII SOLUBILITIES OF SOME INORGANIC COMPOUNDS IN WATER AT VARIOUS TEMPERATURES

The Table gives the number of grams of the *anhydrous* substance which can be dissolved in 100 g of water at the temperature (°C) indicated at the head of the column. Where the formula is followed by an asterisk, the value is expressed in grams of the anhydrous substance in 100 g of the saturated solution at the temperature indicated.

Formula	Molecular weight	0°	10°	20°	30°	40°	50°	60°	70°	80°	90°	100 °C
$AlCl_3,6H_2O$	241.43	30.5	31.0	31.4	31.8	32.1		32.5		32.7		32.9
$Al_2(SO_4)_3,18H_2O$*	666.42	23.8	25.1	26.7	28.8	31.4	34.3	37.2	39.8	42.2	44.7	47.1
$Al_2(SO_4)_3,(NH_4)_2SO_4,24H_2O$	906.65	2.10	4.99	7.74	10.94	14.88	20.10	26.70				
$Al_2(SO_4)_3,K_2SO_4,24H_2O$	948.75	3.0	4.0	5.9	8.4	11.7	17.0	24.8	40.0	71.0	109.0	
NH_4Cl	53.49	29.4	33.3	37.2	41.4	45.8	50.4	55.2	60.2	65.6	71.3	77.3
NH_4NO_3	80.04	118.3		192.4	241.8	297.0	344.0	421.0	499.0	580.0	740.0	871.0
$(NH_4)_2C_2O_4,H_2O$*	142.11	2.31	3.11	4.26	5.74	7.57	9.74	12.25	15.10	18.30	21.84	25.73
$(NH_4)_2HPO_4$	132.06	30.0	38.6	40.8	42.9	45.0	47.2	49.3	51.4	54.2		
$(NH_4)H_2PO_4$	115.03	18.5	22.8	27.2	31.7	36.2		45.2				63.4
$(NH_4)_2SO_4$*	132.13	41.4	42.1	42.9	43.8	44.7	45.8	47.0				
NH_4SCN*	76.12	54.5	59.0	63.0	67.5							
$BaBr_2,2H_2O$	333.18	98	101	104	109	114	118	123	128	135		149
$BaCl_2,2H_2O$	244.28	31.6	33.3	35.7	38.2	40.7	43.6	46.4	49.4	52.4		58.8
$Ba(OH)_2,8H_2O$	315.48	1.67	2.48	3.89	5.59	8.22	13.12	20.94		101.4		
$Ba(NO_3)_2$*	261.35	4.7	6.3	8.3	10.3	12.4	14.6	16.9	19.1	21.4	23.6	25.6
Br_2	159.81	4.22	3.40	3.20	3.13							
$2CdCl_2,5H_2O$*	456.69	47.3			56.9			57.7		58.4		59.6
$3CdSO_4,8H_2O$	769.49	76.48	76.00	76.60		78.54		83.68				
$Ca(CH_3COO)_2,2H_2O$	194.20	37.4	36.0	34.7	33.8	33.2		32.7			31.1	29.7
$CaBr_2,6H_2O$	307.99	125	132	143		213		278		295		312
CaI_2*	293.89	64.6	66.0	67.6	69.0	70.8		74.0		78.0		100.0
$Ca(OH)_2$*	74.09	0.130	0.125		0.109	0.100	0.092		0.045	0.031	0.027	0.011
$CaSO_4,2H_2O$*	172.17				0.064	0.063	0.057				0.059	
$CoCl_2,6H_2O$	237.93	30.3	32.3	34.6	37.4	41.0						
$Co(NO_3)_2,6H_2O$	291.03	45.7		49.5	52.7							
$CoSO_4,7H_2O$*	281.10	20.3	23.4	26.5	29.6	32.8		33.5				
$CoSO_4,(NH_4)_2SO_4,6H_2O$	395.22	6.0	9.5	13.0	17.0	22.0	27.0		40.0	49.0		
$Cu(NO_3)_2,6H_2O$*	295.65	45.5	50.0	55.5								
$CuSO_4,5H_2O$	249.68	14.3	17.4	20.7	25.0	28.5	33.3	40.0		55.0		75.4
$FeSO_4,7H_2O$	278.01	15.7	20.5	26.5	32.9	40.2	48.5	54.9				
$FeSO_4,(NH_4)_2SO_4,6H_2O$	392.13	12.5				33.0	40.0		52.0			
$Pb(CH_3COO)_2,3H_2O$	379.33	19.7	29.3	44.3	69.7	116.0	221.0					
$Pb(NO_3)_2$*	331.21	27.3	31.6	35.2	38.8	41.9		47.8		52.7		57.1

Compound	M											
MgSO$_4$,7H$_2$O*	246.47	18.0	22.0	25.2	28.0	30.8	33.4	35.3	—	—	48.3	—
MgSO$_4$,(NH$_4$)$_2$SO$_4$,6H$_2$O	360.59	11.8	14.6	18.0	21.7	25.9	30.2	35.2	—	—	—	65.7
MnBr$_2$,4H$_2$O*	286.81	56.0	57.6	59.5	61.1	62.8	64.5	66.3	68.0	—	—	—
MnCl$_2$,4H$_2$O	197.90	63.4	68.1	73.9	80.7	88.6	98.2	108.6	—	—	—	—
MnSO$_4$,4H$_2$O	223.06	53.2	—	—	67.8	68.8	—	—	—	—	—	—
HgBr$_2$*	360.40	—	—	0.55	0.65	0.90	1.25	1.65	—	2.70	—	4.70
HgCl$_2$*	271.50	3.5	4.6	6.1	7.3	9.3	—	14.0	16.6	—	23.1	38.0
NiBr$_2$,6H$_2$O*	326.61	53	55	57	58	59	60	—	—	—	—	—
NiCl$_2$,6H$_2$O*	237.71	35.0	37.5	39.1	40.8	43.2	—	—	—	—	—	—
NiI$_2$,6H$_2$O*	420.61	55.4	57.5	59.7	61.7	63.5	64.7	—	—	—	—	—
NiSO$_4$,7H$_2$O	280.87	28.1	33.0	38.4	44.1	48.2	52.8	56.9	61.0	—	—	—
KHCO$_3$*	100.12	18.6	21.8	25.0	28.1	31.3	34.2	37.5	40.6	—	—	—
K$_2$CO$_3$*	138.21	51.3	51.9	52.5	53.2	53.9	54.8	55.9	57.1	58.3	59.6	60.9
KBr	119.00	53.5	59.5	65.2	70.6	75.5	80.2	85.5	90.0	95.0	99.2	100.1
KBrO$_3$	167.00	3.1	4.8	6.9	9.5	13.2	17.5	22.7	—	34.0	—	50.1
K$_4$[Fe(CN)$_6$],3H$_2$O*	422.39	12.5	17.4	22.0	26.0	32.6	—	—	38.2	—	—	—
KSCN	97.18	177	—	217	—	—	—	—	—	—	—	—
KCl	74.55	27.6	31.0	34.0	37.0	40.0	42.6	45.5	48.3	51.1	54.0	56.7
KClO$_3$*	122.55	3.2	4.9	6.8	9.2	12.2	15.6	19.2	23.2	27.3	31.5	36.0
KClO$_4$*	138.55	0.75	1.05	1.65	2.50	3.60	4.90	6.80	9.20	11.8	15.0	18.2
K$_2$Cr$_2$O$_7$*	294.18	4.47	—	10.97	—	20.83	—	31.3	—	42.2	—	50.0
KI	166.00	128	136	144	152	160	168	176	184	192	200	208
KIO$_3$*	214.00	4.4	5.9	7.5	9.3	11.2	13.3	15.5	16.6	—	19.9	24.4
K$_2$SO$_4$	174.25	6.9	8.5	10.0	11.5	12.9	14.2	15.4	16.5	17.6	18.6	19.4
KHSO$_4$*	136.16	26.6	—	32.7	35.4	37.9	40.3	43.3	46.1	—	—	54.9
AgNO$_3$	169.87	55.6	63.3	69.5	74.0	80.2	80.2	85.5	—	—	—	90.0
AgNO$_2$	153.87	0.16	0.22	0.34	0.50	0.72	1.00	1.36	—	—	—	—
Ag$_2$SO$_4$*	311.79	0.57	0.69	0.75	0.88	0.97	1.05	1.14	1.21	1.28	1.34	1.39
NaBr,2H$_2$O*	138.92	44.5	46.0	47.6	49.6	53.8	53.8	54.1	—	—	—	—
NaBrO$_3$*	150.89	—	—	26.7	29.9	32.8	35.5	38.5	—	43.1	—	—
Na(CH$_3$COO),3H$_2$O	136.08	36.3	40.8	46.5	54.5	65.5	—	—	—	—	—	—
NaHCO$_3$	84.01	6.9	8.2	9.6	11.1	12.7	14.5	16.4	—	—	—	—
Na$_2$CO$_3$,10H$_2$O	286.14	7.0	12.5	21.5	28.4	—	—	—	—	—	—	—
Na$_2$C$_2$O$_4$	134.00	2.62	2.96	3.30	3.67	4.01	4.37	4.70	5.05	—	5.40	6.10
NaCl	58.42	35.6	35.7	35.8	36.0	36.3	36.7	37.1	37.5	38.0	38.5	39.1
Na$_2$Cr$_2$O$_7$,2H$_2$O*	298.00	62.0	—	—	—	68.3	70.5	72.9	—	—	79.0	80.6
NaF*	41.99	3.53	3.90	—	4.05	4.21	4.35	4.47	—	4.66	—	4.83
NaI,2H$_2$O	185.92	158.7	168.6	178.7	190.3	205.0	227.8	256.8	294	296	296	302
NaIO$_3$,H$_2$O*	215.91	2.42	4.39	—	9.63	11.71	14.0	16.5	19.0	21.0	22.8	24.8
NaNO$_3$*	84.99	42.2	44.7	46.7	48.7	50.5	52.8	54.9	57.1	59.7	—	—
NaOH	40.00	—	—	109	119	129	145	174	—	—	313	—
Na$_2$SO$_4$,10H$_2$O	322.19	5.0	9.0	19.4	40.8	48.8	46.7	45.3	—	—	43.7	42.5
Na$_2$S$_2$O$_3$,5H$_2$O	248.17	50.2	59.7	70.1	—	—	—	—	81.8	85.9	90.5	100.8

APPENDIX VIII   SOURCES OF ANALYSED SAMPLES

Throughout this book the use of a number of standard analytical samples is recommended in order that practical experience may be gained on substances of known composition.

In the United Kingdom such samples (sometimes known as Ridsdale's Samples), suitable for metallurgical, chemical, and spectrographic analysis, are supplied by:

The National Bureau of Analysed Standards Ltd, Newham Hall, Middlesborough, Yorkshire, England, from whom a detailed list is available.

In the United States of America a similar wide range of standards, listed in NBS Special Publication 260, 1975–6 Edition, can be obtained from:

US Department of Commerce, National Bureau of Standards, Washington DC, 20234, USA.

Elements and compounds of high purity and known composition are marketed by Johnson, Matthey. Each batch has been subjected to spectroscopic analysis and a detailed laboratory report accompanies each batch of material. These are listed in booklet No. 1760, 'Spectrographically Standardised Substances', and are supplied by:

Johnson, Matthey Chemicals Ltd, Hatton Garden, London, EC1, England.

An additional source of non-ferrous metal standards suitable for emission spectroscopy and as chemical standards is the:

British Non-Ferrous Metals Technology Centre, Grove Laboratories, Denchworth Road, Wantage, Oxfordshire, England.

APPENDIX IX   BUFFER SOLUTIONS AND SECONDARY pH STANDARDS

The British standard for the pH scale is a $0.05\,M$ solution of potassium hydrogen–phthalate (British Standard No. 1647: 1950 and 1961) which has a pH of 4.001 at 20 °C. The values at various other temperatures are collected below.

British pH standard: $0.05\,M$-potassium hydrogenphthalate

Temp., °C	pH	Temp., °C	pH	Temp., °C	pH
0	4.011	35	4.020	65	4.105
5	4.005	40	4.031	70	4.121
10	4.001	45	4.045	75	4.140
15	4.000	50	4.061	80	4.161
25	4.005	55	4.080	85	4.185
30	4.011	60	4.091	90	4.211

Subsidiary pH standards at 25 °C include:

	pH
$0.05\,M$-HCl $+0.09\,M$-KCl	2.07
$0.1\,M$-potassium tetroxalate	1.48
$0.1\,M$-potassium dihydrogen citrate	3.72
$0.1\,M$-acetic acid $+0.1\,M$-sodium acetate	4.64
$0.01\,M$-acetic acid $+0.01\,M$-sodium acetate	4.70
$0.01\,M$-KH$_2$PO$_4$ $+0.01\,M$-Na$_2$HPO$_4$	6.85
$0.05\,M$-borax	9.18
$0.025\,M$-NaHCO$_3$ $+0.025\,M$-Na$_2$CO$_3$	10.00
$0.01\,M$-Na$_3$PO$_4$	11.72

The following table covering the pH range 2.6–12.0 (18 °C) is included as an example of a **universal buffer mixture**.

A mixture of 6.008 g of A.R. citric acid, 3.893 g of A.R. potassium dihydrogen phosphate, 1.769 g of A.R. boric acid, and 5.266 g of pure diethylbarbituric acid is dissolved in water and made up to 1 dm^3. The pH values at 18 °C of mixtures of 100 cm^3 of this solution with various volumes (X) of $0.2\,M$-sodium hydroxide solution (free from carbonate) are tabulated below.

pH	X(cm^3)	pH	X(cm^3)	pH	X(cm^3)
2.6	2.0	5.8	36.5	9.0	72.7
2.8	4.3	6.0	38.9	9.2	74.0
3.0	6.4	6.2	41.2	9.4	75.9
3.2	8.3	6.4	43.5	9.6	77.6
3.4	10.1	6.6	46.0	9.8	79.3
3.6	11.8	6.8	48.3	10.0	80.8
3.8	13.7	7.0	50.6	10.2	82.0
4.0	15.5	7.2	52.9	10.4	82.9
4.2	17.6	7.4	55.8	10.6	83.9
4.4	19.9	7.6	58.6	10.8	84.9
4.6	22.4	7.8	61.7	11.0	86.0
4.8	24.8	8.0	63.7	11.2	87.7
5.0	27.1	8.2	65.6	11.4	89.7
5.2	29.5	8.4	67.5	11.6	92.0
5.4	31.8	8.6	69.3	11.8	95.0
5.6	34.2	8.8	71.0	12.0	99.6

The National Bureau of Standards (NBS) pH standards (including notes on the preparation of the buffer solutions) are given in Section **XIV, 16**, Table XIV, 3.

APPENDIX X APPROXIMATE pH VALUES OF SOME COMMON REAGENT SOLUTIONS AT ABOUT ROOM TEMPERATURE

Compound	Molarity	pH
Acid, benzoic	(Saturated)	2.8
Acid, boric	0.1	5.3
Acid, citric	0.1	2.1
Acid, citric	0.01	2.6
Acid, hydrochloric	0.1	1.1
Acid, oxalic	0.1	1.3
Acid, salicylic	(Saturated)	2.4
Acid, succinic	0.1	2.7
Acid, tartaric	0.1	2.0
Ammonia, aqueous	0.1	11.3
Ammonium alum	0.05	4.6
Ammonium chloride	0.1	4.6
Ammonium dihydrogenphosphate	0.1	4.0
Ammonium oxalate	0.1	6.4
Diammonium hydrogenphosphate	0.1	7.9
Ammonium sulphate	0.1	5.5
Sodium tetraborate	0.1	9.2
Calcium hydroxide	(Saturated)	12.4
Potassium acetate	0.1	9.7
Potassium alum	0.1	4.2
Potassium carbonate	0.1	11.5
Potassium dihydrogencitrate	0.1	3.7
Potassium dihydrogencitrate	0.02	3.8
Potassium hydrogencarbonate	0.1	8.2
Potassium dihydrogenphosphate	0.1	4.5
Potassium hydrogenoxalate	0.1	2.7
Sodium acetate	0.1	8.9
Sodium benzoate	0.1	8.0
Sodium carbonate	0.1	11.5
Sodium carbonate	0.01	11.0
Sodium hydrogencarbonate	0.1	8.3
Sodium hydrogensulphate	0.1	1.4
Sodium dihydrogenphosphate	0.1	4.5
Sodium hydroxide	0.1	12.9
Disodium hydrogen phosphate	0.1	9.2
Trisodium phosphate	0.01	11.7
Sulphamic acid	0.01	2.1

The pH values of $0.1 M$-HCl and of $0.1 M$-NaOH solutions at different temperatures ($°C$) are:

Temp.	0.1M-HCl	0.1M NaOH	Temp.	0.1M-HCl	0.1M-NaOH
5°	1.10	13.62	35°	1.10	12.57
10°	1.10	13.43	40°	1.10	12.42
15°	1.10	13.24	45°	1.10	12.28
20°	1.10	13.06	50°	1.10	12.15
25°	1.10	12.88	55°	1.11	12.02
30°	1.10	12.72	60°	1.11	11.90

APPENDIX XI DISSOCIATION CONSTANTS OF SOME ACIDS IN WATER AT 25 °C

Dissociation constants are expressed as $pK_a (= -\log K_a)$.

Acid		pK_a	Acid		pK_a
Aliphatic acids					
Formic		3.75	Succinic	K_1	4.21
Acetic		4.76		K_2	5.64
Propanoic		4.87	Glutaric	K_1	4.34
Butanoic		4.82		K_2	5.27
3-Methyl propanoic		4.85	Adipic	K_1	4.43
Pentanoic		4.84		K_2	5.28
Fluoroacetic		2.58	Methylmalonic	K_1	3.07
Chloroacetic		2.86		K_2	5.87
Bromoacetic		2.90	Ethylmalonic	K_1	2.96
Iodoacetic		3.17		K_2	5.90
Cyanoacetic		2.47	Dimethylmalonic	K_1	3.15
Diethylacetic		4.73		K_2	6.20
Lactic		3.86	Diethylmalonic	K_1	2.15
Pyruvic		2.49		K_2	7.47
Acrylic		4.26	Fumaric	K_1	3.02
Vinylacetic		4.34		K_2	4.38
Tetrolic		2.65	Maleic	K_1	1.92
trans-Crotonic		4.69		K_2	6.23
Furoic		3.17	Tartaric	K_1	3.03
Oxalic	K_1	1.27		K_2	4.37
	K_2	4.27	Citric	K_1	3.13
Malonic	K_1	2.85		K_2	4.76
	K_2	5.70		K_3	6.40
Aromatic acids					
Benzoic		4.21			
Phenylacetic		4.31	2-Benzoylbenzoic		3.54
Sulphanilic		3.23	Phthalic K_1		2.95
Phenoxyacetic		3.17	K_2		5.41
Mandelic		3.41	*cis*-Cinnamic		3.88
1-Naphthoic		3.70	*trans*-Cinnamic		4.44
2-Naphthoic		4.16	Phenol		10.00
1-Naphthylacetic		4.24	1-Nitroso-2-naphthol		7.77
2-Naphthylacetic		4.26	2-Nitroso-1-naphthol		7.38

	ortho (2-)	meta (3-)	para (4-)
Aromatic acids			
Fluorobenzoic	3.27	3.86	4.14
Chlorobenzoic	2.94	3.83	3.98
Bromobenzoic	2.85	3.81	3.97
Iodobenzoic	2.86	3.85	3.93
Hydroxybenzoic	3.00	4.08	4.53
Methoxybenzoic	4.09	4.09	4.47
Nitrobenzoic	2.17	3.49	3.42
Aminobenzoic	4.98	4.79	4.92
Toluic	3.91	4.24	4.34
Chlorophenol	8.48	9.02	9.38
Nitrophenol	7.23	8.40	7.15
Methylphenol (cresol)	10.29	10.09	10.26
Methoxyphenol	9.98	9.65	10.21

Acid		pK$_a$	Acid		pK$_a$
Inorganic acids					
Arsenious		9.22	Nitrous		3.35
Arsenic	K_1	2.30	Phosphoric	K_1	2.12
	K_2	7.08		K_2	7.21
	K_3	9.22		K_3	12.30
Boric		9.24	Phosphorous	K_1	1.8
Carbonic	K_1	6.37		K_2	6.15
	K_2	10.33	Sulphuric	K_2	1.92
Hydrocyanic		9.14	Sulphurous	K_1	1.92
Hydrofluoric		4.77		K_2	7.20
Hydrogen sulphide	K_1	7.00	Thiosulphuric	K_1	1.7
	K_2	14.00		K_2	2.5
Hypochlorous		7.25			

Acidic Dissociation Constants of Some Bases in Water at 25 °C.

The data for bases are expressed as acidic dissociation constants, e.g., for ammonia, the value pK$_a$ = 9.24 is given for the ammonium ion:

$$NH_4{}^+ + H_2O \rightleftharpoons NH_3 + H_3O^+$$

Otherwise expressed, bases are considered from the standpoint of the ionisation of the conjugated acids. The basic dissociation constant for the reaction

$$NH_3 + H_2O \rightleftharpoons NH_4{}^+ + OH^-$$

may then be obtained from the relation:

$$pK_a \text{ (acidic)} + pK_b \text{ (basic)} = pK_w \text{ (water)}$$

where pK$_w$ is 14.00 at 25 °C.* For simplicity, the name of the base will be expressed in the 'basic' form, e.g., ammonia for ammonium ion, propylamine for propylammonium ion, piperidine for piperidinium ion, aniline for anilinium ion, etc., although it is appreciated that this is not strictly correct: no difficulty should be experienced in writing down the correct name, if required.

Base		pK$_a$	Base		pK$_a$
Ammonia		9.24	Hydrazine		7.93
Methylamine		10.64	Hydroxylamine		5.82
Ethylamine		10.63	Benzylamine		9.35
Propylamine		10.57	Aniline		4.58
Butylamine		10.62	o-Toluidine		4.39
Cyclohexylamine		10.64	m-Toluidine		4.68
Dimethylamine		10.77	p-Toluidine		5.09
Diethylamine		10.93	2-Chloroaniline		2.62
Monoethanolamine		9.50	3-Chloroaniline		3.32
Triethanolamine		7.77	4-Chloroaniline		3.81
Trimethylamine		9.80	N-Methylaniline		4.85
Triethylamine		10.72	NN-Dimethylaniline		5.15
tris-(Hydroxymethyl)-			Pyridine		5.17
aminomethane		8.08	2-Methylpyridine		5.97
Piperidine		11.12	3-Methylpyridine		5.68
Ethylenediamine	K_1	7.00	4-Methylpyridine		6.02
	K_2	10.09	Benzidine	K_1	4.97
1,3-Propylenediamine	K_1	8.64		K_2	3.75
	K_2	10.62	1,10-Phenanthroline		4.86
1,4-Butylenediamine	K_1	9.35			
	K_2	10.80			

* The values at 20 °C and 30 °C are 14.17 and 13.83 respectively.

APPENDIX XII   POTENTIALS OF THE COMMON REFERENCE ELECTRODES

Electrode	Potential at 25 °C, volts *vs.* N.H.E.
Hg/Hg_2Cl_2 (sat.), KCl (sat.) [S.C.E.]	+0.244
Hg/Hg_2Cl_2 (sat.), $1.0M$-KCl [N.C.E.]	+0.281
Hg/Hg_2Cl_2 (sat.), $0.10M$-KCl	+0.336
Hg/Hg_2SO_4 (sat.), K_2SO_4 (sat.)	+0.64
Hg/Hg_2SO_4 (sat.), 0.05-$4M$-H_2SO_4	+0.615
Ag/AgCl (sat.), KCl (sat.)	+0.199
Ag/AgCl (sat.), $1.0M$-KCl	+0.227
Ag/AgCl (sat.), $0.10M$-KCl	+0.290

APPENDIX XIII   POLAROGRAPHIC HALF-WAVE POTENTIALS

Ion	Supporting electrolyte	$E_{\frac{1}{2}}$ (volts $vs.$ S.C.E.)
Ba^{2+}	$0.1M$-$N(CH_3)_4Cl$	-1.94
Bi^{3+}	$1M$-HCl	-0.09
	$0.5M$-H_2SO_4	-0.04
	$0.5M$-tartrate $+0.1M$ NaOH	-1.0
	$0.5M$-sodium hydrogentartrate, pH 4.5	-0.23
Cd^{2+}	$0.1M$-KCl	-0.64
	$1M$-$NH_3 + 1M$-NH_4^+	-0.81
	$1M$-HNO_3	-0.59
	$1M$-KI	-0.74
	$1M$-KCN	-1.18
Co^{2+}	$0.1M$-KCl	-1.20
	$0.1M$-pyridine $+0.1M$-pyridinium ion	-1.07
Cu^{2+}	$0.1M$-KCl	$+0.04$
	$1M$-$NH_3 + 1M$-NH_4Cl	-0.24 (1st wave)
		-0.50 (2nd wave)
	$0.5M$-sodium hydrogentartrate, pH 4.5	-0.09
Fe^{2+}	$0.1M$-KCl	-1.3
Fe^{3+}	$0.5M$-tartrate, pH 9.4	-1.20 (1st wave)
		-1.73 (2nd wave)
	$0.1M$-EDTA $+2M$-CH_3COONa	-0.13 (1st wave)
		-1.3 (2nd wave)
K^+	$0.1M$-$N(CH_3)_4OH$ in 50 % ethanol	-2.10
Li^+	$0.1M$-$N(CH_3)_4OH$ in 50 % ethanol	-2.31
Mn^{2+}	$1M$-KCl	-1.51
	$0.2M$-$H_2P_2O_7^{2-}$, pH 2.2	$+0.1$
Na^+	$0.1M$-$N(CH_3)_4Cl$	-2.07
Ni^{2+}	$1M$-KCl	-1.1
	$1M$-KSCN	-0.70
	$1M$-KCN	-1.36
	$1M$-pyridine $+$ HCl, pH 7.0	-0.78
	$1M$-$NH_3 + 0.2M$-NH_4^+	-1.06
O_2	Most buffers, pH 1–10	-0.05 (1st wave)
		-0.9 (2nd wave)
Pb^{2+}	$0.1M$-KCl	-0.40
	$1M$-HNO_3	-0.40
	$1M$-NaOH	-0.75
	$0.5M$-sodium hydrogentartrate, pH 4.5	-0.48
	$0.5M$-tartrate $+0.1M$-NaOH	-0.75
Sn^{2+}	$1M$-HCl	-0.47
Sn^{4+}	$1M$-HCl $+4M$-NH_4^+	-0.25 (1st wave)
		-0.52 (2nd wave)
Zn^{2+}	$0.1M$-KCl	-1.00
	$1M$-NaOH	-1.53
	$1M$-$NH_3 + 1M$-NH_4^+	-1.33
	$0.5M$-tartrate, pH 9	-1.15

APPENDIX XIV TABLES OF ARC 'RAIES ULTIMES' AND PERSISTENT LINES FOR SPECTROGRAPHIC ANALYSIS*

The 'Raies Ultimes' and persistent lines marked in the enlargements of the spectrograms of the R.U. powder (taken on a Hilger large quartz Littrow spectrograph) are collected together in the table and are listed under the elements to which they are due. The wavelengths are recorded in Ångstroms, together with their relative intensities as determined from the original enlargements.

The arc sensitivity of an element may be defined as the logarithm of the ratio of the amount of diluent, such as the base used for the R.U. powder, to the amount of the given element such that when the mixture is excited the strongest line of the element will have an intensity 1. Thus a sensitivity of 6 corresponds to one part per million, and a sensitivity of 1 to one part in 10.

The table also indicates the relative sensitivity, under arc excitation, of the more important lines. Thus 'a' means the most sensitive and 'b' the next in sensitivity. The elements are grouped according to their absolute sensitivities in a table at the end.

	Barium *contd.*	**Cadmium** *contd.*	**Cobalt** *contd.*
Aluminium	4934.09 *(5) a*	2288.02 *(2) a*	2407.25 *(0)*
3961.53 *(7) a*	4554.04 *(15) a*		
3944.03 *(5) a*		**Calcium**	**Copper**
3092.71 *(3) b*	**Beryllium**	4455.89 *(25)*	3273.96 *(15) a*
3082.16 *(3) b*	3321.34 *(2)*	4434.96 *(20)*	3247.51 *(20) a*
	3321.09 *(2)*	4425.44 *(15)*	
Antimony	3131.07 *(2) b*	4318.65 *(12)*	**Fluorine**
3267.50 *(2)*	3130.42 *(3) b*	to	CaF Band Head at λ 5291
3232.50 *(1)*	2650.78 *(0)*	4283.01 *(12)*	
3029.81 *(0)*	2650.47 *(0)*	4226.73 *(50) a*	**Gallium**
2877.92 *(4) a*	2494.73 *(0)*	3968.47 *(40) a*	4172.06 *(5) a*
2769.94 *(1)*	2494.56 *(0)*	3933.67 *(40) a*	2943.64 *(4)*
2598.06 *(4) a*	2348.61 *(5) a*		2874.24 *(4)*
2528.54 *(4) a*		**Carbon**	
2311.47 *(3) a*	**Bismuth**	2478.57 *(2) a*	**Germanium**
	3067.72 *(5) a*		3039.06 *(2) a*
Arsenic	2989.03 *(2) b*	**Cesium**	2754.59 *(2) a*
2898.71 *(2)*	2897.98 *(2) b*	4593.18 *(1) a*	2709.63 *(2) a*
2860.45 *(3) b*		4555.36 *(3) a*	2691.34 *(1)*
2780.20 *(3) b*	**Boron**		2651.58 *(1)*
2745.00 *(2)*	2497.73 *(5) a*	**Chromium**	2651.18 *(2) a*
2492.91 *(2)*	2496.78 *(4) a*	4289.72 *(2)*	2592.54 *(1)*
2456.53 *(2)*		4274.80 *(2)*	
2381.18 *(2)*	**Cadmium**	4254.35 *(2) a*	**Gold**
2370.77 *(2)*	5085.82 *(1)*	3605.33 *(2)*	2675.95 *(2) b*
2369.67 *(2)*	4799.92 *(1)*	3593.49 *(2)*	2427.95 *(2) a*
2349.84 *(4) a*	3610.61 *(3) b*	3578.69 *(3)*	
2288.12 *(3) b*	3467.66 *(2)*		**Indium**
	3466.20 *(3) b*	**Cobalt**	4511.32 *(6) a*
Barium	3403.65 *(2)*	3453.51 *(3) a*	4101.77 *(4) a*
5535.55 *(8) a*	3261.06 *(5) a*	3405.12 *(1) b*	3258.56 *(2)*

* Reproduced by courtesy of the Research Laboratories of the General Electric Company Ltd, Wembley, Middlesex.

Further details will be found in Booklet No. 1762, 'Sensitive Arc Lines of 50 elements including the Use of R.U. Powder in Spectroscopic Analysis', available from Johnson, Matthey & Co. Ltd, Hatton Garden, London, EC1. The R.U. Powder is available from this firm.

Indium *contd.*

3256.09	(5) b
3039.36	(2)

Iridium

3513.65	(1) b
3220.78	(1) b
3133.32	(1) a
2639.71	(0)
2543.97	(0)

Iron

3734.87	(2)
3719.94	(2)
3617.79	(1)
3581.20	(2)
3465.86	(2)
3440.61	(2)
3021.07	(1)
3020.64	(2)
2983.57	(0)
2973.24	(0)
2966.90	(0)
2719.03	(1)
2631.05	(0)
2599.40	(1)
2527.43	(0)
2522.85	(2)
2488.15	(2)
2483.27	(3) a

Lanthanum

4429.90	(1)
4333.73	(1)
4123.23	(2)
4086.71	(2)
3995.75	(1)
3949.11	(4) a
3337.49	(3)

Lead

4057.82	(10) a
3739.95	(2)
3683.47	(8) a
3639.58	(6) a
3572.73	(2)
2873.32	(4)
2833.07	(5) a
2823.19	(2)
2802.00	(4)
2663.17	(3)
2614.18	(4)
2577.26	(1)
2476.38	(2)
2393.79	(1)

Lithium

6707.85	(20) a

Lithium *contd.*

6103.64	(5)
4602.86	(2) b
3232.61	(1)

Magnesium

5183.62	(30)
5172.70	(25)
5167.34	(20)
3838.26	(25)
3832.31	(20)
3829.35	(15)
3096.90	(30)
3093.00	(15)
3091.08	(10)
2852.13	(100) a
2802.70	(30)
2795.53	(30)

Manganese

4034.49	(3) a
4033.07	(3) a
4030.76	(4) a
2798.27	(3)
2794.82	(1)
2605.69	(1)
2593.73	(1)
2576.10	(1)

Mercury

2536.52	(2) a

Molybdenum

3902.96	(4) a
3864.11	(5) a
3798.25	(5) a
3447.12	(1)
3358.12	(0)
3208.83	(1)
3193.97	(3)
3170.35	(3)
3158.17	(2)
3132.59	(3)
3112.12	(0)

Nickel

3524.54	(2) b
3515.05	(0)
3446.26	(1)
3414.77	(2) a
3003.63	(0)
3002.49	(0)

Niobium

4123.81	(1)
4100.92	(1)
4079.73	(2)
4058.94	(3) a

Niobium *contd.*

3358.42	(2)

Osmium

4260.85	(1)
3267.95	(2)
3262.29	(1)
3058.66	(2)
3030.70	(1)
3018.04	(1)
2909.06	(3) a
2838.63	(1)
2488.55	(2)

Palladium

3634.70	(1)
3609.55	(1)
3516.94	(2)
3481.15	(2)
3460.77	(2)
3421.24	(3)
3404.58	(4) a
3242.70	(3)

Phosphorus

2554.93	(1)
2553.28	(2) a
2535.65	(2) a
2534.01	(1)

Platinum

3064.71	(2) a
3042.64	(1)
2997.97	(1)
2929.79	(1)
2830.30	(1)
2719.04	(1)
2705.89	(0)
2702.40	(1)
2659.45	(2) a
2650.86	(1)
2628.03	(0)

Potassium

7698.98	(10) a
7644.91	(10) a
4047.20	(5) b
4044.14	(5) b
3447.70	(1)
3446.72	(1)

Rhodium

4374.80	(1) b
3692.36	(1) b
3657.99	(0)
3596.19	(0)
3502.52	(1)
3434.89	(2) a

Rhodium *contd.*

3396.85	(0)
3323.09	(0)

Ruthenium

3728.03	(2)
3726.93	(2)
3661.35	(1)
3634.92	(1)
3596.18	(0)
3593.02	(1)
3498.94	(4) a
3436.74	(2) a
3428.63	(1)
3428.31	(1)
3417.35	(0)
2874.98	(1)
2735.72	(1)
2721.56	(1)

Rubidium

7947.60	(10) a
7800.23	(10) a
4215.56	(4) b
4201.85	(4) b

Scandium

4246.83	(1)
4023.69	(2) a
4020.40	(1) a
3911.81	(2) a
3907.48	(1) a
3642.79	(0)
3580.93	(1)

Silicon

3905.53	(1)
2881.58	(8) a
2528.52	(4)
2524.12	(4)
2519.21	(4)
2516.12	(5) a
2514.33	(4)
2506.90	(4)
2435.16	(2)

Silver

3382.89	(10) a
3280.68	(10) a

Sodium

5895.92	(30) a
5889.95	(30) a
5688.22	(1)
5682.66	(0)
3302.99	(4) b
3302.32	(4) b

Strontium

4607.33 (*10*) *a*
4215.52 (*6*) *a*
4077.71 (*10*) *a*

Tantalum

3318.84 (*1*)
3311.16 (*1*)
3103.25 (*0*)
2714.67 (*2*) *a*
2656.61 (*1*)
2653.27 (*1*)
2647.47 (*1*)
2646.22 (*1*)

Thallium

5350.46 (*7*) *a*
3775.72 (*6*) *a*
3529.43 (*4*) *b*
3519.24 (*6*) *b*
3229.75 (*1*)
2921.52 (*1*)
2918.32 (*1*)
2767.87 (*3*)
2580.14 (*1*)
2379.69 (*1*)

Tin

3330.59 (*2*)
3262.33 (*4*) *a*
3175.02 (*3*)

Tin *contd.*

3034.12 (*3*)
3009.15 (*1*)
2863.33 (*4*) *a*
2839.99 (*4*) *a*
2706.51 (*3*)
2661.25 (*0*)
2546.55 (*1*)
2429.50 (*2*)
2354.85 (*2*)

Titanium

4536.05 (*1*)
to
4533.24 (*1*)
3998.64 (*2*)
3989.76 (*1*)
3981.76 (*0*)
3958.21 (*1*)
3956.34 (*0*)
3653.50 (*1*)
3642.68 (*1*)
3635.46 (*1*)
3377.59 (*1*)
3372.80 (*2*)
3371.45 (*2*)
3370.44 (*1*)
3361.21 (*2*)
3354.64 (*1*)
3349.41 (*2*)
3349.04 (*1*)
3341.88 (*3*)

Titanium *contd.*

3322.94 (*0*)
3241.99 (*0*)
3239.04 (*1*)
3236.57 (*1*)
3234.52 (*1*)
3199.92 (*1*)

Tungsten

4294.61 (*0*) *b*
4008.75 (*1*) *a*
2946.98 (*0*)
2944.40 (*0*)

Vanadium

4594.11 (*0*)
4460.29 (*1*)
4408.51 (*1*)
to
4379.24 (*2*)
4134.49 (*0*)
4132.02 (*1*)
4128.07 (*2*)
4099.80 (*1*)
3185.40 (*2*)
3183.98 (*2*)
3183.41 (*1*)
3110.71 (*0*)
3102.30 (*1*)
3066.38 (*1*)
3060.46 (*1*)
2924.64 (*1*)

Vanadium *contd.*

2924.03 (*0*)
2923.62 (*0*)
2908.82 (*1*)

Yttrium

4643.70 (*0*)
4374.94 (*1*)
4142.84 (*2*)
4102.38 (*3*) *a*
3982.60 (*0*)
3774.33 (*2*)
3710.29 (*2*)
3620.94 (*1*)
3242.28 (*2*)
3216.68 (*1*)

Zinc

6362.35 (*20*) *b*
4810.53 (*50*) *a*
4722.16 (*50*) *a*
4680.14 (*30*) *a*
3345.93 (*30*) *b*
3345.57 (*30*) *b*
3302.94 (*25*)
3302.59 (*25*)
3282.33 (*20*)

Zirconium

3496.21 (*2*) *a*
3481.15 (*0*)
3438.23 (*2*) *a*
3391.98 (*2*) *a*

Arc Sensitivity between *5* and *6*:
Ag, Co, Cr, Cu, In, Li, Mg,
Na, Ni, Os, Pd, Pt, Rh, Ru.

Arc Sensitivity between *4* and *5*:
Al, Au, Ba, Be, Ca, Fe, Ga, Ge, Hg,
Ir, K, Mn, Mo, Pb, Sc, Sn, Sr, Ti, V.

Arc Sensitivity between *3* and *4*:
B, Bi, Cd, La, Rb, Sb,
Si, Tl, Y, Zn, Zr.

Arc Sensitivity between *2* and *3*:
As, Cs, Nb, P, Ta, W.

APPENDIX XV PERCENTAGE POINTS OF THE *t*-DISTRIBUTION

ϕ	50%	20%	10%	5%	2%	1%	0.5%	0.1%
1	1.00	3.08	6.31	12.71	31.82	63.66	127.32	636.62
2	0.816	1.89	2.92	4.30	6.97	9.92	14.09	31.60
3	0.765	1.64	2.35	3.18	4.54	5.84	7.45	12.92
4	0.741	1.53	2.13	2.78	3.75	4.60	5.60	8.61
5	0.727	1.48	2.02	2.57	3.37	4.03	4.77	6.87
6	0.718	1.44	1.94	2.45	3.14	3.71	4.32	5.96
7	0.711	1.42	1.89	2.37	3.00	3.50	4.03	5.41
8	0.706	1.40	1.86	2.31	2.90	3.36	3.83	5.04
9	0.703	1.38	1.83	2.26	2.82	3.25	3.69	4.78
10	0.700	1.37	1.81	2.23	2.76	3.17	3.58	4.59
11	0.697	1.36	1.80	2.20	2.72	3.11	3.50	4.44
12	0.695	1.36	1.78	2.18	2.68	3.06	3.43	4.32
13	0.694	1.35	1.77	2.16	2.65	3.01	3.37	4.22
14	0.692	1.35	1.76	2.14	2.62	2.98	3.33	4.14
15	0.691	1.34	1.75	2.13	2.60	2.95	3.29	4.07
16	0.690	1.34	1.75	2.12	2.58	2.92	3.25	4.02
17	0.689	1.33	1.74	2.11	2.57	2.90	3.22	3.97
18	0.688	1.33	1.73	2.10	2.55	2.88	3.20	3.92
19	0.688	1.33	1.73	2.09	2.54	2.86	3.17	3.88
20	0.687	1.33	1.73	2.09	2.53	2.85	3.15	3.85
21	0.686	1.32	1.72	2.08	2.52	2.83	3.14	3.82
22	0.686	1.32	1.72	2.07	2.51	2.82	3.12	3.79
23	0.685	1.32	1.71	2.07	2.50	2.81	3.10	3.77
24	0.685	1.32	1.71	2.06	2.49	2.80	3.09	3.75
25	0.684	1.32	1.71	2.06	2.49	2.79	3.08	3.73
26	0.684	1.32	1.71	2.06	2.48	2.78	3.07	3.71
27	0.684	1.31	1.70	2.05	2.47	2.77	3.06	3.69
28	0.683	1.31	1.70	2.05	2.47	2.76	3.05	3.67
29	0.683	1.31	1.70	2.05	2.46	2.76	3.04	3.66
30	0.683	1.31	1.70	2.04	2.46	2.75	3.03	3.65
40	0.681	1.30	1.68	2.02	2.42	2.70	2.97	3.55
60	0.679	1.30	1.67	2.00	2.39	2.66	2.92	3.46
120	0.677	1.29	1.66	1.98	2.36	2.62	2.86	3.37
∞	0.674	1.28	1.65	1.96	2.33	2.58	2.81	3.29

The table gives the percentage of the area under the *two* tails of the *t*-curve, and therefore gives the probability that *t* will exceed the tabular entry in *absolute* value.

Appendices XV, XVI and XVII have been derived, with the permission of the Biometrika Trustees, from the corresponding tables in *Biometrika Tables for Statisticians*, Vol. I, Third Edition (1966), and *Biometrika Tables for Statisticians*, Vol. II, (1972), E. S. Pearson and H. O. Hartley, eds., Cambridge University Press.

APPENDIX XVI F-DISTRIBUTION

Probability level	ϕ_2	ϕ_1 (corresponding to greater mean square)											
		1	2	3	4	5	6	7	8	9	10	15	∞
0.10	1	39.9	49.5	53.6	55.8	57.2	58.2	58.9	59.4	59.9	60.2	61.2	63.3
0.05		161.4	199.5	215.7	224.6	230.2	234.0	236.8	238.9	240.5	241.9	246.0	254.3
0.01		4,052	4,999	5,403	5,625	5,764	5,859	5,928	5,981	6,023	6,056	6,157	6,366
0.10	2	8.53	9.00	9.16	9.24	9.29	9.33	9.35	9.37	9.38	9.39	9.42	9.49
0.05		18.5	19.0	19.2	19.2	19.3	19.3	19.4	19.4	19.4	19.4	19.4	19.5
0.01		98.5	99.0	99.2	99.2	99.3	99.3	99.4	99.4	99.4	99.4	99.4	99.5
0.10	3	5.54	5.46	5.39	5.34	5.31	5.28	5.27	5.25	5.24	5.23	5.20	5.13
0.05		10.1	9.55	9.28	9.12	9.01	8.94	8.89	8.85	8.81	8.79	8.70	8.53
0.01		34.1	30.8	29.5	28.7	28.2	27.9	27.7	27.5	27.3	27.2	26.9	26.1
0.10	4	4.54	4.32	4.19	4.11	4.05	4.01	3.98	3.95	3.94	3.92	3.87	3.76
0.05		7.71	6.94	6.59	6.39	6.26	6.16	6.09	6.04	6.00	5.96	5.86	5.63
0.01		21.2	18.0	16.7	16.0	15.5	15.2	15.0	14.8	14.7	14.5	14.2	13.5
0.10	5	4.06	3.78	3.62	3.52	3.45	3.40	3.37	3.34	3.32	3.30	3.24	3.10
0.05		6.61	5.79	5.41	5.19	5.05	4.95	4.88	4.82	4.77	4.74	4.62	4.36
0.01		16.3	13.3	12.1	11.4	11.0	10.7	10.5	10.3	10.2	10.1	9.72	9.02
0.10	6	3.78	3.46	3.29	3.18	3.11	3.05	3.01	2.98	2.96	2.94	2.87	2.72
0.05		5.99	5.14	4.76	4.53	4.39	4.28	4.21	4.15	4.10	4.06	3.94	3.67
0.01		13.7	10.9	9.78	9.15	8.75	8.47	8.26	8.10	7.98	7.87	7.56	6.88
0.10	7	3.59	3.26	3.07	2.96	2.88	2.83	2.78	2.75	2.72	2.70	2.63	2.47
0.05		5.59	4.74	4.35	4.12	3.97	3.87	3.79	3.73	3.68	3.64	3.51	3.23
0.01		12.2	9.55	8.45	7.85	7.46	7.19	6.99	6.84	6.72	6.62	6.31	5.65
0.10	8	3.46	3.11	2.92	2.81	2.73	2.67	2.62	2.59	2.56	2.54	2.46	2.29
0.05		5.32	4.46	4.07	3.84	3.69	3.58	3.50	3.44	3.39	3.35	3.22	2.93
0.01		11.3	8.65	7.59	7.01	6.63	6.37	6.18	6.03	5.91	5.81	5.52	4.86
0.10	9	3.36	3.01	2.81	2.69	2.61	2.55	2.51	2.47	2.44	2.42	2.34	2.16
0.05		5.12	4.26	3.86	3.63	3.48	3.37	3.29	3.23	3.18	3.14	3.01	2.71
0.01		10.6	8.02	6.99	6.42	6.06	5.80	5.61	5.47	5.35	5.26	4.96	4.31
0.10	10	3.29	2.92	2.73	2.61	2.52	2.46	2.41	2.38	2.35	2.32	2.24	2.06
0.05		4.96	4.10	3.71	3.48	3.33	3.22	3.14	3.07	3.02	2.98	2.85	2.54
0.01		10.0	7.56	6.55	5.99	5.64	5.39	5.20	5.06	4.94	4.85	4.56	3.91
0.10	12	3.18	2.81	2.61	2.48	2.39	2.33	2.28	2.24	2.21	2.19	2.10	1.90
0.05		4.75	3.89	3.49	3.26	3.11	3.00	2.91	2.85	2.80	2.75	2.62	2.30
0.01		9.33	6.93	5.95	5.41	5.06	4.82	4.64	4.50	4.39	4.30	4.01	3.36
0.10	15	3.07	2.70	2.49	2.36	2.27	2.21	2.16	2.12	2.09	2.06	1.97	1.76
0.05		4.54	3.68	3.29	3.06	2.90	2.79	2.71	2.64	2.59	2.54	2.40	2.07
0.01		8.68	6.36	5.42	4.89	4.56	4.32	4.14	4.00	3.89	3.80	3.52	2.87
0.10	16	3.05	2.67	2.46	2.33	2.24	2.18	2.13	2.09	2.06	2.03	1.94	1.72
0.05		4.49	3.63	3.24	3.01	2.85	2.74	2.66	2.59	2.54	2.49	2.35	2.01
0.01		8.53	6.23	5.29	4.77	4.44	4.20	4.03	3.89	3.78	3.69	3.41	2.75
0.10	24	2.93	2.54	2.33	2.19	2.10	2.04	1.98	1.94	1.91	1.88	1.78	1.53
0.05		4.26	3.40	3.01	2.78	2.62	2.51	2.42	2.36	2.30	2.25	2.11	1.73
0.01		7.82	5.61	4.72	4.22	3.90	3.67	3.50	3.36	3.26	3.17	2.89	2.21
0.10	60	2.79	2.39	2.18	2.04	1.95	1.87	1.82	1.77	1.74	1.71	1.60	1.29
0.05		4.00	3.15	2.76	2.53	2.37	2.25	2.17	2.10	2.04	1.99	1.84	1.39
0.01		7.08	4.98	4.13	3.65	3.34	3.12	2.95	2.82	2.72	2.63	2.35	1.60
0.10	∞	2.71	2.30	2.08	1.94	1.85	1.77	1.72	1.67	1.63	1.60	1.49	1.00
0.05		3.84	3.00	2.60	2.37	2.21	2.10	2.01	1.94	1.88	1.83	1.67	1.00
0.01		6.63	4.61	3.78	3.32	3.02	2.80	2.64	2.51	2.41	2.32	2.04	1.00

APPENDIX XVII   PERCENTAGE POINTS OF THE χ^2-DISTRIBUTION

ϕ	99%	97.5%	95%	90%	50%	10%	5%	2.5%	1%	0.1%
1	0.000	0.001	0.004	0.016	0.455	2.71	3.84	5.02	6.63	10.83
2	0.020	0.051	0.103	0.211	1.39	4.61	5.99	7.38	9.21	13.82
3	0.115	0.216	0.352	0.584	2.37	6.25	7.81	9.35	11.34	16.27
4	0.297	0.484	0.711	1.06	3.36	7.78	9.49	11.14	13.28	18.47
5	0.554	0.831	1.15	1.61	4.35	9.24	11.07	12.83	15.09	20.52
6	0.872	1.24	1.64	2.20	5.35	10.64	12.59	14.45	16.81	22.46
7	1.24	1.69	2.17	2.83	6.35	12.02	14.07	16.01	18.48	24.32
8	1.65	2.18	2.73	3.49	7.34	13.36	15.51	17.53	20.09	26.13
9	2.09	2.70	3.33	4.17	8.34	14.68	16.92	19.02	21.67	27.88
10	2.56	3.25	3.94	4.87	9.34	15.99	18.31	20.48	23.21	29.59
11	3.05	3.82	4.57	5.58	10.34	17.28	19.68	21.92	24.72	31.26
12	3.57	4.40	5.23	6.30	11.34	18.55	21.03	23.34	26.22	32.91
13	4.11	5.01	5.89	7.04	12.34	19.81	22.36	24.74	27.69	34.53
14	4.66	5.63	6.57	7.79	13.34	21.06	23.68	26.12	29.14	36.12
15	5.23	6.26	7.26	8.55	14.34	22.31	25.00	27.49	30.58	37.70
16	5.81	6.91	7.96	9.31	15.34	23.54	26.30	28.85	32.00	39.25
17	6.41	7.56	8.67	10.09	16.34	24.77	27.59	30.19	33.41	40.79
18	7.01	8.23	9.39	10.86	17.34	25.99	28.87	31.53	34.81	42.31
19	7.63	8.91	10.12	11.65	18.34	27.20	30.14	32.85	36.19	43.82
20	8.26	9.59	10.85	12.44	19.34	28.41	31.41	34.17	37.57	45.32
21	8.90	10.28	11.59	13.24	20.34	29.62	32.67	35.48	38.93	46.80
22	9.54	10.98	12.34	14.04	21.34	30.81	33.92	36.78	40.29	48.27
23	10.20	11.69	13.09	14.85	22.34	32.01	35.17	38.08	41.64	49.73
24	10.86	12.40	13.85	15.66	23.34	33.20	36.42	39.36	42.98	51.18
25	11.52	13.12	14.61	16.47	24.34	34.38	37.65	40.65	44.31	52.62
26	12.20	13.84	15.38	17.29	25.34	35.56	38.89	41.92	45.64	54.05
27	12.88	14.57	16.15	18.11	26.34	36.74	40.11	43.19	46.96	55.48
28	13.56	15.31	16.93	18.94	27.34	37.92	41.34	44.46	48.28	56.89
29	14.26	16.05	17.71	19.77	28.34	39.09	42.56	45.72	49.59	58.30
30	14.95	16.79	18.49	20.60	29.34	40.26	43.77	46.98	50.89	59.70
40	22.16	24.43	26.51	29.05	39.34	51.81	55.76	59.34	63.69	73.40
50	29.71	32.36	34.76	37.69	49.34	63.17	67.50	71.42	76.15	86.66
60	37.49	40.48	43.19	46.46	59.33	74.40	79.08	83.30	88.38	99.61
70	45.44	48.76	51.74	55.33	69.33	85.53	90.53	95.02	100.42	112.32
80	53.54	57.15	60.39	64.28	79.33	96.58	101.88	106.63	112.33	124.84
90	61.75	65.65	69.13	73.29	89.33	107.57	113.14	118.14	124.12	137.21
100	70.06	74.22	77.93	82.36	99.33	118.50	124.34	129.56	138.81	149.45

APPENDIX XVIII   FOUR-FIGURE LOGARITHMS

	0	1	2	3	4	5	6	7	8	9	Mean differences								
											1	2	3	4	5	6	7	8	9
10	0000	0043	0086	0128	0170	0212	0253	0294	0334	0374	4	8	12	17	21	25	29	33	37
11	0414	0453	0492	0531	0569	0607	0645	0682	0719	0755	4	8	11	15	19	23	26	30	34
12	0792	0828	0864	0899	0934	0969	1004	1038	1072	1106	3	7	10	14	17	21	24	28	31
13	1139	1173	1206	1239	1271	1303	1335	1367	1399	1430	3	6	10	13	16	19	23	26	29
14	1461	1492	1523	1553	1584	1614	1644	1673	1703	1732	3	6	9	12	15	18	21	24	27
15	1761	1790	1818	1847	1875	1903	1931	1959	1987	2014	3	6	8	11	14	17	20	22	25
16	2041	2068	2095	2122	2148	2175	2201	2227	2253	2279	3	5	8	11	13	16	18	21	24
17	2304	2330	2355	2380	2405	2430	2455	2480	2504	2529	2	5	7	10	12	15	17	20	22
18	2553	2577	2601	2625	2648	2672	2695	2718	2742	2765	2	5	7	9	12	14	16	19	21
19	2788	2810	2833	2856	2878	2900	2923	2945	2967	2989	2	4	7	9	11	13	16	18	20
20	3010	3032	3054	3075	3096	3118	3139	3160	3181	3201	2	4	6	8	11	13	15	17	19
21	3222	3243	3263	3284	3304	3324	3345	3365	3385	3404	2	4	6	8	10	12	14	16	18
22	3424	3444	3464	3483	3502	3522	3541	3560	3579	3598	2	4	6	8	10	12	14	15	17
23	3617	3636	3655	3674	3692	3711	3729	3747	3766	3784	2	4	6	7	9	11	13	15	17
24	3802	3820	3838	3856	3874	3892	3909	3927	3945	3962	2	4	5	7	9	11	12	14	16
25	3979	3997	4014	4031	4048	4065	4082	4099	4116	4133	2	3	5	7	9	10	12	14	15
26	4150	4166	4183	4200	4216	4232	4249	4265	4281	4298	2	3	5	7	8	10	11	13	15
27	4314	4330	4346	4362	4378	4393	4409	4425	4440	4456	2	3	5	6	8	9	11	13	14
28	4472	4487	4502	4518	4533	4548	4564	4579	4594	4609	2	3	5	6	8	9	11	12	14
29	4624	4639	4654	4669	4683	4698	4713	4728	4742	4757	1	3	4	6	7	9	10	12	13
30	4771	4786	4800	4814	4829	4843	4857	4871	4886	4900	1	3	4	6	7	9	10	11	13
31	4914	4928	4942	4955	4969	4983	4997	5011	5024	5038	1	3	4	6	7	8	10	11	12
32	5051	5065	5079	5092	5105	5119	5132	5145	5159	5172	1	3	4	5	7	8	9	11	12
33	5185	5198	5211	5224	5237	5250	5263	5276	5289	5302	1	3	4	5	6	8	9	10	12
34	5315	5328	5340	5353	5366	5378	5391	5403	5416	5428	1	3	4	5	6	8	9	10	11
35	5441	5453	5465	5478	5490	5502	5514	5527	5539	5551	1	2	4	5	6	7	9	10	11
36	5563	5575	5587	5599	5611	5623	5635	5647	5658	5670	1	2	4	5	6	7	8	10	11
37	5682	5694	5705	5717	5729	5740	5752	5763	5775	5786	1	2	3	5	6	7	8	9	10
38	5798	5809	5821	5832	5843	5855	5866	5877	5888	5899	1	2	3	5	6	7	8	9	10
39	5911	5922	5933	5944	5955	5966	5977	5988	5999	6010	1	2	3	4	5	7	8	9	10
40	6021	6031	6042	6053	6064	6075	6085	6096	6107	6117	1	2	3	4	5	6	8	9	10
41	6128	6138	6149	6160	6170	6180	6191	6201	6212	6222	1	2	3	4	5	6	7	8	9
42	6232	6243	6253	6263	6274	6284	6294	6304	6314	6325	1	2	3	4	5	6	7	8	9
43	6335	6345	6355	6365	6375	6385	6395	6405	6415	6425	1	2	3	4	5	6	7	8	9
44	6435	6444	6454	6464	6474	6484	6493	6503	6513	6522	1	2	3	4	5	6	7	8	9
45	6532	6542	6551	6561	6571	6580	6590	6599	6609	6618	1	2	3	4	5	6	7	8	9
46	6628	6637	6646	6656	6665	6675	6684	6693	6702	6712	1	2	3	4	5	6	7	7	8
47	6721	6730	6739	6749	6758	6767	6776	6785	6794	6803	1	2	3	4	5	5	6	7	8
48	6812	6821	6830	6839	6848	6857	6866	6875	6884	6893	1	2	3	4	5	5	6	7	8
49	6902	6911	6920	6928	6937	6946	6955	6964	6972	6981	1	2	3	4	4	5	6	7	8
50	6990	6998	7007	7016	7024	7033	7042	7050	7059	7067	1	2	3	3	4	5	6	7	8
51	7076	7084	7093	7101	7110	7118	7126	7135	7143	7152	1	2	3	3	4	5	6	7	8
52	7160	7168	7177	7185	7193	7202	7210	7218	7226	7235	1	2	2	3	4	5	6	7	7
53	7243	7251	7259	7267	7275	7284	7292	7300	7308	7316	1	2	2	3	4	5	6	6	7
54	7324	7332	7340	7348	7356	7364	7372	7380	7388	7396	1	2	2	3	4	5	6	6	7
	0	1	2	3	4	5	6	7	8	9	1	2	3	4	5	6	7	8	9

APPENDIX XVIII   FOUR-FIGURE LOGARITHMS

	0	1	2	3	4	5	6	7	8	9	Mean differences								
											1	2	3	4	5	6	7	8	9
55	7404	7412	7419	7427	7435	7443	7451	7459	7466	7474	1	2	2	3	4	5	5	6	7
56	7482	7490	7497	7505	7513	7520	7528	7536	7543	7551	1	2	2	3	4	5	5	6	7
57	7559	7566	7574	7582	7589	7597	7604	7612	7619	7627	1	2	2	3	4	5	5	6	7
58	7634	7642	7649	7657	7664	7672	7679	7686	7694	7701	1	1	2	3	4	4	5	6	7
59	7709	7716	7723	7731	7738	7745	7752	7760	7767	7774	1	1	2	3	4	4	5	6	7
60	7782	7789	7796	7803	7810	7818	7825	7832	7839	7846	1	1	2	3	4	4	5	6	6
61	7853	7860	7868	7875	7882	7889	7896	7903	7910	7917	1	1	2	3	4	4	5	6	6
62	7924	7931	7938	7945	7952	7959	7966	7973	7980	7987	1	1	2	3	3	4	5	6	6
63	7993	8000	8007	8014	8021	8028	8035	8041	8048	8055	1	1	2	3	3	4	5	5	6
64	8062	8069	8075	8082	8089	8096	8102	8109	8116	8122	1	1	2	3	3	4	5	5	6
65	8129	8136	8142	8149	8156	8162	8169	8176	8182	8189	1	1	2	3	3	4	5	5	6
66	8195	8202	8209	8215	8222	8228	8235	8241	8248	8254	1	1	2	3	3	4	5	5	6
67	8261	8267	8274	8280	8287	8293	8299	8306	8312	8319	1	1	2	3	3	4	5	5	6
68	8325	8331	8338	8344	8351	8357	8363	8370	8376	8382	1	1	2	3	3	4	4	5	6
69	8388	8395	8401	8407	8414	8420	8426	8432	8439	8445	1	1	2	2	3	4	4	5	6
70	8451	8457	8463	8470	8476	8482	8488	8494	8500	8506	1	1	2	2	3	4	4	5	6
71	8513	8519	8525	8531	8537	8543	8549	8555	8561	8567	1	1	2	2	3	4	4	5	5
72	8573	8579	8585	8591	8597	8603	8609	8615	8621	8627	1	1	2	2	3	4	4	5	5
73	8633	8639	8645	8651	8657	8663	8669	8675	8681	8686	1	1	2	2	3	4	4	5	5
74	8692	8698	8704	8710	8716	8722	8727	8733	8739	8745	1	1	2	2	3	4	4	5	5
75	8751	8756	8762	8768	8774	8779	8785	8791	8797	8802	1	1	2	2	3	3	4	5	5
76	8808	8814	8820	8825	8831	8837	8842	8848	8854	8859	1	1	2	2	3	3	4	5	5
77	8865	8871	8876	8882	8887	8893	8899	8904	8910	8915	1	1	2	2	3	3	4	4	5
78	8921	8927	8932	8938	8943	8949	8954	8960	8965	8971	1	1	2	2	3	3	4	4	5
79	8976	8982	8987	8993	8998	9004	9009	9015	9020	9025	1	1	2	2	3	3	4	4	5
80	9031	9036	9042	9047	9053	9058	9063	9069	9074	9079	1	1	2	2	3	3	4	4	5
81	9085	9090	9096	9101	9106	9112	9117	9122	9128	9133	1	1	2	2	3	3	4	4	5
82	9138	9143	9149	9154	9159	9165	9170	9175	9180	9186	1	1	2	2	3	3	4	4	5
83	9191	9196	9201	9206	9212	9217	9222	9227	9232	9238	1	1	2	2	3	3	4	4	5
84	9243	9248	9253	9258	9263	9269	9274	9279	9284	9289	1	1	2	2	3	3	4	4	5
85	9294	9299	9304	9309	9315	9320	9325	9330	9335	9340	1	1	2	2	3	3	4	4	5
86	9345	9350	9355	9360	9365	9370	9375	9380	9385	9390	1	1	2	2	3	3	4	4	5
87	9395	9400	9405	9410	9415	9420	9425	9430	9435	9440	0	1	1	2	2	3	3	4	4
88	9445	9450	9455	9460	9465	9469	9474	9479	9484	9489	0	1	1	2	2	3	3	4	4
89	9494	9499	9504	9509	9513	9518	9523	9528	9533	9538	0	1	1	2	2	3	3	4	4
90	9542	9547	9552	9557	9562	9566	9571	9576	9581	9586	0	1	1	2	2	3	3	4	4
91	9590	9595	9600	9605	9609	9614	9619	9624	9628	9633	0	1	1	2	2	3	3	4	4
92	9638	9643	9647	9652	9657	9661	9666	9671	9675	9680	0	1	1	2	2	3	3	4	4
93	9685	9689	9694	9699	9703	9708	9713	9717	9722	9727	0	1	1	2	2	3	3	4	4
94	9731	9736	9741	9745	9750	9754	9759	9763	9768	9773	0	1	1	2	2	3	3	4	4
95	9777	9782	9786	9791	9795	9800	9805	9809	9814	9818	0	1	1	2	2	3	3	4	4
96	9823	9827	9832	9836	9841	9845	9850	9854	9859	9863	0	1	1	2	2	3	3	4	4
97	9868	9872	9877	9881	9886	9890	9894	9899	9903	9908	0	1	1	2	2	3	3	4	4
98	9912	9917	9921	9926	9930	9934	9939	9943	9948	9952	0	1	1	2	2	3	3	4	4
99	9956	9961	9965	9969	9974	9978	9983	9987	9991	9996	0	1	1	2	2	3	3	3	4
	0	1	2	3	4	5	6	7	8	9	1	2	3	4	5	6	7	8	9

INDEX

The following abbreviations are used:

aa	= atomic absorption	**fl**	= flame emission	**sepn.**	= separation
am	= amperometry	**fu**	= fluorimetric	**soln.**	= solution
ch	= chromatographic	**g**	= gravimetric	**stdn.**	= standardisation
cm	= coulometric	**hf**	= high frequency	**temp.**	= temperature
cn	= conductometric	**p**	= potentiometric	**th**	= thermal
D	= determination	**prep.**	= preparation	**ti**	= titrimetric
eg	= electrogravimetric	**s**	= spectrophotometric	**v**	= voltammetry
em	= emission spectrographic	**se**	= solvent extraction